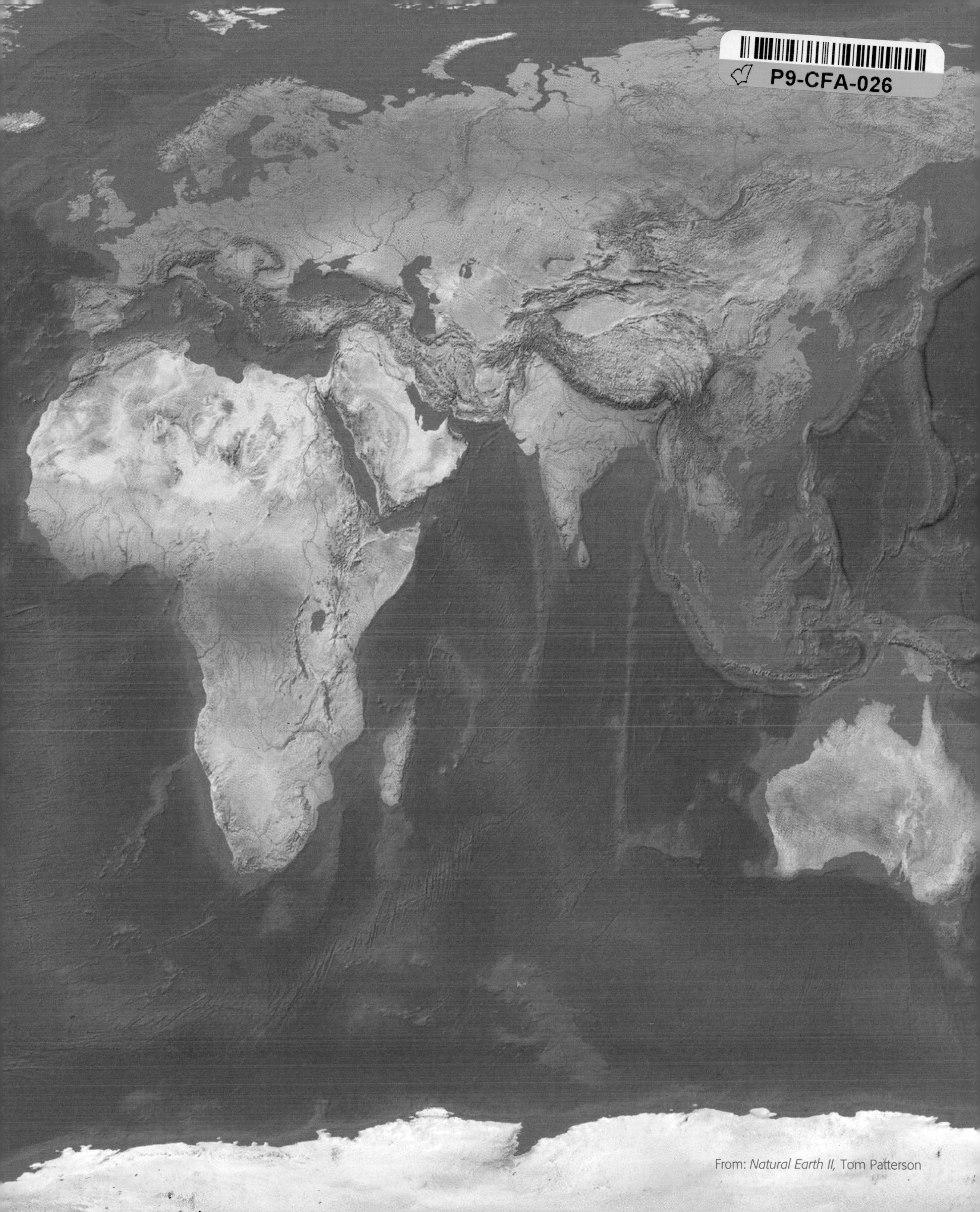
P9-CFA-026

From: *Natural Earth II,* Tom Patterson

PHYSICAL GEOGRAPHY

Ninth Edition

EDITION

9 PHYSICAL GEOGRAPHY

Robert E. Gabler
Emeritus, Western Illinois University

James F. Petersen
Texas State University–San Marcos

L. Michael Trapasso
Western Kentucky University

Dorothy Sack
Ohio University–Athens

Australia • Brazil • Japan • Korea • Mexico • Singapore • Spain • United Kingdom • United States

Physical Geography, Ninth Edition
Robert E. Gabler, James F. Petersen, L. Michael Trapasso, Dorothy Sack

Development Editor: Amy Collins

Assistant Editor: Liana Monari

Editorial Assistant: Paige Leeds

Technology Project Manager: Alexandria Brady

Marketing Communications Manager: Belinda Krohmer

Project Manager, Editorial Production: Hal Humphrey

Art Director: Vernon Boes

Print Buyer: Karen Hunt

Permissions Editor: Roberta Broyer

Production Service: Katy Bastille, Pre-PressPMG

Text Designer: Diane Beasley

Photo Researcher: Terri Wright

Copy Editor: Deborah Bader

Illustrators: Accurate Art, Precision Graphics, Rolin Graphics, Pre-Press PMG

Cover Designer: Cheryl Carrington

Cover Image: Stockbyte/Getty Images

Compositor: Pre-Press PMG

© 2009, 2007 Brooks/Cole, Cengage Learning

ALL RIGHTS RESERVED. No part of this work covered by the copyright herein may be reproduced, transmitted, stored, or used in any form or by any means graphic, electronic, or mechanical, including but not limited to photocopying, recording, scanning, digitizing, taping, Web distribution, information networks, or information storage and retrieval systems, except as permitted under Section 107 or 108 of the 1976 United States Copyright Act, without the prior written permission of the publisher.

For product information and technology assistance, contact us at
Cengage Learning Customer & Sales Support, 1-800-354-9706.

For permission to use material from this text or product,
submit all requests online at **cengage.com/permissions**
Further permissions questions can be e-mailed to
permissionrequest@cengage.com.

Library of Congress Control Number: 2008928956

ISBN-13: 978-0-495-55506-3
ISBN-10: 0-495-55506-1

Brooks/Cole
10 Davis Drive
Belmont, CA 94002-3098
USA

Cengage Learning is a leading provider of customized learning solutions with office locations around the globe, including Singapore, the United Kingdom, Australia, Mexico, Brazil, and Japan. Locate your local office at **international.cengage.com/region**

Cengage Learning products are represented in Canada by Nelson Education, Ltd.

For your course and learning solutions, visit **academic.cengage.com**

Purchase any of our products at your local college store or at our preferred online store **www.ichapters.com**

Printed in Canada
1 2 3 4 5 6 7 12 11 10 09 08

Preface

Earth, our planetary home, is a wondrous life-support system, yet it is also complex and ever-changing. Our planet's environments are robust enough to adapt to many environmental changes, but if certain limits are approached they may be threatened or damaged. Today, the modern technologies that we use tend to insulate us from fully experiencing our environment, so we can become lulled into forgetting about our dependence on Earth's natural systems and resources. Sometimes it is hard to imagine that while you read this you are also moving through space on a living planetary oasis surrounded by the vastness of space—empty and, as far as we know, devoid of life. Understanding our planet, the nature of its environments, and how they operate is as critical as it has ever been for humankind.

For as long as people have existed, the resources provided by their physical environments have been the key to survival. Preindustrial societies, such as those dependent on hunting and gathering or small-scale agriculture, tended to have small populations that exerted relatively little impact on their natural surroundings. In contrast, today's industrialized societies have large populations, demand huge quantities of natural resources, and can influence or cause environmental change, not only on a local scale, but also on a global one. A great concern today is about the potential impacts of changes in global climates, which certainly have been influenced by human activities.

As the world's population has increased, so have the scales, degrees, and cumulative effects of human impacts on the environment. We have polluted the air and water. We have used up tremendous amounts of nonrenewable resources and have altered many natural landscapes without fully assessing the potential consequences. Too often, we have failed to respect the power of Earth's natural forces when constructing our homes and cities or while pursuing our economic activities. In the 21st century, it is now evident that if we continually fail to comprehend Earth's potential and respect its limitations as a human habitat, we may be putting ourselves and future generations at risk. Despite the many differences between our current lifestyles and those of early humans, the ways we use and affect our physical environment provide the keys to our survival.

Today, this important message is gaining acceptance. We understand that Earth does not offer limitless natural resources. The news media have expanded their coverage of environmental characteristics and issues, including human impacts. Many governmental representatives work to enact legislation that will address environmental problems. Scientists and governmental leaders from around the world meet to discuss environmental issues that increasingly cross international boundaries. Humanitarian organizations, funded by governments as well as by private citizens, struggle to alleviate the suffering that results from natural disasters or from environmental degradation. The more we know about Earth and its environments, the more effective we can be in working toward stewardship and preservation.

Geography is a highly regarded subject in most nations of the world, and in recent years it has undergone a renaissance in the United States. National education standards that include physical geography support offering high-quality geography curricula in U.S. elementary, secondary, and postsecondary schools. Employers are increasingly recognizing the value and importance of geographic knowledge, skills, and techniques in the workplace. Physical geography as an applied field makes use of many computer-assisted and space-age technologies, such as geographic information systems (GIS), computer-assisted mapmaking (cartography), the global positioning system (GPS), and satellite image interpretation. At the collegiate level, physical geography offers an introduction to the concerns, ideas, knowledge, and tools that are necessary for further study of our planet. More than ever before, physical geography is being recognized as an ideal science course for general-education students—students who will make decisions that consider human needs and desires, but also environmental limits and possibilities. It is for these students that *Physical Geography* has been written.

Features

Comprehensive View of the Earth System *Physical Geography* introduces all major aspects of the Earth system, identifying physical phenomena and natural processes and stressing their characteristics, relationships, interactions, and distributions. The text covers a wide range of topics, including the atmosphere, the solid Earth, oceans and other water bodies, and the living environments of our planet.

Clear Explanation The text uses an easily understandable, narrative style to explain the origins, development, significance, and distribution of processes, physical features, and events that occur within, on, or above Earth's surface. The writing style is targeted toward rapid comprehension and making the study of physical geography meaningful and enjoyable.

Introduction to the Geographer's Tools Space-age and computer technologies have revolutionized the ways that we can study our planet, its features, its environmental aspects, and its natural processes. A full chapter is devoted to maps and other forms of spatial imagery and data used by geographers. Illustrations throughout the book include images gathered from space, accompanied by interpretations of the environmental aspects that the scenes illustrate. Also included are introductory discussions of techniques currently used by geographers to analyze or display location and environmental aspects of Earth, including remote sensing, geographic information systems, computer-assisted cartography, and the global positioning system.

Focus on Student Interaction The text uses numerous methods to encourage interaction between students, the textbook, and the subject matter of physical geography. The activities at the end of each chapter, which can be completed individually or as a group, are designed to engage students and promote active, rather than passive learning. Questions following the captions of most illustrations prompt students to think beyond the map, graph, diagram, image, or photograph and give further consideration to the topic.

Three Unique Perspectives

Physical geography is a field that seeks to develop an understanding and appreciation of our Earth and its environmental diversity. In approaching this goal, this textbook employs article boxes that illustrate the three major perspectives of physical geography. Through a **spatial science perspective,** physical geography focuses on understanding and explaining the locations, distribution, and spatial interactions of natural phenomena. Physical geography can also be approached from a **physical science perspective,** which applies the knowledge and methods of the natural and physical sciences, for example, by using the scientific method and systems analysis techniques. Through an **environmental science perspective,** physical geographers consider impacts, influences, and interactions among human and natural components of the environment, in other words, how the environment influences human life and how humans affect the environment.

Map Interpretation Series Learning map interpretation skills is a priority in a physical geography course. To meet the needs of students who do not have access to a laboratory setting, this text includes map activities with accompanying explanations, full-color maps printed at their original map scale, satellite images, and interpretation questions. These maps give students an opportunity to develop valuable map-reading skills. In courses that have a lab section, the map interpretation features offer a supplement to lab activities and a link between class lectures, the text, and lab work.

Objectives

Since the first edition, the authors have sought to accomplish four major objectives:

To Meet the Academic Needs of the Student Instructors familiar with the style and content of *Physical Geography* know that this textbook is written specifically for the student, and it is designed to satisfy the major purposes of a liberal education. Students are provided with the knowledge and understanding they need to make informed decisions involving the environments that they will interact with throughout their lives. The text assumes little or no prior background in physical geography or other Earth sciences. Numerous examples from throughout the world are included to illustrate important concepts and help nonscience majors bridge the gap between scientific theory and practical application.

To Strongly Integrate the Illustrations with the Written Text Numerous photographs, maps, satellite images, scientific visualizations, block diagrams, graphs, and line drawings have been carefully chosen to clearly illustrate important concepts in physical geography. The text discussions of concepts often contain repeated references to the illustrations, so students are able to examine in graphic form, as well as mentally visualize, the physical processes and phenomena involved. Some examples of topics that are clearly explained through the integration of visuals and text include map and image interpretation (Chapter 2), the seasons (Chapter 3), the heat energy budget (Chapter 4), surface wind systems (Chapter 5), storms (Chapter 7), soils (Chapter 12), plate tectonics (Chapter 13), rivers (Chapter 17), glaciers (Chapter 19), and coastal processes (Chapter 20).

To Communicate the Nature of Geography The nature of geography and three major perspectives of physical geography (spatial science, physical science, and environmental science) are discussed in Chapter 1. In subsequent chapters, important topics of geography involving all three perspectives are discussed. For example, location is a dominant topic in Chapter 2 and remains an important theme throughout the text. Spatial distributions are stressed as the climatic elements are discussed in Chapters 4 through 6. The changing Earth system is a central focus in Chapter 8. Characteristics of environments constitute Chapters 9 and 10. Spatial interactions are demonstrated in discussions of weather systems (Chapter 7), soils (Chapter 12), and volcanic and tectonic activity (Chapters 13 and 14). Article boxes in every chapter present interesting and important examples of each perspective.

To Fulfill the Major Requirements of Introductory Physical Science College Courses *Physical Geography* offers a full chapter on the tools and methodologies of physical geography. The Earth as a system and the physical processes that are responsible for the location, distribution, and spatial relationships of physical phenomena beneath, at, and above Earth's surface are examined in detail. Scientific method, hypothesis, theory, and explanation are continually stressed. In addition, end-of-chapter questions that involve understanding and interpreting graphs of environmental data (or graphing data for analysis), quantitative transformation or calculation of environmental variables, and/or hands-on map analysis directly support science learning. Models and systems are frequently cited in the discussion of important concepts, and scientific classification is presented in several chapters—some of these topics include air masses, tornadoes, and hurricanes (Chapter 7), climates

(Chapters 8 and 9), biogeography (Chapter 11), soils (Chapter 12), rivers (Chapter 17), and coasts (Chapter 20).

Ninth Edition Revision

Revising *Physical Geography* for a ninth edition involved thoughtful consideration of the input from many reviewers with varied opinions. Not only is our planet ever-changing, but so are the many ways that we study, observe, measure, and analyze Earth's characteristics, environments, and processes. New scientific findings and new ways of communicating those findings are continually being developed. This edition has been revised so that the latest and most important information is presented to those who are studying physical geography. As authors we continually seek to include coverage on physical geographic topics that will spark student interest. We also seek to keep current on recent environmental concerns, findings, and natural hazards by explaining the events, the conditions that led to those events, and how they are related to physical geography. Some recent examples include natural disasters such as deadly mudslides in the Philippines, terrible wildfires in Southern California and Texas, flooding in many areas, and numerous damaging and deadly outbreaks of multiple tornadoes. Hurricane Katrina and the tragic South Asia tsunami continue to be discussed in terms of human impact and new efforts toward avoidance of such tragedies in the future. These events and others are addressed as examples of Earth processes and human–environment interactions.

In addition, we thoroughly revised the text; prepared new graphs, maps, and diagrams; integrated 221 new photographs; and updated information on numerous worldwide environmental events. What follows is a brief review of other major changes made to this ninth edition.

New Co-Author We are privileged to welcome Dorothy Sack of Ohio University as a new co-author. A geomorphologist with a broad background in physical geography and a strong interest in coastal and arid environments, Dorothy's expertise, fresh outlook, and commitment to geographic education have been a valuable asset to this edition.

Chapter Reorganization The number of chapters has been reduced to 20, allowing us to strengthen discussions and improve illustrations while keeping the book at approximately the same length. Previous chapters on the world's oceans and on coastal processes and landforms have been combined into a single chapter. The chapter on atmospheric pressure, winds, and circulation patterns was reorganized to better conform to the scale of weather systems. The global climates and climate change chapter received major revision, and can be used as a standalone chapter for climate discussions in a one-semester class, making the more detailed climate chapters optional. The weathering and mass wasting chapter has been revised with more precise definitions and greater emphasis on the importance of the breakdown of rock matter. The chapter on tectonic forces and landforms that result from them is now organized by direction of the force (compressional, tensional, and shearing) rather than by type of structure (folds and faults). The map and graph interpretation exercises remain at the ends of chapters to avoid interrupting the flow of text discussion, and some improved images or photographs are included in these exercises. These and other organization changes provide increased course flexibility without significantly altering the sequence of topics or compelling instructors to make major changes in syllabi.

New and Revised Text New material has been added on a variety of topics. Great concern has been given to unusual weather conditions and the potential impacts of global warming. In 2007 the National Weather Service (NOAA) adopted a revised version of the Fujita scale for rating tornadoes that will provide better understanding of how damage is related to wind speeds and construction type. Also in 2007 the International Panel on Climate Change released an exhaustive report by hundreds of climate scientists worldwide that examines the evidence for links between human activities and global warming. Both of these important and new weather-related scientific findings and approaches are addressed in this text revision.

Earth systems approaches are reinforced with additional content, illustrations, and examples. The concept of spatial scale in atmospheric processes has been given a stronger emphasis. Sections on the greenhouse effect and global warming have been expanded, and there is a new discussion concerning Near Earth Objects (NEOs). A graph interpretation activity is included that involves the analysis and classification of climatic data and characteristics through use of climographs. In the fluvial chapter, the section on hydrology has been expanded to include the important topic of flood recurrence intervals. Many other sections of the book contain new material, new line art, new photographs, and new feature boxes. These include new regional-spatial examples, and human interactions with the environment (Chapter 1), new examples of vertical exaggeration (Chapter 2), using solar energy (Chapter 3), the urban heat island (Chapter 4), upper air circulation (Chapter 5), tornado chasers (Chapter 7), desertification and deforestation (Chapter 9), soil conservation (Chapter 12), major landslides (Chapter 15), water pollution (Chapter 16), the impact of dams (Chapter 17), and off-road vehicles and deserts (Chapter 18).

New Student Activities The end-of-chapter material has significantly improved with the addition of "Apply & Learn," hands-on activities that ask students to apply concepts and illustrate understanding by drawing a map or image, writing about a specific application of geography, or solving a problem using quantitative analysis. Also new for selected chapters is "Locate & Explore," engaging and informative Google Earth® exercises that reinforce concepts and provide experience using Web-based digital maps and imagery. Read "About Locate & Explore Activities" for more information.

Enhanced Program of Illustrations The illustration program has undergone substantial revision. The new and expanded topics required many new figures and updates to others, including numerous photographs, satellite images, and maps. Two hundred twenty-one figures have been replaced by new photographs and there are 73 new or revised line drawings.

An Increased Focus on Geography as a Discipline The undergraduate students of today include the professional geographers of tomorrow. Several changes in the text provide students with a better appreciation of geography as a discipline worthy of continued study and serious consideration as a career choice. The focus on the applications begins with the definition of geography, the discipline's tools and methodologies, selected topics to illustrate the role of geography as a spatial science, and the practical applications of the discipline, all topics found in Chapter 1.

Physical geography plays a central role in understanding environmental issues, human–environment interactions, and in approaches to solving environmental problems. Spreading the message about the importance and relevance of geography in today's world is essential to the viability and strength of geography in schools and universities. *Physical Geography,* Ninth Edition, seeks to reinforce that message to our students.

About Locate & Explore Activities

Throughout this textbook you will find Locate & Explore activities at the end of many chapters, which require you to use Google Earth®. Google Earth is a virtual globe browser that allows you to interactively display and investigate geographic data from anywhere in the world. To perform these exercises, you should have the latest version of Google Earth installed on your computer. The exercises require you to use some data layers that are included with Google Earth, as well as some additional data layers that you must download. For detailed instructions about using Google Earth, and to download the necessary data, go to academic.cengage.com/earthscience/gabler9e.

Ancillaries

Instructors and students alike will greatly benefit from the comprehensive ancillary package that accompanies this text.

For the Instructor

Class Preparation and Assessment Support

Instructor's Manual with Test Bank and Lab Pack The downloadable manual contains suggestions concerning teaching methodology as well as evaluation resources including course syllabi, listings of main concepts, chapter outlines and notes, answers to review questions, recommended readings, and a complete test item file. The Instructor's Manual also includes answers for the accompanying *Lab Pack*. Available exclusively for download from our password-protected instructor's Web site: academic.cengage.com/earthscience/gabler9e.

ExamView® Computerized Testing Create, deliver, and customize tests and study guides (both print and online) in minutes with this easy-to-use assessment and tutorial system. Preloaded with the *Physical Geography* test bank, *ExamView* offers both a *Quick Test Wizard* and an *Online Test Wizard*. You can build tests of up to 250 questions using as many as 12 question types. *ExamView's* complete word-processing capabilities also allow you to enter an unlimited number of new questions or edit existing questions.

Dynamic Lecture Support

PowerLecture with JoinIn™ A complete all-in-one reference for instructors, the PowerLecture CD contains PowerPoint® slides with lecture outlines, images from the text, stepped art from the text, zoomable art figures from the text, and active figures that interactively demonstrate concepts. Besides providing you with fantastic course presentation material, the PowerLecture CD contains electronic files of the Test Bank and Instructor's Manual, as well as JoinIn, the easiest Audience Response System to use, featuring instant classroom assessment and learning.

Active Earth CD The Active Earth Collection allows you to pick and choose from over 120 earth science animations and active figures, ABC® natural hazard video clips, and in-depth Google Earth® lecture activities, and includes a link to the Earth Science Newsroom. Grab your students' attention by creating your lectures using these dynamic tools.

Laboratory and GIS Support

Lab Pack ISBN: 0-495-56515-6. The perfect lab complement to the text, this *Lab Pack* contains over 50 exercises, varying in length and difficulty, designed to help students achieve a greater understanding and appreciation of physical geography.

GIS Investigations Michelle K. Hall-Wallace, C. Scott Walker, Larry P. Kendall, Christian J. Schaller, and Robert F. Butler of the University of Arizona, Tucson.

The perfect accompaniment to any physical geography course, these four groundbreaking guides tap the power of *ArcView® GIS and ArcGIS®* to explore, manipulate, and analyze large data sets. The guides emphasize the visualization, analysis, and multimedia integration capabilities inherent to GIS and enable students to "learn by doing" with a full complement of GIS capabilities. The guides contain all the software and data sets needed to complete the exercises.

Exploring the Dynamic Earth:
GIS Investigations for the Earth Sciences
ISBN with ArcView CD: 0-534-39138-9
ISBN for use with ArcGIS site license: 0-495-11509-6

Exploring Tropical Cyclones:
GIS Investigations for the Earth Sciences
ISBN with ArcView CD: 0-534-39147-8
ISBN for use with ArcGIS site license: 0-495-11543-6

Exploring Water Resources:
GIS Investigations for the Earth Sciences
ISBN with ArcView CD: 0-534-39156-7
ISBN for use with ArcGIS site license: 0-495-11512-6

Exploring the Ocean Environment:
GIS Investigations for the Earth Sciences
ISBN with ArcView CD: 0-534-42350-7
ISBN for use with ArcGIS site license: 0-495-11506-1

For the Student

Geography Resource Center This password-protected site includes interactive maps, animations, and an array of other discipline-related resources to complement your experience with geography. Go to academic.cengage.com/earthscience/gabler9e to get started.

Acknowledgments

This edition of *Physical Geography* would not have been possible without the encouragement and assistance of editors, friends, and colleagues from throughout the country. Great appreciation is extended to Sarah Gabler; Martha, Emily, and Hannah Petersen; and Greg Nadon and Carolyn Moore for their patience, support, and understanding.

Special thanks go to the splendid freelancers and staff members of Brooks/Cole Cengage Learning. These include Marcus Boggs, Earth Sciences Director; Amy K. Collins, Development Editor; Liana Monari, Assistant Editor; Alexandria Brady and Melinda Newfarmer, Technology Project Managers; Hal Humphrey, Content Project Manager; Diane Beasley, Designer; Terri Wright, Photo Researcher; illustrators Pre-PressPMG, Accurate Art, Precision Graphics, and Rolin Graphis; Katy Bastille, Pre-PressPMG Production Coordinator; Paige Leeds, Editorial Assistant; and Dr. Chris Houser, creator of our wonderful *Locate & Explore* activities.

Colleagues who reviewed the plans and manuscript for this and previous editions include: Peter Blanken, University of Colorado; Brock Brown, Texas State University; J. Michael Daniels, University of Wyoming; Ben Dattilo, University of Nevada, Las Vegas; Leland R. Dexter, Northern Arizona University; James Doerner, University of Northern Colorado; Percy "Doc" Dougherty, Kutztown State University; Tom Feldman, Joliet Junior College; Roberto Garza, San Antonio College; Greg Gaston, University of North Alabama; Perry Hardin, Brigham Young University; David Helgren, San Jose State University; Chris Houser, University of West Florida; Fritz C. Kessler, Frostburg State University; Elizabeth Lawrence, Miles Community College; Jeffrey Lee, Texas Tech University; Michael E. Lewis, University of North Carolina, Greensboro; John Lyman, Bakersfield College; Charles Martin, Kansas State University; Debra Morimoto, Merced College; Andrew Oliphant, San Francisco State University; James R. Powers, Pasadena City College; Joyce Quinn, California State University, Fresno; Colin Thorn, University of Illinois at Urbana-Champaign; Dorothy Sack, Ohio University; George A. Schnell, State University of New York, New Paltz; Peter Siska, Austin Peay State University; Richard W. Smith, Harford Community College; Ray Sumner, Long Beach City College; Michael Talbot, Pima Community College; David L. Weide, University of Nevada, Las Vegas; Thomas Wikle, Oklahoma State University, Stillwater; Amy Wyman, University of Nevada, Las Vegas; Craig ZumBrunnen, University of Washington, Redmond; and Joanna Curran, Richard Earl, and Mark Fonstad, all of Texas State University.

Photos courtesy of: Bill Case and Chris Wilkerson, Utah Geological Survey; Center for Cave and Karst Studies, Western Kentucky University; Hari Eswaran, USDA Natural Resources Conservation Service; Richard Hackney, Western Kentucky University; David Hansen, University of Minnesota; L. Elliot Jones, U.S. Geological Survey; Susan Jones, Nashville, Tennessee; Bob Jorstad, Eastern Illinois University; Carter Keairns, Texas State University; Parris Lyew-Ayee, Oxford University, UK; Dorothy Sack, Ohio University; Anthony G. Taranto Jr., Palisades Interstate Park–New Jersey Section; Justin Wilkinson, Earth Sciences, NASA Johnson Space Center.

The detailed comments and suggestions of all of the above individuals have been instrumental in bringing about the many changes and improvements incorporated in this latest revision of the text. Countless others, both known and unknown, deserve heartfelt thanks for their interest and support over the years.

Despite the painstaking efforts of all reviewers, there will always be questions of content, approach, and opinion associated with the text. The authors wish to make it clear that they accept full responsibility for all that is included in the ninth edition of *Physical Geography*.

Robert E. Gabler
James F. Petersen
L. Michael Trapasso
Dorothy Sack

Foreword to the Student

Why Study Geography?

In this global age, the study of geography is absolutely essential to an educated citizenry of a nation whose influence extends throughout the world. Geography deals with location, and a good sense of where things are, especially in relation to other things in the world, is an invaluable asset whether you are traveling, conducting international business, or sitting at home reading the newspaper.

Geography examines the characteristics of all the various places on Earth and their relationships. Most important in this regard, geography provides special insights into the relationships between humans and their environments. If all the world's people could have one goal in common, it should be to better understand the physical environment and protect it for the generations to come.

Geography provides essential information about the distribution of things and the interconnections of places. The distribution pattern of Earth's volcanoes, for example, provides an excellent indication of where Earth's great crustal plates come in contact with one another; and the violent thunderstorms that plague Illinois on a given day may be directly associated with the low pressure system spawned in Texas two days before. Geography, through a study of regions, provides a focus and a level of generalization that allows people to examine and understand the immensely varied characteristics of Earth.

As you will note when reading Chapter 1, there are many approaches to the study of geography. Some courses are regional in nature; they may include an examination of one or all of the world's political, cultural, economic, or physical regions. Some courses are topical or systematic in nature, dealing with human geography, physical geography, or one of the major subfields of the two.

The great advantage to the study of a general course in physical geography is the permanence of the knowledge learned. Although change is constant and is often sudden and dramatic in the human aspects of geography, alterations of the physical environment on a global scale are exceedingly slow when not influenced by human intervention. Theories and explanations may differ, but the broad patterns of atmospheric and oceanic circulation and of world climates, landforms, soils, natural vegetation, and physical landscapes will be the same tomorrow as they are today.

Keys to Successful Study

Good study habits are essential if you are to master science courses such as physical geography, where the topics, explanations, and terminology are often complex and unfamiliar. To help you succeed in the course in which you are currently enrolled, we offer the following suggestions.

Reading Assignments

- Read the assignments before the material contained therein is covered in class by the instructor.
- Compare what you have read with the instructor's presentation in class. Pay particular attention if the instructor introduces new examples or course content not included in the reading assignment.
- Do not be afraid to ask questions in class and seek a full understanding of material that may have been a problem during your first reading of the assignment.
- Reread the assignment as soon after class as possible, concentrating on those areas that were emphasized in class. Highlight only those items or phrases that you now consider to be important, and skim those sections already mastered.
- Add to your class notes important terms, your own comments, and summarized information from each reading assignment.

Understanding Vocabulary

Mastery of the basic vocabulary often becomes a critical issue in the success or failure of the student in a beginning science course.

- Focus on the terms that appear in boldface type in your reading assignments. Do not overlook any additional terms that the instructor may introduce in class.
- Develop your own definition of each term or phrase and associate it with other terms in physical geography.
- Identify any physical processes associated with the term. Knowing the process helps to define the term.
- Whenever possible, associate terms with location.
- Consider the significance to humans of terms you are defining. Recognizing the significance of terms and phrases can make them relevant and easier to recall.

Learning Earth Locations

A good knowledge of place names and of the relative locations of physical and cultural phenomena on Earth is fundamental to the study of geography.

- Take personal responsibility for learning locations on Earth. Your instructor may identify important physical features and place names, but you must learn their locations for yourself.

- Thoroughly understand latitude, longitude, and the Earth grid. They are fundamental to location on maps as well as on a globe. Practice locating features by their latitude and longitude until you are entirely comfortable using the system.
- Develop a general knowledge of the world political map. The most common way of expressing the location of physical features is by identifying the political unit (state, country, or region) in which it can be found.
- Make liberal use of outline maps. They are the key to learning the names of states and countries and they can be used to learn the locations of specific physical features. Personally placing features correctly on an outline map is often the best way to learn location.
- Cultivate the atlas habit. The atlas does for the individual who encounters place names or the features they represent what the dictionary does for the individual who encounters a new vocabulary word.

Utilizing Textbook Illustrations

The secret to making good use of maps, diagrams, and photographs lies in understanding why the illustration has been included in the text or incorporated as part of your instructor's presentation.

- Concentrate on the instructor's discussion. Taking notes on slides, overhead transparencies, and illustrations will allow you to follow the same line of thought at a later date.
- Study all textbook illustrations on your own. Be sure to note which were the focus of considerable classroom attention. Do not quit your examination of an illustration until it makes sense to you, until you can read the map or graph, or until you can recognize what a diagram or photograph has been selected to explain.
- Hand-copy important diagrams and graphs. Few of us are graphic artists, but you might be surprised at how much better you understand a graph or line drawing after you reproduce it yourself.
- Read the captions of photos and illustrations thoroughly and thoughtfully. If the information is included, be certain to note where a photograph was taken and in what way it is representative. What does it tell you about the region or site being illustrated?
- Attempt to place the principle being illustrated in new situations. Seek other opportunities to test your skills at interpreting similar maps, graphs, and photographs and think of other examples that support the text being illustrated.
- Remember that all illustrations are reference tools, particularly tables, graphs, and diagrams. Refer to them as often as you need to.

Taking Class Notes

The password to a good set of class notes is selectivity. You simply cannot and, indeed, you should not try to write down every word uttered by your classroom instructor.

- Learn to paraphrase. With the exception of specific quotations or definitions, put the instructor's ideas, explanations, and comments into your own words. You will understand them better when you read them over at a later time.
- Be succinct. Never use a sentence when a phrase will do, and never use a phrase when a word will do. Start your recall process with your note-taking by forcing yourself to rebuild an image, an explanation, or a concept from a few words.
- Outline where possible. Preparing an outline helps you to discern the logical organization of information. As you take notes, organize them under main headings and subheadings.
- Take the instructor at his or her word. If the instructor takes the time to make a list, then you should do so too. If he or she writes something on the board, it should be in your notes. If the instructor's voice indicates special concern, take special notes.
- Come to class and take your own notes. Notes trigger the memory, but only if they are your notes.

Doing Well on Tests

Follow these important study techniques to make the most of your time and effort preparing for tests.

- Practice distillation. Do not try to reread but skim the assignments carefully, taking notes in your own words that record as economically as possible the important definitions, descriptions, and explanations. Do the same with any supplementary readings, handouts, and laboratory exercises. It takes practice to use this technique, but it is a lot easier to remember a few key phrases that lead to ever increasing amounts of organized information than it is to memorize all of your notes. And the act of distillation in itself is a splendid memory device.
- Combine and reorganize. Merge all your notes into a coherent study outline.
- Become familiar with the type of questions that will be asked. Knowing whether the questions will be objective, short-answer, essay, or related to diagrams and other illustrations can help in your preparation. Some instructors place old tests on file where you can examine them or will forewarn you of their evaluation styles if you inquire. If not, then turn to former students; there are usually some around the department or residence halls who have already experienced the instructor's tests.
- Anticipate the actual question that will likely be on the test. The really successful students almost seem to be able to predict the test items before they appear. Take your educated guesses and turn them into real questions.
- Try cooperative study. This can best be described as role playing and consists very simply of serving temporarily as the instructor. So go ahead and teach. If you can demonstrate a technique, illustrate an idea, or explain a process or theory to another student so that he or she can understand it, there is little doubt that you can answer test questions over the same material.
- Avoid the "all-nighter." Use the early evening hours the night before the test for a final unhurried review of your study outline. Then get a good night's sleep.

The Importance of Maps

Like graphs, tables, and diagrams, maps are an excellent reference tool. Familiarize yourself with the maps in your textbook in order to better judge when it is appropriate to seek information from these important sources.

Maps are especially useful for comparison purposes and to illustrate relationships or possible associations of things. But the map reader must beware. Only a small portion of the apparent associations of phenomena in space (areal associations) are actually cause-and-effect relationships. In some instances the similarities in distribution are a result of a third factor that has not been mapped. For instance, a map of worldwide volcano distribution is almost exactly congruent with one of incidence of earthquakes, yet volcanoes are not the cause of earthquakes, nor is the obverse true. A third factor, the location of tectonic plate boundaries, explains the first two phenomena.

Finally, remember that the map is the most important statement of the professional geographer. It is useful to all natural and social scientists, engineers, politicians, military planners, road builders, farmers, and countless others, but it is the essential expression of the geographer's primary concern with location, distribution, and spatial interaction.

About Your Textbook

This textbook has been written for you, the student. It has been written so that the text can be read and understood easily. Explanations are as clear, concise, and uncomplicated as possible. Illustrations have been designed to complement the text and to help you visualize the processes, places, and phenomena being discussed. In addition, the authors do not believe it is sufficient to offer you a textbook that simply provides information to pass a course. We urge you to think critically about what you read in the textbook and hear in class.

As you learn about the physical aspects of Earth environments, ask yourself what they mean to you and to your fellow human beings throughout the world. Make an honest attempt to consider how what you are learning in your course relates to the problems and issues of today and tomorrow. Practice using your geographic skills and knowledge in new situations so that you will continue to use them in the years ahead. Your textbook includes several special features that will encourage you to go beyond memorization and reason geographically.

Chapter Activities At the end of each chapter, Consider & Respond and Apply & Learn questions require you to go well beyond routine chapter review. The questions are designed specifically so that you may apply your knowledge of physical geography and on occasion personally respond to critical issues in society today. Locate & Explore activites (found at the end of many chapters) teach you how to use the Google Earth application as an exploratory learning tool. Check with your instructor for answers to the problems.

Caption Questions With almost every illustration and photo in your textbook a caption links the image with the chapter text it supports. Read each caption carefully because it explains the illustration and may also contain new information. Wherever appropriate, questions at the ends of captions have been designed to help you seize the opportunity to consider your own personal reaction to the subject under consideration.

Map Interpretation Series It is a major goal of your textbook to help you become an adept map reader, and the Map Interpretation Series in your text has been designed to help you reach that goal.

Environmental Systems Diagrams Viewing Earth as a system comprising many subsystems is a fundamental concept in physical geography for researchers and instructors alike. The concept is introduced in Chapter 1 and reappears frequently throughout your textbook. The interrelationships and dependencies among the variables or components of Earth systems are so important that a series of special diagrams (see, for example, Figure 6.4) have been included with the text to help you visualize how the systems work. Each diagram depicts the system and its variables and also demonstrates their interdependence and the movement or exchanges that occur within each system. The diagrams are designed to help you understand how human activity can affect the delicate balance that exists within many Earth systems.

Geography Resource Center The Geography Resource Center is a password-protected site that includes interactive maps, animations, and an array of other discipline-related resources to complement your experience with geography and give you additional tools for success in your geography course. Go to academic.cengage.com/earthscience/gabler9e to get started.

As authors of your textbook, we wish you well in your studies. It is our fond hope that you will become better informed about Earth and its varied environments and that you will enjoy the study of physical geography.

Brief Contents

Contents

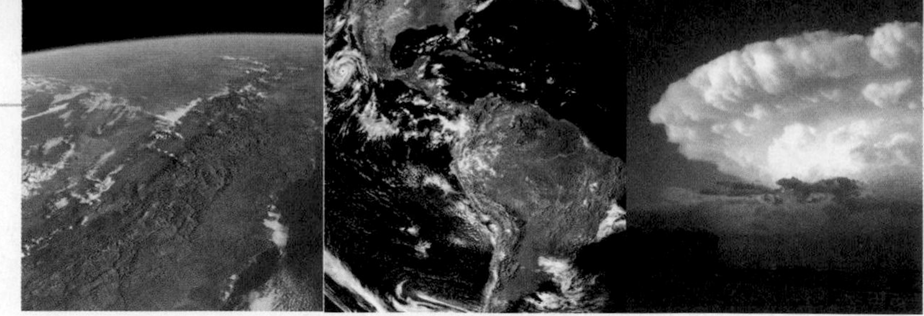

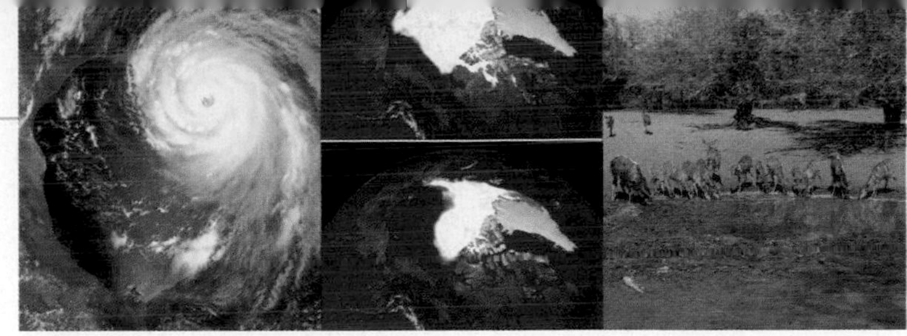

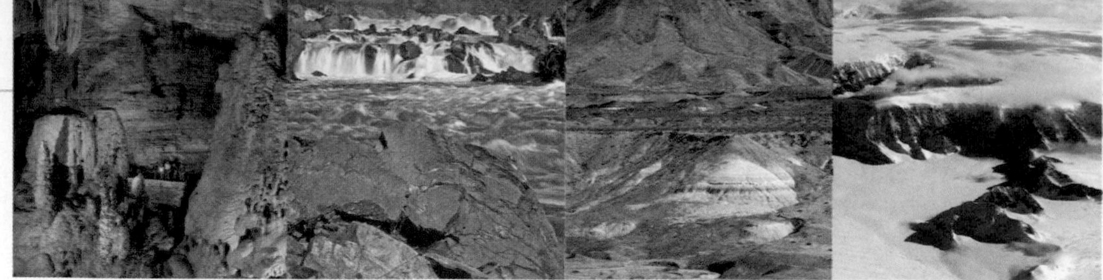

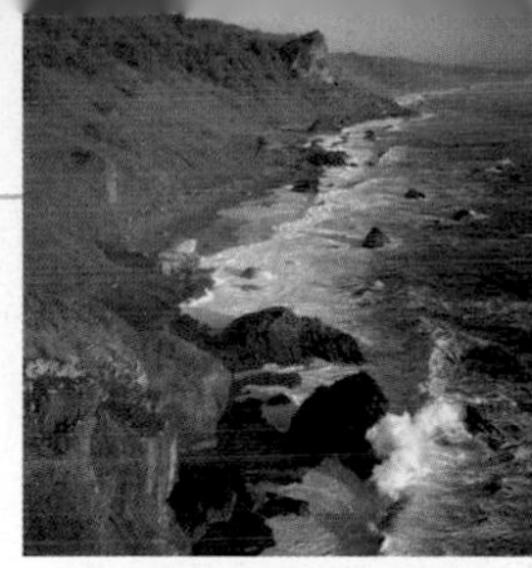

List of Major Maps

The World

The Ocean

The Contiguous United States

North America

Author Biographies

Robert E. Gabler During his nearly five decades of professional experience, Professor Gabler has taught geography at Hunter College, City of New York, Columbia University, and Western Illinois University, in addition to 5 years in public elementary and secondary schools. At times in his career at Western he served as Chairperson of the Geography and Geology Department, Chairperson of the Geography Department, and University Director of International Programs. He received three University Presidential Citations for Teaching Excellence and University Service, served two terms as Chairperson of the Faculty Senate, edited the *Bulletin of the Illinois Geographical Society,* and authored numerous articles in state and national periodicals. He is a Past President of the Illinois Geographical Society, former Director of Coordinators and Past President of the National Council for Geographic Education, and the recipient of the NCGE George J. Miller Distinguished Service Award.

James F. Petersen James F. Petersen is Professor of Geography at Texas State University, in San Marcos, Texas. He is a broadly trained physical geographer with strong interests in geomorphology and Earth Science education. He enjoys writing about topics relating to physical geography for the public, particularly environmental interpretation, and has written a landform guidebook for Enchanted Rock State Natural Area in central Texas and a number of field guides. He is a strong supporter of geographic education, having served as President of the NCGE in 2000 after more than 15 years of service to that organization. He has also written or served as a senior consultant for nationally published educational materials at levels from middle school through university, and has done many workshops for geography teachers. Recently, he contributed the opening chapter in an environmental history of San Antonio that explains the physical geographic setting of central Texas.

L. Michael Trapasso L. Michael Trapasso is Professor of Geography at Western Kentucky University, the Director of the College Heights Weather Station, and a Research Associate with the Kentucky Climate Center. His research interests include human biometeorology, forensic meteorology, and environmental perception. He has received the Ogden College Faculty Excellence Award, and has also received Fulbright and Malone Fellowships to conduct research in various countries. His explorations have extended to all seven continents, and he has written and lectured extensively on these travels. He has also contributed to a lab manual, a climatology textbook, and topical encyclopedias; and he has written and narrated educational television programming concerning weather and climate for the Kentucky Educational Television (KET) Network and WKYU-TV 24, Western Kentucky University Television.

Dorothy Sack Dorothy Sack, Professor of Geography at Ohio University in Athens, Ohio, is a physical geographer who specializes in geomorphology. Her research emphasizes arid region landforms, including geomorphic evidence of paleolakes, which contributes to paleoclimate reconstruction. She has published research results in a variety of professional journals, academic volumes, and Utah Geological Survey reports. She also has research interests and publications on the history of geomorphology and the impact of off-road vehicles. Her work has been funded by the National Geographic Society, NSF, Association of American Geographers (AAG), American Chemical Society, and other organizations. She is active in professional organizations, having served as chairperson of the AAG Geomorphology Specialty Group, and several other offices for the AAG, Geological Society of America, and History of Earth Sciences Society. She enjoys teaching and research, and has received the Outstanding Teacher Award from Ohio University's College of Arts and Sciences.

PHYSICAL GEOGRAPHY

Ninth Edition

Physical Geography: Earth Environments and Systems

1

CHAPTER PREVIEW

Physical geography investigates and seeks to explain the spatial aspects, functions, and characteristics of Earth's physical phenomena.

- Why is geography often called the spatial science?
- Why are the topics of spatial interaction and change important in physical geography?

Although it is closely related to many other sciences, physical geography has its own unique focus and perspectives for studying Earth.

- What are the three major perspectives of physical geography?
- Why is a holistic approach important to understanding physical geography?

The use of models and the analysis of various Earth systems are important research and educational techniques used by geographers.

- What kinds of models may be used to portray Earth, its features, and its physical processes?
- In what ways can systems analysis lead to an understanding of complex environments?

Unlike some other physical sciences, physical geography places a special emphasis on human–environment relationships.

- Why is geography so important in the study of the environmental sciences today?
- Why do ecosystems provide such excellent opportunities for physical geographers to study the interactions between humans and the natural environment?

Every physical environment offers an array of advantages as well as challenges or hazards to the human residents of that location.

- What environmental adaptations are necessary for humans to live in your area?
- What impacts do humans have on the environment where you live?

◀ **Earth's incredible environmental diversity: An oasis of life in the vastness of space.**
Image provided by GeoEye and NASA SeaWiFs Project

Viewed from far enough away to see an entire hemisphere, Earth is both beautiful and intriguing—a life-giving planetary oasis. From this perspective we can begin to appreciate "the big picture," a global view of our planet's physical geography through its display of environmental diversity. Characteristics of the oceans, the atmosphere, the landmasses, and evidence of life as revealed by vegetated regions, are apparent. Looking carefully, we can recognize geographic patterns, shaped by the processes that make our world dynamic and ever-changing. Except for the external addition of energy from the sun, our planet is a self-contained system that has all the requirements to sustain life.

Earth may seem immense and almost limitless from the perspective of humans living on its surface. In contrast, viewing the "big picture" reveals its conspicuous limits and fragility—a spherical island of life surrounded by the vast, dark emptiness of space. However, from our vantage point in space, we cannot comprehend the details of how processes involving air, water, land, and living things interact to create a diverse array of landscapes and environmental conditions on Earth. These distant images display the basic aspects of Earth that make our existence possible, but they only hint at the complexity of our planet. Being aware of "the big picture" is important, but this knowledge should be bolstered by a detailed understanding of how Earth's features and processes interact to develop the extraordinary environmental diversity that exists on our planet. Developing this understanding is the goal of a course in physical geography.

The Study of Geography

Geography is a word that comes from two Greek roots. *Geo-* refers to "Earth," and *-graphy* means "picture or writing." The primary objective of geography is the examination, description, and explanation of Earth—its variability from place to place, how places and features change over time, and the processes responsible for these variations and changes. Geography is often called the **spatial science** because it includes recognizing, analyzing, and explaining the variations, similarities, or differences in phenomena located (or distributed) on Earth's surface. The major geographic organizations in the United States have provided us with a good description of geography.

> Where is something located? Why is it there? How did it get there? How does it interact with other things? Geography is not a collection of arcane information. Rather it is the study of spatial aspects of human existence.
>
> People everywhere need to know about the nature of their world and their place in it. Geography has much more to do with asking questions and solving problems than it does with rote memorization of facts.
>
> So what exactly is geography? It is an integrative discipline that brings together the physical and human dimensions of the world in the study of people, places, and environments. Its subject matter is the Earth's surface and the processes that shape it, the relationships between people and environments, and the connections between people and places.
>
> Geography Education Standards Project, 1994
> *Geography for Life*

Geography is distinctive among the sciences by virtue of its definition and central purpose. Unlike most scientists in related disciplines (for example, biologists, geologists, chemists, economists), who are bound by the phenomena they study, geographers may focus their research on nearly any topic related to the scientific analysis of human or natural processes on Earth (● Fig. 1.1). Geographers generally consider all of the human and natural phenomena that are relevant to a given problem or issue; in other words, they often take a **holistic** approach to understanding aspects of our planet.

Geographers study the physical and/or human characteristics of places, seeking to identify and explain characteristics that two or more locations may have in common as well as why places vary in their geographic attributes. Geographers gather, organize, and analyze many kinds of geographic data and information, yet a unifying factor among them is a focus on explaining spatial locations, distributions, and relationships. They apply a variety of skills, techniques, and tools to the task of answering geographic questions. Geographers also study processes that influenced Earth's landscapes in the past, how they continue to affect them today, how a landscape may change in the future, and the significance or impact of these changes.

Because geography embraces the study of virtually any global phenomena, it is not surprising that the subject has many subdivisions and it is common for geographers to specialize in one or more subfields of the discipline. Geography is also subdivided along academic lines; some geographers are social scientists and some are natural scientists, but most are involved in studying human or natural processes and how they affect our planet, as well as the interactions among these processes. The main subdivision that deals with human activities and the impact of these activities is called cultural or **human geography.** Human geographers are concerned with such subjects as population distributions, cultural patterns, cities and urbanization, industrial and commercial location, natural resource utilization, and transportation networks (● Fig. 1.2). Geographers are interested in how to divide and synthesize areas into meaningful divisions called **regions,** which are areas identified by certain characteristics they contain that make them distinctive and distinguish them from surrounding areas. A

● FIGURE 1.1

When conducting research or examining one of society's many problems, geographers are prepared to consider any information or aspect of a topic that relates to their studies.

What advantage might a geographer have when working with other physical scientists seeking a solution to a problem?

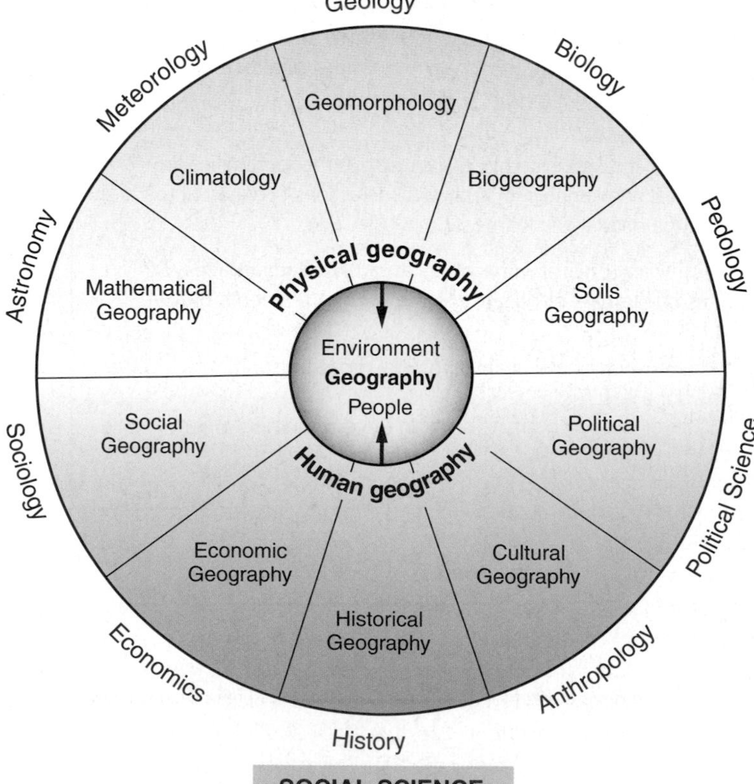

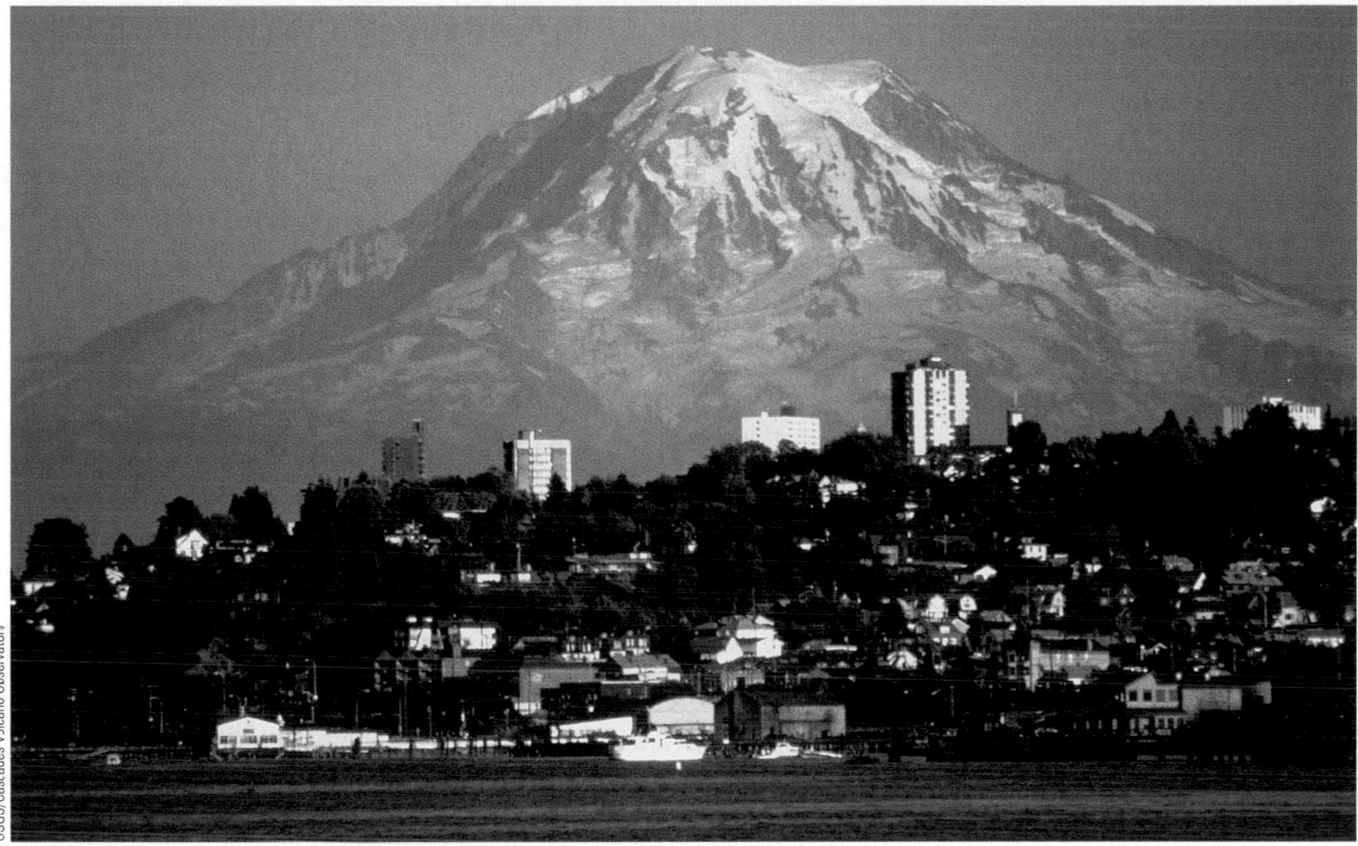
USGS/Cascades Volcano Observatory

• **FIGURE 1.2**
Settlement patterns, economic activities, recreational opportunities, and many aspects of human activities are a function of interactions among geographic factors, both human and physical.
What human geographic characteristics can you interpret from this scene?

region can be defined by characteristics that are physical, human, or a combination of factors. Geographic study that concentrates on both the general physical and human characteristics of a region, such as Canada, the Great Plains, the Caribbean, or the Sahara, is termed **regional geography.**

Physical Geography

Physical geography encompasses the processes and features that make up Earth, including human activities where they interface with the environment. In fact, physical geographers are concerned with nearly all aspects of Earth and can be considered generalists because they are trained to view a natural environment in its entirety, and how it functions as a unit (•Fig. 1.3). However, after completing a broad education in basic physical geography, most physical geographers focus their expertise on advanced study in one or two specialties. For example, *meteorologists* and *climatologists* consider how the interaction of atmospheric components influences weather and climate. Meteorologists are interested in the atmospheric processes that affect daily weather, and they use current data to forecast weather conditions. Climatologists are interested in the averages and extremes of long-term weather data, regional classification of climates, monitoring and understanding climatic change and climatic hazards, and the long-range impact of atmospheric conditions on human activities and the environment.

The study of the nature, development, and modification of landforms is a specialty called *geomorphology*, a major subfield of physical geography. Geomorphologists are interested in understanding and explaining variation in landforms, the processes that produce physical landscapes, and the nature and geometry of Earth's surface features. The factors involved in landform development are as varied as the environments on Earth, and include gravity, running water, stresses in the Earth's crust, flowing ice in glaciers, volcanic activity, and the erosion or deposition of Earth's surface materials. *Biogeographers* examine natural and human-modified environments and the ecological processes that influence their characteristics and distributions, including vegetation change over time. They also study the ranges and patterns of vegetation and animal species, seeking to discover the environmental factors that limit or facilitate their distributions. Many *soil scientists* are geographers, who are involved in mapping and analyzing soil types, determining the suitability of soils for certain uses, such as agriculture, and working to conserve soil as a natural resource.

Copyright and photograph by Dr. Parvinder S. Sethi

• **FIGURE 1.3**
Physical geographers study the elements and processes that affect natural environments. These include rock structures, landforms, soils, vegetation, climate, weather, and human impacts.
What physical geography characteristics can you interpret from this scene?

Finally, because of the critical importance of water to life on Earth, geographers are widely involved in the study of water bodies and their processes, movements, impact, quality, and other characteristics. They may serve as *hydrologists, oceanographers*, or *glaciologists*. Many geographers involved with water studies also function as water resource managers, who work to ensure that lakes, watersheds, springs, and groundwater sources are suitable to meet human or environmental needs, provide an adequate water supply, and are as free of pollution as possible.

Technology, Tools, and Methods

The technologies that physical geographers use in their efforts to learn more about Earth are rapidly changing. The abilities of computer systems to capture, process, model, and display spatial data—functions that can be performed on a personal computer—were only a dream 30 years ago. Today the Internet provides access to information and images on virtually any topic. The amounts of data, information, and imagery available for studying Earth and its environments have exploded. Graphic displays of environmental data and information are becoming more vivid and striking as a result of sophisticated methods of data processing and visual representation. Increased computer power allows the presentation of high-resolution images, three-dimensional scenes, and animated images of Earth features, changes, and processes (• Fig. 1.4).

Continuous satellite imaging of Earth has been ongoing for more than 30 years, which has given us a better perspective on environmental changes as they occur. Using satellite imagery it is possible to monitor changes in a single place over time or to compare different places at a point in time. Using various energy sources to produce images from space, we are able to see, measure, monitor, and map processes and the effects of certain processes including many that are invisible to the naked eye. Satellite technology is being used to determine the precise location of a positioning receiver on Earth's surface, a capability that has many useful applications for geography and mapping. Today, most mapmaking (*cartography*) and many aspects of map analysis are computer-assisted operations, although the ability to visually interpret a map, a landscape, or an environmental image remains an important geographic skill.

Making observations and gathering data in the field are valuable skills for most physical geographers, but they must

Image by R. B. Husar, Washington University; the land layer from the SeaWiFS Project; fire maps from the European Space Agency; the sea surface temperature from the Naval Oceanographic Office's Visualization Laboratory; and cloud layer from SSEC, University of Wisconsin

• **FIGURE 1.4**
Complex computer-generated model of Earth, based on data gathered from satellites.
How does this image compare to the Earth image in the chapter opening?

• **FIGURE 1.5**
A geographer uses computer technology to analyze maps and imagery.
In what ways are computer-generated maps and landscape images helpful in studies of physical geography?

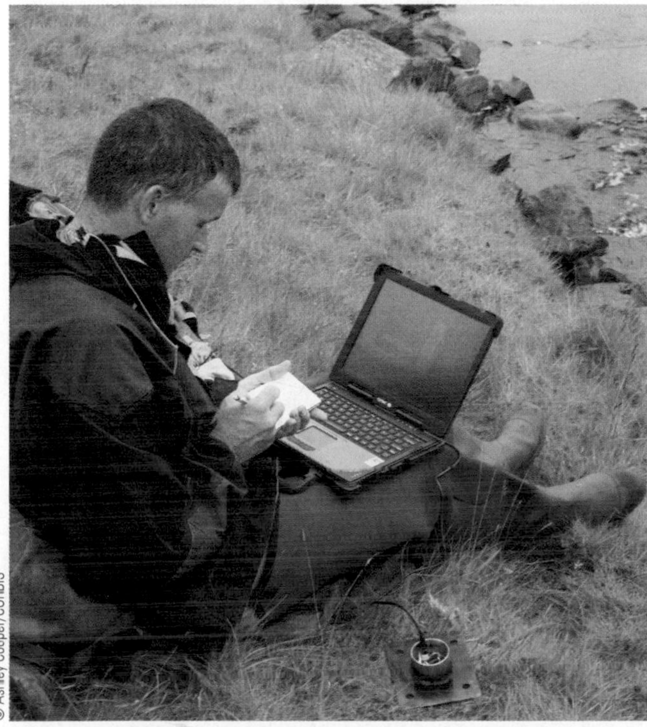

© Ashley Cooper/CORBIS

also keep up with new technologies that support and facilitate traditional fieldwork. Technology may provide maps, images, and data, but a person who is knowledgeable about the geographical aspects of the subject being studied is essential to the processes of analysis and problem solving (• Fig. 1.5). Many geographers are gainfully employed in positions that apply technology to the problems of understanding our planet and its environments, and their numbers are certain to increase in the future.

Major Perspectives in Physical Geography

Your textbook has been designed to demonstrate three major perspectives that physical geography emphasizes: spatial science, physical science, and environmental science. Although the emphasis on each of these perspectives may vary from chapter to chapter, the contributions of all three perspectives to scientific study will be apparent throughout the book. As you read this chapter, take note of how directly each scientific perspective relates to the unique nature of geography as a discipline.

The Spatial Science Perspective

A central role of geography among the sciences is best illustrated by its definition as the ***spatial science*** (the science of Earth space). No other discipline has the specific responsibility for investigating and attempting to explain the spatial aspects of Earth phenomena. Even though physical geographers may have many divergent interests, they share a common goal of understanding and explaining the spatial variation existing on Earth's surface.

How do physical geographers examine Earth from a spatial point of view? What are the spatial questions that physical geographers raise, and what are some of the problems they seek to understand and solve? From among the nearly unlimited number of topics available to physical geographers, we have chosen five to clearly illustrate the role of geography as the spatial science. In keeping with the quote from *Geography for Life*, that geography is about asking questions and solving problems, common study questions have been included for each topic.

Location Geographic knowledge and studies often begin with locational information. The location of a feature usually employs one of two methods: **absolute location,** which is expressed by a coordinate system (or address), or **relative location,** which identifies where a feature exists in relation to something else, usually a fairly well-known location. For example, Pikes Peak, in the Rocky Mountains of Colorado, with an elevation of 4302 meters (14,115 ft), has a location of latitude 38°51' north and longitude 105°03' west. A global address like this is an absolute location. However, another way to report its location would be to state that it is 36 kilometers (22 mi) west of Colorado Springs (• Fig. 1.6). This is an example of relative location (its position in relation to Colorado Springs). Typical spatial

GEOGRAPHY'S SPATIAL SCIENCE PERSPECTIVE

The Regional Concept: Natural and Environmental Regions

The term *region* is familiar to us all, but it has a precise meaning and special significance to geographers. Simply stated, a region is an area that is defined by a certain shared characteristic (or a set of characteristics) existing within its boundaries. Regions are spatial models, just as systems are operational models. Systems help us understand how things work, and regions help us make spatial sense of our world. The concept of a region is a tool for thinking about and analyzing logical divisions of areas based on their geographic characteristics. Just as it helps us to understand Earth by considering smaller parts of its overall system, dividing space into coherent regions helps us understand the arrangement and nature of areas on our planet. Regions can be described based on either human or natural characteristics, or a combination of the two.

Regions can also be divided into subregions. For example, North America is a region, but it can be subdivided into many subregions. Examples of subregions based on natural characteristics include the Atlantic Coastal Plain (similarity of landforms, geology, and locality), the Prairies (ecological type), the Sonoran Desert (climate type, ecological type, and locality), the Pacific Northwest (general locality), and Tornado Alley (region of high potential for these storms).

The regions that physical geographers are mainly interested in are based on *natural* and human–environmental characteristics. The term natural, as used here, means

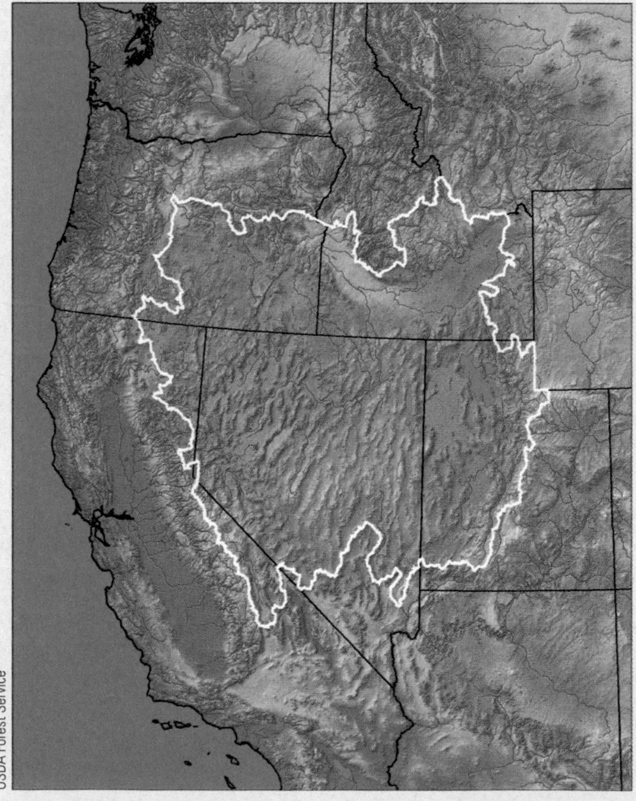

The Great Basin of the Western United States is a landform region that is clearly defined based on important physical geographic characteristic. No rivers flow to the ocean from this arid and semiarid region of mountains and topographic basins. The rivers and streams that exist flow into enclosed basins, where the water evaporates away from temporary lakes, or they flow into lakes like the Great Salt Lake, which has no outlet to the sea. Topographic features called drainage divides (mountain ridges) form the outer edges of the Great Basin, defining and enclosing this natural region. **Using topographic maps of the region, would it be relatively easy to outline the Great Basin?**

primarily related to natural processes and landscape features. However, we recognize that today human activities have an impact on virtually every natural process, and human–environmental regions offer significant opportunities for geographic analysis. Geographers not only study and explain regions, their locations, and their characteristics but also strive to delimit them—to outline their boundaries on a map. An unlimited number of regions can be derived for each of the four major Earth subsystems.

There are three important points to remember about natural and environmental regions. Each of these points has endless applications and adds considerably to the questions that the process of defining regions based on spatial characteristics seeks to answer.

- **Natural regions can change in size and shape over time in response to environmental changes.** These changes can be fast enough to observe as they occur, or so gradual that they require intensive study to detect. An example is desertification, the expansion of desert regions that has occurred in recent years in response to climatic change and human impacts on the land, such as overgrazing, which can form a desertlike landscape. Using images from space, we can see and monitor changes in the area covered by deserts, as well as other natural regions.
- **Boundaries separating different natural or environmental regions tend to be indistinct or transitional, rather than sharp.** For example, on a climate map, lines separating desert from nondesert regions do not imply that extremely arid conditions instantly appear when the line is crossed; rather, if we travel to a desert, it is likely to get progressively more arid as we approach our destination.
- **Regions are spatial models, devised by humans, for geographic analysis, study, and understanding.** Natural or environmental regions, like all regions, are conceptual models that are specifically designed to help us comprehend and organize spatial relationships and geographic distributions. Learning geography is an invitation to think spatially, and regions provide an essential, extremely useful, conceptual framework in that process.

Understanding regions, through an awareness of how areas can be divided into geographically logical units and why it is useful to do so, is essential in geography. Regions help us to understand, reason about, and make sense of, the spatial aspects of our world.

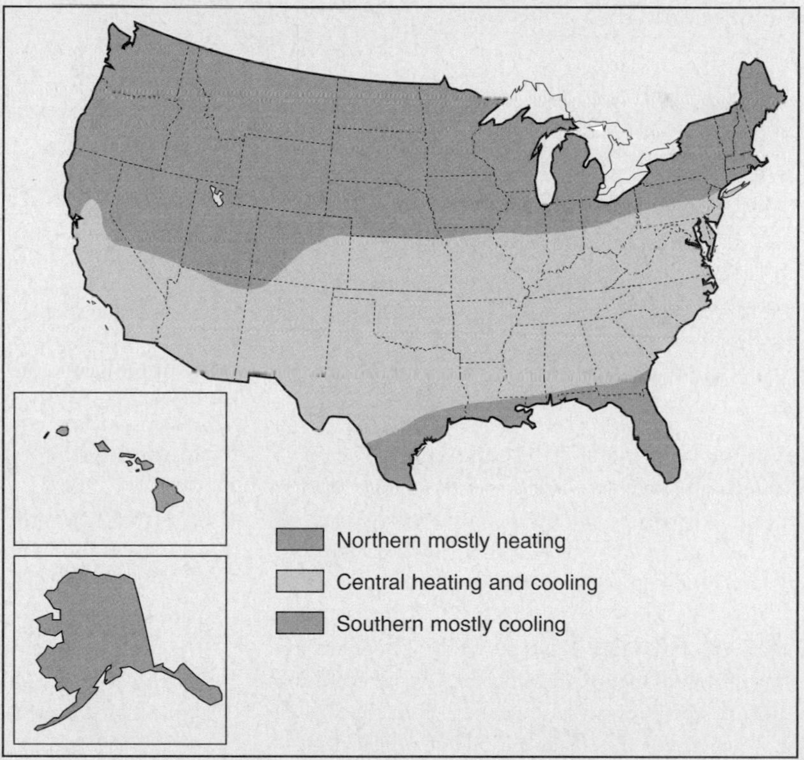

This human–environmental map divides the United States into three regions based on annual heating and cooling needs. Using spatial climatic data, the United States was divided into regions according to their similar home-heating or -cooling requirements. The reddish-brown means that heating is required more often than cooling. The tan region represents roughly equal heating and cooling needs. Blue represents a stronger or greater demand for cooling than for heating. The map is clearly related to climate regions. **Do you think that the boundaries between these regions are as sharply defined in reality as they are on this map? Can you recognize the spatial patterns that you see? Do the shapes of these regions, and the ways that they are related to each other, seem spatially logical?**

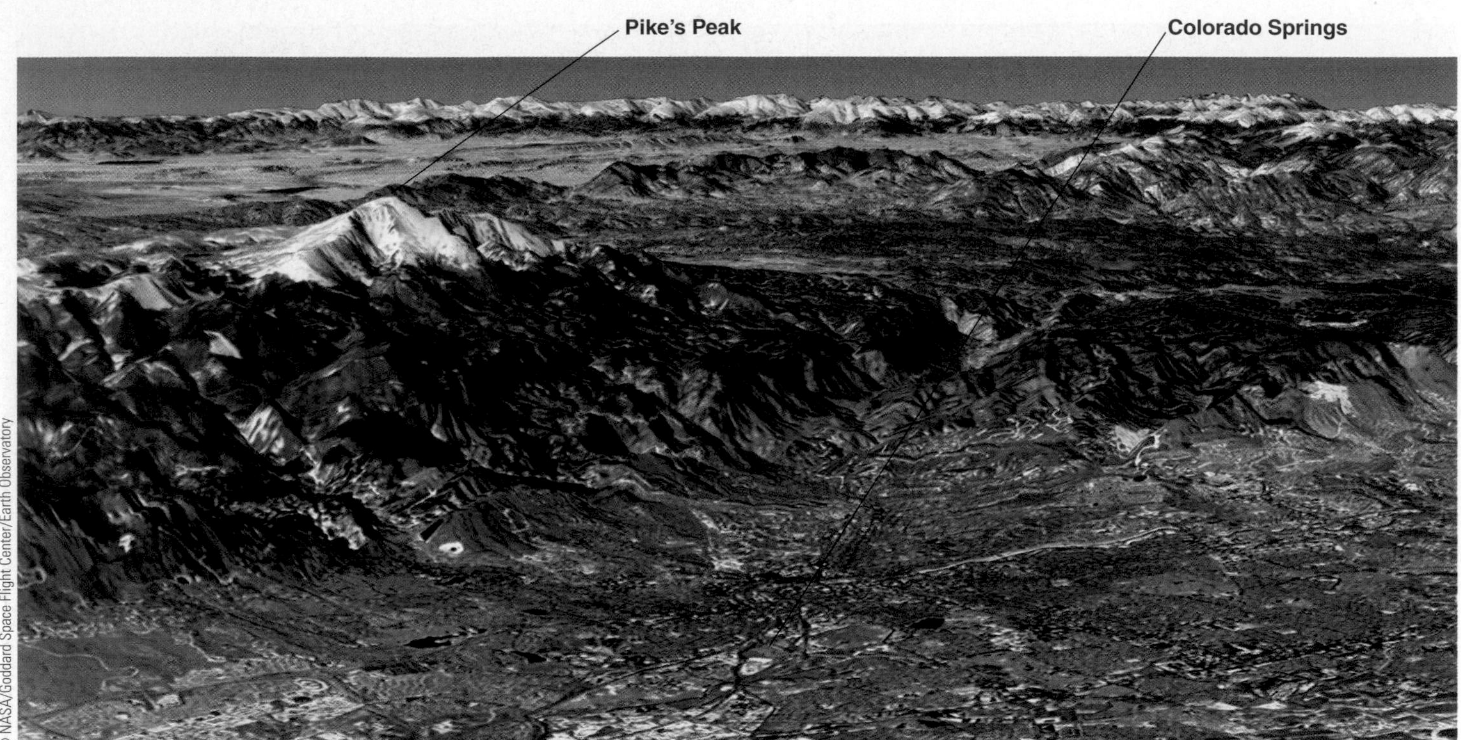

© NASA/Goddard Space Flight Center/Earth Observatory

• **FIGURE 1.6**
A three-dimensional digital model shows the relative location of Pikes Peak to Colorado Springs, Colorado. Because this is a perspective view, the 36-km (22-mi) distance appears to be shorter than its actual ground distance. A satellite image was merged with elevation data gathered by radar from the space shuttle to create this scene.
What can you learn about the physical geographic characteristics of this place from the image?

questions involving location include the following: *Where is a certain type of Earth feature found, and where is it not found? Why is a certain feature located where it is? What methods can we use to locate a feature on Earth? How can we describe its location? What is the most likely or least likely location for a certain Earth feature?*

Characteristics of Places Physical geographers are interested in the environmental features and processes that combine to make a place unique, and they are also interested in the shared characteristics between places. For example, what physical geographic features make the Rocky Mountains appear as they do? Further, how are the Appalachian Mountains different from the Rockies, and what characteristics are common to both of these mountain ranges? Another aspect of the characteristics of places is analyzing the environmental advantages and challenges that exist in a place. Other examples might include: *How does an Australian desert compare to the Sonoran Desert of the southwestern United States? How do the grasslands of the Great Plains of the United States compare to the grasslands of Argentina? What are the environmental conditions at a particular site? How do places on Earth vary in their environments, and why? In what ways are places unique, and in what ways do they share similar characteristics with other places?*

Spatial Distributions and Spatial Patterns

When studying how features are arranged in space, geographers are usually interested in two spatial factors. **Spatial distribution** means the extent of the area or areas where a feature exists. For example, where on Earth do we find the tropical rainforests? What is the distribution of rainfall in the United States on a particular day? Where on Earth do major earthquakes occur? **Spatial pattern** refers to the arrangement of features in space—are they regular or random, clustered together or widely spaced? The distribution of population can be either dense or sparse (• Fig. 1.7). The spatial pattern of earthquakes may be aligned on a map because earthquake faults display similar linear patterns. *Where are certain features abundant, and where are they rare? How are particular factors or elements of physical geography arranged in space, and what spatial patterns exist, if any? What processes are responsible for these distributions or patterns? If a spatial pattern exists, what does it signify?*

Spatial Interaction Few processes on Earth operate in isolation; areas on our planet are interconnected, which means linked to conditions elsewhere on Earth. A condition, an occurrence, or a process in one place generally has an impact on other places. Unfortunately, the exact nature of this **spatial interaction** is often difficult to establish with certainty except after years of study. A cause–effect relationship can often only be suspected because a direct relationship is often difficult to prove. It is much easier to observe that changes seem to be associated with each other, without knowing if one event causes the other or if this result is coincidental.

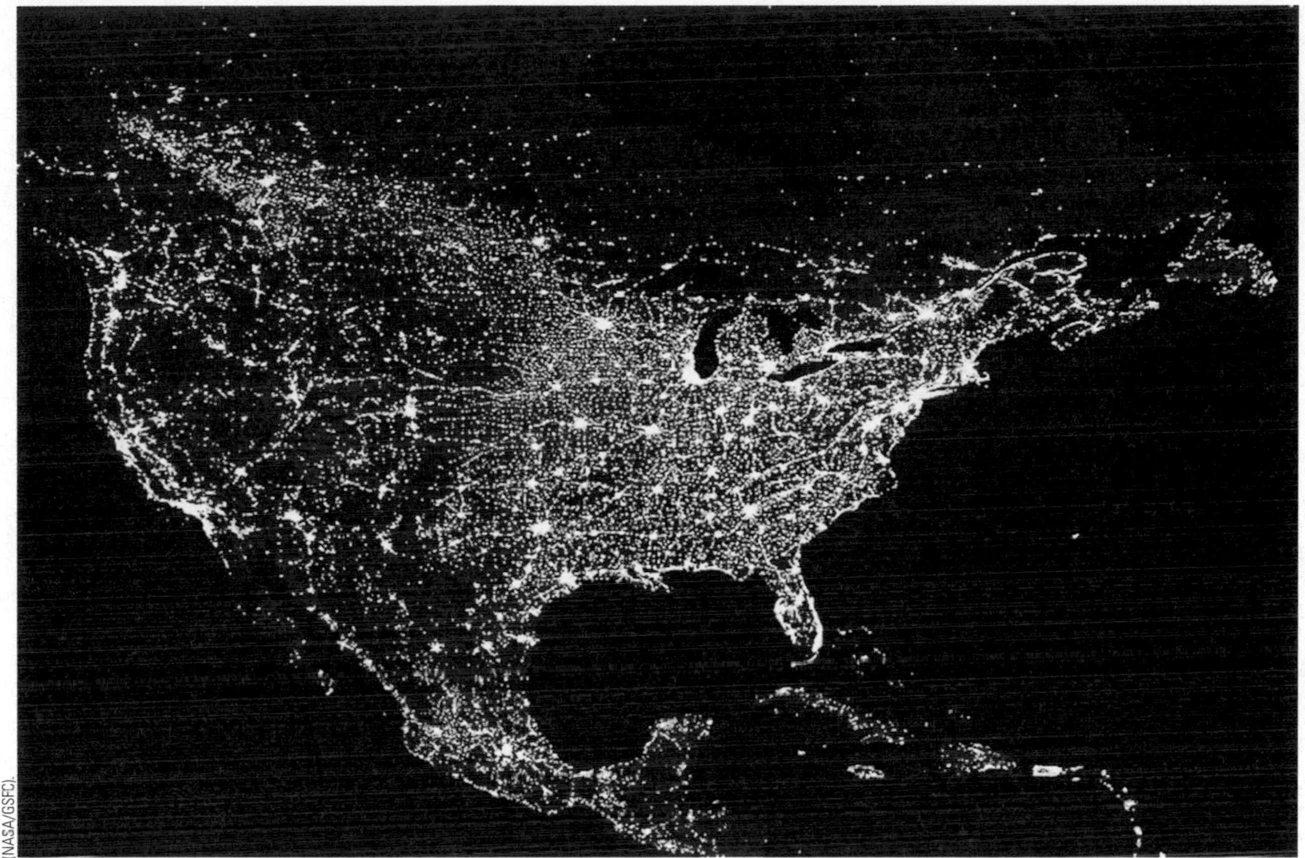

Data courtesy Marc Imhoff (NASA/GSFC) and Christopher Elvidge (NOAA/NGDC). Image by Craig Mayhew (NASA/GSFC) and Robert Simmon (NASA/GSFC).

• **FIGURE 1.7**
A nighttime satellite image provides several good illustrations of distribution and pattern, shown here on most of North America. Spatial distribution is where features are located (or perhaps, absent). Spatial pattern refers to their arrangement. Geographers seek to explain these spatial relationships.
Can you locate and propose possible explanations for two patterns and two distributions in this scene?

For example, the presence of abnormally warm ocean waters off South America's west coast, a condition called El Niño, seems to be related to unusual weather in other parts of the world. Clearing the tropical rainforest may have a widespread impact on world climates. Interconnections are one reason for considering interactions, impacts, and their potential links, at various scales from local, to regional, to global. *What are the relationships among places and features on Earth? How do they affect one another? What important interconnections link the oceans to the atmosphere and the atmosphere to the land surface?*

Ever-Changing Earth Earth's features and landscapes are continuously changing in a spatial context. Weather maps show where and how weather elements change from day to day, over the seasons, and from year to year. Storms, earthquakes, landslides, and stream processes modify the landscape. Coastlines may change position because of storm waves, tsunamis, or changes in sea level. Areas that were once forested have been clear-cut, changing the nature of the environment there. Vegetation and wildlife are becoming reestablished in areas that were devastated by recent volcanic eruptions or wildfires. Desertlike conditions seem to be expanding in many arid regions of the world. Volcanic islands have been created in historic times (• Fig. 1.8), and a new Hawaiian island is now forming beneath the waters of the Pacific Ocean.

World climates have changed throughout Earth's history, with attendant shifts in the distributions of plant and animal life. Today, changes in Earth's climates and environments are complicated by the impact of human activities. Earth and its environments are always changing, although at different time scales so the impact and direction of certain changes can be difficult to ascertain. *How are Earth features changing in ways that can be recorded in a spatial sense? What processes contribute to the change? What is the rate of change? Does change occur in a cycle? Can humans witness this change as it is taking place, or is a long-term study required to recognize the change? Do all places on Earth experience the same levels of change, or is there spatial variation?*

The previous five topics illustrate geography's strong emphasis on the spatial perspective. Learning the relevant questions to ask is the first step toward finding answers and explanations, and it is a major objective of your physical geography course.

Icelandic Ministry for the Environment

• **FIGURE 1.8**
Surtsey, Iceland, is an island in the north Atlantic that did not exist until about 45 years ago when undersea eruptions reached the ocean surface to form this new volcanic island. Since the 1960s when the volcanic eruptions stopped, erosion by waves and other processes have reduced the island by half of its original size.
Once the island formed and cooled, what other environmental changes should slowly begin to take place?

The Physical Science Perspective

As physical geographers apply their expertise to the study of Earth, they observe phenomena, compile data, and seek solutions to problems or the answers to questions that are also of interest to researchers in one or more of the other physical sciences. Physical geographers who specialize in climatology share many ideas and information with atmospheric physicists. Soil geographers study some of the same elements and compounds analyzed by chemists. Biogeographers are concerned about environments that support the same plants and animals that are classified by biologists. However, to whatever questions are raised and whatever problems require a solution, physical geographers bring unique points of view—a spatial perspective and a holistic approach that will carefully consider all Earth phenomena that may be involved. Physical geographers are concerned with the processes that affect Earth's physical environments at scales from global to regional to local. By examining the factors, features, and processes that influence the environment and learning how these elements work together, we can better understand our planet's ever-changing physical geography. We can also appreciate the importance of viewing Earth in its entirety as a constantly functioning system.

The Earth System A **system** is any entity that consists of interrelated parts or components, and the analysis of systems provides physical geographers with ideal opportunities to study these relationships as they affect Earth's features and environments. Earth certainly fits this definition because many continuously changing variables combine to make our home planet, the **Earth system,** function the way that it does. The individual components of a system, termed **variables,** are studied or grouped together because these variables interact with one another as parts of a functioning unit.

A change in one aspect of the Earth system affects other parts, and the impact of these changes can be significant enough to appear in regional or even worldwide patterns, clearly demonstrating the interconnections among these variables. For example, the presence of mountains influences the distribution of rainfall, and variations in rainfall affect the density, type, and variety of vegetation. Plants, moisture, and the underlying rock affect the kind of soil that forms in an area. Characteristics of vegetation and soils influence the runoff of water from the land, leading to completion of the circle, because the amount of runoff is a major factor in stream erosion, which eventually can reduce the height of mountains. Many cycles such as this operate

GEOGRAPHY'S PHYSICAL SCIENCE PERSPECTIVE

The Scientific Method

Science . . . is the systematic and organized inquiry into the natural world and its phenomena. Science is about gaining a deeper and often useful understanding of the world.

Multicultural History of Science web page, Vanderbilt University

The real purpose of the scientific method is to make sure nature hasn't misled you into thinking you know something you don't actually know.

Robert M. Pirsig,
Zen and the Art of Motorcycle Maintenance

Physical geography is a science that focuses on the Earth system, how its components and processes interact, and how and why aspects that affect Earth's surface are spatially arranged, as well as how humans and their environments are interrelated.

To wonder about your environment and attempt to understand it is a fundamental basis of human life. Increasing our awareness, satisfying our curiosity, learning how our world works, and determining how we can best function within it are all parts of a satisfying but never-ending quest for understanding. Without curiosity about the world, supported by making observations, noting relationships and patterns, and applying the knowledge discovered, humans would not have survived beyond their earliest beginnings. Science gives us a method for answering questions and testing ideas by examining evidence, drawing conclusions, and making new discoveries.

The sciences search for new knowledge using a strategy that minimizes the possibility of erroneous conclusions. This highly adaptable process is called the *scientific method.* It is a general framework for research, but it can accommodate an infinite number of topics and strategies for deriving conclusions. Although the scientific method is strongly associated with the physical sciences, it is applicable to nearly all fields of scientific research including studies that involve all three perspectives in physical geography—the physical, environmental, and spatial sciences.

Scientific method generally involves the following steps:

- **Making an observation that requires an explanation.** We may wonder if the observation represents a general pattern or is a "fluke" occurrence. For example, on a trip to the mountains, you notice that it gets colder as you go up in elevation. Is that just a result of conditions on the day you were there, or just the conditions at the location where you were, or is it a relationship that generally occurs everywhere?
- **Restating the observation as a hypothesis.** Here is an example: As we go higher in elevation, the temperature gets cooler (or, as a question, Does it get cooler as we go up in elevation?). The answer may seem obvious, yet it is generally but not always true, depending on environmental conditions that will be discussed in later chapters. Many scientists recommend a strategy called *multiple working hypotheses*, which means that we consider and test many possible hypotheses to discover which one best answers the questions while eliminating other possibilities.
- **Determining a technique for testing the hypothesis and collecting necessary data.** The next step is finding a technique for evaluating data (numerical information) and/or facts that concern that hypothesis. In our example, we would gather temperature and elevation data (taken at about the same time for all data points) for the area we are studying.
- **Applying the technique or strategy to test the validity of the hypothesis.** Here we discover if the hypothesis is supported by adequate evidence, collected under similar conditions to minimize bias. The technique will recommend either acceptance or rejection of the hypothesis.

If the hypothesis is rejected, we can test an alternate hypothesis or modify our existing one and try again, until we discover a hypothesis that is supported by the data. If the test supports the hypothesis, our observation is confirmed, at least for the location and environmental conditions in which our data and information were gathered.

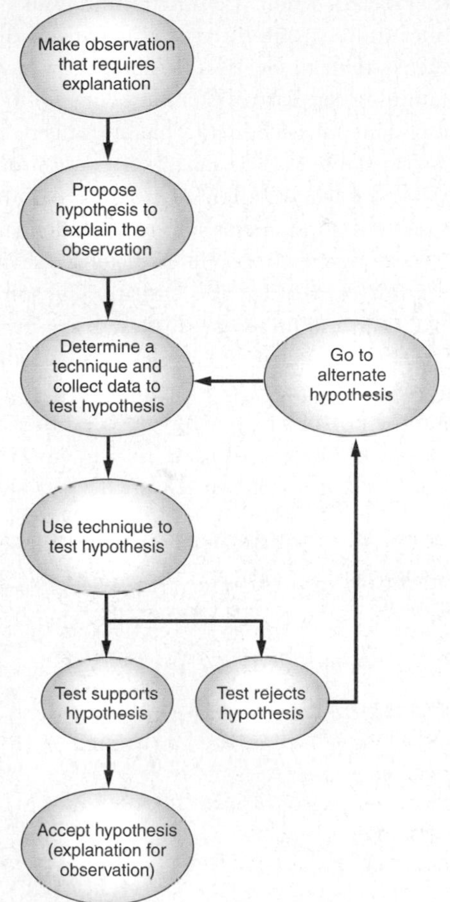

Steps in applying the scientific method.

After similar tests are conducted, if the hypothesis is supported in many places and under other conditions, then the hypothesis may become a theory. Theories are well-tested concepts or relationships that, given specified circumstances, can be used to explain and predict outcomes.

The processes of asking questions, seeking answers, and finding solutions through the scientific method have contributed greatly to human existence, our technologies, and our quality of life. Obviously, there are always more questions to be answered and problems yet to be solved. In fact, new findings typically yield new questions. Human curiosity, along with an intrinsic need for knowledge through observation and experience, has formed the basis for scientific method, an objective, structured approach that leads us toward the primary goal of physical geography—understanding how our world works.

to change our planet, but the Earth system is complex, and these cycles and processes operate at widely varying rates and over widely varying time spans.

Earth's Major Subsystems Systems can be divided into **subsystems**, which are functioning units of a major system and demonstrate strong internal connections (for example, a car has a fuel system, an electrical system, and a suspension system, etc.). Examining the Earth system as being composed of a set of interdependent subsystems is a major concept in understanding the physical sciences. The Earth system comprises four major subsystems (• Fig. 1.9). The **atmosphere** is the gaseous blanket of air that envelops, shields, and insulates Earth. The movements and processes of the atmosphere create the changing conditions that we know as weather and climate. The solid Earth—landforms, rocks, soils, and minerals—makes up the **lithosphere.** The waters of the Earth system—oceans, lakes, rivers, and glaciers—constitute the **hydrosphere.** The fourth major division, the **biosphere,** is composed of all living things: people, other animals, and plants.

It is the nature of these four major subsystems and the interactions among them that create and nurture the conditions necessary for life on Earth. For example, the hydrosphere provides the water supply for life on Earth, including humans, and provides a home environment for aquatic plants and animals. The hydrosphere directly affects the lithosphere as water moving in streams, waves, and currents shapes landforms. It also influences the atmosphere through evaporation, condensation, and the effects of ocean temperatures on climate. The impact and intensity of interactions among Earth's subsystems are not identical everywhere on our planet, and it is these variations that lead to the geographic patterns of environmental diversity.

Many other examples of overlap exist among these four major subsystems of Earth. Soil can be examined as part of the lithosphere, the biosphere, or the hydrosphere, because soils typically contain minerals, organisms, and water (and gases as well). The water stored in plants and animals is part of both the biosphere and the hydrosphere, and the water in clouds is a component of the atmosphere as well as the hydrosphere. The fact that we cannot draw sharp boundaries between these divisions underscores the interrelatedness among various components of the Earth system. However, like a machine, a computer, or the human body, planet Earth is a system that functions well only when all of its parts (and its subsystems) work together harmoniously.

All, Copyright and photograph by Dr. Parvinder S. Sethi; center inset, NASA

• **FIGURE 1.9**
Earth's four major subsystems. Studying Earth as a system is central to understanding changes in our planet's environments and adjusting to or dealing with these changes. Earth consists of many interconnected subsystems.
How do these systems overlap? For example, how does the atmosphere overlap with the hydrosphere, or the biosphere?

Earth Impacts We are aware that the Earth system is *dynamic*, responding to continuous changes in its four major subsystems, and that we can directly observe some of these changes—the seasons, the ocean tides, earthquakes, floods, volcanic eruptions. Other aspects of our planet may take years, or even more than a lifetime, to accumulate enough change so that humans can recognize their impact. Long-term changes in our planet are often difficult to understand or predict with certainty. The evidence must be carefully and scientifically studied to determine what is really occurring and what the potential consequences might be (• Fig. 1.10). Changes of this type include shifts in world climates, drought cycles, the spread of deserts, worldwide rise or fall in sea level, erosion of coastlines, and major changes in river systems. Yet understanding changes in our planet is critical to human existence. We are, after all, a part of the Earth system. Changes in the system may be naturally caused or human induced, or they may result from a combination of these factors. Today, much of the concern about environmental changes, such as the many potential impacts of global warming, centers on the increasing impact that human activities are exerting on Earth's natural systems. To understand our planet, therefore, we must learn about its components and the processes that operate to change or regulate the Earth system. Such knowledge is in the best interest for humankind as they interact with and influence Earth's natural systems, which form the habitat for all living things.

Top, W. C. Alden (USGS), Courtesy of Glacier National Park Archives; Bottom, Blase Reardon (USGS), Courtesy of Glacier National Park Archives

• **FIGURE 1.10**
Photographs taken 92 years apart in Montana's Glacier National Park show that Shepard Glacier, like other glaciers in the park, has dramatically receded during that time. This retreat is in response to climate warming and droughts, which have reduced the amount of snowfall that would form into glacial ice. The U.S. Geological Survey estimates that if this climatic trend continues, the glaciers in the park will disappear by 2030.
What other kinds of environmental change might require long-term observation and recording of evidence?

The Environmental Science Perspective

Today, we regularly hear talk about the environment and ecology and we are concerned about damage to ecosystems caused by human activity. We also hear news reports of disasters caused by humans being exposed to such violent natural processes as earthquakes, floods, tornadoes, or the terrible consequences of the South Asia tsunami in 2004 and Hurricane Katrina in 2005. Newspapers and magazines often devote entire sections to discussions of these and other environmental issues. But what are we really talking about when we use words like *environment, ecology,* or *ecosystem?* In the broadest sense, our **environment** can be defined as our surroundings; it is made up of all physical, social, and cultural aspects of our world that affect our growth, our health, and our way of living.

Environments are also systems because they function through the interrelationships among many variables. Environmental understanding involves giving consideration to a wide variety of factors, characteristics, and processes involving weather, climate, soils, rocks, terrain, plants, animals, water, humans, and how these factors interconnect and interact with each other to produce an environment. The holistic approach of physical geography is an advantage in this understanding, because the potential influence of each of these factors must be considered not only individually, but also in terms of how they affect one another as parts of an environmental system.

Human Impacts Physical geographers are keenly interested in environmental processes and interactions, and they give special attention to the relationships that involve humans and their activities. Much of human existence throughout time has been a product of the adaptations that various cultures have made and the modifications they have imposed on their natural surroundings. Primitive skills and technology generally require people to make greater adjustments in adapting to their environment. The more complex a culture's technology is, the greater the amount of environmental modification. Thus, human–environment interaction is a two-way relationship, with the environment influencing human behavior and humans impacting the environment. Today, meeting the needs of a growing population exerts an ever-increasing pressure on our planet's resources and environments.

Just as humans interact with their environment, so do other living things. The study of relationships between organisms, whether animal or plant, and their environments is a science known as **ecology.** Ecological relationships are complex but naturally balanced "webs of life." Altering the natural ecology of a community of organisms may have negative results (although this is not always so). For example, filling in or polluting coastal marshlands may disrupt the natural ecology of those wetlands. As a result, fish spawning grounds may be destroyed, and the food supply of some marine animals and migratory birds could be depleted. The end product of certain environmental impacts may be the destruction of valuable plant and animal life. Human activities will always affect the environment in some way, but if we understand the factors and processes involved, we can minimize the negative impacts.

The word *ecosystem* is a contraction of *ecological system*. An **ecosystem** is a community of organisms and the relationships of those organisms to each other and to their environment (• Fig. 1.11). An ecosystem is dynamic in that its various parts are always in flux. For instance, plants grow, rain falls, animals eat, and soils develop—all changing the environment of a particular ecosystem. Because each member of the ecosystem belongs to the environment of every other part of that system, a change in one alters the environment for the others. As those components react to the alteration, they in turn continue to transform the environment for the others. A change in the weather, for example, from sunshine to rain, affects plants, soils, and animals. Heavy rain, however, may carry away soils and plant nutrients so that plants may not be able to grow as well and animals, in turn, may not have as much to eat. In contrast, the addition of moisture to the soil may help some plants grow, increasing the amount of shade beneath them and thus keeping other plants from growing.

GEOGRAPHY'S ENVIRONMENTAL SCIENCE PERSPECTIVE

Human–Environment Interactions

As the world population has grown, the effects of human activities on the environment, as well as the impacts of environmental processes on humans, have become topics of increasing concern. There are many circumstances where human–environment relationships have been mutually beneficial, yet two negative aspects of those interactions have gained serious attention in recent years. Certain environmental processes, with little or no warning, can become dangerous to human life and property, and certain human activities threaten to cause major, and possibly irrevocable, damage to Earth environments.

Earth Impacts

The environment becomes a hazard to humans and other life forms when relatively uncommon and extraordinary natural events occur that are associated most directly with the atmosphere, hydrosphere, or lithosphere. Living under the conditions provided by these three subsystems, it is elements of the biosphere, including humans, that suffer the damaging consequences of sporadic natural events of extraordinary intensity. The routine processes of these three subsystems become a problem and spawn environmental hazards for two reasons. First, on occasion and often unpredictably, they operate in an unusually intense or violent fashion. Summer showers may become torrential rains that occur repeatedly for days or even weeks. Ordinary tropical storms gain momentum as they travel over warm ocean waters, and they reach coastlines as full-blown hurricanes, as Hurricane Katrina did in 2005. Molten rock and associated gases from deep beneath Earth move slowly toward the surface and suddenly trigger massive eruptions that literally blow apart volcanic mountains. The 2004 tsunami wave that devastated coastal areas along the Indian Ocean provided an example of the potential for the occasional occurrences of natural processes that far exceed our expectable "norm."

Each of these examples of Earth systems operating in sudden or extraordinary fashion is a noteworthy environmental event, but it does not become an environmental hazard unless people or their properties are affected. Thus, the second reason environmental hazards exist is because people live where potentially catastrophic environmental events may occur.

Why do people live where environmental hazards pose a major threat? Actually, there are many reasons. Some people have no choice. The land they live on could be their land by birthright; it was their family's land for generations. Especially in densely populated developing nations, there may be no other place to go. Other people choose to live in hazardous areas because they believe the advantages outweigh the potential for natural disaster. They are attracted by productive farmland, the natural beauty of a region or building site, or the economic possibilities available at a location. In addition, nearly every populated area of the world is associated with an environmental

USGS Western Coastal and Marine Geology

Environmental impacts on humans: a destructive tsunami. In December of 2004, a powerful undersea earthquake generated a large tsunami, which devastated many coastal areas along the Indian Ocean, particularly in Thailand, Sri Lanka, and Indonesia. Nearly a quarter of a million people were killed, and the homes of about 1.7 million people were destroyed. Here a huge barge was left onshore by the tsunami, which leveled buildings, and stripped the vegetation from the cliffs to a height of 31 meters (102 ft). Some natural-environmental processes, like this one, can be detrimental to humans and their built environment, and others are beneficial. **Can you cite some examples of natural processes that can affect the area where you live?**

hazard or perhaps several hazards. Forested regions are subject to fire; earthquake, landslide, and volcanic activities plague mountain regions; violent storms threaten interior plains; and many coastal regions experience periodic hurricanes or typhoons (the term used for hurricanes that strike Asia).

Human Impacts

Just as the environment can pose an ever-present danger to humans, through their activities, humans can constitute a serious threat to the environment. Issues such as global warming, acid precipitation, deforestation and the extinction of biological species in tropical areas, damage to the ozone layer of the atmosphere, and desertification have risen to the top of agendas when world leaders meet and international conferences are held. Environmental concerns are recurring subjects of magazine and newspaper articles, books, and television programs.

Much environmental damage has resulted from atmospheric pollution associated with industrialization, particularly in support of the wealthy, developed nations. But as population pressures mount and developing nations struggle to industrialize, human activities are exacting an increasing toll on the soils, forests, air, and waters of the developing world as well. Environmental deterioration is a problem of worldwide concern, and solutions must involve international cooperation in order to be successful. As citizens of the world's wealthiest nation, Americans must seriously consider what steps can be taken to counter major environmental threats related to human activities. What are the causes of these threats? Are the threats real and well documented? What can I personally do to help solve environmental problems? With limited resources on Earth, what will we leave for future generations?

Examining environmental issues from the physical geographer's perspective requires that characteristics of both the environment and the humans involved in those issues be given strong consideration. As will become apparent in this study of geography, physical environments are changing constantly, and all too frequently human activities result in negative environmental consequences. In addition, throughout Earth, humans live in constant threat from various and spatially distributed environmental hazards such as earthquake, fire, flood, and storm. The natural processes involved are directly related to the physical environment, but causes and solutions are imbedded in human–environment interactions that include the economic, political, and social characteristics of the cultures involved. The recognition that geography is a holistic discipline—that it includes the study of all phenomena on Earth—requires that physical geographers play a major role in the environmental sciences.

Both, United Nations Environmental Program (UNEP)

Human impacts on the environment: the shrinking Aral Sea. Located in the central Asian desert between Kazakhstan and Uzbekistan, the Aral Sea is an inland lake that does not have an outlet stream. The water that flows in is eventually lost by evaporation to the air. Before the 1960s, rivers flowing out of mountain regions supplied enough water to maintain what was the world's fourth-largest body of inland water. Since that time, diversion of river water for agriculture has caused the Aral Sea to dramatically shrink, and its salinity has increased by 600%. The result has been the disappearance of many species that relied on the lake for survival, along with frequent dust storms, and an economic disaster for the local economy. Without the waters of the lake to moderate temperatures, the winters have become colder, and the summers hotter. Today, efforts are under way to restore at least part of the lake and its environments.

What are some examples of how humans have impacted the environment where you live?

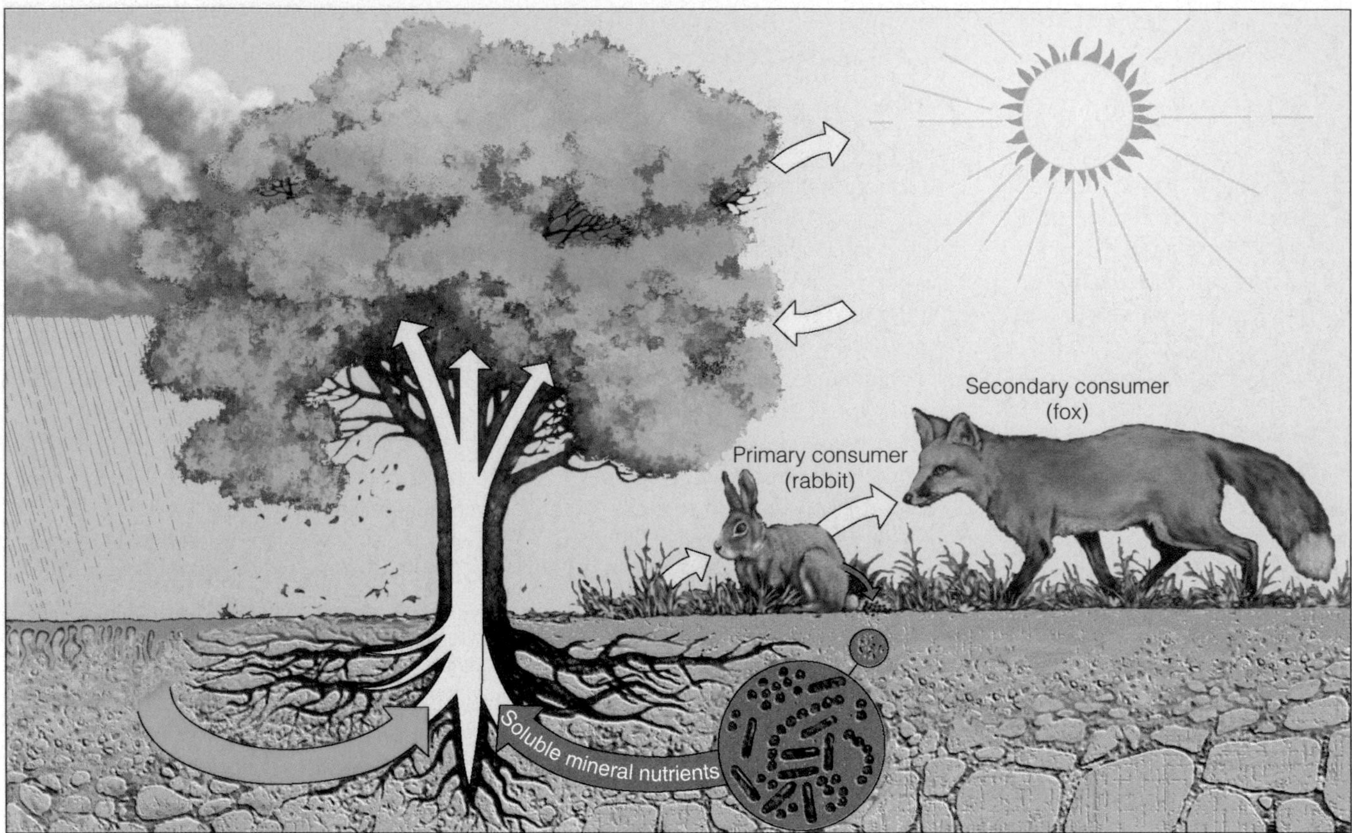

• FIGURE 1.11
Ecosystems are an important aspect of natural environments, which are affected by the interaction of many processes and components.
How do ecosystems illustrate the interactions in the environment?

The ecosystem concept (like other systems models) can be applied on almost any scale from local to global, in a wide variety of geographic locations, and under all environmental conditions that support life. Hence, your backyard, a farm pond, a grass-covered field, a marsh, a forest, or a portion of a desert can be viewed as an ecosystem. Ecosystems exist wherever there is an exchange of materials among living organisms and where there are functional relationships between the organisms and their natural surroundings. Although some ecosystems, such as a lake or a desert oasis, have relatively clear-cut boundaries, the limits of many others are not as precisely defined. Typically, the change from one ecosystem to another is obscure and transitional, occurring gradually over distance.

A Life-Support System

Certainly the most critical and unique attribute of Earth is that it is a **life-support system.** On Earth, natural processes produce an adequate supply of oxygen; the sun interacts with the atmosphere, oceans, and land to maintain tolerable temperatures; and photosynthesis or other continuous cycles of creation provide new food supplies for living things. If a critical part of a life-support system is significantly changed or fails to operate properly, living organisms may no longer be able to survive. Spacecraft can also provide a life-support system for astronauts, but they are dependent on Earth for sustenance, maintenance, and supplies of necessities (● Fig. 1.12). For instance, if all the oxygen in a spacecraft is used up, the crew inside will die. If a spacecraft cannot control the proper temperature range, its occupants may burn or freeze. If food supplies run out, the astronauts will starve. Other than the input of energy from the sun, the Earth system provides the necessary environmental constituents and conditions to permit life, as we know it, to exist.

Earth, then, is made up of a set of interrelated components, operating within systems that are vital and necessary for the existence of all living creatures. About 40 years ago, Buckminster Fuller, a distinguished scientist, philosopher, and inventor, coined the notion of Spaceship Earth—the idea that our planet is a life-support system, transporting us through space. Fuller also thought that knowing how Earth works is important—indeed this knowledge may be required for human survival—but that humans are only slowly learning the processes involved. He compared this information to an operating manual, like the owner's manual for an automobile.

> One of the most interesting things to me about our spaceship is that it is a mechanical vehicle, just as is an automobile. If you own an automobile, you realize that you must put oil and gas into it, and you must put water in the radiator and take care of the car as a whole. You know that you are going to have to keep the machine in good order or it's going to be in trouble and fail to function.
>
> We have not been seeing our Spaceship Earth as an integrally designed machine which to be persistently successful must be comprehended and serviced in total . . . there is one outstandingly important fact regarding Spaceship Earth, and that is that no instruction book came with it.
>
> R. Buckminster Fuller
> *Operating Manual for Spaceship Earth*

NASA

• **FIGURE 1.12**
The International Space Station can function as a life-support system and astronauts can venture out on a spacewalk, but they remain dependent on resources like air, food, and water that are shipped in from Earth.
What do the limited resources on space vehicles suggest about our environmental situation on Earth?

Today, we realize that critical parts of our planet's life-support system, **natural resources,** can be abused, wasted, or exhausted, potentially threatening the function of planet Earth as a human life-support system. A concern is that humans may be rapidly depleting critical natural resources, especially those needed for fuel. Many natural resources on our planet are nonrenewable, meaning that nature will not replace them once they are exhausted. Coal and oil are nonrenewable resources. When nonrenewable resources such as these mineral fuels are gone, the alternative resources may be less desirable or more expensive.

One type of abuse of Earth's resources is **pollution,** an undesirable or unhealthy contamination in an environment resulting from human activities (• Fig. 1.13). We are aware that critical resources, such as air, water, and even land areas, can be polluted to the point where they become unusable or even lethal to some life forms. By polluting the oceans, we may be killing off important fish species, perhaps allowing less desirable species to increase in number. Acid rain, caused by atmospheric

U.S. Environmental Protection Agency

• **FIGURE 1.13**
Pollution of the air, water, and land remains a significant environmental problem.
What pollutants threaten the air and water in your community and what are the probable sources of this pollution?

pollutants from industries and power plants, is damaging forests and killing fish in freshwater lakes. Air pollution has become a serious environmental problem for urban centers throughout the world (● Fig. 1.14). What some people do not realize, however, is that pollutants are often transported by winds and waterways hundreds or even thousands of kilometers from their source. Lead from automobile exhausts has been found in the ice of Antarctica, as has the insecticide DDT. Pollution is a worldwide problem that does not stop at political, or even continental, boundaries.

In modern times, the ability of humans to alter the landscape has been increasing. For example, a century ago the interconnected Kissimmee River–Lake Okeechobee–Everglades ecosystem constituted one of the most productive wetland regions on Earth. But marshlands and slow-moving water stood in the way of urban and agricultural development. Intricate systems of ditches and canals were built, and since 1900, half of the original 1.6 million hectares (4 million acres) of the Everglades has disappeared (● Fig. 1.15). The Kissimmee River was channelized into an arrow-straight ditch, and wetlands along the river were drained. Levees have prevented water in Lake Okeechobee from contributing water flow to the Everglades, and highway construction further disrupted the natural drainage patterns.

Fires have been more frequent and more destructive, and entire biotic communities have been eliminated by lowered water levels. During excessively wet periods, portions of the Everglades are deliberately flooded to prevent drainage canals from overflowing. As a result, animals drown and birds cannot rest and reproduce. South Florida's wading bird population has decreased by 95% in the last hundred years. Without the natural purifying effects of wetland systems, water quality in south Florida has deteriorated; with lower water levels, saltwater encroachment is a serious problem in coastal areas.

Today, backed by government agencies, scientists are struggling to restore south Florida's ailing ecosystems. There are extensive plans to allow the Kissimmee River to flow naturally across its former flood plain, to return agricultural land to wetlands, and to restore water-flow patterns through the Everglades. The problems of south Florida should serve as a useful lesson. Alterations of the natural environment should not be undertaken without serious consideration of all consequences.

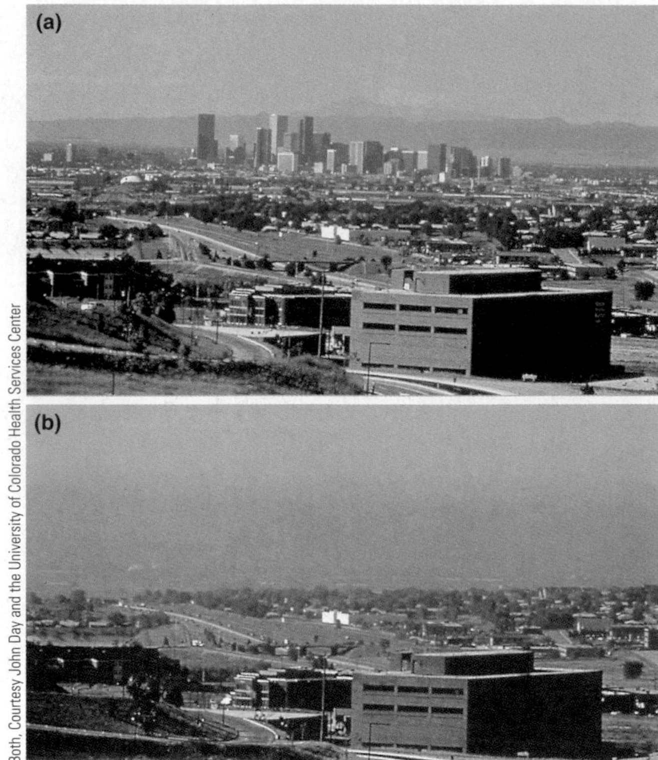

Both, Courtesy John Day and the University of Colorado Health Services Center

● **FIGURE 1.14**
(a) Denver, Colorado, on a clear day, with the Rocky Mountains visible in the background. (b) On a smoggy day from the same location, even the downtown buildings are not visible.
If you were choosing whether to live in a small town, a rural area, or a major city, would pollution affect your decision?

The Human–Environment Equation Despite the wealth of resources available from the air, water, soil, minerals, vegetation, and animal life on Earth, the capacity of our planet to support the growing numbers of humans may have an ultimate limit, a threshold population. Dangerous signs indicate that such a limit may someday be reached. The world population has passed the 6.7 billion mark, and United Nations' estimates indicate more than 9 billion people by 2050 if current growth rates continue. Today, more than half the world's people must tolerate substandard living conditions and insufficient food. A major problem today is the distribution of food supplies, but ultimately, over the long term, the size of the human population cannot exceed the environmental resources necessary to sustain them.

Although our current objective is to study physical geography, we should not ignore the information shown in the World Map of Population Density (inside textbook back cover). The map shows the distribution of people over the land areas of Earth and illustrates an important aspect of the human–environment equation. World population distributions are highly irregular; people have chosen to live and have multiplied rapidly in some places but not in others. One reason for this uneven distribution is the differing capacities of Earth's varied environments to support humans in large numbers. We are learning that, much like life on a spaceship, there are limits to the suitable living space on Earth, and we must use our lands wisely. Usable land is a limited resource (● Fig. 1.16). In our search for livable space, we occasionally construct buildings in locations that are not environmentally safe. Also, we sometimes plant crops in areas that are ill suited to agriculture while at the same time paving over prime farmland for other uses.

The relationships between humans and the environments in which they live will be emphasized throughout this book. Geographers are keenly aware that the nature or behavior of each of the parties in the relationship may have direct effects on the other. However, when considering the human–environment equation and the sustaining of acceptable human living standards for generations to come, it is important to note that environments do not change their nature to accommodate humans. Humans should make greater attempts to alter their behavior to accommodate the limitations and potentials of Earth environments. It has been said that humans are not passengers on Spaceship Earth; rather, they are the crew. This means we have the responsibility to maintain our own habitat. Poised at the interface between Earth and human existence, geography has much to offer in helping us understand

• FIGURE 1.15
(a) As a natural stream channel, the Kissimmee River originally meandered (flowed in broad, sweeping bends) on its floodplain for a 100-mile stretch from Lake Kissimmee to Lake Okeechobee. (b) In the 1960s and early 1970s, the river was artificially straightened, disrupting the previously existing ecosystem at the expense of plants, animals, and human water supplies. As part of a project to restore this habitat, the Kissimmee is today reestablishing its flood plain, wetland environments, and its meandering channel. (c) An ongoing problem is the invasion of weedy plants that cause a serious fire hazard during the dry season. Controlled burns by the U.S. Fish and Wildlife Department are necessary to avoid more catastrophic fires, and to help restore the natural vegetation.
What factors should be considered prior to any attempts to return rivers and wetland habitats to their original condition?

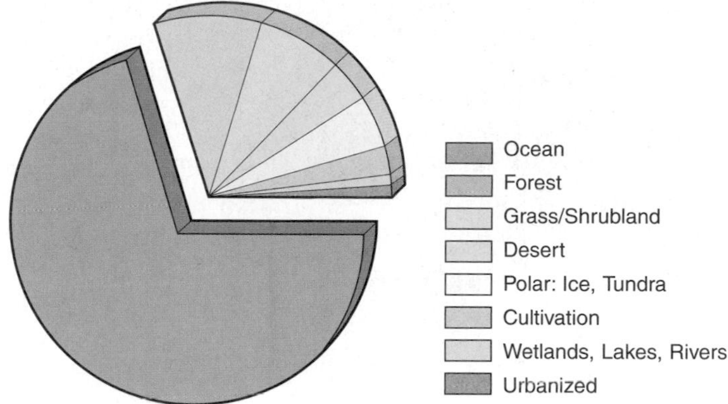

• FIGURE 1.16
The percentages of land and water areas on Earth. Habitable land is a limited resource on our planet.
What options do we have for future settlement of Earth's lands?

the factors involved in meeting this responsibility. Scientific studies directed toward environmental monitoring are helping us learn more about the changes on Earth's surface that are associated with human activities. All citizens of Earth must understand the impact of their actions on the complex environmental systems of our planet.

Models and Systems

As physical geographers work to describe, understand, and explain the often-complex features of planet Earth and its environments, they support these efforts, as other scientists do, by developing representations of the real world called models. A **model** is a useful simplification of a more complex reality that permits prediction, and each model is designed with a specific purpose in mind. As examples, maps and globes are models—representations that provide us with useful information required to meet specific needs. Models are simplified versions of what they depict, devised to convey the most important information about a feature or process without an overwhelming amount of detail. Models are essential to understanding and predicting the way that nature operates, and they vary greatly in their levels of complexity. Today, many models are computer generated because computers can handle great amounts of data and perform the mathematical calculations that are often necessary to construct and display certain types of models.

There are many kinds of models (• Fig. 1.17). **Physical models** are solid three-dimensional representations, such as a world globe or a replica of a mountain. **Pictorial/graphic models** include pictures, maps, graphs, diagrams, and drawings. **Mathematical/statistical models** are used to predict possibilities such as the flooding of rivers or changes in weather conditions that may result from climate change. Words, language, and the definitions of terms or ideas can also serve as models.

Another important type is a **conceptual model**—the mind imagery that we use for understanding our surroundings and experiences. Imagine for a minute (perhaps with your eyes closed) the image that the word *mountain* (or *waterfall, cloud, tornado, beach, forest, desert*) generates in your mind. Can you describe this feature's characteristics in detail? Most likely what you "see" (conceptualize) in your mind is sketchy rather than

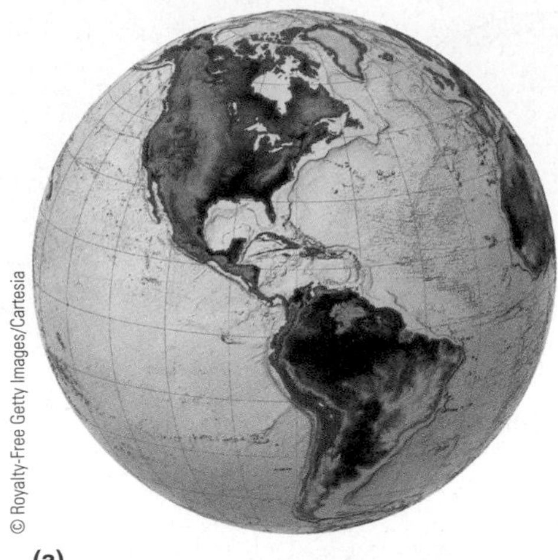
© Royalty-Free Getty Images/Cartesia
(a)

U.S. National Park Service
(b)

EPA, South Florida Water Management Division
(c)

• **FIGURE 1.17**
Models help us understand Earth and its subsystems by focusing our attention on major features or processes.
(a) Globes are physical models that demonstrate many terrestrial characteristics—planetary shape, configuration and distributions of landmasses and oceans, and spatial relationships.
(b) A digital landscape model of the Big Island of Hawaii shows the environment of Hawaii Volcanoes National Park. Computer-generated clouds, shadows, and reflections were added to provide "realism" to the scene. The terrain is faithfully rendered.
(c) This working physical model of the Kissimmee River in Florida was constructed to investigate ways to restore the river. Proposed modifications could be analyzed on this model before work was done on the actual river (see Fig. 1.15). A similar model exists of San Francisco Bay.

detailed, but enough information is there to convey a mental idea of a mountain. This image is a conceptual model. For geographers, a particularly important type of conceptual model is the **mental map**, which we use to think about places, travel routes, and the distribution of features in space. Psychologists have shown in many studies that maps, whether mental or pictorial, are very efficient in conveying a great amount of spatial information that the brain can recognize, store, and access. Try to think of other conceptual models that represent our planet's environments or one of its features. How could we even begin to understand our world without conceptual models, and in terms of spatial understanding, without mental maps?

Systems Theory

If you try to think about Earth in its entirety, or to understand how a part of the Earth system works, often you will discover that there are just too many factors to envision. Our planet is too complex to permit a single model to explain all of its environmental components and how they affect one another. Yet it is often said that to be responsible citizens of Earth, we should "think globally, but act locally." To begin to comprehend Earth as a whole or to understand most of its environmental components, physical geographers use a powerful strategy for analysis called **systems theory**. Systems theory suggests that the way to understand how anything works is to use the following strategy:

1. Clearly define the system that you are studying.
   What are the boundaries (limits) of the system?
2. Break the defined system down into its component parts (variables). The variables in a system are either matter or energy.
   What important parts and processes are involved in this system?
3. Attempt to understand how these variables are related to (or affect, react with, or impact) one another.
   How do the parts interact with one another to make the system work? What will happen in the system if a part changes?

The systems approach is a beneficial tool for studying any level of environmental conditions on Earth. Subsystems, the interacting divisions of the Earth system, are also important to consider. The atmosphere, hydrosphere, lithosphere, and biosphere each function as a subsystem of Earth. The human body is a system (• Fig. 1.18) that is composed of many subsystems (for example, the respiratory system, circulatory system, and digestive system). Subsystems can also be divided into subsystems, and so on.

Geographers often divide the Earth system into smaller subsystems in order to focus their attention on understanding a particular part of the whole. Examples of subsystems examined by physical geographers include the water cycle, climatic systems, storm systems, stream systems, the systematic heating of the atmosphere, and ecosystems. A great advantage of systems analysis is that it can be applied to environments at virtually any spatial scale, from global to microscopic.

How Systems Work

Basically, the world "works" by the movement (or transfer) of matter and energy and the processes involved with these transfers. For example, as shown in • Figure 1.19, sunlight (*energy*) warms (*process*)

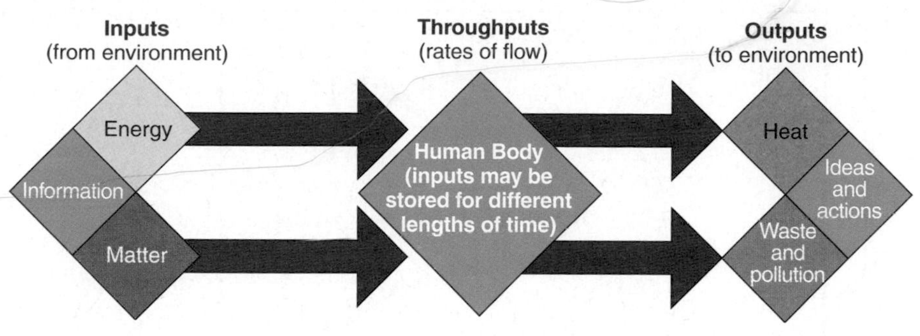

● **FIGURE 1.18**
The human body is an example of a system, with inputs of energy and matter.
What characteristics of the human body as a system are similar to the Earth as a system?

a body of water (*matter*), and the water evaporates (*process*) into the atmosphere. Later, the water condenses (*process*) back into a liquid, and the rain (*matter*) falls (*process*) on the land and runs off (*process*) downslope back to the sea. In a systems model, geographers can trace the movement of energy or matter into the system (**inputs**), their storage in the system and their movements out of the system (**outputs**), as well as the interactions between components within the system.

A **closed system** is one in which no substantial amount of *matter* crosses its boundaries, although *energy* can go in and out of a closed system (● Fig. 1.20). Planet Earth, or the Earth system as a whole, is essentially a closed system. Except for meteorites that reach Earth's surface, the escape of gas molecules or spacecraft from the atmosphere, and a few moon rocks brought back by astronauts, the Earth system is essentially closed to the input or output of matter. The hydrosphere is another good example of a closed system. Water may exist in the system in all three of its states—liquid, gas, or solid ice—and may be transformed from one state to another many times, but there is virtually no gain or loss of water (no output of matter) in the system.

Most Earth subsystems, however, are **open systems** because both energy and matter move freely across subsystem boundaries as inputs and outputs. A stream is an excellent illustration of an open subsystem, in which matter and energy in the form of soil particles, rock fragments, solar energy, and precipitation enter the stream while heat energy dissipates into the atmosphere and the stream bed. Water and sediments leave the stream where it empties into the ocean or some other standing body of water, and precipitation provides an input of water to the stream system.

When we describe Earth as a system or as a complex set of interrelated systems, we are using models to help us organize our thinking about what we are observing. Models also assist us in explaining the processes involved in changing, maintaining, or regulating our planet's life-support systems. Throughout the chapters that follow, we will use the systems concept, as well as many other kinds of models, to help us simplify and illustrate complex features of the physical environment.

Equilibrium in Earth Systems

The parts, or variables, of a system have a tendency to reach a balance with one another and with the external factors that influence that system. If the inputs entering the system are balanced by

● **FIGURE 1.19**
An example of environmental interactions: energy, matter, process. Being aware of energy and matter and the interactive processes that link them is an important part of understanding how environmental systems operate.
Can you think of another environmental system and break it down into its components of energy, matter, and process?

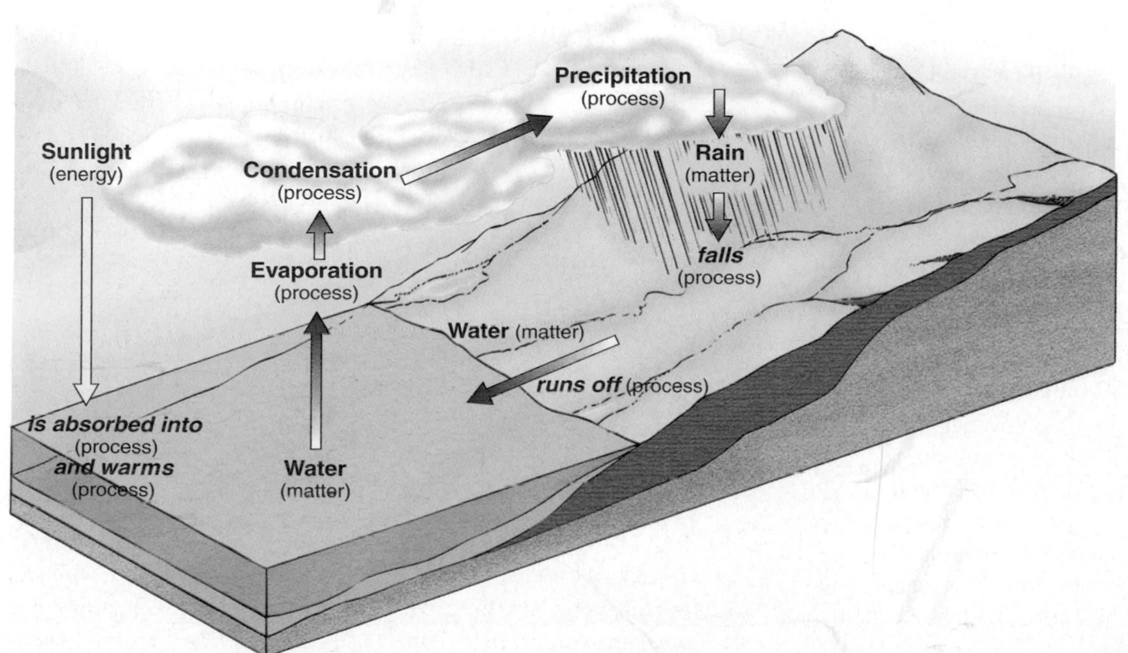

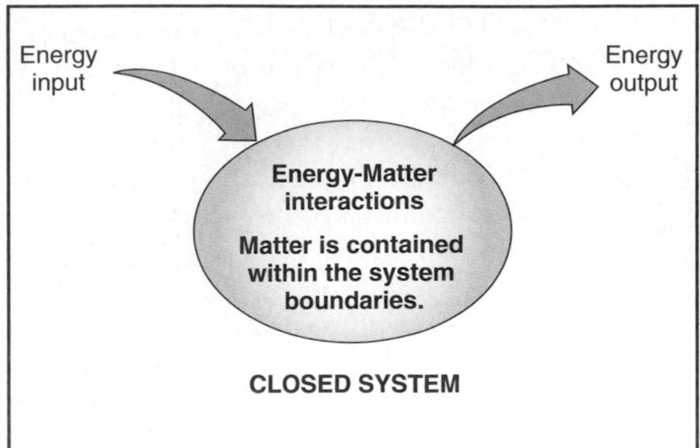

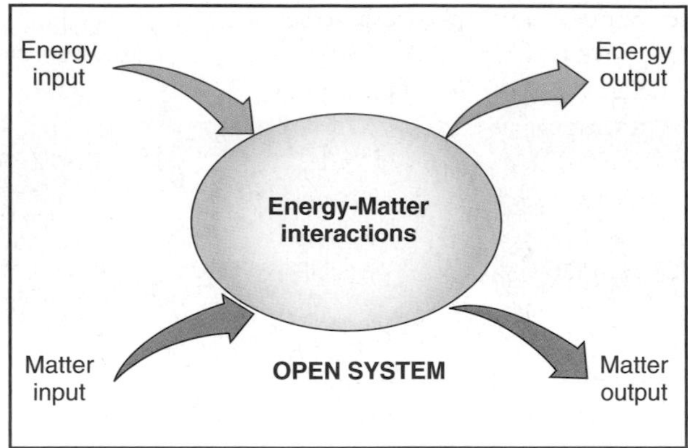

• FIGURE 1.20

Closed systems allow only energy to pass in and out. Open systems involve the inputs and outputs of both energy and matter. Earth is basically a closed system. Solar energy (input) enters the Earth system, and that energy is dissipated (output) to space mainly as heat. External inputs of matter are virtually nil, mainly meteorites and moon rock samples. Except for outgoing space vehicles, equipment, or space "junk," virtually no matter is output from the Earth system. Because Earth is a closed system, humans face limits to their available natural resources. Most subsystems on the planet, however, are open systems, with incoming and outgoing matter and energy. Processes are driven by energy.

Think of an example of an open system, and outline some of the matter–energy inputs and outputs involved in such a system.

outputs, the system is said to have reached a state of **equilibrium.** Most natural systems have a tendency toward stability (equilibrium) regarding environmental systems, and we often hear this called the "balance of nature." What this means is that natural systems have built-in mechanisms that tend to counterbalance, or accommodate, change without changing the system dramatically. Animal populations—deer, for example—will adjust naturally to the food supply of their habitats. If the vegetation on which they browse is sparse because of drought, fire, overpopulation, or human impact, deer may starve, reducing the population. The smaller deer population may enable the vegetation to recover, and in the next season the deer may increase in numbers. Most systems are continually shifting slightly one way or another as a reaction to external conditions. This change within a range of tolerance is called **dynamic equilibrium;** that is, a balance exists but maintaining it requires adjustment to changing conditions, much as tightrope walkers sway back and forth and move their hands up and down to keep their balance. Dynamic equilibrium, however, also means that the balance is not static but in the long term changes may be accumulating. A reservoir contained by a dam is a good example of equilibrium in a system (● Fig. 1.21).

The interactions that cause change or adjustment between parts of a system are called **feedback.** Two kinds of feedback relationships operate in a system. **Negative feedback,** whereby one change tends to offset another, creates a natural counteracting effect that is generally beneficial because it tends to help the system maintain equilibrium (an inverse relationship). Earth subsystems can also exhibit **positive feedback** sequences for a while—that is, changes that reinforce the direction of an initial change (a direct relationship). For example, several times in the past 2 million years, Earth has experienced significant decreases in global temperatures. This cooling of the atmospheric system led to the growth of great ice sheets, glaciers that covered large portions of Earth's surface. The massive ice sheets increased the amount of solar energy that was reflected back to space from Earth's surface, thus increasing the cooling trend and the further growth of the glaciers. The result over a considerable period of time was positive feedback. But ultimately the climate got so cold that evaporation from the oceans decreased and the cover of sea ice expanded, cutting off the supply of moisture to storms that fed snow to the glaciers. The reduction of moisture is an example of what is called a **threshold,** a condition

• FIGURE 1.21

A reservoir serves as an example of dynamic equilibrium in systems. The amount of water coming in may increase or decrease over time, but it must equal the water going out, or the level of the lake will rise or fall. If the input–output balance is not maintained, the lake will get larger or smaller as the reservoir system adjusts to hold more or less water in storage. A state of equilibrium (balance) will always exist between inputs, outputs, and storage in the system.

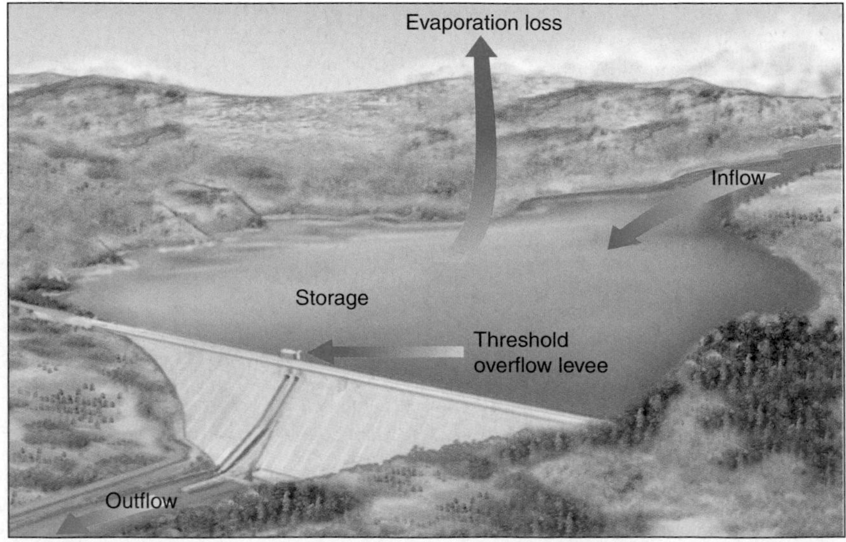

that causes a system to change dramatically, in this case bringing the positive feedback to a halt. The decrease in snowfall caused the glaciers to shrink and the climate began to warm, thus beginning another cycle.

Thresholds are conditions that, if reached or exceeded (or not met), can cause a fundamental change in a system and the way that it behaves. For example, earthquakes will not occur until the built-up stress reaches a threshold level that overcomes the strength of the rocks to resist breaking. Thresholds are common regulators of systems processes. As another example, fertilizing a plant will help it to grow larger and faster. But if more and more fertilizer is added, will this positive feedback relationship continue forever? Too much fertilizer may actually poison the plant and cause it to die. Either exceeding or not meeting certain critical conditions (thresholds) can change a system dramatically. With environmental systems, an important question that we often try to answer is how much change a system can tolerate without becoming drastically or irreversibly altered, particularly if the change has negative consequences.

To further illustrate how feedback works, let's consider a simplified example—a hypothetical scenario of what might happen if human-caused damage to the atmosphere's ozone layer continues unimpeded by human counteraction. ● Figure 1.22 shows a **feedback loop**—a circular set of feedback operations that can be repeated as a cycle. Generally in natural systems, the overall result of a feedback loop is negative feedback because the sequence of changes serves to counteract the direction of change in the initial element. The example is intended to show you how to think about Earth processes and interactions functioning as a system. First, however, we must start with some facts:

1. We know that the ozone layer in the upper atmosphere protects us by blocking harmful ultraviolet (UV) radiation from space, radiation that could otherwise cause harmful skin cancers and cell mutations.
2. We also know that chlorofluorocarbons (CFCs), and some related chemicals that have been widely used in air conditioners, can migrate to the upper atmosphere and cause chemical reactions that destroy ozone.

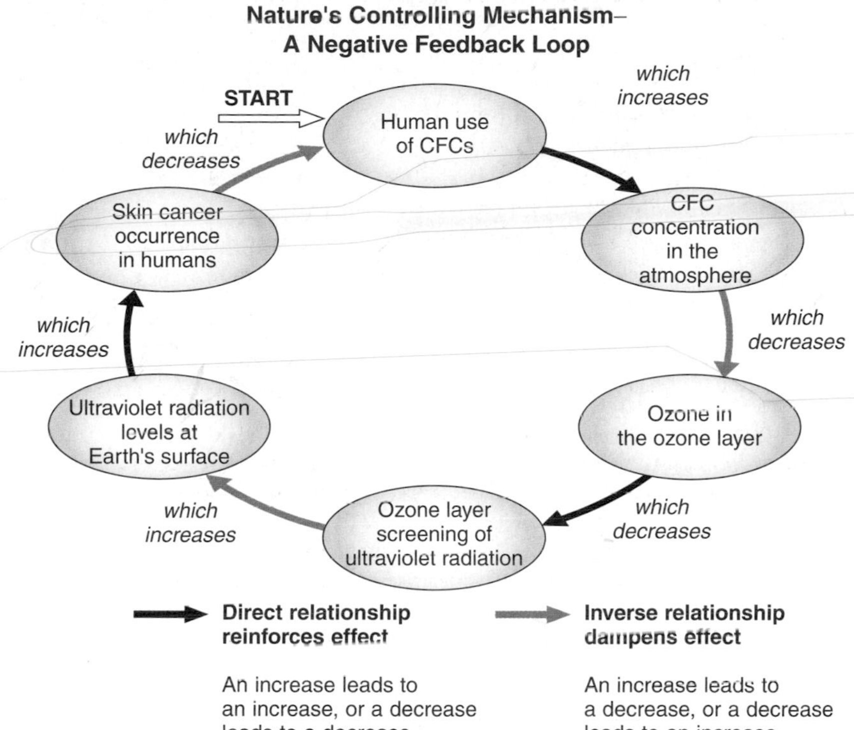

● **FIGURE 1.22**
A negative feedback loop: nature's controlling mechanism. The ozone layer absorbs UV radiation from the sun. If ozone diminishes, more UV radiation will reach the surface. A feedback loop illustrates how negative feedback adds stability to a system. Relationships between two variables (one link to the next in the loop) can be either direct or inverse. A direct relationship means that either an increase or a decrease in the first variable will lead to the same effect on the next. For example, a decrease in ozone leads to a decrease in ozone screening of UV radiation. An inverse relationship means that the change in the first variable will result in an opposite change in the next. For example, an increase in CFCs leads to a decrease in ozone in the ozone layer. After one pass through a negative feedback loop, a shift will occur: the effect on the first variable reverses, thus reversing all subsequent changes in the next cycle. The variables maintain the same relationships, either direct or inverse. Follow a second pass through the feedback loop (reversing the increase or decrease interactions) to understand how this works. Human decision making can play an important role in environmental systems. The last link between skin cancer and human use of CFCs would likely result in people taking actions to reduce the problem.
What might be the potential (and extreme) alternative resulting from a lack of corrective action by humans?

Knowing these facts, keep in mind that this systems example is simplified, and presents an extreme scenario. In fact, strong efforts have been undertaken in the last 25 years or so to minimize or eliminate the use of CFCs in the United States and internationally. Today, new automobiles and trucks are sold with air conditioners that use an "ozone-friendly," non-CFC unit to cool the vehicles' interiors. However, because the replacement refrigerant used in many of these units forms a gas that can contribute to global warming, research and development efforts continue to seek a more benign alternative.

Systems analysis allows us to see how these processes will affect the variables and helps us answer "what if?" questions. For example, if CFCs continue to deplete the ozone layer, what will happen? The feedback loop in Figure 1.22 shows six of the most important factors related to ozone-layer damage by CFCs. Each of these factors is linked by a feedback interaction to the next variable in the loop.

Follow Figure 1.22, starting with the human use of CFCs at the top of the diagram, and trace the feedback links.

1. If the amount of CFCs used by humans increases, the amount of CFCs in the atmosphere will also increase. An increase leads to an increase in the next factor, so this is a *direct* (positive) relationship.
2. *Increasing* the CFCs in the atmosphere will lead to a *decrease* of ozone in the ozone layer. Here an increase leads to a decrease in the next factor, so this is an *inverse* (negative) relationship between atmospheric CFCs and ozone.

3. *Decreasing* the ozone in the upper atmosphere will *decrease* the amount of harmful ultraviolet (UV) radiation that is blocked by the ozone layer. Here a decrease leads to a decrease; this is a *direct* relationship because the decreasing effect is reinforced.
4. *Decreasing* the blocking of harmful UV radiation will cause an *increased* amount of harmful UV radiation at Earth's surface. A decrease leads to an increase, so this is an *inverse* relationship.
5. *Increasing* the level of UV radiation at Earth's surface will cause an *increased* amount of skin cancer in humans, which can be fatal. An increase leads to an increase, so this is a *direct* relationship.
6. *Increasing* skin cancer in humans could lead to policy changes that *decrease* the release of CFCs into the atmosphere, producing negative feedback relative to the initial variable (item 1 above) in the feedback loop.

Finally, some important questions remain: What is likely to happen to the human use of CFCs if the occurrence of skin cancer continues to increase? Will humans act to correct the problem, or not? What would be the potential outcome in each case? Ironically, negative feedback loop operations are beneficial to the environment because they regulate a system through a tendency toward balance. Feedback loops in nature normally do not operate for extended periods on positive feedback because environmental limiting factors (thresholds) act to return the process to a state of equilibrium. What are some other examples of feedback operations in natural systems?

It is essential to remember that systems are models, and so they are not the same as reality. They are products of the human mind and are only one way of looking at the real world. Examining various Earth subsystems helps us understand the natural processes involved in the development of the atmosphere, lithosphere, hydrosphere, and biosphere. Models may even help us simulate past events or predict future change. But we must be careful not to confuse simplified models with the complexities of the real world.

Physical Geography and You

The characteristics of the physical environment affect our everyday lives. The principles, processes, and perspectives of physical geography provide keys that help us be environmentally aware, assess environmental situations, analyze the factors involved, and make informed choices among possible courses of action.

What are the environmental advantages and disadvantages of a particular home site? Should you plant a new lawn before or after the spring rains? What sort of environmental impacts might be expected from a proposed shopping center? What potential impacts of natural hazards—flooding, landslides, earthquakes, hurricanes, and tornadoes—should you be aware of where you live? What can you do to minimize potential damage to your household from a natural hazard? What can you do to ensure that both you and your family are as prepared as possible for the kind of natural hazard that might affect your home?

It is apparent, then, that the study of physical geography and the understanding of the natural environment that it provides are valuable to all of us. Perhaps you have wondered, however, what do people with a background in physical geography do in the workplace? What kinds of jobs do they hold? Physical geography sounds interesting and exciting, but can I make a living at it?

By applying their knowledge, skills, and techniques to real-world problems, physical geographers make major contributions to human well-being and to environmental stewardship. Physical geographers emphasize the Earth system, but also consider the effect of people on that system or the impact that an environment may have on people and the way they live. A knowledge of physical geography can help us analyze and solve environmental problems, such as whether we should continue to build nuclear power plants, allow offshore oil development, or drain coastal marshlands. Each of these questions may generate a different answer depending on the physical geography of the location in question. A recent publication about geography-related jobs by the U.S. Department of Labor stated that people in any career field that deals with maps, location, spatial data, or the environment would benefit from an educational background in geography.

Finally, knowledge of physical geography provides not only opportunities for personal enrichment and possible employment but also a source of perpetual enjoyment. Geography is a visual science, and it is really more than just a subject. Geography is a way of looking at the world and of observing its features. It involves asking questions about the nature of those features as well as appreciating their beauty and complexity. It encourages you to seek explanations, gather information, and use geographic skills, tools, and knowledge to solve problems. Even if you forget many of the facts discussed in this book, you will have been shown new ways to consider, see, and evaluate the world around you. Just as you see a painting differently after an art course, so too will you see sunsets, waves, storms, deserts, valleys, rivers, forests, prairies, and mountains with an "educated eye." You should retain knowledge of geography for life. You will see greater variety in the landscape because you will have been trained to observe Earth differently, with greater awareness and with a deeper understanding.

Chapter 1 Activities

Define & Recall

geography
spatial science
holistic approach
human geography
region
regional geography
physical geography
absolute location
relative location

spatial distribution
spatial pattern
spatial interaction
system
Earth system
variable
subsystem
atmosphere
lithosphere
hydrosphere
biosphere
environment
ecology
ecosystem
life-support system
natural resource
pollution
model
physical model
pictorial/graphic model
mathematical/statistical model
conceptual model
mental map
systems theory
input
output
closed system
open system
equilibrium
dynamic equilibrium
feedback
negative feedback
positive feedback
threshold
feedback loop

Discuss & Review

1. Why can geography be considered both a physical and a social science? What are some of the subfields of physical geography, and what do geographers study in those areas of specialization?
2. Why is geography known as the spatial science? What are some topics that illustrate the role of geography as the spatial science?
3. What does a holistic approach mean in terms of thinking about an environmental problem?
4. How do physical geography's three major perspectives make it unique among the sciences?
5. What are the four major divisions of the Earth system, and how do the divisions interact with one another?
6. How does the study of systems relate to the role of geography as a physical science?
7. How does the examination of human–environment relationships in ecosystems serve to illustrate the role of geography as an environmental science?
8. What is meant by the human–environment equation? Why is the equation falling further out of balance?
9. How do open and closed systems differ? How does feedback affect the dynamic equilibrium of a system?
10. How does negative feedback maintain a tendency toward balance in a system? What is a threshold in a system?

Consider & Respond

1. Give examples from your local area that demonstrate each of the five topics listed concerning spatial science.
2. List some potential sources of pollution in your city or town. Could these kinds of pollution affect your life? What are some potential solutions to this problem?
3. Give one example of an ecosystem in your local area that has been affected by human activity. In your opinion, was the change good or bad? What values are you using in making such a judgment?
4. There are advantages and disadvantages to the use of models and the study of systems by scientists. List and compare the advantages and disadvantages from the point of view of a physical geographer.
5. How can a knowledge of physical geography be of value to you now and in the future? What steps should you take if you wish to seek employment as a physical geographer? What advantages might you have when applying for a job?

Apply & Learn

1. From memory, draw on paper the "mental map" that you envision when you think about the geography of your local natural environment, or neighborhood. Try to maintain some reasonable geographic (spatial) relationships for the sizes of areas and distances between features.
2. Draw a simple, circular feedback loop to illustrate the interactions between components and processes involved in some system that you are familiar with. Label the positive (direct) and negative (inverse) feedback relationships. What threshold conditions exist and how do they enact change?

Representations of Earth

2

CHAPTER PREVIEW

Maps and other graphic representations of Earth are essential to understanding geography.

- In what ways are maps useful in daily life or in the workplace, and why are they essential to geographers?
- Why are maps an important means of communication?

The Earth's shape is generally referred to as spherical.

- Why does Earth's shape deviate from a perfect sphere and by how much?
- A globe is the only way to represent the entire Earth without distortion, so what does this imply about maps?

The geographic grid is a coordinate system for describing and finding locations on Earth.

- How are latitude and longitude associated with navigation and time zones?
- How is the Public Lands Survey System different from and similar to the geographic grid?

Maps, remote sensing, and the global positioning system (GPS) are useful tools for physical geographers.

- How can maps, aerial photographs, and remotely sensed images provide complementary information about a place?
- What advantages do digital images offer to understanding the environment?
- How does the GPS operate to find a location?

Geographic information system (GIS) technology allows the direct comparison and combination of many map information layers.

- Why is it useful to compare the locations and distributions of two or more environmental variables?
- How does a GIS use different map layers to help us understand spatial relationships in environmental systems?

◀ Opposite: The San Francisco Bay Area in a digital "false-color" satellite image of visible and near-infrared light. Healthy vegetation appears red. This image is similar to those taken by a digital camera. The inset is an enlargement of the airport and shows the pixels that make up the image.
NASA/GSFC/METI/ERSDAC/JAROS, and U.S./Japan ASTER Science Team

Knowing where certain features are located and being able to convey that information to others is essential to describing and analyzing aspects of the Earth system. Many of the principles that are used in dealing with locational problems have been known for centuries, but the technologies applied to these tasks are rapidly improving and changing. Computer-assisted and space-age technologies are now widely used for locating, describing, storing, and accessing spatial data. Digital technology has greatly increased the abilities of physical geographers and other scientists to analyze vast amounts of data relatively quickly. Today, computer systems can be used to generate high-quality maps and three-dimensional (3-D) displays of Earth system features that would have been nearly impossible or extremely time-consuming to produce a decade or two ago. Many geographers use these technologies to help them understand environmental concerns, and they also work to improve the capabilities of computer systems to analyze and display spatial information.

Location on Earth

Perhaps as soon as people began to communicate with each other, they also began to develop a language of location, using landscape features as directional cues. Today, we still use landmarks to help us find our way. When ancient peoples began to sail the oceans, they recognized the need for ways of finding directions and describing locations. Long before the first compass was developed, humans understood that the positions of the sun and the stars—rising, setting, or circling in the sky—could provide accurate locational information. Observing relationships between the sun and the stars to find a position on Earth is a basic skill in **navigation,** the science of location and wayfinding. Navigation is basically the process of getting from where you are to where you want to go.

Maps and Mapmaking

The first maps were probably made by early humans who drew locational diagrams on rocks or in the soil. Ancient maps were fundamental to the beginnings of geography as they helped humans communicate spatial thinking and were useful in finding directions (● Fig. 2.1). The earliest known maps were constructed of sticks or were drawn on clay tablets, stone slabs, metal plates, papyrus, linen, or silk. Throughout history maps have become increasingly more common, as a result of the appearance of paper, followed by the printing press, and then the computer. Today, we encounter maps nearly everywhere.

Maps and globes convey spatial information through graphic symbols, a "language of location," that must be understood to appreciate and comprehend the rich store of information that they display (see Appendix B). Although we typically think of maps as being representations of Earth or a part of its surface, maps and globes have now been made to show extraterrestrial features such as the moon and some of the planets.

Cartography is the science and profession of mapmaking. Geographers who specialize in cartography supervise the development of maps and globes to ensure that mapped information and data are accurate and effectively presented. Most cartographers would agree that the primary purpose of a map is to communicate spatial information. In recent years, computer technology has revolutionized cartography.

> Cartographers can now gather spatial data and make maps faster than ever before—within hours—and the accuracy of these maps is excellent. Moreover, digital mapping enables mapmakers to experiment with a map's basic characteristics (for example, scale, projections), to combine and manipulate map data, to transmit entire maps electronically, and to produce unique maps on demand.
>
> United States Geological Survey (USGS)
> *Exploring Maps,* page 1

The changes in map data collection and display that have occurred through the use of computers and digital techniques are dramatic. Information that was once collected manually from ground observations and surveys can now be collected instantly by orbiting satellites that send recorded data back to Earth at the speed of light. Maps that once had to be hand-drawn (● Fig. 2.2) can now be created on a computer and printed in a relatively short amount of time. Although artistic talent is still an advantage, today's cartographers must also be highly skilled users of computer mapping systems, and of course understand the principles of geography, cartography, and map design.

We can all think of reasons why maps are important for conveying spatial information in navigation, recreation, political science, community planning, surveying, history, meteorology, and geology. Many high-tech locational and mapping technologies are now in widespread use by the public, through the Internet and also satellite-based systems that display locations for use in hiking, traveling, and direction finding for all means of transportation. Maps are ever-present in the modern world; they are in newspapers, on television news or weather broadcasts, in our homes, and in our cars. How many maps do you see in a typical day? How many would that equal in a year? How do these maps affect your daily life?

● FIGURE 2.1
In France, cave paintings made between 17,000 and 35,000 years ago apparently depict the migration routes of animals. This view shows detail of stags crossing a river, and experts suggest that some of the artwork represents a rudimentary map with marks that appear to represent locational information. If so, this is the earliest known example of humans recording their spatial knowledge.
Why would prehistoric humans want to record locational information?

©De Sazo/ Photo Researchers, Inc.

Size and Shape of Earth

We were first able to image our planet's shape from space in the 1960s but even as early as 540 BC, ancient Greeks theorized that our planet was a sphere. In 200 BC. Eratosthenes, a philosopher and geographer, estimated Earth's circumference fairly closely to its actual size (how he accomplished this will be illustrated in the next chapter). Earth can generally be considered a sphere, with an equatorial circumference of 39,840 kilometers (24,900 mi), but the centrifugal force associated with Earth's daily *rotation* causes the equatorial region to bulge outward, and slightly flattens the polar regions into a shape that is basically an **oblate spheroid.**

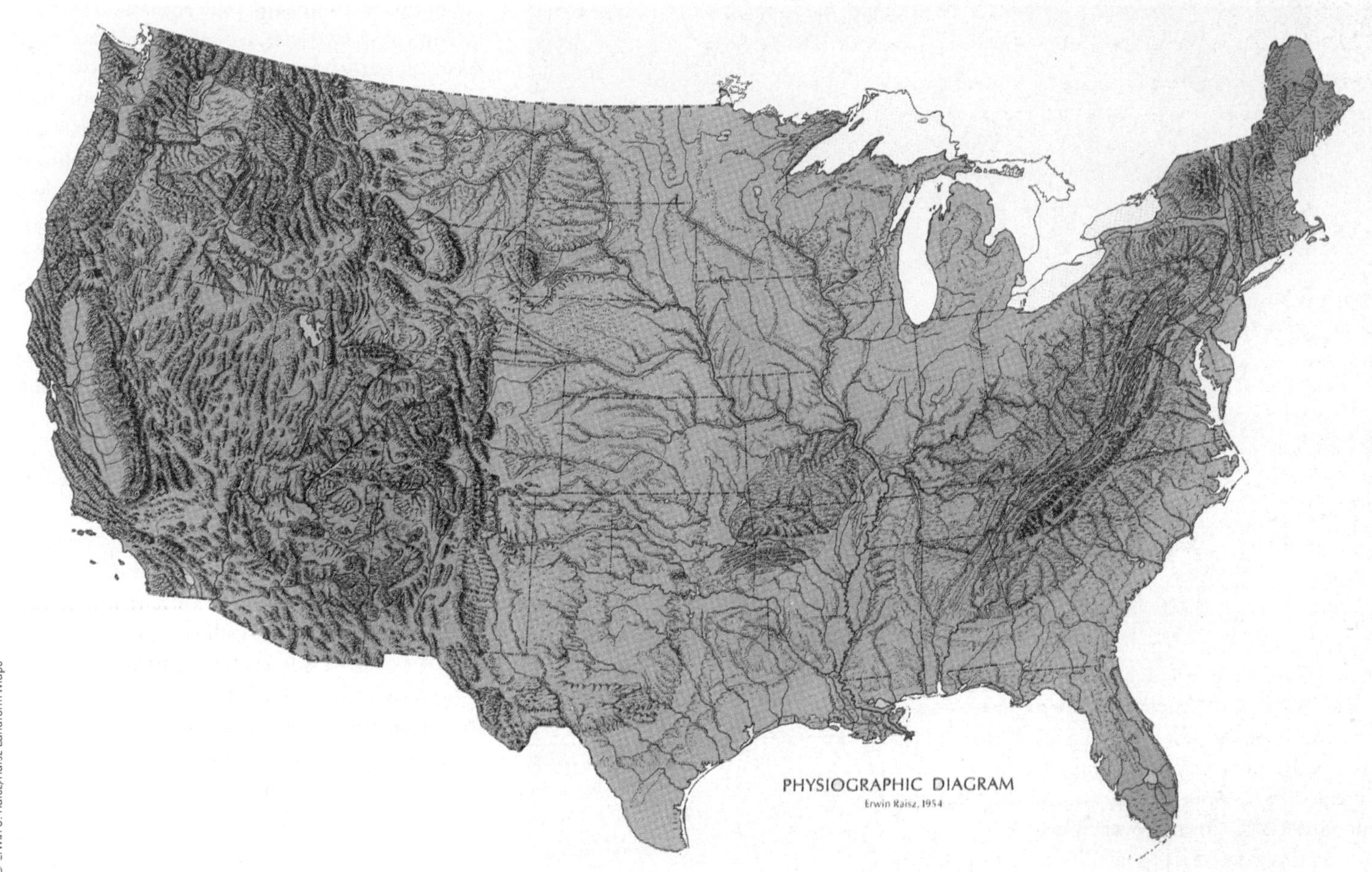

© Erwin J. Raisz/Raisz Landform Maps

• FIGURE 2.2
When maps had to be hand-drawn, artistic talent was required in addition to knowledge of cartographic principles. Mapmaking was a lengthy process, much more difficult than it is today, with computer mapping software and satellite imagery readily available. Erwin Raisz, a famous and talented cartographer, drew this map of U.S. landforms in 1954.
Are maps like this still valuable for learning about landscapes, or are they obsolete?

Our planet's deviations from a true sphere are relatively minor. Earth's diameter at the equator is 12,758 kilometers (7927 mi), while from pole to pole it is 12,714 kilometers (7900 mi). On a globe with a 12-inch diameter (30.5 cm), this difference of 44 kilometers (27 mi) would be about as thick as the wire in a paperclip. Landforms also cause deviations from true sphericity. Mount Everest in the Himalayas is the highest point on Earth at 8850 meters (29,035 ft) above sea level. The lowest point is the Challenger Deep, in the Mariana Trench of the Pacific Ocean southwest of Guam, at 11,033 meters (36,200 ft) below sea level. The difference between these two elevations, 19,883 meters, or just over 12 miles (19.2 km), would also be insignificant when reduced in scale on a 12-inch (30.5 cm) globe.

Earth's variation from a spherical shape is less than one third of 1%, and is not noticeable when viewing Earth from space (• Fig. 2.3). Nevertheless, people working in the very precise fields of navigation, surveying, aeronautics, and cartography must give consideration to Earth's deviations from a perfect sphere.

Globes and Great Circles

As nearly perfect models of our planet, world globes show our planet's shape and accurately display the shapes, sizes, and comparative areas of Earth features, landforms, and water bodies; distances between locations; and true compass directions. Because globes have essentially the same geometric form as Earth, a globe represents geographic features and spatial relationships virtually without distortion. For this reason, if we want to view the entire world, a globe provides the most accurate representation of our planet.

Yet, a globe would not help us find our way on a hiking trail. It would be awkward to carry, and our location would appear as a tiny pinpoint, with very little, if any, local information. We would need a *map* that clearly showed elevations, trails, and rivers and could be folded to carry in a pocket or pack. Being familiar with the characteristics of a globe helps us understand maps and how they are constructed.

NASA

• **FIGURE 2.3**
Earth, photographed from space by Apollo 17 astronauts, showing most of Africa, the surrounding oceans, storm systems in the Southern Hemisphere, and the relative thinness of the atmosphere. Earth's spherical shape is clearly visible; the flattening of the polar regions is too minor to be visible.
What does this suggest about the degree of "sphericity" of Earth?

An imaginary circle drawn in any direction on Earth's surface and whose plane passes through the center of Earth is a **great circle** (• Fig. 2.4a). It is called "great" because this is the largest circle that can be drawn around Earth that connects any two points on the surface. Every great circle divides Earth into equal halves called **hemispheres.** An important example of a great circle is the *circle of illumination,* which divides Earth into light and dark halves—a day hemisphere and a night hemisphere. Great circles are useful to navigation, because any trace along any great circle marks the shortest travel routes between locations on Earth's surface. Any circle on Earth's surface that does not divide the planet into equal halves is called a **small circle** (Fig. 2.4b). The planes of small circles do not pass through the center of Earth.

The shortest route between two places can be located by finding the great circle that connects them. Put a rubber band (or string) around a globe to visualize this spatial relationship. Connect any two cities, such as Beijing and New York, San Francisco and Tokyo, New Orleans and Paris, or Kansas City and Moscow, by stretching a large rubber band around the globe so that it touches both cities and divides the globe in half. The rubber band then marks the shortest route between these two cities. Navigators chart *great circle routes* for aircraft and ships because traveling the shortest distance saves time and fuel. The farther away two points are on Earth, the greater the travel distance savings will be by following the great circle route that connects them.

Latitude and Longitude

Imagine you are traveling by car and you want to visit the Football Hall of Fame in Canton, Ohio. Using the Ohio road map, you look up Canton in the map index and find that it is located at "G-6." The letter G and the number 6 meet in a box marked on the map. In box G-6, you locate Canton (• Fig. 2.5). What you have used is a **coordinate system** of intersecting lines, a system of *grid cells* on the map. Without a locational coordinate system, it would be difficult to describe a location. A problem that a sphere presents, however, is deciding where the starting points should be for a grid system. Without reference points, either natural or arbitrary, a sphere is a geometric form that looks the same from any direction, and has no natural beginning or end points.

Measuring Latitude

The **North Pole** and the **South Pole** provide two natural reference points because they mark the opposite positions of Earth's *rotational* axis, around which it turns in 24 hours. The **equator,** halfway between the poles, forms a great circle that divides the planet into the Northern and Southern Hemispheres. The equator is the reference line for measuring **latitude** in degrees north or degrees south—the equator is 0° latitude.

• **FIGURE 2.4**
Any imaginary geometric plane that passes through Earth's center, thus dividing it into two equal halves, forms a great circle where the plane intersects Earth's surface. This plane can be oriented in any direction as long as it defines two (equal) hemispheres (a). The plane shown in (b) slices the globe into unequal parts, so the line of intersection of such a plane with Earth is a small circle.

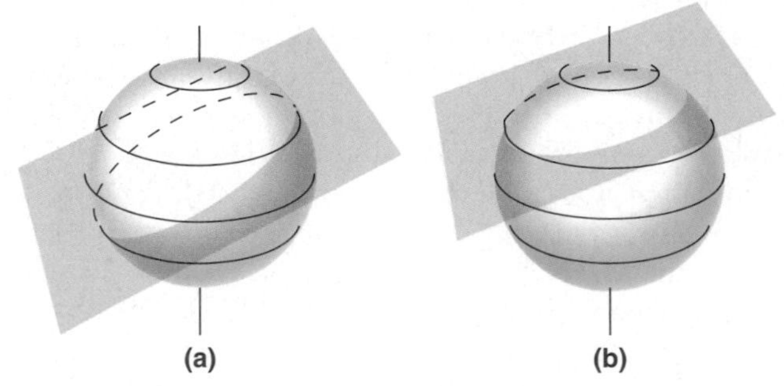

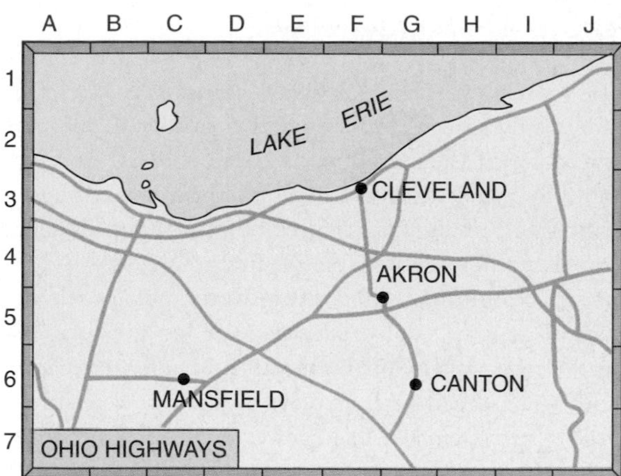

• **FIGURE 2.5**
Using a simple rectangular coordinate system to locate a position. This map employs an alpha-numeric location system, similar to that used on many road maps (and campus maps).
What are the rectangular coordinates of Mansfield? What is at location F-3?

North or south of the equator, the angles and their arcs increase until we reach the North or South Pole at the maximum latitudes of 90° north or 90° south.

To locate the latitude of Los Angeles, for example, imagine two lines that radiate out from the center of Earth. One goes straight to Los Angeles and the other goes to the equator at a point directly south of the city. These two lines form an angle that is the latitudinal distance (in degrees) that Los Angeles lies north of the equator (• Fig. 2.6a). The angle made by these two lines is just over 34°—so the latitude of Los Angeles is about 34°N (north of the equator). Because Earth's circumference is approximately 40,000 kilometers (25,000 mi) and there are 360 degrees in a circle, we can divide (40,000 km/360°) to find that 1° of latitude equals about 111 kilometers (69 mi).

A single degree of latitude covers a relatively large distance, so degrees are further divided into minutes (') and seconds (") of arc. There are 60 minutes of arc in a degree. Actually, Los Angeles is located at 34°03'N (34 degrees, 3 minutes north latitude). We can get even more precise: 1 minute is equal to 60 seconds of arc. We could locate a different position at latitude 23°34' 12"S, which we would read as 23 degrees, 34 minutes, 12 seconds south latitude. A minute of latitude equals 1.85 kilometers (1.15 mi), and a second is about 31 meters (102 ft).

A **sextant** can be used to determine latitude by celestial navigation (• Fig. 2.7). This instrument measures the angle between our *horizon,* the visual boundary line between the sky and Earth, and a celestial body such as the noonday sun or the North Star (Polaris). The latitude of a location, however, is only half of its global address. Los Angeles is located approximately 34° north of the equator, but an infinite number of points exist on the same line of latitude.

Measuring Longitude To accurately describe the location of Los Angeles, we must also determine where it is situated along the line of 34°N latitude. However, to describe an east or west position, we must have a starting line, just as the equator provides our reference for latitude. To find a location east or west, we use longitude lines, which run from pole to pole, each one forming half of a great circle. The global position of the 0° east–west reference line for longitude is arbitrary, but was established by international agreement. The longitude line passing through Greenwich, England (near London), was accepted as the **prime meridian,** or 0° longitude in 1884. **Longitude** is the angular distance east or west of the prime meridian.

• **FIGURE 2.6**
Finding a location by latitude and longitude. (a) The geometric basis for the latitude of Los Angeles, California. Latitude is the angular distance in degrees either north or south of the equator. (b) The geometric basis for the longitude of Los Angeles. Longitude is the angular distance in degrees either east or west of the prime meridian, which passes through Greenwich, England. (c) The location of Los Angeles is 34°N, 118°W.
What is the latitude of the North Pole and does it have a longitude?

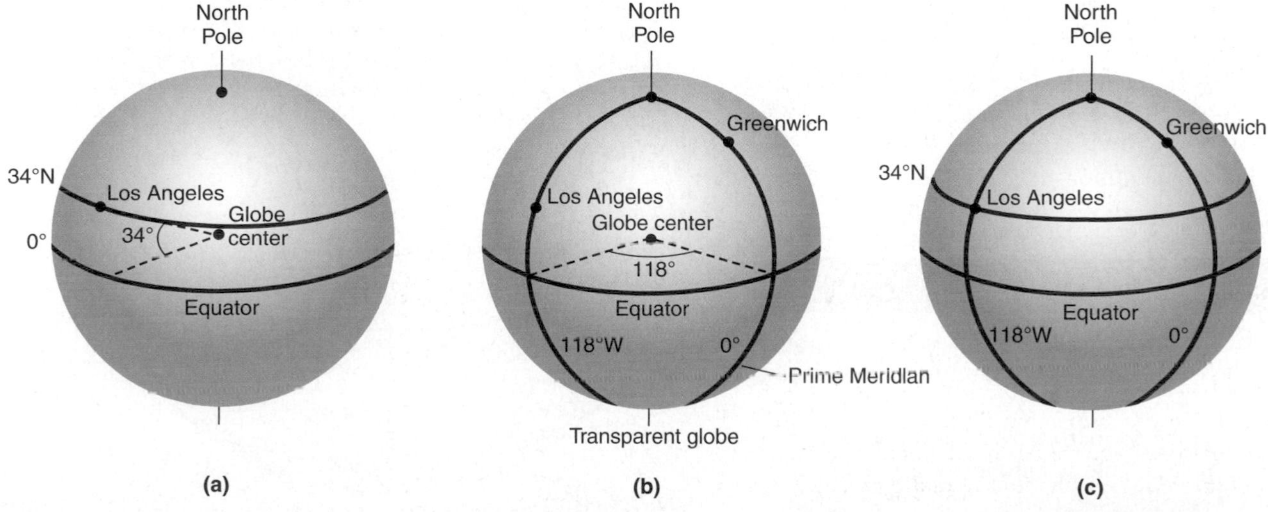

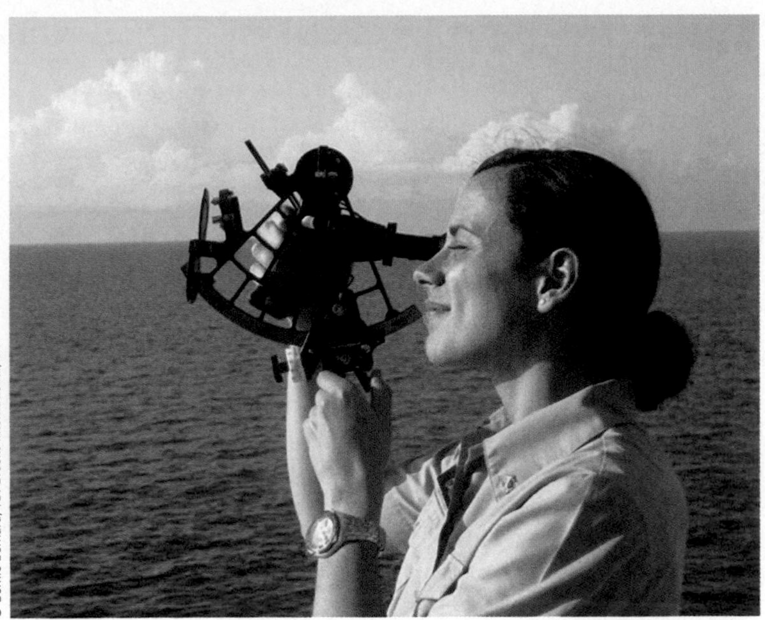
© Bernie Bernard, TDI-Brooks International, Inc.

• FIGURE 2.7
Finding latitude by celestial navigation. A traditional way to determine latitude is by measuring the angle between the horizon and a celestial body with a sextant. Today a satellite-assisted technology called the global positioning system (GPS) supports most air and sea navigation (as well as land travel).
With high-tech location systems like the GPS available, why might understanding how to use a sextant still be important?

Like latitude, longitude is also measured in degrees, minutes, and seconds. Imagine a line drawn from the center of Earth to the point where the north–south running line of longitude that passes through Los Angeles crosses the equator. A second imaginary line will go from the center of Earth to the point where the prime meridian crosses the equator (this location is 0°E or W and 0°N or S). Figure 2.6b shows that these two lines drawn from Earth's center define an angle, the arc of which is the angular distance that Los Angeles lies west of the prime meridian (118°W longitude). Figure 2.6c provides the global address of Los Angeles by latitude and longitude.

Our longitude increases as we go farther east or west from 0° at the prime meridian. Traveling eastward from the prime meridian, we will eventually be halfway around the world from Greenwich, in the middle of the Pacific Ocean at 180°. Longitude is measured in degrees up to a maximum of 180° east or west of the prime meridian. Along the prime meridian (0° E–W) or the 180° meridian, the E–W designation does not matter, and along the Equator (0° N–S), the N–S designation does not matter, and is not needed for indicating location.

The Geographic Grid

Every point on Earth's surface can be located by its latitude north or south of the equator in degrees, and its longitude east or west of the prime meridian in degrees. Lines that run east and west around the globe to mark latitude and lines that run north and south from pole to pole to indicate longitude form the **geographic grid** (• Fig. 2.8).

Parallels and Meridians

The east–west lines marking latitude completely circle the globe, are evenly spaced, and are parallel to the equator and each other. Hence, they are known as **parallels.** The equator is the only parallel that is a great circle; all other lines of latitude are small circles. One degree of latitude equals about 111 kilometers (69 mi) anywhere on Earth.

Lines of longitude, called **meridians,** run north and south, converge at the poles, and measure longitudinal distances east or west of the prime meridian. Each meridian of longitude, when joined with its mate on the opposite side of Earth, forms a great circle. Meridians at any given latitude are evenly spaced, although meridians get closer together as they move poleward from the equator. At the equator, meridians separated by 1° of longitude are about 111 kilometers (69 mi) apart, but at 60°N or 60°S latitude, they are only half that distance apart, about 56 kilometers (35 mi).

Longitude and Time

The relationship between longitude, Earth's *rotation,* and time, was used to establish **time zones.** Until about 125 years ago, each town or area used what was known as *local time.* **Solar noon** was determined by the precise moment in a day when a vertical stake cast its shortest shadow. This meant that the sun had reached its highest angle in the sky for that day at that location—noon—and local clocks were set to that time. Because of Earth's *rotation* on its axis, noon in a town toward the east occurred earlier, and towns to the west experienced noon later.

• FIGURE 2.8
A globelike representation of Earth, which shows the geographic grid with parallels of latitude and meridians of longitude at 15° intervals.
How do parallels and meridians differ?

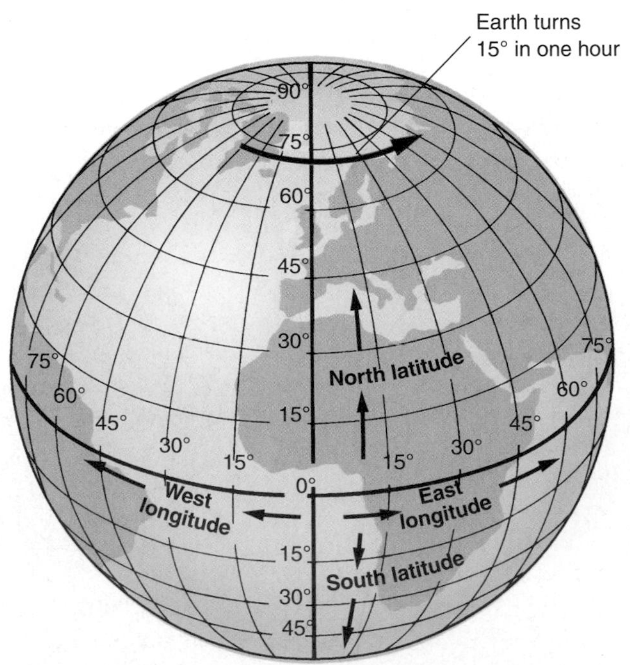

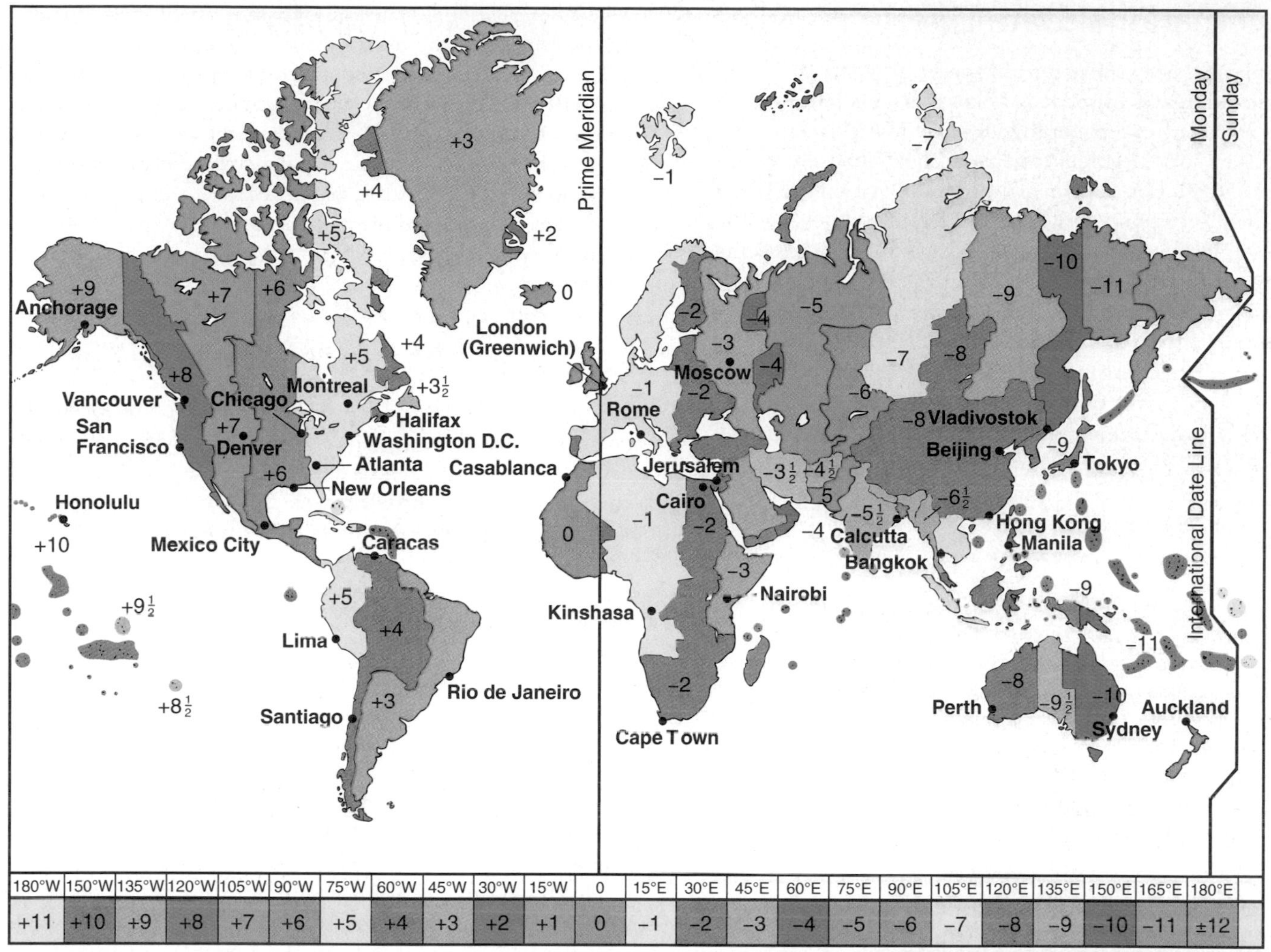

• FIGURE 2.9
World time zones reflect the fact that Earth rotates through 15° of longitude each hour. Thus, time zones are approximately 15° wide. Political boundaries usually prevent the time zones from following a meridian perfectly.
How many hours of difference are there between the time zone where you live and Greenwich, England, and is it earlier in England or later?

By the late 1800s, advances in travel and communication made the use of local time by each community impractical. In 1884 the International Meridian Conference in Washington, D.C., set standardized time zones and established the longitude passing through Greenwich as the prime meridian (0° longitude). Earth was divided into 24 time zones, one for each hour in a day, because Earth turns 15° of longitude in an hour (360° ÷ 24 hours). Ideally, each time zone spans 15° of longitude. The prime meridian is the *central meridian* of its time zone, and the time when solar noon occurs at the prime meridian was established as noon for all places between 7.5°E and 7.5°W of that meridian. The same pattern was followed around the world. Every line of longitude evenly divisible by 15° is the central meridian for a time zone extending 7.5° of longitude on either side. However, as shown in • Figure 2.9, time zone boundaries do not follow meridians exactly. In the United States, time zone boundaries commonly follow state lines. It would be inconvenient and confusing to have a time zone boundary dividing a city or town into two time zones, so jogs in the lines were established to avoid most of these problems.

The time of day at the prime meridian, known as *Greenwich Mean Time* (GMT, but also called Universal Time, UTC, or Zulu Time), is used as a worldwide reference. Times to the east or west can be easily determined by comparing them to GMT. A place 90°E of the prime meridian would be 6 hours later (90° ÷ 15° per hour) while in the Pacific Time Zone of the United States and Canada, whose central meridian is 120°W, the time would be 8 hours earlier than GMT.

For navigation, longitude can be determined with a *chronometer*, an extremely accurate clock. Two chronometers are used, one set on Greenwich time and the other on local time. The number of hours between them, earlier or later, determines longitude (1 hour = 15° of longitude). Until the advent of electronic navigation by ground- and satellite-based systems, the sextant and chronometer were a navigator's basic tools for determining location.

The International Date Line

On the opposite side of Earth from the prime meridian is the **International Date Line.** It is a line that generally follows the 180th meridian, except for jogs to separate Alaska and Siberia and to skirt some Pacific island groups (• Fig. 2.10). At the International Date Line, we turn our calendar back a full day if we are traveling east and forward a full day if we are traveling west. Thus, if we are going east from Tokyo to San Francisco and it is 4:30 p.m. Monday just before we cross the International Date Line, it will be 4:30 p.m. Sunday on the other side. If we are traveling west from Alaska to Siberia and it is 10:00 a.m. Wednesday when we reach the International Date Line, it will be 10:00 a.m. Thursday once we cross it. As a way of remembering this relationship, many world maps and globes have Monday and Sunday (M | S) labeled in that order on the opposite sides of the International Date Line. To find the correct day, you just substitute the current day for Monday or Sunday, and use the same relationship.

• FIGURE 2.10
The International Date Line. The new day officially begins at the International Date Line (IDL) and then sweeps westward around the Earth to disappear when it again reaches the IDL. West of the line is always a day later than east of the line. Maps and globes often have either "Monday | Sunday" or "M | S" shown on opposite sides of the line to indicate the direction of the day change. This is the IDL as it is officially accepted by the United States.
Why does the International Date Line deviate from the 180° meridian in some places?

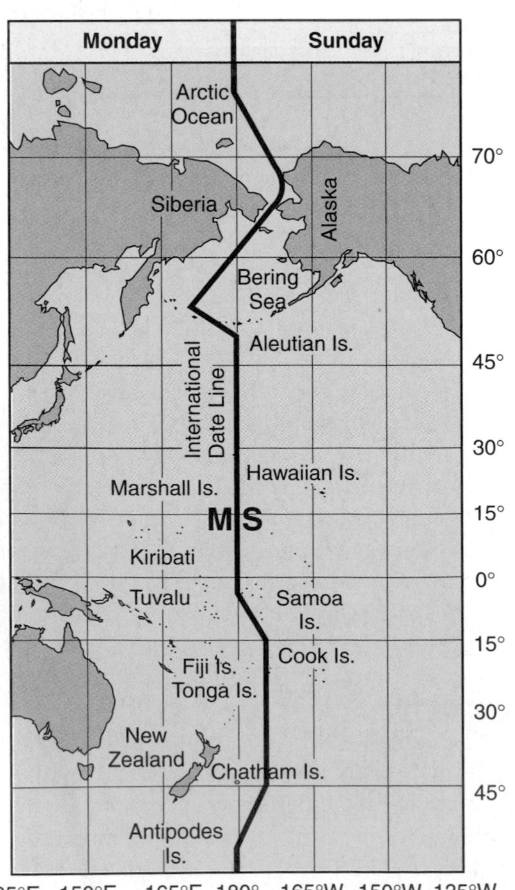

The International Date Line was not established officially until the 1880s, but the need for such a line on Earth to adjust the day was inadvertently discovered by Magellan's crew who, from 1519 to 1521, were the first to circumnavigate the world. Sailing westward from Spain when they returned from their voyage, the crew noticed that one day had apparently been missed in the ship's log. What actually happened is that in going around the world in a westward direction, the crew had experienced one less sunset and one less sunrise than had occurred in Spain during their absence.

The U.S. Public Lands Survey System

The longitude and latitude system was designed to locate the *points* where those lines intersect. A different system is used in much of the United States to define and locate land *areas.* This is the **U.S. Public Lands Survey System,** or the *Township and Range System,* developed for parceling public lands west of Pennsylvania. Lands in the eastern U.S. had already been surveyed into irregular parcels at the time this system was established. The Township and Range System divides land areas into parcels based on north–south lines called **principal meridians** and east–west lines called **base lines** (• Fig. 2.11). Base lines were surveyed along parallels of latitude. The north–south meridians, though perpendicular to the base lines, had to be adjusted (jogged) along their length to accommodate Earth's curvature. If these adjustments were not made, the north–south lines would tend to converge and land parcels defined by this system would be smaller in northern regions of the United States.

The Township and Range System forms a grid of nearly square parcels called townships laid out in horizontal *tiers* north and south of the base lines and in vertical *columns* ranging east and west of the principal meridians. A **township** is a square plot 6 miles on a side (36 sq mi, or 93 sq km). As illustrated in • Figure 2.12,

• FIGURE 2.11
Principal meridians and base lines of the U.S. Public Lands Survey System (Township and Range System).
Why wasn't the Township and Range System applied throughout the eastern United States?

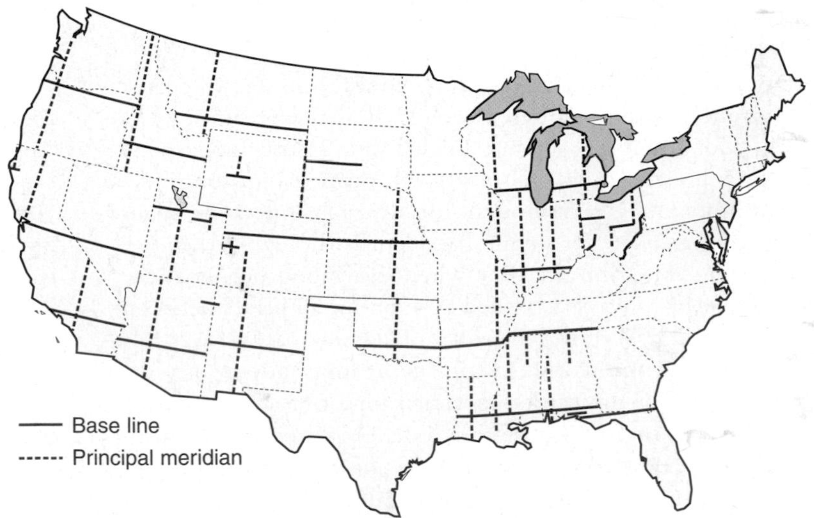

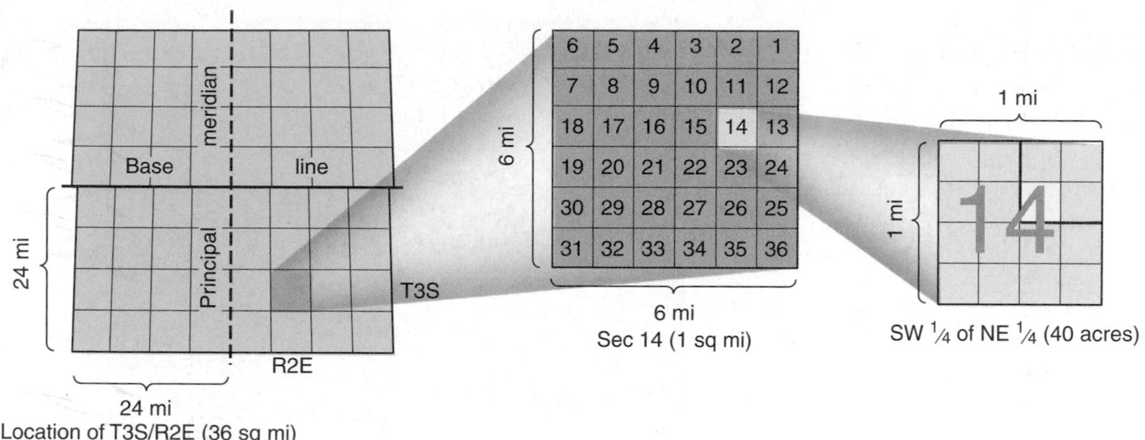

• FIGURE 2.12

The method of location for areas of land according to the Public Lands Survey System.

How would you describe the extreme southeastern 40 acres of section 20 in the middle diagram?

townships are first labeled by their position north or south of a base line; thus, a township in the third tier south of a base line will be labeled Township 3 South, which is abbreviated T3S. However, we must also name a township according to its *range*—its location east or west of the principal meridian for the survey area. Thus, if Township 3 South is in the second range east of the principal meridian, its full location can be given as T3S/R2E (Range 2 East).

The Public Lands Survey System divides townships into 36 **sections** of 1 square mile, or 640 acres (2.6 sq km, or 259 ha). Sections are designated by numbers from 1 to 36 beginning in the northeasternmost section with section 1, snaking back and forth across the township, and ending in the southeast corner with section 36. Sections are divided into four *quarter sections*, named by their location within the section—northeast, northwest, southeast, and southwest, each with 160 acres (65 ha). Quarter sections are also subdivided into four *quarter-quarter sections*, sometimes known as *forties*, each with an area of 40 acres (16.25 ha). These quarter-quarter sections, or 40-acre plots, are also named after their position in the quarter: the northeast, northwest, southeast, and southwest forties. Thus, we can describe the location of the 40-acre tract that is shaded in Figure 2.12 as being in the SW ¼ of the NE ¼ of Sec. 14, T3S/R2E, which we can find if we locate the principal meridian and the base line. The order is consistent from smaller division to larger, and township location is always listed before range (T3S/R2E).

The Township and Range system has exerted an enormous influence on landscapes in many areas of the United States and gives most of the Midwest and West a checkerboard appearance from the air or from space (• Fig. 2.13). Road maps in states that use this survey system strongly reflect its grid, and many roads follow the regular and angular boundaries between square parcels of land.

• FIGURE 2.13

Rectangular field patterns resulting from the Public Lands Survey System in the Midwest and western United States. Note the slight jog in the field pattern to the right of the farm buildings near the lower edge of the photo.

How do you know this photo was not taken in the midwestern United States?

© Grant Heilman/Grant Heilman Photography

The Global Positioning System

The **Global Positioning System (GPS)** is a modern technology for determining a location on Earth. This high-tech system was originally created for military applications but today is being adapted to many public uses, from surveying to navigation. The global positioning system uses radio signals, transmitted by a network of satellites orbiting 17,700 kilometers (11,000 mi) above Earth (● Fig. 2.14). By accessing signals from several satellites, a GPS receiver calculates the distances from those satellites to its location on Earth. GPS is based on the principle of *triangulation,* which means that if we can find the distance to our position, measured from three or more different locations (in this case, satellites), we can determine our location. GPS receivers vary in size, and handheld units are common (● Fig. 2.15). Small GPS receivers are very useful to travelers, hikers, and backpackers who need to keep track of their location. The distances from a receiver to the satellites are calculated by measuring the time it takes for a satellite radio signal, broadcast at the speed of light, to arrive at the receiver. A GPS receiver performs these calculations and displays a locational readout in latitude, longitude, and elevation, or on a map display. Map-based GPS systems—where GPS data is translated to a map display—not only are becoming popular for hikers, but larger units are widely used in vehicles and also on boats and aircraft. With sophisticated GPS equipment and techniques, it is possible to find locational coordinates within small fractions of a meter (● Fig. 2.16).

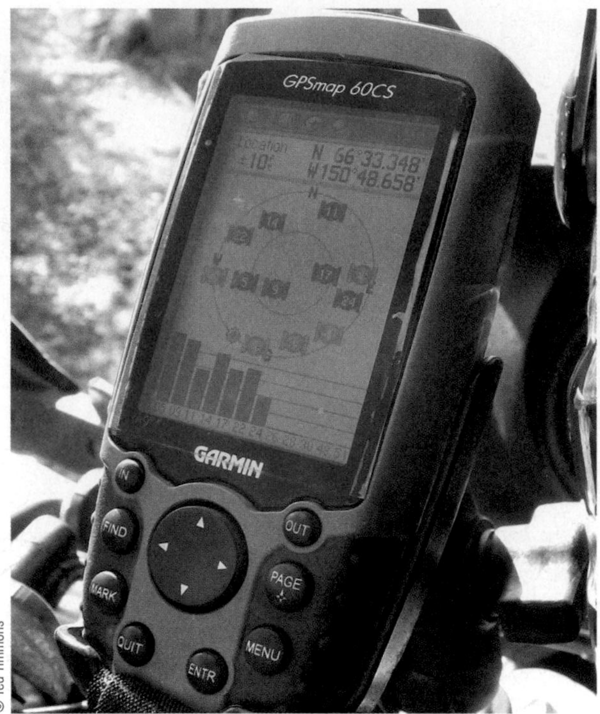

© Ted Timmons

● FIGURE 2.15
A GPS receiver provides a readout of its latitudinal and longitudinal position based on signals from a satellite network. Small handheld units provide an accuracy that is acceptable for many uses, and many can also display locations on a map. This receiver was mounted on a motorcycle for navigation on a trip to Alaska; the latitude shown is at the Arctic Circle. **What other uses can you think of for a small GPS unit like this that displays its longitude and latitude as it moves from place to place?**

● FIGURE 2.14
The global positioning system (GPS) uses signals from a network of satellites to determine a position on Earth. A GPS receiver on the ground calculates the distances from several satellites (a minimum of three) to find its location by longitude, latitude, and elevation. With the distance from three satellites, a position can be located within meters, but with more satellite signals and sophisticated GPS equipment, the position can be located very precisely.

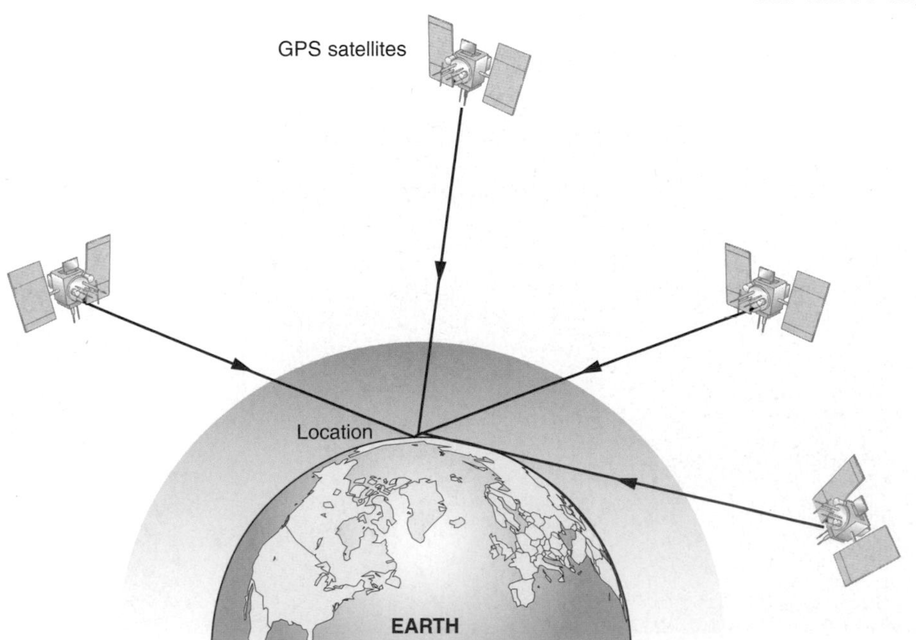

Maps and Map Projections

Maps can be reproduced easily, can depict the entire Earth or show a small area in great detail, are easy to handle and transport, and can be displayed on a computer monitor. There are many different varieties of maps, and they all have qualities that can be either advantageous or problematic, depending on the application. It is impossible for one map to fit all uses. Knowing some basic concepts concerning maps and cartography will greatly enhance a person's ability to effectively use a map, and to select the right map for a particular task.

Advantages of Maps

If a picture is worth a thousand words, then a map is worth a million. Because they are graphic representations and use symbolic language, maps show spatial relationships and portray geographic information with great efficiency. As visual representations, maps supply an enormous amount of information that would take many pages to describe in words (probably less successfully).

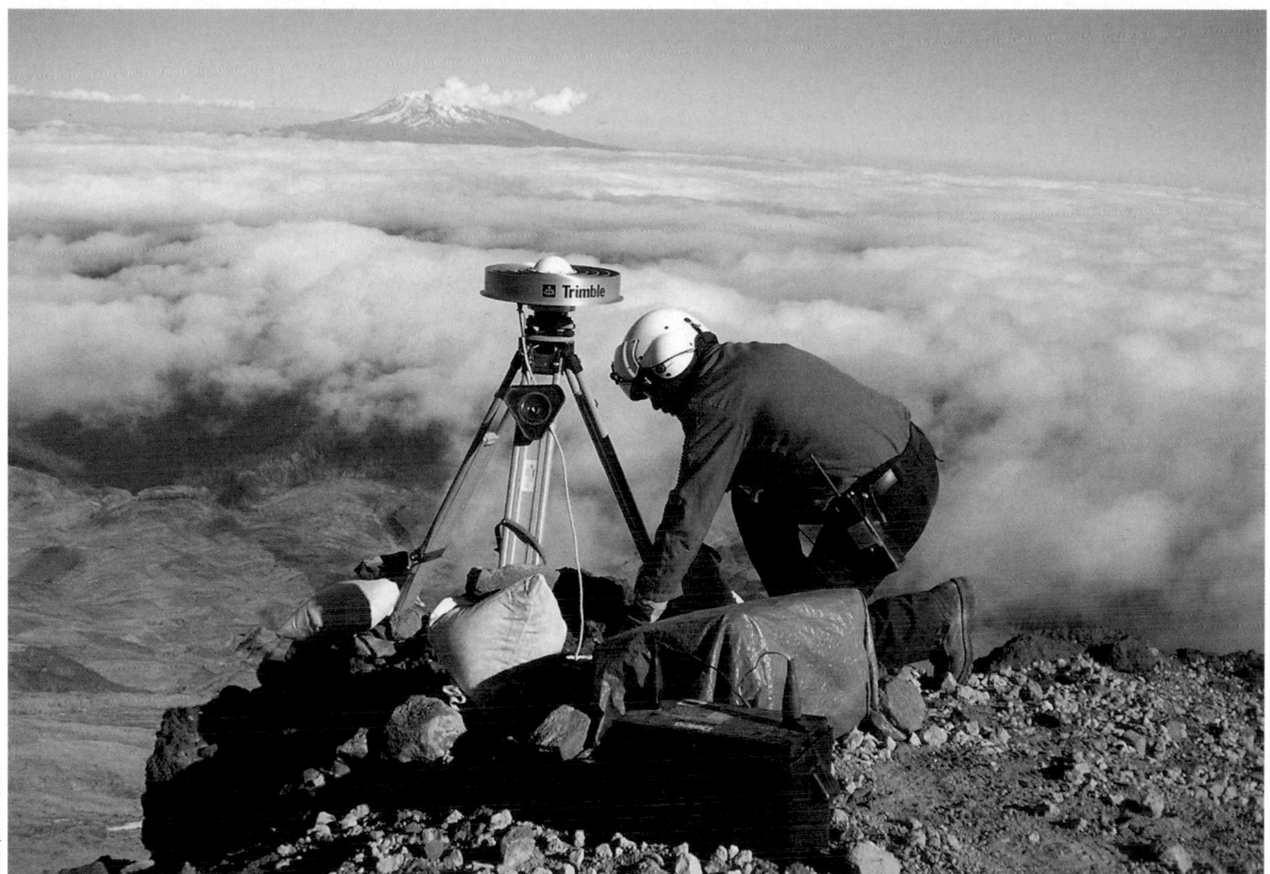

USGS/Mike Poland

● **FIGURE 2.16**
A scientist monitoring volcanoes in Washington State uses a professional GPS system to record a precise location by longitude, latitude, and elevation. Highly accurate land surveying by GPS requires advanced techniques and equipment that is more sophisticated than the typical handheld GPS receiver. This is the view from Mount St. Helens, with Mount Adams, another volcano in the distance.

Imagine trying to tell someone about all of the information that a map of your city, county, state, or campus provides: sizes, areas, distances, directions, street patterns, railroads, bus routes, hospitals, schools, libraries, museums, highway routes, business districts, residential areas, population centers, and so forth. Maps can display true courses for navigation and accurate shapes of Earth features. They can be used to measure areas, or distances, and they can show the best route from one place to another. The potential applications of maps are practically infinite, even "out of this world," because our space programs have produced detailed maps of the moon (● Fig. 2.17) and other extraterrestrial features.

Cartographers can produce maps to illustrate almost any relationship in the environment. For many reasons, whether it is presented on paper, on a computer screen, or in the mind, the map is the geographer's most important tool.

● **FIGURE 2.17**
Lunar Geography. A detailed map of the moon shows a major crater that is 120 kilometers in diameter (75 mi). Even the side of the moon that never faces Earth has been mapped in considerable detail.
How were we able to map the moon in such detail?

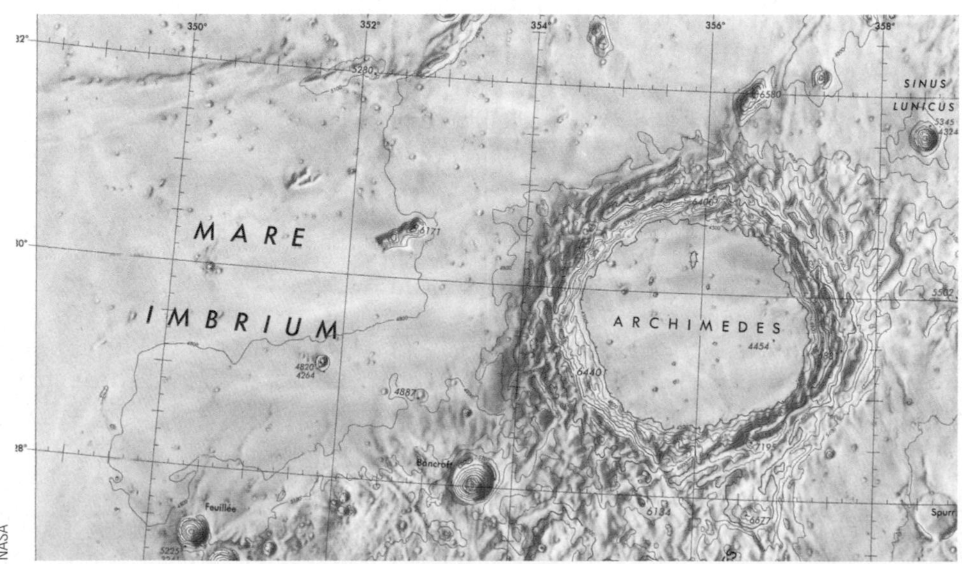

NASA

Limitations of Maps

On a globe, we can directly compare the size, shape, and area of Earth features, and we can measure distance, direction, shortest routes, and true directions. Yet, because of the distortion inherent in maps, we can never compare or measure all of these properties on a single map. It is impossible to present a spherical planet on a flat (two-dimensional) surface and accurately maintain all of its geometric properties. This process has been likened to trying to flatten out an eggshell.

Distortion is an unavoidable problem of representing a sphere on a flat map, but when a map depicts only a small area, the distortion should be negligible. If we use a map of a state park for hiking, the distortion will be too small to affect us. On maps that show large regions or the world, Earth's curvature causes apparent and pronounced distortion. To be skilled map users, we must know which properties a certain map depicts accurately, which features it distorts, and for what purpose a map is best suited. If we are aware of these map characteristics, we can make accurate comparisons and measurements on maps and better understand the information that the map conveys.

Properties of Map Projections

The geographic grid has four important geometric properties: (1) Parallels of latitude are always parallel, (2) parallels are evenly spaced, (3) meridians of longitude converge at the poles, and (4) meridians and parallels always cross at right angles. There are thousands of ways to transfer a spherical grid onto a flat surface to make a **map projection,** but no map projection can maintain all four of these properties at once. Because it is impossible to have all these properties on the same map, cartographers must decide which properties to preserve at the expense of others. Closely examining a map's grid system to determine how these four properties are affected will help us discover areas of greatest and least distortion.

Although maps are not actually made this way, certain projections can be demonstrated by putting a light inside a transparent globe so that the grid lines are projected onto a plane or flat surface (*planar projection*), a cylinder (*cylindrical projection*), or a cone (*conic projection*), geometric forms that are flat or can be cut and flattened out (● Figs. 2.18a–c). Today, map projections are developed mathematically, using computers to fit the geographic grid to a surface. Distortions in the geographic grid that are required to make a map can affect the geometry of several characteristics of the areas and features that a map portrays.

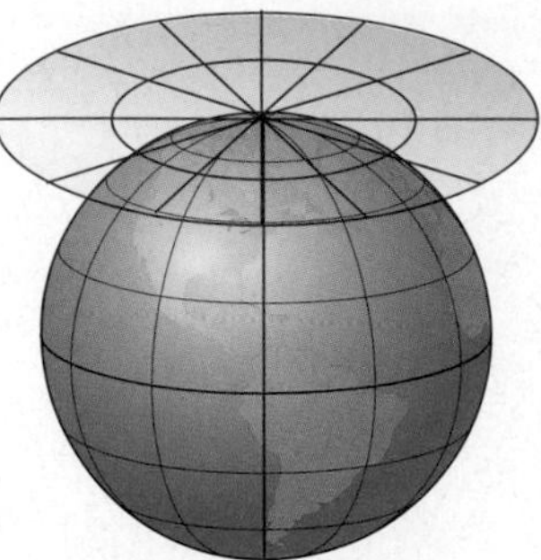

(a) Planar projection

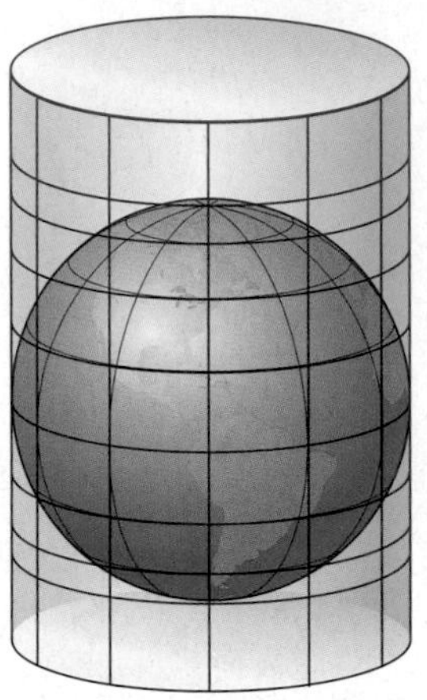

(b) Cylindrical projection

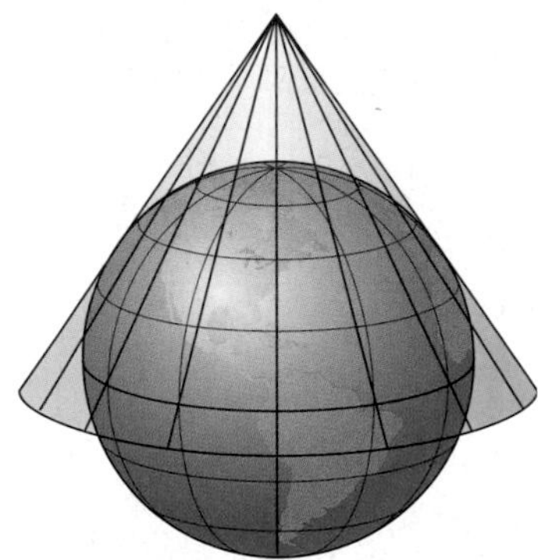

(c) Conical projection

● FIGURE 2.18
The theory behind the development of (a) planar, (b) cylindrical, and (c) conic projections. Although projections are not actually produced this way, they can be demonstrated by projecting light from a transparent globe.
Why do we use different map projections?

Shape

Flat maps cannot depict large regions of Earth without distorting either their shape or their comparative sizes in terms of area. However, using the proper map projection will depict the true shapes of continents, regions, mountain ranges, lakes, islands, and bays. Maps that maintain the correct shapes of areas are **conformal maps.** To preserve the shapes of Earth features on a conformal map, meridians and parallels always cross at right angles just as they do on the globe.

Most of us are familiar with the **Mercator projection** (● Fig. 2.19), commonly used in schools and textbooks, although less so in recent years. The Mercator projection does present correct shapes, so it is a conformal map, but areas away from the equator are exaggerated in size. Because of its widespread use, the Mercator projection's distortions led generations of students to

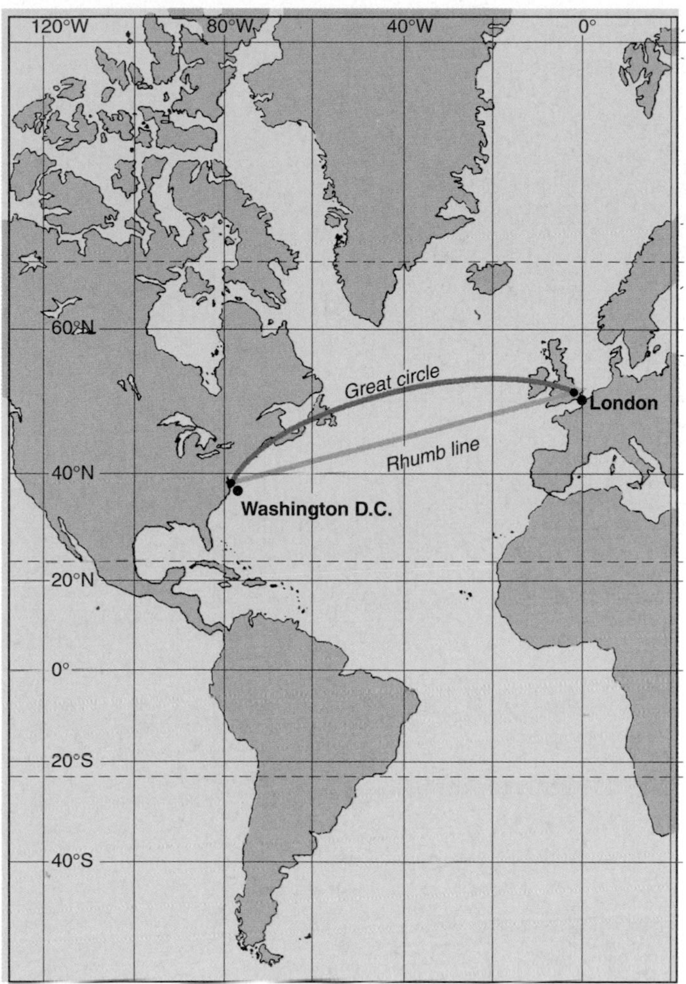

• FIGURE 2.19
The Mercator projection was designed for navigation, but has often been misused as a general-purpose world map. Its most useful property is that lines of constant compass heading, called *rhumb* lines, are straight lines. The Mercator is developed from a cylindrical projection.
Compare the sizes of Greenland and South America on this map to their proportional sizes on a globe. Is the distortion great or small?

believe incorrectly that Greenland is as large as South America. On Mercator's projection, Greenland is shown as being about equal in size to South America (see again Fig. 2.19), but South America is actually about eight times larger.

Area Cartographers are able to create a world map that maintains correct area relationships; that is, areas on the map have the same size proportions to each other as they have in reality. Thus, if we cover any two parts of the map with, let's say, a quarter, no matter where the quarter is placed it will cover equivalent areas on Earth. Maps drawn with this property, called **equal-area maps,** should be used if size comparisons are being made between two or more areas. The property of equal area is also essential when examining spatial distributions. As long as the map displays equal area and a symbol represents the same quantity throughout the map, we can get a good idea of the distribution of any feature—for example, people, churches, cornfields, hog farms, or volcanoes. However, equal-area maps distort the shapes of map features (• Fig. 2.20) because it is impossible to show both equal areas and correct shapes on the same map.

Distance No flat map can maintain a constant distance scale over Earth's entire surface. The scale on a map that depicts a large area cannot be applied equally everywhere on that map. On maps of small areas, however, distance distortions will be minor, and the accuracy will usually be sufficient for most purposes. Maps can be made with the property of **equidistance** in specific instances. That is, on a world map, the equator may have equidistance (a constant scale) along its length, and all meridians may have equidistance, but not the parallels. On another map, all straight lines drawn from the center may have equidistance, but the scale will not be constant unless lines are drawn from the center.

Direction Because the longitude and latitude directions run in straight lines, but curve around the spherical Earth, not all flat maps can show true directions as straight lines. A given map may be able to show true north, south, east, and west, but the directions between those points may not be accurate in terms of the angle between them. So, if we are sailing toward an island, its location may be shown correctly according to its longitude and latitude, but the direction in which we must sail to get there may not be accurately displayed, and we might pass right by it. Maps that show true directions as straight lines are called **azimuthal** projections. These are drawn with a central focus, and all straight lines drawn from that center are true compass directions (• Fig. 2.21).

Examples of Map Projections

All maps based on projections of the geographic grid maintain one aspect of Earth—the property of location. Every place shown on a map must be in its proper location with respect to latitude and longitude. No matter how the arrangement of the global grid is changed by projecting it onto a flat surface, all places must still be located at their correct latitude and longitude.

The Mercator Projection As previously mentioned, one of the best-known world maps is the Mercator projection, a mathematically adjusted cylindrical projection (see again Fig. 2.18b). Meridians appear as parallel lines instead of converging at the poles. Obviously, there is enormous east–west distortion of the high latitudes because the distances between meridians are stretched to the same width that they are at the equator (see again Fig. 2.19). The spacing of parallels on a Mercator projection is also not equal, in contrast to their arrangement on a globe. The resulting grid consists of rectangles that become larger toward the poles. Obviously, this projection does not display equal area, and size distortion increases toward the poles.

Gerhardus Mercator devised this map in 1569 to provide a property that no other world projection has. A straight line drawn anywhere on a Mercator projection is a true compass direction.

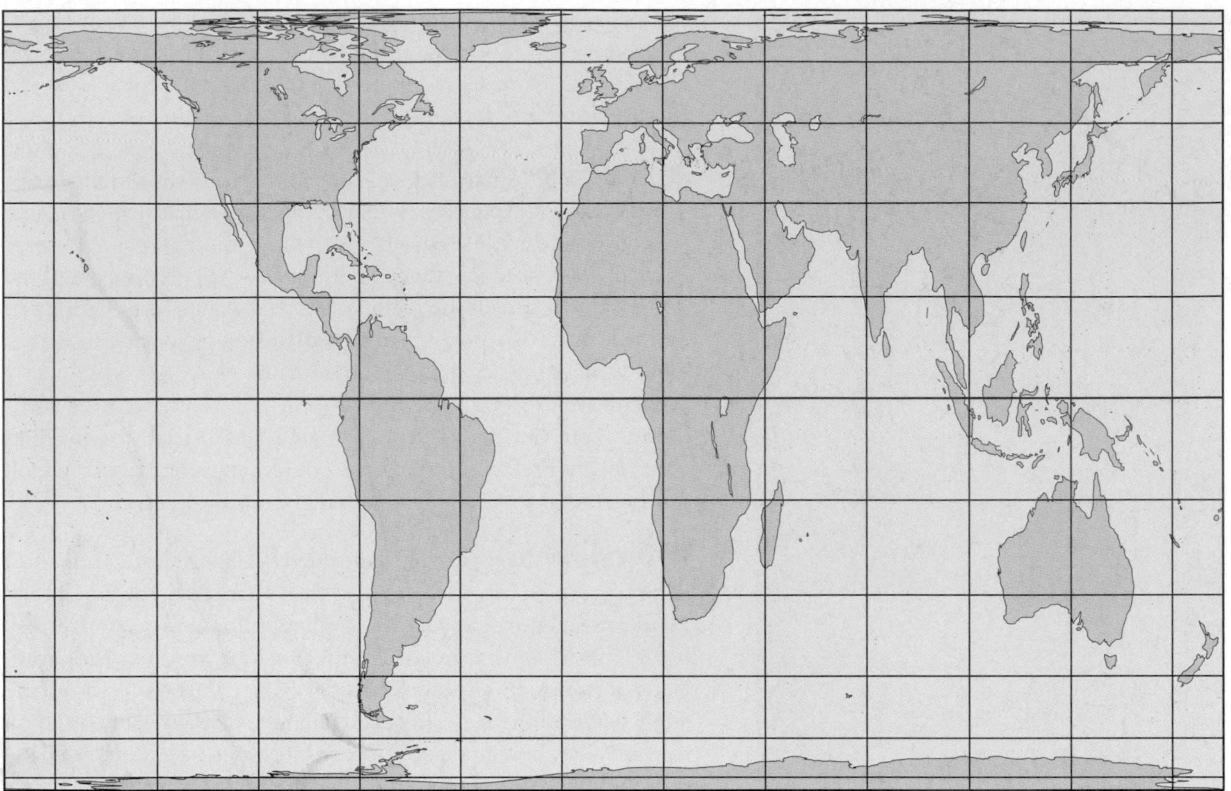

• FIGURE 2.20
An equal-area world projection map. This map preserves area relationships but distorts the shape of landmasses.
Which world map would you prefer, one that preserves area or one that preserves shape, and why?

• FIGURE 2.21
Azimuthal map centered on the North Pole. Although a polar view is the conventional orientation of such a map, it could be centered anywhere on Earth. Azimuthal maps show true directions between all points, but can only show a hemisphere on a single map.

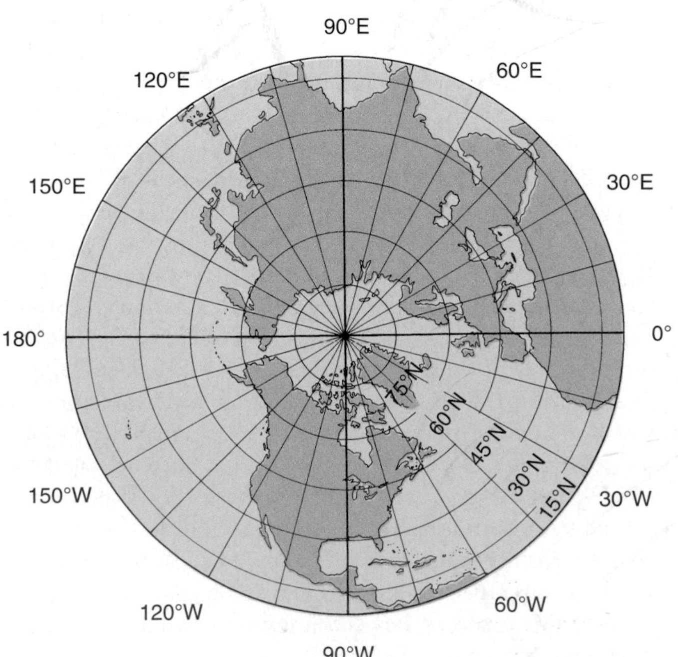

A line of constant direction, called a **rhumb line,** has great value to navigators (see again Fig. 2.19). On Mercator's map, navigators could draw a straight line between their location and the place where they wanted to go, and then follow a constant compass direction to get to their destination.

Gnomonic Projections **Gnomonic projections** are planar projections, made by projecting the grid lines onto a plane, or flat surface (see again Fig. 2.18a). If we put a flat sheet of paper tangent to (touching) the globe at the equator, the grid will be projected with great distortion. Despite their distortion, gnomonic projections (• Figure 2.22) have a valuable characteristic: they are the only maps that display all arcs of great circles as straight lines. Navigators can draw a straight line between their location and where they want to go, and this line will be the shortest route between the two places.

An interesting relationship exists between gnomonic and Mercator projections. Great circles on the Mercator projection appear as curved lines, and rhumb lines appear straight. On the gnomonic projection the situation is reversed—great circles appear as straight lines, and rhumb lines are curves.

Conic Projections Conic projections are used to map middle-latitude regions, such as the United States (other than Alaska and Hawaii), because they portray these latitudes with minimal distortion. In a simple conic projection, a cone is fitted

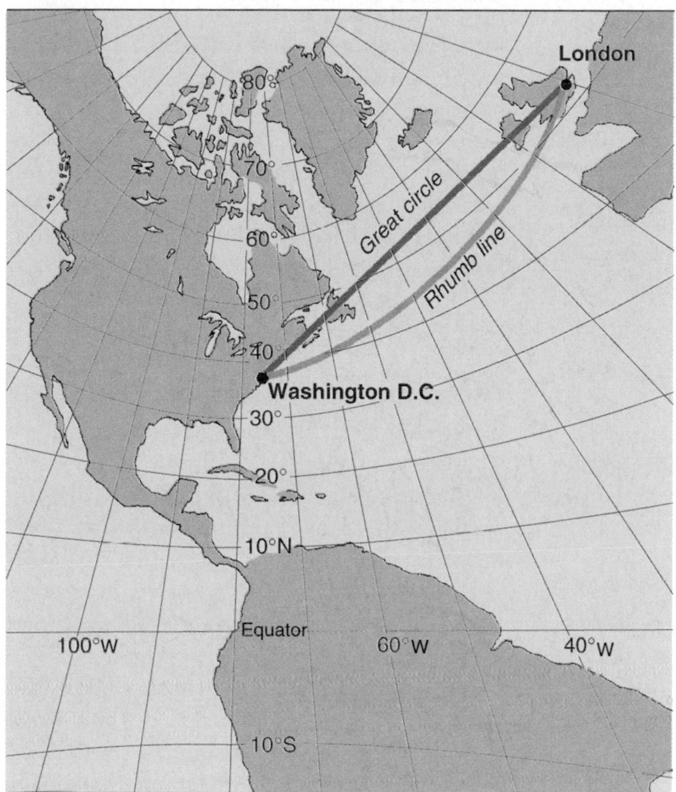

• FIGURE 2.22
The gnomonic projection produces extreme distortion of distances, shapes, and areas. Yet it is valuable for navigation because it is the only projection that shows all great circles as straight lines. It is developed from a planar projection.
Compare this figure with Figure 2.19. How do these two projections differ?

over the globe with its pointed top centered over a pole (see again Fig. 2.18c). Parallels of latitude on a conic projection are concentric arcs that become smaller toward the pole, and meridians appear as straight lines radiating from the pole.

Compromise Projections In developing a world map, one cartographic strategy is to compromise by creating a map that shows both area and shape fairly well but is not really correct for either property. These world maps are **compromise projections** that are neither conformal nor equal area, but an effort is made to balance distortion to produce an "accurate looking" global map (● Fig. 2.23a). An *interrupted projection* can also be used to reduce the distortion of landmasses (Fig. 2.23b) by moving much of the distortion to the oceanic regions. If our interest was centered on the world ocean, however, the projection could be interrupted in the continental areas to minimize distortion of the ocean basins.

Map Basics

Maps not only contain spatial information and data that the map was designed to illustrate, but they also display essential information about the map itself. This information and certain graphic features (often in the margins) are intended to facilitate using and understanding the map. Among these items are the map title, date, legend, scale, and direction.

Title A map should have a title that tells what area is depicted and what subject the map concerns. For example, a hiking and camping map for Yellowstone National Park should have a title like "Yellowstone National Park: Trails and Camp Sites." Most maps should also indicate when they were published and the date to which its information applies. For instance, a population map of the United States should tell when the census was taken, to let us know if the map information is current, or outdated, or whether the map is intended to show historical data.

Legend A map should also have a **legend**—a key to symbols used on the map. For example, if one dot represents 1000 people or the symbol of a pine tree represents a roadside park, the legend should explain this information. If color shading is used on the map to represent elevations, different climatic regions, or other factors, then a key to the color coding should be provided. Map symbols can be designed to represent virtually any feature (see Appendix B).

Scale Obviously, maps depict features smaller than they actually are. If the map used for measuring sizes or distances, or if the size of the area represented might be unclear to the map user, it is essential to know the map scale (● Fig. 2.24). A map **scale** is an expression of the relationship between a distance on Earth and the same distance as it appears on the map. Knowing the map scale is essential for accurately measuring distances and for determining areas. Map scales can be conveyed in three basic ways.

A **verbal scale** is a statement on the map that indicates, for example, "1 centimeter to 100 kilometers" (1 cm represents 100 km) or "1 inch to 1 mile" (1 in on the map represents 1 mi on the ground). Stating a verbal scale tends to be how most of us would refer to a map scale in conversation. A verbal scale, however, will no longer be correct if the original map is reduced or enlarged. When stating a verbal scale it is acceptable to use different map units (centimeters, inches) to represent another measure of true length it represents (kilometers, miles).

A **representative fraction (RF)** scale is a ratio between a unit of distance on the map to the distance that unit represents in reality (expressed in the same units). Because a ratio is also a fraction, units of measure, being the same in the numerator and denominator, cancel each other out. An RF scale is therefore free of units of measurement and can be used with any unit of linear measurement—meters, or centimeters, feet, inches—as long as the same unit is used on both sides of the ratio. As an example, a map may have an RF scale of 1:63,360, which can also be expressed 1/63,360. This RF scale can mean that 1 inch on the map represents 63,360 inches on the ground. It also means that 1 centimeter on the map represents 63,360 centimeters on the ground. Knowing that 1 inch on the map represents 63,360 inches on the ground may be difficult to conceptualize unless we realize that 63,360 inches is equal to 1 mile. Thus, the representative fraction 1:63,360 means the map has the same scale as a map with a verbal scale of 1 inch to 1 mile.

A **graphic scale,** or **bar scale,** is useful for making distance measurements on a map. Graphic scales are graduated lines (or bars)

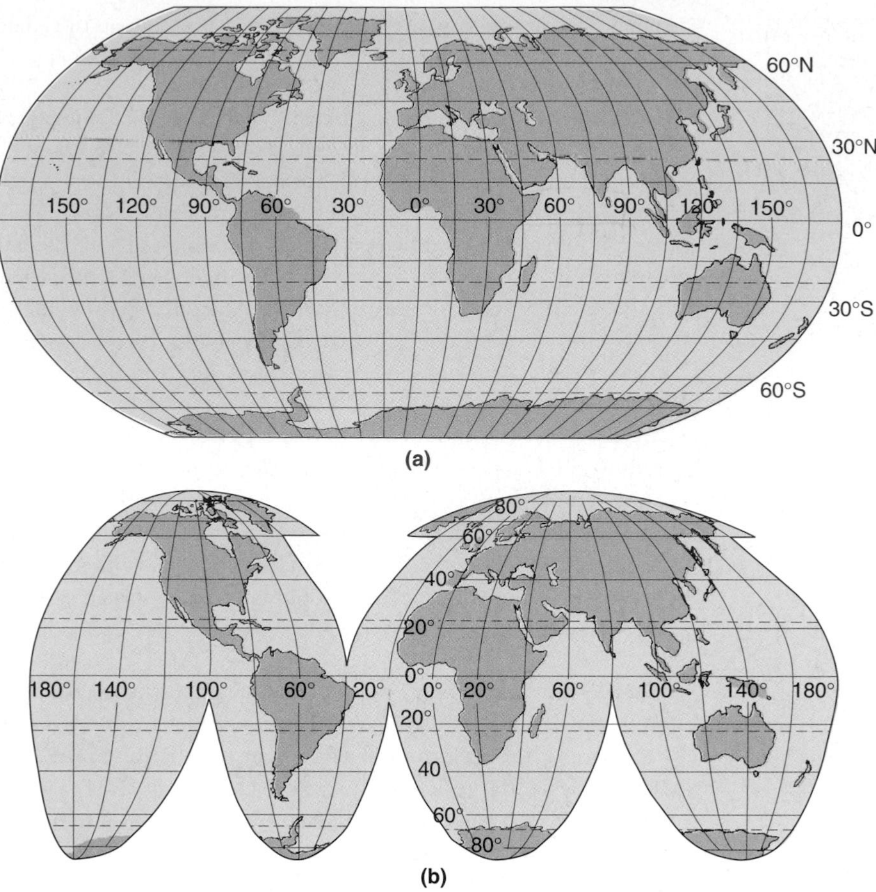

• FIGURE 2.23

The Robinson projection (a) is considered a compromise projection because it departs from equal area to better depict the shape of the continents, but seeks to show both area and shape reasonably well, although neither are truly accurate. Distortion in projections can be also reduced by interruption (b)—that is, by having a central meridian for each segment of the map.

Compare the distortion of these maps with the Mercator projection (Fig. 2.19). What is a disadvantage of (b) in terms of usage?

marked with map distances that are proportional to distances on the Earth. To use a graphic scale, take a straight edge of a piece of paper, and mark the distance between any two points on the map. Then use the graphic scale to find the equivalent distance on Earth's surface. Graphic scales have two major advantages:

1. It is easy to determine distances on the map, because the graphic scale can be used like a ruler to make measurements.
2. They are applicable even if the map is reduced or enlarged, because the graphic scale (on the map) will also change proportionally in size. This is particularly useful because maps can be reproduced or copied easily in a reduced or enlarged scale using computers or photocopiers. The map and the graphic scale, however, must be enlarged or reduced together (the same amount) for the graphic scale to be applicable.

Maps are often described as being of small, medium, or large scale (• Fig. 2.25). *Small-scale* maps show large areas in a relatively small size, include little detail, and have large denominators in their representative fractions. *Large-scale* maps show small areas of Earth's surface in greater detail and have smaller denominators in their representative fractions. To avoid confusion, remember that 1/2 is a *larger* fraction than 1/100, and small scale means that Earth features are shown very small. A large-scale map would show the same features larger. Maps with representative fractions larger than 1:25,000 are large scale. Medium-scale maps have representative fractions between 1:25,000 and 1:250,000. Small-scale maps have representative fractions less than 1:250,000. This classification follows the guidelines of the U.S. Geological Survey (USGS), publisher of many maps for the federal government and for public use.

Direction The orientation and geometry of the geographic grid give us an indication of direction because parallels of latitude are east–west lines and meridians of longitude run directly north-south. Many maps have an arrow pointing to north as displayed on the map. A north arrow may indicate either *true north* or *magnetic north*—or two north arrows may be given, one for true north and one for magnetic north.

Earth has a magnetic field that makes the planet act like a giant bar magnet, with a magnetic north pole and a magnetic south pole, each with opposite charges. Although the magnetic poles shift position slightly over time, they are located in the Arctic and Antarctic regions and do not coincide with the geographic poles. Aligning itself with Earth's magnetic field, the north-seeking end of a compass needle points toward the magnetic north pole. If we know the **magnetic declination,** the angular difference between magnetic north and true geographic north, we can compensate for this dif-

• FIGURE 2.24

Map scales. A *verbal scale* states the relationship between a map measurement and the corresponding distance that it represents on the Earth. Verbal scales generally mix units (centimeters/ kilometer or inches/mile). A *representative fraction (RF)* scale is a ratio between a distance on a map (1 unit) and its actual length on the ground (here, 100,000 units). An RF scale requires that measurements be in the same units both on the map and on the ground. A *graphic scale* is a device used for measuring distances on the map in terms of distances on the ground.

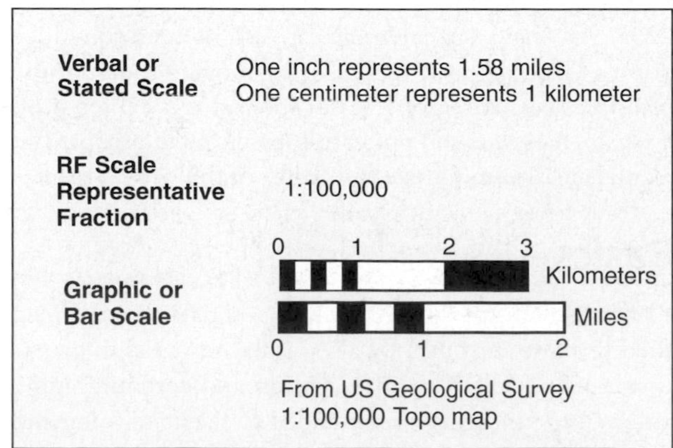

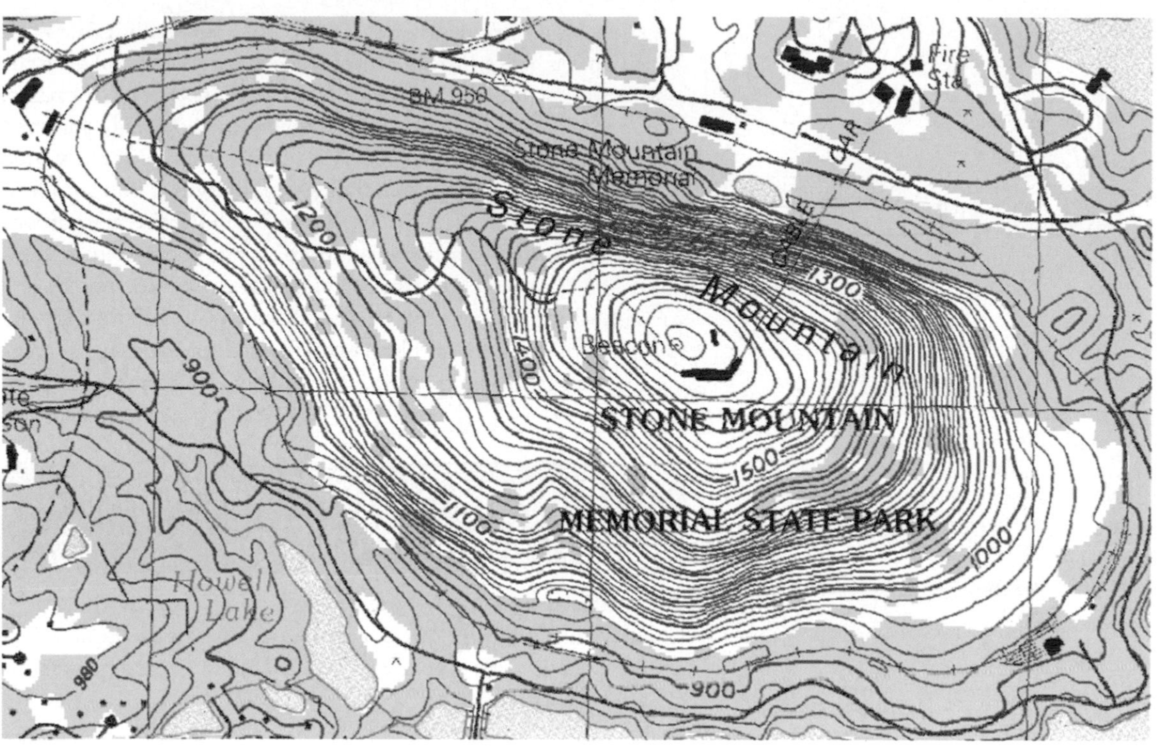

(a) 1:24,000 large-scale map

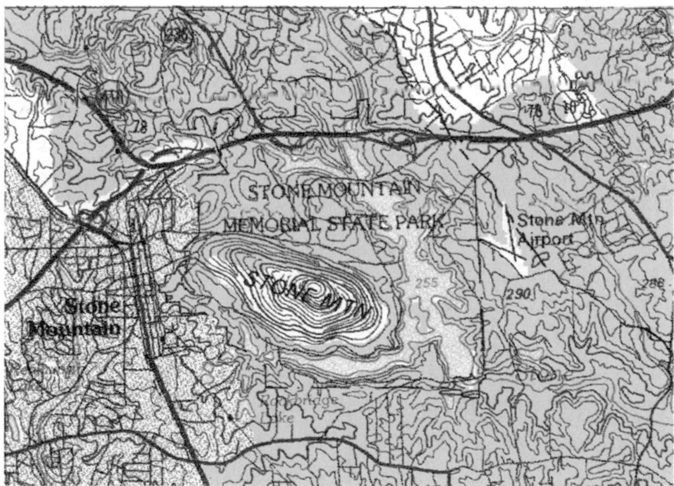

(b) 1:100,000 small-scale map

• FIGURE 2.25

Map scales: Larger versus smaller. The designations *small scale* and *large scale* are related to a map's *representative fraction* (RF) scale. These maps of Stone Mountain Georgia illustrate two scales: (a) 1:24,000 (larger scale) and (b) 1:100,000 (smaller scale). It is important to remember that an RF scale is a *fraction* that represents the proportion between a length on the map and the true distance it represents on the ground. One centimeter on the map would equal the number of centimeters in the denominator of the RF on the ground.

Which number is smaller—1/24,000 or 1/100,000? Which scale map shows more land area—the larger-scale map or the smaller-scale map?

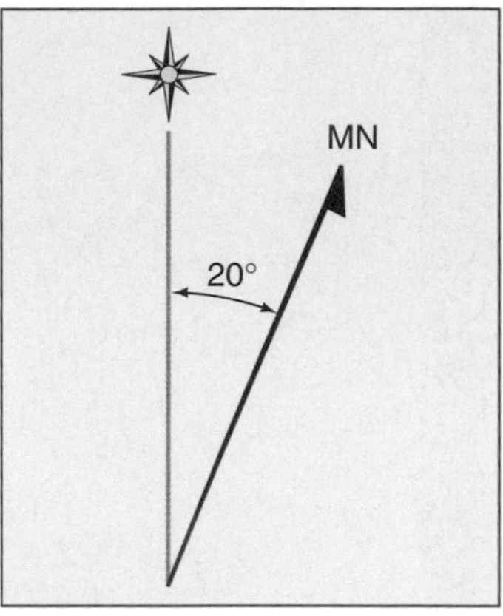

• FIGURE 2.26
Map symbol showing true north, symbolized by a star representing Polaris (the North Star), and magnetic north, symbolized by an arrow. The example indicates 20°E magnetic declination.
In what circumstances would we need to know the magnetic declination of our location?

ference (• Fig. 2.26). Thus, if our compass points north and we know that the magnetic declination for our location is 20°E, we can adjust our course knowing that our compass is pointing 20°E of true north. To do this, we should turn 20°W from the direction indicated by our compass in order to face true north. Magnetic declination varies from place to place and also changes through time. For this reason, magnetic declination maps are revised periodically, so using a recent map is very important. A map of magnetic declination is called an *isogonic map* (• Fig. 2.27), and *isogonic lines* connect locations that have equal declination.

Compass directions can be given by either the azimuth system or the bearing system (see Appendix B). In the **azimuth** system, direction is given in degrees of a full circle (360°) clockwise from north. That is, if we imagine the 360° of a circle with north at 0° (and at 360°) and read the degrees clockwise, we can describe a direction by its number of degrees away from north. For instance, straight east would have an azimuth of 90°, and due south would be 180°. The **bearing** system divides compass directions into four quadrants of 90° (N, E, S, W), each numbered by directions in degrees away from either north or south. Using this system, an azimuth of 20° would be north, 20° east (20° east of due north), and an azimuth of 210° would be south, 30° west (30° west of due south). Both azimuths and bearings are used for mapping, surveying, and navigation for both military and civilian purposes.

Displaying Spatial Data and Information on Maps

Thematic maps are designed to focus attention on the spatial extent and distribution of one feature (or a few related ones). Examples include maps of climate, vegetation, soils, earthquake epicenters, or tornadoes.

Discrete and Continuous Data

There are two major types of spatial data, discrete and continuous. **Discrete data** means that either the phenomenon is located at a particular place or it is not—for example, hot springs, tropical rainforests, rivers, tornado paths, or earthquake faults. Discrete data are represented on maps by point, area, or line symbols to show their locations and distributions (• Figs. 2.28a–c). *Regions* are discrete areas that exhibit a common characteristic or set of characteristics within their

• FIGURE 2.27
Isogonic map of the conterminous United States, showing the magnetic declination that must be added (west declination) or subtracted (east declination) from a compass reading to determine true directions.
What is the magnetic declination of your hometown to the nearest degree?

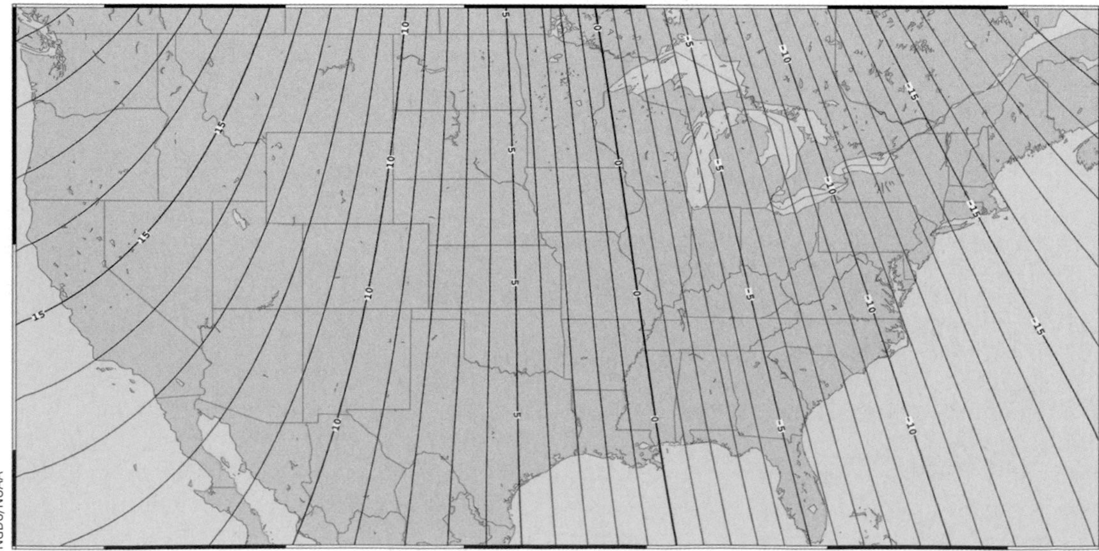

NGDC/NOAA

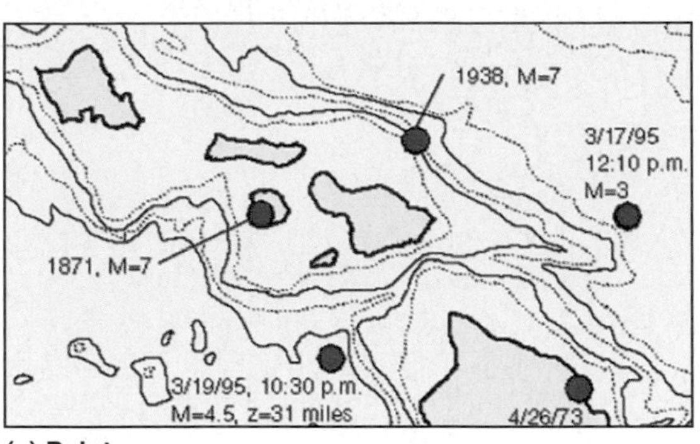

(a) Points

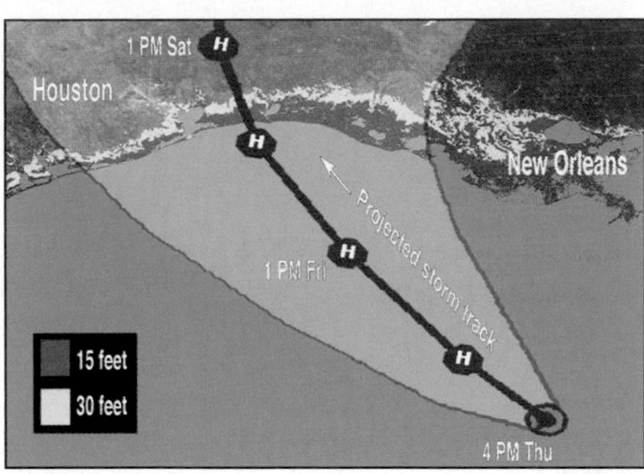

(b) Lines

NASA/JPL/NGA

(c) Areas

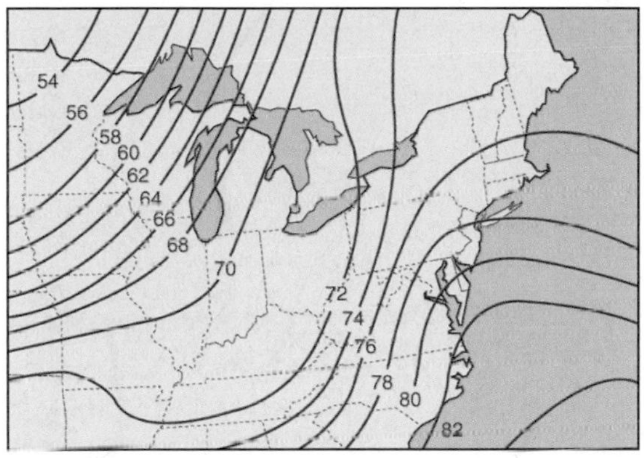

(d) Continuous variable

• FIGURE 2.28
Discrete and continuous spatial data (variables). *Discrete variables* represent features that are present at certain locations but do not exist everywhere. The locations, distributions, and patterns of discrete features are of great interest in understanding spatial relationships. Discrete variables can be (a) *points* representing, for example, locations of large earthquakes in Hawaii (or places where lightning has struck or locations of water-pollution sources), (b) *lines* as in the path taken by Hurricane Rita (or river channels, tornado paths, or earthquake fault lines), (c) *areas* like the land burned by a wildfire (or clear-cuts in a forest, or the area where an earthquake was felt). *A continuous variable* means that every location has a certain measurable characteristic; for example, everywhere on Earth has an elevation, even if it is zero (at sea level) or below (a negative value). Changes in a continuous variable over an area can be represented by isolines, shading, or colors, or with a 3-D appearance. The map (d) shows the continuous distribution of temperature variation in part of eastern North America.
Can you name other environmental examples of discrete and continuous variables?

boundaries, and are typically represented by different colors or shading to differentiate one region from another. Physical geographic regions include areas of similar soil, climate, vegetation, landform type, or many other characteristics (see the world and regional maps throughout this book).

Continuous data means that a measurable numerical value for a certain characteristic exists everywhere on Earth (or within the area of interest displayed); for example, every location on Earth has a measurable elevation (or temperature, or air pressure, or population density). The distribution of continuous data is often shown using **isolines**—lines on a map that connect points with the same numerical value (Fig. 2.28d). Isolines that we will be using later on include *isotherms,* which connect points of equal temperature; *isobars,* which connect points of equal barometric pressure; *isobaths* (also called bathymetric contours), which connect points with equal water depth; and *isohyets,* which connect points receiving equal amounts of precipitation.

Topographic Maps

Topographic contour lines are isolines that connect points on a map that are at the same elevation above mean sea level (or below sea level such as in Death Valley, California). For example, if we walk around a hill along the 1200-foot contour line shown on the map, we would always be 1200 feet above sea level, maintaining

GEOGRAPHY'S SPATIAL SCIENCE PERSPECTIVE

Using Vertical Exaggeration to Portray Topography

Most maps present a landscape as if viewed from directly overhead, looking straight down. This perspective is sometimes referred to as a *map view* or *plan view* (like architectural house plans). Measurements of length and distance are accurate, as long as the area depicted is not so large that Earth's curvature becomes a major factor. Topographic maps, for example, show spatial relationships in two dimensions (length and width on the map, called x and y coordinates in mathematical Cartesian terms). Illustrating terrain, as represented by differences in elevation, requires some sort of symbol to display elevational data on the map. Topographic maps use contour lines, which can also be enhanced by relief shading (see the Map Interpretation, Volcanic Landforms, in Chapter 14 for an example).

For many purposes, though, a side view, or an oblique view, of what the terrain looks like (also called *perspective*) helps us visualize the landscape (see Figs. 2.34 and 2.35). Block diagrams, 3-D models of Earth's surface, are very useful for showing the general layout of topography from a perspective view. They provide a perspective with which most of us are familiar, similar to looking out an airplane window or from a high vantage point. Block diagrams are excellent for illustrating 3-D relationships in a landscape scene, and information about the subsurface can be included. But such diagrams are not intended for making accurate measurements, and many block diagrams represent hypothetical or stylized, rather than actual, landscapes.

A topographic *profile* illustrates the shape of a land surface as if viewed directly from the side. It is basically a graph of elevation changes over distance along a transect line. Elevation and distance information collected from a topographic map or from other elevation data in spatial form can be used to draw a topographic profile. Topographic profiles show the terrain. If the geology of the subsurface is represented as well, such profiles are called *geologic cross sections*.

Block diagrams, profiles, and cross sections are typically drawn in a manner that stretches the vertical presentation of the features being depicted. This makes mountains appear taller than they are in comparison to the landscape, the valleys deeper, the terrain more rugged, and the slopes steeper. The main reason why vertical exaggeration is used is that it helps make subtle changes in the terrain more noticeable. In addition, land surfaces are really much flatter than most people think they are. In fact, cartographers have worked with psychologists to determine what degree of vertical exaggeration makes a profile or block diagram appear most "natural" to people viewing a presentation of elevation differences in a landscape. For technical applications, most profiles and block diagrams will indicate how much the vertical presenta-

USGS/ digital elevation model by Steve Schilling; geo-referenced by Frank Trusdell

Anatahan Island in a natural-scale presentation, without vertical exaggeration (compare to Fig. 2.31).

a constant elevation and walking on a level line. Contour lines are an excellent means for showing the elevation changes and the form of the land surface on a map. The arrangement, spacing, and shapes of the contours give a map reader a mental image of what the topography (the "lay of the land") is like (● Fig. 2.29).

● Figure 2.30 illustrates how contour lines portray the land surface. The bottom portion of the diagram is a simple contour map of an asymmetrical hill. Note that the elevation difference between adjacent contour lines on this map is 20 feet. The constant difference in elevation between adjacent contour lines is called the **contour interval.**

If we hiked from point A to point B, what kind of terrain would we cover? We start from sea level point A and immediately begin to climb. We cross the 20-foot contour line, then the 40-foot, the 60-foot, and, near the top of our hill, the 80-foot contour level. After walking over a relatively broad summit that is above 80 feet but not as high as 100 feet (or we would cross another contour line), we once again cross the 80-foot contour line, which means we must be starting down. During our descent, we cross each lower level in turn until we arrive back at sea level (point B).

In the top portion of Figure 2.30, a **profile** (side view) helps us to visualize the topography we covered in our walk.

tion has been stretched, so that there is no misunderstanding. Two times vertical exaggeration means that the feature is presented two times higher than it really is, but the horizontal scale is correct. Note that the image of Anatahan in Figure 2.31 has three times vertical exaggeration; that is, the mountains appear to be three times as high and steep as they really are. Compare that presentation to the natural scale (not vertically exaggerated) version shown here. This is how the island and the seafloor actually look in terms of slope steepness and relief.

To illustrate why vertical exaggeration is used, look at the three profiles of a volcano in the Hawaiian Islands. Which do you think shows the true, natural-scale profile of this volcanic mountain? Which one "looks" the most natural to you? What is the true shape of this volcano? After making a guess, check below for the answer and the degree of vertical exaggeration in each of the three profiles. Note that this is a huge volcano—the profile extends horizontally for 100 kilometers.

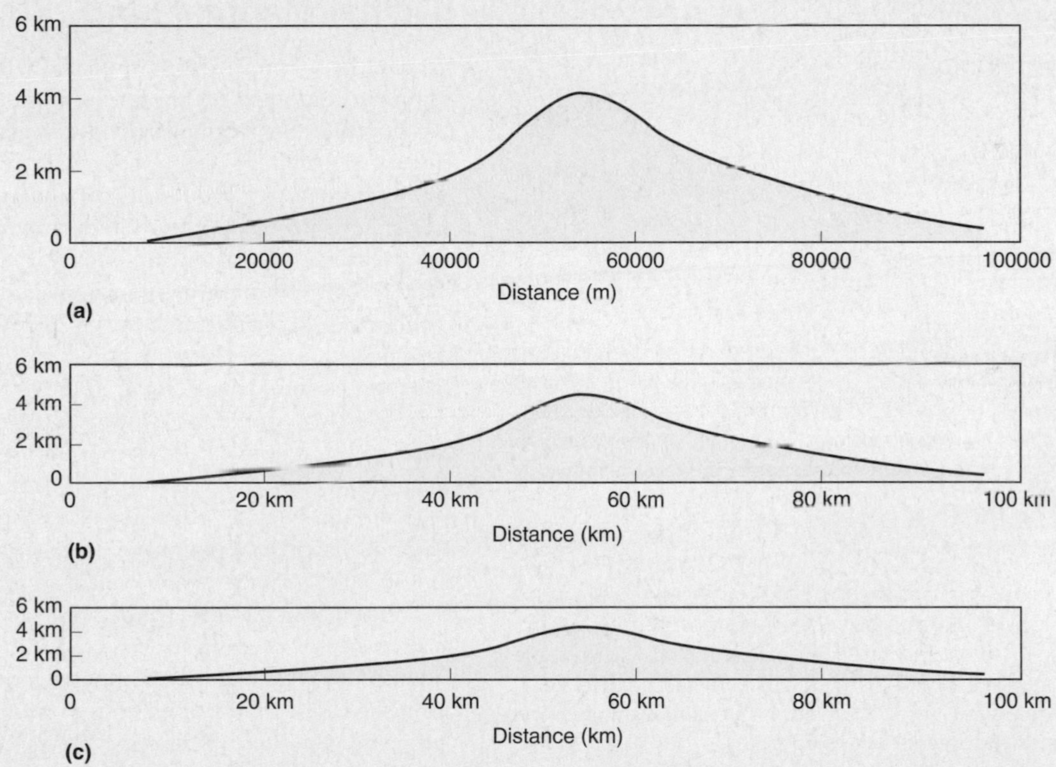

Profiles of Mauna Kea, Hawaii (data from NASA): (a) 4X vertical exaggeration; (b) 2X vertical exaggeration; (c) natural-scale profile—no vertical exaggeration.

We can see why the trip up the mountain was more difficult than the trip down. Closely spaced contour lines near point A represent a steeper slope than the more widely spaced contour lines near point B. Actually, we have discovered something that is true of all isoline maps: The closer together the lines are on the map, the steeper the **gradient** (the greater the rate of vertical change per unit of horizontal distance). When studying a contour map, we should understand that the slope between contours almost always changes gradually, and it is unlikely that the land drops off in steps downslope as the contour lines might suggest.

Topographic maps use symbols to show many other features in addition to elevations (see Appendix B)—for instance, water bodies such as streams, lakes, rivers, and oceans or cultural features such as towns, cities, bridges, and railroads. The USGS produces topographic maps of the United States at several different scales. Some of these maps—1:24,000, 1:62,500, and 1:250,000—use English units for their contour intervals. Many recent maps are produced at scales of 1:25,000 and 1:100,000 and use metric units. Contour maps that show undersea topography are called *bathymetric charts,* and in the United States, they are produced by the National Ocean Service.

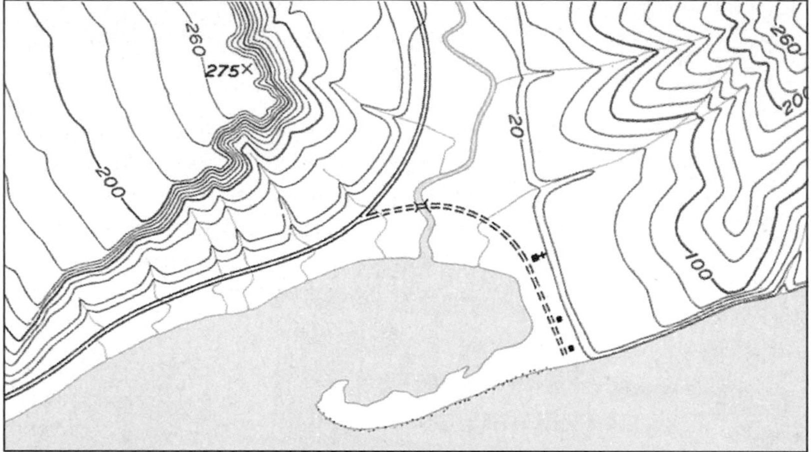

• FIGURE 2.29
(Top) A view of a river valley and surrounding hills, shown on a shaded-relief diagram. Note that a river flows into a bay partly enclosed by a sand spit. The hill on the right has a rounded surface form, but the one on the left forms a cliff along the edge of an inclined but flat upland. (Bottom) The same features represented on a contour map. **If you had only a topographic map, could you visualize the terrain shown in the shaded-relief diagram?**

• FIGURE 2.30
A topographic profile and contour map. Topographic contours connect points of equal elevation relative to mean sea level. The upper part of the figure shows the topographic profile (side view) of an island. Horizontal lines mark 20-foot intervals of elevation above sea level. The lower part of the figure shows how these contour lines look in map view. **Study this figure and the maps in Figure 2.25. What is the relationship between the spacing of contour lines and steepness of slope?**

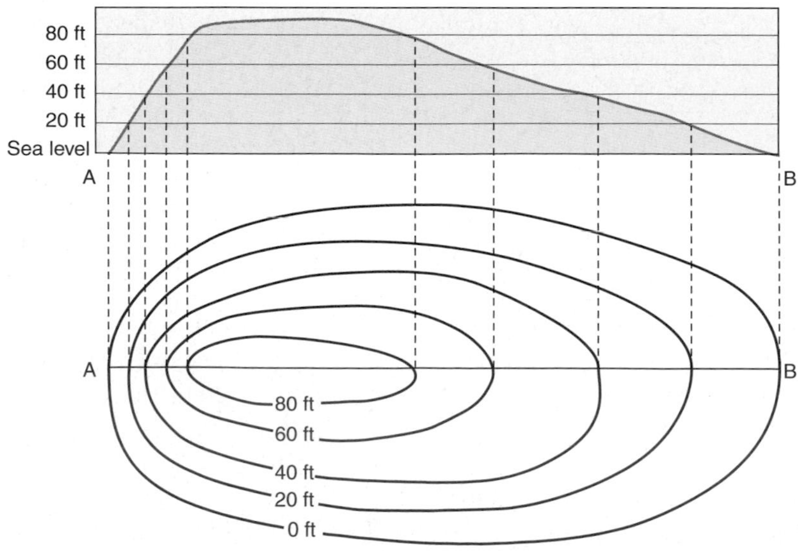

Modern Mapping Technology

Cartography has undergone a technological revolution, from slow, manual methods to an automated and interactive process, using computer systems to store, analyze, and retrieve data, and to draw the final map. For most mapping projects, computer systems are faster, more efficient, and less expensive than the hand-drawn cartographic techniques they have replaced. However, it is still important to understand basic cartographic principles to make a good map. A computer mapping system will draw only what an operator instructs it to draw.

Digital Mapmaking

Today, the vast majority of professionally made maps employ computer technologies, because computer systems offer many important advantages to mapmaking. In digital form, maps can be easily revised because they do not have to be manually redrawn with each revision or major change. Map data, stored in a computer, can be displayed on a monitor and corrected, changed, or improved by the mapmaker. Hundreds of millions of data bits, representing elevations, depths, temperatures, or populations, can be stored in a digital database, accessed, and displayed on a map. The database for a map may also include information on coastlines, political boundaries, city locations, river systems, map projections, and coordinate systems. More than 100 million bits of information are stored and thousands of bits of data plotted to make a typical digitally produced USGS topographic map.

Mapmakers can *tile* together adjacent maps to view a large area or zoom in to see detail on a small area. In addition to scale changes, computer maps allow users to make easy metric conversions as well as changes in projections, contour intervals, symbols, colors, and directions of view (rotating the orientation). The ability to interact with a map and make on-the-spot modifications is essential when representing changing phenomena such as weather systems, air pollution, ocean currents, volcanic eruptions, and forest fires. Digital maps can be instantly disseminated and shared via the Internet, which is a great advantage when spatial information is rapidly changing, or it is important to communicate mapped information as soon as possible.

Digital elevation models (DEMs), computer-generated, 3-D views of topography, are particularly useful to physical geographers, geologists, civil engineers, and other scientists (• Fig. 2.31). A DEM is useful for displaying topography in a way that simulates a 3-D view. Digital elevation data can be used to make many types of terrain displays and maps, including color-scaled contour maps, where areas between contours are assigned a certain color, conventional contour maps, and

USGS/ digital elevation model by Steve Schilling; geo-referenced by Frank Trusdell

• **FIGURE 2.31**
A digital elevation model (DEM) of Anatahan Island (145° 40' E, 16° 22" N), and the surrounding Pacific Ocean floor has been presented in 3-D and colorized according to elevation and sea-floor depth relative to sea level. The vertical scale has been stretched three times compared to the horizontal scale. Refer to the box on vertical exaggeration to see a natural-scale image.
The vertical scale bar represents a distance of 3800 meters, so taking the vertical exaggeration into account, what horizontal distance would the same scale bar length represent in meters?

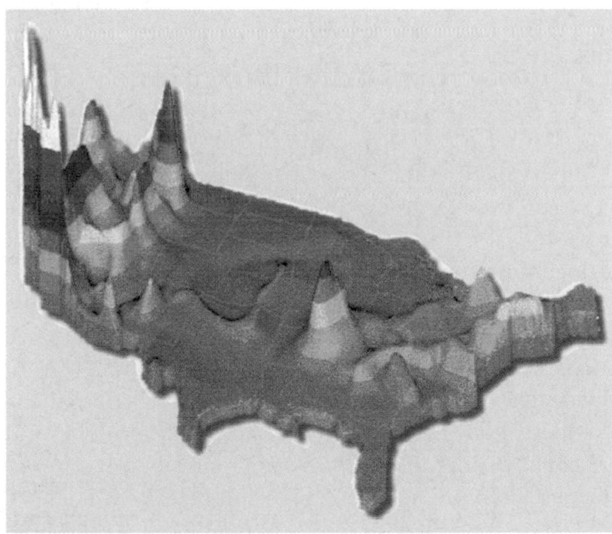

USGS

• **FIGURE 2.32**
Earthquake hazard in the conterminous United States: a continuous variable displayed as a continuous surface in 3-D perspective. Here it is easy to develop a mental map of how potential earthquake danger varies across this part of the United States.
Are you surprised by any of the locations that are shown to have substantial earthquake hazard?

shaded-relief maps. Digital terrain models may be designed to show **vertical exaggeration** by stretching the vertical scale of the display to enhance the relief of an area, as seen in Figure 2.31, where the ocean depth and the island's topography have been increased by a factor of three. This means that the vertical scale is three times larger than the horizontal scale (see the feature on vertical exaggeration in this chapter). Actually, any geographic factor represented by continuous data can be displayed either as a two-dimensional contour map or as a 3-D surface to enhance the visibility of the spatial variation that it conveys (• Fig. 2.32).

Geographic Information Systems

A **geographic information system (GIS)** is an incredibly versatile innovation for map analysis that stores spatial databases, supports spatial data analysis, and facilitates the production of digital maps. A GIS is a computer-based technology that assists the user in the entry, analysis, manipulation, and display of geographic information derived from combining any number of digital *map layers,* each composed of a specific *thematic map* (• Fig. 2.33). A GIS can be used to make the scale and map projection of these map layers identical, thus allowing the information from several or all layers to be combined into new and more meaningful composite maps. GIS is especially useful to geographers as they work to address problems that require large amounts of spatial data from a variety of sources.

What a GIS Does Imagine that you are in a giant map library with thousands of paper maps, all of the same area, but each map shows a different aspect of the same location: one map shows roads, another highways, another trails, another rivers (or soils, or vegetation, or slopes, or rainfall, and so on—the possibilities are limitless). The maps were originally produced at many different sizes, scales, and projections (including some maps that do not preserve shape or area). These cartographic factors will make it very difficult to visually overlay and compare the spatial information among these different maps. You also have digital terrain models and satellite images that you would like to compare to the maps. Further, because few aspects of the environment involve only one factor or exist in spatial isolation, you want to be able to combine a selection of these geographic aspects on a single composite map. You have a spatial-geographic problem, and to solve that problem, you need a way to make several representations of a part of Earth directly comparable. What you need is a GIS and the knowledge of how to use this system.

Data and Attribute Entry The first step is to enter map and image data into a computer system. Each data set is input and stored as a layer of spatial information that represents an individual thematic map layer as a separate digital file (see again Fig. 2.33). Another step is **geocoding,** which is entering and locating spatial data and information in relation to grid coordinates such as latitude and

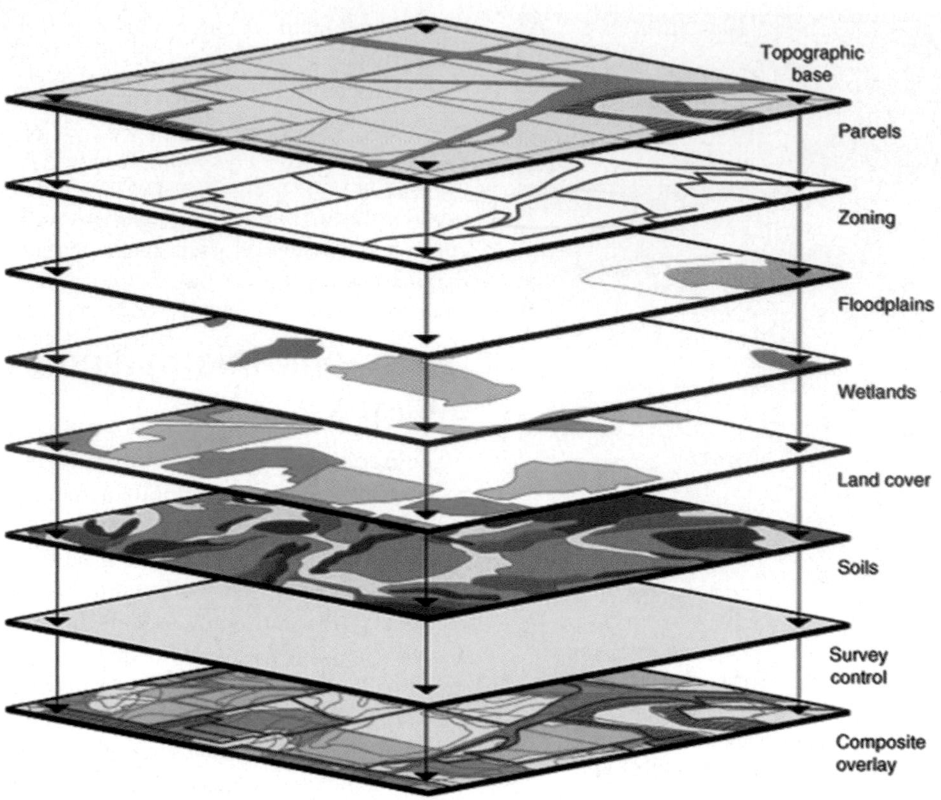

• FIGURE 2.33
Geographic information systems store different information and data as individual map layers. GIS technology is widely used in geographic and environmental studies in which several different variables need to be assessed and compared spatially to solve a problem.
Can you think of other applications for geographic information systems?

longitude. Further, a list of *attributes* (specific feature characteristics, such as the lengths and names of rivers) is attached to each map layer and can be easily viewed.

Registration and Display A GIS can display any layer or any combination of layers, geometrically registered (fitted) to any map projection and at any scale that you specify. The maps, images, and data sets can now be directly compared at the same size, based on grid coordinates arranged on the same map projection and map scale. A GIS can digitally *overlay* any set of thematic map layers that are needed. If you want to see the locations of *homes* on a *river floodplain,* a GIS can quickly create a useful map by retrieving, combining, and displaying the *home* and *floodplain* map layers simultaneously. If you want to see *earthquake faults* and *artificially landfilled areas* in relation to locations of *fire stations* and *police stations,* that composite map will require four layers, but that is no problem for a GIS.

Visualization Models Also referred to as *visualizations,* **visualization models** are computer-generated image models designed to illustrate and explain complex processes and features. Many visualizations are presented as 3-D images and/or as animations. For example, the Earth image shown in the first chapter (see again Fig. 1.4) is a visualization model. Visualization models can combine and present several components of the Earth system in stunning 3-D views, based on actual environmental data and satellite images or air photos. An example is shown in • Figure 2.34 where a satellite image and a DEM are layers in a GIS that can be combined in a 3-D view to produce a landscape visualization model, the Rocky Mountain front at Salt Lake City, based on real image and elevation data. This process is called *draping* (like draping cloth over some object) but the scale and the perspective are accurately registered among the map layers. Visualization models help us understand and conceptualize many environmental processes and features.

Today, the products and techniques of cartography are very different from their beginning forms and they continue to be improved, but the goal of making a representation of Earth remains the same—to effectively communicate geographic and spatial knowledge in a visual format. An example of a digital landscape visualization produced by combining elevation data and a satellite image is shown in • Figure 2.35.

GIS in the Workplace A simple example will help to illustrate the utility of a GIS. Suppose you are a geographer working for the Natural Resources Conservation Service. Your current problem is to control erosion along the banks of a reservoir. You know that erosion is a function of many environmental variables, including soil types, slope steepness, vegetation characteristics, and others. Using a GIS, you would enter map data for each of these variables as a separate layer. You could analyze these variables individually, or you could integrate information from individual layers (soils, slope, vegetation, and so on) to identify the locations most susceptible to erosion. Your resources and personnel could then be directed toward controlling erosion in those target areas. In physical geography and the Earth sciences, GIS is being used to analyze potential coastal flooding from sea-level rise, areas in need of habitat restoration, flood hazard potential, and earthquake distributions, just to list a few examples. The spatial analysis capabilities of a GIS are nearly unlimited and are applicable in almost any career field.

Many geographers are employed in careers that apply GIS technology. The capacity of a GIS to integrate and analyze a wide variety of geographic information, from census data to landform characteristics, makes it useful to both human and physical geographers. With nearly unlimited applications in geography and other disciplines, GIS will continue to be an important tool for understanding our environment and making important decisions based on spatial information.

NASA
(a)

USGS
(b)

NASA/JPL/NIMA
(c)

• **FIGURE 2.34**
A GIS can include (a) digital landscape images from satellites or aircraft, and also (b) digital elevation models, and combine them to make (c) a 3-D model of a landscape, one type of visualization model. This digital model of Salt Lake City, Utah, and the Rocky Mountain front was made by draping a satellite image over a 3-D presentation of the land surface. Digital models like this can be rotated on a computer screen to be viewed from any angle or direction. The examples here are enlarged to show the pixel resolution.

Remote Sensing of the Environment

Remote sensing is the collection of information and data about distant objects or environments. Remote sensing involves the gathering and interpretation of aerial and space imagery, images that have many maplike qualities. Using remote sensing systems, we can also detect objects and scenes that are not visible to humans and can display them on images that we can visually interpret.

Remote sensing is commonly divided into photographic techniques and digital imaging, which may use equipment similar to digital cameras or employ more sophisticated technologies. Today, with the recent widespread use of digital cameras, the use

NASA/JPL/NIMA

• **FIGURE 2.35**
The physical environment of Cape Town, South Africa, is presented in this landscape visualization. Satellite imagery and elevation data were combined to produce this scene, and computer enhanced to show a 3-D perspective. The topography is vertically exaggerated by a factor of two to enhance the terrain. This image, unlike the example in Figure 2.34, is not greatly enlarged so the pixels are less visible.
Does the terrain in this landscape look vertically exaggerated to you, or does the scene look fairly natural?

of the term "photograph" is changing in common usage, but in technical terms, **photographs** are made by using cameras to record a picture on *film*. Digital cameras or image scanners produce a **digital image**—an image that is converted into numerical data. Most images returned from space are digital, because digital data can be easily broadcast back to Earth. Digital imagery also offers the advantage of computer-assisted data processing, image enhancement, interpretation, and image sharing, and can provide a landscape image as a thematic layer in a GIS. Digital images consist of **pixels,** a term that is short for "picture element," the smallest area resolved in a digital picture (as seen in the enlarged inset of the San Francisco International Airport in the chapter-opening image). A key factor in digital images is **spatial resolution,** expressed as how small an area (on the Earth) each pixel represents—for example, 15–30 meters for a satellite image of a city or small region. Satellites that image an entire hemisphere at once, or large continental areas, use resolutions that are much more coarse, to produce a more generalized scene. A digital image is similar to a mosaic, made up of grid cells with varying colors or tones that form a picture. Each cell (pixel) has a locational address within the grid and a value that represents the brightness or color of the picture area that the pixel represents. The digital values in an array of grid cells (pixels) are translated into an image by computer technology.

Digital cameras for personal use express resolution in *megapixels,* or how many million pixels make an image. The more megapixels a camera or digital scanner can image, the better the resolution and the sharper the image will be, but this also depends on how much an image is enlarged or reduced in size, while maintaining the same spatial resolution. If the pixel size is small enough on the finished image, the mosaic effect will be either barely noticeable or invisible to the human eye.

Aerial Photography and Image Interpretation

Aerial photographs have provided us with "bird's-eye" views of our environment via kites and balloons even before airplanes were invented, but aircraft led to a tremendous increase in the availabil-

USDA

(a)

USGS

(b)

• FIGURE 2.36

(a) Oblique photos provide a "natural view," like looking out of an airplane window. This oblique aerial photograph in natural color shows farmland, countryside, and forest. (b) Vertical photos provide a maplike view that is more useful for mapping and making measurements (as in this view of Tampa Bay, Florida).

What are the benefits of an oblique view, compared to a vertical view?

ity of aerial photography (• Figs. 2.36a and 2.36b). Both air photos and digital images may be *oblique* (Fig. 2.36a), taken at an angle other than perpendicular to Earth's surface, or *vertical* (Fig. 2.36b), looking straight down. Image interpreters use aerial photographs and digital imagery to examine and describe relationships among objects on Earth's surface. A device called a *stereoscope* allows overlapped pairs of images (typically aerial photos) taken from different positions to allow viewing of features in three dimensions.

Near-infrared (NIR) energy, light energy at wavelengths that are too long for our eyes to see, cuts though atmospheric haze better than visual light does. Natural-color photographs taken from very high altitudes or from space, tend to have low contrast and can appear hazy (• Fig. 2.37a). Photographs and digital images that use NIR tend to provide very clear images when taken from high altitude or space. Color NIR photographs and digital images are sometimes referred to as "false color" pictures, because on NIR, healthy grasses, trees, and most plants will show up as bright red, rather than green (Fig. 2.37b and see again the chapter opening satellite image). Near-infrared photographs and images have many applications for environmental study, particularly for water resources, vegetation, and crops. An incorrect, but widely held, notion of NIR techniques is that they image heat, or temperature variations. Near-infrared energy is light, images as is *reflected* off of surfaces, and not *radiated* heat energy.

• FIGURE 2.37

A comparison of a natural-color photograph (a) to the same scene in false-color near-infrared. (b) Red tones indicate vegetation; dark blue—clear, deep water; and light blue—shallow or muddy water. This is a wetlands area on the coast of Louisiana.

If you were asked to make a map of vegetation or water features, which image would you prefer to use and why?

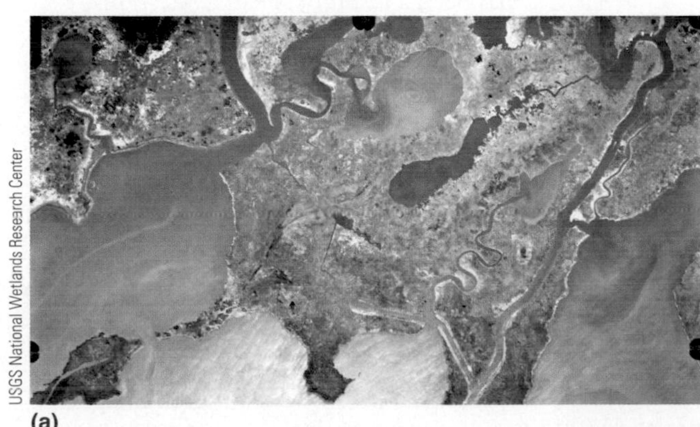
USGS National Wetlands Research Center

(a)

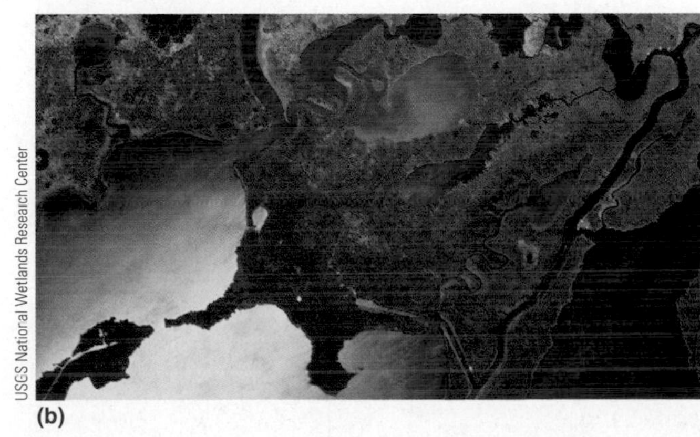
USGS National Wetlands Research Center

(b)

Specialized Remote Sensing Techniques

Many different remote sensing systems are in use, each designed for specific imaging applications. Remote sensing may use UV light, visible light, NIR light, thermal infrared energy (heat), lasers, and microwaves (radar) to produce images.

Thermal infrared (TIR) images show patterns of heat and temperature instead of light and can be taken either day or night by TIR sensors. Spatially recorded heat patterns are digitally converted into a visual image. TIR images record temperature differences, and can detect features that are hot or cold compared to their surroundings. Hot objects show up in light tones, and cool objects will be dark, but typically a computer is used to emphasize heat differences by colorizing the image. Some TIR applications include finding volcanic hot spots and geothermal sites, locating forest fires through dense smoke, finding leaks in building insulation or in pipelines, and detecting thermal pollution in lakes and rivers.

Weather satellites also use thermal infrared imaging for understanding certain atmospheric conditions. We have all seen these TIR images on television when the meteorologist says, "Let's see what the weather satellite shows." Clouds are depicted in black on the original thermal image because they are colder than their background, the surface of Earth below. Because we don't like to see black clouds, the image tones are reversed, like a photo negative, so that the clouds appear white. These images may also be colorized to show cloud heights, because clouds are progressively colder at higher altitudes (● Fig. 2.38).

Radar (*RA*dio *D*etection *A*nd *R*anging) transmits radio waves and produces an image by reading the energy signals that are reflected back. Radar systems can operate day or night and can see through clouds.

There are several kinds of **imaging radar** systems that sense the surface (topography, rock, water, ice, sand dunes, and so forth) by converting radar reflections into a maplike image. **Side-Looking Airborne Radar (SLAR)** was designed to image areas located to the side of an aircraft. Imaging radar generally does not "see" trees (depending on the system), so it makes an image of the land surface rather than a crown of trees. Excellent for mapping terrain and water features (● Fig. 2.39), SLAR is used most often to map remote, inhospitable, inaccessible, cloud-covered regions, or heavily forested areas.

Radar is also used to monitor and track thunderstorms, hurricanes, and tornadoes (● Fig. 2.40). **Weather radar** systems produce maplike images of precipitation. Radar penetrates clouds (day or night) but reflects off of raindrops and other precipitation, producing a signal on the radar screen. Precipitation patterns are typically the kind of weather radar image that we see on television. The latest systems include *Doppler radar,* which can determine precipitation patterns, direction of movement, and how fast a storm is approaching (much as police radar measures vehicle speed).

Sonar (*SO*und *NA*vigation and *R*anging) uses the reflection of emitted sound waves to probe the ocean depths. Much of our understanding of sea floor topography, and mapping of the sea floor, has been a result of sonar applications.

• FIGURE 2.38
Thermal infrared weather images show patterns of heat and cold. This is part of the southeastern United States beamed back from a U.S. weather satellite called GOES (Geostationary Operational Environmental Satellite). Original thermal images are black and white, but in this image the stormy areas have been colorized. Reds, oranges, and yellows show where the storm is most intense and blues less intense.
Why are the storm patterns on weather images like this useful to us?

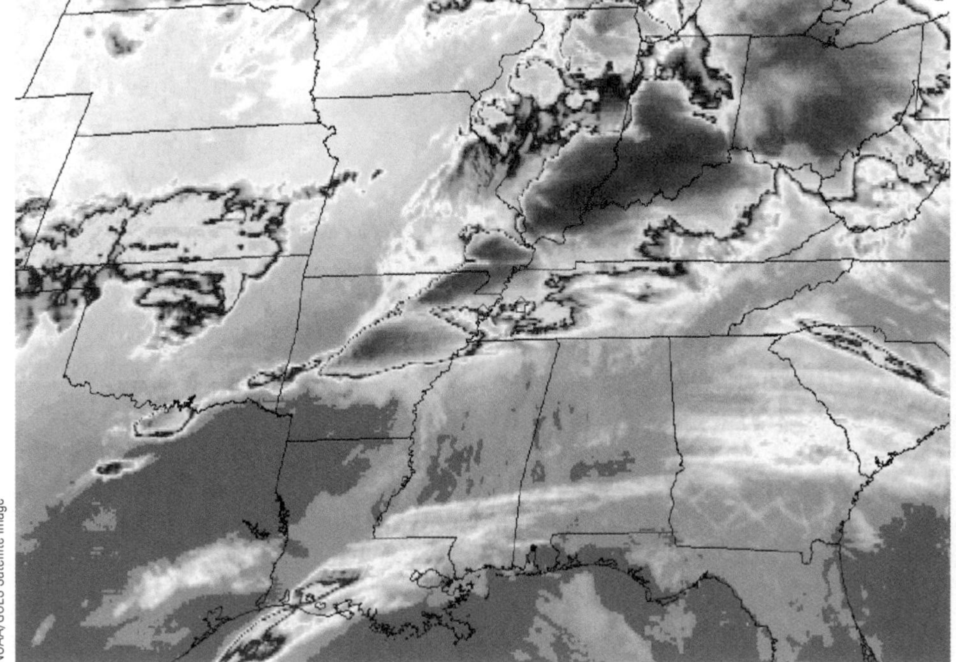

NOAA/GOES Satellite Image

Multispectral Remote Sensing Applications

Multispectral remote sensing means using and comparing more than one type of image of the same place, whether taken from space or not (for example, radar and TIR images, or NIR and normal color photos). Common on satellites, *multispectral scanners* produce digital images by sensing many kinds of energy simultaneously that are relayed to receiving stations to be stored as separate image files. Each

• **FIGURE 2.39**
Imaging radar reflections produce an image of a landscape. Radar reflections are affected by many factors, particularly the surface materials, as well as steepness and orientation of the terrain. This radar scene taken from Earth orbit shows the topography near Sunbury, Pennsylvania, where the West Branch River flows into the Susquehanna River (north is at the top of the image). Parallel ridges, separated by linear valleys, form the Appalachian Mountains in Pennsylvania. River bridges provide a sense of scale.

part of the energy spectrum yields different information about aspects of the environment. The separate images, just like thematic map layers in a GIS can be combined later, depending on which ones are needed for analysis.

Many types of images can be generated from multispectral data, but the most familiar is the **color composite image** (• Fig. 2.41). Blending three images of the same location, by overlaying pixel data from three different *wavelengths* of reflected light, creates a digital color composite. A common color composite image resembles a false-color NIR photograph, with a color assignment that resembles color NIR photos (see again the chapter-opening image). On a standard NIR color composite, red is healthy vegetation, pinkish tones may represent vegetation that is under stress; barren areas show up as white or brown; clear, deep-water bodies are dark blue; and muddy water appears light blue. Clouds and snow are bright white. Urbanized areas are blue-gray. Although the colors are visually important, the greatest benefit of digital multispectral imagery is that computers can be used to identify, classify, and map (in a first approximation) these kinds of areas automatically, based on color and tone differences. These digital images can be input as thematic layers for integration into a GIS, and geographic information systems are often closely linked to the analysis of remotely sensed images.

The use of digital technologies in mapping and imaging our planet and its features continues to provide us with data and information that contribute to our understanding of the Earth system. Through continuous monitoring of the Earth system, global, regional, and even local changes can be detected and mapped. Geographic information systems have the capability to match and combine thematic layers of any sort, instantly accessing any combination of layers that we need to solve complex spatial problems.

Maps and various kinds of representations of Earth continue to be essential tools for geographers and other scientists, whether they are on paper, displayed on a computer monitor, hand drawn in the field, or stored as a mental image. Digital mapping, GPS, GIS, and remote sensing have revolutionized the field of geography, but the fundamental principles concerning maps and cartography remain basically unchanged.

• **FIGURE 2.40**
NEXRAD radar image of thunderstorms associated with a storm front shows detail of a severe storm with a hook-shaped pattern that is associated with tornadoes. Colors show rainfall intensity: green—light rainfall, yellow—moderate, and orange-red—heavy. The radar has also picked up reflections from huge groups of flying bats, among the millions that live in this region.
How are weather and imaging radar scenes different in terms of what they record about the environment?

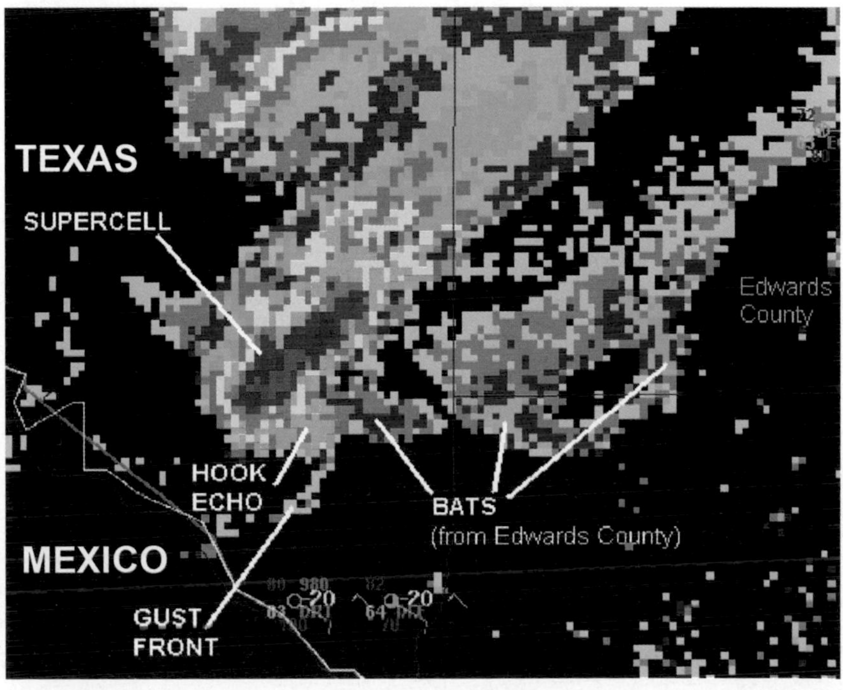

GEOGRAPHY'S PHYSICAL SCIENCE PERSPECTIVE

Polar versus Geostationary Satellite Orbits

Satellite systems that return images from orbit are designed to produce many different kinds of Earth imagery. Some of these differences are related to the type of orbit the satellite system is using while scanning the surface. There are two distinctively different types of orbits: the *polar orbit* and the *geostationary orbit* (sometimes called a *geosynchronous orbit*), each with a different purpose.

The polar orbit was developed first; as its name implies, the satellite orbits Earth from pole to pole. This orbit has some distinct advantages. It is typically a low orbit for a satellite, usually varying in altitude from 700 kilometers (435 mi) to 900 kilometers (560 mi). At this height, but also depending on the equipment used, a polar-orbiting satellite can produce clear, close-up images of Earth. However, at this distance the satellite must move at a fast orbital velocity to overcome the gravitational pull of Earth. This velocity can vary, but for polar orbiters it averages around 27,400 kilometers/hour (17,000 mph), traveling completely around Earth in about 90 minutes. While the satellite orbits from pole to pole, Earth rotates on its axis below, so each orbit views a different path along the surface. Thus, polar orbits will at times cover the dark side of the planet. To adjust for this, a slightly modified polar orbit was developed, called a *sun synchronous orbit.* If the polar orbit is tilted a few degrees off the vertical, then it can remain over the sunlit part of the globe at all times. Most modern polar-orbiting satellites are sun synchronous (a near-polar orbit).

The geostationary orbit, developed later, offered some innovations in satellite image gathering. A geostationary orbit must have three characteristics: (1) it must move in the same direction as Earth's rotation; (2) it must orbit exactly over the equator; and (3) the orbit must be perfectly circular. The altitude of the orbit must be also exact, at 35,900 kilometers (22,300 mi). At this greater height, the orbital velocity is less than that for a polar orbit—11,120 kilometers/hour (6900 mph). When these conditions are met, the satellite's orbit is perfectly synchronized with Earth's rotation, and the satellite is always located over the same spot above Earth. This orbit offers some advantages. First, at its great distance, a geostationary satellite can view an entire Earth hemisphere in one image (that is, the half it is always facing—a companion satellite images the other hemisphere). Another great advantage is that geostationary satellites can send back a continuous stream of images for monitoring changes in our atmosphere and oceans. A film loop of successive geostationary images is what we see on TV weather broadcasts when we see motion in the atmosphere. Geostationary satellite images give us broad regional presentations of an entire hemisphere at once. Near polar–orbiting satellites take image after image in a swath and rely on Earth's rotation to cover much of the planet, over a time span of about a week and a half.

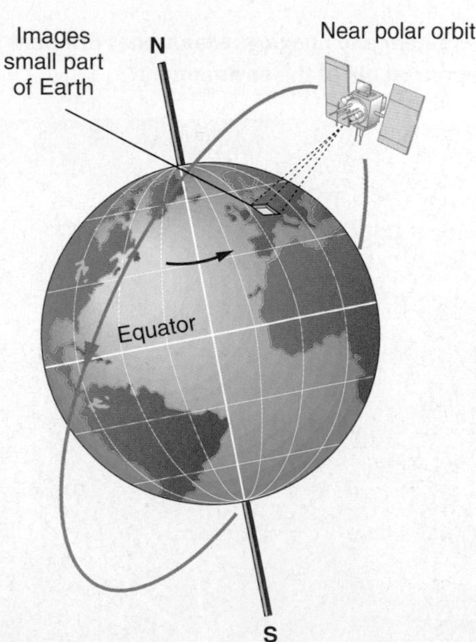

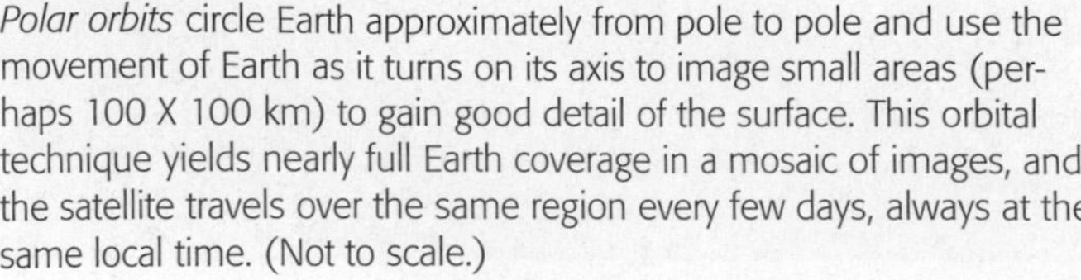
Polar orbits circle Earth approximately from pole to pole and use the movement of Earth as it turns on its axis to image small areas (perhaps 100 X 100 km) to gain good detail of the surface. This orbital technique yields nearly full Earth coverage in a mosaic of images, and the satellite travels over the same region every few days, always at the same local time. (Not to scale.)

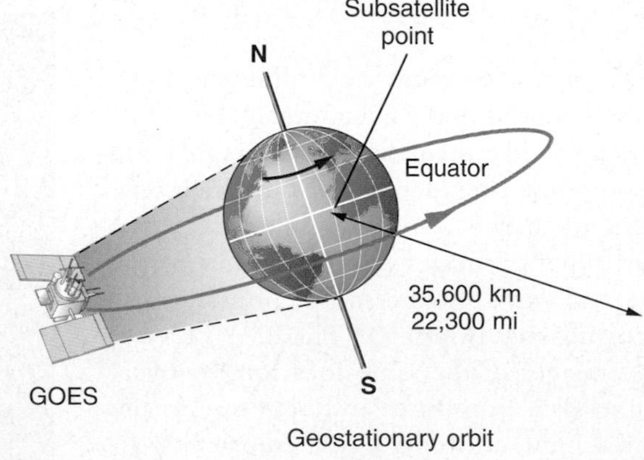

Geostationary orbits are used with satellites orbiting above the equator at a speed that is synchronized with Earth rotation so that the satellite can image the same location continuously. Many weather satellites use this orbit at a height that will permit imaging an entire hemisphere of Earth. (Not to scale.)

• **FIGURE 2.41**
This color composite satellite image of the New Orleans area along the Mississippi River in Louisiana was taken 5 years before the disastrous impact of Hurricane Katrina, which devastated much of the city in 2005.
What features and geographic patterns can you recognize on this natural-color image?

Chapter 2 Activities

Define & Recall

navigation
cartography
oblate spheroid
great circle
hemisphere
small circle
coordinate system
North Pole
South Pole
equator
latitude
sextant
prime meridian
longitude
geographic grid
parallel
meridian
time zone
solar noon
International Date Line
U.S Public Lands Survey System
principal meridian
base line
township
section
global positioning system (GPS)
map projection

conformal map
Mercator projection
equal-area map
equidistance
azimuthal map
rhumb line
gnomonic projection
compromise projection
legend
scale
verbal scale
representative fraction (RF scale)
graphic (bar) scale
magnetic declination
azimuth
bearing
thematic map
discrete data
continuous data
isoline
topographic contour line
contour interval
profile
gradient
digital elevation model (DEM)
vertical exaggeration
geographic information system (GIS)
geocoding
visualization models
remote sensing
aerial photograph
digital image
pixel
resolution (spatial resolution)
near-infrared (NIR)
thermal infrared (TIR)
radar
imaging radar
side-looking airborne radar
weather radar
sonar
multispectral remote sensing
color composite image

Discuss & Review

1. Why is a great circle useful for navigation?
2. What are the latitude and longitude coordinates of the place (town, city) where you live?
3. Approximately how precise in meters could you be if you tried to locate a building in your city to the nearest second of latitude and longitude? Using a GPS?
4. What time zone are you in? What is the time difference between Greenwich time and your time zone?
5. If you fly across the Pacific Ocean from the United States to Japan, how will the International Date Line affect you?
6. How has the use of the Public Lands Survey System affected the landscape of the United States? Has your local area been affected by its use? How?
7. Why is it impossible for maps to provide a completely accurate representation of Earth's surface? What is the difference between a conformal map and an equal-area map?
8. What is the difference between an RF and a verbal map scale?
9. What does a small-scale map show in comparison with a large-scale map?
10. What does the concept of *thematic map layers* mean in a geographic information system?
11. What specific advantages do computers offer to the map-making process?
12. What is the difference between a photograph and a digital image?
13. What does a weather radar image show in order to help us understand weather patterns?

Consider & Respond

1. Select a place within the United States that you would most like to visit for a vacation. You have with you a highway map, a USGS topographic map, and a satellite image of the area. What kinds of information could you get from one of these sources that is not displayed on the other two? What spatial information do they share (visible on all three)?
2. If you were an applied geographer and wanted to use a geographic information system to build an information database about the environment of a park (pick a state or national park near you), what are the five most important layers of mapped information that you would want to have? What combinations of two or more layers would be particularly important to your purpose?

Apply & Learn

1. If it is 2:00 a.m. Tuesday in New York (EST), what time and day is it in California (PST)? What time is it in London (GMT)? What is the date and time in Sydney, Australia (151° East)?
2. A topographic profile has a linear scale of 1:2400, and a vertical scale of 1 inch equals 100 feet. How many feet does 1 inch equal on the linear scale? If there is vertical exaggeration, what is it?
3. If 10 centimeters (3.94 in.) on a map equal 1 kilometer (3281 ft) on the ground, what is the RF scale of the map? You can round the answer to the nearest thousand. This is the formula to use for scale conversions of this kind: Map distance/Earth distance = 1/Representative Fraction Denominator.

Locate & Explore

Note: Please read the section of the Preface titled "About Locate & Explore Activities" before beginning these exercises.

1. The coordinate system used on a globe is latitude and longitude, representing angular distance (in degrees) north and south of the equator, and angular distance east and west of the prime meridian that passes through Greenwich, England. Using the Search window in Google Earth, fly to the heart of the following cities and identify the latitude and longitude. Measure the latitude and longitude using decimal degrees with two decimal places (ex: 41.89 N as opposed to 41°88'54.32" N). Make sure that you correctly note whether the latitude is North (N or +) or South (S or –) of the equator and whether the longitude is East (E or +) or West (W or –) of the prime meridian.

   Tip: Change the latitude/longitude setting in the Tools > Options dialog box.

   a. London, England
   b. Paris, France
   c. New York City
   d. San Francisco, California
   e. Buenos Aires, Argentina
   f. Cape Town, South Africa
   g. Moscow, Russia
   h. Beijing, China
   i. Sydney, Australia
   j. Your home town

   Now reverse the latitude and longitude of your home town and note where you are in the world.
2. To find your location on the surface of the earth you can use a global positioning system (GPS) device, which gives location in latitude and longitude. Enter the following coordinates into Google Earth to identify the location:

   a. 41.89 N, 12.492 E
   b. 33.857 S, 151.215 E
   c. 29.975 N, 31.135 E
   d. 90.0 N, 0 E
   e. 90.0 S, 90.0 W
   f. 27.175 N, 78.042 E
   g. 27.99 N, 86.92 E
   h. 40.822 N, 14.425 E
   i. 48.858 N, 2.295 E

   Tip: Use the zoom, tilt, rotate, and elevation exaggeration functions of Google Earth to help view and interpret the landform object shown in the browser.
3. When looking at a topographic profile or using the terrain feature in Google Earth you can control the elevation exaggeration, which is calculated as the ratio of the units on the horizontal (x) axis to the units on the vertical (z) axis. In Google Earth you can adjust the elevation exaggeration between 0.5 and 3, thereby making subtle objects more noticeable. (Go to Tools > Options to set the elevation exaggeration.) In Google Earth, turn on the Elevation Exaggeration Layer and then Fly to Mount Everest, the Nebraska Sand Hills, and the Goosenecks of the San Juan River. Adjust the exaggeration from .5 to 3 and notice the change in the terrain. In which of these landscapes is a higher level of vertical exaggeration most useful in interpreting the natural terrain?

   Tip: Use the zoom, tilt, and rotate functions of Google Earth to help view and interpret the landform object shown in the browser.
4. Using Google Earth, open the Stone Mountain Topographic Layer and draw a topographic profile (similar to Fig. 2.30 in your text) from Point A to Point B along the line shown. Use the contour lines and Google Earth's elevation exaggeration and tilt features to decipher the landform.

   Tip: Adjust the transparency of the topographic layer to see the image below. Turn off unnecessary layers for better visibility.

Map Interpretation

TOPOGRAPHIC MAPS

The Map

A topographic map is a widely used tool for graphically depicting variations in elevation within an area. A contour line connects points of equal elevation above some reference datum, usually mean sea level. A vast storehouse of information about the relief and the terrain can be interpreted from these maps by understanding the spacing and configuration of contours. For example, elevations of mountains and valleys, steepness of slopes, and the direction of stream flow can be determined by studying a topographic map. In addition to contour lines, many standard symbols are used on topographic maps to represent mapped features, data, and information (a guide to these symbols is in Appendix B).

The elevation difference represented by adjacent contour lines depends on the map scale and the relief in the mapped area, and is called the contour interval. Contour intervals on topographic maps are typically in elevation measurements divisible by ten. In mountainous areas wider intervals are needed to keep the contours from crowding and visually merging together. A flatter locality may require a smaller contour interval to display subtle relief features. It is good practice to note both the map scale and the contour interval when first examining a topographic map.

Keep in mind several important rules when interpreting contours:

- Closely spaced contours indicate a steep slope, and widely spaced contours indicate a gentle slope.
- Evenly spaced contours indicate a uniform slope.
- Closed contour lines represent a hill or a depression.
- Contour lines never cross but may converge along a vertical cliff.
- A contour line will bend upstream when it crosses a valley.

Interpreting the Map

1. What is the contour interval on this map?
2. The map scale is 1:24,000. One inch on the map represents how many feet on the Earth's surface?
3. What is the highest elevation on the map? Where is it located?
4. What is the lowest elevation on the map? Where is it located?
5. Note the mountain ridge between Boat and Emerald Canyons (C-4). Is it steeper on its east side or its west side? What led you to your conclusion?
6. In what direction does the stream in Boat Canyon flow? What led you to your conclusion?
7. The aerial photograph below depicts a portion of the topographic map on the opposite page. What area of the air photo does the map depict? How well do the contours represent the physical features seen on the air photo?
8. Identify some cultural features on the map. Describe the symbols used to depict these features. The map shown is older than the aerial photograph. Can you identify some cultural features on the aerial photograph not depicted on the contour map?

© Bruce Perry, Department of Geological Sciences, CSU Long Beach

Aerial photograph of the coast at Laguna Beach, California.

Opposite:
Laguna Beach, California
Scale 1:24,000
Contour interval = 20 feet
U.S. Geological Survey

Opposite: © Bruce Perry, Department of Geological Sciences, CSU Long Beach

A
B
C
D
E
1
2
3
4
5
6
Muddy
Reef Hill
626
Center
844
Canyon
Moro
Reservoirs
El Morro
Sch
Trailer
Park
BM 18
LANE
Emerald
Canyon
Boat
Canyon
Guna
719
Fire
Sta
Park
WT
Water
Tank
Irvine Bowl
Park
Laguna
133
Abalone Point
BM 97
BM 86
Emerald Bay
HILLCREST
DR
COAST
HIGHWAY
BM
86
MYRTLE
ST
HIGH
DR
ASTER
ST
BROADWAY
Sewage
Disposal
City
Hall
Fire Sta
Two Rock Point
Crescent
Bay
Twin Points
Heisler Park
Art
Gallery
Recreation Point
BM 17
FOREST
AVE
Thurston
Sch
PO
High
Sch
LAGUNA BEACH
BM 44
Cheneys Point
THALIA
COAST
CATALINA
BM 56
G U L F

Earth–Sun Relationships and Solar Energy

3

CHAPTER PREVIEW

The sun is the original and ultimate source of the energy that drives the various components of the Earth system.

- How does the sun's energy reach Earth?
- How does this energy affect the Earth system?

The types of energy emitted by the sun are represented in the electromagnetic spectrum.

- What bands of this spectrum control heating and cooling in the Earth energy system?
- What bands of this spectrum affect humans directly?

The regular movements of Earth, termed *rotation* and *revolution*, are the fundamental elements of Earth–sun relationships, which initially control the dynamics of our atmosphere and the phenomena related to it.

- Why is this one of the most important understandings in physical geography?
- What other understandings follow from this concept?

The relationship of Earth's axis to the plane of Earth's orbit is the key to an explanation of seasons on Earth.

- How does it operate in conjunction with Earth's revolution to produce seasons?
- How does it influence variations in the amounts of insolation reaching different portions of Earth's surface?

On specific dates through the year, incoming sun angles strike certain lines of latitude that divide Earth into large horizontal zones.

- How many of these zones exist?
- How are they named?

◀ **Opposite: Our sun, the ultimate energy source for Earth–atmosphere systems.**

Courtesy of SOHO/[instrument] consortium. SOHO is a project of international cooperation between ESA and NASA.

With a radius 110 times that of Earth and a mass 330,000 times greater, the sun reigns as the center of our solar system. The gravitational pull of this fierce, stormy ball of gas holds Earth in orbit, and its emissions power the Earth–atmosphere systems on which our lives depend. As the source of almost all the energy in our world, it holds the key to many of our questions about Earth and sky.

Everyone has wondered about environmental changes that take place throughout the year and from place to place over Earth's surface. Perhaps when you were young, you wondered why it got so much warmer in summer than in winter and why some days were long whereas those in other seasons were much shorter. These questions and many like them are probably as old as the earliest human thoughts, and the answers to them help provide us with an understanding of the physical geography of our world.

Physical geographers' concerns take them beyond planet Earth to a consideration of the sun and Earth's position in the solar system. Geographers examine the relationship between the sun and Earth to explain such earthly phenomena as the alternating periods of light and dark that we know as day and night. Other relationships between Earth and sun also help explain seasonal variations in climate. Although a knowledge of solar dynamics is not within the realm of physical geography, an examination of Earth's relationship to its ultimate energy source is vital to understanding the environments that support life as we know it.

The Solar System and Beyond

If you look at the night sky on a clear night, all the stars that you can see are a part of a single collection of stars called the Milky Way Galaxy. A **galaxy** (● Fig. 3.1) is an enormous island in the universe—an almost incomprehensible cluster of stars, dust, and gases. Our sun is one of billions of stars that comprise the Milky Way Galaxy. In turn, the observable universe contains billions of other galaxies.

Distances within the universe are so vast that it is necessary to use a large unit of measure termed a **light-year** (the distance that light travels in 1 year). A light-year is equal to 6 trillion miles. Light travels at the amazing speed of 298,000 kilometers per second (186,000 mi/sec). Thus, in 1 second, light could travel seven times around the circumference of Earth. Although that may seem like a great distance, the *closest* star to Earth, other than the sun, is Proxima Centauri at 4.2 light-years away, and the *closest* galaxy to ours is the Canis Major Dwarf Galaxy, at 25,000 light-years away.

● FIGURE 3.1
This image shows one of the billions of galaxies that make up our vast universe. This galaxy, referred to as Galaxy M81, is 12 million light-years away from Earth, and has a spiral shape just like our Milky Way Galaxy.

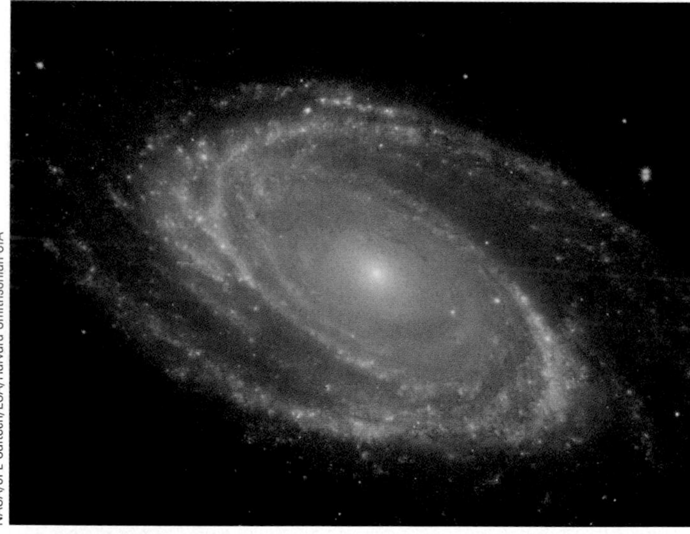

NASA/JPL-Caltech/ESA/Harvard-Smithsonian CfA

The Solar System

The sun is the center of our solar system. A **solar system** can be defined as all the heavenly bodies surrounding a particular star because of the star's dominant mass and gravity. **Gravity** is the attractive force one body has for another. The greater the **mass** or amount of matter a body has, the greater the gravitational pull it will exert on other bodies. The principal celestial bodies in our sun's system are the eight major planets. A **planet,** as defined by the International Astronomical Union in 2006, is a celestial body in orbit around the sun, with sufficient gravitational attraction to overcome rigid body forces and assume a nearly spherical shape, and has cleared the neighborhood around its orbit (● Fig. 3.2). Under this new definition Pluto, formerly our ninth planet, has been reclassified as a *dwarf planet.* It is generally agreed that Pluto is a large body captured from the Kuiper Belt (a disk-shaped region containing small icy bodies that lies past the orbit of Neptune) by the gravitational pull of the sun.

Our solar system also includes approximately 138 **satellites** (like Earth's moon, these bodies orbit the planets) and numerous **asteroids,** which are small solar-system bodies with a diameter of less than 500 miles (800 km), as well as **comets** and **meteors.** A comet is made up of a head—a collection of solid fragments held together by ice—and a tail, sometimes millions of miles long, composed of gases (● Fig. 3.3). Meteors are small, stonelike or metallic bodies that, when entering Earth's atmosphere, burn and often appear as a streak of light, or "shooting star." A meteor that survives the fall through the atmosphere and strikes Earth's surface is called a **meteorite.**

The Planets

The four planets closest to the sun (Mercury, Venus, Earth, and Mars) are called the **terrestrial planets.** They are relatively small, warmed by their proximity to the sun, and composed of rock and metal. They all have solid surfaces that exhibit records of geological forces in the form of craters, mountains, and volcanoes. The last four planets (Jupiter, Saturn, Uranus, and Neptune) are much larger and composed primarily of lighter ices, liquids, and gases. These planets are termed the **giant planets,** or **gas planets.** Although they have solid cores at their centers, they are more like huge balls of gas and liquid with no solid surface on which to walk.

The eight major planets that revolve around the sun have several phenomena in common. From a point far out in space above the sun's "north pole," they would all appear to move around the sun in the

● FIGURE 3.2
The solar system, showing the sun and planets in their proper order according to their distance from the sun. The approximate size relationships between the individual planets are shown. However, the planetary orbits are much condensed, and the scale of the sun and planets is greatly exaggerated. The planets would be much too small to be visible at the scale of the orbits shown.
What happened to Pluto?

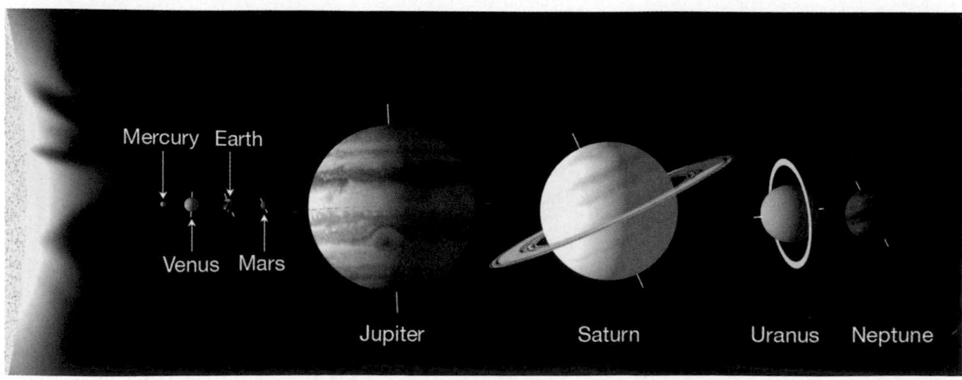

© Dr. Richard Hackney, Western Kentucky University

© Dr. Richard Hackney, Western Kentucky University

• FIGURE 3.3
The comet Hale–Bopp shows a split tail because two different types of icy material are emitting different jets of gasses.

TABLE 3.1
Comparison of the Planets

Name	Distance from Sun (AU)*	Revolution Period (yr)	Diameter (km)	Mass (10^{23} kg)	Density (g/cm³)
Mercury	0.39	0.24	4,878	3.3	5.4
Venus	0.72	0.62	12,102	48.7	5.3
Earth	1.00	1.00	12,756	59.8	5.5
Mars	1.52	1.88	6,787	6.4	3.9
Jupiter	5.20	11.86	142,984	18,991	1.3
Saturn	9.54	29.46	120,536	5,686	0.7
Uranus	19.18	84.07	51,118	866	1.2
Neptune	30.06	164.82	49,660	1,030	1.6

*An AU or (astronomical unit) is the distance from Earth to the sun.

same counterclockwise direction. Their orbits follow an elliptical, almost circular, path. All planets also rotate, or spin, on their own axes. With the exception of Venus and Uranus, all rotate in the same counterclockwise direction. All the planet's orbits lie close to the same plane (the plane of the ecliptic) passing through the sun's equator. All planets have an atmospheric layer of gases with the exception of Mercury, which is not dense or heavy enough for its gravity to hold appreciable amounts of gases (Table 3.1).

The Earth–Sun System

Earth receives about 1/2,000,000,000 (one two billionth) of the radiation given off by the sun, but even this tiny amount drives the biological and physical characteristics of Earth's surface. Other bodies in the solar system receive some of the sun's radiant energy, but the vast proportion of it travels out through space unimpeded. The sun's energy is the most important factor determining environmental conditions on Earth. With the exception of geothermal heat sources (such as volcanic eruptions and geyser springs) and heat emitted by radioactive minerals, the sun remains the source of all the energy for Earth and atmospheric systems.

The intimate and life-producing relationship between Earth and sun is the result of the amount and distribution of radiant energy received from the sun. Such factors as our planet's size, its distance from the sun, its atmosphere, the movement of Earth around the sun, and the planet's rotation on an axis all affect the amount of radiant energy that Earth receives. Though some processes of our physical environment result from Earth forces not related to the sun, these processes would have little relevance were it not for the life-giving, life-sustaining energy of the sun.

Earth revolves around the sun at an average distance of 150 million kilometers (93 million mi). The sun's size and its distance from us challenge our comprehension. About 130 million Earths could fit inside the sun, and a plane flying at 500 miles per hour would take 21 years to reach the sun.

GEOGRAPHY'S ENVIRONMENTAL SCIENCE PERSPECTIVE

Passive Solar Energy, an Ancient and Basic Concept

When we think of using solar energy to heat buildings or to produce electrical power, we look toward complex and developing technology to find the way. Indeed, photovoltaic (PV) cells that convert sunlight into electricity, and solar thermal towers that convert water into steam to drive electrical turbines, are excellent ways to harness this inexhaustible energy source. However, long before these were invented, people wished to be comfortable in their homes, and used a passive form of solar energy. This can easily be done, assuming you know the sun angles that affect your home.

The concept is very simple; flood your home with solar energy in the wintertime. This makes better use of indoor sunlight and adds more heat during the cold season. Then, restrict the amount of insolation entering the home in summer; this keeps the interior cooler during the hottest months, while still allowing daylight to illuminate the interior. These days, environmentally conscious home designers can do this by adjusting the number and placement of windows in the home and controlling the length and angle of the eaves (or roof overhang).

This concept is very old. The Cliff Palace in Mesa Verde, Colorado, is a wonderful example of an 800-year-old cliff-dwelling. Here the cave roof and overhang perform the same service as the environmentally conscious home design. More direct sunlight enters the structures during the winter and indirect sunlight enters during the summer.

How did these people know about sun angles? For millennia ancient pagan cultures worshipped the Sun God, and their astronomers observed and calculated sun angles. Ancient cultures in China, as well as the ancient Egyptians, Greeks, and Romans, designed their architecture to best utilize solar energy. In the Americas, the ancient Maya, Incas, Aztecs, and North American tribes used their knowledge of sun angles as guides to erect buildings, temples, and pyramids to their chief god—the sun. This is not a new concept.

Using this knowledge, we can form a simple rule to help us to save money on future heating and air-conditioning costs. Keep your home or apartment shaded in the summer and sunlit in the winter. Window curtains, shades, and blinds can do a lot more to save on energy bills than you think!

National Park Service Geologic Resources/D. Luchsinger

The Cliff Palace at Mesa Verde in Colorado shows that early Native Americans understood the use of passive solar energy in locating their cliff dwellings under natural overhangs.

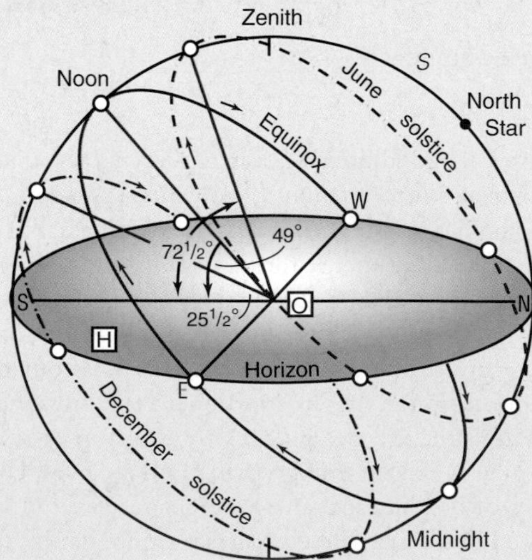

Sun path diagrams help us to find seasonal and daily changes in alignment of the sun, relative to the horizon, at a particular latitude.

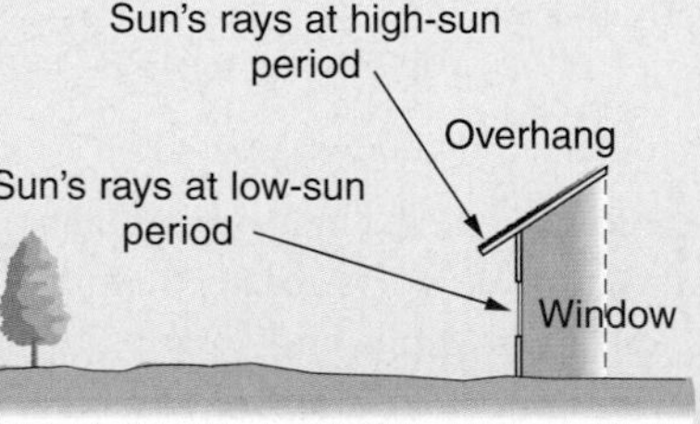

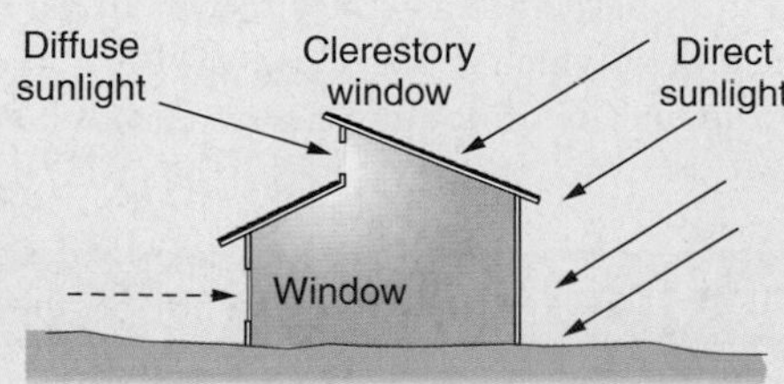

Modern house designs take seasonal changes in sun angles into account. The top diagram shows the maximum and minimum sun angles experienced by a hypothetical location at 40°N latitude. The bottom diagram shows how the home can be designed for maximum passive solar efficiency.

As far as we know with certainty, within our solar system, only on Earth has the energy from the sun been used to create life—to create something that can grow, develop, reproduce, and eventually die. Yet there remains a possibility of life, or at least the basic organic building blocks, on Mars and perhaps even on one or two of the moons of Saturn. What fascinates scientists, geographers, and philosophers alike, however, is the likelihood that millions of planets like Earth in the universe may have developed life-forms that might be more sophisticated than humans.

The Sun and Its Energy

The sun, like all other stars in the universe, is a self-luminous sphere of gases that emits radiant energy. A slightly less than average-sized star, our sun is the only self-luminous body in our solar system and is the source of almost all the light and heat for the surfaces of the various celestial bodies in our planetary system. The energy emitted by the sun comes from **fusion (thermonuclear) reactions** that take place at its core. There, under extremely high pressure, and temperatures that exceed 15,000,000°C (27,000,000°F), two hydrogen atoms fuse together to form one helium atom in a process similar to that of a hydrogen bomb explosion (• Fig. 3.4a and b). This fusion reaction releases tremendous amounts of energy that radiate out from the core to just below the solar surface where countless convective currents act like a pot of boiling water (hotter gas rising and cooler gas sinking). These convective currents give the sun that "grainy appearance" seen in special X-ray imagery (see chapter-opening image). The *photosphere* (sphere of light) is what the human eye sees as the surface of the sun, and is the densest layer. It has an estimated temperature of between 5500°C and 6100°C (10,000°F and 11,000°F). The *chromosphere* (sphere of color) is a thin layer of gases above the photosphere and appears red in color. Lastly, the *corona* (or crown) is the outermost layer of the sun's atmosphere. Its constantly changing shape is caused by charged particles trapped by the sun's magnetic field (see again the chapter-opening image). Charged particles (mainly protons and electrons) from the corona can flow along the sun's magnetic field lines millions of miles into space as **solar wind.** Unlike solar radiation, which moves at the speed of light, solar wind travels about 400 kilometers (640 mi) per second and takes more than 4 days to reach our planet. When these solar winds reach Earth, they are prevented from harming the surface by Earth's magnetic field and are confined to the upper atmosphere (• Fig. 3.5). During these times, they can disrupt radio and television communication and may disable orbiting satellites. The

• FIGURE 3.5
Earth's magnetic field protects the surface from the harmful effects of solar wind.

Courtesy of NASA / Marshall Space Flight Center

• FIGURE 3.4
(a) The fireball explosion of a hydrogen bomb is created by thermonuclear fusion. (b) This same reaction powers the sun.
What elements drive a fusion reaction?

© US Navy/Photo Researchers, Inc.

(a)

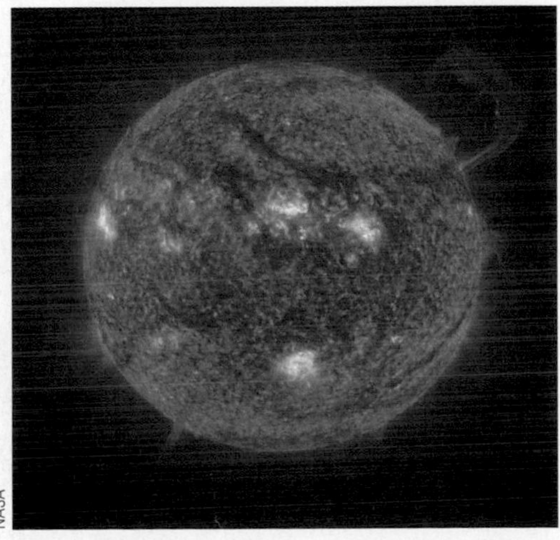

NASA

(b)

magnetic field of Earth tends to direct this solar wind into the outer atmosphere in the regions around our planet's magnetic poles (● Fig. 3.6). When this happens, the solar wind energizes the ions in the outer atmosphere, and results in an amazing light show known as the **auroras.** The Aurora Borealis, known as the *northern lights* (● Fig. 3.7), and the Aurora Australis, called the *southern lights,* happen simultaneously in the northern latitudes and southern latitudes.

The intensity of solar winds is influenced by the best-known solar feature, **sunspots.** Visible on the photosphere, these dark regions are about 1500°C–2000°C cooler than the surrounding temperature (● Fig. 3.8). Galileo began recording sunspots back in the 1600s, and for many years they have been used to indicate solar activity. Sunspots seem to observe an 11-year cycle from one maximum (where 100 or more may

● FIGURE 3.6

Solar wind, directed toward the magnetic poles, forms ring-shaped auroras, over and around the poles in each hemisphere. This is the northern aurora, Aurora Borealis.

What is the aurora in the Southern Hemisphere called?

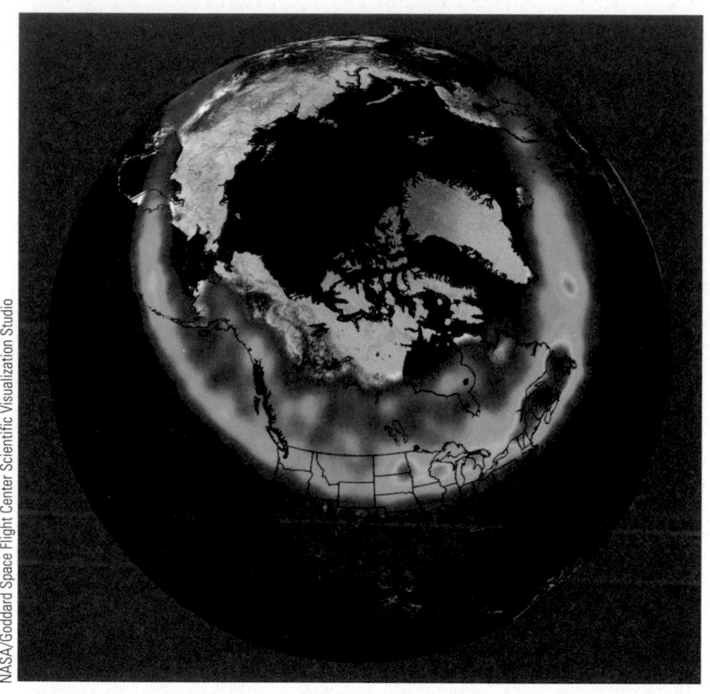

NASA/Goddard Space Flight Center Scientific Visualization Studio

● FIGURE 3.8

Sunspots as they appear on the solar surface. Insets show the area of the sun that is illustrated here and the relative size of Earth.

About how many Earth diameters can fit east to west across this sunspot?

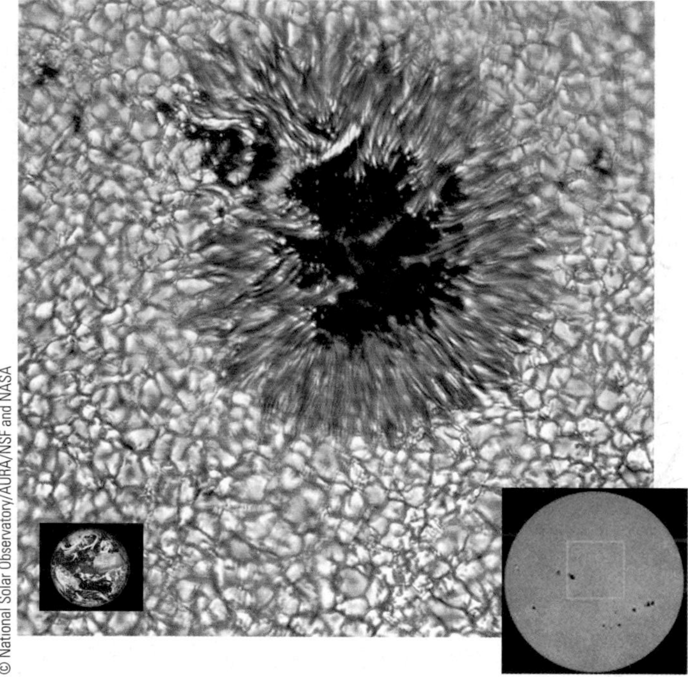

© National Solar Observatory/AURA/NSF and NASA

● FIGURE 3.7

(a) Solar wind and the ions in Earth's atmosphere interact to produce the Aurora Borealis in the Northern Hemisphere. (b) The record-setting solar activity of November 2004 caused the Aurora Borealis to be seen as far south as Houston, Texas. This photo (b) was taken near Bowling Green, Kentucky, at 37°N latitude.

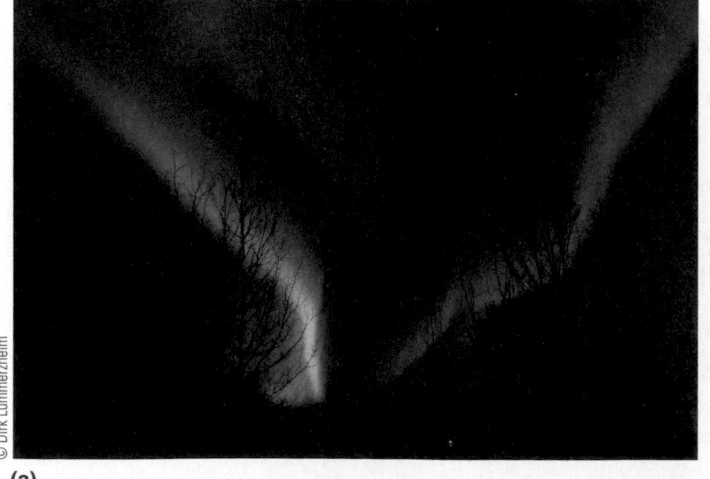

© Dirk Lummerzheim

(a)

© Dr. Richard Hackney, Western Kentucky University

(b)

be visible) to the next. Our next cycle (Number 24) is now on the rise and should peak around the year 2012. An individual sunspot may last for less than a day or as much as 6 months. Just how sunspots might affect Earth's atmosphere is still a matter of controversy. Proving direct connections between sunspot numbers and weather or climate is difficult, but such relationships have been suggested.

Solar Energy and Atmospheric Dynamics

As we have previously noted, our sun is the major source of energy, either directly or indirectly, for the entire Earth system. Earth does receive very small proportions of energy from other stars and from the interior of Earth itself (volcanoes and geysers provide certain amounts of heat energy); however, when compared with the amount received from the sun, these other sources are insignificant.

Energy is emitted by the sun in the form of **electromagnetic energy,** which travels at the speed of light in a spectrum of varying wavelengths (• Fig. 3.9). It takes about 8.3 minutes for this energy to reach Earth. Approximately 9% of solar energy is made up of *gamma rays, X-rays,* and *ultraviolet radiation,* all of which are shorter in wavelength than visible light. These wavelengths cannot be seen but can affect other tissues of the human body. Thus, absorbing too many X-rays can be dangerous, and excessive ultraviolet waves give us sunburned skin and are a primary cause of skin cancer. About 41% of the solar spectrum comes in the form of visible light rays, where each color is distinguishable by its specific wavelength band. However there are large bands of the electromagnetic spectrum not visible to the human eye. About 49% of the sun's radiant energy exists in wavelengths that are longer than visible light rays. Although these wavelengths are invisible, they can sometimes be sensed by human skin. The shorter wavelengths of infrared, known as *near infrared,* are harmless to living organisms. Longer waves in the *far infrared* part of the spectrum, also called *thermal infrared,* can be felt as heat. The last 1% of solar radiation falls into the band regions of microwave, television, and radio wavelengths.

Collectively, gamma rays, X-rays, ultraviolet rays, visible light, and near infrared are considered to have shorter wavelengths and are known as **shortwave radiation.** Starting from thermal infrared, the longer wavelengths of energy are considered **longwave radiation.** Through our advances in technology, we have learned to harness some electromagnetic wavelength bands for our own uses. There are many examples. In the field of communications, we employ radio waves, microwaves, and television signals; in diagnostic health care, we utilize X-rays. In the fields of remote sensing and national defense, visible light is necessary for photography and visible satellite imagery; we also use radar, which uses microwaves (to detect weather patterns and aircraft), and thermal infrared sensors (for heat imagery and heat-seeking weaponry).

The sun radiates energy into space at an almost steady rate. At its outer edge, Earth's atmosphere intercepts an amount of energy slightly less than 2 calories per square centimeter per minute. A **calorie** is the amount of *energy* required to raise the temperature of 1 gram of water 1°C. This can also be expressed in units of power—in this case, around 1370 watts per square meter. The rate of a planet's receipt of solar energy is known as the **solar constant** and has been measured with great precision outside Earth's atmosphere by orbiting satellites. The atmosphere affects the amount of solar radiation received on the surface of Earth because some energy is absorbed by clouds, some is reflected (bounced off), and some is refracted (bent). If we could remove the atmosphere from Earth, we would find that the solar energy striking the surface at a particular location for a particular time would be a constant value determined by the latitude of the location.

Of course, the measured value of the solar constant varies with distance from the sun as the same amount of energy radiates out into larger areas. For example, if we measured the solar constant for the planet Mercury, it would be much higher than that for Earth. When Earth is closest to the sun in its orbit, its solar constant is slightly higher than the yearly average, and when it is farthest away, the solar constant is slightly lower than average. However, this difference does not have a significant effect on Earth's temperatures. When Earth is farthest from the sun in July and the solar constant is lowest because of the distance from the sun, the Northern Hemisphere is in the midst of a summer with temperatures that are not significantly different from those in the Southern Hemisphere 6 months later. The solar constant also varies slightly with changes in activity on the sun; during intense sunspot or solar storm activity, for example, the solar constant will be slightly higher than usual. However, these variations are not even as great as those caused by Earth's elliptical orbit.

• FIGURE 3.9

Radiation from the sun travels toward Earth in a wide spectrum of wavelengths, which are measured in micrometers (μm) (1 μm equals one millionth of a meter). Visible light occurs at wavelengths of approximately 0.4–0.7 micrometers. Solar radiation is considered shortwave radiation (less than 4.0 μm), whereas terrestrial (Earth) radiation is of long wavelengths (more than 4.0 μm).
Are radio signals considered longwave or shortwave radiation?

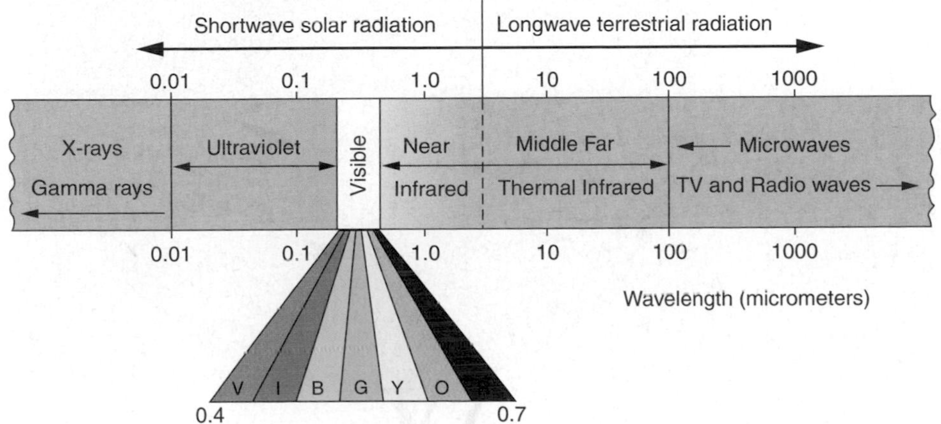

Movements of Earth

Earth has three basic movements: **galactic movement, rotation,** and **revolution.** The first of these is the movement of Earth with the sun and the rest of the solar system in an orbit around the center of the Milky Way Galaxy. This movement has limited effect on the changing environments of Earth and is generally the concern of astronomers rather than geographers. The other two movements of Earth, rotation on its axis and revolution around the sun, are of vital interest to the physical geographer. The consequences of these movements are the phenomena of day and night, variations in the length of day, and the changing seasons.

Rotation Rotation refers to the spin of Earth on its axis, an imaginary line extending from the North Pole to the South Pole. Earth rotates on its axis at a uniform rate, making one complete turn with respect to the sun in 24 hours.

Earth turns in an eastward direction (● Fig. 3.10). The sun "rises" in the east and appears to move westward across the sky, but it is actually Earth, not the sun, that is moving, rotating toward the morning sun (that is, toward the east).

Earth, then, rotates in a direction opposite to the apparent movement of the sun, moon, and stars across the sky. If we look down on a globe from above the North Pole, the direction of rotation is counterclockwise. This eastward direction of rotation not only defines the movement of the zone of daylight on Earth's surface but also helps define the circulatory movements of the atmosphere and oceans.

The velocity of rotation at the Earth's surface varies with the distance of a given place from the equator (the imaginary circle around Earth halfway between the two poles). All points on the globe take 24 hours to make one complete rotation (360°). Thus, the *angular velocity* for all locations on Earth's surface is the same—360° per 24 hours, or 15° per hour. However, the *linear velocity* depends on the distance (not the angle) covered during that 24 hours. The linear velocity at the poles is zero. You can see this by spinning a globe with a postage stamp affixed to the North Pole. The stamp rotates 360° but covers no distance and therefore has no linear velocity. If you place the stamp anywhere between the North and South Poles, however, it will cover a measurable distance during one rotation of the globe. The greatest linear velocity is found at the equator, where the distance traveled by a point in 24 hours is largest. At Kampala, Uganda, near the equator, the velocity is about 460 meters (1500 ft) per second, or approximately 1660 kilometers (1038 mi) per hour (● Fig. 3.11). In comparison, at St. Petersburg, Russia (60°N latitude), where the distance traveled during one complete rotation of Earth is about half that at the equator, Earth rotates about 830 kilometers (519 miles) per hour.

We are unaware of the speed of rotation because (1) the angular velocity is constant for each place on Earth's surface, (2) the atmosphere rotates with Earth, and (3) there are no nearby objects, either stationary or moving at a different rate with respect to Earth, to which we can compare Earth's movement. Without such references, we cannot perceive the speed of rotation.

Rotation accounts for our alternating days and nights. This can be demonstrated by shining a light at a globe while rotating the globe slowly toward the east. You can see that half the sphere is always illuminated while the other half is not and that new points are continually moving into the illuminated section of the globe while others are moving into the darkened sector. This corresponds to Earth's rotation and the sun's energy striking Earth. While one half of Earth receives the light and energy of solar radiation, the other half is in darkness. As noted in Chapter 2, the great circle separating day from night is known as the **circle of illumination** (● Fig. 3.12).

● FIGURE 3.10
Earth turns around a tilted axis as it follows its orbit around the sun. Earth's rotation is from west to east, making the stationary sun appear to rise in the east and set in the west.

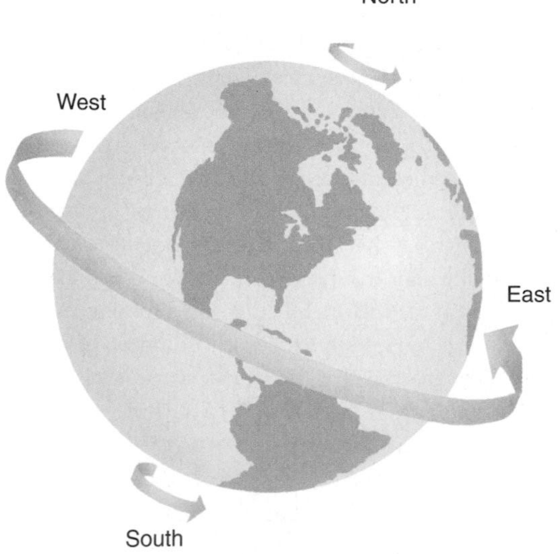

● FIGURE 3.11
The speed of rotation of Earth varies with the distance from the equator. **How much faster does a point on the equator move than a point at 60°N latitude?**

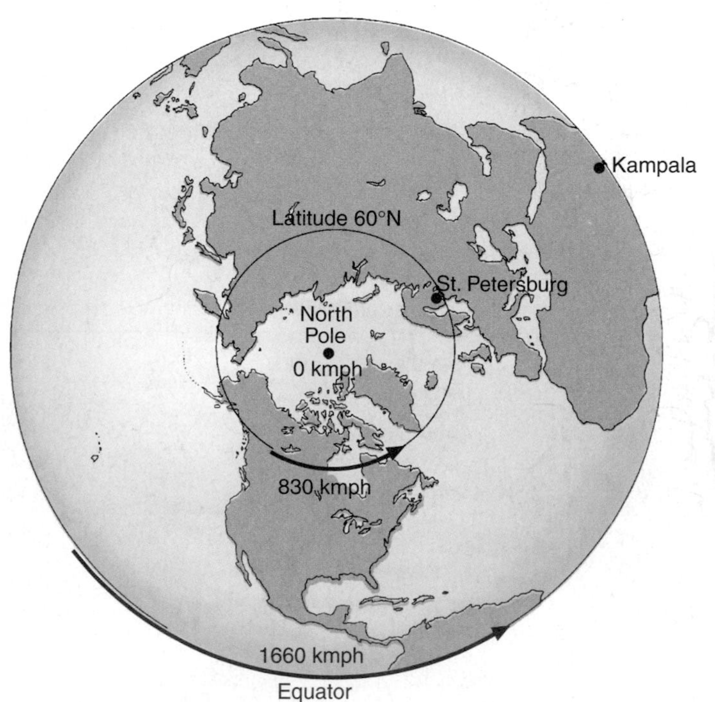

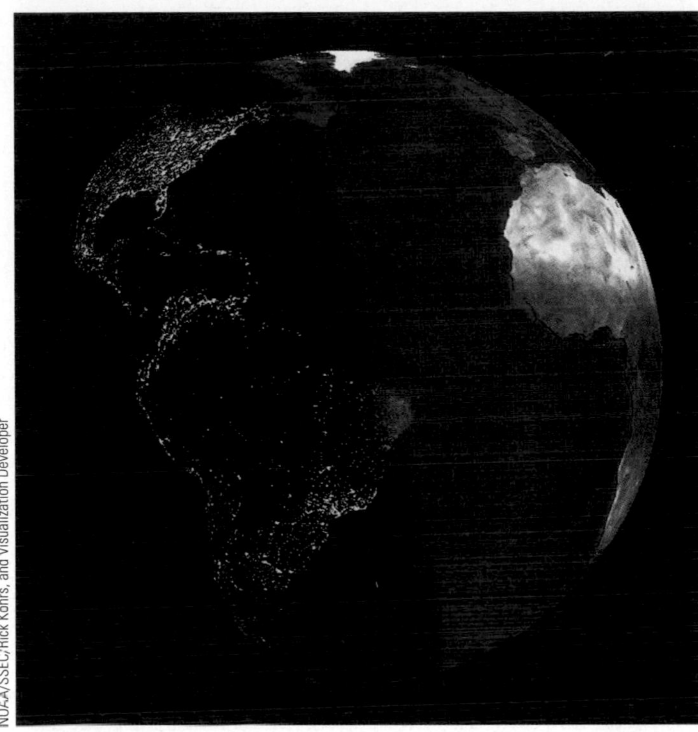

• **FIGURE 3.12**
The circle of illumination, which separates day from night, is clearly seen on this image of Earth.
Which way is the circle of illumination moving across Earth's surface?

Revolution While Earth rotates on its axis, it also revolves around the sun in a slightly elliptical orbit at an average distance from the sun of about 150 million kilometers (93 million mi) (• Fig. 3.13). On about January 3, Earth is closest to the sun and is said to be at **perihelion** (from Greek: *peri,* close to; *helios,* sun); its distance from the sun then is approximately 147.5 million kilometers. At around July 4, Earth is about 152.5 million kilometers from the sun. It is then that Earth has reached its farthest point from the sun and is said to be at **aphelion** (Greek: *ap,* away; *helios,* sun). Five million kilometers is relatively insignificant in space, and these varying distances from Earth to the sun only minimally affect (about 3.5% difference) the receipt of energy on Earth. Hence, they have little relationship to the seasons.

The period of time that Earth takes to make one revolution around the sun determines the length of 1 year. Earth makes 365¼° rotations on its axis during the time it takes to complete one revolution of the sun; therefore, a year is said to have 365¼° days. Because of the difficulty of dealing with a fraction of a day, it was decided that a year would have 365 days, and every fourth year, called *leap year,* an extra day would be added as February 29.

Plane of the Ecliptic, Inclination, and Parallelism In its orbit around the sun, Earth moves in a constant plane, known as the **plane of the ecliptic.** Earth's equator is tilted at an angle of 23½° from the plane of the ecliptic, causing Earth's axis to be tilted 23½° from a line perpendicular to the plane (• Fig. 3.14). In addition to this constant **angle of inclination,** Earth's axis maintains another characteristic called **parallelism.** As Earth revolves around the sun, Earth's axis remains parallel to its former positions. That is, at every position in Earth's orbit, the axis remains pointed toward the same spot in the sky. For the North Pole, that spot is close to the star that we call the North Star, or Polaris. Thus, Earth's axis is fixed with respect to the stars outside our solar system but not with respect to the sun (see again the axis representation in Fig. 3.10).

Before continuing, it should be noted that, although the patterns of Earth rotation and revolution are considered constant in our current discussion, the two movements are subject to change. Earth's axis wobbles through time and will not always remain at an angle of exactly 23½° from perpendicular to the plane of the ecliptic. Moreover, Earth's orbit around the sun will change from more circular to more elliptical through periods that can be accurately determined. These and other cyclical changes were calculated and compared by Milutin Milankovitch, a Serbian astronomer during the 1940s, as a possible explanation for the ice ages. Since then the *Milankovitch Cycles* have often been used when climatologists attempt to explain climatic variations. These variations will be discussed in more detail along with other theories of climatic change in Chapter 8.

• **FIGURE 3.13**
An oblique view of Earth's elliptical orbit around the sun. Earth is closest to the sun at perihelion and farthest away at aphelion. Note that in the Northern Hemisphere summer (July), Earth is farther from the sun than at any other time of the year.
When is Earth closest to the sun?

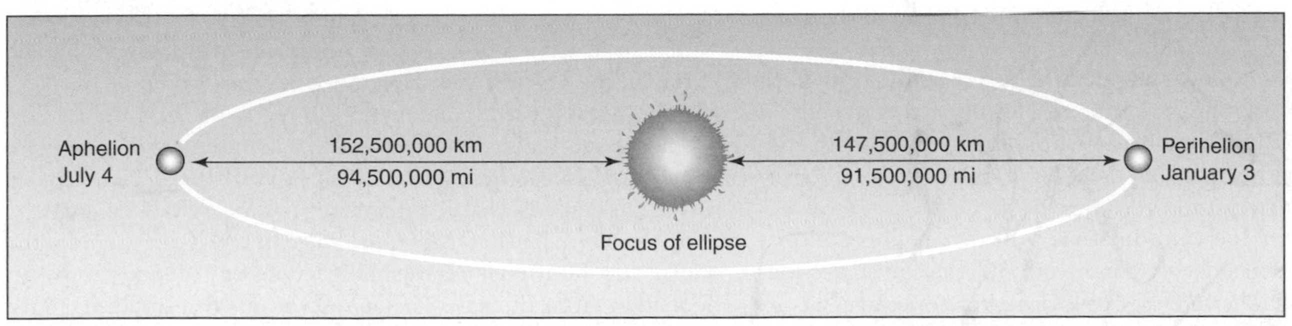

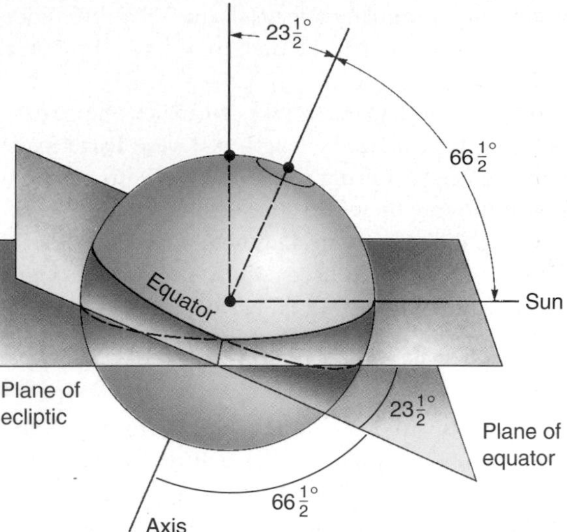

• **FIGURE 3.14**
The plane of the ecliptic is defined by the orbit of Earth around the sun. The 23½° inclination of Earth's rotational axis causes the plane of the equator to cut across the plane of the ecliptic.
How many degrees is Earth's axis tilted from the vertical?

Sun Angle, Duration, and Insolation

Understanding Earth's relationships with the sun leads us directly into a discussion of how the intensity of the sun's rays varies from place to place throughout the year and into an examination of the seasonal changes on Earth. Solar radiation received by the Earth system, known as **insolation** (for ***in***coming ***sol***ar radi***ation***), is the main source of energy on our planet. The seasonal variations in temperature that we experience are due primarily to fluctuations in insolation.

What causes these variations in insolation and brings about seasonal changes? It is true that Earth's atmosphere affects the amount of insolation received. Heavy cloud cover, for instance, will keep more solar radiation from reaching Earth's surface than will a clear sky. However, cloud cover is irregular and unpredictable, and it affects total insolation to only a minor degree over long periods of time.

The real answer to the question of what causes variations in insolation lies with two major phenomena that vary regularly for a given position on Earth as our planet rotates on its axis and revolves around the sun: the duration of daylight and the angle of the solar rays. The amount of daylight controls the duration of solar radiation, and the angle of the sun's rays directly affects the intensity of the solar radiation received. Together, the intensity and the duration of radiation are the major factors that affect the amount of insolation available at any location on Earth's surface.

Therefore, a location on Earth will receive more insolation if (1) the sun shines more directly, (2) the sun shines longer, or (3) both. The intensity of solar radiation received at any one time varies from place to place because Earth presents a spherical surface to insolation. Therefore, only one line of latitude on the Earth's rotating surface can receive radiation at right angles, while the rest receive varying oblique (sharp) angles (• Fig. 3.15a). As we can see from Figure 3.15b and c, solar energy that strikes Earth at a nearly vertical angle renders more intense energy but covers less area than an equal amount striking Earth at an oblique angle.

The intensity of insolation received at any given latitude can be found using *Lambert's Law,* named for Johann Lambert, an 18th-century German scientist. Lambert developed a formula by which the intensity of insolation can be calculated using the sun's zenith angle (that is, the sun angle deviating from 90° directly overhead). Using Lambert's Law, one can identify, based on latitude, where greater or lesser solar radiation is received on Earth's surface. • Figure 3.16 shows the intensity of total solar energy received at various latitudes, when the most direct radiation (from 90° angle rays) strikes directly on the equator.

In addition, the atmospheric gases act to diminish, to some extent, the amount of insolation that reaches Earth's surface. Because oblique rays must pass through a greater distance of atmosphere than vertical rays, more insolation will be lost in the process. In 1854, German scientist and mathematician August Beer established a relationship to calculate the amount of solar energy lost as it comes through our atmospheric gases. *Beer's Law,* as it's called, is strongly affected by the thickness of the atmosphere through which the energy must pass.

Since no insolation is received at night, the duration of solar energy is related to the length of daylight received at a particular point on Earth (Table 3.2). Obviously, the longer the period of daylight, the greater the amount of solar radiation that will be received at that location. As we will see in our next section, periods of daylight vary in length through the seasons of the year, as well as from place to place, on Earth's surface.

The Seasons

Many people assume that the seasons must be caused by the changing distance between Earth and the sun during Earth's yearly revolution. As noted earlier, the change in this distance is very small. Further, for people in the Northern Hemisphere, Earth is actually closest to the sun in January and farthest away in July (see again Fig. 3.13). This is exactly opposite of that hemisphere's seasonal variations. As we will see, seasons are caused by the 23½° tilt of Earth's equator to the plane of the ecliptic (see again Fig. 3.14) and the parallelism of the axis that is maintained as Earth orbits the sun. About June 21, Earth is in a position in its orbit so that the northern tip of its axis is inclined toward the sun at an angle of 23½°. In other words, the plane of the ecliptic (the 90° sun angle) is directly on 23½° N latitude. This day during Earth's orbit is called the summer **solstice** (from Latin: *sol,* sun; *sistere,* to stand) in the Northern Hemisphere. We can best see what is happening if we refer to • Figure 3.17, position A. In that diagram, we can see that the Northern and Southern Hemispheres receive unequal amounts of light from the sun. That is, as we imagine rotating Earth

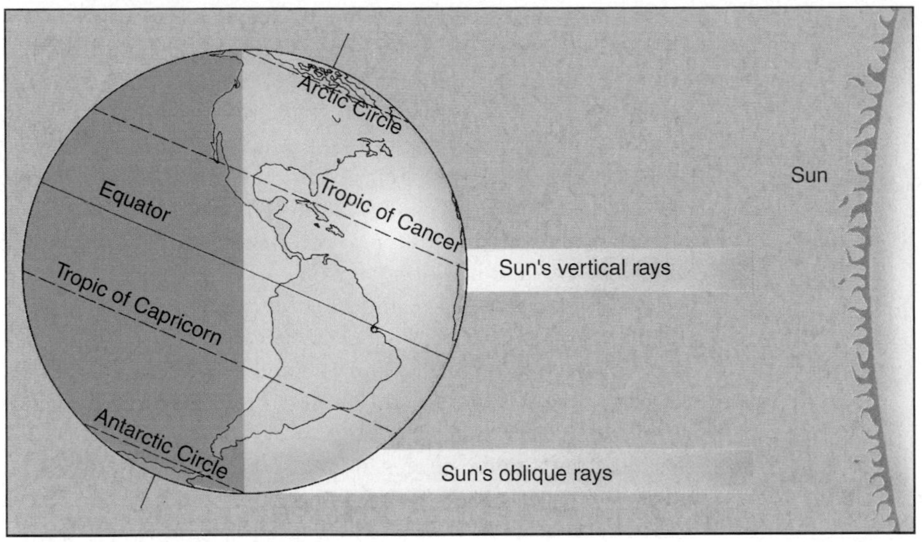

(a)

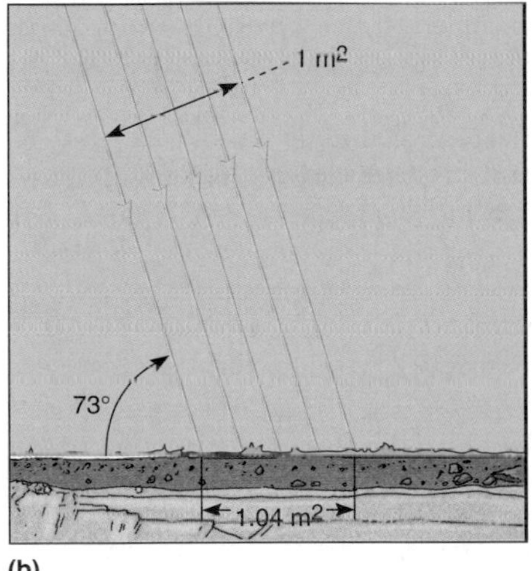

(b)

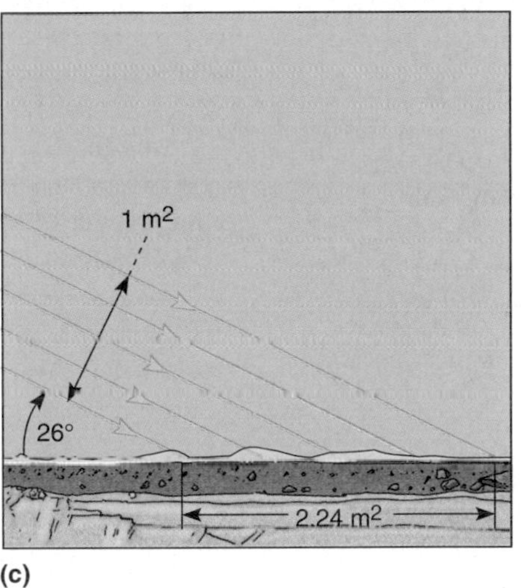

(c)

• FIGURE 3.15

(a) The angle at which the sun's rays strike Earth's surface determines the amount of solar energy received per unit of surface area. This amount in turn affects the seasons. The diagram represents the June condition, when solar radiation strikes the surface perpendicularly on the Tropic of Cancer, creating summer conditions in the Northern Hemisphere. In the Southern Hemisphere, the sun's rays are more oblique and spread over larger areas, thus receiving less energy per unit of area, making this the winter hemisphere. **How would a similar figure of Earth–sun relationships in December differ?** The sun's rays in summer (b) and winter (c). In summer the sun appears high in the sky, and its rays hit Earth more directly, spreading out less. In winter the sun appears low in the sky, and its rays spread out over a much wider area, becoming less effective at heating the ground.

under these conditions, a larger portion of the Northern Hemisphere than the Southern Hemisphere remains in daylight. Conversely, a larger portion of the Southern Hemisphere than the Northern Hemisphere remains in darkness. Thus, a person living at Repulse Bay, Canada, north of the Arctic Circle, experiences a full 24 hours of daylight at the June solstice. On the same day, someone living in New York City will experience a longer period of daylight than of darkness. However, someone living in Buenos Aires, Argentina, will have a longer period of darkness than daylight on that day. This day is called the winter solstice in the Southern Hemisphere. Thus, June 21 is the longest day, with the highest sun angles of the year in the Northern Hemisphere, and the shortest day, with the lowest sun angles of the year, in the Southern Hemisphere.

Now let's imagine the movement of Earth from its position at the June solstice toward a position a quarter of a year later, in September. As Earth moves toward that new position, we can imagine the changes that will be taking place in our three cities. In Repulse Bay, there will be an increasing amount of darkness through July, August, and September. In New York, sunset will be

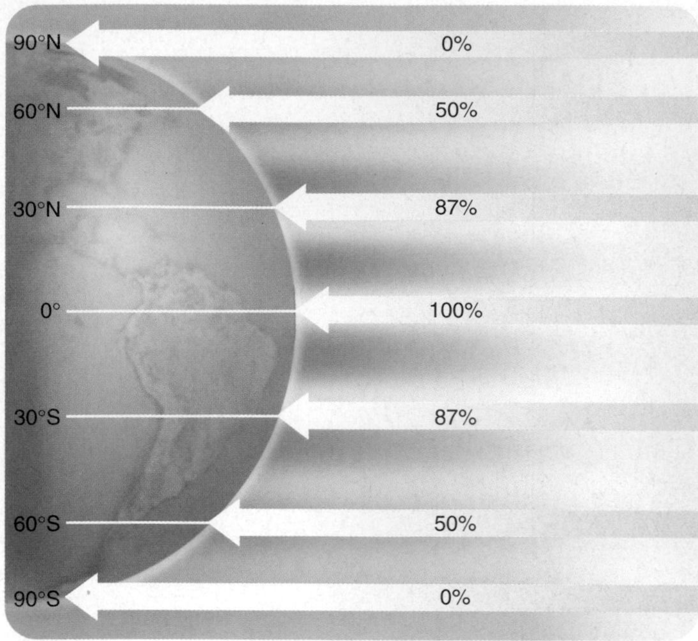

• FIGURE 3.16
The percentage of incoming solar radiation (insolation) striking various latitudes during an equinox date according to Lambert's Law. **How much less solar energy is received at 60° latitude than that received at the equator?**

arriving earlier. In Buenos Aires, the situation will be reversed; as Earth moves toward its position in September, the periods of daylight in the Southern Hemisphere will begin to get longer, the nights shorter.

Finally, on or about September 22, Earth will reach a position known as an **equinox** (Latin: *aequus,* equal; *nox,* night). On this date (the autumnal equinox in the Northern Hemisphere), day and night will be of equal length at all locations on Earth. Thus, on the equinox, conditions are identical for both hemispheres. As you can see in • Figure 3.18, position B, Earth's axis points neither toward nor away from the sun (imagine the axis is pointed at the reader); the circle of illumination passes through both poles, and it cuts Earth in half along its axis.

Imagine again the revolution and rotation of Earth while moving from around September 22 toward a new position another quarter of a year later in December. We can see that in Repulse Bay the nights will be getting longer until, on the winter solstice, which occurs on or about December 21, this northern town will experience 24 hours of darkness (Fig. 3.17, position C). The only natural light at all in Repulse Bay will be a faint glow at noon refracted from the sun below the horizon. In New York, too, the days will get shorter, and the sun will set earlier. Again, we can see that in Buenos Aires the situation is reversed. Around December 21, that city will experience its summer solstice; conditions will be much as they were in New York City in June.

Moving from late December through another quarter of a year to late March, Repulse Bay will have longer periods of daylight, as will New York, while in Buenos Aires the nights will be getting longer. Then, on or about March 20, Earth will again be in an equinox position (the vernal equinox in the Northern Hemisphere) similar to the one in September (Fig. 3.18, position D). Again, days and nights will be equal all over Earth (12 hours each).

TABLE 3.2
Duration of Daylight for Certain Latitudes

Length of Day (Northern Hemisphere) (read down)			
LATITUDE (IN DEGREES)	**MAR. 20/SEPT. 22**	**JUNE 21**	**DEC. 21**
0.0	12 hr	12 hr	12 hr
10.0	12 hr	12 hr 35 min	11 hr 25 min
20.0	12 hr	13 hr 12 min	10 hr 48 min
23.5	12 hr	13 hr 35 min	10 hr 41 min
30.0	12 hr	13 hr 56 min	10 hr 4 min
40.0	12 hr	14 hr 52 min	9 hr 8 min
50.0	12 hr	16 hr 18 min	7 hr 42 min
60.0	12 hr	18 hr 27 min	5 hr 33 min
66.5	12 hr	24 hr	0 hr
70.0	12 hr	24 hr	0 hr
80.0	12 hr	24 hr	0 hr
90.0	12 hr	24 hr	0 hr
LATITUDE	**MAR. 20/SEPT. 22**	**DEC. 21**	**JUNE 21**
Length of Day (Southern Hemisphere) (read up)			

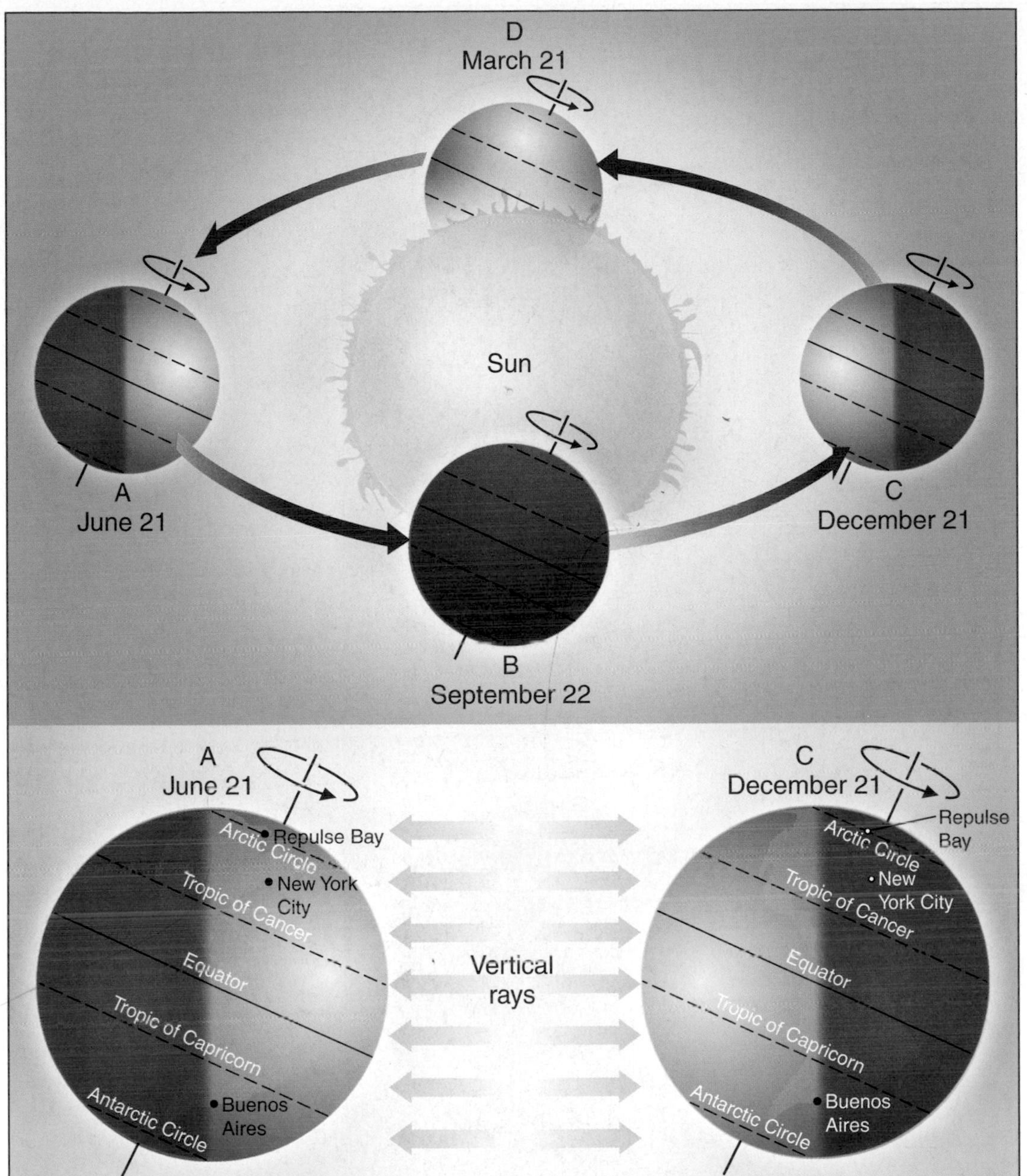

• **FIGURE 3.17**
The geometric relationships between Earth and the sun during the June and December solstices. Note the differing day lengths at the summer and winter solstices in the Northern and Southern Hemispheres.

Finally, moving through another quarter of the year toward the June solstice where we began, Repulse Bay and New York City are both experiencing longer periods of daylight than darkness. The sun is setting earlier in Buenos Aires until, on or about June 21, Repulse Bay and New York City will have their longest day of the year and Buenos Aires its shortest. Further, we can see that around June 21, a point on the Antarctic Circle in the Southern Hemisphere will experience a winter solstice similar to that which Repulse Bay had around December 21 (Fig. 3.17, position A). There will be no daylight in 24 hours, except what appears at noon as a glow of twilight in the sky.

Lines on Earth Delimiting Solar Energy

Looking at the diagrams of Earth in its various positions as it revolves around the sun, we can see that the angle of inclination is important. On June 21, the plane of the ecliptic is directly on 23½°N latitude. The sun's rays can reach 23½° beyond the North Pole, bathing it in sunlight. The **Arctic Circle,** an imaginary line drawn around Earth 23½° from the North Pole (or 66½° north of the equator) marks this limit. We can see from the diagram that all points on or north of the Arctic Circle will experience no darkness on the June solstice and that all points south of the

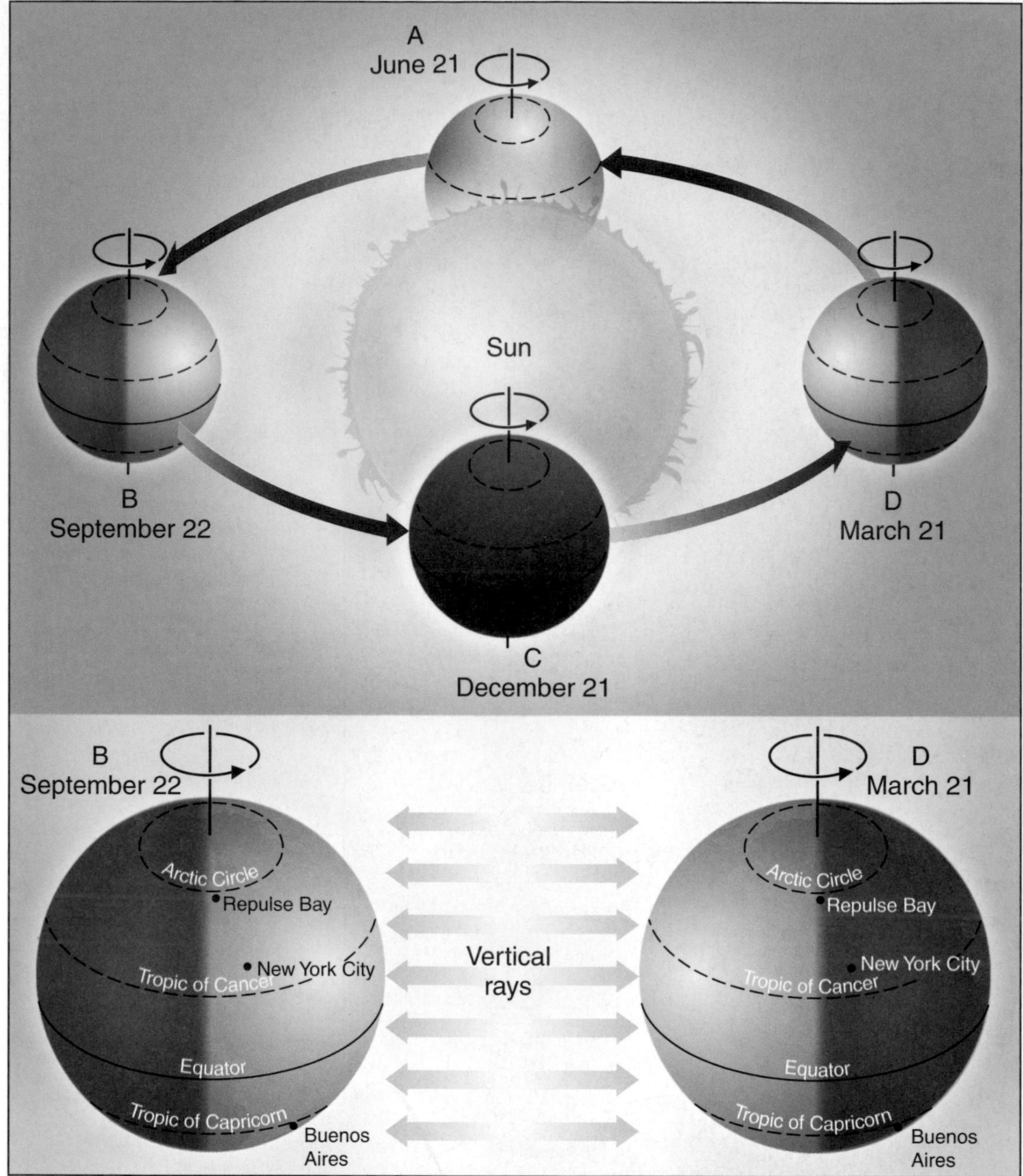

• FIGURE 3.18
The geometric relationships between Earth and the sun at the March and September equinoxes. Daylight and darkness periods are 12 hours everywhere because the circle of illumination crosses the equator at right angles and cuts through both poles.
If Earth were not inclined on its axis, would there still be latitudinal temperature variations? Would there be seasons?

Arctic Circle will have some darkness on that day. The **Antarctic Circle** in the Southern Hemisphere (23½° north of the South Pole, or 66½° south of the equator) marks a similar limit.

Furthermore, it can be seen from the diagrams that the sun's **vertical (direct) rays** (rays that strike Earth's surface at right angles) also shift position in relation to the poles and the equator as Earth revolves around the sun. At the time of the June solstice, the sun's rays are vertical, or directly overhead, at noon at 23½° *north* of the equator. This imaginary line around Earth marks the northernmost position at which the solar rays will ever be directly overhead during a full revolution of our planet around the sun. The imaginary line marking this limit is called the **Tropic of Cancer** (23½°N latitude). Six months later, at the time of the December solstice, the solar rays are vertical, and the noon sun is directly overhead 23½° *south* of the equator. The imaginary line marking this limit is known as the **Tropic of Capricorn** (23½°S latitude). At the times of the March and September equinoxes, the vertical solar rays will strike directly

GEOGRAPHY'S PHYSICAL SCIENCE PERSPECTIVE

Using the Sun's Rays to Measure the Spherical Earth—2200 Years Ago

About 240 BC in Egypt, Eratosthenes, a Greek philosopher and geographer, observed that the noonday sun's angle above the horizon changed along with the seasons. Knowing that our planet was spherical, he used geometry and solar observations to make a remarkably accurate estimate of Earth's circumference. A librarian in Alexandria, he read an account of a water well in Syene (today Aswan, Egypt), located to the south about 800 kilometers (500 mi) on the Nile River. On June 21 (summer solstice), this account stated, the sun's rays reached the bottom of the well and illuminated the water. Because the well was vertical, this meant that the sun was directly overhead on that day. Syene was also located very near the Tropic of Cancer, the latitude of the subsolar point on that date.

Eratosthenes had made many observations of the sun's angle over the year, so he knew that the sun's rays were never vertical in Alexandria, and at noon on that day in June a vertical column near the library formed a shadow. Measuring the angle between the column and a line from the column top to the shadow's edge, he found that the sun's angle was 7.2° away from vertical.

Assuming that the sun's rays strike Earth's spherical surface in a parallel fashion, Eratosthenes knew that Alexandria was located 7.2° north of Syene. Dividing the number of degrees in a circle (360°) by 7.2°, he calculated that the two cities were separated by 1/50 of Earth's circumference. The distance between Syene and Alexandria was 5000 stades, with a stade being the distance around the running track at a stadium. Therefore, 5000 stades times 50 meant that Earth must be 250,000 stades in circumference.

Unfortunately in ancient times, stades of different lengths were being used in different regions, and it is not certain which distance Eratosthenes used. A commonly cited stade length is about 0.157 kilometers (515 ft), and in using this measure, the resulting distance estimate would be 39,250 kilometers (24,388 mi). This distance is very close to the actual circumference of the great circle that would connect Alexandria and Syene.

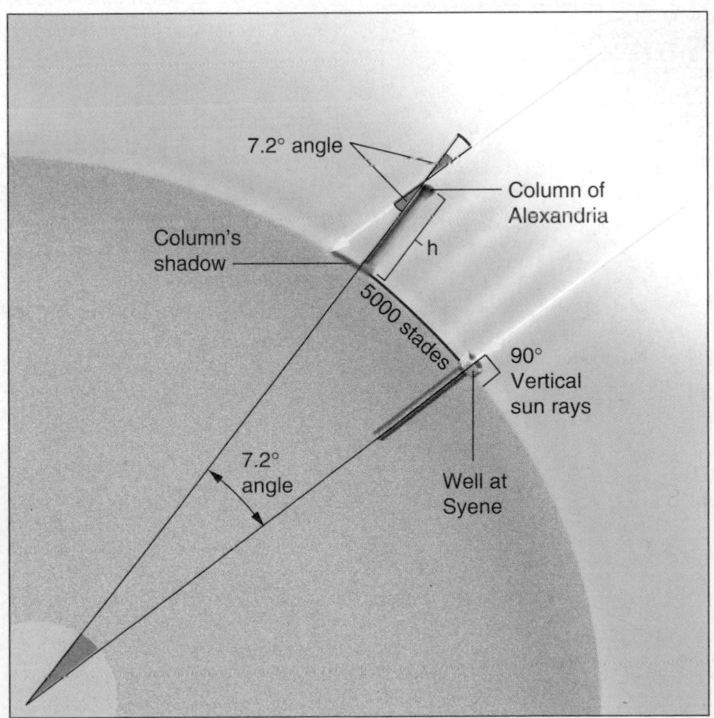

By observing the noon sun angle cast by a column where he lived and knowing that no shadow was cast on that same day in Syene to the south, Eratosthenes used geometry to estimate the Earth's circumference. On a spherical Earth, a 7.2° difference in angle also meant that Syene was 7.2° south in latitude from Alexandria, or 1/50 of Earth's circumference.

at the equator; the noon sun is directly overhead at all points on that line (0° latitude).

Note also that on any day of the year the sun's rays will strike Earth at a 90° angle at only one position, either on or between the two tropics. All other positions that day will receive the sun's rays at an angle of less than 90° (or will receive no sunlight at all).

The Analemma

The latitude at which the noon sun is directly overhead is also known as the sun's **declination.** Thus, if the sun appears directly overhead at 18°S latitude, the sun's declination is 18°S. A figure called an **analemma,** which is often drawn on globes as a big-bottomed "figure 8," shows the declination of the sun throughout the year. A modified analemma is presented in ● Figure 3.19. Thus, if you would like to know where the sun will be directly overhead on April 25, you can look on the analemma and see that it will be at 13°N. The analemma actually charts the passage of the direct rays of the sun over the 47° of latitude that they cover during a year.

● FIGURE 3.19
An analemma is used to find the solar declination (latitudinal position) of the vertical noon sun for each day of the year.
What is the declination of the sun on October 30th?

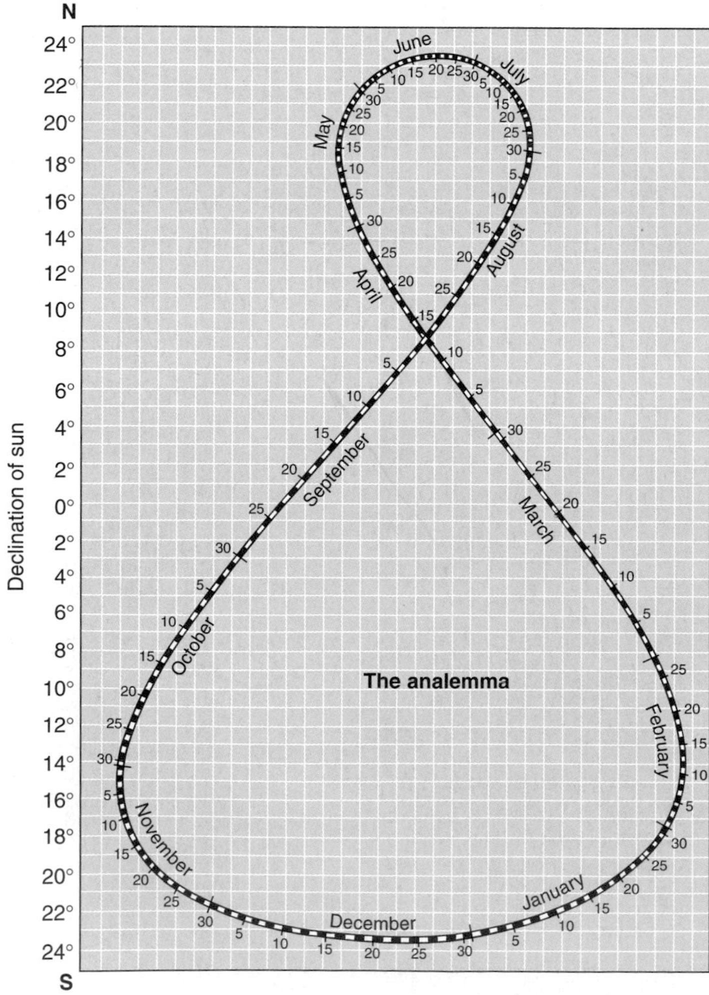

Variations of Insolation with Latitude

Neglecting for the moment the influence of the atmosphere on variations in insolation during a 24-hour period, the amount of energy received by the surface begins after daybreak and increases as Earth rotates toward the time of solar noon. A place will receive its greatest insolation at solar noon when the sun has reached its zenith, or highest point in the sky, for that day. The amount of insolation then decreases as the sun angle lowers toward the next period of darkness. Obviously, at any location, no insolation is received during the darkness hours.

We also know that the amount of daily insolation received at any one location on Earth varies with latitude (see again Fig. 3.16). The seasonal limits of the most direct insolation are used to determine recognizable zones on Earth. Three distinct patterns occur in the distribution of the seasonal receipt of solar energy in each hemisphere. These patterns serve as the basis for recognizing six latitudinal zones, or bands, of insolation and temperature that circle Earth (● Fig. 3.20).

If we look first at the Northern Hemisphere, we may take the Tropic of Cancer and the Arctic Circle as the dividing lines for three of these distinctive zones. The area between the equator and the Tropic of Cancer can be called the *north tropical zone.* Here, insolation is always high but is greatest at the time of the year that the sun is directly overhead at noon. This occurs twice a year, and these dates vary according to latitude (see again Fig. 3.19). The *north middle-latitude zone* is the wide band between the Tropic of Cancer and the Arctic Circle. In this belt, insolation is greatest on the June solstice when the sun reaches its highest noon

● FIGURE 3.20
The equator, the Tropics of Cancer and Capricorn, and the Arctic and Antarctic Circles define six latitudinal zones that have distinctive insolation characteristics.
Which zone(s) would have the least annual variation in insolation? Why?

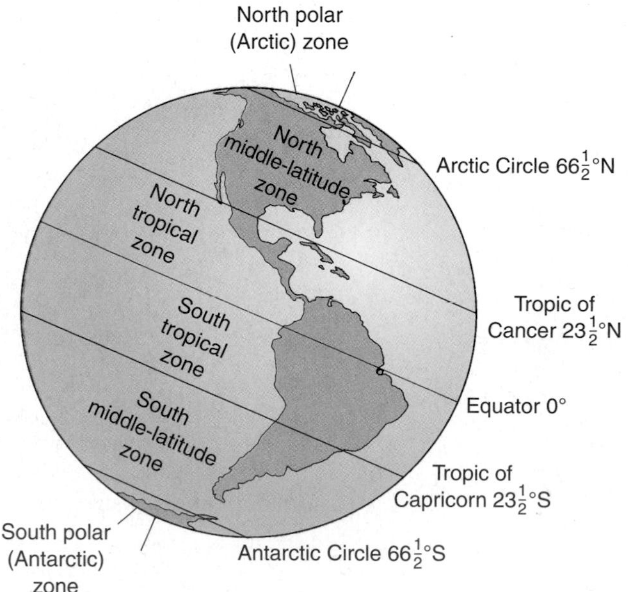

angle and the period of daylight is longest. Insolation is least at the December solstice when the sun is lowest in the sky and the period of daylight the shortest. The north *polar zone,* or *Arctic zone,* extends from the Arctic Circle to the pole. In this region, as in the middle-latitude zone, insolation is greatest at the June solstice, but it ceases during the period that the sun's rays are blocked entirely by the tilt of Earth's axis. This period lasts for 6 months at the North Pole but is as short as 1 day directly on the Arctic Circle.

Similarly, there is a *south tropical zone,* a *south middle-latitude zone,* and a *south polar zone,* or *Antarctic zone,* all separated by the Tropic of Capricorn and the Antarctic Circle in the Southern Hemisphere. These areas get their greatest amounts of insolation at opposite times of the year from the northern zones.

Despite various patterns in the amount of insolation received in these zones, we can make some generalizations. For example, total annual insolation at the top of the atmosphere over a particular latitude remains nearly constant from year to year (the solar constant). Furthermore, annual insolation tends to decrease from lower latitudes to higher latitudes (Lambert's Law together with Beer's Law). The closer to the poles a place is located, the greater will be its seasonal variations caused by fluctuations in insolation.

The amount of insolation received by the Earth system is an important concept in understanding atmospheric dynamics and the distribution of climate, soils, and vegetation across the globe. Such climatic elements as temperature, precipitation, and winds are controlled in part by the amount of insolation received by Earth. People depend on certain levels of insolation for physical comfort, and plant life is especially sensitive to the amount of available insolation. You may have noticed plants that have wilted in too much sunlight or that have grown brown in a dark corner away from a window. Over a longer period of time, deciduous plants have an annual cycle of budding, flowering, leafing, and losing their leaves. This cycle is apparently determined by the fluctuations of increasing and decreasing solar radiation that mark the changing seasons. Even animals respond to seasonal changes. Some animals hibernate; many North American birds fly south toward warmer weather as winter approaches; and many animals breed at such a time that their offspring will be born in the spring, when warm weather is approaching.

Ancient civilizations around the world (from China to Mexico) realized the incredible influence of the solar energy and many societies worshiped the sun, as chief among the pagan gods (● Fig. 3.21). And of course they would—the sun was vital to their survival and they knew it! Most humans do not worship the sun as a god any longer, but it is extremely important to understand and appreciate its role as Earth's ultimate source of energy.

● FIGURE 3.21

El Castillo, a Mayan Pyramid at Chichén Itzá, Mexico, was oriented to the annual change in sun angle, as the Maya worshiped the sun. Each side has 91 steps—the number of days separating the solstice and equinox days. On the vernal equinox, the afternoon sun casts a shadow that makes it appear like a giant snake is crawling down the pyramid's north side.

Why would ancient civilizations worship the sun?

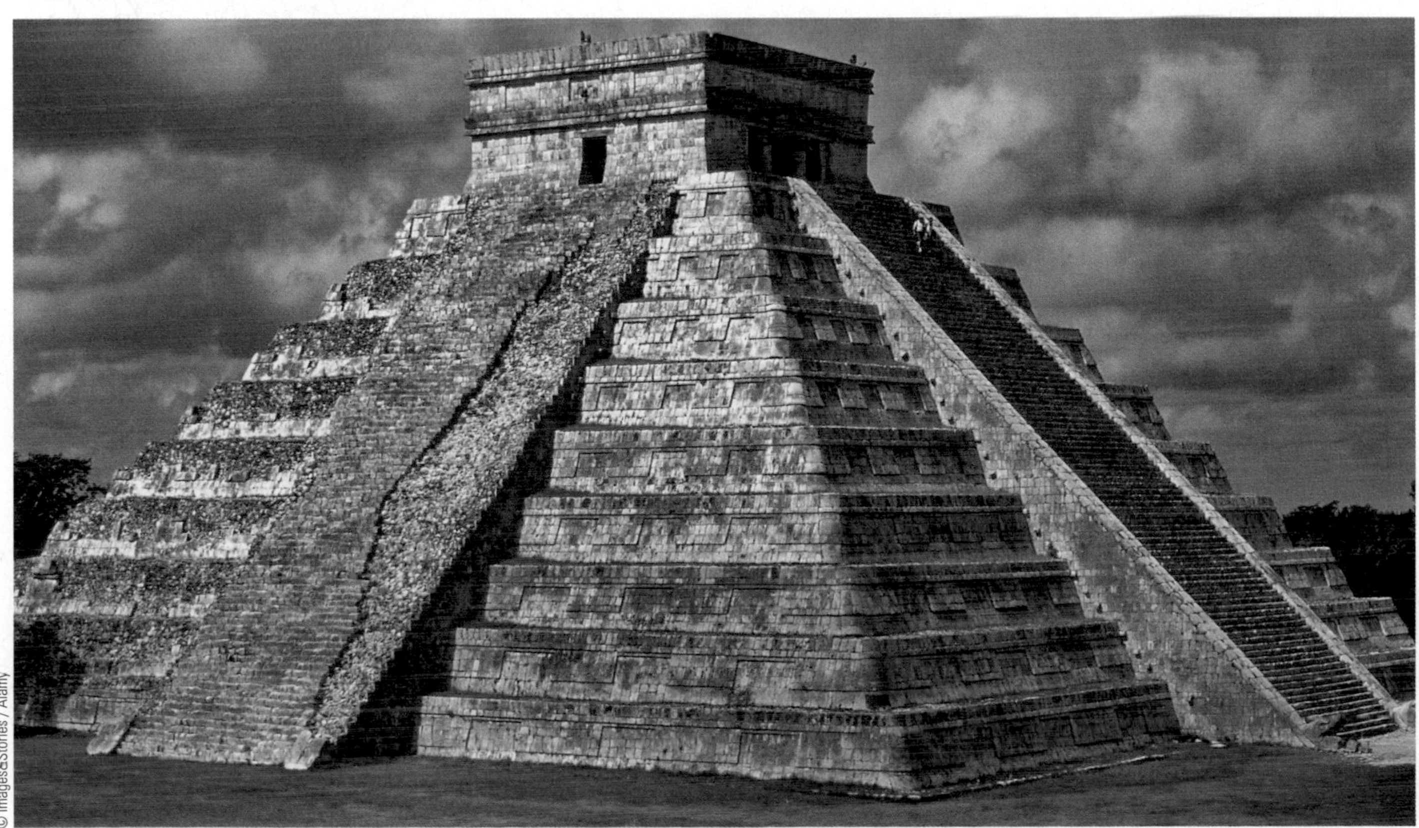

© Images&Stories / Alamy

Chapter 3 Activities

Define & Recall

galaxy
light-year
solar system
gravity
mass
planet
satellite
asteroid
comet
meteor
meteorite
terrestrial planet
giant planet (gas planet)
fusion (thermonuclear) reaction
solar wind
aurora
sunspot
electromagnetic energy
shortwave radiation
longwave radiation
calorie
solar constant
galactic movement
rotation
revolution
circle of illumination
perihelion
aphelion
plane of the ecliptic
angle of inclination
parallelism
insolation
solstice
equinox
Arctic Circle
Antarctic Circle
vertical (direct) rays
Tropic of Cancer
Tropic of Capricorn
declination
analemma

Discuss & Review

1. What is a solar system? What bodies constitute our solar system?
2. How is the energy emitted from the sun produced?
3. Name the terrestrial planets. What do they have in common? Name the giant planets. What do they have in common?
4. Which planets are capable of maintaining a gaseous atmosphere?
5. The electromagnetic spectrum displays various types of energy by their wavelengths. Where is the division between longwave and shortwave energy? In what ways do humans use electromagnetic energy?
6. Is the amount of solar energy reaching Earth's outer atmosphere constant? What might make it change?
7. Describe briefly how Earth's rotation and revolution affect life on Earth.
8. If the sun is closest to Earth on January 3, why isn't winter in the Northern Hemisphere warmer than winter in the Southern Hemisphere?
9. Identify the two major factors that cause regular variation in insolation throughout the year. How do they combine to cause the seasons?

Consider & Respond

1. Given what you know of the sun's relation to life on Earth, explain why the solstices and equinoxes have been so important to cultures all over the world.
2. Use the discussion of solar angle, including Figure 3.15, to explain why we can look directly at the sun at sunrise and sunset but not at the noon hour.
3. Describe in your own words the relationship between insolation and latitude.
4. Use the analemma presented in Figure 3.19 to determine the latitude where the noon sun will be directly overhead on February 12, July 30, November 2, December 30.

Apply & Learn

1. Imagine you are at the equator on March 20. The noon sun would be directly overhead. However, for every degree of latitude that you travel to the north or south, the noon solar angle would decrease by the same amount. For example, if you travel to 40°N latitude, the solar angle would be 50°.

   Explain this relationship.

   Develop a formula or set of instructions to generalize this relationship.

   What would be the solar angle at 40°N on June 21? On December 21?

   Lambert's Law can be expressed as: $I = I_0 \text{ Cos } \gamma$

   Where: I = Intensity of solar radiation received at the surface.
   I_0 = Intensity of solar radiation received from a 90° angle.
   Cos = Cosine of γ
   γ = The sun's zenith angle

   The zenith angle is measured from 90° straight above, down to where the sun's position in the sky.

2. Using Lambert's Law, calculate how intense is the insolation on September 22 at the following locations:

   0° latitude
   27°S latitude
   40°N latitude
   65°S latitude
   83°N latitude

The Atmosphere, Temperature, and the Heat Budget

4

CHAPTER PREVIEW

Our planet's atmosphere is essential to life as we know it here on Earth.

- How is this true?
- How should this fact affect human behavior?

Earth has an energy budget with a multiplicity of inputs and outputs (exchanges) that ultimately remain in balance despite recurring deficits and surpluses from time to time and from place to place.

- How is the budget concept useful to an understanding of atmospheric heating and cooling?
- How can we tell that the budget remains in balance?

Water plays a very important role in the exchanges of energy that fuel atmospheric dynamics.

- What characteristics of water are responsible for its importance in energy exchange?
- In what ways is water involved in the heating of the atmosphere?

As a direct result of differences in insolation and the mechanics of atmospheric heating, air temperature varies over time and both horizontally and vertically through space.

- What are the most obvious variations?
- How do temperatures stay within the ranges suitable for life if there are such great differences in the amounts of insolation received?

Atmospheric elements are affected by atmospheric controls to produce weather and climate.

- How do the elements differ from the controls?
- How does weather differ from climate?

◀ Opposite: The atmosphere is a very thin layer of gases held to the surface by Earth's gravity. This view from the International Space Station is looking toward the south along the Andes Mountain and the coast of Chile. Part of Argentina can be seen to the east of the Andes.
NASA

Water and oxygen are vital for animals and humans to survive. Plant life requires carbon dioxide as well as a sufficient water supply. Most living things we know cannot survive extreme temperatures, nor can they live long if exposed to large doses of harmful radiation. It is the atmosphere, the envelope of air that surrounds Earth, that supplies most of the oxygen and carbon dioxide and that helps maintain a constant level of water and radiation in all Earth systems.

Though actually a thin film of air, the atmosphere serves as an insulator, maintaining the viable temperatures we find on Earth. Without the atmosphere, Earth would experience temperature extremes of as much as 260°C (500°F) between day and night. The atmosphere also serves as a shield, blocking out much of the sun's ultraviolet (UV) radiation and protecting us from meteor showers. The atmosphere is also described as an ocean of air surrounding Earth. This description reminds us of the currents and circulation of the atmosphere—its dynamic movements—which create the changing conditions on Earth that we know as weather.

For comparison, we can look at our moon—a celestial body with virtually no atmosphere—in order to see the importance of our own atmosphere. Most obviously, a person standing on the moon without a space suit would immediately die for lack of oxygen. Our lunar astronauts recorded temperatures of up to 204°C (400°F) on the hot, sunlit side of the moon, and, on the dark side, temperatures approaching −121°C (−250°F). These temperature extremes would certainly kill an unprotected human.

The next thing any astronaut on the moon would notice is the "unearthly" silence. On Earth, we hear sounds because sound waves move by vibrating the molecules in the air. Because the moon has no atmosphere and no molecules to carry the sound waves, the lunar visitor cannot hear any sounds; only radio communications are possible. Also, because there is no atmosphere, an astronaut cannot fly aircraft or helicopters, and it would be fatal to try to use a parachute. In addition, lack of atmosphere means no protection from the bombardment of meteors that fly through space and collide with the moon. Nearing Earth, most meteors burn up before reaching the surface because of the friction they encounter while moving through the atmosphere. Without an atmosphere for protection, the ultraviolet rays of the sun would also burn a visitor to the moon. On Earth, we are protected to a large degree from UV radiation because the ozone layer of the upper atmosphere absorbs the major portion of this harmful radiation.

We can see that, in contrast to our stark, lifeless moon, Earth presents a hospitable environment for life almost solely because of its atmosphere. All living things are adapted to its presence. For example, many plants reproduce by pollen and spores that are carried by winds. Birds can fly only because of the air. The water cycle of Earth is maintained through the atmosphere, as are the heat and radiation "budgets." The atmosphere diffuses sunlight as well, giving us our blue skies and the fantastic reds, pinks, oranges, and purples of sunrise and sunset. Without this diffusion, the sky would appear black, as it does from the moon (● Fig. 4.1).

● FIGURE 4.1
Without an atmosphere, the moon's environment would be deadly to an unprotected astronaut.
How do astronauts communicate with each other on the moon?

NASA Glenn Research Center (NASA-GRC)

Further, the atmosphere provides a means by which the systems of Earth attempt to reach equilibrium. Changes in weather are ultimately the result of the atmospheric effects that equalize temperature and pressure differences on Earth's surface by transferring heat and moisture through atmospheric and oceanic circulation systems.

Characteristics of the Atmosphere

The atmosphere extends to approximately 480 kilometers (300 mi) above Earth's surface. Its density decreases rapidly with altitude; in fact, 97% of the air is concentrated in the first 25 kilometers (16 mi) or so. Because air has mass, the atmosphere exerts pressure on Earth's surface. At sea level, this pressure is about 1034 grams per square centimeter (14.7 lb/sq in.), but the higher the elevation, the lower is the atmospheric pressure. In Chapter 5, we will examine the relationship between atmospheric pressure and elevation in more detail.

Composition of the Atmosphere

The atmosphere is composed of numerous gases (Table 4.1). Most of these gases remain in the same proportions regardless of the density of the atmosphere. A bit more than 78% of the atmosphere's volume is made up of nitrogen, and nearly 21% consists of oxygen. Argon comprises most of the remaining 1%. The percentage of carbon dioxide in the atmosphere has risen through time, but is a little less than 0.04% by volume. There are traces of other gases as well: ozone, hydrogen, neon, xenon, helium, methane, nitrous oxide, krypton, and others.

Nitrogen, Oxygen, Argon, and Carbon Dioxide

Of these four most abundant gases that make up the atmosphere, nitrogen gas (N_2) makes up the largest proportion of air. It is a very important element supporting plant growth and will be discussed in more detail in Chapter 11. In addition, some of the other gases in the atmosphere are vital to the development and maintenance of life on Earth. One of the most important of the atmospheric gases is of course oxygen (O_2), which humans and all other animals use to breathe and oxidize (burn) the food that they eat. *Oxidation,* which is technically the chemical combination of oxygen with other materials to create new products, occurs in situations outside animal life as well. Rapid oxidation takes place, for instance, when we burn fossil fuels or wood and thus release large amounts of heat energy. The decay of certain rocks or organic debris and the development of rust are examples of slow oxidation. All these processes depend on the presence of oxygen in the atmosphere. The third most abundant gas in our atmosphere is Argon (Ar). It is not a chemically active gas and therefore neither helps nor hinders life on Earth.

Carbon dioxide (CO_2), the fourth most abundant gas, is involved in the system known as the *carbon cycle.* Plants, through a process known as **photosynthesis,** use sunlight

TABLE 4.1
Composition of the Atmosphere Near Earth's Surface

Permanent Gases			Variable Gases			
GAS	SYMBOL	PERCENT (BY VOLUME) DRY AIR	GAS (AND PARTICLES)	SYMBOL	PERCENT (BY VOLUME)	PARTS PER MILLION (PPM)*
Nitrogen	N_2	78.08	Water vapor	H_2O	0 to 4	
Oxygen	O_2	20.95	Carbon dioxide	CO_2	0.038	380*
Argon	Ar	0.93	Methane	CH_2	0.00017	1.7
Neon	Ne	0.0018	Nitrous oxide	N_2O	0.00003	0.3
Helium	He	0.0005	Ozone	O_3	0.000004	0.04†
Hydrogen	N_2	0.00006	Particles (dust, soot, etc.)		0.000001	0.01–0.15
Xenon	X_2	0.000009	Chlorofluorocarbons (CFCs)		0.00000002	0.0002

*For CO_2, 380 parts per million means that out of every million air molecules, 380 are CO_2 molecules.

†Stratospheric values at altitudes between 11 km and 50 km are about 5 to 12 ppm.

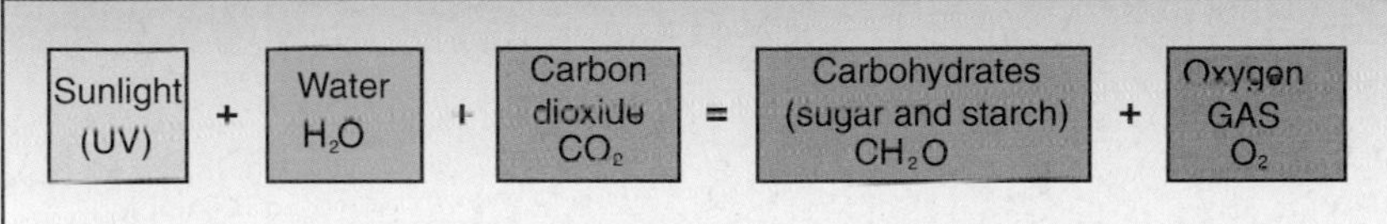

• FIGURE 4.2
The equation of photosynthesis shows how solar energy (mainly UV radiation) is used by plants to manufacture sugars and starches from atmospheric carbon dioxide and water, liberating oxygen in the process. The stored food energy is then eaten by animals, which also breathe the oxygen released by photosynthesis.

(mainly ultraviolet radiation) as the driving force to combine carbon dioxide and water to produce carbohydrates (sugars and starches), in which energy, derived originally from the sun, is stored and used by vegetation (• Fig. 4.2). Oxygen is given off as a by-product. Animals then use the oxygen to oxidize the carbohydrates, releasing the stored energy. A by-product of this process in animals is the release of carbon dioxide, which completes the cycle when it is in turn used by plants in photosynthesis.

Water Vapor, Liquids, Particulates, and Aerosols

Water vapor is always mixed in some proportion with the dry air of the lower part of the atmosphere; it is the most variable of the atmospheric gases and can range from 0.02% by volume in a cold, dry climate to more than 4% in the humid tropics. The percentage of water vapor in the air will be discussed later under the broad topic of humidity, but it is important to note here that the variations in this percentage over time and place are an important consideration in the examination and comparison of climates.

Water vapor also absorbs heat in the lower atmosphere and so prevents its rapid escape from Earth. Thus, like carbon dioxide, water vapor plays a large role in the insulating action of the atmosphere. In addition to gaseous water vapor, liquid water also exists in the atmosphere as rain and as fine droplets in clouds, mist, and fog. Solid water is found in the atmosphere in the form of ice crystals, snow, sleet, and hail.

Particulates are solids suspended in the atmosphere, and *aerosols* also include tiny liquid droplets and/or ice crystals composed of chemicals other than water. For example, sulfur dioxide ice crystals (SO_2) and other aerosols are also found in our atmosphere. Particulates can be considered as aerosols, but aerosols are not necessarily particulate matter. Particulates can be pollutants from transportation and industry, but the majority are natural particles and aerosols that have always existed in our atmosphere (• Fig. 4.3). Particles such as dust, smoke, pollen and spores, volcanic emissions, bacteria, and salts from ocean spray can all play an important role in absorption of energy and in the formation of raindrops.

• FIGURE 4.3
Volcanic eruptions, like this one at Mount St. Helens in Washington State, add a variety of gases, particulates, aerosols, and water vapor into our atmosphere.
What other ways are particles added to the atmosphere?

USGS

Atmospheric Environmental Issues

Two gases in our atmosphere play significant roles in important environmental issues. One is carbon dioxide, a gas that is directly involved in an apparent slow but steady rise in global temperatures. The other is ozone, which comprises a layer in the upper atmosphere and protects Earth from excessive UV radiation but is endangered by other gases associated with industrialization.

Carbon Dioxide and the Greenhouse Effect We are all familiar with what happens to the inside of a parked car on a sunny street if all the windows are left closed. Shortwave radiation (mainly visible light) from the sun can penetrate the glass windows with ease (● Fig. 4.4). When the insolation strikes the interior of the car, it is absorbed and heats the exposed surfaces. Energy, emitted from the surfaces as longwave radiation (mainly heat), cannot escape through the glass as freely. The result is that the interior of the vehicle gets hotter throughout the day. Many drivers recognize this as a blast of hot air as the car door opens. In extreme cases, windows in some cars have cracked under the heat as a result of *thermal expansion* (the property of all materials to expand when heated). A more serious effect is when temperatures become so great in automobiles with closed windows as to pose a deadly threat to small children and pets left behind.

Termed the **greenhouse effect,** this is the primary reason for the moderate temperatures observed on Earth. A greenhouse (plant beds surrounded by a glass structure) will behave like the closed vehicle parked in the sun. Insolation (shortwave radiation) can flow through the glass roof and walls of the greenhouse unimpeded and help the plants inside to thrive, even in a cold outdoor environment. However, the resulting heat energy (longwave radiation) within the greenhouse cannot escape as rapidly as insolation coming in and thus the interior of the greenhouse becomes warmer (● Fig. 4.5).

Like the glass of a greenhouse, carbon dioxide and water vapor (and other so-called greenhouse gases) in the atmosphere are largely transparent to incoming shortwave solar radiation, but can impede the escape of longwave radiation by absorbing it and then radiating it back to Earth. For example, carbon dioxide emits about half of its absorbed heat energy back to Earth's surface. Of course, although the results are similar, the processes involving the glass of the car or greenhouse on the one hand and the atmosphere on the other are significantly different. The heat of a closed car, or a greenhouse, increases because the air is trapped and cannot circulate with the outside air. Our atmosphere is free to circulate, but is selective as to which wavelengths of energy it will transmit. Though other analogies have been sought to explain the unequal exchange of radiation wavelengths in our atmosphere, for now the term "greenhouse effect" is still acceptable. The greenhouse effect in Earth's atmosphere is not a bad thing, for, without any greenhouse gases in the atmosphere, Earth's surface would be too cold to sustain human life. The greenhouse process helps maintain the warmth of the planet and is a factor in Earth's *heat energy budget* (discussed later in this chapter).

However, a serious environmental issue arises when increasing concentrations of greenhouse gases cause measurable increases in worldwide temperatures. Since the Industrial Revolution, human

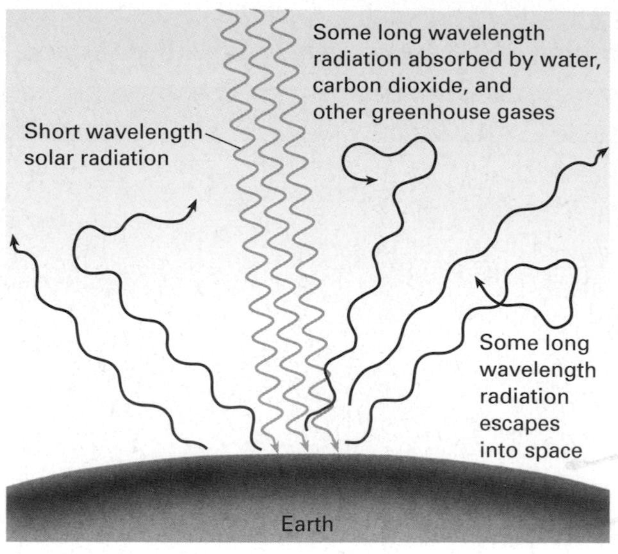

(a)

Short waves from sun
Long waves heat car interior
Selected waves escape

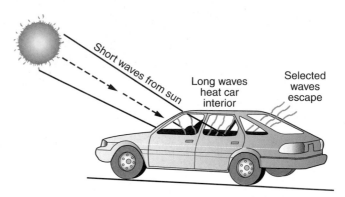

(b)

● FIGURE 4.4

(a) Greenhouse gases in our atmosphere allow short-wavelength solar radiation (sunlight) to penetrate Earth's atmosphere relatively unhampered, while some of the long-wavelength radiation (heat) is kept from escaping into outer space. (b) A similar sort of heat buildup occurs in a closed car. The penetration of shortwave radiation through the car windows is plentiful, but the glass prevents some of the longwave radiation to escape. **How might you prevent your car interior from becoming so hot on a summer day?**

beings have been adding more and more carbon dioxide to the atmosphere through their burning of fossil (carbon) fuels. At the same time, Earth has undergone massive *deforestation* (the removal of forests and other vegetation, including prime agricultural lands, for urban, commercial, and industrial development). Vegetation uses large amounts of carbon dioxide in *photosynthesis* (see again Fig. 4.2), and removal of the vegetation permits more carbon dioxide to remain in the atmosphere. ● Figure 4.6 shows how these two human activities have worked together to increase the amount of carbon dioxide in the atmosphere through time. Because carbon dioxide absorbs the longwave radiation from Earth's surface, restricting its escape to space, the rising amounts of carbon dioxide in the atmosphere increase the greenhouse effect and help produce a global rise in temperatures.

• FIGURE 4.5
Greenhouses used by horticulturalists at Western Kentucky University. Notice the shaded roof on the greenhouse in the background, which coupled with open windows and exhaust fans bears witness to the fact that greenhouses can actually get too hot for the plants.

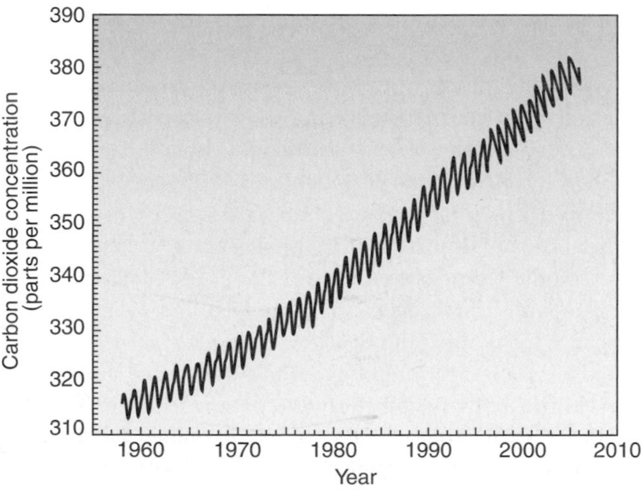

• FIGURE 4.6
Since 1958 these measurements of atmospheric carbon dioxide recorded at Mauna Loa, Hawaii, have shown an upward trend.
Why do you suppose the line zigzags each year?

At the present time, numerous researchers worldwide are closely monitoring the trends and amounts of change associated with Earth's temperatures and are watching for physical manifestations of greenhouse warming. This issue will be discussed in more detail with other causes of climate change, in Chapter 8.

The Ozone Layer Another vital gas in Earth's atmosphere is **ozone.** The ozone molecule (O_3) is related to the oxygen molecule (O_2), except it is made up of three oxygen atoms whereas oxygen gas consists of only two. Ozone is formed in the upper atmosphere when an oxygen molecule is split into two oxygen atoms (O) by the sun's ultraviolet radiation. Then the free unstable atoms join two oxygen gas molecules to form two molecules of ozone gas consisting of three oxygen atoms each:

(1) $2O_2 + 2O^- \rightarrow 2O_3$ (using UV radiation)

In the lower atmosphere, ozone is formed by electrical discharges (like high-tension power lines and lightning strokes) as well as incoming shortwave solar radiation. It is a toxic pollutant and a major component of urban smog, which can cause sore and watery eyes, soreness in the throat and sinuses, and difficulty in breathing. Near the surface of Earth, ozone is a menace and can only hurt lifeforms. However, in the upper atmosphere, ozone is essential to both terrestrial and marine life. Ozone is vital to living organisms because it is capable of absorbing large amounts of the sun's UV radiation.

In the upper atmosphere, UV radiation is consumed as it breaks the chemical bonds of ozone (O_3) to form an oxygen gas molecule (O_2) and an oxygen atom (O).

(2) $2O_3 \rightarrow 2O_2 + 2O$ (using UV radiation)

Then more UV radiation is consumed to recombine the oxygen gas and the oxygen atom back into ozone, as in Formula 1.

This process is repeated over and over again, thereby involving large amounts of UV energy that would otherwise reach Earth's surface. The chemistry of Formulas 1 and 2, repeating back and forth, creates a very efficient UV filter.

Without the ozone layer of the upper atmosphere, excessive UV radiation reaching Earth would severely burn human skin, increase the incidence of skin cancer and optical cataracts, destroy certain microscopic forms of marine life, and damage plants. Throughout the globe, UV radiation is, at least, responsible for painful sunburns or sensible suntans, depending on individual skin tolerance and exposure time.

For many years, there has been concern that human activity, especially the addition of chlorofluorocarbons (CFCs) and nitrogen oxides (NO_x) to the atmosphere, may permanently damage Earth's fragile ozone layer. CFCs, known commercially as Freon, have been used extensively in refrigeration and air-conditioning. Through time refrigeration has become ingrained in modern societies around the world, and the use of CFCs has become a necessary chemical in our lives. Developing an ozone-friendly refrigeration agent should be a top priority in all countries worldwide.

The ozone destruction reaction works in the following way: chlorine atoms (Cl), found in CFCs, split off and enter the stratosphere. There they bond to oxygen atoms (O) to form chlorine monoxide and oxygen gas.

(3) $$Cl + O_3 \rightarrow ClO + O_2$$

These oxygen atoms have now bonded with the chlorine and cannot be used to replenish the original amount of ozone (O_3) as in Formula 1. This and other chemical reactions attack the ozone layer, and threaten our natural UV filter. Nitrogen oxide compounds (NO_x), emitted with automobile and jet engine exhaust, also have the ability to enter the stratosphere and destroy our ozone shield.

The small proportion of UV radiation that the ozone layer allows to reach Earth does serve useful purposes. For instance, it has a vital function in the process of photosynthesis (see again Fig. 4.2). It is important in the production of certain vitamins (especially vitamin D), it can help treat certain types of skin disorders, and it helps the growth of some beneficial viruses and bacteria. However, increasing amounts of UV radiation reaching the surface can become a serious problem for Earth's environments, and the ozone layer must be protected from the pollutants that threaten its existence.

Ozone "Hole" First of all, there is no hole in our protective ozone layer. From Earth's surface to the outer reaches of the atmosphere, there is always some ozone present (keeping in mind ozone is only a trace gas). There have been years when ozone was missing from specific levels above the ground, but there is always some ozone above us. If all the ozone in our atmosphere were forced down to sea level, the atmospheric pressure there would compress the ozone into a worldwide layer ranging from 3 to 4 millimeters (<1/16 of an inch) thick. So the ozone hole is really an area where the amount of ozone is considerably less than it should be. ● Figure 4.7 displays an elliptical area centered over the South Pole (the area of maximum ozone loss).

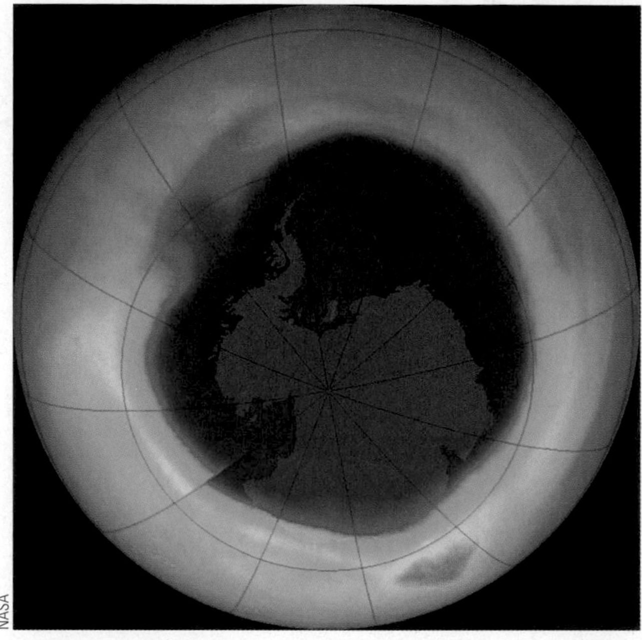

● FIGURE 4.7
For decades, satellite sensors have produced images of the ozone hole (shaded in purple) over Antarctica. This shows the spatial extent of the ozone hole in September of 2007.
What are the potential effects of ozone depletion on the world's human population?

Atmospheric ozone levels are measured in Dobson Units (du), established by G. M. B. Dobson along with his Dobson Spectrometer in the late 1920s. A range of 300 to 400 du indicates a sufficient amount of ozone to prevent damage to Earth's life-forms. To understand these units better, consider that 100 du equals 1 millimeter of thickness that ozone would have at sea level. Ozone measured inside the "hole" has dropped as low as 95 du in recent years and the area of the ozone-deficient atmosphere (a more accurate description than a "hole") has exceeded that of the North American continent.

Ozone-destroying pollutants enter a *stratospheric circulation pattern* high in the atmosphere that transports them over the tropics, over the middle latitudes, and toward the poles. Near the South Pole, ozone with its destroyers (CFCs and NO_x) become trapped together in what is called the *Southern Hemisphere polar vortex*. This is a closed circulation pattern of extremely cold air circling around the dark South Pole during the winter months (June, July, and August). During the South Polar spring (September, October, and November), incoming solar radiation starts to dissolve the vortex, and at the same time energizes the ozone-destroying Formula 3. Usually the ozone hole reaches its greatest extent each year within the first two weeks of October. Formula 3 works most effectively on ice crystal surfaces, and *polar stratospheric clouds* (ice clouds in the stratosphere) provide the perfect laboratory. When taking into account the stratospheric circulation, the Southern Hemisphere polar vortex, and the polar stratospheric clouds, the area of maximum ozone destruction would first occur at the poles, and then would appear like a swirling elliptical region like those seen in Total Ozone Mapping Spectrometer (TOMS) satellite images (see again Fig. 4.7).

Vertical Layers of the Atmosphere

Though people function primarily in the lowest level of the atmosphere, there are times, such as when we fly in aircraft or visit mountainous terrain, when we leave our normal altitude. The thinner atmosphere at these higher altitudes may affect us if we are not accustomed to it. Visitors to Inca ruins in the Andes or high-altitude Himalayan climbers may experience altitude sickness, and even skiers in the Rockies near mile-high Denver may need time to adjust. The air at these levels is much *thinner* than most of us are used to; by this we mean there is more empty space between air molecules and thus less oxygen and other gases in each breath of air.

As one travels from the surface to outer space, the atmosphere undergoes various changes, and it is necessary to look at the vertical layers that exist with Earth's atmospheric envelope. There are several systems used to divide the atmosphere into vertical layers. One system uses temperature and rates of temperature changes. Another uses the changes in the content of the gases in the atmosphere, and yet a third deals with the functions of these various layers.

System of Layering by Temperature Characteristics The atmosphere can be divided into four layers according to differences in temperature and rates of temperature change (• Fig. 4.8). The first of these layers, lying closest to Earth's surface, is the **troposphere** (from Greek: *tropo,* turn—the turning or mixing zone), which extends about 8–16 kilometers (5–10 mi) above Earth. Its thickness, which tends to vary seasonally, is least at the poles and greatest at the equator. It is within the troposphere that people live and work, plants grow, and virtually all Earth's weather and climate take place.

The troposphere has two distinct characteristics that differentiate it from other layers of the atmosphere. One is that the water vapor and particulates of the atmosphere are concentrated in this one layer; they are rarely found in the atmospheric layers above the troposphere. The other characteristic of this layer is that temperature normally decreases with increased altitude. The average rate at which temperatures within the troposphere decrease with altitude is called the **normal lapse rate** (or the environmental lapse rate); it amounts to 6.5°C per 1000 meters (3.6°F/1000 ft).

The altitude at which the temperature ceases to drop with increased altitude is called the *tropopause*. It is the boundary that separates the troposphere from the **stratosphere**—the second layer of the atmosphere. The temperature of the lower part of the stratosphere remains fairly constant (about −57°C, or −70°F) to an altitude of about 32 kilometers (20 mi). It is in the stratosphere that we find the ozone layer that does so much to protect life on Earth from the sun's UV radiation. As the ozone layer absorbs UV radiation, this absorbed energy results in the release of heat, and thus temperatures increase in the upper parts of the stratosphere.

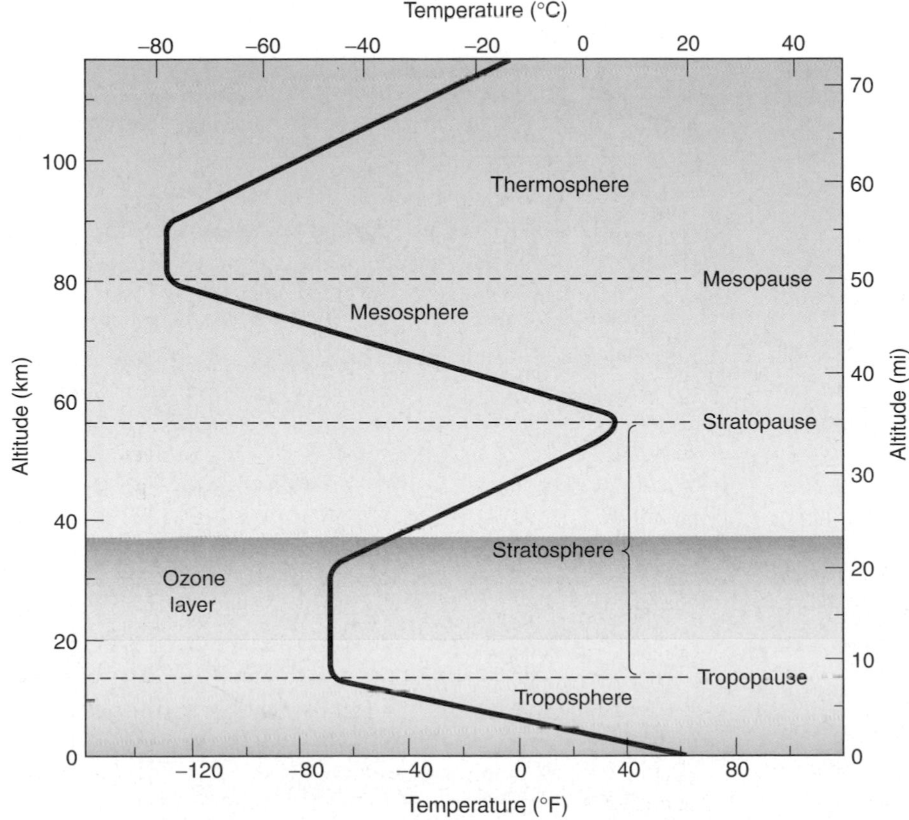

• FIGURE 4.8
Vertical temperature changes in Earth's atmosphere are the basis for its subdivision into the troposphere, stratosphere, mesosphere, and thermosphere.
At what altitude is our atmosphere the coldest?

Some water is available in the stratosphere, but it appears as stratospheric ice clouds. These thin veils of ice clouds have no effect on weather as we experience it. Temperatures at the *stratopause* (another boundary), which is about 50 kilometers (30 mi) above Earth, are about the same as temperatures found on Earth's surface, although little of that heat can be transferred because the air is so thin.

Above the stratopause is the **mesosphere,** in which temperatures tend to drop with increased altitude; the *mesopause* (the last boundary) separates the mesosphere from the **thermosphere,** where temperatures increase until they approach 1100°C (2000°F) at noon. Again, the air is so thin at this altitude that there is practically a vacuum and little heat can be transferred.

System of Layering by Functional Characteristics Astronomers, geographers, and communications experts sometimes use a different method of layering the atmosphere, one based on the protective function these layers provide. In this system, the atmosphere is divided into two distinct layers, the lowest of which is the **ozonosphere.** This layer lies approximately between 15 and 50 kilometers (10–30 mi) above the surface. The ozonosphere is another name for the ozone layer mentioned previously. Again, ozone effectively filters the UV energy from the sun and gives off heat energy instead. As we have noted, although ozone is a toxic pollutant at Earth's surface, aloft, it serves a vital function for Earth's life systems.

From about 60–400 kilometers (40–250 mi) above the surface lies the layer known as the **ionosphere.** This name denotes the ionization of molecules and atoms that occurs in this layer, mostly as a result of UV rays, X-rays, and gamma radiation. Ionization refers to the process whereby atoms are changed to ions through the removal or addition of electrons, giving them an electrical charge. The ionosphere in turn helps shield Earth from the harmful shortwave forms of radiation. This electrically charged layer also aids in transmitting communication and broadcast signals to distant regions of Earth. It is in the ionosphere that the auroras (defined in Chapter 3) occur. The ionosphere gradually gives way to interplanetary space (● Fig. 4.9).

System of Layering by Chemical Composition

Atmospheric chemists and physicists are at times concerned with the actual chemical makeup of the atmosphere. To this end, there is one more system to divide the atmosphere into two vertical layers. The first is termed the **homosphere** (from Greek: *homo,* same throughout). This layer begins on the surface and extends to an altitude of 80 kilometers (50 mi). In this layer, the gases in our atmosphere maintain the same percent by volume as those listed in Table 4.1. There are a few areas of concentration of specific gases, like the water vapor near Earth's surface and the ozone layer aloft, but for the most part the mixture is homogeneous. Other than rapid decreases in pressure and density while ascending through this layer, this is essentially the same air that we breathe on the surface.

From an altitude of about 80 kilometers (50 mi) and reaching into the vacuum of outer space lies the **heterosphere** (from Greek: *hetero,* different). In this layer, atmospheric gases are no longer evenly mixed but begin to separate out into distinct sublayers of concentration. This separation of gases is caused by Earth's gravity in which heavier gases are pulled closer to the surface and the lighter gases drift farther outward. The regions of concentration and their corresponding gases occur in the following order: nitrogen gas (N_2) is the heaviest and therefore the lowermost, followed by atomic oxygen (O), then by helium (He), and finally atomic hydrogen (H)—the lightest element that concentrates at the outermost region (● Fig. 4.10).

It is interesting to note that, when watching television news reports from the International Space Station or NASA shuttle missions, the astronauts who we see are still in the atmosphere. The background appears black, making it look like interplanetary space, but in reality these missions still take place within the realm of the outer atmosphere. One should also keep in mind that these different layering systems can focus on the same regions. For example, note that the thermosphere, the ionosphere, and the heterosphere all occupy the same altitudes above Earth—that is, from 80 kilometers (50 mi) and outward. The names are different because of the criteria used in the differing systems.

Effects of the Atmosphere on Solar Radiation

As the sun's energy passes through Earth's atmosphere, more than half of its intensity is lost through various processes. In addition, as discussed in Chapter 3, the amount of insolation actually received at a particular location depends on not only the processes involved

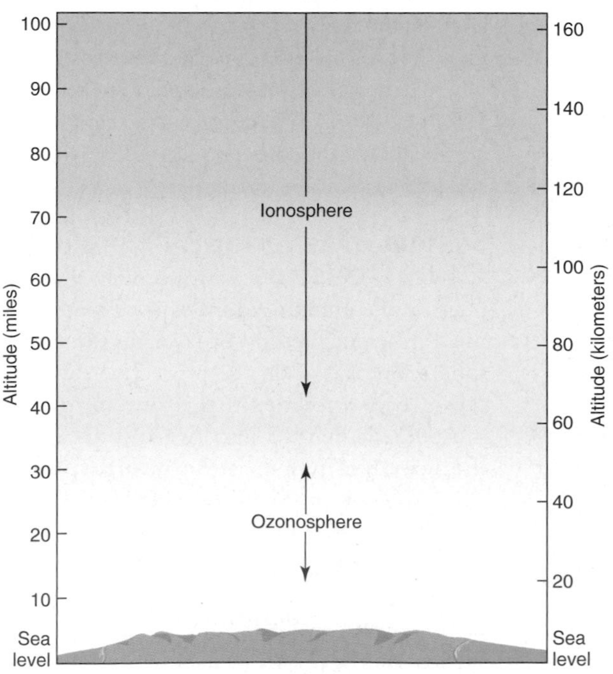

● **FIGURE 4.9**
Vertical changes in Earth's atmosphere based on functions of the gases cause the atmosphere to be subdivided into the ozonosphere and the ionosphere.
How do these layers protect life on Earth?

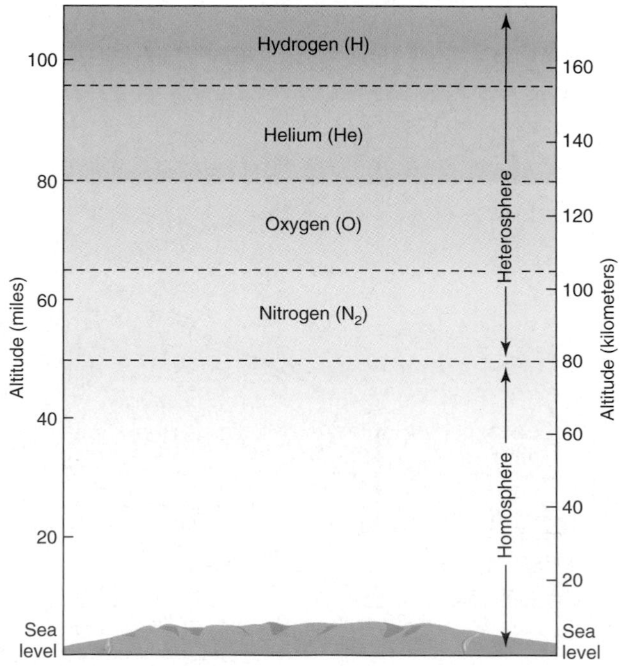

● **FIGURE 4.10**
The changes in chemical composition of the gases allow for the subdivision of the atmosphere into the homosphere and the heterosphere.
What gas in the atmosphere is the last to reach into outer space?

but also latitude, time of day, time of year, and atmospheric thickness (all of which are related to the angle at which the sun's rays strike Earth). The transparency of the atmosphere (or the amount of cloud cover, moisture, carbon dioxide, and solid particles in the air) also plays a vital role.

When the sun's energy passes through the atmosphere, several things happen to it (the following figures represent approximate averages for entire Earth; at any one location or time, they may differ): (1) 26% of the energy is reflected directly back to space by clouds and the ground; (2) 8% is *scattered* by minute atmospheric particles and returned to space as diffuse radiation; (3) 19% is *absorbed* by the ozone layer and water vapor in the clouds of the atmosphere; (4) 20% reaches Earth's surface as diffuse radiation after being scattered; (5) 27% reaches Earth's surface as direct radiation (● Fig. 4.11). In other words, on a worldwide average, 47% of the incoming solar radiation eventually reaches the surface, 19% is retained in the atmosphere, and 34% is returned to space. Because Earth's energy budget is in equilibrium, the 47% received at the surface is ultimately returned to the atmosphere by processes that we now examine.

Water as Heat Energy

As it penetrates our atmosphere, some of the incoming solar radiation is involved in several energy exchanges. One such exchange involves how water in the Earth system is altered from one state to another. Water is the only substance that can exist in all three states of matter—as a solid, a liquid, and a gas—within the normal temperature range of Earth's surface. In the atmosphere, water exists as a clear, odorless gas called *water vapor.* It is also a *liquid* in the atmosphere (as clouds, fog, and rain), in the oceans, and in other water bodies on and beneath the surface of Earth. Liquid water is also found within vegetation and animals. Finally, water can be found as a *solid* in snow and ice in the atmosphere, as well as on and under the surface of the colder parts of Earth.

Not only does water exist in all three states of matter, but it can change from one state to another, as illustrated in ● Figure 4.12. In doing so, it becomes involved in the heat energy system of Earth. The molecules of a gas move faster than do those of a liquid. Thus, during the process of *condensation,* when water vapor changes to liquid water, its molecules slow down and some of their

● **FIGURE 4.11**
Environmental Systems: Earth's radiation budget From one year to the next, Earth's overall average temperature varies very little. This fact indicates that a long-term global balance, or equilibrium, must exist between the energy received and emitted by the Earth system. Note in the diagram that only 47% of the incoming solar energy reaches and is absorbed by Earth's surface. Eventually, all the energy gained by the atmosphere is lost to space. However, the radiation budget is a dynamic one. In other words, alterations to one element affect the other elements. As a result, there is growing concern that one of the elements, human activity, will cause the atmosphere to absorb more Earth emitted energy, thus raising global temperatures.

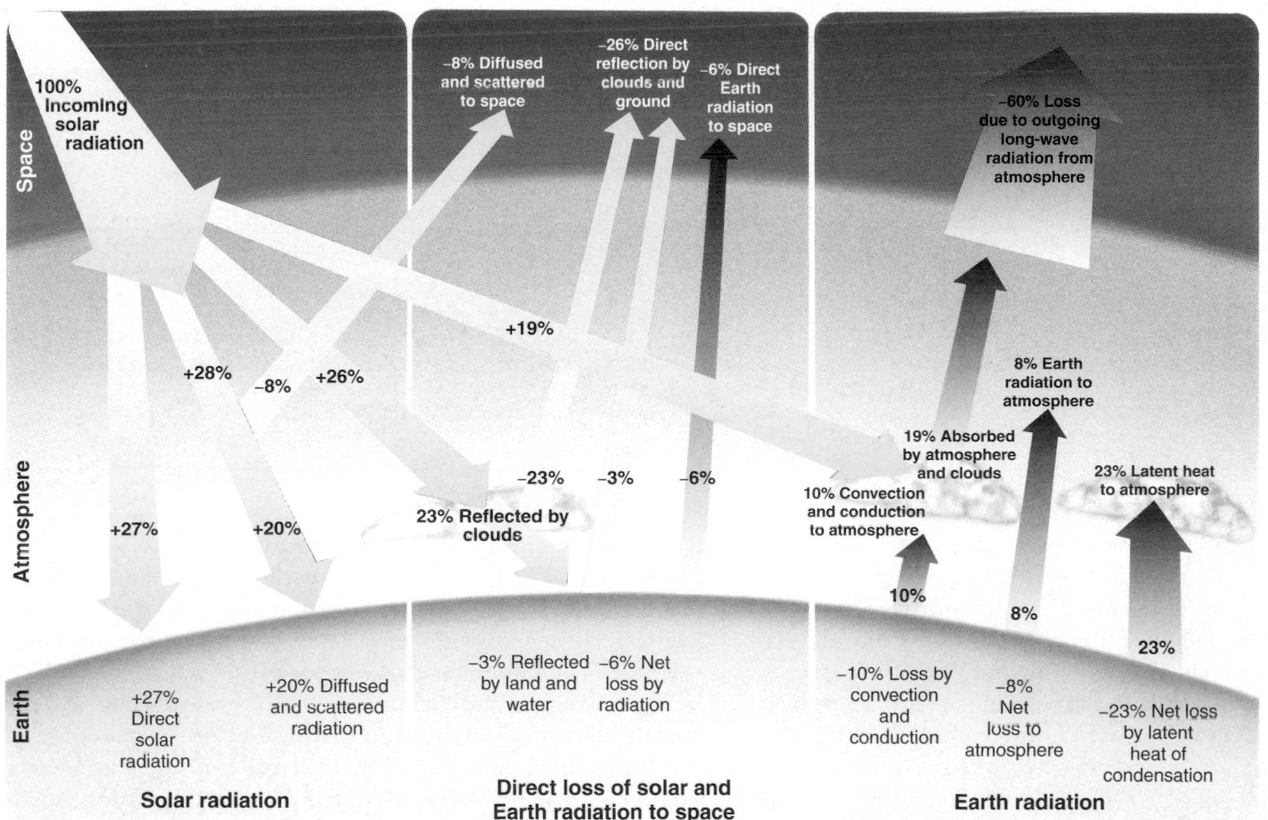

GEOGRAPHY'S PHYSICAL SCIENCE PERSPECTIVE

Colors of the Atmosphere

When we look to the skies, we are used to seeing certain colors: the brilliant blue of a clear day, the vibrant reds and oranges of a sunset, clouds ranging from pure white to ominous gray-black, and sometimes even a colorful rainbow. These are wondrous colors, and we sometimes take them for granted. Why they exist, however, can be explained by the concept of *atmospheric scattering*. This process is explained in the text, but let's examine how it affects the colors of our atmosphere.

It is important to first realize that as sunlight is scattered in our atmosphere, certain wavelengths are scattered more than others. Each wavelength of visible light corresponds to one of the colors of the visible light spectrum. For example, a wavelength of 0.45–0.50 micrometers (1 micrometer = 1/1000 of a millimeter) corresponds to the color blue, and 0.66–0.70 micrometers to the color red. The wavelengths that are scattered, then enter the human eye, are transmitted to the human brain, and then translated into the colors. In other words, the wavelengths that are scattered are the colors that we see.

In 1871, Lord Rayleigh, a radiation scientist, explained that our atmospheric gases tend to scatter the shorter wavelengths of visible light. In other words, *Rayleigh scattering* makes a clear sky appear blue. In 1908, another radiation specialist, Gustav Mie, explained that oblique (or sharp) sun angles coming through the atmosphere tend to cause the scattering of the longer wavelengths of visible light. In other words, Mie scattering explains why sunsets bring about a red sky. Another form of scattering, called *indiscriminant scattering,* is caused by tiny water droplets and ice crystals in our atmosphere that tend to scatter all the wavelengths of visible light with equal intensity. When all the wavelengths are scattered equally, white light emerges, and this explains the white clouds we see. (Keep in mind that all clouds are white. When we see dark gray to black storm clouds, we are actually seeing white clouds plus shadows cast by the height and thickness of these massive clouds.)

Rainbows are much more complicated, and their full explanation involves a number of trigonometric functions. Rainbows are caused by water droplets suspended in the atmosphere. The droplets act like tiny glass prisms, which take the incoming white sunlight and project the colors of the spectrum. Key elements necessary to observe a rainbow involve the sun angle, water droplet density and orientation, and the observer angle. When you see a rainbow from a moving vehicle, it is only a matter of time until the angles change and the rainbow disappears.

Incidentally, the colors described above are those distinguishable by the human eye in regard to Earth's atmosphere. Humans visiting another planet with an atmosphere may never see the color blue at all. Furthermore, animals (those that are not color-blind) may see colors other than those described here.

M. Trapasso

Sunset over the North Pacific Ocean.

M. Trapasso

A complete rainbow over Easter Island in the Pacific Ocean.

energy is released into the environment, about 590 calories per gram (cal/g). The molecules of a solid move even more slowly than those of a liquid, so during the process of *freezing,* when water changes to ice, additional energy is released into the environment, this time 80 calories per gram. When the process is reversed, heat must be added to the ice. Thus, *melting* ice requires the addition of 80 calories per gram to the ice, from the surrounding environment. Further, *evaporation* requires the addition of 590 calories per gram be added to the liquid water, from the environment. This added energy is stored in the water as *latent* (or hidden) *heat.* The **latent heat of fusion** refers to the 80 calories per gram released into the environment when water freezes into ice; to melt ice into water, this heat energy comes from the environment. The **latent heat of evaporation** (590 cal/g) is added to the water, from the environment, to form water vapor, and the **latent heat of condensation** (also 590 cal/g) is removed from the water vapor, and released into the environment as it condenses into liquid water. The last is **latent heat of sublimation** at 670

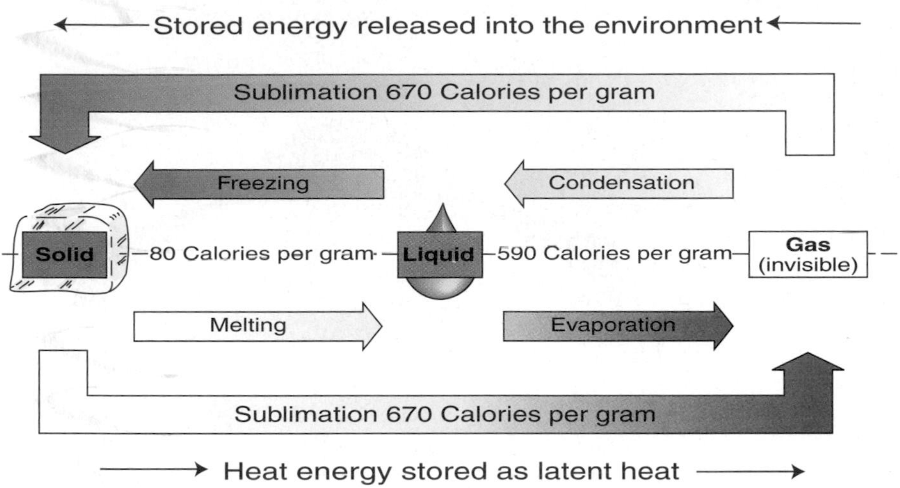

● **FIGURE 4.12**
The three physical states of water and the energy exchanges between them. This is a simple diagram to read, if you consider where the heat energy comes from and where it goes. For example, when water freezes into ice, heat energy must come out of the water as it freezes, and thus enters the environment. To evaporate water into vapor, the heat must go into the water for it to evaporate, and therefore must come from the surrounding environment.
Why do you suppose that some of the energy in these exchanges is referred to as "latent heat"?

calories per gram (the addition of 590 cal/g + 80 cal/g). *Sublimation* is the process where ice turns to vapor or vapor turns to ice without going through the intermediate liquid phase. Snowflakes and frost are formed by sublimation.

Some of these energy exchanges can be easily demonstrated. For example, if you hold an ice cube in your hand, your hand feels cold because the heat removed from your hand (the surrounding environment in this case) is needed to melt the ice. We are cooled by evaporating perspiration from our skin because heat must be absorbed by the evaporating perspiration, from our skin (the surrounding environment in this case), thereby lowering skin temperature.

Heating the Atmosphere

The 19% of direct solar radiation that is retained by the atmosphere is "locked up" in the clouds and the ozone layer and thus is not available to heat the troposphere. Other sources must be found to explain the creation of atmospheric warmth. The explanation lies in the 47% of incoming solar energy reaching Earth's surface (on both land and water) and in the transfer of heat energy from Earth back to the atmosphere. This is accomplished through such physical processes as (1) radiation, (2) conduction, (3) convection (along with the related phenomenon, advection), and (4) the latent heat of condensation (● Fig. 4.13).

Processes of Heat Energy Transfer

Radiation The process by which electromagnetic energy is transferred from the sun to Earth is called **radiation.** We should be aware that all objects emit electromagnetic radiation. The characteristics of that radiation depend on the temperature of the radiating body. In general, the warmer the object, the more energy it will emit, and the shorter are the wavelengths of peak emission. Because the sun's absolute temperature is 20 times that of Earth, we can predict that the sun will emit more energy, and at shorter wavelengths, than Earth. This is borne out by the facts: The energy output per square meter by the sun is approximately 160,000 times that of Earth! Further, the majority of solar energy is emitted at wavelengths shorter than 4.0 micrometers, whereas most of Earth's energy is radiated at wavelengths much longer than 4.0 micrometers (see again Fig. 3.9). Thus, *shortwave radiation* from the sun reaches Earth and heats its surface, which, being cooler than the sun, gives off energy in the form of longwaves. It is this *longwave radiation* from Earth's surface that heats the lower layers of the atmosphere and accounts for the heat of the day.

Conduction The means by which heat is transferred from one part of a body to another or between two touching objects is called **conduction.** Heat flows from the warmer to the cooler part of a body in order to equalize temperature. Conduction actually occurs as heat is passed from one molecule to another in chainlike fashion. It is conduction that makes the bottom of your soup bowl too hot to touch.

Atmospheric conduction occurs at the interface of (zone of contact between) the atmosphere and Earth's surface. However, it is actually a minor method of heat transfer in terms of warming the atmosphere because it affects only the layers of air closest to the surface. This is because air is a poor conductor of heat. In fact, air is just the opposite of a good conductor; it is a good insulator. This property of air is why a layer of air is sometimes put between two panes of glass to help insulate the window. Air is also used as a layer of insulation in sleeping bags and cold-weather parkas. In fact, if air were a good conductor of heat, our kitchens would become unbearable every time we turned on the stove or oven.

Convection In the atmosphere, as pockets of air near the surface are heated, they expand in volume, become less dense than the surrounding air, and therefore rise. This vertical transfer of heat through the atmosphere is called **convection;** it is the same type of process by which boiling water circulates in a pot on the stove. The water near the bottom is heated first, becoming lighter and less dense as it is heated. As this water tends to rise, colder, denser surface water flows down to replace it. As this new water is warmed, it too flows upward while additional colder water moves downward. This movement within the fluid is called a *convective current.* These currents set into motion by the heating of a fluid (liquid or gas) make up a *convectional system.* Such systems account for much of the vertical transfer of heat within the atmosphere and the oceans and are a major cause of clouds and precipitation.

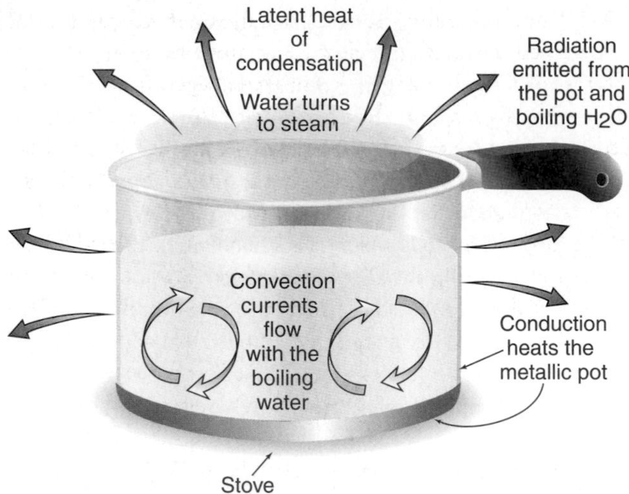

• FIGURE 4.13

Mechanisms of heat transfer. Conduction occurs when heat travels from molecule to molecule, warming the metallic pot. Convection is shown by the hotter water flowing upward, and the cooler water sinking downward, forming the "convective current" in the boiling water. Radiation is displayed as heat emitted outward from the boiling pot. Lastly, heat of condensation is released as water vapor turns back into steam.

How might we add advection to this small system?

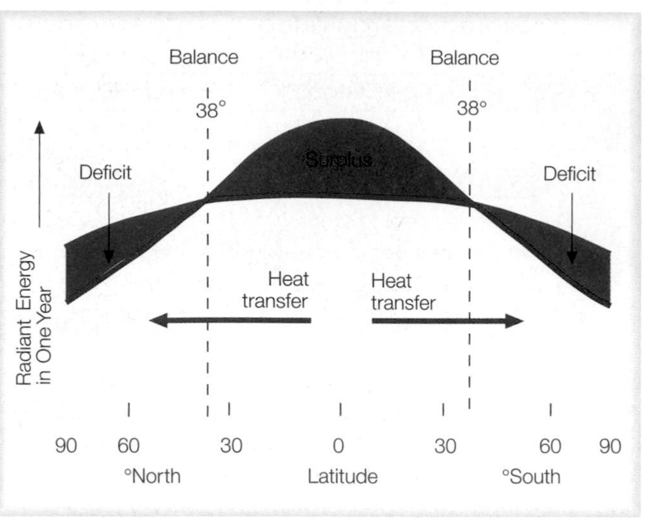

• FIGURE 4.14

Latitudinal variation in the energy budget. Low latitudes receive more insolation than they lose by reradiation and have an energy surplus. High latitudes receive less energy than they lose outward and therefore have an energy deficit.

How do you think the surplus energy in the low latitudes is transferred to higher latitudes?

Advection **Advection** is the term applied to horizontal heat transfer. There are two major advection agents within the Earth–atmosphere system: winds and ocean currents. Both agents help transfer energy horizontally between the equatorial and polar regions, thus maintaining the energy balance in the Earth–atmosphere system (• Fig. 4.14).

Latent Heat of Condensation As we have seen, when water evaporates, a significant amount of energy is stored in the water vapor as latent heat (see again Fig. 4.12). This water vapor is then transported by advection or convection to new locations where condensation takes place and the stored energy is released. This process plays a major role in the transfer of energy within the Earth system: The *latent heat of evaporation* helps cool the atmosphere while the *latent heat of condensation* helps warm the atmosphere and, in addition, is a source of energy for storm activity. The power of all severe weather is supplied by the latent heat of condensation.

The Heat Energy Budget

The Heat Energy Budget at Earth's Surface Now that we know the various means of heat transfer, we are in a position to examine what happens to the 47% of solar energy that reaches Earth's surface (see again Fig. 4.11). Approximately 14% of this energy is emitted by Earth in the form of longwave radiation. This 14% includes a net loss of 6% (of the total) directly to outer space, and the other 8% is captured by the atmosphere. In addition, there is a net transfer back to the lower atmosphere (by conduction and convection) of 10 of the 47% that reached Earth. The remaining 23% returns to the atmosphere through the release of latent heat of condensation. Thus, 47% of the sun's original insolation that reached Earth's surface is all returned to other segments of the system. There has been no long-term gain or loss. Therefore, at Earth's surface, the heat energy budget is in balance.

Examination of the heat energy budget of Earth's surface helps us understand the *open energy system* that is involved in the heating of the atmosphere. The *input* in the system is the incoming shortwave solar radiation that reaches Earth's surface; this is balanced by the *output* of longwave terrestrial (Earth's) radiation back to the atmosphere and to space. As these functions adjust to remain in balance, we can say that the overall heat budget of Earth's surface is in a state of *dynamic equilibrium*.

Of course, it should be noted that the percentages mentioned earlier represent an oversimplification in that they refer to *net* losses that occur over a long period of time. In the shorter term, heat may be passed from Earth to the atmosphere and then back to Earth in a chain of cycles before it is finally released into space. The absorption and reflection of incoming solar radiation, and the emission of outgoing terrestrial radiation, can all be affected by surface ground cover and any human activity that may change the surface cover.

The Heat Energy Budget in the Atmosphere At one time or another, about 60% of the solar energy intercepted by the Earth system is temporarily retained by the atmosphere. This includes 19% of direct solar radiation absorbed by the clouds and the ozone layer, 8% emitted by *longwave radiation* from the Earth's surface, 10% transferred from the surface by *conduction* and *convection,* and 23% released by the *latent heat of condensation*. Some of this energy is recycled back to the surface for short periods of time, but eventually all of it is lost into outer space as more solar energy is received. Hence, just as was the case at Earth's surface, the heat energy budget in the atmosphere is in balance over long periods of time—a dynamically stable system. However, many scientists believe that an imbalance in the heat energy budget, with possible negative effects, could develop due to the *greenhouse effect*.

Variations in the Heat Energy Budget Remember that the figures we have seen for the heat energy budget are averages for the whole Earth over many years. For any *particular* location, the heat energy budget is most likely not balanced. Some places have a surplus of incoming solar energy over outgoing energy loss, and others have a deficit. The main causes of these variations are differences in latitude and seasonal fluctuations.

As we have noted previously, the amount of insolation received is directly related to latitude (see again Fig. 4.14). In the tropical zones, where insolation is high throughout the year, more solar energy is received at Earth's surface and in the atmosphere than can be emitted back into space. In the Arctic and Antarctic zones, on the other hand, there is so little insolation during the winter, when Earth is still emitting longwave radiation, that there is a large deficit for the year. Locations in the middle-latitude zones have lower deficits or surpluses, but only at about latitude 38° is the budget balanced. If it were not for the heat transfers within the atmosphere and the oceans, the tropical zones would get hotter and the polar zones would get colder through time.

At any location, the heat energy budget varies throughout the year according to the seasons, with a tendency toward a surplus in the summer or high-sun season and a tendency toward a deficit 6 months later. Seasonal differences may be small near the equator, but they are great in the middle-latitude and polar zones.

Air Temperature

Temperature and Heat

Although heat and temperature are highly related, they are not the same. **Heat** is a form of energy—the total kinetic energy of all the atoms that make up a substance. All substances are made up of molecules that are constantly in motion (vibrating and colliding) and therefore possess kinetic energy—the energy of motion. This energy is manifested as heat. **Temperature,** on the other hand, is the average kinetic energy of the individual molecules of a substance. When something is heated, its atoms vibrate faster, and its temperature increases. It is important to remember that the amount of heat energy depends on the mass of the substance under discussion, whereas the temperature refers to the energy of individual molecules. Thus, a burning match has a high temperature but minimal heat energy; the oceans have moderate temperatures but high heat energy content.

Temperature Scales

Three different scales are generally used in measuring temperature. The one with which Americans are most familiar is the **Fahrenheit scale,** devised in 1714 by Daniel Fahrenheit, a German scientist. On this scale, the temperature at which water boils at sea level is 212°F, and the temperature at which water freezes is 32°F. This scale is used in the English system of measurements.

The **Celsius scale** (also called the **centigrade scale**) was devised in 1742 by Anders Celsius, a Swedish astronomer. It is part of the metric system. The temperature at which water freezes at sea level on this scale was arbitrarily set at 0°C, and the temperature at which water boils was designated as 100°C.

The Celsius scale is used nearly everywhere but in the United States, and even in the United States, the Celsius scale is used by the majority of the scientific community. By this time, you have undoubtedly noticed that throughout this book comparable figures in both the Celsius and Fahrenheit scales are given side by side for temperatures. Similarly, whenever important figures for distance, area, weight, or speed are given, we use the metric system followed by the English system. Appendix A at the back of your text may be used for comparison and conversion between the two systems.

● Figure 4.15 can help you compare the Fahrenheit and Celsius systems as you encounter temperature figures outside this

● FIGURE 4.15

The Fahrenheit and Celsius temperature scales. The scales are aligned to permit direct conversion of readings from one to the other.

When it is 70°F, what is the temperature in Celsius degrees?

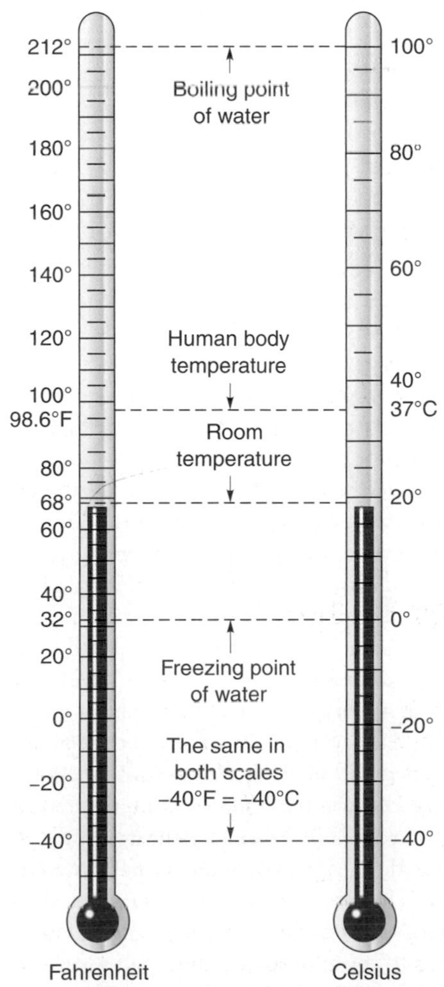

book. In addition, the following formulas can be used for conversion from Fahrenheit to Celsius or vice versa:

$$C = (F - 32) \div 1.8$$

$$F = (C \times 1.8) + 32$$

The third temperature scale, used primarily by scientists, is the **Kelvin scale.** Lord Kelvin, a British radiation scientist, felt that negative temperatures were not proper and should not be used. In his mind, no temperature should ever go below zero. His scale is based on the fact that the temperature of a gas is related to the molecular movement within the gas. As the temperature of a gas is reduced, the molecular motion within the gas slows. There is a temperature at which all molecular motion stops and no further cooling is possible. This temperature, approximately –273°C, is termed *absolute zero.* The Kelvin scale uses absolute zero as its starting point. Thus, 0°K equals –273°C. Conversion of Celsius to Kelvin is expressed by the following formula:

$$K = C + 273$$

Short-Term Variations in Temperature

Local changes in atmospheric temperature can have a number of causes. These are related to the mechanics of the receipt and dissipation of energy from the sun and to various properties of Earth's surface and the atmosphere.

The Daily Effects of Insolation

As we noted earlier, the amount of insolation at any particular location varies both throughout the year (annually) and throughout the day (diurnally). Annual fluctuations are associated with the sun's changing declination and hence with the seasons. Diurnal changes are related to the rotation of Earth about its axis. Each day, insolation receipt begins at sunrise, reaches its maximum at noon (local solar time), and returns to zero at sunset.

Although insolation is greatest at noon, you may have noticed that temperatures usually do not reach their maximum until 2–4 p.m. (● Fig. 4.16). This is because the insolation received by Earth from sunrise until the afternoon hours exceeds the energy being lost through Earth radiation. Hence, during that period, as Earth and atmosphere continue to gain energy, temperatures normally show a gradual increase. Sometime around 3–4 p.m., when outgoing Earth radiation begins to exceed insolation, temperatures start to fall. The daily lag of Earth radiation and temperature behind insolation is accounted for by the time it takes for Earth's surface to be heated to its maximum and for this energy to be radiated to the atmosphere.

Insolation receipt ends at sunset, but energy that has been stored in Earth's surface layer during the day continues to be lost throughout the night and the ability to heat the atmosphere decreases. The lowest temperatures occur around dawn, when the maximum amount of energy has been emitted and before replenishment from the sun can occur. Thus, if we disregard other factors for the moment, we can see that there is a predictable hourly change in temperature called the **daily march of temperature.** There is a gentle decline from midafternoon until dawn and a rapid increase in the 8 hours or so from dawn until the next maximum is reached.

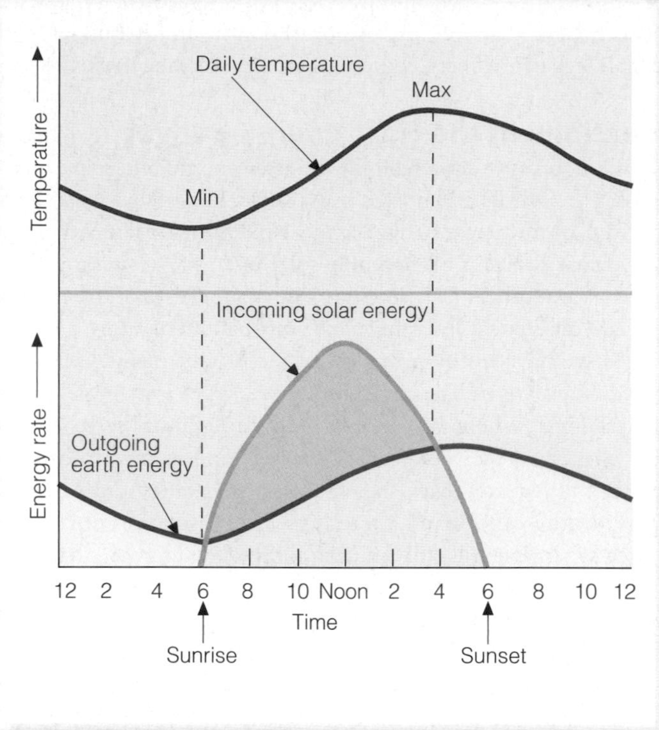

● FIGURE 4.16
Diurnal changes in air temperature are controlled by insolation and outgoing Earth radiation. Where incoming energy exceeds outgoing energy (orange), the air temperature rises. Where outgoing energy exceeds incoming energy (blue), air temperature drops.
Why does temperature rise even after solar energy declines?

Cloud Cover

The extent of cloud cover is another factor that affects the temperature of Earth's surface and the atmosphere (● Fig. 4.17). Weather satellites have shown that, at any time, about 50% of Earth is covered by clouds (● Fig. 4.18). This is important because a heavy cloud cover can reduce the amount of insolation a place receives, thereby causing daytime temperatures to be lower on a cloudy day. On the other hand, we also have the greenhouse effect, in which clouds, composed in large part of water droplets, are capable of absorbing heat energy radiating from Earth. Clouds therefore keep temperatures near Earth's surface warmer than they would otherwise be, especially at night. The general effect of cloud cover, then, is to moderate temperature by lowering the potential maximum and raising the potential minimum temperatures. In other words, cloud cover makes for cooler days and warmer nights.

Differential Heating of Land and Water

For reasons we will later explain in detail, bodies of water heat and cool more slowly than the land. The air above Earth's surface is heated or cooled in part by what is beneath it. Therefore, temperatures over bodies of water or on land subjected to ocean winds (**maritime** locations) tend to be more moderate than those of land-locked places at the same latitude. Thus, the greater the **continentality** of a location (the distance removed from a large body of water), the less its temperature pattern will be modified.

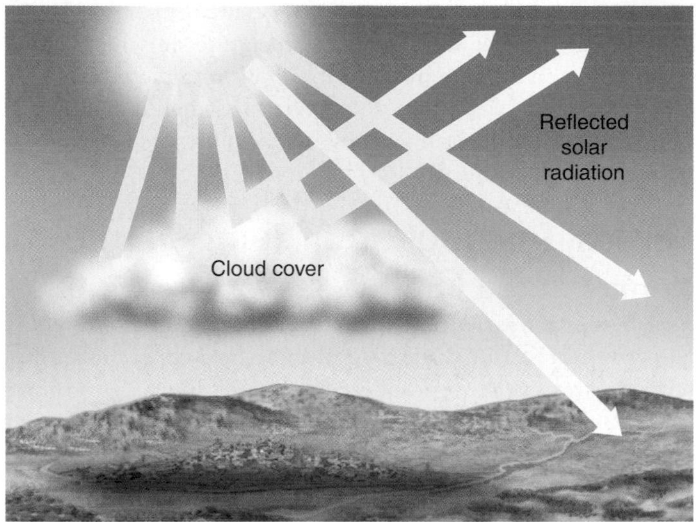

(a) Day

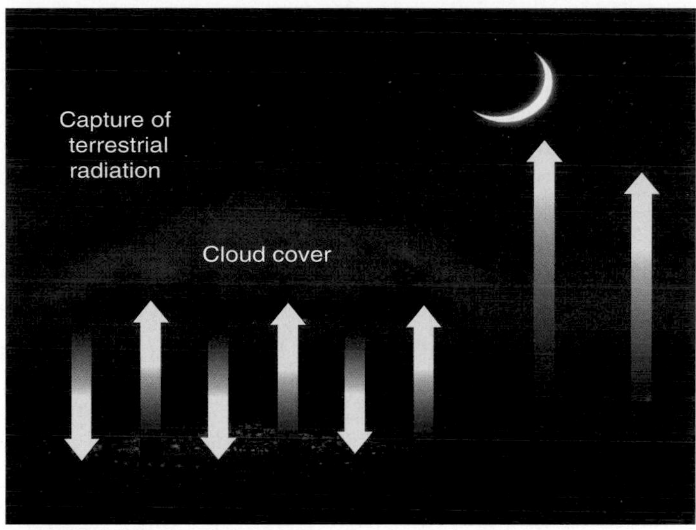

(b) Night

• FIGURE 4.17
The effect of cloud cover on temperatures. (a) By intercepting insolation, clouds produce lower air temperatures during the day. (b) By trapping longwave radiation from Earth, clouds increase air temperatures at night. The overall effect is a great reduction in the diurnal temperature range.
Why do desert regions have large diurnal variations in temperature?

NASA Goddard Space Flight Center Image: Reto Stöckli. Enhancements: Robert Simmon. Data/support: MODIS Land Group, Atmosphere Group, Ocean Group, and Science Data Support Team. Additional data: USGS EROS Data Center.

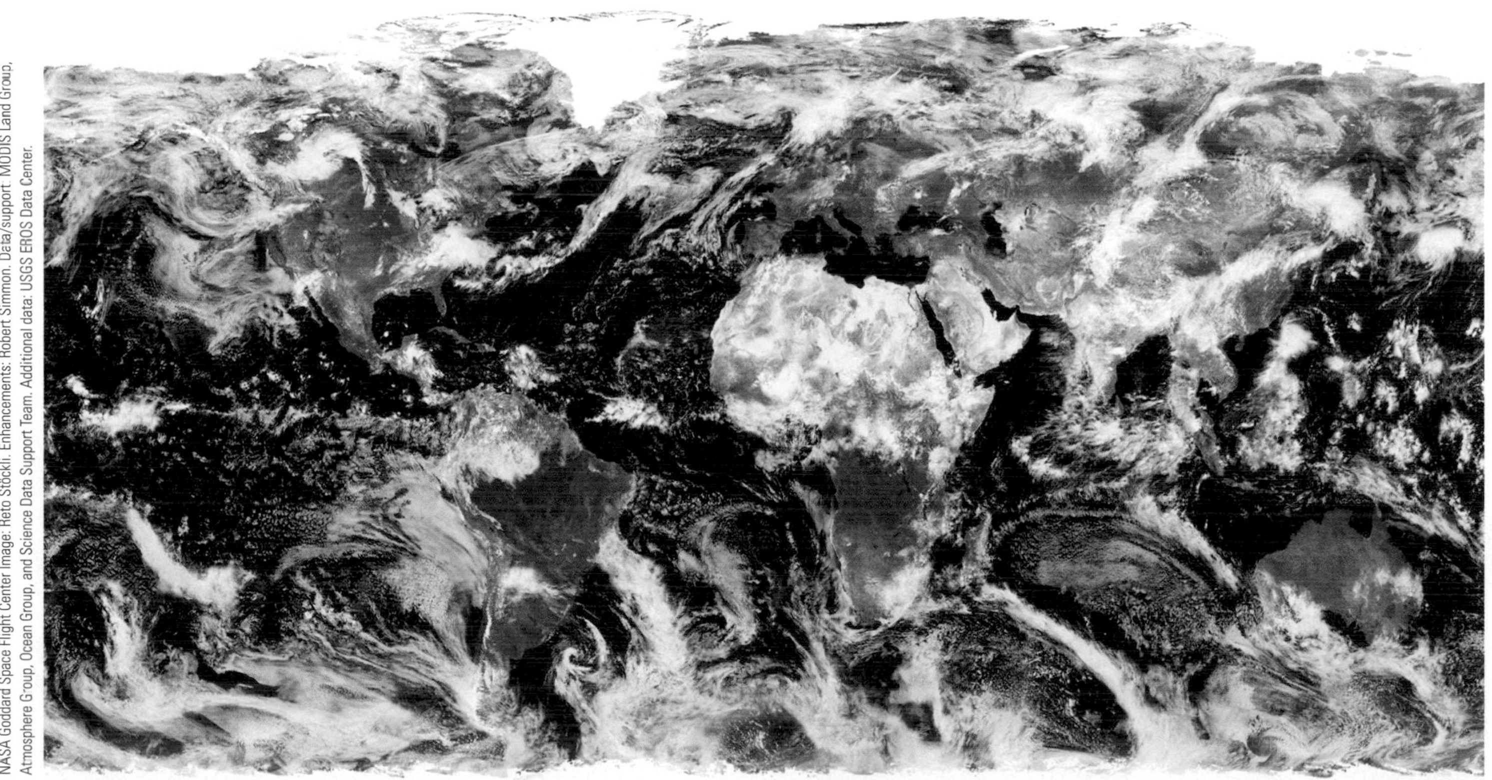

• FIGURE 4.18
This composite of several satellite images shows a variety of cloud cover and storm systems across Earth.
In general which are the cloudiest latitude zones and which are the zones with the clearest skies?

Reflection The capacity of a surface to reflect the sun's energy is called its **albedo;** a surface with a high albedo has a high percentage of reflection. The more solar energy that is reflected back into space by Earth's surface, the less that is absorbed for heating the atmosphere. Temperatures will be higher at a given location if its surface has a low albedo rather than a high albedo.

As you may know from experience, snow and ice are good reflectors; they have an albedo of 90–95%. Or put another way, only 5–10% of the incoming solar radiation is absorbed by the snow and ice. This is one reason why glaciers on high mountains do not melt away in the summer or why there may still be snow on the ground on a sunny day in the spring: Most of the solar energy is reflected away. A forest, on the other hand, has an albedo of only 10–15% (or 85–90% absorption), which is good for the trees because they need solar energy for photosynthesis. The albedo of cloud cover varies, from 40 to 80%, according to the thickness of the clouds. The high albedo of many clouds is why much solar radiation is reflected directly back into space by the atmosphere.

The albedo of water varies greatly, depending on the depth of the water body and the angle of the sun's rays. If the angle of the sun's rays is high, smooth water will reflect little. In fact, if the sun is vertical over a calm ocean, the albedo will be only about 2%. However, a low sun angle, such as just before sunset, causes an albedo of more than 90% from the same ocean surface (• Fig. 4.19). Likewise, a snow surface in winter, when solar angles are lower, can reflect up to 95% of the energy striking it, and skiers must constantly be aware of the danger of severe sunburns and possible snow blindness from reflected solar radiation. In a similar fashion, the high albedo of sand causes the sides of sunbathers' legs to burn faster when they lie on the beach.

Horizontal Air Movement We have already seen that advection is the major mode of horizontal transfer of heat and energy over Earth's surface. Any movement of air due to the wind, whether on a large or small scale, can have a significant short-term effect on the temperatures of a given location. Thus, wind blowing from an ocean to land will generally bring cooler temperatures in summer and warmer temperatures in winter. Large quantities of air moving from polar regions into the middle latitudes can cause sharp drops in temperature, whereas air moving poleward will usually bring warmer temperatures.

• FIGURE 4.19
At low sun angles, water reflects most of the solar radiation that strikes it.
Why is it so difficult to assign one albedo value to a water surface?

M. Trapasso

Vertical Distribution of Temperature

Normal Lapse Rates We have learned that Earth's atmosphere is primarily heated from the ground up as a result of longwave terrestrial radiation, conduction, and convection. Thus, temperatures in the troposphere are usually highest at ground level and decrease with increasing altitude. As noted earlier in the chapter, this decrease in the free air of approximately 6.5°C per 1000 meters (3.6°F/1000 ft) is known as the *normal lapse rate.*

The lapse rate at a particular place can vary for a variety of reasons. Lower lapse rates can exist if denser and colder air is drained into a valley from a higher elevation or if advectional winds bring air in from a cooler region at the same altitude. In each case, the surface is cooled. On the other hand, if the surface is heated strongly by the sun's rays on a hot summer afternoon, the air near Earth will be disproportionately warm, and the lapse rate will increase. Fluctuations in lapse rates due to abnormal temperature conditions at various altitudes can play an important role in the weather a place may have on a given day.

Temperature Inversions Under certain circumstances, the normal observed decrease of temperature with increased altitude might be reversed; temperature may actually *increase* for several hundred meters. This is called a **temperature inversion.**

Some inversions take place 1000 or 2000 meters above the surface of Earth where a layer of warmer air interrupts the normal decrease in temperature with altitude (• Fig. 4.20). Such inversions tend to stabilize the air, causing less turbulence and discouraging both precipitation and the development of storms. Upper air inversions may occur when air settles slowly from the upper atmosphere. Such air is compressed as it sinks and rises in temperature, becoming more stable and less buoyant. Inversions caused by descending air are common at about 30–35° north and south latitudes.

An upper air inversion common to the coastal area of California results when cool marine air blowing in from the Pacific Ocean moves under stable, warmer, and lighter air aloft created by subsidence and compression. Such an inversion layer tends to maintain itself; that is, the cold underlying air is heavier and cannot rise through the warmer air above. Not only does the cold air resist rising or moving, but pollutants, such as smoke, dust particles, and automobile exhaust, created at Earth's surface, also fail to disperse. They therefore accumulate in the lower atmosphere. This situation is particularly acute in the Los Angeles area, which is a basin surrounded by higher mountainous areas (• Fig. 4.21). Cooler air blows into the basin from the ocean and then cannot escape either horizontally, because of the landform barriers, or vertically, because of the inversion.

Some of the most noticeable temperature inversions are those that occur near the surface when Earth cools the lowest layer of air through conduction and radiation (• Fig. 4.22). In this situation, the

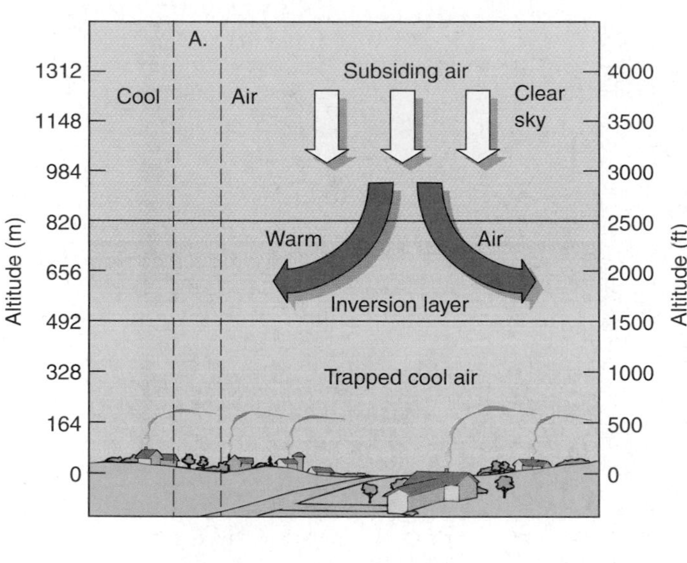

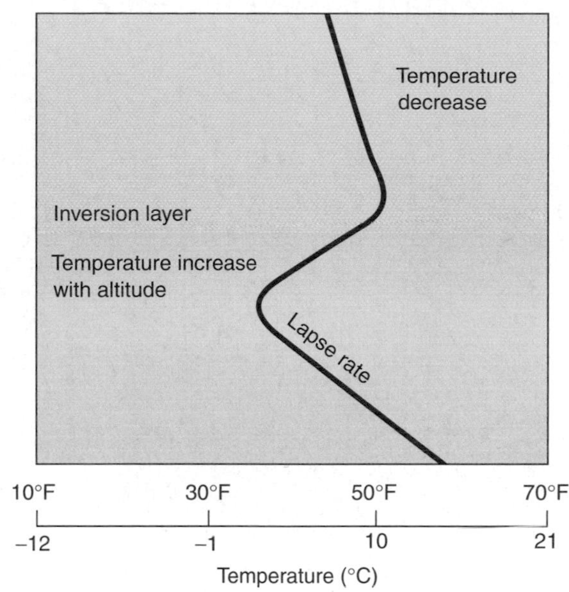

• **FIGURE 4.20**
(Left) Temperature inversion caused by subsidence of air. (Right) Lapse rate associated with the column of air (A) in the left-hand drawing.
Why is the pattern (to the right) called a temperature inversion?

coldest air is nearest the surface and the temperature rises with altitude. Inversions near the surface most often occur on clear nights in the middle latitudes. They may be enhanced by snow cover or the recent advection of cool, dry air into the area. Such conditions produce extremely rapid cooling of Earth's surface at night as it loses the day's insolation through radiation. Then the layers of the atmosphere that are closest to Earth are cooled by radiation and conduction more than those at higher altitudes. Calm air conditions near the surface help produce and partially result from these temperature inversions.

Surface Inversions: Fog and Frost Fog and frost will be discussed again in Chapter 6, but they often occur as the result of a surface inversion. Especially where Earth's surface is hilly, cold, dense surface air will tend to flow downslope and accumulate in the lower valleys. The colder air on the valley floors and other low-lying areas sometimes produces fog or, in more extreme cases, a killing frost. Farmers use a variety of methods to prevent such frosts from destroying their crops. For example, fruit trees in California are often

• **FIGURE 4.21**
Conditions producing smog-trapping inversion in the Los Angeles area.
Why is the air clear above the inversion?

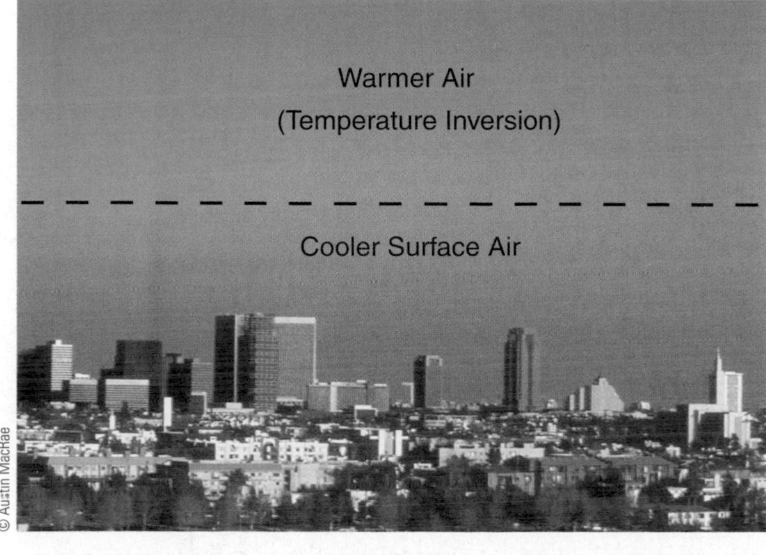

• **FIGURE 4.22**
Temperature inversion caused by the rapid cooling of the air above the cold surface of Earth at night.
What is the significance of an inversion?

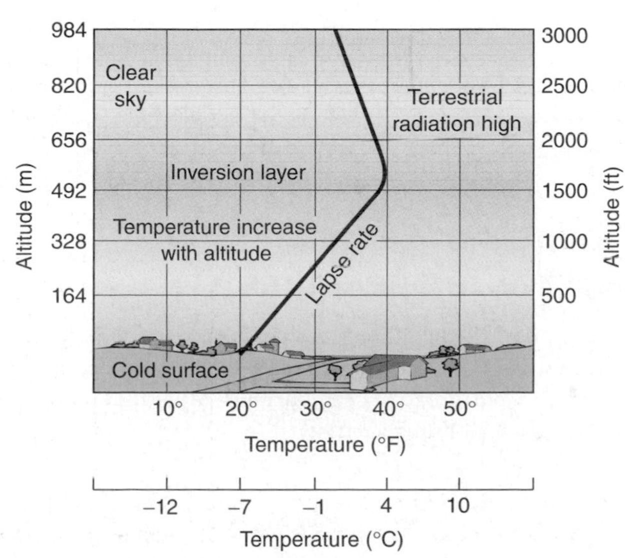

© R. Sager

(a)

© D. Falconer/Photolink/Getty Images

(b)

• **FIGURE 4.23**
(a) Fans (propellers) are used to protect Washington apple orchards from frost. (b) Smudge pots were an older method of trying to keep surface temperatures above freezing.

planted on the warmer hillsides instead of in the valleys. Farmers may also put blankets of straw, cloth, or some other insulator over their plants. This prevents the escape of Earth's heat radiation to outer space and thereby keeps the plants warmer. Large fans and helicopters are sometimes used in an effort to mix surface layers and disturb the inversion (• Fig. 4.23a). Huge orchard heaters that warm the air can also be used to disturb the temperature layers. Smudge pots, an older method of preventing frost, pour smoke into the air, which provides a blanket of insulation much like blankets of cloth or straw (Fig. 4.23b). However, smudge pots have declined in favor because of their air-pollution potential.

Controls of Earth's Surface Temperatures

Variations in temperatures over Earth's surface are caused by several **controls.** The major controls are (1) latitude, (2) land and water distribution, (3) ocean currents, (4) altitude, (5) landform barriers, and (6) human activity.

Latitude Latitude is the most important control of temperature variation involved in weather and climate. Recall that, because of the inclination and parallelism of Earth's axis as it revolves around the sun (Chapter 3), there are distinct patterns in the latitudinal distribution of the seasonal and annual receipt of solar energy over Earth's surface. This has a direct effect on temperatures. In general, annual insolation tends to decrease from lower latitudes to higher latitudes (see again Fig. 3.16). Table 4.2 shows the average annual temperatures for several locations in the Northern Hemisphere. We can see that, responding to insolation (with one exception), a poleward decrease in temperature is true for these locations. The exception is near the equator itself. Because of the heavy cloud cover in equatorial regions, annual temperatures there tend to be lower than at places slightly to the north or south, where skies are clearer.

Another very simple way to see this general trend of decreasing temperatures as we move toward the poles is to think about the kinds of clothes we would take along for 1 month—say, January—if we were to visit Ciudad Bolivar, Venezuela; Raleigh, North Carolina; or Point Barrow, Alaska (see Table 4.2).

Land and Water Distribution Not only do the oceans and seas of Earth serve as storehouses of water for the whole system, but they also store tremendous amounts of heat energy. Their widespread distribution makes them an important atmospheric control that does much to modify the atmospheric elements. All things heat and cool at different rates. This is especially true

TABLE 4.2
Average Annual Temperature

Location	Latitude	(°C)	(°F)
Libreville, Gabon	0°23'N	26.5	80
Ciudad Bolivar, Venezuela	8°19'N	27.5	82
Bombay, India	8°58'N	26.5	80
Amoy, China	24°26'N	22.0	72
Raleigh, North Carolina	35°50'N	18.0	66
Bordeaux, France	44°50'N	12.5	55
Goose Bay, Labrador, Canada	53°19'N	−1.0	31
Markova, Russia	64°45'N	−9.0	15
Point Barrow, Alaska	71°18'N	−12.0	10
Mould Bay, NWT, Canada	76°17'N	−17.5	0

when comparing land to water, in that land heats and cools faster than water. There are three reasons for this phenomenon. First, the *specific heat* of water is greater than that of land. Specific heat refers to the amount of heat necessary to raise the temperature of 1 gram of any substance 1°C. Water, with a specific heat of 1 calorie/gram degree C, must absorb more heat energy than land with specific heat values of about 0.2 calories/gram degree C, to be raised the same number of degrees in temperature.

Second, water is *transparent* and solar energy can penetrate through the surface into the layers below, whereas in *opaque* materials like soil and rock, solar energy is concentrated on the surface. Thus, a given unit of heat energy will spread through a greater volume of water than land. Third, because liquid water circulates and mixes, it can transfer heat to deeper layers within its mass. The result is that as summer changes to winter, the land cools more rapidly than bodies of water, and as winter becomes summer, the land heats more rapidly. Because the air gets much of its heat from the surface with which it is in contact, the differential heating of land and water surfaces produces inequalities in the temperature of the atmosphere above these two surfaces.

The mean temperature in Seattle, Washington, in July is 18°C (64°F), while the mean temperature during the same month in Minneapolis, Minnesota, is 21°C (70°F). Because the two cities are at similar latitudes, their annual pattern and receipt of solar energy are also similar. Therefore, their different temperatures in July must be related to a control other than latitude. Much of this difference in temperature can be attributed to the fact that Seattle is near the Pacific coast, whereas Minneapolis is in the heart of a large continent, far from the moderating influence of an ocean. Seattle stays cooler than Minneapolis in the summer because the surrounding water warms up slowly, keeping the air relatively cool. Minneapolis, on the other hand, is in the center of a large landmass that warms very quickly. In the winter, the opposite is true. Seattle is warmed by the water while Minneapolis is not. The mean temperature in January is 4.5°C (40°F) in Seattle and −15.5°C (4°F) in Minneapolis.

Not only do water and land heat and cool at different rates, but so do various land surface materials. Soil, forest, grass, and rock surfaces all heat and cool differentially and thus vary the temperatures of the overlying air.

Ocean Currents Surface ocean currents are large movements of water pushed by the winds. They may flow from a place of warm temperatures to one of cooler temperatures and vice versa. These movements result, as we saw in Chapter 1, from the attempt of Earth systems to reach a balance—in this instance, a balance of temperature and density.

The rotation of Earth affects the movements of the winds, which in turn affect the movement of the ocean currents. In general, the currents move in a clockwise direction in the Northern Hemisphere and in a counterclockwise direction in the Southern Hemisphere (● Fig. 4.24). Because the temperature of the ocean greatly affects the temperature of the air above it, an ocean current that moves warm equatorial water toward the poles (a warm current) or cold polar water toward the equator (a cold current) can significantly modify the air temperatures of those locations. If the currents pass close to land and are accompanied by ocean breezes, they can have a significant impact on the coastal climate.

The Gulf Stream, with its extension, the North Atlantic Drift, is an example of an ocean current that moves warm water poleward. This warm water keeps the coasts of Great Britain, Iceland, and Norway ice free in wintertime and moderates the climates of nearby land areas (● Fig. 4.25). We can see the effects of the Gulf Stream if we compare the winter conditions of the British Isles with those of Labrador in northeastern Canada. Though both are at the same latitude, the average temperature in Glasgow, Scotland, in January is 4°C (39°F), while during the same month it is −21.5°C (−7°F) in Nain, Labrador.

The California Current off the West Coast of the United States helps moderate the climate of that coast as it brings cold water south. As the current swings southwest away from the coast of central California, cold bottom water is drawn to the surface, causing further chilling of the air above. San Francisco's cool summers (July average: 14°C, or 58°F) show the effect of this current.

Altitude As we have seen, temperatures within the troposphere decrease with increasing altitude. In Southern California, you can find snow for skiing if you go to an altitude of 2400–3000 meters (8000–10,000 ft). Mount Kenya, 5199 meters (17,058 ft) high and located at the equator, is still cold enough to have glaciers. Anyone who has hiked upward 500, 1000, or

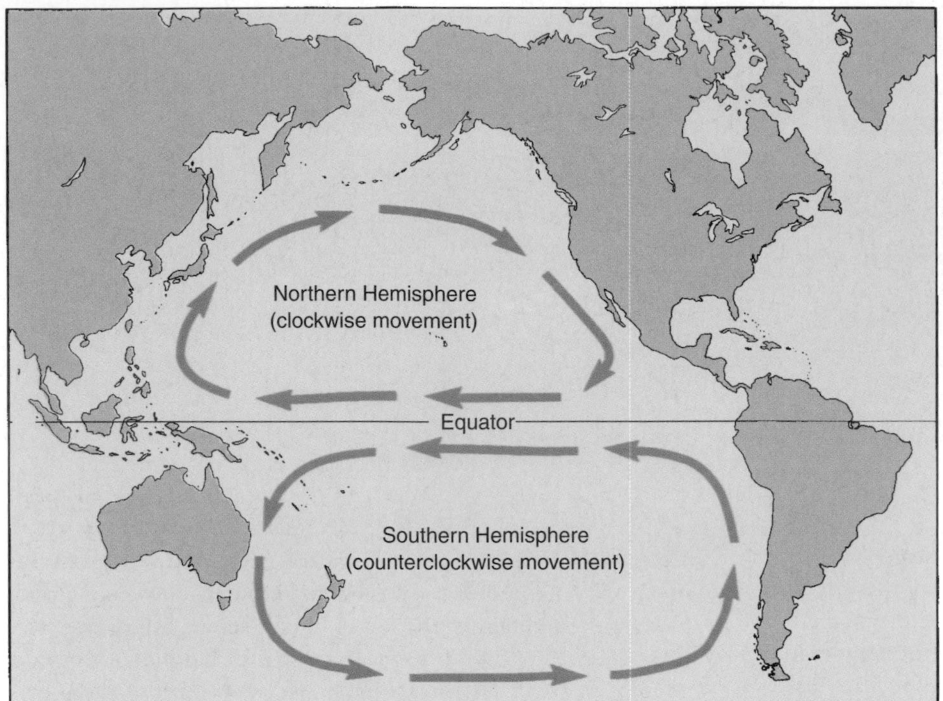

• **FIGURE 4.24**
A highly simplified map of currents in the Pacific Ocean to show their basic rotary pattern. The major currents move clockwise in the Northern Hemisphere and counterclockwise in the Southern Hemisphere. A similar pattern exists in the Atlantic.
What path would a hurricane forming off western Africa take as it approached the United States?

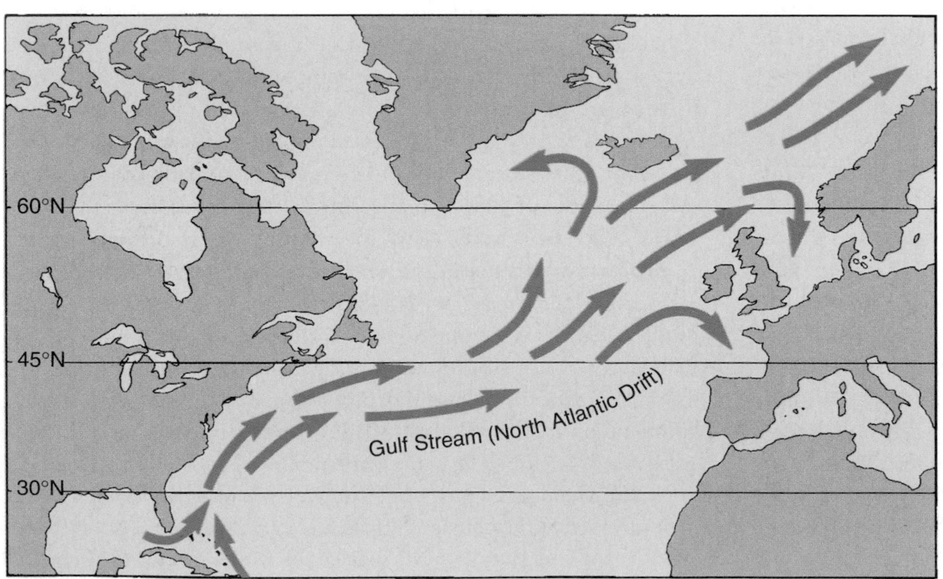

• **FIGURE 4.25**
The Gulf Stream (the North Atlantic Drift farther eastward) is a warm current that greatly moderates the climate of northern Europe. Use this figure and the information gained in Figure 4.24 to discuss the route sailing ships would follow from the United States to England and back.

1500 meters in midsummer has experienced a decline in temperature with increasing altitude. Even if it is hot on the valley floor, you may need a sweater once you climb a few thousand meters (• Fig. 4.26). The city of Quito, Ecuador, only 1° south of the equator, has an average temperature of only 13°C (55°F) because it is located at an altitude of about 2900 meters (9500 ft). This concept will be discussed again when dealing with highland climates in Chapter 10.

Landform Barriers Landform barriers, especially large mountain ranges, can block movements of air from one place to another and thus affect the temperatures of an area. For example, the Himalayas keep cold, wintertime Asiatic air out of India, giving the Indian subcontinent a year-round tropical climate. Mountain orientation can create some significant differences as well. In North America, for example, southern slopes face the sun and tend to be warmer than the shady north-facing slopes. Snowcaps on the south-facing slopes may have less snow and may exist at a higher elevation. North-facing slopes usually have more snow, and it extends to lower elevations.

Human Activities Human beings, too, may be considered "controls" of temperature. In addition, human activities like destroying forests, draining swamps, or creating large reservoirs can significantly affect local climatic patterns and, possibly, world temperature patterns as well. The building and expansion of cities around the world have created pockets of warm temperatures that are known as *urban heat islands*. In each of the examples above, human activities have changed the surface landscape and surface cover, which affect, among other things, the surface *albedo* and available moisture for *latent heat exchanges.*

Temperature Distribution at Earth's Surface

Displaying the distribution of temperatures over the surface of the Earth requires a mapping device called **isotherms.** Isotherms (from Greek: *isos,* equal; *therm,* heat) are defined as lines that connect points of equal temperature. When constructing isothermal maps showing temperature distribution, we need to account for elevation by adjusting temperature readings to what they would be at sea level. This adjustment means adding 6.5°C for every 1000 meters of elevation (the normal lapse rate). The rate of temperature change on an isothermal map is called the **temperature gradient.** Closely

National Park Service (NPS Photo)

• **FIGURE 4.26**
Snow-capped mountains show the visual evidence that temperatures decrease with altitude. This mountain is in Grand Teton National Park in Wyoming, named after the dramatic range of jagged peaks, such as this one.
At what rate per 1000 meters do temperatures decrease with height in the troposphere?

spaced isotherms indicate a steep temperature gradient (a rapid temperature change over a shorter distance), and widely spaced lines indicate a weak one (a slight temperature change over a longer distance).

• Figure 4.27 and • Figure 4.28 show the horizontal distribution of temperatures for Earth at two critical times, during January and July, when the seasonal extremes of high and low temperatures are most obvious in the Northern and Southern Hemispheres. The easiest feature to recognize on both maps is the general orientation of the isotherms; they run nearly east-west around Earth, as do the parallels of latitude.

A more detailed study of Figures 4.27 and 4.28 and a comparison of the two maps reveal some additional important features. The highest temperatures in January are in the Southern Hemisphere; in July, they are in the Northern Hemisphere. Comparing the latitudes of Portugal and Southern Australia can demonstrate this point. Note on the July map that Portugal in the Northern Hemisphere is nearly on the 20°C isotherm, whereas in Southern Australia in the Southern Hemisphere the average July temperature is around 10°C, even though the two locations are approximately the same distance from the equator. The temperature differences between the two hemispheres are again a product of insolation, this time changing as the sun shifts north and south across the equator between its positions at the two solstices.

Note that the greatest deviation from the east–west trend of temperatures occurs where the isotherms leave large landmasses to cross the oceans. As the isotherms leave the land, they usually bend rather sharply toward the pole in the hemisphere experiencing winter and toward the equator in the summer hemisphere. This behavior of the isotherms is a direct reaction to the differential heating and cooling of land and water. The continents are hotter than the oceans in the summer and colder in the winter. Other interesting features on the January and July maps can be mentioned briefly. Note that the isotherms poleward of 40° latitude are much more regular in their east–west orientation in the Southern than in the Northern Hemisphere. This is because in the Southern Hemisphere (often called the "water hemisphere") there is little land south of 40°S latitude to produce land and water contrasts. Note also that the temperature gradients are much steeper in winter than in summer in both hemispheres. The reason for this can be understood when you recall that the tropical zones have high temperatures throughout the year, whereas the polar zones have large seasonal differences. Hence, the difference in temperature between tropical and polar zones is much greater in winter than in summer.

As a final point, observe the especially sharp swing of the isotherms off the coasts of eastern North America, southwestern South America, and southwestern Africa in January and off Southern California in July. In these locations, the normal bending of the isotherms due to land–water differences is augmented by the presence of warm or cool ocean currents.

Annual March of Temperature

Isothermal maps are commonly plotted for January and July because there is a lag of about 30–40 days from the solstices, when the amount of insolation is at a minimum or maximum (depending on the hemisphere), to the time minimum or maximum temperatures are reached. This **annual lag of temperature** behind insolation is similar to the daily lag of temperature explained previously. It is a result of the changing relationship between incoming solar radiation and outgoing Earth radiation.

Temperatures continue to rise for a month or more after the summer solstice because insolation continues to exceed Earth's radiation loss. Temperatures continue to fall after the winter solstice until the increase in insolation finally matches Earth's radiation. In short, the lag exists because it takes time for Earth to heat or cool and for those temperature changes to be transferred to the atmosphere.

The annual changes of temperature for a location can be plotted in a graph. The mean temperature for each month in a place such as Peoria, Illinois, or Sydney, Australia, is recorded and a line drawn connecting the 12 temperatures (• Fig. 4.29). The mean monthly temperature is the average of the daily mean temperatures recorded at a weather station during a month. The daily mean temperature is the average of the high and low temperatures for a 24-hour period. The curve that connects the 12 monthly temperatures depicts the **annual march of temperature** and

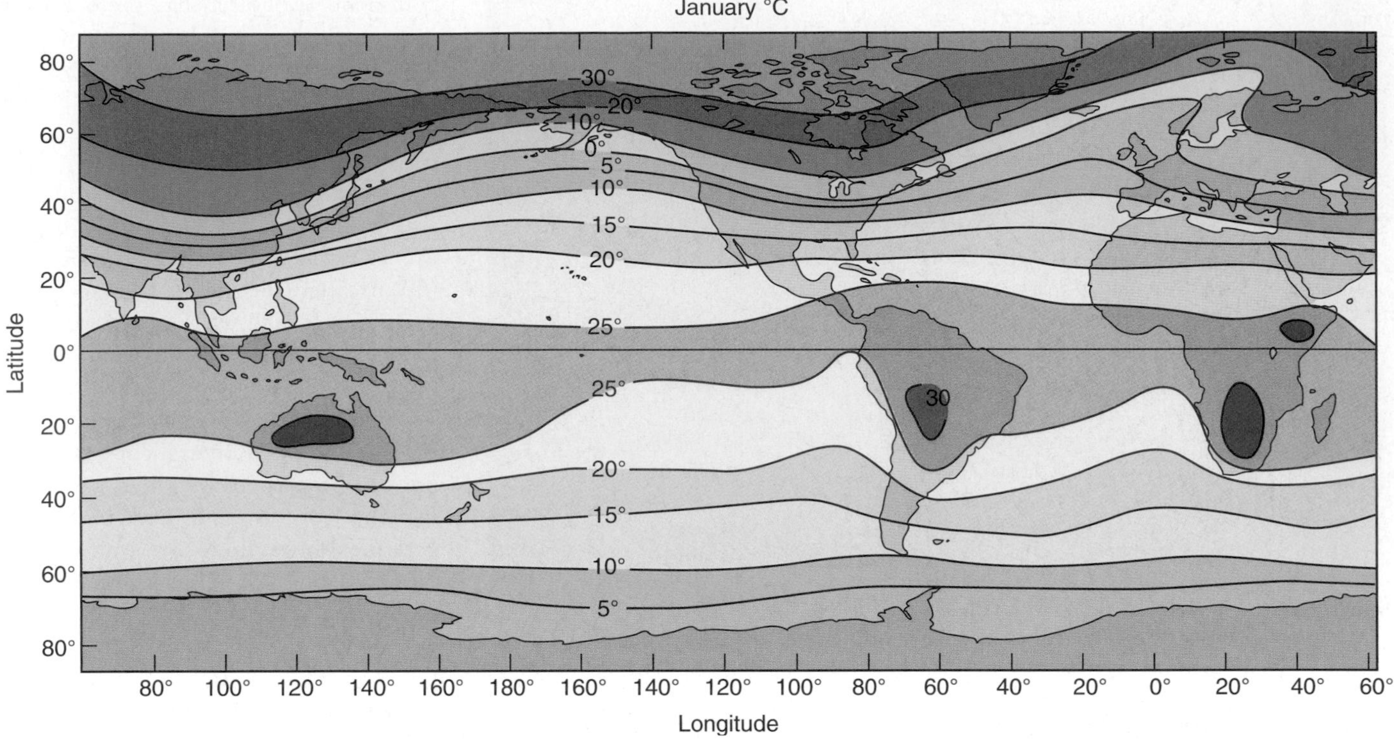

• FIGURE 4.27
Average sea-level temperatures in January (°C).

• FIGURE 4.28
Average sea-level temperatures in July (°C).
Observe the temperature gradients between the equator and northern Canada in January and July. Which is greater? Why?

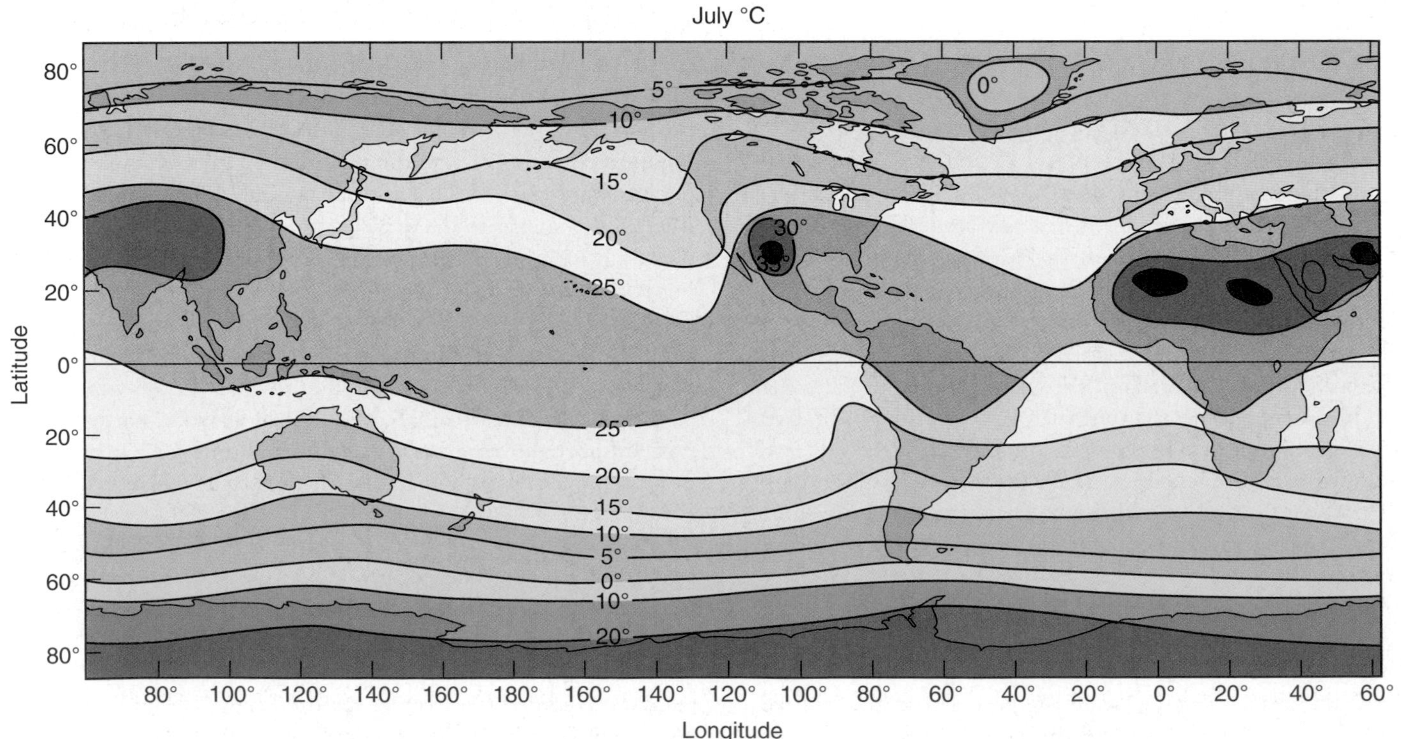

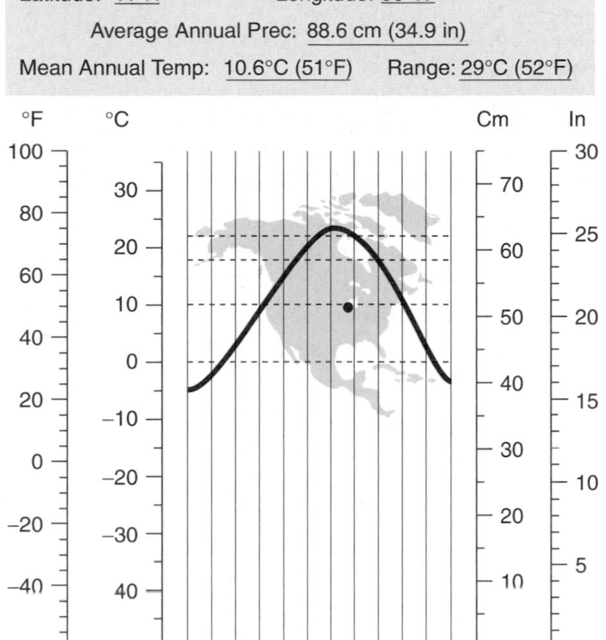

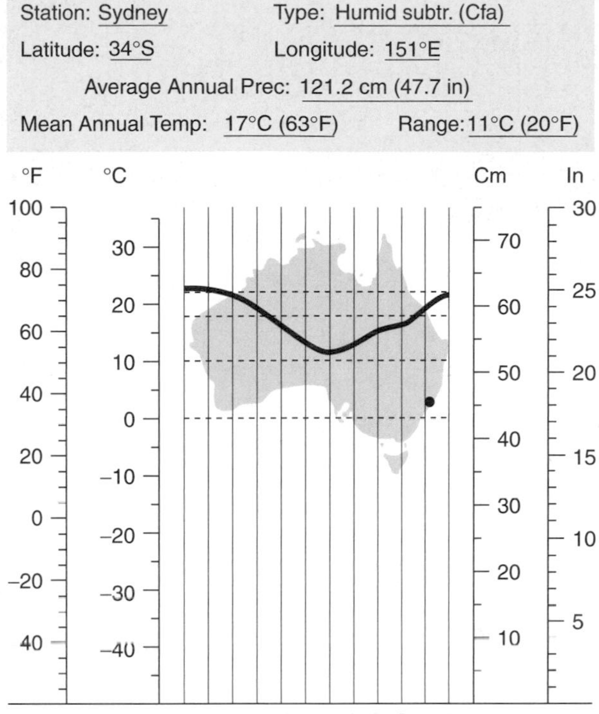

• **FIGURE 4.29**
The annual march of temperature at Peoria, Illinois, and Sydney, Australia.
Why do these two locations have opposite temperature curves?

shows both the decrease in solar radiation, as reflected by a decrease in temperature, from midsummer to midwinter and the increase in temperature from midwinter to midsummer caused by the increase in solar radiation.

Weather and Climate

We frequently use the words *weather* and *climate* in general conversation, but it is important in a physical science course to carefully distinguish between the terms. **Weather** refers to the condition of atmospheric elements at a given time and for a specific area. That area could be as large as the New York metropolitan area or as small and specific as a weather observation station. It is the lowest layer, the troposphere, that exhibits Earth's weather and is of the greatest interest to physical geographers and weather forecasters who survey the changing conditions of the atmosphere in the field of study known as **meteorology.**

Many observations of the weather of a place over a period of at least 30 years provide us with a description of its climate. **Climate** describes an area's average weather, but it also includes those common deviations from the normal or average that are likely to occur and the processes that make them so. Climates also include extreme situations, which can be very significant. Thus, we could describe the climate of the southeastern United States in terms of average temperatures and precipitation through a year, but we would also have to include mention of the likelihood of events such as hurricanes and snowstorms during certain periods of the year. **Climatology** is the study of the varieties of climates, both past and present, found on our planet and their distribution over its surface.

Weather and climate are of prime interest to the physical geographer because they affect and are interrelated with all of Earth's environments. The changing conditions of atmospheric elements such as temperature, rainfall, and wind affect soils and vegetation, modify landforms, cause flooding of towns and farms, and in a multitude of other ways, influence the function of systems in each aspect of Earth's physical environments.

Five basic **elements** of the atmosphere serve as the "ingredients" of weather and climate: (1) solar energy (or insolation), (2) temperature, (3) pressure, (4) wind, and (5) precipitation. We must examine these elements in order to understand and categorize weather and climate. Thus, a weather forecast will generally include the present temperature, the probable temperature range, a description of the cloud cover, the chance of precipitation, the speed and direction of the winds, and air pressure.

In Chapter 3, we noted that the amount of solar energy received at one place on Earth's surface varies during a day and throughout the year. The amount of insolation a place receives is the most important weather element; the other four elements depend in part on the intensity and duration of solar energy.

GEOGRAPHY'S SPATIAL SCIENCE PERSPECTIVE

The Urban Heat Island

In 1820 Sir Luke Howard (who also created our cloud classification system, discussed in Chapter 6) documented that temperatures in London, England, varied from those of the surrounding countryside. This is an early reference to what has become known as an *urban heat island*. This is a measurable pocket of warm air produced by a large urban area. This comes as no surprise, since cities represent areas where human activity is concentrated and large densities of population live and work. Research in the United States began in the mid-1950s, but serious urban climate monitoring started in the early 1970s. The Metropolitan Meteorological Experiment (METROMEX) in St. Louis, Missouri, became the first extensive research project on urban climate. Since those times, urban heat island has been the standard title for this phenomenon.

Several factors mandate that the temperatures of cities are generally 2–10°F (1–6°C) warmer than the surrounding rural areas.

1. Energy use: People live and work in temperature-controlled environments (utilizing both heating and air-conditioning). In addition, lighting fixtures, appliances (major and minor), other machinery, and engines are employed in vast numbers in the city. In a rural area, there would be only a small fraction of this energy use.
2. Automobiles: In large urban areas, millions of automobile engines drive within and through major cities. In the country, there are fewer vehicles (trucks and farm machinery) that function each day.
3. Industry: Industrial parks surrounding some urban areas may contain factories, which operate large machinery, and use great amounts of energy. Few industries would stray far from major urban areas and exist on the countryside.
4. Human populations: The human body with a core temperature of 37°C (98.6°F) adds heat (through conduction and advection) to the atmosphere. Respiration (especially exhaling) also adds both heat and humidity to the air as well. Normally a few individuals would not produce a strong effect, but in cities where millions of people reside, this heat accumulates. The low population densities of rural regions make this factor insignificant.

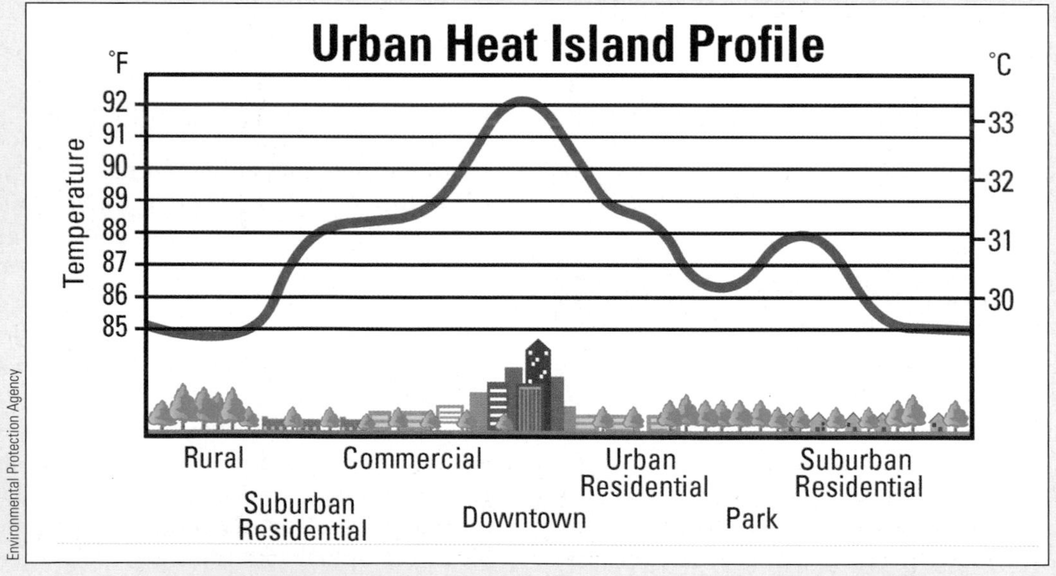

A temperature profile of the urban "heat island" shows the increase in temperature with increasing urbanization.

The temperature of the atmosphere at a given place on or near the surface of Earth is largely a function of the insolation received at that location. It is also influenced by many other factors, such as land and water distribution and altitude. Unless there is some form of precipitation occurring, the temperature of the air may be the first element of weather that we describe when someone asks us what it is like outside.

However, if it is raining or foggy or snowing, we will probably notice and mention that condition first. We are less aware of the amount of water vapor or moisture in the air (except in very arid or humid areas). However, moisture in the air is a vital

5. Materials and buildings: The building material of cities (concrete, asphalt, glass, and metal) can heat up quickly (with low specific heat and low albedo values) during the day, as opposed to the grass, trees, and cropland of the surrounding countryside. Though the urban materials also cool down quickly at night, the heat is often trapped within the *urban canyons* (streets with tall skyscrapers on both sides). In some cases, urban heat islands appear to be more noticeable when analyzing overnight low temperatures.
6. Pollution levels: Particulate and aerosol levels are higher in cities than the surrounding countryside. A dome of pollution and haze can form over large cities. Some of these pollutants may reflect incoming solar radiation, and shade the city somewhat by day. But this pollution may also act to absorb Earth's heat trying to escape, and reradiate it back to the surface, especially at night.
7. Water on the surface: In the country, the evaporation from lakes, ponds, streams, or any standing water can help cool the surroundings using the latent heat of evaporation (590 cal/g). Urban areas are designed to drain surface water quickly and efficiently with gutters, drains, and sewers. With few if any bodies of water, the city cannot avail itself of this heat loss mechanism.

It has been estimated that, by the year 2025, about 80% of the world's population will reside in an urban area. As world population grows and urban development increases, these seven factors will intensify as well, and the urban heat island will directly affect more people than the climatology of the region in general.

NASA/Marshall Space Flight Center

The urban heat island effect in Sacramento, California. The thermal image (right) shows the heat retention of buildings and densely urbanized areas at night. The hottest areas are in red, less hot areas in yellow or green, and cooler areas in blue, like the Sacramento River, parks, and neighborhoods (which are cooled by the presence of trees). Comparing sections of the thermal image to the aerial photograph (left) illustrates the relationships between built-up areas, vegetated areas, water bodies, and their relative temperatures at night.

weather element in the atmosphere, and its variations play an important role in the likelihood of precipitation (● Fig. 4.30).

We all know that weather varies. Because it is the momentary state of the atmosphere at a given location, it varies in both time and place. There are even variations in the amount that weather varies. In some places and at some times of the year, the weather changes almost daily. In other places, there may be weeks of uninterrupted sunshine, blue skies, and moderate temperatures followed by weeks of persistent rain. A few places experience only minor differences in the weather throughout the year. The language of the original people of Hawaii is said to have no word for weather because conditions there varied so little.

Los Alamos National Laboratory/Nancy W. Ambrosiano

(a)

M. Trapasso

(b)

• FIGURE 4.30

(a) Weather, such as a snowy day, is a short-term meteorological event. (b) These glaciers in Alaska, however, come about as the result of a long-term colder climate.

Can you make the same weather versus climate comparisons using rainfall?

Complexity of Earth's Energy Systems

This chapter was designed to show the variations of Earth's energy systems and dynamic balances. These variations are the result of complex interrelationships between the characteristics of Earth and its atmosphere and the energy gained and lost by Earth's environments. The variations are both horizontal across the surface and vertical through our atmosphere. Further, they vary on both daily and seasonal time frames.

Variations of Earth's energy systems impose both diurnal and annual rhythms on our agricultural activities, recreational pursuits, clothing styles, architecture, and energy bills. Human activities are constantly influenced by temperature changes, which reflect the input–output patterns of Earth's energy systems.

Chapter 4 Activities

Define & Recall

photosynthesis
greenhouse effect
ozone
troposphere
normal lapse rate
stratosphere
mesosphere
thermosphere
ozonosphere
ionosphere
homosphere
heterosphere
latent heat of fusion
latent heat of evaporation
latent heat of condensation
latent heat of sublimation
radiation
conduction
convection
advection
heat
temperature
Fahrenheit scale
Celsius (centigrade) scale
Kelvin scale
daily march of temperature
maritime
continentality
albedo
temperature inversion
control (temperature)
isotherm
temperature gradient
annual lag of temperature
annual march of temperature
weather
meteorology
climate
climatology
element (atmosphere)

Discuss & Review

1. What function does ozone play in the support of life on Earth? Where and how is ozone formed?
2. How does Earth's atmosphere affect incoming solar radiation (insolation)? By what processes is insolation prevented from reaching Earth's surface? What percentages are involved in a generalized situation? What percentages reach the surface, and by what processes?
3. Discuss the role of water in energy exchange. What characteristics of water make it so important?
4. How is the atmosphere heated from Earth's surface?
5. What is meant by Earth's heat energy budget? List and define the important energy exchanges that keep it in balance.
6. At what time of day does insolation reach its maximum? Its minimum? Compare this to the daily temperature maximum and minimum.
7. What is a temperature inversion? Give several reasons why temperature inversions occur.
8. Would you expect an area like Seattle, Washington, to have a milder or a harsher winter than Grand Forks, North Dakota? Why?
9. Describe the behavior of the isotherms in Figures 4.27 and 4.28. What factors cause the greatest deviation from an east–west trend? What factors cause the greatest differences between the January and July maps?
10. What is the difference between meteorology and climatology?
11. What are the basic characteristics that we call *atmospheric elements* of weather and climate?
12. What factors cause variation in the elements of weather and climate?

Consider & Respond

1. Convert the following temperatures to Fahrenheit: 20°C, 30°C, and 15°C; and these to Celsius: 60°F, 15°F, and 90°F.
2. Refer to Figure 4.11. List the major means by which the atmosphere gains heat and loses heat.
3. What are the major weather and climate controls that operate in your area?
4. The normal lapse rate is 6.5°C/1000 meters. If the surface temperature is 25°C, what is the air temperature at 10,000 meters above Earth's surface? Convert your answer to degrees Fahrenheit.
5. Refer to Figures 4.27 and 4.28. What location on Earth's surface exhibits the greatest annual range of temperature? Why?

Apply & Learn

1. With respect to incoming solar radiation, how are albedo and absorption related? (a) Develop a mathematical relationship between these two processes. (b) If the albedo of a grassy lawn is 23% and that of a black-top driveway is 4%, what is the difference in the absorption between these two surfaces? (c) Albedo and absorption can also be expressed as decimal values. Convert your answers to (a) and (b) into decimal form.
2. A skier is at the base of a mountain and the temperature is 68°F. How high up the mountain side (in meters and feet) might the skier expect to find snow on the slope?

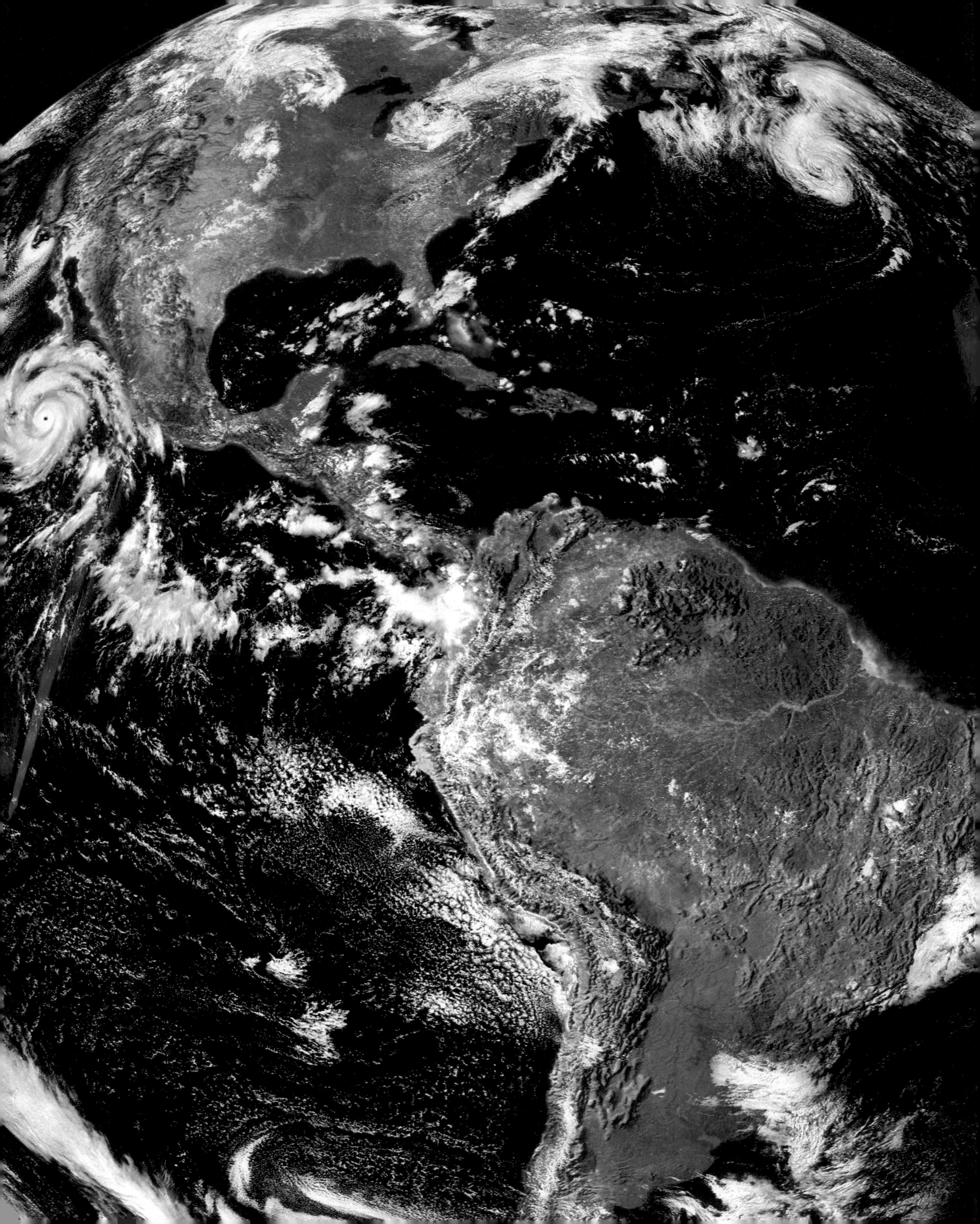

Atmospheric Pressure, Winds, and Circulation Patterns

5

CHAPTER PREVIEW

Latitudinal differences in temperature (as a result of differential receipt of insolation) provide a partial explanation for latitudinal differences in pressure.

- What is the relationship between temperature and pressure?
- Why is this only a partial explanation?

The fact that land heats and cools more rapidly than water is of significance not only to world patterns of temperature but also to world patterns of pressure, winds, and precipitation.

- How can you explain this fact?
- What effect does this fact have on world patterns?

Planetary (global) wind systems in association with global pressure patterns play a major role in global circulation.

- What are the six major planetary (global) wind belts or zones, and what are their chief characteristics?
- Why do the wind belts migrate with the seasons?

Upper air winds and atmospheric circulation play a major role in controlling surface weather and climatic conditions.

- What is upper air circulation like?
- How do ocean currents affect atmospheric conditions of land areas?

El Niños can have a devastating impact on our global weather.

- What is an El Niño?
- How does it influence global weather?

◀ Opposite: The swirling circulation patterns seen in Earth's atmosphere are created by changes in pressure and winds.
NASA/GSFC

An individual gas molecule weighs almost nothing; however, the atmosphere as a whole has considerable weight and exerts an average pressure of 1034 grams per square centimeter (14.7 lb/sq in.) on Earth's surface. The reason why people are not crushed by this atmospheric pressure is that we have air and water inside us—in our blood, tissues, and cells—exerting an equal outward pressure that balances the inward pressure of the atmosphere. Atmospheric pressure is important because variation in pressure within the Earth–atmosphere system creates our atmospheric circulation and thus plays a major role in determining our weather and climate. It is the differences in *atmospheric pressure* that create our *winds*. Further, the movement of the winds drives our *ocean currents,* and thus atmospheric pressure works its way into several of Earth's systems.

In 1643, Evangelista Torricelli, a student of Galileo, performed an experiment that was the basis for the invention of the *mercury barometer,* an instrument that measures atmospheric (also called barometric) pressure. Torricelli took a tube filled with mercury and inverted it in an open pan of mercury. The mercury inside the tube fell until it was at a height of about 76 centimeters (29.92 in.)

above the mercury in the pan, leaving a vacuum bubble at the closed end of the tube (• Fig. 5.1). At this point, the pressure exerted by the atmosphere on the open pan of mercury was equal to the pressure from the mercury trying to drain from the tube. Torricelli observed that as the air pressure increased, it pushed the mercury up higher into the tube, increasing the height of the mercury until the pressure exerted by the mercury (under the pull of gravity) would equal the pressure of the air. On the other hand, as the air pressure decreased, the mercury level in the column dropped.

In the strictest sense, a mercury barometer does not actually measure the pressure exerted by the atmosphere on Earth's surface, but instead measures the *response* to that pressure. That is, when the atmosphere exerts a specific pressure, the mercury will respond by rising to a specific height (• Fig. 5.2). Meteorologists usually prefer to work with actual pressure units. The unit most often used is the millibar (mb). Standard sea-level pressure of 1013.2 millibars will cause the mercury to rise 76 centimeters (29.92 in.).

Our study of the atmospheric elements that combine to produce weather and climate has to this point focused on the fundamental influence of solar energy on the global distributional patterns of temperature. The unequal receipt of insolation by latitude over Earth's surface produces temperature patterns that vary from the equator to the poles. In this chapter, we learn that these temperature differences are one of the major causes of the development of patterns of higher and lower pressure that also vary with latitude. In addition, we examine patterns of another kind—patterns of movement or, more properly, circulation, in which both energy and matter travel cyclically through Earth subsystems.

Geographers are particularly interested in circulation patterns because they illustrate spatial interaction, one of geography's major themes introduced in Chapter 1. Patterns of movement between one place and another reveal that the two places have a relationship and prompt geographers to seek both the nature and effect of that relationship. It is also important to understand the causes of the spatial interaction taking place. As we examine the circulation patterns featured later in this chapter, you should make a special effort once again to trace each pattern back to the fundamental influence of solar energy.

• FIGURE 5.1
A simple mercury barometer. Standard sea-level pressure of 1013.2 millibars will cause the mercury to rise 76 centimeters (29.92 in.) in the tube.
When air pressure increases, what happens to the mercury in the tube?

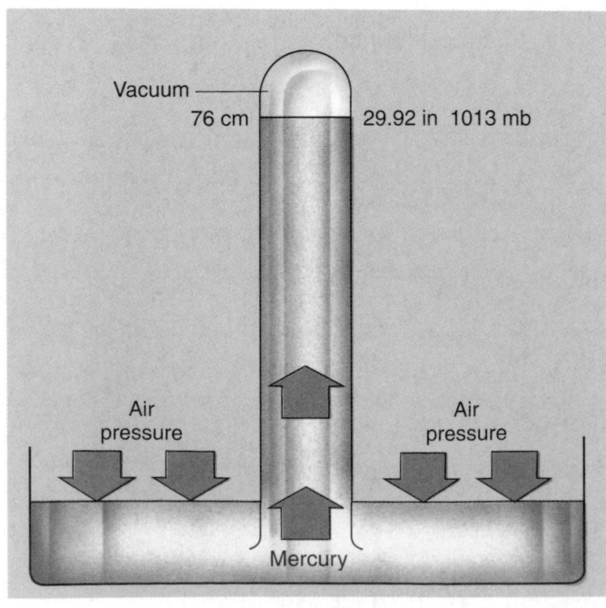

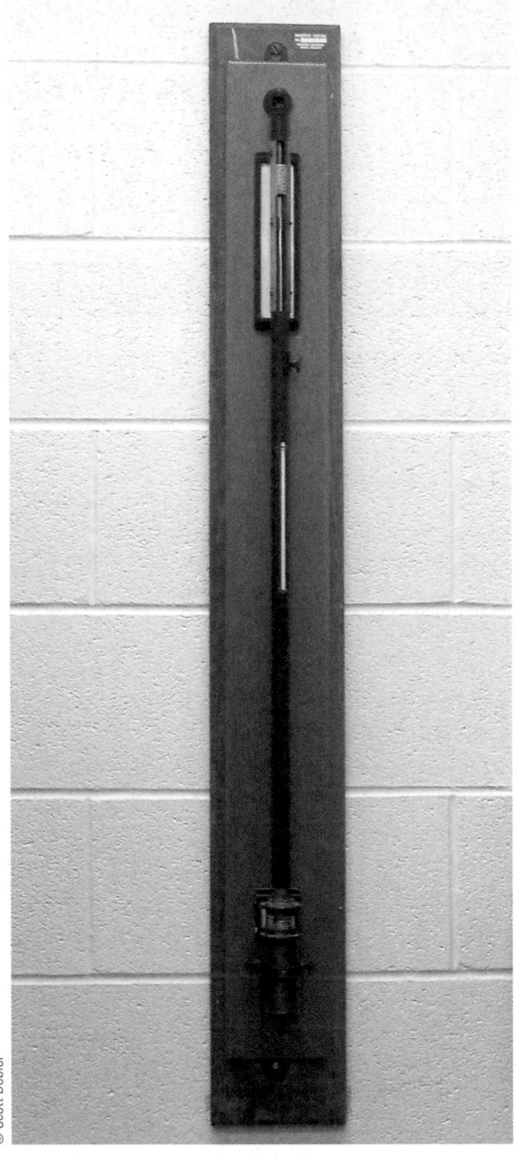

• FIGURE 5.2
This mercury barometer is bolted to the wall of the College Heights Weather Station in Bowling Green, Kentucky.
Why must this instrument be so tall to work properly?

Variations in Atmospheric Pressure

Vertical Variations in Pressure

Imagine a pileup of football players during a game. The player on the bottom gets squeezed more than a player near the top because he has the weight of all the others on top of him. Similarly, air pressure decreases with elevation, for the higher we go, the more diffused, and more widely spaced the air molecules become. The increased intermolecular space results in lower air density and lower air pressure (● Fig. 5.3). In fact, at the top of Mount Everest (elevation 8848 m, or 29,028 ft), the air pressure is only about one third the pressure at sea level.

Humans are usually not sensitive to small, everyday variations in air pressure. However, when we climb or fly to altitudes significantly above sea level, we become aware of the effects of air pressure on our system. When jet aircraft fly at 10,000 meters (33,000 ft), they have to be pressurized and nearly airtight so that a near-sea-level pressure can be maintained. Even then, the pressurization may not work perfectly, so our ears may pop as they adjust to a rapid change in pressure when ascending or descending. Hiking or skiing at heights that are a few thousand meters in elevation will affect us if we are used to the air pressure at sea level. The reduced air pressure means less oxygen is contained in each breath of air. Thus, we sometimes find that we get out of breath far more easily at high elevations until our bodies adjust to the reduced air pressure and corresponding drop in oxygen level.

● FIGURE 5.3

Both air pressure and air density decrease rapidly with increasing altitude. **By approximately how much does density drop between 0 and 100 km?**

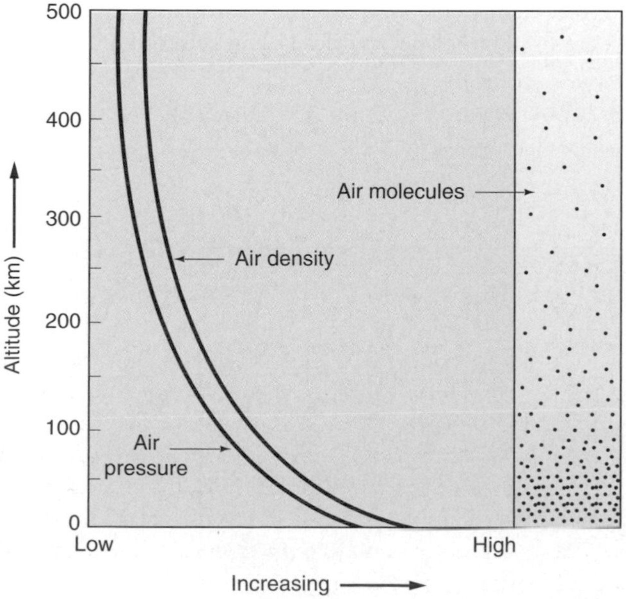

Changes in air pressure are not solely related to altitude. At Earth's surface, small but important variations in pressure are related to the intensity of insolation, the general movement of global circulation, and local humidity and precipitation. Consequently, a change in air pressure at a given locality often indicates a change in the weather. Weather systems themselves can be classified by the structure and tendency toward change of their pressure.

Horizontal Variations in Pressure

The causes of horizontal variation in air pressure are grouped into two types: thermal (determined by temperature) and dynamic (related to motion of the atmosphere).

We look at the simpler thermal type first. In Chapter 4, we saw that Earth is heated unevenly because of unequal distribution of insolation, differential heating of land and water surfaces, and different albedos of surfaces. One of the basic laws of gases is that the pressure and density of a given gas vary inversely with temperature. Thus, during the day, as Earth's surface heats the air in contact with it, the air expands in volume and decreases in density. Such air has a tendency to rise as its density decreases. When the warmed air rises, there is less air near the surface, with a consequent decrease in surface pressure. The equator is an area where such low pressure occurs regularly.

In an area with cold air, there is an increase in density and a decrease in volume. This causes the air to sink and pressure to increase. The poles are areas where such high pressures occur regularly. Thus, the constant low pressure in the equatorial zone and the high pressure at the poles are thermally induced.

From this we might expect a gradual increase in pressure from the equator to the poles to accompany the gradual decrease in average annual temperature. However, actual readings taken at Earth's surface indicate that pressure does not increase in a regular fashion poleward from the equator. Instead, there are regions of high pressure in the subtropics and regions of low pressure in the subpolar regions. The dynamic causes of these zones, or *belts,* of high and low pressure are more complex than the thermal causes.

These dynamic causes are related to the rotation of Earth and the broad patterns of circulation. For example, as air rises steadily at the equator, it moves toward the poles. Earth's rotation, however, causes the poleward-flowing air to drift to the east. In fact, by the time it is over the subtropical regions, the air is flowing from west to east. This bending of the flow as it moves poleward impedes the northward movement and causes the air to pile up over the subtropics, which results in increased pressure at Earth's surface there.

With high pressure over the polar and subtropical regions, dynamically induced areas of low pressure are created between them, in the subpolar region. As a result, air sinks into and flows from the highs to the lows, where it enters and rises. Thus, both the subtropical and subpolar pressure regions are dynamically induced. This example describes horizontal pressure variations on a global scale. We concentrate on this scale later in this chapter.

Basic Pressure Systems

Before we begin our discussion of circulation patterns leading up to the global scale, we must start by describing the two basic types of pressure systems: the **low,** or **cyclone,** and the **high,** or **anticyclone.** These are represented by the capital letters **L** and **H** that we commonly see on TV, newspaper, and official weather maps.

A low, or cyclone, is an area where air is ascending. As air moves upward away from the surface, it relieves pressure from that surface. In this case, barometer readings will begin to fall. A high, or anticyclone, is just the opposite. In a high, air is descending toward the surface and thus barometer readings will begin to rise, indicating an increased pressure on the surface. Lows and highs are illustrated in • Figure 5.4.

Convergent and Divergent Circulation

As we have just seen, winds blow toward the center of a cyclone and can be said to *converge* toward it. Hence, a cyclone is a closed pressure system whose center serves as the focus for **convergent wind circulation.** The winds of an anticyclone blow away from the center of high pressure and are said to be *diverging.* In the case of an anticyclone, the center of the system serves as the source for **divergent wind circulation.** Figure 5.4 shows converging and diverging winds moving in straight paths. This is not a true picture of reality. In fact, winds moving out of a high and into a low do so in a spiraling motion created by another force, which we cover in the chapter section on wind.

Mapping Pressure Distribution

Geographers and meteorologists can best study pressure systems when they are mapped. In mapping air pressure, we reduce all pressures to what they would be at sea level, just as we changed temperature to sea level in order to eliminate altitude as a factor. The adjustment to sea level is especially important for atmospheric pressure because the variations due to altitude are far greater than those due to atmospheric dynamics and would tend to mask the more meteorologically important regional differences.

Isobars (from Greek: *isos,* equal; *baros,* weight) are lines drawn on maps to connect places of equal pressure. When the isobars appear close together, they portray a significant difference in pressure between places, hence a strong **pressure gradient.** When the isobars are far apart, a weak pressure gradient is indicated. When depicted on a map, high and low pressure cells are outlined by concentric isobars that form a closed system around centers of high or low pressure.

Wind

Wind is the horizontal movement of air in response to differences in pressure. Winds are the means by which the atmosphere attempts to balance the uneven distribution of pressure over Earth's surface. The movements of the wind also play a major role in correcting the imbalances in radiational heating and cooling that occur over Earth's surface. On average, locations below 38° latitude receive more radiant energy than they lose, whereas locations poleward of 38° lose more than they gain (see again Fig. 4.14). Our global wind system transports energy poleward to help maintain an energy balance. The global wind system also gives rise to the ocean currents, which are another significant factor in equalizing the energy imbalance. Thus, without winds and their associated ocean currents, the equatorial regions would get hotter and the polar regions colder through time.

Besides serving a vital function in the advectional (horizontal) transport of heat energy, winds also transport water vapor from the air above bodies of water, where it has evaporated, to land surfaces, where it condenses and precipitates. This allows greater precipitation over land surfaces than could otherwise occur. In addition, winds exert influence on the rate of evaporation itself. Furthermore, as we become more aware of and concerned about the effect that the burning of fossil fuels has on our atmosphere, we look for alternate energy sources. Natural sources such as water, solar energy, and wind become increasingly attractive alternatives to fossil fuels. They are clean, abundant, and renewable.

• FIGURE 5.4
Winds converge and ascend in cyclones (low pressure centers) and descend and diverge from anticyclones (high pressure centers).
How is temperature related to the density of air?

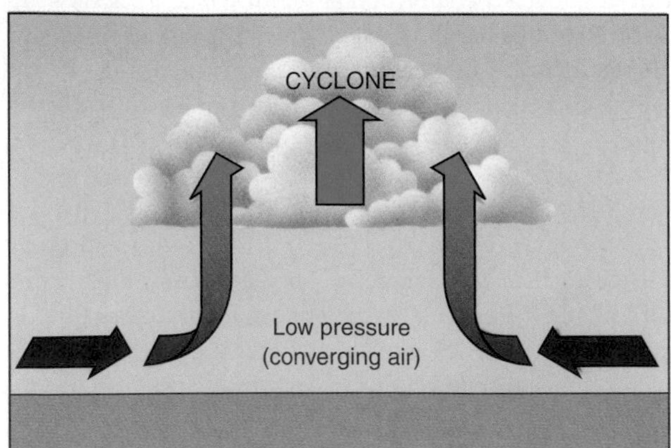

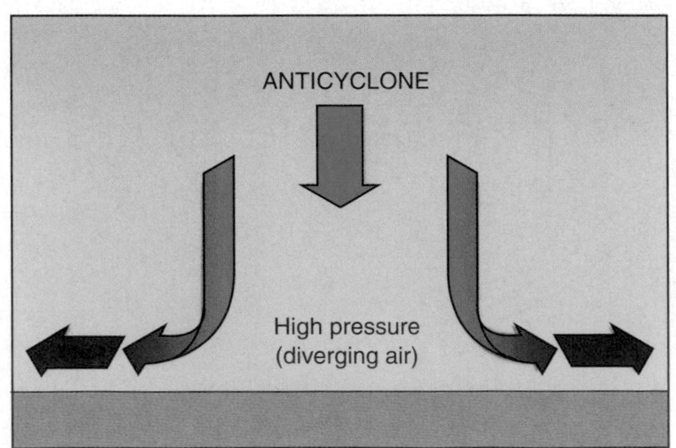

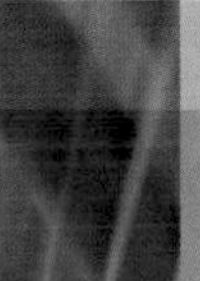

GEOGRAPHY'S ENVIRONMENTAL SCIENCE PERSPECTIVE

Harnessing the Wind

For centuries, windmills provided the power to pump water and grind grain in rural areas throughout the world. But the widespread availability of inexpensive electricity changed the role of most windmills to that of a nostalgic tourist attraction. Should we then conclude that energy from the wind is only a footnote in the history of power? In no way is that a reasonable assumption. The mounting needs for electricity and increasing problems from atmospheric pollution associated with fossil fuels must be taken into consideration.

Wind power is an inexhaustible source of clean energy. Although the cost of electrical energy produced by the wind depends on favorable sites for the location of wind turbines, wind power is already cost competitive with power produced from fossil fuels. One expert calls wind generation the fastest-growing electricity-producing technology in the world. During the last decade, power production from the wind increased more than 25%. Much of the growth was in Europe, where most of the world's 17,000 megawatts of wind power are generated. As examples, 13% of Denmark's power and more than 20% of the power in the Netherlands, Spain, and Germany is supplied by the wind.

Two criteria are more important than others in the location of wind turbines. The site must have persistent strong winds, and it must be in an already developed region so that the power from the turbines can be linked directly to an existing electrical grid system. Although individual wind turbines (such as those located on farms scattered throughout the Midwest and Great Plains of the United States) can be found producing electricity, most wind power is generated from *wind farms*. These are long rows, or more concentrated groups, of as many as 50 or more turbines. Each turbine can economically extract up to 60% of the wind's energy at minimum wind speeds of 20 kilometers (12 mi) per hour, although higher wind speeds are desirable. Because the power generated is proportional to the cube of the wind speed, a doubling of wind velocity increases energy production eight times.

Although North America currently lags far behind Europe in the production of energy from the wind, the continent has great potential. Excellent sites for the location of wind farms exist throughout the open plains of North America's interior and along its coasts from the Maritime Provinces of Canada to Texas and from California to the Pacific Northwest. In addition, the newest wind-power technology places wind farms out of sight and sound in offshore locations that avoid navigation routes and marine-life sanctuaries. And North America has some of the largest coastlines in the world with major adjacent power needs. The sites are available, the technology has been developed, the costs are competitive, and the resolve to shift from fossil fuels is growing. Is it not time for power from the winds to come to North America?

Copyright and photograph by Dr. Parvinder S. Sethi

Fields of windmills, like this one in Southern California that is used to generate electricity, are called wind farms.

US Army Corps of Engineers/Julie Stone

Windmills like this, used to pump well water on ranches and farms, are in semiarid and arid regions of North America.

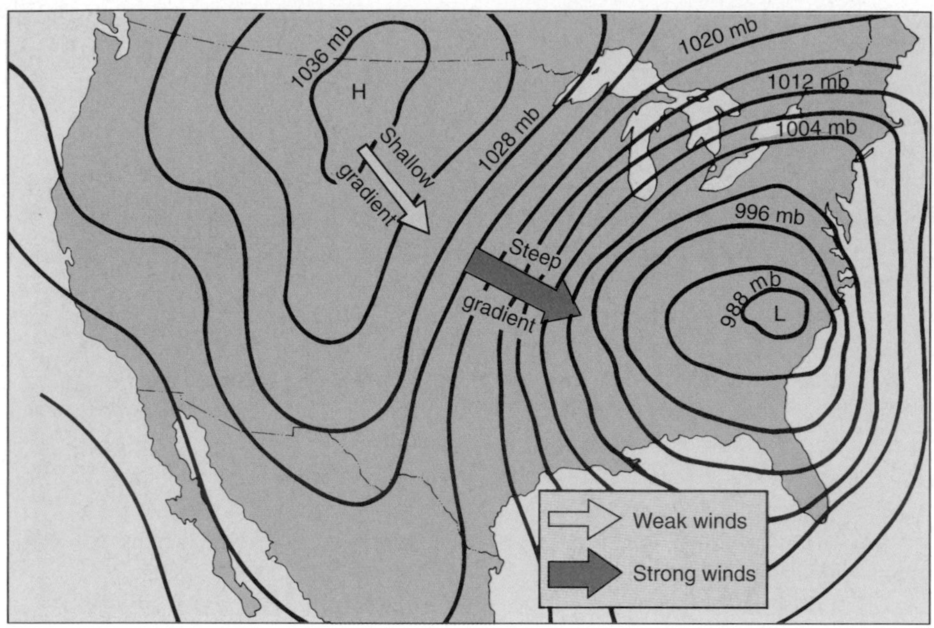

• FIGURE 5.5
The relationship of wind to the pressure gradient: The steeper the pressure gradient, the stronger will be the resulting wind.
Where else on this figure (other than the area indicated) would winds be strong?

Pressure Gradients and Winds

Winds vary widely in velocity, duration, and direction. Much of their strength depends on the size or strength of the pressure gradient to which they are responding. As we noted previously, *pressure gradient* is the term applied to the rate of change of atmospheric pressure between two points (at the same elevation). The greater this change—that is, the steeper the pressure gradient—the greater will be the wind response (• Fig. 5.5). Winds tend to flow down a pressure gradient from high pressure to low pressure, just as water flows down a slope from a high point to a low one. A useful little rhyme, "Winds always blow, from high to low," will always remind you of the direction of surface winds. The steeper the pressure gradients involved, the faster and stronger will be the winds. Yet wind does not flow directly from high to low, as we might expect, because other factors also affect the direction of wind.

The Coriolis Effect and Wind

Two factors, both related to our Earth's rotation, greatly influence wind direction. First, our fixed-grid system of latitude and longitude is constantly rotating. Thus, our frame of reference for tracking the path of any free-moving object—whether it is an aircraft, a missile, or the wind—is constantly changing its position. Second, the speed of rotation of Earth's surface increases as we move equatorward and decreases as we move toward the poles (see again Fig. 3.11). Thus, to use our previous example, someone in St. Petersburg (60°N latitude), where the distance around a parallel of latitude is about half that at the equator, moves at about 840 kilometers per hour (525 mph) as Earth rotates, while someone in Kampala, Uganda, near the equator, moves at about 1680 kilometers per hour (1050 mph).

Because of these Earth rotation factors, anything moving horizontally appears to be deflected to the right of the direction in which it is traveling in the Northern Hemisphere and to the left in the Southern Hemisphere. This apparent deflection is termed the **Coriolis effect.** The degree of deflection, or curvature, is a function of the speed of the object in motion and the latitudinal location of the object. The higher the latitude, the greater will be the Coriolis effect (• Fig. 5.6). In fact, not only does the Coriolis effect decrease at lower latitudes, but it does not exist at the equator. Also, the faster the object is moving, the greater will be the apparent deflection, and the greater the distance something must travel, the greater will be the Coriolis effect.

As we have said, anything that moves horizontally over Earth's surface exhibits the Coriolis effect. Thus, both the atmosphere and the oceans are deflected in their

• FIGURE 5.6
Schematic illustration of the apparent deflection (Coriolis effect) of an object caused by Earth's rotation when an object (or the wind) moves north, south, east, or west in both hemispheres.
If no Coriolis effect exists at the equator, where would the maximum Coriolis effect be located?

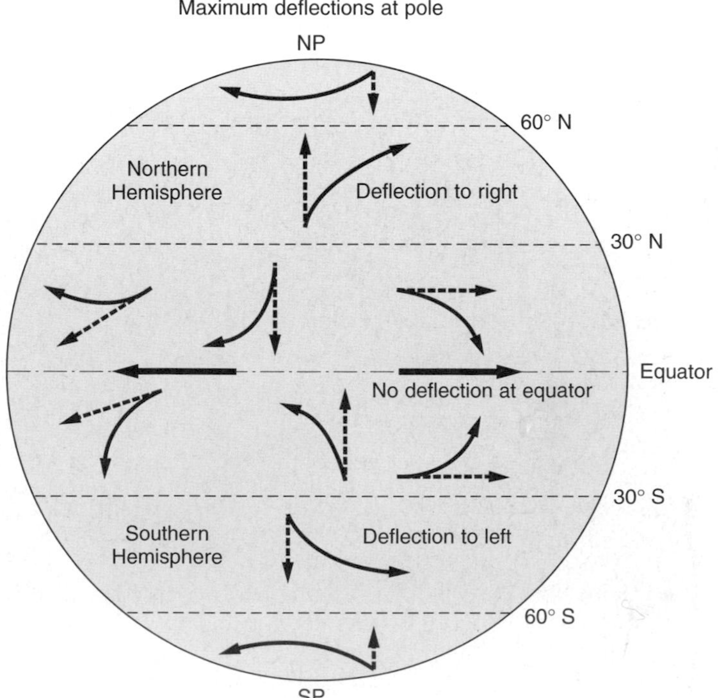

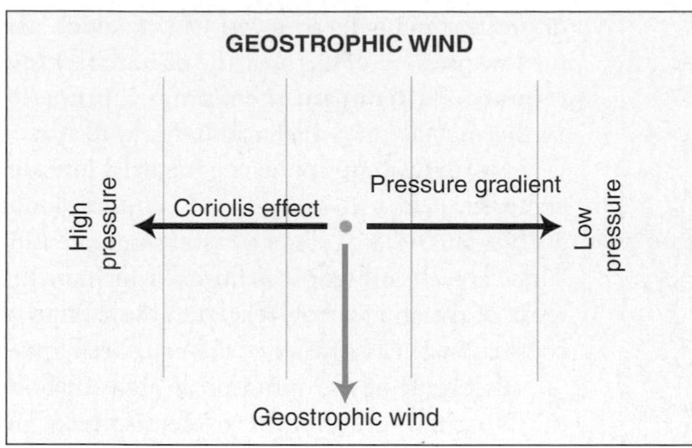

• **FIGURE 5.7**
This Northern Hemisphere example illustrates that in a geostrophic wind, the Coriolis effect causes it to veer to the right until the pressure gradient and Coriolis effect reach an equilibrium and the wind flows between (and parallel to) the isobars.

movements. Winds in the Northern Hemisphere moving across a gradient from high to low pressure are apparently deflected to the right of their expected path (and to the left in the Southern Hemisphere). In addition, when considering winds at Earth's surface, we must take into account another force. This force, **friction,** interacts with the pressure gradient and the Coriolis effect.

Friction and Wind

Above Earth's surface, frictional drag is of little consequence to wind development. At this level, the wind starts down the pressure gradient and turns 90° in response to the Coriolis effect. At this point, the pressure gradient is balanced by the Coriolis effect, and the wind, termed a **geostrophic wind,** flows parallel to the isobars (• Fig. 5.7).

However, at or near Earth's surface (up to about 1000 m above the surface), frictional drag is important because it reduces the wind speed. A reduced wind speed in turn reduces the Coriolis effect, but the pressure gradient is not affected. With the pressure gradient and Coriolis effect no longer in balance, the wind does not flow between the isobars like its upper-level counterpart. Instead, a surface wind flows obliquely (about a 30° angle) across the isobars toward an area of low pressure.

Wind Terminology

Winds are named after their source. Thus, a wind that comes out of the northeast is called a northeast wind. One coming from the south, even though going toward the north, is called a south (or southerly) wind. It is helpful for students to use the phrase "out of" when describing a wind direction. That phrase will help students to keep the correct direction. For example, if the winds are blowing to the south, then by saying, "the winds are out of the north," automatically makes the student think about the direction of the wind's origin.

Windward refers to the direction from which the wind blows. The side of something that faces the direction from which the wind is coming is called the *windward* side. Thus, a windward slope is the side of a mountain against which the wind blows (• Fig. 5.8). **Leeward,** on the other hand, means the direction toward which the wind is blowing. Thus, when the winds are coming out of the west, the *leeward* slope of a mountain would be the east slope. We know that winds can blow from any direction, yet in some places winds may tend to blow more from one direction than any other. We speak of these as the **prevailing winds.**

Cyclones, Anticyclones, and Winds

Imagine a high pressure cell (anticyclone) in the Northern Hemisphere in which the air is moving from the center in all directions down pressure gradients. As it moves, the air will be deflected to the right, no matter which direction it was originally going. Therefore, the wind moving out of an anticyclone in the Northern Hemisphere will move from the center of high pressure in a clockwise spiral (• Fig. 5.9).

Air tends to move down pressure gradients from all directions toward the center of a low pressure area (cyclone). However, because the air is apparently deflected to the right in the Northern Hemisphere, the winds move into the cyclone in a counterclockwise spiral. Because all objects including air and water are apparently deflected to the left in the Southern Hemisphere, spirals there are reversed. Thus, in the Southern Hemisphere, winds moving away from an anticyclone do so in a counterclockwise spiral, and winds moving into a cyclone move in a clockwise spiral.

• **FIGURE 5.8**
Illustration of the meaning of windward (facing into the wind) and leeward (facing away from the wind).
How might vegetation differ on the windward and leeward sides of an island?

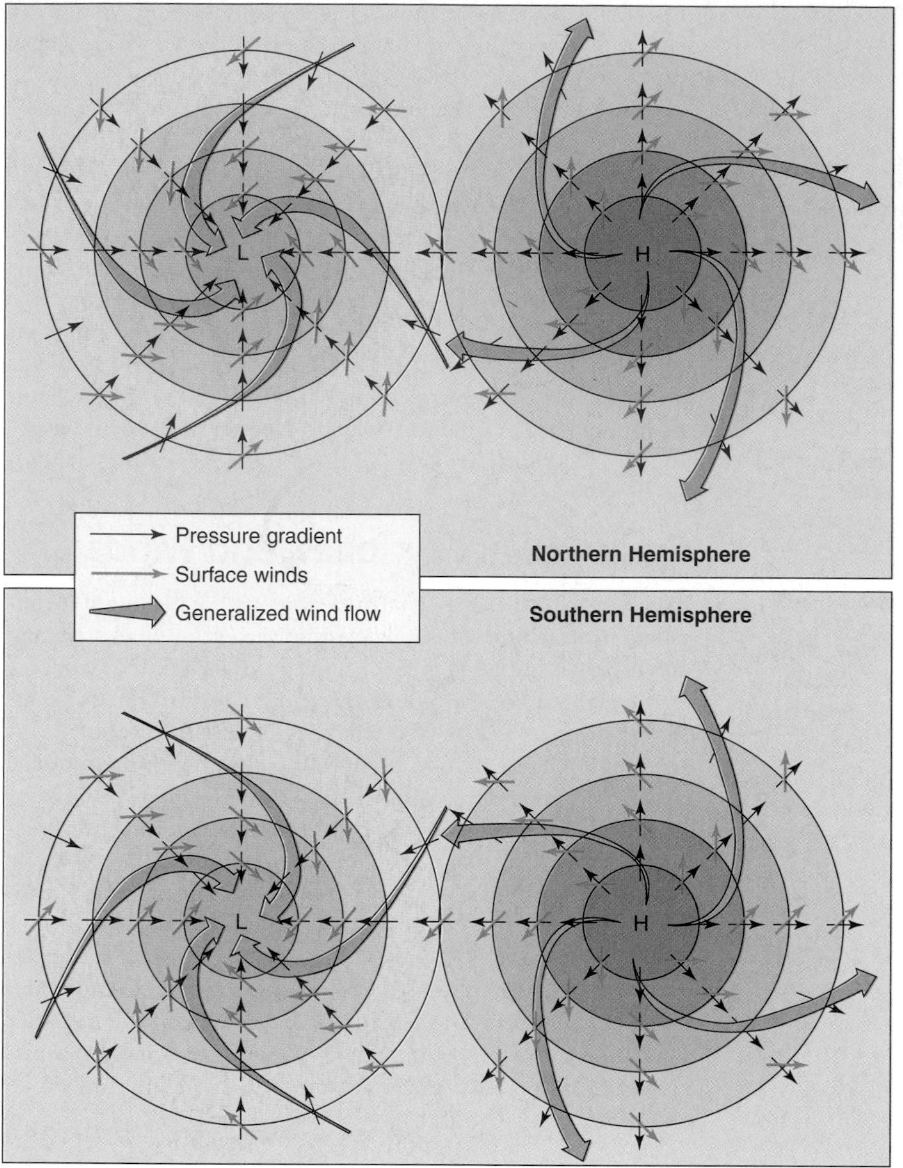

• FIGURE 5.9
Movement of surface winds associated with low pressure centers (cyclones) and high pressure centers (anticyclones) in the Northern and Southern Hemispheres. Note that the surface winds are to the right of the pressure gradient in the Northern Hemisphere and to the left of the pressure gradient in the Southern Hemisphere.
What do you think might happen to the diverging air of an anticyclone if there is a cyclone nearby?

Global Pressure Belts

Idealized Global Pressure Belts

Using what we have learned about pressure on Earth's surface, we can construct a theoretical model of the pressure belts of the world (• Fig. 5.10). Later, we see how real conditions depart from our model and examine why these differences occur.

Centered approximately over the equator in our model is a belt of low pressure, or a **trough.** Because this is the region on Earth of greatest annual heating, we can conclude that the low pressure of this area, the **equatorial low (equatorial trough),** is determined primarily by thermal factors, which cause the air to rise.

North and south of the equatorial low and centered on the so-called horse latitudes, about 30°N and 30°S, are cells of relatively high pressure. These are the **subtropical highs,** which are the result of dynamic factors related to the sinking of convectional cells initiated at the equatorial low.

Poleward of the subtropical highs in both the Northern and Southern Hemispheres are large belts of low pressure that extend through the upper-middle latitudes. Pressure decreases through these **subpolar lows** until about 65° latitude. Again, dynamic factors play a role in the existence of subpolar lows.

In the polar regions are high pressure systems called the **polar highs.** The extremely cold temperatures and consequent sinking of the dense polar air in those regions create the higher pressures found there.

This system of pressure belts that we have just developed is a generalized picture. Just as temperatures change from month to month, day to day, and hour to hour, so do pressures vary through time at any one place. Our long-term global model disguises these smaller changes, but it does give an idea of broad pressure patterns on the surface of Earth.

The Global Pattern of Atmospheric Pressure

As our idealized model suggests, the atmosphere tends to form belts of high and low pressure along east–west axes in areas where there are no large bodies of land. These belts are arranged by latitude and generally maintain their bandlike pattern. However, where there are continental landmasses, belts of pressure are broken and tend to form cellular pressure systems. The landmasses affect the development of belts of atmospheric pressure in several ways. Most influential is the effect of the differential heating of land and water surfaces. In addition, landmasses affect the movement of air and consequently the development of pressure systems through friction with their surfaces. Landform barriers such as mountain ranges also block the movement of air and thereby affect atmospheric pressure.

Seasonal Variations in the Pattern

In general, the global atmospheric pressure belts shift northward in July and southward in January, following the changing position of the sun's direct rays as they migrate between the Tropics of Cancer and Capricorn. Thus, there are thermally induced sea-

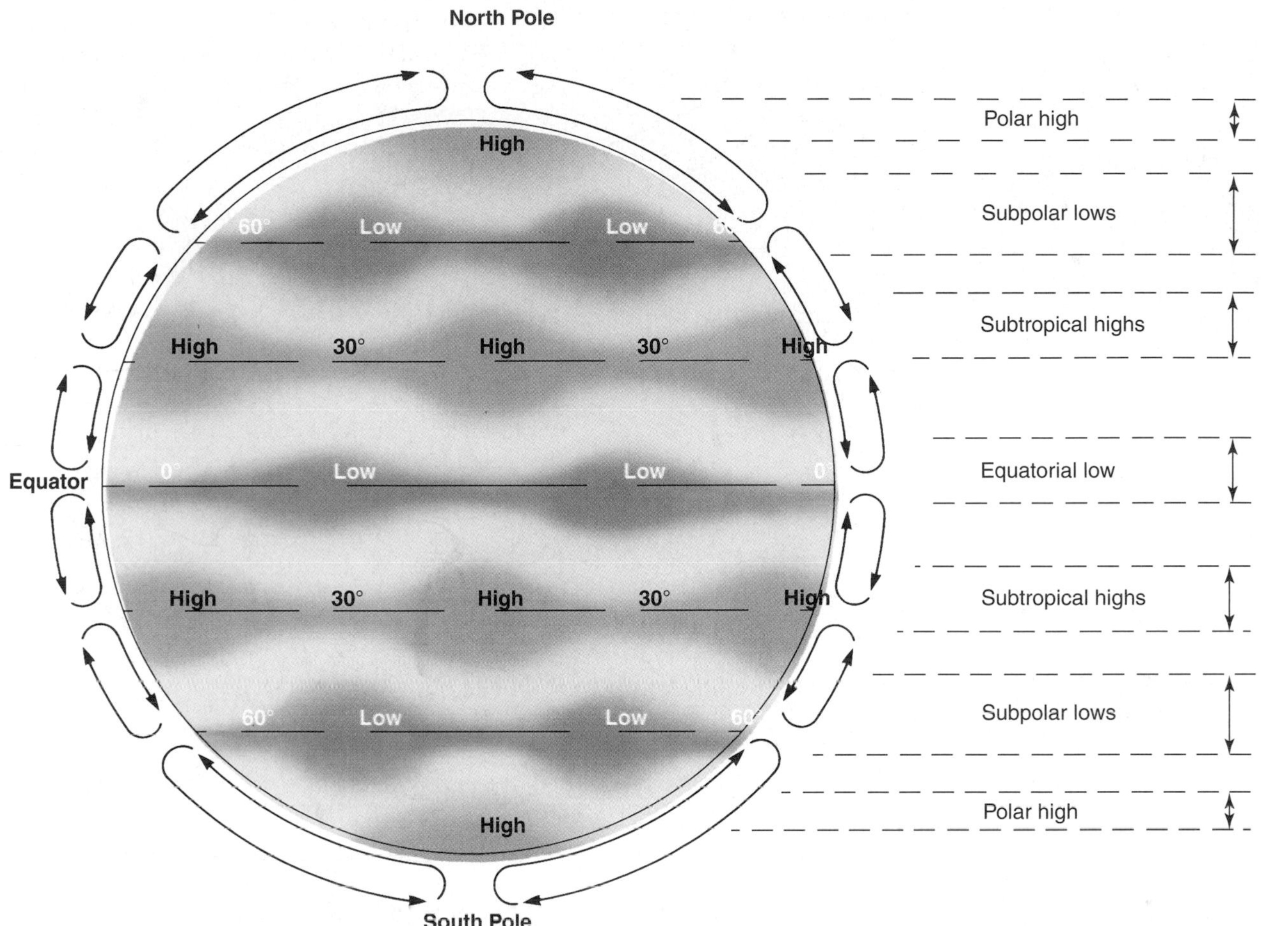

• FIGURE 5.10
Idealized world pressure belts. Note the arrows on the perimeter of the globe that illustrate the cross-sectional flow associated with the surface pressure belts.
Why do most of these pressure belts come in pairs?

sonal variations in the pressure patterns, as seen in • Figures 5.11a and b. These seasonal variations tend to be small at low latitudes, where there is little temperature variation, and large at high latitudes, where there is an increasing contrast in length of daylight and angle of the sun's rays. Furthermore, landmasses tend to alter the general pattern of seasonal variation. This is an especially important factor in the Northern Hemisphere, where land accounts for 40% of the total Earth surface, as opposed to less than 20% in the Southern Hemisphere.

January Because continents cool more quickly than the oceans, their temperatures will be lower in winter than those of the surrounding seas. Figure 5.11a shows that in the middle latitudes of the Northern Hemisphere this variation leads to the development of cells of high pressure over the land areas. In contrast, the subpolar lows develop over the oceans because they are comparatively warmer. Over eastern Asia, there is a strongly developed anticyclone during the winter months that is known as the **Siberian High.** Its equivalent in North America, known as the **Canadian High,** is not nearly so well developed because the North American landmass is considerably smaller than the Eurasian continent.

In addition to the Canadian High and the Siberian High, two low pressure centers develop: one in the North Atlantic, called the **Icelandic Low,** and the other in the North Pacific, called the **Aleutian Low.** The air in them has relatively lower pressure than either the subtropical or the polar high systems. Consequently, air moves toward these low pressure areas from both north and south. Such low pressure regions are associated with cloudy, unstable weather and are a major source of winter storms, whereas high pressure areas are associated with clear, blue-sky days; calm, starry nights; and cold, stable weather. Therefore, during the winter months, cloudy and sometimes dangerously stormy weather tends to be associated with the two oceanic lows and clear weather with the continental highs.

We can also see that the polar high in the Northern Hemisphere is well developed. This development is due primarily to thermal factors because January is the coldest time of the year. The subpolar lows have developed into the Aleutian and Icelandic cells described earlier. At the same time, the subtropical highs of the Northern Hemisphere appear slightly south of their average annual position because of the migration of the sun toward the Tropic of Capricorn. The equatorial trough also appears centered south of its average annual position over the geographic equator.

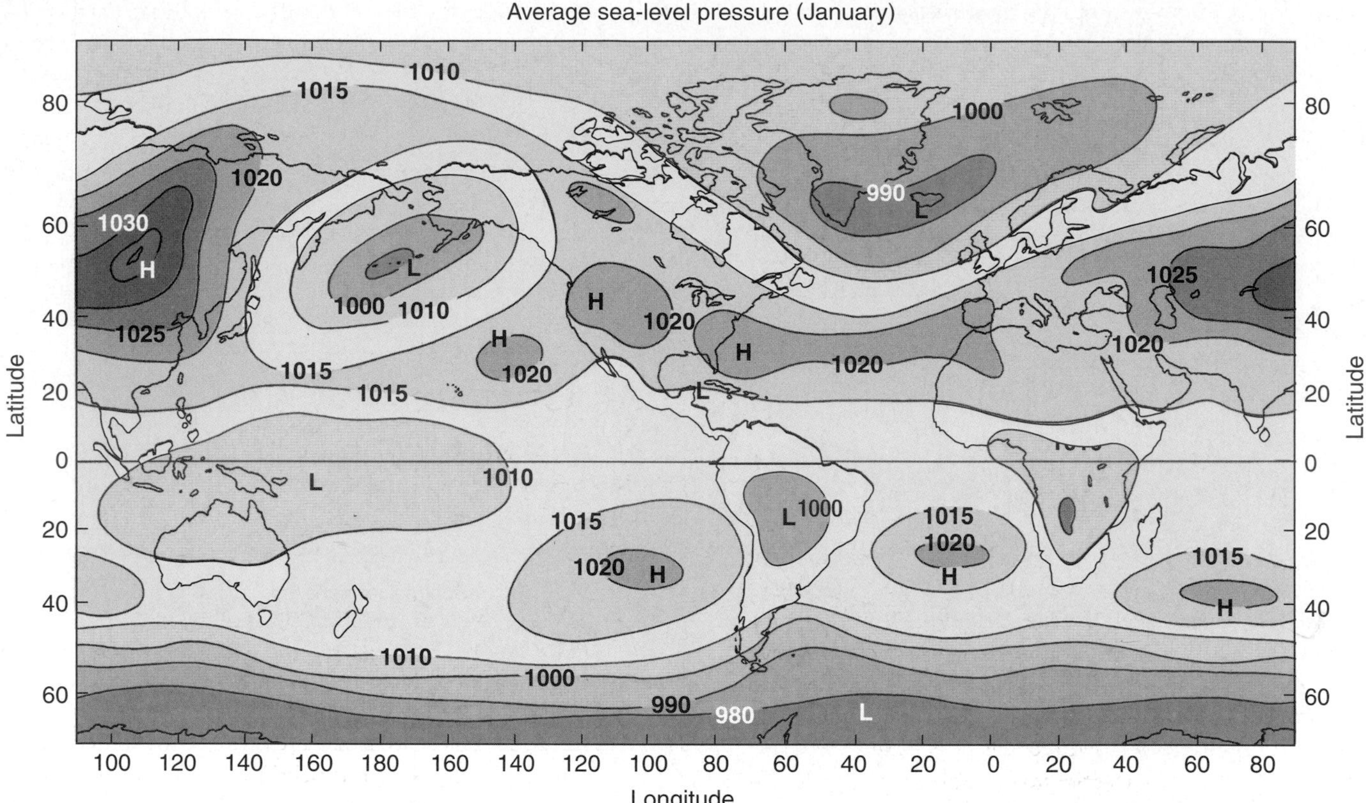

(a)

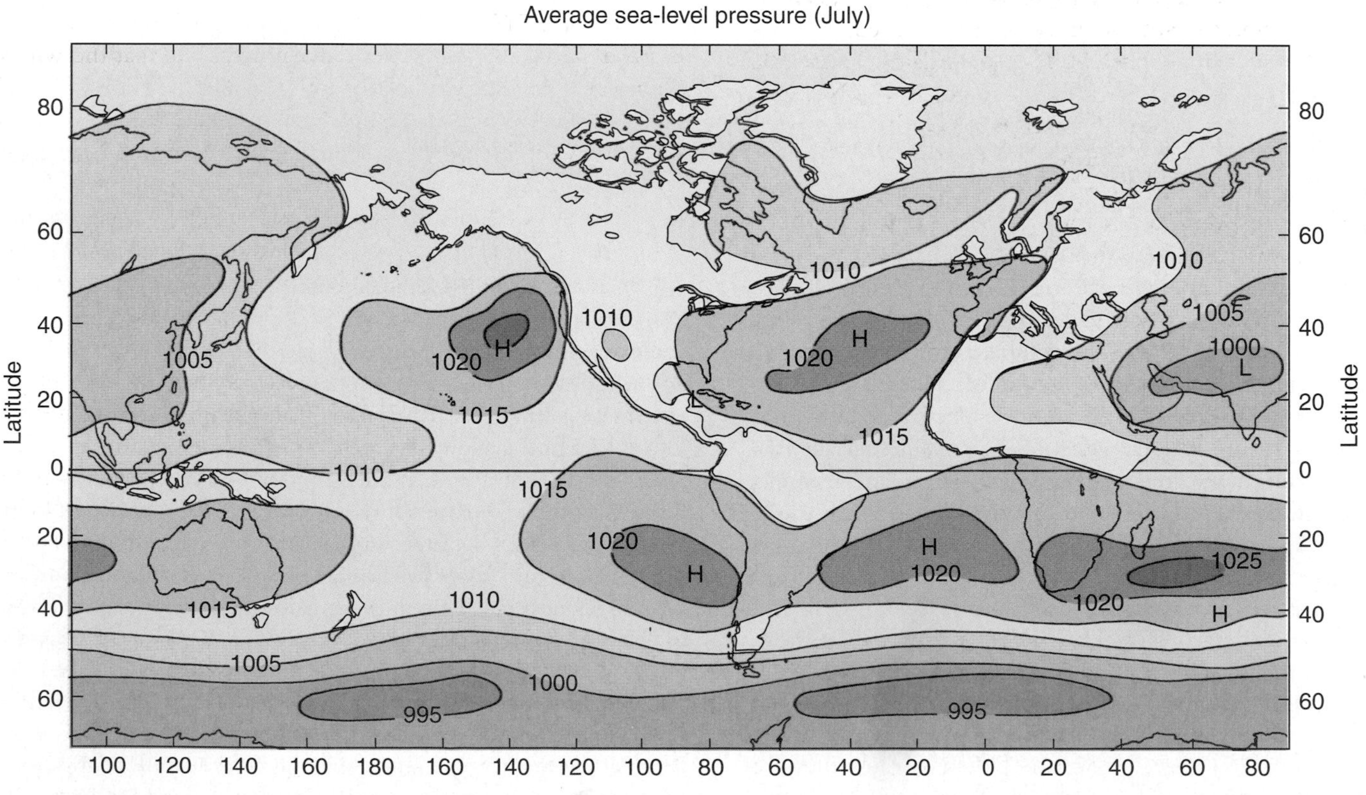

(b)

• FIGURE 5.11

(a) Average sea-level pressure (in millibars) in January. (b) Average sea-level pressure (in millibars) in July. **What is the difference between the January and July average sea-level pressures at your location? Why do they vary?**

In January in the Southern Hemisphere, the subtropical belt of high pressure appears as three cells centered over the oceans because the belt of high pressure has been interrupted by the continental landmasses where temperatures are much higher and pressure tends to be lower than over the oceans. Because there is virtually no land between 45°S and 70°S latitude, the subpolar low circles Earth as a belt of low pressure and is not divided into cells by any landmasses. There is little seasonal change in this belt of low pressure other than in January (summer in the Southern Hemisphere), when it lies a few degrees north of its July position.

July The anticyclone over the North Pole is greatly weakened during the summer months in the Northern Hemisphere, primarily because of the lengthy (24-hour days) heating of the oceans and landmasses in that region (Fig. 5.11b). The Aleutian and Icelandic Lows nearly disappear from the oceans, while the landmasses, which developed high pressure cells during the cold winter months, have extensive low pressure cells slightly to the south during the summer. In Asia, a low pressure system develops, but it is divided into two separate cells by the Himalayas. The low pressure cell over northwest India is so strong that it combines with the equatorial trough, which has moved north of its position 6 months earlier. The subtropical highs of the Northern Hemisphere are more highly developed over the oceans than over the landmasses. In addition, they migrate northward and are highly influential factors in the climate of landmasses nearby. In the Pacific, this subtropical high is termed the **Pacific High;** this system of pressure plays an important role in moderating the temperatures of the West Coast of the United States. In the Atlantic Ocean, the corresponding cell of high pressure is known as the **Bermuda High** to North Americans and as the **Azores High** to Europeans and West Africans. As we have already mentioned, the equatorial trough of low pressure moves north in July, following the migration of the sun's vertical rays, and the subtropical highs of the Southern Hemisphere lie slightly north of their January locations.

In examining pressure systems at Earth's surface, we have seen that there are essentially seven belts of pressure (two polar highs, two subpolar lows, two subtropical highs, and one equatorial low), which are broken into cells of pressure in some places primarily because of the influence of certain large landmasses. We have also seen that these belts and cells vary in size, intensity, and location with the seasons and with the migration of the sun's vertical rays over Earth's surface. Since these global-scale pressure systems migrate by latitude with the position of the direct sun angle, they are sometimes referred to as *semipermanent pressure systems* because they are never permanently fixed in the same location.

Global Surface Wind Systems

The planetary, or global, wind system is a response to the global pressure patterns and also plays a role in the maintenance of those same pressures. This wind system, which is the major means of transport for energy and moisture through Earth's atmosphere, can be examined in an idealized state. To do so, however, we must ignore the influences of landmasses and seasonal variations in solar energy. By assuming, for the sake of discussion, that Earth has a homogeneous surface and that there are no seasonal variations in the amount of solar energy received at different latitudes, we can examine a theoretical model of the atmosphere's planetary circulation. Such an understanding will help explain specific features of climate such as the rain and snow of the Sierra Nevada and Cascade Mountains and the existence of arid regions farther to the east. It will also account for the movement of great surface currents in our oceans that are driven by this atmospheric engine.

Idealized Model of Atmospheric Circulation

Because winds are caused by pressure differences, various types of winds are associated with different kinds of pressure systems. Therefore, a system of global winds can be demonstrated using the model of pressures that we previously developed (see again Fig. 5.10).

The characteristics of convergence and divergence are very important to our understanding of global wind patterns. Surface air diverges from zones of high pressure and converges on areas of low pressure. We also know that, because of the pressure gradient, surface winds always blow from high pressure to low pressure.

Knowing that surface winds originate in areas of high pressure and taking into account the global system of pressure cells, we can develop our model of the wind systems of the world (● Fig. 5.12). This model takes into account differential heating, Earth rotation, and atmospheric dynamics. Note that the winds do not blow in a straight north–south line. The variation is due of course to the Coriolis effect, which causes an apparent deflection to the right in the Northern Hemisphere and to the left in the Southern Hemisphere.

Our idealized model of global atmospheric circulation includes six wind belts, or zones, in addition to the seven pressure zones that we have previously identified. Two wind belts, one in each hemisphere, are located where winds move out of the polar highs and down the pressure gradients toward the subpolar lows. As these winds are deflected to the right in the Northern Hemisphere and to the left in the Southern, they become the **polar easterlies.**

The remaining four wind belts are closely associated with the divergent winds of the subtropical highs. In each hemisphere, winds flow out of the poleward portions of these highs toward the subpolar lows. Because of their general movement from the west, the winds of the upper-middle latitudes are labeled the **westerlies.** The winds blowing from the highs toward the equator have been called the **trade winds.** Because of the Coriolis effect, they are the **northeast trades** in the Northern Hemisphere and the **southeast trades** south of the equator.

Our model does not conform exactly to actual conditions. First, as we know, the vertical rays of the sun do not stay precisely over the equator but migrate as far north as the Tropic of Cancer in June and south to the Tropic of Capricorn in December. Therefore, the pressure systems, and consequently the winds, must move to adjust to the change in the position of the sun. Then, as we have already discovered, the existence of the continents, especially in

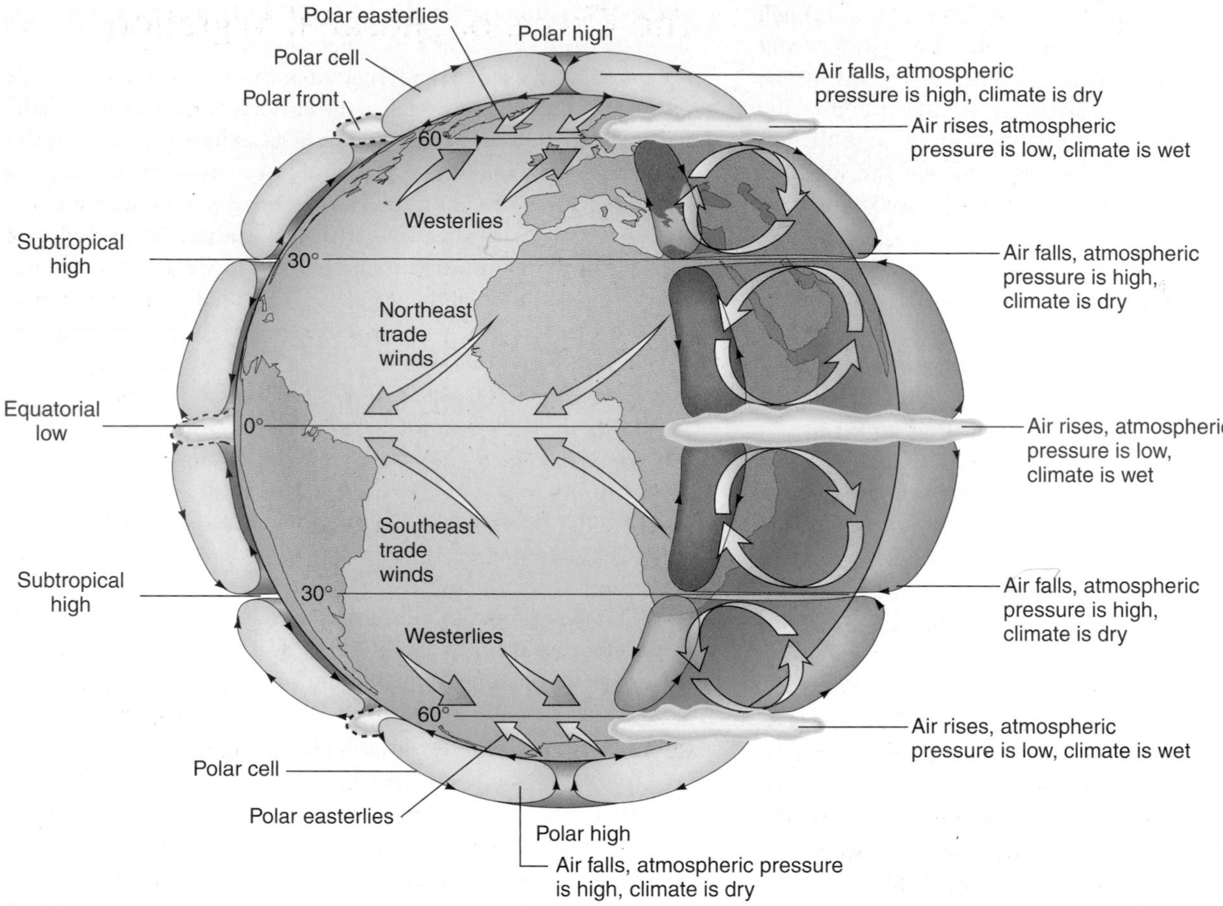

• FIGURE 5.12
The general circulation of Earth's atmosphere.

the Northern Hemisphere, causes longitudinal pressure differentials that affect the zones of high and low pressure.

Conditions within Latitudinal Zones

Trade Winds A good place to begin our examination of winds and associated weather patterns as they actually occur is in the vicinity of the subtropical highs. On Earth's surface, the trade winds, which blow out of the subtropical highs toward the equatorial trough in both the Northern and Southern Hemispheres, can be identified between latitudes 5° and 25°. Because of the Coriolis effect, the northern trades move away from the subtropical high in a clockwise direction out of the northeast. In the Southern Hemisphere, the trades diverge out of the subtropical high toward the equatorial trough from the southeast, as their movement is counterclockwise. Because the trades tend to blow out of the east, they are also known as the **tropical easterlies.**

The trade winds tend to be constant, steady winds, consistent in their direction. This is most true when they cross the eastern sides of the oceans (near the eastern portion of the subtropical high). The area of the trades varies somewhat during the solar year, moving north and south a few degrees of latitude with the sun. Near their source in the subtropical highs, the weather of the trades is clear and dry, but after crossing large expanses of ocean, the trades have a high potential for stormy weather.

Early Spanish sea captains depended on the northeast trade winds to drive their galleons to destinations in Central and South America in search of gold, spices, and new lands. Going eastward toward home, navigators usually tried to plot a course using the westerlies to the north. The trade winds are one of the reasons that the Hawaiian Islands are so popular with tourists; the steady winds help keep temperatures pleasant, even though Hawaii is located south of the Tropic of Cancer.

Doldrums Where the trade winds converge in the equatorial trough (or tropical low) lies a zone of calm and weak winds of no prevailing direction. Here the air, which is very moist and heated by the sun, tends to expand and rise, maintaining the low pressure of the area. These winds, which are roughly between 5°N and 5°S, are generally known as the **doldrums.** This area is called the **intertropical convergence zone (ITCZ),** or the "equatorial belt of variable winds and calms." Because of the converging moist air and high potential for rainfall in the doldrums, this region coincides with the world's latitudinal belt of heaviest precipitation and most persistent cloud cover.

Old sailing ships often remained becalmed in the doldrums for days at a time. It is interesting to note that the word *doldrums* in the English language means a bored or depressed state of mind. The sailors were in the doldrums in more ways than one.

Subtropical Highs The areas of subtropical high pressure, generally located between latitudes 25° and 35°N and S, and from which winds blow poleward to become the westerlies and equatorward as the trade winds, are often called the subtropical belts of variable winds, or the "horse latitudes." This name comes from the occasional need by the Spanish conquistadors to eat their horses or throw them overboard in order to conserve drinking water and lighten the weight when their ships were becalmed in these latitudes. The subtropical highs are areas, like the doldrums, in which there are no strong prevailing winds. However, unlike the doldrums, which are characterized by convergence, rising air, and heavy rainfall, the subtropical highs are areas of sinking and settling air from higher altitudes, which tend to build up the atmospheric pressure. Weather conditions are typically clear, sunny, and rainless, especially over the eastern portions of the oceans where the high pressure cells are strongest.

Westerlies The winds that flow poleward out of the subtropical high pressure cells in the Northern Hemisphere are deflected to the right and thus blow from the southwest. Those in the Southern Hemisphere are deflected to the left and blow out of the northwest. Thus, these winds have been correctly labeled the westerlies. They tend to be less consistent in direction than the trades, but they are usually stronger winds and may be associated with stormy weather. The westerlies occur between about 35° and 65°N and S latitudes. In the Southern Hemisphere, where there is less land than in the Northern Hemisphere to affect the development of winds, the westerlies attain their greatest consistency and strength. Much of Canada and most of the United States—except Florida, Hawaii, and Alaska—are under the influence of the westerlies.

Polar Winds Accurate observations of pressure and wind are sparse in the two polar regions; therefore, we must rely on remotely sensed information (mainly by weather satellite imagery). Our best estimate is that pressures are consistently high throughout the year at the poles and that prevailing easterly winds blow from the polar regions to the subpolar low pressure systems.

Polar Front Despite our limited knowledge of the wind systems of the polar regions, we do know that the winds can be highly variable, blowing at times with great speed and intensity. When the cold air flowing out of the polar regions and the warmer air moving in the path of the westerlies meet, they do so like two warring armies: One does not absorb the other. Instead, the denser, heavier cold air pushes the warm air upward, forcing it to rise rapidly. The line along which these two great wind systems battle is appropriately known as the **polar front.** The weather that results from the meeting of the cold polar air and the warmer air from the subtropics can be very stormy. In fact, most of the storms that move slowly through the middle latitudes in the path of the prevailing westerlies are born at the polar front.

The Effects of Seasonal Migration

Just as insolation, temperature, and pressure systems migrate north and south as Earth revolves around the sun, Earth's wind systems also migrate with the seasons. During the summer months in the Northern Hemisphere, maximum insolation is received north of the equator. This condition causes the pressure belts to move north as well, and the wind belts of both hemispheres shift accordingly. Six months later, when maximum heating is taking place south of the equator, the various wind systems have migrated south in response to the migration of the pressure systems. Thus, seasonal variation in wind and pressure conditions is one important way in which actual atmospheric circulation differs from our idealized model.

The seasonal migration will most affect those regions near the boundary zone between two wind or pressure systems. During the winter months, such a region will be subject to the impact of one system. Then, as summer approaches, that system will migrate poleward and the next equatorward system will move in to influence the region. Two such zones in each hemisphere have a major effect on climate. The first lies between latitudes 5° and 15°, where the wet equatorial low of the high-sun season (summer) alternates with the dry subtropical high and trade winds of the low-sun season (winter). The second occurs between 30° and 40°, where the subtropical high dominates in summer but is replaced by the wetter westerlies in winter.

California is an example of a region located within a zone of transition between two wind or pressure systems (● Fig. 5.13). During the winter, this region is under the influence of the westerlies blowing out of the Pacific High. These winds, turbulent and full of moisture from the ocean, bring winter rains and storms to "sunny" California. As summer approaches, however, the subtropical high and its associated westerlies move north. As California comes under the influence of the calm and steady high pressure system, it experiences again the climate for which it is famous: day after day of warm, clear, blue skies. This alternation of moist winters and dry summers is typical of the western sides of all landmasses between 30° and 40° latitude.

Longitudinal Differences in Winds

We have seen that there are sizable latitudinal differences in pressure and winds. In addition, there are significant longitudinal variations, especially in the zone of the subtropical highs.

As was previously noted, the subtropical high pressure cells, which are generally centered over the oceans, are much stronger on their eastern sides than on their western sides. Thus, over the eastern portions of the oceans (west coasts of the continents) in the subtropics, subsidence and divergence are especially noticeable. The above-surface temperature inversions so typical of anticyclonic circulation are close to the surface, and the air is calm and clear. The air moving equatorward from this portion of the high produces the classic picture of the steady trade winds with clear, dry weather.

Over the western portions of the oceans (eastern sides of the continents), conditions are markedly different. In its passage

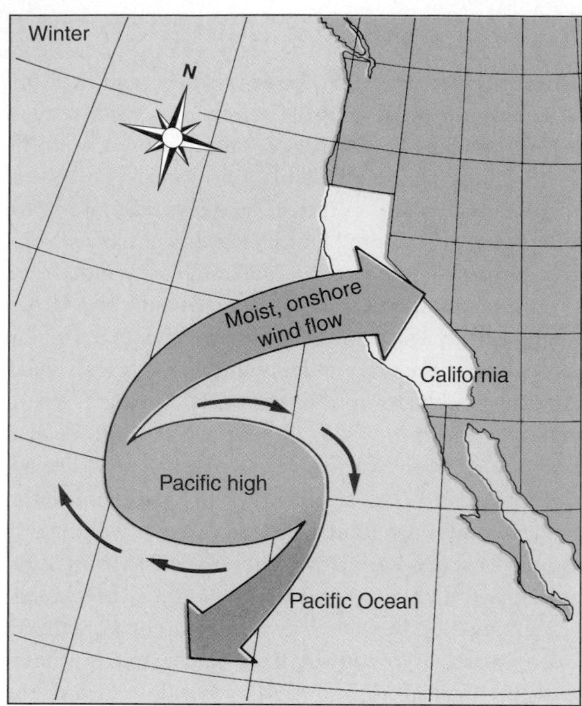

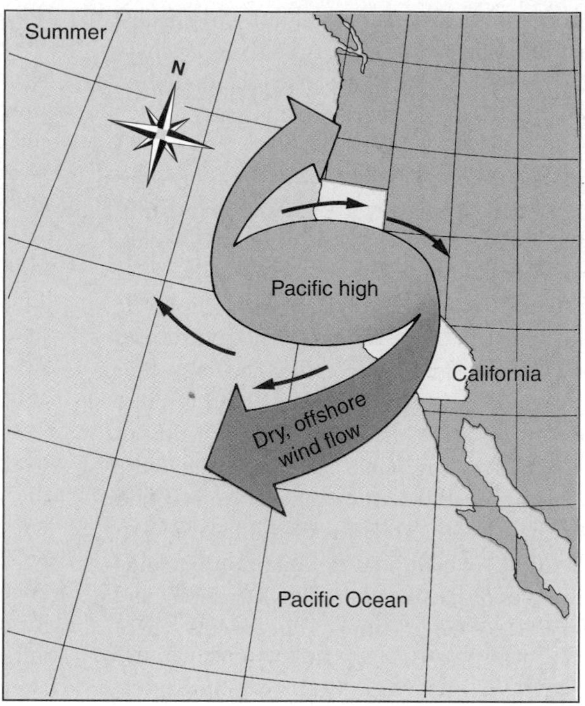

• FIGURE 5.13

Winter and summer positions of the Pacific anticyclone in relation to California. In the winter, the anticyclone lies well to the south and feeds the westerlies that bring the cyclonic storms and rain from the North Pacific to California. The influence of the anticyclone dominates during the summer. The high pressure blocks cyclonic storms and produces warm, sunny, and dry conditions.

In what ways would the seasonal migration of the Pacific anticyclone affect agriculture in California?

• FIGURE 5.14

Circulation pattern in a Northern Hemisphere subtropical anticyclone. Subsidence of air is strongest in the eastern part of the anticyclone, producing calm air and arid conditions over adjacent land areas. The southern margin of the anticyclone feeds the persistent northeast trade winds.

What wind system is fed by the northern margin?

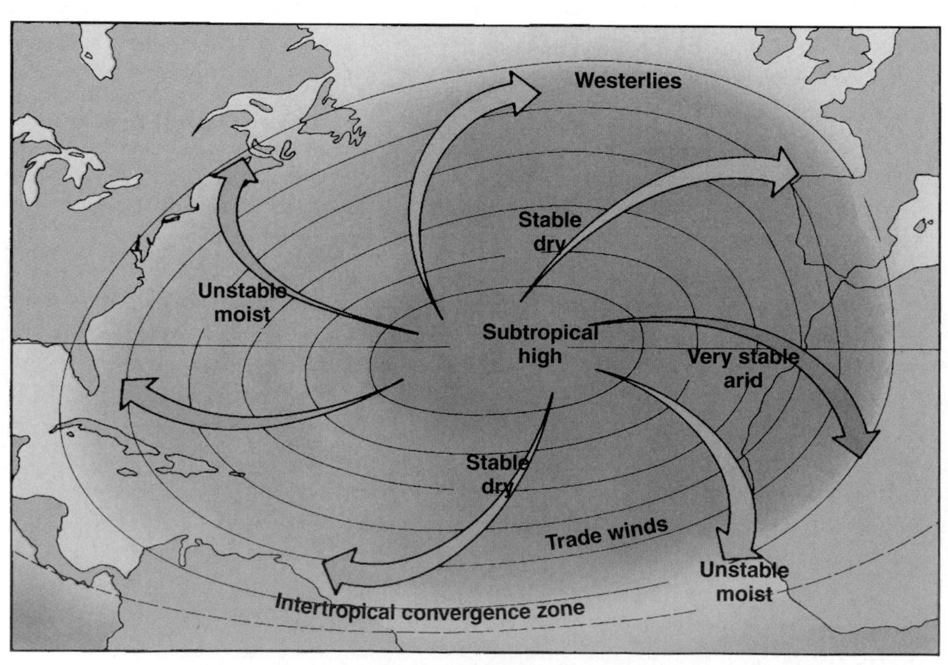

over the ocean, the diverging air is gradually warmed and moistened; turbulent and stormy weather conditions are likely to develop. As indicated in ● Figure 5.14, wind movement in the western portions of the anticyclones may actually be poleward and directed toward landmasses. Hence, the trade winds in these areas are especially weak or nonexistent much of the year.

As we have pointed out in discussing Figures 5.11 and 5.12, there are great land–sea contrasts in temperature and pressure throughout the year farther toward the poles, especially in the Northern Hemisphere. In the cold continental winters, the land is associated with pressures that are higher than those over the oceans, and thus there are strong, cold winds from the land to the sea. In the summer, the situation changes, with relatively low pressure existing over the continents because of higher temperatures. Wind directions are thus greatly affected, and the pattern is reversed so that winds flow from the sea toward the land.

Upper Air Winds and Jet Streams

Thus far, we have closely examined the wind patterns near Earth's surface. Of equal, or even greater, importance is the flow of air above Earth's surface—in particular, the flow of air at altitudes above 5000 meters (16,500 ft), and higher in the upper troposphere. The formation, movement, and decay of surface cyclones and anticyclones in the middle latitudes depend to a great extent on the flow of air high above Earth's surface.

The circulation of the upper air winds is a far less complex phenomenon than surface wind circulation. In the upper troposphere, an average westerly flow, the *upper air westerlies,* is maintained poleward of about 15°–20° latitude in both hemispheres. Because of the reduced frictional drag, the upper air westerlies move much more rapidly than their surface counterparts. Between 15° and 20°N and S latitudes are the *upper air easterlies,* which can be considered the high-altitude extension of the trade winds. The flow of the upper air winds became very apparent during World War II when high-altitude bombers moving eastward were found to cover similar distances faster than those flying westward. Pilots had encountered the upper air westerlies, or perhaps even the **jet streams**—very strong air currents embedded within the upper air westerlies.

The upper air westerlies form as a response to the temperature difference between warm tropical air and cold polar air. The air in the equatorial latitudes is warmed, rises convectively to high altitudes, and then flows toward the polar regions. At first this seems to contradict our previous statement, relative to surface winds, that air flows from cold areas (high pressure) toward warm areas (low pressure). This apparent discrepancy disappears, however, if you recall that the pressure gradient, down which the flow takes place, must be assessed between two points *at the same elevation.* A column of cold air will exert a higher pressure at Earth's surface than a column of warm air. Consequently, the pressure gradient established at Earth's surface will result in a flow from the cooler air toward the warmer air. However, cold air is denser and more compact than warm air. Thus, pressure decreases with height more rapidly in cold air than in warm air. As a result, at a specific height above Earth's surface, a lower pressure will be encountered above cold surface air than above warm surface air. This will result in a flow (pressure gradient) from the warmer surface air toward the colder surface air at that height. ● Figure 5.15 illustrates this concept.

Returning to our real-world situation, as the upper air winds flow from the equator toward the poles (down the pressure gradient), they are turned eastward because of the Coriolis effect. The net result is a broad circumpolar flow of westerly winds throughout most of the upper atmosphere (● Fig. 5.16). Because the upper air westerlies form in response to the thermal gradient between tropical and polar areas, it is not surprising that they are strongest in winter (the low-sun season), when the thermal contrast is greatest. During the summer (the high-sun season), when the contrast in temperature over the hemisphere is much reduced, the upper air westerlies move more slowly.

The temperature gradient between tropical and polar air, especially in winter, is not uniform but rather is concentrated where the warm tropical air meets cold polar air. This boundary, called the polar front, with its stronger pressure gradient, marks the location of the **polar front jet stream.** Ranging from 40 to 160 kilometers (25–100 mi) in width and up to 2 or 3 kilometers (1–2 mi) in depth, the polar front jet stream can be thought of as a faster, internal current of air within the upper air westerlies. While the polar front jet stream flows over the middle latitudes, another westerly **subtropical jet stream** flows above the sinking air of the subtropical highs in the lower-middle latitudes. Like the upper air westerlies, both jets are best developed in winter when hemispherical temperatures exhibit their steepest gradient (● Fig. 5.17). During the summer, both jets weaken in intensity. The subtropical jet stream frequently disappears completely, and the polar front jet tends to migrate northward.

We can now go one step further and combine our knowledge of the circulation of the upper air and surface to yield a more realistic portrayal of the vertical circulation pattern of our atmosphere (● Fig. 5.18). In general, the upper air westerlies and the associated polar jet stream flow in a fairly smooth pattern (● Fig. 5.19a). At times, however, the upper air westerlies develop oscillations, termed *long waves,* or **Rossby waves,** after the Swedish meteorologist Carl Rossby who first proposed and then proved

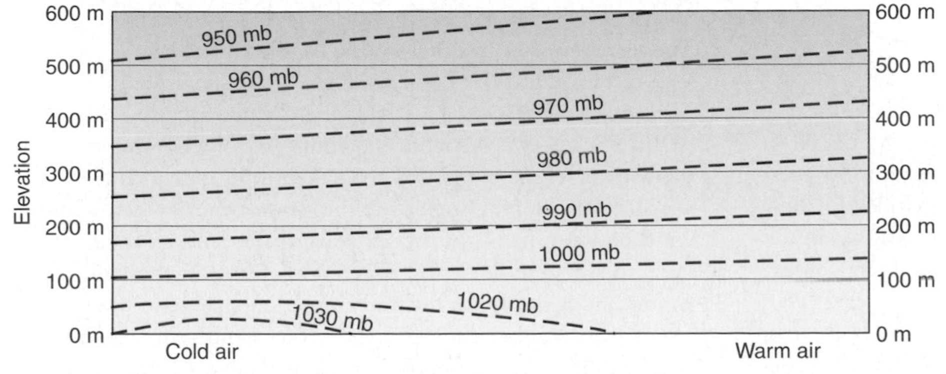

● FIGURE 5.15
Variation of pressure surfaces with height. Note that the horizontal pressure gradient is from cold to warm air at the surface and in the opposite direction at higher elevations (such as 400 m).
In what direction would the winds flow at 300 meters?

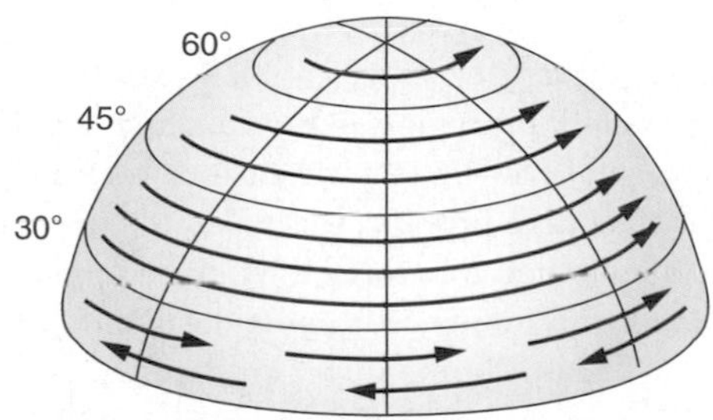

● FIGURE 5.16
The upper air westerlies form a broad circumpolar flow throughout most of the upper atmosphere.

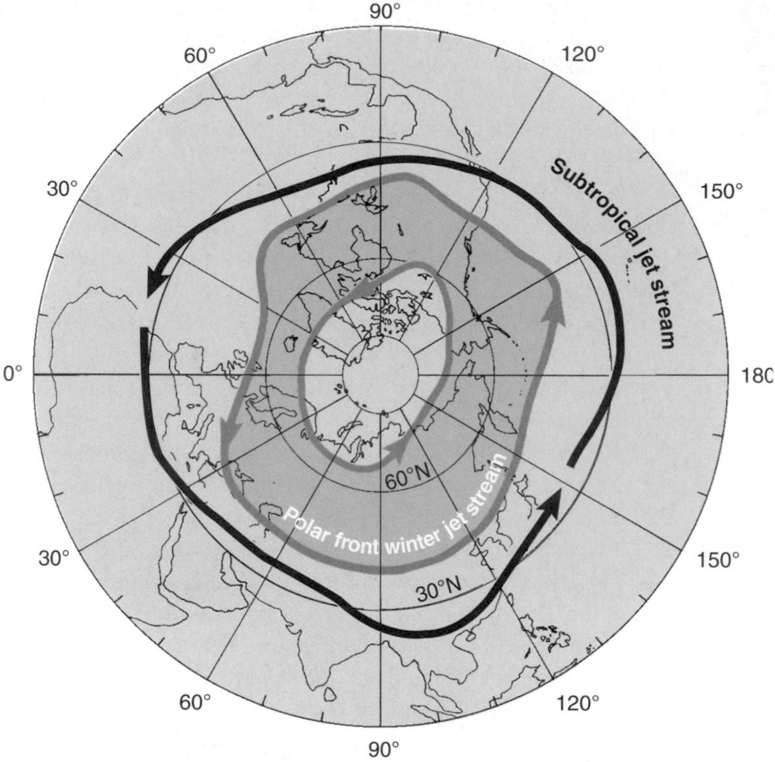

• **FIGURE 5.17**
Approximate location of the subtropical jet stream and area of activity of the polar front jet stream (shaded) in the Northern Hemisphere winter.
Which jet stream is most likely to affect your home state?

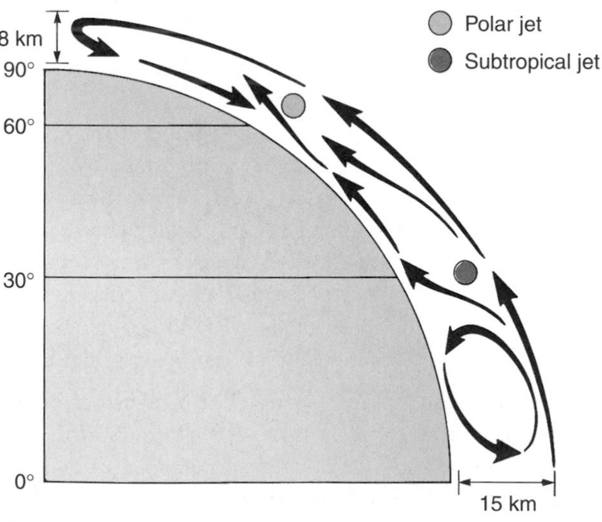

• **FIGURE 5.18**
A more realistic schematic cross section of the average circulation in the atmosphere.

their existence (• Fig. 5.19b). Rossby waves result in cold polar air pushing into the lower latitudes and forming *troughs* of low pressure, while warm tropical air moves into higher latitudes, forming *ridges* of high pressure. It is when the upper air circulation is in this configuration that surface weather is most influenced. We will examine this influence in more detail in Chapter 7.

Eventually, the upper air oscillations become so extreme that the "tongues" of displaced air are cut off, forming upper air cells of warm and cold air (Fig. 5.19c and d). This process helps maintain a net poleward flow of energy from equatorial and tropical areas. The cells eventually dissipate, and the pattern returns to normal (see again Fig. 5.19a). The complete cycle takes approximately 4–8 weeks. Although it is not completely clear why the upper atmosphere goes into these oscillating patterns, we are currently gaining additional insights. One possible cause is variation in ocean-surface temperatures. If the oceans in, say, the northern Pacific or near the equator become unusually warm or cold (for example, El Niño or La Niña, discussed later in this chapter), this apparently triggers oscillations, which continue until the ocean-surface temperature returns to normal. Other causes are being investigated at this time.

In addition to this influence on weather, jet streams are important to study for other reasons. They can carry pollutants, such as radioactive wastes or volcanic dust, over great distances and at relatively rapid rates. It was the polar jet stream that carried ash from the Mount St. Helens eruption (in 1980) eastward across the United States and Southern Canada. Nuclear fallout from the Chernobyl incident in the former Soviet Union could be monitored in succeeding days as it crossed the Pacific, and later the United States, in the jet streams. Pilots flying eastward—for example, from North America to Europe—take advantage of the jet stream, so the flying times in this direction may be significantly shorter than those in the reverse direction.

Subglobal Surface Wind Systems

As we have seen, winds develop whenever differential heating causes differences in pressure. The global wind system is a response to the constant temperature imbalance between tropical and polar regions. On a smaller, or subglobal, scale, additional wind systems develop. We begin with a discussion of monsoon winds, which are continental in size and develop in response to the seasonal variations in temperature and pressure. Last, on the smallest scale are local winds, which develop in response to the diurnal (daily) variation in heating and its local effects upon pressure and winds.

Monsoon Winds

The term *monsoon* comes from the Arabic word *mausim,* meaning season. This word has been used by Arab sailors for many centuries to describe seasonal changes in wind direction across the Arabian Sea between Arabia and India. As a meteorological term, **monsoon** refers to the directional shifting of winds from one season to the next. Usually, the monsoon occurs when a humid wind blowing from the ocean toward the land in the summer shifts to a dry, cooler wind blowing seaward off the land in the winter, and it involves a full 180° direction change in the wind.

The monsoon is most characteristic of southern Asia although it occurs on other continents as well. As the large landmass of Asia

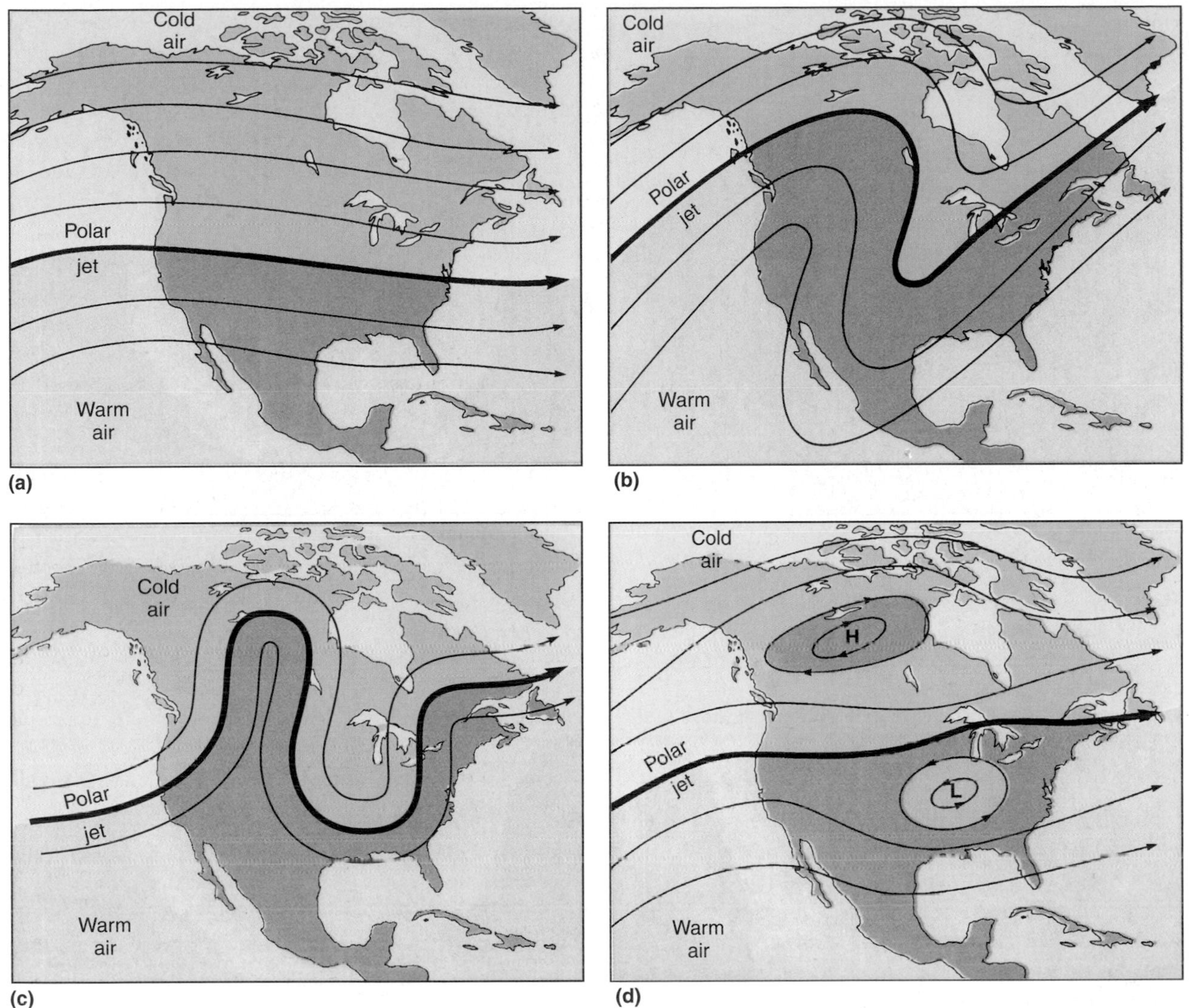

• FIGURE 5.19

Development and dissipation of Rossby waves in the upper air westerlies. (a) A fairly smooth flow prevails. (b) Rossby waves form, with a ridge of warm air extending into Canada and a trough of cold air extending down to Texas. (c) The trough and ridge may begin to turn back on themselves. (d) The trough and ridge are cut off and will eventually dissipate. The flow will then return to a pattern similar to (a).

How are Rossby waves closely associated with the changeable weather of the central and eastern United States?

cools more quickly than the surrounding oceans, the continent develops a strong center of high pressure from which there must be an outflow of air in winter (• Fig. 5.20). This outflow blows across much land toward the tropical low before reaching the oceans. It brings cold, dry air south.

In summer the Asian continent heats quickly and develops a large low pressure center. This development is reinforced by a poleward shift of the warm, moist tropical air to a position over southern Asia. Warm, moist air from the oceans is attracted into this low. Though full of water vapor, this air does not in itself cause the wet summers with which the monsoon is associated. However, any turbulence or landform barrier that makes this moist air rise and, as a result, cool down will bring about precipitation. This precipitation is particularly noticeable in the foothills of the Himalayas, the western Ghats of India, and the Annamese Highlands of Vietnam.

In the lower latitudes, a monsoonal shift in winds can come about as a reaction to the migration of the direct rays of the sun. For example, the winds of the equatorial zone migrate during the summer months northward toward the southern coast of Asia, bringing with them warm, moist, turbulent air. The winds of the Southern Hemisphere also migrate north with the sun, some crossing the equator. They also bring warm, moist air (from their trip over the ocean) to the southern and especially the southeastern coasts of India. In the winter months, the equatorial and tropical winds migrate south, leaving southern Asia under the influence of the dry, calm winds of the tropical Northern Hemisphere. Asia and northern Australia are true monsoon areas, with a full 180° wind shift with changes from summer to winter. Other regions, like the southern United States and West Africa, have "monsoonal tendencies," but are not monsoons in the true meaning of the term.

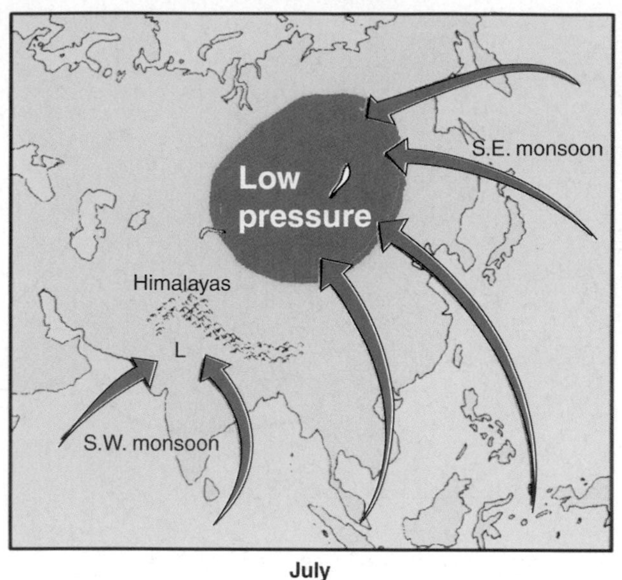

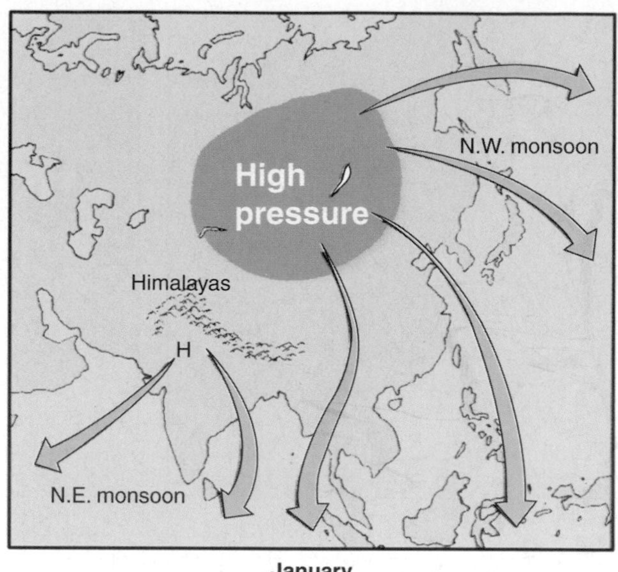

• FIGURE 5.20
Seasonal changes in surface wind direction that create the Asiatic monsoon system. The "burst" of the "wet monsoon," or the sudden onshore flow of tropical humid air in July, is apparently triggered by changes in the upper air circulation, resulting in heavy precipitation. The offshore flow of dry continental air in winter creates the "dry monsoon" and drought conditions in southern Asia.
How do the seasonal changes of wind direction in Asia differ from those of the southern United States?

The phenomenon of monsoon winds and their characteristic seasonal shifting cannot be fully explained by the differential heating of land and water, however, or by the seasonal shifting of tropical and subtropical wind belts. Some aspects of the monsoon system—for example, its "burst" or sudden transition between dry and wet in southern Asia—must have other causes. Meteorologists looking for a more complete explanation of the monsoon are examining the role played by the jet stream and other wind movements of the upper atmosphere.

Local Winds

Earlier, we discussed the major circulation patterns of Earth's atmosphere. This knowledge is vital to understanding the climate regions of Earth and the fundamental climatic differences between those regions. Yet we are all aware that there are winds that affect weather on a far smaller scale. These *local winds* are often a response to local landform configurations and add further complexity to the problem of understanding the dynamics of weather.

Chinooks and Other Warming Winds One type of local wind is known by several names in different parts of the world—for example, **Chinook** in the Rocky Mountain area and **foehn** (pronounced "fern") in the Alps. Chinook-type winds occur when air originating elsewhere must pass over a mountain range. As these winds flow down the leeward slope after crossing the mountains, the air is compressed and heated at a greater rate than it was cooled when it ascended the windward slope (• Fig. 5.21). Thus, the air enters the valley below as warm, dry winds. The rapid temperature rise brought about by such winds has been known to damage crops, increase forest-fire hazard, and set off avalanches.

An especially hot and dry wind is the **Santa Ana** of Southern California. It forms when high pressure develops over the interior desert regions of Southern California. The clockwise circulation of the high drives the air of the desert southwest over the mountains of eastern California, accentuating the dry conditions as the air moves down the western slopes. The hot, dry Santa Ana winds are notorious for fanning forest and brush fires, which plague the southwestern United States, especially in California.

Drainage Winds Also known as **katabatic winds, drainage winds** are local to mountainous regions and can occur only under calm, clear conditions. Cold, dense air will accumulate in a high valley, plateau, or snowfield within a mountainous area. Because the cold air is very dense, it tends to flow downward, escaping through passes and pouring out onto the land below. Drainage winds can be extremely cold and strong, especially when they result from cold air accumulating over ice sheets such as Greenland and Antarctica. These winds are known by many local names; for example, on the Adriatic coast, they are called the *bora;* in France, the *mistral;* and in Alaska, the *Taku.*

Land Breeze–Sea Breeze The **land breeze–sea breeze cycle** is a diurnal (daily) one in which the differential heating of land and water again plays a role (• Fig. 5.22). During the day, when the land—and consequently the air above it—is heated more quickly and to a higher temperature than the nearby ocean (sea or large lake), the air above the land expands and rises.

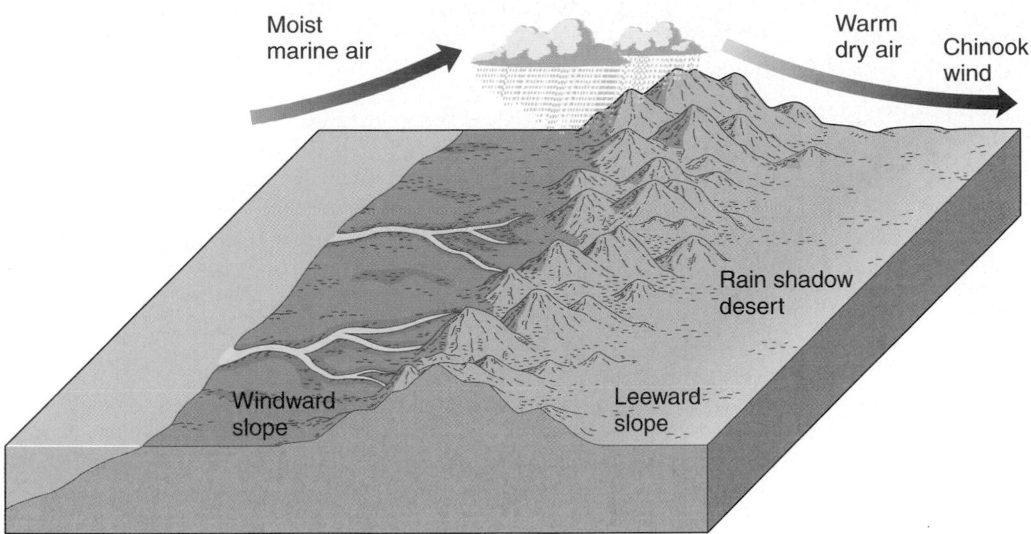

• **FIGURE 5.21**
Chinook (or foehn) winds result when air ascends a mountain barrier, becoming cooler as it expands and losing some of its moisture through condensation and precipitation. As the air descends the leeward side of the range, its relative humidity becomes lower as the air compresses and warms. This produces the relatively warm, dry conditions with which foehn winds are associated.
The term Chinook, a type of foehn wind, means "snow eater." Can you offer an explanation for how this name came about?

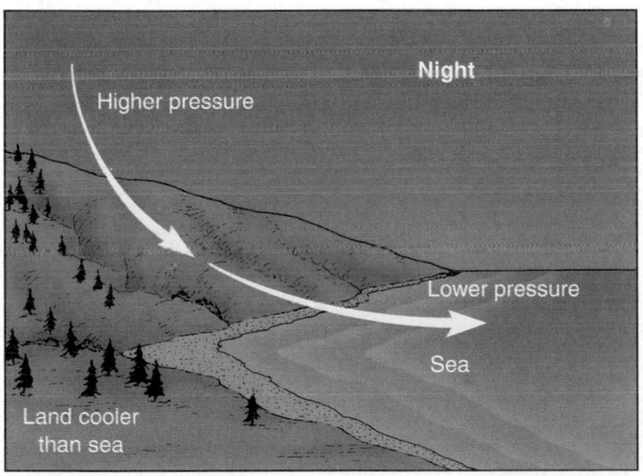

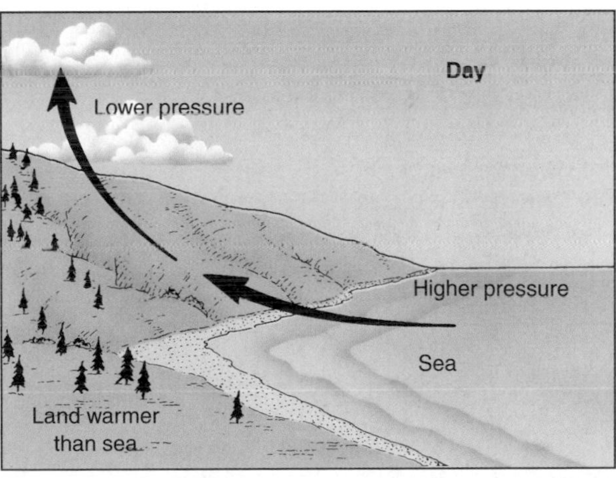

• **FIGURE 5.22**
Land and sea breezes. This day-to-night reversal of winds is a consequence of the different rates of heating and cooling of land and water areas. The land becomes warmer than the sea during the day and colder than the sea at night; the air flows from the cooler to the warmer area.
What is the impact on daytime coastal temperatures of the land and sea breeze?

This process creates a local area of low pressure, and the rising air is replaced by the denser, cooler air from over the ocean. Thus, a sea breeze of cool, moist air blows in over the land during the day. This sea breeze helps explain why seashores are so popular in summer; cooling winds help alleviate the heat. At times, however, sea breezes are responsible for afternoon cloud cover and light rain, spoiling an otherwise sunny day at the shore. These winds can mean a 5°C–9°C (9°F–16°F) reduction in temperature along the coast, as well as a lesser influence on land perhaps as far from the sea as 15–50 kilometers (9–30 mi). During hot summer days, such winds cool cities like Chicago, Milwaukee, and Los Angeles. At night, the land and the air above it cool more quickly and to a lower temperature than the nearby water body and the air above it. Consequently, the pressure builds higher over the land and air flows out toward the lower pressure over the water, creating a land breeze. For thousands of years, sailboats have left their coasts at dawn, when there is still a land breeze, and have returned with the sea breeze of the late afternoon.

GEOGRAPHY'S SPATIAL SCIENCE PERSPECTIVE

The Santa Ana Winds and Fire

Wildfires require three factors to occur: *oxygen, fuel,* and an *ignition source.* The conditions for all three factors vary geographically, so their spatial distributions are not equal everywhere. In locations where all three factors have the potential to exist, the danger from wildfires is high. Oxygen in the atmosphere is constant, but winds, which supply more oxygen as a fire consumes it, vary with location, weather, and terrain. High winds cause fires to spread faster and make them difficult to extinguish. Fuel in wild land fires is usually supplied by dry vegetative litter (leaves, branches, and dry annual grasses). Certain environments have more of this fuel than others. Dense vegetation tends to support the spread of fires. Growing vegetation can also become desiccated—dried out by transpiration losses during a drought or an annual dry season. In addition, once a fire becomes large, extreme heat in the areas where it is spreading causes vegetation along the edges of the burning area to lose its moisture through evaporation. Ignition sources are the means by which a fire is started. Lightning and human causes, such as campfires and trash fires, provide the main ignition sources for wildfires.

Southern California offers a regional example of how conditions related to these three factors combine with the local physical geography to create an environment that is conducive to wildfire hazard. This is also a region where many people live in forested or scrub-covered locales or along the urban–wild land fringe—areas that are very susceptible to fire. High pressure, warm weather, and low relative humidity dominate the Mediterranean climate of Southern California's coastal region for much of the year. When these conditions occur, this region experiences high fire potential because of the warm dry air and the vegetation that has dried out during the arid summer season.

The most dangerous circumstances for wildfires in Southern California occur when high winds are sweeping the region. When a strong cell of high pressure forms east of Southern California, the clockwise (anticyclonic) circulation directs winds from the north and east toward the coast. These warm, dry winds (called Santa Ana winds) blow down from nearby high-desert regions, becoming adiabatically warmer and drier as they descend into the coastal lowlands. The Santa Ana winds are most common in fall and winter, and wind speeds can be 50–90 kilometers per hour (30–50 mph) with stronger local wind gusts reaching 160 kilometers per hour (100 mph). Just like using a bellows or blowing on a campfire to get it started, the Santa Ana winds produce fire weather that can cause the spread of a wildfire to be extremely rapid after ignition. Most people take great care during these times to avoid or strictly control any activities that could cause a fire to start, but occasionally accidents, acts of arson, or lightning strikes ignite a wildfire. Given the physical geography of the Los Angeles region, when the Santa Ana winds are blowing, the fire danger is especially extreme.

Ironically, although the Santa Ana winds create dangerous fire conditions, they also provide some benefits to local residents because the winds tend to blow air pollutants offshore and out of the urban region. In addition, because they are strong winds flowing opposite to the direction of ocean waves, experienced surfers can enjoy higher than normal waves during those periods when Santa Ana winds are present.

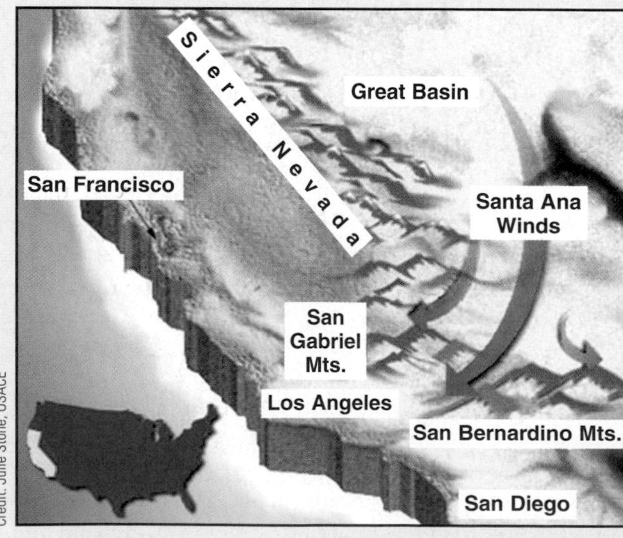

Credit: Julie Stone, USACE

Geographic setting and wind direction for Santa Ana winds.

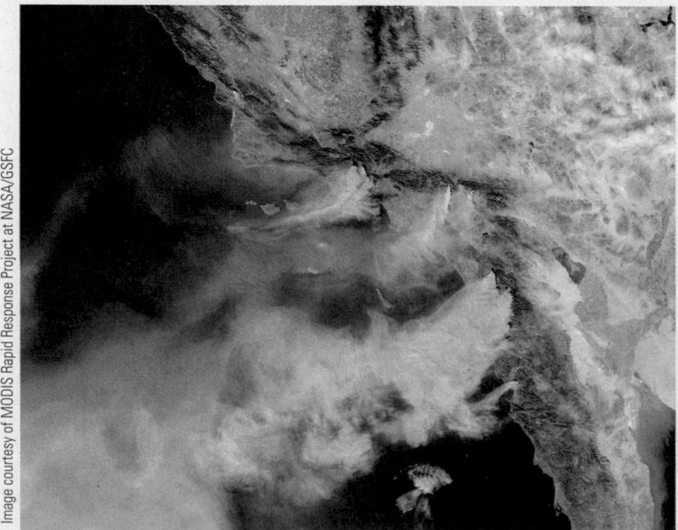

Image courtesy of MODIS Rapid Response Project at NASA/GSFC

This satellite image shows strong Santa Ana winds from the northeast fanning wildfires in Southern California and blowing the smoke offshore for many kilometers.

because there is little land poleward of 40°S, the West Wind Drift (or Antarctic Circumpolar Drift) circles Earth as a cool current across all three major oceans almost without interruption. It is cooled by the influence of the Antarctic ice sheet.

In general then, warm currents move poleward as they carry tropical waters into the cooler waters of higher latitudes, as in the case of the Gulf Stream or the Brazil Current. Cool currents deflect water equatorward, as in the California Current and the Humboldt Current. Warm currents tend to have a humidifying and warming effect on the east coasts of continents along which they flow, whereas cool currents tend to have a drying and cooling effect on the west coasts of the landmasses. The contact between the atmosphere and ocean currents is one reason why subtropical highs have a strong side and a weak side. Subtropical highs on the west coast of continents are in contact with cold ocean currents, which cool the air and make the eastern side of a subtropical high more stable and stronger. On the east coasts of continents, contacts with warm ocean currents cause the western sides of subtropical highs to be less stable and weaker.

The general circulation just described is consistent throughout the year, although the position of the currents follows seasonal shifts in atmospheric circulation. In addition, in the North Indian Ocean, the direction of circulation reverses seasonally according to the monsoon winds.

The cold currents along west coasts in subtropical latitudes are frequently reinforced by **upwelling.** As the trade winds in these latitudes drive the surface waters offshore, the wind's frictional drag on the ocean surface displaces the water to the west. As surface waters are dragged away, deeper, colder water rises to the surface to replace them. This upwelling of cold waters adds to the strength and effect of the California, Humboldt (Peru), Canary, and Benguela Currents.

El Niño

As you can see in Figure 5.25, the cold Humboldt Current flows equatorward along the coasts of Ecuador and Peru. When the current approaches the equator, the westward-flowing trade winds cause upwelling of nutrient-rich cold water along the coast. Fishing, especially for anchovies, is a major local industry.

Every year usually during the months of November and December, a weak warm countercurrent replaces the normally cold coastal waters. Without the upwelling of nutrients from below to feed the fish, fishing comes to a standstill. Fishermen in this region have known of the phenomenon for hundreds of years. In fact, this is the time of year they traditionally set aside to tend to their equipment and await the return of cold water. The residents of the region have given this phenomenon the name **El Niño,** which is Spanish for "The Child," because it occurs about the time of the celebration of the birth of the Christ Child.

The warm-water current usually lasts for 2 months or less, but occasionally the disruption to the normal flow lasts for many months. In these situations, water temperatures are raised not just along the coast but for thousands of kilometers offshore (● Fig. 5.26). Over the past decade, the term *El Niño* has come to describe these exceptionally strong episodes and not the annual event. During the past 50 years, approximately 18 years qualify as having El Niño conditions (with sea-surface temperatures 0.5°C higher, or warmer, than normal for 6 consecutive months).

● FIGURE 5.26

These enhanced satellite images show a significant El Niño (left) and La Niña (right) episodes in the Tropical Pacific. The red and white shades display the warmer sea surface temperatures, while the blues and purples mark areas of cooler temperatures.

From what continent does an El Niño originate?

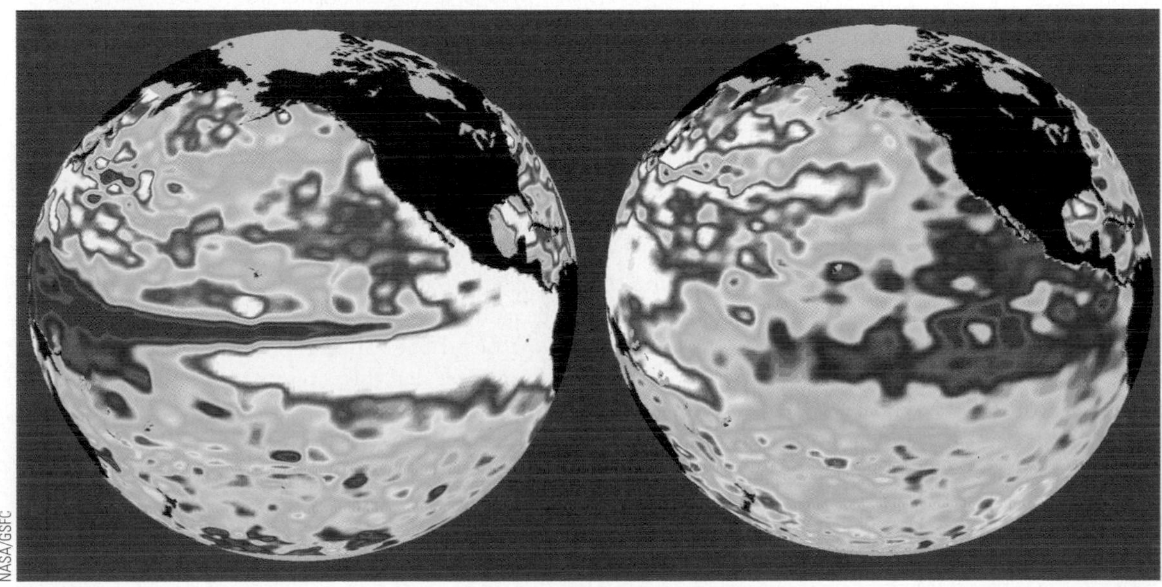

Not only do the El Niños affect the temperature of the equatorial Pacific, but the strongest of them also impact worldwide weather.

El Niño and the Southern Oscillation

To completely understand the processes that interact to produce an El Niño requires that we study conditions all across the Pacific, not just in the waters off South America. In the 1920s, Sir Gilbert Walker, a British scientist, discovered a connection between surface-pressure readings at weather stations on the eastern and western sides of the Pacific. He noted that a rise in pressure in the eastern Pacific is usually accompanied by a fall in pressure in the western Pacific and vice versa. He called this seesaw pattern the **Southern Oscillation.** The link between El Niño and the Southern Oscillation is so great that they are often referred to jointly as ENSO (*El Niño/Southern Oscillation*). These days the atmospheric pressure values from Darwin, Australia, are compared to those recorded on the Island of Tahiti, and the relationship between these two values defines the Southern Oscillation.

During a typical year, the eastern Pacific has a higher pressure than the western Pacific. This east-to-west pressure gradient enhances the trade winds over the equatorial Pacific waters. This results in a surface current that moves from east to west at the equator. The western Pacific develops a thick, warm layer of water while the eastern Pacific has the cold Humboldt Current enhanced by upwelling.

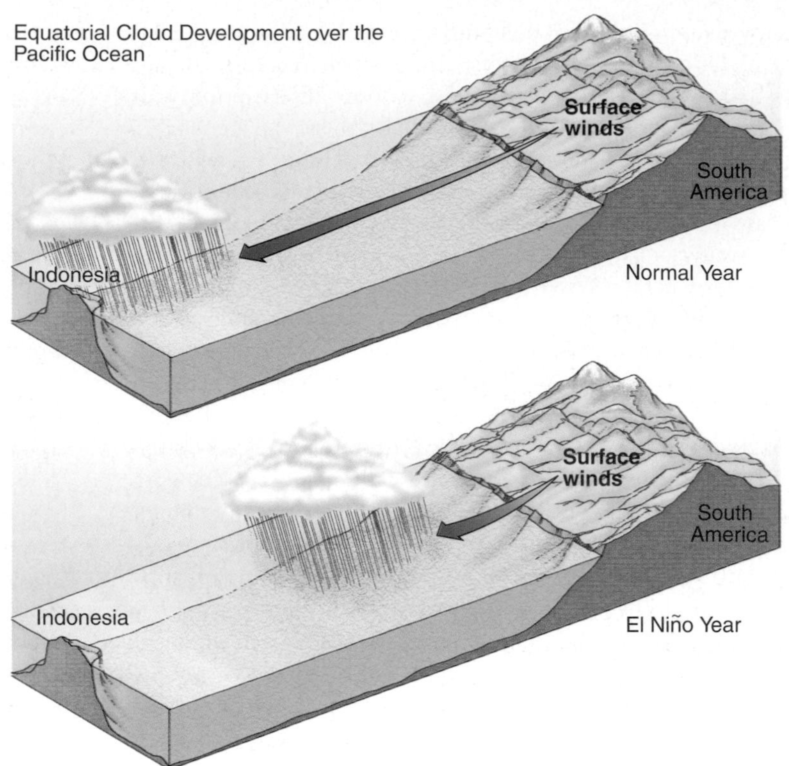

• **FIGURE 5.27**
During El Niño, the easterly surface winds weaken and retreat to the eastern Pacific, allowing the central Pacific to warm and the rain area to migrate eastward.
Near what country or countries does El Nino begin?

Then, for unknown reasons, the Southern Oscillation swings in the opposite direction, dramatically changing the usual conditions described above, with pressure increasing in the western Pacific and decreasing in the eastern Pacific. This change in the pressure gradient causes the trade winds to weaken or, in some cases, to reverse. This causes the warm water in the western Pacific to flow eastward, increasing sea-surface temperatures in the central and eastern Pacific. This eastward shift signals the beginning of El Niño.

In contrast, at times and for reasons we do not fully know, the trade winds will intensify. These more powerful trade winds will cause even stronger upwelling than usual to occur. As a result, sea-surface temperatures will be even colder than normal. This condition is known as **La Niña** (in Spanish, "Little Girl," but scientifically simply the opposite of El Niño). La Niña episodes will at times, but not always, bring about the opposite effects of an El Niño episode (see again Fig. 5.26).

El Niño and Global Weather

Cold ocean waters impede cloud formation. Thus, under normal conditions, clouds tend to develop over the warm waters of the western Pacific but not over the cold waters of the eastern Pacific. However, during an El Niño, when warm water migrates eastward, clouds develop over the entire equatorial region of the Pacific (• Fig. 5.27). These clouds can build to heights of 18,000 meters (59,000 ft). Clouds of this magnitude can disrupt the high-altitude wind flow above the equator. As we have seen, a change in the upper air wind flow in one portion of the atmosphere will trigger wind flow changes in other portions of the atmosphere. Alterations in the upper air winds result in alterations to surface weather.

Scientists have tried to document as many past El Niño events as possible by piecing together bits of historical evidence, such as sea-surface temperature records, daily observations of atmospheric pressure and rainfall, fisheries' records from South America, and even the writings of Spanish colonists living along the coasts of Peru and Ecuador dating back to the 15th century. Additional evidence comes from the growth patterns of coral and trees in the region. Researchers are constantly discovering new techniques to identify El Niños through history.

Based on this historical evidence, we know that El Niños have occurred as far back as records go. One disturbing fact is that they appear to be occurring more often. Records indicate that during the 16th century, an El Niño occurred, on average, every 6 years. Evidence gathered over the past few decades indicates that El Niños are now occurring, on average, every 2.2 years. Even more alarming is the fact that they appear to be getting stronger. The record-setting El Niño of 1982–1983 was recently surpassed by the one in 1997–1998.

The 1997–1998 El Niño brought copious and damaging rainfall to the southern United States, from California to Florida. Snowstorms in the northeast portion of the United States were more frequent and stronger than in most years. The warm El Niño winters fueled Hurricane Linda, which devastated the western coast of Mexico. Linda was the strongest hurricane ever recorded in the eastern Pacific.

In recent years, scientists have become better able to monitor and forecast El Niño and La Niña events. An elaborate network of ocean-anchored weather buoys plus satellite observations provide an enormous amount of data that can be analyzed by computer to help predict the formation and strength of El Niño and La Niña events.

North Atlantic Oscillation

Our improved observation skills have led to the discovery of the **North Atlantic Oscillation (NAO)**—a relationship between the Azores (subtropical) High and the Icelandic (subpolar) Low. The east-to-west, seesaw motion of the Icelandic Low and the Azores High control the strength of the westerly winds and the direction of storm tracks across the North Atlantic. There are two recognizable phases associated with the established NAO index.

A positive NAO index phase is identified by higher than average pressure in the Azores High and lower than average pressure in the Icelandic Low. The increased pressure difference between the two systems results in stronger winter storms, occurring more often and following a more northerly track (● Fig. 5.28a). This promotes warm and wet winters in Europe, but cold, dry winters in Canada and Greenland. The eastern United States may experience a mild and wet winter. The negative NAO index phase occurs with a weak Azores High and higher pressure in the Icelandic Low. The smaller pressure gradient between these two systems will weaken the westerlies resulting in fewer and weaker winter storms (Fig. 5.28b). Northern Europe will experience cold air with moist air moving into the Mediterranean. The East Coast of the United States will experience more cold air and snowy winters. This index varies from year to year but also has a tendency to stay in one phase for periods lasting several years in a row.

The North Atlantic Oscillation (NAO) is not as well understood as ENSO. Truly, both oscillations require more research in the future if scientists are to better understand how these ocean phenomena affect weather and climate. Will scientists ever be able to predict the occurrence of such phenomena as ENSO or the NAO? No one can answer that question, but as our technology improves, our forecasting ability will also increase. We have made tremendous progress: In the past few decades, we have come to recognize the close association between the atmosphere and hydrosphere as well as to better understand the complex relationship between these Earth systems.

This chapter began with an examination of the behavior of atmospheric gases as they respond to solar radiation and other dynamic forces. This information enabled a definition and thorough discussion of global pressure systems and their accompanying winds. This discussion in turn permitted a description of atmospheric circulation patterns on the global and subglobal scale. Once again, we can recognize the interactions among Earth's systems. Earth's radiation budget helps create movements in our atmosphere, which in turn help drive ocean circulation, which in turn creates feedback with the atmosphere:

Solar radiation → Atmosphere → Hydrosphere →
Back to the atmosphere

In following chapters, we will examine the role of the atmospheric systems in controlling variations in weather and climate and, later, weather and climate systems as they affect surface landforms.

● FIGURE 5.28
Positions of the pressure systems and winds involved with the (a) positive and (b) negative phases of the North Atlantic Oscillation (NAO).
Which two pressure systems are used to establish the NAO phases?

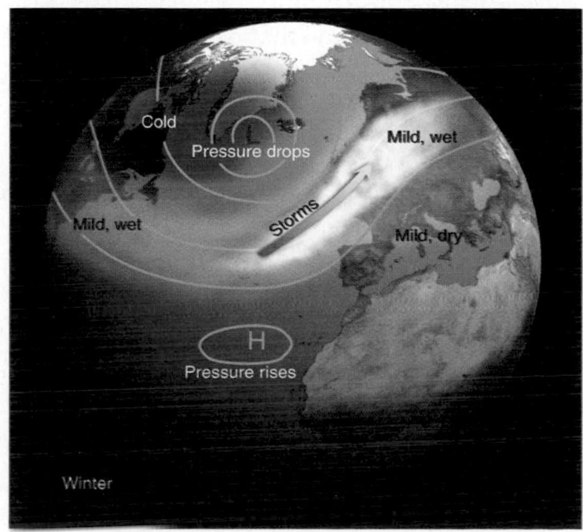

(a) Positive phase

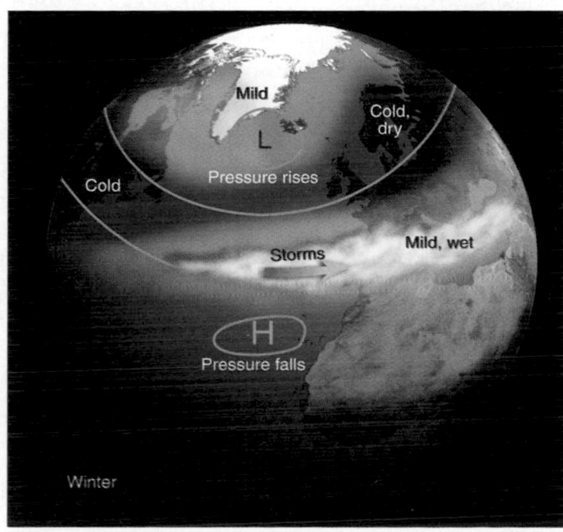

(b) Negative phase

Chapter 5 Activities

Define & Recall

cyclone (low)
anticyclone (high)
convergent wind circulation
divergent wind circulation
isobar
pressure gradient
wind
Coriolis effect
friction
geostrophic wind
windward
leeward
prevailing wind
trough
equatorial low (equatorial trough)
subtropical high
subpolar low
polar high
Siberian High
Canadian High
Icelandic Low
Aleutian Low
Pacific High
Bermuda High (Azores High)
polar easterlies
westerlies
trade winds
northeast trades
southeast trades
tropical easterlies
doldrums
intertropical convergence zone (ITCZ)
polar front
jet stream
polar front jet stream
subtropical jet stream
Rossby wave
monsoon
Chinook
foehn
Santa Ana
drainage wind (katabatic wind)
land breeze–sea breeze cycle
mountain breeze–valley breeze cycle
gyre
upwelling
El Niño
Southern Oscillation
La Niña
North Atlantic Oscillation (NAO)

Discuss & Review

1. What is atmospheric pressure at sea level? How do you suppose Earth's gravity is related to atmospheric pressure?
2. Horizontal variations in air pressure are caused by thermal or dynamic factors. How do these two factors differ?
3. What kind of pressure (high or low) would you expect to find in the center of an anticyclone? Describe and diagram the wind patterns of anticyclones in the Northern and Southern Hemispheres.
4. What is the circulation pattern around a center of low pressure (cyclone) in the Northern Hemisphere? In the Southern Hemisphere? Draw diagrams to illustrate these circulation patterns.
5. Explain how water and land surfaces affect the pressure overhead during summer and winter. How does this relate to the afternoon sea breeze?
6. How do landmasses affect the development of belts of atmospheric pressure over Earth's surface?
7. Why do Earth's wind systems and pressure belts migrate with the seasons?
8. How are the land breeze–sea breeze and monsoon circulations similar? How are they different?
9. What effect on valley farms could a strong drainage wind have?
10. What are monsoons? What causes them? Name some nations that are concerned with the arrival of the "wet monsoon."
11. What is the relationship between ocean currents and global surface wind systems? How does the gyre in the Northern Hemisphere differ from the one in the Southern Hemisphere?
12. Where are the major warm and cool ocean currents located in respect to Earth's continents? Which currents have the greatest effect on North America?
13. What is an El Niño? What are some impacts that it has on global weather?

Consider & Respond

1. Look at the January (Fig. 5.11a) and July (Fig. 5.11b) maps of average sea-level pressure. Answer the following questions:
   a. Why is the subtropical high pressure belt more continuous (linear, not cellular) in the Southern Hemisphere than in the Northern Hemisphere in July?
   b. During July, what area of the United States exhibits the lowest average pressure? Why?
2. How did the trade winds of the Atlantic Ocean affect the sailing routes of early traders?
3. Describe the movements of the upper air. How have pilots applied their experience of the upper air to their flying patterns?
4. Is the polar front jet stream stronger in the summer or the winter? Why?
5. What effect would foehn-type winds have on farming, forestry, and ski resorts?

Apply & Learn

1. The amount of power that can be generated by wind is determined by the equation

$$p = \tfrac{1}{2}D \times S^3$$

where P is the power in watts, D is the density, and S is the wind speed in meters per second (m/sec). Because $D = 1.293\ \text{kg/m}^3$, we can rewrite the equation as

$$P = 0.65 \times S^3$$

   a. How much power (in watts) is generated by the following wind speeds: 2 meters per second, 6 meters per second, 10 meters per second, 12 meters per second?
   b. Because wind power increases significantly with increased wind speed, very windy areas are ideal locations for "wind farms." Cities A and B both have average wind speeds of 6 meters per second. However, city A tends to have very consistent winds; in city B, half of its winds tend to be at 2 meters per second and the other half at 10 meters per second. Which site would be the better location for a wind generation plant?
2. Atmospheric pressure decreases at the rate of 0.036 millibar per foot as one ascends through the lower portion of the atmosphere.
   a. The Sears Tower in Chicago, Illinois, is one of the world's tallest buildings at 1450 feet. If the street-level pressure is 1020.4 millibars, what is the pressure at the top of the Sears Tower?
   b. If the difference in atmospheric pressure between the top and ground floor of an office building is 13.5 millibars, how tall is the building?
   c. A single story of a building is 12 feet. You enter an elevator on the top floor of the building and want to descend five floors. The elevator has no floor markings—only a barometer. If the initial reading was 1003.2 millibars, at what pressure reading would you want to get off?

Moisture, Condensation, and Precipitation

6

CHAPTER PREVIEW

The hydrologic cycle involves the circulation of water throughout all the major Earth spheres; it is therefore fundamental to the nature and operation of the entire Earth system.

- How does the hydrologic cycle involve all of Earth's spheres?
- What are the major stages of the hydrologic cycle?

Although water may evaporate from all Earth surfaces and transpiration may add considerable moisture to the air, the oceans are the most important source of water vapor in the atmosphere.

- What portion (latitudes) of the oceans would have the highest evaporation rates, and why?
- What times of year would evaporation be at its greatest?

There is only one way for significant condensation, cloud formation, and precipitation to occur: Air must be forced to rise so that sufficient adiabatic cooling will take place.

- Why does adiabatic cooling take place in rising air?
- Why might the rate of cooling differ in wet and dry air?

The geographic distribution of precipitation over Earth can be explained by either one or a combination of the following: frontal, cyclonic, orographic, or convectional precipitation.

- Which is most important in your community?
- Which is most closely related to landforms?

As a general rule, the less precipitation a region receives, the greater will be the variability of precipitation in that region from year to year.

- What is the effect of this variability on human beings?
- How might this rule affect a rainforest or a desert?

◀ **Opposite: This large thunderstorm demonstrates an important and violent transfer of moisture from the atmosphere to Earth's surface.**
NOAA/NWS/Greg Lundeen

Water is vital to all life on Earth. Although some living organisms can survive without air, nothing can survive without water. Water is necessary for photosynthesis, cell growth, protein formation, soil formation, and the absorption of nutrients by plants and animals.

Water affects Earth's surface in innumerable ways. The structure of the water molecule is such that water can dissolve an enormous number of substances—so many in fact that it has been called the *universal solvent.* Because water acts as a solvent for so many substances, it is almost never found in a pure state. Even rainwater is filled with impurities picked up in the atmosphere. Indeed, without these impurities to condense around, neither clouds nor precipitation could occur. In addition, rainwater usually contains some dissolved carbon dioxide from the air. Therefore, rain is a very weak form of carbonic acid. We will see later (Chapter 16) that this fact affects how water shapes certain landforms. The weak acidity of rainwater should not be confused with the environmentally damaging *acid rain,* which is at least ten times more acidic.

Not only can water dissolve and transport many minerals, it also can transport solid particles in suspension. These characteristics make water a unique transportation system for Earth. Water supplies nutrients that would not otherwise be available to plants. Water carries minerals and nutrients down streams, through the soil, through the openings in subsurface rocks, and through plants and animals. It deposits solid matter on stream floodplains, in river deltas, and on the ocean floor.

The surface tension of water and the behavior of water molecules to draw together make possible **capillary action**—the ability of water to pull itself upward through small openings against gravity. Capillary action also permits transport of dissolved material in an upward direction—that is, against the pull of gravity. Capillary action moves water into the stems and leaves of plants—even to the topmost needles of the great California redwoods and the top leaves of the rainforest trees. Capillary action is also important in the movement of blood through our bodies. Without it, many of our cells could not receive the necessary nutrients carried by the blood.

Another important and highly unusual property of water is that it expands when it freezes. Most substances contract when cooled and expand when heated. Water follows these rules until it is cooled below 4°C (39°F); then it begins to expand. Ice is therefore less dense than water and consequently will float on water, as do ice floes and icebergs (● Fig. 6.1).

Finally, compared with solids, water is slow to heat and slow to cool. Therefore, as we learned in Chapter 4, large bodies of water on Earth act as reservoirs of heat during winter and have a cooling effect in summer. This moderating effect on temperature can be seen in the vicinity of lakes as well as on seacoasts.

Earth's water—the *hydrosphere* (from Latin: *hydros,* water)—is found in all three states: as a liquid in rivers, lakes, oceans, and rain; as a solid in the form of snow and ice; and as a vapor in our atmosphere. Even the water temporarily stored in living things can be considered part of the hydrosphere. About 73% of Earth's surface is covered by water, with the largest proportion contained within the world's oceans (● Fig. 6.2). In all, the total water content of the Earth system, whether liquid, solid, or vapor, is about 1.33 billion cubic kilometers (326 million cu mi). Although water cycles in and out of the atmosphere, lithosphere, and biosphere, the total amount of water in the hydrosphere remains constant (● Fig. 6.3).

● FIGURE 6.1
Huge icebergs in the ocean near Antarctica float like the ice in your beverage glass.
If the ice floating in a beverage glass melts completely before you can drink from it, will the liquid level rise, fall, or remain the same as before? Why?

M. Trapasso

The Hydrologic Cycle

The circulation of water from one part of the overall Earth system to another is known as the **hydrologic cycle.** The air contains water vapor that has entered the atmosphere through evaporation from Earth's surface. When water vapor condenses and falls as precipitation, several things may happen to it. First, it may go directly into a body of water where it is immediately available for evaporation back into the atmosphere. Alternatively, it may fall onto the land where it may run off the surface to form streams, ponds, or lakes. Or it may be absorbed into the ground where it can either be contained by the soil or flow through open spaces, called *interstices,* that exist in loose sand, gravel, silt, and voids in solid rock. Ultimately, much of the water in or on Earth's surface reaches the oceans. Some water that reaches the surface as snow becomes a part of massive ice over Greenland and Antarctica as well as in high mountain glaciers, while some snow remains frozen until spring thaw and then reenters the hydrologic cycle as a liquid. Other water is used by plants and animals and temporarily becomes a part of living things. Thus, there are six storage areas for water in the hydrologic cycle: the atmosphere, the oceans, bodies of fresh water on the surface, plants and animals, open spaces beneath Earth's surface, and glacial ice.

Liquid water is returned to the atmosphere as a gas through evaporation. Water evaporates from all bodies of water on Earth, from plants and animals, and from soils; it can even evaporate from falling precipitation. Once the water is an atmospheric gas again, the cycle can be repeated.

The hydrologic cycle is basically one of evaporation, condensation, and precipitation, and to some extent transportation because these processes are in constant motion. The hydrologic cycle is a *closed system;* that is, water may appear in the system in all three major states—as a liquid, vapor, or solid—and may be transferred from one state to another, but there is no gain or loss of water by the system. This is unlike the radiation budget depicted in Figure 4.11, which is an *open system* where energy can flow both into and out of the system. Although the hydrologic cycle is a closed system, it is not static but exceedingly dynamic. The percentage of water associated with any one component of the system changes constantly over time and place. For example, during the last ice age, evaporation and precipitation were greatly reduced. Also, some changes are human induced; the cutting down of a forest or the damming of a river will cause adjustments among the components.

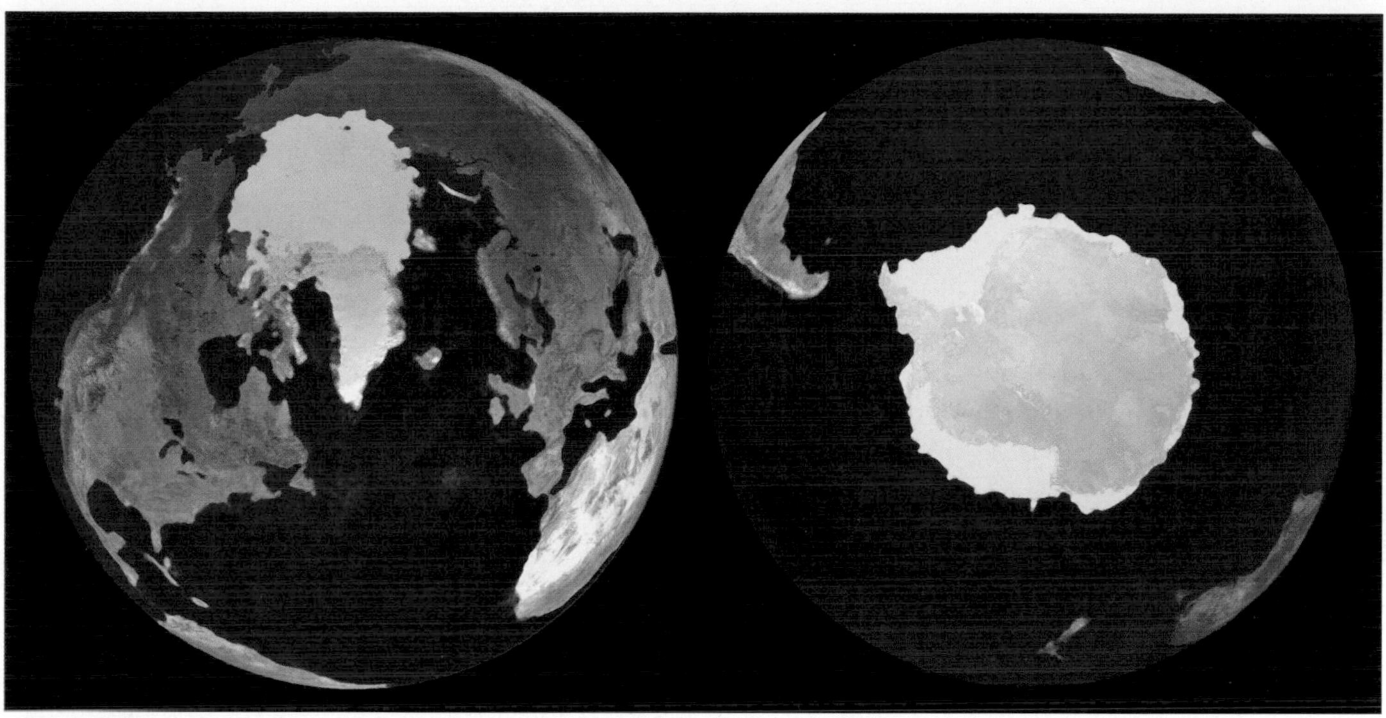

• **FIGURE 6.2**
Most of Earth's fresh water is locked in glacial ice, as seen in this view of the Northern and Southern Hemispheres.
Can you distinguish between the Greenland and Antarctic glaciers and the seasonal (pack ice and ice shelves) that float on the ocean surface?

• **FIGURE 6.3**
This illustration of Earth's water sources emphasizes that the vast majority of water in the hydrosphere is salt water, stored in the world's oceans. The bulk of the supply of fresh water is relatively unavailable because it is stored in polar ice sheets.
How might global warming or cooling alter this figure?

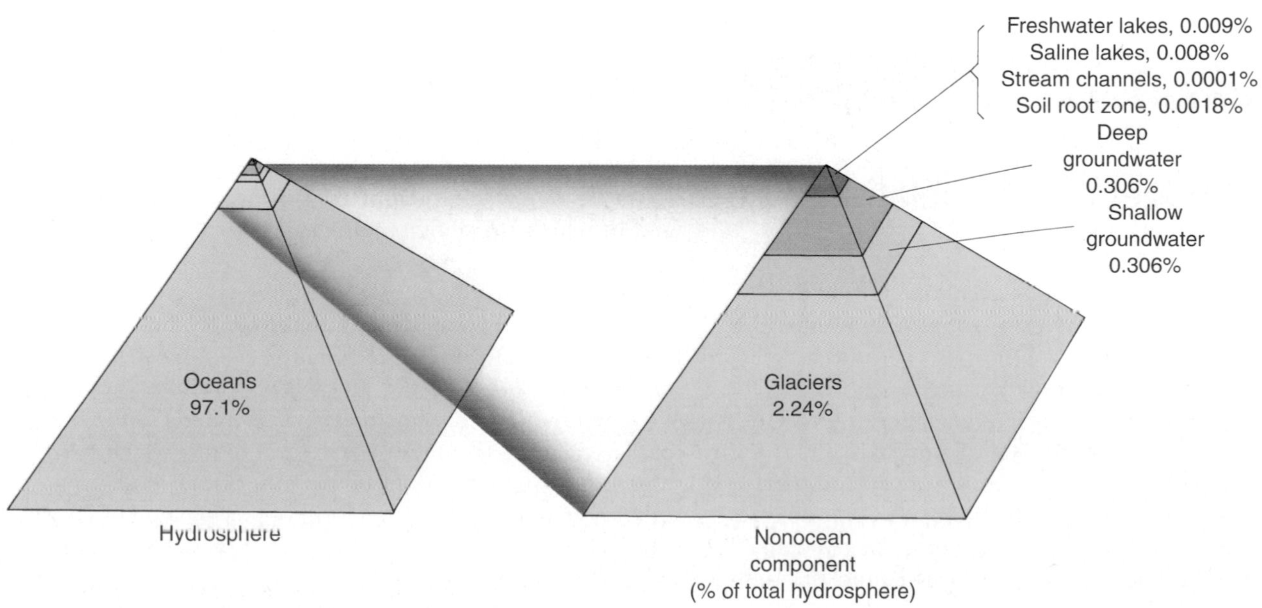

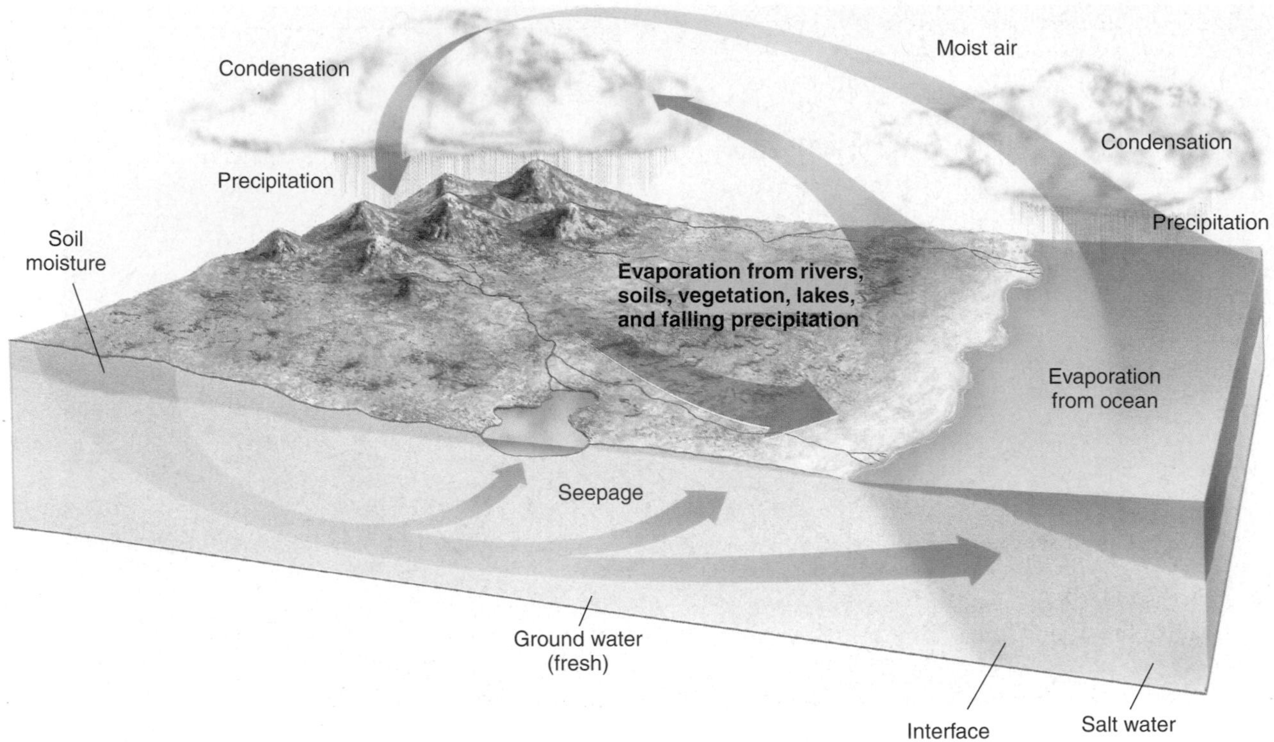

• **FIGURE 6.4**
Environmental Systems: The Hydrologic System The hydrologic system is concerned with the circulation of water from one part of the Earth system to another. The subsystem of the hydrologic system illustrated in this diagram is referred to as the hydrologic cycle. Largely through condensation, precipitation, and evaporation, water is cycled endlessly between the atmosphere, the soil, subsurface storage, lakes and streams, plants and animals, glacial ice, and the principal reservoir—the oceans.
Is this a closed system or an open system? Why?

The hydrologic cycle is one of the most important subsystems of the larger Earth system. It is linked to numerous other subsystems that rely on water as an agent of movement. For example, it plays a major role in the redistribution of energy over Earth's surface. (• Fig. 6.4) provides a schematic illustration of the circulation of water in the hydrologic cycle.

Water in the Atmosphere

The Water Budget and Its Relation to the Heat Budget

We are most familiar with and most often take notice of water in its liquid form, as it pours from a tap or as it exists in fine droplets within clouds or fog. Water also exists as a tasteless, odorless, transparent gas known as water vapor, which is mixed with the other gases of the atmosphere in varying proportions. Water vapor is found within approximately the first 5500 meters (18,000 ft) of the troposphere and makes up a small but highly variable percentage of the atmosphere by volume. Atmospheric water is the source of all condensation and precipitation. Through these processes and through evaporation, water plays a significant role as Earth's temperature regulator and modifier. In addition, as we noted in Chapter 4, water vapor in the atmosphere absorbs and reflects a significant portion of both incoming solar energy and outgoing Earth radiation. By preventing great losses of heat from Earth's surface, water vapor helps maintain the moderate range of temperature found on this planet.

As we stated previously, Earth's hydrosphere is a *closed system;* that is, water is neither received from outside the Earth system nor given off from it. Thus, an increase in water within one subsystem must be accounted for by a loss in another. Put another way, we say that the Earth system operates with a *water budget,* in which the total quantity of water remains the same and in which the deficits must balance the gains throughout the entire system.

We know that the atmosphere gives up a great deal of water, most obviously by condensation into clouds, fog, and dew and through several forms of precipitation (rain, snow, hail, sleet). If the quantity of water in the atmosphere remains at the same level through time, the atmosphere must be absorbing from other parts of the system an amount of water equal to that which it is giving up. During 1 minute, over 1 billion tons of water are given up by the atmosphere through some form of precipitation or condensation, while another billion tons are evaporated and absorbed as water vapor by the atmosphere.

If we look again at our discussion of the heat energy budget in Chapter 4, we can see that a part of that budget is the latent heat of condensation. Of course, this energy is originally derived from the sun. The sun's energy is used in evaporation and is then stored in the molecules of water vapor, to be released only during condensation. Although the heat transfers involved in evaporation and condensation within the total heat energy budget are proportionately small, the actual energy is significant. Imagine the amount of energy released every minute when a billion tons of water condense out of the atmosphere. It is this vast storehouse of energy, the latent heat of condensation, that provides the major source of power for Earth's storms: hurricanes, tornadoes, and thunderstorms.

There are limits to the amount of water vapor that can be held by any parcel of air. A very important determinant of the amount of water vapor that can be held by the air is temperature. The warmer air is, the greater the quantity of water vapor it can hold. Therefore, we can make a generalization that air in the polar regions can hold far less water vapor (approximately 0.2% by volume) than the hot air of the tropics and equatorial regions of Earth, where the air can contain as much as 5% by volume.

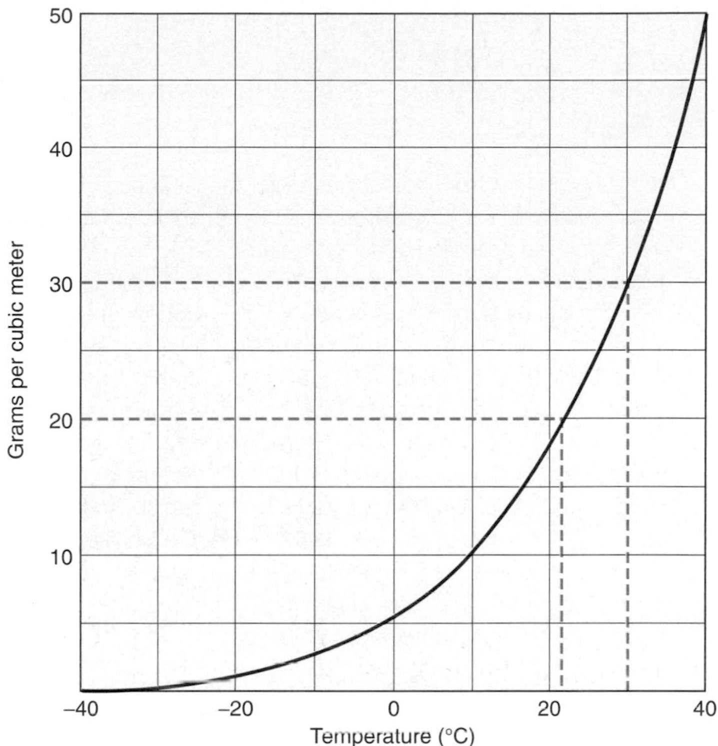

• FIGURE 6.5
The graph shows the maximum amount of water vapor that can be contained in a cubic meter of air over a wide range of temperatures.
Compare the change in capacity when the air temperature is raised from 0°C to 10°C with a change from 20°C to 30°C. What does this indicate about the relationship between temperature and capacity?

Saturation and Dew Point

When air of a given temperature holds all the water vapor that it possibly can, it is said to be in a state of **saturation** and to have reached its **capacity.** If a constant temperature is maintained in a quantity of air, there will come a point, as more water vapor is added, when the air will be saturated and unable to hold any more water vapor. For example, when you take a shower, the air in the room becomes increasingly humid until a point is reached at which the air cannot contain more water. Then, excess water vapor condenses onto the colder mirrors and walls.

We know that the capacity of air to hold water vapor varies with temperature. In fact, as we can see in (• Fig. 6.5), this capacity of air to contain moisture increases with rising temperatures. Some examples will help illustrate the relationship between temperature and water vapor capacity. If we assume that a parcel of air at 30°C is saturated, then it will contain 30 grams of water vapor in each cubic meter of air (30 g/m^3). Now suppose the temperature of the air increases to 40°C *without* increasing the water vapor content. The parcel is no longer saturated because air at 40°C can hold more than 30 grams per cubic meter of water vapor (actually, 50 g/m^3). Conversely, if we decrease the temperature of saturated air from 30°C (which contains 30 g/m^3 of water vapor) to 20°C (which has a water vapor capacity of only 17 g/m^3), some 13 grams of the water vapor will condense out of the air because of the reduced capacity.

It is also evident that if an unsaturated parcel of air is cooled, it will eventually reach a temperature at which the air will become saturated. This critical temperature is known as the **dew point**—the temperature at which condensation takes place. For example, we know that if a parcel of air at 30°C contains 20 grams per cubic meter of water vapor, it is not saturated because it can hold 30 grams per cubic meter. However, if we cool that parcel of air to 21°C, it would become saturated because the capacity of air at 21°C is 20 grams per cubic meter. Thus, that parcel of air at 30°C has a dew point of 21°C. It is the cooling of air to below its *dew point temperature* that brings about the condensation that must precede any precipitation.

Because the capacity of air to hold water vapor increases with rising temperatures, air in the equatorial regions has a higher dew point than air in the polar regions. Thus, because the atmosphere can hold more water in the equatorial regions, there is greater potential for large quantities of precipitation than in the polar regions. Likewise, in the middle latitudes, summer months, because of their higher temperatures, have more potential for large-scale precipitation than do winter months.

Humidity

The amount of water vapor in the air at any one time and place is called **humidity.** There are three common ways to express the humidity content of the air. Each method provides information that contributes to our discussion of weather and climate.

Absolute and Specific Humidity

Absolute humidity is the measure of the mass of water vapor that exists within a given volume of air. It is expressed either in the metric system as the number of grams per cubic meter (g/m^3) or in the English system as grains per cubic foot (gr/ft^3). **Specific humidity** is the mass of

GEOGRAPHY'S SPATIAL SCIENCE PERSPECTIVE

The Wettest and Driest Places on Earth

Though there may be disagreement as to which exact locations on Earth hold the distinction of being the wettest and driest, there are two geographical locations that lay claim to these titles.

Mount Waialeale (Hawaiian: for rippling waters) stands at 1569 meters (5148 ft) above sea level and is the second-highest peak on the island of Kauai in the Hawaiian Island Chain. This location averages 11.68 meters (460 in.) of rain per year, with a record of 17.34 meters (683 in.) in 1982.

There are several reasons why this peak consistently accumulates huge amounts of rainfall: (1) Being northernmost of the inhabited Hawaiian Islands allows some exposure to winter time frontal systems; (2) its shape is nearly conical and thus all sides are exposed to the moisture-laden winds; and (3) its very steep cliffs cause the humid air to rise dramatically—over 1000 meters (3000 ft) in less than 1/2 mile distance. This allows the rainfall to concentrate more on one spot instead of being spread out over a gentler mountain slope.

Though some frontal activity brings about winter rainfall, most of it comes in the form of orographic precipitation. The trade winds flow mainly out of the east, but any direction of the winds will waft the moisture-laden ocean air up the sides of this mountain. The precipitation falls mainly in the form of light rain, and drizzle. The area around this mountain peak is perennially covered in fog and mist. The rainfall is rarely heavy, but it is consistent. The small amounts of rain each day accumulate through the year to reach these huge totals.

The driest spot on Earth is located in northern Chile, South America. Two locations can claim some fame in this discussion. Arica, Chile, claims the lowest average annual precipitation at 0.5 millimeters (considered a *trace*) of precipitation. However the location shown in the image below is located between Arica and the city of Iquique, Chile. This valley nested in the Andes mountain range has not received rain according to anyone's living memory. No data has been collected, because as far as anyone knows, it has never rained there. No doubt, if it ever rains at this location, it will be recorded and likely to be celebrated as well! This location contains no vegetation and is likely not to contain many living microbes either.

Northern Chile is dominated by two mechanisms that keep it rain free. The first is the dominance of a subtropical high pressure system. This semipermanent anticyclone with its stabilizing force remains in close proximity to this location for long periods of time. Second, this valley is on the leeward (rain shadow) side of the Andes Mountains. Thus, with both of these processes working together, this location simply does not receive precipitation.

Although both the wettest and driest locations are unique in their own ways, by applying our knowledge of pressure systems, winds, and precipitation processes (Chapters 5 and 6) it is not difficult to understand how these geographical distinctions can exist.

M. Trapasso

The wettest place on Earth is Mt. Waialeale on the Hawaiian island of Kauai.

M. Trapasso

The driest spot on Earth is a valley in northern Chile.

water vapor (given in grams) per mass of air (given in kilograms). Obviously, both are measures of the actual amount of water vapor in the air. Because most water vapor gets into the air through the evaporation of water from Earth's surface, it stands to reason that absolute and specific humidity will decrease with height from Earth.

We have also learned that air is compressed as it sinks and expands as it rises. Thus, a given parcel of air changes its volume as it moves vertically, but there may be no change in the amount of water vapor in that quantity of air. We can see, then, that absolute humidity, although it measures the amount of water vapor, can vary as a result of the vertical movement changing the volume of the air parcel. Specific humidity, on the other hand, changes *only* as the quantity of the water vapor changes. For this reason, when assessing the changes of water vapor content in large masses of air, which often have vertical movement, specific humidity is the preferred measurement among geographers and meteorologists.

Relative Humidity Probably the best-known means of describing the content of water vapor in the atmosphere—the one commonly given on television and radio weather reports—is **relative humidity.** It is simply the ratio between the amount of water vapor in air of a given temperature and the maximum amount of vapor that the air could hold at that temperature; it is reported as a percentage that expresses how close the air is to saturation. This method of describing humidity has its strengths and weaknesses. Its strength lies in the ease with which it is communicated. Most people can understand the concept of "percent" and are unlikely to understand "grams per cubic meter." However, relative humidity percentages can vary widely through the day, even when the water vapor content of the air remains the same. The unsteady nature of relative humidity is its weakness.

If the temperature and absolute humidity of an air parcel are known, its relative humidity can be determined by using Figure 6.5. For instance, if we know that a parcel of air has a temperature of 30°C and an absolute humidity of 20 grams per cubic meter, we can look at the graph and determine that if it were saturated its absolute humidity would be 30 grams per cubic meter. To determine relative humidity, all we do is divide 20 grams (actual content) by 30 grams (content at capacity) and multiply by 100 (to get an answer in percentage):

$$(20 \text{ grams} \div 30 \text{ grams}) \times 100 = 67\%$$

The relative humidity in this case is 67%. In other words, the air is holding only two thirds of the water vapor it could contain at 30°C; it is at only 67% of its capacity.

Two important factors are involved in the horizontal distribution and variation of relative humidity. One of these is the availability of moisture. For example, the air above bodies of water is apt to contain more moisture than similar air over land surfaces because there is simply more water available for evaporation. Conversely, the air overlying a region like the central Sahara Desert is usually very dry because it is far from the oceans and little water is available to be evaporated. The second factor in the horizontal variation of relative humidity is temperature. In regions of higher temperature, relative humidity for the same amount of water vapor will be lower than it would be in a cooler region.

At any one point in the atmosphere, relative humidity varies if the amount of water vapor increases as a result of evaporation *or* if the temperature increases or decreases. Thus, although the quantity of water vapor may not change through a day, the relative humidity will vary with the daily temperature cycle. As air temperature increases from around sunrise to its maximum in midafternoon, the relative humidity decreases as the air becomes capable of holding greater and greater quantities of water vapor. Then, as the air becomes cooler, decreasing toward its minimum temperature around sunrise, the relative humidity increases (● Fig. 6.6).

● FIGURE 6.6
These graphs illustrate the relationship between air temperature (top) and relative humidity (bottom) through a week in May in Bowling Green, Kentucky.
What do you notice about the relationship between these two lines?

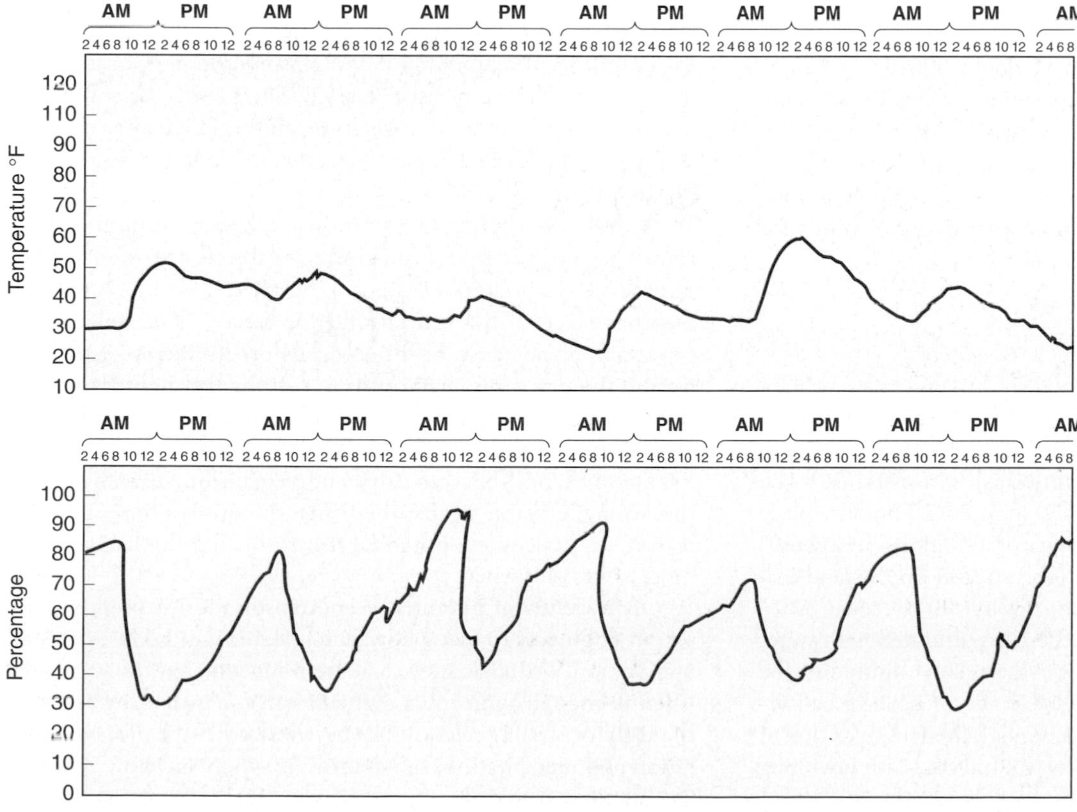

Relative humidity affects our comfort through its relationship to the rate of evaporation. Perspiration evaporates into the air, leaving behind a salty residue, which you can taste if you lick your lips after perspiring a great deal. Evaporation is a cooling process because the heat used to change the perspiration to water vapor (which becomes locked in the water vapor as latent heat) is subtracted from your skin. This is the reason that, on a hot August day when the temperature approaches 35°C (95°F), you will be far more uncomfortable in Atlanta, Georgia, where the relative humidity is 90%, than in Tucson, Arizona, where it may be only 15% at the same temperature. At 15%, your perspiration will be evaporated at a faster rate at the lower relative humidity, and you will benefit from the resultant cooling effects. When the relative humidity is 90%, the air is nearly saturated, far less evaporation can take place, and less heat is drawn from your skin.

Sources of Atmospheric Moisture

In our earlier discussion of the hydrologic cycle, we saw that the atmosphere receives water vapor through the process of evaporation. Water evaporates into the atmosphere from many different places, most important of which are the surfaces of Earth's bodies of water. Water also evaporates from wet ground surfaces and soils, from droplets of moisture on vegetation, from city pavements, building roofs, cars, and other surfaces, and even from falling precipitation.

Vegetation provides another source of water vapor. Plants give up water in a complex process known as **transpiration,** which can be a significant source of atmospheric moisture. A mature oak tree, for instance, can give off 400 liters (105 gal) of water per day, and a cornfield may add 11,000–15,000 liters (2900–4000 gal) of water to the atmosphere per day for each acre under cultivation. In some parts of the world—notably tropical rainforests of heavy, lush vegetation—transpiration accounts for a significant amount of atmospheric humidity. Together, evaporation and transpiration, or **evapotranspiration,** accounts for virtually all the water vapor in the air.

Rate of Evaporation

The rate of evaporation is affected by several factors. First, it is affected by the amount and temperature of accessible water. Thus, as Table 6.1 shows, the rate of evapotranspiration tends to be greater over the oceans than over the continents. The only place this generalization is not true is in equatorial regions between 0° and 10°N and S, where the vegetation is so lush on the land that transpiration provides a large amount of water for the air.

Second is the degree to which the air is saturated with water vapor. The drier the air and the lower the relative humidity, the greater the rate of evaporation can be. Some of us have had direct experience with this principle. Compare the length of time it takes your bathing suit to dry on a hot, humid day with how long it takes on a day when the air is dry.

Third is the wind, which affects the rate of evaporation. If there is no wind, the air that overlies a water surface will approach saturation as more and more molecules of liquid water change to water vapor. Once saturation is reached, evaporation will cease. However, if there is a wind, it will blow the saturated or nearly saturated air away from the evaporating surface, replacing it with air of lower humidity. This allows evaporation to continue as long as the wind keeps blowing saturated air away and bringing in drier air. Anyone who has gone swimming on a windy day has experienced the chilling effects of rapid evaporation.

Temperature affects the rate of evaporation by affecting the first and second factors above. As air temperature increases, so does the temperature of the water at the evaporation source. Such increases in temperature ensure that more energy is available to the water molecules for their escape from a liquid state to a gaseous one. Consequently, more molecules can make the transition. Also as the temperature of the air increases, its capacity to contain moisture also increases. As the air gets warmer, it is energized, the molecules of air separate farther apart, and air density decreases. With more energy and wider spacing between the air molecules, additional water molecules can enter the atmosphere, thus increasing evaporation.

Potential Evapotranspiration

So far, we have discussed actual evaporation and transpiration (evapotranspiration). However, geographers and meteorologists are also concerned with **potential evapotranspiration** (● Fig. 6.7). This term refers to the idealized conditions in an area under which there would be sufficient moisture for all possible evapotranspiration to occur. Various formulas have been derived for estimating the potential evapotranspiration at a location because it is difficult to measure directly. These formulas commonly use temperature, latitude, vegetation, and soil character (permeability, water-retention ability) as factors that could affect the potential evapotranspiration.

In places where precipitation exceeds potential evapotranspiration, there is a surplus of water for storage in the ground and in bodies of water, and water can even be exported to other places if transportation (like canals) is feasible. When potential evapotranspiration exceeds precipitation, as it does during the dry summer months in California, then there is no water available for storage; in fact, the water stored during previous rainy months evaporates quickly into the warm, dry air (● Fig. 6.8). Soil becomes dry and vegetation turns brown as the available water is absorbed into the atmosphere. For this reason, fires become a potential hazard during the late summer months in California.

Knowledge of potential evapotranspiration is used by irrigation engineers to learn how much water will be lost through evaporation. With that, they can determine whether the water that is left is enough to justify a drainage canal. Farmers, by assessing the daily or weekly relationship between potential evapotranspiration and precipitation, can determine when and how much to irrigate their crops.

TABLE 6.1
Distribution of Actual Mean Evapotranspiration

	Latitude					
Zone	**60°–50°**	**50°–40°**	**40°–30°**	**30°–20°**	**20°–10°**	**10°–0°**
Northern Hemisphere						
Continents	36.6 cm (14.2 in.)	33.0 (13.0)	38.0 (15.0)	50.0 (19.7)	79.0 (31.1)	115.0 (45.3)
Oceans	40.0 (15.7)	70.0 (27.6)	96.0 (37.8)	115.0 (45.3)	120.0 (47.2)	100.0 (39.4)
Mean	38.0 (15.0)	51.0 (20.1)	71.0 (28.0)	91.0 (35.8)	109.0 (42.9)	103.0 (40.6)
Southern Hemisphere						
Continents	20.0 cm (7.9 in.)	NA	51.0 (20.1)	41.0 (16.1)	90.0 (35.4)	122.0 (48.0)
Oceans	23.0 (9.1)	58.0 (22.8)	89.0 (35.0)	112.0 (44.1)	119.0 (47.2)	114.0 (44.9)
Mean	22.5 (8.8)	NA	NA	99.0 (39.0)	113.0 (44.5)	116.0 (45.7)

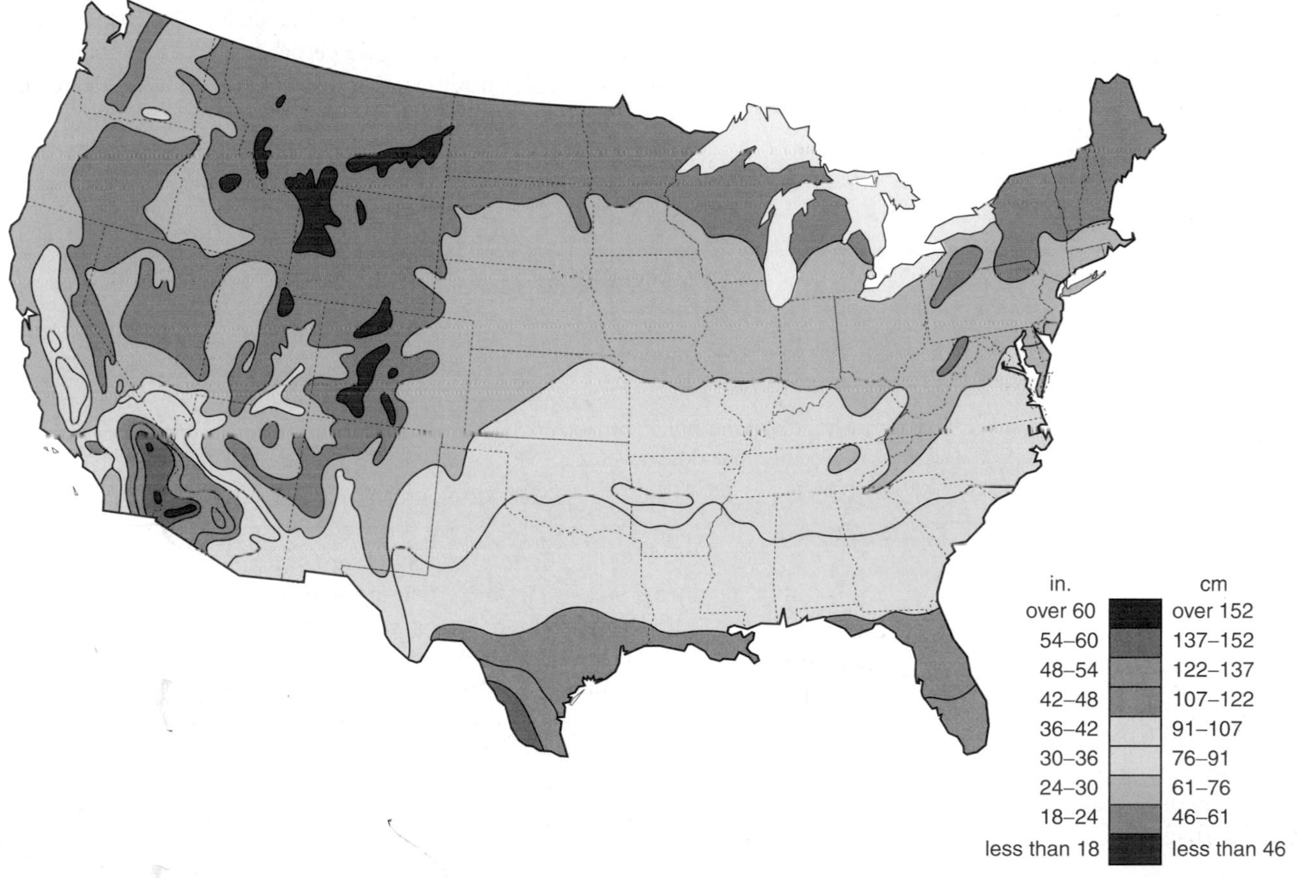

• **FIGURE 6.7**
Annual potential evapotranspiration for the contiguous 48 states.
Why is potential evapotranspiration so high in the desert southwest?

Condensation

Condensation is the process by which a gas is changed to a liquid. In our present discussion of atmospheric moisture and precipitation, condensation refers to the change of water vapor to liquid water.

Condensation occurs when air saturated with water vapor is cooled. Viewed in another way, we can say that if we lower the temperature of air until it has a relative humidity of 100% (the air has reached the dew point), condensation will occur with additional cooling. It follows, then, that condensation depends on (1) the relative humidity of the air and (2) the degree of cooling. In the arid air of Death Valley, California, a great amount of cooling must take place before the dew point is reached. In contrast, on a humid summer afternoon in Biloxi, Mississippi, a minimal amount of cooling will bring on condensation.

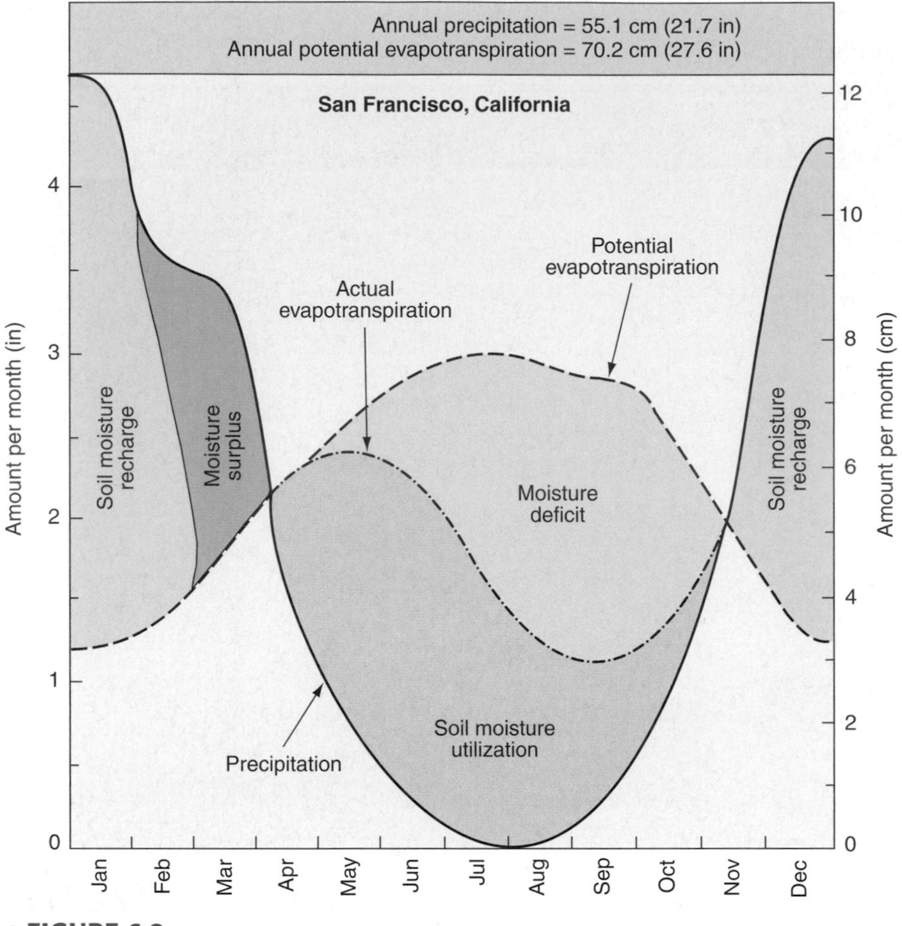

• **FIGURE 6.8**
This is an example of the *Thornthwaite water budget system,* which "keeps score" of the balance between water input by precipitation and water loss to evaporation and transpiration, permitting month-by-month estimates of both runoff and soil moisture.
When would irrigation be necessary at this site?

This is the principle behind the formation of droplets of water on the side of a glass containing a cold drink on a warm afternoon: The temperature of the air is lowered when it comes in contact with the cold glass. Consequently, the air's capacity to hold water vapor is diminished. If air touching the glass is cooled sufficiently, its relative humidity will reach 100%. Any cooling beyond that point will result in condensation in the form of water droplets on the glass.

Condensation Nuclei

For atmospheric condensation to occur, one other factor is necessary: the presence of **condensation nuclei.** These are minute particles in the atmosphere that provide a surface upon which condensation can take place. Condensation nuclei are most often sea-salt particles in the air from the evaporation of seawater. They also can be particles of dust, smoke, pollen, or volcanic material. Some particles are said to be more *hygroscopic* (the property to attract moisture) than others. More commonly, these are chemical particles that are the by-products of industrialization. The condensation that takes place on such chemical nuclei is often corrosive and dangerous to human health; when it is, we know it as *smog* (an invented term derived by combining the words *smoke* and *fog*).

Theoretically, if all particles were removed from a volume of air, we could cool that air below its dew point without condensation occurring. Conversely, if there is a superabundance of particles, condensation may take place at relative humidity less than 100%. For example, ocean fogs, which are an accumulation of condensation droplets formed on sea-salt particles, can form when the relative humidity is as low as 92%.

In nature, condensation appears in a number of forms. Fog, clouds, and dew are all the results of condensation of water vapor in the atmosphere. The type of condensation produced depends on a number of factors, including the cooling process itself. The cooling that produces condensation in one form or another can occur as a result of radiational cooling, through advection, through convection, or through a combination of these processes.

Fog

Fogs and clouds appear when water vapor condenses on nuclei and a large number of these droplets form a mass. Not being transparent to light in the way that water vapor is, these masses of condensed water droplets appear to us as fog or clouds, in any of a number of shapes and forms, usually in shades of white or gray.

In the water budget and the hydrologic cycle, fog is a minor form of condensation. Yet, in certain areas of the world, it has important climatic effects. The "drip factor" helps sustain vegetation along desert coastlines where fog occurs. Fog also plays havoc with our modern transportation systems. Navigation on the seas is made more difficult by fog, and air travel can be greatly impeded. In fact, fog sometimes causes major airports to shut down until visibility improves. Highway travel is also greatly hampered by heavy fogs, which can lead to huge, chain-reaction pileups of cars.

Radiation Fog Radiational cooling can lead to **radiation fog,** also called **temperature-inversion fog.** This kind of fog is likely to occur on a cold, clear, calm night, usually in the middle latitudes. These conditions allow for maximum outgoing radiation from the ground with no incoming radiation. The ground gets colder during the night as it gives up more of the heat that it has received during the day. As time passes, the air directly above is cooled by conduction through contact with the cold ground. Because the cold surface can cool only the lower few meters of the atmosphere, an inversion is created in which cold air at the surface is overlain by warmer air above. If this cold layer of air at the surface is cooled to a temperature below its dew point, then condensation will occur, usually in the form of a low-lying fog. However, wind strong enough to disturb this inversion layer can prevent the formation of the fog by not allowing the air to stay at the surface long enough to become cooled below its dew point.

The chances of a temperature-inversion fog occurring are increased by certain types of land formations. In valleys and depressions, cold air accumulates through air drainage. During a cold night, this air can be cooled below its dew point, and a fog forms like a pond in the bottom of the valley (● Fig. 6.9a). It is common in mountainous areas to see an early morning radiation fog on the valley floor while snow-capped mountaintops shine against a clear blue sky. Radiation fog has a diurnal cycle. It forms during the night and is usually the densest around sunrise when temperatures are lowest. It then "burns off" during the day when the heat from the sun slowly penetrates the fog and warms the surface. Earth in turn warms the air directly above it, increasing its temperature and consequently its capacity to hold water vapor. This greater capacity allows the fog to evaporate into the air. As Earth's heat penetrates to higher layers of air, the fog continues to burn off—from the ground up!

Radiation fog often forms even more densely in industrial areas where the high concentration of chemical particles in the air provides abundant condensation nuclei. Such a fog is usually thicker and denser than "natural" radiation fogs and less easily dissipated by wind or sun.

Advection Fog Another common type of fog is **advection fog,** which occurs through the movement of warm, moist air over a colder surface, either land or water. When the warm air is cooled below its dew point through heat loss by radiation and conduction from the colder surface below, condensation occurs in the form of fog. Advection fog is usually less localized than radiation fog. It is also less likely to have a diurnal cycle, though if not too thick, it can be burned off early in the day to return again in the late afternoon or early evening. More common, however, is the persistent advection fog that spreads itself over a large area for days at a time. Advection fog is a major reason why ski resorts are forced to close. Warm, moist air moving over the cold snow causes the dense fog.

Advection fog forms over land during the winter months in middle latitudes. It forms, for example, in the United States when warm, moist air from the Gulf of Mexico flows northward over the cold, frozen, and sometimes snow-covered upper Mississippi Valley.

During the summer months, advection fog may form over large lakes or over the oceans. Formation over lakes occurs when warm continental air flows over a colder water surface, such as when a warm air mass passes over the cool surface of Lake Michigan. An advection fog also can be formed when a warm air mass moves over a cold ocean current and the air is cooled sufficiently to bring about condensation. This variety of advection fog is known as a sea fog. Such a situation accounts for fogs along the West Coast of the United States. During the summer months, the Pacific subtropical high moves north with the sun, and winds flow out of the high toward the coast where

© Richard Hamilton Smith/ CORBIS

(a)

© Craig Aurness/ CORBIS

(b)

M. Trapasso

(c)

● **FIGURE 6.9**
Types of fog. (a) Radiation fog forms when cold air drains into a valley. (b) Advection (sea) fog is caused by warm, moist air passing over colder water or a colder coastal surface. (c) Upslope fog is caused by moist air adiabatically cooling as it rises up a mountain slope.
What unique problems might coastal residents face as a result of fog?

they pass over the cold California Current. When condensation occurs, fogs form that flow in over the shore, pushed from behind by the eastward movement of air and pulled by the low pressure of the warmer land (Fig. 6.9b). Advection fogs also occur in New England, especially along the coasts of Maine and the Canadian Maritime Provinces, when warm, moist air from above the Gulf Stream flows north over the colder waters of the Labrador Current. Advection fog over the Grand Banks off Newfoundland has long been a hazard for cod fishermen there.

Upslope Fog Another type of fog clings to windward sides of mountain slopes and is known as **upslope fog.** Its appearance is sometimes the source of geographic place names—for example, the Great Smoky Mountains where this type of fog is quite common. During early morning hours in middle-latitude locations, a light, moist breeze may ascend a slope and cool to the dew point, leaving a blanket of fog behind (Fig. 6.9c). In tropical rainforest regions, mountain slopes may be covered in a misty fog any time of day. Because the atmosphere is much more humid, reaching the dew point is easier.

Other Minor Forms of Condensation

Dew, which is made up of tiny droplets of water, is formed by the condensation of water vapor at or near the surface of Earth. Dew collects on surfaces that are good radiators of heat (such as your car or blades of grass). These good radiators give up large amounts of heat during the hours of darkness. When the air comes in contact with these cold surfaces, it cools; if it is cooled to the dew point, droplets of water will form as beads on the surface. When the temperature of the air is below 0°C (32°F), **white frost** forms. It is important to note that frost is not frozen dew but instead represents a sublimation process—water vapor changing directly to the frozen state (see again Fig. 4.12).

Sometimes, under very still conditions with low air pressure, though air temperatures may be below 0°C (32°F), the liquid droplets that make up clouds or fogs are not frozen into solid particles. When such *supercooled* water droplets come in contact with a surface, such as the edge of an airplane wing or a tree branch or a window, ice crystals are created on that surface in a formation known as **rime.** Icing on the wings, nose, and tail of an airplane is extremely hazardous and has been the cause of aviation disasters through time.

Clouds

Clouds are the most common form of condensation and are important for several reasons. First, they are the source of all precipitation. **Precipitation** is made up of condensed water particles, either liquid or solid, that fall to Earth. Obviously, not all clouds result in precipitation, but we cannot have precipitation without the formation of a cloud first. Clouds also serve an important function in the heat energy budget. We have already noted that clouds absorb some of the incoming solar energy. They also reflect some of that energy back to space and scatter and diffuse other wavelengths of the incoming energy before it strikes Earth as diffuse radiation. In addition, clouds absorb some of Earth's radiation so that it is not lost to space and then reradiate it back to the surface. Finally, clouds are a beautiful and ever-changing aspect of our environment. The colors of the sky and the variations in the shapes and hues of clouds have provided us all with a beautiful backdrop to the natural scenery here on Earth.

Cloud Forms Clouds are composed of billions of tiny water droplets and/or ice crystals so small (some measured in 1000ths of a millimeter) that they can remain suspended in the atmosphere. As you will learn in this and later chapters, clouds are a manifestation of various movements in our atmosphere. Clouds appear white or in shades of gray, even deep gray approaching black. They differ in color depending on how thick or dense they are and if the sun is shining on the surface that we see. The thicker a cloud is, the more sunlight it is able to absorb and thus block from our view, and the darker it will appear. Clouds also seem dark when we are seeing their shaded side instead of their sunlit side.

In 1803 an Englishman, Sir Luke Howard, proposed the first cloud classification scheme, which has been modified through time into the current system. Cloud names may (but not always) consist of two parts. The first part of the name refers to the cloud's height: low-level clouds, below 2000 meters (6500 ft), are called **strato;** middle-level clouds, from 2000 to 6000 meters (6500–19,700 ft), are named **alto;** and high-level clouds, above 6000 meters (19,700 ft), are termed **cirro.**

The second part of the name concerns the morphology, or shape, of the clouds. The three basic shapes are termed *cirrus, stratus,* and *cumulus.* Classification systems categorize these cloud formations into many subtypes; however, most subtypes are variations of the three basic shapes. ● Figure 6.10 illustrates the appearance and the general heights of common clouds; ● Figure 6.11 provides a pictorial summary of the major cloud types.

Cirrus clouds (from Latin: *cirrus,* a lock or wisp of hair) form at very high altitudes, normally 6000–10,000 meters (19,800–36,300 ft), and are made up of ice crystals rather than droplets of water. They are thin, stringy, white clouds that trail like feathers across the sky. When associated with fair weather, cirrus clouds are scattered white patches in a clear blue sky.

Stratus clouds (from Latin: *stratus,* layer) can appear anywhere from near the surface of Earth to almost 6000 meters (19,800 ft). The variations of stratus clouds are based in part on their altitude. The basic characteristic of stratus clouds is their horizontal sheet-like appearance, lying in layers with fairly uniform thickness. The horizontal configuration indicates that they form in stable atmospheric conditions, which inhibit vertical development.

Often stratus clouds cover the entire sky with a gray cloud layer. It is stratus clouds that make up the dull, gray, overcast sky common to winter days in much of the midwestern and eastern United States. The stratus cloud formation may overlie an area for days, and any precipitation will be light but steady and persistent.

Cumulus clouds (from Latin: *cumulus,* heap or pile) develop vertically rather than forming the more horizontal structures of the cirrus and stratus forms. Cumulus are massive piles of clouds, rounded or cauliflower in appearance, usually with a flat base, which can be

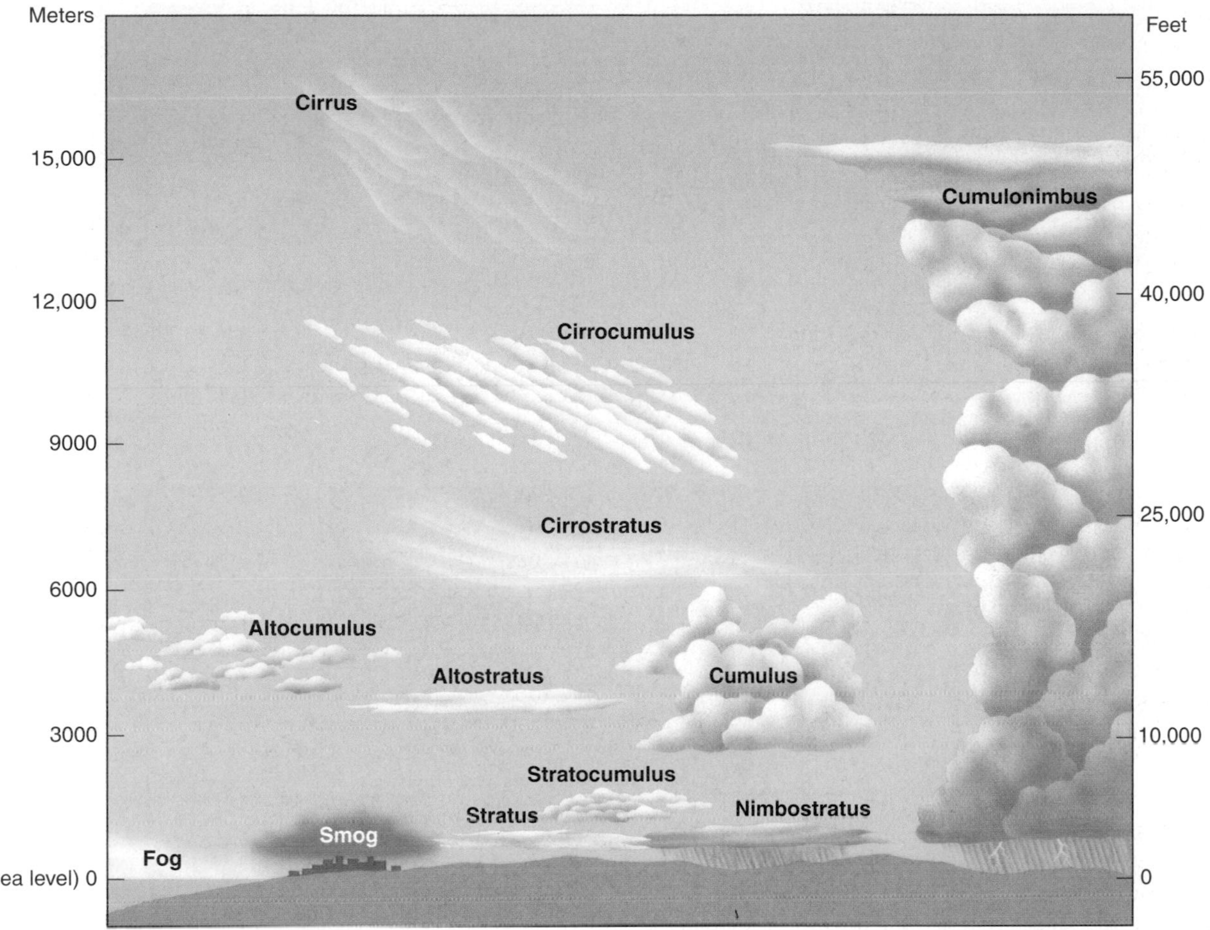

• **FIGURE 6.10**
Cloud classification scheme. Clouds are named based on their height and their form.
Observe this figure and Figure 6.11; what cloud type is present in your area today?

anywhere from 500 to 12,000 meters (1650–39,600 ft) above sea level. From this base, they pile up into great rounded structures, often with tops like cauliflowers. The cumulus cloud is the visible evidence of an unstable atmosphere; its base is the point where condensation has begun in a column of air as it moves upward.

Examine Figures 6.10 and 6.11 to familiarize yourself with the basic cloud types and their names. Keep in mind that some cloud shapes exist in all three levels—for example, *stratocumulus* (*strato* = low level + *cumulus* = a rounded shape), *altocumulus,* and *cirrocumulus*. These three share the similar rounded or cauliflower appearance of cumulus clouds, which can exist at all three levels. You may notice that *altostratus* (*alto* = middle level + *stratus* = layered shape) and *cirrostratus* have two-part names, but low-level layered clouds are called *stratus* only. Lastly, thin, stringy *cirrus* clouds are found only as high-level clouds, so the term *cirro* (meaning high-level cloud) is not necessary here.

Other terms used in describing clouds are **nimbo** or **nimbus,** meaning precipitation (rain is falling). Thus, the *nimbostratus* cloud may bring a long-lasting drizzle, and the *cumulonimbus* is the thunderstorm cloud. This latter cloud has a flat top, called an anvil head, as well as a relatively flat base, and it becomes darker as it grows higher and thicker and thus blocks the incoming sunlight. The cumulonimbus is the source of many atmospheric concerns including high-speed winds, torrential rain, flash flooding, thunder, lightning, hail, and possibly tornadoes. This type of cloud can develop in several different ways as we will soon discuss.

Adiabatic Heating and Cooling The cooling process that leads to cloud formation is quite different from that associated with the other condensation forms that we have already examined. The cooling process that produces fog, frost, and dew is either radiation or advection. On the other hand, clouds usually develop from a cooling process that results when a parcel of air on Earth's surface is lifted into the atmosphere.

The rising parcel of air will expand as it encounters decreasing atmospheric pressure with height. This expansion allows the air molecules to spread out, which causes the parcel's temperature to decrease. This is known as **adiabatic cooling** and occurs at the constant lapse rate of approximately 10°C per 1000 meters (5.6°F/1000 ft). By the same token, air descending through the atmosphere is compressed by the increasing pressure and undergoes **adiabatic heating** of the same magnitude.

• FIGURE 6.11
Types of clouds.

However, the rising and cooling parcel of air will eventually reach its dew point—the temperature at which water vapor begins to condense out, forming cloud droplets. From this point on, the adiabatic cooling of the rising parcel will decrease as latent energy released by the condensation process is added to the air. To differentiate between these two adiabatic cooling rates, we refer to the precondensation rate (10°C/1000 m) as the **dry adiabatic lapse rate** and the lower, postcondensation rate as the **wet adiabatic lapse rate.** The latter rate averages 5°C per 1000 meters (3.2°F/1000 ft) but varies according to the amount of water vapor that condenses out of the air.

A rising air parcel will cool at one of these two adiabatic rates. Which rate is in operation depends on whether condensation is (wet adiabatic rate) or is not (dry adiabatic rate) occurring.

NOAA/NWS

Cirrus

© John Cunningham/Visuals Unlimited

Cumulonimbus

©Martin Miller/ Visuals Unlimited

Nimbostratus

M. Trapasso

Cumulus

• **FIGURE 6.11** *(continued)*

On the other hand, the warming temperatures of descending air allow it to hold greater quantities of water vapor. In other words, as the air temperature rises farther above the dew point, condensation will not occur, so the heat of condensation will not affect the rate of rise in temperature. Thus, the temperature of air that is descending and being compressed always increases at the dry adiabatic rate.

It is important to note that adiabatic temperature changes are the result of changes in volume and do not involve the addition or subtraction of heat from external sources.

It is also extremely important to differentiate between the *environmental lapse rate* and *adiabatic lapse rates.* In Chapter 4, we found that in general the temperature of our atmosphere decreases with increasing height above Earth's surface; this is known as the environmental lapse rate, or the normal lapse rate. Although it averages 6.5°C per 1000 meters (3.6°F/1000 ft), this rate is quite variable and must be measured through the use of meteorological instruments sent aloft. Whereas the environmental lapse rate reflects nothing more than the vertical temperature structure of the atmosphere, the adiabatic lapse rates are concerned with temperature changes as a parcel of air moves through the atmospheric layers (• Fig. 6.12).

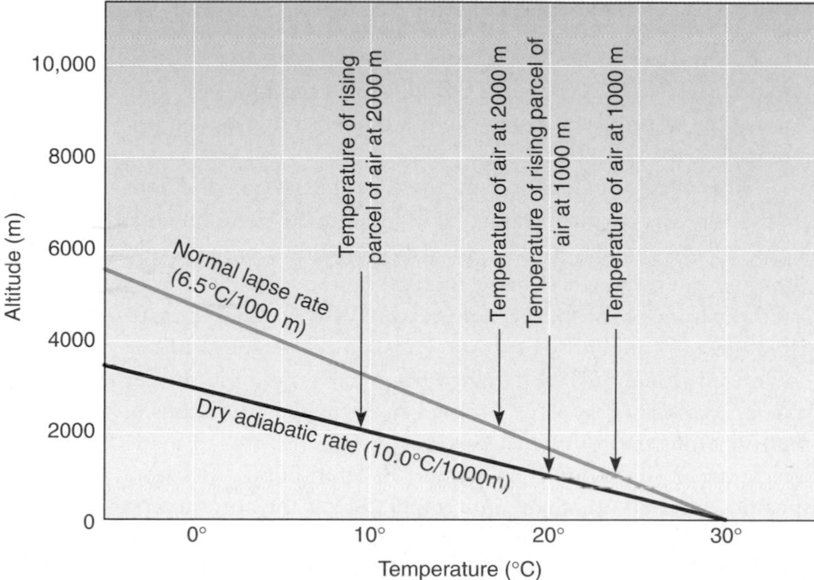

• **FIGURE 6.12**

Comparison of the dry adiabatic lapse rate and the environmental lapse rate. The environmental lapse rate is the average vertical change in temperature. Air displaced upward will cool (at the dry adiabatic rate) because of expansion.

In this example, using the environmental lapse rate, what is the temperature of the layer of air at 2000 meters?

Stability and Instability Although adiabatic cooling results in the development of clouds, the various forms of clouds are related to differing degrees of vertical air movement. Some clouds are associated with rapidly rising, buoyant air, whereas other forms result when air resists vertical movement.

An air parcel will rise of its own accord as long as it is warmer than the surrounding layer of air. When it reaches a layer of the atmosphere that is the same temperature as itself, it will stop rising. Thus, an air parcel warmer than the surrounding atmospheric air will rise and is said to be *unstable*. On the other hand, an air parcel that is colder than the surrounding atmospheric air will resist any upward movement and will likely sink to lower levels. Then the air is said to be *stable*.

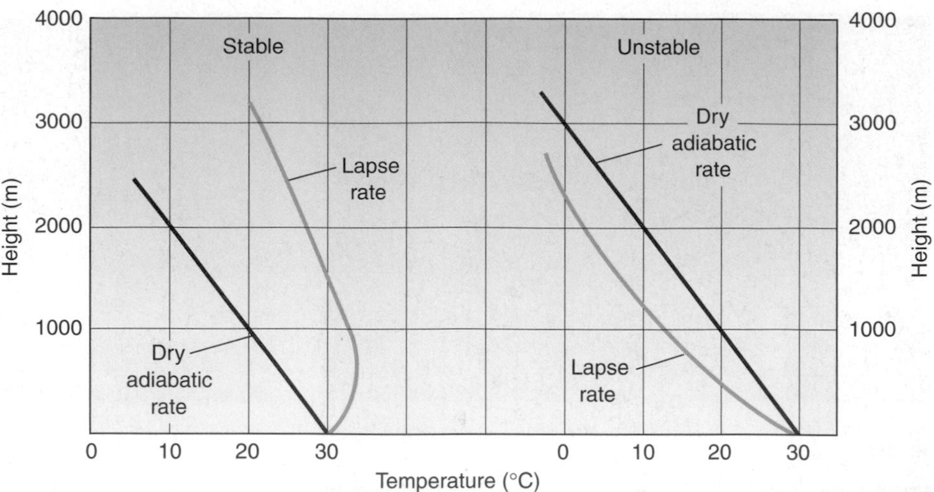

• **FIGURE 6.13**
Relationship between lapse rates and air mass stability. When air is forced to rise, it cools adiabatically. Whether it continues to rise or resists vertical motion depends on whether adiabatic cooling is less rapid or more rapid than the prevailing vertical temperature lapse rate. If the adiabatic cooling rate exceeds the lapse rate, the lifted air will be colder than its surroundings and will tend to sink when the lifting force is removed. If the adiabatic cooling rate is less than the lapse rate, the lifted air will be warmer than its surroundings and will be buoyant, continuing to rise even after the original lifting force is removed.
In these examples, what would be the temperature of the lifted air if it rose to 2000 meters?

Determining the stability or instability of an air parcel involves nothing more than asking the question, If an air parcel were lifted to a specific elevation (cooling at an adiabatic lapse rate), would it be warmer, colder, or the same temperature as the atmospheric air (determined by the environmental lapse rate at that time) at that same elevation?

If the air parcel is warmer than the atmospheric air at the selected elevation, then the parcel would be unstable and would continue to rise, because warmer air is less dense and therefore buoyant. Thus, under conditions of **instability,** the environmental lapse rate must be *greater than* the adiabatic lapse rate in operation. For example, if the environmental lapse rate is 12°C per 1000 meters and the ground temperature is 30°C, then the atmospheric air temperature at 2000 meters would be 6°C. On the other hand, an air parcel (assuming that no condensation occurs) lifted to 2000 meters would have a temperature of 10°C. Because the air parcel is warmer than the atmospheric air around it, it is unstable and will continue to rise (• Fig. 6.13).

Now let's assume that it is another day and all the conditions are the same, except that measurements indicate the environmental lapse rate on this day is 2°C per 1000 meters. Consequently, although our air parcel if lifted to 2000 meters would still have a temperature of 10°C, the temperature of the atmosphere at 2000 meters would now be 26°C. Thus, the air parcel would be colder and would sink back toward Earth as a result of its greater density (see again Fig. 6.13). As you can see, under conditions of **stability,** the environmental lapse rate is *less than* the adiabatic lapse rate in operation. If an air parcel, upon being lifted to a specific elevation, has the same temperature as the atmospheric air surrounding it, it is neither stable nor unstable. Instead, it is considered *neutral;* it will neither rise nor sink but will remain at that elevation.

Whether an air parcel will be stable or unstable is related to the amount of cooling and heating of air at Earth's surface. With cooling of the air through radiation and conduction on a cool, clear night, air near the surface will be relatively close in temperature to that aloft, and the environmental lapse rate will be low, thus enhancing stability. With the rapid heating of the surface on a hot summer day, there will be a very steep environmental lapse rate because the air near the surface is so much warmer than that above, and instability will be enhanced.

Pressure zones can also be related to atmospheric stability. In areas of high pressure, stability is maintained by the slow subsiding air from aloft. In low pressure regions, on the other hand, instability is promoted by the tendency for air to converge and then rise.

Precipitation Processes

Condensed droplets within cloud formations stay in the air and do not fall to Earth because of their tiny size (0.02 mm, or less than 1000th of an inch), their general buoyancy, and the upward movement of the air within the cloud. These droplets of condensation are so minute that they are kept floating in the cloud formation; their mass and the consequent pull of gravity are insufficient to overcome the buoyant effects of air and the vertical currents, or updrafts, within the clouds. • Figure 6.14 shows the relative sizes of a condensation nucleus, a cloud droplet, and a raindrop. It takes about a million cloud droplets to form one raindrop.

Precipitation occurs when the droplets of water, ice, or frozen water vapor grow and develop masses too great to be held aloft. They then fall to Earth as rain, snow, sleet, or hail. The form that precipitation takes depends largely on the method of formation and the temperature during formation. Among the many theories

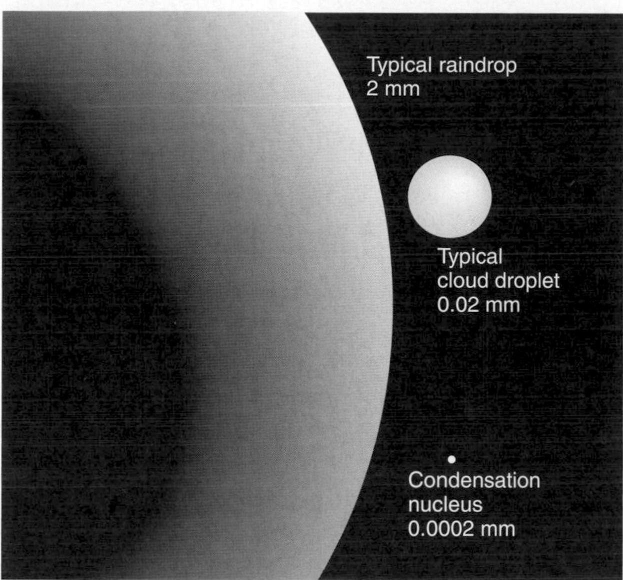

• FIGURE 6.14
The relative sizes of raindrops, cloud droplets, and condensation nuclei.
If the diameter of a raindrop is 100 times larger than a cloud droplet, why does it take a million cloud droplets to produce one raindrop?

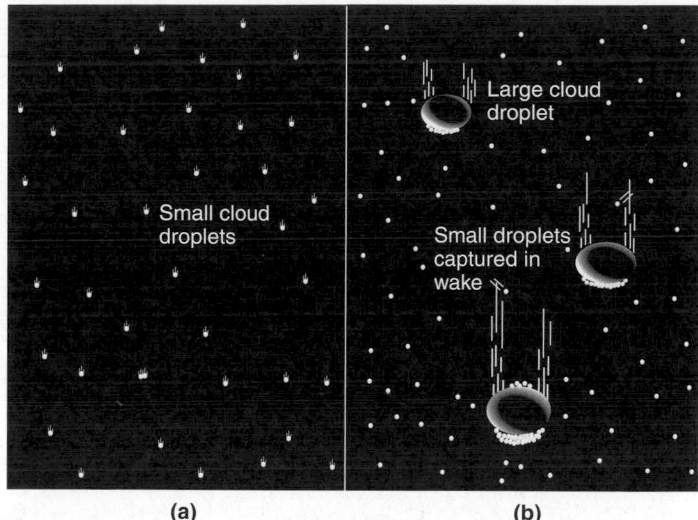

• FIGURE 6.15
Collision and coalescence. (a) In a warm cloud consisting of small cloud droplets of uniform size, the droplets are less likely to collide because they are falling very slowly and at about the same speed. (b) In a cloud of different-sized droplets, some droplets fall more rapidly and can overtake and capture some of the smaller droplets.
Why do these tiny droplets fall at different speeds?

that try to explain the formation of precipitation, the **collision–coalescence process** for warm clouds in low latitudes and the **Bergeron (or ice crystal) Process** for cold clouds at higher latitudes are the most widely accepted.

Precipitation in the lower latitudes of the tropics and in warm clouds is likely to form by the collision–coalescence process. The collision–coalescence process is one in which the name itself describes the process. By nature, water is quite cohesive (able to stick to itself). When water droplets are colliding in the circulation of the cloud, they tend to coalesce (or grow together). This is especially true as the water droplets begin to fall toward the ground. In falling, the larger droplets overtake the smaller, more buoyant droplets and capture them to form even larger raindrops. The mass of these growing raindrops eventually overcomes the updrafts of the cloud and fall to Earth, under the pull of gravity. This process occurs in the warm section of clouds where all the moisture exists as liquid water (• Fig. 6.15).

At higher latitudes, storm clouds can possess three distinctive layers. The lowermost is a warm layer of liquid water. Here the temperatures are above the freezing point of 0°C (32°F). Above this is the second layer composed of some ice crystals but mainly **supercooled water** (liquid water that exists at a temperature below 0°C). In the uppermost layer of these tall clouds, when temperatures are lower than or equal to –40°C (–40°F), ice crystals will dominate (• Fig. 6.16). It is in relation to these layered clouds that Scandinavian meteorologist Tor Bergeron presented a more complex explanation.

The Bergeron (or ice crystal) Process begins at great heights in the ice crystal and supercooled water layers of the clouds. Here, the supercooled water has a tendency to freeze on any available surface. (It is for this reason that aircraft flying through middle- to high-latitude thunderstorms run the risk of severe icing and invite disaster.) The ice crystals mixed in with the supercooled water in the highest layers of the clouds can become *freezing nuclei* and form the centers of growing ice crystals. (Essentially, this is the process that can also create snow.) As the supercooled water continues to freeze onto these frozen nuclei, their masses grow until gravity begins to pull them toward Earth. As this frozen precipitation enters the lower layer of the clouds, the above-freezing temperatures there melt the ice crystals into liquid rain before they hit the ground. Therefore, according to Bergeron, rain in these clouds begins as frozen precipitation and melts into a liquid before reaching Earth.

As the melted precipitation falls through the lower, warmer section of the cloud, the collision–coalescence process may take over and cause the raindrops to grow even larger as they descend toward the surface.

Major Forms of Precipitation

Rain, consisting of droplets of liquid water, is by far the most common form of precipitation. Raindrops vary in size but are generally about 2–5 millimeters (approximately 0.1–0.25 in.) in diameter (see again Fig. 6.14). As we all know, rain can come in many ways: as a brief afternoon shower, a steady rainfall, or the deluge of a tropical rainstorm. When the temperature of an air mass is only slightly below the dew point, the raindrops may be very small (about 0.5 mm or less in diameter) and close together. The result is a fine mist called **drizzle.** Drizzle is so light that it is greatly affected by the direction of air currents and the variability of winds. Consequently, drizzle seldom falls vertically.

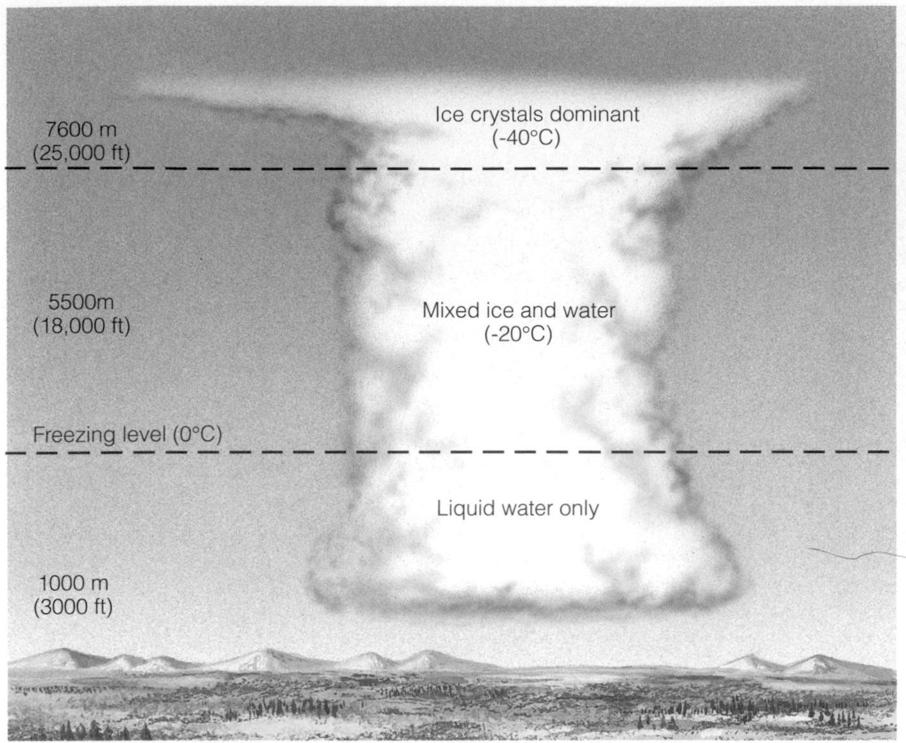

• **FIGURE 6.16**
The distribution of water, supercooled water, and ice crystals in a high-latitude storm cloud according to the Bergeron Process theory.
What is the difference between water and supercooled water?

Snow is the second most common form of precipitation. When water vapor is frozen directly into a solid without first passing through a stage as liquid water (or sublimation), it forms minute ice crystals around the freezing nuclei (of the Bergeron Process). These crystals characteristically appear as six-sided, symmetric shapes. Combinations of these ice-crystal shapes make up the intricate patterns of snowflakes. Snow will reach the ground if the entire cloud and the air beneath the cloud maintain below-freezing temperatures.

Sleet is frozen rain, formed when rain, in falling to Earth, passes through a relatively thick layer of cold air near the surface and freezes. The result is the creation of small, solid particles of clear or milky ice. In English-speaking countries outside the United States, sleet refers not to this phenomenon of frozen rain but rather to a mixture of rain and snow.

Hail is a less common form of precipitation than the three just described. It occurs most often during the spring and summer months and is the result of thunderstorm activity. Hail appears as rounded lumps of ice, called hailstones, which can vary in size from 5 millimeters (0.2 in.) in diameter and up to sizes larger than a baseball (• Fig. 6.17). The world record is a hailstone 30 centimeters (12 in.) in diameter that fell in Australia. Hailstones dropping from the sky can be highly destructive to crops and other vegetation, as well as to cars and buildings. Though primarily a property destroyer, hailstones have been known to kill animals and humans.

Hail forms when ice crystals are lifted by strong updrafts in a cumulonimbus (thunderstorm) cloud. Then, as these ice crystals circulate around the storm cloud, supercooled water droplets attach themselves and are frozen as a layer. Sometimes these pellets are lifted up into the cold layer of air and then dropped again and again. The resulting hailstone, made up of concentric layers of ice, has a frosty, opaque appearance when it finally breaks out of the strong updrafts of the cloud formation and falls to Earth. The larger the hailstone, the more times it is cycled through the freezing process and accumulated additional frozen layers.

On occasion, a raindrop can form and have a temperature below 0°C (32°F). This will occur when there is a shallow layer of below-freezing temperatures all the way to the ground so that the liquid rain can reach a supercooled state. These supercooled droplets will freeze the instant they fall onto a surface that is also at a below-freezing temperature. The resulting icy covering on trees, plants, and telephone and power lines is known as **freezing rain** (or **glaze**). People usually call the rain and its blanket of ice an "ice storm" (• Fig. 6.18). Because of the weight of ice, glazing can break off large branches of trees, bringing down telephone and power lines. It can also make roads practically impassable. A small counterbalance against the negative effects of glazing is the beauty of the natural landscape after an ice storm. Sunlight catches on the ice, reflecting and making a diamond-like surface covering the most ordinary weeds and tree branches.

• **FIGURE 6.17**
Hailstones can be the size of golf balls, or even larger.
What gives them their spherical appearance?

NOAA

FEMA Photo/Michael Raphael

• **FIGURE 6.18**
An ice storm can cover a city with a dangerous glazing of ice.
Why are power failures a common occurrence with ice storms?

Factors Necessary for Precipitation

Three factors are necessary for the formation of any type of precipitation on Earth. The first is the presence of *moist air* on the surface. This air obviously represents the source of moisture (for the precipitation) and energy (in the form of latent heat of condensation). Second are the *condensation nuclei* around which the water vapor can condense, discussed earlier in this chapter. Third is a *mechanism of uplift*. These **uplift mechanisms** are responsible for forcing the air higher into the atmosphere so that it can cool down (by the dry adiabatic rate) to the dew point. These uplift mechanisms are vital to the process of precipitation.

A parcel of air can be forced to rise in four major ways. All the precipitation that falls anywhere on Earth can be traced back to one of these four uplift mechanisms (• Fig. 6.19):

- **Convectional precipitation** results from the displacement of warm air upward in a convectional system.
- **Frontal precipitation** takes place when a warm air mass rises after encountering a colder, denser air mass.
- **Cyclonic** (or **convergence) precipitation** occurs when air converges upon and is lifted up into a low pressure system.
- **Orographic precipitation** results when a moving air mass encounters a land barrier, usually a mountain, and must rise above it in order to pass.

Convectional Precipitation The simple explanation of convection is that when air is heated near the surface it expands, becomes lighter, and rises. It is then displaced by the cooler, denser air around it to complete the convection cycle. The important factor in convection for our discussion of precipitation is that the heated air rises and thus fulfills the one essential criterion for significant condensation and, ultimately, precipitation.

To enlarge our understanding of convectional precipitation, let's apply what we have learned about instability and stability. • Figure 6.20 illustrates two different cases in which air rises due to convection. In both, the lapse rate in the free atmosphere is the same; it is especially high during the first few thousand meters but slows after that (as on a hot summer day).

In the first case (Fig. 6.20a), the air parcel is not very humid, and thus the dry adiabatic rate applies throughout its ascent. By the time the air reaches 3000 meters (9900 ft), its temperature and density are the same as those of the surrounding atmospheric air. At this point, convectional lifting stops.

In the second case (Fig. 6.20b), we have introduced the latent heat of condensation. Here again, the unsaturated rising column of air cools at the dry adiabatic rate of 10°C per 1000 meters (5.6°F/1000 ft) for the first 1000 meters (3300 ft). However, because the air parcel is humid, the rising air column soon reaches the dew point, condensation takes place, and cumulus clouds begin to form. As condensation occurs, the heat locked up in the water vapor is released and heats the moving parcel of air, retarding the adiabatic rate of cooling so that the rising air is now cooling at the wet adiabatic rate (5°C/1000 meters). Hence, the temperature of the rising air parcel remains warmer than that of the atmospheric layer it is passing through, and the air parcel will continue to rise on its own. In this case, which incorporates the latent heat of condensation, we have massive condensation, towering cumulus clouds, and a thunderstorm potential.

Convectional precipitation is most common in the humid equatorial and tropical areas that receive much of the sun's energy and in summer in the middle latitudes. Though differential heating of land surfaces plays an important role in convectional precipitation, it is not the sole factor. Other factors, such as surface topography and atmospheric dynamics associated with the upper air winds, may provide the initial upward lift for air that is potentially unstable. Once condensation begins in a convectional column, additional energy is available from the latent heat of condensation for further lifting.

This convectional lifting can result in the heavy precipitation, thunder, lightning, and tornadoes of spring and summer afternoon thunderstorms. When the convectional currents are strong in the characteristic cumulonimbus clouds, hail can result.

Frontal Precipitation The zones of contact between relatively warm and relatively cold bodies of air are known as **fronts.** When two large bodies of air that differ in density, humidity, and temperature meet, the warmer one is lifted above the colder. When this happens, the major criterion for large-scale condensation and precipitation is once again met. Frontal precipitation thus occurs as the moisture-laden warm air rises above the front caused by contact with the cold air. Continuous frontal precipitation has caused some devastating floods through time.

To fully understand fronts, we must examine what causes unlike bodies of air to come together and what happens when they do. This will be discussed in Chapter 7, where we will take a more detailed look at frontal disturbances and precipitation.

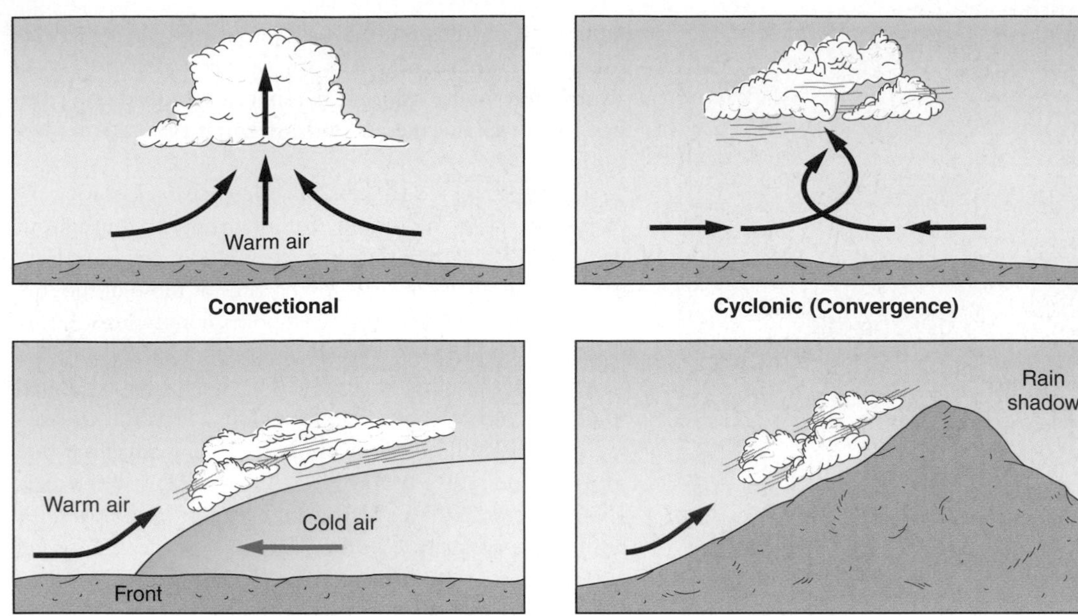

• **FIGURE 6.19**
The principal cause of precipitation is upward movement of moist air resulting from convectional, frontal, cyclonic, or orographic lifting.
What kind of air movement is common to all four diagrams?

Cyclonic (Convergence) Precipitation The third mechanism, the cyclonic (also known as convergence), was first introduced in Chapter 5 (see again Fig. 5.4). When air enters a low pressure system, or *cyclone,* it does so (in a counterclockwise fashion in the Northern Hemisphere) from all directions. When air *converges* on a low pressure system, it has little option but to rise. Therefore, clouds and possible precipitation are common around the center of a cyclone.

Orographic Precipitation As was the case with convectional rainfall, orographic rainfall has a simple definition and a somewhat more complex explanation. When land barriers—such as mountain ranges, hilly regions, or even the escarpments (steep edges) of plateaus or tablelands—lie in the path of prevailing winds, large portions of the atmosphere are forced to rise above these barriers. This fills the one main criterion for significant precipitation—that large masses of air are cooled by ascent and expansion until large-scale condensation takes place. The resultant precipitation is termed *orographic* (from Greek: *oros,* mountains). As long as the air parcel rising up the mountainside remains stable (cooling at a greater rate than the environmental lapse rate), any resulting cloud cover will be a type of stratus cloud. However, the situation can be complicated by the same circumstances illustrated in Figure 6.20b. A potentially unstable air parcel may need only the initial lift provided by the orographic barrier to set it in motion. In this case, it will continue to rise of its own accord (no longer forced) as it seeks air of its own temperature and density. Once the land barrier provides the initial thrust, it has performed its function as a lifting mechanism.

Because the air deposits most of its moisture on the windward side of a mountain, there will normally be a great deal less precipitation on the leeward side; on this side, the air will be much drier and the dew point consequently much lower. Also, as air descends the leeward slope, its temperature warms (at the dry adiabatic rate), and condensation ceases. The leeward side of the mountain is thus said to be in the **rain shadow** (• Fig. 6.21a). Just as being in the shade, or in shadow, means that you are not receiving any direct sun, so being in the rain shadow means that you do not receive much rain. If you live near a mountain range, you can see the effects of orographic precipitation and the rain shadow in the pattern of vegetation (Figs. 6.21b and c). The windward side of the mountains (say, the Sierra Nevada in California) will be heavily forested and thick with vegetation. The opposite slopes in the rain shadow will usually be drier and the cover of vegetation sparser.

Distribution of Precipitation

The precipitation a region receives can be described in different ways. We can look at average annual precipitation to get an overall picture of the amount of moisture that a region gets during a year. We can also look at its number of raindays—days on which 1.0 millimeter (0.01 in.) or more of rain is received during a 24-hour period. Less than this amount is known as a **trace** of rain. If we divide the number of raindays in a month or year by the total number of days in that period, the resulting figure represents the probability of rain. Such a measure is important to farmers and to ski or summer resort owners whose incomes may depend on precipitation or the lack of it.

We can also look at the average monthly precipitation. This provides a picture of the seasonal variations in precipitation (• Fig. 6.22). For instance, in describing the climate of the west

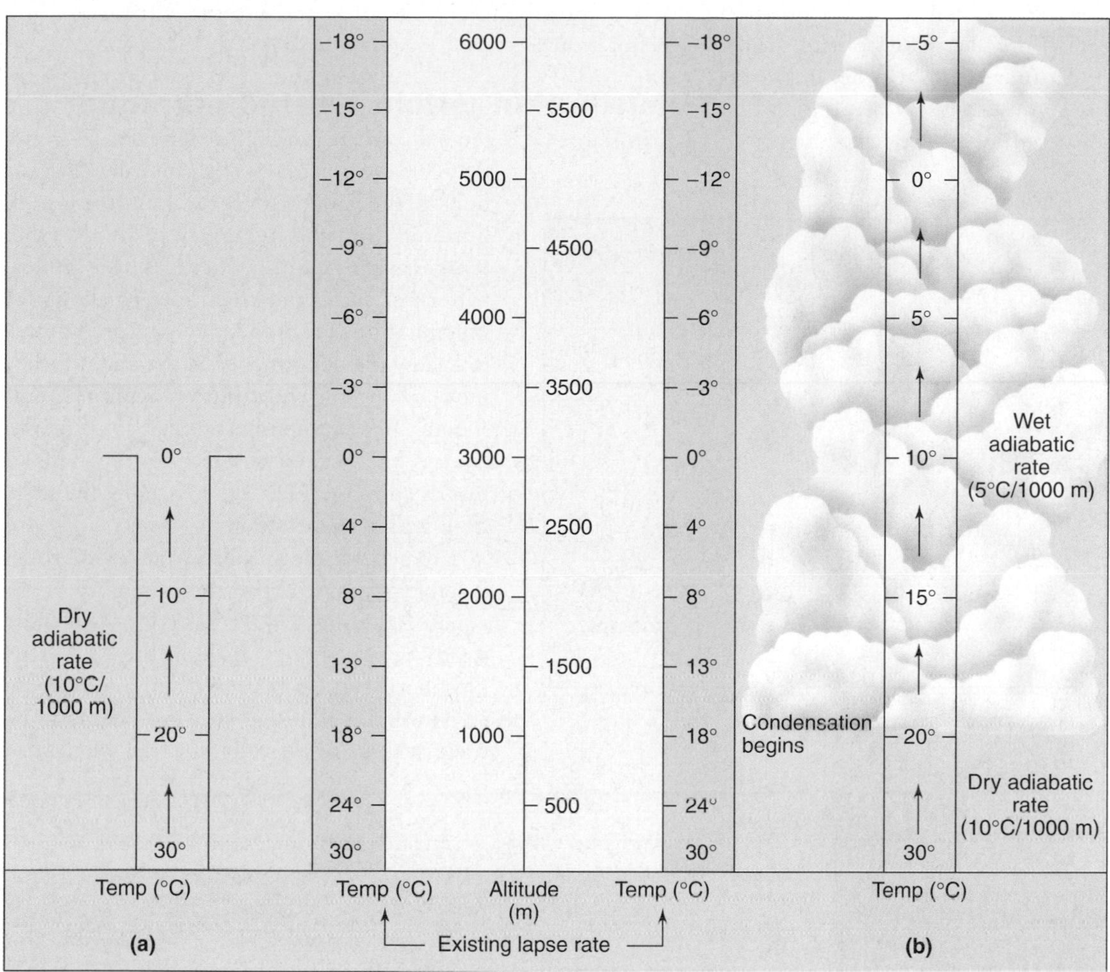

• FIGURE 6.20

Effect of humidity on air mass stability. (a) Warm, dry air rises and cools at the dry adiabatic rate, soon becoming the same temperature as the surrounding air, at which point convectional uplift terminates. Because the rising dry air did not cool to its dew point temperature by the time that convectional lifting ended, no cloud formed. (b) Rising warm, moist air soon cools to its dew point temperature. The upward-moving air subsequently cools at the wet adiabatic rate, which keeps the air warmer than the surrounding atmosphere so that the uplift continues. Only when all moisture is removed by condensation will the air cool rapidly enough at the dry adiabatic rate to become stable.

What would be necessary for the cloud in (b) to stop its upward growth at 4500 meters?

coast of California, average annual precipitation would not give the full story because this figure would not show the distinct wet and dry seasons that characterize this region.

Horizontal Distribution of Precipitation • Figure 6.23 shows average annual precipitation for the world's continents. We can see that there is great variability in the distribution of precipitation over Earth's surface. Although there is a zonal distribution of precipitation related to latitude, this distribution is obviously not the only factor involved in the amount of precipitation an area receives.

The likelihood and amount of precipitation are based on two factors. First, precipitation depends on the degree of lifting that occurs in air of a particular region. This lifting, as we have already seen, may be due to the collision of different air masses (frontal), to the convergence of air into a low pressure system (cyclonic or convergence), to differential heating of Earth's surface (convection), to the lifting that results when an air mass encounters a rise in Earth's surface (orographic), or to a combination of these processes. The second factor affecting the likelihood of precipitation depends on the internal characteristics of the air itself, including its degree of instability, its temperature, and its humidity.

Because higher temperatures, as we have seen, allow air masses to hold greater amounts of water vapor and because, conversely, cold air masses can hold less water vapor, we can expect a general decrease of precipitation from the equator to the poles that is related to the unequal zonal distribution of incoming solar energy discussed in Chapter 3.

However, if we look again at Figure 6.23, we see a great deal of variability in average annual precipitation beyond the general pattern of a decrease with increased latitude. In the following discussion, we examine some of these variations and give the reasons for them. We also apply what we have already learned about temperature, pressure systems, wind belts, and precipitation.

Distribution within Latitudinal Zones The equatorial zone is generally an area of high precipitation—more than 200 centimeters (79 in.) annually—largely due to the zone's high temperatures, high humidity, and the instability of its air. High temperatures and instability lead to a general pattern of rising air, which in turn allows for precipitation. This tendency is strongly reinforced by the convergence of the trades as they move toward the equator from opposite hemispheres. In fact, the intertropical convergence zone is one of the two great zones where air masses converge. (The other is along the polar front within the westerlies.)

In general, the air of the trade wind zones is stable compared with the instability of the equatorial zone. Under the control of these steady winds, there is little in the way of atmospheric disturbances to lead to convergent or convectional lifting. However, because the trade winds are basically easterly, when they

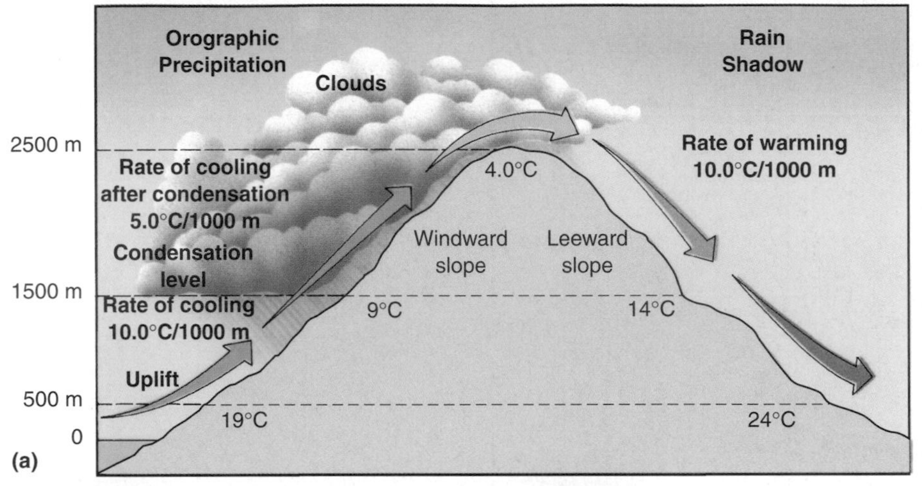

• **FIGURE 6.21**
Orographic precipitation and the rain-shadow effect.
(a) Orographic uplift over the windward (western) slope of the Sierras produces condensation, cloud formation, and precipitation, resulting in (b) dense stands of forest. (c) Semi-arid or rain-shadow conditions occur on the leeward (eastern) slope of the Sierras.
Can you identify a mountain range in Eurasia in which the leeward side of that range is in the rain shadow?

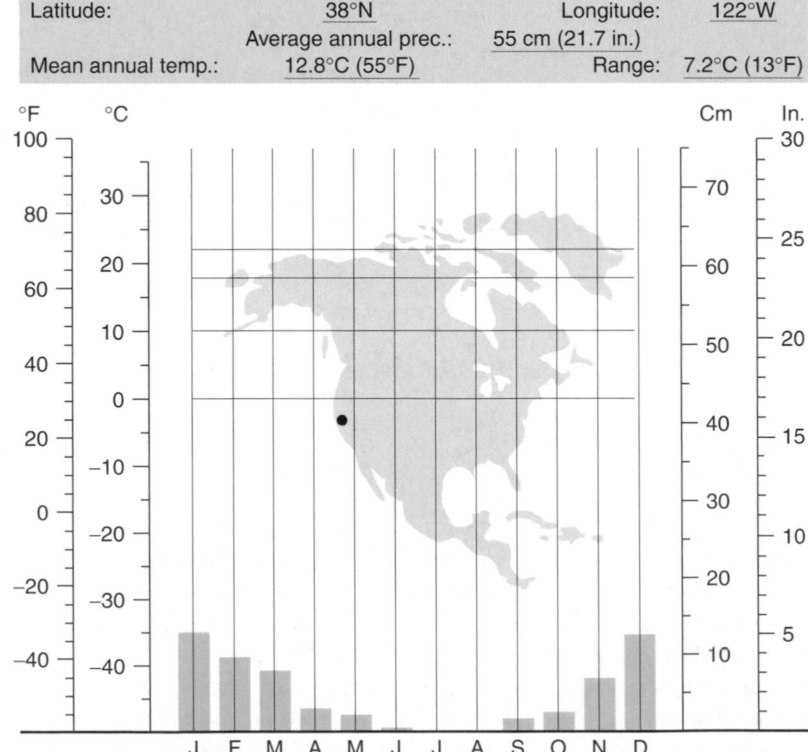

• **FIGURE 6.22**
Average monthly precipitation in San Francisco, California, is represented by colored bars along the bottom of the graph. A graph of monthly precipitation figures like this one gives a much more accurate picture than the annual precipitation total, which does not tell us that nearly all the precipitation occurs in only half of the year.
How would this rainfall pattern affect agriculture?

GEOGRAPHY'S PHYSICAL SCIENCE PERSPECTIVE

The Lifting Condensation Level (LCL)

When you look at clouds in our atmosphere, it is often quite easy to see their relatively flat bases. Cloud tops may appear quite irregular, but cloud bases are often flat. Even if the cloud bases do not seem flat, it will be obvious that the clouds you see all seem to be formed at the same level above the surface. This level represents the altitude to which the air must be lifted (and cooled at the dry adiabatic rate) before saturation is reached. Any additional lifting and clouds will form and build upward. Therefore, the height at which clouds form from lifting is called the lifting condensation level (LCL) and can be estimated by the equation:

$$\text{LCL (in meters)} = 125 \text{ meters} \times (\text{Celsius temperature} - \text{Celsius dew point})$$

For example, if the surface temperature is 7.2°C (45°F) and the dew point temperature is 4.4°C (40°F), then the LCL is estimated at 350 meters (1148 ft) above the surface.

Caution: Keep in mind that different layers of clouds may exist at the same time. Low, middle, and high clouds as defined in this chapter may all appear on the same afternoon. These clouds may have formed in other regions and be only passing overhead. The formula presented here is best used with the lowest level of cloud cover that appears overhead.

M. Trapasso

The stratocumulus clouds (bottom layer) show the lifting condensation level (LCL).

move onshore along east coasts or islands with high elevations, they bring moisture from the oceans with them. Thus, within the trade wind belt, continental east coasts tend to be wetter than continental west coasts.

In fact, where the air of the equatorial and trade wind regions—with its high temperatures and vast amounts of moisture—moves onshore from the ocean and meets a landform barrier, record rainfalls can be measured. The windward slope of Mount Waialeale on Kauai, Hawaii, at approximately 22°N latitude, holds the world's record for greatest average annual rainfall—1168 centimeters (460 in.).

Moving poleward from the trade wind belts, we enter the zones of subtropical high pressure where the air is subsiding. As it sinks lower, it is warmed adiabatically, increasing its moisture-holding capacity and consequently reducing the amount of precipitation in this area. In fact, if we look at Figure 6.23, which shows average annual precipitation on a latitudinal basis, we can see a dip in precipitation level corresponding to the latitude of the subtropical high pressure cells. These areas of subtropical high pressure are in fact where we find most of the great deserts of the world: in northern and southern Africa, Arabia, North America, and Australia. The exceptions to this subtropical aridity occur along the eastern sides of the landmasses where, as we have already noted, the subtropical high pressure cells are weak and wind direction is often onshore. This exception is especially true of regions affected by the monsoons.

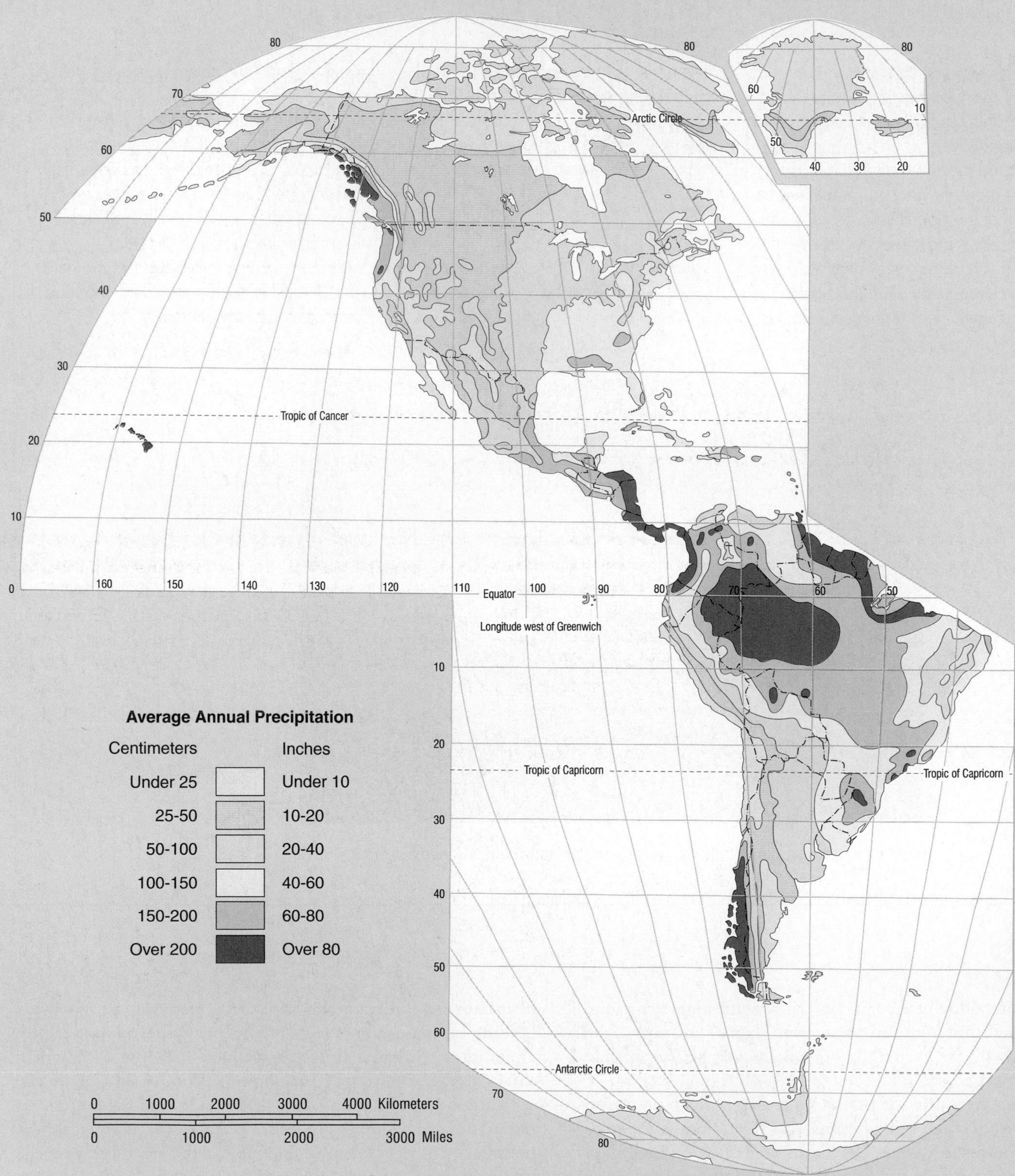

• FIGURE 6.23
World map of average annual precipitation.
In general, where on Earth's surface does the heaviest rainfall occur? Why?

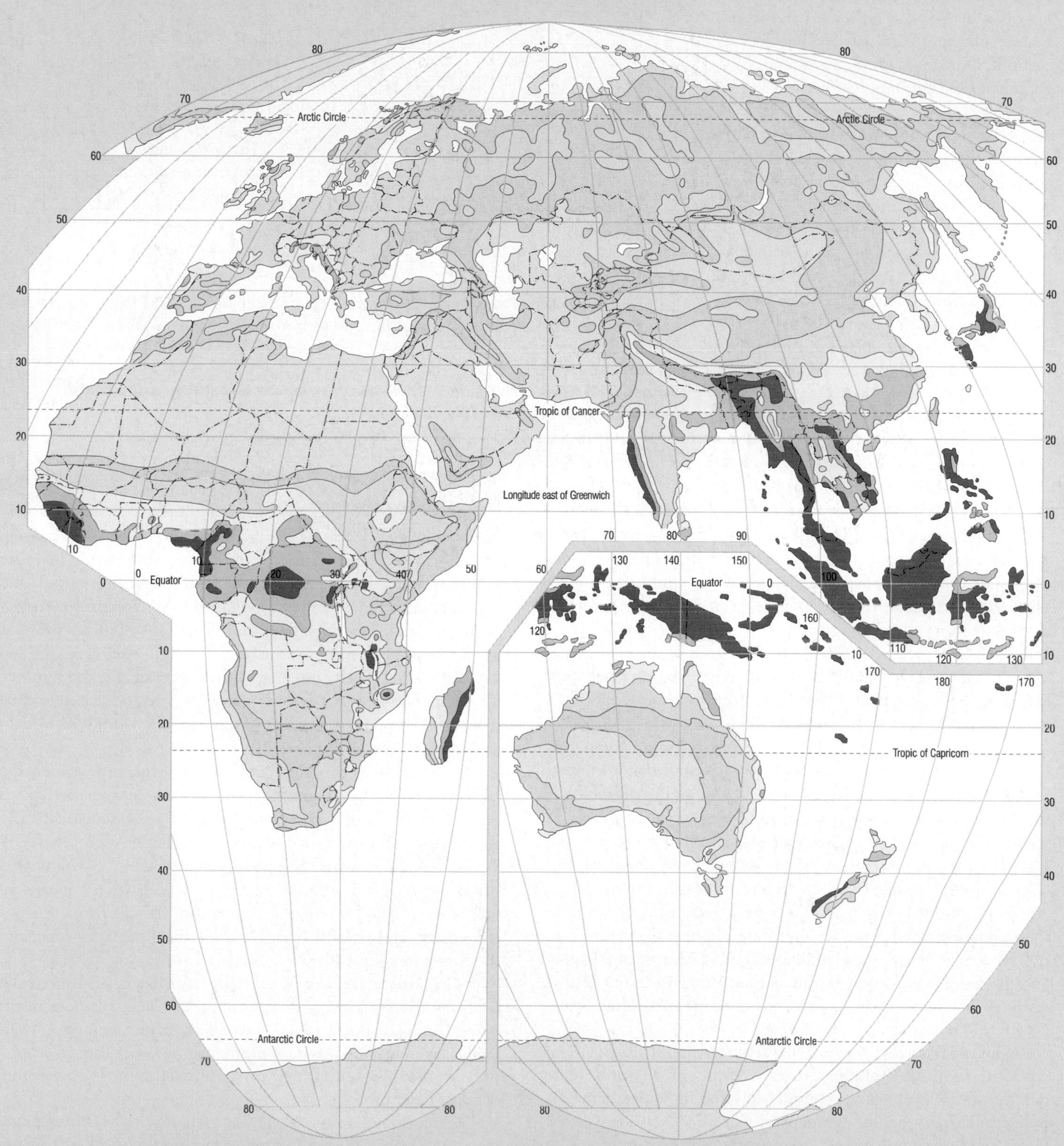

A Western Paragraphic Projection
developed at Western Illinois University

In the zones of the westerlies, from about 35° to 65°N and S latitude, precipitation occurs largely as a result of the meeting of cold, dry polar air masses and warm, humid subtropical air masses along the polar front. Thus, there is much cyclonic (convergence) and frontal precipitation in this zone.

Naturally, the continental interiors of the middle latitudes are drier than the coasts because they are farther away from the oceans. Furthermore, where air in the prevailing westerlies is forced to rise, as it is when it crosses the Cascades and Sierra Nevada of the Pacific Northwest and California, respectively, especially during the winter months, there is heavy orographic precipitation. Thus, in the middle latitudes, continental west coasts tend to be wet, and precipitation decreases with movement eastward toward continental interiors. Along eastern coasts within the westerlies, precipitation usually increases once again because of proximity to humid air from the oceans.

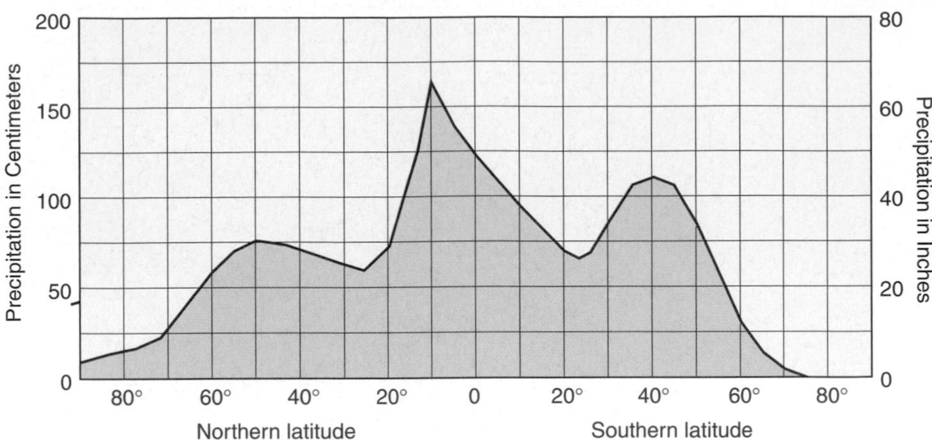

• FIGURE 6.24

Latitudinal distribution of average annual precipitation. This figure illustrates the four distinctive precipitation zones: high precipitation caused by convergence of air in the tropics and in the middle latitudes along the polar front and low precipitation caused by subsidence and divergence of air in the subtropical and polar regions.

Compare this figure with Figure 5.11. What is the relationship between world rainfall patterns and world pressure distribution?

In the United States, the interior lowlands are not as dry as we might expect within the prevailing westerlies. This is because of the great amount of frontal activity resulting from the conflicting northward and southward movements of polar and subtropical air. If there were a high east–west mountain range extending from central Texas to northern Florida, the lowlands of the continental United States north of that range would be much drier than they actually are because they would be cut off from moist air originating in the Gulf of Mexico.

Also characteristic of the belt of the westerlies are desert areas that occur in the rain shadows of prevailing winds that are forced to rise over mountain ranges. This effect is in part responsible for the development and maintenance of California's Death Valley, as well as the desert zone of eastern California and Nevada in the United States, the mountain-ringed deserts of eastern Asia, and Argentina's Patagonian Desert, which is in the rain shadow of the Andes. Note in • Figure 6.24 that there is greater precipitation in the middle latitudes of the Southern Hemisphere than there is in the Northern Hemisphere middle latitudes. This occurs largely because there is a lot more ocean and less landmass in the Southern Hemisphere westerlies than in the corresponding zone of the Northern Hemisphere.

Moving poleward, we find that temperatures decrease, along with the moisture-holding capacity of the air. The low temperatures also lead to low evaporation rates. In addition, the air in the polar regions shows a general pattern of subsidence that yields areas of high pressure. This settling of the air in the polar regions is the opposite of the lifting needed for precipitation. All these factors combine to cause low precipitation values in the polar zones.

Variability of Precipitation

The rainfall depicted in Figure 6.23 is an annual average. It should be remembered, however, that in many parts of the world there are significant variations in precipitation, both within any one year and between years. For example, areas like the Mediterranean region, California, Chile, South Africa, and Western Australia, which are on the west sides of the continents and roughly between 30° and 40° latitude, get much more rain in the winter than in the summer. There are also areas between 10° and 20° latitude that get much more of their precipitation in the summer (high-sun season) than in the winter (low-sun season).

Rainfall totals can change markedly from one year to the next, and tragically for many of the world's people, the drier a place is on the average, the greater will be the statistical variability in its precipitation (compare • Fig. 6.25 with Fig. 6.23). To make matters worse for people in dry areas, a year with a particularly high amount of rainfall may be balanced with several years of below-average precipitation. This situation has occurred recently in West Africa's Sahel, the Russian steppe, and the American Great Plains.

Thus, there are years of drought and years of flood, each bringing its own kind of disaster upon the land. Farmers, resort owners, construction workers, and others whose economic well-being depends in one way or another on weather can determine only a probability of rainfall on an annual, monthly, or even a seasonal basis.

Meteorologists cannot predict rainfall with 100% accuracy. This inability is due to the many factors involved in causing precipitation—temperature, available moisture, atmospheric disturbances, landform barriers, frontal activity, air mass movement, upper air winds, and differential surface heating, among

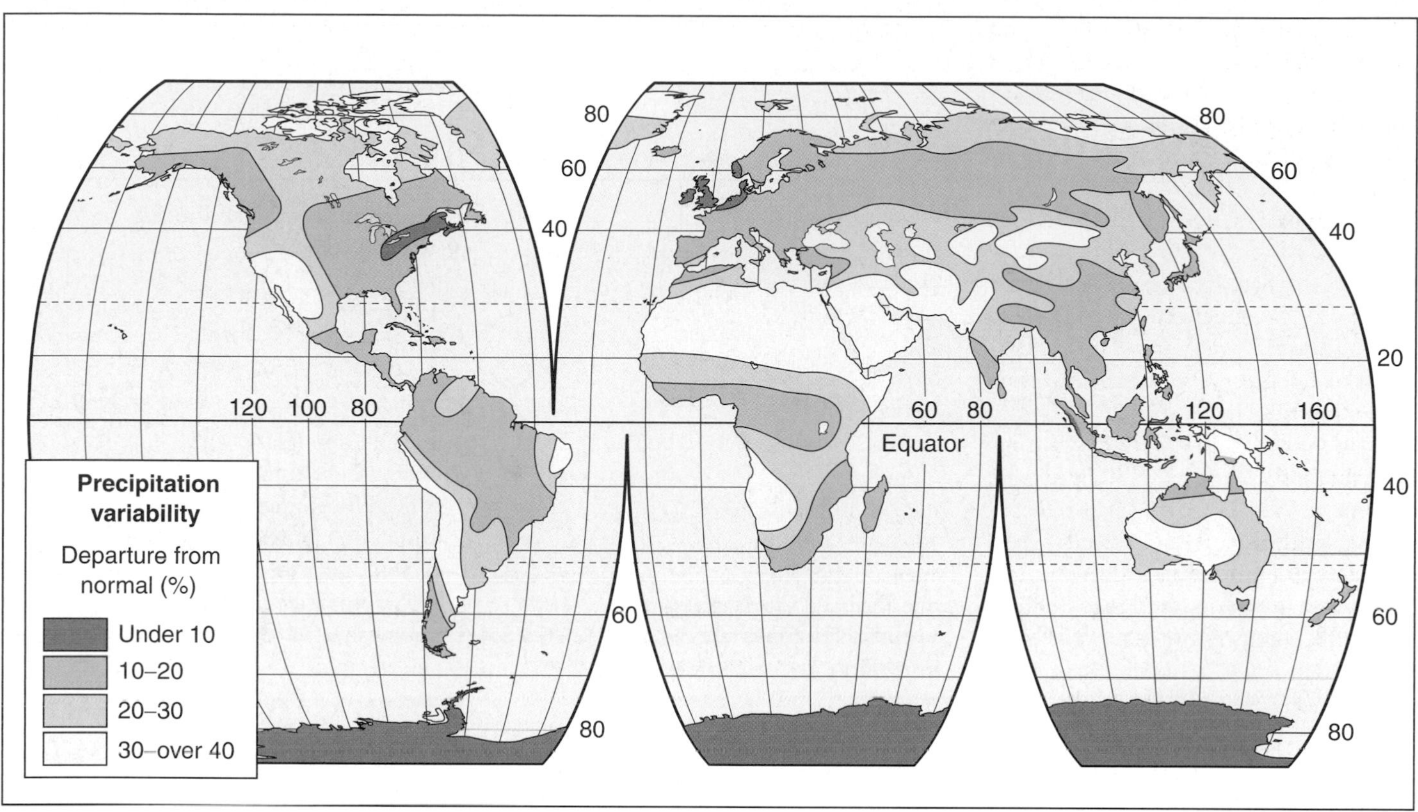

• FIGURE 6.25
World map of precipitation variability. The greatest variability in year-to-year precipitation totals occurs in the dry regions, accentuating the critical problem of moisture supply in those parts of the world.
Compare this figure with Figure 6.23. What are some of the similarities and differences?

others. In addition, the interaction of these factors in the development of precipitation is very complex and not completely understood.

Study of the hydrologic system helps scientists understand how changes in one subsystem can greatly impact other subsystems. For example, it is theorized that changes in the radiation budget of the Earth system can produce changes in the hydrologic cycle that in turn can explain changes in glacial subsystems and bring about continental glaciation (an ice age). If the energy available to heat Earth's atmosphere were reduced, the colder temperatures would affect the hydrologic cycle. A colder atmosphere would store less water vapor. Thus, there would be less precipitation, but the colder temperatures would result in more snowfall. With time, glacial ice would increase, and sea level would drop. The high albedo of the ice would further reduce the energy available to heat our atmosphere, which would allow glacial ice to increase and eventually result in continental ice sheets covering large portions of Earth.

Even without a dramatic example like the one described above, it should already become apparent that our planet's hydrologic system permeates all other major Earth subsystems; so much so, that it is extremely difficult to include all discussions concerning water in a single chapter, all its own. Instead, in future chapters, the vital role of water will be emphasized as each of the other subsystems is examined in detail.

Chapter 6 Activities

Define & Recall

capillary action
hydrologic cycle
saturation
capacity
dew point
humidity
absolute humidity
specific humidity
relative humidity
transpiration
evapotranspiration
potential evapotranspiration
condensation nuclei
radiation fog (temperature-inversion fog)
advection fog
upslope fog
dew
white frost
rime
precipitation
strato
alto
cirro
cirrus
stratus
cumulus
nimbus (nimbo)
adiabatic cooling
adiabatic heating
dry adiabatic lapse rate
wet adiabatic lapse rate
instability
stability
collision–coalescence process
Bergeron (ice crystal) Process
supercooled water
rain
drizzle
snow
sleet
hail
freezing rain (glaze)
uplift mechanism
convectional precipitation
frontal precipitation
cyclonic (convergence) precipitation
orographic precipitation
front
rain shadow
trace

Discuss & Review

1. How is the hydrologic cycle related to Earth's water budget?
2. What is the difference between absolute and specific humidity? What is relative humidity?
3. Imagine that you are deciding when, in your daily schedule, to water the garden. What time of day would be best for conserving water? Why?
4. What factors must be taken into consideration when calculating potential evapotranspiration? How is evapotranspiration related to the water budget of a region?
5. What factors affect the formation of temperature-inversion fogs?
6. What causes adiabatic cooling? Differentiate between the environmental lapse rate and adiabatic lapse rates.
7. How and why does the wet adiabatic lapse rate differ from the dry adiabatic lapse rate?
8. How is atmospheric stability related to the adiabatic lapse rates?
9. What atmospheric conditions are necessary for precipitation to occur?
10. Find out how many inches of precipitation have fallen in your area this year. Is that average or unusually high or low?
11. Compare and contrast convectional, orographic, cyclonic (convergence), and frontal precipitation.
12. How is rainfall variability related to total annual rainfall? How might this relationship be considered a double problem for people?

Consider & Respond

1. Refer to Figure 6.5.
   a. What is the water vapor capacity of air at 0°C? 20°C? 30°C?
   b. If a parcel of air at 30°C has an absolute humidity (actual water vapor content) of 20.5 grams per cubic meter, what is the parcel's relative humidity?
   c. If the relative humidity of a parcel of air is 33% and the air temperature is 15°C, what is the absolute humidity of the air in grams per cubic meter?
   d. A major concern of northern-climate residents is the low relative humidity within their homes during the winter. Low relative humidity is not healthy, and it has an adverse effect on the homes' furnishings. The problem results when cold air, which can hold little water vapor, is brought indoors and heated up. The following example will illustrate the problem: Assume that the air outside is 5°C and has a relative humidity of 60%. What is the actual water vapor content of this air? If it is brought indoors (through the doors, windows, and cracks in the home) and heated to 20°C, with no increase in water vapor content, what is the new relative humidity?

2. Recall that as a parcel of air rises, it expands and cools. The rate of cooling, termed the dry adiabatic lapse rate, is 10°C per 1000 meters. (A descending parcel of air will always warm at this rate.) In addition, the dew point temperature decreases about 2°C per 1000 meters within a rising parcel of air. At the height at which the dew point temperature is reached, condensation begins and, as discussed in the text, the wet adiabatic lapse rate of 5°C per 1000 meters becomes operational. When the wet adiabatic lapse rate is in operation, the dew point temperature will be the same as the air temperature. When a parcel descends through the atmosphere, its dew point temperature increases 2°C per 1000 meters.

The height at which condensation begins, termed the lifting condensation level (LCL), can be determined by using the formula found in *Geography's Physical Science Perspective: The Lifting Condensation Level (LCL)*, page 163.

a. A parcel of air has a temperature of 25°C and a dew point temperature of 14°C. What is the height of the LCL? If that parcel were to rise to 4000 meters, what would be its temperature?
A parcel of air at 6000 meters has a temperature of −5°C and a dew point of −10°C. If it descended to 2000 meters, what would be its temperature and dew point temperature?

Apply & Learn

1. Using the data set below, for each month of the year, calculate: (a) A running total of precipitation month to month. (b) The departure from the mean value (in surplus or deficit) for each month. (c) The annual departure from the mean value (in surplus or deficit) for the whole year.
How does the year begin with respect to surplus or deficit? How does the year end?
Which month accumulated the greatest deficit for the year?
Which month is the first to show a surplus?

Month	Recorded Rainfall	Mean Rainfall
January	4.94″	4.94″
February	2.40″	4.10″
March	3.02″	5.26″
April	3.21″	4.36″
May	3.69″	4.22″
June	2.69″	4.17″
July	5.08″	4.28″
August	5.41″	3.62″
September	7.89″	3.19″
October	4.10″	2.68″
November	4.27″	3.72″

2. Using the data set below, answer the following questions: (a) Following the procedure described in the section Distribution of Precipitation, page 160, calculate the probability of precipitation for each month of the year. (b) Which month has the highest probability? Which has the lowest probability? (c) What is the average probability for rainfall for the year?

Month	Number of days with >1mm (0.01 in.) of rain
January	12
February	10
March	7
April	11
May	16
June	11
July	8
August	9
September	10
October	10
November	11
December	9

Locate & Explore

Note: Please read the About Locate & Explore Activities section of the Preface before beginning these exercises.

1. Using Google Earth, Fly to the Aral Sea (45.2°N, 59.9°E) on the Uzbekistan–Kazakhstan border. Once you arrive at your coordinates, zoom out to view the extent of the Aral Sea, which was once one of the largest lakes in the world. As a result of irrigation projects and stream diversions, much of the water flowing into the lake was cut off and the lake has been significantly reduced in size. You can see the outline of the lake before it was reduced in size. Assuming that the lake can be characterized as a rectangle (area = length × width), what has been the change in the lake's area (in square miles and as a percentage of the original lake area)?
Tip: Use the ruler tool to measure the width and length.

Air Masses and Weather Systems

7

CHAPTER PREVIEW

The movement of relatively large bodies of air (air masses) is responsible for the transportation of distinguishable characteristics of temperature and humidity to regions far from their original sources.

- How is this important to the operation of Earth systems?
- How might air masses be modified?

The meeting of the leading edges of two unlike air masses occurs along a sloping surface of discontinuity called a front.

- How do air masses differ?
- Why are fronts important in explaining middle-latitude weather?

The major explanation for the variable and nearly unpredictable weather of the middle latitudes may be found in the irregular migration of relatively short-lived low pressure systems (cyclones) in the path of the prevailing westerlies.

- Why do cyclones play such a significant role?
- What are the human consequences of variable and unpredictable weather?

Tropical cyclones and extratropical cyclones are among the largest weather systems in the world. These two weather systems are known by other names.

- What are they?
- How do they differ?

Meteorology is an inexact science, and there is much yet to be learned about the behavior of air masses, fronts, and pressure systems. We should therefore anticipate that weather forecasting will remain a complicated art.

- How accurate is weather prediction?
- How successful are humans at altering the weather?

◄ **Opposite: An enhanced satellite image of Hurricane Katrina as it swirls toward the New Orleans area.**
Image courtesy of MODIS Rapid Response Project at NASA/GSFC

If we are to understand the types of weather that rule the middle latitudes, we must first come to grips with the vital parts of our basic weather systems. In the previous four chapters, we have looked at the elements of the atmosphere and investigated some of the controls that act upon those elements, causing them to vary from place to place and through time. However, even more is involved in the examination of weather. We have not yet looked at storms (atmospheric disturbances)—their types and characteristics, their origin, and their development. Weather systems that produce storms of various scales will be discussed in this chapter. Storms are an important part of the weather story. They help illustrate the interactions among the weather elements. Further, they represent a major means of energy exchange within the atmosphere.

Air Masses

Before we begin to study weather systems, we should understand the nature and significance of air masses. In themselves, air masses provide a straightforward way of looking at the weather. An **air mass** is a large body of air, at times subcontinental in size, that moves over Earth's surface with distinguishable characteristics. An air mass is relatively homogeneous in temperature and humidity; that is, at approximately the same altitude within the air mass, the temperature and humidity will be similar. As a result of this temperature and moisture uniformity, the density of air will be much the same throughout any one level within an air mass. Of course, because an air mass may extend over 20 or 30 degrees of latitude, we can expect some slight variations due to changes in sun angle and its corresponding insolation, which are significant over that distance. Changes caused by contact with differing land and ocean surfaces also affect the characteristics of air masses.

The similar characteristics of temperature and humidity within an air mass are determined by the nature of its **source region**—the place where the air mass originates. Only a few areas on Earth make good source regions. For the air mass to have similar characteristics throughout, the source region must have a nearly homogeneous surface. For example, it can be a desert, an ice sheet, or an ocean body, but not a combination of surfaces. In addition, the air mass must have sufficient time to acquire the characteristics of the source region. Hence, gently settling, slowly diverging air will mimic a source region, whereas converging, rising air will not.

Air masses are identified by a simple letter code. The first is always a lowercase letter. There are two choices: The letter *m,* for maritime, means the air mass originates over the sea and is therefore relatively moist. The letter *c,* for continental, means the air mass originates over land and is therefore relatively dry. The second letter is always a capital. These help to locate the latitude of the source region. *E* stands for Equatorial; this air is very warm. The letter *T* identifies a Tropical origin and is therefore warm air. A *P* represents Polar; this air can be quite cold. Lastly, an *A* identifies Arctic air, which is very cold. These six letters can be combined to give us the classification of air masses first described in 1928 and still used today: **Maritime Equatorial (*mE*), Maritime Tropical (*mT*), Continental Tropical (*cT*), Continental Polar (*cP*), Maritime Polar (*mP*),** and **Continental Arctic (*cA*).** These six types are described more fully in Table 7.1. From now on, we will use the symbols rather than the full names as we discuss each type of air mass.

Modification and Stability of Air Masses

As a result of the general circulation patterns within the atmosphere, air masses do not remain stationary over their source regions indefinitely. When an air mass begins to move over Earth's surface along a path known as a trajectory, for the most part it retains its distinct and homogeneous characteristics. However, modification does occur as the air mass gains or loses some of its thermal energy and moisture content to the surface below. Although this modification is generally slight, the gain or loss of thermal energy can make an air mass more stable or unstable.

An air mass is further classified by whether it is warmer or colder than the surface over which it travels because this has a bearing on its stability. If an air mass is colder than the surface over which it passes, then the surface will heat the air mass from below. This will in turn increase the environmental lapse rate, enhancing the prospect of instability. To describe such a situation, the letter *k* (from German: *kalt,* cold) is added to the other letters that symbolize the air mass. For example, an *mT* air mass originating over the Gulf of Mexico in summer that moves onshore over warmer land would be denoted *mTk.* Such an air mass is often unstable and can produce copious convective precipitation. On the other hand, this same *mT* air mass moving onshore during the winter would be warmer than the land surface. Consequently, the air mass would be cooled from below, decreasing its environmental lapse rate, which enhances the prospect of stability. We describe this situation with the letter *w* (from German: *warm,* warm), and the air mass would be denoted *mTw.* In this case, stratiform (lighter), not convective (heavier), precipitation is most likely.

The modification of air masses can also involve moisture content. For example, during the early-winter to midwinter seasons, cold, dry *cP* or *cA* air from Canada can move southeastward across the Great Lakes region. While passing over the lakes, this air mass can pick up moisture, thus increasing its humidity level. This modified *cP* or *cA* air reaches the frigid land on the leeward shores of the Great Lakes and precipitates, at times, large amounts of *lake-effect snows.* These snowfall areas may appear as *snow belts* or bands of snow, extending downwind from the lakes. The chances for lake-effect snow events diminish in late winter as the surfaces of the lakes freeze, thus cutting off the moisture supply to the air masses flowing across them.

North American Air Masses

Most of us are familiar with the weather in at least one region of the United States or Canada; therefore, in this chapter we will concentrate on the air masses of North America and their effects on weather. What we learn will be applicable to the rest of the world, and as we examine climate regions in some of the following chapters, we will be able to understand that weather everywhere is most often affected by the movements of air masses. Especially in middle-latitude regions, the majority of atmospheric disturbances result from the confrontations of different air masses.

Five types of air masses (*cA, cP, mP, mT,* and *cT*) influence the weather of North America, some more than others. Air masses assume characteristics of their source regions (● Fig. 7.1). Consequently, as the source regions change with the seasons, primarily because of changing insolation, the air masses also will vary.

Continental Arctic Air Masses The frigid, frozen surface of the Arctic Ocean and the land surface of northern Canada and Alaska serve as source regions for this air mass. It is extremely cold, very dry, and very stable. Though it will affect parts of Canada, even during the winter when this air mass is best developed, it seldom travels far enough south to affect the United States. However, on those few occasions when it does extend down into the midwestern and southeastern United States, its impact is awesome. Record-setting

TABLE 7.1
Types of Air Masses

Source	Region	Usual Characteristics at Source	Accompanying Weather
Maritime Equatorial (*mE*)	Equatorial oceans	Ascending air, very high moisture content	High temperature and humidity, heavy rainfall; never reaches the United States
Maritime Tropical (*mT*)	Tropical and subtropical oceans	Subsiding air; fairly stable but some instability on western side of oceans; warm and humid	High temperatures and humidity, cumulus clouds, convectional rain in summer; mild temperatures, overcast skies, fog, drizzle, and occasional snowfall in winter; heavy precipitation along *mT/cP* fronts in all seasons
Continental Tropical (*cT*)	Deserts and dry plateaus of subtropical latitudes	Subsiding air aloft; generally stable but some local instability at surface; hot and very dry	High temperatures, low humidity, clear skies, rare precipitation
Maritime Polar (*mP*)	Oceans between 40° and 60° latitude	Ascending air and general instability, especially in winter; mild and moist	Mild temperatures, high humidity; overcast skies and frequent fogs and precipitation, especially during winter; clear skies and fair weather common in summer; heavy orographic precipitation, including snow, in mountainous areas
Continental Polar (*cP*)	Plains and plateaus of subpolar and polar latitudes	Subsiding and stable air, especially in winter; cold and dry	Cool (summer) to very cold (winter) temperatures, low humidity; clear skies except along fronts; heavy precipitation, including winter snow, along *cP/mT* fronts
Continental Arctic (*cA*)	Arctic Ocean, Greenland, and Antarctica	Subsiding very stable air; very cold and very dry	Seldom reaches United States, but when it does, bitter cold, subzero temperatures, clear skies, often calm conditions

• FIGURE 7.1
Source regions of North American air masses. Air mass movements import the temperature and moisture characteristics of these source regions into distant areas.
Use Table 7.1 and this figure to determine which air masses affect your location. Are there seasonal variations?

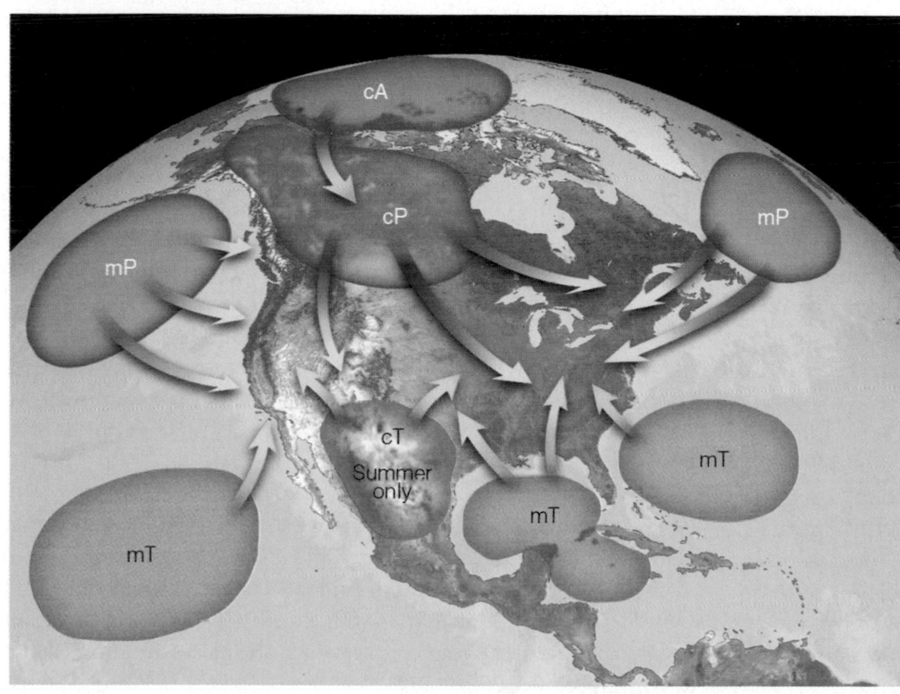

cold temperatures often result. If the *cA* air mass remains in the Midwest for an extended period, vegetation unaccustomed to the extreme cold can be severely damaged or killed.

Continental Polar Air Masses

At its source in north-central North America, a *cP* air mass is cold, dry, and stable because it is warmer than the surface beneath it; the weather of a *cP* air mass is cold, crisp, and clear. Because there are no east–west landform barriers in North America, *cP* air can migrate south across Canada and the United States. A tongue of *cP* air can sometimes reach as far south as the Gulf of Mexico or Florida. When winter *cP* air extends into the United States, its temperature and humidity are raised only slightly. The movement of such an air mass into the Midwest and South brings with it a cold wave characterized by colder-than-average temperatures and clear, dry air and can cause freezing temperatures as far south as Florida and Texas.

The general westerly direction of atmospheric circulation in the middle latitudes rarely allows a *cP* air mass to break through the great western mountain ranges to the West

Coast of the United States. When such an air mass does reach the Washington, Oregon, and California coasts, it brings with it unusual freezing temperatures that do great damage to agriculture.

Maritime Polar Air Masses During winter months, the oceans tend to be warmer than the land, so an *mP* air mass tends to be warmer than its counterpart on land (the *cP* air mass). Much *mP* air is originally cold, dry *cP* air that has moved to a position over the ocean. There, it is modified by the warmer water and collects heat and moisture. Thus, *mP* air is cold (although not as cold as *cP* air) and damp, with a tendency toward instability. The northern Pacific Ocean serves as the source region for *mP* air masses, which, because of the general westerly circulation of the atmosphere in the middle latitudes, affect the weather of the northwestern United States and southwestern Canada. When this *mP* air meets an uplift mechanism (such as a mass of colder, denser air or coastal mountain ranges), the result is usually very cloudy weather with a great deal of precipitation. An *mP* air mass may still be the source of many midwestern snowstorms even after crossing the western mountain ranges.

Generally, an *mP* air mass that develops over the northern Atlantic Ocean does not affect the weather of the United States because such an air mass tends to flow eastward toward Europe. However, on some occasions, there may be a reversal of the dominant wind direction accompanying a low pressure system, and New England can be made miserable by the cool, damp winds, rain, and snow of a weather system called a *nor'easter.* A nor'easter may, at times, bring serious winter storm conditions to our New England states.

Maritime Tropical Air Masses The Gulf of Mexico and subtropical Atlantic and Pacific Oceans serve as source regions for *mT* air masses that have a great influence on the weather of the United States and at times southeastern Canada. During winter, the waters are warm, and the air above is warm, and moist. As the warm, moist air moves northward up the Mississippi lowlands, it travels over increasingly cooler land surfaces. The lower layers of air are chilled, and dense advection fog often results. When it reaches the *cP* air migrating southward from Canada, the warm *mT* air is forced to rise over the colder, drier *cP* air, and significant precipitation can occur.

The longer days and more intense insolation of summer months modify an *mT* air mass at the source region by increasing its temperature and moisture content. However, during summer, the land is warmer than the nearby waters, and as the *mT* air mass moves onto the land, the instability of the air mass increases. This air mass is a factor in the formation of great thunderstorms and convective precipitation on hot, humid days, and it is also responsible for much of the hot, humid weather of the southeastern and eastern United States.

Maritime tropical air masses also form over the Pacific Ocean in the subtropical latitudes. These air masses tend to be slightly cooler than those that form over the Gulf of Mexico and the Atlantic, partly because of their passage over the cooler California Current. A Pacific *mT* air mass is also more stable because of the strong subsidence associated with the eastern portion of the Pacific subtropical high. This air mass contributes to the dry summers of Southern California and occasionally brings moisture in winter as it rises over the mountains of the Pacific Coast.

Continental Tropical Air Masses A fifth type of air mass may affect North America, but it is the least important to the weather of the United States and Canada. This is the *cT* air mass that develops over large, homogeneous land surfaces in the subtropics. The Sahara Desert of North Africa is a prime example of a source region for this type of air mass. The weather typical of the *cT* air mass is usually very hot and dry, with clear skies and major heating from the sun during daytime.

In North America, there is little land in the correct latitudes to serve as a source region for a *cT* air mass of any significant proportion. A small *cT* air mass can form over the deserts of the southwestern United States and central Mexico in the summer. In the source region, a *cT* air mass provides hot, dry, clear weather. When it moves eastward, however, it is usually greatly modified as it comes in contact with larger and stronger air masses of different temperature, humidity, and density values. At times, *cT* air from Mexico and Texas meets with *mT* air from the Gulf of Mexico. This boundary is known as a *dry line.* Here, the drier air is denser and will lift the moister air over it. This mechanism of uplift may act as a trigger for precipitation episodes and perhaps thunderstorm activity.

Fronts

We have seen that air masses migrate with the general circulation of the atmosphere. Over the United States, which is influenced primarily by the westerlies, there is a general eastward flow of the air masses. In addition, air masses tend to diverge from areas of high pressure and converge toward areas of low pressure. This tendency means that the tropical and polar air masses, formed within systems of divergence, tend to flow toward areas of convergence within the United States. As previously noted, an important feature of an air mass is that it maintains the primary characteristics first imparted to it by its source region, although some slight modification may occur during its migration.

When air masses differ, they do so primarily in their temperature and in their moisture content, which in turn affect the air masses' density and atmospheric pressure. As we saw in Chapter 6, when different air masses come together, they do not mix easily but instead come in contact along sloping boundaries called *fronts.* Although usually depicted on maps as a one-dimensional boundary line separating two different air masses, a front is actually a three-dimensional surface with length, width, and height. To emphasize this concept, a front is sometimes referred to as a **surface of discontinuity.** This surface of discontinuity is a zone that can cover an area from 2 to 3 kilometers (1–2 mi) wide to as wide as 150 kilometers (90 mi). Hence, it is more accurate to speak of a frontal zone rather than a frontal line.

The sloping surface of a front is created as the warmer and lighter of the two contrasting air masses is lifted or rises above the cooler and denser air mass. Such rising, known as *frontal uplift,* is a major source of precipitation in middle-latitude countries like the United States and Canada (as well as middle-latitude European and Asian countries) where contrasting air masses are most likely to converge.

The steepness of the frontal surface is governed primarily by the degree of difference between the two converging air masses. When there is a sharp difference between the two air masses, as when an *mT* air mass of high temperature and moisture content meets a *cP* air mass with its cold, dry characteristics, the slope of the frontal surface will be steep. With a steep slope, there will be greater frontal uplift. Provided other conditions (for example, temperature and moisture content) are equal, a steep slope with its greater frontal uplift will produce heavier precipitation than will a gentler slope.

Fronts are also differentiated by determining whether the colder air mass is moving in on the warmer one, or vice versa. The weather that occurs along a front also depends on which air mass is the "aggressor."

Cold Front

A **cold front** occurs when a cold air mass actively moves in on a warmer air mass and pushes it upward. The colder air, denser and heavier than the warm air it is displacing, stays at the surface while forcing the warmer air to rise. As we can see in • Figure 7.2, a cold front usually results in a relatively steep slope in which the warm air may rise 1 meter vertically for every 40–80 meters of horizontal distance. If the warm air mass is unstable and has a high moisture content, heavy precipitation can result, sometimes in the form of violent thunderstorms. A **squall line** may result when several storms align themselves on (or in advance of) a cold front. In any case, cold fronts are usually associated with strong weather disturbances or sharp changes in temperature, air pressure, and wind.

• **FIGURE 7.2**
Cross section of a cold front. Cold fronts generally move rapidly, with a blunt forward edge that drives adjacent warmer air upward. This can produce violent precipitation from the warmer air.

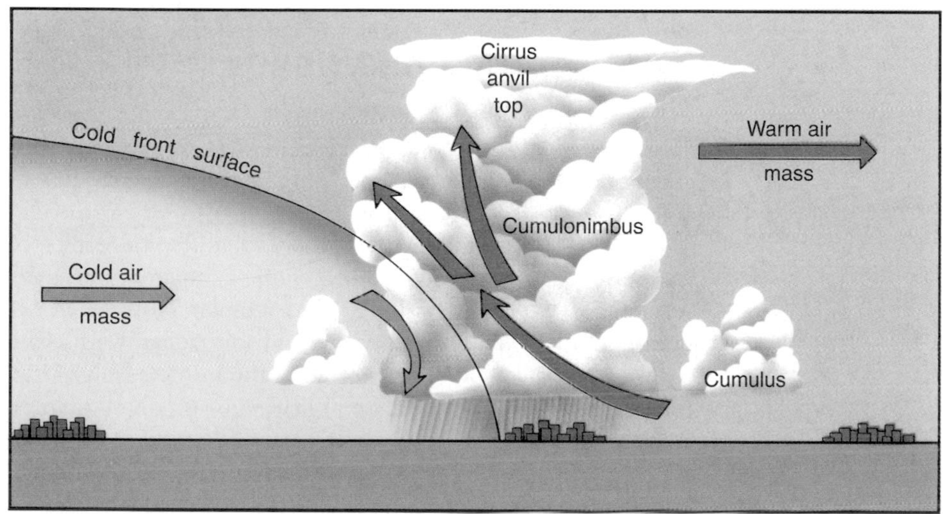

• **FIGURE 7.3**
Cross section of a warm front. Warm fronts advance more slowly than cold fronts and replace rather than displace cold air by sliding upward over it. The gentle rise of the warm air produces stratus clouds and gentle rain.
Compare Figures 7.2 and 7.3. How are they different? How are they similar?

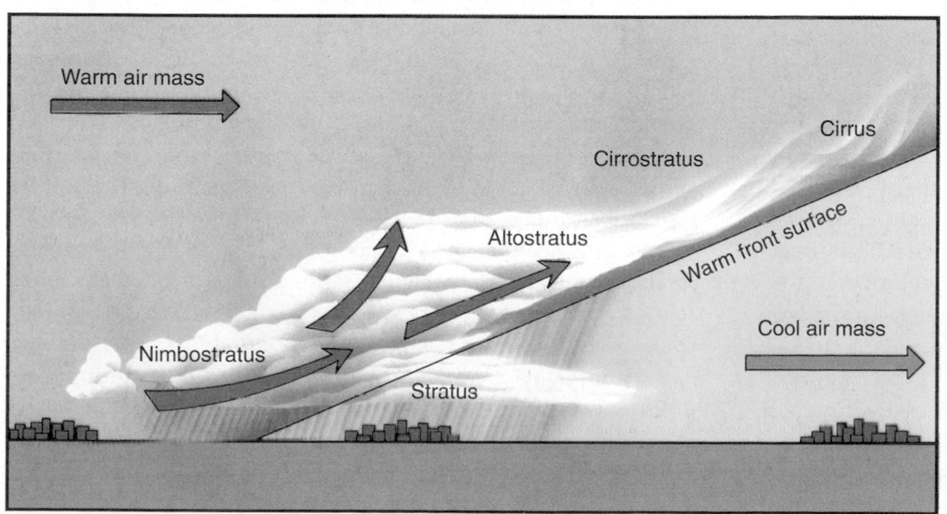

Warm Front

When a warmer air mass is the aggressor and invades a region occupied by a colder air mass, a **warm front** results. At a warm front, the warmer air, as it slowly pushes against the cold air, also rises over the colder, denser air mass, which again stays in contact with Earth's surface. The slope of the surface of discontinuity that results is usually far gentler than that occurring in a cold front. In fact, the warm air may rise only 1 meter vertically for every 100 or even 200 meters of horizontal distance. Thus, the frontal uplift that develops will not be as great as that occurring along a cold front. The result is that the warmer weather associated with the passage of a warm front tends to be less violent and the changes less abrupt than those associated with cold fronts.

If we look at • Figure 7.3, we can see why the advancing warm front affects the weather of areas ahead of the actual surface location of the frontal zone. Changes in the weather from approaching fronts can sometimes be indicated by the series of cloud types that precede them.

Stationary and Occluded Fronts

When two air masses have converged and formed a frontal boundary but then neither moves, we have a situation known as a *quasi-stationary* or, as it is more commonly called, a **stationary front.** Locations under the influence of a stationary front are apt to experience clouds, drizzle, and rain (or possible thunderstorms) for several days. In fact, a stationary front and its accompanying

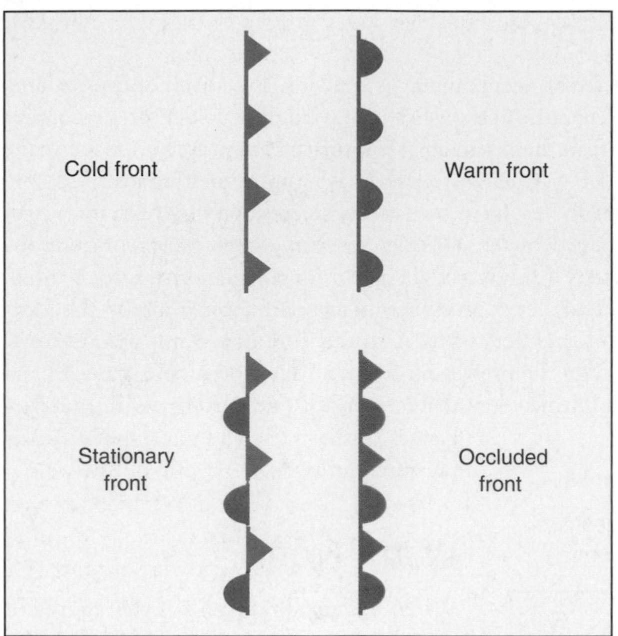

• **FIGURE 7.4**
The four major frontal symbols used on weather maps.

weather will remain until either the contrasts between the two air masses are reduced or the circulation of the atmosphere finally causes one of the air masses to move. If a stationary front holds a position for a length of time, then regional flooding is likely to occur.

An **occluded front** occurs when a faster-moving cold front overtakes a warm front, pushing all of the warm air aloft. This frontal situation usually occurs in the latter stages of a middle-latitude cyclone, which will be discussed next. Map symbols for the four frontal types are shown in • Figure 7.4.

Atmospheric Disturbances

Middle Latitude Cyclones

Embedded within the wind belts of the general atmospheric circulation (see Chapter 5) are secondary circulations. These are made up of storms and other atmospheric disturbances. We use the term **atmospheric disturbance** because it is more general than a storm and includes variations in the secondary circulation of the atmosphere that cannot be correctly classified as storms.

Partly because our primary interest is in the weather of North American, we concentrate on an examination of **middle-latitude cyclones,** sometimes known as **extratropical cyclones.**

Shortly after World War I, Norwegian meteorologists Jacob Bjerknes and Halvor Solberg put forth the *polar front theory,* which provided insight into the development, movement, and dissipation of middle-latitude storms. They recognized the middle latitudes as an area of convergence where unlike air masses, such as cold polar air and warm subtropical air, commonly meet at a boundary called the *polar front*. Though the polar front may be a continuous boundary circling the entire globe, it is most often fragmented into several individual line segments. Furthermore, the polar front tends to move north and south with the seasons and is apt to be stronger in winter than in summer. It is along this wavy polar front that the upper air westerlies (see again Figs. 5.16 and 5.17), also known as the *polar front jet stream,* develop and flow.

Middle-latitude storms develop at the front and then travel along it. These migrating storms, with their opposing cold, dry polar air and warm, humid tropical air, can cause significant variation in the day-to-day weather of the locations over which they pass. It is not unusual in some parts of the United States and Canada for people to go to bed at the end of a beautiful warm day in early spring and wake up to falling snow the next morning. Such variability is common for middle-latitude weather, especially during certain times of the year when the weather changes from a period of cold, clear, dry days to a period of snow, only to be followed by one or two more moderate but humid days.

Cyclones and Anticyclones

Nature, Size, and Appearance on Maps We have previously distinguished cyclones and anticyclones according to differences in pressure and wind direction. Also, when studying maps of world pressure distribution, we identified large areas of semipermanent cyclonic and anticyclonic circulation in Earth's atmosphere (the subtropical high, for example). Now, when examining middle-latitude atmospheric disturbances, we use the terms *cyclone* and *anticyclone* to describe the moving cells of low and high pressure, respectively, that drift with varying regularity in the path of the prevailing westerly winds. As systems of higher pressure, anticyclones are usually characterized by clear skies, gentle winds, and a general lack of precipitation. As centers for converging, rising air, cyclones create the storms of the middle latitudes, with associated fronts of various types.

As we know from experience, no middle-latitude cyclonic storm is ever exactly like any other. The storms vary in their intensity, their longevity, their speed, the strength of their winds, their amount and type of cloud cover, the quantity and kind of their precipitation, and the surface area they affect.

Because there are an endless variety of cyclones, we describe "model cyclones" in the following discussions. Not every storm will act in the way we describe, but certain generalizations are helpful in understanding middle-latitude cyclones.

A cyclone has a low pressure center; thus, winds tend to converge toward that center in an attempt to equalize pressure. If we visualize air moving in toward the center of the low pressure system, we can see that the air that is already at the center must be displaced upward. Incoming *mT* air spirals upward, and the lifting (convergence uplift also known as cyclonic uplift) that occurs in a cyclone results in clouds and precipitation.

Anticyclones are high pressure systems in which atmospheric pressure decreases toward the outer limits of the system. Visualizing an anticyclone, or high, we can see that air in the center of the system must be subsiding, in turn displacing surface air outward, away from the center of the system. Hence, an anticyclone has diverging winds. In addition, an anticyclone tends to be a fair-weather system; the subsiding air in its center increases in temperature and stability, reducing the opportunity for condensation.

We should note here that the pressures we are referring to in these two systems are relative. What is important is that in a cyclone, pressure decreases toward the center, and in an anticyclone, pressure increases toward the center. Furthermore, the intensities of the winds involved in these systems depend on the steepness of the *pressure gradients* (the change in pressure over a horizontal distance) involved. Thus, if there is a steep pressure gradient in a cyclone, with the pressure much lower at the center than at the outer portions of the system, the winds will converge toward the center with considerable velocity.

The situation is easier to visualize if we imagine these pressure systems as landforms. A cyclone is shaped like a basin (● Fig. 7.5). If we are filling the basin with water, we know that the water will flow in faster the steeper the sides and the deeper the depression. If we visualize an anticyclone as a hill or mountain, then we can also see that just as water flowing down the sides of such landforms will flow faster with increased height and steepness, so will the air blowing out of an area of very high pressure move rapidly.

On a surface weather map, cyclones and anticyclones are depicted by concentric isobars of increasing pressure toward the center of a high and of decreasing pressure toward the center of a low. Usually a high will cover a larger area than a low, but both pressure systems are capable of covering and affecting extensive areas. There are times when nearly the entire midwestern United States is under the influence of the same system. The average diameter of an anticyclone is about 1500 kilometers (900 mi); that of a cyclone is about 1000 kilometers (600 mi).

● FIGURE 7.5

The horizontal and vertical structure of pressure systems. Close spacing of isobars around a cyclone or anticyclone indicates a steep pressure gradient that will produce strong winds. Wide spacing of isobars indicates a weaker system.

Where would be the strongest winds in this figure? Where would be the weakest winds?

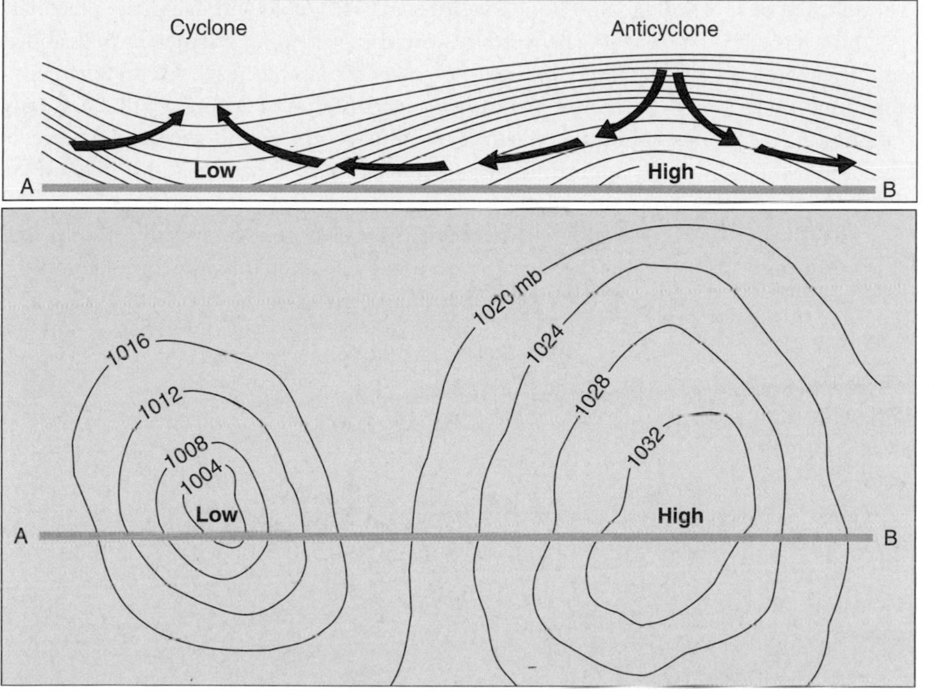

General Movement

The cyclones and anticyclones of the middle latitudes are steered, or guided, along a path reflecting the configuration and speed of the upper air westerlies (or the jet stream). The upper air flow can be quite variable with wild oscillations. However, a general west-to-east pattern does prevail. Consequently, people in most of the eastern United States look at the weather occurring to the west to see what they might expect in the next few days. Most storms that develop in the Great Plains or Far West move across the United States during a period of a few days at an average speed of about 36 kilometers per hour (23 mph) and then travel on into the North Atlantic before occluding.

Although neither cyclones nor anticyclones develop in exactly the same places at the same times each year, they do tend to develop in certain areas or regions more frequently than in others. They also follow the same general paths, known as **storm tracks** (● Fig. 7.6). These storm tracks vary

● FIGURE 7.6

Common storm tracks for the United States. Virtually all cyclonic storms move from west to east in the prevailing westerlies and swing northeastward across the Atlantic coast. Storm tracks originating in the Gulf of Mexico represent tropical hurricanes.

What storm tracks influence your location?

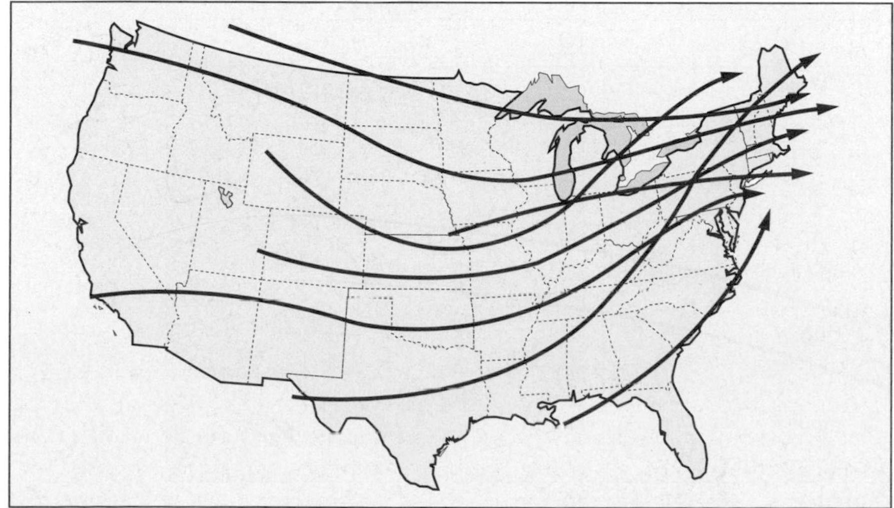

with the seasons. In addition, because the temperature variations between the air masses are stronger during the winter months, the atmospheric disturbances that develop in the middle latitudes during those months are greater in number and intensity.

Cyclones Now let's look more closely at cyclones—their origin, development, and characteristics. Warm and cold air masses meet at the polar front where most cyclones develop. These two contrasting air masses do not merge but may move in opposite directions along the frontal zone. Although there may be some slight uplift of the warmer air along the edge of the denser, colder air, the uplift will not be significant. There may be some cloudiness and precipitation along such a frontal zone, though not of storm caliber.

For reasons not completely understood but certainly related to the wind flow in the upper troposphere, a wavelike kink may develop along the polar front. This is the initial step in the formation of a fully grown middle-latitude cyclone (• Fig. 7.7). At this bend in the polar front, we now have warm air pushing poleward (a warm front) and cold air pushing equatorward (a cold front), with a center of low pressure at the location where the two fronts are joined.

As the contrasting air masses jockey for position, the clouds and precipitation that exist along the fronts are greatly intensified, and the area affected by the storm is much greater. Along the warm front, precipitation will be more widespread but less intense than along the cold front. One factor that can vary the kind of precipitation occurring at the warm front is the stability of the warm air mass. If it is stable, then its uplifting over the cold air mass may cause only a fine drizzle or a light, powdery snow if the temperatures are cold enough. On the other hand, if the warm air mass is moist and unstable, the uplifting may set off heavier precipitation. As you can see by referring again to Figure 7.3, the precipitation that falls at the warm front may *appear* to be coming from the colder air. Though weather may feel cold and damp, the precipitation is actually coming from the overriding warmer air mass above, then falling through the colder air mass to reach Earth's surface.

Because a cold front usually moves faster, it will eventually overtake the warm front. This produces the situation we previously identified as an *occluded front*. Because additional warm, moist air will not be lifted after occlusion, condensation and the release of latent heat energy will diminish, and the system will soon die. Occlusions are usually accompanied by rain and are the major process by which middle-latitude cyclones dissipate.

• FIGURE 7.7
Environmental Systems: Middle-Latitude Cyclonic Systems Stages in the development of a middle-latitude cyclone. Each view represents the development somewhat eastward of the preceding view as the cyclone travels along its storm track. Note the occlusion in (e).
In (c), where would you expect rain to develop? Why?

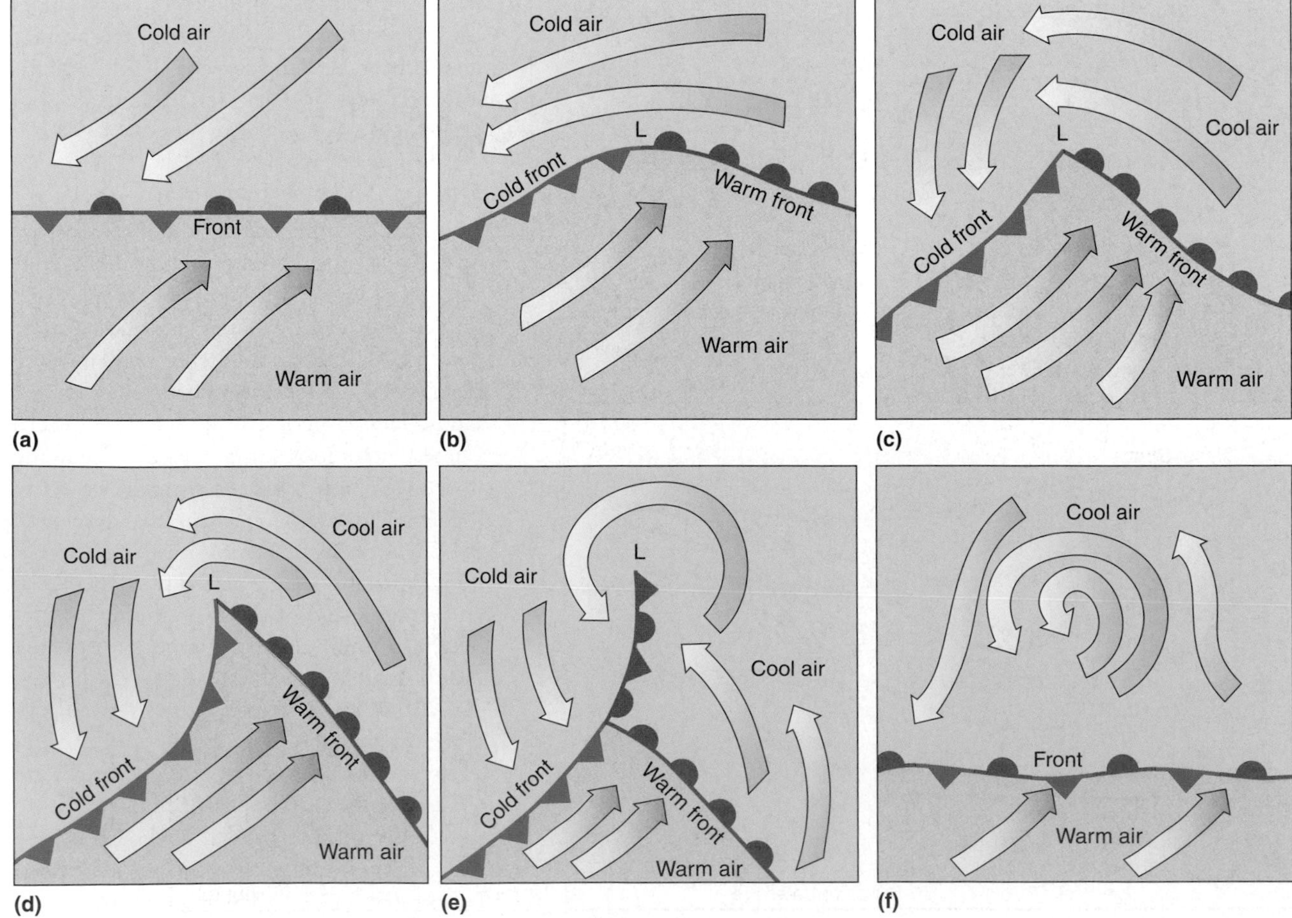

Cyclones and Local Weather Different parts of a middle-latitude cyclone exhibit different weather. Therefore, the weather that a location experiences at a particular time depends on which portion of the middle-latitude cyclone is over the location. Also, because the entire cyclonic system tends to travel from west to east, a specific sequence of weather can be expected at a given location as the cyclone passes over that location.

Let's assume that it is late spring. A cyclonic storm has originated in the southeast corner of Nebraska and is following a track (see again Figure 7.6) across northern Illinois, northern Indiana, northern Ohio, through Pennsylvania, and finally out over the Atlantic Ocean. A view of this storm on a weather map, at a specific time in its journey, is presented in • Figure 7.8a. Figure 7.8b shows a cross-sectional view north of the center of the cyclone, and Figure 7.8c shows a cross-sectional view south of the center of the cyclone. As the storm continues eastward, at 33–50 kilometers per hour (20–30 mph), the sequence of weather will be different for Detroit, where the warm and cold fronts will pass just to the south, than for Pittsburgh, where both fronts will pass overhead. To illustrate this point, let's examine, element by element, the variation in weather that will occur in Pittsburgh, with reference, where appropriate, to the differences that occur in Detroit as the cyclonic system moves east.

As we have previously stated, atmospheric temperature and pressure are closely related. As temperature increases, air expands and pressure decreases. Therefore, these two elements are discussed together. Because a cyclonic storm is composed of two dissimilar air masses, there are usually significant temperature contrasts. The sector of warm, humid *mT* air between the two fronts of the cyclone is usually considerably warmer than the cold *cP* air surrounding it. The temperature contrast is accentuated in the winter when the source region for *cP* air is the cold cell of high pressure normally found in Canada at that time of year. During the summer, the contrast between these air masses is greatly reduced.

As a consequence of the temperature difference, the atmospheric pressure in the warm sector is considerably lower than the atmospheric pressure in the cold air behind the cold front. Far in advance of the warm front, the pressure is also high, but as the warm front (see again Figure 7.3) approaches, increasingly more cold air is replaced by uplifted warm air, thus steadily reducing the surface pressure.

• FIGURE 7.8
Environmental Systems: Middle-Latitude Cyclonic Systems This diagram models a middle-latitude cyclone positioned over the Midwest as the system moves eastward: (a) a map view of the weather system; (b) a cross section along line AB north of the center of low pressure; (c) a cross section along line CD south of the center of low pressure.

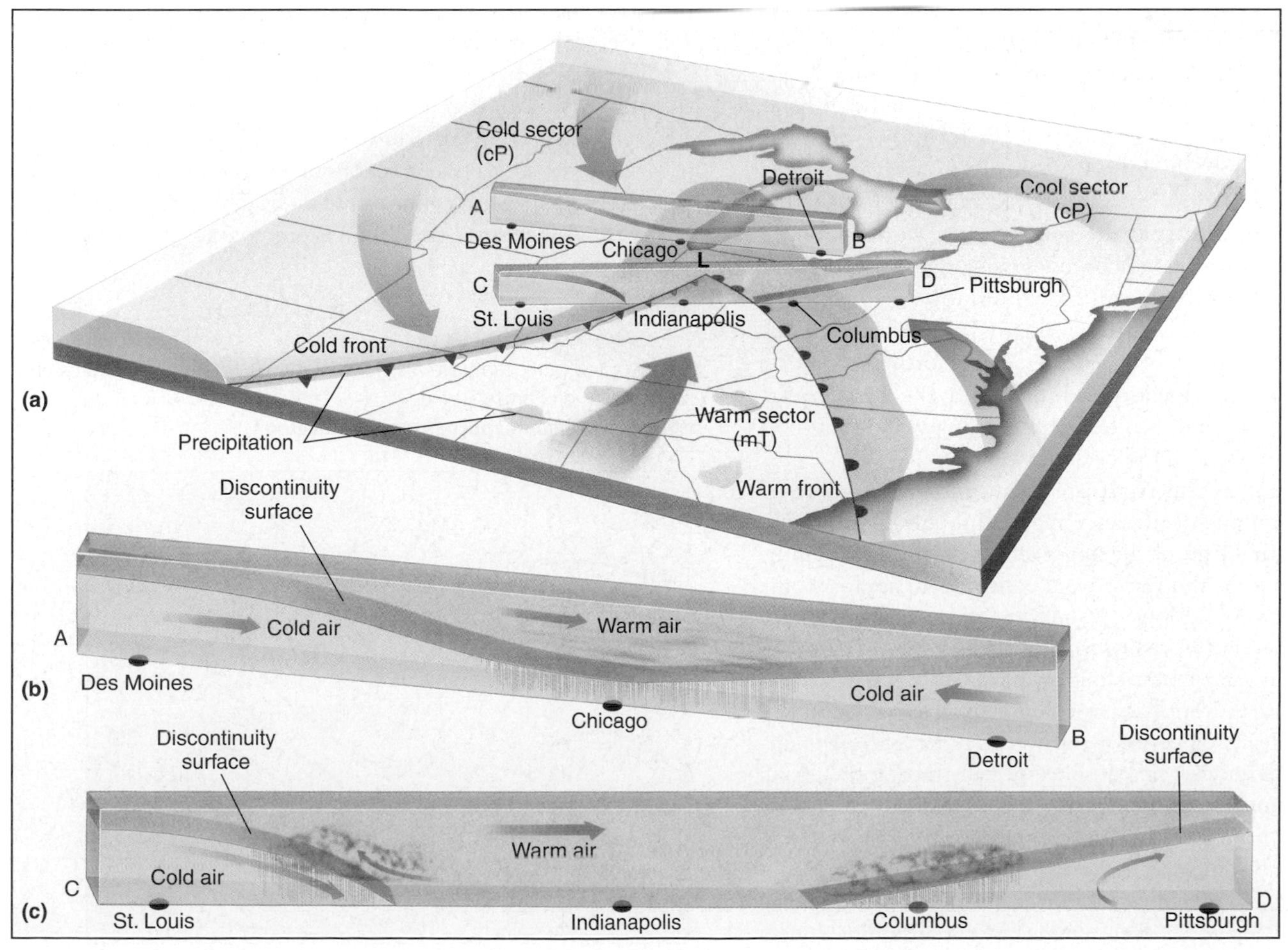

Therefore, as the warm front of this late-spring cyclonic storm approaches Pittsburgh, the pressure will decrease. After the warm front passes through Pittsburgh where the temperature may have been 8°C (46°F) or more, the pressure will stop falling, and the temperature may rise up to 18°–20°C (64°F–69°F) as *mT* air invades the area. At this point, Indianapolis has already experienced the passage of the warm front. After the cold front passes, the pressure will rise rapidly and the temperature will drop. In this late-spring storm, the *cP* air temperature behind the cold front might be 2°C–5°C (35°F–40°F). Detroit, which is to the north of the center of the cyclone, will miss the warm air sector entirely and therefore will experience a slight increase in pressure and a temperature change from cool to cold as the cyclone moves to the east.

Changes in wind direction are one signal of the approach and passing of a cyclonic storm. Because a cyclone is a center of low pressure, winds flow counterclockwise toward its center. Also, winds are caused by differences in pressure. Therefore, the winds associated with a cyclonic storm are stronger in winter when the pressure (and temperature) differences between air masses are greatest.

In our example, Pittsburgh is located to the south and east of the center of low pressure and ahead of the warm front, and it is experiencing winds from the southeast. As the entire cyclonic system moves east, the winds in Pittsburgh will shift to the south-southwest after the warm front passes. Indianapolis is currently in this position. After the cold front passes, the winds in Pittsburgh will be out of the north-northwest. St. Louis has already experienced the passage of the cold front and currently has winds from the northwest. The changing direction of wind, clockwise around the compass from east to southeast to south to southwest to west and northwest, is called a **veering wind shift** and indicates that your position is south of the center of a low. On the other hand, Detroit, which is also experiencing winds from the southeast, will undergo a completely different sequence of directional wind changes as the cyclonic storm moves eastward. Detroit's winds will shift to the northeast as the center of the storm passes to the south. Chicago has just undergone this shift. Finally, after the storm has passed, the winds will blow from the northwest. Des Moines, to the west of the storm, currently has northwest winds. Such a change of wind direction, from east to northeast to north to northwest, is called a **backing wind shift,** as the wind "backs" counterclockwise around the compass. A backing wind shift indicates that you are north of the cyclone's center.

The type and intensity of precipitation and cloud cover also vary as a cyclonic disturbance moves through a location. In Pittsburgh, the first sign of the approaching warm front will be high cirrus clouds. As the warm front continues to approach, the clouds will thicken and lower. When the warm front is within 150–300 kilometers (90–180 mi) of Pittsburgh, light rain and drizzle may begin, and stratus clouds will blanket the sky.

After the warm front has passed, precipitation will stop and the skies will clear. However, if the warm, moist *mT* air is unstable, convective showers may result, especially during the spring and summer months with their high afternoon sun angles.

As the cold front passes, warm air in its path will be forced to move aloft rapidly. This may mean that there will be a cold, hard rain, but the band of precipitation normally will not be very wide because of the steep angle of the surface of discontinuity along a cold front. In our example, the cold front and the band of precipitation have just passed St. Louis.

Thus, Pittsburgh can expect three zones of precipitation as the cyclonic system passes over its location: (1) a broad area of cold showers and drizzle in advance of the warm front; (2) a zone within the moist, subtropical air from the south where scattered convectional showers can occur; and (3) a narrow band of hard rainfall associated with the cold front (Fig. 7.8c). However, locations to the north of the center of the cyclonic storm, such as Detroit, will usually experience a single, broad band of light rains resulting from the lifting of warm air above cold air from the north (Fig. 7.8b). In winter, the precipitation is likely to be snow, especially in locations just to the northwest of the center of the storm, where the humid *mT* air overlies extremely cold *cP* air.

As you can see, different portions of a middle-latitude cyclone are accompanied by different weather. If we know where the cyclone will pass relative to our location, we can make a fairly accurate forecast of what our weather will be like as the storm moves east along its track (see Map Interpretation: Weather Maps).

Cyclones and the Upper Air Flow The upper air wind flow greatly influences our surface weather. We have already discussed the role of these upper air winds in the steering of surface storm systems. Another less obvious influence of the upper air flow is related to the undulating, wavelike flow so often exhibited by the upper air. As the air moves its way through these waves, it undergoes divergence or convergence because of the atmospheric dynamics associated with curved flow. This upper air convergence and divergence greatly influence the surface storms below.

The region between a ridge and the next downwind trough (A–B in • Fig. 7.9) is an area of upper-level convergence. In our atmosphere, an action taken in one part of the atmosphere is compensated for by an opposite reaction somewhere else. In this case, the upper air convergence is compensated for by divergence at the surface. In this area, anticyclonic circulation is promoted as the air is pushed

• FIGURE 7.9
Waves in the polar front jet stream. The upper air wind pattern, such as that depicted here, can have a significant influence on temperatures and precipitation on Earth's surface.
Where would you expect storms to develop?

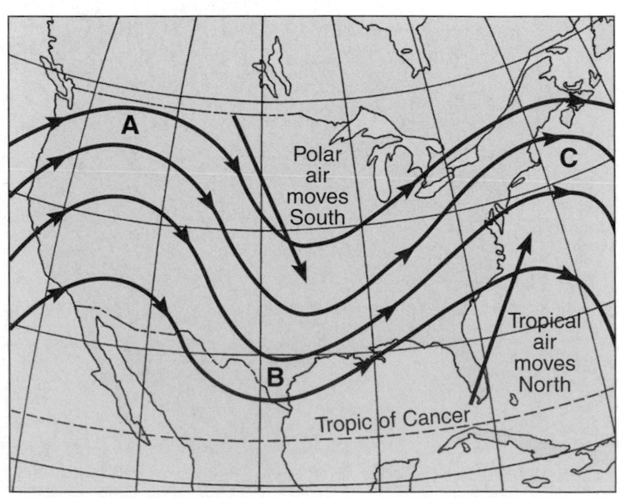

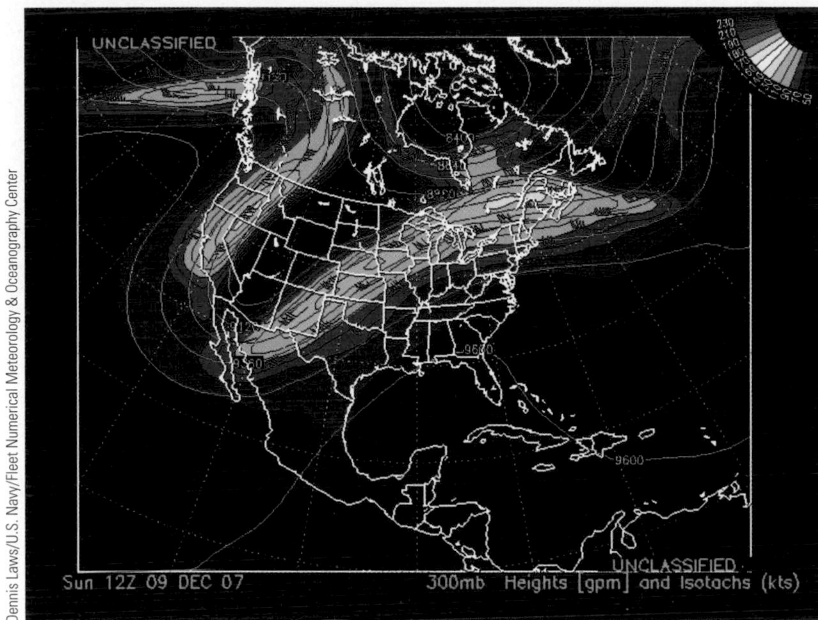

Dennis Laws/U.S. Navy/Fleet Numerical Meteorology & Oceanography Center

• **FIGURE 7.10**
Polar front jet stream analysis. This wind analysis map was produced at an altitude of 300 mb (approximately 10,000 m or 33,000 ft above sea level). At this height the long waves of the jet stream can more easily be seen. In this true winds depiction, the troughs and ridges are not as smooth and regular as they are in theory. Knots (kts), or nautical miles per hour, are a little faster than statute miles per hour. **Which country does most of this pattern occupy?**

downward. This pattern will inhibit the formation of a middle-latitude storm or cause an existing storm to weaken or even dissipate. On the other hand, the region between a trough and the next downwind ridge (B–C in Fig. 7.9) is an area of upper-level divergence, which in turn is compensated for by surface convergence. This is an area where air is drawn upward and cyclonic circulation is encouraged. Convergence at the surface will certainly enhance the prospects of storm development or strengthen an already existing storm.

In addition to storm development or dissipation, upper air flow will have an impact on temperatures as well. If we assume that our "average" upper air flow is from west to east, then any deviation from that pattern will cause either colder air from the north or warmer air from the south to be advected into an area. Thus, after the atmosphere has been in a wavelike pattern for a few days, the areas in the vicinity of a trough (area B in Fig. 7.9) will be colder than normal as polar air from higher latitudes is brought into that area. Just the opposite occurs at locations near a ridge (area C in Fig. 7.9). In this case, warmer air from more southerly latitudes than would be the case with west-to-east flow is advected into the area near the ridge. • Figure 7.10 shows that in reality, the jet stream curves with less regularity. Comparing Figures 7.9 and 7.10 you can see the difference between the theoretical and the real waves in the polar front jet stream.

Anticyclones Just as cyclones are centers of low pressure that are typified by the convergence and uplift of air, so anticyclones are cells of high pressure in which air descends and diverges. The subsidence of air in the center of an anticyclone encourages stability as the air is warmed adiabatically while sinking toward the surface. Consequently, the air can hold additional moisture as its capacity increases with increasing temperatures. The weather resulting from the influence of an anticyclone is often clear, with no rainfall. There are, however, certain conditions under which there can be some precipitation within a high pressure system. When such a system passes near or crosses a large body of water, the resulting evaporation can cause variations in humidity significant enough to result in some precipitation.

There are two sources for the relatively high pressures that are associated with anticyclones in the middle latitudes of North America. Some anticyclones move into the middle latitudes form northern Canada and the Arctic Ocean in what are called outbreaks of cold Arctic air. These outbreaks can be quite extensive, covering much of the midwestern and eastern United States. The temperatures in an anticyclone that has developed in a *cA* air mass can be markedly lower than those expected for any given time of year. They may be far below freezing in the winter. Other anticyclones are generated in zones of high pressure in the subtropics. When they move across the United States toward the north and northeast, they bring waves of hot, clear weather in summer and unseasonably warm days in the winter months.

Hurricanes

Though their overall diameter may be less than that of a *middle latitude cyclone* with its extended *fronts,* hurricanes are essentially the largest storms on Earth. Hurricanes are severe *tropical cyclones* that receive a great deal of attention from scientists and laypeople alike, primarily because of their tremendous destructive powers (• Fig. 7.11).

• **FIGURE 7.11**
Damage incurred by Hurricane Andrew in 1992. Until the hurricanes of 2004 and 2005, the damage by Hurricane Andrew, as shown in this photo, was the costliest in U.S. history.

NOAA 7.15: CIMSS at UW-Madison, produced by the Tropical Cyclones team

Abundant, even torrential, rains and winds often exceeding 160 kilometers per hour (100 mph) characterize hurricanes. Though hurricanes develop over the oceans, their paths at times do take them over islands and coastal lands. The results can be devastating destruction of property and loss of life. It is not just the rains and winds that cause damage. Accompanying the hurricane are unusually high seas, called **storm surges,** which can flood entire coastal communities (● Fig. 7.12).

A **hurricane** is a circular, cyclonic system with wind speeds in excess of 118 kilometers per hour (74 mph). It has a diameter of 160–640 kilometers (100–400 mi). Extending upward to heights of 12–14 kilometers (40,000–45,000 ft), the hurricane is a towering column of spiraling air (● Fig. 7.13). At its base, air is sucked in by the very low pressure at the center and then spirals inward. Once within the hurricane structure, air rises rapidly to the top and spirals outward. This rapid upward movement of moisture-laden air produces enormous amounts of rain. Furthermore, the release of latent heat energy provides the power to drive the storm.

At the center of the hurricane lies the eye of the storm, an area of calm, clear, usually warm and humid, but rainless air.

● FIGURE 7.12
As a hurricane moves ashore, a storm surge combines with the normal high tide to create a storm tide. This mound of water, topped by battering waves, moves ashore along an area of the coastline as much as 100 miles wide. The combination of the storm surge, battering waves, and high winds is deadly.
Why is the timing of landfall so critical to coastal areas?

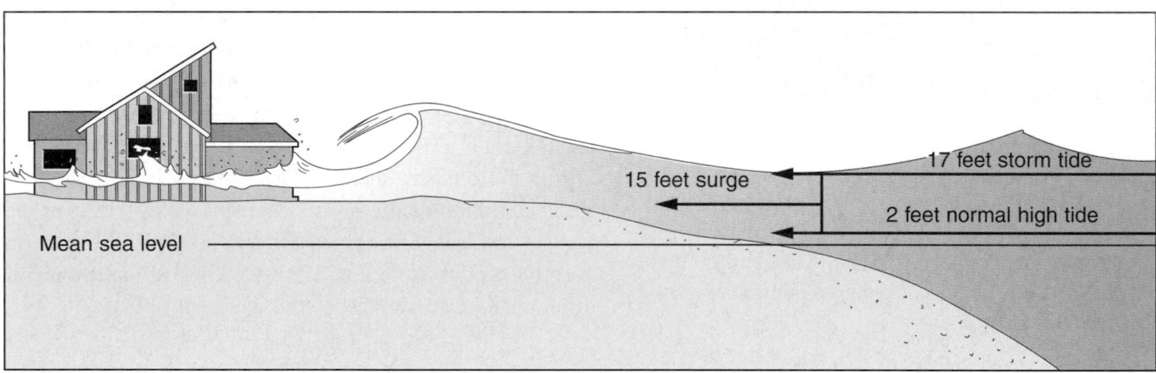

● FIGURE 7.13
In this cross section of a hurricane, circulation patterns show inflow of air in the spiraling arms of the cyclonic system, rising air in the towering circular wall cloud, and outflow in the upper atmosphere. Subsidence of air in the storm's center produces the distinctive calm, cloudless "eye" of the hurricane.
Why is this so?

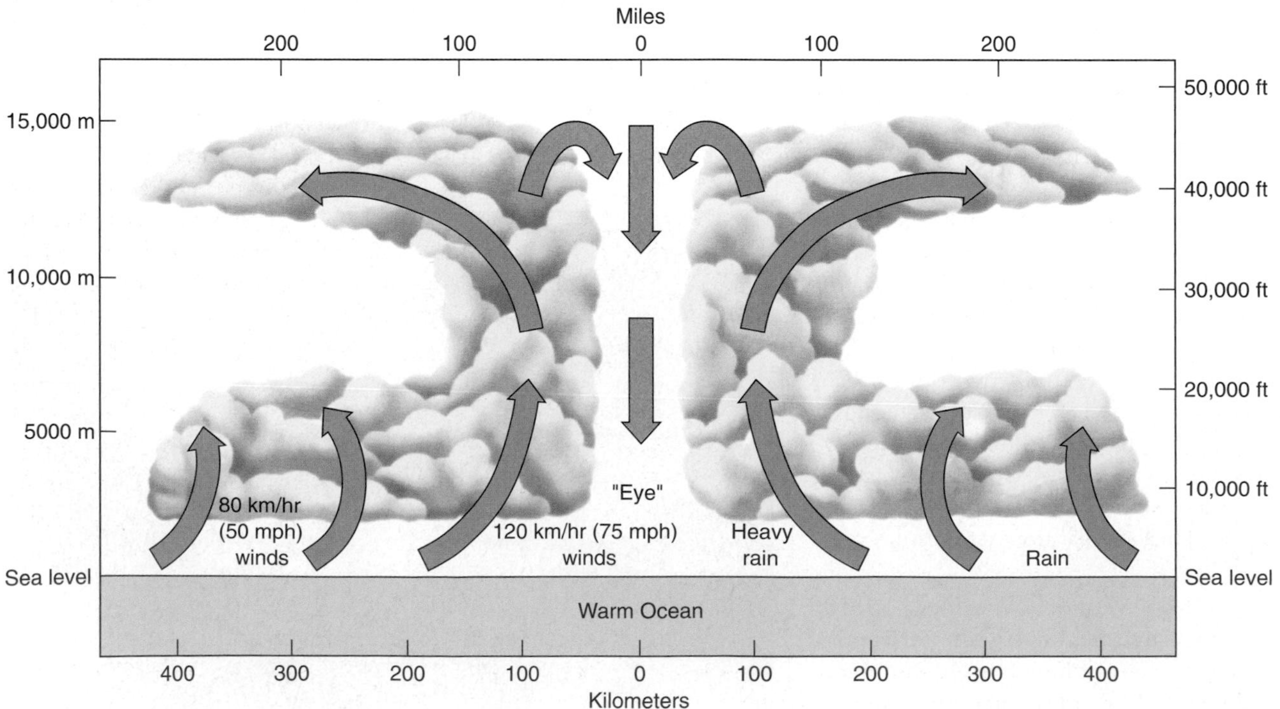

Sailors traveling through the eye have been surprised to see birds flying there. Unable to leave the eye because of the strong winds surrounding it, these birds will often alight on the passing ship as a resting spot.

Hurricanes have very strong pressure gradients because of the extreme low pressure at their centers. The strong pressure gradients in turn cause the powerful winds of the hurricane. In contrast to the middle-latitude cyclone, a hurricane is formed from a single air mass and does not have the different temperature sectors like a frontal system. Rather, a hurricane has a fairly even, circular temperature distribution, which we might have expected due to its circular winds.

The Saffir–Simpson Hurricane Scale provides a means of classifying hurricane intensity and potential damage by assigning a number from 1 to 5 based on a combination of central pressure, wind speed, and the height of the storm surge (Table 7.2).

Although a great deal of time, effort, and money has been spent on studying the development, growth, and paths of hurricanes, much is still not known. For example, it is not yet possible to predict the path of a hurricane, even though it can be tracked with radar and studied by planes and weather satellites. In addition, meteorologists can list factors favorable for the development of a hurricane but cannot say that in a certain situation a hurricane will definitely develop and travel along a particular path. As with tornadoes, there are also "Hurricane Alleys," or areas where their development is more likely to occur (● Fig. 7.14).

Among the factors leading to hurricane development are a warm ocean surface of about 25°C (77°F) and warm, moist overlying air. These factors are probably the reasons why hurricanes occur most often in the late summer and early fall when air masses have maximum humidity and ocean surface temperatures are highest. Also, the Coriolis effect must be sufficient to support the rapid spiraling of the hurricane. Therefore, hurricanes neither develop nor survive in the equatorial zone from about 8°S to 8°N, for there the Coriolis effect is far too weak. Hurricanes begin as weak tropical disturbances such as the easterly wave (described later in this chapter) and in fact will not develop without the impetus of such a disturbance. It is further suspected that some sort of turbulence in the upper air may play a part in a hurricane's initial development.

Names are assigned to storms once they reach *tropical storm* status, with wind speeds between 62 and 118 kilometers per hour (39–74 mph). Each year the names are selected from a different alphabetical list of alternating

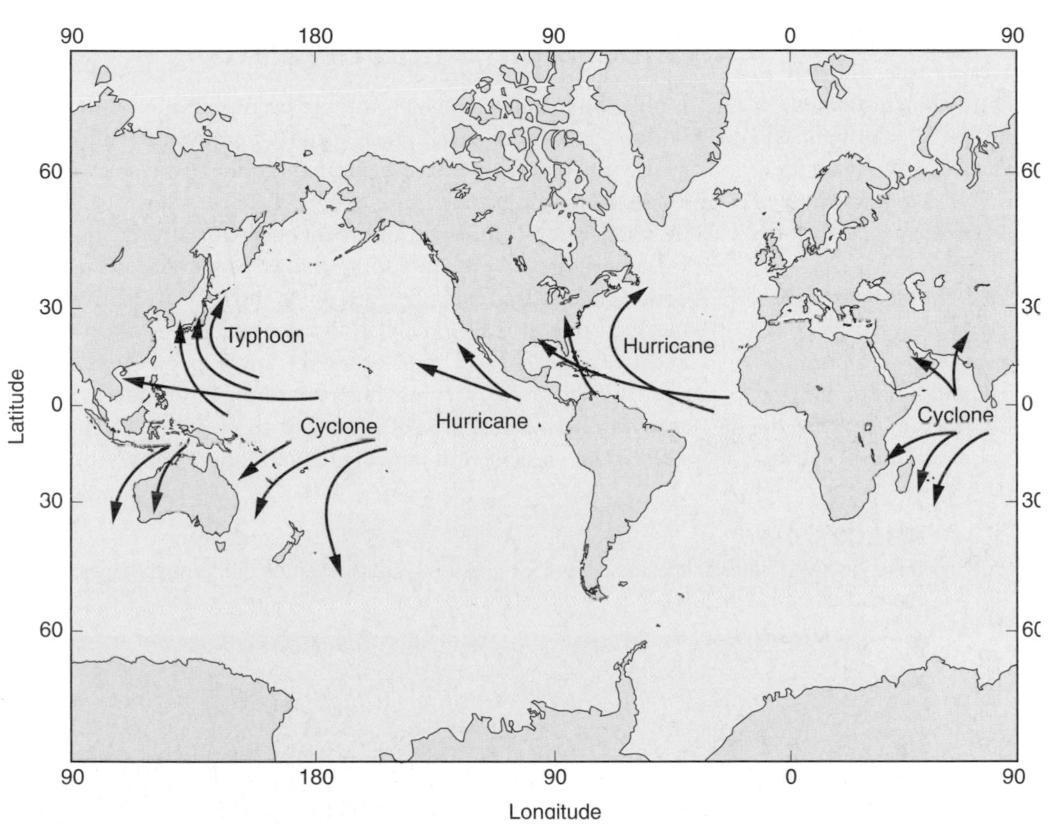

● FIGURE 7.14
A world map showing major "Hurricane Alleys."
Which coastlines seem unaffected by these tracks?

TABLE 7.2
Saffir–Simpson Hurricane Scale

Scale Number	Central Pressure	Wind Speed		Storm Surge		Damage
(CATEGORY)	(MILLIBARS)	(KPH)	(MPH)	(METERS)	(FEET)	
1	980	119–153	74–95	1.2–1.5	4–5	Minimal
2	965–979	154–177	96–110	1.6–2.4	6–8	Moderate
3	945–964	178–209	111–130	2.5–3.6	9–12	Extensive
4	920–944	210–250	131–155	3.7–5.4	13–18	Extreme
5	<920	>250	>155	>5.4	>18	Catastrophic

female and male names—one list for the North Atlantic and one for the North Pacific. If a hurricane is especially destructive and becomes a part of recorded history, its name is retired and never used again.

Hurricanes do not last long over land because their source of moisture (and consequently their source of energy) is cut off. Also, friction with the land surface produces a drag on the whole system. North Atlantic hurricanes that move first toward the west with the trade winds and then north and northeast as they intrude into the westerlies become polar cyclones if they remain over the colder region of the ocean and eventually die out. Over land, they will also become simple cyclonic storms, but even when they have lost some of their power, hurricanes can still do great damage.

Hurricanes can occur over most subtropical and tropical oceans and seas; until recently, the South Atlantic was considered the exception, though it was not understood why. However, on March, 26, 2004, tropical Cyclone Catarina was the first to attack the southern coast of Brazil, much to the amazement of atmospheric scientists all over the world! In the South Pacific and Australia, hurricanes are called *tropical cyclones* (or willy-willies). Near the Philippines, they are known as *bagyos,* but in most of East Asia they are called **typhoons.** In the Bay of Bengal, they are referred to as *cyclones.*

The year 2004 was certainly one for the record books. Typhoon Tokage struck the Japanese coast near Tokyo with significant loss of life. A total of ten tropical cyclones pounded Japan in 2004 alone. In our own Caribbean and Gulf of Mexico, three hurricanes, Charley, Frances, and Jeanne, directly hit Florida. A fourth, Ivan, whose center struck Gulf Shores, Mississippi, caused devastation in Florida's western panhandle (• Fig.7.15). The damages from these storms are estimated to reach $23 billion. This exceeds the cost of Hurricane Andrew in 1992, which at the time was called the costliest natural disaster in U.S. history and caused more than $20 billion in damage.

However, for the United States the worst was yet to come. In 2005, Hurricane Katrina, with 225 kilometer per hour (140 mph) winds and *storm surges* rising over 16 feet in height, struck the Gulf of Mexico coasts of Louisiana, Mississippi, and Alabama. The storm surges from Katrina breached the levee systems designed to protect the city of New Orleans. When the levees gave way to the surging ocean waves, New Orleans (much of which is below sea level to begin with) was flooded. This caused even more massive destruction, from which the city is still trying to recover. Katrina was responsible for the deaths of hundreds of people, and the areas destroyed by the winds and floodwaters will have long-term recovery costs estimated to be as much as $200 billion. In contrast, the 2006 hurricane season was unusual because no hurricanes made landfall in the United States; the last time that happened was during the 2001 hurricane season. On May 9th 2007, the formation of Subtropical Storm Andrea marked a premature beginning to the official hurricane season, which normally runs from June 1 through November 30 in the Atlantic Ocean. The exact number and severity of tropical cyclones can vary quite dramatically from one year to the next. • Figure 7.16 maps 54 years worth of hurricane landfall sites for the United States.

Some people have suggested that we seek ways to control these destructive storms. On the other hand, hurricanes are a major source of rainfall and an important means of transferring energy within Earth's systems away from the tropics. Eliminating them might cause unwanted and unforeseen climate changes.

Snow Storms and Blizzards

Unlike hurricanes, snow-producing events are obviously found within the middle and higher latitudes because they are associated with colder winter temperatures. However, the areas affected by these storms may be quite extensive. These snow events must be triggered by the same uplift mechanisms already discussed for other types of precipitation—that is, orographic, frontal, and convergence (cyclonic). Convection is more of a warm weather mechanism and is less likely to be involved in snow-producing events. In middle- to high-latitude winters, people experience snowfall events of varying severity. They can come as a *snow shower* or *snow flurry,* a brief period of snowfall in which intensity can be variable and may change rapidly. A **snowstorm** is a

• FIGURE 7.15
This montage of satellite images shows the remarkable similarity of the hurricane tracks as they approached Florida in 2004.

CIMSS at UW-Madison, produced by the Tropical Cyclones team

GEOGRAPHY'S SPATIAL SCIENCE PERSPECTIVE

Hurricane Paths and Landfall Probability Maps

Hurricanes (called typhoons in Asia) are generated over tropical or subtropical oceans and build strength as they move over regions of warm ocean water. Ships and aircraft regularly avoid hurricane paths by navigating away from these huge violent storms. People living in the path of an oncoming hurricane try to prepare their belongings, homes, and other structures, and may have to evacuate if the hazard potential of the impending storm is great enough. Landfall refers to the location where the eye of the storm encounters the coastline. Storm surges present the most dangerous hazard associated with hurricanes, where the ocean violently washes over and floods low-lying coastal areas. In 1900, 6000 residents of Galveston Island in Texas were killed when a hurricane pushed a 7-meter-high (23-ft-high) wall of water over the island. Much of the city was destroyed by this storm, the worst natural disaster ever to occur in the United States in terms of lives lost.

Today we have sophisticated technology for tracking and evaluating tropical storms. Computer models, developed from maps of the behavior of past storms, are used to indicate a hurricane's most likely path and landfall location, as well as the chance that it may strike the coast at other locations. The nearer a storm is to the coast, the more accurate the predicted landfall site should be, but in some cases a hurricane may begin to move in a completely different direction. In general, hurricanes that originate in the North Atlantic Ocean tend to move westward toward North America and then turn northward along the Atlantic or Gulf Coasts.

Nature still remains unpredictable, so potential landfall sites are shown on probability maps, which show the degree of likelihood for the hurricane path. These maps help local authorities and residents decide what course of action is best to take in preparing for the approach of a hurricane. A 90% probability means that nine times out of ten storms under similar regional weather conditions have moved onshore in the direction indicated by that level on the map. A 60% probability means that six of ten hurricanes moved as indicated, and so forth. Regions where the hurricane is considered likely to move next are represented on the map by color shadings that correspond to varying degrees of probability for the storm path. In recent years, the National Weather Service has worked hard to develop new computer models that will yield better predictions of hurricane paths, intensities, and landfall areas. If you live in a coastal area affected by hurricanes and tropical disturbances, understanding these maps of landfall probability may be very important to your safety and your ability to prepare for a coming storm.

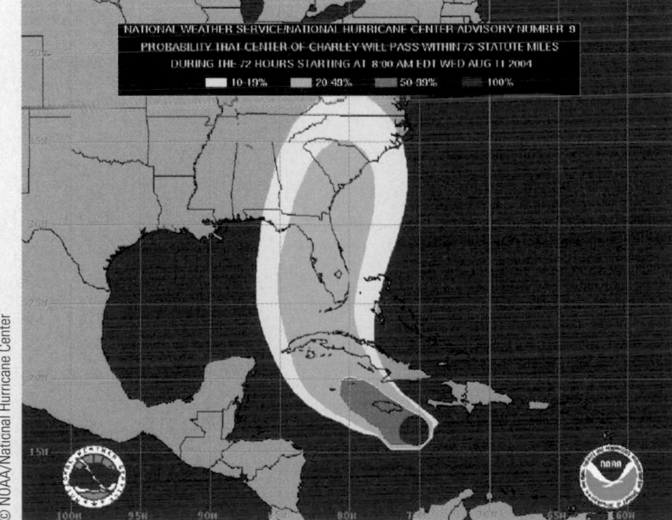

© NOAA/National Hurricane Center

Landfall probability map for Hurricane Charley on Wednesday, August 11, 2004, showing the most likely place for landfall at the time the map was produced. This was 2 days before the hurricane struck Florida's west coast.

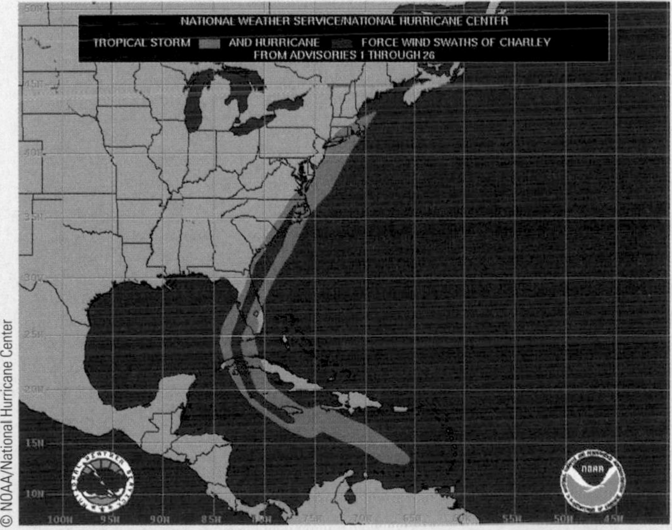

© NOAA/National Hurricane Center

This map, produced after the storm had dissipated, shows the actual path of Hurricane Charley and its severity. In this case, the landfall probability map proved fairly accurate. From landfall along the Gulf Coast, Hurricane Charley crossed the Florida peninsula, was downgraded to tropical storm status, but then struck Atlantic coastal areas of the Carolinas with strong winds, heavy rain, and coastal flooding.

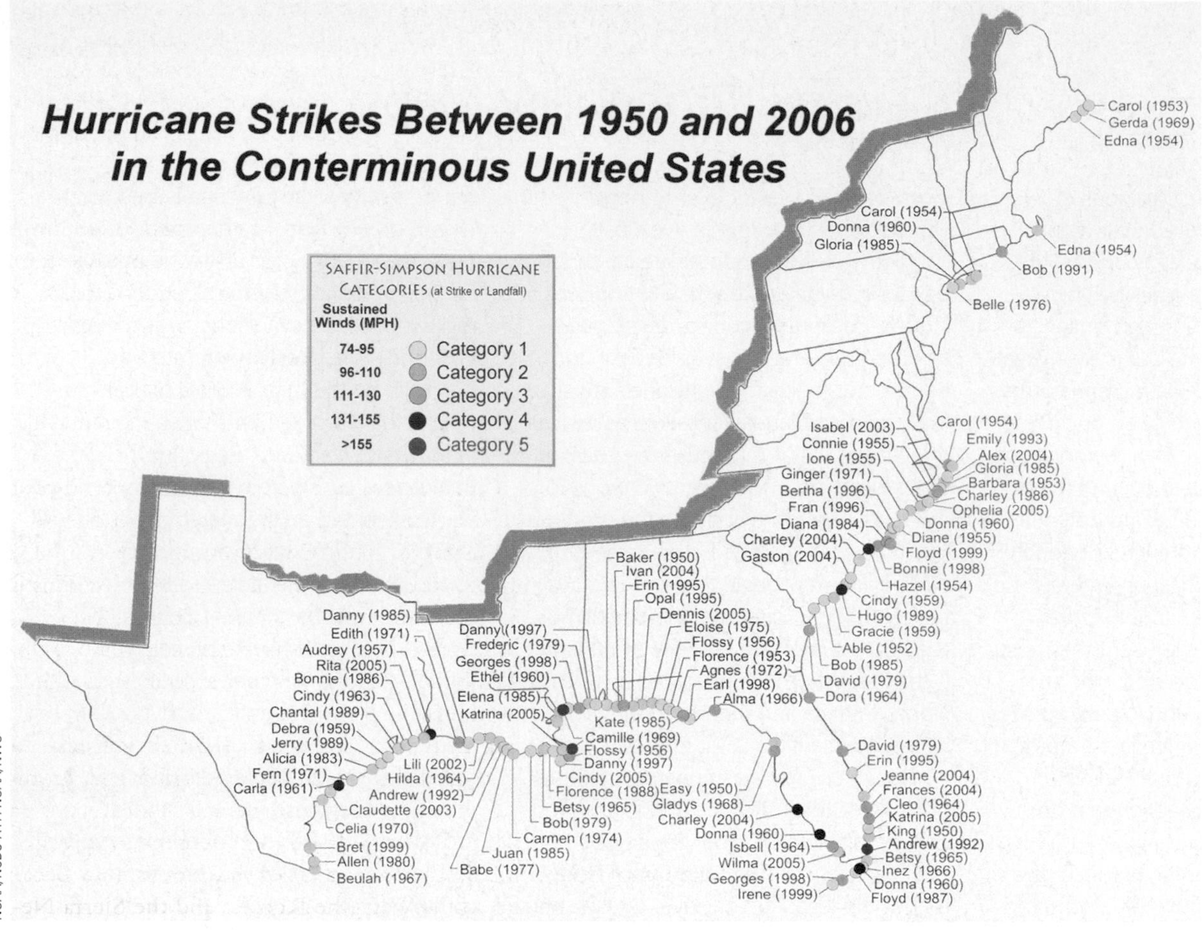

NOAA/NCDC 7.17: NOAA/NWS

• FIGURE 7.16
This map shows hurricane landfall sites along the U.S. East and Gulf Coasts. The sites are labeled by storm name, the year, and the Saffir–Simpson Category.
What areas of our coasts seem to have escaped landfall, so far?

storm where frozen precipitation falls in the form of snow and is much more severe. Some snowstorms create enough turbulence to create lightning discharges, a phenomenon once thought to be impossible.

A **blizzard** is the most severe weather event. It is characterized as a heavy snowstorm accompanied by strong winds. At wind speeds of about 55 kilometers per hour (35 mph) or greater, a blizzard can reduce visibility to zero due to falling and blowing snow. Here, the term *whiteout* can apply. Visibility is reduced so that all a person can see is white, and an individual can totally lose track of distance and direction. This is especially dangerous for people using any mode of transportation. Airport closings and traffic accidents are common during blizzards (● Fig. 7.17).

• FIGURE 7.17
A weather station in central Illinois during a blizzard in February of 2000. With winds gusting to 45 miles per hour, greatly reduced visibility was one effect of this blizzard.
How far would you estimate the visibility to be in this neighborhood?

NOAA/NWS

Thunderstorms

Thunderstorms are common local storms of the middle and lower latitudes. Very simply, a thunderstorm is a storm accompanied by thunder and lightning. Lightning is an intense discharge of electricity. For lightning to occur, positive and negative electrical charges must be generated within a cloud. It is believed that the intense friction of the air on moving ice particles within a cumulonimbus cloud generates these charges. Usually, but not always, a clustering of positive charges tends to occur in the upper portion of the cloud, with negative charges clustering in the lower portion. When the potential difference between these charges becomes large enough to overcome the natural insulating effect of the air, a lightning flash, or discharge, takes place. These discharges, which often involve over 1 million volts, can occur within the cloud, between two clouds, or from cloud to ground. The air immediately around the discharge is momentarily heated to temperatures in excess of 25,000°C (45,000°F), which is about four times hotter than the surface of the sun! The heated air expands explosively, creating the shock wave we call thunder (• Fig. 7.18).

• FIGURE 7.18
A cross-sectional view of a thunderstorm showing the distribution of electrical charges.
Where do you place a lightning bolt in this diagram?

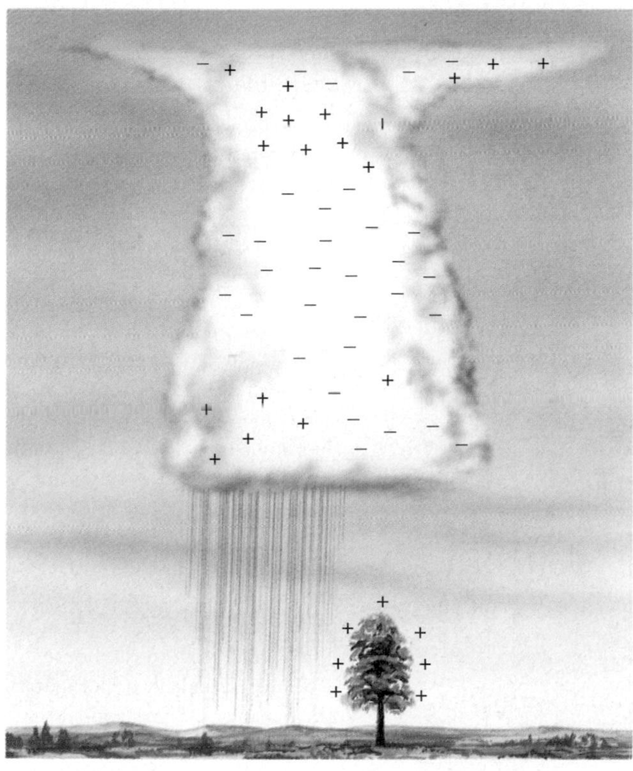

• FIGURE 7.19
Thermal convection and orographic uplift.
What are the other mechanisms of uplift?

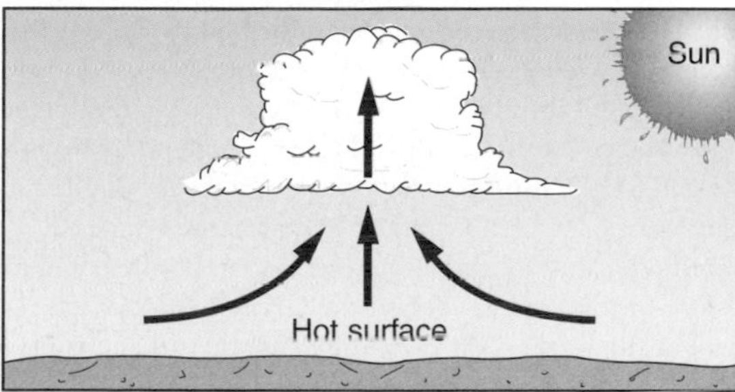

(a) Convectional uplift

(b) Orographic uplift

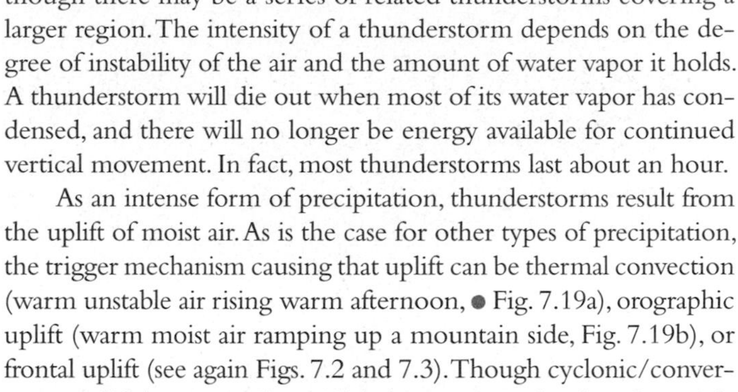

Thunderstorms usually cover a small area of a few miles although there may be a series of related thunderstorms covering a larger region. The intensity of a thunderstorm depends on the degree of instability of the air and the amount of water vapor it holds. A thunderstorm will die out when most of its water vapor has condensed, and there will no longer be energy available for continued vertical movement. In fact, most thunderstorms last about an hour.

As an intense form of precipitation, thunderstorms result from the uplift of moist air. As is the case for other types of precipitation, the trigger mechanism causing that uplift can be thermal convection (warm unstable air rising warm afternoon, • Fig. 7.19a), orographic uplift (warm moist air ramping up a mountain side, Fig. 7.19b), or frontal uplift (see again Figs. 7.2 and 7.3). Though cyclonic/convergence uplift (see again Fig. 6.19) helps to create clouds and precipitation, it is less effective in triggering severe thunderstorms.

Convective thunderstorms are most common in lower latitudes during the warmer months of the year and during the warmer hours of the day. It is apparent, then, that the amount of solar heating affects the development of thunderstorms. This is true because the intense heating of the surface steepens the environmental lapse rate, which in turn leads to increased instability of the air, allowing for greater moisture-holding capacity and adding to the buoyancy of the air.

Orographic thunderstorms occur when air is forced to rise over land barriers, providing the necessary initial trigger action leading to the development of thunderstorm cells. Thunderstorms of orographic origin play a large role in the tremendous precipitation of South and Southeast Asia. In North America, they occur over the mountains in the West (the Rockies and the Sierra Nevada), especially during summer afternoons when heating of south-facing slopes increases the air's instability. For this reason, pilots of small plans try to avoid flying in the mountains during summer afternoons for fear of getting caught in the turbulence of a thunderstorm.

Frontal thunderstorms are often associated with cold fronts where a cooler air mass forces a warmer air mass to rise. This action can bring about the strong, vertical updrafts necessary for precipitation. In fact, at times a cold front is immediately preceded by a line of thunderstorms (a squall line) resulting from such frontal uplift (see again Fig. 7.2).

As we mentioned in the discussion of precipitation types in Chapter 6, hail can be a product of thunderstorms when the vertical updrafts of the cells are sufficiently intense to carry water droplets repeatedly into a freezing layer of air. Fortunately, since thunderstorms are primarily associated with warm weather areas, only a very small percentage of storms around the world produce

NOAA/NWS

• **FIGURE 7.20**
A powerful F4 tornado in central Illinois during July of 2004.

Greg Henshall / FEMA 7.23: NOAA

• **FIGURE 7.21**
Terrible destruction caused by a F5 tornado at Greensburg, Kansas, on May 16, 2007.

hail. In fact, hail seldom occurs in thunderstorms in the lower latitudes. In the United States, there is little hail along the Gulf of Mexico where thunderstorms are most common.

Tornadoes

Tornadoes are the most violent storms on the face of Earth (• Fig. 7.20 and • Fig.7.21). They can occur almost anywhere but are far more common in the interior of North America than any other country in the world. In fact, Oklahoma and Kansas lie in the path of so many "twisters" that together they are sometimes referred to as "Tornado Alley."

Systematic government documentation of tornado activity, such as that depicted in • Figures 7.22a and 7.22b, began in 1875. Accounts of tornadoes occurring prior to 1875 must be tracked down through other sources. These accounts, though often unverifiable and vague, do offer interesting and informative insights into our forebears' perceptions of tornadoes. The accounts below describe a tornado that killed several people as it swept across several counties in western Illinois on May 21, 1859.

It was "a violent storm or hurricane [which] did immense damage to houses, barns, fences, and also caused some destruction of life." It was described as having a "frightful . . . balloon or funnel shape, and appeared . . . peculiarly bright and luminous, not at all black or dark in any of its parts, except its base or bottom." A vivid account of what surely must be related to the output of static electricity associated with a tornado is given in this account of the same tornado as it swept across Morgan county: "Mr. Cowell was plowing his field . . . He saw the frightful cloud approaching . . . and at once attempted to drive his horses and plow to the house . . . The horses suddenly took fright . . . their manes and tails and all their hair 'stood right out straight' as he expressed it, and . . . the iron in the harness . . . and plow, in his language 'seemed all covered with fire.' He felt a violent pulling of his own hair which left 'his head sore for some days' and the hair itself rigid and inflexible." In addition, although unconfirmed by others, Mr. Cowell was one of the few individuals to have a tornado pass directly over him and live to tell about it. He described the light in the center of the tornado as being "so brilliant that he could not endure it with his eyes open, and for the most part kept them shut . . . Yet [inside the tornado] there was no wind, no thunder and no noise whatever. . ." Another interesting feature of this same tornado can be attributed to the low pressure of the vortex: "When the terrific whirl struck . . . [it] stripped all of the feathers off from the hens and turkeys, as perfectly clean as if picked for the table. Some, though badly plucked, and made entirely blind, still lived." Such a bizarre occurrence prob-

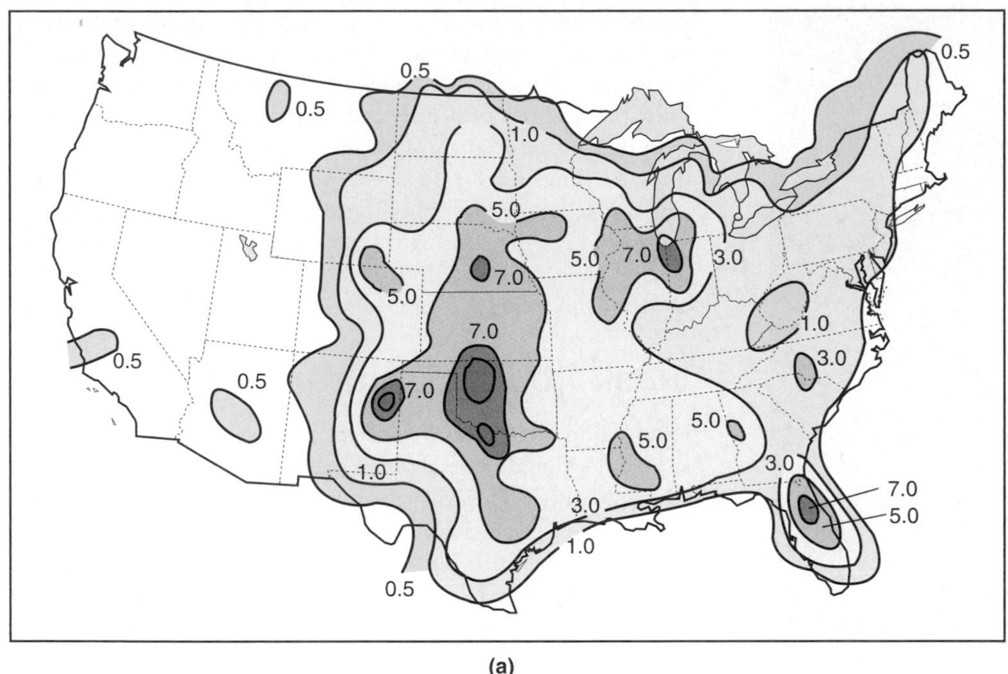

(a)

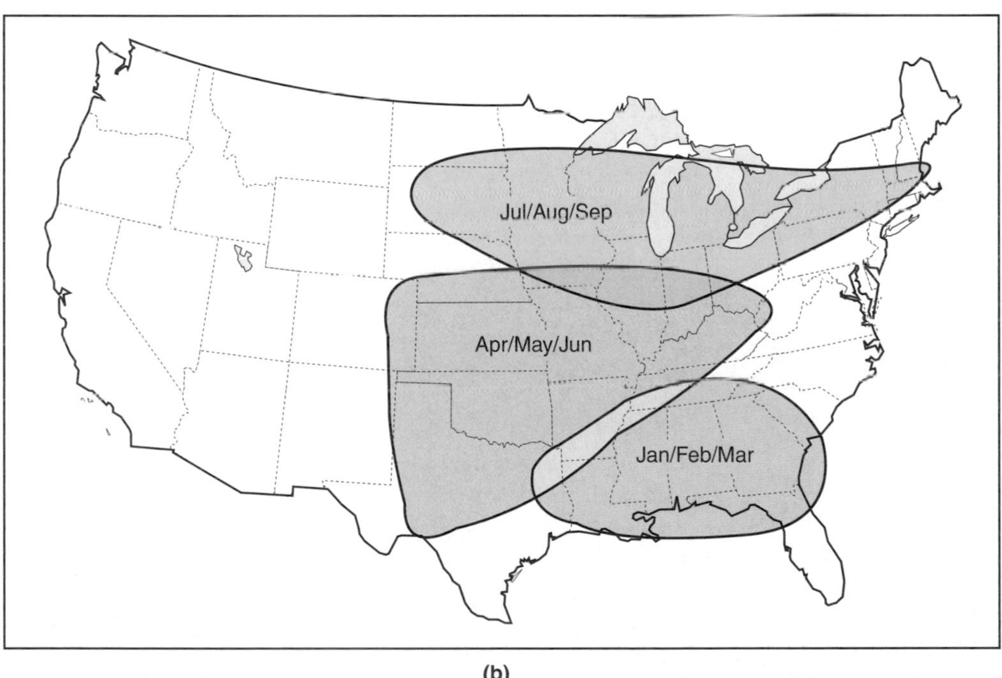

(b)

• **FIGURE 7.22**

(a) Average annual number of tornadoes per 26,000 square kilometers (10,000 sq mi). (b) Seasonal march of peak tornado activity.

How do tornadoes affect your geographic area?

ably resulted when the hollow quills of the feathers expanded so suddenly—as the low-pressure vortex moved over the area—that the birds' feathers "exploded."

Transactions of the Illinois Natural
History Society
Phillips Bros., 1861

A **tornado** is actually a small, intense cyclonic storm of very low pressure, violent updrafts, and converging winds of enormous contrast. Fortunately, they are small and short lived. Even in Tornado Alley, a tornado is likely to strike a given locale only once in 250 years.

Although only 1% of all thunderstorms produce a tornado, 80% of all tornadoes are associated with thunderstorms and middle-latitude cyclones. The remaining 20% of tornadoes are spawned by hurricanes that make landfall. In the past decade, over 1000 tornadoes have occurred each year in the United States, most of them from March to July in the late afternoon or early evening in the central part of the country.

Because of their small size and limited life span, tornadoes are extremely difficult to detect and forecast. However, relatively new radar technology, **Doppler radar,** improves tornado detection and forecasting significantly. Doppler radar has more power concentrated in a narrower beam than previous radar units. This allows meteorologists to assess storms in much greater detail (• Fig. 7.23). Even more important is a Doppler radar's ability to measure wind speeds flowing toward the radar site and wind speeds blowing away from the radar site (that is, the Doppler effect). When the energy emitted by radar strikes precipitation, a small portion is scattered back to the radar. Depending on whether the precipitation is moving toward or away from the radar site, the wavelength of the returned energy is either compressed or elongated. The faster the winds flow, the greater will be the change in wavelength. Previous radars could not measure this change; however, Doppler radar does and uses it to estimate the wind circulation and rotation within the storm. In fact, Doppler radar is so sensitive that it can detect the wind pattern in clear air by detecting the backscattered energy from clouds, pollution, insects, and so forth. It allows meteorologists to see the formation of a tornado, thus increasing the warning time to the public. Doppler radar also permits the detection around airports of clear air turbulence (CAT), a major factor in airline accidents. The U.S. government, through the NEXRAD program (NEXt-generation weather RADar), has installed 141 Weather Service Doppler Radar (WSR-88D) sites across the country.

NOAA

• FIGURE 7.23
Doppler radar displayed a strong cold front triggering several squall lines (in red) and severe weather on November 15, 2005.
How many squall lines can you see on this image?

Doppler radar observations indicate that most tornadoes (63%) are fairly weak with wind speeds of 162 kilometers per hour (100 mph) or less. About 35% of tornadoes can be classified as strong, with wind speeds reaching 320 kilometers per hour (200 mph). Nearly 70% of all tornado fatalities result from violent tornadoes. Although very rare (only 2% of all tornadoes reach this stage), these may last for hours and have wind speeds in excess of 480 kilometers per hour (300 mph).

Before Doppler radar, wind speeds within a tornado could not be measured directly. Therefore, tornado intensity was estimated from the damage produced by the storm. The late T. Theodore Fujita, a former professor at the University of Chicago, originated a scale of tornado intensity. The scale is termed the Fujita Intensity Scale or, more commonly, the F-scale (Table 7.3). In 2007 the National Weather Service adopted a refined and modified version of the original F-scale, based on new data and observations that were not available to Fujita. This is the **Enhanced Fujita Scale,** the EF-scale, and the main difference is in how wind speeds are estimated based on damage observations, and changes in the wind speeds for the new EF-scale.

A tornado first appears as a swirling, twisting funnel cloud that moves across the landscape at 35–51 kilometers per hour (22–32 mph). Its narrow end may be only 100 meters (330 ft) across. The funnel cloud becomes a tornado when its narrow end is in contact with the ground where the greatest damage is done often along a linear track (• Figs. 7.24a and b). Above the ground, the end can swirl and twist, but little or nothing is done to the ground below. The color of a tornado can be milky white to black, depending on the amount and direction of sunlight and the type of debris being picked up by the storm as it travels across the land. Although most tornado damage is caused by the violent winds, most tornado injuries and deaths result from flying debris. The small size and short duration of a tornado greatly limit the number of deaths caused by tornadoes. In fact, more people die from lightning strikes each year than from tornadoes. At times however, severe storms may spawn a **tornado outbreak,** where multiple tornadoes are produced by the same system. The worst outbreak in recorded history occurred on April 3–4, 1974, when 148 twisters touched down in 13 states, injuring almost 5500 people and killing 330 others.

Weak Tropical Disturbances

Until World War II, the weather of tropical regions was described as hot and humid, generally fair, but basically pretty monotonous. The only tropical disturbance given any attention was the tropical cyclone (also called a *hurricane* or, in other parts of the world, a *typhoon*), a spectacular but relatively uncommon storm that affects only islands, coastal lands, and ships at sea.

Even a few decades ago, an aura of mystery remained about the weather of the tropics. One reason for this lack of information was that the few weather stations located in tropical areas were widely scattered and often poorly equipped. As a result, it was difficult to understand completely the passing weather disturbances in the tropics.

Largely through satellite technology and computer analysis, it is now known that a variety of weak atmospheric disturbances affects the weather and relieves the monotony, although it is likely that the full number of these disturbances has not yet been recognized. The primary impact of these weak tropical disturbances on the weather of tropical regions is not on the temperature but rather on the cloud cover and the amount of precipitation. Temperatures in the tropics are largely unaffected during the passage of a tropical storm, except that as the cloud cover increases, temperature extremes are reduced.

Easterly Wave The best known of the weak tropical disturbances is the **easterly wave.** It shows up in • Figure 7.25 as a trough-shaped, weak, low pressure region that is generally aligned on an approximate north–south axis. Traveling slowly in the trade winds belt from east to west, it is preceded by fair, dry weather and followed by cloudy, showery weather. This occurs because air tends to converge into the low from its rear, or the east, causing lifting and convectional showers. The resulting divergence and subsidence to the west account for the fair weather. Meteorologists believe that this type of disturbance can on occasion develop into a tropical cyclone (or hurricane).

Polar Outbreak Occasionally, an outbreak of polar air may follow a low into the subtropics and tropics. Such an outbreak would of course be preceded by the squalls, clouds, and rain associated with a cold front. Following, however, would be a period of cool, clear, fair weather, as the modified polar air influences are felt. On rare occasions near the equator in the Brazilian Amazon, such an Antarctic outburst, known locally as a *friagem,* can bring freezing temperatures and widespread damage to vegetation. Farther to the south, near São Paulo, the coffee crop can be ruined, causing coffee prices in North America to rise.

TABLE 7.3
The Fujita Tornado Intensity Scale and the Enhanced Fujita Scale

	Wind Speed			Wind Speed		
F-SCALE	**KPH**	**MPH**	**EF-SCALE**	**KPH**	**MPH**	**EXPECTED DAMAGE**
F-0	<116	<72	EF-0	<138	<86	**Light Damage** Damage to chimneys and billboards; broken branches; shallow-rooted trees pushed over
F-1	116–180	72–112	EF-1	138–177	86–110	**Moderate Damage** Surfaces peeled off roofs; mobile homes pushed off foundations or overturned; exterior doors blown off; windows broken; moving autos pushed off the road
F-2	181–253	113–157	EF-2	178–217	111–135	**Considerable Damage** Roofs torn off houses; mobile homes demolished; boxcars pushed over; large trees snapped or uprooted; light-object missiles generated
F-3	254–332	158–206	EF-3	218–265	136–165	**Severe Damage** Roofs and some walls torn off well-constructed houses; trains overturned; most trees in forest uprooted; heavy cars lifted off ground and thrown
F-4	333–419	207–260	EF-4	266–322	166–200	**Devastating Damage** Well constructed houses leveled; structures with weak foundations blown some distance; cars thrown and large missiles generated
F-5	>419	>260	EF-5	>322	>200	**Incredible Damage** Strong frame houses lifted off foundations and carried considerable distance to disintegrate; automobile-sized missiles fly through the air farther than 100 meters; trees debarked; incredible phenomena occur

• FIGURE 7.24
(a) The destructive track of a powerful tornado is visible on this satellite image as a linear swath of damage across the landscape of La Plata, Maryland. (b) Only the foundation and basement are left of this home that was struck by the devastating La Plata tornado in April of 2002.

Image courtesy Lawrence Ong, EO-1 Mission Science Office, NASA/GSFC

(a)

Image courtesy Lawrence Ong, EO-1 Mission Science Office, NASA/GSFC

(b)

GEOGRAPHY'S PHYSICAL SCIENCE PERSPECTIVE

Tornado Chasers and Tornado Spotters

It is amazing how many students choose to study meteorology because they want to chase tornadoes. This was especially true after the 1996 movie *Twister.* If this type of activity has ever interested you, you should know a few things from the start. First of all, there is a distinct difference between a *tornado spotter* and a *tornado chaser.*

Tornado spotters are trained by the National Weather Service (NWS) to serve their communities by watching and warning for severe weather. When severe weather is approaching, Doppler radar may not tell the complete story about what is happening on the ground. Sometimes Doppler radar only shows where a tornado may begin to form, and certain categories of tornadoes may begin before radar "tornado-signature" is even detected. A spotter in the field can solve these problems by pinpointing the tornado touchdown and tracking the storm at a safe distance.

NWS professionals often conduct these training sessions for state Emergency Management Agencies and amateur radio groups. Amateur radio operators (known as "hams") are well suited for this important service. A good spotter must know what to look for *and* be able to communicate a warning back to the NWS, so ham-radio spotters fit the bill nicely. These volunteers form groups known as SKYWARN networks and perform a significant public service.

A tornado chaser, on the other hand, is not necessarily out to warn others of danger. Some of them are simply thrill seekers engaging in an exciting hobby. Others do it as a part of their job: They may be scientists collecting data for research, and some are professional photographers, photojournalists, and news reporters. Chasers are usually more mobile than spotters and will drive hundreds of miles to encounter a tornado. More often then not, a tornado chase results in a "bust," meaning a failed trip. Experienced tornado chasers may go on 50 or 60 trips over several years without seeing a tornado. Some spend their vacation time driving cross-country, hoping to see a tornado—only to go home disappointed.

Incidentally, neither of these activities is an actual paying job. People get involved for any number of reasons, including scientific field study, storm photography, self-education, news coverage, or the adrenaline rush. Some financial gain may be possible, by selling photos and videos or collecting a stipend from a research grant, but in general neither tornado spotting nor tornado chasing is a career in itself.

There are an estimated 1000 tornado chasers in the United States. Some of their professions include engineers, store owners, pilots, roofers, students, postal workers, teachers, and (of course) meteorologists. Their average age is about 35 years, but ages range from 18 to 65. Women comprise about 2% of this group. Many tornado chasers have a college education. Though most live in the Great Plains or Midwest (where tornadoes are more frequent), tornado chasers now reside in all the lower 48 states. Regardless of who they are and where they live, most of them have one thing in common: a working knowledge of meteorology. Armed with that knowledge, they have the best chance of witnessing a tornado; otherwise, "shooting in the dark" will only disappoint the uninformed tornado chaser with bust after bust.

NOAA

A specialized van, equipped by the National Weather Service with sophisticated sensors for detecting tornadoes and hazardous thunderstorms, stands by as NASA prepares to launch the space shuttle.

© Ian Giammanco

Scientists who "chase" tornadoes seek to observe them from as near as is safely possible, in order to gain important information to help us understand how these storms function. With a large tornado in the background, this vehicle is mounted with special equipment for tornado tracking and data gathering.

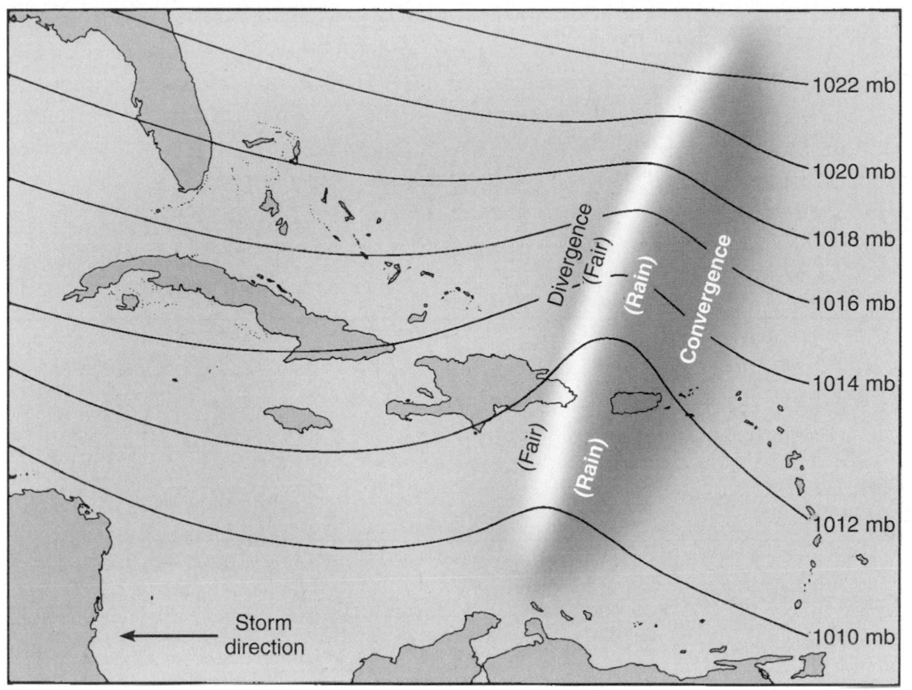

• FIGURE 7.25
A typical easterly wave in the tropics. Note that the isobars (and resulting winds) do not close in a circle but merely make a poleward "kink," indicating a low pressure trough rather than a closed cell. The resulting weather is a consequence of convergence of air coming into the trough, producing rains, and divergence of air coming out of the trough, producing clear skies.
Why do easterly waves move toward the west?

• FIGURE 7.26
This image from the GOES East satellite shows large areas on two continents and the adjacent ocean areas.
Can these kinds of images help us forecast tropical storms and hurricanes?

NASA/NOAA GOES and NOAA AVHRR

Weather Forecasting

Weather forecasting, at least in principle, is fairly straightforward. Meteorological observations are made, collected, and mapped to depict the current state of the atmosphere. From this information, the probable movement, as well as any anticipated growth or decay, of the current weather systems is projected for a specific amount of time into the future.

When a forecast goes wrong—which we all know occurs—it is usually either because limited or erroneous information has been collected and processed in the first place or, more likely, because errors have been made in anticipating the path or growth of the storm systems. Little errors will compound themselves over time. For example, a few degrees' shift in a storm's path may result in an error of a few miles in the projected location of that storm in a 2-hour forecast. However, this same few-degree error may result in a projected locational error of hundreds of miles in a 48-hour forecast. Consequently, the farther into the future one tries to forecast, the greater the chance of error.

Although forecasts are not perfect, they are much better today than in the past. Much of this improvement can be attributed to our current sophisticated technology and equipment. Increased knowledge and surveillance of the upper atmosphere have improved the accuracy of weather prediction. Weather satellites have helped tremendously by providing meteorologists with a better understanding of weather and weather systems. They have been of particular value to forecasters on the West Coast of the United States (• Fig. 7.26). Before the advent of weather satellites, these forecasters had to rely on information relayed from ships, leaving enormous areas of the Pacific unobserved. Thus, forecasters were often caught off guard by unexpected weather events.

In addition, high-speed computers allow rapid processing and mapping of observed weather conditions. Computers also allow the processing of numerical forecasts, which are based on the solution of physical equations that govern our atmosphere. Numerical forecasts and long-term forecasts based on solving statistical relationships and equations would not be possible without computers. In fact, computers now play such an important role in forecasting that some of the world's largest and fastest computers are used to forecast the weather.

Though forecasters now possess a great deal of knowledge and a variety of highly sophisticated devices that were previously unavailable, these devices are not foolproof. Understanding some of the problems the weather forecaster faces may make us more understanding when a forecast fails. No one can promise a sunny day. Nor can anyone say that it will definitely rain tomorrow, for no one can truly predict the future. The weather forecaster combines science and art, fact and interpretation, data and intuition, to come up with some probabilities about future weather conditions.

In this chapter, we combined some of the elements of the atmosphere learned in previous chapters to explain our major weather systems. We have noted how temperature and moisture differences characterize air masses and their leading edges, called fronts. With their additional pressure and wind components, air masses and fronts form the recognizable weather systems that all of us deal with from day to day. These systems may be relatively small, like tornadoes, or cover large areas, like middle-latitude cyclones. But most important, these weather systems affect our lives, whether in small ways, like an inconvenient forecast, or in devastating ways, as in life-threatening storms.

Chapter 7 Activities

Define & Recall

air mass
source region
Maritime Equatorial (*mE*)
Maritime Tropical (*mT*)
Continental Tropical (*cT*)
Continental Polar (*cP*)
Maritime Polar (*mP*)
Continental Arctic (*cA*)
surface of discontinuity
cold front
squall line
warm front
stationary front
occluded front
atmospheric disturbance
middle-latitude cyclone (extratropical cyclone)
storm track
veering wind shift
backing wind shift
storm surge
hurricane
typhoon
snowstorm
blizzard
convective thunderstorm
orographic thunderstorm
frontal thunderstorm
tornado
Doppler radar
Enhanced Fujita Scale
tornado outbreak
easterly wave

Discuss & Review

1. What is an air mass?
2. Do all areas on Earth produce air masses? Why or why not?
3. What letter symbols are used to identify air masses? How are these combined? What air masses influence the weather of North America? Where and at what time of the year are they most effective?
4. Use Table 7.1 and Figure 7.1 to find out what kinds of air masses are most likely to affect your local area. How do they affect weather in your area?
5. Why do you suppose air masses can be classified by whether they develop over water or over land?
6. What kind of air mass forms over the southwestern United States in summer? Have you ever experienced weather in such an air mass? What was it like? What kind of weather might you expect to experience if such an air mass met an *mP* air mass?
7. What is a front? How does it occur?
8. Compare warm and cold fronts. How do they differ in duration and precipitation characteristics?
9. What kind of weather often results from a stationary front? What kinds of forces tend to break up stationary fronts?
10. How does the westerly circulation of winds affect air masses in your area? What kinds of weather result?
11. Draw a diagram of a mature (fully developed) middle-latitude cyclone that includes the center of the low with several isobars, the warm front, the cold front, wind direction arrows, appropriate labeling of warm and cold air masses, and zones of precipitation.
12. If a wind changes to a clockwise direction, what is the shift called? Where does it locate you in relation to the center of a low pressure system? Explain why this happens.
13. How does the configuration of the upper air wind patterns play a role in the surface weather conditions?
14. Describe the sequence of weather events over a 48-hour period in St. Louis, Missouri, if a typical low pressure system (cyclone) passes 300 kilometers (180 mi) north of that location in the spring.
15. List three major causes of thunderstorms. How might the storms that develop from each of these causes differ?

Consider & Respond

1. Collect a three-day series of weather maps from your local newspapers. Based on the migration of high and low pressure systems during that period, discuss the likely pattern of the upper air winds.
2. Look at Figure 7.8a. Assume you are driving from point A to point B. Describe the changes in weather (temperature, wind speed and direction, barometric pressure, precipitation, and cloud cover) you would encounter on your trip. Do the same analysis for a trip from point C to point D.
3. The location of the polar front changes with seasons. Why? In what way is Figure 7.22b related to the seasonal migration of the polar front?
4. Redraw Figure 7.7 so that it depicts a Southern Hemisphere example.
5. List the ideal conditions for the development of a hurricane.

Apply & Learn

1. Wind speeds are sometimes given in knots (nautical miles per hour) instead of statute miles per hours as most people use. A nautical mile (used mainly by sailors and pilots) is equal to 6080 feet, slightly longer than the statute mile at 5280 feet; therefore, a knot is a little faster than 1 statute mile per hour (mph). The conversion from miles per hour to knots is:

   $$\text{KNOTS} = \text{MPH} \times 0.87$$

   Using Table 7.2, convert the ranges of wind speeds for Hurricane Categories 1 through 5 from miles per hour to knots.
2. A cold front squall line moving 35 miles per hour has spawned tornadoes at 9:00 a.m. near Memphis, Tennessee. The remainder of Tennessee is covered by an unstable *mT* air mass. Make a statement about what and when we might expect to happen in Nashville, Tennessee (some 200 miles away).

Locate & Explore

Note: Please read the About Locate & Explore Activities section of the Preface before beginning these exercises.

1. Using Google Earth and the weather layer provided by Google Earth (in the Layers window), track the daily variation in weather (temperature, pressure, wind speed, wind direction, dewpoint, precipitation, and cloud type) at any weather station in the following cities. Can you see any relationship in the weather between these cities? Can you use the weather in Omaha to predict the weather later in the week in Louisville or Washington?
   Portland, Oregon (45.53°N, 122.69°W)
   Casper, Wyoming (42.86°N, 106.29°W)
   Omaha, Nebraska (41.25°N, 95.88°W)
   Louisville, Kentucky (38.27°N, 85.74°W)
   Washington, D.C. (38.93°N, 77.03°W)
2. Using Google Earth and the weather layer provided by Google Earth (in the Layers window), look at the different weather systems across the United States and their associated cloud type. Over the next week, go outside every day at the same time and look at the sky. Record the cloud type in your area, using Figure 6.10 from your text as a guide. How do the cloud type and general weather conditions for your area (rain, sun, wind, and so on) relate to the regional weather systems (low pressures, high pressures, fronts, and so on) that are currently tracking across the United States?
3. Using Google Earth and the Wind Vector Layer for Hurricane Katrina, describe how the wind speed and direction changed in New Orleans, Louisiana (30.0°N, 90.1°W) and Biloxi, Mississippi (30.4°N, 88.9°W) as the storm passed. These layers show the speed of the wind (background color) and the direction (small arrows) every 2 hours as the storm passed. Provide an explanation for the observed wind speed and direction based on distance of each city from the center of the storm and whether the city was east or west of the storm.
   Tip: Create an *x-y* scatterplot in Excel using time on the *x*-axis and wind speed on the *y*-axis.

Map Interpretation

WEATHER MAPS

One of the most widely used weather maps is a surface weather map. These maps, which portray meteorological conditions over a large area at a given moment in time, are important for current weather descriptions and forecasting. Simultaneous observations of meteorological data are recorded at weather stations across the United States (and worldwide). This information is electronically relayed to the National Centers for Environmental Prediction near Washington, D.C., where the surface data are analyzed and mapped.

Meteorologists at the Centers then use the individual pieces of information to depict the general weather picture over a larger area. For example, isobars (lines of equal atmospheric pressure) are drawn to reveal the locations of cyclones (L) and anticyclones (H), and to indicate frontal boundaries. Areas that were receiving precipitation at the time the map depicts are shaded in green so that these areas are highlighted. The end result is a map of surface weather conditions that can be used to forecast changes in weather patterns. This map is accompanied by a satellite image that was taken on the same date.

1. Isobars are lines of equal atmospheric pressure expressed in millibars. What is the interval (in millibars) between adjacent isobars on this weather map?
2. What kind of front is passing through central Florida at this time?
3. Which Canadian high pressure system is stronger—the one located over British Columbia or the one near Newfoundland?
4. Which state is free of precipitation at this time: Nebraska, Connecticut, Mississippi, or Kentucky?
5. What kind of front is located over Nevada and Utah?
6. Does the surface map accurately depict the cloud cover indicated on the accompanying satellite image?
7. Can you find the low pressure systems on the satellite image?
8. Is there any cloud cover over West Virginia? How can you tell?
9. Do the locations of the fronts and areas of precipitation depicted on the map agree with the idealized relationship represented in Figure 7.8 (Middle-Latitude Cyclonic Systems)?
10. On the map, what kind of frontal symbol lies off the U.S. coast between New Jersey and Connecticut?
11. Looking at the satellite image, comment about the cold frontal symbols extending out to the northeast from Florida, and extending out to the southeast from the New England coast.
12. Are both of the low pressure systems depicted on this map occluding?

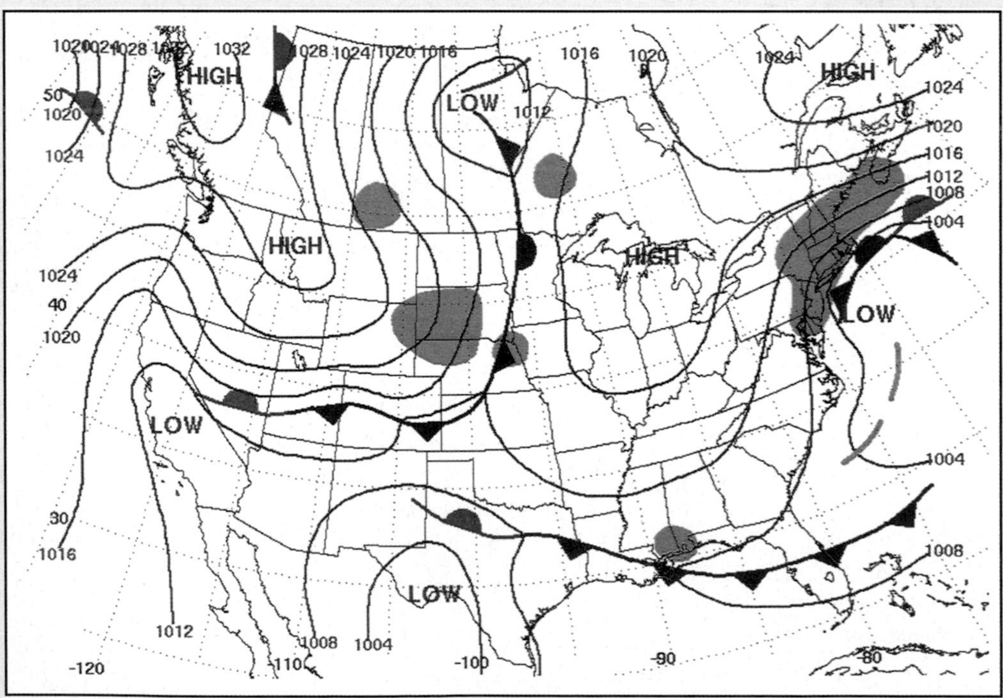

This surface weather map illustrates the spatial distribution of quantitative weather elements (air pressure, temperatures, wind speed, wind directions) as well as the locations of fronts and areas of precipitation. Isobars define high and low pressure cells, and the kinds of fronts are also identified.

Opposite: A satellite image of the atmospheric conditions shown on the accompanying weather map.

NOAA

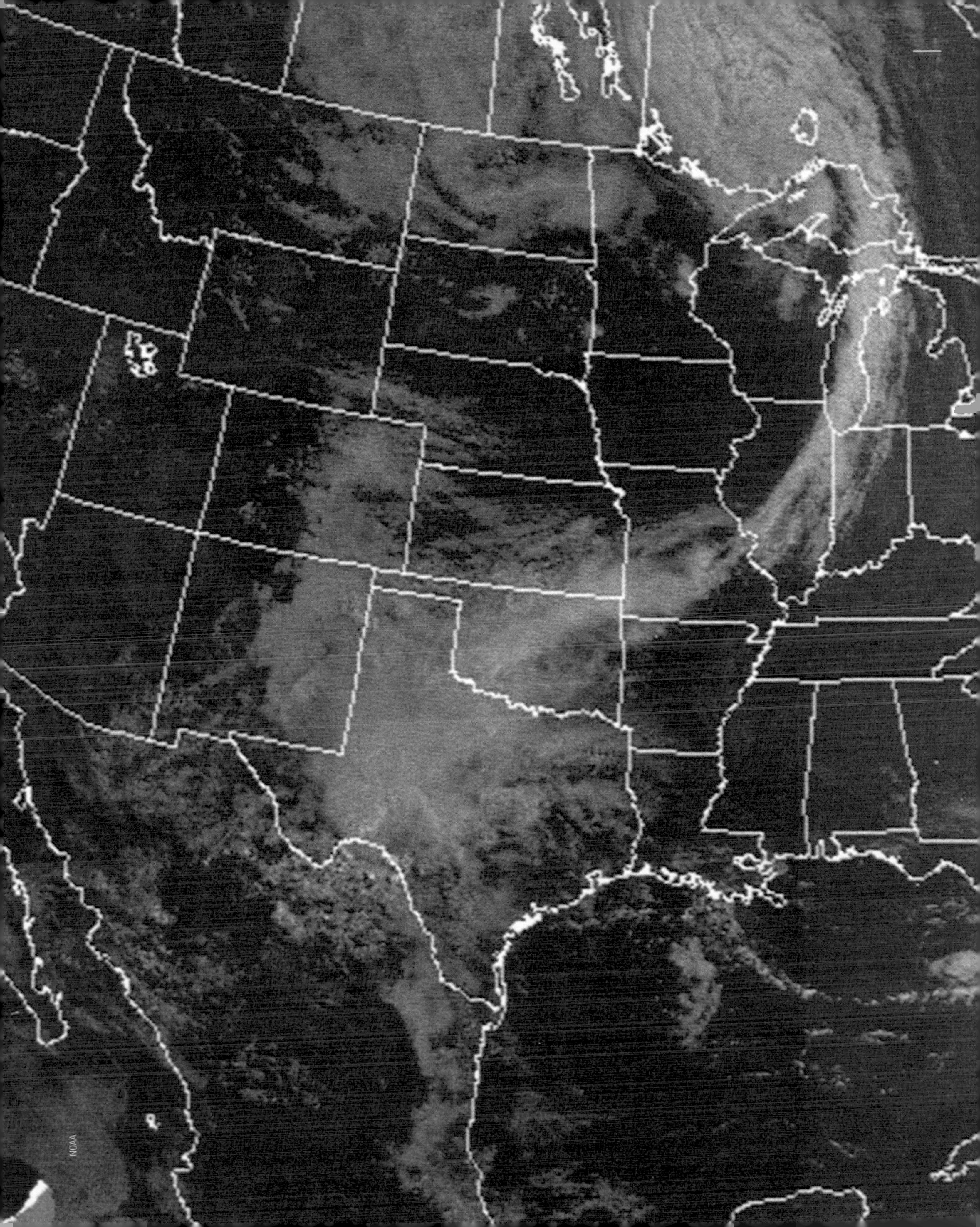
NOAA

Global Climates and Climate Change

8

CHAPTER PREVIEW

Atmospheric elements vary greatly from place to place on Earth, so scientists have classified climates by combining places with similar climatic statistics into a manageable number of climate regions.

- What two atmospheric elements are most often used when classifying climates?
- Why is there a need for more than one system of climate classification?

On a global, or macro, scale, the Köppen system of climate classification is one of the most widely used.

- What are some of the advantages and disadvantages of the Köppen system?
- What are some advantages of the Thornthwaite climate system, and how does it differ from the Köppen system?

Geographers use regions for much the same reasons that scholars in other disciplines use arbitrary systems for the organization of information—to create an orderly presentation of diverse phenomena.

- How does a geographer identify and define a region?
- What type of phenomena can be organized into region?

Even when applying the best scientific methods and the most modern technology, predicting future climate remains a difficult process, involving multiple lines of evidence.

- What methods and technology are most commonly used?
- What climate changes are most likely in the immediate and more distant future?

Why global warming is occurring is a complex issue, so it has taken an international effort by many scientists to estimate degree of influence of the factors that are most likely responsible.

- What are some of the major factors that can contribute to global warming?
- How are human activities most likely involved?

◄ Opposite: In the last decade, the cover of Arctic sea ice has continued to shrink in response to warming climatic conditions. September is the month when the ice melts back to its smallest areal extent. These photos compare the September ice extent in 1977 (top) to 2007 (bottom).
NASA/Goddard Space Flight Center Scientific Visualization Studio

If someone asked you "What's the weather like where you live?" how might you respond? Would you talk about the storm that occurred last week or say that winters are very mild where you reside? You may find that a question dealing with local atmospheric conditions is difficult to answer. Is the question referring to *weather* or *climate*? It is essential that you can distinguish between the two. Weather and climate were defined briefly in Chapter 4. In Chapter 7, we discussed the fundamentals of weather. In this chapter, we begin the study of climate in much greater detail.

Unlike weather, which describes the state of the atmosphere over short periods of time, climatic analysis relies heavily on averages, expected variations, and statistical probabilities involving data accumulated for the atmospheric elements over periods of many years. Climatic descriptions include such things as averages, extremes, and patterns of change for temperature, precipitation, pressure, sunshine, wind velocity and direction, and other weather elements throughout the year.

In the first part of this chapter, we will introduce the characteristics and classification of modern climates. Because climate can be defined at different scales, from a single hillside to a region as large as the Sahara stretching across much of northern Africa, two systems of describing and classifying climates are discussed. The Thornthwaite system is introduced because it is one of the most effective systems available to various scientists for the classification of climates on a more local scale. The Köppen system, however, has been widely adopted by physical geographers and other scientists; in a modified version, it will be the basis for the worldwide regional study of present-day climates in Chapters 9 and 10.

The remainder of this chapter focuses on climate change. Climates in the past were not the same as they are today, and there is every reason to believe that future climates will be different as well. It is now widely recognized that humans may also alter Earth's climate.

For decades scientists have realized that Earth has experienced major climate shifts during its history. It was believed that these shifts were gradual and could not be detected by humans during their lifetimes. However, recent research reveals that climate has shifted repeatedly between extremes over some exceedingly short intervals. Moreover, the research has revealed that climate during the most recent 10,000 years has been extraordinarily stable compared to similar intervals in the past.

To predict future climates, it is critical that we examine the details of past climate changes, including both the magnitude and rates of prehistoric climate change. Earth has experienced both ice ages and lengthy periods that were warmer than today. These fluctuations serve as indicators of the natural variability of climate in the absence of significant human impact. Using knowledge of present and past climates, as well as models of how and why climate changes, we conclude this chapter with some predictions of future climate trends.

Classifying Climates

Knowledge that climate varies from region to region dates to ancient times. The early Greeks (such as Aristotle, circa 350 BC) classified the known world into Torrid, Temperate, and Frigid zones based on their relative warmth. It was also recognized that these zones varied systematically with latitude and that the flora and fauna reflected these changes as well. With the further exploration of the world, naturalists noticed that the distribution of climates could be explained using factors such as sun angles, prevailing winds, elevation, and proximity to large water bodies.

The two weather variables used most often as indicators of climate are temperature and precipitation. To classify climates accurately, climatologists require a minimum of 30 years of data to describe the climate of an area. The invention of an instrument to reliably measure temperature—the thermometer—dates only to Galileo in the early 17th century. European settlement of and sporadic collection of temperature and precipitation data from distant colonies began in the 1700s but was not routine until the mid-19th century. This was soon followed in the early 20th century by some of the first attempts to classify global climates using actual temperature and precipitation data.

As we have seen in earlier chapters, temperature and precipitation vary greatly over Earth's surface. Climatologists have worked to reduce the infinite number of worldwide variations in atmospheric elements to a comprehensible number of groups by combining elements with similar statistics (● Fig. 8.1). That is, they can classify climates strictly on the basis of atmospheric elements, ignoring the causes of those variations (such as the frequency of air mass movements). This type of classification, based on statistical and mathematical parameters or physical characteristics, is called an **empirical classification.** A classification based on the causes, or *genesis,* of climate variation is known as a **genetic classification.**

Ordering the vast wealth of available climatic data into descriptions of major climatic groups, on either an empirical or a genetic basis, enables geographers to concentrate on the larger-scale causes of climatic differentiation. In addition, they can examine exceptions to the general relationships, the causes of which are often one or more of the other atmospheric controls. Finally, differentiating climates helps explain the distribution of other climate-related phenomena of importance to humans.

Despite its value, climate classification is not without its problems. Climate is a generalization about observed facts and data based on the averages and probabilities of weather. It does not describe a real weather situation; instead, it presents a composite weather picture. Within such a generalization, it is impossible to include the many variations that actually exist. Thus, classification systems must sometimes be adjusted to changes in climate. On a global scale, generalizations, simplifications, and compromises are made to distinguish among climate types and regions.

The Thornthwaite System

One system for classifying climates concentrates on a local scale. This system is especially useful for soil scientists, water resources specialists, and agriculturalists. For example, for a farmer interested in growing a specific crop in a particular area, a system classifying large regions of Earth is inadequate. Identifying the major vegetation type of the region and the annual range of both temperature and precipitation does not provide a farmer with information concerning the amounts and timing of annual soil moisture surpluses or deficits. From an agricultural perspective, it is much more important to know that moisture will be available in the growing season, whether it comes directly in the form of precipitation or from the soil.

Developed by an American climatologist, C. Warren Thornthwaite, the **Thornthwaite system** establishes moisture availability at the subregional scale (● Fig. 8.2). It is the system preferred by those examining climates on a local scale. Development of detailed climate classification systems such as the Thornthwaite system became possible only after temperature and precipitation data were widely collected at numerous locations beginning in the latter half of the 19th century.

The Thornthwaite system is based on the concept of **potential evapotranspiration (potential ET),** which approximates the water use of plants with an unlimited water supply. (Evapotranspiration, discussed in Chapter 6, is a combination of evaporation and transpiration, or water loss through vegetation.) Potential ET is a theoretical value that increases with increasing temperature, winds, and length of daylight and decreases with increasing hu-

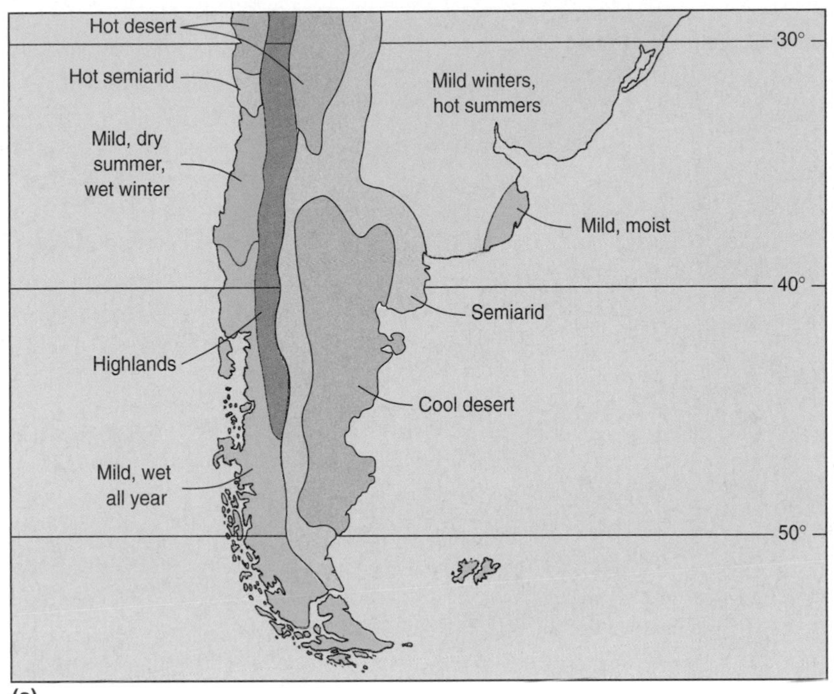

(a)

M. Trapasso
(b)

M. Trapasso
(c)

FIGURE 8.1
(a) This map shows the diversity of climates possible in a relatively small area, including portions of Chile, Argentina, Uruguay, and Brazil. The climates range from dry to wet and from hot to cold, with many possible combinations of temperature and moisture characteristics. (b) The Argentine Patagonian Steppe. (c) A meadow in the Argentine, Tierra del Fuego.
What can you suggest as the causes for the major climate changes as you follow the 40°S latitude line from west to east across South America?

midity. In contrast, **actual evapotranspiration (actual ET)** reflects actual water use by plants. This water can be supplied during the dry season by soil moisture if the soil is saturated, the climate is relatively cool, and/or the day lengths are short. Thus, measurements of actual ET relative to potential ET and available soil moisture are the determining factors for most vegetation and crop growth. Figure 6.8 shows a visual representation of the Thornthwaite system as it applies to the San Francisco, California area.

The Thornthwaite system recognizes three climate zones based on potential ET values: low-latitude climates, with potential ET greater than 130 centimeters (51 in.); middle-latitude climates, with potential ET less than 130 but greater than 52.5 centimeters (20.5 in.); and high-latitude climates, with potential ET less than 52.5 centimeters. Climate zones may be subdivided based on how long and by how much actual ET is below potential ET. Moist climates have either a surplus or a minor deficit of less than 15 centimeters (6 in.). Dry climates have an annual deficit greater than 15 centimeters.

Thornthwaite's original equations for potential ET were based on analyses of data collected in the midwestern and eastern United States. The method was subsequently used with less success in other parts of the world. Over the past few decades, many attempts have been made to improve the accuracy of the Thornthwaite system for regions outside the United States.

The Köppen System

The most widely used climate classification is based on regional temperature and precipitation patterns. It is referred to as the **Köppen system** after the German botanist and climatologist who developed it. Wladimir Köppen recognized that major vegetation associations reflect the area's climate. Hence, his climate regions were formulated to coincide with well-defined vegetation regions, and each climate region was described by the natural vegetation most often found there. Evidence of the strong influence of Köppen's system is seen in the wide usage of his climatic terminology, even in nonscientific literature (for example, steppe climate, tundra climate, rainforest climate).

Advantages and Limitations of the Köppen System Not only are temperature and precipitation two of the easiest weather elements to measure, but they are also measured more often and in more parts of the world than any other variables. By using temperature and precipitation statistics to define his boundaries, Köppen was able to develop precise definitions for each climate region, eliminating the imprecision that can develop in verbal and sometimes in genetic classifications.

Moreover, temperature and precipitation are the most important and effective weather elements. Variations caused by the atmospheric controls will show up most obviously in temperature and

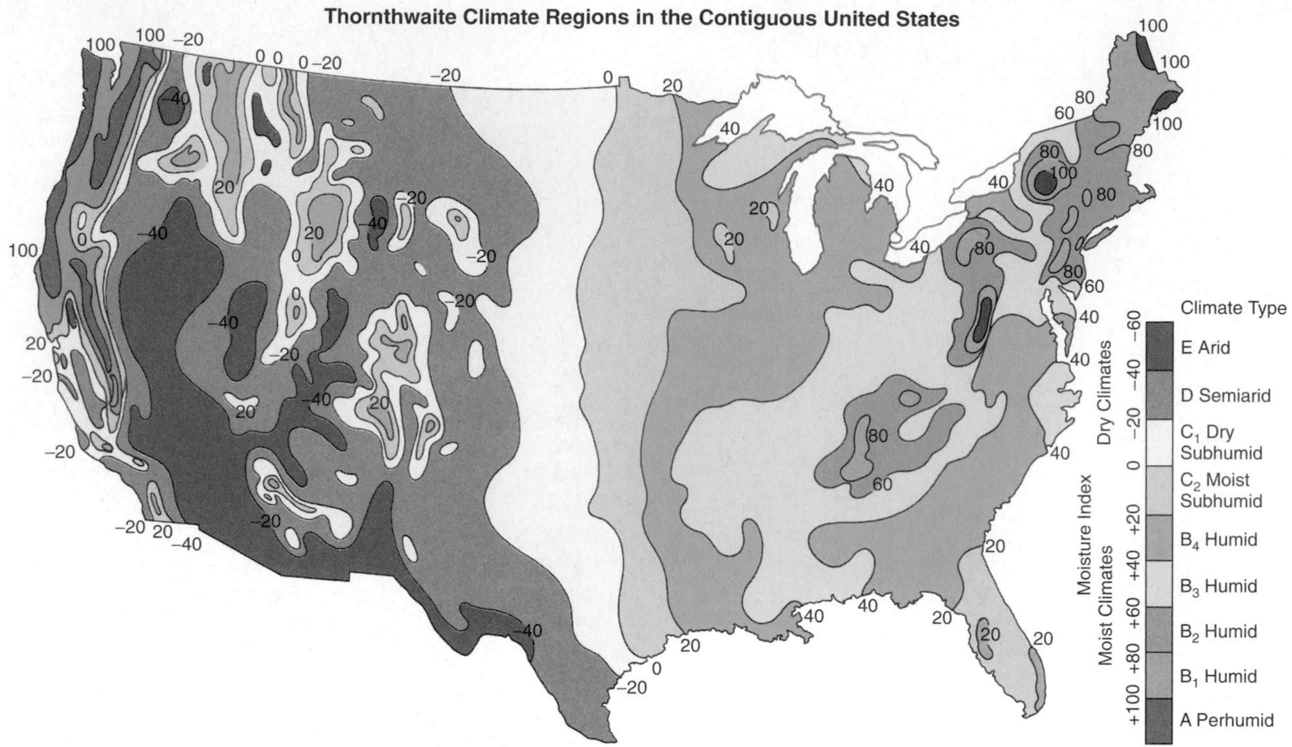

• FIGURE 8.2
Thornthwaite climate regions in the contiguous United States are based on the relationship between precipitation (P) and potential evapotranspiration (ET). The moisture index (MI) for a region is determined by this simple equation:

$$\text{MI} = 100 \times \frac{\text{P} - \text{Potential ET}}{\text{Potential ET}}$$

Where precipitation exceeds potential ET, the index is positive; where potential ET exceeds precipitation, the index is negative.
What are the moisture index and Thornthwaite climate type for coastal California?

precipitation statistics. At the same time, temperature and precipitation are the weather elements that most directly affect humans, other animals, vegetation, soils, and the form of the landscape.

Köppen's climate boundaries were designed to define the vegetation regions. Thus, Köppen's climate boundaries reflect "vegetation lines." For example, the Köppen classification uses the 10°C (50°F) monthly isotherm because of its relevance to the timberline—the line beyond which it is too cold for trees to thrive. For this reason, Köppen defined the treeless polar climates as those areas where the mean temperature of the warmest month is below 10°C. Clearly, if climates are divided according to associated vegetation types and if the division is based on the atmospheric elements of temperature and precipitation, then the result will be a visible association of vegetation with climate types. The relationship with the visible world in Köppen's climate classification system is one of its most appealing features to geographers and other scientists.

There are of course limitations to Köppen's system. For example, Köppen considered only average monthly temperature and precipitation in making his climate classifications. These two elements permit estimates of precipitation effectiveness but do not measure it with enough precision to permit comparison from one specific locality to another. In addition, for the purposes of generalization and simplification, Köppen ignored winds, cloud cover, intensity of precipitation, humidity, and daily temperature extremes—much, in fact, of what makes local weather and climate distinctive.

Simplified Köppen Classification The Köppen system, as modified by later climatologists, divides the world into six major climate categories. The first four are based on the annual range of temperatures: humid tropical climates (*A*), humid mesothermal (mild winter) climates (*C*), humid microthermal (severe winter) climates (*D*), and polar climates (*E*). Another category, the arid and semiarid climates (*BW* and *BS*), identifies regions that are characteristically dry based on both temperature and precipitation values. Because plants need more moisture to survive as the temperature increases, the arid and semiarid climates include regions where the temperatures range from cold to very hot. The final category, highland climates (*H*), identifies mountainous regions

GEOGRAPHY'S PHYSICAL SCIENCE PERSPECTIVE

Using Climographs

As stated at the beginning of this chapter, weather and climate are different ways of looking at how our atmosphere affects various locations on Earth. *Weather* deals with the state of the atmosphere at one point in time, or in the short term. It describes what is going on outside today or in the next few days. *Climate* deals with the conditions of the atmosphere in the long term, in other words, how the atmosphere behaves in a particular area through the months and years. Usually, a minimum record of 30 years is required to establish what an area's climate might bring. Therefore a climate can be described as, but is not restricted to, a compilation of average values used to summarize atmospheric conditions of an area.

It is possible to summarize the nature of the climate at any point on Earth in graph form, as shown in the accompanying figure. Given information on mean monthly temperature and rainfall, we can express the nature of the changes in these two elements throughout the year simply by plotting their values as points above or below (in the case of temperature) a zero line. To make the pattern of the monthly temperature changes clearer, we can connect the monthly values with a continuous line, producing an annual temperature curve. Monthly precipitation amounts are usually shown as bars reaching to various heights above the line of zero precipitation. Such a display of a location's climate is called a **climograph**. To read the graph, one must relate the temperature curve to the values given along the left side and the precipitation amounts to the scale on the right. Other information may also be displayed, depending on the type of climograph used.

The climograph shown here represents the type that we use in Chapters 9 and 10. This climograph can be used to determine the Köppen classification of the station as well as to show its specific temperature and rainfall measures. The climate classification abbreviations relating to all climographs are found in Table 8.1.

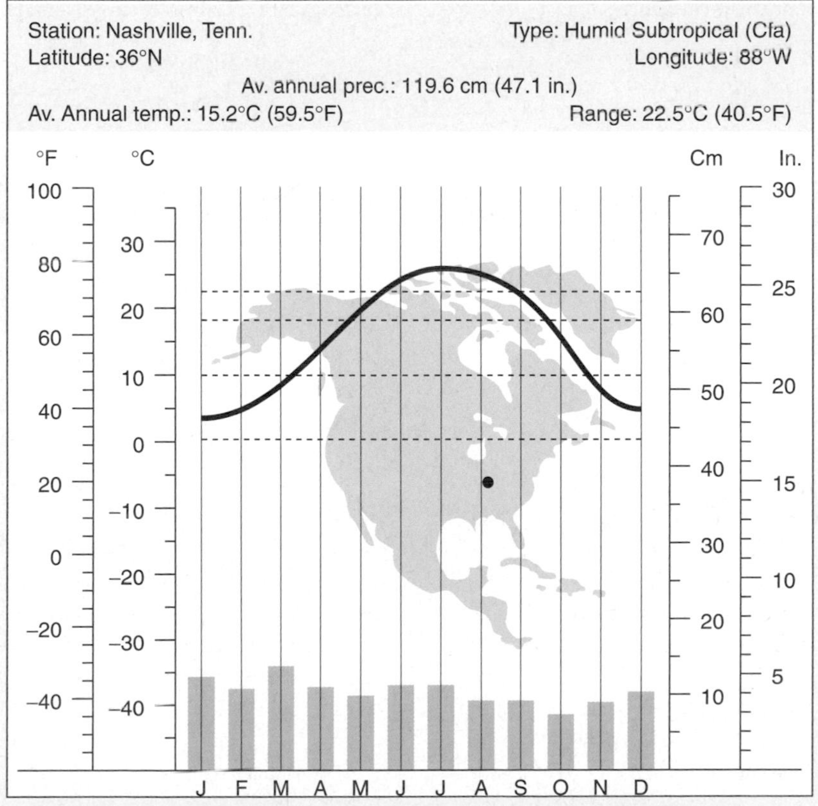

A standard climograph showing average monthly temperature (curve) and rainfall (bars). The horizontal index lines at 0°C (32°F), 10°C (50°F), 18°C (64.4°F), and 22°C (71.6°F) are the Köppen temperature parameters by which the station is classified.

where vegetation and climate vary rapidly as a result of changes in elevation and exposure.

Within each of the first five major categories, individual climate types and subtypes are differentiated from one another by specific parameters of temperature and precipitation. Table 1 in the "Graph Interpretation" exercise, pages 226–229, outlines the letter designations and procedures for determining the types and subtypes of the Köppen classification system. This table can be used with any Köppen climate type presented in this chapter as well as Chapters 9 and 10.

The Distribution of Climate Types

Five of the six major climate categories of the Köppen classification include enough differences in the ranges, total amounts, and seasonality of temperature and precipitation to produce the 13 distinctive climate types listed in Table 8.1. The tropical and arid climate types are discussed in some detail in the next chapter; the mesothermal, microthermal, and polar climates are presented in Chapter 10, along with a brief coverage of undifferentiated highland climates.

Tropical (*A*) Climates

Near the equator we find high temperatures year-round because the noon sun is never far from 90° (directly overhead). Humid climates of this type with no winter season are Köppen's **tropical climates.** As his boundary for tropical climates, Köppen chose 18°C (64.4°F) for the average temperature of the coldest month because it closely coincides with the geographic limit of certain tropical palms.

Table 8.1 shows that there are three humid tropical climates, reflecting major differences in the amount and distribution of rainfall within the tropical regions. Tropical climates extend poleward to 30° latitude or higher in the continent's interior but to lower latitudes near the coasts because of the moderating influence of the oceans on coastal temperatures.

Regions near the equator are influenced by the *intertropical convergence zone (ITCZ).* However, the convergent and rising air of the ITCZ, which brings rain to the tropics, is not anchored in one place; instead, it follows the 90° sun angle (see again "The Analemma," Chapter 3), migrating with the seasons. Within 5°–10° latitude of the equator, rainfall occurs year-round because the ITCZ moves through twice a year and is never far away (● Fig. 8.3). Poleward of this zone, the stabilizing influence of *subtropical high pressure systems* causes precipitation to become seasonal. When the ITCZ is over the region during the high-sun period (summer), there is adequate rainfall. However, during the low-sun period (winter), the subtropical highs and trade winds invade the area, bringing clear, dry weather.

We find the tropical rainforest climate *(Af)* in the equatorial region flanked both north and south by the dry-winter tropical savanna *(Aw)* climate (● Fig. 8.4). Finally, along coasts facing the strong, moisture-laden inflow of air associated with the summer monsoon, we find the tropical monsoon *(Am)* climate (● Fig. 8.5). Note the equatorial regions of Africa and South America shown in ● Figure 8.6. The atmospheric processes that produce the various tropical (*A*) climates are discussed in Chapter 9.

TABLE 8.1
Simplified Köppen Climate Classes

Climates	Climograph Abbreviation
Humid Tropical Climates (*A*)	
Tropical Rainforest Climate	Tropical Rf.
Tropical Monsoon Climate	Tropical Mon.
Tropical Savanna Climate	Tropical Sav.
Arid Climates (*B*)	
Steppe Climate	Low-lat./Mid-lat. Steppe
Desert Climate	Low-lat./Mid-lat. Desert
Humid Mesothermal (Mild Winter) Climates (*C*)	
Mediterranean Climate	Medit.
Humid Subtropical Climate	Humid Subt.
Marine West Coast Climate	Marine W.C.
Humid Microthermal (Severe Winter) Climates (*D*)	
Humid Continental, Hot-Summer Climate	Humid Cont. H.S.
Humid Continental, Mild-Summer Climate	Humid Cont. M.S.
Subarctic Climate	Subarctic
Polar Climates (*E*)	
Tundra Climate	Tundra
Ice-sheet Climate	Ice-sheet
Highland Climates (*H*) Various climates based on elevation differences.	No single climograph can depict these varied (or various) climates

R. Gabler

• **FIGURE 8.3**
Tropical rainforest climate: island of Jamaica.

R. Gabler

• **FIGURE 8.4**
Tropical savanna climate: East African high plains.

R. Gabler

• **FIGURE 8.5**
Tropical monsoon climate: Himalayan foothills, West Bengal, India.

Polar (*E*) Climates Just as the tropical climates lack winters (cold periods), the polar climates—at least statistically—lack summers. **Polar climates,** as defined by Köppen, are areas in which no month has an average temperature exceeding 10°C (50°F). Poleward of this temperature boundary, trees cannot survive. The 10°C isotherm for the warmest month more or less coincides with the Arctic Circle, poleward of which the sun does not rise above the horizon in midwinter and though the length of day increases during polar summers, the insolation strikes at a low angle.

The polar climates are subdivided into tundra and ice-sheet climates. The ice-sheet *(EF)* climate (• Fig. 8.7) has no month with an average temperature above 0°C (32°F). The Tundra *(ET)* climate (• Fig. 8.8) occurs where at least 1 month averages above 0°C (32°F). Look at the far northern regions of Eurasia, North America, and Antarctica in Figure 8.6. The processes creating the polar (*E*) climates are explained in Chapter 10.

Mesothermal (*C*) and Microthermal (*D*) Climates Except where arid climates intervene, the lands between the tropical and polar climates are occupied by the transitional middle-latitude mesothermal and microthermal climates. As they are neither tropical nor polar, the mild and severe winter climates must have at least 1 month averaging below 18°C (64.4°F) and 1 month averaging above 10°C (50°F). Although both middle-latitude climate categories have distinct temperature seasons, the **microthermal climates** have severe winters with at least 1 month averaging below freezing. Once again, vegetation reflects the climatic differences. In the severe-winter climates, all broadleaf and even some species of needle-leaf trees defoliate naturally during the winter (generally, needle-leaf trees do not defoliate in winter) because soil water is temporarily frozen and unavailable. Much of the natural vegetation of the mild-winter **mesothermal climates** retains its foliage throughout the year because liquid water is always present in the soil. The line separating mild from severe winters usually lies in the vicinity of the 40th parallel.

A number of important internal differences within the mesothermal and microthermal climate groups produce individual climate types based on precipitation patterns or seasonal temperature contrasts. The Mediterranean, or dry summer, mesothermal (*Csa, Csb*) climate (like Southern California or southern Spain in Figure 8.6) appears along west coasts between 30° and 40° latitude (• Fig. 8.9). On the east coasts, in generally the same latitudes, the humid subtropical climate is found (• Fig. 8.10). This type of climate is found in regions like the southeastern United States and southeastern China in Figure 8.6.

The distinction between the humid subtropical (*Cfa*) and marine west coast (*Cfb, Cfc*) climates illustrates a second important criterion for the internal subdivisions of middle-latitude climates: seasonal contrasts. Both mesothermal climates have year-round precipitation, but humid subtropical summer temperatures are much higher than those in the marine west coast climate. Therefore, summers are hot. In contrast, the mild summers of the marine west coast climate, located poleward of the Mediterranean climate along continental west coasts, often extend beyond 60° latitude

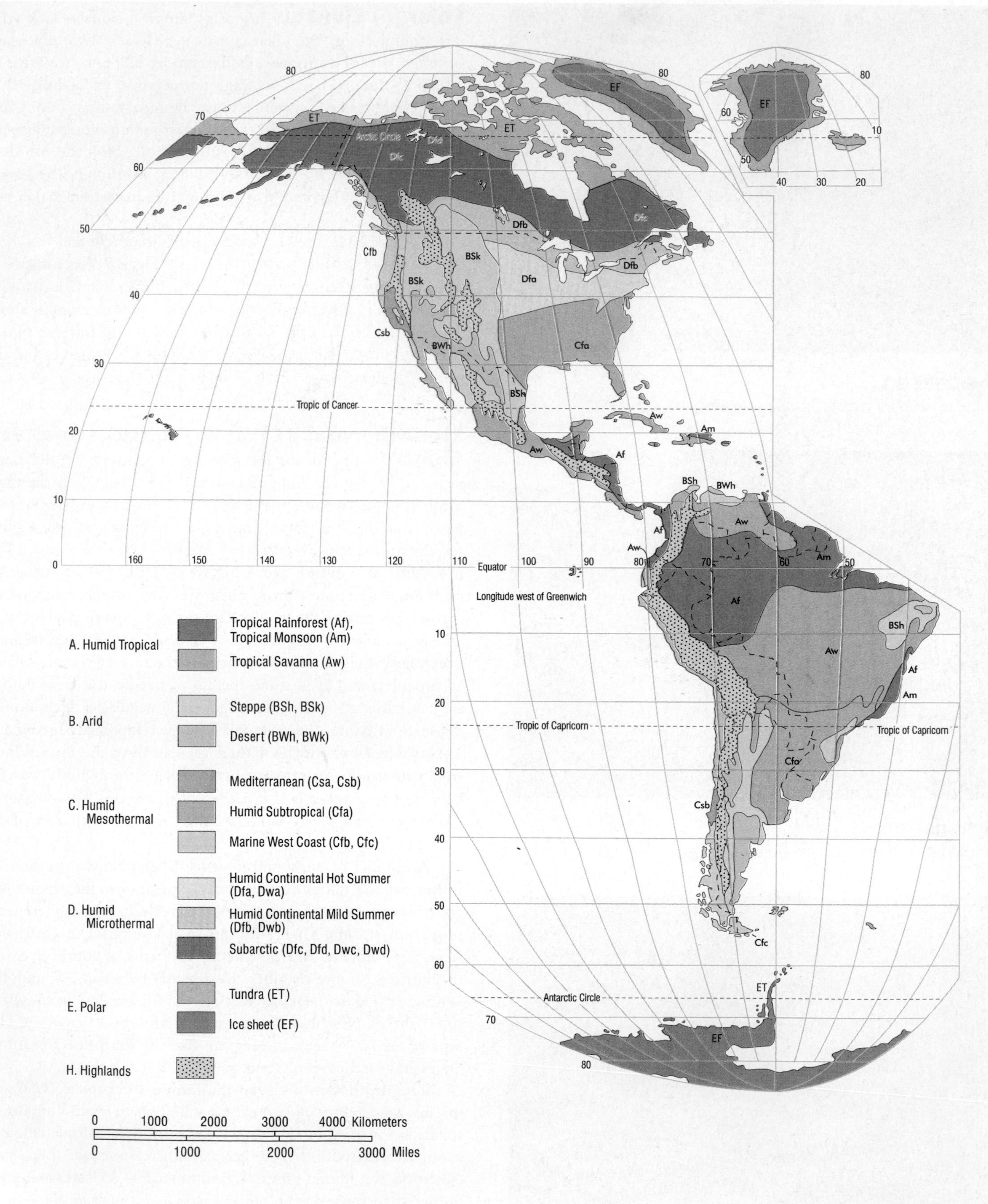

• FIGURE 8.6
World map of climates in the modified Köppen classification system.

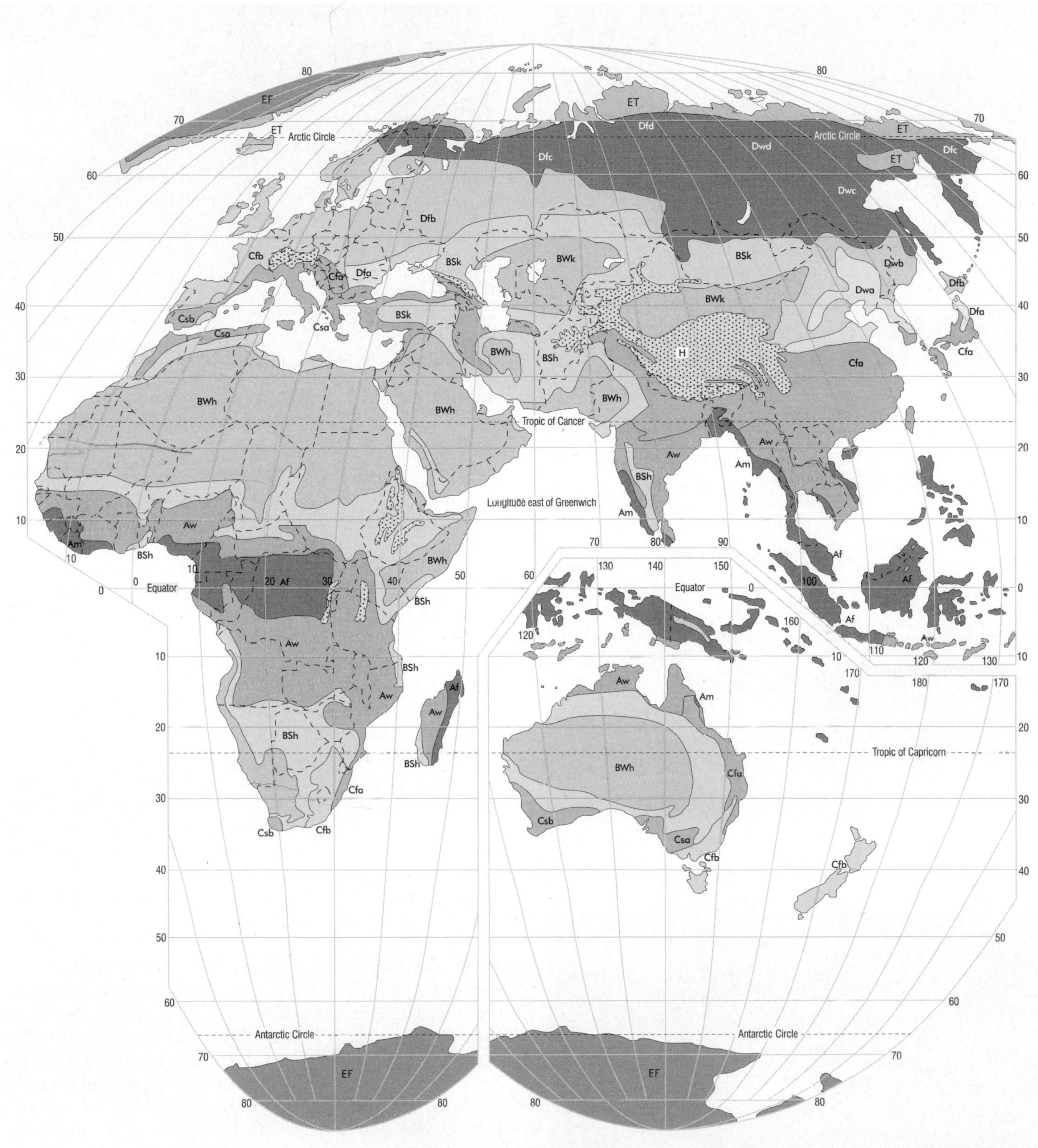

A Western Paragraphic Projection developed at Western Illinois University

J. Petersen

• **FIGURE 8.7**
Polar ice-sheet climate: glaciers near the southern coast of Greenland.

Alaska Image Library/USFWS

• **FIGURE 8.8**
Tundra climate: caribou in the Alaskan tundra during the summer.

R. Gabler

• **FIGURE 8.9**
Mediterranean mesothermal climate: village in southern Spain.

R. Gabler

• **FIGURE 8.10**
Humid subtropical climate: grapefruit grove in central Florida.

(• Fig. 8.11). Some examples of marine west coast climates, shown in Figure 8.6, are along the northwest coast of the United States and extending into Canada, the west coast of Europe, and the British Isles.

Another example of internal differences is found among microthermal (*D*) climates, which usually receive year-round precipitation associated with middle-latitude cyclones traveling along the polar front. Internal subdivision into climate types is based on summers that become shorter and cooler and winters that become longer and more severe with increasing latitude and continentality. Microthermal climates are found exclusively in the Northern Hemisphere (• Fig. 8.12) because there is no land in the Southern Hemisphere latitudes that would normally be occupied by these climate types. In the Northern Hemisphere, these climates progress poleward through the humid continental, hot-summer (*Dfa, Dwa*) climate, to the humid continental, mild-summer (*Dfb, Dwb*) climate, and finally, to the subarctic (*Dfc, Dfd, Dwc, and Dwd*) climate (• Fig. 8.13). Microclimate regions can be seen in the eastern United States and Canada or northward through eastern Europe in Figure 8.6.

Arid (*B*) Climates Climates that are dominated by year-round moisture deficiency are called **arid climates.** These climates will penetrate deep into the continent, interrupting the latitudinal zonation of climates that would otherwise exist. The definition of climatic aridity is that precipitation received is less than potential ET. Aridity does not depend solely on the amount of precipitation received; potential ET rates and temperature must also be taken into account. In a low-latitude climate with relatively high temperatures, the potential ET rate is greater than in a colder, higher-latitude climate. As a result, more rain must fall in the lower latitudes to produce the same effects (on vegetation) that smaller amounts of precipitation produce in areas with lower temperatures and, consequently, lower potential ET rates. Potential ET rates also decrease with altitude, which helps to explain why higher altitude (highland) climates are distinguished separately.

R. Gabler

• **FIGURE 8.11**
Marine west coast climate: North Sea coast of Scotland.

Arid climates are concentrated in a zone from about 15°N and S to about 30°N and S latitude along the western coasts, expanding much farther poleward over the heart of each landmass. The correspondence between the arid climates and the belt of subtropical high pressure systems is quite unmistakable (like in the southwestern United States, central Australia, and north Africa in Fig. 8.6), and the poleward expansion is a consequence of remoteness from the oceanic moisture supply.

In desert (*BW*) climates, the annual amount of precipitation is less than half the annual potential ET (• Fig. 8.14). Bordering the deserts are steppe (*BS*) climates—semiarid climates that are transitional between the extreme aridity of the deserts and the moisture surplus of the humid climates (• Fig. 8.15). The definition of the steppe climate is an area where annual precipitation is less than potential ET but more than half the potential ET. *B* climates and the processes that create them are discussed in more detail in Chapter 9.

R. Gabler

• **FIGURE 8.12**
Microthermal, severe winter climate: winter in Illinois.

R. Gabler

• **FIGURE 8.14**
Desert climate: Sonoran Desert of Arizona.

Alaska Image Library/USFWS

• **FIGURE 8.13**
Microthermal, subarctic climate: these trees endure a long winter in Alaska.

R. Gabler

• **FIGURE 8.15**
Steppe climate: Sand Hills of Nebraska.

Natural Resources Conservation Service/Gene Alexander

• **FIGURE 8.16**
Highland climate: Uncompahare National Forest, Colorado.

Highland (*H*) Climates The pattern of climates and extent of aridity are affected by irregularities in Earth's surface, such as the presence of deep gulfs, interior seas, or significant highlands. The climatic patterns of Europe and North America are quite different because of such variations.

Highlands can channel air mass movements and create abrupt climatic divides. Their own microclimates form an intricate pattern related to elevation, cloud cover, and exposure (• Fig. 8.16). One significant effect of highlands aligned at right angles to the prevailing wind direction is the creation of arid regions extending tens to hundreds of kilometers leeward. Look at the mountain ranges in Figure 8.6: the Rockies, the Andes, the Alps, and the Himalayas show *H* climates. These undifferentiated **highland climates** are discussed in more detail in Chapter 10.

Climate Regions

Each of our modified Köppen climate types is defined by specific parameters for monthly averages of temperature and precipitation; thus, it is possible to draw boundaries between these types on a world map. The areas within these boundaries are examples of one type of world region. The term **region,** as used by geographers, refers to an area that has recognizably similar internal characteristics that are distinct from those of other areas. A region may be described on any basis that unifies it and differentiates it from others.

As we examine the climate regions of the world in the chapters that follow, you should make frequent reference to the map of world climate regions (see again Fig. 8.6). It shows the patterns of Earth's climates as they are distributed over each continent. However, a word of caution is in order. On a map of climate regions, distinct lines separate one region from another. Obviously, the lines do not mark points where there are abrupt changes in temperature or precipitation conditions. Rather, the lines signify **zones of transition** between different climate regions. Furthermore, these zones or boundaries between regions are based on monthly and annual averages and may shift as temperature and moisture statistics change over the years.

The actual transition from one climate region to another is gradual, except in cases in which the change is brought about by an unusual climate control such as a mountain barrier. It would be more accurate to depict climate regions and their zones of transition on a map by showing one color fading into another. Always keep in mind, as we describe Earth's climates, that it is the core areas of the regions that best exhibit the characteristics that distinguish one climate from another.

Now, let's look more closely at Figure 8.6. One thing that is immediately noticeable is the change in climate with latitude. This is especially apparent in North America when we examine the East Coast of the United States moving north into Canada. Here the sun angles and length of day play an important role. We can also see that similar climates usually appear in similar latitudes and/or in similar locations with respect to landmasses, ocean currents, or topography. These climate patterns emphasize the close relationship among climate, the weather elements, and the climate controls. These elements and controls are more fully correlated with their corresponding climates in Chapters 9 and 10. There is an order to Earth's atmospheric conditions and so also to its climate regions.

A striking variation in these global climate patterns becomes apparent when we compare the Northern and Southern Hemispheres. The Southern Hemisphere lacks the large landmasses of the Northern Hemisphere; thus, no climates in the higher latitudes (in land regions) can be classified as humid microthermal, and only one small peninsula of Antarctica can be said to have a tundra climate.

Scale and Climate

Climate can be measured at different scales (macro, meso, or micro). The climate of a large (macro) region, such as the Sahara, may be described correctly as hot and dry. Climate can also be described at mesoscale levels; for example, the climate of coastal Southern California is sunny and warm, with dry summers and wet winters. Finally, climate can be described at local scales, such as on the slopes of a single hill. This is termed a **microclimate.**

At the microclimate level, many factors will cause the climate to differ from nearby areas. For example, in the United States and other regions north of the Tropic of Cancer, south-facing slopes tend to be warmer and drier than north-facing slopes because they receive more sunlight (• Fig. 8.17). This variable is referred to as *slope aspect*—the direction a mountain slope faces in respect to the sun's rays. Microclimatic differences such as slope aspect can cause significant differences in vegetation and soil moisture. In what is sometimes called *topoclimates,* tall mountains often possess vertical zones of vegetation that reflect changes in the microclimates as one ascends from the base of the mountain (which may be surrounded by a tropical-type vegetation) to higher slopes with middle latitude–type vegetation to the summit covered with ice and snow.

Human activities can influence microclimates as well. Recent research indicates that the construction of a large reservoir leads to greater annual precipitation immediately downwind of this impounded water. This occurs because the lake supplies addi-

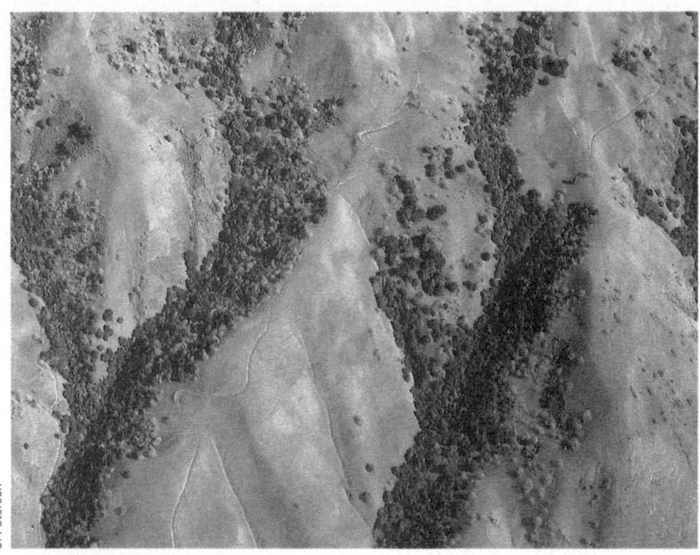

• **FIGURE 8.17**
This aerial photograph, facing eastward over valleys in the Coast Ranges of California, illustrates the significance of slope aspect. South-facing slopes on the left sides of valleys receive direct rays of the sun, and are hotter and drier than the more shaded north-facing slopes. The south-facing slopes support grasses and only a few trees, while the shaded, north-facing slopes are tree covered.
Why do the differing angles that the sun's rays strike the two opposite slopes affect temperatures?

tional water vapor to passing storms, which intensifies the rainfall or snows immediately downwind of the lake. These microclimatic effects are similar to the *lake-effect snows* that occur downwind of the Great Lakes in the early winter when the lakes are not frozen (discussed in Chapter 7). Another example of human impact on microclimates is the urban heat-island effect (discussed in Chapter 4), which leads to changes in temperature (urban centers tend to be warmer than their outlying rural areas), rainfall, wind speeds, and many other phenomena.

Climates of the Past

To try to predict future climates, it is critical to understand the magnitude and frequency of previous climate changes. Knowledge that Earth experienced major climate changes in the past is not new. In 1837, Louis Agassiz, a European naturalist, proposed that Earth had experienced major periods of *glaciation,* periods known as ice ages, when large areas of the continents were covered by huge sheets of ice. He presented evidence that glaciers (flowing ice) had once covered most of England, northern Europe, and Asia, as well as the foothill regions of the Alps. Agassiz arrived in the United States in 1846 and found similar evidence of widespread glaciation throughout North America.

• **FIGURE 8.18**
This map identifies the extensive areas of Canada and the northern United States that were covered by moving sheets of ice as recently as 18,000 years ago.
Why does the ice move in various directions in different regions of the continent?

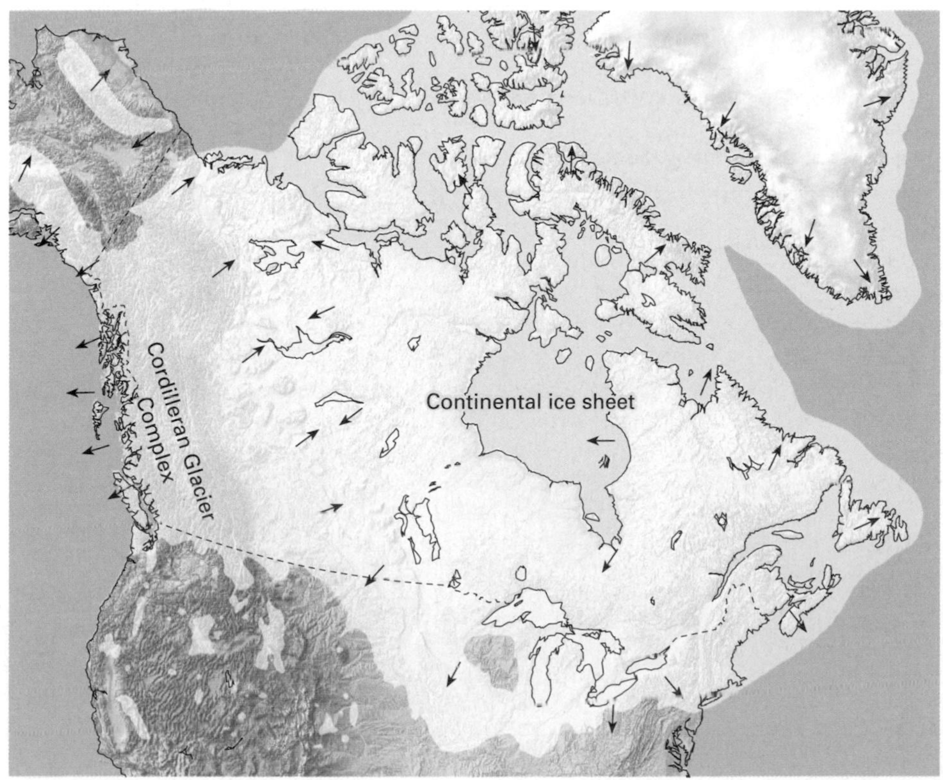

The Ice Ages

Until the 1960s, it was widely believed that Earth had experienced four major glacial advances followed by warmer interglacial periods. These glacial cycles occurred during the geologic epoch known as the *Pleistocene* (from about 1.6 to 2.0 million years up to 10,000 years ago). In Europe, these glacial epochs were termed the Günz (oldest), Mindel, Riss, and Würm. Likewise in North America, evidence of four glacial periods was recognized; these were termed the Nebraskan (oldest), Kansan, Illinoian, and Wisconsinan glaciations (• Fig. 8.18).

A major problem with studying the advance and retreat of glaciers on land is that each subsequent advance of the glaciers tends to destroy, bury, or greatly disrupt the sedimentary evidence of the previous glacial period. The evidence of the fourfold record of glacial advances was largely recognized on the basis of glacial deposits lying beyond the limit of the more recent glaciations. Evidence of "average" glacial advances that were subsequently overridden by more recent glaciers was rarely recognized.

Before the advent of radiometric-dating techniques (mineral and organic material can be dated by measuring the extent to which radioactive elements in the material have decayed through time), the timing of the glacial advances in both Europe and the United States was only crudely known. For example, estimates of the age of the last interglacial period were based on the rates at which Niagara Falls had eroded headward after the areas were first exposed when the glaciers retreated. Calculations ranging from 8000 to 30,000 years ago were produced.

Modern Research

Two major advances in scientific knowledge about climate change occurred in the 1950s. First, radiometric techniques, such as radiocarbon dating, that measured the absolute ages of landforms produced by the glaciers, began to be widely used. Radiocarbon dating conclusively showed that the last ice advance peaked about 18,000 years ago. This ice sheet covered essentially all of Canada and the northern United States, extending down to the Ohio and Missouri Rivers. It flowed over modern-day city sites such as Boston, New York, Indianapolis, and Des Moines.

The second major discovery was that evidence of detailed climate changes has been recorded in the sediments on the ocean floors. Unlike the continental record, the deep-sea sedimentary record had not been disrupted by subsequent glacial advances. Rather, the slow, continuous sediment record provides a complete history of climate changes during the past several million years. The most important discovery of the deep-sea record is that Earth has experienced numerous major glacial advances during the Pleistocene, not just the four that had been identified previously. Today, the names of only two of the North American glacial periods, the Illinoian and Wisconsinan, have been retained.

Because the deep-sea sedimentary record is so important to climate-change studies, it is important to understand how the record is deciphered. The deep-sea mud contains the microscopic record of innumerable surface-dwelling marine animals that built tiny shells for protection. When they died, these tiny shells sank to the seafloor, forming the layers of mud. Different species thrive in different surface-water temperatures; therefore, the stratigraphic record of the tiny fossils produces a detailed history of water-temperature fluctuations.

These tiny seashells are composed of calcium carbonate ($CaCO_3$); therefore, the analyses also record the oxygen composition of the seawater in which they were formed. One common measurement technique for determining oxygen composition is known as **oxygen-isotope analysis** (further explained in the next section). Modern seawater has a fixed ratio of the two oxygen isotopes. The O_{18}/O_{16} ratios will indicate changes in ocean temperatures relating to glacial cycles. A review of the oxygen-isotope record indicates that the last glacial advance about 18,000 years ago was only one of many major glacial advances during the past 2.4 million years. Evidence has suggested there may have been as many as 28 glacial-type climatic episodes.

Today, climatologists are aware that the present climate is but a short interval of relative stability in a time of major climate shifts. Moreover, the modern climate epoch, known as the *Holocene* (10,000 years ago to the present), is a time of extraordinarily stable, warm temperatures compared to most of the last 2.4 million years (● Fig. 8.19). Based on the deep-sea record, it appears that global climates tend to rest at one of two extremes: a very cold interval characterized by major glaciers and lower sea levels and shorter intervals between the glacial advances marked by unusually warm temperatures and high sea levels. With the realization that global climates have changed dramatically numerous times, two obvious questions arise: What causes global climate to change, and how quickly does global climate change from one extreme to the other?

● FIGURE 8.19

Analyses of oxygen-isotope ratios in ice cores taken from the glacial ice of Antarctica and Greenland provide evidence of surprising shifts of climate over short periods of time.

Has the general trend of temperatures on Earth been warmer or colder during the Holocene?

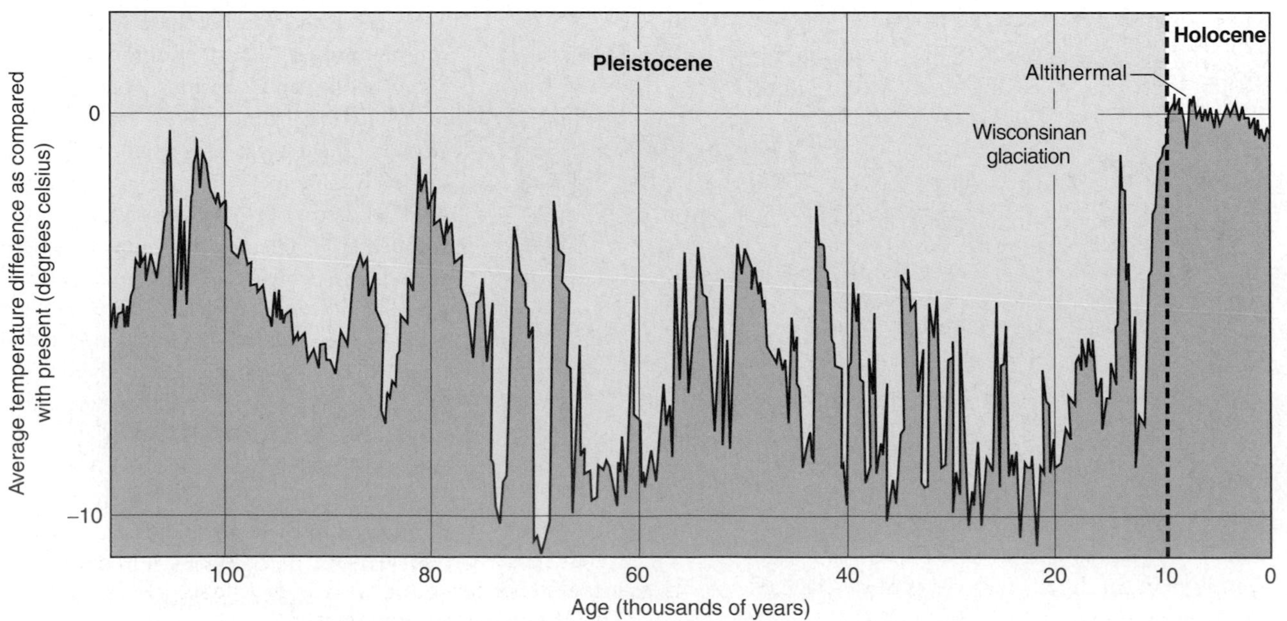

Methods for Revealing Climates of the Past

There are many methods used to uncover clues to the climates of the past. They are too numerous to mention in this one chapter. However, there are a few reliable methods that have been used for many years in reconstructing paleoclimates (ancient climates). Radiocarbon dating, mentioned earlier, is a means of determining how old an object (which contained carbon) may be. This is a very helpful tool but in itself does not indicate the climate of the past. *Oxygen-isotope analysis,* on the other hand, is a means of reconstructing paleoclimates.

To understand this method, it is helpful to review some basic definitions used in physics and chemistry. *Isotopes* are defined as atoms with the same atomic number but different atomic mass. The atomic number is equal to the positive charge of the nucleus—essentially, the number of *protons* in the nucleus. The atomic mass (or atomic weight) is equal to the number of *protons* and *neutrons* that comprise the nucleus of the atom. *Electrons,* which orbit the atom, are negatively charged particles that possess no appreciable mass (or weight). In a neutral atom, the number of electrons should equal the number of protons. When an electron is gained or lost to the atom, then a net (–) or (+) charge, respectively, will result, and the particle is then classified as an *ion*.

When dealing with isotopes, the atomic number (proton number) must remain the same (giving the atom its identity in the Periodic Table of Elements), but the number of neutrons can vary. In the example of oxygen isotopes, the atomic number is always 8. This is necessary to identify the atom as oxygen. However, the atom may contain 8 neutrons (O^{16}, the lighter isotope) or 10 neutrons (O^{18}, the heavier isotope). In oxygen-isotope analysis, the ratio of O^{18} to O^{16} is measured and compared to normal values. We have already discussed the O^{18}/O^{16} ratios of seafloor ($CaCO_3$) sediment; however, when dealing with yearly layers within Greenland and Antarctic ice cores, we must use them in a different way. When water evaporates from the ocean, slightly more of the O^{16} than O^{18} evaporates because water containing the lighter-weight oxygen evaporates more readily. During an ice age, the evaporated water is stored in the form of glacial ice rather than returned to the oceans and the O^{18}/O^{16} ratio in the ocean changes slightly to reflect the O^{16}-enriched water being stored in the glaciers. In this way O^{18}/O^{16} ratios can help reconstruct climates of the past from glacial ice layers (● Fig. 8.20).

Through time, paleoclimatologists have discovered new and different ways to determine climates of the past. The oxygen-isotope analysis is one of the most widely accepted methods, but there are other long-established methods as well. Two are worth a brief discussion; they are dendrochronology and palynology.

Dendrochronology (or tree-ring dating) has been used for decades. This analysis calls for the examination of tree rings exposed by cores taken through the middle of certain species of trees. The core (small enough so as not to harm the tree) will reveal each yearly tree ring. The rings are counted back through time to establish a time scale for the analysis. Each ring, by its thickness, color, and texture, can reveal the climate conditions (temperature and precipitation characteristics) during that particular year of the tree's growth. Thus, a short-term climate record can be determined through careful examination of tree rings (● Fig. 8.21).

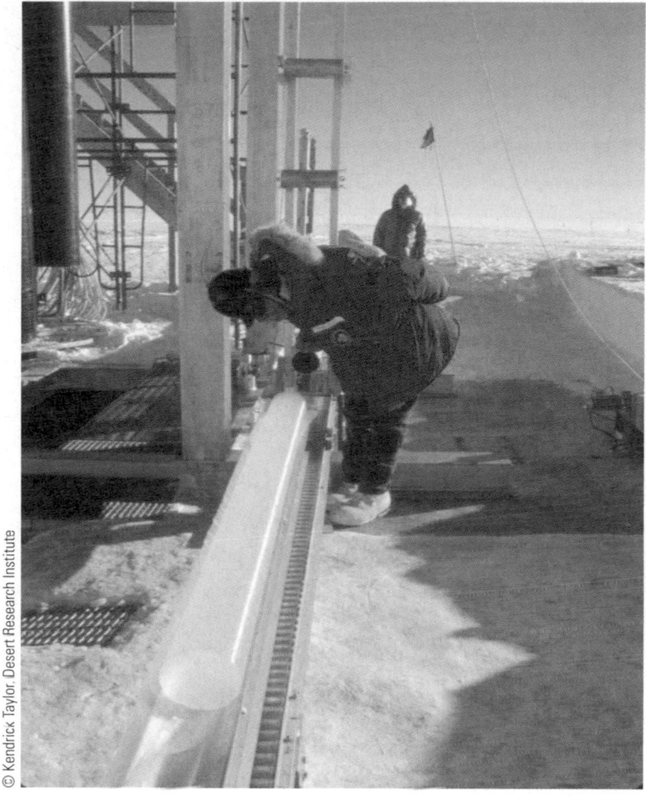

© Kendrick Taylor, Desert Research Institute

● **FIGURE 8.20**
Paleoclimates can be reconstructed using O^{18}/O^{16} ratios found in ice cores gathered from Greenland and Antarctica.
How can they distinguish the age of the various layers of ice?

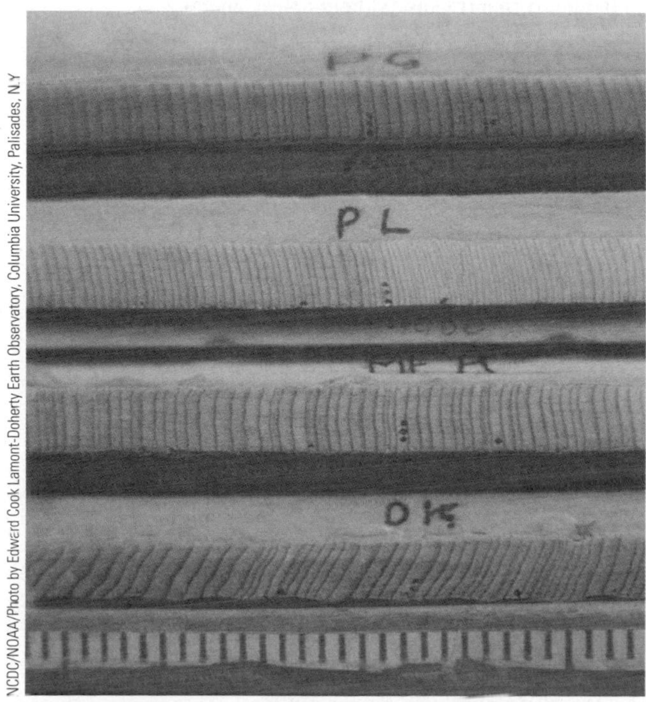

NCDC/NOAA/Photo by Edward Cook Lamont-Doherty Earth Observatory, Columbia University, Palisades, N.Y.

● **FIGURE 8.21**
The thickness, color, and texture of tree rings indicate the type of climate in existence during that particular growing season.
Where are the oldest tree rings found?

Wind

Core through bog deposits

Bog

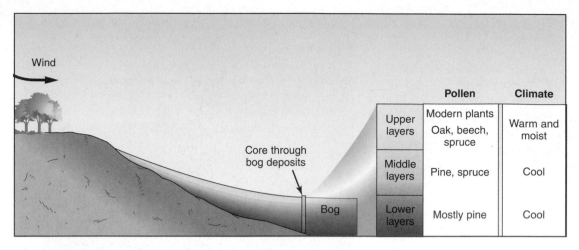

	Pollen	Climate
Upper layers	Modern plants Oak, beech, spruce	Warm and moist
Middle layers	Pine, spruce	Cool
Lower layers	Mostly pine	Cool

• **FIGURE 8.22**
This simple diagram shows the use of palynology to reconstruct paleoclimates.
What problems might occur with wind-blown pollen samples?

Palynology (or pollen-analysis dating) is also a well-established way of reconstructing past climates and environments. Though pollen samples can be recovered from a variety of environments, lakes and organic bogs are the best places to extract the samples to be analyzed. A core is drilled and removed to show the layers of sediment and organic material all the way to the bottom layer of the lake or bog. Then each layer with organic material can be radiocarbon dated to identify its age. Afterward, all the pollen is removed from each individual layer of the core and analyzed to identify the various tree or other plant types present. Thus, pollen is used to identify the numbers, types, and relative distributions of the trees and plants from which the pollen came (• Fig. 8.22 and • Fig. 8.23). It is then left to the paleoclimatologists and botanists to determine what type of climate would be required to sustain a forest or other environment of the type described by the analysis of each layer in the core.

• **FIGURE 8.23**
Pollen is used to identify the type of vegetation found in a particular layer of bog material.
How can botanists (plant experts) help paleoclimatologists reconstruct climates of the past?

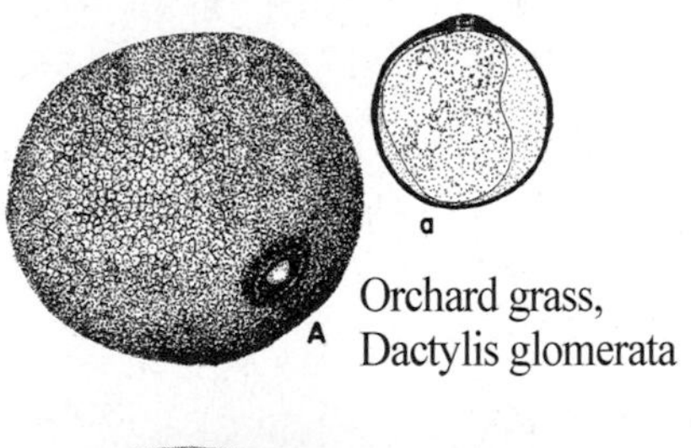

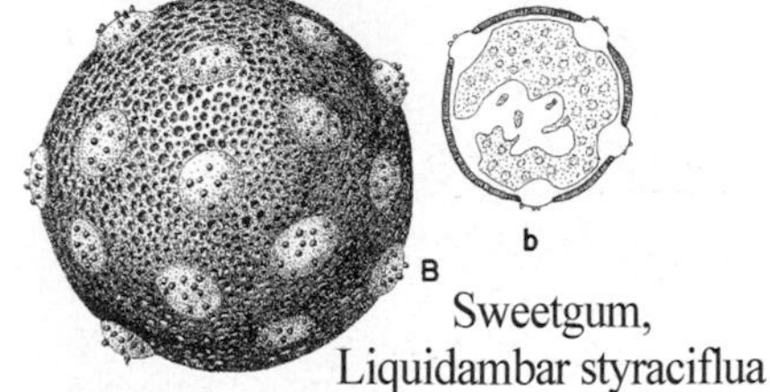

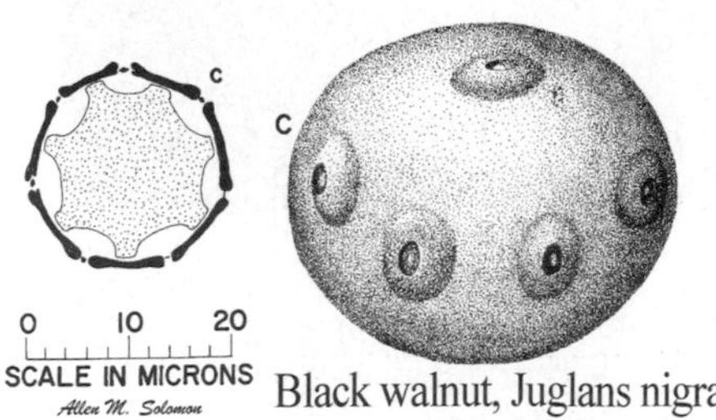

Drawn by Allen M. Solomon, Coos Bay, Oregon

Rates of Climate Change

Throughout most of Canada and into the United States, glaciers covered large areas north of the Missouri and Ohio Rivers 18,000 years ago. In the west, freshwater lakes more than 500 feet deep covered much of Utah and Nevada. However, the United States was mostly glacier free and the western lake basins were dry by about 9000 years ago. Abundant evidence has even been found that the climate about 7000 years ago (a time known as the **Altithermal**) was warmer than today (see again Fig. 8.19).

For glaciers several thousand feet thick to melt completely and for deep lakes to evaporate, a substantial increase in insolation is required over a few thousand years. Where did so much extra energy come from?

To answer questions about such rapid rates of climate change requires a more detailed record of climate than the deep-sea sediments can provide. This is because the deep-sea sedimentary record is extraordinarily slow—a few centimeters of sea mud accumulates in a thousand years. Rapid shifts in climate during periods of a few hundred years are not recorded clearly in the seafloor sediments. This problem has been solved by coring the thick glaciers covering Antarctica and Greenland. Glacial ice records yearly amounts of snowfall and is much more likely to provide short-term evidence of climate changes. *Oxygen isotope analysis* is utilized again, this time with the glacial ice of Antarctica and, most recently, Greenland. These analyses have revealed a detailed record of climate changes during the past 250,000 years (see again Fig. 8.19).

A surprising discovery of the ice-sheet analyses is the speed at which climate changes. Rather than changing gradually from glacial to interglacial conditions over thousands of years, the ice record indicates that the shifts can occur in a few years or decades. Thus, whatever is most responsible for major climate changes can develop rapidly. This probably requires a *positive feedback system,* which means, as explained in Chapter 1, that a change in one variable will cause changes in other variables that magnify the amount of original change. For example, most glaciers have high albedos, reflecting significant amounts of sunlight back to space. However, if the ice sheets retreat for whatever reason, low-albedo land begins to absorb more insolation, increasing the amount of energy available to melt the ice. Thus, the more ice that melts, the more energy is available to melt the ice further, magnifying the initial glacial retreat.

In contrast, a *negative feedback system,* where changes in one of the variables induce the system to remain stable, also affects the likelihood or rate of climate change. For example, increasing global temperatures cause evaporation rates to increase. The more water that evaporates from the ocean surface, the more clouds will form. The more clouds that exist, the more insolation is reflected back to space, cooling Earth's surface. (A counterargument to this effect is that clouds also operate as a greenhouse blanket, trapping heat in the lower atmosphere.) Thus, for climate changes to occur rapidly, negative feedback cycles such as this one must be overwhelmed by positive feedback cycles.

Causes of Climate Change

Although theories about the causes of climate change are numerous, they can be organized into five broad categories: (1) astronomical variations in Earth's orbit; (2) changes in Earth's atmosphere; (3) changes in oceanic circulation; (4) changes in landmasses; and (5) asteroid and comet impacts.

Orbital Variations

Astronomers have detected slow changes in Earth's orbit that affect the distance between the sun and Earth as well as the deviation of Earth's axis on the plane of the ecliptic. These orbital cycles produce regular changes in the amount of solar energy that reaches Earth. The longest is known as the **eccentricity cycle,** which is a 100,000-year variation in the shape of Earth's orbit around the sun. In simple terms, Earth's orbit changes from an ellipse (oval), to a more circular orbit, and then back, affecting Earth–sun distance. More elliptical orbits seem to be associated with warm periods and more circular orbits may correspond to ice ages.

A second cycle, termed the **obliquity cycle,** represents a 41,000-year variation in the tilt of Earth's axis from a maximum 24.5° to a minimum of 22.0° and then back. The more Earth is tilted, the greater is the seasonality at middle and high latitudes. Therefore, less tilt should bring cooler summers to the polar regions and less melting of ice sheets, which may promote an ice age.

Finally, a **precession cycle** has been recognized with a periodicity of 21,000 years. The precession cycle determines the time of year that perihelion occurs. Today, Earth is closest to the sun on January 3 and, as a result, receives about 3.5% greater insolation than the average in January. When aphelion occurs on January 3 in about 10,500 years, the Northern Hemisphere winters should be somewhat colder (● Fig. 8.24).

These cycles operate collectively, and the combined effect of the three cycles can be calculated. The first person to examine all three of these cycles in detail was the mathematician Milutin

● FIGURE 8.24
Milankovitch calculated the periodicity for (a) eccentricity, (b) obliquity, and (c) and (d) precession.
What effect should these changes in receipt of insolation have on global climates?

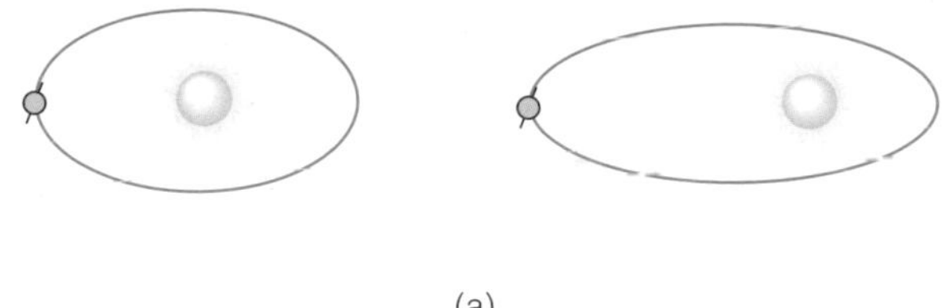

(a)

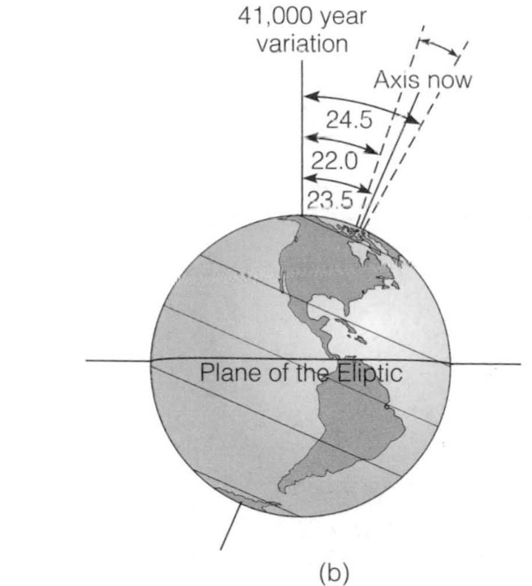

(b)

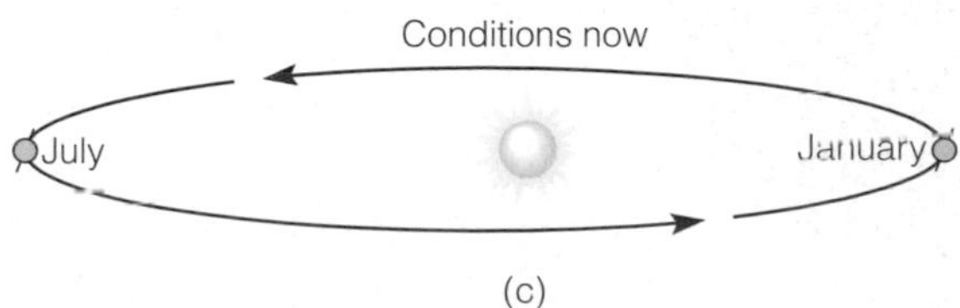

(c)

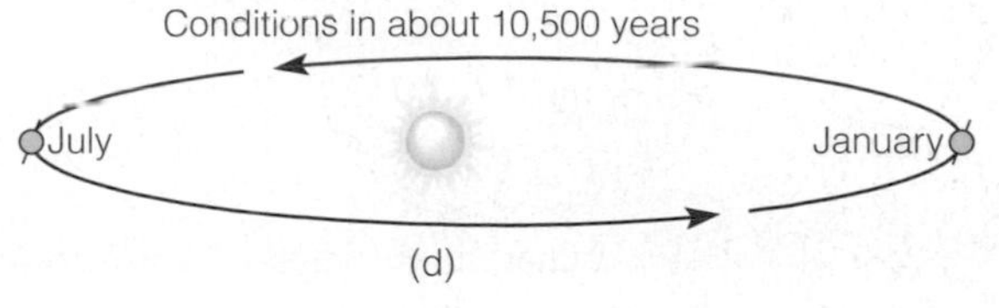

(d)

Milankovitch, who completed the complex mathematical calculations and showed how these changes in Earth's orbit would affect insolation. Milankovitch's calculations indicated that numerous glacial cycles should occur during 1 million-year intervals.

By the late 1970s, most paleoclimate (ancient climate) scientists were convinced that an unusually good correlation existed between the deep-sea record and Milankovitch's predictions. This suggests that the primary driving force behind glacial cycles is regular orbital variations, and it indicates that long-term climate cycles are entirely predictable! Unfortunately for humans, the Milankovitch theory indicates that the warm Holocene interglacial will soon end and that Earth is destined to experience full glacial conditions (glacial ice possibly as far south as the Ohio and Missouri Rivers) in about 20,000 years.

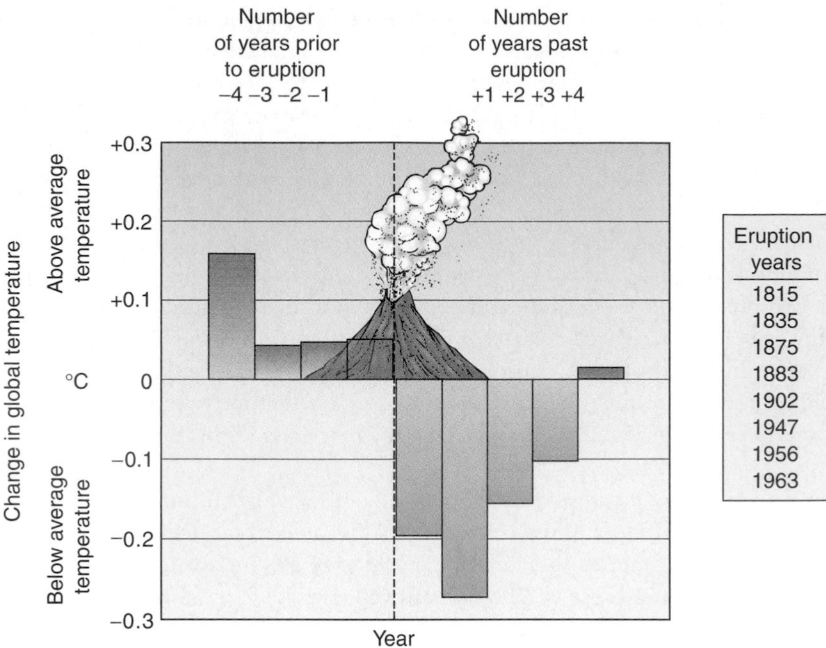

• **FIGURE 8.26**
Examination of global temperatures within 4 years before and after major volcanic eruptions provides compelling evidence that volcanic activity can have a direct effect upon the amounts of insolation reaching Earth's surface.
At what period after an eruption year does the effect seem the greatest?

Changes in Earth's Atmosphere

Many theories attribute climate changes to variations in atmospheric dust levels. The primary villain is volcanic activity, which pumps enormous quantities of particulates and aerosols (especially sulfur dioxide) into the stratosphere, where strong winds spread it around the world. The dust reduces the amount of insolation reaching Earth's surface for periods of 1–3 years (• Fig. 8.25 and • Fig. 8.26).

Volcanic Activity The climatic cooling effect of volcanic activity is unquestioned; all of the coldest years on record over the past two centuries have occurred in the year following a major eruption. Following the massive eruption of Tambora (in Indonesia) in 1815, 1816 was known as "the year without a summer." Killing frosts in July ruined crops in New England and Europe, resulting in famines. Several decades later, following the massive eruption of Krakatoa (also in Indonesia) in 1883, temperatures decreased significantly during 1884. Although no 20th-century eruptions have approached the magnitude of these two, the 1991 eruption of Mount Pinatubo (in the Philippine Islands) produced a substantial respite of cool conditions in an otherwise continuous series of record warm years (• Fig. 8.27).

• **FIGURE 8.25**
Volcanic activity at Mount St. Helens in Washington State pumps gases and particulates into the atmosphere. The volcanic peak of Mount Rainier, a potentially active volcano, is in the background.
Besides affecting the climate, what other hazards result from volcanic explosions?

USGS/Jim Vallance

Atmospheric Gases Another phenomenon closely correlated with average global temperatures is the composition of atmospheric gases. Scientists have known for many years that carbon dioxide (CO_2) acts as a "**greenhouse gas.**" There is no question that CO_2 is transparent to incoming shortwave radiation and blocks outgoing longwave radiation, similar to the effect of the glass panes in a greenhouse or in your automobile on a sunny day (refer again to the greenhouse discussion in Chapter 4). Thus, as the atmospheric content of greenhouse gases rises, so will the amount of heat trapped in the lower atmosphere.

Captured in the glacial ice of Antarctica and Greenland are air bubbles containing minor samples of the atmosphere that existed at the time that the ice formed. One of the important discoveries of the ice-core projects is that prehistoric atmospheric CO_2 levels increased during interglacial periods and decreased during major glacial advances.

The fact that average global temperatures and CO_2 levels are so closely correlated suggests that Earth will experience record warmth as the atmospheric level of CO_2 increases. The present level of approximately 380 parts per million of CO_2 is already higher than at any time in the past million years.

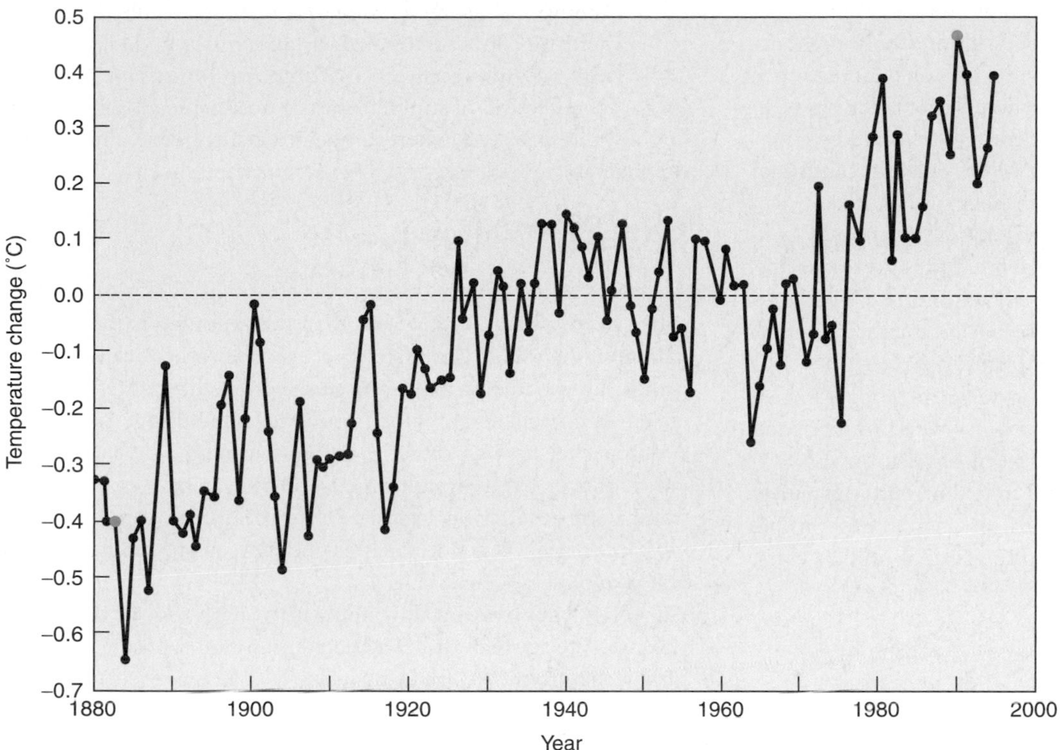

• **FIGURE 8.27**
This graph shows the gradual warming trend in global temperatures since 1880. It also documents the sharp reversal of the trend and the cooling of temperatures after the eruptions of Krakatoa in 1883 and Mount Pinatubo in 1991.
Is volcanic activity responsible for all of these temperature reversals?

Carbon dioxide is not the only greenhouse gas. Molecule for molecule, methane (CH_4) is more than 20 times more effective than CO_2 as a greenhouse gas but is considered less important because the atmospheric concentrations and the length of time the molecules of gas remain in the atmosphere (residence time) is much smaller. Garbage dump emissions and termite mounds both produce substantial quantities of CH_4. But a much more important source of atmospheric methane may come from the tundra regions or the deep sea. If warming the tundra or ocean water indeed releases large amounts of methane as is theorized, the resulting positive feedback cycle of warming could be enormous.

Other greenhouse gases include CFCs (chlorofluorocarbons) and N_2O (nitrous oxide). The relative greenhouse contribution of common greenhouse gases and their average residence times in the atmosphere are presented in • Figure 8.28.

Changes in the Ocean

Oceans cover over 70% of Earth's surface. Their enormous volume and high heat capacity make the oceans the single largest buffer against changes in Earth's climate. Whenever changes occur in oceanic temperatures, chemistry, or circulation, significant changes in global climate are certain to follow.

Surface oceanic currents are driven mostly by winds. However, a much slower circulation deep below the surface moves large volumes of water between the oceans. A major driving force of the deep circulation appears to be differences in water buoyancy caused by differences in salinity (salt content). Where surface evaporation is rapid, the rising salinity content causes the seawater density to increase, inducing subsidence. On the other

• **FIGURE 8.28**
Gases other than carbon dioxide released to the atmosphere by human activity contribute approximately 40% to the greenhouse effect. The figures in parentheses indicate the average number of years that the different gases remain in the atmosphere and contribute to temperature change.
Which gas has the longest residence time?

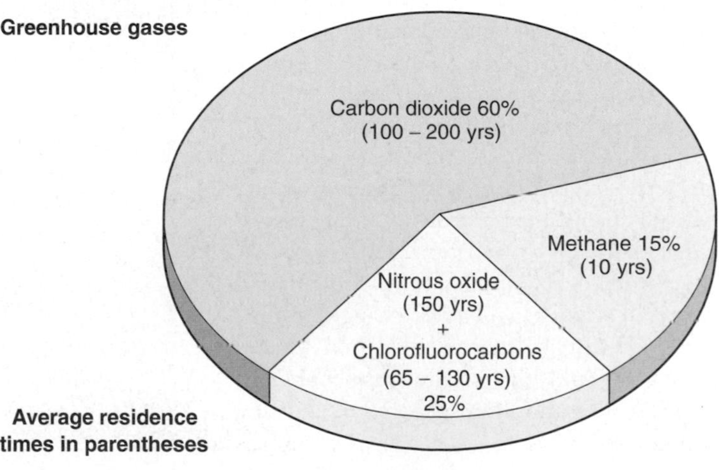

hand, when major influxes of freshwater flow from adjacent continents or concentrations of melting icebergs flood into the oceans, the salinity is reduced, thereby increasing the buoyancy of the water. When the surface water is buoyant, deep-water circulation slows. In many cases, the freshwater influx is immediately followed by a major flow of warm surface waters into the North Atlantic, causing an abrupt warming of the Northern Hemisphere. Subsurface ocean currents are also affected by water temperature. Extremely cold Arctic and Antarctic waters are quite dense and tend to subside, whereas tropical water is warmer and may tend to rise. Therefore, salinity and temperature taken together bring about rather complex subsurface flows deep within our ocean basins.

In modern times, short-term changes in Pacific circulation are primarily responsible for El Niño and La Niña events (discussed in Chapter 5). The onset of El Niño/Southern Oscillation (ENSO) climatic events is both rapid and global in extent, and it is widely believed that changes in oceanic circulation may be responsible for similar rapid climate changes during the last 2.4 million years.

Changes in Landmasses

The fourth category of climate change theories involves changes in Earth's surface to explain lengthy periods of cold climates. A number of ice ages, some with multiple glacial advances, occurred during Earth's history. To explain some of the previous glacial periods, scientists have proposed several factors that might be responsible. For example, one characteristic that all of these glacial periods have in common with the Pleistocene is the presence of a continent in polar latitudes. Polar continents permit glaciers to accumulate on land, which results in lowered sea levels and consequent global effects.

Another geologic factor sometimes invoked as a cause of climate change is the formation, disappearance, or movement of a landmass that restricts oceanic or atmospheric circulation. For example, eruptions of volcanoes and the formation of the Isthmus of Panama severed the connection between the Atlantic and Pacific, thereby closing a pathway of significant ocean circulation. This redirection of ocean water created the Gulf Stream Current/North Atlantic Drift (see again Fig. 5.25). Another example is the uplift of the Himalayas, altering atmospheric flows, and monsoonal effects in Asia. Both of these events, and several other significant changes, immediately predate the onset of the modern series of glaciations. Which events caused climate changes and which are simply coincidences has yet to be determined.

Shifting in landmasses (in both latitude and altitude) will affect changes in the types and distributions of vegetation. These changes would further affect atmospheric composition and atmospheric circulation patterns.

A final group of theories involve changes in albedo, caused either by major snow accumulations on high-latitude landmasses or by large oceanic ice sheets drifting into lower latitudes. The increased reflection of sunlight starts a positive feedback cycle of cooling that may end when the polar oceans freeze, shutting off the primary moisture source for the polar ice sheets.

Impact Events

As described in Chapter 3, *asteroids* are small, rocky, or metallic solar system bodies, usually less than 800 kilometers (500 mi) in diameter. They may break apart into smaller pieces called meteoroids. These objects orbit our sun, along with comets. *Comets* are small objects of rocky or iron material held together by ice. A comet's ice will vaporize in sunlight, leaving a distinguishable tail of dust or gas. Through time these objects have struck Earth, some with devastating impact. There is no doubt such impacts will occur again. It is not a matter of if, but when, another impact will take place (● Fig. 8.29).

On a daily basis Earth is bombarded with tons of this material; most are so small that they burn up in our atmosphere before hitting the ground. At night these small objects are seen as shooting stars. Objects that are smaller than 40 meters in diameter will incinerate with the friction encountered in our atmosphere. Objects ranging from 40 meters to about 1 kilometer in diameter can do tremendous damage on a local scale when they reach Earth's surface. This size impact can be expected every 100 years or so. The last occurred in 1908 near Tunguska, Siberia, and devastated a huge area of the Siberian wilderness, with a blast estimated at 15 megatons (a megaton explosion is equal to 1 million tons of TNT).

Every few hundred thousand years or so, an object with a diameter of greater than 1.6 kilometers (1 mi or greater) will

● FIGURE 8.29
Barringer Crater, Arizona, shows the results of an impact with an iron-nickel meteorite of about 50 meters (165 ft) in diameter.

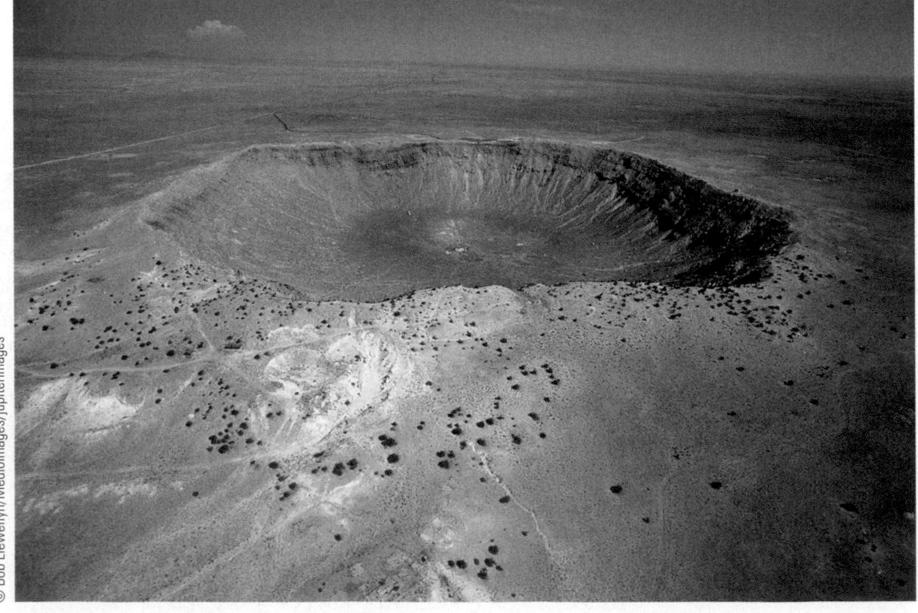

© Bob Llewellyn/MedioImages/jupiterimages

GEOGRAPHY'S SPATIAL SCIENCE PERSPECTIVE

Climate Change and Its Impact on Coastlines

When we look at a map or a globe, one of the pieces of geographic information that we see is so obvious and basic that we often take it for granted, perhaps failing to recognize that it is spatial information. This is the location of coastlines—the boundaries between land and ocean regions. One of the most basic aspects of our planet that a world map shows us is where landmasses exist and where the oceans are located, as well as the generally familiar shape of these major Earth features. But maps of our planet today only show where the coastline is currently located. We know that sea level has changed over time and that it rose 20–30 centimeters (8–12 in.) during the 20th century. The hydrologic system on Earth is a closed system because the total amount of water (as a gas, liquid, and solid) on our planet is fixed. When the climate supports more glacial ice, sea level falls. When world climates experience a warming tendency, sea level rises. More ice in glaciers means less water in the oceans, and vice versa. If global warming trends continue at their present rate, the U.S. Environmental Protection Agency (EPA) predicts that sea level will rise 31 centimeters (1 ft) in the next 25–50 years. That amount of sea-level rise will cause problems for low-lying coastal areas; the populations of some coral islands in the Pacific are already concerned, as their homelands are barely above the high-tide level.

Looking at a map of world population distribution and comparing it to a world physical map shows a strong link between settlement density and coastal areas. For low-lying coastal regions, sea-level rise is a major concern, and the gentler the slope of the coast is, the farther inland the inundation would be with every increment of sea-level rise. Scientists at the United States Geological Survey (USGS) have determined that if all the glaciers on Earth were to melt, sea level would rise 80 meters (263 ft) and that a 10-meter (33-ft) rise would displace 25% of the U.S. population.

Before the ice ages of the Pleistocene, worldwide climates were generally warmer; through much of Earth's history, no glaciers have existed on the planet. During times like those, when Earth was ice free, sea level would have been at a maximum, and that might occur again in the distant future under similar climatic conditions. At the time when glaciers were most extensive, during the maximum advance of Pleistocene glaciers, sea level fell to about 100 meters (330 ft) below today's level. Maps that create the positions of coastlines and the shape of continents during times of major environmental change show how temporary and vulnerable coastal areas can be.

The accompanying maps show the present coastline (dark green), the pre-Pleistocene maximum rise of sea level (light green), the Pleistocene drop in sea level (light blue), and the impact on North American coastlines. It may seem odd to think of the coastlines shown on a world map as temporary, but because coasts can shift over time, a future world map could look quite different from the map we know today.

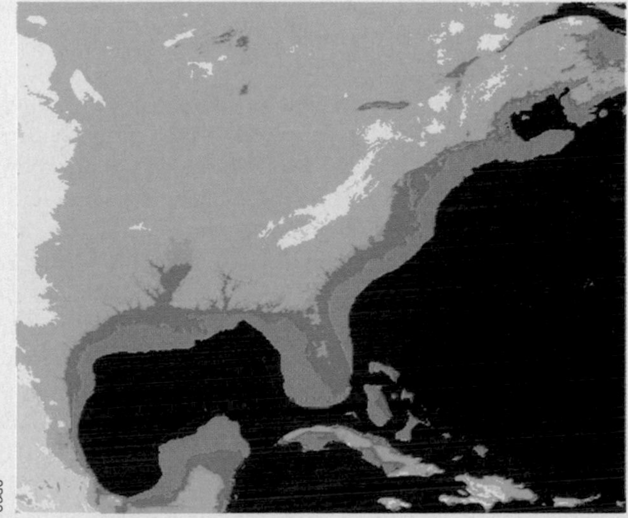
USGS

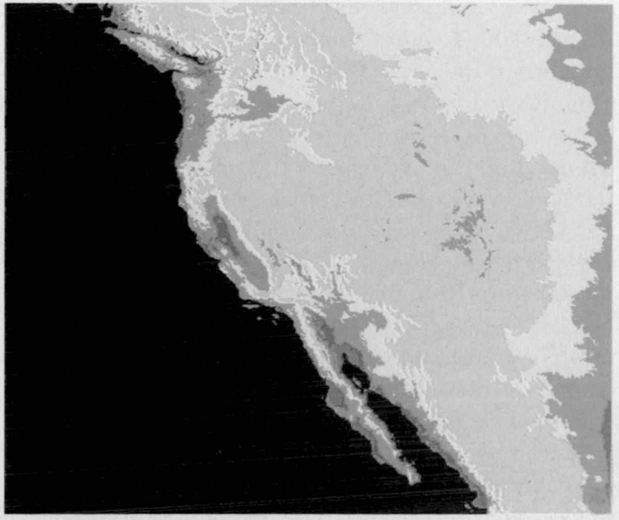
USGS

These maps show how changes in sea level would affect the coastline. Dark green shows our coastlines as they appear today. Light green shows the coastline if all glaciers on Earth were to melt. Light blue shows the coastline if glaciers expanded to the level of maximum extent during the Pleistocene.

impact Earth, producing severe environmental damage and climate change on a global scale. The power of the blasts from such impacts could equal a million megatons of energy. The likely affect would be an "impact winter," characterized by skies darkened with particulates, thereby blocking insolation and causing a drastic drop in temperatures. Firestorms would result from heated impact debris raining back down on Earth, and large amounts of acid rain would precipitate as well. A catastrophe of this kind would result in loss of crops worldwide, followed by starvation and disease. The largest known impacts on Earth in its history, like the one that may have contributed to the extinction of the dinosaurs 65 million years ago, have been estimated to be about 15 kilometers (about 10 mi) in diameter and may have exploded with a force of 100 million megatons.

A vigilant group of professional and amateur astronomers are constantly watching the night skies for **Near Earth Objects (NEOs)** or objects whose trajectory may bring them into a collision course with our planet. If we know far enough in advance that an NEO will impact Earth, we may be able to avert a disaster by modifying the object's course so as to avoid a collision. At this point however, whether or not anything can be done about such a collision remains mere speculation.

Predicting Future Climates

With so many variables potentially responsible for climate change, reliably predicting future climate is an exceedingly difficult proposition at best. The primary problem in climate prediction is posed by natural variability. ● Figure 8.30 displays the frequency and magnitudes of climate changes that have occurred naturally over the past 150,000 years. Although the Holocene has been the most stable interval of the whole period, a detailed examination of the Holocene record reveals a wide range of climates. For example, a long interval of climates, hotter than today's climate, occurred during the *Altithermal*. This interval was characterized by the dominance of grasslands in the Sahara and severe droughts on the Great Plains. Other warm intervals occurred during the Bronze Age, during the second half of the Roman Empire, and in medieval times. An unusually cold interval began with the eruption of Santorini (the site of a civilization that some believe was the basis for the Atlantis myth) in the Aegean Sea. Other cold periods occurred during the Dark Ages and again beginning about 1150 to 1460 in the North Atlantic and 1560 to 1850 in continental Europe and North America. These last episodes collectively have been termed the **Little Ice Age.** The Little Ice Age had major impacts on civilizations—from the isolation of the Greenland settlements established during the medieval warm period to the abandonment of the Colorado Plateau region by the Anasazi cultures. An important point to remember is that, with the exception of the cold interval that began with the eruption of Santorini, climatologists do not know what variables changed to cause each of these major climate fluctuations.

Attempts to predict future climates are complicated further by the operation of many feedback cycles. Simply increasing the amount of heat that is trapped by gases in the lower atmosphere may or may not result in long-term warming. Negative feedback processes such as increased cloud formation and increased plant uptake of CO_2 may operate to counteract the warming. However, warming of the oceans and tundra may release additional greenhouse gases, setting into motion some significant positive feedback cycles. Which feedback mechanisms will dominate is not certain; therefore, all predictions must be tentative.

There have been numerous attempts to simulate the variables that affect climate. *General Circulation Models* (GCMs) are complex computer simulations based on the relationships among weather and climate variables discussed throughout this book: sun angles, temperature, evaporation rates, land versus water effects, energy transfers, and so on. Some of the variables, and relationships between them, are at best difficult to model, or left out of the models altogether. The complexity and the usefulness of GCMs are both increasing rapidly. Although improvement on GCMs continues, they are not infallible. However, these circulation models appear to do a good job in predicting how conditions will change in specific regions as the Earth warms or cools, and they have added new insights into how some climatic variables interact.

Based on the record of climate changes during the past, only one thing can be concluded about future climates: they will change. Looking into the distant future, the Milankovitch cycles indicate that another glacial cycle is probably on the way. The most rapid cooling should occur between 3000 and 7000 years from now. In the near term, however, global warming is most likely. The rise of greenhouse gases such as carbon dioxide and methane, the widespread destruction of vegetation, and the feedback cycles that will most likely result are bound to increase the average global temperature for the foreseeable future. An average increase of 1°C (nearly 2°F) would be equivalent to the change that has occurred since the end of the Little Ice Age in about 1850. A 2°C warming would be greater than anything that has happened in the Holocene, including the Altithermal. A 3°C warming would exceed anything that has happened in the past million years. Current estimates and the most reliable GCMs predict a 1°C–3.5°C (2°–6°F) warming in the 21st century.

It is clear that not all areas will be affected equally. One of the most important effects is expected to be a more vigorous hydrologic cycle, fueled largely by increases in evaporation from the ocean. Intense rainfalls will be more likely in many regions, as will droughts in other regions such as the Great Plains. Temperatures will rise most in the polar regions, mainly during the winter months. As a result of the warming, sea levels will rise, mostly because of the thermal expansion of ocean water and melting ice sheets. By 2100, sea levels should be between 15 and 95 centimeters (0.5–3.1 ft) higher than today. In addition, the ranges of tropical diseases will expand toward higher latitudes, tree lines will rise, and many alpine glaciers will continue to disappear (● Fig. 8.31).

However, some greenhouse effects may be beneficial to humans. Growing seasons in the high latitudes should increase in length. The increase in atmospheric CO_2 will help some crops such as wheat, rice, and soybeans grow larger faster. In the United States, a 1°C increase in average temperature should decrease heating bills

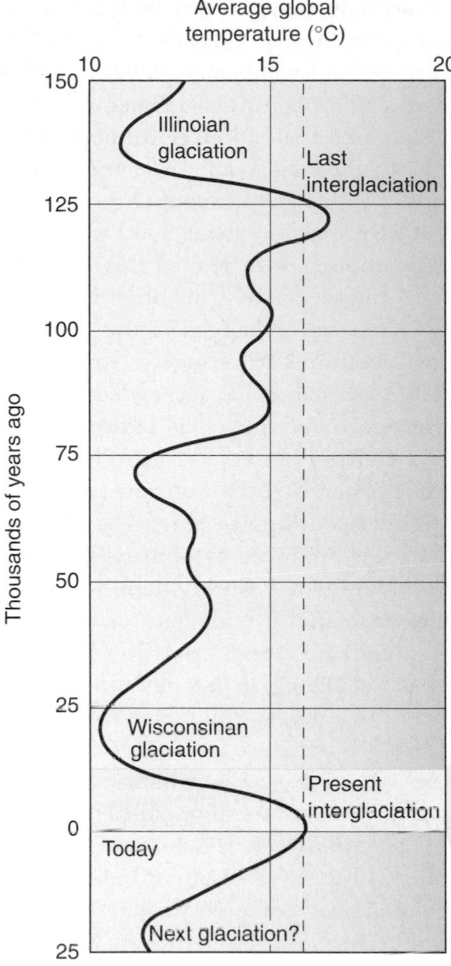

From Skinner & Porter, *Physical Geology.*

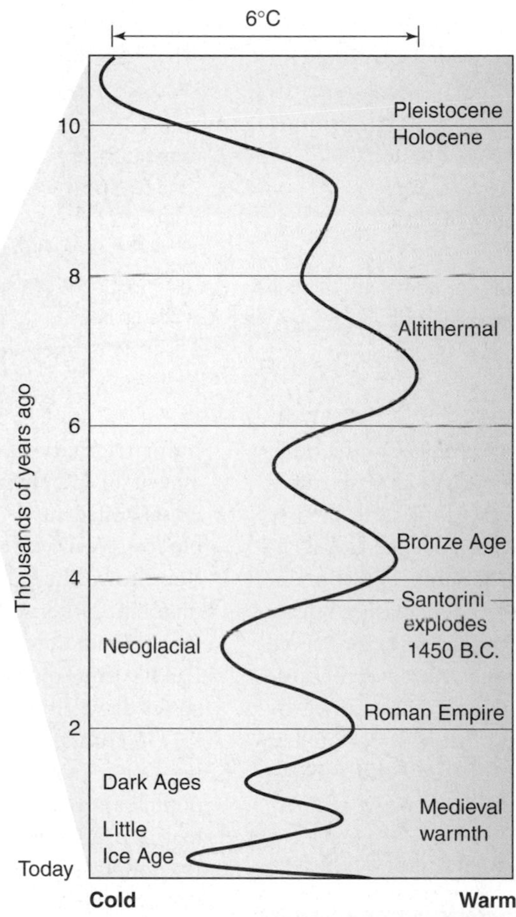

After Imbrie & Imbrie, *Ice Ages: Solving the Mystery,* Enslow Publishers, Short Hills, NJ, p. 179.

• FIGURE 8.30
This figure shows the broad climate trends of the past 150,000 years, with significant details for the Holocene. Climatologists have been remarkably successful in dating recent climate change, but predicting future climates remains difficult.
Why is this so?

by about 11%. The benefits as well as the detriments of a changing climate are still speculative. In general, throwing a climate out of the status quo is liable to cost us more money as we try to adapt to the new changes.

The vast majority of scientists believe that the **global warming** is already occurring. Eleven of the 12 hottest years on record have occurred since 1995, and each subsequent year usually sets a new record. Average annual global temperatures have already risen between 0.3°C and 0.6°C (0.5°F–1.1°F), and sea level has risen between 10 and 25 centimeters (4–10 in.) during the past 100 years. Given the long residence times of many greenhouse gases (see again Fig. 8.28) and the heat capacity of the oceans,

M. Trapasso

• FIGURE 8.31
With few exceptions, mountain glaciers worldwide are retreating. This valley is leading to the terminus of the Franz Josef Glacier in New Zealand. In 1865 this entire valley was filled with glacial ice.
Is this a sign of global warming?

some warming is inevitable and will likely continue. On the other hand, the lesson of this chapter is that short-term climate trends and some longer-term climate trends are very difficult to predict, because of the many uncertain factors that can influence global climates. Major volcanic eruptions, changing oceanic circulation, or human impacts (such as the ongoing effects of increasing release of greenhouse gases, massive deforestation, and urbanization) could significantly disrupt climate trends at any time. Moreover, because sudden, major shifts in climate systems that were not caused by human activity are abundant in the record of the last 2.4 million years, predicting climate will always be a process with varying degrees of uncertainty.

The Issue of Global Warming

Global warming and concerns about climate change seem to be discussed everywhere today, in television programs, movies, and the news media. Scientists, environmentalists, politicians, and celebrities are speaking out about global warming. Although it is well known that global temperatures are rising, there is some resistance to accepting the conclusion that humans are the major cause. Individuals on both sides of the question accuse each other of doing "junk science" to support their positions. No doubt this is, and will continue to be, a difficult issue to reconcile, because of political and economic agendas. But overwhelmingly, scientists who have studied the effects, impacts, and potential influenced of global warming have concluded that human activities have had a significant impact on the warming trends of the last 100 years or more. Further, they are concerned that the human activities that influence global warming are increasing at unprecedented rates, driven by increasing industrialization, global population growth, and the use of natural resources. To understand the issues, let us summarize a few aspects of what we know about global warming.

It is important to realize that climates change—and they have many times in Earth's history. They have changed in the past and they will continue to change in the future. Since the last major ice age ended, 10,000 years or so ago, and the lesser "Little Ice Age" about a century ago, the climate has been warming. There is little controversy about that trend. According to extensive research and data sets produced by climate researchers from around the world, the most recent decades show global temperatures warming at an accelerated rate. A controversy comes about when trying to establish the major causes of this global warming, and to estimate the degrees of influence for each of these factors. Are natural processes or human activities causing this upward trend in global temperatures? The controversy is not about whether human activity has an impact on the atmosphere and all other aspects of the physical environment. Some uncertainty, however, exists concerning to what extent human activities are responsible for global warming and to what extent humans can slow the trend toward rising temperature There are some scientists (and many more nonscientists) who feel that it is incorrect to assume that human activities are responsible for significant changes in Earth's atmospheric temperatures. They believe that natural processes drive all major climate changes.

In contrast, the vast majority of climate scientists contend that humans are a more significant force on our planet. With a growing population of about 6.6 billion in number, these scientists contend that humans have the ability to effect serious atmospheric change. By polluting both the atmosphere and the hydrosphere, creating massive amounts of solid waste, destroying forests, and damming rivers, humans can disrupt the natural environment and seriously alter climate.

In an effort to understand global warming, including its influencing factors, as well as its current and future impacts, the United Nations and the Intergovernmental Panel on Climate Change (IPCC) have cooperated to gather as much relevant information and data as possible. The IPCC is a worldwide group of distinguished atmospheric scientists. In 2007, after years of research and study, a series of comprehensive reports on global warming by the IPCC was released that involved more than 800 climate scientists from 130 countries. These scientists studied multiple lines of evidence worldwide, from tree-ring and ice-core data, to glacial retreat and sea-level rise, to changes in the atmosphere, to changes in weather phenomena, and they strongly considered the many potential influences of both natural processes and human activities. The conclusion of the IPCC was that it is "*very likely*" (>90% probability) that emissions of greenhouse gases from *anthropogenic* (human-induced) activities have caused " . . . most of the observed increase in globally-averaged temperatures since the mid-20th century." They also state that in the last 50 years, the influence of Earth–sun relationships and volcanic activity would *likely* have caused a cooling trend. The results of computer models generated by the IPCC show how observed temperatures have increased in the last 100 years, compared to the predicted impact of natural influences alone, and compared to a combination of human and natural factors. The best fit is the one that includes human influence in global warming (• Fig. 8.32).

The IPCC has summarized their findings thus: "Today, the time for doubt has passed. The IPCC has unequivocally affirmed the warming of our climate system, and linked it directly to human activities." The IPCC goes on to state that dealing with the environmental changes associated with global warming and/or working to minimize human impacts on climate change will be an important concern worldwide in the coming years.

Determining an appropriate course of action based on these findings could be complicated. If humans hope to halt or reduce the rate of global warming, and return to, or maintain, a more optimum climate, consensus must be built regarding some important questions. For example, what is the optimum climate, and who decides what levels of temperature and precipitation constitute the optimum climate? Further, the impact of global warming and in fact, of all major climatic change, will always vary among different geographic locations and climatic regions (see again Fig. 8.32). With the wide variety of environments on Earth, some geographic regions would benefit from a warmer climate and other areas will bear significant negative impacts (for example, the world's heavily populated coastal regions as sea level rises). We are not able to adjust our atmosphere as easily as we can set a thermostat in our homes. The question then becomes, "How should we approach this issue?"

Recommendations for the Future

Whether or not you believe humans are a major cause of global warming and its ramifications, the following recommendations should be followed if humans are to be successful stewards of planet Earth. If humans are a *major* driving force, and we do little or nothing about reducing the human activities that contribute

• FIGURE 8.32

Using the best computer models available to evaluate global climate change, the International Panel on Climate Change has found that only the models that include increasing human releases of greenhouse gases (shown in pink) fit the temperature trends that have been observed in the last century (shown by the black lines). The blue tones estimate the ranges of what the temperatures would be without human impacts on global warming.

On what continent has the observed temperature fluctuated the most during this time period, and which one the least?

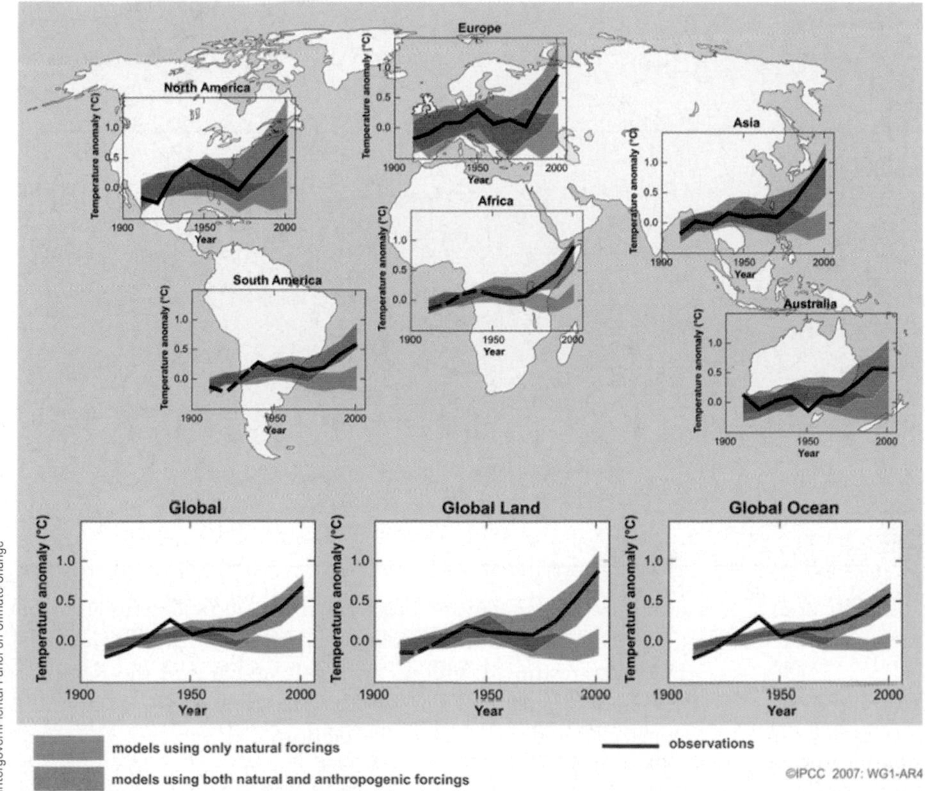

Intergovernmental Panel on Climate Change

to global warming, the environmental consequences will be quite serious. However, if we work to reduce our impacts on climate and the environment, both humankind and Earth, our life-support system, will be better off, no matter what climatic scenario we face in the future.

On a Global Scale To the extent possible, the nations of the world should devote serious research and monetary resources to: (1) Developing alternative sources of energy. Whatever the effects of burning fossil fuels on global temperatures, it also pollutes the air humans breathe, making it dangerous to human health. Energy from solar radiation, wind, geothermal heat, ocean tides, biofuels, hydroelectric generation, even nuclear reactors, helps keep the atmosphere cleaner for generations to come. (2) Curtailing, or better managing, our energy usage. With Earth's growing human population, the developed nations' high rates of consumption, and the increasing industrialization in developing nations, energy demands are increasing and will continue to grow in the foreseeable future. How much energy we use and how we can conserve energy must be major considerations. (3) Recycling our waste materials. Currently, human populations are consuming our nonrenewable resources at rates that cannot be sustained. Recycled resources will ease the strain on those that are vanishing at such a rapid rate, and will save the energy needed to create new resources. (4) Curtailing deforestation. This very destructive process should be restricted everywhere on Earth. Forest vegetation is a primary agent in the removal of CO_2 from our atmosphere through the process of photosynthesis. (5) Desalinizing ocean water (removing salt from seawater). Making this process easier and less costly should be a major research effort all over the world. Arid climates create deserts, but irrigation can turn desert regions into productive lands. It is ironic that millions of people are starving in drought-stricken regions on our planet, which is mostly covered by water.

On a Personal Scale There are simple things we can do every day that can help reduce our negative impact on Earth's fragile environment. The following are just a few suggestions: (1) Use car pools and mass transportation; drive smaller cars; drive less often and at reduced speeds. (2) Use more energy-efficient lighting and appliances and turn them off when not in use. (3) Set thermostats to use less energy to cool your home in summer and warm it in the winter. (4) Recycle materials (metals, glass, plastic, paper, and others) as often as possible. (5) Consciously protect your own physical environment and remind others around you to follow your example.

One of the few things all humans have in common, regardless of age, sex, race, religion, or nationality, is that we all occupy Earth together. It is our responsibility to care for the planet that sustains us. We must work toward the proper care of our world for our own descendants and for the generations throughout the world who follow after us.

Chapter 8 Activities

Define & Recall

empirical classification
genetic classification
Thornthwaite system
potential evapotranspiration (potential ET)
actual evapotranspiration (actual ET)
Köppen system
climograph
tropical climate
polar climate
microthermal climate
mesothermal climate
arid climate
highland climate
region
zone of transition
microclimate
oxygen-isotope analysis
dendrochronology
palynology
Altithermal
eccentricity cycle
obliquity cycle
precession cycle
greenhouse gases
Near Earth Objects (NEOs)
Little Ice Age
global warming

Discuss & Review

1. Why is it important to study the nature and possible causes of past climates when attempting to predict future climate change?
2. Why are temperature and precipitation the two atmospheric elements most widely used as the sources of statistics for climate classification? How are these two elements used in the Köppen system to identify six major climate categories?
3. What are the advantages and disadvantages of the Köppen system for geographers? Why are the Köppen climate boundaries often referred to as "vegetation lines"?

4. What is a climograph? What seasonal and annual patterns of climate does it reflect?
5. How does the Thornthwaite system of climate classification differ from the Köppen system? What are the advantages of the Thornthwaite system?
6. Why is the occurrence, frequency, and dating of glacial advances and retreats so important to the study of past climates? How has modern research changed earlier theories of glacial coverage and associated climate change during the Pleistocene?
7. How have scientists been able to document the rapid shifts of climates that have occurred during the latter part of the Pleistocene?
8. What are the major possible causes of global climate change? What contribution did the mathematician Milankovitch make to theories regarding glaciation?
9. What is the evidence that volcanic activity can affect global temperatures? How does this occur?
10. What effects can changes in the amounts of CO_2 and other greenhouse gases in the atmosphere have on global temperatures? How can past changes in amounts of CO_2 be determined?
11. How might changes in Earth's oceans and landmasses affect global climates?
12. What is the primary difficulty for any climatologist who attempts to predict future climates?
13. What changes are likely to occur in Earth's major subsystems if global warming continues for the near term, as most scientists believe?

Consider & Respond

1. Study the definitions for the individual Köppen climate types described in the "Graph Interpretation" exercise. Why do you think Köppen and later climatologists who modified the system selected the particular temperature and precipitation parameters that separate the individual types from one another?
2. Examine the climograph in the "Using Climographs" box on page 203. During what month does Nashville experience the greatest precipitation? What major change would immediately identify this graph as representing a Southern Hemisphere location? What do the four horizontal dashed lines represent?
3. Review Figures 8.19 and 8.30. State in your own words the general conclusions you would draw from a study of these two figures.
4. After studying Chapters 7 and 8 in your textbook, which do you believe is more important to you now and in the future—the subject of weather or of climate? Defend your answer.

Apply & Learn

1. Using the climograph for Nashville, Tennessee, found in the "Using Climographs" box on page 203, find an average temperature reading for every month of the year on the line graph. Then calculate the mean temperature for the year and the annual temperature range. How close are the data you calculated compared to those given at the top of the climograph?
2. Using the precipitation data presented in the bar graph on the Nashville climograph, derive a precipitation value for every month of the year, and then calculate an annual average precipitation value and an annual precipitation range. How do your calculations compare with those printed on the climograph? What can you say about the distribution of Nashville's precipitation through the year?

Locate & Explore

Note: Please read the About Locate & Explore Activities section of the Preface before beginning these exercises.

1. Using Google Earth, fly to Lake Chad (13.42°N, 14.01°E). Once you arrive at your coordinates, zoom out to view the extent of the lake. Lake Chad was once the largest lake in Africa, but ongoing drought has significantly reduced the lake in area. Since the lake is shallow, small changes in the discharge of the Chari River lead to large changes in lake area. Assuming that the lake can be characterized as rectangles (area = length × width), what has been the change in area (in square miles and as a percentage) from the original lake boundary to the lake today?

   Tip: Use the ruler tool to measure the width and length.

Graph Interpretation

THE KÖPPEN CLIMATE CLASSIFICATION SYSTEM

The key to understanding any system of classification is found by personally practicing use of the system. This is one reason why the Consider & Respond review sections of both Chapters 9 and 10 are based on the classification of data from sites selected throughout the world. Correctly classifying these sites in the modified Köppen system may seem complicated at first, but you will find that, after applying the system to a few of the sample locations, the determination of a correct letter symbol and associated climate name for any other site data should be routine.

Before you begin, take the time to familiarize yourself with Table 1. You will note that there are precise definitions in regard to temperature or precipitation that identify a site as one of the five major climate categories in the Köppen system (*A*, tropical; *B*, arid; *C*, mesothermal; *D*, microthermal; *E*, polar). Furthermore, you will note that the additional letters required to identify the actual climate type also have precise definitions or are determined by the use of the graphs. In other words, Table 1 is all you need to classify a site if monthly and annual means of precipitation and temperature are available. Table 1 should be used in a systematic fashion to determine first the major climate category and, once that is determined, the second and third letter symbols (if needed) that complete the classification. As you begin to classify, it is strongly recommended that you use the following procedure. (After a few examples you may find that you can omit some steps with a glance at the statistics.)

TABLE 1
Simplified Köppen Classification of Climates

First Letter	Second Letter	Third Letter
E Warmest month less than 10°C (50°F) POLAR CLIMATES *ET*–Tundra *EF*–Ice Sheet	*T* Warmest month between 10°C (50°F) and 0°C (32°F) *F* *Warmest* month below 0°C (32°F)	NO THIRD LETTER (with polar climates) SUMMERLESS
B Arid or semiarid climates ARID CLIMATES BS–Steppe BW–Desert	*S* Semiarid climate (see Graph 1) *W* Arid climate (see Graph 1)	*h* Mean annual temperature greater than 18°C (64.4°F) *k* Mean annual temperature

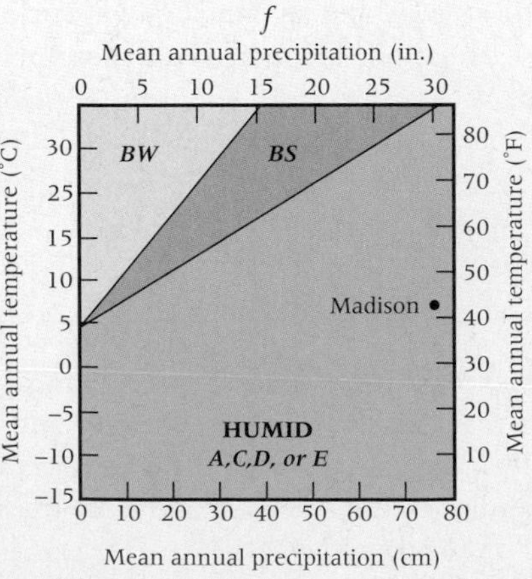

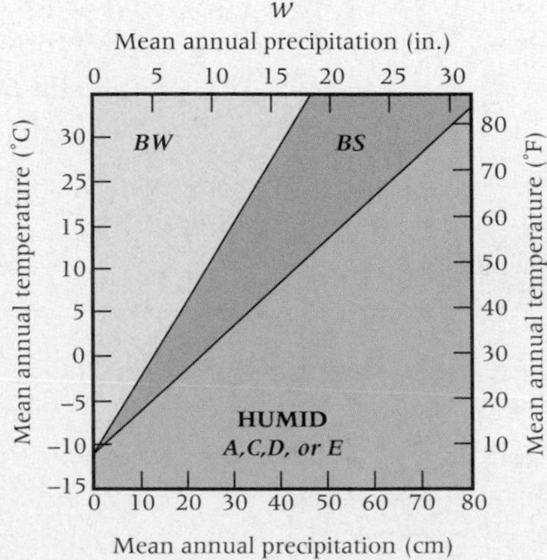

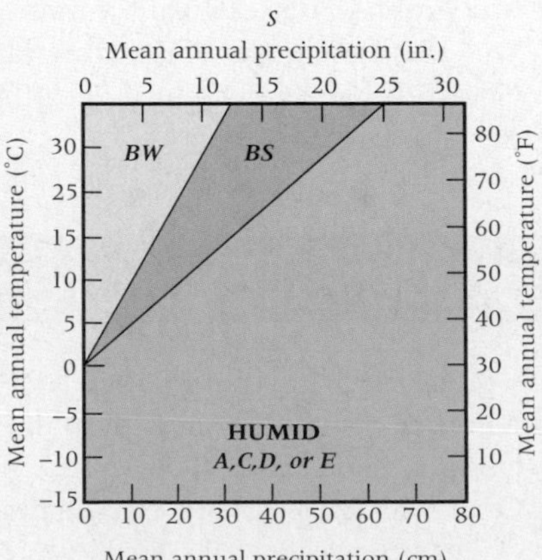

Graph 1 Humid/Dry Climate Boundaries

TABLE 1
Simplified Köppen Classification of Climates *(Continued)*

First Letter	Second Letter	Third Letter
A Coolest month greater than 18°C (64.4°F) TROPICAL CLIMATES *Am*—Tropical monsoon *Aw*—Tropical savanna *Af*—Tropical rainforest	*f* Driest month has at least 6 cm (2.4 in.) of precipitation *m* Seasonally, excessively moist (see Graph 2) *w* Dry winter, wet summer (see Graph 2)	NO THIRD LETTER (with tropical climates) WINTERLESS
C Coldest month between 18°C (64.4°F) and 0°C (32°F); at least one month over 10°C (50°F) MESOTHERMAL CLIMATES *Csa, Csb*—Mediterranean *Cfa, Cwa*—Humid subtropical *Cfb, Cfc*—Marine west coast	*s* (DRY SUMMER) Driest month in the summer half of the year, with less than 3 cm (1.2 in.) of precipitation and less than one third of the wettest winter month	*a* Warmest month above 22°C (71.6°F) *b* Warmest month below 22°C (71.6°F), with at least four months above 10°C (50°F)
D Coldest month less than 0°C (32°F); at least one month over 10°C (50°F) MICROTHERMAL CLIMATES *Dfa, Dwa—Humid continental, hot summer* *Dfb, Dwb—Humid continental, mild summer* *Dfc, Dwc, Dfd, Dwd—Subarctic*	*w* (DRY WINTER) Driest month in the winter half of the year, with less than one tenth the precipitation of the wettest summer month *f* (ALWAYS MOIST) Does not meet conditions for *s* or *w* above	*c* Warmest month below 22°C (71.6°F), with one to three months above 10°C (50°F) *d* Same as *c*, but coldest month is below −38°C (−36.4°F)

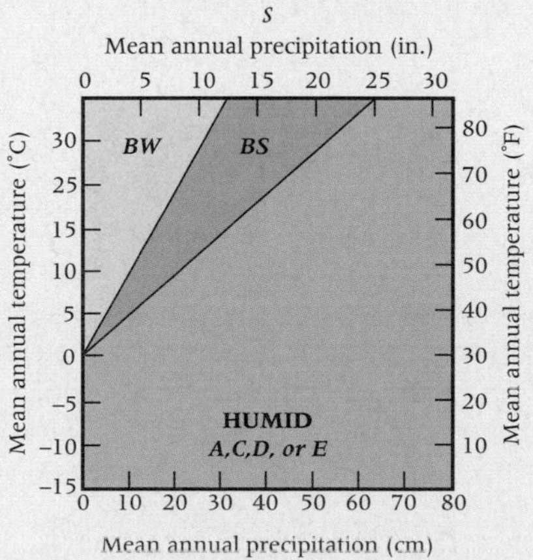

Graph 2 Rainfall of driest month (cm)

Step 1. Ask: Is this a polar climate (*E*)? Is the warmest month less than 10°C (50°F)? If so, is the warmest month between 10°C (50°F) and 0°C (32°F) (*ET*) or below 0°C (32°F) (*EF*)? If not, move on to:

Step 2. Ask: Is there a seasonal concentration of precipitation? Examine the monthly precipitation data for the driest and wettest summer and winter months for the site. Take careful note of temperature data as well because you must determine whether the site is located in the Northern or Southern Hemisphere. (April to September are summer months in the Northern Hemisphere but winter months in the Southern Hemisphere. Similarly, the Northern Hemisphere winter months of October to March are summer south of the equator.) As the table indicates, a site has a dry summer (*s*) if the driest month in summer has less than 3 centimeters (1.2 in.) of precipitation and less than one third of the precipitation of the wettest winter month. It has a dry winter (*w*) if the driest month in winter has less than one tenth the precipitation of the wettest summer month. If the site has neither a dry summer nor a dry winter, it is classified as having an even distribution of precipitation (*f*). Move on to:

Step 3. Ask: Is this an arid climate (*B*)? Use one of the small graphs (included in Graph 1) to decide. Based on your answer in Step 2, select one of the small graphs and compare mean annual temperature with mean annual precipitation. The graph will indicate whether the site is an arid (*B*) climate or not. If it is, the graph will indicate which one (*BW* or *BS*). You should further classify the site by adding *h* if the mean annual temperature is above 18°C (64.4°F) and *k* if it is below. If the site is neither *BW* nor *BS*, it is a humid climate (*A*, *C*, or *D*). Move on to:

Step 4a. Ask: Is this a tropical climate (*A*)? The site has a tropical climate if the temperature of the coolest month is higher than 18°C (64.4°F). If so, use Graph 2 in Table 1 to determine which tropical climate the site represents. (Note that there are no additional lowercase letters required.) If not, move on to:

Step 4b. Ask: Which major middle-latitude climate group does that site represent, mesothermal (*C*) or microthermal (*D*)? If the temperature of the coldest month is between 18°C (64.4°F) and 0°C (32°F), the site has a mesothermal climate. If is below 0°C (32°F), it has a microthermal climate. Once you have answered the question, move on to:

Step 5. Ask: What was the distribution of precipitation? This was determined back in Step 2. Add *s*, *w*, or *f* for a *C* climate or *w* or *f* for a *D* climate to the letter symbol for the climate. Then, move on to:

Step 6. Ask: What is needed to express the details of seasonal temperature for the site? Refer again to Table 1 and the definitions for the letter symbols. Add *a*, *b*, or *c* for the mesothermal (*C*) climates or *a*, *b*, *c*, or *d* for the microthermal (*D*) climates, and you have completed the classification of your climate. However, note that you may not have come this far because you might have completed your classification at Steps 1, 3, or 4a.

We should now be ready to try out the use of Table 1, following the steps we have recommended. Data for Madison, Wisconsin, is presented below for our example.

	J	F	M	A	M	J	J	A	S	O	N	D	Year
T (°C)	−8	−7	−1	7	13	19	21	21	16	10	2	−6	7
P (cm)	3.3	2.5	4.8	6.9	8.6	11.0	9.6	7.9	8.6	5.6	4.8	3.8	77.0

The correct answer is derived below:

Step 1. We must determine whether or not our site has an *E* climate. Because Madison has several months averaging above 10°C, it does not have an *E* climate.

Step 2. We must determine if there is a seasonal concentration of precipitation. Because Madison is driest in winter, we compare the 2.5 centimeters of February precipitation with the precipitation of June (1/10 of 11.0 cm, or 1.1 cm) and conclude that Madison has neither a dry summer nor a dry winter but instead has an even distribution of precipitation (*f*). [*Note:* The 2.5 cm of February precipitation is not less than 1/10 (1.1 cm) of June precipitation.]

Step 3. Next we assess, through the use of Graphs 1(*f*), 1(*w*), or 1(*s*), whether our site is an arid climate (*BW*, *BS*) or a humid climate (*A*, *C*, or *D*). Because we have previously determined that Madison has an even distribution of precipitation, we will use Graph 1(*f*). Based on Madison's mean annual precipitation (77.0 cm) and mean annual temperature (7°C), we conclude that Madison is a humid climate (*A*, *C*, or *D*).

Step 4. Now we must assess which humid climate type Madison falls under. Because the coldest month (−8°C) is below 18°C, Madison does *not* have an *A* climate. Although the warmest month (21°C) is above 10°C, the coldest month

(−8°C) is not between 0°C and 18°C, so Madison does *not* have a *C* climate. Because the warmest month (21°C) is above 10°C and the coldest month is below 0°C, Madison *does* have a *D* climate.

Step 5. Because Madison has a *D* climate, the second letter will be *w* or *f*. Because precipitation in the driest month of winter (2.5 cm) is not less that one tenth of the amount of the wettest summer month (1/10 × 11.0 cm = 1.1 cm), Madison does *not* have a *Dw* climate. Madison therefore has a *Df* climate.

Step 6. Because Madison is a *Df* climate, the third letter will be *a, b, c*, or *d*. Because the average temperature of the warmest month (21°C) is not above 22°C, Madison does *not* have a *Dfa* climate. Because the average temperature of the warmest month is below 22°C, with at least 4 months above 10°C, Madison is a *Dfb* climate.

Low-Latitude and Arid Climate Regions

9

CHAPTER PREVIEW

Although the humid tropical climates are all characterized by high temperatures throughout the year, they exhibit significant differences based on either the amounts or the distribution of the precipitation they receive.

- What temperature parameter do these climates have in common?
- How do they differ from one another on the basis of precipitation?

Except for a few unusual circumstances, the tropical rainforest climate regions are among the least populated areas of the world, despite being coincident with the belt of heaviest rainfall, insolation, and vegetative growth.

- Why are there so few people in these regions?
- In what ways are these regions valuable to humankind?

Although rainfall is seasonal in both tropical savanna and tropical monsoon climate regions, the differences in total rainfall between the two climates cause major dissimilarities in their environments, resource characteristics, and human use.

- What are the chief dissimilarities?
- How is rainfall responsible?

Earth's arid regions are created by various processes that are found in a wide range of latitudes across the globe.

- Where are the most extensive arid regions found?
- What processes create these different regions?

Knowledge of the location of the world's deserts is similar to an understanding of the distribution of the world's steppe regions.

- What is the association between deserts and steppes?
- How do the regions differ?

◄ Opposite: In tropical climate regions that experience a lengthy dry season, like this site in East Africa, waterholes are important resources for sustaining the wildlife population.

© Jeremy Woodhouse/ Getty Images

The odds are overwhelming that within your lifetime, if you have not already done so, you will travel. You may travel extensively either within or beyond the borders of North America to destinations far from where you are living today. You may be moving to a new home or place of employment. You may be traveling on business or simply for pleasure. Whatever the reason, it is likely that you will ask the question almost every other traveler asks: "I wonder what the weather will be like?"

Realistically, as you have learned from reading previous chapters, the question should probably be, "I wonder what the climate is like?" The constant variability associated with weather in many areas of the world makes it difficult to predict. However, the long-term averages and ranges upon which climate is based allow geographers to provide the traveler with a general idea of the atmospheric conditions likely to be experienced at specific locations throughout the world during different times of the year.

Of course, there are many reasons other than travel why knowledge of climate and its variation over Earth's surface is a valuable asset. An understanding of climate in other areas of the world helps us understand the adaptations to atmospheric conditions that have been made by the people who live there. We can better appreciate some of their economic activities and certain aspects of their cultures.

In addition, the climate of any place on Earth has a dominant effect on native vegetation and animal life. It influences the rate and manner by which rock material is destroyed and soil is formed. It is a contributing factor in the way landforms are reduced and physical landscapes are sculptured. In short, knowledge of climates provides endless clues to not only atmospheric conditions but also numerous other aspects of the physical environment.

In this chapter and the next, you will be provided with a broad descriptive survey of world climates: their locations, distributions, general characteristics, formation processes, associated features, and related human activities. The information contained in these two chapters can serve as a valuable knowledge base as you prepare for the future. Throughout these chapters, we use the modified version of the Köppen climate classification that was introduced in Chapter 8. It is interesting to note that each of the climates discussed can be found within North America, so even if you never travel beyond this continent's borders, you will still find the discussions of climates valuable preparation as you move about your own country.

Humid Tropical Climate Regions

We have already learned a good deal about the climate regions of the humid tropics through our preliminary discussions of these climate types in Chapter 8. In addition, we can review the location of these climates in relation to other climate regions through regular examination of the world map of climates (see again Fig. 8.6). It now remains for us to identify the major characteristics of each humid tropical climate type in turn, along with its associated world regions.

Study of • Figure 9.1 and a careful reading of Table 9.1 will provide the locations of the humid tropical climates and a preview of the significant facts associated with them. The table also reminds us that, although each of the three humid tropical climates has high average temperatures throughout the year, they differ greatly in the amount and distribution of precipitation.

Tropical Rainforest Climate

The **tropical rainforest climate** probably comes most readily to mind when someone says the word *tropical*. Hot and wet throughout the year, the tropical rainforest climate has been the stage for many stories of both fact and fiction. One cannot easily forget the life-and-death struggle with the elements portrayed by Humphrey Bogart and Katharine Hepburn in the classic film *The African Queen*. More recent films about the Vietnam War also depict the difficulties of moving and fighting in such formidable environments. Upon visiting this type of climate, one would easily feel the high temperatures, oppressive humidity, and the frequent heavy rains, which sustain the massive vegetative growth for which it is known (• Fig. 9.2).

• FIGURE 9.1
Index map of humid tropical climates.

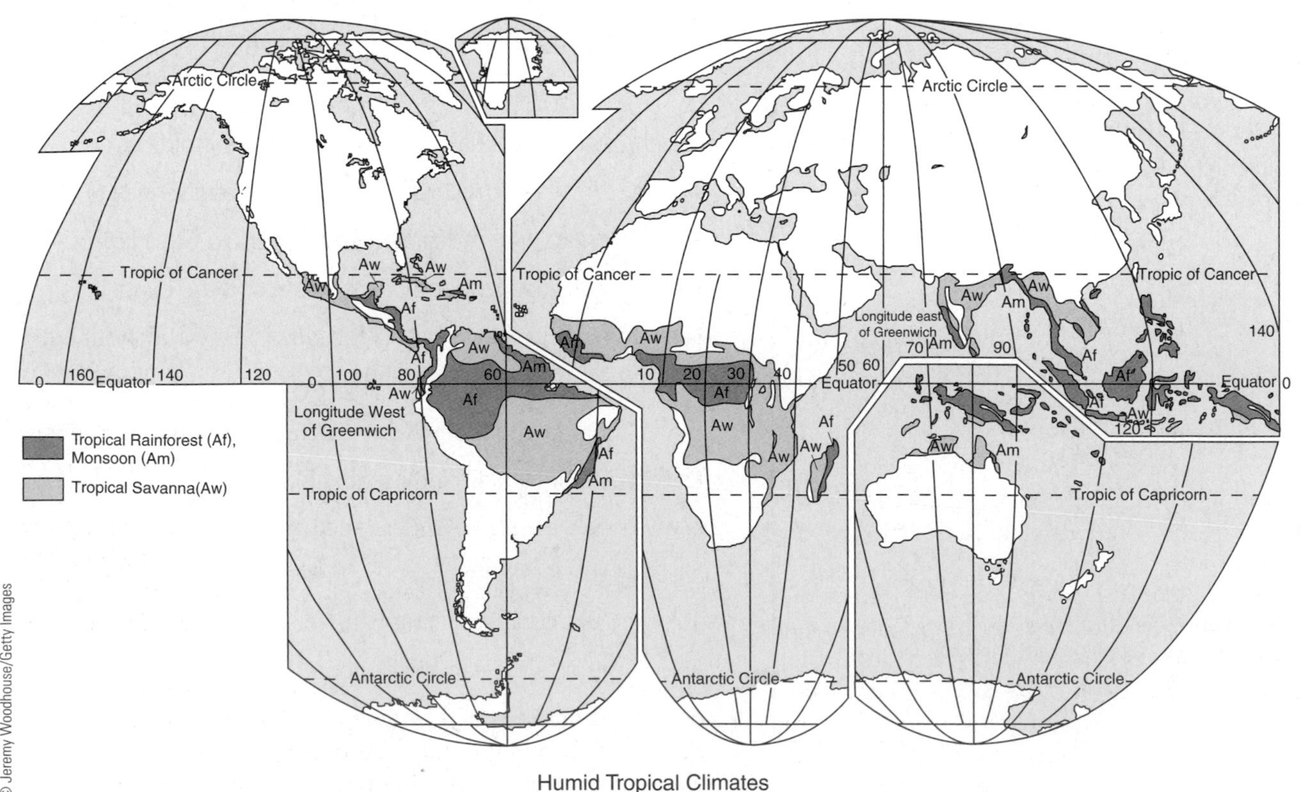

© Jeremy Woodhouse/Getty Images

TABLE 9.1
The Humid Tropical Climates

Name and Description	Controlling Factors	Geographic Distribution	Distinguishing Characteristics	Related Features
Tropical Rainforest Coolest month above 18°C (64.4°F); driest month with at least 6 cm (2.4 in.) of precipitation	High year-round insolation and precipitation of doldrums (ITCZ); rising air along trade wind coasts	Amazon R. Basin, Congo R. Basin, east coast of Central America, east coast of Brazil, east coast of Madagascar, Malaysia, Indonesia, Philippines	Constant high temperatures; equal length of day and night; lowest (2°C–3°C/3°F–5°F) annual temperature ranges; evenly distributed heavy precipitation; high amount of cloud cover and humidity	Tropical rainforest vegetation (selva); jungle where light penetrates; tropical iron-rich soils; climbing and flying animals, reptiles, and insects; slash-and-burn agriculture
Tropical Monsoon Coolest month above 18°C (64.4°F); one or more months with less than 6 cm (2.4 in.) of precipitation; excessively wet during rainy season	Summer onshore and winter offshore air movement related to shifting ITCZ and changing pressure conditions over large landmasses; also transitional between rainforest and savanna	Coastal areas of southwest India, Sri Lanka, Bangladesh, Myanmar, southwest Africa, Guyana, Surinam, French Guiana, northeast and southeast Brazil	Heavy high-sun rainfall (especially with orographic lifting), short low-sun drought; 2°C–6°C (3°F–10°F) annual temperature range, highest temperature just prior to rainy season	Forest vegetation with fewer species than tropical rainforest; grading to jungle and thorn forest in drier margins; iron-rich soils; rainforest animals with larger leaf-eaters and carnivores near savannas; paddy rice agriculture
Tropical Savanna Coolest month above 18°C (64.4°F); wet during high-sun season, dry during lower-sun season	Alternation between high-sun doldrums (ITCZ) and low-sun subtropical highs and trades caused by shifting winds and pressure belts	Northern and eastern India, interior Myanmar and Indo-Chinese Peninsula; northern Australia; borderlands of Congo R., south central Africa; llanos of Venezuela, campos of Brazil; western Central America, south Florida, and Caribbean Islands	Distinct high-sun wet and low-sun dry seasons; rainfall averaging 75–150 cm (30–60 in.); highest temperature ranges for humid tropical climates	Grasslands with scattered, drought-resistant trees, scrub, and thorn bushes; poor soils for farming, grazing more common; large herbivores, carnivores, and scavengers

Constant Heat and Humidity Most weather stations in the tropical rainforest climate regions record average monthly temperatures of 25°C (77°F) or more (● Fig. 9.3). Because these regions are usually located within 5° or 10° of the equator, the sun's noon rays are always close to being directly overhead. Days and nights are of almost equal length, and the amount of insolation received remains nearly constant throughout the year. Consequently, no appreciable temperature variations can be linked to the sun angle and therefore be considered seasonal. In other words, the concept of summer and winter as being hot and cold seasons, respectively, does not exist here.

The **annual temperature range**—the difference between the average temperatures of the warmest and coolest months of the year—reflects the consistently high angle of the sun's rays. As indicated in Figure 9.3, the annual range is seldom more than 2°C or 3°C (4°F or 5°F). In fact, at Ocean Island in the central Pacific, the annual range is 0°C because of the additional moderating influence of the ocean on the nearly uniform pattern of insolation.

One of the most interesting features of the tropical rainforest climate is that the **daily (diurnal) temperature ranges**—the differences between the highest and lowest temperatures during the day—are usually far greater than the annual range. Highs of 30°C–35°C (86°F–95°F) and lows of 20°C–24°C (68°F–75°F) produce daily ranges of 10°C–15°C (18°F–27°F). However, the drop in temperature at night is small comfort. The high humidity causes even the cooler evenings to seem oppressive. (Recall, water vapor is a greenhouse gas and helps retain the heat energy.)

The climographs of Figure 9.3 illustrate that significant variations in precipitation can occur even within rainforest regions. Although most rainforest locations receive more than

Don Deering, NASA/LBA Project

(a)

Jim Ross/NASA/DRFC

(b)

• FIGURE 9.2

(a) The typical vegetation in a tropical rainforest climate forms a cover of trees, growing at different heights to make a multilayered treetop canopy. This is the rainforest canopy in the Amazon region of Brazil. (b) Tall and massive hardwood trees with distinctive buttressed trunks thrive in a climate that is hot and wet all year, but shady on the forest floor.
How many tree layers can you see in (a)?

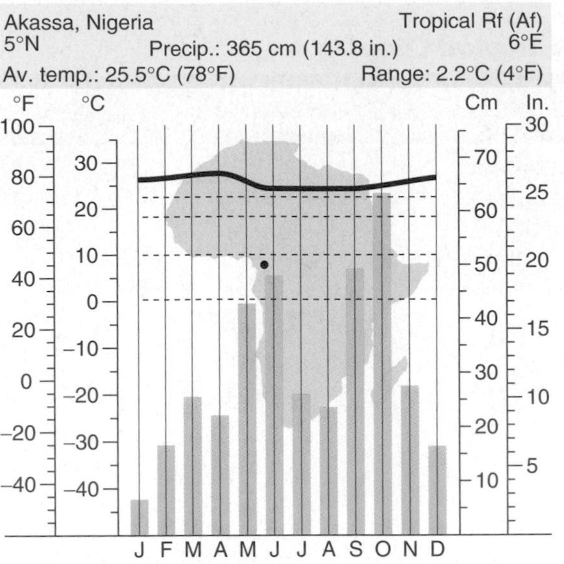

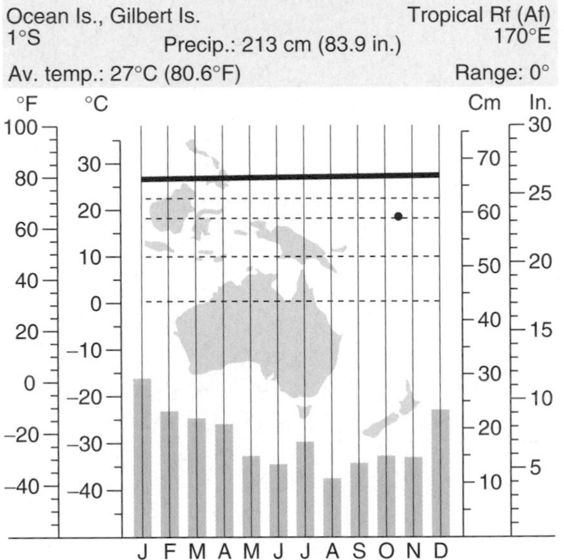

• FIGURE 9.3

Climographs for tropical rainforest climate stations.
Why is it difficult, without looking at the climograph keys, to determine whether each station is located in the Northern or Southern Hemisphere?

200 centimeters (80 in.) a year of precipitation and the average is in the neighborhood of 250 centimeters (100 in.), some locations record an annual precipitation of more than 500 centimeters (200 in.). Ocean locations, near the greatest source of moisture, tend to receive the most rain. Mount Waialeale, in the Hawaiian Islands, receives a yearly average of 1168 centimeters (460 in.), making it the wettest spot on Earth. As a group, climate stations in the humid tropics experience much higher annual totals than typical humid middle-latitude stations. Compare, for example, the 365 centimeters in Akassa, Nigeria, with the average 112 centimeters received annually in Portland, Oregon, or the 61 centimeters received in London, England.

We should recall that the heavy precipitation of the tropical rainforest climate is associated with the warm, humid air of the doldrums and the unstable conditions along the ITCZ (intertropical convergence zone). Both convection and convergence serve as uplift mechanisms, causing the moist air to rise and condense and resulting in the heavy rains that are characteristic of this climate. These processes are enhanced on the east coasts of continents where warm ocean currents allow humid tropical climates to extend farther poleward. On west coasts where cold ocean currents flow, these mechanisms of uplift are somewhat inhibited. There is heavy cloud cover during the warmer, daylight hours when convection is at its peak, although the nights and early mornings can be quite clear. Variations in rainfall can usually be traced to the ITCZ and its low pressure cells of varying strength. Many tropical rainforest locations (Akassa, for example) exhibit two maximum precipitation periods during the year, one during each appearance of the ITCZ as it follows the migration of the sun's direct rays (recall, the sun crosses the equator on the equinox days in March 21 and September 22). In addition, although no season can be called dry, during some months it may rain on only 15 or 20 days.

Cloud Forests Highland areas near a seacoast both in the tropical regions like Costa Rica (• Fig. 9.4) and in the middle latitudes (coastal Washington State) may contain **cloud forests.** Here, moisture-laden maritime air is lifted up the windward slopes of mountains. This orographic precipitation may not fall as heavy rain showers or thunderstorms but rather as an almost constant misty fog. Through time, this cloudy environment can precipitate enough moisture to qualify as a rainforest, but without the oppressive heat found in the rainforests of the low-altitude tropics. One advantage of the cloud forests is a scarcity of flying insects that cannot survive in the colder temperatures.

A Delicate Balance The most common vegetation of tropical rainforest climate regions is multistoried, broadleaf evergreen forest made up of many species whose tops form a thick, almost continuous canopy cover that blocks out much of the sun's light. This type of rainforest is sometimes called a **selva.** Within the selva, there is usually little undergrowth on the forest floor because sunlight cannot penetrate enough to support much low-growing vegetation (• Fig. 9.5). When a tree dies in a selva, and the new opening in the canopy allows for sunlight to enter, another tree will immediately fill that void and use the insolation. In a rainforest, sunlight is more important to vegetative growth than rainfall.

The relationship between the soils beneath the selva and the vegetation that the soils support is so close that there exists a nearly perfect ecological balance between the two, threatened only by people's efforts to earn a living from the soil. The trees of the selva supply the tropical soils with the nutrients that the trees themselves need for growth. As leaves, flowers, and branches fall to the ground or as roots die, the numerous soil-dwelling animals and bacteria act on them, transforming the forest litter into organic matter with vital nutrients. However, if the trees are removed, there is no replenishment of these nutrients and no natural barrier (forest litter and root systems) to prevent large amounts of rain from percolating through the soil. This percolating water can dissolve and remove nutrients and minerals from the topsoil. The intense activities of microorganisms, worms, termites, ants, and other insects cause rapid deterioration of the remaining organic debris, and soon all that remains is an infertile mixture of insoluble manganese, aluminum, and iron compounds. In essence, without the vegetation to protect and feed the soils, rainforest soils are quite barren.

In recent years, there has been large-scale harvesting of the tropical rainforests by the lumber industry, and land has been cleared for agriculture and livestock production, especially in the Amazon River basin. Such deforestation can have a significant and, unfortunately, permanent impact on the delicate balance that exists among Earth's systems.

Environmental conditions vary from place to place within climate regions; therefore, the typical rainforest situation that we have just described does not apply everywhere in the tropical rainforest climate. Some regions are covered by true jungle, a term often misused when describing the rainforest. **Jungle** is a dense tangle of vines and smaller trees that develops where

• FIGURE 9.4
A cloud forest in the highlands of Costa Rica.
Can a constant, misty rain drop as much water as less frequent rainstorms?

M. Trapasso

• FIGURE 9.5
On the forest floor, a tropical rainforest offers considerable open space.
Why is this so?

M. Trapasso

GEOGRAPHY'S ENVIRONMENTAL SCIENCE PERSPECTIVE

The Amazon Rainforest

Currently, an area of Earth's virgin tropical forest somewhat larger than a football field is being destroyed every second. During a recent decade, the rate of deforestation doubled, and in the Amazon Basin, where more than half of the rainforest resources are located, the rate nearly tripled. Simple mathematics indicates that even at the present rate of deforestation, the Amazon rainforest will virtually disappear in 150 years.

From the point of view of a developing country, there are basic economic reasons for clearing the Amazon rainforest, and Peru, Ecuador, Colombia, and Brazil (where most of the forest is located) are developing countries. Tropical timber sales provide short-term income to finance national growth and repay staggering debts owed to foreign banks. In addition, Brazil, in particular, has viewed the Amazon rainforest as a frontier land available for agricultural development and for resettlement of the poor from overcrowded urban areas.

No one questions that the decisions concerning the future of the Amazon rainforest rest with the governments of nations that control these resources. So why has the rainforest become a serious international issue? The answer is found in the concerns of physical geographers and other environmental scientists throughout the world.

When the tropical rainforests are removed, the hydrologic cycle and energy budgets in the previously forested areas are dramatically altered, often irreversibly. In tropical rainforests, the canopy shades the forest floor, thus helping to keep it cooler. In addition, the huge mass of vegetation provides a tremendous amount of water vapor to the atmosphere through transpiration. The water vapor condenses to form clouds, which in turn provide rainfall to nourish the forest. With the forests removed, transpiration is diminished, which leads to less cloud cover and less rainfall. With fewer clouds and no forest canopy, more solar energy reaches Earth's surface.

The unfortunate outcome is that areas that are deforested soon become hotter and drier, and any ecosystem in place is seriously damaged or destroyed. Once the rainforests are removed, the soils lose their source of plant nutrients, and this precludes the growth of any significant crops or plants. In addition, the rainforest, which was in harmony with the soil, cannot

©Gregory Dimijian/Photo Researchers, Inc.

A section of Amazon rainforest cleared by slash-and-burn techniques for potential farming or grazing.

©Dan Guravich/Photo Researchers, Inc.

Cattle grazing along the Rio Salimoes in Brazil in an area of former rainforest.

direct sunlight does reach the ground, as in clearings and along streams (● Fig. 9.6). Other regions have soils that remain fertile or have bedrock that is chemically basic and provides the soils above with a constant supply of soluble nutrients through the natural weathering processes. Examples of the former region are found along major river floodplains; examples of the latter are the volcanic regions of Indonesia and the limestone areas of Malaysia and Vietnam. Only in such regions of continuous soil fertility can agriculture be intensive and continuous enough to support population centers in the tropical rainforest climate.

Human Activities Throughout much of the tropical rainforest climate, humans are far outnumbered by other forms of animal life. Though there are few large animals of any kind, a great variety of smaller tree-dwelling and aquatic species live in the rainforest. Small predatory cats, birds, monkeys, bats, alligators, crocodiles, snakes, and amphibians such as frogs of many varieties abound. Animals that can fly or climb into the food-rich leaf canopy have become the dominant animals in this world of trees.

Most common of all, though, are the insects. Mosquitoes, ants, termites, flies, beetles, grasshoppers, butterflies, and bees live

reestablish itself. Thus, the multitude of flora and fauna species indigenous to the rainforest is lost forever. It is impossible to calculate the true cost of this reduction in biodiversity (the total number of different plant and animal species in the Earth system).

The lost species may have held secrets to increased food production; a cure for AIDS, cancer, or other health problems; or a base for better insecticides that do not harm the environment. Similar services to humanity already have been provided by tropical forest species.

Tropical deforestation is also threatening the natural chemistry of the atmosphere. Rainforests are a major source of the atmospheric oxygen so essential to all animal life. And deforestation encourages global warming by enhancing the greenhouse effect because forests act as a major reservoir of carbon dioxide. It has been estimated that forest clearing since the mid-1800s has contributed more than 130 billion tons of carbon to the atmosphere, more than two thirds as much as has been added by the burning of coal, oil, and natural gases combined.

What can be done? The reasons for tropical deforestation and the solutions to the problem may be economic, but the issues are extremely complex. It is not sufficient for the rest of the world to point out to governments of tropical nations that their forests are a major key to human survival. It is unacceptable for scientists and politicians from nations where barely one fourth of the original forests remain to insist that the citizens of the tropics cease cutting trees and establish forest plantations on deforested land. These are desired outcomes, but it is first the responsibility of all the world's people to help resolve the serious economic and social problems that have prevented most tropical nations from considering their forests as a sustainable resource.

NASA/Robert Simmon, ASTER

A satellite image of an area near Rodonia, Brazil, shows an example of deforestation in the region, between September of 2000 (left) and the same month in 2006 (right). The area shown is about 4.8 by 3.2 kilometers (3 × 2 mi).

everywhere in the rainforest. Insects can breed continuously in this climate without danger from cold or drought.

Besides the insects, there are genuine health hazards for human inhabitants of the tropical rainforest. Not only does the oppressive, sultry weather impose uncomfortable living conditions, but also any open wound would heal more slowly in the steamy environment. This climate also allows a variety of parasites and disease-carrying insects to threaten human survival. Malaria, yellow fever, dengue fever, and sleeping sickness are all insect-borne (sometimes fatal) diseases of the tropics and uncommon in the middle latitudes.

Whenever native populations have existed in the rainforest, subsistence hunting and gathering of fruits, berries, small animals, and fish have been important. Since the introduction of agriculture, land has been cleared, and crops such as manioc, yams, beans, maize (corn), bananas, and sugarcane have been grown. It has been the practice to cut down the smaller trees, burn the resulting debris, and plant the desired crops. With the forest gone, this kind of farming is possible for only 2 or 3 years before the soil is completely exhausted of its small supply of nutrients and the surrounding area is depleted of game. At this point, the native population

M. Trapasso

• **FIGURE 9.6**
A jungle along the Usumacinta River on the Mexico–Guatemala border.
Why is the vegetation so dense here, when it is more open inside the forest at ground level?

moves to another area of forest to begin the practice over again. This kind of subsistence agriculture is known as **slash-and-burn** or simply **shifting cultivation.** Its impact on the close ecological balance between soil and forest is obvious in many rainforest regions. Sometimes the damage done to the system is irreparable, and only jungle, thorn bushes, or scrub vegetation will return to the cleared areas (• Fig. 9.7).

In terms of numbers of people supported, the most important agricultural use of the tropical rainforest climate is the wet-field (paddy) rice agriculture on the river floodplains of southeastern Asia. However, this type of agriculture is best developed in the monsoon variant of this climate. Commercial plantation agriculture is also significant. The principal plantation crops are rubber, sugarcane, and cacao, all of which originally grew with abundance in the forests of the Amazon Basin but are now of greatest importance in other rainforest regions—rubber in Malaysia and Indonesia, sugarcane and cacao in West Africa and the Caribbean area.

• **FIGURE 9.7**
An example of subsistence slash-and-burn agriculture: preparing an area for planting in Ecuador.
Would you expect shifting cultivation to be on the increase or decrease in tropical rainforests?

M. Trapasso

Tropical Monsoon Climate

We associate the **tropical monsoon climate** most closely with the peninsula lands of Southeast Asia. Here the alternating circulation of air (from sea to land in summer and from land to sea in winter) is strongly related to the shifting of the ITCZ. During the summer, the ITCZ moves north into the Indian subcontinent and adjoining lands to latitudes of 20° or 25°. This is due in part to the attracting force of the deep low pressure system of the Asian continent. However, as we have previously noted, the mechanism is complex and involves changes in the upper air flow as well as in surface currents. Several months later, the moisture-laden summer monsoon is replaced by an outflow of dry air from the massive Siberian high pressure system that develops in the winter season over central Asia. By this time, the ITCZ has shifted to its southernmost position (see again Fig. 5.20).

Figure 9.1 and • Figure 9.8 confirm that climate regions outside of Asia fit the simplified Köppen classification of tropical monsoon as well. A modified version of the monsoonal wind shift occurs at Freetown, Sierra Leone, in Africa, but the climate there might also be described as transitional between the constantly wet rainforest climate and the sharply seasonal wet and dry conditions of the tropical savanna.

Distinctions between Rainforest and Monsoon

Whatever the factors are that produce tropical monsoon climate regions, these regions have strong similarities to those classified as tropical rainforest. In fact, although their core regions are distinctly different, the two climates are often intermixed over *zones of transition*. A major reason for the similarity between monsoon and rainforest climates is that a monsoon area has enough precipitation to allow continuous vegetative growth with no dormant period during the year. Rains are so abundant and intense and the dry season so short that the soils usually do not dry out completely. As a result, this climate and its soils support a plant cover much like that of the tropical rainforests.

However, there are clear distinctions between rainforest and monsoon climate regions. The most important distinction of course concerns precipitation, including both distribution and amount. The monsoon climate has a short dry season, whereas the rainforest does not. Perhaps even more interesting, the average rainfall in monsoon regions varies more widely from place to place. It usually totals between 150 and 400 centimeters (60 to 150 in.) and may be massive where the onshore monsoon winds are forced to rise over mountain barriers. Mahabaleshwar, altitude 1362 meters (4467 ft), on the windward side of India's Western Ghats, averages more than 630 centimeters (250 in.) of rain during the 5 months of the summer monsoon.

The annual march of temperature of the monsoon climate differs appreciably from the monotony of the rainforest climate. The heavy cloud cover of the rainy monsoon reduces insolation and temperatures during that time of year. During the period of clear skies just prior to the onslaught of the rains, higher temperatures are recorded.

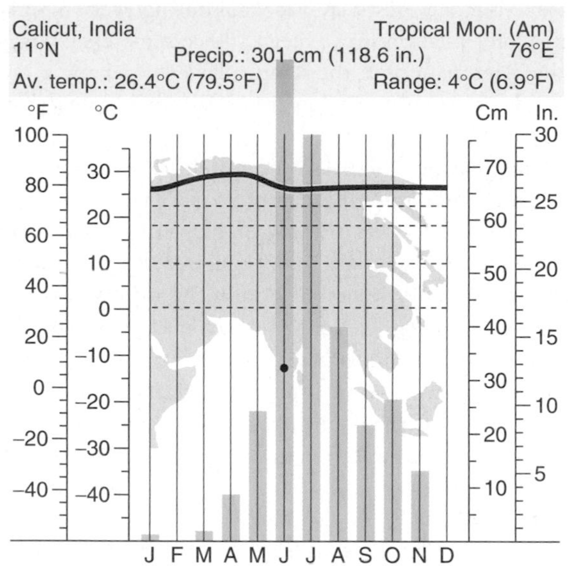

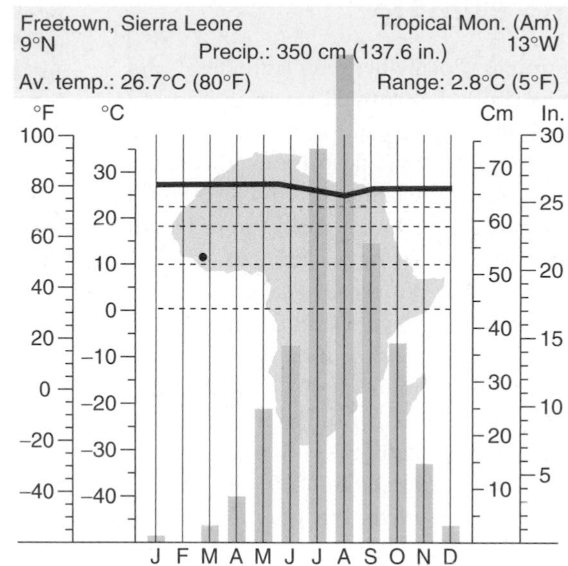

• FIGURE 9.8

Climographs for tropical monsoon climate stations.

What is the approximate range of precipitation between the highest and lowest months?

As a result, the annual temperature range in a monsoon climate is 2°C–6°C (compared with 2°C–3°C in the tropical rainforest).

Some additional distinctions between monsoon and rainforest regions can be found in vegetation and animal life. Toward the wetter margins, the tropical monsoon forest resembles the tropical rainforest, but fewer species are present and certain ones become dominant. The seasonality of rainfall in the monsoon narrows the range of species that will prosper. Toward the drier margins of the climate, the trees grow farther apart, and the monsoon forest often gives way to jungle or a dwarfed thorn forest. The composition of the animal kingdom here also changes. The climbing and flying species that dominate the forest are joined by larger, hooved leaf eaters and by larger carnivores such as the famous tigers of Bengal.

Effects of Seasonal Change The seasonal precipitation of the tropical monsoon climate is of major importance for economic reasons, especially to the people of Southeast Asia and India. Most of the people living in those areas are farmers, and their major crop is rice, which is the staple food for millions of Asians. Rice is most often an irrigated crop, so the monsoon rains are very important to its growth. Harvesting, on the other hand, must be done during the dry season (• Fig. 9.9).

Each year, an adequate food supply for much of South and Southeast Asia depends on the arrival and departure of the monsoon rains. The difference between famine and survival for many people in these regions is very much associated with the climate.

• FIGURE 9.9

Crop selection and agricultural production must be adjusted to the (a) wet and (b) dry seasons throughout Southeast Asia.

Would it be beneficial to the people of Southeast Asia if the traditional rice farming methods were replaced by mechanized rice agriculture as practiced in the United States?

Lou Linwei/Alamy

(a)

©Armstrong Roberts, robertstock.com

(b)

Tropical Savanna Climate

Located well within the tropics (usually between latitudes 5° and 20° on either side of the equator), the **tropical savanna climate** has much in common with the tropical rainforest and monsoon. The sun's vertical rays at noon are never far from overhead, the receipt of solar energy is nearly at a maximum, and temperatures remain constantly high. Days and nights are of nearly equal length throughout the year, as they are in other tropical regions.

However, as previously noted, its distinct seasonal precipitation pattern identifies the tropical savanna. As the latitudinal wind and pressure belts shift with the direct angle of the sun, savanna regions are under the influence of the rain-producing ITCZ (doldrums) for part of the year and the rain-suppressing subtropical highs for the other part. In fact, the poleward limits of the savanna climate are approximately the poleward limits of migration of the ITCZ, and the equatorward limits of this climate are the equatorward limits of movement by the subtropical high pressure systems.

As you can see in Figure 9.1 and Table 9.1, the greatest areas of savanna climate are found peripheral to the rainforest climates of Central and South America and Africa. Lesser but still important savanna regions occur in India, peninsular Southeast Asia, and Australia. In some instances, the climate extends poleward of the tropics, as it does in the southernmost portion of Florida.

Transitional Features of the Savanna

Of particular interest to the geographer is the transitional nature of the tropical savanna. Often situated between the humid rainforest climate on one side and the rain-deficient steppe climate on the other, the savanna experiences some of the characteristics of both. During the rainy, high-sun season, atmospheric conditions resemble those of the rainforest, whereas the low-sun season can be as dry as nearby arid lands are all year. The gradational nature of the climate causes precipitation patterns to vary considerably (● Fig. 9.10). Savanna locations close to the rainforest may have rain during every month, and total annual precipitation may exceed 180 centimeters (70 in.). In contrast, the drier margins of the savanna have longer and more intensive periods of drought and lower annual rainfalls, less than 100 centimeters (40 in.).

Other characteristics of the savanna help demonstrate its transitional nature. The higher temperatures just prior to the arrival of the ITCZ produce annual temperature ranges 3°C–6°C (5°F–11°F) wider than those of the rainforest, but still not as wide as those of the steppe and desert. Although the typical savanna vegetation (known as *llanos* in Venezuela, *campos* in Brazil, and *pampas* in Argentina) is a mixture of grassland and trees, scrub, and thorn bushes, there is considerable variation. Near the equatorward margins of these climates, grasses are taller, and trees, where they exist, grow fairly close together (● Fig. 9.11).

● FIGURE 9.11

The pampas of Argentina after some rainfall display lush tropical savanna vegetation.

Why do you think Argentina is a major exporter of beef cattle?

M. Trapasso

● FIGURE 9.10

Climographs for tropical savanna climate stations.

Consider the differences in climate and human use of the environment between Key West and Kano. Which are more important in the geography of the two places, the physical or the human factors?

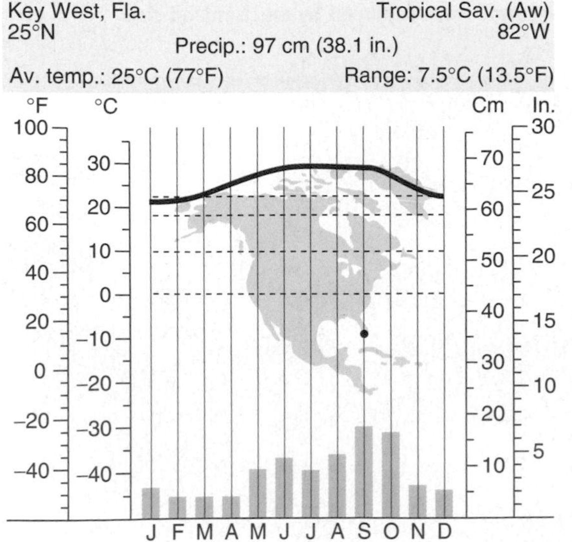

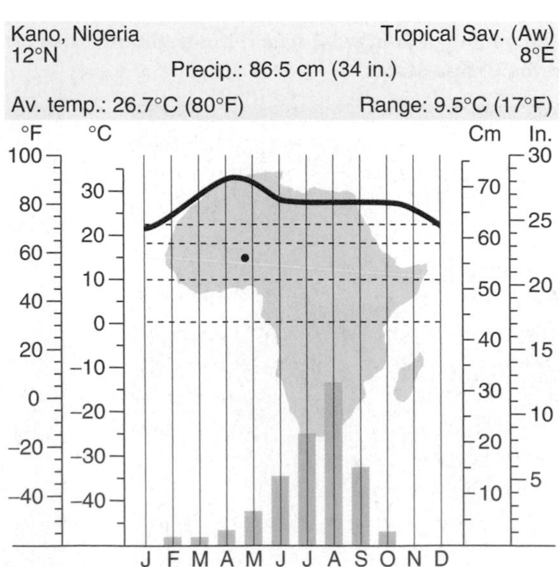

Toward the drier, poleward margins, trees are more widely scattered and smaller, and the grasses are shorter. Soils, too, are affected by the climatic gradation as the iron-rich reddish soils of the wetter sections are replaced by darker-colored, more organic-rich soils in the drier regions.

Both vegetation and soils have made special adaptations to the alternating wet–dry seasons of the savanna. During the wet (high-sun) period, the grasslands are green, and the trees are covered with foliage. During the dry (low-sun) period, the grass turns brown, dry, and lifeless, and most of the trees lose their leaves as an aid in reducing moisture loss through transpiration. The trees develop deep roots that can reach down to water in the soil during the dry season. They are also fire resistant, an advantage for survival in the savanna where the grasses may burn during the winter drought.

Savanna Potential Conditions within tropical savanna regions are not well suited to agriculture although many of our domesticated grasses (grains) are presumed to have grown wild there. Rainfall is far less predictable than in the rainforest or even the monsoon climate. For example, Nairobi, Kenya, has an average rainfall of 86 centimeters (34 in.). Yet from year to year, the amount of rain received may vary from 50 to 150 centimeters (20–60 in.). As a rule, the drier the savanna station, the more unreliable the rainfall becomes. However, the rains are essential for human and animal survival in savanna regions. When they are late or deficient, as they have been in West Africa in recent years, severe drought and famine result. On the other hand, when the rains last longer than usual or are excessive, they can cause major floods, often followed by outbreaks of disease.

Savanna soils (except in areas of recent stream deposits) also limit productivity. During the rains of the wet season, they may become gummy; during the dry season, they are hard and almost impenetrable. Consequently, people in the savannas have often found the soils better suited to grazing than to farming. The Masai, a tribe of cattle herders and fierce warriors of East Africa, are world-famous examples (● Fig. 9.12). However, even animal husbandry has its problems. Many savanna regions make poor pasturelands, at least during the dry part of the year.

The savannas of Africa have exhibited the greatest potential of the world's savanna regions. They have been veritable zoological gardens for the larger tropical animals, to such an extent that the popularity of classic photo safaris has made the African savannas a major center for tourism. The grasslands support many different herbivores (plant eaters), such as the elephant, rhinoceros, giraffe, zebra, and wildebeest (● Fig. 9.13). The herbivores in turn are eaten by the carnivores (flesh eaters), such as the lion, leopard, and cheetah. Lastly, scavengers, such as hyenas, jackals, and vultures, devour what remains of the carnivore's kill. During the dry season, the herbivores find grasses and water along stream banks and forest margins and at isolated water holes. The carnivores follow the herbivores to the water, and a few human hunters and scavengers still follow them both.

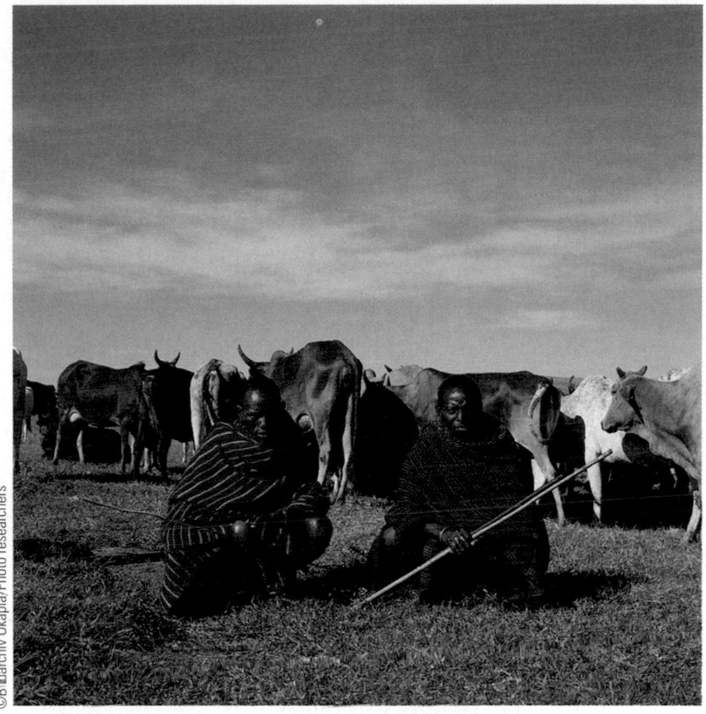
©Bildarchiv Okapia/Photo researchers

● FIGURE 9.12
The grasslands of the East African savanna are well suited to supporting large numbers of grazing animals. These Masai cattle herders in Kenya count their wealth by the numbers of animals they own.
What environmental problems may be created as cattle herds grow?

● FIGURE 9.13
Giraffes have always been a majestic sight in the savanna climate of East Africa.
How is a giraffe's height so well adapted to the savanna environment?

M. Trapasso

Arid Climate Regions

Arid climate regions in the simplified Köppen system are widely distributed over Earth's surface. A brief study of • Figure 9.14 confirms that they are found from the vicinity of the equator to more than 50°N and S latitude. There are two major concentrations of desert lands, and each illustrates one of the important causes of climatic aridity. The first is centered on the Tropics of Cancer and Capricorn (23½°N and S latitudes) and extends 10°–15° poleward and equatorward from there. This region contains the most extensive areas of arid climates in the world. The second is located poleward of the first and occupies continental interiors, particularly in the Northern Hemisphere.

The concentration of deserts in the vicinity of the two tropic lines is directly related to the subtropical high pressure systems. Although the boundaries of the subtropical highs may migrate north and south with the direct rays of the sun, their influence remains constant in these latitudes. We have already learned that the subsidence and divergence of air associated with these systems is strongest along the eastern portions of the oceans (recall, cold ocean currents off the western coasts of continents help stabilize the atmosphere). Hence, the clear weather and dry conditions of the subtropical high pressure extend inland from the western coasts of each landmass in the subtropics. The Atacama, Namib, and Kalahari Deserts and the desert of Baja California are restricted in their development by the small size of the landmass or by landform barriers to the interior. However, the western portion of North Africa and the Middle East comprises the greatest stretch of desert in the world and includes the Sahara, Arabian, and Thar Deserts. Similarly, the Australian Desert occupies most of the interior of the Australian continent.

The second concentration of deserts is located within continental interiors remote from moisture-carrying winds. Such arid lands include the vast cold-winter deserts of inner Asia and the Great Basin of the western United States. The dry conditions of the latter region extend northward into the Columbia Plateau and southward into the Colorado Plateau and are increased by the mountain barriers that restrict the movement of rain-bearing air masses from the Pacific. Similar *rain-shadow* conditions help to explain the Patagonia Desert of Argentina and the arid lands of western China.

Both wind direction and ocean currents can accentuate aridity in coastal regions. When prevailing winds blow parallel to a coastline instead of onshore, desert conditions are likely to occur because little moisture is brought inland. This seems to be the case in eastern Africa and in northeastern Brazil. Where a cold current flows next to a coastal desert, foggy conditions may develop. Warm, moist air from the ocean may be cooled to its dew point as it passes over the cooler current. A temperature inversion is created, increasing stability and preventing the upward movement of air required for precipitation. The unique, fog-shrouded coastal deserts in Chile (the Atacama), southwest Africa (the Namib), and Baja California have the lowest precipitation of any regions on Earth.

Figure 9.14 shows deserts of the world to be core areas of aridity, usually surrounded by the slightly moister steppe regions. Hence, our explanations for the location of deserts hold

• FIGURE 9.14
A map of the world's arid lands.
What does a comparison of this map with the Map of World Population Density (inside back cover) suggest?

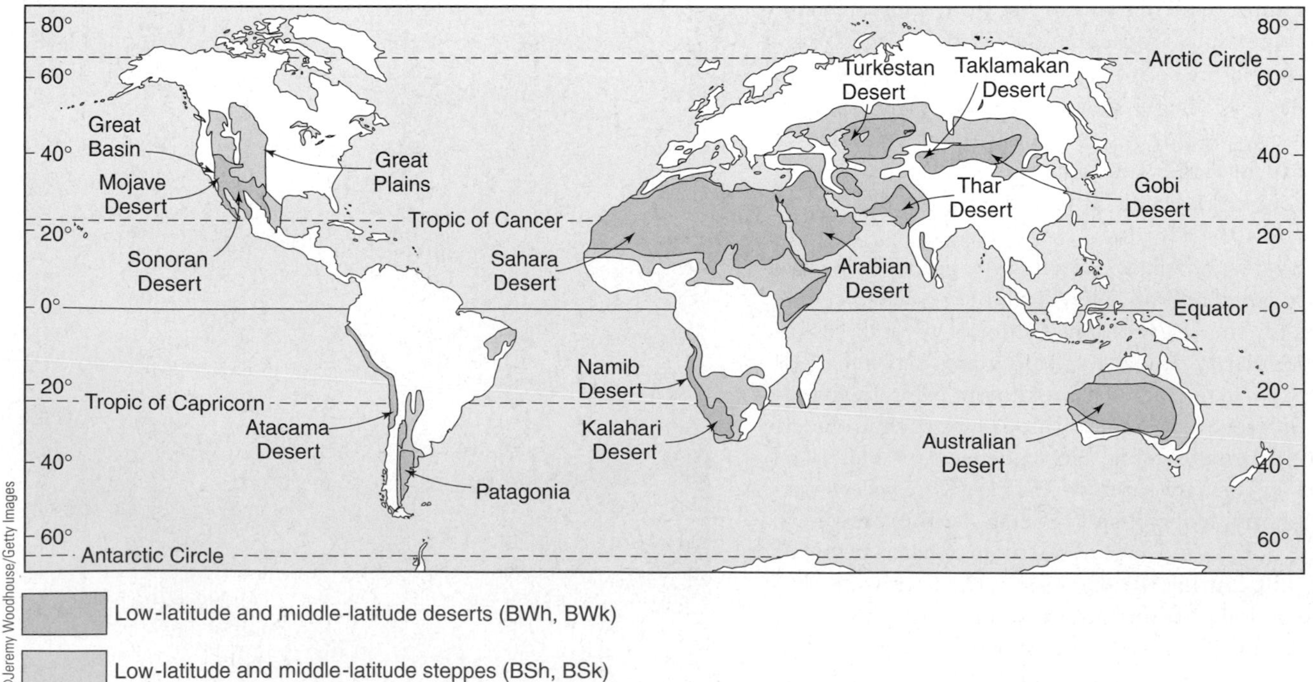

©Jeremy Woodhouse/Getty Images

true for the steppes as well. The steppe climates either are sub-humid borderlands of the humid tropical, mesothermal, and microthermal climates or are transitional between these climates and the deserts. As previously noted, we classify both steppe and desert on the basis of the relation between precipitation and potential ET (evapotranspiration, see again Chapter 6 and Fig. 6.8). In the **desert climate,** the amount of precipitation received is less than half the potential ET. In the **steppe climate,** the precipitation is more than half but less than the total potential ET.

The criterion for determining whether a climate is desert, steppe, or humid is *precipitation effectiveness.* The amount of precipitation actually available for use by plants and animals is the *effective precipitation*. Precipitation effectiveness is related to temperature. At higher temperatures, it takes more precipitation to have the same effect on vegetation and soils than at lower temperatures. The result is that areas with higher temperatures that promote greater ET can receive more precipitation than cooler regions and yet have a more arid climate.

Because of the temperature influence, precipitation effectiveness depends on the season in which an arid region's meager precipitation is concentrated. Obviously, precipitation received during the low-sun period will be more effective than that received during the high-sun period when temperatures are higher because less will be lost through ET. The simplified Köppen graphs based on the concept of precipitation effectiveness are included in the "Graph Interpretation" exercise (at the end of Chapter 8) and may be used to determine whether a particular location has a desert, steppe, or humid climate.

Desert Climates

The deserts of the world extend through such a wide range of latitudes that the simplified Köppen system recognizes two major subdivisions. The first are low-latitude deserts where temperatures are relatively high year-round and frost is absent or infrequent even along poleward margins; the second are middle-latitude deserts, which have distinct seasons, including below-freezing temperatures during winter (Table 9.2). However, the significant characteristic of all deserts is their aridity. The relative unimportance of temperature is emphasized by the small number of occasions on which we will distinguish between low-latitude and middle-latitude deserts in the discussion that follows.

Land of Extremes By definition, deserts are associated with a minimum of precipitation, but they represent the extremes in other atmospheric conditions as well. With few clouds and little water vapor in the air, as much as 90% of insolation reaches Earth in desert regions. This is why the highest insolation and highest temperatures are recorded in low-latitude desert areas and not in the more humid tropical climates that are closer to the equator. Again because the desert air contains so little moisture (recall, water vapor is a greenhouse gas), and with little or no cloud cover, there is little atmospheric effect, and much of the energy received by Earth during the day is radiated back to the atmosphere at night. Consequently, night temperatures in the desert drop far below their daytime highs. This excessive heating and cooling give low-latitude deserts the greatest diurnal temperature ranges in the world, and middle-latitude deserts are not far behind. In the spring and fall, these ranges may be

TABLE 9.2
The Arid Climates

Name and Description	Controlling Factors	Geographic Distribution	Distinguishing Characteristics	Related Features
Desert Precipitation less than half of potential evapotranspiration; mean annual temperature above 18°C (64.4°F) (low-lat.), below (mid-lat.)	Descending, diverging circulation of subtropical highs; continentality often linked with rainshadow location	Coastal Chile and Peru, southern Argentina, southwest Africa, central Australia, Baja California and interior Mexico, North Africa, Arabia, Iran, Pakistan and western India (low-lat.); inner Asia and western United States (mid-lat.)	Aridity; low relative humidity; irregular and unreliable rainfall; highest percentage of sunshine; highest diurnal temperature range; highest daytime temperatures; windy conditions	Xerophytic vegetation; often barren, rocky, or sandy surface; desert soils; excessive salinity; usually small, nocturnal burrowing animals; nomadic herding
Steppe Precipitation more than half but less than potential evapotranspiration mean annual temperature above 18°C (64.4°F) (low-lat.), below (mid-lat.)	Same as deserts; usually transitional between deserts and humid climates	Peripheral to deserts, especially in Argentina, northern and southern Africa, Australia, central and southwest Asia, and western United States	Semiarid conditions, annual rainfall distribution similar to nearest humid climate; temperatures vary with latitude, elevation, and continentality	Dry savanna (tropics) or short grass vegetation; highly fertile black and brown soils; grazing animals in vast herds; predators and small animals; ranching and dry farming

GEOGRAPHY'S SPATIAL SCIENCE PERSPECTIVE

Desertification

Although desertification is sometimes confused with drought, the two terms describe distinctly different processes. Drought—a longer than normal period of little or no rainfall—is a naturally recurring climatic event. It is especially common in arid and semiarid regions but can occur in subhumid and sometimes even humid climates. As a general rule, the drier a climate is, the greater are the variability of rainfall and the risk of drought.

Desertification, by contrast, is the natural process of desert expansion caused by climatic change but accelerated by human activities. It is a more serious situation than drought because it involves long-term environmental and human consequences. Desertification expands the margins of the desert when rare rains cause gully erosion, sheet erosion, and loss of soil. It also increases wind erosion, causing dust storms and sand dune movement into grassland and farmland areas. Desertification is pronounced in regions of the world where humans have accelerated the expansion of desert climate and landform features into former grassland and woodland regions. Although climate change may be the trigger, the process is accelerated by deforestation, overcultivation, soil salinization due to irrigation, and overgrazing by cattle, sheep, and goats.

Desertification is not new. Archeological evidence from Israel and Jordan indicates that as far back as 4000 BC early farming communities may have destroyed the soil and deforested the hills, causing desertification. Recent research into ancient environmental catastrophes has shown a similar pattern of denudation of the hilly landscape of Greece as early as 3000 BC. Today, evidence of desertification is visible in areas of Spain that exhibit deep gully erosion, in northwestern India as the Thar Desert expands into Rajasthan's farming areas, and throughout much of the Middle East, northern China, and Africa. Along with the threat to the human population, desertification endangers habitats for wildlife.

It was not until the 1970s, however, that desertification became well known, as television revealed starving and suffering citizens of the nations of the African Sahel. It showed bone-thin cattle trying to find a blade of grass in a barren landscape. The TV also revealed villages being invaded by sand dunes. The Sahel is the semiarid zone bordering the southern margin of the Sahara. It extends across northern Africa from Mauritania on the Atlantic coast to Somalia on the Indian Ocean. The term *desertification* was popularized at a U.N. conference dealing with problems like those of the Sahel, and most people associate the term with the continuing plight of the people in the region.

The United Nations Environment Program (UNEP) includes a cost estimate of up to $20 billion annually for 20 years in order to successfully fight worldwide desertification. In 1994, 87 nations signed the Desertification Convention in Paris. When ratified by 50 nations, this treaty will budget funds to help protect the fertility of lands that are at the greatest risk of desertification. It will take the support of all the world's nations for antidesertification programs to be successful. Only a major international effort can deal with a natural hazard that causes such large-scale environmental deterioration and human suffering.

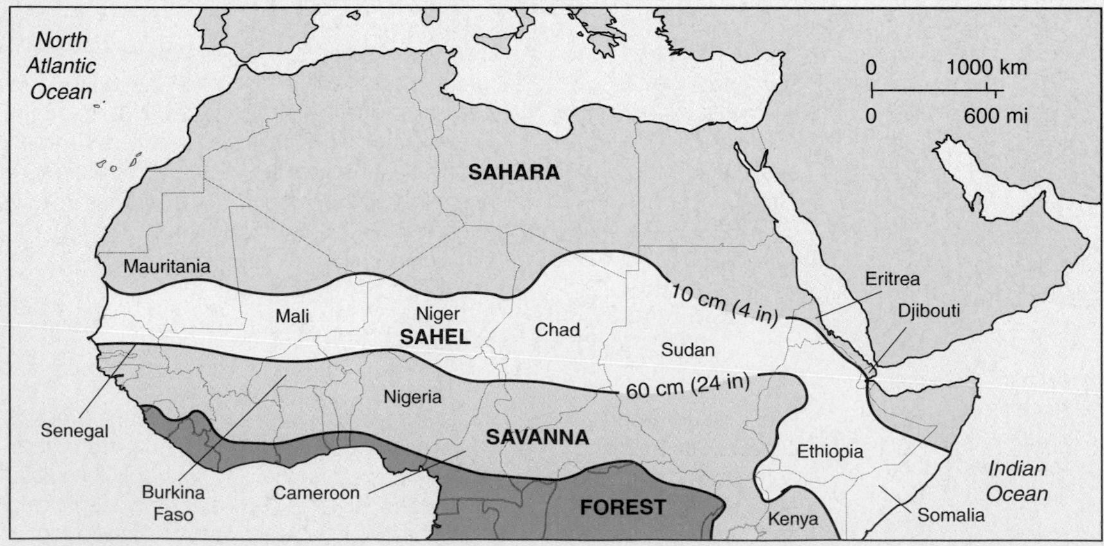

The Sahel region of Africa, shown here in a light-tan color, is the transition zone between the extreme aridity of the Sahara and the tropical humid areas of Africa (in green tones). In recent years, the Sahel has experienced desertification through climate change and overuse of this marginal land by human activities.

UNEP Sudan

Dune sand encroaches on agricultural land in Sudan, near the Nile River in the East African part of the Sahel region. Desertification reduces the amount of land that is directly usable for agriculture and grazing.

UNEP Sudan

This pond in the Sudan, built to impound water, has dried up completely even after more rainfall had occurred than has been typical in recent years.

as great as 40°C (72°F) in a day. More common diurnal temperature ranges in deserts are 22°C–28°C (40°F–50°F).

The sun's rays are so intense in the clear, dry desert air that temperatures in shade are much lower than those a few steps away in direct sunlight. (Keep in mind that all temperatures for meteorological statistics are recorded in the shade.) Khartoum, Sudan (in the Sahara), has an *average* annual temperature of 29.5°C (85°F), which is a *shade* temperature. Temperatures in the bright desert sun under cloudless skies at Khartoum are often 43°C (110°F) or more. Soil temperatures rise close to 95°C (200°F) in midsummer in the Mojave Desert of Southern California.

During low-sun or winter months, deserts experience colder temperatures than more humid areas at the same latitude, and in summer they experience hotter temperatures. Just as with the high diurnal ranges in deserts, these high annual temperature ranges can be attributed to the lack of moisture in the air.

Annual temperature ranges are usually greater in middle-latitude deserts, such as the Gobi in Asia, than in low-latitude deserts because of the colder winters experienced at higher latitudes. Compare, for example, the climograph for Aswan in south central Egypt—at 24°N, a low-latitude desert location—with the climograph for Turtkul, Uzbekistan—at 41°N, a middle-latitude desert location (• Fig. 9.15). The annual range for Aswan is 17°C (31°F); in Turtkul, it is 34°C (61°F).

©Jonathan Blair/CORBIS

• FIGURE 9.16
A rainstorm in the Mojave Desert of California produces a double rainbow.
What environmental clues suggest that rainfall is an infrequent event?

Precipitation in the desert climate is irregular and unreliable, but when it comes, it may arrive in an enormous cloudburst, bringing more precipitation in a single rainfall than has been recorded in years (• Fig. 9.16). Recall from Chapter 6 that the *variability of precipitation* is greater in regions where precipitation totals are lowest. This happened in the extreme at the port of Walvis Bay, on the coast of the Namib Desert, a cold-current coastal desert of southwest Africa. The equivalent of 10 years' rain was re-

• FIGURE 9.15
Climographs for desert climate stations.
If you consider the serious limitations of desert climates, how do you explain why some people choose to live in desert regions?

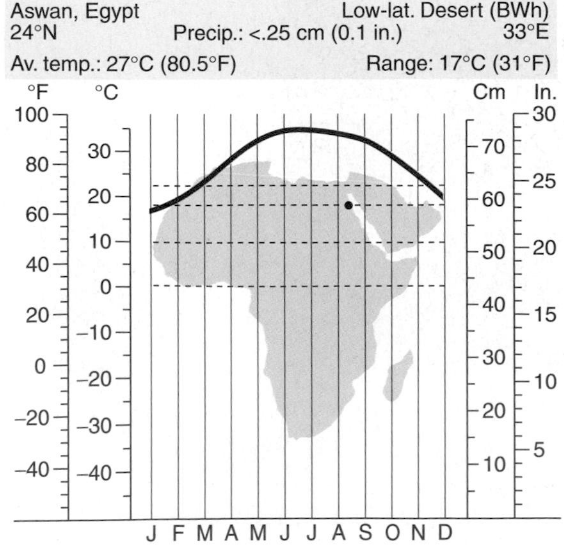

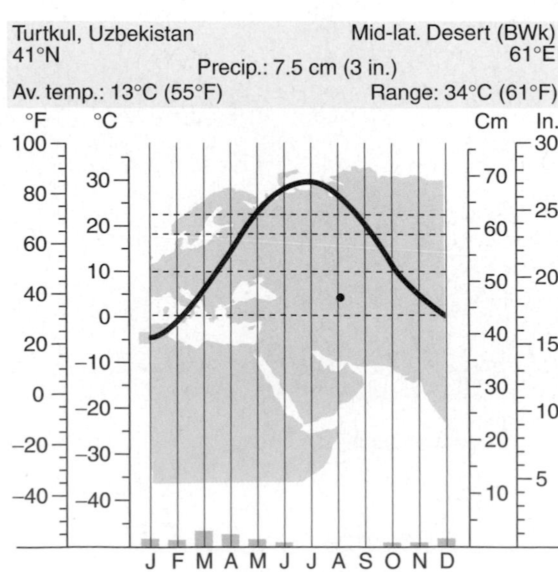

• **FIGURE 9.17**
A dust storm invades the city of Khartoum in Sudan.
How might storms like these affect human health? What about machinery and vehicles?

ceived in one night when a freak storm dumped 3.2 centimeters (1.3 in.) of rain.

The actual amount of water vapor in the air may be high in desert areas. However, the hot daytime temperatures increase the capacity of the air to hold moisture so that the relative humidity during daylight hours is quite low (10–30%). Desert nights are a different story. Radiation of energy is rapid in the clear air. As temperature drops, relative humidity increases and the formation of dew in the cool hours of early morning is common. Where measurements have been made, the amount of dew formed has sometimes considerably exceeded the annual rainfall for that location. It has been suggested, in fact, that dew may be of great importance to plant and animal life in the desert. Studies are now being carried out on the use of dew as a moisture source for certain crops, thereby minimizing the need for large-scale irrigation.

The convection currents set up by the intense heating of the land during the day help make the desert a windy place. In addition, the sparseness of vegetation and the absence of topographic interruptions in some deserts allow winds to sweep across these arid lands unimpeded. Sand and dust are carried by the desert winds, lowering visibility and irritating eyes and throats (• Fig. 9.17).

Adaptations by Plants and Animals

Deserts tend to have sparse vegetation, and large tracts may be barren bedrock, sand, or gravel. The plants that do exist are **xerophytic,** or adapted to extreme drought. They may have thick bark, thorns, little foliage, and waxy leaves, all of which reduce loss of water by transpiration. Another characteristic adaptation is the storage of moisture in stem or leaf cells, as in the cactus (saguaro and barrel cactus, prickly pear) (• Fig. 9.18). Some plants, such as creosote bush, mesquite, and acacias, have deep root systems to reach water; others, such as the Joshua tree, spread their roots widely near the surface for their moisture supply.

Even humans, the most adaptable of animals, find the desert environment a lasting challenge. For the most part, people have

• **FIGURE 9.18**
(a) Vegetation adapted to the arid conditions in the Atacama Desert in Chile. (b) After recent rainfall the desert landscape here in Organ Pipe National Park in Arizona becomes greener and comes into bloom.
What physical characteristics of cacti help them to survive the heat, drought, and evaporation rates of the desert?

(a)

(b)

M. Trapasso

• **FIGURE 9.19**
This oasis supplies enough water to support a small suburb of Ica, in southern Peru. **What can you see in the background that proves this is a desert environment?**

been hunters and gatherers, nomadic herders, and subsistence farmers wherever there was a water supply from wells, oases, or exotic streams (streams bearing water from outside the region), such as the Nile, Tigris, Euphrates, Indus, and Colorado Rivers (• Fig. 9.19). Desert people have learned to adjust their habits to the environment. For example, they wear loose clothing to protect themselves from the burning rays of the sun and to prevent moisture loss by evaporation from the skin. At night, when the temperatures drop, the clothing keeps them warm by insulating and minimizing the loss of body heat.

Permanent agriculture has been established in desert regions all around the world wherever river or well water is available. Some produce mainly subsistence crops, but others have become significant producers of commercial crops for export.

Steppe Climates

Further study of Figure 9.14 and Table 9.2 provides a reminder that the distribution of the world's steppe lands is closely related to the location of deserts. Both moisture-deficient climate types share the controlling factors of continentality, rain-shadow location, the subtropical high pressure systems, or some combination of the three. The transitional nature of the steppes may make them seem like better-watered deserts at one time and like slightly subhumid versions of their humid-climate neighbors at another. Herein lies the major problem of steppe regions: How and to what extent should these variable and unpredictable climate regimes be used by humans?

Similarities to Deserts We are already aware that steppe regions are differentiated from deserts by their greater precipitation. Whereas most low-latitude desert locations receive fewer than 25 centimeters (10 in.) of rain annually, low-latitude steppe regions usually receive between 25 and 50 centimeters (10–20 in.). However, the similarities between deserts and steppes are often greater than the differences. In both climates, the potential ET exceeds the precipitation. As in the deserts, precipitation in steppe regions is unpredictable and varies widely in total amount from year to year. Annual rainfall differs significantly from place to place within both desert and steppe regions, and vegetation varies accordingly.

To be more specific, both the general precipitation pattern and the nature of the vegetation of a steppe region are usually closely related to the more humid climate immediately adjacent to it. Thus, when the steppe is located between desert and tropical savanna, the steppe's rains come with the high-sun season. Next to a Mediterranean climate, the steppe receives primarily winter precipitation. Similarly, the short, shallow-rooted grasses most commonly associated with the steppe climate occur in the areas of transition from mesothermal and microthermal climates to desert. However, in the areas of transition from tropical savanna to desert, the vegetation is the dry savanna type, including scrub tree and bush growth, which becomes more stunted and sparse toward the drier margins until the typical desert shrub–vegetation type is dominant. • Figure 9.20 can be described as a *tropical scrub* or *tropical thorn forest,* and can be a hostile environment for human activities.

Both low-latitude and middle-latitude steppe varieties are identified by mean annual temperature (Table 9.2). As in the desert,

• **FIGURE 9.20**
This tropical scrub or tropical thorn forest can be a hostile environment for humans and many animals, but these African warthogs thrive there. **Why is it important not to wander off the well-worn paths and trails in this environment?**

UNEP Sudan

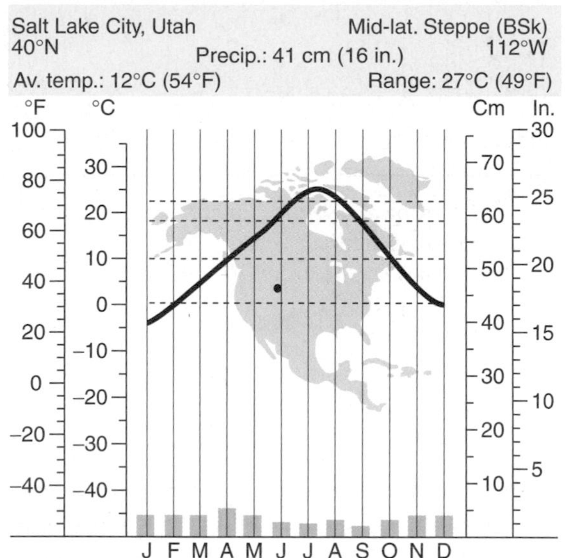

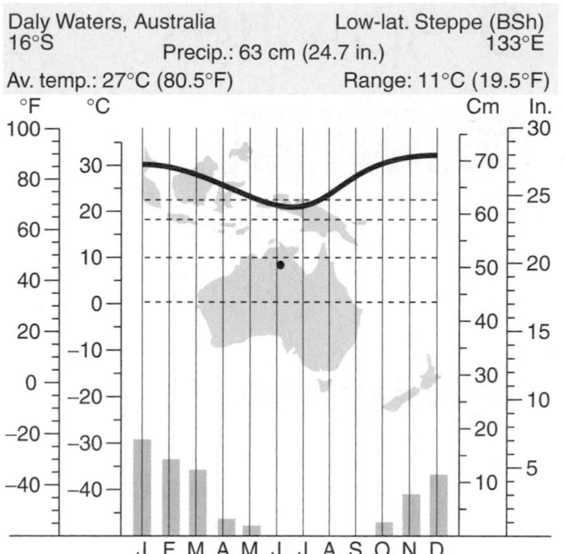

• FIGURE 9.21
Climographs for steppe climate stations.
What are the major differences between the climograph data for Salt Lake City and Daly Waters? What are the causes of these differences?

steppe temperatures vary throughout the climate type with latitude, distance from the sea, and elevation. The climographs of • Figure 9.21 demonstrate that although summer temperatures are high in all steppe regions, the differences in winter temperatures can produce annual ranges in middle-latitude steppes that are two or three times as great as those in low-latitude steppes.

A Dangerous Appeal Although the surface cover is often incomplete, in more humid regions of the middle-latitude steppes the grasses have been excellent for pasture. In North America, this was the realm of the bison and antelope; in Africa, it was the domain of wildebeests and zebra. Steppe soils are usually high in organic matter and soluble minerals. Attributes such as these have attracted farmers and herders alike to the rich grasslands—but not without penalty to both humans and land.

The climate is dangerous for agriculture, even in the middle-latitude steppe, and people take a sizable risk when they attempt to farm. Although dry-farmed wheat and drought-resistant barley and sorghum can be successfully raised because both farming methods and crops are adapted to the environment, the use of techniques employed in more humid regions can lead to serious problems. During dry cycles, crops fail year after year, and with the land stripped of its natural sod, the soil is exposed to wind erosion. Even using the grasses for grazing domesticated animals is not always the answer, for overgrazing can just as quickly create "Dust Bowl" conditions (as occurred during the extreme droughts of the 1930s in the southwestern United States).

The difficulties in making steppe regions more productive point out again the sensitive ecological balance of Earth's systems. The natural rains in the steppe are usually sufficient to support a vegetation cover of short grasses that can feed the roaming herbivores that graze on them. The herbivore population in turn is kept in check by the carnivores who prey on them (• Fig. 9.22). When people enter the scene, however, sending out more animals to graze, plowing the land, or merely killing off the predators, the ecological balance is tipped, and sometimes the results are disastrous.

The *A* and *B* climates discussed in this chapter represent some of the greatest extremes in atmospheric elements worldwide. These extremes vary from the heavy precipitation in tropical rainforest and tropical monsoon climates to the extreme dryness in our subtropical desert regions. Daily temperatures vary greatest in the desert regions of the middle latitudes, and the variability of precipitation is always greatest in the driest regions of Earth. *A* and *B* climates may be harsh and delicate at the same time. These climates make survival difficult for the human populations that inhabit them, and in turn, the human intervention can have devastating effects on these environments.

• FIGURE 9.22
This small herd of zebra has more to worry about than an attack by a predator.
What other types of dangers might threaten this environment?

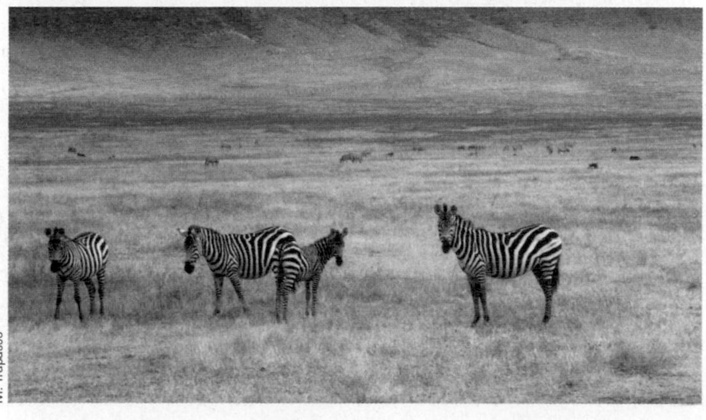

Chapter 9 Activities

Define & Recall

tropical rainforest climate
annual temperature range
daily (diurnal) temperature range
cloud forest
selva
jungle
slash-and-burn (shifting cultivation)
tropical monsoon climate
tropical savanna climate
desert climate
steppe climate
xerophytic

Discuss & Review

1. Explain the role of insolation in the existence of three different tropical environments.
2. Describe the three tropical (*A*) climates.
3. How do ocean currents affect tropical and arid climates?
4. What are the controlling factors that explain the tropical rainforest climate? Give a brief verbal description of this climate type.
5. Describe the delicate balance between vegetative growth and soil fertility in a tropical rainforest environment.
6. What is the difference between a rainforest and a jungle? What factor differentiates these two forest types?
7. Where do you find the greatest difference in temperature between direct sunlight and shade, the rainforest or the savanna? Why are these two so different?
8. Explain the seasonal precipitation pattern of the tropical savanna climate. State some of the transitional features of this climate. How have vegetation and soils adapted to the wet–dry seasons?
9. What conditions give rise to desert climates?
10. How do steppes differ from deserts? Why might human use of steppe regions in some ways be more hazardous than use of deserts?
11. What exceptions are there to the general rule that the pattern of climate regions is produced by Earth–sun relationships? How do the Northern and Southern Hemispheres differ in climate patterns, and why?
12. Do each of the following for the listed climates: tropical rainforest, tropical monsoon, tropical savanna, desert, and steppe.
    a. Identify the climate from a set of data or a climograph provided in the chapter indicating average monthly temperature and precipitation for a representative station within a region typical of that climate.
    b. Match the climate type with a written statement that includes one or more of the following: the statistical parameters of the climate in the modified Köppen classification; the particular climate controls (controlling factors) that produce the climate; and the geographic distributions of the climate, stated in terms of physical or political location.
    c. Distinguish between the important subtypes (if any) of each climate by identifying the characteristics that separate them from one another.

Consider & Respond

1. Based on the classification scheme presented in the "Graph Interpretation" exercise (end of Chapter 8), classify the following climate stations from the data provided.

		J	F	M	A	M	J	J	A	S	O	N	D	Yr.
a.	Temp. (°C)	20	23	27	31	34	34	34	33	33	31	26	21	29
	Precip. (cm)	0.0	0.0	0.0	0.0	0.2	0.1	1.1	0.2	0.0	0.0	0.0	0.0	1.6
b.	Temp. (°C)	21	24	28	31	30	28	26	25	26	27	25	22	26
	Precip. (cm)	0.0	0.0	0.2	0.8	7.1	11.9	20.9	31.1	13.7	1.4	0.0	0.0	87.2
c.	Temp. (°C)	19	20	21	23	26	27	28	28	27	26	22	20	24
	Precip. (cm)	5.1	4.8	5.8	9.9	16.3	18.0	17.0	17.0	24.0	20.0	7.1	3.0	149.0
d.	Temp. (°C)	2	4	8	13	18	24	26	24	21	14	7	3	14
	Precip. (cm)	1.0	1.0	1.0	1.3	2.0	1.5	3.0	3.3	2.3	2.0	1.0	1.3	20.6
e.	Temp. (°C)	27	26	27	27	27	27	27	27	27	27	27	27	27
	Precip. (cm)	31.8	35.8	35.8	32.0	25.9	17.0	15.0	11.2	8.9	8.4	6.6	15.5	243.8
f.	Temp. (°C)	13	17	17	19	22	24	26	26	26	24	19	15	20
	Precip. (cm)	6.6	2.0	2.0	0.5	0.3	0.0	0.0	0.0	0.3	1.8	4.6	6.6	26.7

2. The following six locations are represented by the data in the previous table, although not in the order listed: Albuquerque, New Mexico; Belém, Brazil; Benghazi, Libya; Faya, Chad; Kano, Nigeria; Miami, Florida. Use an atlas and your knowledge of climates to match the climatic data with the locations.
3. Benghazi, Libya, and Albuquerque, New Mexico, both exhibit steppe climates, but the cause (or control) of their dry conditions is quite different. Describe the primary factor responsible for the dry conditions at each location.
4. Tropical *B* climates are located more poleward than *A* climates, yet their daytime high temperatures are often higher. Why?
5. Kano, Nigeria, has copious amounts of rainfall during the summer season. There are two sources, or causes, of this rainfall. What are they?

Apply & Learn

	Jan	Feb	Mar	Apr	May	Jun	Jul	Aug	Sep	Oct	Nov	Dec
Average Max. Temperature (F)	66.2	73.4	81.1	89.7	99.5	109.2	115.6	113.8	105.8	92.9	76.3	65.2
Average Min. Temperature (F)	39.3	46.0	53.8	61.5	71.5	80.6	87.2	85.3	75.2	61.6	47.8	38.1
Average Total Precipitation (in)	0.32	0.50	0.31	0.13	0.07	0.05	0.13	0.14	0.19	0.10	0.19	0.16

1. The data set above shows measurements collected from Death Valley, California, between April 1, 1961 and December 31, 2005. For this data set, calculate the mean annual maximum temperature, and annual maximum temperature range; the mean annual minimum temperature, and annual minimum temperature range; the mean annual precipitation values, and the annual precipitation range.
   The highest temperature ever recorded in the United States was set here in July 1913, at 134° F. How much higher was that record from the mean maximum temperature for July in the above data set? Is July typically the month with the highest maximum temperatures?
2. The Amazon rainforest occupies some 5.5 million square kilometers. Between May 2000 and August 2005, Brazil lost more than 132,000 square kilometers of forest—an area larger than Greece. Assuming a constant rate* of deforestation at approximately 132,000 square kilometers per 5-year time period, how many years will it take before the Amazon rainforest is completely destroyed?
   *Actually, the destruction of the Amazon rainforest is occurring at an accelerating rate.

Middle-Latitude, Polar, and Highland Climate Regions

10

CHAPTER PREVIEW

The locations of the Mediterranean and the humid subtropical climates on opposite sides of continents in the lower-middle latitudes provide convincing evidence of the dominance of the subtropical highs as the climatic control of these latitudes.

- In what ways are the subtropical highs responsible for these contrasting climates?
- How are the oceans involved?

Some geographers refer to the marine west coast climate as the "temperate oceanic," or simply the "marine" climate.

- Why do we call this climate the marine west coast climate?
- Why do all these names emphasize maritime terms?

Although humid continental hot-summer and mild-summer regions have more in common than not, there are distinct differences in natural and cultivated vegetation between the two climates, particularly in North America.

- What are these differences, and why do they exist?
- How do people deal with these differences?

The climatic and related physical characteristics that distinguish the subarctic climate and the two polar climates are accompanied by human utilization patterns significantly different from those found in the humid continental climates.

- In what ways does the physical environment in each of these high-latitude climates affect human use patterns?
- How delicate are these climates?

Highlands are occupied by a complex pattern of widely varying microclimates far too intricate to be shown on anything but large-scale maps.

- What factors are most important in explaining the existence of a particular highland microclimate?
- What are the major peculiarities of mountain climates?

◀ **Opposite: Forests surround a small glacial lake in Alaska.**
M. Trapasso

In this chapter, we further explain the world climates as they dominate higher latitudes and higher altitudes. As we have noted, the tropical climates exhibit the constant characteristic of heat, whereas polar climates exhibit the constant characteristic of cold. The arid climates have inadequate precipitation in common. But what is the constant characteristic of the mesothermal and microthermal middle-latitude climates, which we are about to examine? If there is one, then it is the oxymoron "constant inconsistency." Each of the middle-latitude climates is dominated by the changing of the seasons and the variability of atmospheric conditions associated with migrating air masses or cyclonic activity along the polar front.

Middle-Latitude Climates

If change is constant in the mesothermal and microthermal climate groups, then degree of change is what distinguishes one climate type from another. In one or another of the middle-latitude climates, summers vary from hot to cool and winters from mild to extremely cold. Some climates receive adequate monthly precipitation year-round; others experience winter drought or, even more challenging to humans who live there, dry months during the normal summer growing season. Despite the changing atmospheric conditions of middle-latitude climates, their regions are home to the majority of Earth's people, and they are major factors in some of Earth's most attractive, interesting, and productive physical environments.

Humid Mesothermal Climate Regions

When we use the term *mesothermal* (from Greek: *mesos,* middle) in describing climates, we are usually referring to the moderate temperatures that characterize such regions. However, we could also be referring to their middle position between those climates that have high temperature throughout the year and those that experience severe cold. By definition, the mesothermal climates experience seasonality, with distinct summers and winters that distinguish them from the humid tropics. Their summers are long and their winters mild, and this separates them climatically from the microthermal climates, which lie poleward.

The three distinct mesothermal climates introduced in Chapter 8 are displayed in Table 10.1. In all three, the annual precipitation exceeds the annual potential evapotranspiration, but the Mediterranean climate has a lengthy period of precipitation deficit in the summer season that distinguishes it from the humid subtropical and the marine west coast climates. The latter two are further differentiated by the fact that the humid subtropical regions have hot summers, whereas the marine west coast regions experience mild summers.

Mediterranean Climate

The **Mediterranean climate** (*Csa, Csb*) is one of the best examples for organizing a study of the environment or developing an understanding of world regions, based on climate classification. Such a climate appears with remarkable regularity in the vicinity of 30° to 40° latitude along the west coasts of each landmass (● Fig. 10.1). The alternating controls of subtropical high pressure in summer and westerly wind movement in winter are so predictable that all Mediterranean lands have notably similar and easily recognized temperature and precipitation characteristics (● Fig. 10.2). The special appearance, combinations, and climatic adaptations of Mediterranean vegetation not only are unusual but also are clearly distinguishable from those of other climates. Agricultural practices, crops, recreational activities, and architectural styles all exhibit strong similarities within Mediterranean lands.

● FIGURE 10.1
Index map of humid mesothermal climates.

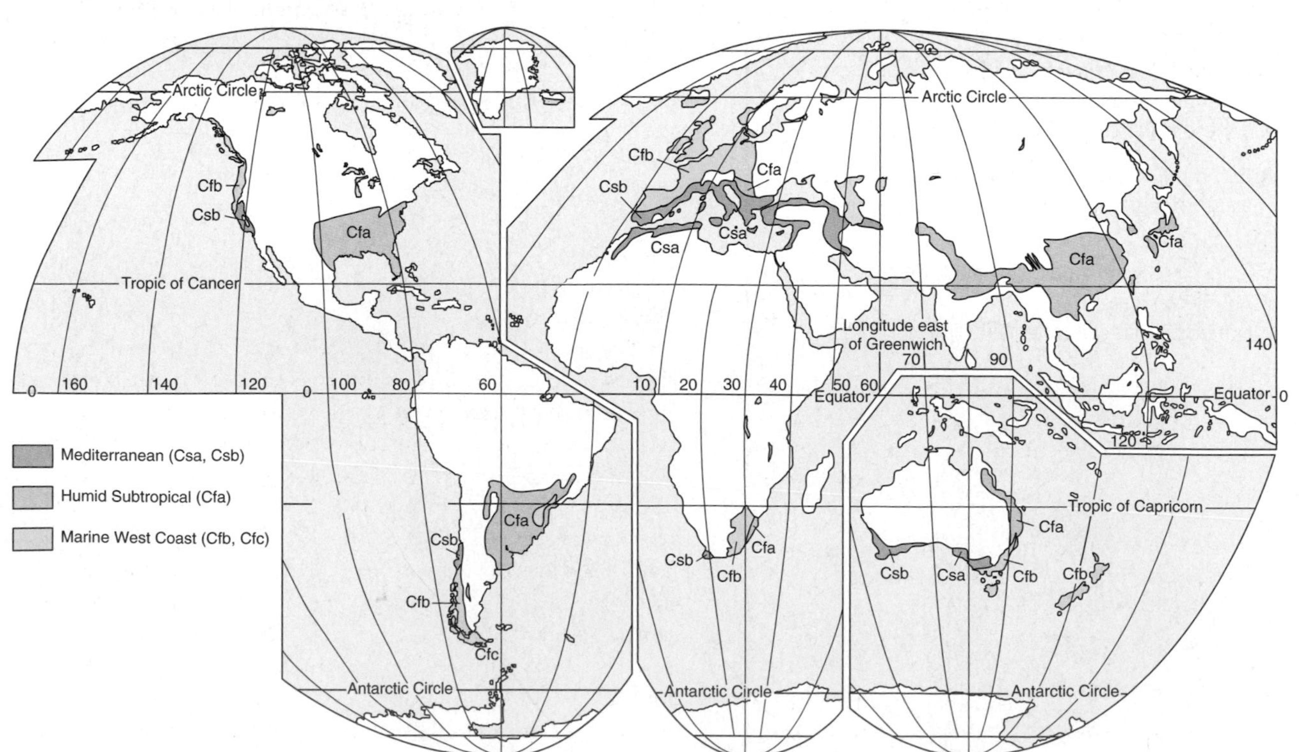

TABLE 10.1
The Mesothermal Climates

Name and Description	Controlling Factors	Geographic Distribution	Distinguishing Characteristics	Related Features
Mediterranean Warmest month above 10°C (50°F); coldest month between 18°C (64.4°F) and 0°C (32°F); summer drought; hot summers (inland), mild summers (coastal)	West coast location between 30° and 40°N and S latitudes; alternation between subtropical highs in summer and westerlies in winter	Central California; central Chile; Mediterranean Sea borderlands, Iranian highlands; Cape Town area of South Africa; southern and southwestern Australia	Mild, moist winters and hot, dry summers inland with cooler, often foggy coasts; high percentage of sunshine; high summer diurnal temperature range; frost danger	Sclerophyllous vegetation; low, tough brush (chaparral); scrub woodlands; varied soils, erosion in Old World regions; winter-sown grains, olives, grapes, vegetables, citrus, irrigation
Humid Subtropical Warmest month above 10°C (50°F); coldest month between 18°C (64.4°F) and 0°C (32°F); hot summers; generally year-round precipitation, winter drought (Asia)	East coast location between 20° and 40°N and S latitudes; humid onshore (monsoonal) air movement in summer, cyclonic storms in winter	Southeastern United States; southeastern South America; coastal southeast South Africa and eastern Australia; eastern Asia from northern India through south China to southern Japan	High humidity; summers like humid tropics; frost with polar air masses in winter; precipitation 62–250 cm (25–100 in.), decreasing inland; monsoon influence in Asia	Mixed forests, some grasslands, pines in sandy areas; soils productive with regular fertilization; rice, wheat, corn, cotton, tobacco, sugarcane, citrus
Marine West Coast Warmest month above 10°C (50°F); coldest month between 18°C (64.4°F) and 0°C (32°F); year-round precipitation; mild to cool summers	West coast location under the year-round influence of the westerlies; warm ocean currents along some coasts	Coastal Oregon, Washington, British Columbia, and southern Alaska; southern Chile; interior South Africa; southeast Australia and New Zealand; northwest Europe	Mild winters, mild summers, low annual temperature range; heavy cloud cover, high humidity; frequent cyclonic storms, with prolonged rain, drizzle, or fog; 3- to 4-month frost period	Naturally forested, green year-round; soils require fertilization; root crops, deciduous fruits, winter wheat, rye, pasture and grazing animals; coastal fisheries

Warm, Dry Summers; Mild, Moist Winters The major characteristics of the Mediterranean climate are a dry summer; a mild, moist winter; and abundant sunshine (90% of possible sunshine in summer and as much as 50–60% even during the rainy winter season). Summers are warm throughout the climate, but there are enough differences between the monthly temperatures in coastal and interior locations to recognize two distinct subtypes. The moderate-summer subtype (*Csb*) has the lower summer temperatures associated with coastal fogs and a strong maritime influence. The hot-summer subtype (*Csa*) is located further inland and reflects an increased influence of *continentality.* The inland version has higher summer and daytime temperatures and slightly cooler winter and nighttime temperatures than its coastal counterpart, and it has greater annual and diurnal temperature ranges as well. Compare, for example, the annual range for Red Bluff, California, an inland station, with that of San Francisco, a coastal station about 240 kilometers (150 mi) farther south (see again Fig. 10.2).

Whichever the subtype, Mediterranean summers clearly show the influence of the subtropical highs. Weeks go by without a sign of rain, and evapotranspiration rates are high. Effective precipitation is lower than actual precipitation, and the summer drought is as intense as that of the desert. Days are warm to hot, skies are blue and clear, and sunshine is abundant. The high percentage of insolation coupled with nearly vertical rays of the noon sun may drive daytime temperatures as high as 30°C–38°C (86°F–100°F), except where moderated by a strong ocean breeze or coastal fog.

Fog is common throughout the year in coastal locations and is especially noticeable during the summer. As moist maritime air moves onshore, it passes over the cold ocean currents that typically parallel west coasts in Mediterranean latitudes. The air is cooled, condensation takes place, and fog regularly creeps in

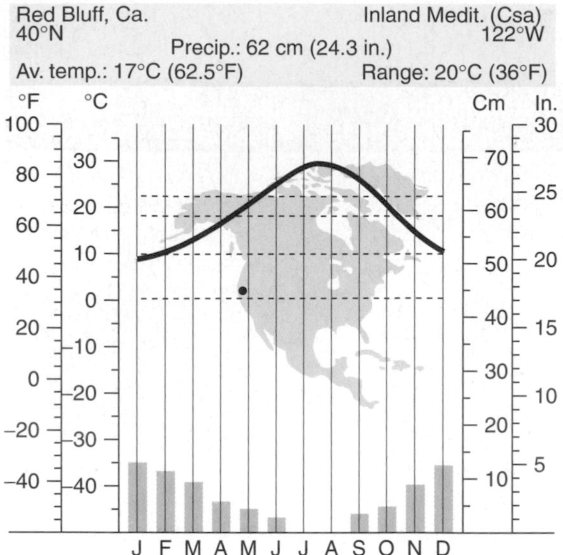

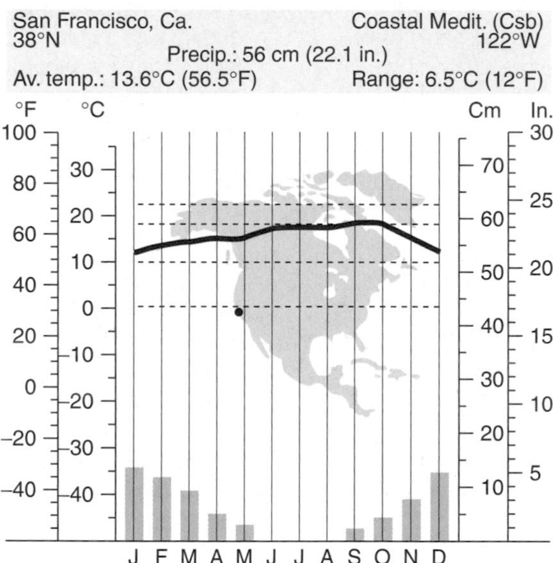

• FIGURE 10.2
Climographs for Mediterranean climate stations.
In what way do these climographs differ? What causes the differences?

during the late afternoon, remains through the night, and "burns off" during the morning hours. As in the desert, radiation loss is rapid at night, and even in summer, nighttime temperatures are commonly only 10°C–15°C (50°F–60°F).

Winter is the rainy season in the Mediterranean climate. The average annual rainfall in these regions is usually between 35 and 75 centimeters (15–30 in.), with 75% or more of the total rain falling during the winter months. The precipitation results primarily from the cyclonic storms and frontal systems common with the westerlies. Annual amounts increase with elevation and decrease with increased distance from the ocean. Only because the rain comes during the cooler months, when evapotranspiration rates are lower, is there sufficient precipitation to make this a humid climate.

Despite the rain during the winter season, there are often many days of fine, mild weather. Insolation is still usually above 50%, and the average temperature of the coldest month rarely falls below 4°C–10°C (40°F–50°F). Frost is uncommon and, because of its rarity, many less hardy tropical varieties of fruits and vegetables are grown in these regions. However, when frost does occur, it can do great damage.

Special Adaptations The summer drought, not frost, is the great challenge to vegetation in Mediterranean regions. The natural vegetation reflects the wet–dry seasonal pattern of the climate. During the rainy season, the land is covered with lush, green grasses that turn golden and then brown under the summer drought. Only with winter and the return of the rains does the landscape become green again. Much of the natural vegetation is **sclerophyllous** (hard-leafed) and drought resistant. Like xerophytes, these plants have tough surfaces, shiny, thick leaves that resist moisture loss, and deep roots to help combat aridity.

One of the most familiar plant communities is made up of many low, scrubby bushes that grow together in a thick tangle. In the western United States, this is called **chaparral** (• Fig. 10.3). Chaparral areas have the potential for frequent wild fires that occur in this dry brush typically during the summer and fall. The fires help perpetuate the chaparral because the associated heat is required to open seedpods and allow many chaparral species to reproduce. People often remove chaparral as a preventive measure against fires, but the removal can have disastrous results because the chaparral acts as a check against erosion of soils during the rainy season. With the chaparral removed, soils wash or slide down hillsides during the heavy rains of winter, frequently taking homes with them.

• FIGURE 10.3
Typical remnant chaparral vegetation as found in southern Spain.
Why would you expect to find little "native" vegetation left in Old World Mediterranean lands?

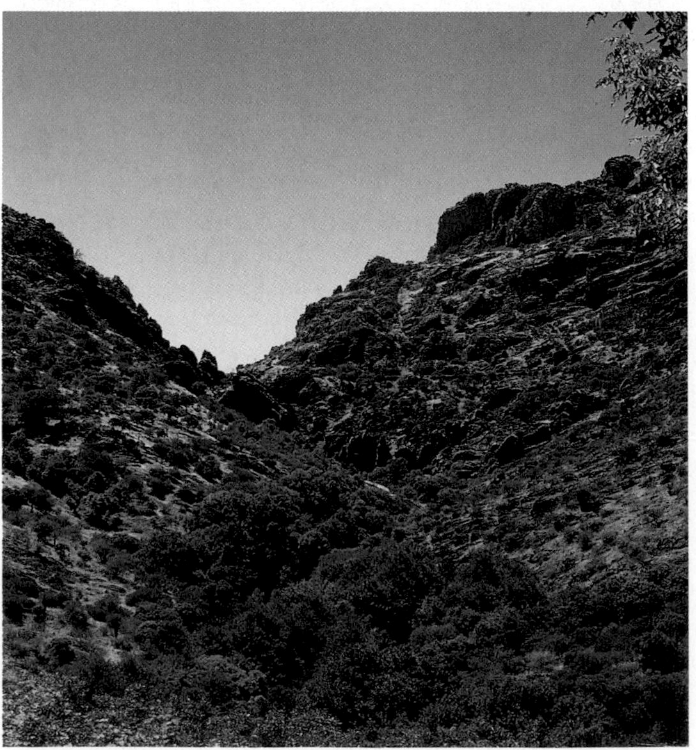
R. Gabler

Brush similar to the chaparral of California is called *mallee* in Australia, *mattoral* in Chile, and *maquis* in France. In fact, the "maquis" French Underground that fought against Nazi occupation in World War II literally meant "underbrush."

Trees in the Mediterranean climate also respond to the seasonally wet and dry moisture conditions. Because of their drought-resistant qualities, the needle-leaf pines are among the more common species. Groves of deciduous and evergreen oaks appear in depressions where moisture collects and on the shady north sides of hills where evapotranspiration rates are lower. Where the summer drought is more distinct, the scrub and woodlands open up to parklands of grasses and scattered oak trees. Even the great redwood forests of northern California probably could not survive without the heavy fogs that regularly invade the coastal lands in summer (• Fig. 10.4).

In response to the dry summers, most soils in Mediterranean regions are high in soluble nutrients, but the potential for agriculture differs widely from one region to another. The lime-rich soils around the Mediterranean Sea originally were highly productive, but destructive agricultural practices and overgrazing over thousands of years of human use have caused serious erosion problems. Today, bare white limestone is widely visible on hillsides in Spain, Italy, Greece, Crete, Syria, Lebanon, Israel, and Jordan.

• FIGURE 10.4

California redwoods (*Sequoia sempervirens*) may reach heights of 100 meters (330 ft) and live for thousands of years. They grow in cool, shady, and often foggy areas. This ancient tree survives despite the tunnel that was cut many years ago.

Why is it considered unusual to find redwoods growing in a Mediterranean climate?

M. Trapasso

In the Mediterranean regions of California, the soils are dense clays, gluelike when wet and hard as concrete when dry. During the Spanish period in California, the clay was formed into adobe bricks and used for building material. These clay soils are hazardous on slopes because they absorb so much water during the wet season that they can become mudflows destructive to roads and homes in areas where vegetation has been removed by fire or human interference.

In all Mediterranean regions, the most productive areas are the lowlands covered by stream deposits. Here farmers have made special adaptations to climatic conditions. There is sufficient rainfall in the cool season to permit fall planting and spring harvesting of winter wheat and barley. These grasses originally grew wild in the eastern Mediterranean region. Grapevines, fig and olive trees, and the cork oak, which undoubtedly were also native to Mediterranean lands, are especially well adapted to the dry summers because of their deep roots and thick, well-insulated stems or bark (• Fig. 10.5). Where water for irrigation is available, an incredible diversity of crops may be seen. These include, in addition to those already mentioned, oranges, lemons, limes, melons, dates, rice, cotton, deciduous fruits, various types of nuts, and countless vegetables. California, blessed both with fertile valleys for growing fruits, vegetables, and flowers and with snow meltwater for irrigation, is probably the most agriculturally productive of the Mediterranean regions.

Even the houses of these regions show people's adaptation to the climate. Usually white or pastel in color, they gleam in the brilliant sunshine against clear blue skies. Many have shuttered windows to reduce incoming sunlight and to keep the houses cool in summer. On the other hand, much less attention is paid to keeping homes warm during the cooler winter months. This is true even in the United States, where many Midwesterners and Easterners are surprised at the lack of insulation in California homes and the small number and size of heating devices.

• FIGURE 10.5

Oranges adapt well to a Mediterranean climate. This vineyard grove is located in California. These are not tall trees.

What factors of this climate might limit their growth?

M. Trapasso

Humid Subtropical Climate

The **humid subtropical climate** (*Cfa*) extends inland from continental east coasts between 15° and 20° and 40°N and S latitude (see again Fig. 10.1 and Table 10.1). Thus, it is located within approximately the same latitudes and in a similar transitional position as the Mediterranean climate, but on the eastern instead of the western continental margins. There is ample evidence of this climatic transition. Summers in the humid subtropics are similar to the humid tropical climates farther equatorward. When the noon sun is nearly overhead, these regions are subject to the importation of moist tropical air masses. High temperatures, high relative humidity, and frequent convectional showers are all characteristics that they share with the tropical climates. In contrast, during the winter months, when the pressure and wind belts have shifted equatorward, the humid subtropical regions are more commonly under the influence of the cyclonic systems of the continental middle latitudes. Polar air masses can bring colder temperatures and occasional frost.

Comparison with the Mediterranean Climate

Like the inland version of the Mediterranean climate, the humid subtropical climate has mild winters and hot summers. But it has no dry season. Whereas the Mediterranean lands are under the drought-producing eastern flank of the subtropical high pressure systems, the humid subtropical regions are located on the weak western sides of the subtropical highs. Subsidence and stability are greatly reduced or absent, even during the summer months. Here again, ocean temperatures play a significant role. The warm ocean currents that are commonly found along continental east coasts in these latitudes also moderate the winter temperatures and warm the lower atmosphere, thus increasing lapse rates, which enhances instability. Furthermore, a modified monsoon effect (especially in Asia and to some extent in the southern United States) increases summer precipitation as the moist, unstable tropical air is drawn in over the land.

As might be expected from the year-round rainfall, average annual precipitation for humid subtropical locations usually exceeds that for Mediterranean stations and may vary more widely as well. Humid subtropical regions receive anywhere from 60 to 250 centimeters (25–100 in.) a year. Precipitation generally decreases inland toward continental interiors and away from the oceanic sources of moisture. It is not surprising that these regions are noticeably drier the closer they are to steppe regions inland toward their western margins.

Both the Mediterranean and humid subtropical climates receive winter moisture from cyclonic storms, which travel along the polar front. As we have noted, the great contrast occurs in the summer when the humid subtropics receive substantial precipitation from convectional showers, supplemented in certain regions by a modified monsoon effect. In addition, because of the shift in the sun and wind belts during the summer months, the humid subtropical climates are subject to tropical storms, some of which develop into hurricanes (or typhoons), especially in late summer. These three factors—the modified monsoon effect, convectional activity, and tropical storms—combine in most of these regions to produce a precipitation maximum in late summer. The climographs for New Orleans, Louisiana, and Brisbane, Australia, illustrate these effects (• Fig. 10.6).

A subtype of the humid subtropical climate is found most often on the Asian continent, where the monsoon effect is most pronounced because of the magnitude of the seasonal pressure changes over this immense landmass. There, the low-sun period, or winter season, is noticeably drier than the high-sun period. High pressure over the continent blocks the importation of moist air so that some months receive less than 3 centimeters (1.2 in.) of precipitation.

Temperatures in the humid subtropics are much like those of the Mediterranean regions. Annual ranges are similar, despite a greater variation among climatic stations in the humid subtropical climate, primarily because the climate covers a far larger land area. Mediterranean stations record higher summer day-

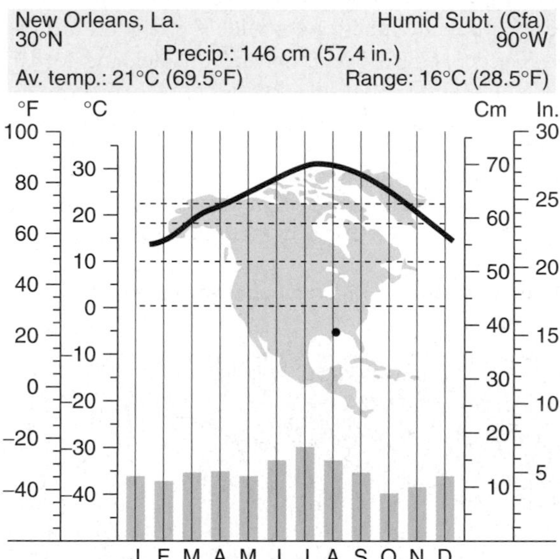

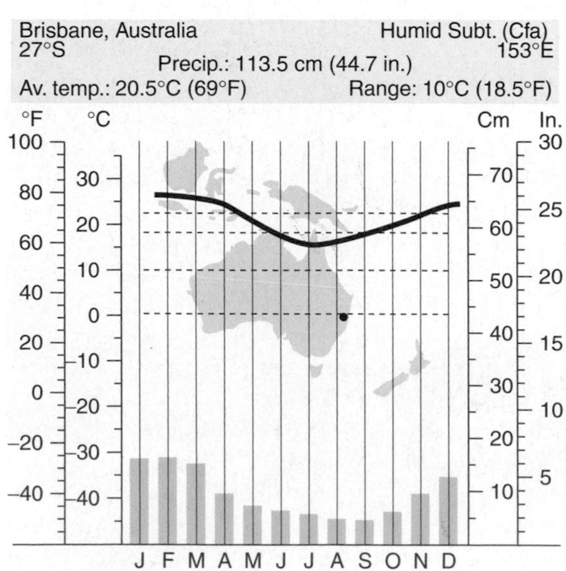

• FIGURE 10.6
Climographs for humid subtropical climate stations.
What hemispheric characteristics are shown in these graphs?

time temperatures, but summer months in both climates average around 25°C (77°F), increasing to as much as 32°C (90°F) as maritime influence decreases inland. Winter months in both climates average around 7°C–14°C (45°F–57°F). Frost is a similar problem. The long growing season in the warmest humid subtropical regions enables farmers to grow such delicate crops as oranges, grapefruit, and lemons, but, as in the Mediterranean climates, farmers must be prepared with various means to protect their more sensitive crops from the danger of freezing. The growers of citrus crops in Florida are concentrated in the Central Lake District to take advantage of the moderating influence of nearby bodies of water.

Variation in humidity greatly affects the effective temperatures—the temperatures we feel—in the humid subtropical and Mediterranean climates. The summer temperatures in humid subtropical regions feel far warmer than they are because of the high humidity there. In fact, summers in this climate are oppressively hot, sultry, and uncomfortable. Nor is there the relief of lower night temperatures, as in the Mediterranean regions. The high humidity of the humid subtropical climate prevents much radiative loss of heat at night. Consequently, the air remains hot and sticky. Diurnal temperature ranges, in winter as well as in summer, are far smaller in the humid subtropical than in the drier Mediterranean regions. Despite the relatively mild temperatures, humid subtropical winters seem cold and damp, again because of the high humidity.

A Productive Climate Vegetation generally thrives in humid subtropical regions, with their abundant rainfall, high temperatures, and long growing season. The wetter portions support forests of broad-leaf deciduous trees, pine forests on sandy soils, and mixed forests (● Fig. 10.7). In the drier interiors near the steppe regions, forests give way to grasslands, which require less moisture. There is an abundant and varied fauna. A few of the common species are deer, bears, foxes, rabbits, squirrels, opossums, raccoons, skunks, and birds of many sizes and species. Bird life in lake and marsh areas is incredibly rich. Alligators inhabit the American swamps in North Carolina, Georgia, and Florida.

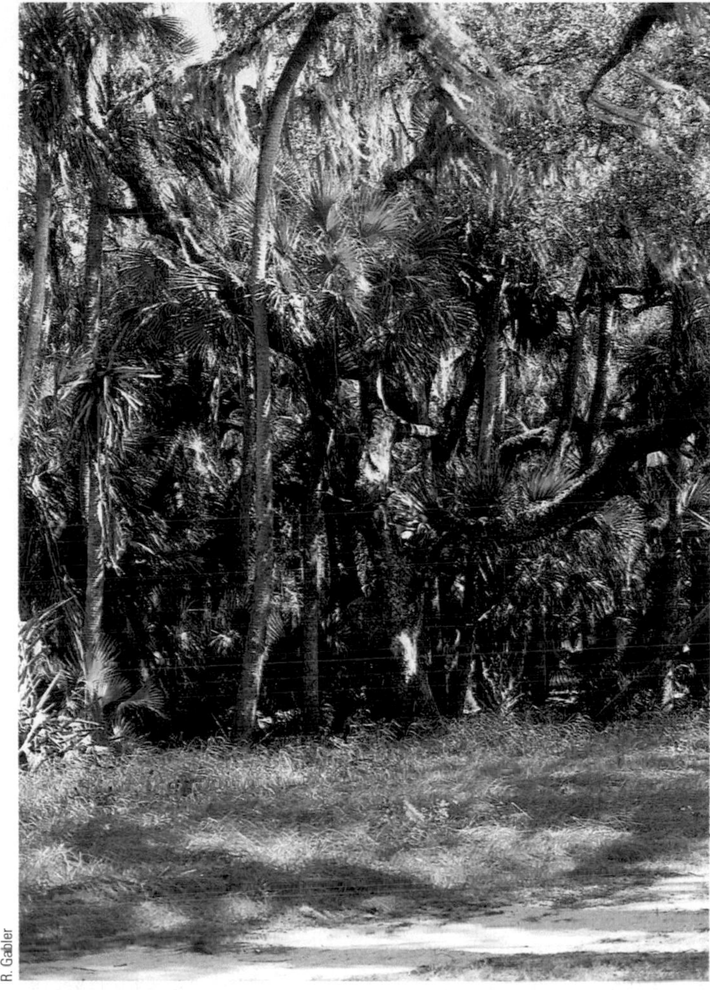
R. Gabler

● FIGURE 10.7
Forest vegetation similar to this in Myakka State Park originally covered much of humid subtropical central Florida.
How has the physical landscape of central Florida been changed by human occupancy?

As in the tropics, soils tend to have limited fertility because of rapid removal of soluble nutrients. However, there are exceptions in drier grassland areas, such as the pampas of Argentina and Uruguay, which constitute South America's "bread basket." Whatever the soil resource, the humid subtropical regions are of enormous agricultural value because of their favorable temperature and moisture characteristics. They have been used intensively for both subsistence crops, such as rice and wheat in Asia, and commercial crops, such as cotton and tobacco in the United States. When we consider that this climate (with its monsoon phase) is characteristic of south China as well as of the most densely populated portions of both India and Japan, we realize that this climate regime contains and feeds far more human beings than any other type (● Fig. 10.8). The care with which agriculture has been practiced and the soil resource conserved over thousands of years of intensive use in eastern Asia is in sharp contrast to the agricultural exploitation of the past 200 years in the corresponding area of the United States. The traditional system of cotton and tobacco farming, in particular, devastated the land by exhausting the soil and triggering massive sheet

● FIGURE 10.8
The terraced fields in this humid subtropical climate region near Wakayama, Japan, are ideally suited for rice production.
Why is rice a preferred crop in Japan?

©Robert Essel, NYC/CORBIS

and gully erosion. Over much of the old cotton and tobacco belt, extending from the Carolinas through Georgia and Alabama to the Mississippi delta, all topsoil or a significant part of it has been lost, and we see only the red clay subsoil and occasionally bare rock where crops formerly flourished. In the remaining areas, practices have had to change to conserve the soil that is left. Heavy applications of fertilizer, scientific crop rotation, and careful tilling of the land are now the rule.

Where forests still form the major natural vegetation, forest products, such as lumber, pulpwood, and turpentine, are important commercially (• Fig. 10.9). The long-leaf and slash pines of the southeastern United States are a source of lumber, as well as the resinous products of the pine tree (pitch, tar, resin, and turpentine). The absence of temperature and moisture limitations strongly favors forest growth. In Georgia, for example, trees may grow two to four times faster than in colder regions such as New England. This means that trees can be planted and harvested in much less time than in cooler forested regions, offering distinct commercial advantages.

The fact that living things thrive in the humid subtropical climate also presents certain problems, because parasites and disease-carrying insects also thrive along with other lifeforms. African "killer bees," for example, have migrated through Mexico and now pose a new threat to humans and livestock in the southern United States.

Agriculture and lumbering are not the only important industries in the humid subtropical climates. In the southeastern United States, livestock raising has been greatly increasing in importance. Despite the commercial advantages of this climate, people often find it an uncomfortable one in which to live. The development and spread of air-conditioning helps mitigate this problem. Where the ocean offers relief from the summer heat, as in Florida, the humid subtropical climate is an attractive recreation and retirement region. The beauty of its more unusual features, such as its cypress swamps and forests draped with Spanish moss, has to be experienced to be fully appreciated.

• FIGURE 10.9
A commercial tree farm in Alabama. Note that the pine trees are planted in orderly rows to expedite cultivation and harvest.
Why are tree farms common in the U.S. Southeast and Pacific Northwest but not in New England or the upper Midwest?

USDA/NRCS Alabama

Marine West Coast Climate

Proximity to the sea and prevailing onshore winds make the **marine west coast climate** (*Cfb, Cfc*) one of the most temperate in the world. Thus, it is sometimes known as the *temperate oceanic climate.* Found in those middle-latitude regions (between 40° and 65°) that are continuously influenced by the westerlies, the marine west coast climate receives ample precipitation throughout the year. However, unlike the humid subtropical climate just discussed, it has mild to cool summers. The climograph for Bordeaux, France, in • Figure 10.10 is representative of the mild-summer marine west coast climate (*Cfb*), and the climograph for Reykjavik, Iceland, represents the cool-summer variety (*Cfc*).

• FIGURE 10.10
Climographs for marine west coast climate stations.
How do you explain why Reykjavik has a lower temperature range than Bordeaux?

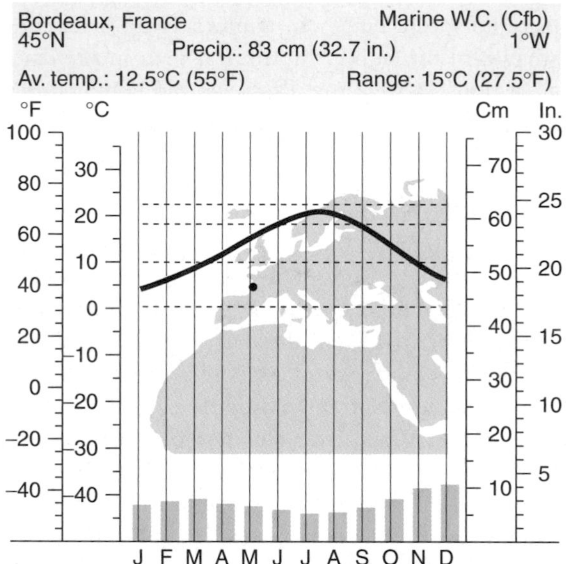

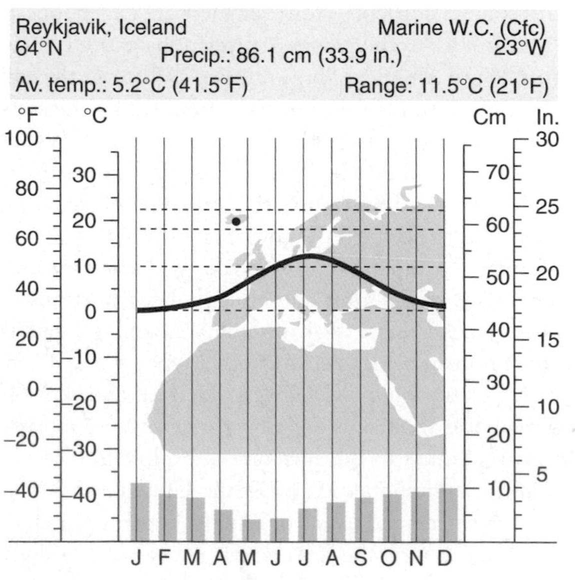

Oceanic Influences As they travel onshore, the westerlies carry with them the moderating marine influence on temperature, as well as much moisture. In addition, warm ocean currents, such as the North Atlantic Drift, bathe some of the coastal lands in the latitudes of the marine west coast climate, further moderating climatic conditions and accentuating humidity. This latter influence is particularly noticeable in Europe where the marine west coast climate extends along the coast of Norway to beyond the Arctic Circle (see again Fig. 10.1).

In this climate zone, the marine influence is so strong that temperatures decrease little with poleward movement. Thus, the influence of the oceans is even stronger than latitude in determining these temperatures. This is obvious when we examine isotherms in the areas of marine west coast climates on a map of world temperatures (see again Figs. 4.27 and 4.28). Wherever the marine west coast climate prevails, the isotherms swing poleward, parallel to the coast, clearly demonstrating the dominant *marine influence*.

Another result of the ocean's moderating effect is that the annual temperature ranges in the marine west coast climates are relatively small, considering the latitude. For an illustration of this, compare the monthly temperature graphs for Portland, Oregon, and Eau Claire, Wisconsin (• Fig. 10.11). Though these two cities are at the same latitude, the annual range at Portland is 15.5°C (28°F), while at Eau Claire it is 31.5°C (57°F). The moderating effect of the ocean on Portland's temperatures is clearly contrasted to the effect of *continentality* on the temperatures in Eau Claire.

Diurnal temperature ranges are also smaller than they are in other climate regions at similar latitudes and in more arid climates. Heavy cloud cover and high humidity in both summer and winter diminish daytime heating and prevent much radiational cooling at night. Consequently, the difference between the daily maximum and minimum temperatures is small.

Of course, these climographs and climate statistics represent averages, which can be misleading. Marine west coast climate regions experience the unpredictable weather conditions associated with the polar front. Occasional invasions of a tropical air mass in summer or a polar air mass in winter can move against the general westerly flow of air in these latitudes and produce surprising results. For example, under just such weather conditions, temperatures in Seattle, Washington, have reached a high of 38°C (100°F) and a low of –19°C (–3°F).

Despite the insulating effect of cloudy skies and high-moisture content of the air, slowing heat loss at night, frost is a significant factor in the marine west coast climate. It occurs more often, may last longer, and is more intense than in other mesothermal regions. The growing season is limited to 8 months or less, but even during the months when freezing temperatures may occur, only half the nights or fewer may experience them. The possibility of frost and the frequency of its occurrence increase inland far more rapidly than they do poleward, once more illustrating the importance of the marine influence.

As final evidence of oceanic influences, study the distribution of the marine west coast climate in Figure 10.1. Where mountain barriers prevent the movement of maritime air inland, this climate is restricted to a narrow coastal strip, as in the Pacific Northwest of North America and in Chile. Where the land is surrounded by water, as in New Zealand, or where the air masses move across broad plains, as in much of northwestern Europe, the climate extends well into the interior of the landmass.

• FIGURE 10.11
Effect of maritime influence on climates of two stations at the same latitude. Portland, Oregon, exemplifies the maritime influences dominating marine west coast climates. Eau Claire, Wisconsin, shows the effect of location in the continental interior. The difference in temperature ranges for the two stations is significant, but note also the interesting differences in precipitation distribution.
How do you explain these differences?

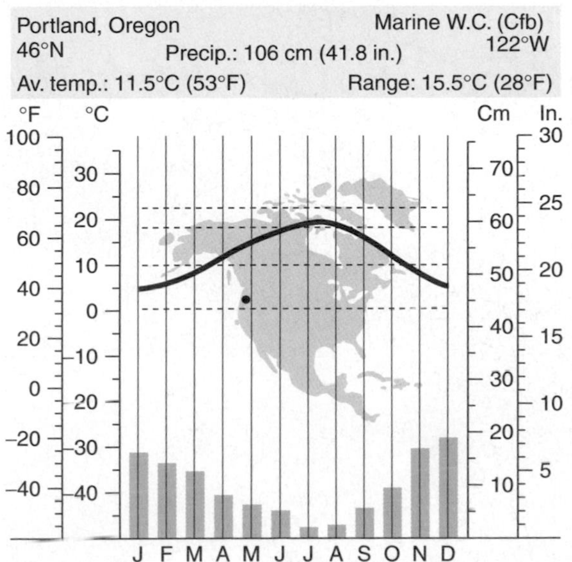

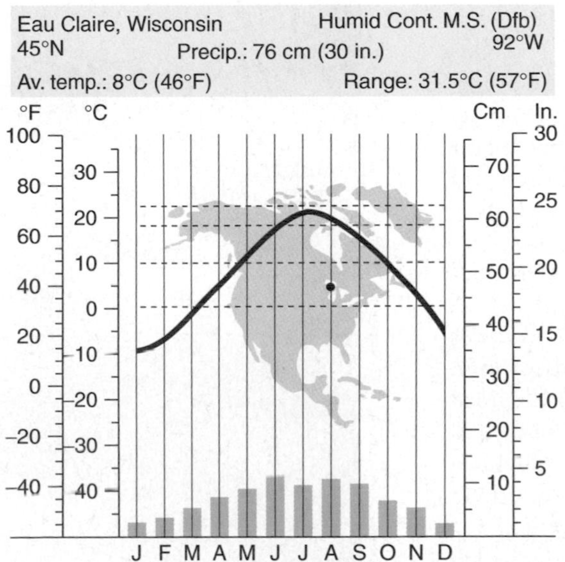

Clouds and Precipitation The marine west coast has a justly deserved reputation as one of the cloudiest, foggiest, rainiest, and stormiest climates in the world. This is particularly true during the winter season. Rain or drizzle may last off and on for days, though the amount of rain received is small for the number of rainy days recorded. Even when rain is not falling, the weather is apt to be cloudy or foggy. Advection fog may be especially common and long lasting in the winter months when air masses pass over warm ocean currents and pick up considerable moisture, which is then condensed as a fog when the air masses move over colder land. The cyclonic storms and frontal systems are also strongest in the winter when the subtropical highs have shifted equatorward. Conspicuous winter maximums in rainfall occur near the coasts and near boundaries with the Mediterranean climate. However, farther inland a summer maximum may occur.

Though all parts of this climate type receive ample precipitation, there is much greater site-to-site variation in precipitation averages than in temperature statistics. Precipitation tends to decrease very gradually as one moves inland, away from the oceanic source of moisture. It also decreases equatorward, especially during summer months, as the influence of the subtropical highs increases and the influence of the westerlies decreases. This can bring about periods of beautiful, clear weather, something rarely associated with this climate but not uncommon in our Pacific Northwest.

The most important factor in the amount of precipitation received is local topography. When a mountain barrier such as the Cascades in the Pacific Northwest or the Andes in Chile parallels the coast, abundant precipitation, both cyclonic and orographic, falls on the windward side of the mountains. Valdivia, Chile, located windward of the Andes, receives an average of 267 centimeters (105 in.) of precipitation a year. A similar location in Canada, Henderson Lake, British Columbia, averages 666 centimeters (262 in.) of rain a year, the highest figure for the entire North American continent. During the colder Pleistocene Epoch, these high precipitation amounts, falling largely as winter snow, produced large mountain glaciers. In many cases, these came down to the sea, excavating deep troughs that now appear as elongated inlets or fjords. Fjord coasts are present in Norway, British Columbia, Chile, and New Zealand—all areas of marine west coast climates today (● Fig. 10.12). In contrast, where there are lowlands and no major landforms of high elevation, precipitation is spread more evenly over a wide area, and the amount received at individual stations is more moderate, around 50–75 centimeters (20–30 in.) annually. This is the situation in much of the Northern European Plain, extending from western France to eastern Poland.

Two aspects of precipitation are directly related to the moderate temperatures of this climate regime. Snow falls infrequently and, when it does, it melts or turns to slush as soon as it hits the ground. Snow is especially rare in lowland regions of this climate zone. Paris averages only 14 snow days a year; London, 13; and Seattle, 10. In addition, thunderstorms and convectional showers are uncommon although they occur occasionally. Even in summer, surface heating is rarely sufficient to produce the towering cumulonimbus clouds.

Resource Potential There is little doubt that this climate offers certain advantages for agriculture. The small annual temperature ranges, mild winters, long growing seasons, and abundant precipitation all favor plant growth. Many crops, such as wheat, barley, and rye, can be grown farther poleward than in more continental regions. Although the soils common to these regions are not naturally rich in soluble nutrients, highly successful agriculture is possible with the application of natural or commercial fertilizers (● Fig. 10.13). Root crops (such as potatoes, beets, and turnips), deciduous fruits (such as apples and pears), berries, and grapes join the grains previously mentioned as important farm products. Grass in particular requires little sunshine, and pastures are always lush. The greenness of Ireland—the Emerald Isle—is evidence of these favorable conditions, as is the abundance of herds of beef and dairy cattle.

The magnificent forests that form the natural vegetation of the marine west coast regions have been a readily available resource.

● FIGURE 10.12
The scenic fjords of coastal Norway, shown here, were produced by glacial erosion during the Pleistocene ice advance.
In what other areas of the world are fjords common?

©Iconotec Royalty Free Photograph/Fotosearch

R. Gabler

● **FIGURE 10.13**
Reliable precipitation makes a diversified type of agriculture possible in the marine west coast climate areas, with emphasis on grain, orchard crops, vineyards, vegetables, and dairying. The village in the photograph is Iphofen, Germany.
Although the climate is similar, why would a photo taken in a marine west coast agricultural region of the United States depict a scene that is significantly different from this one?

Some of the finest stands of commercial timber in the world are found along the Pacific coast of North America where pines, firs, and spruces abound, commonly exceeding 30 meters (100 ft) in height. Europe and the British Isles were once heavily forested, but most of those forests (even the famous Sherwood Forest of Robin Hood fame) have been cut down for building material and have been replaced by agricultural lands and urbanization.

Humid Microthermal Climate Regions

Our definition of humid microthermal includes temperatures high enough during part of the year to have a recognizable summer and cold enough 6 months later to have a distinct winter. In between are two periods, called spring and fall, when all life, and especially vegetation, makes preparations for the temperature extremes. Thus, in this section, we talk about climate regions that clearly display four readily identifiable seasons.

Seasonality is not the only reason we often use the word *variable* when describing the humid microthermal climates. As ● Figure 10.14 indicates, these climates are generally located between 35°N and 75°N on the North American and Eurasian landmasses. (Since there are no large landmasses at similar latitudes in the Southern Hemisphere, these climate types exist only in the Northern Hemisphere.) Thus, they share the westerlies and the storms of the polar front with the marine west coast climate. However, their position in the continental interiors and in high latitudes prevents them from experiencing the moderating influence of the oceans. In fact, the dominance of continentality in these climates is best demonstrated by the fact that they do not exist in the Southern Hemisphere where there are no large landmasses in the appropriate latitudes.

The recognition of three separate microthermal climates is based mainly on latitude and the resulting differences in the length and severity of the seasons (Table 10.2). Winters tend to be longer and colder toward the poleward margins because of latitude and toward interiors throughout the microthermal climates because of the continental influence. Summers inland are also inclined to be hotter, but they become progressively shorter as the winter season lengthens poleward. Thus, the three microthermal climates can be defined as humid continental, hot summer (*Dfa, Dwa*); humid continental, mild summer (*Dfb, Dwb*); and subarctic (*Dfc, Dfd, Dwc, Dwd*), with a cool summer and, in extreme cases, a long, bitterly cold winter.

All microthermal climates have several features in common. By definition, they all experience a surplus of precipitation over potential evapotranspiration, and they have year-round precipitation. An exception to this rule lies in an area of Asia where the Siberian High causes winter droughts to occur. The greater frequency of maritime tropical air masses in summer and continental polar air masses in winter, combined with the monsoonal effect and strong summer convection, produce a precipitation maximum in the summer. Although the length of time that snow remains

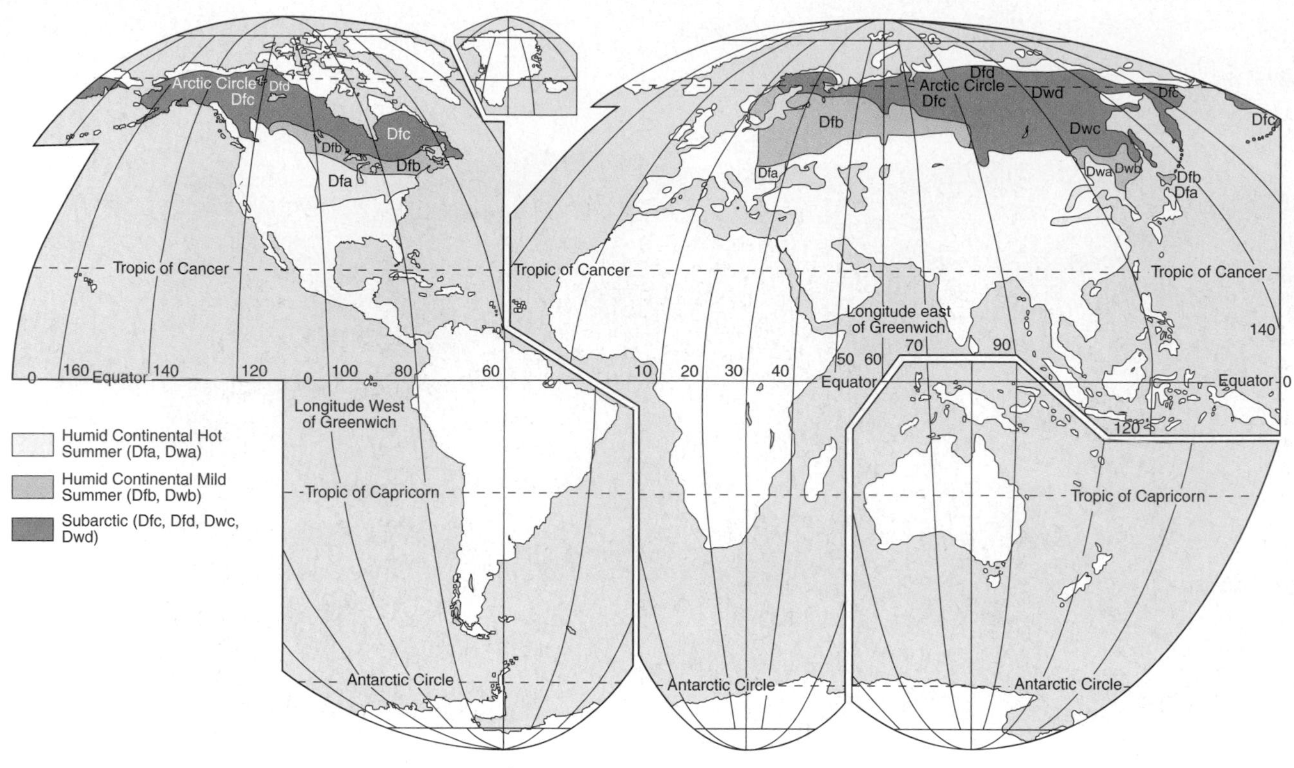

• FIGURE 10.14
Index map of humid microthermal climates.

• FIGURE 10.15
Map of the contiguous United States, showing average annual number of days of snow cover.
What areas of the United States average the greatest number of days of snow cover?

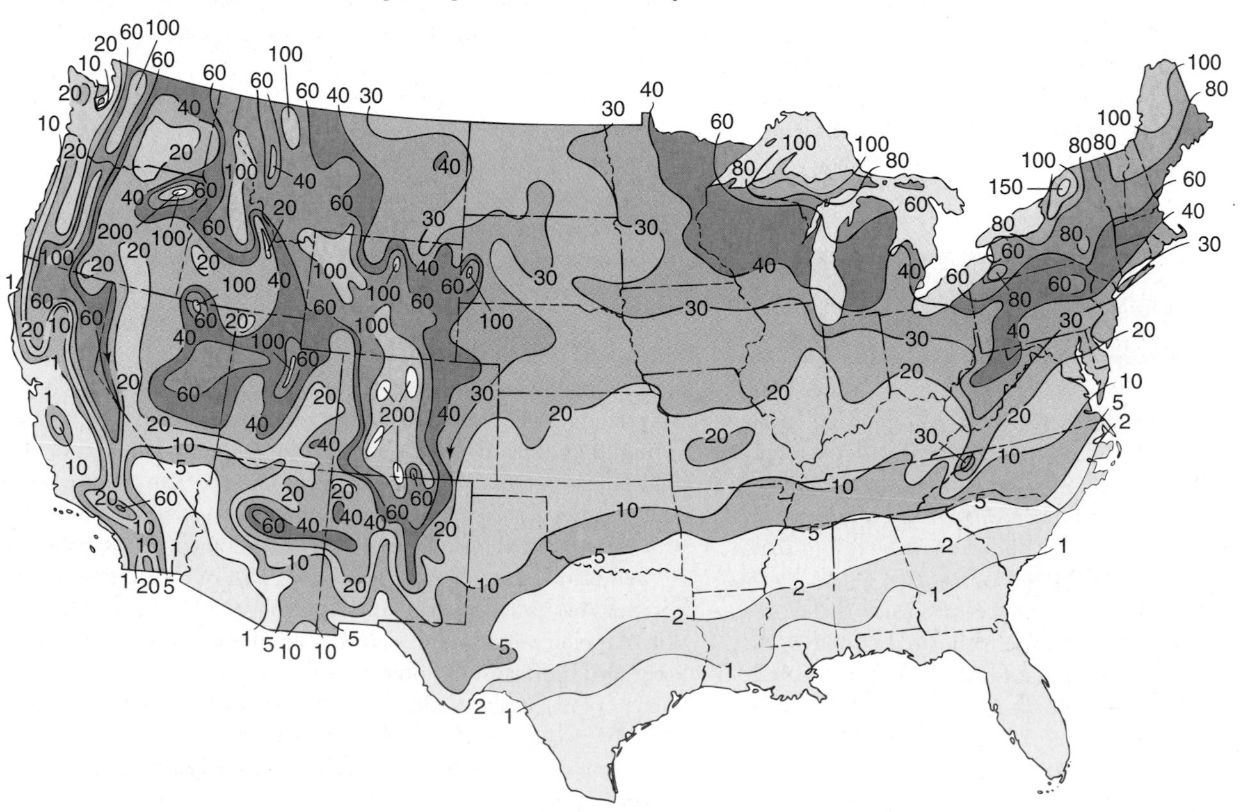

TABLE 10.2
The Microthermal Climates

Name and Description	Controlling Factors	Geographic Distribution	Distinguishing Characteristics	Related Features
Humid Continental, Hot Summer				
Warmest month above 10°C (50°F); coldest month below 0°C (32°F); hot summers; usually year-round precipitation, winter drought (Asia)	Location in the lower middle latitudes (35°–45°); cyclonic storms along the polar front; prevailing westerlies; continentality; polar anticyclone in winter (Asia)	Eastern and midwestern U.S. from Atlantic coast to 100°W longitude; east central Europe; northern and northeast (Manchuria) China, northern Korea, and Honshu (Japan)	Hot, often humid summers; occasional winter cold waves; rather large annual temperature ranges; weather variability; precipitation 50–115 cm (20–45 in.), decreasing inland and poleward; 140- to 200-day growing season	Broad-leaf deciduous and mixed forest; moderately fertile soils with fertilization in wetter areas; highly fertile grassland and prairie soils in drier areas; "corn belt," soybeans, hay, oats, winter wheat
Humid Continental, Mild Summer				
Warmest month above 10°C (50°F); coldest month below 0°C (32°F); mild summers; usually year-round precipitation, winter drought (Asia)	Location in the middle latitudes (45°–55°); cyclonic storms along the polar front; prevailing westerlies; continentality; polar anticyclone in winter (Asia)	New England, the Great Lakes region, and south central Canada; southeastern Scandinavia; eastern Europe, west central Asia; eastern Manchuria (China) and Hokkaido (Japan)	Moderate summers; long winters with frequent spells of clear, cold weather; large annual temperature ranges; variable weather with less total precipitation than farther south; 90- to 130-day growing season	Mixed or coniferous forest; moderately fertile soils with fertilization in wetter areas; highly fertile grassland and prairie soils in drier areas; spring wheat, corn for fodder, root crops, hay, dairying
Subarctic				
Warmest month above 10°C (50°F); coldest month below 0°C (32°F); cool summers, cold winters poleward; usually year-round precipitation, winter drought (Asia)	Location in the higher middle latitudes (50°–70°); westerlies in summer, strong polar anticyclone in winter (Asia); occasional cyclonic storms; extreme continentality	Northern North America from Newfoundland to Alaska; northern Eurasia from Scandinavia through most of Siberia to the Bering Sea and the Sea of Okhotsk	Brief, cool summers; long, bitterly cold winters; largest annual temperature ranges; lowest temperatures outside Antarctica; low precipitation, 20–50 cm (10–20 in.); unreliable 50- to 80-day growing season; permafrost common	

on the ground increases poleward and toward the continental interior (● Fig. 10.15), all three microthermal climates experience significant snow cover. This decreases the effectiveness of insolation and helps explain their cold winter temperatures. Finally, the unpredictable and variable nature of the weather is especially apparent in the humid microthermal climates.

With these generalizations in mind, let's compare the microthermal climates with the mesothermal climates that we have just examined. Regions with microthermal climates have more severe winters, a lasting snow cover, shorter summers, shorter growing seasons, shorter frost-free seasons, a truer four-season development, lower nighttime temperatures, greater average annual temperature ranges, lower relative humidity, and much more variable weather than do the mesothermal climate regions.

Humid Continental, Hot-Summer Climate

Unlike the other two microthermal climates, the **humid continental, hot-summer climate** (*Dfa, Dwa*) is relatively limited in its distribution on the Eurasian landmass (see Table 10.2). This is unfortunate for the people of Europe and Asia because it has by far the greatest agricultural potential and is the most productive of the microthermal climates. In the United States, this climate is distributed over a wide area that begins with the eastern seaboard of New York, New Jersey, and southern New England and stretches continuously across the heartland of the eastern United States to encompass much of the American Midwest. It is one of the most densely populated, highly developed, and agriculturally productive regions in the world.

In terms of environmental conditions, the hot-summer variety of microthermal climate has some obvious advantages over its poleward counterparts. Its higher summer temperatures and longer growing season permit farmers to produce a wide variety of crops. Those lands within the hot-summer region that were covered by ice sheets are far enough equatorward that there has been sufficient time for most negative effects of continental glaciation to be removed, and primarily positive effects remain. Soils are inclined to be more fertile, especially under forest cover, where the typical soil-forming processes are not as extreme and where deciduous trees are more common than the acid-associated pine. Of course, some advantages are matched by liabilities. The lower fuel bills of winter in the humid continental, hot-summer climate are often more than offset by the cost of air-conditioning during the long, hot summers not found in other microthermal climates.

Internal Variations From place to place within the humid continental, hot-summer climate, there are significant differences in temperature characteristics. The length of the growing season is directly related to latitude, varying from 200 days equatorward to as little as 140 days along poleward margins of the climate. In addition, the degree of continentality can have an effect on both summer and winter temperatures and, as a result, on temperature range. Ranges are consistently large, but they become progressively larger toward continental interiors. Especially near the coasts in this climate region, temperatures may be modified by a slight marine influence so that temperatures are milder, in both summer and winter, than those at inland locations at comparable latitudes. Large lakes may cause a similar effect. Even the size of the continent exerts an influence. Galesburg, Illinois, a typical station in the United States, has a significantly lower temperature range than Shenyang, northeast China (Manchuria), which is located at almost the same latitude but which experiences the greater seasonal contrasts of the Eurasian landmass (• Fig. 10.16).

The amount and distribution of precipitation are also variable from station to station. The total precipitation received decreases both poleward and inland (• Fig. 10.17). A move in either direction is a move away from the source regions of warm maritime air masses that provide much of the moisture for cyclonic storms and convectional showers. This decrease can be seen in the average annual precipitation figures for the following cities, all at a latitude of about 40°N: New York (longitude 74°W), 115 centimeters (45 in.); Indianapolis, Indiana (86°W), 100 centimeters (40 in.); Hannibal, Missouri (92°W), 90 centimeters (35 in.); and Grand Island, Nebraska (98°W), 60 centimeters (24 in.). Most stations have a precipitation maximum in summer when the warm, moist air masses dominate. In certain regions of Asia, not only does a summer maximum of precipitation exist, but also the monsoon circulation inhibits winter precipitation to such a degree that they experience winter drought (see again Shenyang, Fig. 10.16).

As might be expected, vegetation and soils vary with the climatic elements, especially precipitation. In the wetter regions, forests and forest soils predominate. At one time, in certain sections of the American Midwest, tall prairie grasses grew where precipitation was insufficient to support forests; in all the drier portions of the climate, grasslands are the natural vegetation. The soils that developed under these grasslands are among the richest in the world.

Seasonal Changes The four seasons are highly developed in the humid microthermal, hot summer climate. Each is distinct from the other three and has a character all its own. The winter is cold and often snowy; the spring is warmer, with frequent showers that produce flowers, budding leaves, and green grasses; the hot, humid summer brings occasional violent thunderstorms; and the fall has periods of both clear and rainy weather, with mild days and frosty nights in which the green leaves of summer turn to colorful reds, oranges, yellows, and browns before falling to the ground.

• FIGURE 10.16
Climographs for humid continental, hot-summer climate stations.
What are the reasons for the differences in temperature and precipitation between the two stations?

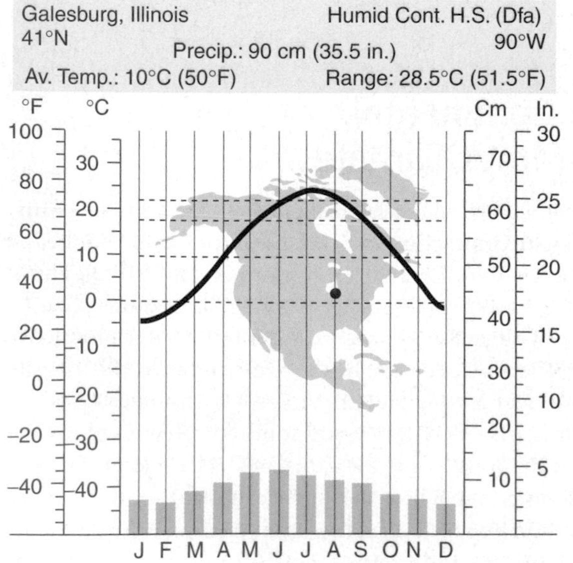

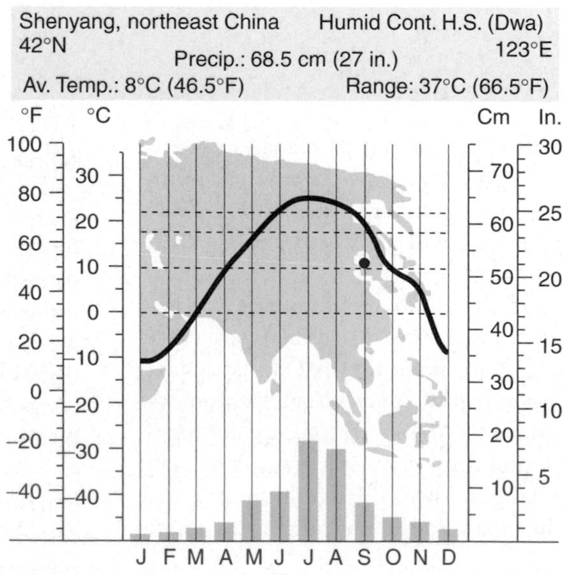

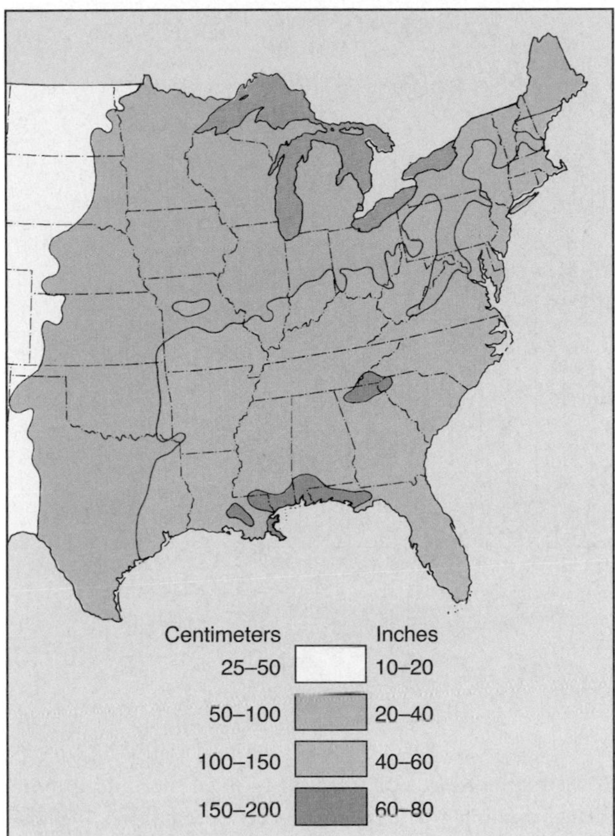

● **FIGURE 10.17**
Decreases of precipitation inland from coastal regions are clearly evident in this map of the eastern United States.
Why does precipitation decrease inland and poleward?

Of course, the most significant differences are between summer and winter. In all regions, the summers are long, humid, and hot. The centers of the migrating low pressure systems usually pass poleward of these regions, and they are dominated by tropical maritime air. So-called hot spells can go on day and night for a week or more, with only temporary relief available from convectional thunderstorms or an occasional cold (cool) front. Asia, in particular, experiences the heavy summer precipitation associated with the monsoonal effect. Conditions are usually ideal for vigorous vegetative growth. The summer heat and humidity are also the ideal formula for insects; mosquitoes, flies, gnats, and bugs of all kinds abound.

Winters are not as severe as those farther poleward, but average January temperatures are usually between −5°C and 0°C (23°F–32°F) or below. Once again, the averages tell only part of the story. There is invariably a prolonged invasion of cold, dry, arctic air once or twice during the winter. This often occurs just after a storm has passed and the ground is covered with snow. The sky remains clear and blue for days at a time; the temperatures will stay near −18°C (0°F) and may occasionally dip to 30°C or 35°C (20°F or 30°F) *below zero* at night. The ground remains frozen for long periods, and snow cover may be present for several days, even weeks, at a time. However, these characteristics do not last continuously because the greatest frequency of cyclones occurs during winter and sudden weather contrasts are common. Cold air precedes warmer air, and thaw follows freeze. Vegetation remains dormant throughout the winter season but bursts into life again with the return of consistently warmer temperatures. Throughout its early growth, vegetation is in constant danger of late-spring frost.

As should be apparent from this description, the atmospheric changes within seasons are just as significant as those between seasons. The humid continental, hot-summer climate is the classic example of variable middle-latitude weather. This is the domain of the polar front. Cyclonic storms are born as tropical air masses move northward and confront polar air masses migrating to the south. The daily weather in these regions is dominated by days of stormy frontal activity followed by the clear conditions of a following anticyclone. Above the land is a battlefield in which storms mark the struggles of air masses for dominance, and as in battle, the conflict occurs at the *front*. The general circulation of the atmosphere in these latitudes continuously carries the cyclones and anticyclones toward the east along the polar front. When the polar front is most directly over these regions, as it is in winter and spring, one storm and its associated fronts seem to follow directly behind another with such speed and regularity that the only safe weather prediction is that the weather will change (see again Chapter 7).

Humid Continental, Mild-Summer Climate

If you review the relative distributions of the humid continental, hot-summer and mild-summer climates in Figure 10.14, the close relationship between the two is unmistakable. Where one is found, the other is found as well; in each situation, the mild-summer climate invariably lies adjacent to and poleward from the hot-summer climate.

In most instances, the **humid continental, mild-summer climate** (*Dfb, Dwb*) is a more continental or severe-winter version of its equatorward counterpart. It is characterized by distinct seasonality. There is significant climatic variation, particularly with respect to precipitation, from place to place within the climate. Variable weather is the rule, and storms along the polar front provide most of the precipitation within this climate type. Of course, there are differences between the neighboring climates. These are especially apparent when we examine certain aspects of temperature, growing season, vegetation, and human activity.

Mild-Summer–Hot-Summer Comparison In the microthermal climates, precipitation tends to decrease poleward; therefore, the humid continental, mild-summer climate tends to have less precipitation than the hot-summer regions closer to the equator. Precipitation continues to decrease throughout this climate type toward the poleward margins and from the coasts toward the arid continental interiors. As in its hot-summer counterpart, the monsoon effect in the mild-summer climate is strong enough in Asia to produce a dry-winter season (see Vladivostok, ● Fig. 10.18).

Winters in the mild-summer climate are more severe and longer than in its neighbor to the south. Summers, on the other hand, are not as long or hot. The combination of more severe winters and shorter summers makes for a growing season of

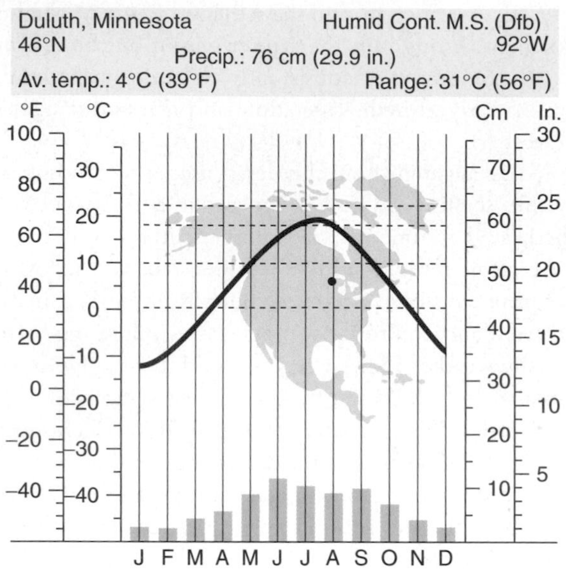

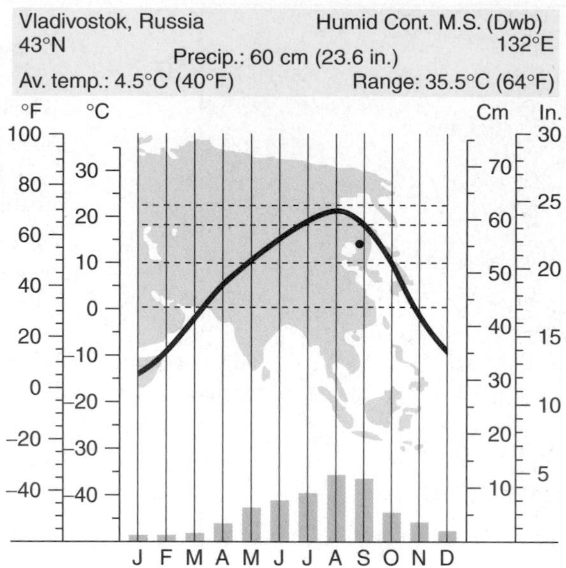

• FIGURE 10.18

Climographs for humid continental, mild-summer climate stations.

What characteristics of these climographs distinguish them from the climographs of Figure 10.16?

between 90 and 130 days, which is 1–3 months shorter than in the hot-summer climate. In addition, although overall precipitation totals—50 to 100 centimeters (20–40 in.)—are generally lower, snowfall is greater, and snow cover is both thicker and longer lasting (• Fig. 10.19).

The humid continental, mild-summer regions exhibit seasonal changes just as clearly as the hot-summer regions. Annual temperature ranges are generally larger. Vigorous polar and tropical air mass interaction makes weather change a common occurrence. However, the more poleward position of the mild summer climate brings about a greater dominance of the colder air masses and explains why temperature variability is not as abrupt or as great as it is farther south. Under normal conditions, tropical air is strongly modified by the time it reaches mild-summer regions; even in the high-sun season, intrusions of warm, humid air rarely last more than a few days at a time. By contrast, winter invasions of cold arctic air periodically bring several successive days or weeks of clear skies and frigid temperatures.

As in the hot-summer climate, the wetter regions of the mild-summer climate are associated with natural forest vegetation. However, many trees common in the hot-summer climate, such as oaks, hickories, and maples, find it difficult to compete with firs, pines, and spruces toward the colder, polar margins of these regions. Fortunately for agriculture, northward extensions of the grasslands and the rich soils that accompany them in the hot-summer regions are found in the drier portions of the mild-summer climate.

• FIGURE 10.19

People, animals, and plants living in the humid continental, mild-summer regions have learned to cope with abundant snow, which may be present continuously for long periods of time. Here is a typical winter scene near Buffalo, New York.

The snow presents a variety of problems for some people living in humid continental, mild-summer regions. Can you name a few?

©Annie Griffiths Belt/CORBIS

Human Activity in the Humid Continental Climates

Perhaps the greatest contrast between the hot-summer and mild-summer humid continental regions is exhibited in agriculture. Despite the unpredictability of the weather, the humid continental hot-summer agricultural regions are among the finest in the world. The favorable combination of long, hot summers, ample rainfall, and highly fertile soils has made the American Midwest a leading producer of corn, beef cattle, and hogs. Soybeans, which are native to similar climate regions in northern China, are now second to corn throughout the Midwest as feed for animals and as a raw material for the food-processing, plastics, and vegetable oil industries. Wheat, barley, and other grains are especially important in European and Asian regions, and winters are sufficiently mild that fall-sown varieties can be raised in the United States. In the mild-summer climate, on the other hand, a shorter growing season imposes certain limitations on agriculture and restricts the crops that can be grown. Farmers rely more on quick-ripening varieties, grazing animals, orchard products, and root crops. Dairy products—milk, cheese, butter, cream—are mainstays in the economies of Wisconsin, New York, and northern New England. The moderating effect of the Great Lakes or other water bodies permits the growth of deciduous fruits, such as apples, plums, and cherries.

The length of the growing season is the most obvious reason for the differences in agriculture between the two humid continental climates, but there is another climate-related reason as well. The great ice sheets of the Pleistocene Epoch have had significant but different effects on mild-summer and hot-summer regions, especially in North America. In the hot-summer regions, the ice sheets thinned and receded, releasing the enormous load of soil and solid rock debris they had stripped off the lands nearer to their centers of origin. The material was laid down in a blanket hundreds of feet thick in the areas of maximum glacial advance. As the ice retreated northward, less and less debris was deposited, much of it flushed away by meltwater streams. The more southerly, hot-summer region consequently has an undulating topography underlain by thick masses of glacial debris. The soils formed on this debris are well developed and fertile, and plant nutrients are more likely to be evenly distributed because steep slopes are lacking. The more northerly, mild-summer region, on the other hand, mainly shows the effects of glacial erosion (● Fig. 10.20). Rockbound lakes and marshy lowlands alternate with ice-scoured rock hills. Soils are either thin and stony or waterlogged. Because of its lower agricultural potential, large sections of this area remain in forest.

However, because of its wilderness character and the abundance of lakes in basins produced by glacial erosion, recreational possibilities in a mild-summer region far exceed those of a more subdued hot-summer region. Minnesota calls itself the "Land of 10,000 Lakes," and in New York and New England, lakes, rough mountains, and forest combine to produce some of the most spectacular scenery east of the Rocky Mountains (● Fig. 10.21).

Copyright and photograph by Dr. Parvinder S. Sethi

● FIGURE 10.20
Although the effect of glacial action farther south was to deposit material, here in New Hampshire, we see an area of glacial erosion—Newfound Lake, a former glacial trough.
How might this area be considered an economic resource?

● FIGURE 10.21
Gooseberry Falls in Minnesota are on the Gooseberry River, which flows into Lake Superior, the largest of the Great Lakes.
What can you say about the relative abundance of water resources of states like New York and Minnesota?

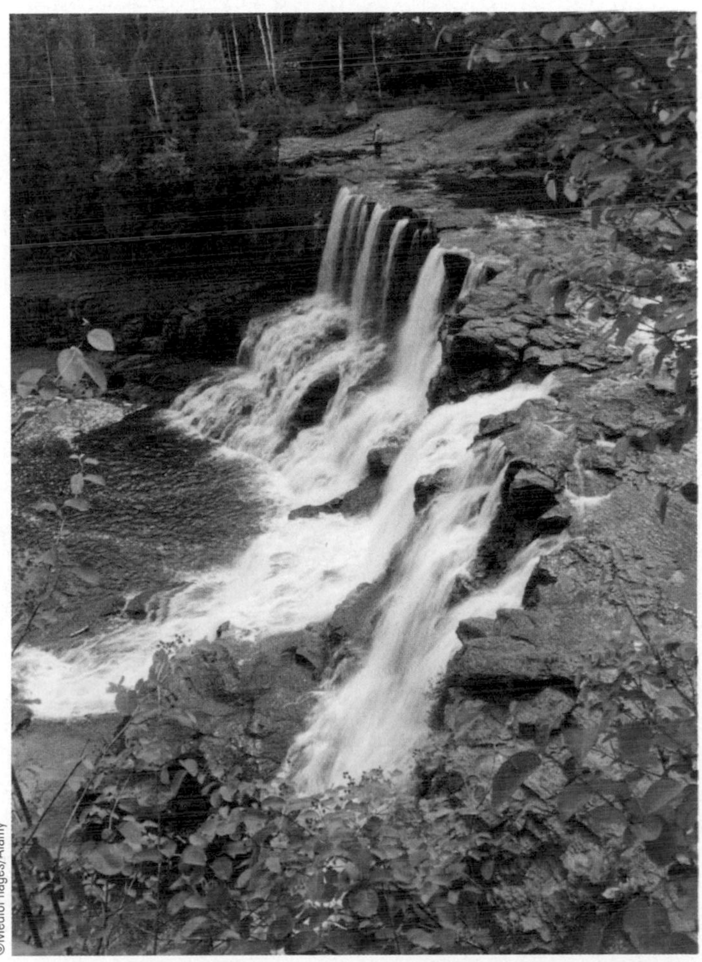

©Medioimages/Alamy

Subarctic Climate

The **subarctic climate** (*Dfc, Dfd, Dwc, Dwd*) is the farthest poleward and most extreme of the microthermal climates. By definition, it has at least 1 month with an average temperature above 10°C (50°F), and its poleward limit roughly coincides with the 10°C isotherm for the warmest month of the year. As you may recall from our earlier discussion of the simplified Köppen system, forests cannot survive where at least 1 month does not have an average temperature over 10°C. Thus, the poleward boundary of the subarctic climate is the poleward limit of forest growth as well.

As Figure 10.14 indicates, the subarctic climate, like the other microthermal climates, is found exclusively in the

GEOGRAPHY'S PHYSICAL SCIENCE PERSPECTIVE

Effective Temperatures

Effective temperatures (formerly known as *sensible temperatures*) are temperatures as they might be experienced by a person at rest, in ordinary clothing, in a motionless atmosphere. In other words, at any given time, how comfortable does a temperature feel to the individual experiencing it? This temperature value cannot be obtained by simply reading a thermometer.

Several factors come into play when considering effective temperatures. These factors can be divided into two categories: atmospheric factors and human factors. Of the atmospheric factors, four are most important. First is the actual temperature; obviously, thermometers will help distinguish between cold and warm days. Second is humidity; because the evaporation of sweat is a cooling process for the human body, humid days feel warmer than dry days. Third is wind speed; winds not only carry heat from the body but can also accelerate the evaporation of sweat. Fourth is the percentage of clear sky; shady areas are cooler than sunny areas.

Of the many human factors, the following stand out as important. First is respiration; breathing in a lungful of cold air

Heat index (or apparent temperature)

Air temperature (°F) \ Relative humidity (%)	0	5	10	15	20	25	30	35	40	45	50	55	60	65	70	75	80	85	90	95	100
140	125																				
135	120	128																			
130	117	122	131																		
125	111	116	123	131	141																
120	107	111	116	123	130	139	148														
115	103	107	111	115	120	127	135	143	151												
110	99	102	105	108	112	117	123	130	137	143	150										
105	95	97	100	102	105	109	113	118	123	129	135	142	149								
100	91	93	95	97	99	101	104	107	110	115	120	126	132	138	144						
95	87	88	90	91	93	94	96	98	101	104	107	110	114	119	124	130	136				
90	83	84	85	86	87	88	90	91	93	95	96	98	100	102	106	109	113	117	122		
85	78	79	80	81	82	83	84	85	86	87	88	89	90	91	93	95	97	99	102	105	108
80	73	74	75	76	77	77	78	79	79	80	81	81	82	83	85	86	86	87	88	89	91
75	69	69	70	71	72	72	73	73	74	74	75	75	76	76	77	77	78	78	79	79	80
70	64	64	65	65	66	66	67	67	68	68	69	69	70	70	70	70	71	71	71	71	72

A chart of effective temperatures can be used to determine the heat index, which combines temperature and humidity conditions to find an apparent temperature.

Northern Hemisphere. It covers vast areas of subpolar Eurasia and North America. Conditions vary widely over such great distances. Extremely severe winter regions are located along the polar margins or deep in the interior of the Asian landmass, and climate subtypes with winter drought are found in association with the Siberian high and its clear skies, bitter cold, and strong subsidence of air over interior Asia during winter. Other subarctic regions experience less severe winters or year-round precipitation.

Further study of Figure 10.14 suggests two additional observations. First, ocean currents tend to influence the distribution of the subarctic climate. Along the west coasts of the continents, especially in North America, the warm ocean currents modify temperatures sufficiently to permit the marine west coast climate to extend into latitudes normally occupied by the subarctic and to cause the subarctic to be found well beyond the Arctic Circle. Along east coasts, where cold ocean currents help reduce winter temperatures, the subarctic is situated farther south. Second, the development of the subarctic climate is not as extensive in North America as it is in Eurasia. This is because (1) the Eurasian continent is a larger landmass, which increases the effect of continentality, and (2) the large water surface of Hudson Bay in Canada provides a modifying marine influence inland, which tends to counter the effect of continentality there.

The Effects of High Latitude and Continentality Subarctic regions experience short, cool summers and long, bitterly cold winters (● Fig. 10.22). The rapid heating and cooling

will make one feel colder. Second is perspiration; the evaporative-cooling process is quite efficient for the human body, but it differs from one individual to another. Third is the amount of activity involved; physical work or playing a physical sport can heat the body rapidly. Fourth is the amount of exposed skin; tank tops versus sweatshirts can make a world of difference.

Effective temperatures are established by considering the interplay between these two sets of factors. For example, the well-known heat index (also called apparent temperature) takes into account the temperature and humidity of a summer day and calculates how it might feel. The equally well-known wind chill index considers both temperature and wind speed to establish how cold it might feel on a winter day. Keep in mind that these are only theoretical values. No one can predict exactly how comfortable a particular person will feel on a given day. However, these temperature indices can help guide us when dealing with seasonally extreme days.

Wind Chill Chart

Temperature (°F)

Wind (mph) / Calm	40	35	30	25	20	15	10	5	0	−5	−10	−15	−20	−25	−30	−35	−40	−45
5	36	31	25	19	13	7	1	-5	-11	-16	-22	-28	-34	-40	-46	-52	-57	-63
10	34	27	21	15	9	3	-4	-10	-16	-22	-28	-35	-41	-47	-53	-59	-66	-72
15	32	25	19	13	6	0	-7	-13	-19	-26	-32	-39	-45	-51	-58	-64	-71	-77
20	30	24	17	11	4	-2	-9	-15	-22	-29	-35	-42	-48	-55	-61	-68	-74	-81
25	29	23	16	9	3	-4	-11	-17	-24	-31	-37	-44	-51	-58	-64	-71	-78	-84
30	28	22	15	8	1	-5	-12	-19	-26	-33	-39	-46	-53	-60	-67	-73	-80	-87
35	28	21	14	7	0	-7	-14	-21	-27	-34	-41	-48	-55	-62	-69	-76	-82	-89
40	27	20	13	6	-1	-8	-15	-22	-29	-36	-43	-50	-57	-64	-71	-78	-84	-91
45	26	19	12	5	-2	-9	-16	-23	-30	-37	-44	-51	-58	-65	-72	-79	-86	-93
50	26	19	12	4	-3	-10	-17	-24	-31	-38	-45	-52	-60	-67	-74	-81	-88	-95
55	25	18	11	4	-3	-11	-18	-25	-32	-39	-46	-54	-61	-68	-75	-82	-89	-97
60	25	17	10	3	-4	-11	-19	-26	-33	-40	-48	-55	-62	-69	-76	-84	-91	-98

Frostbite occurs in 15 minutes or less

A chart of effective temperatures can be used to determine wind chill, which shows what the combination of cold temperatures and wind speed will make the temperature outside feel like.

associated with continental interiors in the higher latitudes allow little time for the in-between seasons of spring and fall. At Eagle, Alaska, a station in the Klondike region of the Yukon River Valley, the temperature climbs 8°C–10°C (15°F–20°F) per month as summer approaches and drops just as rapidly prior to the next winter season. At Verkhoyansk in Siberia, the change between the seasonal extremes is even more rapid, averaging 15°C–20°C (30°F–40°F) per month.

Because of the high latitudes of these regions, summer days are quite long, and nights are short. The noon sun is as high in the sky during a subarctic summer as during a subtropical winter. The combination of a moderately high angle of the sun's rays and many hours of daylight means that some subarctic locations receive as much insolation at the time of the summer solstice as the equator does. As a result, temperatures during the 1–3 months of the subarctic summer usually average 10°C–15°C (50°F–60°F), and on some days they may even approach 30°C (86°F). Thus, the brief summer in the subarctic climate is often pleasantly warm, even hot, on some days.

The winter season in the subarctic is bitter, intense, and lasts for as long as 8 months. Eagle, Alaska, has 8 months with average temperatures below freezing. In the Siberian subarctic, the January temperatures regularly *average* −40°C to −50°C (−40°F to −60°F). The coldest temperatures in the Northern Hemisphere—officially, −68°C (−90°F) at both Verkhoyansk and Oymyakon; unofficially, −78°C (−108°F) at Oymyakon—have been recorded there. In addition, the winter nights, with an average 18–20 hours of darkness extending well into one's

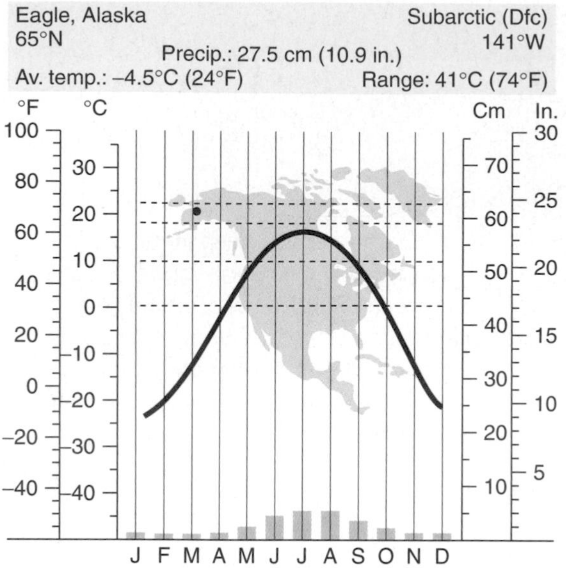

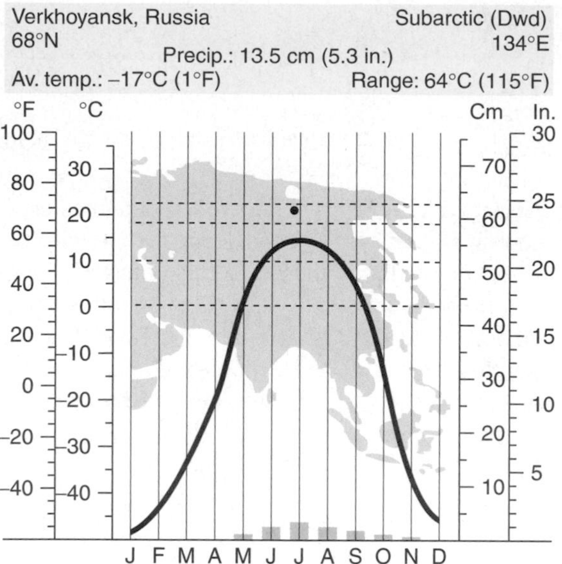

• FIGURE 10.22
Climographs of the subarctic climate stations.
Why would people settle in such severe-winter climate regions?

working hours, can be mentally depressing and can increase the impression of climatic severity.

As a direct result of the intense heating and cooling of the land, the subarctic has the largest annual temperature ranges of any climate. Average annual ranges in equatorward margins of the climate vary from near 40°C (72°F) to more than 45°C (80°F). The exceptions are near western coasts, where warm ocean currents and the marine influence may significantly modify winter temperatures. Average annual ranges for poleward stations are even greater. The climograph for Verkhoyansk, which indicates a range of 64°C (115°F), is an extreme example.

As with subarctic temperature, latitude and continentality influence subarctic precipitation. These climate controls combine to limit annual precipitation amounts to less than 50 centimeters (20 in.) for most regions and to 25 centimeters (10 in.) or less in northern and interior locations. The low temperatures in the subarctic reduce the moisture-holding capacity of the air, thus minimizing precipitation during the occasional passage of cyclonic storms. Location toward the center of large landmasses or near lee coasts increases distance from oceanic sources of moisture. Finally, the higher latitudes occupied by the subarctic climate are dominated by the polar anticyclone, especially in the winter season. The subsidence and divergence of air in the polar anticyclone limit the opportunity for lifting and hence for precipitation in the subarctic regions. This high pressure system also blocks the entry of moist air from warmer areas to the south.

Subarctic precipitation is cyclonic or frontal, and because the polar anticyclone is weaker and farther north during the warmer summer months, more precipitation comes during that season. The meager winter precipitation falls as fine, dry snow. Though there is not as much snowfall as in less severe climates, the temperatures remain so cold for so long that the snow cover lasts for as long as 7 or 8 months. During this time, there is almost no melting of snow, especially in the dark shadows of the forest.

A Limiting Environment The climatic restrictions of subarctic regions place distinct limitations on plant and animal life and on human activity. The characteristic vegetation is coniferous forest, adapted to the severe temperatures; the physiologic drought associated with frozen soil water; and the infertile soils. Seemingly endless tracts of spruce, fir, and pine thrive over enormous areas, untouched by humans (● Fig. 10.23). In Russia, the forest is called the *taiga* (or *boreal forests* in other regions), and this name is sometimes given to the subarctic climate type itself.

The brief summers and long, cold winters severely limit the growth of vegetation in subarctic regions. Even the trees

• FIGURE 10.23
Seemingly endless tracts of taiga (boreal forest) are typical of the vegetation throughout much of the American and Canadian subarctic. This photo was taken in Alaska's Wrangell-St. Elias National Park and Preserve.
Why are these kinds of virgin forests currently of little economic value?

National Park Service—Wrangell-St. Elias

are shorter and more slender than comparable species in less severe climate regimes. There is little hope for agriculture. The growing season averages 50–75 days, and frost may occur even during June, July, or August. Thus, in some years, a subarctic location may have no truly frost-free season. Although scientists are working to develop plant species that can take advantage of the long hours of daylight in summer, only minimal success has been achieved, in southern parts of the climate, with certain vegetables such as cabbage and root crops such as potatoes.

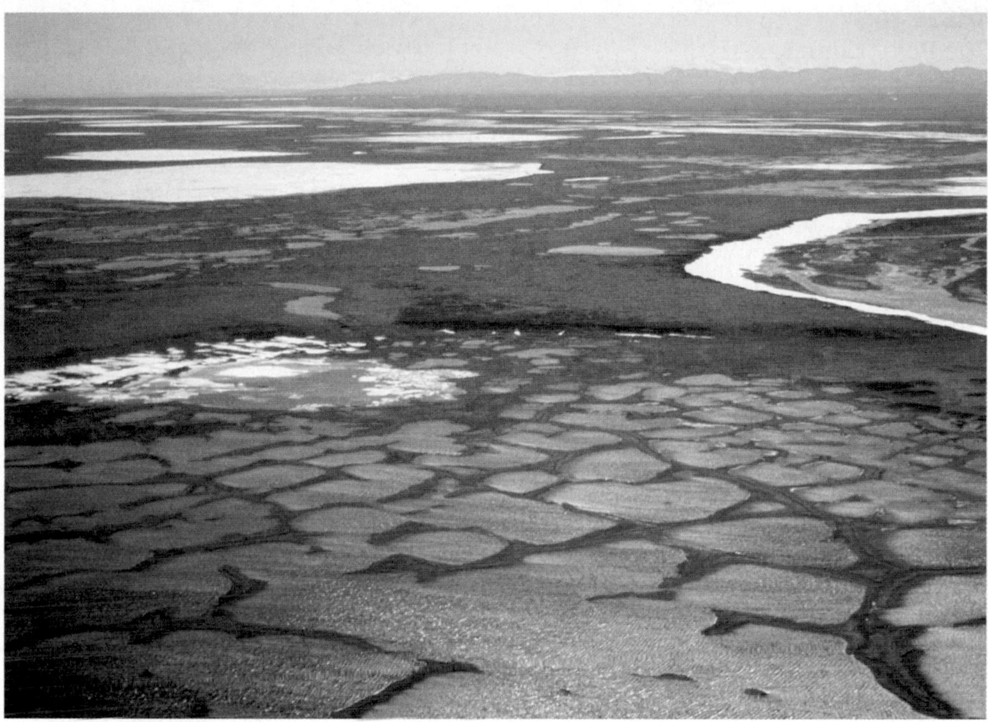

Alaska Image Library/USFWS

• **FIGURE 10.24**
Frost polygons, or patterned ground, as seen in the Arctic National Wildlife Refuge of Alaska. Repeated freezing and thawing cause the soil to produce polygonal shapes. When the ground freezes, it expands, and it shrinks when it thaws.
Why are the edges heaved upward?

A particularly vexing problem to people in subarctic (as well as tundra) regions is **permafrost,** a permanently frozen layer of subsoil and underlying rock that may extend to a depth of 300 meters (1000 ft) or more in the northernmost sections of the climate. Permafrost is present over much of the subarctic climate, but it varies greatly in thickness and is often discontinuous. Where it occurs, the land is frozen completely from the surface down in winter. The warm temperatures of spring and summer cause the top few feet to thaw out, but because the land beneath this thawed top layer remains frozen, water cannot percolate downward, and the thawed soil becomes sodden with moisture, especially in spring, when there is an abundant supply of water from the melting snow. Permafrost poses a problem to agriculture by preventing proper soil drainage. The seasonal freeze and thaw of the surface layer above the permafrost also poses special problems for construction engineers. The cycle causes repeated expansion and contraction, heaving the surface up and then letting it sag down. The effects of this cycle break up roads, force buried pipelines out of the ground, cause walls and bridge piers to collapse, and even swivel buildings off their foundations. Landscapes called **patterned ground,** or **frost polygons,** are commonly found in regions of subarctic yearly freezing and thawing (• Fig. 10.24).

With agriculture a questionable occupation at best, there is little economic incentive to draw humans to subarctic regions. Logging is unimportant because of small tree size. Even use of the vast forests for paper, pulp, and wood products is restricted by their interior location far from world markets. Miners occasionally exploit rich ore deposits; others, many of them native to the subarctic regions, pursue hunting, trapping, and fishing of the relatively limited wildlife. Fur-bearing animals such as mink, fox, wolf, ermine, otter, and muskrat are of greatest value.

Polar Climate Regions

The polar climates are the last of Köppen's humid climate subdivisions to be differentiated on the basis of temperature. These climate regions are situated at the greatest distance from the equator, and they owe their existence primarily to the low annual amounts of insolation they receive. No polar station experiences a month with average temperatures as high as 10°C (50°F), and hence all are without a warm summer (Table 10.3). Trees cannot survive in such a regimen. In the regions where at least 1 month averages above 0°C (32°F), they are replaced by tundra vegetation. Elsewhere, the surface is covered by great expanses of frozen ice. Thus, there are two polar climate types, tundra (*ET*) and ice sheet (*EF*).

There are two important points to keep in mind in the discussion of polar climate regions. First, these regions have a large net annual radiation loss; that is, they give up much more radiation or energy than they receive from the sun during a year, resulting in a major radiation deficiency. The transfer of heat from lower to higher latitudes to make up this deficiency is the driving force of the general atmospheric circulation. Without this compensating poleward transfer of heat from the lower latitudes, the polar regions would become too cold to permit any form of life, and the equatorial regions would heat to temperatures no organism could survive.

A second and equally important characteristic of polar climates is the unique pattern of day and night. At the poles, 6 months of relative darkness, caused when the sun is positioned

TABLE 10.3
The Polar Climates

Name and Description	Controlling Factors	Geographic Distribution	Distinguishing Characteristics	Related Features
Tundra				
Warmest month between 0°C (32°F) and 10°C (50°F); precipitation exceeds potential evapotranspiration	Location in the high latitudes; subsidence and divergence of the polar anticyclone; proximity to coasts	Arctic Ocean borderlands of North America, Greenland, and Eurasia; Antarctic Peninsula; some polar islands	At least 9 months average below freezing; low evaporation; precipitation usually below 25.5 cm (10 in.); coastal fog; strong winds	Tundra vegetation; tundra soils; permafrost; swamps and bogs during melting period; life most common in nearby seas; Inuit; mineral and oil resources; defense industry
Ice Sheet				
Warmest month below 0°C (32°F); precipitation exceeds potential evaporation	Location in the high latitudes and interior of landmasses; year-round influence of the polar anticyclone; ice cover; elevation	Antarctica; interior Greenland; permanently frozen portions of the Arctic Ocean and associated islands	Summerless; all months average below freezing; world's coldest temperature; extremely meager precipitation in the form of snow, evaporation even less; gale-force winds	Ice- and snow-covered surface; no vegetation; no exposed soils; only sea life or aquatic birds; scientific exploration

below the horizon, alternate with 6 months of daylight during which the sun is above the horizon. Even when the sun is above the horizon, however, the sun's rays are at a sharply oblique angle, and little insolation is received for the number of hours of daylight. Moving outward from the poles, the lengths of periods of continuous winter night and continuous summer day decrease rapidly from 6 months at the poles to 24 hours at the Arctic and Antarctic Circles (66 1/2°N and S). Here the 24-hour night or day occurs only at the winter and summer solstices, respectively.

Tundra Climate

Compare the location of the **tundra climate** (*ET*) with that of the subarctic climate in Figure 8.6. You can see that although the tundra climate is situated closer to the poles, it is also along the periphery of landmasses, and, with the exception of the Antarctic Peninsula, it is everywhere adjacent to the Arctic Ocean. Even though temperature ranges in the tundra are large, they are not as large as in the subarctic because of the maritime influence. Winter temperatures in particular are not as severe in the tundra as they are inland (● Fig. 10.25).

It almost seems inappropriate to call the unpleasantly chilly and damp conditions of the tundra's warmer season "summer." Temperatures average around 4°C (40°F) to 10°C (50°F) for the warmest month, and frosts occur regularly. The air does warm sufficiently to melt the thin snow cover and the ice on small bodies of water, but this only causes marshes, swamps, and bogs to form across the land because drainage is blocked by permafrost (● Fig. 10.26). Clouds of black flies, mosquitoes, and gnats swarm in this soggy landscape, known as **muskeg** in Canada and Alaska. A bright note in the landscape is provided by the enormous number of migratory birds that nest in the arctic regions in summer and feed on the insects. However, as soon as the shrinking daylight hours of autumn approach, these birds depart for warmer climates.

Winters are cold and seem to last forever, especially in tundra locations where the sun may be below the horizon for days at a time. The climograph for Barrow, Alaska, illustrates the low temperatures of this climate. Note that average monthly temperatures are *below freezing* 9 months of the year. The average annual temperature is −12°C (10°F). The low-growing tundra vegetation survives despite the forbidding environment. It consists of lichens, mosses, sedges, flowering herbaceous plants, small shrubs, and grasses. In particular, the plants have adjusted to the conditions associated with nearly universal permafrost (● Fig. 10.27).

The tundra regions exhibit several other significant climatic characteristics. Diurnal temperature ranges are small because insolation is uniformly high during the long summer days and uniformly low during the long winter nights. Precipitation is generally low, except in eastern Canada and Greenland, because of exceedingly low absolute humidity and the influence of the polar anticyclone. Icy winds sweep across the open land surface and are an added factor in eliminating the trees that might impede their progress. Coastal fog is characteristic in marine locations, where cool polar maritime air drifts onshore and is chilled below the dew point by contact with the even colder land.

Ice-Sheet Climate

The **ice-sheet climate** (*EF*) is the most severe and restrictive climate on Earth. As Table 10.3 indicates, it covers large areas in both the Northern and Southern Hemispheres, a total of about 16 million square kilometers (6 million sq mi)—nearly the same

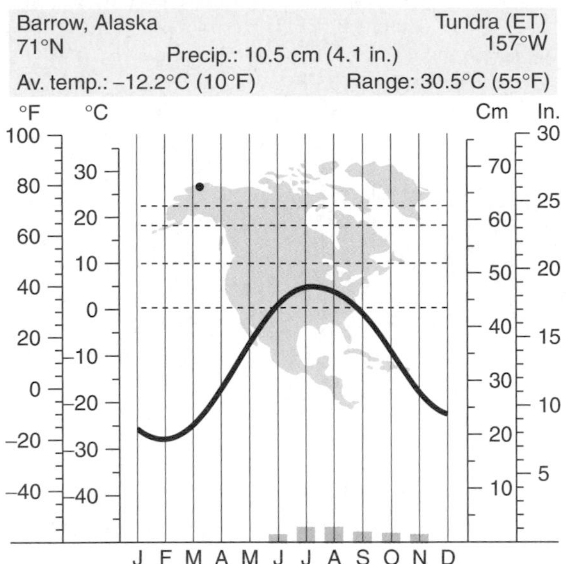

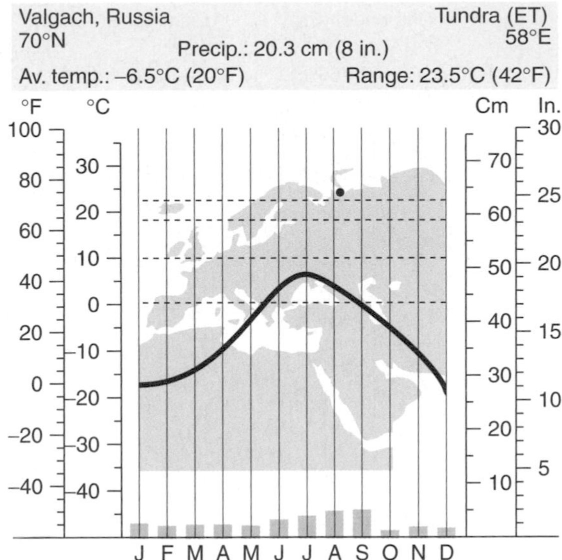

• FIGURE 10.25
Climographs for tundra climate stations.
Why is it not surprising that both stations are located in the Northern Hemisphere?

Alaska Image Library/USFWS

• FIGURE 10.26
Permafrost regions, such as this area at the base of the Alaska Range, become almost impenetrable swampland during the brief Alaskan summer. Travel over land is feasible only in the winter season.
What is the preferred means of travel in the summer?

M. Trapasso

• FIGURE 10.27
One of the Hecho Islands off the Antarctic Peninsula displays thick, bright patches of moss as its most complex vegetation.
What weather elements might help to form this stark landscape?

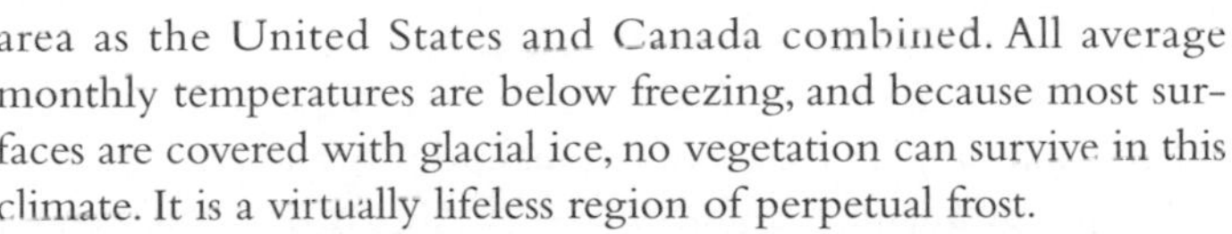
area as the United States and Canada combined. All average monthly temperatures are below freezing, and because most surfaces are covered with glacial ice, no vegetation can survive in this climate. It is a virtually lifeless region of perpetual frost.

Antarctica is the coldest place on Earth (although Siberia sometimes has longer and more severe periods of cold in winter). The world's coldest temperature, –88°C (–127°F), was recorded at Vostok, Antarctica. Consider the climographs for Little America, Antarctica, and Eismitte, Greenland, for a fuller picture of the cold ice-sheet temperatures (● Fig. 10.28).

The primary reason for the low temperatures of ice-sheet climates is the minimal insolation received in these regions. Not only is little or no insolation received during half the year, but also the sun's radiant energy that is received arrives at sharply oblique angles. In addition, the perpetual snow and ice cover of this climate reflects nearly all incoming radiation. A further factor, in both Greenland and Antarctica, is elevation. The ice sheets covering both regions rise more than 3000 meters (10,000 ft) above sea level (● Fig. 10.29). Naturally, this elevation contributes to the cold temperatures.

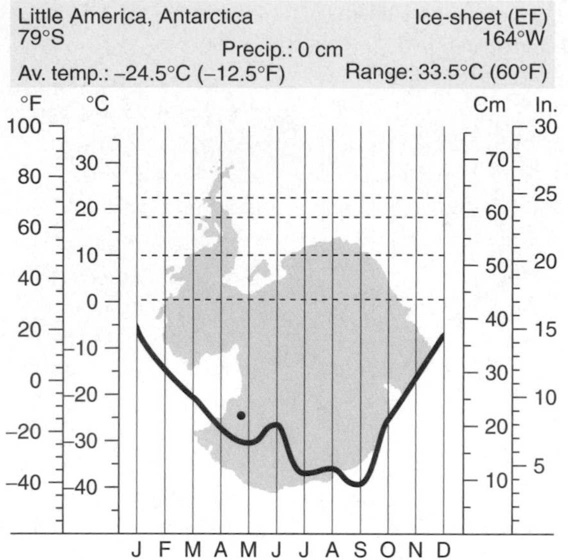

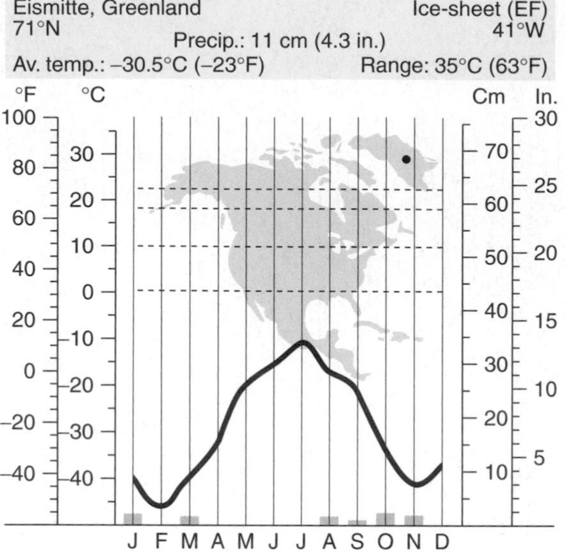

• FIGURE 10.28

Climographs for ice-sheet climate stations.

If you were to accept an offer for an all-expense-paid trip to visit either Greenland or Antarctica, which would you choose, and why would you go?

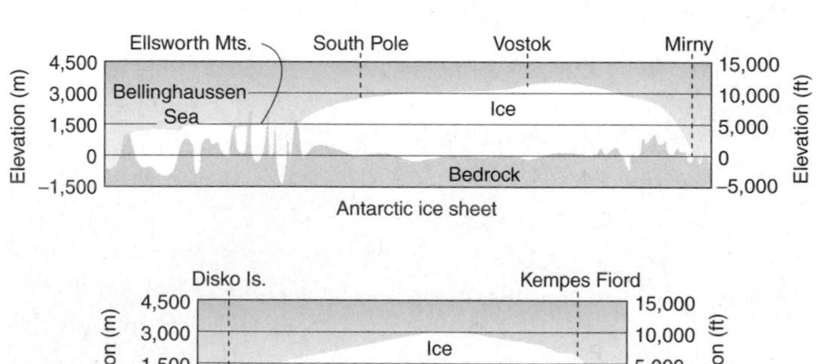

• FIGURE 10.29

The Antarctic ice sheet resulting from the ice-sheet climate of the south polar regions is as much as 4000 meters (13,200 ft) thick locally. Where it is thinnest, it is floated by seawater to produce an ice shelf. The smaller Greenland ice sheet is about 3000 meters (10,000 ft) thick.

What reasons might be given for the fact that more land in Greenland than in Antarctica is free of glacial ice?

The polar anticyclone severely limits precipitation in the ice-sheet climate to the fine, dry snow associated with occasional cyclonic storms. Precipitation is so meager in this climate that regions within this regime are sometimes incorrectly referred to as "polar deserts." However, because of the exceedingly low evaporation rates (evapotranspiration rates are not considered in this climate because there is no plant life to account for any transpiration) associated with the severely cold temperatures, precipitation still exceeds potential evaporation, and the climate can be classified as humid. The annual precipitation surplus produces glaciers, which export snowfall similar to the way rivers export rainfall.

The strong and persistent polar winds are another staple of the harsh ice-sheet climate. Mawson Base, Antarctica, for example, has approximately 340 days a year with gale-force winds of 15 meters per second (33 mph) or more. *Katabatic winds*, which are caused by the downslope drainage of heavy cold air accumulated over ice sheets, are common along the edges of the polar ice. The winds of these regions can result in whiteouts—periods of zero visibility due to blowing fine snow and ice crystals.

Human Activity in Polar Regions

The climatic severity that limits animal life in polar regions to a few scattered species in the tundra is just as restrictive on human settlement. The celebrated Lapps of northern Europe migrate with their reindeer to the tundra from the adjacent forest during warmer months. They join the musk ox, arctic hare, fox, wolf, and polar bear that manage to make a home there despite the prohibitive environment. Only the Inuit (Eskimos) of Alaska, northern Canada, and Greenland have in the past succeeded in developing a year-round lifestyle adapted to the tundra regime. Yet even this group relies less on the resources of the tundra than on the large variety of fish and sea mammals, such as cod, salmon, halibut, seal, walrus, and whale, that occupy the adjacent seas.

As their communication with the rest of the world has increased and they have become acquainted with alternative lifestyles, the permanent population of Inuit living in the tundra has greatly diminished, and life for those remaining has changed drastically. Some have gained new economic security through employment at defense installations or at sites where they join other skilled workers from outside the region to exploit mineral or energy resources. However, the new population centers based

M. Trapasso

• **FIGURE 10.30**

Mile-marker 0 on the Alaska pipeline, near Prudhoe Bay. This is one very profitable venture for humans in the North Slope oil fields. However in 1989, the *Exxon Valdez* spilled 11 million gallons of crude oil into Prince William Sound, Alaska.

Considering the vulnerability of Alaska's physical environment, should development of the North Slope oil fields have been permitted?

• **FIGURE 10.31**

Greenland's ice sheet covers about 85% of its surface. Here we see ice swallowing up the landscape.

What kind of activities might bring individuals from other regions to an ice-sheet climate?

M. Trapasso

on the construction and maintenance of radar and missile defense stations or, as in the case of Alaska's North Slope, on the production and transportation of oil, cannot be considered permanent (• Fig. 10.30). Workers depend on other regions for support and often inhabit this region only temporarily.

The ice-sheet climate cannot serve as a home for humans or other animals. Even the penguins, gulls, leopard seals, and polar bears are coastal inhabitants. It is without question the harshest, most restrictive, most nearly lifeless climate zone on Earth (• Fig. 10.31). Yet, especially in Antarctica, it is of strategic importance and of great scientific interest. Scientists study the oxygen-isotope ratios of the Antarctic ice cores and the gas bubbles trapped within them to help them reconstruct past climates. They have also noted that ozone concentrations have decreased over Antarctica every fall for several decades. The result is a hole in the ozone layer above Antarctica that is as large as the North American continent. This decrease in the ozone layer is a major environmental concern. Antarctica's strategic value is so widely recognized that the world's nations have voluntarily given up claims to territorial rights on the continent in exchange for cooperative scientific exploration on behalf of all humankind.

Highland Climate Regions

As we saw in Chapter 4, temperature decreases with increasing altitude at the rate of about 6.5°C per 1000 meters (3.6°F per 1000 ft). Thus, you might suspect that highland regions exhibit broad zones of climate based on changes in temperature with elevation that roughly correspond to Köppen's climate zones based on change of temperature with latitude (• Fig. 10.32). This is indeed the case, with one important exception: Seasons only exist in highlands if they also exist in the nearby lowland regions. For example, although zones of increasingly cooler temperature occur at progressively higher elevations in the tropical climate regions, the seasonal changes of Köppen's middle-latitude climates are not present.

Elevation is only one of several controls of highland climates; **exposure** is another. Just as some continental coasts face the prevailing wind, so do some mountain slopes; others are lee slopes or are sheltered behind higher topography. The nature of the wind, its temperature, and its moisture content depend on whether the mountain is (1) in a coastal location or deep in a continental interior and (2) at a high or low latitude within or beyond the reaches of cyclonic storms and monsoon circulation. In the middle

M. Trapasso

(a)

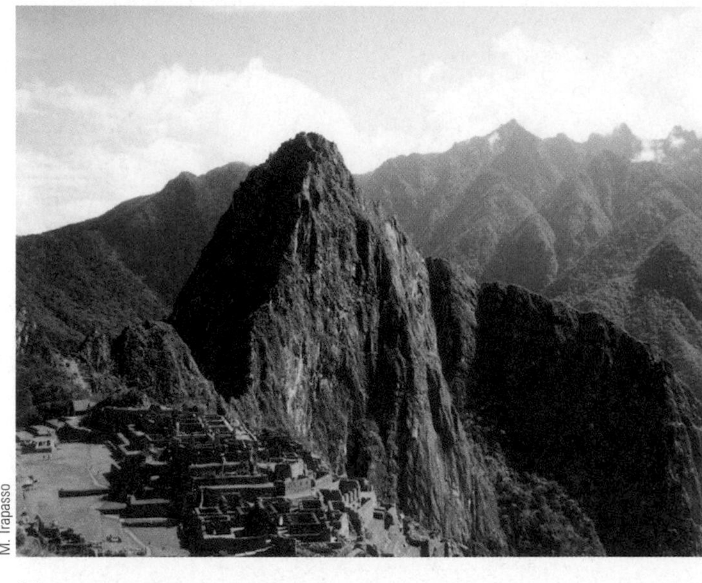
M. Trapasso

(b)

• **FIGURE 10.32**
Note the similarities between (a) Slovar, Norway, 68°N latitude (an Arctic location), at sea level, and (b) Machu Picchu, 13°S latitude (a tropical location), at 2590 meters (8500 feet) above sea level.
In what ways might high latitudes be similar to high altitudes?

and high latitudes, mountain slopes and valley walls that face the equator receive the direct rays of the sun and are warm; poleward-facing slopes are shadowed and cool. West-facing slopes feel the hot afternoon sun, whereas east-facing slopes are sunlit only in the cool of the morning. This factor, known as **slope aspect,** affects where people live in the mountains and where particular crops will do best. The higher one rises in the mountains, the more important direct sunlight is as a source of warmth and energy for plant and animal life processes.

Complexity is the hallmark of highland climates. Every mountain range of significance is composed of a mosaic of climates far too intricate to differentiate on a world map or even on a map of a single continent. Highland climates are therefore undifferentiated, signifying climate complexity. Highland climates are indicated on Figure 8.6 wherever there is marked local variation in climate as a consequence of elevation, exposure, and slope aspect. We can see that these regions are distributed widely over Earth but are particularly concentrated in Asia, central Europe, and western North and South America.

The areas of highland climate on the world map are cool, moist islands in the midst of the zonal climates that dominate the areas around them. Consequently, highland areas are also biotic islands, supporting a flora and fauna adapted to cooler and wetter conditions than those of the surrounding lowlands. This coolness is part of the highland charm, particularly where mountains rise cloaked with forests above arid plains, as do the Canadian Rocky Mountains and California's Sierra Nevada.

Highlands stimulate moisture condensation and precipitation by forcing moving air masses to rise over them (• Fig. 10.33). Where mountain slopes are rocky and forest free, their surfaces grow warm during the day, causing upward convection, which often produces afternoon thundershowers. Mountains receive abundant precipitation and are the source area for multitudes of streams that join to form the great rivers of all of the continents.

There are few streams of significance whose headwaters do not lie in rugged highlands. Much of the stream flow on all continents is produced by the summer melting of mountain snowfields. Thus, the mountains not only draw moisture from the atmosphere but also store much of it in a form that gradually releases it throughout summer droughts when water is most needed for irrigation and for municipal and domestic use.

Peculiarities of Mountain Climates

A general characteristic of mountain weather is its variability from hour to hour as well as from place to place. Strong orographic flow over mountains often causes clouds to form very quickly, leading to thunderstorms and longer rains that do not affect surrounding cloud-free lowlands. Where the cloud cover is diminished, diurnal temperature ranges over mountains are far greater than those over lowlands. Because mountains penetrate upward beyond the densest part of the atmosphere, the greenhouse effect is less developed there than anywhere else on Earth. The thinner layer of low-density air above a mountain site does not greatly impede insolation, thus allowing surfaces to warm dramatically during the daytime. By the same token, the atmosphere in these areas does little to impede longwave radiation loss at night. Consequently, air temperatures overnight are cooler than the elevation alone would indicate. Because the atmospheric shield is thinnest at high elevations, plants, animals, and humans receive proportionately more of the sun's shortwave radiation at high altitudes. Ultraviolet radiation is particularly noticeable; severe sunburn is one of the real hazards of a day in the high country.

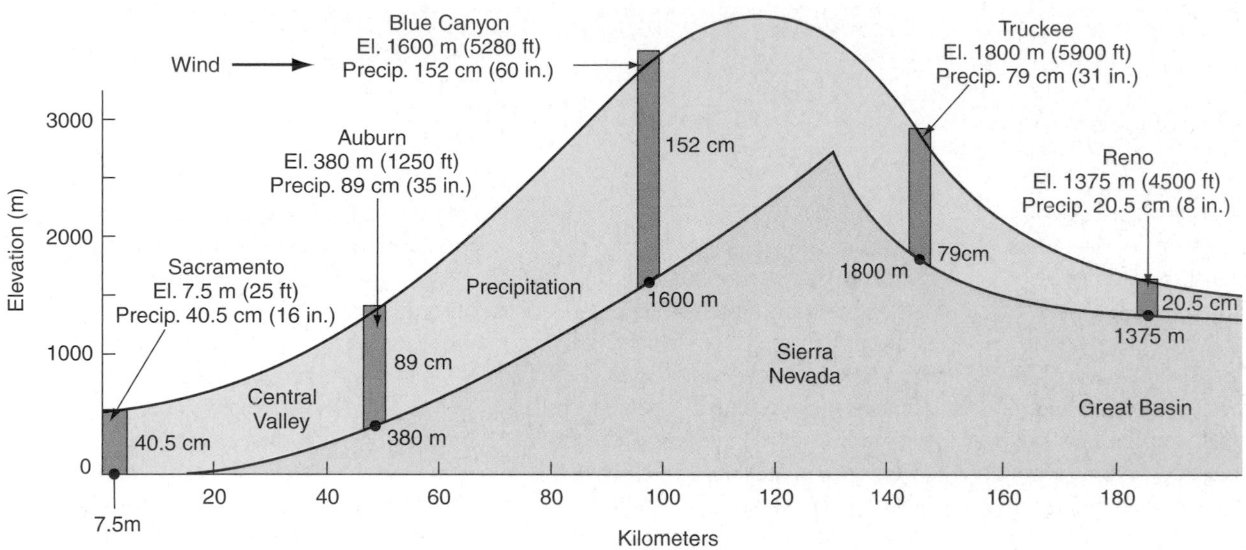

• FIGURE 10.33
Variation in precipitation caused by uplift of air crossing the Sierra Nevada range of California from west to east. The maximum precipitation occurs on the windward slope because air in the summit region is too cool to retain a large supply of moisture. Note the strong rain shadow to the lee that gives Reno a desert climate.
Taking into consideration the locations of the recording stations, during what season of the year does the maximum precipitation on the windward slope occur?

In the middle and high latitudes, mountains rise from mesothermal and microthermal climates into tundra and snow-covered zones. The lower slopes of mountains are commonly forested with conifers, which become more stunted as one moves upward, until the last dwarfed tree is passed at the **tree line**. This is the line beyond which low winter temperatures and severe wind stress eliminate all forms of vegetation except those that grow low to the ground, where they can be protected by a blanket of snow (• Fig. 10.34). Where mountains are high enough, snow or ice permanently covers the land surface. The line above which summer melting is insufficient to remove all of the preceding winter's snowfall is called the **snow line**.

In tropical mountain regions, the vertical zonation of climate is even more pronounced. Both tree line and snow line occur at higher elevations than in middle latitudes. Any seasonal change is mainly restricted to rainfall; temperatures are stable year-round, regardless of elevation. Each climate zone has its own particular association of natural vegetation and has given rise to a distinctive crop combination where agriculture is

• FIGURE 10.34
As in this example in the Colorado Rocky Mountains, the last tree species found at the tree line are stunted, prostrate forms, which often produce an elfin forest. Where the trees are especially gnarled and misshapen by wind stress, the vegetation is called ***krummholz*** (crooked wood).
What do you see in the photograph that indicates prevailing wind direction?

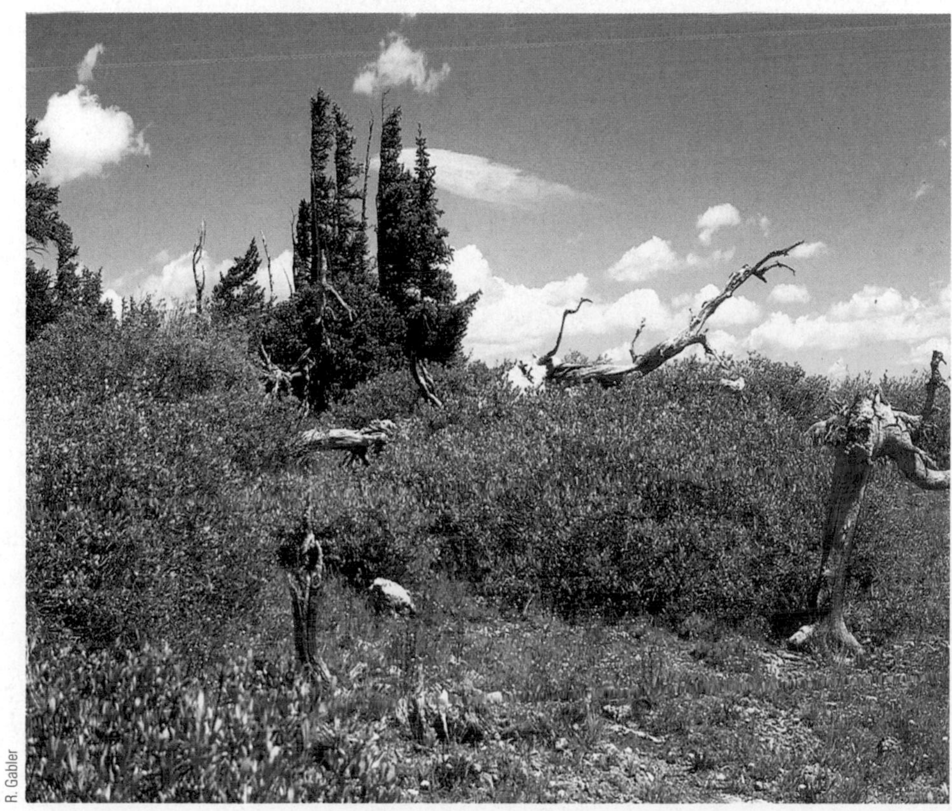

R. Gabler

GEOGRAPHY'S ENVIRONMENTAL SCIENCE PERSPECTIVE

The Effects of Altitude on the Human Body

The changes in climate, soils, and vegetation as an individual climbs into higher altitudes are discussed elsewhere in the text. However, it is also interesting to note some of the ways in which high altitudes affect the human body. Besides food and water, the metabolism of a human being also depends on the consumption of oxygen. A lessening of the normal amount of oxygen intake can have some profound physiological effects. One way to look at the amount of oxygen available for human respiration is to consider something called the *partial pressure of oxygen gas* (pO_2). Partial pressure in this case refers to the portion of total atmospheric pressure attributed to oxygen alone. At sea level, atmospheric pressure is 1013.2 millibars, and the pO_2 is about 212 millibars. In other words, 212 of the 1013.2 millibars are attributed to oxygen gas. At an altitude of 10 kilometers (6.2 mi) the pO_2 drops to only 55 millibars! Even at moderate altitudes, the effects of *hypoxia* (oxygen starvation) can cause headaches and nausea. Above 6 kilometers (3.7 mi), this lack of oxygen can seriously affect the brain. Since the body's need for oxygen does not alter with major changes in altitude, any significant drop in pO_2 with height can cause severe body stress. Pressure-controlled cabins of high-flying aircraft and the use of oxygen by mountain climbers provide dramatic examples of ways to meet the vital need for oxygen.

Fortunately, the human body can acclimatize to moderate changes in altitude, but it takes time (days to sometimes weeks, depending on the altitude). During this time, however, one may experience some of the following symptoms: sleeplessness, headaches, loss of weight, thyroid deficiency, increased excitability, muscle pain, gastrointestinal disturbances, swelling of the lungs, severe infections, psychological and mental disturbances, and others. At the very least, when visiting higher altitude locations, most people can expect headaches (perhaps leading to nausea), a drop in physical endurance (climbing stairs will become more difficult), and hampered mental function (solutions to simple problems may be difficult to grasp). There is little need to worry, though; humans do acclimatize and these symptoms will pass.

It is also important to remember that, in addition to hypoxia, higher altitudes may cause increased susceptibility to severe sunburn. If you climb to a higher altitude, this means that more of Earth's atmosphere is below you. Correspondingly, there is less of the protective atmosphere above you to filter harmful ultraviolet radiation. Keeping latitude constant, sunburn of exposed skin is more likely on a high-altitude mountainside than on a sea-level beach.

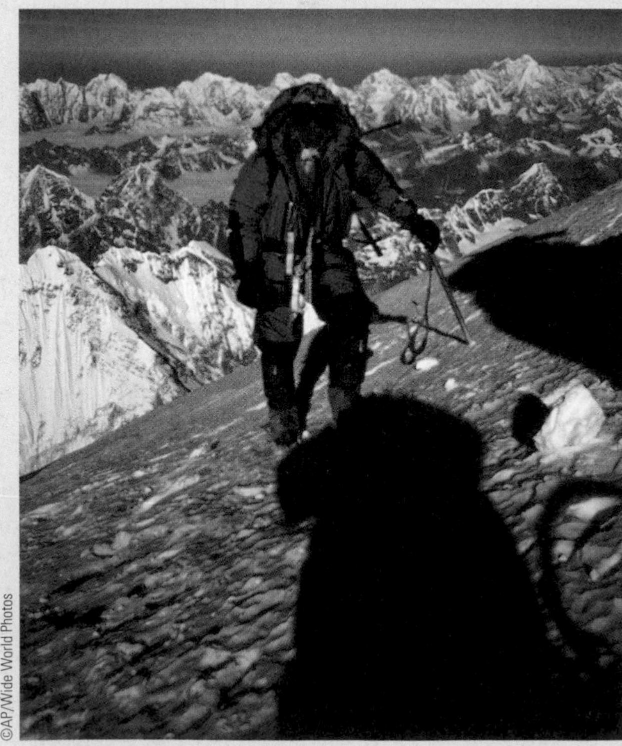
©AP/Wide World Photos

Mountain climbers carry an oxygen supply to help with high altitudes.

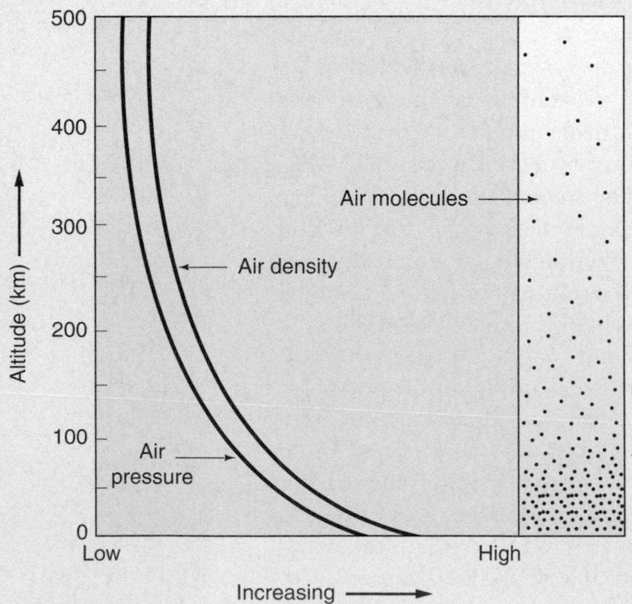

Both air pressure and air density decrease rapidly with increasing altitude.

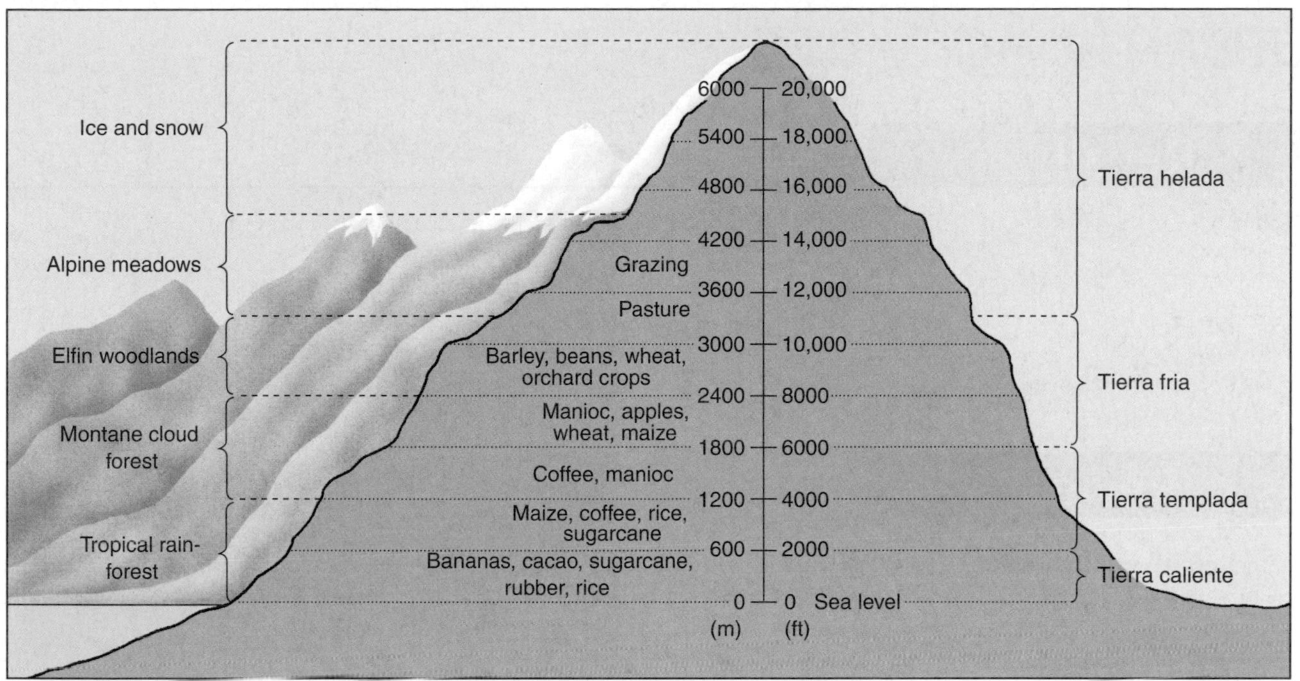

• FIGURE 10.35

Natural vegetation, vertical climate zones, and agricultural products in tropical mountains. Note that this example extends from tropical life zones to the zone of permanent snow and ice. There is little seasonal temperature change in tropical mountains, which allows life-forms sensitive to low temperatures to survive at relatively high elevations.

When Europeans first settled in the highlands of tropical South America, in which vertical climate zone did they prefer to live?

practiced (• Fig. 10.35). In South America, four vertical climate zones are recognized: *tierra caliente* (hot lands), *tierra templada* (temperate lands), *tierra fría* (cool lands), and *tierra helada* (frozen lands).

Highland Climates and Human Activity

In middle-latitude highlands, soils are poor, the growing season is short, and the winter snow cover is heavy in the conifer zone, which dominates the lower and middle mountain slopes. Therefore, little agriculture is practiced, and permanent settlements in the mountains are few. However, as the winter snow melts off the high ground just below the bare rocky peaks, grass springs into life, and humans drive herds of cattle and flocks of sheep and goats up from the warmer valleys. The high pastures are lush throughout the summer, but in early fall they are once again vacated by the animals and their keepers, who return to the valleys. This seasonal movement of herds and herders between alpine pastures and villages in the valleys, termed *transhumance,* was once common in the European highlands (the Alps, Pyrenees, Carpathian Mountains, and mountains throughout Scandinavia) and is still practiced there on a reduced scale.

Otherwise, the middle-latitude highlands serve mainly as sources of timber and of minerals formed by the same geologic forces that elevated the mountains and as arenas for recreation—both summer and winter. Recreational use of the highlands is a relatively recent phenomenon, resulting both from new interest in mountain areas and from new access routes by road, rail, and air that did not exist a century ago.

In contrast to poleward mountain regions, tropical highlands may actually experience more favorable climatic conditions and are often a greater attraction to human settlement (from ancient to modern times) than adjacent lowlands. In fact, large permanent populations are supported throughout the tropics where topography and soil favor agriculture in the vertical climate zones. Highland climates are at such a premium in many areas that steep mountain slopes have been extensively terraced to produce level land for agricultural use. Spectacular agricultural terraces can be seen in Peru, Yemen, the Philippines, and many other tropical highlands. Where the climate is appropriate and population pressure is high, people have created a topography to suit their needs, although they have had to hack it out of mountainsides.

This chapter has discussed the major climates that dominate the middle latitudes of Earth. It is in these latitudes that most of the human population exists. These climates range from some of the most productive to some of the harshest climates on the planet. Again, vegetation, animal life, and human activities have all had to adapt to the specific climatic conditions. And once again, changes such as global warming and destruction of our ozone shield threaten these climates and the life-forms they support.

Chapter 10 Activities

Define & Recall

Mediterranean climate
sclerophyllous
chaparral
humid subtropical climate
marine west coast climate
humid continental, hot-summer climate
humid continental, mild-summer climate
subarctic climate
permafrost
patterned ground (frost polygons)
tundra climate
muskeg
ice-sheet climate
exposure
slope aspect
tree line
snow line

Discuss & Review

1. Summarize the special adaptations of vegetation and soils in Mediterranean regions.
2. Compare the humid subtropical and Mediterranean climates. What are their most obvious similarities and differences?
3. What factors combine to cause a precipitation maximum in late summer in most of the humid subtropical regions?
4. How are temperature, precipitation, and geographic distribution of marine west coast regions linked to the controlling factors for this climate?
5. Explain why the microthermal climates are limited to the Northern Hemisphere.
6. List several features that all humid microthermal climates have in common. How do these features differ from those displayed by the humid mesothermal climates?
7. Describe the relationship between vegetation and climate in the humid continental, mild-summer regions.
8. Refer to Figure 10.22. Using the climographs for Eagle, Alaska, and Verkhoyansk, Russia, describe the temperature patterns of the subarctic regions.
9. What factors limit precipitation in the subarctic regions?
10. Identify and compare the climate factors that strongly influence the tundra and ice-sheet regions. How do these controlling factors affect the distribution of these climates?
11. What kind of plant and animal life can survive in the polar climates? What special adaptations must this life make to the harsh conditions of these regions?
12. How do elevation, exposure, and slope aspect affect the microclimates of highland regions? What are the major climatic differences between highland regions and nearby lowlands?
13. Do each of the following for these climates: Mediterranean; humid subtropical; marine west coast; humid continental, hot-summer; humid continental, mild-summer; subarctic; tundra; ice-sheet.
    a. Identify the climate from a set of data or a climograph indicating average monthly temperature and precipitation for a representative station within a region of that climate (use one of the climographs in this chapter).
    b. Match the climate type with a written statement that includes one or more of the following: the statistical parameters of the climate in the modified Köppen classification; the particular climate controls (controlling factors) that produce the climate; the geographic distribution of the climate as stated in terms of physical or political location; the unique climate characteristics or combination of characteristics that distinguishes the climate from others; types of plants, animals, and soils associated with the climate; and the human utilization typical of the climate.
    c. Distinguish between the important subtypes (if any) of each climate by identifying the characteristics that separate them from one another.

Consider & Respond

1. Based on the classification scheme presented in the "Graph Interpretation" exercise at the end of Chapter 8, classify the following climate stations from the data provided.

		J	F	M	A	M	J	J	A	S	O	N	D	Yr
a.	Temp. (°C)	–42	–27	–40	–31	–20	15	–11	–18	–22	–36	–43	–39	–30
	Precip. (cm)	0.3	0.3	0.5	0.3	0.5	0.8	2.0	1.8	0.8	0.3	0.8	0.5	8.6
b.	Temp. (°C)	3	3	5	8	19	13	15	14	13	10	7	5	9
	Precip. (cm)	4.8	3.6	3.3	3.3	4.8	4.6	8.9	9.1	4.8	5.1	6.1	7.4	65.8
c.	Temp. (°C)	23	23	22	19	16	14	13	13	14	17	19	22	18
	Precip. (cm)	0.8	1.0	2.0	4.3	13.0	18.0	17.0	14.5	8.6	5.6	2.0	1.3	88.1
d.	Temp. (°C)	–27	–28	–26	–18	–8	1	4	3	–1	–8	–18	–24	12
	Precip. (cm)	0.5	0.5	0.3	0.3	0.3	1.0	2.0	2.3	1.5	1.3	0.5	0.5	10.9
e.	Temp. (°C)	–4	–2	5	14	20	24	26	25	20	13	3	–2	12
	Precip. (cm)	0.5	0.5	0.8	1.8	3.6	7.9	24.4	14.2	5.8	1.5	1.0	0.3	62.2
f.	Temp. (°C)	9	9	9	10	12	13	14	14	14	12	11	9	11
	Precip. (cm)	17.0	14.8	13.3	6.8	5.5	1.9	0.3	0.3	1.6	8.1	11.7	17.0	97.6
g.	Temp. (°C)	–3	–2	2	9	16	21	24	23	19	13	4	–2	11
	Precip. (cm)	4.8	4.1	6.9	7.6	9.4	10.4	8.6	8.1	6.9	7.1	5.6	4.8	84.8
h.	Temp. (°C)	0	0	4	9	16	21	24	23	20	14	8	2	12
	Precip. (cm)	8.1	7.4	10.7	8.9	9.4	8.6	10.2	12.7	10.7	8.1	8.9	8.1	111.5

2. The data in the previous table represent the following eight locations, although not in this order: Beijing, China; Point Barrow, Alaska; Chicago, Illinois; Eismitte, Greenland; Eureka, California; Edinburgh, Scotland; New York, New York; Perth, Australia. Use an atlas and your knowledge of climates to match the climatic data with the locations.
3. Eureka, California, Chicago, and New York City are located within a few degrees latitude of one another, yet they represent three distinctly different climate types. Discuss these differences and identify the primary cause, or source, of the differences.
4. The precipitation recorded at Albuquerque, New Mexico (see Consider and Respond, Chapter 9), is almost twice that recorded at Point Barrow, Alaska, yet Albuquerque is considered a dry climate and Point Barrow a humid climate. Why?
5. Why is the *Dw* climate type found only in Asia?
6. *Csa* climates are dry during the summer and *Cfa* climates are wet. Why?

Apply & Learn

Using the data charts in the Geography's Physical Science Perspective: Effective Temperatures box:

1. Determine the Heat Index values for the following data:
   a. 95° F, and 70% relative humidity
   b. 85° F, and 90% relative humidity
   c. 80° F, and 80% relative humidity
   d. 75° F, and 100% relative humidity
   e. 70° F, and 50% relative humidity
   f. 100° F, and 0% relative humidity
2. Determine the wind chill index values for the following data:
   a. 40°F, and 35 mph winds
   b. 20°F, and 20 mph winds
   c. 15°F, and 35 mph winds
   d. –15°F, and 40 mph winds
   e. –30°F, and 30 mph winds
   f. –40°F, and calm winds

Biogeography

11

CHAPTER PREVIEW

Plants, animals, and the environments in which they live are interdependent, each affecting the others.

- In what ways are plants as a group and animals as a group mutually dependent on one another?
- In what ways do humans have a much greater impact on ecosystems than all other life-forms?

As the basic producers, autotrophs are generally considered to be the most important component of an ecosystem.

- What are the differences between autotrophs and heterotrophs?
- In what ways are the autotrophs affected by the other components?

Although other environmental controls may be more important on a local scale, climate has the greatest influence over ecosystems on a worldwide basis.

- Which climatic factors have the greatest effect on plants and animals?
- How does climate influence the shape and size of animals and their appendages?

Earth's major terrestrial ecosystems (biomes) are classified on the basis of the dominant vegetation types that occupy the ecosystems.

- What are these vegetation types?
- Why not base the classification on animals?

Freely floating plants called phytoplankton are the most important link in the ocean food chain.

- What role do they play?
- What function in relation to the atmosphere do phytoplankton perform?

◀ **Opposite: With increasing altitude, trees and bushes become stunted, and tundra grasses and mosses slowly transition to bare rock and soil.**
US Forest Service/Mark Muir

As was pointed out in Chapter 1, biogeographers are those physical geographers who specialize in the study of natural and human-modified environments and the ecological processes that influence each environment's nature and distribution. Along with ecologists from a host of other science disciplines, biogeographers often focus their research on **ecosystems**—communities of organisms that function together in an interdependent relationship with the environments that they occupy.

In almost all respects, the study of ecosystems provides the ideal opportunity to demonstrate the multiple perspectives of physical geography among the sciences: the spatial science perspective, the physical science perspective, and the environmental science perspective. Biogeographers examine the locations, distribution, spatial patterns, and spatial interactions of all plant and animal life. They delineate the boundaries and study the characteristics of the ecosystems that they identify, and also monitor the flow of energy and material through each system. In addition, they pay particular attention to the impact of humans on each ecosystem's living and nonliving physical environment.

Animal life would not exist without plants as basic food, and most plants could not survive without some animals. Together, plants and animals must adapt to their physical environment.

Humans alone have the intelligence and the capacity to alter, either carelessly or deliberately, the plant–animal–physical environment relationship.

What lasting effects will increased industrialization have on the water that animals in the 21st century will drink or on the atmosphere in which plants will grow? What will be the consequences for life-forms if concrete and steel replace additional square kilometers of forest? What could happen to marine life if toxic waste accumulates in the world's oceans? The understanding that comes from the study of ecosystems by biogeographers can provide answers to these questions and can help humans learn to work with, and not against, nature to sustain and improve life on planet Earth as we know it today.

Organization within Ecosystems

It can be said that ecology is an old science. The great voyages of exploration that began in the 15th century carried colonists and adventurers to uncharted lands with exotic environments. The more scholarly observers within each group made careful note of the flora and fauna found in each new part of the world. It soon became apparent that certain plants and animals were found together and that they bore a direct relationship to the climate in which they lived. As information about various world environments became more reliable and readily available, early biologists began to study plant communities and classify vegetation types. As the relationships of animals to these plant communities were recognized, naturalists in the early 20th century began dividing Earth's life-forms into *biotic associations.* Recently, the functional relationships of plants, animals, and their physical environment have been the primary focus of attention, and the concept of the ecosystem has become widely used.

Our definition of an ecosystem is both broad and flexible. The term can be used in reference to the Earth system in its entirety (the ecosphere) or to any group of organisms occupying a given area and functioning together with their nonliving environment. An ecosystem may be large or small, marine or terrestrial (on land), short lived or long lasting (• Fig. 11.1). It may even be an artificial ecosystem, such as a farmer's field. When a farmer plants crops, spreads fertilizer, practices weed control, and sprays insecticides, a new ecosystem is created, but this does not alter the fact that plants and animals are living together in an interdependent relationship with the soil, rainfall, temperatures, sunshine, and other factors that constitute the physical environment (• Fig. 11.2).

As noted in Chapter 1, ecosystems are *open systems.* There is free movement of both energy and materials into and out of these systems. They are usually so closely related to nearby ecosystems or so integrated with the larger ecosystems of which they are a part that they are not isolated in nature or readily delimited. Nevertheless, the concept of the ecosystem is a valuable model for examining the structure and function of life on Earth.

R. Gabler

• FIGURE 11.1
This woodland ecosystem in New Hampshire on the slopes of Mount Cardigan demonstrates the close relationship between living organisms and their nonliving environment.
Why might it be difficult for a biogeographer to determine boundaries for this ecosystem?

USDA/Peggy Greb

• FIGURE 11.2
Hybrid seed, fertilizers, insecticides, and, on occasion, irrigation systems may be used by the farmer to ensure the success of this artificial ecosystem in the Corn Belt.
How does the role of humans in an artificial ecosystem differ from that in a natural one?

Major Components

Ecosystems are many and varied, but the typical ecosystem has four basic components. The first of these is the nonliving, or **abiotic,** part of the system. This is the physical environment in which the plants and animals of the system live. In an aquatic ecosystem (a pond, for example), the abiotic component would include such inorganic substances as calcium, mineral salts, oxygen, carbon dioxide, and water. Some of these would be dissolved in the water, but the majority would lie at the bottom as sediments—a natural reservoir of nutrients for both plants and animals. In a terrestrial ecosystem, the abiotic component provides life-supporting elements and compounds in the soil, groundwater, and atmosphere.

The second, and perhaps most important, component of an ecosystem consists of the basic producers, or **autotrophs** (meaning "self-nourished"). Plants, the most important autotrophs, are essential to virtually all life on Earth because they are capable of using energy from sunlight to convert water and carbon dioxide into organic molecules through the process known as photosynthesis (see again Chapter 4). The sugars, fats, and proteins produced by plants through photosynthesis are the foundation for the food supply that supports other forms of life. It should be noted that some bacteria are also capable of photosynthesis and hence are classed as autotrophs along with plants. Sulfur-dependent organisms that dwell at ocean-bottom thermal vents are also classified as autotrophs.

A third component of most ecosystems consists of consumers, or **heterotrophs** (meaning "other-nourished"). These are animals that survive by eating plants or other animals. Heterotrophs are classified on the basis of their feeding habits. **Herbivores** eat only living plant material; **carnivores** eat other animals; **omnivores** feed on both plants and animals. Animals make an essential contribution to the Earth ecosystem of which they are a part. They use oxygen in their respiration and return as an end product to the atmosphere the carbon dioxide that is required for photosynthesis by plants. They can influence soil development through their digging and trampling activities, and those activities in turn may affect local plant distributions.

We might assume that plants, animals, and a supporting environment are all that are required for a functioning ecosystem, but such is not the case. Without the fourth component of ecosystems, the decomposers, plant growth would soon come to a halt. The **decomposers,** or **detritivores,** feed on dead plant and animal material and waste products. They promote decay and return mineral nutrients to the soil and sea in a form that plants can use.

Trophic Structure

From the discussion of the autotrophs and heterotrophs, it becomes apparent that there is a definite arrangement of the major components of an ecosystem. The components form a sequence in their levels of eating: Herbivores eat plants, carnivores may eat herbivores or other carnivores, and decomposers feed on dead plants and animals and their waste products. The pattern of feeding in an ecosystem is called the **trophic structure,** and the sequence of levels in the feeding pattern is referred to as a **food chain.** The simplest food chain would include only plants and decomposers. However, the chain usually includes at least four steps—for example, grass–field mouse–owl–fungi (plants–herbivore–carnivore–decomposer). More complex food chains may include six or more levels as carnivores feed on other carnivores—for example, zooplankton eat plants, small fish eat zooplankton, larger fish eat small fish, bears eat larger fish, and decomposers consume the bear after it dies.

Organisms within a food chain are often identified by their **trophic level,** or the number of steps they are removed from the autotrophs or plants in a food chain (Table 11.1). Plants occupy the first trophic level, herbivores the second, carnivores feeding on herbivores the third, and so forth until the last level, the decomposers, is reached. Omnivores may belong to several trophic levels because they eat both plants and animals.

TABLE 11.1
Trophic Structure of Ecosystems

Ecosystem Component	Trophic Level	Examples
Autotroph	First	Trees, shrubs, grass
	Second	Locust, rabbit, field mouse, deer, cow, bear
Heterotroph	Third	Praying mantis, owl, hawk, coyote, wolf, bear
	Fourth, etc.	Bobcat, wolf, hawk, bear
Decomposer	Last	Fungi, bacteria

In reality, linear food chains do not operate in isolation; they overlap and interact to form a feeding mosaic within an ecosystem called a **food web** (● Fig. 11.3). Both food chains and food webs merit careful study because they can be used to trace the movement of food and energy from one level to another in the ecosystem.

Nutrient Cycles

Some biologists and ecologists find it helpful to separate the trophic structure into specific nutrient cycles. There are several such cycles, which at times intertwine and help explain the routing for most of the nutrients through our ecosystems. Cycles have been developed for water, carbon, nitrogen, oxygen, sulfur, and phosphorous. Some of these cycles may be familiar to you. Parts of the oxygen and carbon cycles were discussed in various sections of Chapter 4. The water (hydrologic) cycle was highlighted in Chapter 6. Though each of these cycles can be singled out individually, ● Figure 11.4 shows a summary diagram, which incorporates the major processes involved in these cycles. Knowledge of chemical nutrient cycles is essential to an understanding of energy flow in ecosystems.

Energy Flow

When physical geographers study ecosystems, they trace the flow of energy through the system just as they do when they study energy flow in other systems, such as streams or glaciers. Just as in other systems, the laws of thermodynamics apply to ecosystems. For example, as the first law of thermodynamics states, energy cannot be created or destroyed; it can only be changed from one form to another. Energy comes to the ecosystem in the form of sunlight, which is used by plants in photosynthesis. This energy is stored in the system in the form of organic material in plants and animals. It flows through the system along food chains and webs from one trophic level to the next. It is finally released from the system when oxygen is combined with the chemical compounds

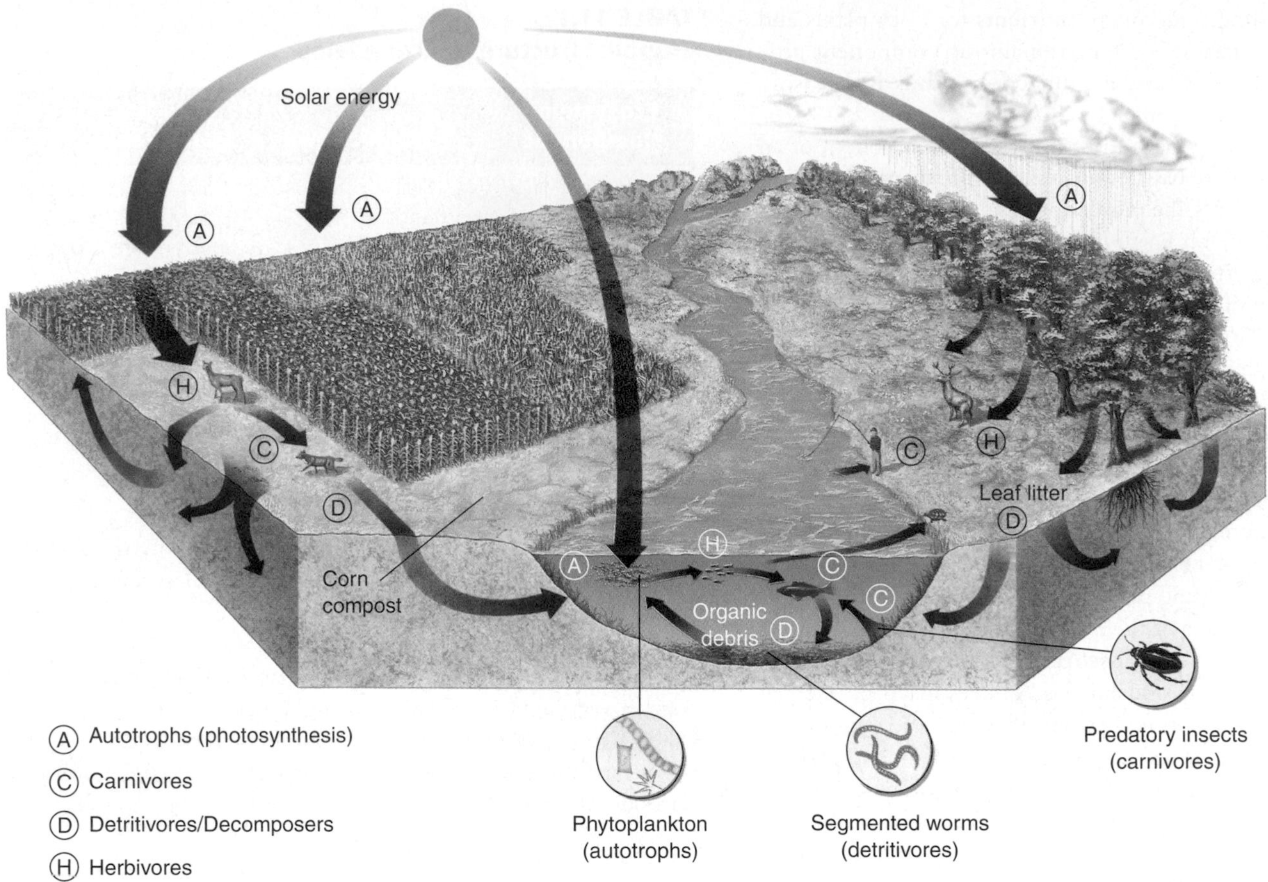

• **FIGURE 11.3**
Environmental Systems: Ecosystems are worthwhile subjects of study by physical geographers. They clearly illustrate the interdependence of the variables in systems, especially the close relationships between the living components of systems (the biosphere) and the nonliving or abiotic components in systems (the atmosphere, hydrosphere, and lithosphere).
Can you trace a trophic structure through this diagram?

of the organic material through the process of oxidation. Respiration, which involves the combination of oxygen with chemical compounds in living cells and can occur at any trophic level, is the major form of oxidation. Fire is yet another form.

The total amount of living material in an ecosystem is referred to as the **biomass.** Because the energy of a system is stored in the biomass, scientists measure the biomass at each trophic level to trace the energy flow through the system. They usually find that the biomass decreases with each successive trophic level (• Fig. 11.5). There are a number of explanations for this, each involving a loss of energy. The first instance occurs between trophic levels. The second law of thermodynamics states that whenever energy is transformed from one state to another there will be a loss of energy through heat. Hence, when an organism at one trophic level feeds on an organism at another, not all of the food energy is used. Some is lost to the system. Additional energy is lost through respiration and movement. At each successive trophic level, the amount of energy required is greater. A deer may graze in a limited area, but the wolf that preys on the deer must hunt over a much larger territory. Whatever the reason for energy loss, it follows that as the flow of energy decreases with each successive trophic level, the biomass also decreases. This principle also applies to agriculture. A great deal more biomass (and food energy) is available in a field of corn than there is in the cattle that eat the corn.

Productivity

Productivity in an ecosystem is defined as the rate at which new organic material is created at a particular trophic level. **Primary productivity** refers to the formation of new organic matter through photosynthesis by autotrophs; **secondary productivity** refers to the rate of formation of new organic material at the heterotroph level.

Primary Productivity

Just how efficient are plants at producing new organic matter through photosynthesis? The answer to this question depends on a number of variables. Photosynthesis requires sunlight, the amount of which depends on the length of day and the angle of the sun's rays, which in turn differ widely with latitude. Photosynthesis is also affected by factors such as soil moisture, temperature, the availability of mineral nutrients, the carbon dioxide content of the atmosphere, and the age and species of the individual plants.

Most studies of productivity in ecosystems have been concerned with measuring the net biomass at the autotroph level

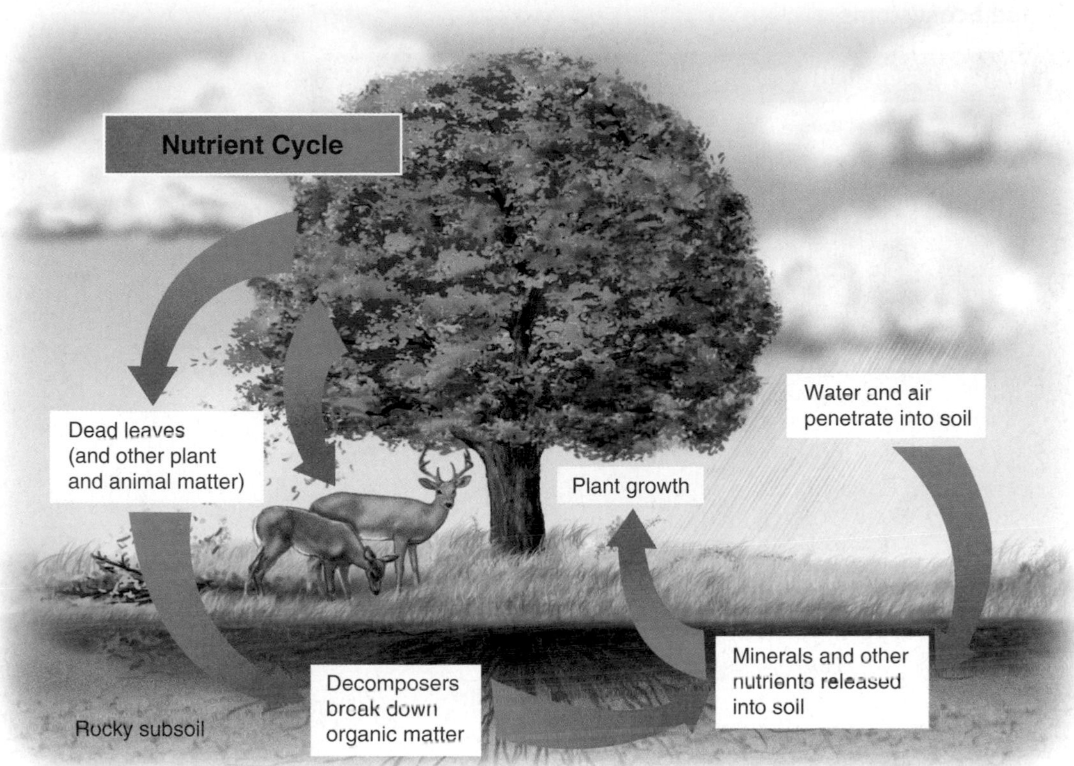

• FIGURE 11.4
This simplified diagram displays the processes used by the nutrient cycles to travel through the ecosystem.
What kinds of processes are taking place beneath the soil surface?

• FIGURE 11.5
Trophic pyramids showing biomass of organisms at various trophic levels in two contrasting ecosystems. Dry weight is used to measure biomass because the proportion of water to total mass differs from one organism to another.
How can you explain the exceptionally large loss of biomass between the first and second trophic levels of the tropical forest ecosystem?

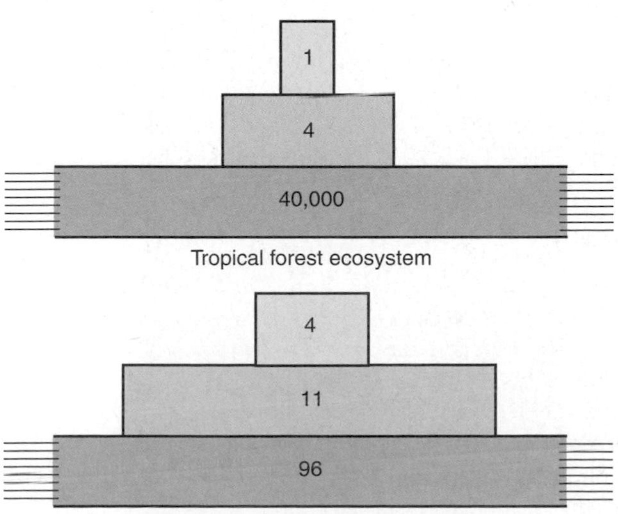

(Table 11.2). Wherever figures have been compiled on the efficiency of photosynthesis, the efficiency has been surprisingly low. Most studies indicate that less than 5% of the available sunlight is used to produce new biomass in ecosystems. For Earth as a whole, the figure is probably less than 1%. Nonetheless, the net primary productivity of the ecosphere is enormous. It is estimated to be in the range of 170 billion metric tonnes (a metric tonne is about 10% greater than a U.S. ton) of organic matter annually. Even though oceans cover approximately 70% of Earth's surface, slightly more than two thirds of net annual productivity is from terrestrial ecosystems and less than one third is from marine ecosystems. Perhaps even more surprising is the fact that humans consume less than 1% of Earth's primary productivity as plant food. However, humans also use biomass in a variety of other ways—for example, we use lumber for construction and paper production, and biomass energy for feedstock and as fodder for range animals.

Table 11.2 illustrates the wide range of net primary productivity displayed by various ecosystems. The latitudinal control of insolation and the subsequent effect on photosynthesis can be easily recognized when comparing figures for terrestrial ecosystems. There is a noticeable decrease in terrestrial productivity from tropical ecosystems to those in middle and higher latitudes. Even the tropical savannas, which are dominated by grasses, produce more biomass in a year than the boreal forests, which are found in the colder climates. Today, satellites monitor Earth's biological productivity and give us a global perspective on our biosphere (• Fig. 11.6).

TABLE 11.2
Net Primary Productivity of Selected Ecosystems

	Net Primary Productivity, gm² per year	
Type of Ecosystem	**Normal Range**	**Mean**
Tropical rainforest	1000–3500	2200
Tropical seasonal forest	1000–2500	1600
Middle-latitude evergreen forest	600–2500	1300
Middle-latitude deciduous forest	600–2500	1200
Boreal forest (taiga)	400–2000	800
Woodland and shrubland	250–1200	700
Savanna	200–2000	900
Middle-latitude grassland	200–1500	600
Tundra and alpine	10–400	140
Desert and semidesert scrub	10–250	90
Extreme desert, rock, sand, and ice	0–10	3
Cultivated land	100–3500	650
Swamp and marsh	800–3500	2000
Lake and stream	100–1500	250
Open ocean	2–400	125
Upwelling zones	400–1000	500
Continental shelf	200–600	360
Algal beds and reefs	500–4000	2500
Estuaries	200–3500	1500

Source: R. H. Whittaker, *Communities and Ecosystems* (2nd ed.). New York: Macmillan, 1975.

• FIGURE 11.6
Worldwide vegetation patterns revealed through a color index derived from environmental satellite observations. Compare this image with the world map of natural vegetation in Figure 11.23.
What color on this map represents desert vegetation?

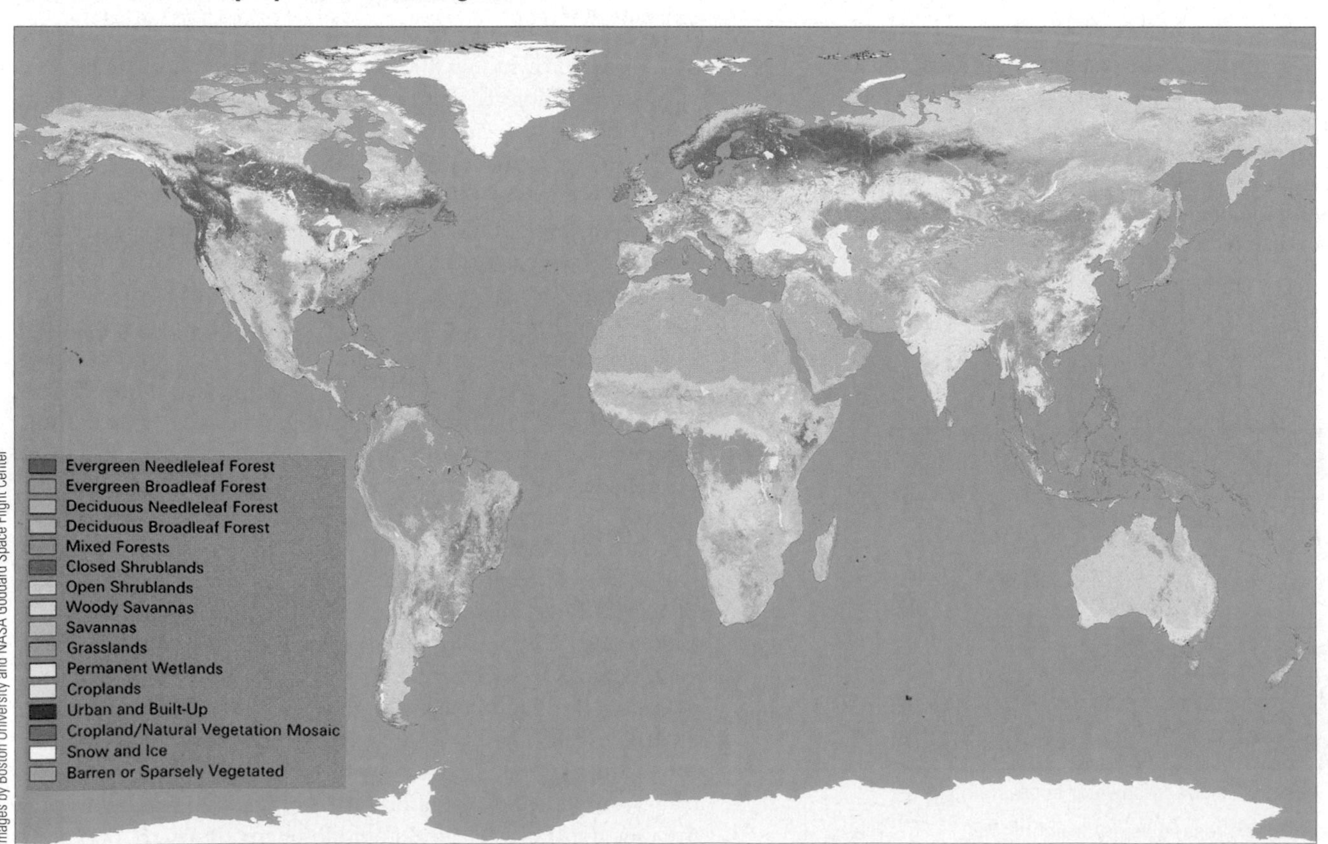

Images by Boston University and NASA Goddard Space Flight Center

The reasons for differences among aquatic, or water-controlled, ecosystems are not quite as apparent. Swamps and marshes are especially well supplied with plant nutrients and therefore have a relatively large biomass at the first trophic level. On the other hand, depth of water has the greatest impact on ocean ecosystems. Most nutrients in the open ocean sink to the bottom, beyond the depth where sunlight can penetrate and make photosynthesis possible. Hence, the most productive marine ecosystems are found in the sunlit, shallow waters of estuaries, continental shelves, or coral reefs or in areas where ocean upwelling carries nutrients nearer to the surface (● Fig. 11.7).

Some artificial ecosystems associated with agriculture can be fairly productive when compared with the natural ecosystems they have replaced. This is especially true in the warmer latitudes where farmers may raise two or more crops in a year or in arid lands where irrigation supplies the water essential to growth. However, Table 11.2 indicates that mean productivity for cultivated land does not approach that of forested land and is just about the same as that of middle-latitude grasslands. Most quantitative studies have shown that agricultural ecosystems are significantly less productive than natural systems in the same environment.

● FIGURE 11.7
The nutrient-rich waters weave through the Great Barrier Reef off Mackay, Australia.
Why are the most productive marine ecosystems found in the shallow waters bordering the world's continents?

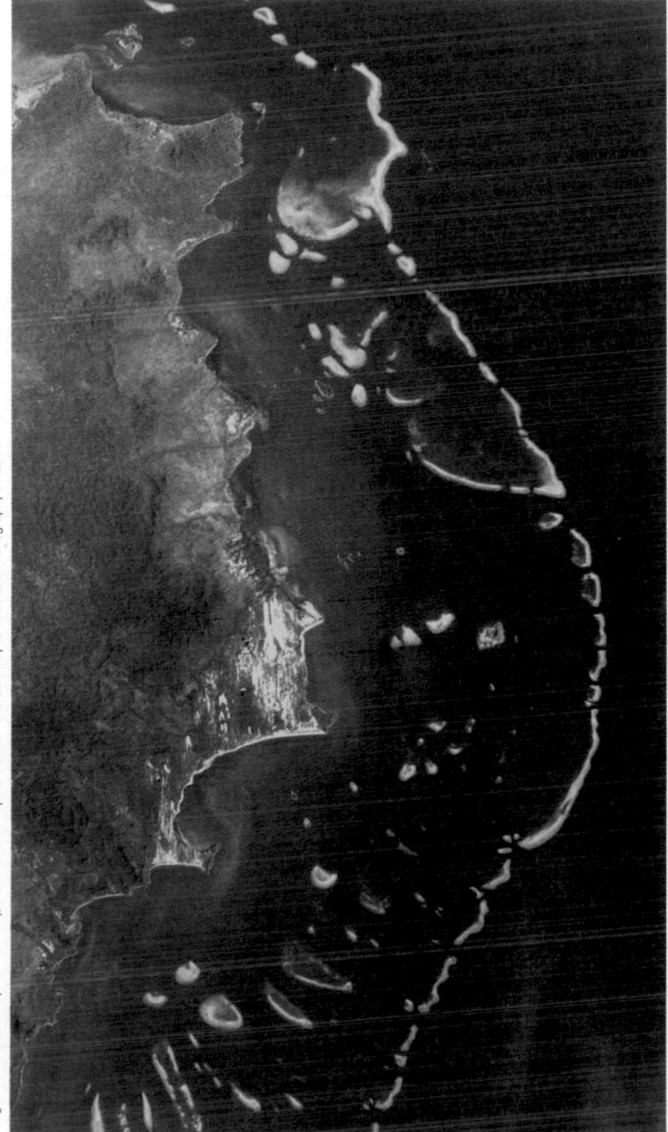

Image Science and Analysis Laboratory, NASA-Johnson Space Center. "The Gateway to Astronaut Photography of Earth."

Secondary Productivity As we have seen, secondary productivity results from the conversion of plant materials to animal substances. We have also noted that the ecological efficiency, or the rate of energy transfer from one trophic level to the next, is low (● Fig. 11.8). The efficiency of transfer from autotrophs to heterotrophs varies widely from one ecosystem to another. The amount of net primary productivity actually eaten by herbivores may range from as high as 15% in some grassland areas to as low as 1 or 2% in certain forested regions. In ocean ecosystems, the figure may be much higher, but there is a greater loss during the digestion process. Once the food is eaten, energy loss through respiration or body movement reduces secondary productivity to a small fraction of the biomass available as net primary productivity.

● FIGURE 11.8
Productivity at the autotroph, herbivore, and carnivore trophic levels as measured in a Tennessee field. The figures represent productivity for 1 square meter of field in 1 year. Note the extremely small proportion of primary productivity that reaches the carnivore level of the food chain.
In this example, which group—carnivores or herbivores—is more efficient (produces the greater percentage of energy available at the trophic level immediately below it in the food chain)?

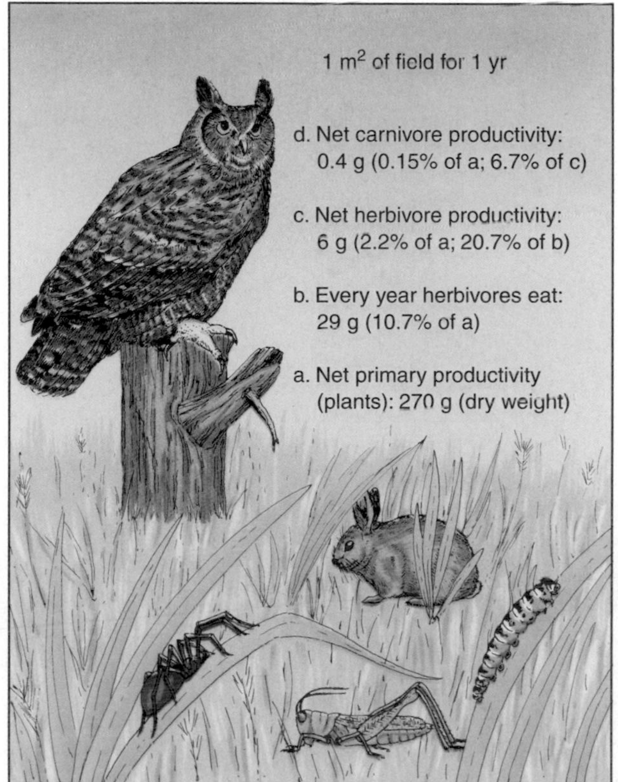

Most authorities consider 10% to be a reasonable estimate of ecological efficiency for both herbivores and carnivores. If both herbivores and carnivores have ecological efficiencies of only 10%, the ratio of biomass at the first trophic level to biomass of carnivores at the third trophic level is several thousand to one. It obviously requires a huge biomass at the autotroph level to support one animal that eats only meat. As human populations grow at increasing rates and agricultural production lags behind, it is indeed fortunate that human beings are omnivores and can adopt a more vegetarian diet (● Fig. 11.9).

● FIGURE 11.9
The triangles illustrate the advantages of a vegetarian diet as we experience another century of rapid population growth. It is fortunate that humans are omnivores and can choose to eat grain products. The same 1350 kilograms of grain that will support, if converted to meat, only 1 person will support 22 people if cattle or other animals are omitted from the food chain.
In what areas of the world today do grain products constitute nearly all of the total food supply?

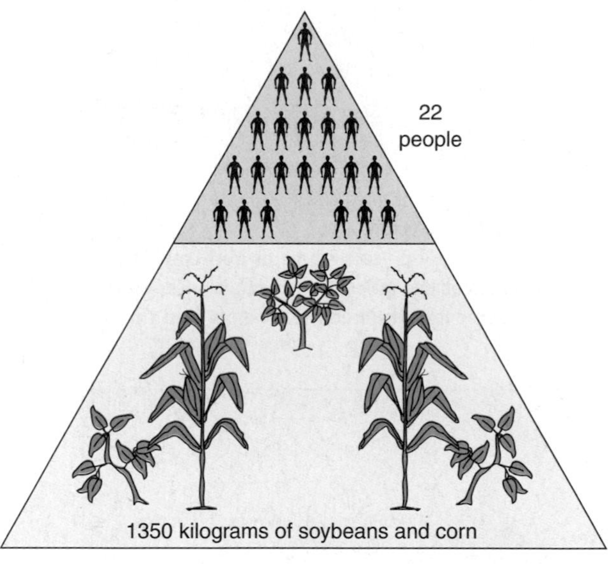

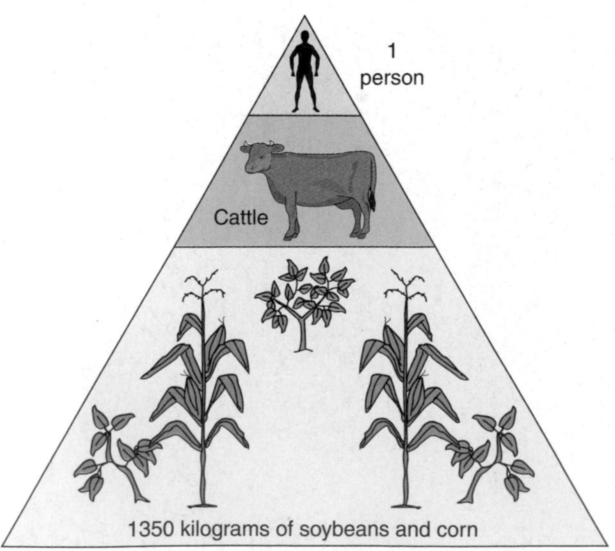

Ecological Niche

There are a surprising number of species in each ecosystem, except for those ecosystems severely restricted by adverse environmental conditions. Yet each organism performs a specific role in the system and lives in a certain location, described as its **habitat.** The combination of role and habitat for a particular species is referred to as its **ecological niche.** A number of factors influence the ecological niche of an organism. Some species are **generalists** and can survive on a wide variety of food. The North American brown, or grizzly, bear, as an omnivore, will eat berries, honey, and fish. On the other hand, the koala of Australia is a specialist and eats only the leaves of certain eucalyptus trees. Specialists do well when their particular food supply is abundant, but they cannot adapt to changing environmental conditions. The generalists are in the majority in most ecosystems because their broader ecological niche allows survival on alternative food supplies.

Some generalists among species occupy an ecological niche in one ecosystem that is quite different from the niche they occupy in another. As food supply varies with habitat, so varies the ecological niche. Humans are the extreme example of the generalist: In some parts of Earth, they are carnivores; in some parts, herbivores; and in some parts, omnivores. It is also true that different species may occupy the same ecological niche in habitats that are similar but located in separate ecosystems.

Succession and Climax Communities

Up to this point, we have been discussing ecosystems in general terms. In the remainder of this chapter, we note that it is the species that occupy the ecosystem that give the ecosystem its character. At least for terrestrial ecosystems, it is the autotrophs—the plant species at the first trophic level—that most easily distinguish one ecosystem from another. All other species in an ecosystem depend on the autotrophs for food, and the association of all living organisms determines the energy flow and the trophic structure of the ecosystem. It should also be noted that the species that occupy the first trophic level are greatly influenced by climate; again we see the interconnections between the major Earth subsystems.

If the plants that comprise the biomass of the first trophic level are allowed to develop naturally without obvious interference from or modification by humans, the resulting association of plants is called **natural vegetation.** These plant associations, or **plant communities,** are compatible because each species within the community has different requirements in relation to major environmental factors such as light, moisture, and mineral nutrients. If two species within a community were to compete, one would eventually eliminate the other. The species forming a community at any specific place and time will be an aggregation of those that together can adapt to the prevailing environmental conditions.

Succession

Natural vegetation of a particular location develops in a sequence of stages involving different plant communities. This developmental process, known as **succession,** usually begins with a relatively simple plant community. Two types of succession, *primary* and *secondary,* are recognized. In primary succession, a bare substrate is the beginning point. No soil or seedbed exists at this point. A *pioneer* community invades the bare substrate (whether it be volcanic lava, glacially deposited sediment, or a bare beach, among others) and begins to alter the environment. As a result, the species structure of the ecosystem does not remain constant. In time, the alterations of the environment become sufficient to allow a new plant community (a community that could not have survived under the original conditions) to appear and eventually to dominate the original community. The process continues with each succeeding community rendering further changes to the environment. Because of the initial absence of soil, primary succession can take hundreds or even a few thousand years. Secondary succession begins when some natural process, such as a forest fire, tornado, or landslide, has destroyed or damaged a great deal of the existing vegetation. Ecologists refer to this process as **gap** creation. Even with such damage, however, soil still typically exists, and seeds may be lying dormant in that soil ready to invade the newly opened gap. Secondary succession, therefore, can occur much more quickly than primary succession.

A common form of secondary succession associated with agriculture in the southeastern United States is depicted in • Figure 11.10. After agriculture has ceased, weeds and grasses are the first vegetative types to adapt to the somewhat adverse conditions associated with bare fields. These low-growing plants will stabilize the topsoil, add organic matter, and in general pave the way for the development of hardwood brush such as sassafras, persimmon, and sweet gum. During the brush stage, the soil will become richer in nutrients and organic matter, and its ability to retain water will increase. These conditions encourage the development of pine forests, the next stage in this vegetative evolutionary process. Pine forests thrive in the newly created environment and will eventually dwarf and dominate the weeds, grasses, and brush that preceded them.

Ironically, the dominance of the pine forest leads to its demise. Pine trees require much sunlight if their seeds are to germinate. When competing with low-lying brush, grasses, and weedy annuals, there is no problem in getting enough sunlight, but once the pines dominate the landscape, their seeds will not germinate in the shade and litter that their dense foliage creates. Thus, pines eventually will give way to hardwoods, such as oak and hickory, whose seeds can germinate under those conditions. These seeds may have been present and dormant in the soil, or blown into the forest, or carried into it by animals. In this example, then, the end result is an oak–hickory forest. If the changes continue uninterrupted, it is estimated that the complete succession will take 100–200 years. In other ecosystems, such as a tropical rainforest, this complete process may take many hundreds of years.

The Climax Community

The theory of plant succession was introduced early in the 20th century. However, some of the original ideas have undergone considerable modification. Succession was considered to be an orderly process that included various *predictable* steps or phases and ended with a dominant vegetative cover that would remain in balance with the environment until disturbed by human activity—or until there were major changes (for example, a climate change) in the

• FIGURE 11.10

A common plant succession in the southeastern United States. Each succeeding vegetation type alters the environment in such a way that species having more stringent environmental requirements can develop.
Why would plant succession be quite different in another region of the United States?

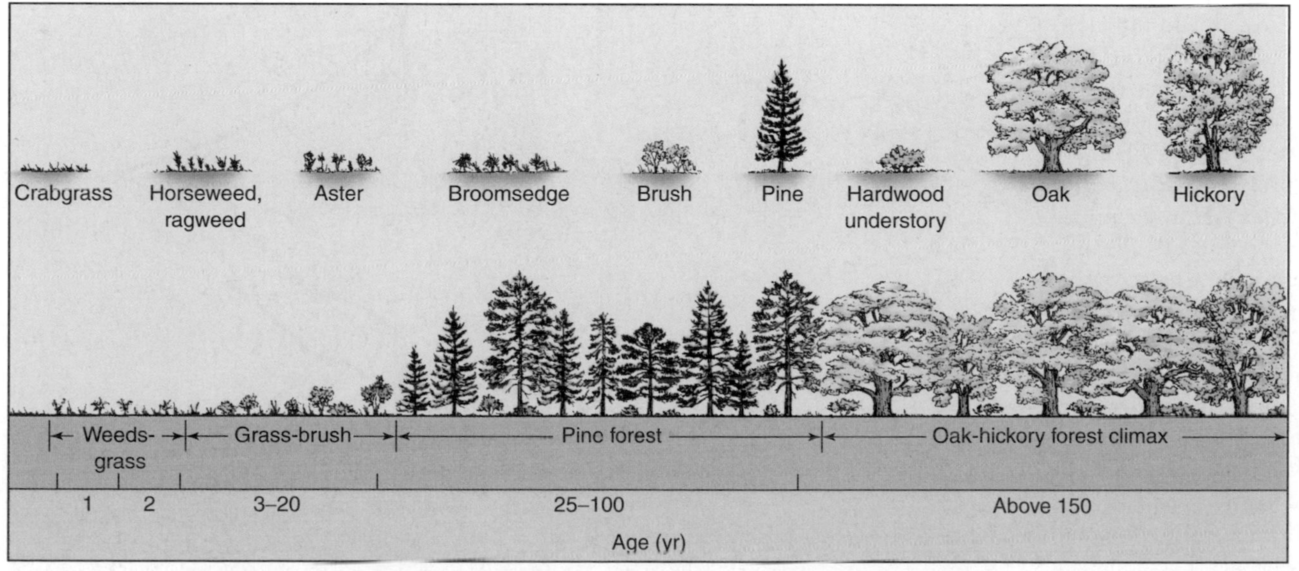

environment. The final step in the process of succession has been referred to as the **climax community.** It was generally agreed that such a community was self-perpetuating and had reached a state of equilibrium or stability with the environment. In our illustration of plant succession in the southeastern United States, the oak–hickory forest would be considered the climax community.

Succession is still a useful model in the study of ecosystems. Why, then, have some of the original ideas been challenged, and what is the most recent thought on the subject? For one thing, early proponents of succession emphasized the predictable nature of the theory. One plant community would follow another in regular order as the species structure of the ecosystem evolved. But it has been demonstrated on many occasions that changes in ecosystems do not occur in such a rigid fashion. More often than not, the movement of species into an area to form a new community occurs in a random fashion and may be largely a function of chance events.

Many scientists today no longer believe that only one type of climax vegetation is possible for each major climate region of the world. Some suggest that one of several different climax communities might develop within a given area, influenced not only by climate but also by drainage conditions, nutrients, soil, or topography. The dynamic nature of climate is now also much more fully understood than it was when the theories of succession and climax were developed, so it is now seen that by the time the species structure of a plant community has adjusted to new climatic conditions, the climate may change again. Because of the dynamic nature of each habitat, no one climax community can exist in equilibrium with the environment for an indefinite period of time. Many biogeographers and ecologists today view plant communities and their ecosystems as a *landscape* that is an expression of all the various environmental factors functioning together. They view the landscape of an area as a **mosaic** of interlocking parts, much like the tiles in a mosaic artwork. In a pine forest, for example, other plants also exist, and some areas do not support pine trees. The dominant area of the mosaic—that is, the pine forest—is referred to as the **matrix.** Gaps within the matrix, resulting from areas of different soil conditions or gaps created by human-created or natural processes, are referred to as **patches** within the matrix (• Fig. 11.11 and • Fig. 11.12). Relatively linear features cutting across the mosaic, including natural features such as rivers and human-created structures such as roads, fence lines, power lines, and hedgerows, are termed **corridors** (• Figs. 11.13 and 11.14). Each particular habitat is unique and constantly changing, and resultant plant and animal communities must constantly adjust to these changes. The dominant environmental influence is climate.

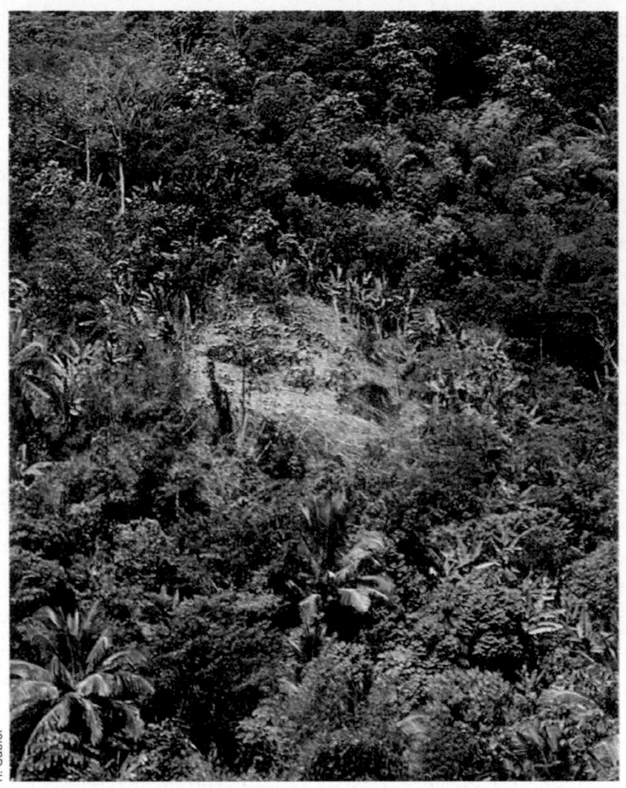

R. Gabler

• **FIGURE 11.11**
This patch in the tropical rainforest matrix of Jamaica is the result of land cleared for shifting (slash-and-burn) cultivation.
What types of human activity might be responsible for patches in the matrix of a middle-latitude forest?

• **FIGURE 11.12**
This patch in the middle-latitude forest near Vancouver, British Columbia, Canada, was the result of a 160 kilometer per hour (100 mph) wind storm.
Why do you suppose the jagged tree trunks were sawed flat?

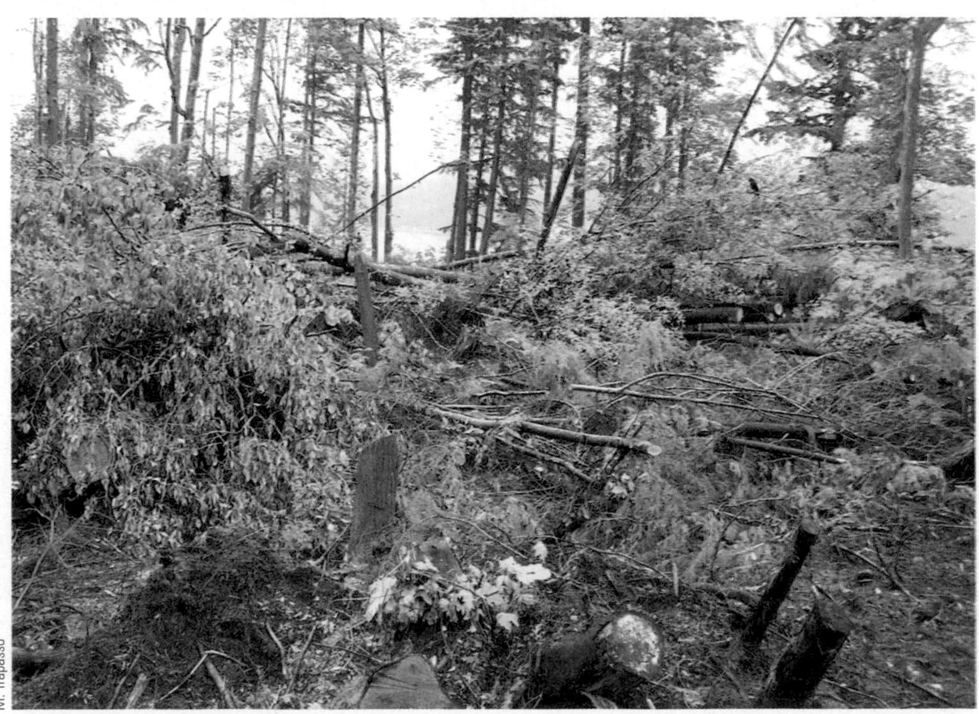

M. Trapasso

As discussed in Chapter 8, climates are also changing. Climate changes occur over both relatively short time periods—on the scales of decades and centuries—and much longer times, measured in millennia. They may be subtle, or they may be sufficiently drastic to create ice ages or warm periods between ice ages. Plant and animal communities must be able to respond to these ongoing changes, or they will not survive. Many modern biogeographers are deeply involved in the reconstruction of the vegetation communities of past climate periods, through the examination of evidence such as tree rings, pollen, insect fossils, and plant fossils. By determining how past climate changes affected and induced change in Earth's ecosystems, biogeographers hope to be able to suggest how future changes may develop as climate continues to change.

© April Baher, CBNERRVA/NOAA National Estuarine Research Reserve Collection

• **FIGURE 11.13**
This aerial view of Taskinas Creek, Virginia, shows riparian corridors passing through a forest matrix on the Atlantic coastal plain.
How does a corridor differ from a patch?

• **FIGURE 11.14**
This view near Franklin, Tennessee, shows a human-made power line corridor.
Why is it important to clear this particular type of corridor?

Jamison R. Gorrell

Environmental Controls

The plant and animal species occupying a particular ecosystem at a given time are those that are most successful in adjusting to the unique environment that constitutes their habitat. Each species has a range within which it can adapt to environmental factors. For example, some plants can survive under a wide range of temperature conditions, whereas others have narrow temperature requirements. Biogeographers and ecologists refer to this characteristic as an organism's range of **tolerance** for a particular environmental condition. The ranges of tolerance for a species will determine where on Earth that species may be found, and species with wide ranges of tolerance will be the most widely distributed. The *ecological optimum* refers to the environmental conditions under which a species will flourish (• Fig. 11.15). As a species moves away from its ecological optimum or as one moves away from the geographic core of a plant or animal community, the environmental conditions become increasingly difficult for that species or community to survive. At the same time, those conditions may be more amenable for another species or community. The **ecotone** is the overlap, or zone of transition, between two plant or animal communities (see again Fig. 11.15).

Climate has the greatest influence over natural vegetation when we observe plant communities on a worldwide basis. The major types of terrestrial ecosystems, or **biomes,** are each associated with specific ranges of temperature and critical precipitation characteristics such as annual amounts and seasonal distribution. Climate influences leaf shape and size in trees and determines if trees can even exist in a region. At the local scale, however, other environmental factors can be as important as climate. A plant's range of tolerance for the acidity of the soil, the drainage of the land, or the salinity of the water may be the critical environmental factor in determining whether that plant is a part of the ecosystem. The discussion that follows serves to illustrate how the major environmental factors influence the organization and structure of ecosystems.

GEOGRAPHY'S ENVIRONMENTAL SCIENCE PERSPECTIVE

The Theory of Island Biogeography

Biogeographers are intrigued by the life-forms and the diversity of species that exist on islands—areas that are isolated from larger landmasses by ocean environments that are not inhabited by these life-forms. How can there be land plants and animals living on an island surrounded by a wide expanse of sea? How did this assortment of flora and fauna become established and flourish on these often distant and geologically recent terrains (volcanic islands, for example, which were barren after they formed)? The farther an island is from the nearest landmasses, the more difficult it is for species to migrate to the island and to establish a viable population there.

The seeds of some plants are carried by the wind, by birds, or by currents to islands, and germinate to develop the vegetative environments on these isolated landmasses. Many other plant and animal species living on islands were introduced to these remote locations by humans as they migrated to these islands. But why did the species adapt and survive?

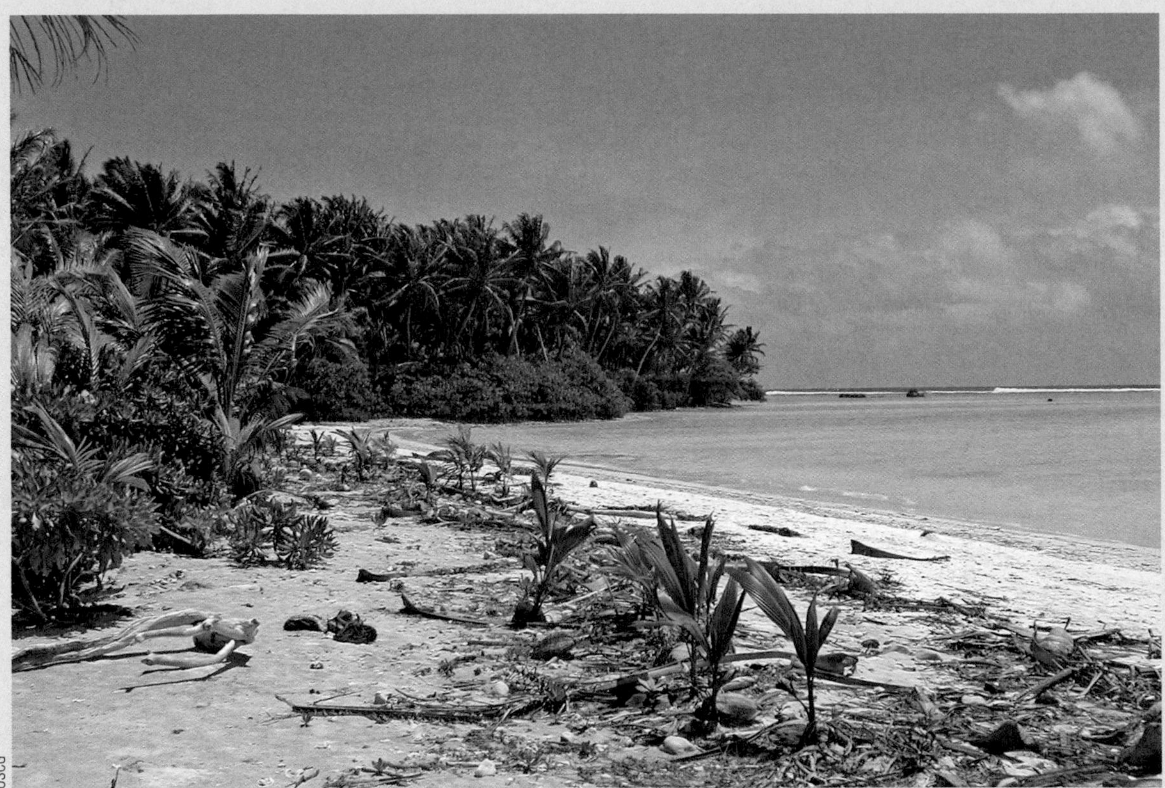

USGS

This beach scene is on Palmyra Island, in the Pacific Ocean, which is one of the most remote islands in the world. The photo illustrates how palm trees become established on tropical islands. Coconuts are the seeds for coconut palm trees and they can float and drift hundreds or thousands of kilometers from one island to another. Waves deposit them on the beach, and buried on the shoreline, the palms sprout and grow. Note the coconuts, recently sprouted palms, and the fully grown coconut palm trees. Only a few other plant species grow here on this very small island. Greater species diversity can be found on islands that are larger and not so distant from major landmasses.

The theory of island biogeography offers an explanation for how natural factors interact to affect successful colonization or extinction of species that initially come to live on an island. The theory considers the degree of isolation of the island (the distance from a mainland source of migrating species), the size of the island, and the number of species living on an island. Generally, the diversity of life-forms on islands is low compared to mainland locations with a similar climate and other environmental characteristics. Low diversity of species typically also means that the floral and faunal populations of that place exist in an environmentally sensitive location. Many extinctions have occurred on islands because of the introduction of some factor that made the habitat nonviable for that species to survive.

Several major factors affect the species diversity on islands (as long as other environmental conditions such as climate are comparable):

1. The farther an island is from the area from which species must migrate, the lower the species diversity. Islands nearer to large landmasses tend to have higher diversity than those that are more distant.
2. The larger the island, the greater the species diversity. This is partly because larger islands tend to offer a wider variety of environments to colonizing organisms than smaller islands do. Larger islands also offer more space for species to occupy.
3. The species diversity of an island results from an equilibrium between the rates of extinction of species on the island and the colonization of species. If the island's extinction rate is higher, only a few hardy species will live there; if the extinction rate is lower compared to the colonization rate, then more species will thrive, and the diversity will be higher.

The theory of island biogeography has also been useful in understanding the ecology and biota of many other kinds of isolated environments, such as high mountain areas that stand, much like islands, above surrounding deserts. In those regions, plants and animals adapted to cool wet environments live in isolation from similar populations on nearby mountains, separated by inhospitable arid environments.

Chris Simpson/Getty Images

The volcanic peak known as Tunupa rises from the vast Uyuni Salt Flats, in Bolivia. Tunupa has a variety of ecosystems unique to its slopes. Isolated mountains, surrounded by arid environments, also tend to fit the concept of island biogeography.

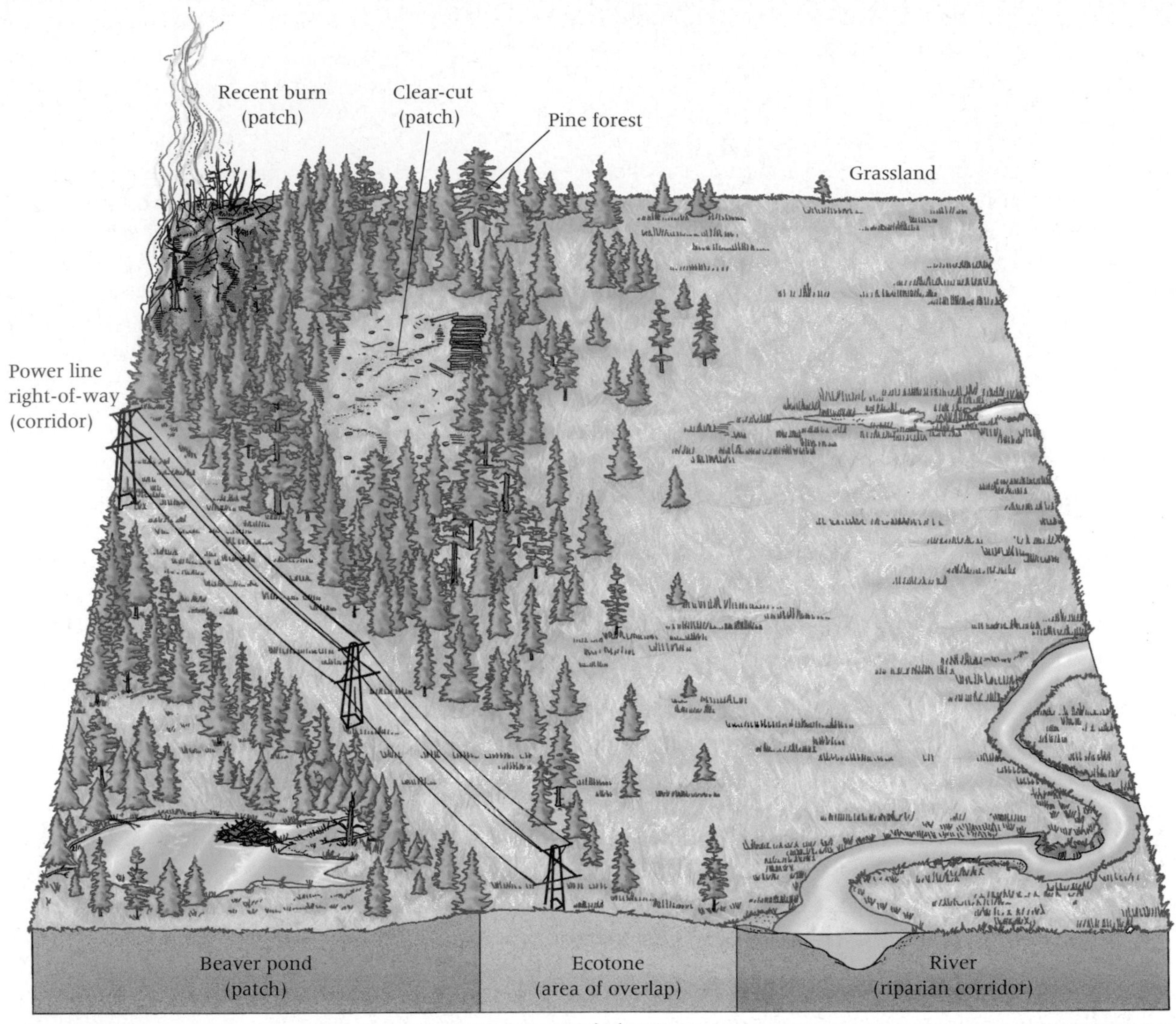

• FIGURE 11.15

The concepts of *ecotone, ecological optimum, range of tolerance, mosaic, matrix, patch,* and *corridor* are illustrated in the diagram.

What effect would a change to drier climatic conditions throughout the area have on the relative sizes of the two ecosystems as well as the position of the ecotone?

Climatic Factors

Of all the various climatic factors that influence the ecosystem, sunlight conditions are often the most critical. Sunlight is the vital source of energy for photosynthesis in plants and a control of life patterns for animals. The competition for light makes forest trees grow taller and limits growth on the forest floor to plants, such as ferns, that can tolerate shade conditions. Leaf sizes, shapes, and even colors may reflect this variation in light reception, with large leaves in areas of limited light reception. The *quality* of light is important, especially in mountain areas, where plant growth may be severely retarded by excess ultraviolet radiation. This radiation does significant damage in the thin air at higher elevations but is effectively screened out by the denser atmosphere at lower elevations. Light *intensity* affects the rate of photosynthesis and hence the rate of primary productivity in an ecosystem. The more intense light of the low latitudes produces a higher energy input and greater biomass in the tropical forest than does the less intense light of the higher latitudes in Arctic regions. The *duration* of daylight, in association with the changing seasons, has a profound effect on the flowering of plants, the activity patterns of insects, and the migration and mating habits of animals.

A second important climatic control of ecosystems is the availability of water. Virtually all organisms require water to survive. Plants require water for germination, growth, and reproduction, and most plant nutrients are dissolved in soil water before

R. Gabler

• **FIGURE 11.16**
Mangrove thicket along the Gulf of Mexico coast of southern Florida.
How might this vegetation type have influenced the routes that were followed by the Spanish adventurers who first explored Florida?

they can be absorbed by plants. Marine plants are adapted to living completely in water (seaweed); some plants, such as mangrove (• Fig. 11.16) and bald cypress (• Figs. 11.17a and 11.17b), rise from coastal marshes and inland swamps; others thrive in the constantly wet rainforest. Certain tropical plants drop their leaves and become dormant during dry seasons; others store water received during periods of rain in order to survive seasons of drought. Desert plants, such as cacti, are especially adapted to obtaining and storing water when it is available while minimizing their water loss from transpiration.

Animals, too, are severely restricted when water is in short supply. In arid regions, animals must make special adaptations to environmental conditions. Many become inactive during the hottest and driest seasons, and most leave their burrows or the shade of plants and rocks only at night. Others, like the camel, can travel for great distances and live for extended periods without a water supply.

• **FIGURE 11.17**
(a) This stand of bald cypress trees is located in extreme southern Illinois, at the poleward limit of growth for this type of vegetation. (b) Extensive cypress forests exist in swampy areas of the southeastern United States. These are cypress "knees," root extensions that stick up above the water to supply air to the trees.
What Köppen climatic type does the site in Illinois represent?

R. Gabler

(a)

National Park Service Digital Archive

(b)

Organisms are affected less by temperature variations than by sunlight and water availability. Many plants can tolerate a wide range of temperatures although each species has optimum conditions for germination, growth, and reproduction. These functions, however, can be impeded by temperature extremes. Temperatures may also have indirect effects on vegetation. For example, high temperatures will lower the relative humidity, thus increasing transpiration. If a plant's root system cannot extract enough moisture from the soil to meet this increase in transpiration, the plant will wilt and eventually die.

Because of their mobility, animals are not as dependent on the vagaries of climate as are plants. Despite the great advantage afforded by mobility, however, animals are nevertheless subject to climatic stress. The geographic distribution of some groups of animals reflects this sensitivity to climate. Cold-blooded animals are, for example, more widespread in warmer climates and more restricted in colder climates. Some warm-blooded animals develop a layer of fat or fur and are able to shiver to protect themselves against the cold. In hot periods, they may sweat, shed fur, or lick their fur in an attempt to stay cool. In extremely cold or arid regions, animals may hibernate. During hibernation, the body temperature of the animal changes roughly in response to outside and ground temperatures. Cold-blooded animals such as the desert rattlesnake move in and out of shade in response to temperature change. Warm-blooded animals may migrate great distances out of environmentally harsh areas.

Some warm-blooded animals exhibit an interesting linkage between body shape and size and variations in average environmental temperature. These adaptations have been described by biologists as *Bergmann's Rule* and *Allen's Rule.* Bergmann's Rule states that, within a warm-blooded species, the body size of the subspecies usually increases with the decreasing mean temperature of its habitat; Allen's Rule notes that, in warm-blooded species, the relative size of exposed portions of the body decreases with the decrease of mean temperature. These rules essentially boil down to the fact that members of the same species living in colder climates eventually evolve shorter or smaller appendages (ears, noses, arms, legs, and so on) than their relatives in warmer climates (Allen's Rule) and that in cold climates body size will be larger with more mass to provide the body heat needed for survival and for protection of the main trunk of the body where vital organs are located (Bergmann's Rule). In cold climates, small appendages are advantageous because they reduce the amount of exposed area subject to temperature loss, frostbite, and cellular disruption (• Fig. 11.18). In warm climates, large body sizes are not necessary for protection of internal organs, but long limbs, noses, and ears allow for heat dissipation in addition to that provided by panting or fur licking.

Although most significant in areas such as deserts, polar regions, coastal zones, and highlands, wind can also serve as a climatic control. Wind may cause direct injury to vegetation or may have an indirect effect by increasing the rate of evapotranspiration. To prevent water loss in the areas of severe wind stress, plants will twist and grow close to the ground to minimize the degree of their exposure (• Fig. 11.19). During severe winters, they are better off buried by snow than exposed to icy gales. In some coastal regions, the shoreline may be devoid of trees or other tall plants; where trees do grow, they are often misshapen or swept bare of leaves and branches on their windward sides.

Soil and Topography

In terrestrial ecosystems, the soils in which plants grow supply much of the moisture and minerals that are transformed into plant tissues. Soil variations are among the most conspicuous influences on plant distribution and often produce sharp boundaries in vegetation type. This is partly a consequence of the varying chemical requirements of different plant species and partly a reflection of other factors such as soil texture. In a particular area, clay soil may retain too much moisture for certain plants, whereas sandy soil retains too little. It is well known that pines thrive in sandy soils, grasses in clays, cranberries in acid soils, and wheat and chili peppers in alkaline soils. The subject of soils and their influence on vegetation will be explored more thoroughly in Chapter 12.

In the discussion of highland climate regions in Chapter 10, we learned that topography influences ecosystems indirectly by providing many microclimates within a relatively small area. Plant communities vary significantly from place to place in highland areas in response to the differing nature of the climatic conditions. Some plants thrive on the sunny south-facing slopes of highland

• FIGURE 11.18

The Arctic polar bear provides an excellent example of both Allen's and Bergmann's Rules. **What warm-climate animal might you suggest as a prime example of Allen's Rule?**

Alaska Image Library/USFWS

R. Gabler

• **FIGURE 11.19**
Krummholz vegetation at the upper reaches of the subalpine zone on Pennsylvania Mountain in the Mosquito Range of the Colorado Rockies. The healthy green vegetation has been covered by snow much of the year and has been protected from the bitterly cold temperatures associated with gale-force winter winds. Note the flag trees, which give a clear indication of wind direction.
What type of vegetation would be found at elevations higher than the one depicted in this photograph?

areas in the Northern Hemisphere; others survive on the colder, shaded, north-facing slopes. The steepness and shape of a slope also affect the amount of time water is present there before draining downslope.

Luxuriant forests tower above the well-watered windward sides of mountain ranges such as the Sierra Nevada and Cascades, but semiarid grasslands and sparse forests cover the leeward sides. Spatial variations in precipitation drainage and resulting vegetation differences would not exist were it not for the presence of the topographic barrier inducing orographic uplift. Each major increase in elevation also produces a different mixture of plant species that can tolerate the lower ranges of temperature found at the higher elevation.

Natural Catastrophes

The distributions of plants and animals are also affected by a diversity of natural processes frequently termed *catastrophes.* It should be noted, however, that this term is applied from a strictly human perspective. What may be catastrophic to a human, such as a hurricane, forest fire, landslide, tsunami, or avalanche (• Fig. 11.20), is simply a natural process operating to produce openings (gaps) in the prevailing vegetative mosaic of a region. The resulting successional processes, whether primary or secondary in nature, produce a variety of patch habitats within the broader regional matrix of vegetation. The study of natural catastrophes and the resulting patch dynamics among the plant and animal residents of an area is a topic of strong interest and ongoing research in modern landscape biogeography.

Biotic Factors

Although they might tend to be overlooked as environmental controls, other plants and animals may be the critical factors in determining whether a given organism is a part of an ecosystem. Some interactions between organisms may be beneficial to both species involved; this is called a **symbiotic relationship.** However other relationships may be of a more *competitive nature* and may have an adverse effect on one or both. Because most ecosystems are suitable to a wide variety of plants and animals, there is always competition between species and among members of a given species to determine which organisms will survive. The greatest competition occurs between species that occupy the same ecological niche, especially during the earliest stages of life cycles when organisms are most vulnerable. Among plants, the greatest competition is for light. Those trees that become dominant in the forest are those that grow the tallest and partially shade the plants growing beneath them.

• **FIGURE 11.20**
The vegetation mosaic of this area in Glacier National Park, Montana, is coniferous forest, but frequent snow avalanches keep rigid-stemmed conifers from invading the patch of low shrubs and grasses.
Why are there so many broken tree stumps in the foreground?

USGS/P. Carrara

Other competition occurs underground, where the roots compete for soil water and plant nutrients.

Interactions between plants and animals and competition both within and among animal species also have significant effects on the nature of an ecosystem. Animals are often helpful to plants during pollination or the dispersal of seeds, and plants are the basic food supply for many animals. The simple act of grazing may help determine the species that make up a plant community. During dry periods, herbivores may be forced to graze an area more closely than usual, with the result that the taller plants are quickly grazed out. Plants that grow close to the ground, that are unpalatable, or that have the strongest root development are the ones that survive. Hence, grazing is a part of the natural selection process, but serious overgrazing rarely results under natural conditions because wild animal populations increase or decrease with the available food supply. To be more precise, the number of animals of a given species will fluctuate between the maximum number that can be supported when its food supply is greatest and the minimum number required for reproduction of the species. For most animals, predators are also an important control of numbers. Fortunately, when the predator's favorite species is scarce, it will seek an alternative species for its food supply.

UNEP-Sudan

• **FIGURE 11.21**
Overgrazing is a major cause of desertification here at this location in Sudan. The environment normally would have been a grassy savanna area with scattered trees.
What are some of the other causes of desertification?

Human Impact on Ecosystems

Throughout human history, we have modified the natural development of ecosystems. Except in regions too remote to be altered significantly by civilization, humans have eliminated much of Earth's natural vegetation. Farming, fire, grazing of domesticated animals, deforestation and afforestation, road building, urban development, dam building and irrigation, raising and lowering of water tables, mining, and the filling in or draining of wetlands are just a few ways in which humans have modified the plant communities around them. Overgrazing by domesticated animals can seriously harm marginal environments in semiarid climates. Trampling and compaction of the soil by grazing herbivores may reduce the soil's ability to absorb moisture, leading to increased surface runoff of precipitation. In turn, the decreased absorption and increased runoff may respectively lead to *land degradation* and gully erosion.

It should be noted that ecosystems are not the only victims as humans alter natural environments; the changes can often produce long-term negative effects on humans themselves. The desertification of large tracts of semiarid portions of East Africa has resulted periodically in widespread famine in countries such as Ethiopia and Somalia (• Fig. 11.21). Elsewhere, the continuing destruction of wetlands not only eliminates valuable plant and animal communities but also often seriously threatens the quality and reliability of the water supplies for the people who drained the land.

We have in fact so changed the vegetation in some parts of the world that we can characterize classes of cultivated vegetation cared for by humans—for example, flowers, shrubs, and grasses to decorate our living areas and grains, vegetables, and fruits that we raise for our own food and to feed the animals that we eat. Our focus in the remainder of this chapter, however, is on the major ecosystems of Earth as we assume they would appear without human modification. Human impact upon and adaptation to the various vegetation types are described in more detail along with the appropriate climate types in Chapters 9 and 10.

Classification of Terrestrial Ecosystems

Classifying the geography of plant communities is no easy task, as their distribution is a complex phenomenon influenced by a variety of factors. However, plant communities are among the most visible of natural phenomena, so they can be easily observed and categorized on the basis of form and structure. Of course, the composition of the natural vegetation changes from place to place

GEOGRAPHY'S SPATIAL SCIENCE PERSPECTIVE

Introduction and Diffusion of an Exotic Species—Fire Ants

Exotic species are plants or animals that have been introduced to a new environment from where they originated, usually by human activities. Some exotics are regarded as beneficial; many landscaping plants used in North America were brought in from other continents or regions and planted as decorative foliage. Problems have occurred, however, with many other exotics, introduced either purposefully or inadvertently, that have adapted to their new environment to the detriment of the native plants or animals. If the exotic is unable to adapt to its new home, then it dies off and will not be a problem. Many exotics, however, thrive at the expense of native populations. One example of a destructive and harmful exotic is the imported red fire ant, which causes a painful bite and kills off many native insect and other populations (often beneficial insects) in the areas they invade. Fire ants came to the United States in the 1930s from Brazil, as "hitchhikers" on a ship that docked at Mobile, Alabama. Since that time, the geographic distribution of fire ant colonies has been closely documented as they have spread throughout the southern states from their point of introduction.

A map of the expansion of fire ants outward from this location illustrates the rapid impact that the introduction of a species to a new environment can have. This spread of fire ants is an example of spatial diffusion—the expansion of a distribution over an area. Climatic factors will influence or limit the eventual distribution of fire ant populations in the United States, but as of now they are still expanding their geographic range. Cold to the north and aridity to the west are the limiting environmental factors, but these ants also have been discovered in Southern California and in settled areas of the arid West, where lawn or agricultural irrigation provides adequate moisture for their survival. The fire ants may have been spread to these locations by potted nursery plants or turf grasses shipped in from areas already infested with this pest.

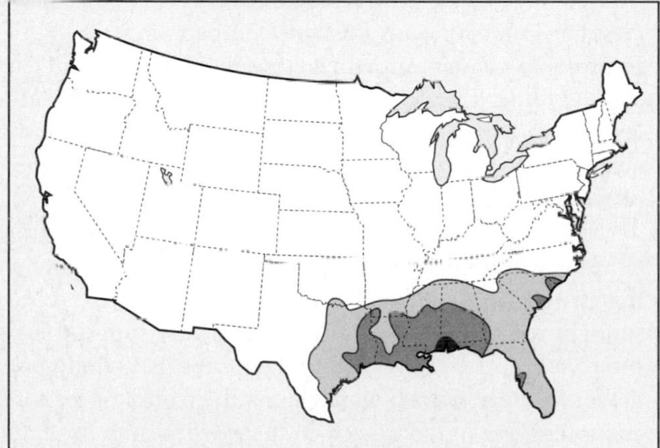

Invasion of Fire Ants 1931–1998.

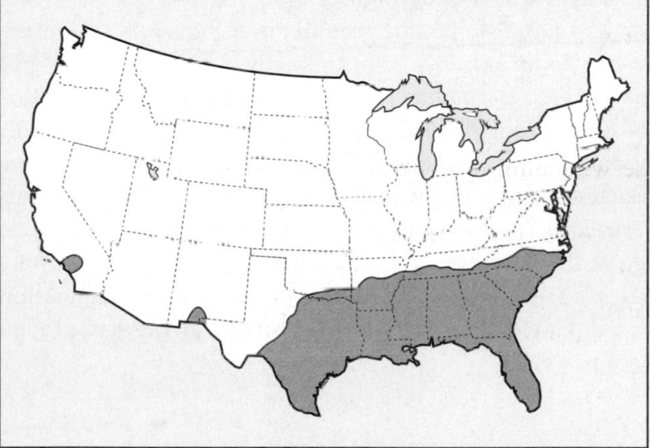

Invasion of Fire Ants 2001. Note the small incursions in the arid southern regions of California and New Mexico.

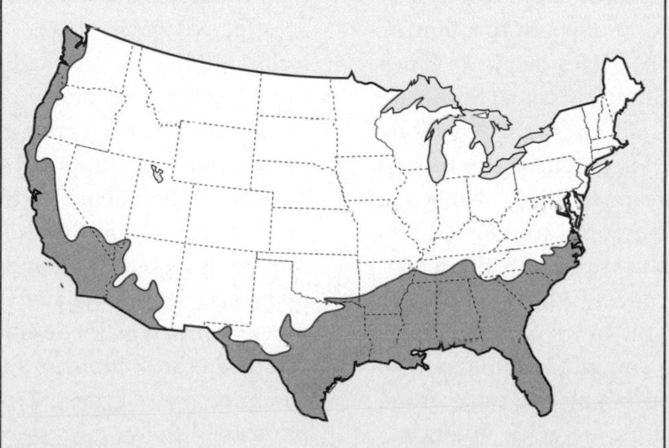

Possible Future Fire Ant Infestation Area. Suburban and agricultural irrigation provides adequate moisture for the potential spread of this invasive exotic species. However, winter temperature colder than 12°C (10°F) is a limiting factor.

in a transitional manner, just as temperature and rainfall do, and although distinctly different types are apparent, there may be broad transition zones (ecotones) between them. Nevertheless, over the world there are distinctive recurring plant communities, indicating a consistent botanical response to systematic controls that are essentially climatic. It is the dominant vegetation of these plant communities that we recognize when we classify Earth's major terrestrial ecosystems (biomes).

All of Earth's terrestrial ecosystems can be categorized into one of four easily recognized types: forest, grassland, desert, and tundra (● Figs. 11.22a, 11.22b, 11.22c, and 11.22d). Because vegetation adapted to cold climates may occur not only in high-latitude regions but also at high elevations at any latitude, biogeographers often refer to the last of the major types as arctic and alpine tundra. However, the forests of the equatorial lowlands are an entirely different world from those of Siberia or of New England, and the original grasslands of Kansas bore little resemblance to those of the Sudan or Kenya. Hence, the four major types of ecosystems can be subdivided into distinctive biomes, each of which is an association of plants and animals of many different species. Pure stands of particular trees, shrubs, or even grasses are extremely rare and are limited to small areas having peculiar soil or drainage conditions.

Earth's major biomes are mapped in ● Figure 11.23 on the basis of the dominant associations of natural vegetation that give each its distinctive character and appearance. The direct influence of climate on the distribution of these biomes is immediately apparent. Temperature (or latitudinal effect on temperature and insolation) and the availability of moisture are the key factors in determining the location of major biomes on the world scale (● Fig. 11.24). Because climatic elements are those that affect vegetation the most, detailed descriptions of major vegetation types are discussed with their corresponding climate types found in Chapters 9 and 10.

Forest Biomes

Forests are easily recognized as associations of large, woody, perennial tree species, generally several times the height of a human, and with a more or less closed canopy of leaves overhead. They vary enormously in density and physical appearance. Some are evergreen and either needle- or broad-leaf; others are deciduous, dropping their leaves to reduce moisture losses during dry seasons or when soil water is frozen. Forests are found only where the annual moisture balance is positive—where moisture availability considerably exceeds potential evapotranspiration in the growing season. Thus, they occur in the tropics, where either the ITCZ (intertropical convergence zone) or the monsoonal circulation brings plentiful rainfall, and in the middle latitudes, where precipitation is associated with cyclones along the polar front, with summer convectional rainfall, or with orographic uplift.

Tropical and middle-latitude forests have evolved different characteristics in response to the nature of the physical limitations in each area. In general, tropical forests have developed in less restrictive forest environments. Temperatures are always high, though not extreme, in the humid tropics, encouraging rapid and luxuriant growth. Middle-latitude forests, on the other hand, must adapt to combat either seasonal cold (ranging from occasional frosts to subzero temperatures) or seasonal drought (which may occur at the worst possible time for vegetative processes).

Tropical Forests

The forests of the tropics are far from uniform in appearance and composition. They grade poleward from the equatorial rainforests, which support Earth's greatest biomass, to the last scattering of low trees that overlook seemingly endless expanses of tall grass or desert shrubs on the tropical margins. We have subdivided the tropical forests into three distinct biomes: the tropical rainforest, the monsoon rainforest, and other tropical forest types, primarily thornbush and scrub. Of course, there are gradations (ecotones) between the different types, as well as distinctive variations that are found in individual localities only.

Tropical Rainforest In the equatorial lowlands dominated by Köppen's tropical rainforest climate, the only physical limitation for vegetation growth is competition between adjacent species. The competition is for light. Temperatures are high enough to promote constant growth, and water is always sufficient. Thus, we find forests consisting of an amazing number of broad-leaf evergreen tree species of rather similar appearance because special adaptations are not required. A cross section of the forest often reveals concentrations of leaf canopies at several different levels. The trees composing the distinctive individual tiers have similar light requirements—lower than those of the higher tiers but higher than those of the lower tiers (● Fig. 11.25). Little or no sunlight reaches the forest floor, which may support ferns but is often rather sparsely vegetated. The forest is literally bound together by vines, **lianas,** which climb the trunks of the forest trees and intertwine in the canopy in their own search for light. Aerial plants may cover the limbs of the forest giants, deriving nutrients from the water and the plant debris that falls from higher levels. Light and variable wind conditions in the rainforest (associated with the tropical belt of the doldrums) preclude wind from being an effective agent of seed and pollen dispersal, so large colorful fruits and flowers, designed to attract animals that will unwittingly carry out seed and pollen dispersal, prevail.

The forest trees commonly depend on widely flared or buttressed bases for support because their root systems are shallow. This is a consequence of the richness of the surface soil and the poverty of its lower levels. The rainforest vegetation and soil are intimately associated. The forest litter is quickly decomposed, its nutrients released and almost immediately reabsorbed by the forest root systems, which consequently remain near the surface. In this way, a rainforest biomass and the available soil nutrients maintain an almost *closed system*. Tropical soils that maintain the amazing biomass of the rainforest are fertile only as long as the forest remains undisturbed. Clearing the forest interrupts the crucial cycling of nutrients between the vegetation and the soil; the copious amounts of water percolating through the soil leach away its soluble constituents, leaving behind only inert iron and aluminum

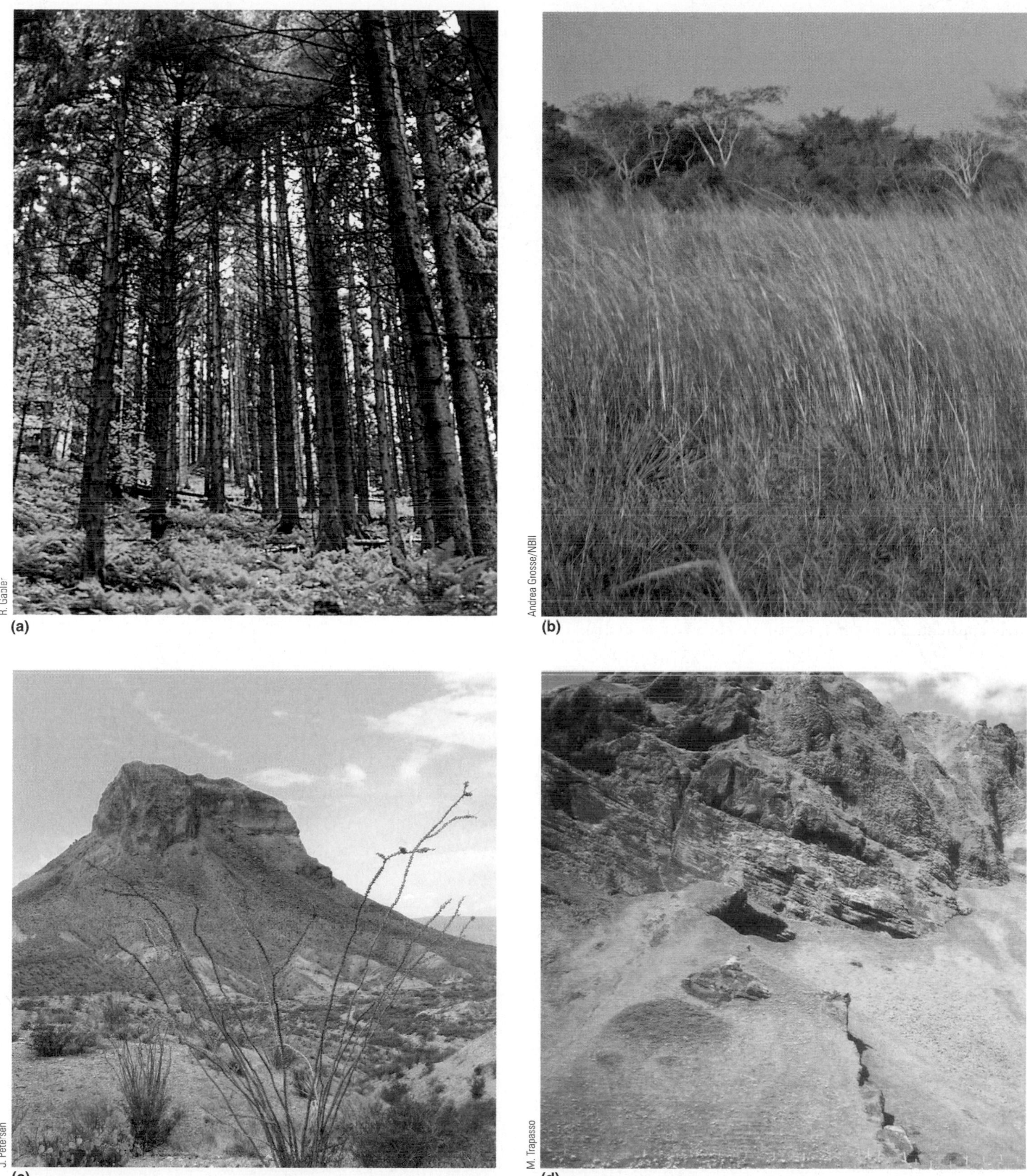

• FIGURE 11.22
The four major types of Earth biomes: (a) forest biome near Dornbirn, Austria; (b) grassland biome in Paraguay; (c) desert biome in Big Bend National Park, Texas; (d) tundra biome in coastal Greenland.
Which one of these images shows an excellent outcome of an ecotone?

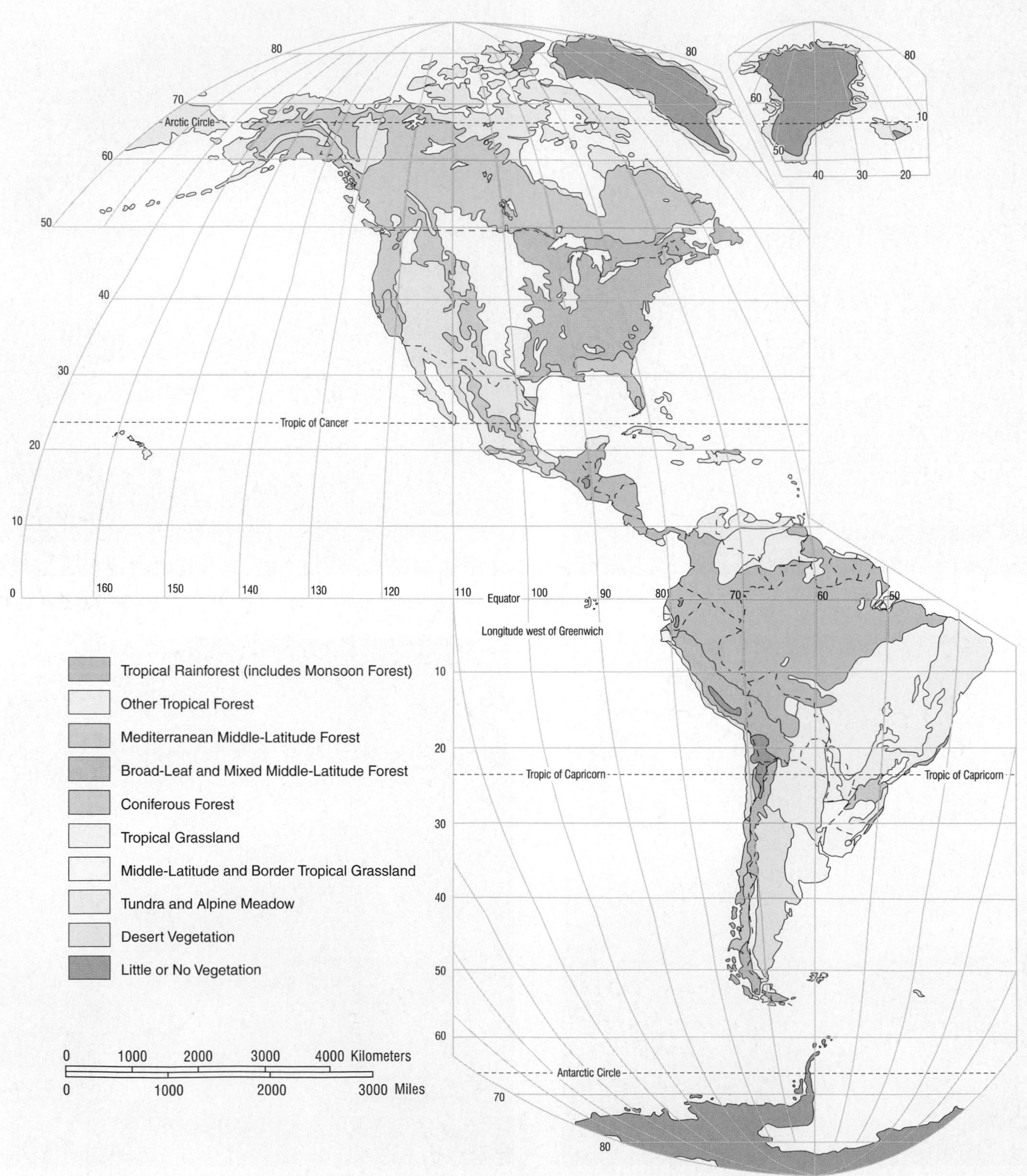

Arctic Circle
Tropic of Cancer
Equator
Longitude west of Greenwich
Tropic of Capricorn
Antarctic Circle
Tropical Rainforest (includes Monsoon Forest)
Other Tropical Forest
Mediterranean Middle-Latitude Forest
Broad-Leaf and Mixed Middle-Latitude Forest
Coniferous Forest
Tropical Grassland
Middle-Latitude and Border Tropical Grassland
Tundra and Alpine Meadow
Desert Vegetation
Little or No Vegetation
0 1000 2000 3000 4000 Kilometers
0 1000 2000 3000 Miles

• FIGURE 11.23
World map of natural vegetation.

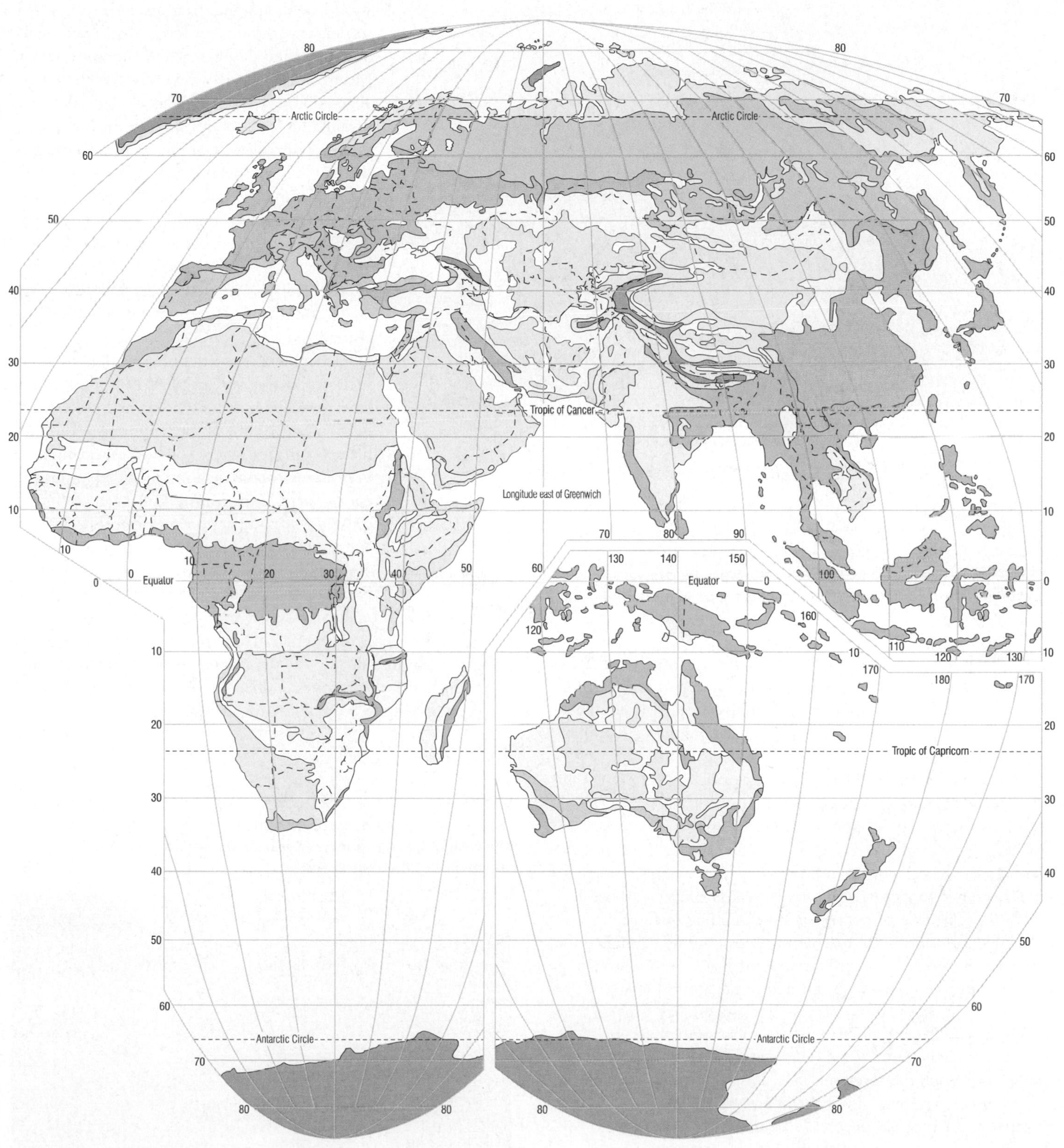

A Western Paragraphic Projection
developed at Western Illinois University

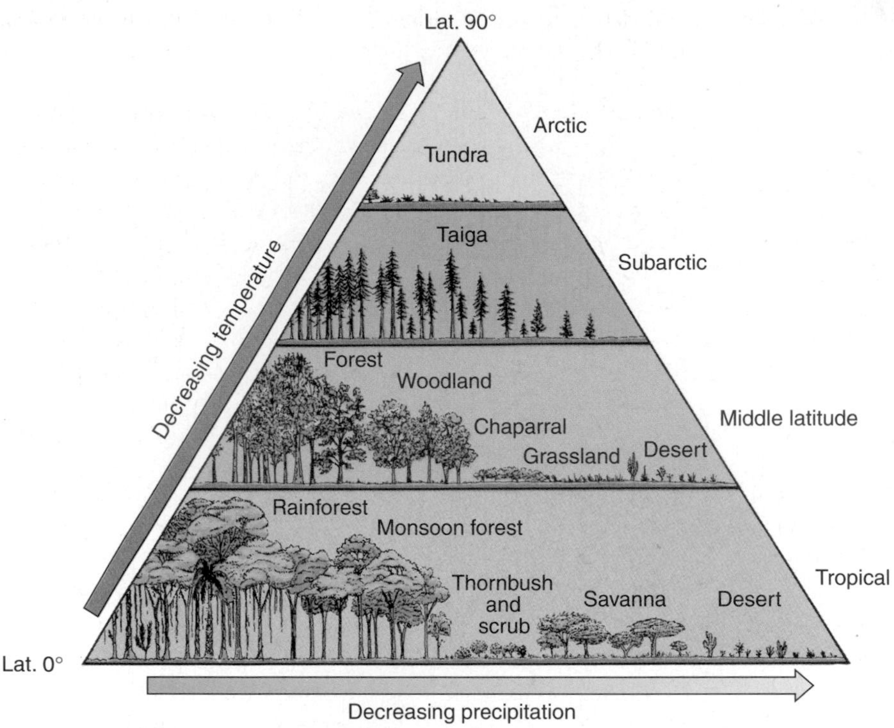

• FIGURE 11.24
This schematic diagram shows distribution of Earth's major biomes as they are related to temperature (latitude) and the availability of moisture. Within the tropics and middle latitudes, there are distinctly different biomes as total biomass decreases with decreasing precipitation. **What major biome dominates the wetter margins of all latitudes but the Arctic?**

oxides that cannot support forest growth. The present rate of clearing threatens to wipe out the worldwide tropical rainforests within the foreseeable future. The largest remaining areas of unmodified rainforests are in the upper Amazon Basin where they cover hundreds of thousands of square kilometers.

Because of the darkness and extensive root systems present on the forest floor, animals of the tropical rainforest are primarily arboreal. A wide variety of species of tree-dwelling monkeys and lemurs, snakes, tree frogs, birds, and insects characterize the rainforest. Even the large herbivorous and carnivorous mammals—such as sloths, ocelots, and jaguars, respectively—are primarily arboreal.

Within the areas of rainforest and extending outward beyond its limits along streams are patches or strips of jungle. Jungle consists of an almost impenetrable tangle of vegetation contrasting strongly with the relatively open nature (at ground level) of the true rainforest. It is often composed of secondary growth that quickly invades the rainforest where a clearing has allowed light to penetrate to the forest floor. Jungle commonly extends into the drier areas beyond the forest margins along the courses of streams. There it forms a gallery of vegetative growth closing over the watercourse and hence has been called **galleria forest.**

Monsoon Rainforest In areas of monsoonal circulation, there is an alternation between the dry monsoon season, when the dominant flow of air is from the land to the sea, and the wet monsoon season, when the atmospheric circulation reverses, bringing moist air onshore along tropical coasts. The wet monsoon season rainfall may be very high, even hundreds of centimeters where air is forced upward by topographic barriers. In any case, it is sufficient to produce a forest that, once established, remains despite the dry monsoon season. Monsoon forests may have discernible tiers of vegetation related to the varying light demands of different species, and they are included with the tropical rainforest in Figure 11.23. However, the number of species is less than in the true rainforest, and the overall height and density of vegetation are also somewhat less. Some of the species are evergreen, but many are deciduous.

Other Tropical Forests, Thornbush, and Scrub Where seasonal drought has precluded the development of true rainforest or where soil characteristics prevent the growth of such vegetation, variant types of tropical forests have developed. These tend to be found on the subtropical margins of the rainforests and on old plateau surfaces where soils are especially poor in nutrients. The vegetation included in this category varies enormously but is generally low growing in comparison to rainforest, without any semblance of a tiered structure, and is denser at ground level. It is commonly thorny, indicating defensive adaptation against browsing animals, and it shows resistance to drought in that it is generally deciduous, dropping its leaves to conserve moisture during the dry winter season. Ordinarily, grass is present beneath the trees and shrubs. As we move away from the equatorial zone, we find the trees more widely spaced and the

• FIGURE 11.25
A tropical rainforest on the island of St. Croix, U.S. Virgin Islands. The dense nature of the forest canopy effectively conceals the vast number of different evergreen tree species and relatively open forest floor. **How might this rainforest differ from the rainforests of the Pacific Northwest of the United States?**

Sean Linehan/NOAA, NGS, Remote Sensing

grassy areas becoming dominant. Along tropical coastlines, a specially adapted plant community, known as mangrove, thrives (see again Fig. 11.16). Here trees are able to grow in salt water.

Middle-Latitude Forests

The forest biomes of the middle latitudes differ from those of the tropics because the dominant trees have evolved mechanisms to withstand periods of water deprivation due to low temperatures and annual variations in precipitation. Evergreen and deciduous plants are present, equipped to cope with seasonal extremes not encountered in tropical latitudes.

Mediterranean Sclerophyllous Woodland Surrounding the Mediterranean Sea and on the southwest coasts of the continents between approximately 30° and 40°N and S latitude, we have seen that a distinctive climate exists—Köppen's mesothermal hot- and dry-summer type (Mediterranean). Here annual temperature variations are moderate, and freezing temperatures are rare. However, little or no rainfall occurs during the warmest months, and plants must be drought resistant. This requirement has resulted in the evolution of distinctive vegetation that is relatively low growing, with small, hard-surfaced leaves and roots that probe deeply for water. The leaves must be capable of photosynthesis with minimum transpiration of moisture. The general look of the vegetation is a thick scrub plant community, called chaparral in the western United States and *maquis* in the Mediterranean region (see again Fig. 10.3). Wherever moisture is concentrated in depressions or on the cooler north-facing hill slopes, deciduous and evergreen oaks occur in groves (● Fig. 11.26). Drought-resistant needle-leaf trees, especially pines, are also part of the overall vegetation association. Thus, the vegetation is a mosaic related to site characteristics and microclimate. Nevertheless, the similarity of the natural vegetative cover in such widely separated areas as Spain, Turkey, and California is astonishing. (Note the location of Mediterranean middle-latitude forest, Fig. 11.23.)

● FIGURE 11.26
The distinctive sclerophyllous evergreen vegetation type encountered wherever hot, dry summers alternate with rainy winters. Oaks commonly occupy relatively damp sites such as the gullies and windward slopes seen here in Southern California.
What are the general characteristics of sclerophyllous vegetation?

R. Gabler

Broad-Leaf Deciduous Forest The humid regions of the middle latitudes experience a seasonal rhythm dominated by warm tropical air in the summer and invasions of cold polar air in the winter. To avoid frost damage during the colder winters and to survive periods of total moisture deprivation when the ground is frozen, trees whose leaves have large transpiring surfaces drop these leaves and become dormant, coming to life and producing new leaves only when the danger period is past. A large variety of trees have evolved this mechanism; certain oaks, hickory, chestnut, beech, and maples are common examples. The seasonal rhythms produce some beautiful scenes, particularly during the periods of transition between dormancy and activity, with the sprouting of new leaves in the spring and the brilliant coloration of the fall as chemical substances draw back into the plant for winter storage (● Fig. 11.27).

The trees of the deciduous forest may be almost as tall as those of the tropical rainforests and, like them, produce a closed canopy of leaves overhead or, in the cold season, an interlaced network of bare branches. However, lacking a multistoried structure and having lower density as a whole, the middle-latitude deciduous forests allow much more light to reach ground level. Forests of this type are the natural vegetation in much of western Europe, eastern Asia, and eastern North America. To the north and south, they merge with mixed forests composed of broad-leaf deciduous trees and conifers. (Broad-leaf forests and mixed forests are combined in Fig. 11.23.) Both the broad-leaf deciduous and mixed forests have been largely logged off or cleared for agricultural land, and the original vegetation of these regions is rarely seen.

Broad-Leaf Evergreen Forest Beyond the tropics, broad-leaf evergreen forests, where the trees remain active throughout the year, are mainly found in certain Southern Hemisphere locations. Here the mild maritime influence is strong enough to prevent either dangerous seasonal droughts or severely low winter temperatures. Southeastern Australia and portions of New Zealand, South Africa, and southern Chile are the principal areas of this type. In the Northern Hemisphere, broad-leaf evergreen forest may once have been significant in eastern Asia, but it has long since been cleared for cultivation. Limited areas occur in the United States in Florida and along the Gulf Coast as a belt of evergreen oaks and magnolias.

Mixed Forest Poleward and equatorward, the broad-leaf deciduous forests in North America, Europe, and Asia gradually merge into mixed forests, including needle-leaf coniferous trees, normally pines. In general, where conditions permit the growth of broad-leaf deciduous trees, coniferous trees cannot compete successfully with them. Thus, in mixed forests, the conifers, which are actually more adaptable to soil and moisture deficiencies, are found in the less hospitable sites: in sandy areas, on acid soils, or where the soil itself is thin. The northern mixed forests reflect the transition to colder climates with increasing latitude; eventually, conifers become dominant in this direction. The southern mixed forests are more problematic in origin. In the United States, they are transitional to pine forests situated on sandy soils of the coastal plain. In Eurasia, they coincide with highlands dominated by conifers during a stage

(a)

(b)

(c)

• **FIGURE 11.27**
The appearance of hardwood forests in middle-latitude regions with cold winters changes dramatically with the seasons. The green leaves of summer (a) change to reds, golds, and browns in fall (b) and drop to the forest floor in winter (c). Leaf dropping in areas of cold winters, such as this example in western Illinois, is a means of minimizing transpiration and moisture loss when the soil water is frozen.
What length of growing season (frost-free period) is associated with the climate of western Illinois?

in the plant succession that began with the change of climate environments at the end of the ice ages, some 10,000 years ago.

Coniferous Forest The coniferous forests occupy the frontiers of tree growth. They survive where most of the broadleaf species cannot endure the climatic severity and impoverished soils. The hard, narrow needles of coniferous species transpire much less moisture than do broad leaves so that needle-leaf species can tolerate conditions of physiologic drought (unavailability of moisture because of excessive soil permeability, a dry season, or frozen soil water) without defoliation. Pines, in particular, also demand little from the soil in the form of soluble plant nutrients, especially basic elements such as calcium, magnesium, sodium, and potassium. Thus, they grow in sandy places and where the soil is acid in character. As a whole, conifers are particularly well adapted to regions having long, severe winters combined with summers warm enough for vigorous plant growth. Because all but a few exceptions retain their leaves (needles) throughout the year, they are ready to begin photosynthesis as soon as temperatures permit without having to produce a new set of leaves to do the work.

Thus, we find a great band of coniferous forests (the **boreal forests,** or **taiga**) dominated by spruce and fir species, with pines on sandy soils, sweeping the full breadth of North America and Eurasia northward of the 50th parallel of latitude, approximately occupying the region of Köppen's subarctic climate (see again Fig. 8.6). Conifers differ from other trees in that their seeds are not enclosed in a case or fruit but are carried naked on cones. All are needle leaf and drought resistant, but a few are not evergreen. Thus, a large portion of eastern Siberia is dominated by larch, which produces a mix of deciduous, coniferous forest. In this area,

January mean temperatures may be –35°C to –51°C (–30°F to –60°F). This is the most severe winter climate in which trees can maintain themselves, and even needle-leaf foliage must be shed for the vegetation to survive. Hardy broad-leaf deciduous birch trees share this extreme climate with the indomitable larches.

Extensive coniferous forests are not confined to high-latitude areas of short summers. Higher elevations in middle-latitude mountains of the Northern Hemisphere have forests of pine, hemlock, and fir (● Fig. 11.28), with subalpine larch and specially adapted pine species characterizing the harshest and highest forest sites. The forests along the sandy coastal plain of the eastern United States are there in part because of the sandy soils, but they may also reflect a stage in plant succession that in time will lead to domination by broad-leaf types. Similarly, a temporary stage in the postglacial vegetation succession of the Great Lakes area included magnificent forests of white pine and hemlock that were completely logged off during the late 19th century.

● FIGURE 11.28
As this scene in the Gifford Pinchot National Forest in Washington State indicates, evergreen coniferous forest is the characteristic vegetation of higher-elevation regions in the middle latitudes. The needle-leaf trees are well adapted to the physiologic drought of the winter season, which is longer and more severe at higher elevations. Here the broad-leaf trees are in fall color prior to losing their leaves during the cold season.
Why are needle-leaf trees better adapted to physiologic drought than broad-leaf trees?

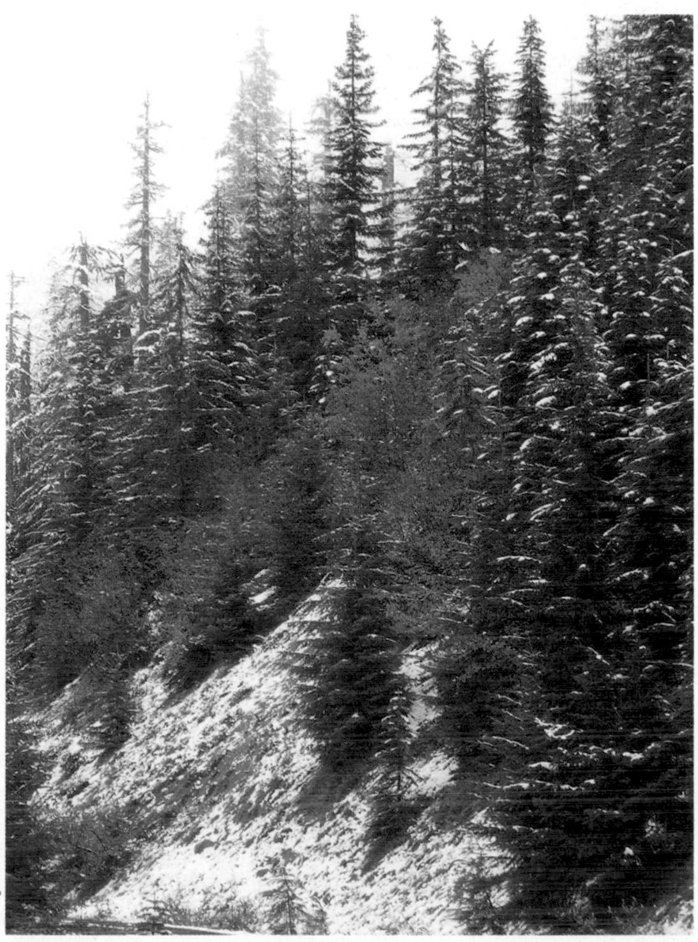

Aldo Aguilar/USFS

A more maritime coniferous forest occupies the West Coast of North America extending from southern Alaska to central California. It is made up of sequoias, Douglas fir, cedar, hemlock, and, farther north, Sitka spruce. Many of the California sequoias are thousands of years old and more than 100 meters (330 ft) high. The southern regions experience summer drought, and farther north, sandy, acidic, or coarse-textured soils dominate.

Grassland Biomes

Grasses, like conifers, appear in a variety of settings and are part of many diverse plant communities. They are in fact an initial form in most plant successions. However, there are enormous, continuous expanses of grasslands on Earth. In general, it is thought that grasses are dominant only where trees and shrubs cannot maintain themselves because of either excessive or deficient moisture in the soil. On the global scale, grassland biomes are located in continental interiors where most, if not all, of the precipitation falls in the summer. Two great geographic realms of grasslands are generally recognized: the tropical and the middle-latitude grasslands. However, it is difficult to define grasslands of either type using any specific climatic parameter, and geographers suspect that human interference with the natural vegetation has caused expansion of grasslands into forests in both the tropical and middle latitudes.

Tropical Savanna Grasslands

The tropical grassland biome differs from grassland biomes of the middle latitudes in that it ordinarily includes a scattering of trees; this is implied in the term **savanna** (● Fig. 11.29). In fact, the demarcation between tropical scrub forest and savanna is seldom

● FIGURE 11.29
The savanna biome of eastern Africa is a classic landscape of grasses and scattered trees (background). In the foreground, vultures (detritivores) eat the carcass of a zebra (herbivore) left behind by a lion (carnivore).
What organisms will take over when the vultures are satisfied?

M. Trapasso

a clear one. The savanna grasses tend to be tall and coarse with bare ground visible between the individual tufts. The related tree species generally are low-growing and wide-crowned forms, having both drought- and fire-resisting qualities, indicating that fires frequently sweep the savannas during the drought season. Savannas occur under a variety of temperature and rainfall conditions, but generally fall within the limits of Köppen's tropical savanna type. They commonly occur on red-colored soils, leached of all but iron and aluminum oxides, which become bricklike when dried. They likewise coincide with areas in which the level of the water table (the zone below which all soil and rock pore space is saturated by water) fluctuates dramatically. The up-and-down movement of the water table may in itself inhibit forest development, and it is no doubt a factor in the peculiar chemical nature of savanna soils. Large migratory herds of grazing animals and associated predators, responding to the periodically abundant grasses followed by seasonal drought, characterized the savanna prior to widespread human disruption.

Middle-Latitude Grasslands

The middle-latitude grasslands occupy the zone of transition between the middle-latitude deserts and forests. On their dry margins, they pass gradually into deserts in Eurasia and are cut off westward by mountains in North America. However, on their humid side, they terminate rather abruptly against the forest margin, again raising questions as to whether their limits are natural or have been created by human activities, particularly the intentional use of fire to drive game animals. The middle-latitude grasslands of North America, like the African savannas, formerly supported enormous herds of grazing animals—in this case, antelope and bison, which were the principal means of support of the American Plains Indians.

Like the savannas, the middle-latitude grasslands were diverse in appearance. They, too, consisted of varying associations of plant species that were never uniform in composition. In North America, the grasses were as much as 3 meters (10 ft) tall in the more humid sections, as in Iowa, Indiana, and Illinois, but only 15 centimeters (6 in.) high on the dry margins from New Mexico to western Canada. Thus, the middle-latitude grassland biomes are usually divided into tall-grass and short-grass prairie, often with a zone of mixture recognized between them. Unlike growth in the tropical savannas, the germination and growth of middle-latitude grasses are attuned to the melting of winter snows, followed by summer rainfall. Whether the grasses are annuals that complete their life cycle in one growing season or perennials that grow from year to year, they are dormant in the winter season. Also, unlike in the savannas, the soils beneath these grasslands are extremely rich in organic matter and soluble nutrients. As a consequence, most of the middle-latitude grasslands have been completely transformed by agricultural activity. Their wild grasses have been replaced by domesticated varieties—wheat, corn, and barley—and they have become the "breadbaskets" of the world.

Tall-Grass Prairie

The **tall-grass prairies** (● Fig. 11.30a) were an impressive sight; in some better-watered areas, they made up endless seas of grass moving in the breeze, reaching higher than a horse's back. Flowering plants were conspicuous, adding to the effect. Unfortunately, this tall-grass prairie scene, which inspired much vivid description by those first encountering it, is barely visible anywhere today. The tall-grass prairies, which once reached continuously from Alberta to Texas, have been almost completely destroyed. Compared to their once vast extent, today only a few tall-grass prairie areas remain in government-protected preserves. Their tough sod, formed by the dense grass root network, defeated the first wooden plows, which had served well enough in breaking up the forest soils. But the steel plow, invented in the 1830s, subdued the sod and was aided by the introduction of subsurface tile for draining the nearly flat uplands and by the simultaneous appearance of well-digging machinery and barbed wire. These four innovations transformed the tall-grass prairie from grazing land to cropland.

In North America, the tall-grass prairie pushed as far eastward as Lake Michigan. Why trees did not invade the prairie in this relatively humid area remains an unanswered question. Farther west, shallow-rooted grass cover is fully understandable because the lower soil levels, to which tree roots must penetrate for adequate support and sustenance, are bone dry. In such areas, trees can survive only along streams or where depressions collect water.

In Eurasia, tall-grass prairies were found on a large scale in a discontinuous belt from Hungary, through Ukraine, Russia, and central Asia, to northern China. The grasslands are known in South America as the pampas of Uruguay and Argentina, and in South Africa as the Veldt. Today, all of these areas have been changed by agriculture. The factor that seems to account best for the tall-grass prairie—precipitation that is both moderate and variable in amount from year to year—is the principal hazard in the use of these regions as farmland. However, this hazard becomes much greater in the areas of short-grass prairie.

Short-Grass Prairie

West of the 100th meridian in the United States and extending across Eurasia from the Black Sea to northern China, roughly coinciding with the areas of Köppen's middle-latitude steppe climate, are vast, nearly level grasslands composed of a mixture of tall and short grass, with short grass becoming dominant in the direction of lower annual precipitation totals (Fig. 11.30b), and tall-grass prairie in semiarid regions that have higher annual precipitation (compare to Fig. 11.30a). On the Great Plains between the Rocky Mountains and the tall-grass prairies, the **short-grass prairie** zone more or less coincides with the zone in which moisture rarely penetrates more than 60 centimeters (2 ft) into the soil, so the subsoil is permanently dry. Moving toward the drier areas, the grassland vegetation association dwindles in diversity and, more conspicuously, in height to less than 30 centimeters (1 ft). This is a consequence of reduction in numbers of tall-growing species and greater abundance of shallow-rooted and lower-growing types. The total amount of ground cover also declines toward the drier margins as the deeply rooted, sod-forming grasses of the prairie grassland give way to bunchgrass species (so called because, instead of forming a con-

• **FIGURE 11.30**

(a) Spring wildflowers bloom in Tallgrass Prairie National Preserve, Kansas, one of the remaining protected areas of that type of prairie; the grasses will grow taller in later spring and summer. (b) Short-grass prairie vegetation, with bison grazing in Wind Cave National Park in South Dakota. Although the tall-grass prairies have been almost completely transformed by humans, vast areas of short-grass prairie remain because of their low and unpredictable precipitation, which makes them poorly suited for agriculture.

Both develop under a semiarid climate, but which of these two types of prairie grassland receives a higher annual precipitation?

Photo courtesy of Tallgrass Prairie National Preserve

(a)

Photo courtesy of Tallgrass Prairie National Preserve

(b)

tinuous grass cover, they occur in isolated clumps or bunches). In their natural conditions, the short-grass prairies of North America and Eurasia supported higher densities of grazing animals—bison and antelope in the former, wild horses in the latter—than did the tall-grass prairies. Indeed, it is suspected that the specific plant association of the short-grass regions may have been a consequence of overgrazing under natural conditions. The short-grass prairies cannot be cultivated without the use of irrigation or dry farming methods, so they remain primarily the domain of wide-ranging grazing animals; however, today's animals are domesticated cattle, not the thundering self-sufficient herds of wild species that formerly made these plains one of Earth's marvels.

Desert

Eventually, lack of precipitation can become too severe even for the hardy grasses. Where evapotranspiration demands greatly exceed available moisture throughout the year, as in Köppen's desert climates, either special forms of plant life have evolved or the surface is bare. Plants that actively combat low precipitation are equipped to probe deeply or widely for moisture, to reduce moisture losses to the minimum, or to store moisture when it is available. Other plants evade drought by merely lying dormant, perhaps for years, until enough moisture is available to ensure successful growth and reproduction. The desert biome is recognized by the presence of plants that are either drought resisting or drought evading (• Fig. 11.31). In extremely dry deserts, only a few plants can survive, and ground cover is much less common. In the driest parts of the Atacama Desert of Chile it is so arid that there is no vegetation at all in some locations (• Fig. 11.32).

Plants that have evolved mechanisms to combat drought are known as *xerophytes*. They are perennial shrubs whose root systems below ground are much more extensive than their visible parts or that have evolved tiny leaves with a waxy covering to combat transpiration. They may have leaves that are needlelike or trunks and limbs that photosynthesize like leaves or that have expandable tissues or accordion-like stems to store water when it is plentiful (the succulent cacti). They may be plants that can tolerate excessively saline water or shrubs that shed their leaves until sufficient moisture is available for new leaf growth. The nonxerophytic vegetation consists mainly of short-lived annuals that germinate and hurry through their complete life cycle of growth—leaf production, flowering, and seed dispersal—in a matter of weeks when triggered by moisture availability. Like other species, these ephemeral plants also require days of a

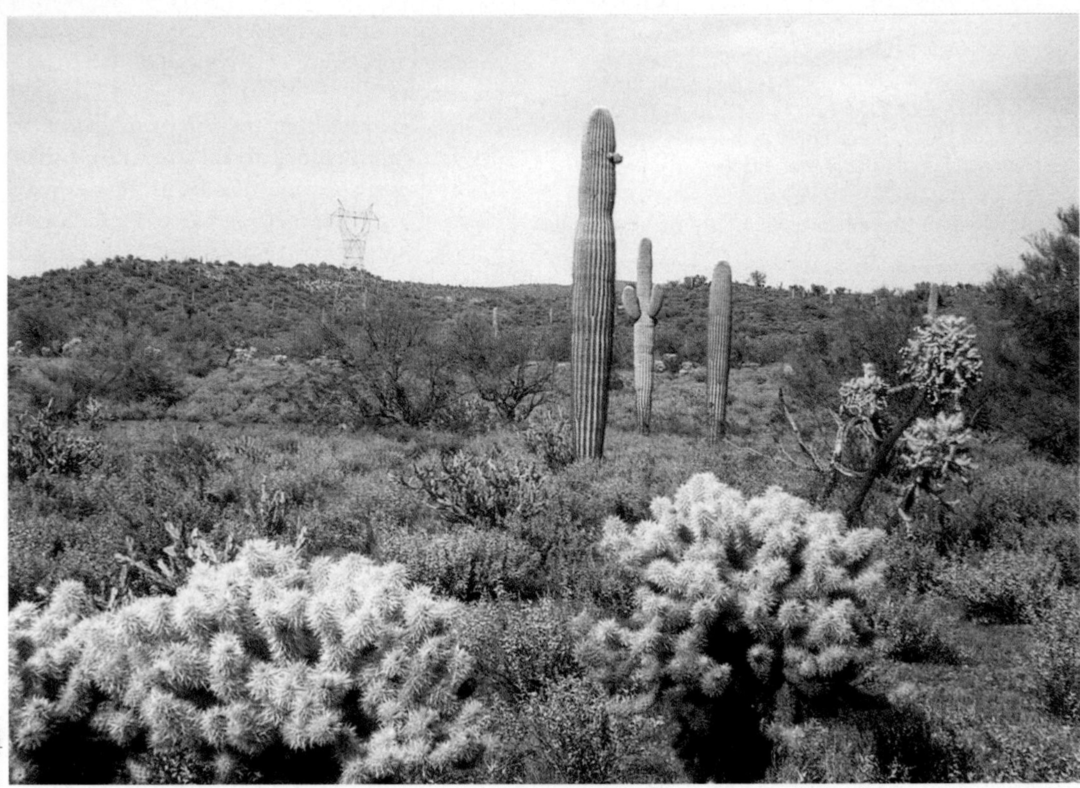

M. Trapasso

• **FIGURE 11.31**
A host of drought-resistant plants can be seen across parts of Arizona.
What drought-resistant adaptations can be easily observed?

Julio L. Betancourt, U.S. Geological Survey

• **FIGURE 11.32**
The absence of vegetation shown here in the Atacama Desert in Chile is a clear indication of the extremely low rainfall and high evaporation rates experienced in this region. The Atacama Desert is the driest region in the world.
Are there any reasons why humans might be found in such desolate regions?

certain length, so they appear only in particular months; therefore, the month-to-month and year-to-year variation in form and appearance of desert vegetation is enormous. Animals of the deserts are primarily nocturnal to avoid the searing heat of the daytime, and many have evolved long ears, noses, legs, and tails that allow for greater blood circulation and cooling. The similar life-forms and habits of the different plant and animal species found in the deserts of widely separated continents are a remarkable display of repeated evolution to ensure survival in similar climatic settings.

Arctic and Alpine Tundra

Proceeding upward in elevation and poleward in latitude, we finally come to regions in which the growing season is too brief to permit tree growth. Nearing the poles, we enter a vast realm dominated by subfreezing temperatures and thin snow cover much of the year, so the ground is frozen to depths of hundreds of meters. Only the top 36–60 centimeters (15–25 in.) thaw during the short summer interval. Still, vegetation survives here and in fact forms a nearly complete cover over the surface. Such vegetation must be equipped to tolerate frozen subsoil (permafrost), icy winds, low sun angles, summer frosts, and soil that is waterlogged during the short growing season. The result is **tundra**—a mixture of grasses, flowering herbs, sedges, mosses, lichens, and occasional low-growing shrubs. Most of the plants are perennials that produce buds close to or beneath the soil surface, protected from the wind. Many of the plants show xerophytic adaptation—such as small, hard leaves—in response to extreme physiologic drought resulting from wind stress. This is particularly true in areas of alpine tundra where extended periods of water-logging of soils are uncommon. The effect of wind is evident from the fact that the less exposed valleys within the tundra region are often occupied by coniferous woodlands.

In a band of varying width reaching across northern Alaska, Canada, Scandinavia, and northern Russia, several types of tundra are recognized: *bush tundra,* consisting of dwarf willow, birch, and alder, which grow along the edge of the coniferous forest; *grass tundra,* which is hummocky and water soaked during the summer (● Fig. 11.33); and *desert tundra,* in which expanses of bare rocks may be covered by colorful lichens. In a few ice-free valleys of Antarctica, only desert tundra occurs.

M. Trapasso

● **FIGURE 11.33**
A variety of tundra grasses grow during the Arctic summer in Alaska. The photograph was taken near the Alaskan Oil Pipeline at Prudhoe Bay.
What environmental concerns might be connected with human impacts on the tundra?

Alpine conditions are not exactly like those in the Arctic latitudes. The deeper snow cover of the high mountains prevents the development of permafrost, and the summer sun results in considerably more evaporation. However, many high areas are swept clean of snow by wind and are thus extremely exposed, supporting only desert tundra. Microclimate becomes an important control of vegetation because of the varying exposures to sun and wind. Nevertheless, the short growing season and severe wind stress produce an overall plant community similar to that in the Arctic regions (● Fig. 11.34).

● **FIGURE 11.34**
Close examination of the tundra biome of the mountainous western United States reveals a mixture of grasses, sedges, rushes, and wildflowers.
Why does all vegetation, including the occasional shrub or stunted tree, grow so close to the ground?

R. Gabler

Animals of the tundra obviously must cope with long periods of extreme cold as well as darkness. Many animals hibernate through the long Arctic winters (or the cold and windy alpine winters); others, such as the caribou of Alaska and Canada, migrate into the boreal forest to escape the extreme cold exacerbated by Arctic winds. Year-round residents, such as the polar bear and musk ox, must have extremely large fat reserves around their large chests, in addition to extremely efficient fur. Many insects, such as the ubiquitous mosquito, cope with the extreme climate by emerging from eggs in the spring, maturing and laying eggs, and dying within one short Arctic summer. Enough eggs survive, buried under insulating snow, to continue the cycle in the following year.

Marine Ecosystems

The living organisms of the ocean can be divided into three groups according to where or how they live in the ocean. The first group is called **plankton.** Plankton is made up mostly of the ocean's smallest—usually microscopic—plants (**phytoplankton**) and animals (**zooplankton**). These tiny plants and animals float freely with the movements of ocean water, are true "drifters," and form the basis of the oceanic food chain. The second group is composed of the animals that swim in the water. This group, called **nekton,** includes fish, squid, marine reptiles, and marine mammals such as whales and seals (• Fig. 11.35). The third group is composed of the plants and animals that live on the ocean floor. This group is called the **benthos.** It includes corals, sponges, and many algae; such burrowing or crawling animals as the barnacle, crab, lobster, and oyster; and attached plants such as turtle grass and kelp.

Life in the ocean depends on the sun's energy and on the nutrients available in the water. Phytoplankton are the most important link in the ocean food chain. At the base of the marine food web, phytoplankton take the dissolved nutrients in the water and, through the process of photosynthesis, produce oxygen and foods needed by zooplankton and by the smallest nekton. Phytoplankton are the only food source for these animals, which in turn form the food source for larger carnivorous fish and marine mammals. These in turn are prey for still larger animals. For example, tiny shrimplike creatures known as *krill* are nicknamed the "power food of the Antarctic." These crustaceans feed on plankton and then become the main food source for birds, penguins, seals, and whales.

• FIGURE 11.35
These elephant seals basking in the sun rely on food sources related to the nutrient-rich and cool upwelled waters along the California coast.
To which group of ocean organisms do these elephant seals belong?

NOAA NMFS SWFSC PRD

The marine food chain just described is part of a full food cycle in the ocean because, through the excretions of animals and the decomposition of both plants and animals in the ocean, chemical nutrients are returned to the water and are again made available for transformation by phytoplankton into usable foods. Phytoplankton also play an essential role in the production of oxygen for Earth's atmosphere. In fact, the greatest concern of scientists who study the annual Antarctic "ozone hole" (see Chapter 4) is that excessive ultraviolet radiation will destroy or diminish the phytoplankton in the Antarctic and other oceans.

The uneven distribution of nutrients in the oceans and the fact that sunlight can penetrate only to a depth of about 120 meters (400 ft), depending on the clarity of the water, means that the distribution of marine organisms is also variable. Most organisms are concentrated in the upper layers of the ocean where the most solar energy is available. In deep waters, where the ocean floor lies below the level to which sunlight penetrates, the benthos organisms depend on whatever nutrients and plant and animal detritus filter down to them. For this reason, benthos animals are scarce in deep and dark ocean waters. They are most common in shallow waters near coasts, such as coral reefs and tide pools, where there is sunlight and a rich supply of nutrients and where phytoplankton and zooplankton are abundant as well.

The waters of the continental shelf have the highest concentration of marine life. The supply of chemical nutrients is greater in waters near the continents where nutrients are washed into the sea from rivers. Marine organisms are also concentrated where deeper waters rise (upwelling) to the surface layers where sunlight is available. Such vertical exchanges are sometimes the result of variations in salinity or density. A similar situation occurs where convection causes bottom layers of water to rise and mix with top layers, as is the case in cold polar waters and in middle-latitude waters during the colder winter months. Marine life is also abundant in areas where there is a mixing of cold and warm ocean currents, as there is off the northeastern coast of the United States (• Fig. 11.36).

During the 1977 dives of the manned submersible vessel *Alvin* off the East Pacific Rise, scientists for the first time observed abundant sea life on the floor of the ocean at depths of more than 2500 meters (8100 ft). It was previously presumed that these cold (2°C/36°F), dark waters were a virtual biological desert. However, the undersea volcanic mountain range produces vents of warm, mineral-rich waters, which nourish bacteria and large colonies of crabs, clams, mussels, and giant 3-meter (10-ft) red tube worms. The discovery of this deep-ocean ecosystem, which exists without the benefit of sunlight, has caused scientists to rethink old theories about the ocean and its chemistry. This unusual "chemosynthetic" vent community has been observed more recently at several other oceanic ridge sites in the Pacific and Atlantic Oceans.

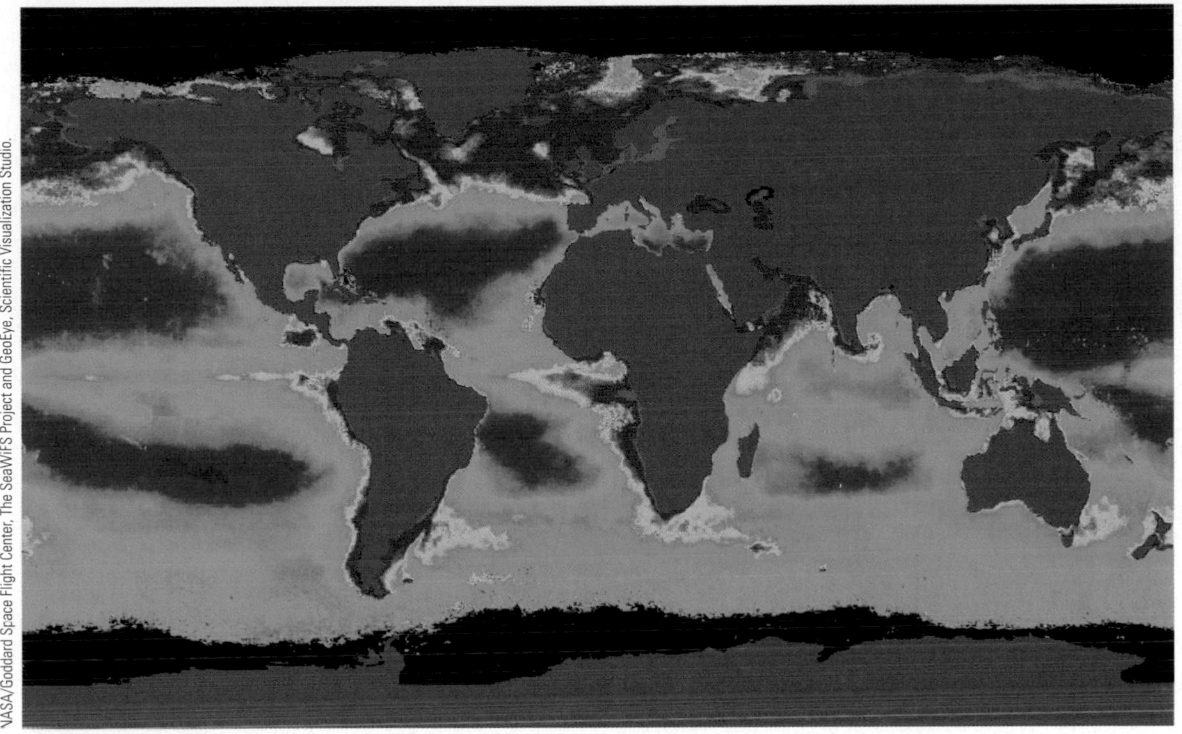

NASA/Goddard Space Flight Center, The SeaWiFS Project and GeoEye, Scientific Visualization Studio.

• FIGURE 11.36
The distribution of chlorophyll-producing marine plankton can be mapped on satellite images. Blue to light blue, to green to yellow to red indicates increasingly higher levels of chlorophyll concentration.
What are two spatial observations that you can make about the geographic locations that have high plankton concentrations?

The Resilience of Life-Forms

Throughout this chapter, an overarching theme becomes evident: Earth's life-forms—both *autotrophs* (plant life) and *heterotrophs* (animal life)—are extremely resilient. Temperature, precipitation, sunlight, and wind are among the most important of the climatic factors that influence the biogeography of our planet. However, it is important to realize that these factors often vary from one year to the next. Most species will adapt to annual fluctuations. Furthermore, climate regions do not need to be vast; microclimatic areas can form a pocket of favorable conditions in which certain life-forms thrive. Even some of the most inhospitable environments are not completely devoid of life; and where one life-form fails to adapt, another more adaptive life-form will survive.

As climates change through time—and they inevitably will—flora and fauna will either adjust to the new climatic conditions, migrate to other areas where more favorable conditions exist, or risk becoming extinct. In the future, mapping the geography of changes in vegetation distribution and animal migration will be an effective way to document the impact of changing climates on the planet's diverse environments.

Chapter 11 Activities

Define & Recall

ecosystem
abiotic
autotroph
heterotroph
herbivore
carnivore
omnivore
decomposer (detritivore)
trophic structure
food chain
trophic level
food web
biomass
primary productivity
secondary productivity
habitat
ecological niche
generalist
natural vegetation
plant community
succession
gap
climax community
mosaic
matrix
patch
corridor
tolerance
ecotone
biome
symbiotic relationship
liana
galleria forest
boreal forest (taiga)
savanna
tall-grass prairie
short-grass prairie
tundra
plankton
phytoplankton
zooplankton
nekton
benthos

Discuss & Review

1. What are some of the reasons why the study of ecosystems is important in the world today? Give several examples of natural ecosystems within easy driving distance of your own residence.
2. What are the four basic components of an ecosystem?
3. What factors are most critical in affecting the net primary productivity of a terrestrial ecosystem?
4. In what ways has the original theory of succession been modified?
5. How do the terms *mosaic, matrix, patch, corridor,* and *ecotone* relate to each other and to a vegetation landscape?
6. What are the four major types of Earth biomes? What important climate characteristics are related to each of the major biomes?
7. Describe a true tropical rainforest. How does such a forest differ from jungle?
8. What are the distinctive features of chaparral vegetation? What climate conditions are associated with chaparral?
9. What conditions of climate or soil might be anticipated for each of the following in the middle latitudes: broad-leaf deciduous forest; broad-leaf evergreen forest; needle-leaf coniferous forest?
10. How have xerophytes adapted to desert climate conditions?
11. What are the chief characteristics of tundra vegetation? What adaptations to climate does tundra vegetation make?
12. The highest concentrations of marine life (see Fig. 11.36) are found in which parts of the oceans? Why?

Consider & Respond

1. Refer to Table 11.2. Based on the means, which is more important for terrestrial ecosystem productivity between the equator and the polar regions: annual temperatures or the availability of water? Use examples from the table to explain your choice.
2. What broad groups might you use if you were to classify Earth's vegetation on the basis of latitudinal zones? Why, do you suppose, did your textbook authors choose not to do this, electing instead to identify broad groups based on major terrestrial ecosystems (biomes)?
3. Refer to Figure 11.23, which shows considerable variation in the natural vegetation of the United States between 30° and 40° north latitude. Describe the broad changes in vegetation that occur within that latitudinal band as you move from the East Coast to the West Coast of the United States.
4. Examine the climate variation (see Fig. 8.6) within the same latitudinal band described in Question 3. Does there appear to be a relationship between natural vegetation and climate? If so, describe this relationship.

Apply & Learn

1. Kudzu is a climbing, woody, perennial vine that originates from Japan. From 1935 to 1950, farmers in the southeastern United States were encouraged to plant it to help reduce soil erosion. Without natural enemies, this vine began to grow out of control. Since 1953, it was identified as a pest weed and efforts to eradicate this invader continue to this day. Today, Kudzu inhabits about 30,000 square kilometers of the southeastern United States. From 1935 to the present, what is the spread of this pest in square kilometers per year?

2. Research has shown that the harvest dates for vineyards in Switzerland vary systematically with the local mean temperatures between April and August. With every 1°C rise in temperature, the harvest date for the grapes moves back by 12 days. Assuming an average harvest date of September 15, what would be the expected harvest date, if the April–August mean temperature was: 1°C above normal, 1.25°C below normal, 1.5°C above normal, and 1°F above normal?

Soils and Soil Development 12

CHAPTER PREVIEW

Soil is an outstanding example of the interrelationships among Earth's subsystems.

- Why is soil such a good example of the interaction and integration of subsystems?
- Why should soil be considered an open system?

Soil water is the means by which plants receive dissolved nutrients that are essential for growth.

- Why is gravitational water such an effective agent of solution?
- How is capillary water important during periods of drought?

Soil fertility depends on many factors, and a soil that is fertile for one vegetation type may not be for another.

- What factors determine a soil's fertility, and how is vegetation involved?
- How are acidity and alkalinity related to soil fertility?

On a global scale, climate exerts a major influence on the formation and characteristics of soils.

- How do temperature, precipitation, and moisture regimes affect the development of soils?
- Why is the regional distribution of soils similar to regions of climate and vegetation?

Soils are among the world's most critical and widely abused yet least understood natural resources.

- How are soils abused or neglected as a resource?
- What can be done to conserve soils?

◄ **Opposite: Scientists work to understand and control erosional losses and other problems that threaten our precious natural resource–soil.**
USDA/ARS/Photo by Jack Dykinga

What are the four most important natural constituents that permit life as we know it to exist on Earth? Many people if asked that question would reply, "air, water, and sunlight," right away, but they might have to think harder and longer about their fourth answer. Most people give little attention to that fourth natural resource, but it is essential for their life on Earth, and it lies right below their feet. *Soil* is that critical resource. The soil mantle that covers most land surfaces is indispensable, but fragile, and threatened by erosion, pollution, or being covered over by the human-built environment. Soil provides nutrients that directly or indirectly support much of life on Earth.

Soil is a dynamic natural body capable of supporting a vegetative cover. It contains chemical solutions, gases, organic refuse, flora, and fauna. The physical, chemical, and biological processes that take place among the components of a soil are integral parts of its dynamic character. Soil responds to climatic conditions (especially temperature and moisture), to the land surface configuration, to vegetative cover and composition, and to animal activity.

Soil has been called "the skin of the Earth." The condition and nature of a soil reflects both the ancient environments under which it formed, and today's environmental conditions. A soil functions as an environmental system, adapting, reflecting, and responding to a great variety of natural and human-influenced processes. Soil is an exceptional example of the integration, interdependence, and overlap among Earth's subsystems because the characteristics of a soil reflect the atmospheric, hydrologic, lithologic, and biotic conditions under which it developed (● Fig. 12.1). In fact, because soils integrate these major subsystems so well, they are sometimes considered a separate system called the *pedosphere* (from Greek: *pedon,* ground).

Soil is also home to numerous living organisms, forming the environments in which they live, both above and below the ground surface. The life-forms that live in or on a soil play significant roles in the development and characteristics of a soil, and through human population growth and expanding civilizations, potentially negative impacts on soils have increased dramatically.

How a soil develops and its resulting characteristics depend on a great number of factors. But when soils are viewed on a world regional scale, a strong and significant influence is climate. The relationships between soils, climate types, and their associated environments were considered in Chapters 9 and 10.

Major Soil Components

What is soil actually made of? What does a shovel contain when it scoops up a load of soil? What soil characteristics support and influence variations in Earth's vegetational environments? Soils contain four major components, and there are many processes that act on these components. The four major components of soil are inorganic materials, soil water, soil air, and organic matter (● Fig. 12.2).

Inorganic Materials

Soils contain varying amounts of insoluble materials—rock fragments and minerals that will not readily dissolve in water. Soils also contain soluble minerals, which supply dissolved chemicals held in solution. Most minerals found in soils are combinations of the common elements of Earth's surface rocks: silicon, aluminum, oxygen, and iron. Some of these constituents occur as solid chemical compounds, and others are found in the air and water that are also vital components of a soil. Soils sustain Earth's land

● FIGURE 12.1
The intertwined links between soil and the major Earth subsystems.
Why is soil considered to be such an integrator of Earth systems?

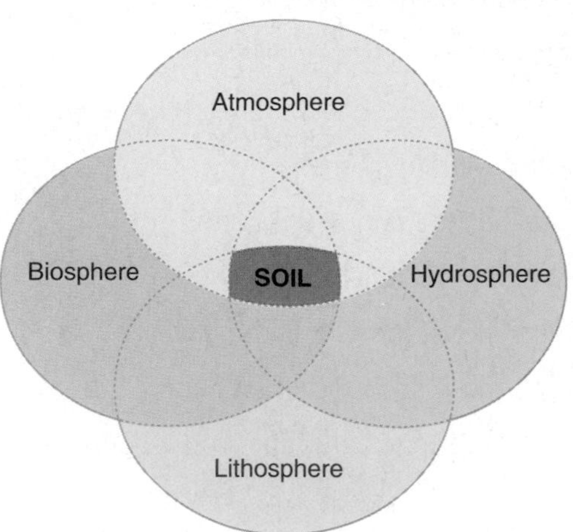

● FIGURE 12.2
The four major components of soil. Soil contains a complex assemblage of inorganic minerals and rocks, along with water, air, and organic matter. The interaction among these components and the proportion of each are important factors in the development of a soil.
How do each of these soil components contribute to making a soil suitable to support plant life?

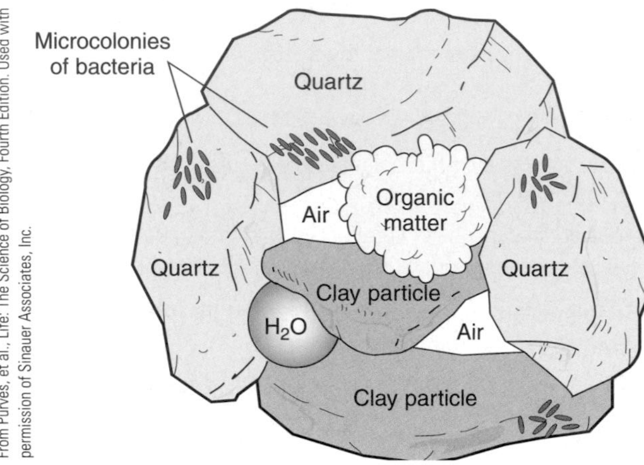

From Purves, et al., Life: The Science of Biology, Fourth Edition. Used with permission of Sinauer Associates, Inc.

● FIGURE 12.3
Fertilizers increase the productivity of soils. This farmer is adding nitrogen fertilizer into the soil on this Iowa farm.
Why can soil fertilizer be either useful or detrimental when it is introduced into the soil system?

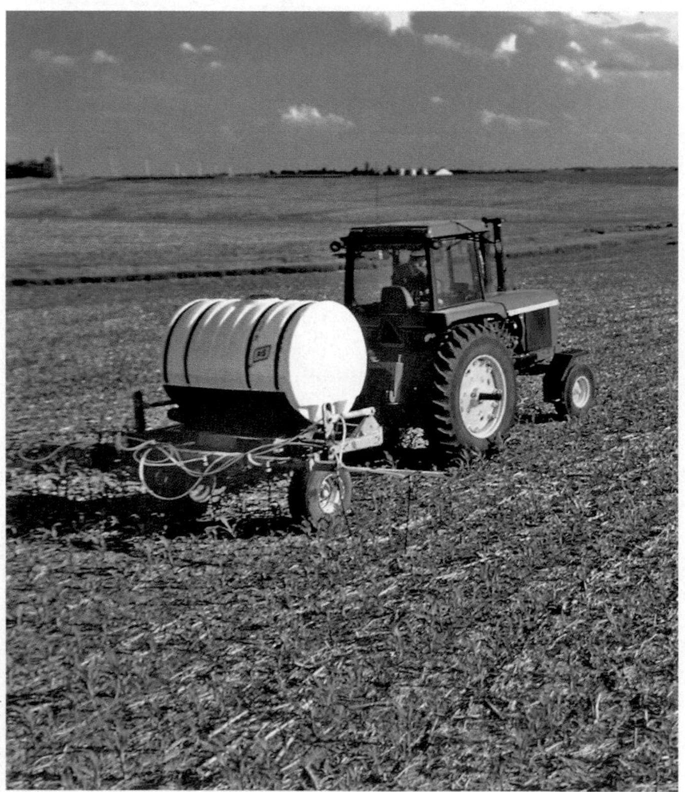

USDA/NRCS/Lynn Betts

ecosystems by providing a great variety of necessary chemical elements and compounds to life-forms. Carbon, hydrogen, nitrogen, sodium, potassium, zinc, copper, iodine, and compounds of these elements are important in soils. The chemical constituents of a soil typically come from many sources—the breakdown (*weathering*) of underlying rocks, deposits of loose sediments, and minerals dissolved in water. Organic activities help to disintegrate rocks, create new chemical compounds, and release gases into the soil.

Plants need many chemical substances for growth, and having a knowledge of a soil's mineral and chemical content is necessary for determining its potential productivity. **Soil fertilization** is the process of adding nutrients or other constituents in order to meet the soil conditions that certain plants require (• Fig. 12.3).

Soil Water

The original source of soil water is precipitation. When precipitation falls on the land, the water that is not evaporated away is either absorbed into the ground or by vegetation, or it runs downslope. Soil water is both an ingredient and a catalyst for chemical reactions that sustain life and influence soil development. Water also provides nutrients in a form that can be extracted by vegetation. As water moves through a soil it washes over and through various soil components, dissolving some of these materials and carrying them through the soil. Soil water is not pure, but is a solution that contains soluble nutrients. Plants need air, water, and minerals to function, live, and grow, and they depend on soil for much of these necessities. A soil functions as an *open system*. Matter and energy flow into and out of a soil, and they are also held in storage (• Fig. 12.4). Understanding these flows—inputs and outputs, the components and processes involved, and how they vary from soil to soil—is a key to appreciating the complexities of this natural resource.

The water in a soil is found in several different circumstances (see again Fig. 12.4). Soil water adheres to soil particles and soil clumps by surface tension (the property that causes small water droplets to form rounded beads instead of spreading out in a thin film). This soil water, called **capillary water,** serves as a stored water supply for plants. Capillary water can move in all directions through soil because it migrates from areas with more water to areas with less. Thus, during dry periods, when there is no gravitational water flowing through the soil, capillary water can move upward or horizontally to supply plant roots with moisture and dissolved nutrients.

• FIGURE 12.4

The interrelationships between a soil and the environmental factors that influence soil development. Soil is an example of an open system because it receives inputs of matter and energy, stores part of these inputs, and outputs matter and energy. Note the inputs and outputs on this diagram.

What are some examples of energy and matter that flow into and out of the soil system?

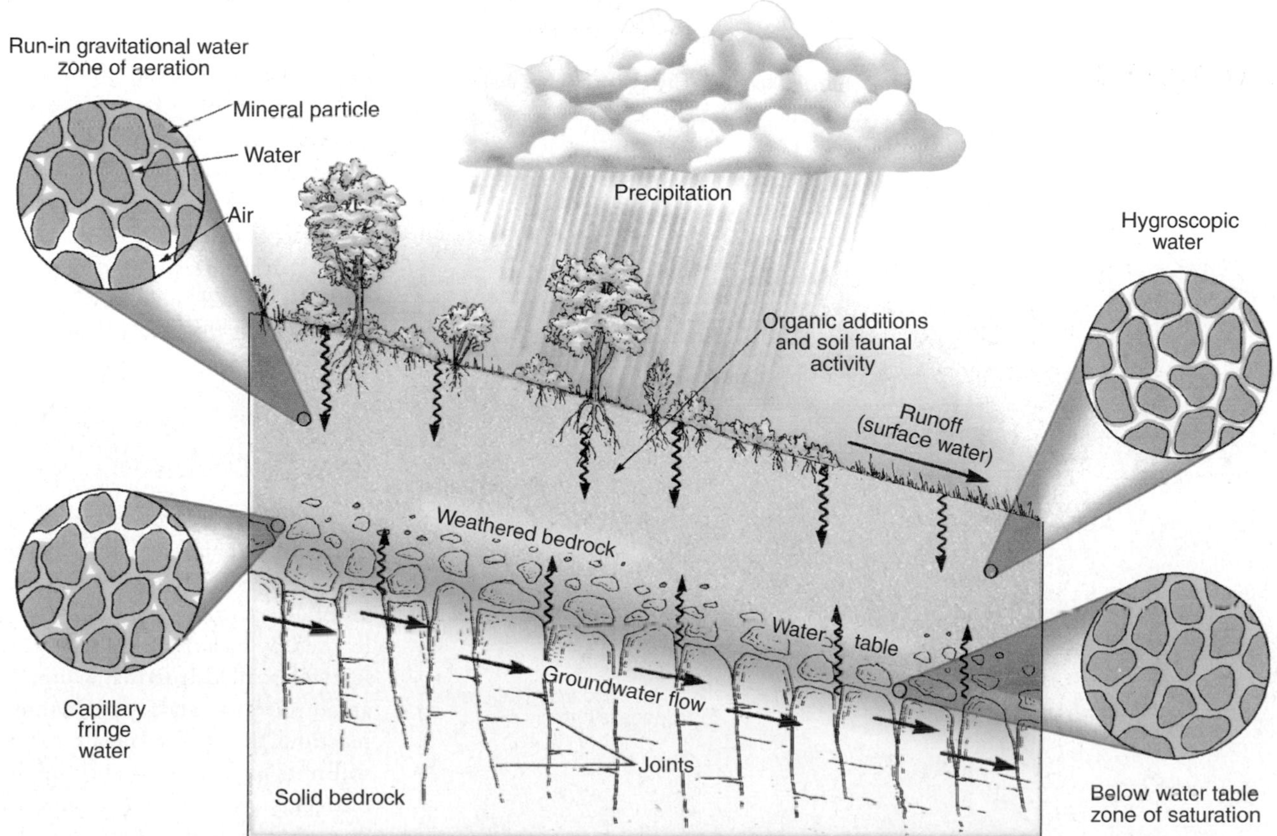

Capillary water migrates upward and moves minerals from the subsoil toward the surface. If this capillary water evaporates away, the formerly dissolved minerals remain, generally as alkaline or saline deposits in the topsoil. High concentrations of certain mineral deposits, like these, can be detrimental to plants and animals existing in the soil. Lime (calcium carbonate) deposited by evaporating soil water can build up to produce a cementlike layer, called *caliche,* which like a clay hardpan prevents the downward percolation of water.

Soil water is also found as a very thin film, invisible to the naked eye, that is bound to the surfaces of soil particles by strong electrical forces. This is **hygroscopic water,** which does not move through the soil, and it also does not supply plants with the moisture that they need.

Water that percolates down through a soil, under the force of gravity, is called **gravitational water.** Gravitational water moves downward through voids between soil particles and toward the *water table*—the level below which all available spaces are filled with water. The quantity of gravitational water in a soil is related to several conditions, including the amount of precipitation, the time since it fell, evaporation rates, the space available for water storage, and how easily the water can move through the soil.

Gravitational water performs several functions in a soil (● Fig. 12.5). As gravitational water percolates downward, it dissolves soluble minerals and carries them into deeper levels of the soil, perhaps to the zone where all open spaces are saturated. Depleting nutrients in the soil by the through flow of water is called **leaching.** In regions of heavy rainfall, leaching is common and can be intense, robbing a topsoil of all but the insoluble substances.

Gravitational water moving down through a soil also takes with it the finer particles (clay and silt) from the upper soil layers. This downward removal of soil components by water is called **eluviation.** As gravitational water percolates downward, it deposits the fine materials that were removed from the topsoil at a lower level in the soil. Deposition by water in the subsoil is called **illuviation.** Gravitational water also mixes soil particles as it moves them downward. One result of eluviation is that the texture of a topsoil tends to become coarser as the fine particles are removed. Consequently, the topsoil's ability to retain water is reduced. Illuviation may eventually cause the subsoil to become dense and compact, forming a clay **hardpan.**

Leaching and eluviation both strongly influence the characteristic layered changes with depth, or **stratification.** Fine particles and substances dissolved from the upper soil are deposited in lower levels, which become dense and may be strongly colored by accumulated iron compounds.

Soil Air

Much of a soil—in some cases, approaching 50%—consists of spaces between soil particles and between *clumps* (aggregates of soil particles). Voids that are not filled with water contain air or certain gases. Compared to the composition of the lower atmosphere, the air in a soil is likely to have less oxygen, more carbon dioxide, and a fairly high relative humidity because of the presence of capillary and hygroscopic water.

For most microorganisms and plants that live in the ground, soil air supplies oxygen and carbon dioxide necessary for life processes. The problem with a water-saturated soil is not necessarily excess water but, if all pore spaces are filled with water, there is no air supply. The lack of air is why many plants find it difficult to survive in water-saturated soils.

Organic Matter

Soil contains organic matter in addition to minerals, gases, and water. The decayed remains of plant and animal materials, partially transformed by bacterial action, are collectively called **humus.** Humus is an important catalyst in chemical reactions that help plants to extract soil nutrients. Humus also supplies nutrients and minerals to the soil. Soils that contain humus are quite

● FIGURE 12.5
Water plays several important roles in the processes that affect soil development. Water is important in moving nutrients and particles vertically, both up and down, in a soil.
How does deposition by capillary water differ from deposition (illuviation) by gravitational water?

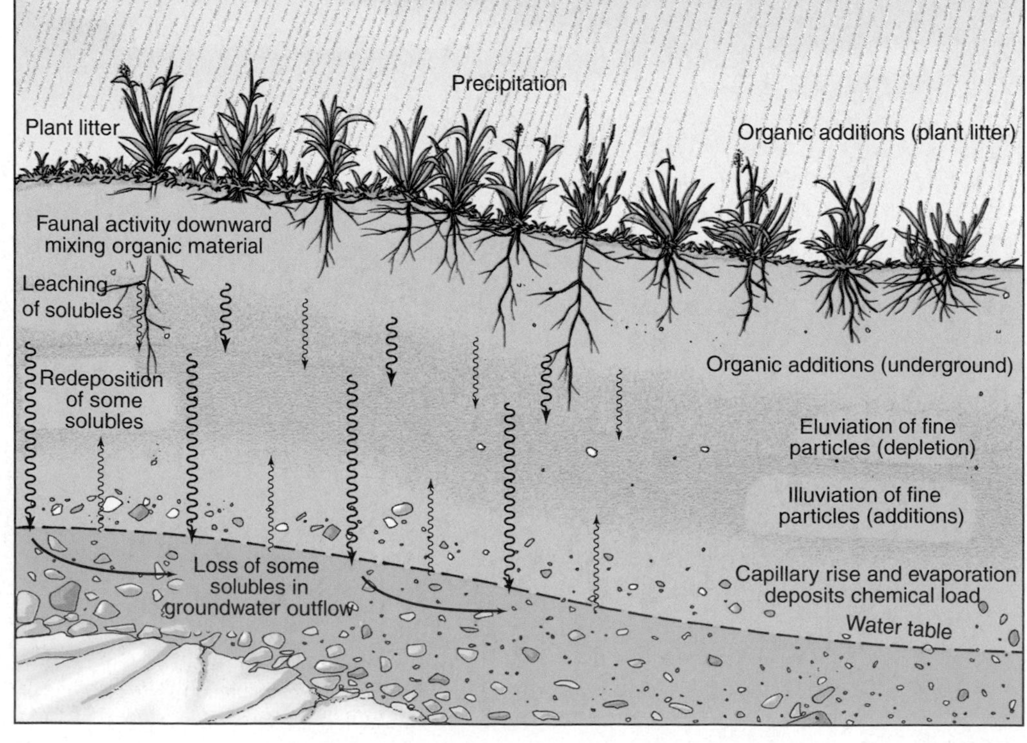

workable and have a good capacity to retain water. Humus also provides an abundant food source for microscopic soil organisms.

Most soils are actually environments that teem with life, ranging from microscopic bacteria and fungi, to earthworms, rodents, and other burrowers. Animals contribute to the soil development and enrichment by creating humus from plant litter. They also mix organic material deeper into the soil, and move inorganic fragments toward the surface. In addition, the functions of plants and their root systems are integral parts of the soil-forming system.

Soils vary at local, regional, and global scales. Particularly strong relationships exist between a soil and the vegetation and climate at its location. For example, soils in middle-latitude grasslands normally have a very high proportion of organic matter; those in deserts are thin, and rich in minerals left behind by evaporating water, like lime and salts; and tropical soils typically have a high content of iron and aluminum oxides. Knowing a soil's water, mineral, and organic components and their proportions can help us determine its productivity and what the best use for that particular soil might be.

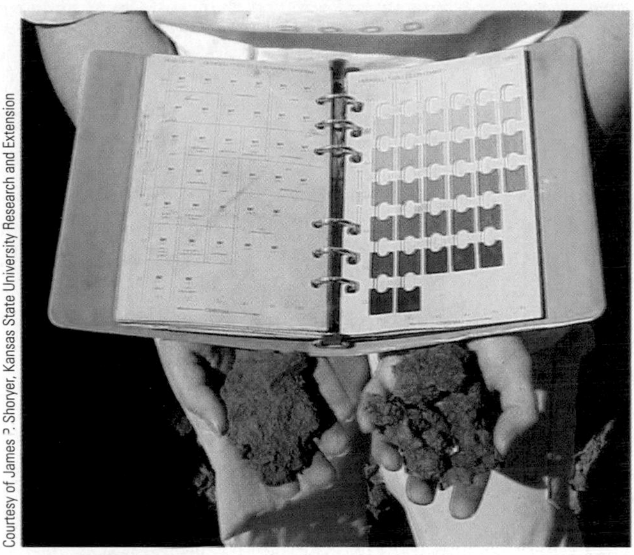

Courtesy of James P. Shroyer, Kansas State University Research and Extension

• **FIGURE 12.6**
Determining soil color. A standardized classification system is used to determine precise color by comparing the soil to the color samples found in Munsell soil color books.
In general, how would you describe the color of the soils where you live?

Characteristics of Soil

Several soil properties that can be readily tested or examined are used to describe and differentiate soil types. The most important properties include color, texture, structure, acidity or alkalinity, and capacity to hold and transmit water and air.

Color

Color is the most visible soil characteristic, but it might not be the most important attribute. Most people are aware of how soils vary in color from place to place. For example, the well-known red clay soils of Georgia are not far from Alabama's belt of black soils. Soils vary in color from black to brown to red, yellow, gray, and near-white. A soil's color is generally related to its physical and chemical characteristics. When describing soils in the field or samples in the laboratory, soil scientists use a book of standardized colors to clearly and precisely identify this coloration (• Fig. 12.6).

Decomposed organic matter is black or brown, so soils with high humus contents tend to be dark. If the humus content of soil decreases because of either low organic activity or loss of organics through leaching, soil colors typically fade to light brown or gray. Soils rich in humus are usually very fertile. For this reason, dark brown or black soils are often referred to as *rich*. However, this is not always true because some black or dark brown soils have little or no humus, but are dark because of other soil-forming factors.

Soils that are red or yellow typically indicate the presence of iron. In moist climates, a light gray or white soil indicates that iron has been leached out, leaving oxides of silicon and aluminum; in dry climates, the same color typically indicates a high proportion of calcium or salts.

Soil colors provide useful clues to the physical and chemical characteristics of soils and make the job of recognizing different soil types easier. But color alone does not answer all the important questions about a soil's qualities or fertility.

Texture

Soil texture refers to the particle sizes (or distribution of sizes) that make up a soil (• Fig. 12.7). In **clayey** soils, the dominant size is **clay** particles, defined as having diameters of less than 0.002 millimeter (soil scientists universally use the metric system). In **silty** soils, the dominant **silt** particles are defined as being between 0.002 and 0.05 millimeter. **Sandy** soils have mostly **sand**-sized particles, with diameters between 0.05 and 2.0 millimeters. Rocks larger than 2.0 millimeters are regarded as pebbles, gravel, or rock fragments, and technically are not soil particles.

The proportion of particle sizes determines a soil's texture. For example, a soil composed of 50% silt-sized particles, 45% clay, and 5% sand would be identified as a silty clay. A triangular graph (• Fig. 12.8) is used to discern different classes of soil texture based on the plot of percentages for each **soil grade** (as sand, silt, and clay are called) within each class. Point A within the silty clay class represents the example just given. A second soil sample (B) that is 20% silt, 30% clay, and 50% sand would be referred to as a sandy clay loam. **Loam** soils, which occupy the central areas of the triangular graph, are soils with a mix of the three grades (sizes) of soil particles without any size being greatly dominant. It is interesting to note that loam soils are generally best suited for supporting vegetation growth.

Soil texture helps determine a soil's capacity to retain moisture and air that are necessary for plant growth. Soils with a higher proportion of larger particles tend to be well aerated and allow water to **infiltrate** (seep through) the soil quickly—sometimes so quickly that plants are unable to use the water. Clay soils present the opposite problem because they retard water movement, becoming waterlogged and deficient in air. Aeration of the soil is an important process in cultivation, and plowing a soil opens its structure and increases its air content.

GEOGRAPHY'S PHYSICAL SCIENCE PERSPECTIVE

Basic Soil Analysis

After studying just one chapter on soils, no one would expect any introductory student to be able to do a detailed soil analysis. However, there are some basic observations that anyone can perform to better understand a few properties of a local soil. No equipment is required to make analyses; only visual and hands-on examinations are necessary.

Soil Color

Soil color can hold clues to the composition and/or the formation processes of that soil. Figure 12.6 shows the Munsell color book, a standard guide for matching and recognizing precise colors of soil types. The book includes common soil colors, but each color can also appear in a wide variety of tones.

Red: Reddish soil usually indicates that oxidation has been an active process—oxygen has chemically reacted with the soil minerals. Red also indicates that iron is in the soil. Just like rusting iron, many iron-rich minerals turn red when oxidized. The formula for this process is $FeO + O_2 \rightarrow Fe_2O_3$ (ferrous oxide + oxygen becomes hematite, a reddish iron oxide).

Blue/Silver/Gray: These tones mean that the soil has likely been reduced; in other words, oxygen has been removed from the soil.

$Fe_2O_3 \rightarrow FeO + O_2$ (the previous formula in reverse)

White: Usually denotes that calcium carbonate ($CaCO_3$) or salts (such as NaCl) may be present in the soil.

Black: A very dark color may indicate a high amount of organic material present in the soil.

More sophisticated field or laboratory analyses are required for absolute identification, but these examples will allow good working hypotheses for soil characteristics represented by colors.

Soil Texture

The particle sizes in a soil determine its texture. Soil texture is a property that you can feel, and your fingers can help in the analysis. Sand-sized particles can be easily recognized because they feel gritty to the touch. Wetting the soil and working it with your hands can help in this process. If the sample is not gritty but rather is smooth to the touch, then the soil contains silt or clay. If the sample feels sticky and you can squeeze a small soil sample into a ribbon (like with modeling clay), then clay-sized particles are abundant. Actual percentages of particle sizes in a soil sample are best established in a laboratory.

Soil Structure

The shape of clumps that a soil makes when it is broken apart is called structure and can be examined by breaking up a handful of soil. The peds (or small clumps of soil) may take on some distinctive shapes. Though the peds may form a variety of shapes, some of the more common are granular (denoting a presence of sand) and platy (showing a presence of clay). Other soil structures, such as blocky, columnar, or prismatic, are shown in Figure 12.9.

Although these simple procedures will not yield a complete analysis of a soil sample, they can certainly be the first steps in the process. It is interesting to note that pedologists (soil scientists) while in the field perform many of these same procedures.

© Jeff Vanuga/USDA Natural Resources Conservation Service

Soil analysis in the field is a hands-on process.

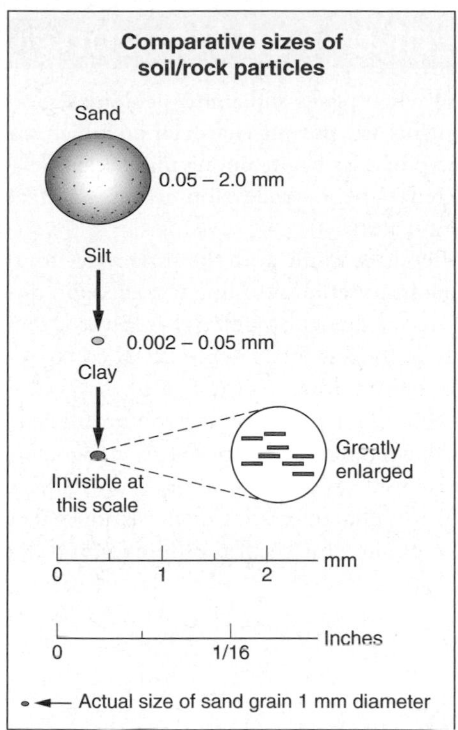

• FIGURE 12.7
Particle sizes in soil. *Sand, silt,* and *clay* are terms that refer to the size of these particles for scientific and engineering purposes. Here, the sizes of each can be compared. Clay particles are tiny, sheetlike particles that cannot be seen.

Structure

In most soils, particles clump into distinctive masses known as **soil peds,** which give a soil a distinctive structure. Soil structure influences a soil's **porosity**—the amount of space that may contain fluids. Soil structure also influences **permeability**—the rate at which fluids such as water can pass through. Permeability is usually greatest in sandy soils, and porosity is usually greatest in clayey soils. Both of these factors control soil drainage as well as the available moisture in a soil. Soils with similar textures may have different structures, and vice versa.

Soil structure can be influenced by outside factors such as moisture regime and the nutrient cycles that plants use to interchange chemicals with the soil, keeping certain ones in the system while others are leached away. We have all seen the structural change in soils that occurs when soils are wet compared to when they are dry. Human activities also influence soil structure through cultivation, irrigation, and fertilization. Fertilizers, as well as lime or decayed organic debris, encourage clumping of soil particles and the maintenance of clumps. Excess sodium and magnesium have the opposite effect, causing clay soils to become a sticky muck when wet and like concrete when dry. The absence of smaller particles typically hinders the development of a well-defined soil structure.

Scientists classify soil structures according to their form. These range from columns, prisms, and angular blocks, to nutlike spheroids, laminated plates, crumbs, and granules (• Fig. 12.9). Soils with massive and fine structures tend to be less useful than aggregates of intermediate size and stability, which permit good drainage and aeration.

Acidity and Alkalinity

An important aspect of soil chemistry is a soil's departure from neutrality toward either acidity or alkalinity (baseness). Levels of acidity or alkalinity are measured on the **pH scale** of 0 to 14. A pH reading indicates the concentration of reactive hydrogen ions present. The pH scale is logarithmic, meaning that each change in a whole pH number represents a tenfold change. It is also an inverted scale—a lower pH means a greater amount of hydrogen ions present (higher acidity). Low pH values indicate an acid soil, and high pH indicates alkaline conditions (• Fig. 12.10).

• FIGURE 12.8
The texture of a soil can be represented by a plotting a point on this diagram. Texture is determined by sieving the soil to determine the percentage of particles falling into the size ranges for clay, silt, and sand. Note that each of the three axes of the triangle is in a different color and the line colors also correspond (clay-red, silt-blue, sand-green).
What would a soil that contains 40% sand, 40% silt, and 20% clay be classified as?

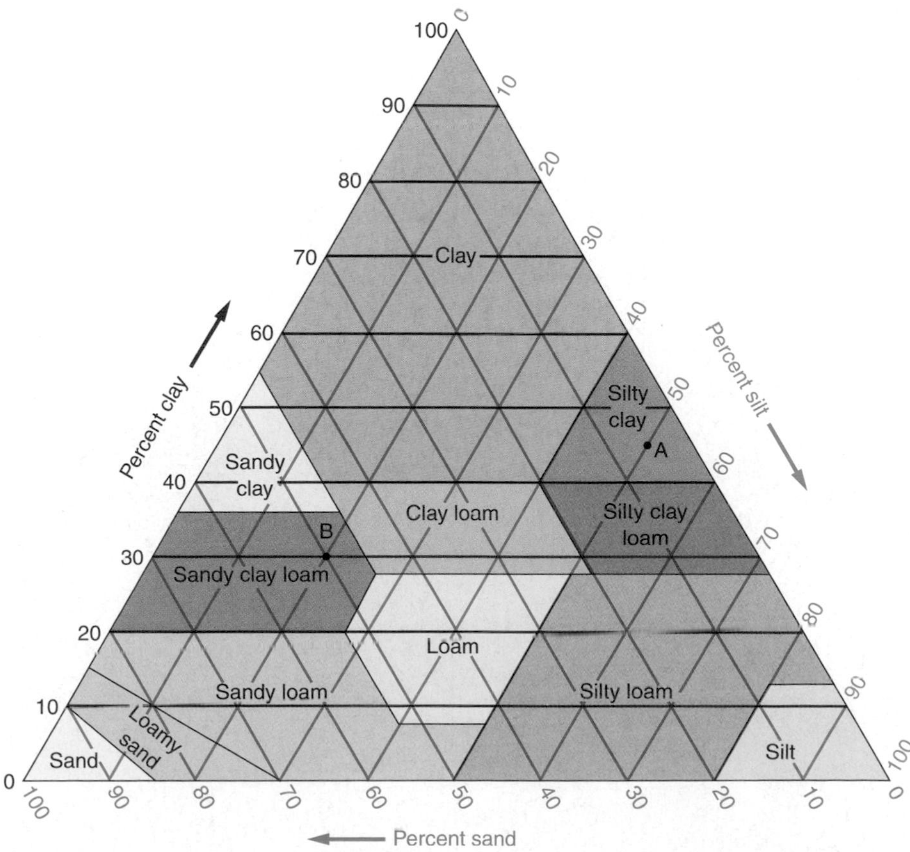

Soil acidity or alkalinity helps determine the available nutrients that affect plant growth. Plants absorb nutrients that are dissolved in liquid. However, soil water that lacks some degree of acidity has little ability to dissolve these nutrients. As a result, even though nutrients are in the soil, plants may not have access to them.

Most complex plants will grow only in soils with levels between pH 4 and pH 10, although the optimum pH for vegetation growth varies with the plant species. Around the world, vegetation has evolved in and adapted to a variety of climates and soil environments, both of which can affect soil pH. Certain species tolerate alkaline soils, and others thrive under more acid conditions.

Leaching caused by high rainfall gradually replaces soil elements such as sodium (Na), potassium (K), magnesium (Mg), and calcium (Ca) with hydrogen. Falling rain picks up atmospheric carbon dioxide and becomes slightly acidic: $H_2O + CO_2 = H_2CO_3$ (carbonic acid), so desert soils tend to be alkaline and soils in humid regions tend to be acidic (• Fig. 12.11). The humus content in the soils of humid areas also contributes to higher soil acidity.

To correct soil alkalinity, common in the arid regions, and to make the soil more productive, the soil can be flushed with irrigation water. Strongly acidic soils are also detrimental to plant growth. In acidic soils, soil moisture dissolves nutrients, but they may be leached away before plant roots can absorb them. Soil acidity can generally be corrected by adding lime to the soil. In addition to affecting plant growth, soil acidity or alkalinity also affects microorganisms in the soil. Microorganisms are highly sensitive to a soil's pH, and each species has an optimum environmental setting.

• FIGURE 12.9

This guide to classifying the structure of a soil on the basis of soil peds can be used to help determine other characteristics of a soil.

How does soil structure affect a soil's usefulness or suitability for agriculture?

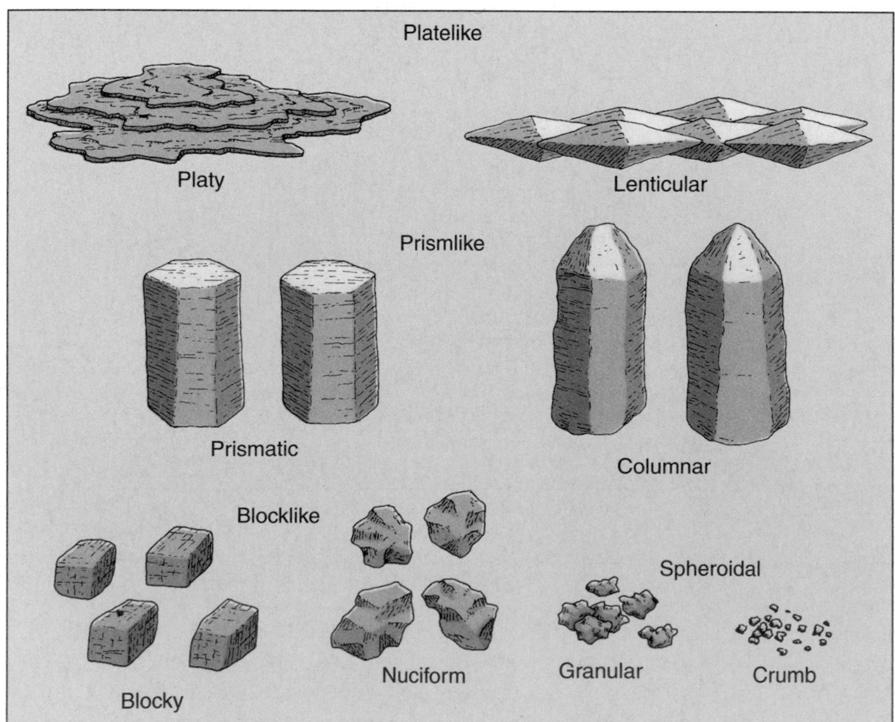

Development of Soil Horizons

Soil development begins when plants and animals colonize rocks or deposits of rock fragments, the **parent material** on which soil will form. Once organic processes begin among mineral particles or rock fragments, differences begin to develop from the surface down through the parent material.

Initially, vertical differences result from the surface accumulation of organic litter and the removal of fine particles and dissolved minerals from upper layers by percolating water that deposits these materials at a lower level. The vertical cross section of a soil from the surface down to the parent material is called a **soil profile** (• Fig. 12.12). Examining soil profiles and the vertical differences they contain is important to recognizing different soil types and how those soils developed. As climate, vegetation, animal life, and characteristics of the land surface affect soil formation over time, this vertical differentiation becomes more and more apparent.

Soil Horizons

Within their soil profiles, well-developed soils typically exhibit several distinct layers, called **soil horizons,** that are distinguished by their physical and chemical properties. Soils are classified largely on the differences in their horizons and in the processes responsible for those differences (• Fig. 12.13). Soil horizons are designated by a set of letters that refer to their composition, dominant process, and/or position in the soil profile.

At the surface, but only in locations where there is a sufficient cover of decomposed vegetation litter, there will be an O *horizon*. This is a layer of organic debris and humus; the "O" designation refers to this horizon's high organic content. Immediately below is the *A horizon,* commonly referred to as "topsoil." In general, the *A* horizon is dark because it contains decomposed organic matter. Beneath the *A* horizon, certain soils have a lighter-colored *E horizon,* named for the action of strong eluvial processes. Below this is a zone of accumulation, the *B horizon,* where much of the materials removed from the *A* and *E* horizons are deposited. Except in soils with a high organic content that has been mixed vertically, the *B* horizon generally has little humus. The *C horizon* is the weathered parent material from which the soil has developed—either fragments of the bedrock, or deposits of rock materials that were transported to the site by water, wind, glacial, or other surface process.

The lowest layer, sometimes called the *R horizon,* is unchanged parent material, either bedrock or transported deposits of rock fragments. Certain horizons in some soils may not be as well developed as others, and some horizons may be missing altogether. Because soils

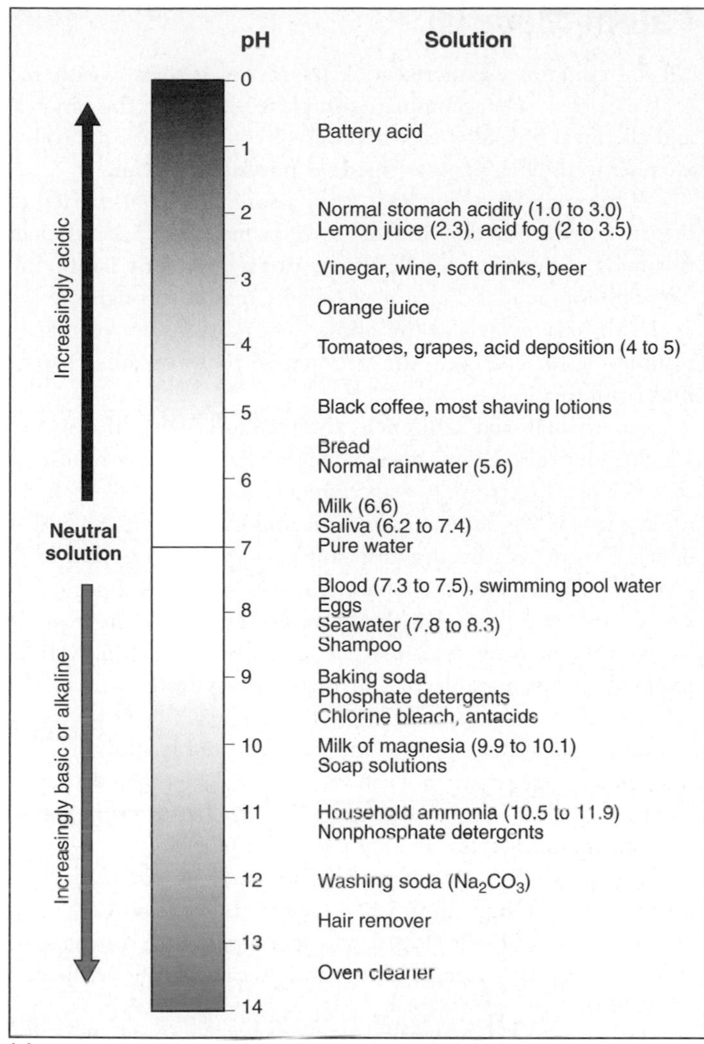

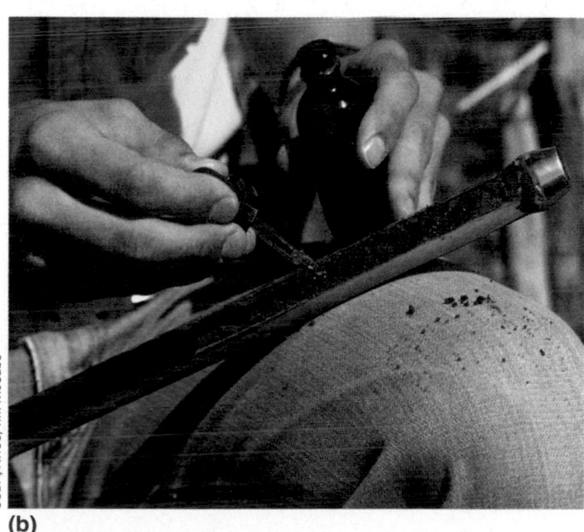

USDA/NRCS/Tim McCabe

• FIGURE 12.10
(a) The pH scale of acidity, neutrality, and alkalinity. The degree of acidity or of alkalinity, called pH, can be easily understood when numbers on the scale are linked to common substances. Low pH means acidic, and high pH means alkaline; a reading of 7 is neutral. (b) Alkalinity in a soil can be tested in the field with drops of a dilute acid. If the soil fizzes in the acid solution, alkalinity is typically high.

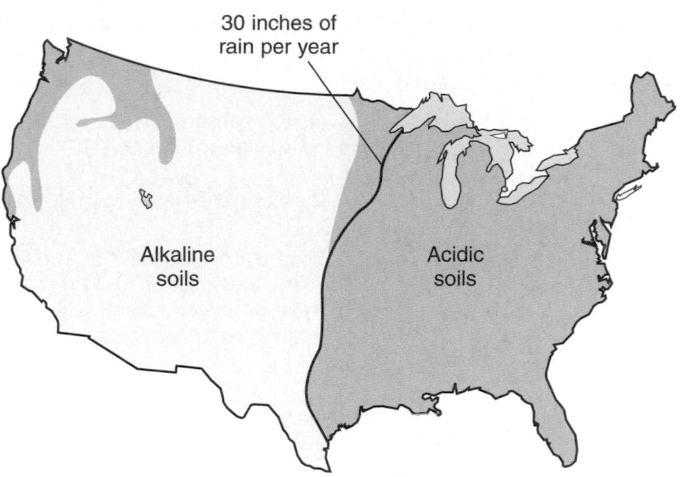

• FIGURE 12.11
The distribution of alkaline and acidic soils in the United States is generally related to climate. Soils in the East tend to be acidic and those in the West, alkaline. The dividing line corresponds fairly well with the 30-inch annual precipitation isohyet.
Other than climate, what environmental factors might cause this east–west variation, and why are some places west of the 30-inch line acidic?

© Hari Eswaran, USDA/NRCS

• FIGURE 12.12
A soil profile is examined by digging a pit with vertical walls to clearly show variations in color, structure, composition, and other characteristics that occur with depth. This soil is in a grassland region of northern Minnesota.

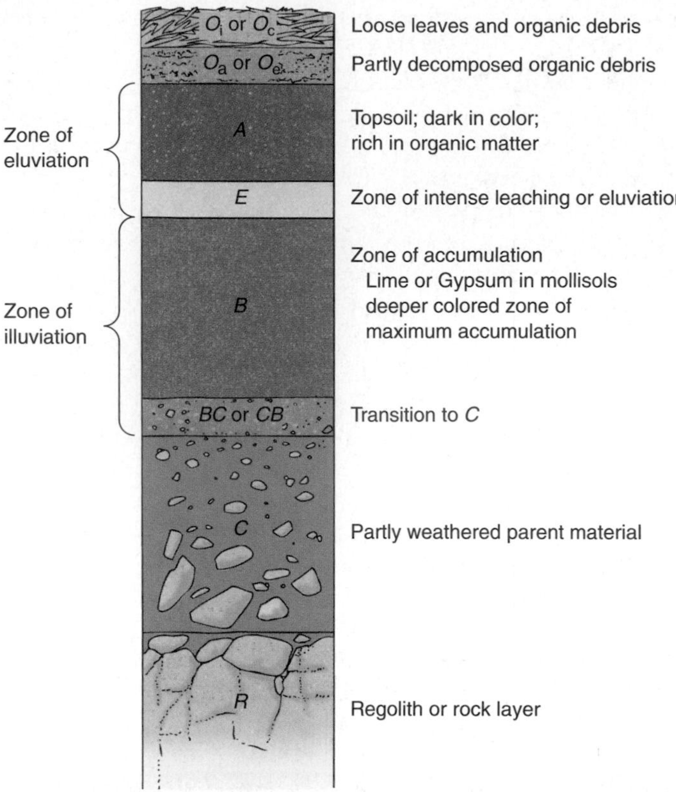

• **FIGURE 12.13**
Soils are categorized by the degree of development and the physical characteristics of their horizons. *Regolith* is a generic term for broken bedrock fragments at or very near the surface.
Which soil profiles shown in Figures 12.28–12.31 display horizons that are easy to recognize?

and the processes that form them vary widely and can be transitional between horizons, the horizon boundaries may be either sharp or gradual. Variations in color and texture within a horizon are also not unusual.

Factors Affecting Soil Formation

Because of the great variety among the components of soils and the processes that affected them, no two soils are identical in all of their characteristics. One important factor is rock *weathering,* which refers to the many natural processes that break down rocks into smaller fragments (weathering will be discussed in detail in Chapter 15). Chemical reactions can cause rocks and minerals to decompose and physical processes also cause the breakup of rocks. Just as statues, monuments, and buildings become "weather-beaten" over time, rocks exposed to the elements eventually break up and decompose.

Hans Jenny, a distinguished soil scientist, observed that soil development was a function of climate, organic matter, relief, parent material, and time—factors that are easy to remember by their initials arranged in the following order: **Cl, O, R, P, T.** Among these factors, parent material is distinctive because it is the raw material. The other factors influence the type of soil that forms from the parent material.

Parent Material

All soil contains weathered rock fragments. If these weathered rock particles have accumulated in place—through the physical and chemical breakdown of bedrock directly beneath the soil—we refer to the fragments as **residual parent material.**

If the rock fragments that form a soil have been carried to the site and deposited by streams, waves, winds, gravity, or glaciers, this mass of deposits is called **transported parent material.** The development and action of organic matter through the life cycles of organisms and the climatic conditions are primarily responsible for changing the fragmented rocks or other parent material into a soil.

Parent material influences the characteristics of a soil in varying degrees. Some parent materials, such as a *sandstone* that contains extremely hard and resistant sand-sized fragments, are far less subject to weathering than others. Soils that develop from weathering-resistant rocks tend to have a high level of similarity to their parent materials. If the bedrock is easily weathered, the soils that develop tend to be more similar to soils in other regions that have a similar climate than to those of comparable parent materials, which formed in a different climate.

On a global basis, climate and the associated plant communities produce greater variations in soil characteristics than do parent materials. Soil differences that are related to variations in parent material are most visible on a local level.

The longer a soil develops, the influence of parent material on its characteristics diminishes. Given the same soil-forming conditions, recently developed soils will show more similarity to its parent material, compared to a soil that has developed over a long time.

Many of the chemicals and nutrients in a soil reflect the composition of its parent material (• Fig. 12.14). For example, calcium-deficient parent materials will produce soils that are low in calcium, and its natural fauna and plant cover will be types that require little calcium. Likewise, a parent material with a high aluminum content will produce a soil that is rich in aluminum. In fact, the ore of aluminum is bauxite, found in tropical soils where it has been concentrated by intense leaching away of the other bases.

The particle sizes that result from the breakdown of parent material are a prime determinant of a soil's texture and structure. A rock material such as sandstone, which contains little clay and weathers into relatively coarse fragments, will produce a soil of coarse texture. Parent materials are also an important influence on the availability of air and water to a soil's living population.

Organic Activity

Plants and animals affect soil development in many ways. The life processes of plants growing in a soil are as important as its microorganisms—the microscopic plants and animals that live in a soil.

Generally, a dense vegetative cover protects a soil from being eroded away by running water or wind. Forests form a protective canopy and produce surface litter, which keeps rain from

R. Gabler

• **FIGURE 12.14**
Despite strong leaching under a wet tropical climate, Hawaiian soils remain high in nutrients because their parent material is of recent volcanic origin.
What other parent materials provide the basis for continuously fertile soils in wet tropical climates?

beating directly on the soil and increases the proportion of rainwater entering the soil rather than running off its surface. Variations in vegetation species and density of cover can also affect the evapotranspiration rates. A sparse vegetative cover will allow greater evaporation of soil moisture and dense vegetation tends to maintain soil moisture.

The characteristics of a plant community affect the nutrient cycles that are involved in soil development (• Fig. 12.15). As plants die and decompose, or leaves fall to the ground, nutrients are returned to the soil. Soils, however, can become impoverished if soluble nutrients that are not used by plants are lost through leaching. The roots of plants help to break up the soil structure, making it more porous, and roots also absorb water and nutrients from the soil.

Leaves, bark, branches, flowers, and root networks contribute to the organic composition of soil, through litter and through the remains of dead plants. The organic content of soil depends on its associated plant life. For example, a grass-covered prairie supplies much more organic matter than the thin vegetative cover of desert regions. There is some question, however, as to whether forests or grasslands (with their thick root networks and annual life cycle) furnish the soil with greater organic content. Many of the world's grassland regions, like the North American prairies, provide some of the world's most fertile soils for cultivation in part because of the high amount of organic matter that a grass cover generates.

In terms of their contribution to soil formation, bacteria are perhaps the most important microorganisms that live in soils. Bacteria break down organic matter, humus, and the debris of living things into organic and inorganic components, allowing the formation of new organic compounds that promote plant growth. It has been suggested that the number of bacteria, fungi, and other microscopic plants and animals living in a soil may be 1 billion per gram (a fifth of a teaspoon) of soil. The activities and remains of these microorganisms, minute though they are individually, add considerably to the organic content of a soil.

Earthworms, nematodes, ants, termites, wood lice, centipedes, burrowing rodents, snails, and slugs also stir up the soil, mixing mineral components from lower levels with organic components from the upper portion. Earthworms contribute greatly to soil development because they take soil in, pass it through their digestive tracts, and excrete it in casts. The process not only helps mix the soil but also changes the texture, structure, and chemical qualities of the soil. In the late 1800s, Charles Darwin estimated that earthworm casts produced in a year would equal as much as 10–15 tons per acre. As for the number of earthworms, a study suggested that the total weight of earthworms beneath a pasture in New Zealand equaled the weight of the sheep grazing above them.

Climate

Chapters 9 and 10 demonstrated that, on a world regional scale, climate is a major factor in soil formation. Of course, if the climate is the same in a region where the soils vary, other factors must be responsible for the local variation. Soil differences that are apparent at a local level tend to reflect the influence of factors such as parent materials, land surface configurations, vegetation types, and time.

Temperature directly affects soil microorganism activity, which in turn affects the decomposition rates of organic matter. In hot equatorial regions, intense activities by soil microorganisms preclude thick accumulations of organic debris or humus. • Figure 12.16 shows that the amounts of organic matter and humus in a soil increase toward the middle latitudes and away from polar regions and the tropics. In the mesothermal and microthermal climates (*C* and *D*), microorganism activity is slow enough to allow decaying organic matter and humus to accumulate in rich layers. Moving poleward into colder regions, retarded microorganism activity and limited plant growth tends to result in thin accumulations of organic matter.

Chemical activity increases and decreases directly with temperature, given equal availability of moisture. As a result, parent materials of soils in hot, humid equatorial regions are altered to a far greater degree by chemical means than are parent materials in colder zones.

Temperature affects soil indirectly through its influence on vegetation associations that are adapted to certain climatic regimes. Soils generally reflect the character of plant cover because of nutrient cycles that tend to keep both vegetation and soil in chemical equilibrium. The combined effects of vegetative cover and the climatic regime tend to produce soil profiles and characteristics that tend to share certain characteristics among different regions that have similar climates and vegetation associations (● Fig. 12.17).

Moisture conditions affect the development and character of soils more directly than any other climatic factor. Without precipitation, and the soil water it provides, terrestrial plant life would be impossible. Ample precipitation supports plant growth that can greatly increase the organic content and thereby the fertility of a soil. However, extremely high rainfall will cause leaching of nutrients, and a relatively infertile soil.

Gravitational and capillary water have pronounced effects on soil development, structure, texture, and color. Precipitation is the original source of soil water (disregarding the minor contribution of dew), and the amount of precipitation received affects leaching, eluviation, and illuviation and thereby rates of soil formation and horizon development. The evaporation rate is a very important factor as well. Salt and gypsum deposits from the upward migration of capillary water are more extensive in hot, dry regions—such as the southwestern United States where evaporation rates are high—than in colder, dry regions (see again Fig. 12.17).

● FIGURE 12.15
The nutrient cycle in a forest. Trees take up nutrients from the soil through soil water absorbed into their root systems. Nutrients are supplied by the breakdown of rocks and minerals, as well as by leaf and other organic litter.

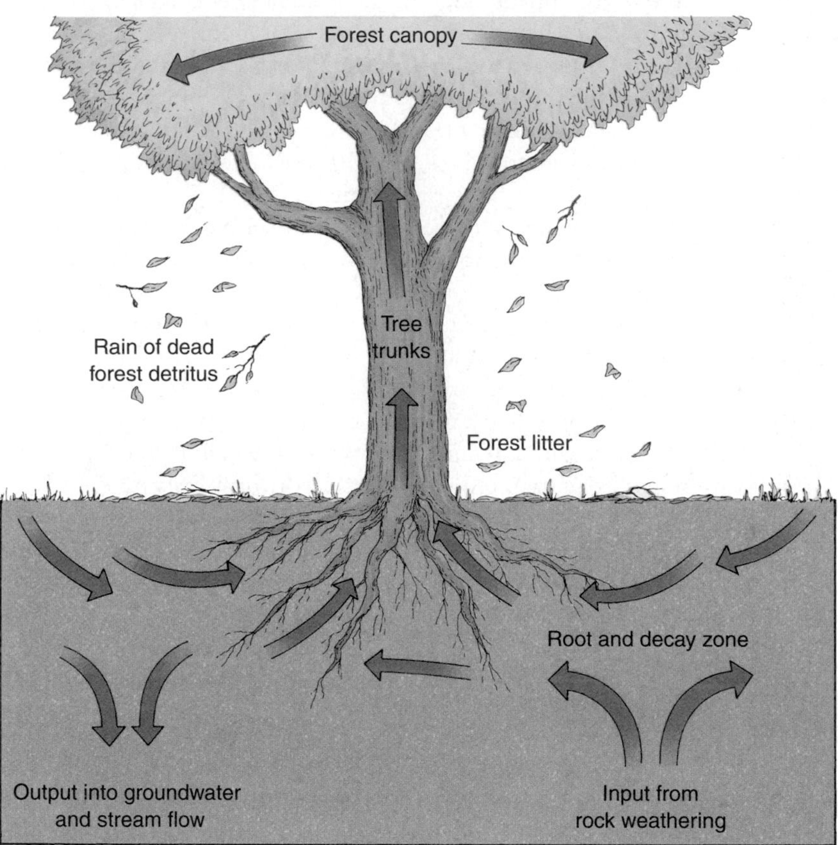

● FIGURE 12.16
The relationship of temperature to production and destruction of organic matter in the soil. **What range of mean annual temperatures is most favorable for the accumulation of humus?**

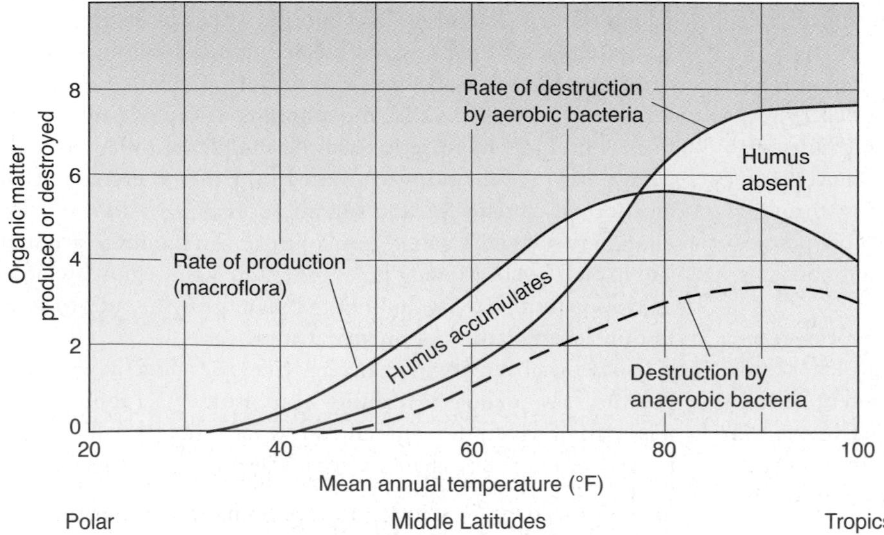

Land Surface Configuration

The slope of the land, its relief, and its *aspect* (the direction it faces) all influence soil development. Steep slopes are generally better drained than gentler ones, and they are also subject to rapid runoff of surface water. As a consequence, there is less infiltration of water on steeper slopes, which inhibits soil development, sometimes to the extent that there will be no soil. In addition, rapid runoff on steep slopes can erode surfaces as fast or faster than soil can develop on them. On gentler slopes, where there is less runoff and higher infiltration, more water is available for soil development and to support vegetation growth, so erosion is not as intense. In fact, erosion rates in areas of gently rolling hills may be just enough to offset the development of soils. Well-developed soils typically form on land that is flat or has a gentle slope.

Slope aspect has a direct effect on microclimates in areas outside of the equatorial tropics. North-facing slopes in the middle and high latitudes of the Northern Hemisphere have microclimates that are cooler and wetter than those on south-facing exposures, which receive the sun's rays at a steeper angle and are therefore warmer and drier. Local variations

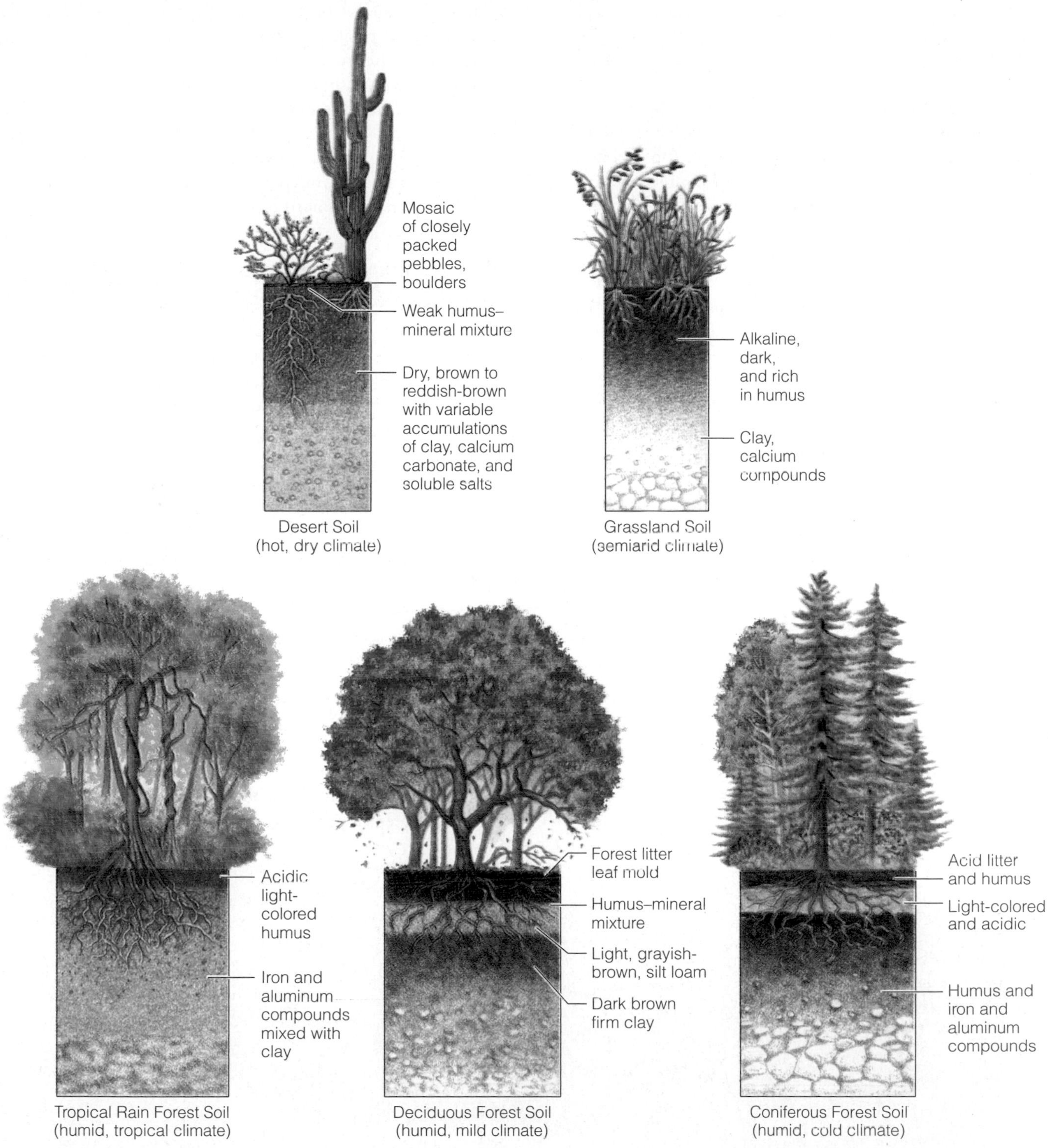

• **FIGURE 12.17**
Idealized diagrams of five different soil profiles illustrate the effects of climate and vegetation on the development of soils and their horizons.
Which two environments produce the most humus and which two produce the least?

in soil depth, texture, and profile development result directly from microclimate differences.

Topography, through its effects on vegetation, indirectly influences soil development. Steep slopes prevent the formation of a soil that would support abundant vegetation, and a modest plant cover yields less organic debris for the soil.

Time

Soils have a tendency to develop toward a state of equilibrium with their environment. A soil is sometimes called "mature" when it has reached such a condition of equilibrium. Young soils are still in the process of developing toward being in equilibrium with their environmental conditions. Mature soils have well-developed horizons that indicate the conditions under which they formed. Young or "immature" soils typically have poorly developed horizons or perhaps none at all (●Fig. 12.18).

Another effect of time is that, as soils develop, their influence of their parent material decreases and they increasingly reflect their climate and vegetative environments. On a global scale, climate typically has the greatest influence on soils, provided sufficient time has passed for the soils to become well developed.

The importance of time in soil formation is especially clear in soils developed on transported parent materials. Depositional surfaces are in many cases quite recent in geologic terms and have not been exposed to weathering long enough for a mature soil to develop. Deposition occurs in a variety of settings: on river floodplains where the accumulating sediment is known as *alluvium;* downwind from dry areas where dust settles out of the atmosphere to form blankets of wind-deposited silts, called *loess;* and in volcanic regions showered by ash and covered by lava. Ten thousand years ago, glaciers withdrew from vast areas, leaving behind jumbled deposits of rocks, sand, silt, and clay.

Because of the great number and variability of materials and processes involved in the formation of soils, there is no fixed amount of time that it takes for a soil to become mature. The Natural Resources Conservation Service, however, estimates that it takes about 500 years to develop 1 inch of soil in the agricultural regions of the United States. Generally, though, it takes thousands of years for a soil to reach maturity.

Soil-Forming Regimes

The characteristics that make major soil types distinctive and different from one another result from their **soil-forming regimes,** which vary mainly because of differences in climate and vegetation. At the broadest scale of generalization, climate differences produce three primary soil-forming regimes: laterization, podzolization, and calcification.

●FIGURE 12.18
The time that a soil has been developing is important to its composition and physical character. Given enough time and the proper environmental conditions, soils will become more maturely developed with a deeper profile and stronger horizon development.
What major changes occur as the soil illustrated here becomes better developed over time?

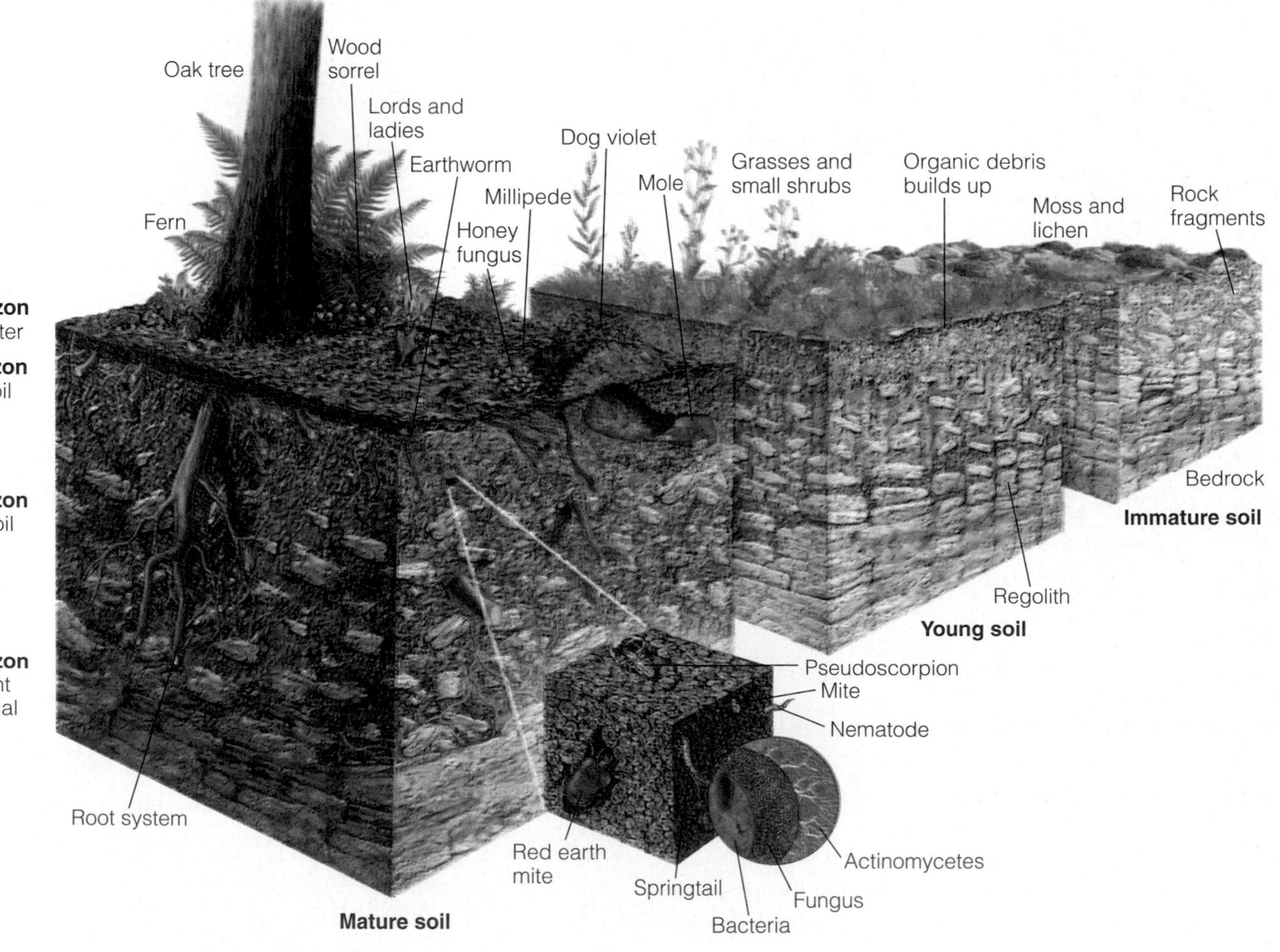

From Derek Elsom, Earth, 1992. Copyright © 1992 by Marshall Editions Developments Limited. New York: Macmillan. Used by permission.

Laterization

Laterization is a soil-forming regime that occurs in humid tropical and subtropical climates as a result of high temperatures and abundant precipitation. These climatic environments encourage rapid breakdown of rocks and decomposition of nearly all minerals. A soil of this type is known as **laterite,** and these soils are generally reddish in color from iron oxides; the term *laterite* means "brick-like." In tropical areas laterite is quarried for building material (● Fig. 12.19).

Despite the dense vegetation that is typical of these climate regions, little humus is incorporated into the soil because the plant litter decomposes so rapidly. Laterites do not have an *O* horizon, and the *A* horizon loses fine soil particles as well as most minerals and bases except for iron and aluminum compounds, which are insoluble primarily because of the absence of organic acids (● Fig. 12.20). As a result, the topsoil is reddish, coarse textured, and tends to be porous. In contrast to the *A* horizon, the *B* horizon in a lateritic soil has a heavy concentration of illuviated materials.

In the tropical forests, soluble nutrients released by weathering are quickly absorbed by vegetation, which eventually returns them to the soil where they are reabsorbed by plants. This rapid cycling of nutrients prevents the total leaching away of bases, leaving the soil only moderately acidic. Removal of vegetation permits total leaching of bases, resulting in the formation of crusts of iron and aluminum compounds (laterites), as well as accelerated erosion of the *A* horizon.

Laterization is a year-round process because of the small seasonal variations in temperature or soil moisture in the humid tropics. This continuous activity and strong weathering of parent material cause some tropical soils to develop to depths of as much as 8 meters (25 ft) or more.

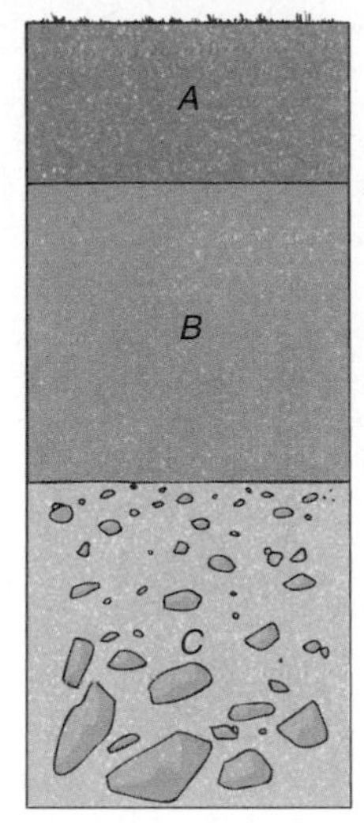

● FIGURE 12.20
Soil profile horizons in laterite, one of three major soil-forming regimes. Laterization is a soil development process that occurs in tropical and equatorial zones that experience warm temperatures year round and wet climates.

● FIGURE 12.19
Laterite cut for building stone and stacked along a village road in the state of Orissa, India.
Why is building with brick or stone rather than wood so important in heavily populated, less developed nations such as India?

R. Gabler

Podzolization

Podzolization occurs mainly in the high middle latitudes where the climate is moist with short, cool summers and long, severe winters. The coniferous forests of these climate regions are an integral part of the podzolization process.

Where temperatures are low much of the year, microorganism activity is reduced enough that humus does accumulate; however, because of the small number of animals living in the soil, there is little mixing of humus below the surface. Leaching and eluviation by acidic solutions remove the soluble bases and aluminum and iron compounds from the *A* horizon (● Fig. 12.21). The remaining silica gives a distinctive ash-gray color to the *E* horizon (*podzol* is derived from a Russian word meaning "ashy"). The needles that coniferous trees drop are chemically acidic and contribute to the soil acidity. It is difficult to determine whether the soil is acidic because of the vegetative cover or whether the vegetative cover is adapted to the acidic soil.

Podzolization can take place outside the typical cold, moist climate regions if the parent material is highly acidic—for example, on the sandy areas common along the East Coast of the United States. The pine forests that grow in such acidic conditions return acids to the soil, promoting the process of podzolization.

Calcification

A third distinctive soil-forming regime is called **calcification.** In contrast to both laterization and podzolization, which require humid climates, calcification occurs in regions where evapotranspiration significantly exceeds precipitation. Calcification is important in the climate regions where moisture penetration is shallow. The subsoil is typically too dry to support tree growth,

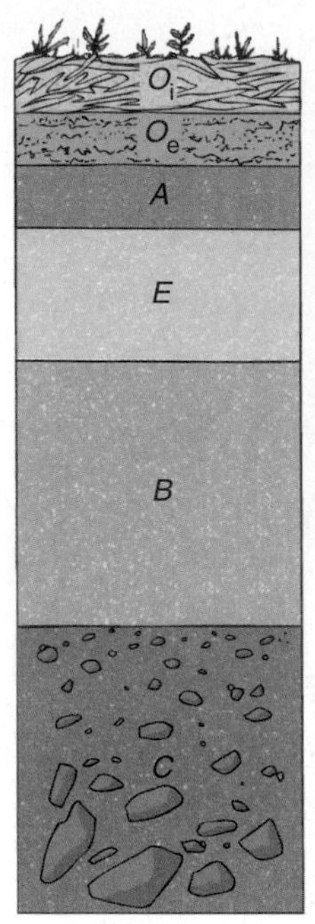

• FIGURE 12.21
Soil profile horizons in a podzol soil, another of the three major soil-forming regimes. Podzolization typically occurs under cool, wet climates in regions of coniferous trees or in boggy environments, and is a very acidic soil type.

and shallow-rooted grass or shrubs are the primary forms of vegetation. Calcification is enhanced as grasses use calcium, drawing it up from lower soil layers and returning it to the soil when the annual grasses die. Grasses and their dense root networks provide large amounts of organic matter, which is mixed deep into the soil by burrowing animals. Middle-latitude grassland soils are rich in bases and in humus and are the world's most productive agricultural soils. The deserts of the American West generally have no humus, and the rise of capillary water can leave not only calcium carbonate but also sodium chloride (salt) at the surface.

In many areas of low precipitation, the air is often loaded with alkali dusts such as calcium carbonate ($CaCO_3$). When calm conditions prevail or when it rains, the dust settles and accumulates in the soil. The rainfall produces an amount of soil water that is just sufficient to translocate these materials to the *B* horizon (• Fig. 12.22). Over hundreds to thousands of years, the $CaCO_3$-enriched dust concentrates in the *B* horizon, forming hard layers of *caliche*. Much thicker accumulations called *calcretes* (• Fig. 12.23) form by the upward (capillary) movement of dissolved calcium in groundwater when the water table is near the surface.

Regimes of Local Importance

Two additional localized soil-forming regimes merit attention. Both characterize areas with poor drainage although they occur under very different climate conditions. The first, **salinization,** or the concentration of salts in the soil, is often detrimental to plant growth. Salinization occurs in stream valleys, interior basins, and other low-lying areas, particularly in arid regions with high groundwater tables. The high groundwater levels can be the result of water from adjacent mountain ranges, stream flow originating in humid regions, or a wet–dry seasonal precipitation regime (• Fig. 12.24). Salinization can also be a consequence of intensive irrigation under arid conditions. Rapid evaporation leaves

• FIGURE 12.22
Soil profile horizons in a calcified soil, formed by the third major soil-forming regime. Calcification is a soil development process that is most prominent in cool to hot subhumid or semiarid climate regions, particularly in grassland regions, but also in deserts.

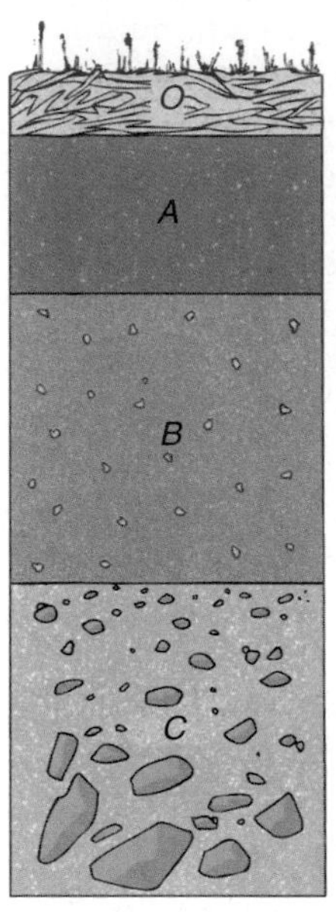

• FIGURE 12.23
This shallow-rooted grass is growing in a thin layer of topsoil over a much thicker layer of calcium carbonate called calcrete. **What precipitation characteristics are associated with the calcification soil-forming process?**

© State of Victoria, Department of Primary Industries, 1997

GEOGRAPHY'S ENVIRONMENTAL SCIENCE PERSPECTIVE

Soil Resources Are Limited and Threatened: How Much Good Soil Is There on Earth?

A widely used analogy for understanding the amount of soil that exists on Earth uses an apple to demonstrate that soil, which is adequate for the world's agricultural needs, is a rather limited natural resource. If the apple is cut into quarters, three of those pieces (75%) represent the water bodies on Earth and should be put away. The piece that remains represents the Earth's land area (25%) where soil *could* exist. Half of the remaining piece should be cut off and put away, representing rocky desert, polar, or high mountain regions, where soil does not exist or will not grow much because of harsh climatic conditions. Now there is an eighth (12.5%) of the apple remaining. This eighth of the apple should then be cut into four pieces, and three of them put away as they represent areas with local conditions that preclude agriculture (too rocky, steep, wet, and so on) and the areas that we have already covered over with cities, towns, and roads. The remaining apple slice (about 3% of Earth) represents areas with good soils for agriculture to feed the world's population.

Soil Degradation and Soil Loss

Today, even areas with good soils are under continuing threat. It may take 200–1000 years or more to develop 2.5 centimeters (1 in.) of soil, but through erosion by water or wind, thousands of years of soil development may be lost in a season or in days. Land degradation by humans is a major cause of soil loss, but impacts from changing climates, particularly desertification, also play a role. Overgrazing, deforestation, overuse of the land, and poor agricultural practices are the major human-related causes of soil loss. The United Nations estimates that, just through erosion, the world loses between 5 and 7 million hectares (12.36–17.30 million acres) of farmland every year. This is an area equal to the size of West Virginia.

In addition to the degradation of soil that is ongoing, farmlands with excellent soils are being taken out of production by urbanization/suburbanization. Many farmlands are attractive to land developers, because they are already cleared, the soils are good for lawns and landscaping, and cities and suburbs are expanding into surrounding agricultural lands.

Our planet is losing its arable soil, while the population is growing, a trend that cannot continue forever. One estimate is that soils are being lost at a rate 17 times faster than the rate at which they form. Many government and private agencies around the world are working to educate people about the critically important problem of soil degradation, and to support solutions to minimize human impacts on soils. The problems associated with soil degradation are globally significant, but the solutions are proper conservation and management at local levels.

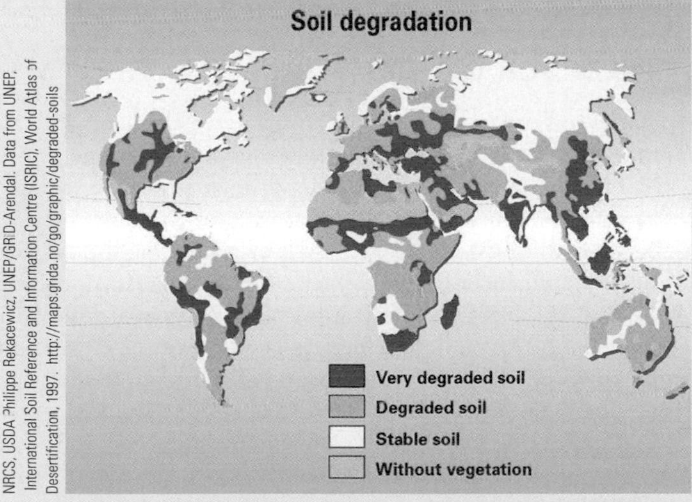

NRCS, USDA Philippe Rekacewicz, UNEP/GRID-Arendal. Data from UNEP, International Soil Reference and Information Centre (ISRIC), World Atlas of Desertification, 1997. http://maps.grida.no/go/graphic/degraded-soils

A world map done by the United Nations shows global patterns—the geography—of soil degradation by various degrees.
What major geographic factors explain the areas with the highest levels of degradation, and what explains areas with the lowest impact on soils?

USDA/NRCS/Lynn Betts

These prime farmlands in Iowa are being converted into suburbs as populations grow and towns expand.
Are there agricultural lands surrounding the place where you live that are currently being converted into housing, commercial, or industrial areas?

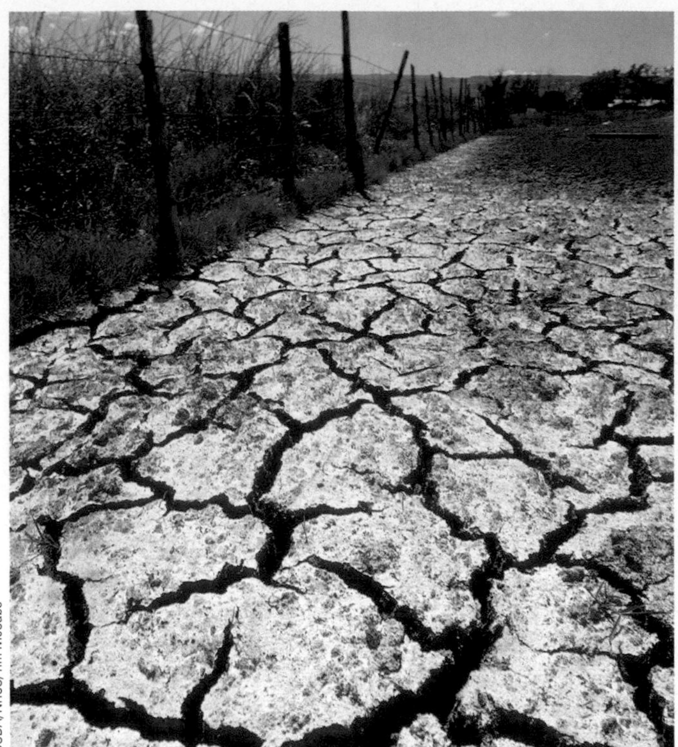

USDA/NRCS/Tim McCabe

• **FIGURE 12.24**
Salinization is indicated by these white deposits on this field in Colorado. Surface salinity has resulted from the upward capillary movement of water and evaporation at the surface causing deposits of salt. The soil cracks also indicate shrinkage caused by evaporative drying of the soil. **What negative soil effects can result when humans practice irrigated agriculture in regions that experience great evaporation rates?**

behind a high concentration of soluble salts and may destroy a soil's agricultural productivity. An extreme example of salinization exists in certain areas of the Middle East, where thousands of years of irrigated agriculture in the desert have made the soils too saline to cultivate today.

Another localized soil regime, **gleization,** occurs in poorly drained areas under cold, wet environmental conditions. Gley soils, as they are called, are typically associated with peat bogs where the soil has an accumulation of humus overlying a blue-gray layer of thick, gummy, water-saturated clay. In poorly drained regions that were formerly glaciated, such as northern Russia, Ireland, Scotland, and Scandinavia, peat has long been harvested and used as a source of energy.

Soil Classification

Soils, like climates, can be classified by their characteristics and mapped by their spatial distributions. In the United States the Soil Survey Division of the Natural Resources Conservation Service (NRCS), a branch of the Department of Agriculture, is responsible for soil classification (termed **soil taxonomy**). As with any classification system, the methods and categories are continually being updated and refined.

Soil classifications are published in **soil surveys,** books that outline and describe the kinds of soils in a region and include maps that show the distribution of soil types, usually at the county level. These documents, available for most parts of the United States, are useful references for factors such as soil fertility, irrigation, and drainage.

The NRCS Soil Classification System

The NRCS soil classification system is based on the development and composition of soil horizons. The largest division in the classification of soils is the **soil order,** of which 12 are recognized by the NRCS. To provide greater detail, soil orders can also be subdivided into suborders, great groups, subgroups, families, and series. More than 10,000 soil series have been recognized in the United States.

The NRCS system of soil types uses names derived from root words of classical languages such as Latin, Arabic, and Greek to refer to the different soil categories. The names, like the system, are precise and consistent and were chosen to describe the characteristics that distinguish one soil from another. Some soil orders reflect regional climate conditions; however, other soil orders reflect the recency or type of parent material, so the distribution of these soils does not conform to climate regions.

When examining a soil for classification under the NRCS system, particular attention is paid to characteristic horizons and textures. Some of these horizons are below the surface (**subsurface horizons**); others, called **epipedons,** are surface layers that usually exhibit a dark shading associated with organic material (humus). Examples of some of the more common horizons, illustrating how names were chosen to represent actual soil properties, are found in Table 12.1.

NRCS Soil Orders

The 12 soil orders are based on a variety of characteristics and processes that can be recognized by examining a soil and its profile. The soil descriptions that follow are based on the sequence shown in • Figure 12.25, which illustrates the links between climate and soils. • Figure 12.26 is a map showing the distribution of dominant soil orders in the United States. Global soils based on the NRCS classification of soil orders are shown in • Figure 12.27. Frequent comparison of Figure 12.27 with the map of world climates in Figure 8.6 will illustrate the relationships between the global distributions of soil and climate.

Entisols are soils that have undergone little or no soil development and lack horizons, because they have only recently begun to form (• Fig. 12.28a). They are often associated with the continuing erosion of sloping land in mountainous regions or with the frequent deposition of alluvium by flooding, or in areas of windblown sand.

Inceptisols are young soils with weak horizon development (Fig. 12.28b). The processes of *A* horizon depletion (eluviation) and *B* horizon deposition (illuviation) are just beginning, usually because of a very cold climate, repeated flood-related deposition, or a high rate of soil erosion. In the United

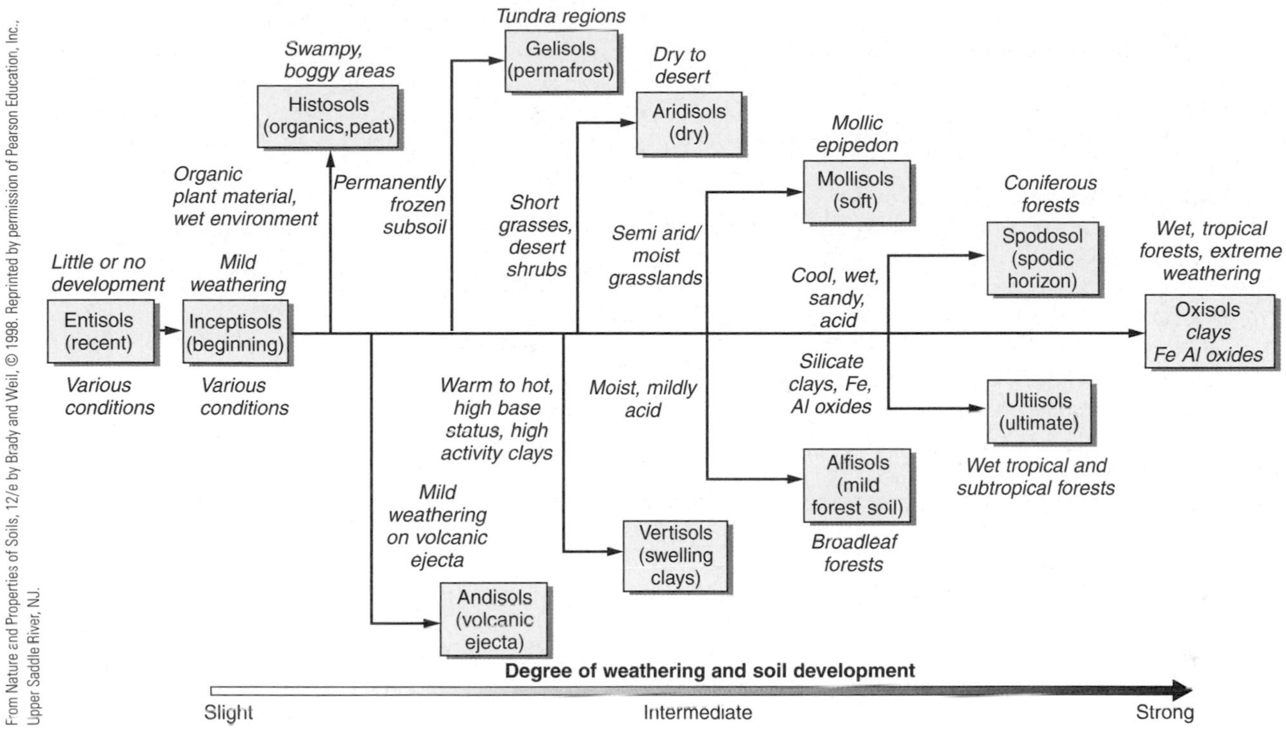

From Nature and Properties of Soils, 12/e by Brady and Weil, © 1998. Reprinted by permission of Pearson Education, Inc., Upper Saddle River, NJ.

• FIGURE 12.25

The NRCS soil orders. The soil orders of the NRCS can be linked to the parent materials, climate, and vegetation of the region in which they formed. The linkages form a treelike pattern, as seen here.

How is degree of weathering related to climatic characteristics?

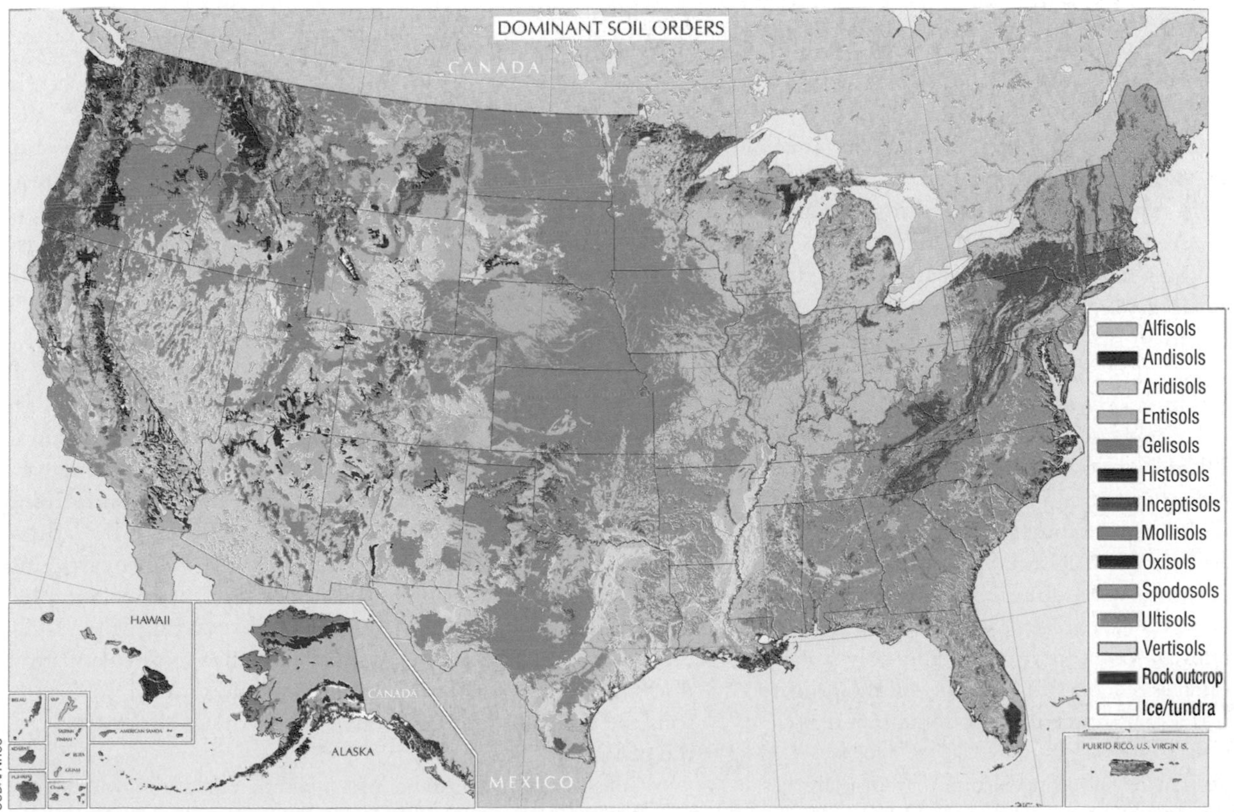

USDA/NRCS

• FIGURE 12.26

The distribution of soils in the United States according to the National Resource Conservation Service's classification.

What kind of soil dominates the place where you live, according to this map?

TABLE 12.1
Common Soil Horizons (NRCS Soil Classification System)*

Oxic horizon (from *oxygen*)
Subsurface horizon, in low-elevation tropical and subtropical climates, that contains oxides of iron and aluminum.

Argillic horizon (from Latin: *argilla*, clay)
Layer formed beneath the *A* horizon by illuviation that contains a high content of accumulated clays.

Ochric epipedon (from Greek: *ochros*, pale)
A surface horizon that is light in color and either very low in organic matter, or very thin.

Albic horizon (from Latin: *albus*, white)
An *A*2 horizon, sandy and light-colored due to the removal of clay and iron oxides, that is above a spodic horizon.

Spodic horizon (from Greek: *spodos*, wood ash)
Beneath an *A*2 horizon, this layer is dark-colored from illuviated humus, and oxides of aluminum and/or iron.

Mollic epipedon (from Latin: *mollis*, soft)
A dark-colored surface layer with a high content of basic substances (calcium, magnesium, potassium).

Calcic horizon (from *calcium*)
A subsurface horizon that is rich in accumulated calcium carbonate or magnesium carbonate.

Salic horizon (from *salt*)
A soil layer, common in desert basins, that is at least 6 inches thick and contains at least 2% salt.

Gypsic horizon (from *gypsum*)
A subsurface soil horizon that is rich in accumulated calcium sulfate (gypsum).

* This table includes only some of the more common horizons.

States, Inceptisols are most common in Alaska, the lower Mississippi River floodplain, and the western Appalachians. Globally, Inceptisols are especially important along the lower portions of the great river systems of South Asia, such as the Ganges-Brahmaputra, the Irrawaddy, and the Mekong. In these areas, sediments associated with periodic flooding constantly enrich Inceptisols, and form the basis for agriculture that supports millions of people.

Histosols develop in poorly drained areas, such as swamps, meadows, or bogs, as a product of gleization (Fig. 12.28c). They are largely composed of decomposing plant material. The waterlogged soil conditions deprive bacteria of the oxygen necessary to decompose the organic matter. Although Histosols may be found in low areas with poor drainage at all latitudes, they are most common in tundra areas or in recently glaciated, high-latitude locations such as Scandinavia, Canada, Ireland, and Scotland. Histosols in the subpolar latitudes are commonly acidic and only suitable for special bog crops such as cranberries. Histosols are the primary source of peat, which is a fuel source in some regions and also is used in landscaping.

Andisols are soils that develop on volcanic parent materials, usually *volcanic ash*, the dust-sized particles emitted by volcanoes (● Fig. 12.29a). Because many of these soils are replenished by eruptions, they are often fertile. Intensive agriculture atop Andisols supports dense populations in the Philippines, Indonesia, and the West Indies. In the United States, Andisols are most common on the slopes of and downwind from volcanoes in the Pacific Northwest and to a lesser extent in Hawaii and Alaska.

Gelisols are soils that experience frequent freezing and thawing of the ground, above *permafrost*, permanently frozen subsoil (Fig. 12.29b). When soil freezes, the ice that forms takes up 9% more space than the liquid water that it replaces. To accommodate the increased space taken up by ice, the soil and the particles in it are pushed upward and outward, away from forming ice cores. When the surface soil thaws, gravity pulls the waterlogged ground back downward. Repeated cycles of freezing and thawing mix and churn the upper soil in a process called *cryoturbation*—mixing (*turbation*) related to freezing (*cryo*). Only the upper part of the soil undergoes freeze–thaw cycles. Permafrost does not permit soil water to percolate downward, so Gelisol soils are typically water saturated when they are not frozen at the surface.

Gelisols occur in tundra and subarctic climate regions where soil development tends to be slow because chemical processes operate slowly in cold environments. This soil type is found in north and central Alaska and Canada, in Siberia, and in high-altitude tundra areas.

Aridisols are soils of desert regions that develop primarily under conditions where precipitation is much less than half of potential evaporation (Fig. 12.29c). Consequently, most Aridisols reflect the calcification process. Where groundwater tables are high, evidence of salinization may also be present.

Although Aridisols tend to have weak horizon development because of limited water movement in the soil, there is often a subsurface accumulation of calcium carbonate (calcic horizon), salt (salic horizon), or calcium sulfate (gypsic horizon). Soil humus is minimal because vegetation is sparse in deserts; therefore, Aridisols are often light in color. Aridisols are usually alkaline, but because few nutrients have been leached, they can support productive agriculture if irrigated to reduce the pH and salinity. Geographically, Aridisols are the most common soils on Earth because deserts cover such a large portion of the land surface.

Vertisols are typically found in regions of strong seasonality of precipitation such as the tropical wet and dry climates (● Fig. 12.30a). In the United States, they are most common where the parent materials produce clay-rich soils. The combination of clayey soils in a wet and dry climate leads to the drying of the soil and consequent shrinkage that forms deep cracks during the dry season, followed by expansion of the soil during the wet season. The constant shrink–swell process disrupts horizon formation to the point that soil scientists often describe Vertisols as "self-plowing" soils. Vertical soil movement may damage highways, sidewalks, foundations, and basements that are built on shrink–swell soils. Vertisols are dark-colored, are high in bases, and contain considerable organic material derived from the grasslands or savanna vegetation with which they are normally associated. Although they harden when dry and become sticky and difficult to cultivate when swollen with moisture, Vertisols can be agriculturally productive.

Mollisols are most closely associated with grassland regions and are among the best soils for sustained agriculture (Fig. 12.30b). Because they are located in semiarid climates, Mollisols are not heavily leached, and they have a generous supply of bases, especially calcium. The characteristic horizon of a Mollisol is a thick, dark-colored surface layer rich in organic matter from the decay of abundant root material. Grasslands and associated Mollisols served as the grazing lands for countless herds of antelope, bison, and horses. Before the invention of the steel plow, the thick root material of grasses made this soil nearly uncultivable in the United States and thus led to the widespread public image of the Great Plains as a "Great American Desert."

In regions of adequate precipitation, such as the tall-grass prairies of the American Midwest, the combination of soils and climate is unexcelled for agriculture. In areas of lesser precipitation, periodic drought is a constant threat, and the temptation of fertile soils was the downfall of many farmers prior to the advent of center pivot irrigation.

Alfisols occur in a wide variety of climate settings. They are characterized by a subsurface clay horizon (argillic *B* horizon), a medium to high base supply, and a light-colored ochric epipedon (Fig. 12.30c). The five suborders of Alfisols reflect climate types and exemplify the hierarchical nature of the classification system: *Aqualfs* are seasonally wet and can be found in mesothermal areas such as Louisiana, Mississippi, and Florida; *Boralfs* are found in moist, microthermal climates such as Montana, Wyoming, and Minnesota; *Udalfs* are common in both microthermal and mesothermal climates that are moist enough to support agriculture without irrigation, such as Wisconsin, Ohio, and Tennessee; *Ustalfs* are found in mesothermal climates that are intermittently dry, such as Texas and New Mexico; and *Xeralfs* are found in California's Mediterranean climate, which is characterized by wet winters and long, dry summers.

Because of their abundant bases, Alfisols can be very productive agriculturally if local deficiencies are corrected: irrigation for the dry suborders, properly drained fields for the wet suborders.

Spodosols are most closely associated with the podzolization soil-forming process. They are readily identified by their strong horizon development (● Fig. 12.31a). There is often a white or light-gray *E* horizon (albic horizon) covered with a thin, black layer of partially decomposed humus and underlain by a colorful *B* horizon enriched in relocated iron and aluminum compounds (spodic horizon).

Spodosols are generally low in bases and form in porous substrates such as glacial drift or beach sands. In New England and Michigan, Spodosols are also acidic. In these regions, as well as in similar regions in northern Russia, Scandinavia, and Poland, only a few types of agricultural plants, such as cucumbers and potatoes, can tolerate the microthermal climates and sandy, acidic soils. Consequently, the cuisine of these regions directly reflects the Spodosols that dominate the areas.

Ultisols, like Spodosols, are also low in bases because they develop in moist or wet regions. Ultisols are characterized by a subsurface clay horizon (argillic horizon) and are often yellow or red because of residual iron and aluminum oxides in the *A* horizon (Fig. 12.31b). In North America, the Ultisols are most closely associated with the southeastern United States. When first cleared of forests, these soils can be agriculturally productive for several decades. But a combination of high rainfall with the associated runoff and erosion from the fields decreases the natural fertility of the soils. Ultisols remain productive only with the continuous application of fertilizers. Today, forests cover many former cotton and tobacco fields of the southeastern United States because of a reduction in soil fertility and extensive soil erosion (see again Fig. 10.9).

Oxisols have developed over long periods of time in tropical regions with high temperatures and heavy annual rainfall. They are almost entirely leached of soluble bases and are characterized by a thick development of iron and aluminum oxides (Fig. 12.31c). The soil consists mainly of minerals that resist weathering (for example, quartz, clays, hydrated oxides). Oxisols are most closely associated with the humid tropics, but they also extend into savanna and tropical thorn forest regions. In the United States, Oxisols are present only in Hawaii. Oxisols are dominated by laterization and retain their natural fertility only as long as the soils and forest cover maintain their delicate equilibrium. The bases in the tropical

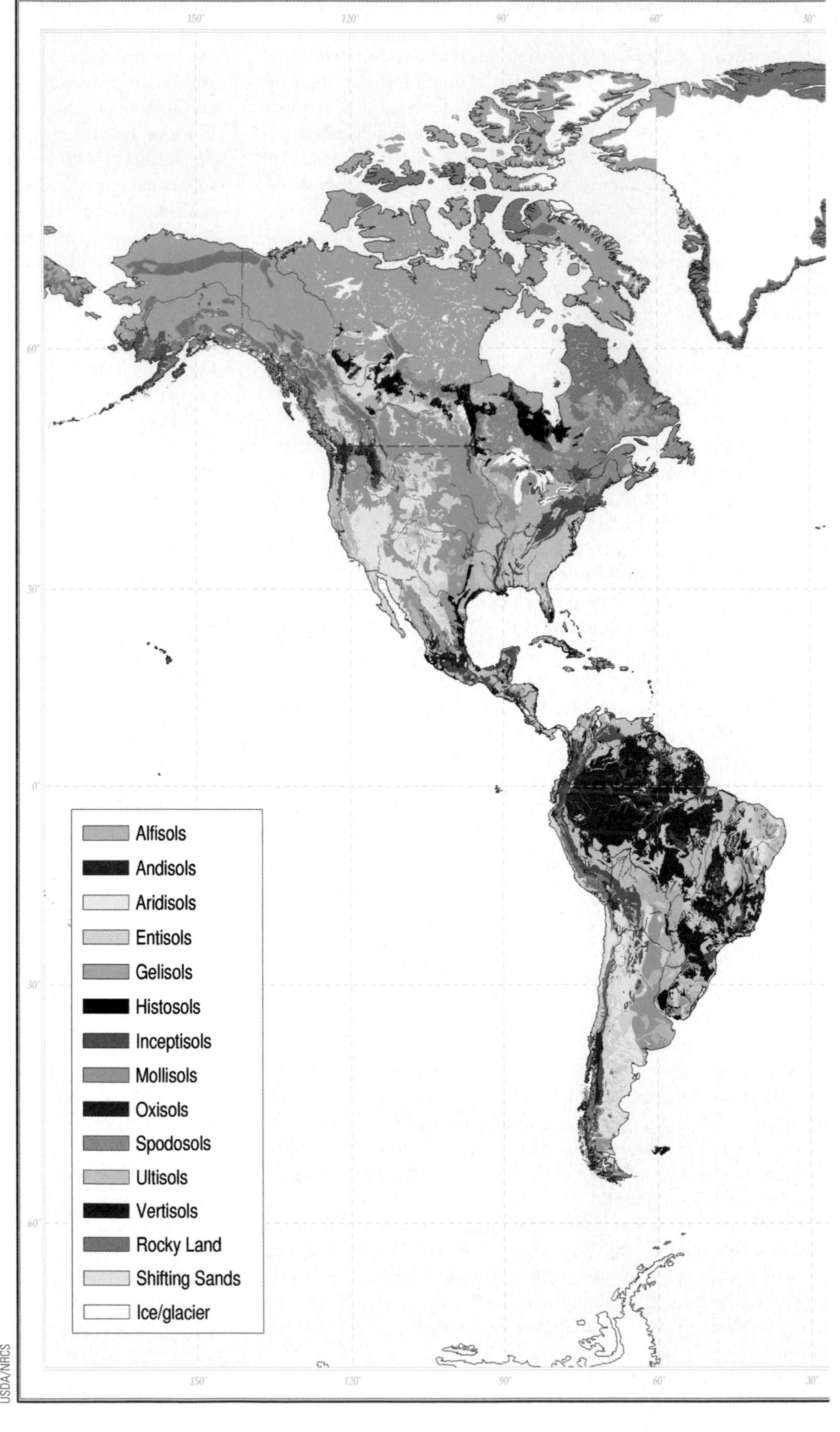

• **FIGURE 12.27**
The global distribution of soils by NRCS soil orders.
How do these patterns resemble the spatial distribution of world climates?

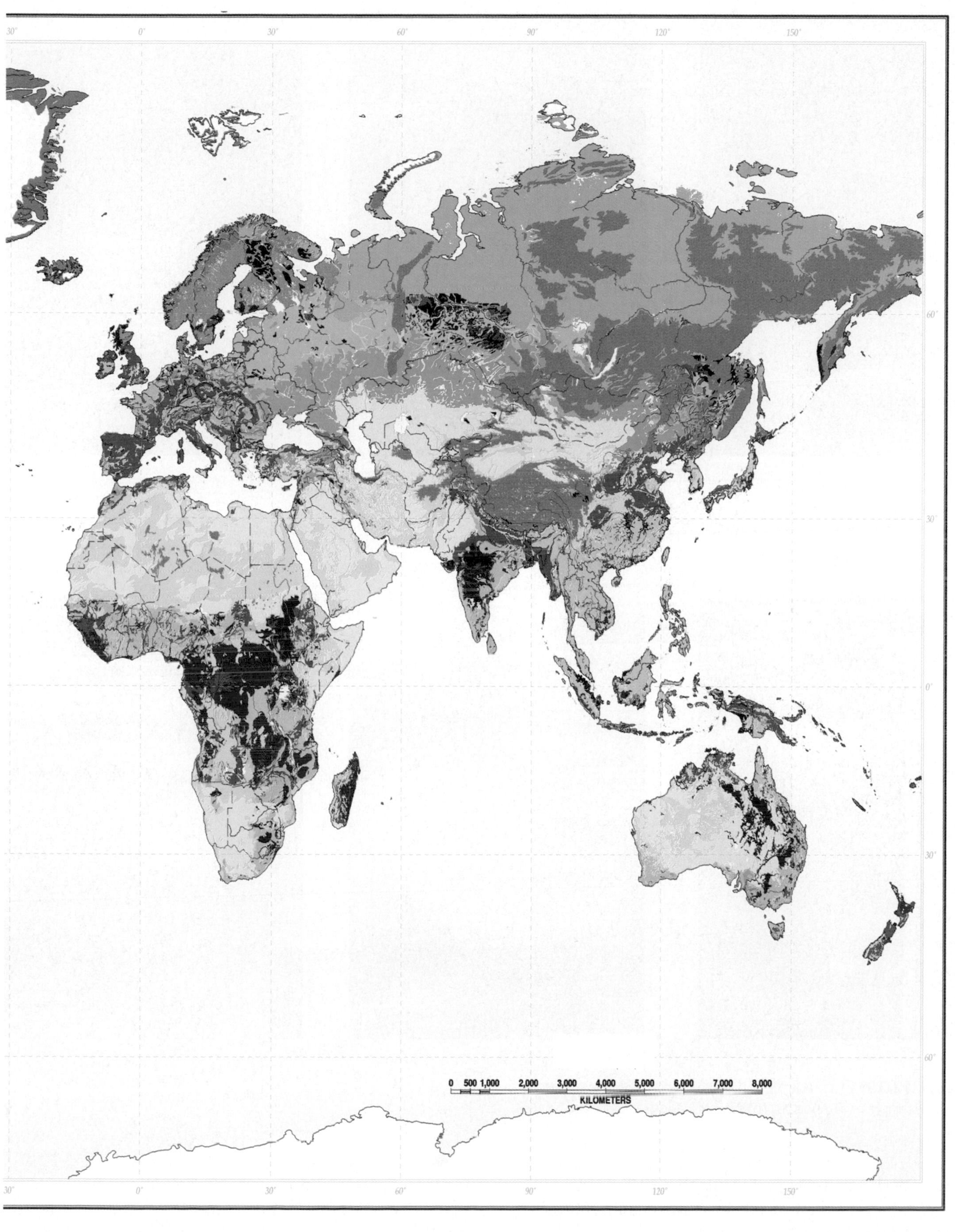

30°
0°
30°
60°
90°
120°
150°
60°
30°
0°
30°
60°
0 500 1,000 2,000 3,000 4,000 5,000 6,000 7,000 8,000
KILOMETERS

© Hari Eswaran, USDA/NRCS
(a)

© Hari Eswaran, USDA/NRCS
(b)

© Hari Eswaran, USDA/NRCS
(c)

• FIGURE 12.28
Soil profile examples: (a) Entisols, (b) Inceptisols, and (c) Histosols.

© Hari Eswaran, USDA/NRCS
(a)

© Hari Eswaran, USDA/NRCS
(b)

© Hari Eswaran, USDA/NRCS
(c)

• FIGURE 12.29
Soil profile examples: (a) Andisols, (b) Gelisols, and (c) Aridisols.

rainforests are stored mainly in the vegetation. When a tree dies, epiphytes and insects must recycle the bases rapidly before the heavy rainfall leaches them from the system.

The burning of vegetation associated with slash-and-burn agriculture in rainforests releases the nutrients necessary for crop growth but quickly results in their loss from the ecosystem. Many tropical Oxisols that once supported lush forests are now heavily dissected by erosion and only support a combination of weeds, shrubs, and grasses.

© Hari Eswaran, USDA/NRCS

(a)

© Hari Eswaran, USDA/NRCS

(b)

© Hari Eswaran, USDA/NRCS

(c)

• FIGURE 12.30
Soil profile examples: (a) Vertisols, (b) Mollisols, and (c) Alfisols.

© Hari Eswaran, USDA/NRCS

(a)

© Hari Eswaran, USDA/NRCS

(b)

© Hari Eswaran, USDA/NRCS

(c)

• FIGURE 12.31
Soil profile examples: (a) Spodosols, (b) Ultisols, and (c) Oxisols.

Soil as a Critical Natural Resource

Regardless of their composition, origin, or state of development, Earth's soils remain one of our most important and vulnerable resources. The word *fertility*, so often associated with soils, has a meaning that takes into consideration the usefulness of a soil to humans. Soils are fertile in respect to their effectiveness in producing specific vegetation types or associations. Some soils may be fertile for corn and others for potatoes. Other soils retain their fertility only as long as they remain in delicate equilibrium with their vegetative cover.

It is clearly the responsibility of all of us who enjoy the agricultural products of farms, ranches, and orchards, as well as appreciate the natural beauty of Earth's diverse biomes,

USDA/NRCS/Lynn Betts

• FIGURE 12.32
Gully erosion on farmlands is a significant problem that can often be avoided or overcome by proper agricultural practices. Gullying, if unchecked, can alter the landscape to the point that the original productivity of the land cannot be regained.
What could have been done to prevent the kind of soil loss shown in this example?

USDA/NRCS/Lynn Betts

• FIGURE 12.33
This farm's contour farming techniques, and the use of buffer zones between fields and along the water course, are excellent examples of soil conservation methods.
What other soil conservation practices are often used to preserve the soil resource?

to help protect our valuable soils. Soil erosion, soil depletion, and the mismanagement of land are problems that we should have great concern about in the world today (• Fig. 12.32). We should also be aware that these problems have reasonable solutions (• Fig. 12.33). Conserving soils and maintaining soil fertility are critical challenges that are essential to maintaining natural environments, as well as supporting life on Earth today and for the future.

Chapter 12 Activities

Define & Recall

soil
soil fertilization
capillary water
hygroscopic water
gravitational water
leaching
eluviation
illuviation
hardpan
stratification
humus
soil texture
clayey
clay
silty
silt
sandy
sand
soil grade
loam
infiltrate
soil ped
porosity
permeability
pH scale
parent material
soil profile
soil horizon
Cl, O, R, P, T
residual parent material
transported parent material
soil-forming regime
laterization
laterite
podzolization
calcification
salinization
gleization
soil taxonomy
soil survey
soil order
subsurface horizon
epipedon
Entisol
Inceptisol
Histosol
Andisol
Gelisol
Aridisol
Vertisol
Mollisol
Alfisol
Spodosol
Ultisol
Oxisol

Discuss & Review

1. Why is soil an outstanding example of the integration and interaction among Earth's subsystems?
2. Describe the different circumstances in which water is found in soil.
3. Under what conditions does leaching take place? What is the effect of leaching on the soil and, consequently, on the vegetation that it supports?
4. How can capillary water contribute to the formation of caliche? What is the effect of caliche on drainage?
5. How is humus formed? What relation does humus have to soil fertility?
6. How is texture used to classify soils? Describe the ways scientists have classified soil structure.
7. What pH range indicates soil suitable for most complex plants?
8. What are the general characteristics of each horizon in a soil profile? How are soil profiles important to scientists?
9. What factors are involved in the formation of soils? Which is most important on a global scale?
10. How does transported parent material differ from residual parent material? List those factors that help determine how much effect the parent material will have on the soil.
11. What are the most important effects of parent material on soil?
12. How does the presence of earthworms and other burrowing animals affect soil?
13. Describe the various ways in which temperature and precipitation are related to soil formation.
14. Describe the three major soil-forming regimes.

Consider & Respond

1. Refer to Figure 12.13 and associated pages in the text.
   a. What horizons make up the zone of eluviation?
   b. What are two processes that occur in the zone of eluviation?
   c. The various *B* horizons are in what zone?
   d. Weathered parent material is the major constituent of what horizon?
   e. Partly decomposed organic debris makes up which horizon?
2. Refer to Table 12.1.
   a. What materials accumulate in an argillic horizon?
   b. Which would generally be better suited for agriculture—a soil with an ochric epipedon or a mollic epipedon? Why?
   c. What name would be given to a 7-inch-thick horizon that contained at least 2% salt?
3. Where would you rank soils in terms of importance among a nation's environmental resources?
4. Give your opinion of the overall value of soils in the United States and the extent to which these soils are preserved and protected.

Apply & Learn

1. Refer to Figure 12.8. Using the texture triangle, determine the textures of the following soil samples.

	Sand	Silt	Clay
a.	35%	45%	20%
b.	75%	15%	10%
c.	10%	60%	30%
d.	5%	45%	50%

   What are the percentages of sand, silt, and clay of the following soil textures? (*Note:* Answers may vary, but they should total 100%.)
   e. Sandy clay
   f. Silty loam
2. Obtain a small sample of soil (a handful or so) and try to discover its texture by using the following method: Wet the soil a bit and work it in your hand.

I. **First, if you can form a ribbon of soil by kneading it with your fingers:**
   If you can form a ribbon that is long relatively strong and flexible: the soil is a clay.
   If you can form a ribbon that is weak and breaks easily, and the soil can be rolled into a coherent ball: the soil is a clay loam.
   If the clay loam looks powdery when dry: the soil is a silty clay loam.
   If the clay loam has a gritty feel with visible sand: it is a sandy clay loam.

II. **Second, if you cannot form a ribbon because the soil breaks up:**
   If damp soil breaks up easily, but is gritty, yet still sticky enough to make a ball: the soil is a loam.
   If it feels gritty and sand can be seen, and the ball breaks up easily in your hands: the soil is a sandy loam.

III. **Third, if the soil is very loose with visible grains of about the same size:**
   If you can see grains in the soil, and the mass in your hand breaks up easily: the soil is a sand.

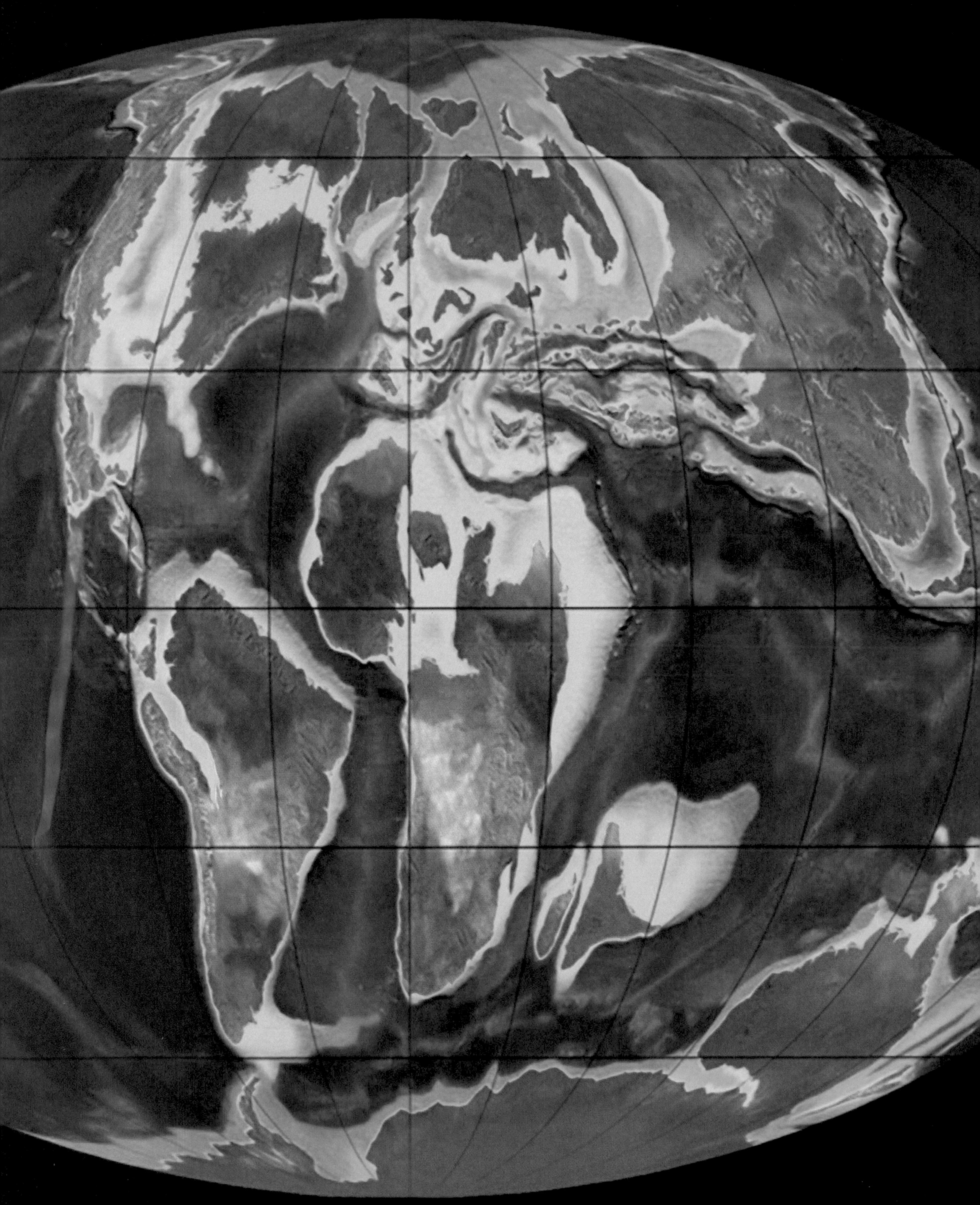

Earth Structure, Earth Materials, and Plate Tectonics 13

CHAPTER PREVIEW

The size and locations of Earth's continents, oceans, and other major landform features have changed significantly through geologic time.

- What are some observations from Earth's surface that have helped scientists explain the processes that lead to these changes?
- How might the movement of these large-scale features affect global climates and environments?

Earth's interior is not uniform, but consists of a few distinct layers.

- What are the main characteristics of each layer?
- How have scientists determined the existence, composition, and nature of these layers?

Only eight chemical elements account for almost 99% of Earth's crust by weight, and the most common minerals are combinations of these same elements.

- What does this suggest concerning the most common rocks in Earth's crust?
- What does this suggest about mineral classification?

The major types of rocks originate under very different circumstances.

- How do the processes of rock formation influence the geographical distribution of the major categories of rock?
- What other factors might influence where different rocks are found today on Earth?

The theory of plate tectonics is accepted by scientists while the older theory of continental drift was not.

- What are the differences and similarities in the two theories?
- How is the history of establishing the theory of plate tectonics illustrative of the scientific method?

◀ **Opposite: Ninety million years ago, Earth had a very different distribution of land and water than it does now. With no ice sheets at the poles, sea level was higher than it is today. In North America, much of the Great Plains as well as the Gulf of Mexico and Atlantic seaboards lay beneath shallow seas.**
R. Blakey, Northern Arizona University

If we could travel back in time to view Earth as it was 90 million years ago, in addition to seeing now-extinct life-forms, including dinosaurs, we would observe a spatial distribution of oceans and landmasses that varies considerably from that which currently exists on our planet. The shapes of the oceans and continents, their locations, and their orientations relative to the poles and equator were very different from what they are today. A vast inland sea cut across what is now the heartland of North America. Dinosaurs left their footprints in large trackways on floodplains of rivers that flowed out of the early Rocky Mountains. Forests grew above the present Arctic Circle. Grasses did not yet exist. The dramatic differences between then and now in climate, soils, and the biosphere, as well as in the shape, size, and distribution of mountain ranges and water bodies, require scientific explanation. These differences also demonstrate the interrelatedness of the four major subsystems of Earth: the atmosphere, hydrosphere, biosphere, and lithosphere.

Like the atmosphere, biosphere, and hydrosphere, that large part of the Earth system that lies beneath our feet undergoes change as a result of flows of energy and matter. Over geologic time, the processes acting within the lithosphere have had a dramatic impact on the geographical distribution of major Earth features and environments. Understanding the structure, materials, and processes of the lithosphere helps explain in scientific terms the past and present distribution of Earth's surface relief (landmasses, ocean basins, mountain ranges, and trenches). This knowledge also helps explain the variability in, and distribution of, rocks and related resources. Large-scale surface relief features and Earth materials of which they are composed provide the foundation or "backdrop" that the surface landforming processes modify. Knowledge of this rocky foundation is thus key to understanding present and future conditions, as well as potential hazards, in our physical environment.

Earth's Planetary Structure

The science of physical geography predominantly focuses on that part of the Earth system that lies at the interface of the atmosphere, hydrosphere, biosphere, and lithosphere, and these come together at and very near Earth's surface. Still, basic knowledge of our planet's internal structure, properties, and processes is needed to understand many aspects of Earth's surface characteristics and landforms.

From low-density gas molecules in the outermost layer of the atmosphere to high-density iron and nickel at the center of the planet, all of the gas, liquid, and solid matter comprising the Earth system is held in this system by gravitational attraction. Sir Isaac Newton taught us that the degree to which particles are drawn to each other by gravity depends on the mass of each particle, which is commonly expressed in units of kilograms, grams, milligrams, and so on. The gravitational force of attraction is greater for objects that have a larger mass than for those with a smaller mass. Because the size of an object influences its mass, it is convenient to use density, which is mass per unit volume (grams per cubic centimeter, for example), to compare masses of equal volumes of various types of material. Earth materials with the greatest density have the greatest gravitational force of attraction, and as a result they have tended to concentrate close together at and near the center of Earth.

The inside of Earth as a whole is primarily composed of solids, the densest of the three states of matter. Most of the liquid water, which is not as dense as solids, lies at the surface of Earth thousands of kilometers above the densest substances that are deep inside the planet. Gases, with an even lower density, have the weakest gravitational attractive force and thus are held relatively loosely around Earth as the atmospheric envelope, rather than within Earth or on its surface. Traveling outward from the center of the Earth system, there exists a density continuum (spectrum) that extends from the densest materials at the center of the planet to the least dense substances at the outer edge of the atmosphere. In previous chapters we have learned a great deal about Earth's atmosphere as well as the biosphere and hydrosphere. In this chapter we begin our study of the solid, or rock, portion of Earth, the lithosphere. As we have already found with the atmosphere and biosphere, the hydrosphere also interacts intimately with the lithosphere, particularly at and near Earth's surface.

Earth has a radius of about 6400 kilometers (4000 mi). Through direct means by mining and drilling we have been able to penetrate and examine directly only an extremely small part of that distance. The lure of gold has taken people (specifically miners) to a depth of 3.5 kilometers (2.2 mi) in South Africa, while drilling for oil and gas has taken machinery to a depth of about 12 kilometers (7.5 mi). These explorations have been helpful in providing information about the solid Earth's outermost layers, but they have just barely scratched the planet's surface. Scientists are continually working to understand the interior of Earth better. Extending scientific knowledge about the structure, composition, and processes operating within Earth helps us learn more about such lithospheric phenomena as earthquakes, volcanic eruptions, the formation of mineral deposits, and the origins of continents. It can even help us learn more about the origin of the planet itself.

Most of what we know about Earth's internal structure and composition has been deduced through indirect means by various forms of remote sensing. Thus far, the most important evidence that scientists have used to gain indirect knowledge of Earth's interior is the behavior of various shock waves, called **seismic waves,** as they travel through the planet. Scientists generate some of these shock waves artificially through controlled explosions, but they mainly use evidence derived by tracking natural earthquake waves as they travel through Earth. A sensitive instrument called a **seismograph** (• Fig. 13.1) can record seismic waves from an earthquake even when the earthquake is centered thousands of kilometers away from the seismograph's location.

Earthquakes produce two major types of seismic waves that travel at different speeds through varying types and densities of material. Of the two types, P (primary) waves travel faster and are the first to arrive at a recording seismograph. S (secondary) waves travel more slowly than P waves, thus they arrive at the seismograph later. Studies of numerous seismograph records show that P and S waves are either refracted (bent) or reflected where they pass through boundaries of major density change between different zones within Earth's interior. Both types of waves speed up in denser material and slow down when passing through material that is less dense. P waves pass through all types of material, including liquids and gases. P waves traveling through the atmosphere are responsible for the rumbling sound that we hear during an earthquake. S waves, on the other hand, can only move through solids; they do not travel through fluids (liquids or air) (• Fig. 13.2a). By considering these and other properties of P and S waves and analyzing data on worldwide patterns of travel of earthquake waves collected over decades, scientists have been able to

• FIGURE 13.1
Seismographs record earthquake waves for scientific study. This seismograph trace shows multiple earthquakes associated with volcanic eruptions.

USGS Volcano Hazards Program

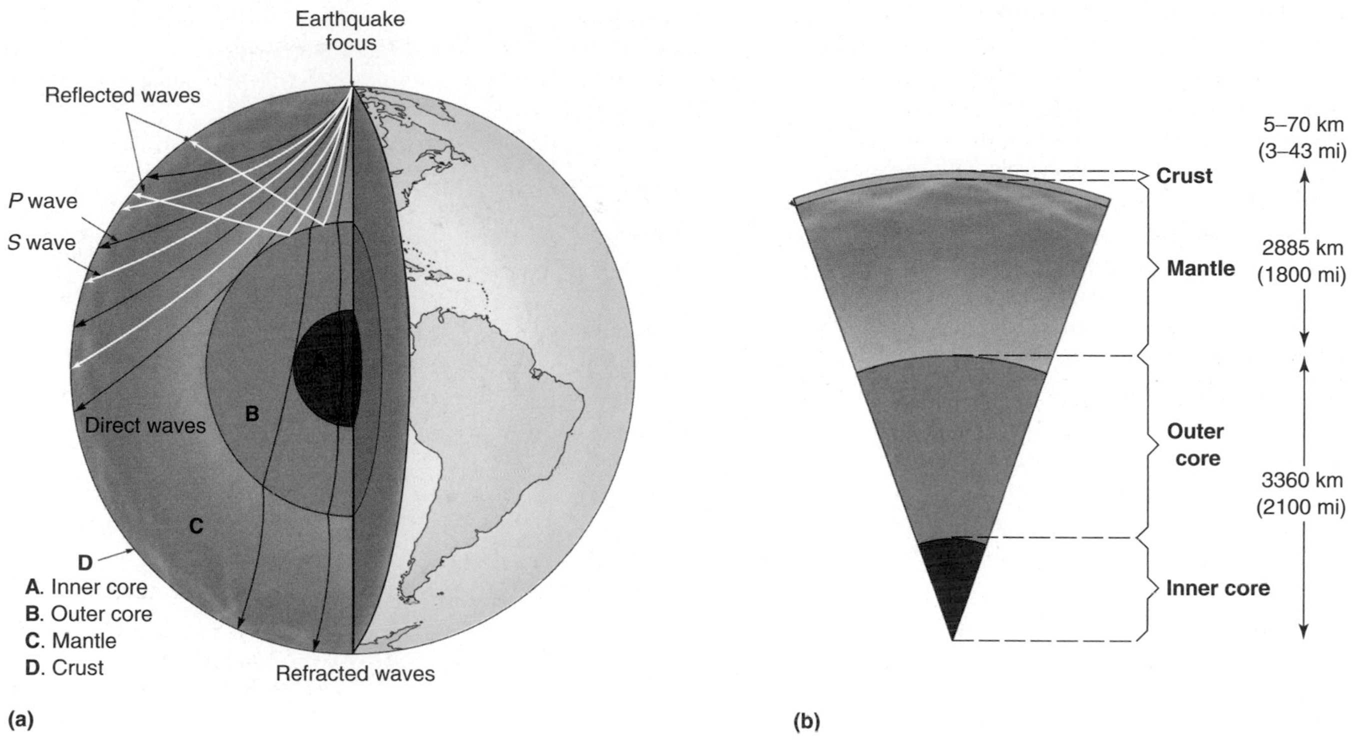

• **FIGURE 13.2**
The internal structure of Earth as revealed by seismic waves. (a) The presence of internal variations is shown by the refraction of P (primary) seismic waves and the inability of S (secondary) waves to pass through Earth's liquid outer core. (b) Cross section through Earth's internal structural zones.
How does the thickness of the crust compare with the thickness of the mantle?

develop a general model of Earth's interior. This information, supplemented by studies of Earth's magnetism and gravitational pull, reveals a series of layers, or zones, in Earth's internal structure. These principal zones, from the center of Earth to the surface, are the core, mantle, and crust (Fig. 13.2b).

Earth's Core

Earth's innermost section, the **core,** contains one third of Earth's mass and has a radius of about 3360 kilometers (2100 mi), which is larger than the planet Mars. Earth's core is under enormous pressure—several million times atmospheric pressure at sea level (more than 7 billion millibars). Scientists have deduced that the core is composed primarily of iron and nickel, and consists of two distinct sections, the inner core and the outer core.

Earth's **inner core** has a radius of about 960 kilometers (600 mi). The speed of P and S waves traveling through the inner core shows that it is a solid with a very high material density of about 13 grams per cubic centimeter (0.5 lbs/in^3). The **outer core** forms a 2400 kilometer (1500 mi) thick band around the inner core. Rock matter at the top of the outer core has a density of about 10 grams per cubic centimeter (0.4 lbs/in^3). Because the outer core blocks the passage of seismic S waves, Earth scientists know that the outer core is molten (melted/liquid rock matter). The high density of both sections of Earth's core supports the notion that they are composed of iron and nickel.

Why is Earth's outer core molten while the inner core is solid? The answer involves the fact that the melting point of mineral matter depends not only on temperature but also on pressure. When rock matter is under higher pressure it melts at a higher temperature than when it is experiencing a lower pressure. The material of the inner core is under higher pressure than the rock matter in the outer core. As a result of this great pressure, the material in the inner core remains solid despite its high temperature. Temperatures in the outer core are lower than in the inner core, but pressures are lower there as well, and this causes the outer core material to exist in the molten state. Internal temperatures are estimated to be 6900°C (12,400°F) at the very center of Earth decreasing to 4800°C (8600°F) at the top of the outer core.

Earth's Mantle

With a thickness of approximately 2885 kilometers (1800 mi) and representing nearly two thirds of Earth's mass, the **mantle** is the largest of Earth's interior zones. Earthquake waves that pass through the mantle indicate that it is composed of solid rocky material, in contrast to the molten outer core that lies beneath it. It is also less dense than the core, with values ranging from 3.3 to 5.5 grams per cubic centimeter (0.1–0.2 lbs/in^3). Although most of the mantle is solid, material near the top of the mantle especially displays characteristics of a **plastic solid,** meaning that the solid rock material can deform and flow very slowly, in this case at rates of a few centimeters per year (an inch or two per year). Scientists agree that the mantle consists of silicate rocks (high in silicon and oxygen) that also contain significant amounts of iron and magnesium.

The mantle is composed of various layers distinguished by different characteristics of strength and rigidity. Of special interest to us are the two uppermost layers. The outermost layer of the mantle, with an average thickness of about 100 kilometers (60 mi), behaves like an **elastic solid.** This contrasts sharply with the plastic solid behavior of the material in the next lower mantle layer. Elastic solids are rigid and brittle. They do not flow, but instead withstand a certain amount of applied stress (force per unit area) with little deformation until a threshold limit of stress is reached. At the threshold value, elastic solids fail by fracturing. The outermost layer of the mantle has a chemical composition like the rest of the mantle, but it responds to applied stress like the overlying Earth layer, the crust. Together, the uppermost mantle layer and the crust form a unit called the **lithosphere.** The term *lithosphere* has traditionally been used to describe the entire solid Earth (as discussed in Chapter 1 and earlier in this chapter). In recent decades, however, the term *lithosphere* has also been used in a separate sense to refer to the brittle outer shell of Earth, including the crust and this rigid, uppermost mantle layer (• Fig. 13.3).

Extending down from the base of the lithosphere about 600 kilometers (375 mi) farther into the mantle is the **asthenosphere** (from Greek: *asthenias,* without strength), a thick layer of plastic mantle material. The material in the asthenosphere can flow both vertically and horizontally, dragging segments of the overlying, rigid lithosphere along with it. Many Earth scientists now believe that the energy for **tectonic forces,** large-scale forces that break and deform Earth's crust, sometimes resulting in earthquakes and often responsible for mountain building, comes from movement within the plastic asthenosphere. Movement in the asthenosphere, in turn, is produced by thermal convection currents that occur in the rest of the mantle below the asthenosphere, and which are driven by heat from decaying radioactive materials in the planet's interior.

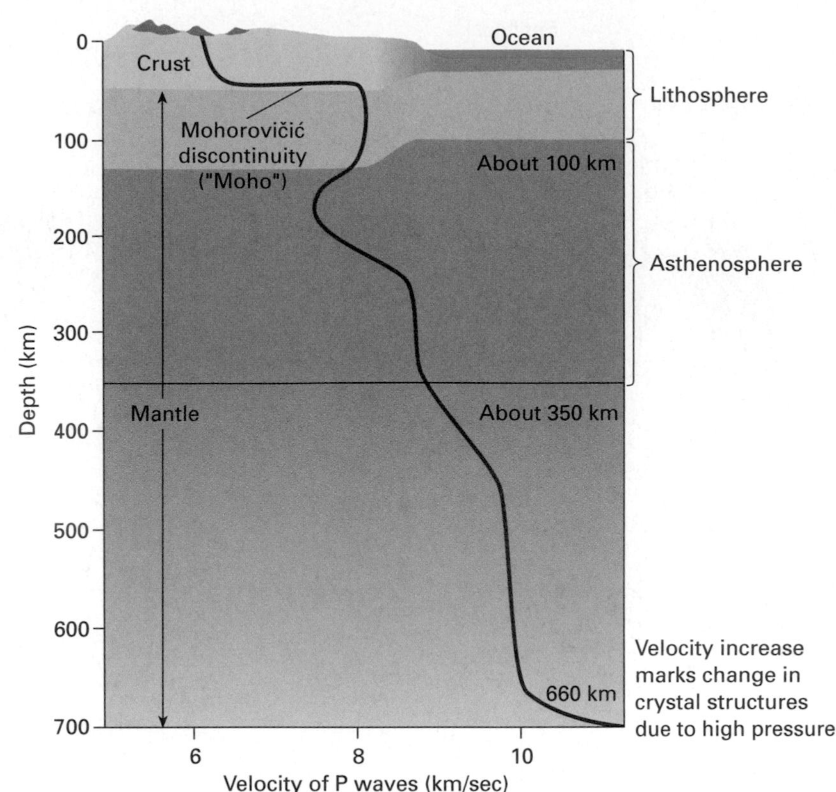

• FIGURE 13.4
Velocity of seismic P (primary) waves in the crust and upper mantle. Velocity increases with depth as material density increases. The Mohorovičić discontinuity is marked by an abrupt increase in seismic wave velocity. (From Thompson & Turk, *Earth Science and the Environment,* 2007)

• FIGURE 13.3
A cross section of the lithosphere, asthenosphere, and lower mantle. The lithosphere is the solid outer part of Earth and includes the crust and solid uppermost mantle, down to the asthenosphere.

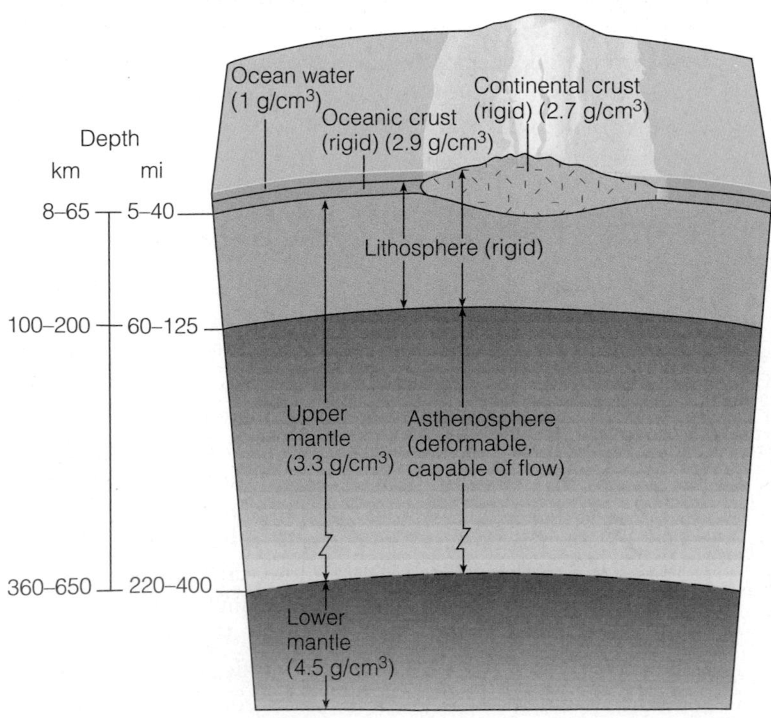

The interface between the mantle and the overlying crust is marked by a significant change of density, called a discontinuity, which is indicated by an abrupt increase in the velocity of seismic waves as they travel down through this internal boundary (• Fig. 13.4). Scientists call this zone the **Mohorovičić discontinuity,** or **Moho** for short, after the Croatian geophysicist who first detected it in 1909. The Moho does not lie at a constant depth but generally mirrors the surface topography, being deepest under mountain ranges where the crust is thick and rising to within 8 kilometers (5 mi) of the ocean floor (see again Fig. 13.3). No geologic drilling has yet penetrated through the Moho into the mantle, but an international scientific partnership, called the Integrated Ocean Drilling Program, is working on such a project. Rock samples eventually retrieved from cores drilled through the Moho will add greatly to our understanding of the composition and structure of Earth's lithosphere.

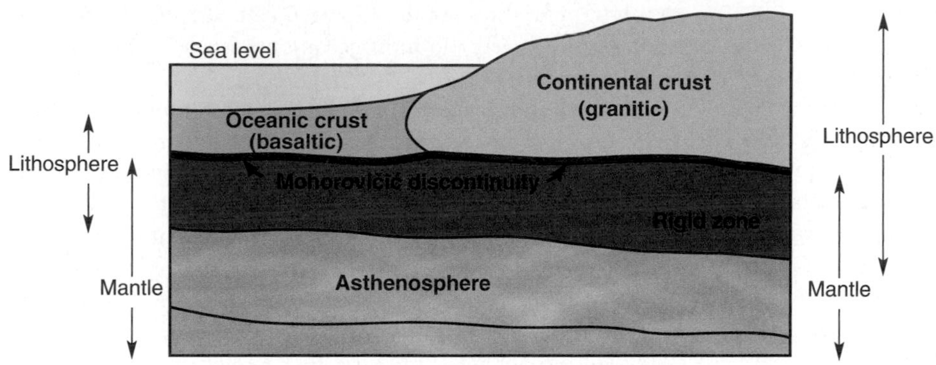

• FIGURE 13.5
Earth has two distinct types of crust, oceanic and continental. The crust and the uppermost part of the mantle, which is rigid like the crust, make up the lithosphere. Below the lithosphere, but still in the upper part of the mantle, is the plastic asthenosphere.

Earth's Crust

Earth's solid and rocky exterior is the **crust,** which is composed of a great variety of rocks that respond in diverse ways and at varying rates to Earth-shaping processes. The crust is the only portion of the lithosphere of which Earth scientists have direct knowledge, yet its related surface materials form only about 1% of Earth's planetary mass. Earth's crust forms the exterior of the lithosphere and is of primary importance in understanding surface processes and landforms. Earth's deep interior components, the core and mantle, are of concern to physical geographers primarily because they are responsible for and can help explain changes in the lithosphere, particularly the crust, which forms the ocean floors and continents.

The density of Earth's crust is significantly lower than that of the core and mantle, and ranges from 2.7 to 3.0 grams per cubic centimeter. The crust is also extremely thin in comparison to the size of the planet. The two kinds of Earth crust, oceanic and continental, are distinguished by their location, composition, and thickness (• Fig. 13.5). Crustal thickness varies from 3 to 5 kilometers (1.9–3 mi) in the ocean basins to as much as 70 kilometers (43 mi) under some continental mountain systems. The average thickness of continental crust is about 32 to 40 kilometers (20–25 mi). The crust is relatively cold, rigid, and brittle compared to the mantle. It responds to stress by fracturing, wrinkling, and raising or lowering rocks into upwarps and downwarps.

Oceanic crust is composed of heavy, dark-colored, iron-rich rocks that are also high in silicon (Si) and magnesium (Mg). Its *basaltic* composition is described more fully in the next section. Compared to continental crust, oceanic crust is quite thin because its density (3.0 g/cm^3) is greater than that of continental crust (2.7 g/cm^3). Forming the vast, deep ocean floors as well as lava flows on all of the continents, basaltic rocks are the most common rocks on Earth.

Continental crust comprises the major landmasses on Earth that are exposed to the atmosphere. It is less dense (2.7 g/cm^3) and much thicker than oceanic crust. Where continental crust extends to very high elevations, such as in mountain ranges, it also descends to great depths below the surface. Continental crust contains more light-colored rocks than oceanic crust does, and can be regarded as *granitic* in composition. The nature of granite and other common rocks are discussed next.

Minerals and Rocks

Minerals are the building blocks of rocks. A mineral is an inorganic, naturally occurring substance represented by a distinct chemical formula and a specific crystalline form. A **rock,** in contrast, is an aggregate (collection) of various types of minerals or an aggregate of multiple individual pieces (grains) of the same kind of mineral. In other words, a rock is not one single, uniform crystal. The most common elements found in Earth's crust, and therefore in the minerals and rocks that make up the crust, are oxygen and silicon, followed by aluminum and iron, and the bases: calcium, sodium, potassium, and magnesium. As you can see in Table 13.1, these eight most common chemical elements, out of the more than 100 known, account for almost 99% of Earth's crust by weight. The most common minerals are combinations of these eight elements.

TABLE 13.1
Most Common Elements in Earth's Crust

Element	Percentage of Earth's Crust by Weight
Oxygen (O)	46.60
Silicon (Si)	27.72
Aluminum (Al)	8.13
Iron (Fe)	5.00
Calcium (Ca)	3.63
Sodium (Na)	2.83
Potassium (K)	2.70
Magnesium (Mg)	2.09
Total	98.70

Source: J. Green, "Geotechnical Table of the Elements for 1953," *Bulletin of the Geological Society of America 64* (1953).

Minerals

Every mineral has distinctive and recognizable physical characteristics that aid in its identification. Some of these characteristics include hardness, luster, cleavage, tendency to fracture, and specific gravity (weight per unit volume). Luster describes the shininess of the mineral. Cleavage signifies how the mineral tends to break along uniform planes, while fracture specifies the form of irregular breaks in the mineral. Minerals are crystalline in nature, although some crystals may only be evident when viewed through a microscope. Mineral crystals display consistent geometric shapes that express their molecular structure (• Fig. 13.6). Halite, for example, which is used as table salt, is a soft mineral that has the specific chemical formula NaCl and a cubic crystalline shape. Quartz, calcite, fluorite, talc, topaz, and diamond are just a few examples of other minerals.

Chemical bonds hold together the atoms and molecules that compose a mineral. The strength and nature of these chemical bonds affect the resistance and hardness of minerals and of the rocks that they form. Minerals with weak internal bonds undergo chemical alteration most easily. Charged particles, that is, *ions,* that form part of a molecule in a mineral may leave or be traded for other substances, generally weakening the mineral structure and forming the chemical basis of *rock weathering.*

Minerals can be categorized into groups based on their chemical composition. Certain elements, particularly silicon, oxygen, and carbon, combine readily with many other elements. As a result, the most common mineral groups are silicates, oxides, and carbonates. Calcite ($CaCO_3$), for example, is a relatively soft but widespread mineral that consists of one atom of calcium (Ca) linked together with a carbonate molecule (CO_3), which consists of one atom of carbon (C) plus three atoms of oxygen (O). The silicates, however, are by far the largest and most common mineral group, comprising 92% of Earth's crust.

• FIGURE 13.6
Crystals of the mineral halite (common salt) formed in rocks from an ancient evaporating sea. The geometrical arrangement of atoms determines the crystal form of a mineral.

J. Petersen

Oxygen and silicon (Si) are the two most common elements in Earth's crust and frequently combine together to form SiO_2, which is called *silica.* **Silicate** minerals are compounds of oxygen and silicon that also include one or more metals and/or bases. They are generally created when molten rock matter containing these elements cools and solidifies, causing the crystallization of different minerals at successively lower temperatures and pressures. Dark, heavy, iron-rich silicate minerals crystallize first (at high temperatures), and light-colored, lower density, iron-poor minerals crystallize later at cooler temperatures. The order of mineral crystallization parallels their relative chemical stability in a rock, with silicate minerals that crystallize later tending to be more stable and more resistant to breakdown. In rocks composed of a variety of silicate minerals, the dark, heavy minerals are the first to decompose, while the iron-poor minerals decompose later. Silica in its crystalline form is the mineral quartz, which has a distinctive prismatic crystalline shape. Because quartz is one of the last silicate minerals to form from solidifying molten rock matter, it is a relatively hard and resistant mineral.

Rocks

Although a few rock types are composed of many particles of a single mineral, most rocks consist of several minerals (• Fig. 13.7). Each constituent mineral in a rock remains separate and retains its own distinctive characteristics. The properties of the rock as a whole are a composite of those of its various mineral constituents. The number of rock-forming minerals that are common is limited, but they combine through a multitude of processes to produce an enormous variety of rock types (refer to Appendix C for information and pictures of common rocks mentioned in the

• FIGURE 13.7
Most rocks consist of several minerals. This piece of cut and polished granite displays intergrown crystals of differing composition, color, and size that give the rock a distinctive appearance.
How does a rock differ from a mineral?

J. Petersen

text). Rocks are the fundamental building materials of the lithosphere. They are lifted, pushed down, and deformed by large-scale tectonic forces originating in the lower mantle and asthenosphere. At the surface, rocks are weathered and eroded, to be deposited as sediment elsewhere.

A mass of solid rock that has not been weathered is called **bedrock.** Bedrock may be exposed at the surface of Earth or it may be overlain by a cover of broken and decomposed rock fragments, called **regolith.** Soil may or may not have formed on the regolith (see again Fig. 12.13). On steep slopes, regolith may be absent and bedrock exposed if running water, gravity, or some other surface process removed the weathered rock fragments. A mass of exposed bedrock is often referred to as an **outcrop** (• Fig. 13.8).

Geologists distinguish three major categories of rocks based on mode of formation. These rock types are igneous, sedimentary, and metamorphic.

Igneous Rocks When molten rock material cools and solidifies it becomes an **igneous rock.** Molten rock matter below Earth's surface is called **magma,** whereas molten rock material at the surface is known specifically as **lava** (• Fig. 13.9). Lava, therefore, is the only form of molten rock matter that we can see. Lava erupts from volcanoes or fissures in the crust at temperatures as high as 1090°C (2000°F). There are two major categories of igneous rocks: extrusive and intrusive.

Molten material that solidifies at Earth's surface creates **extrusive igneous rock,** also called volcanic rock. Extrusive igneous rock, therefore, is created from lava. Very explosive eruptions of molten rock material can cause the accumulation of fragments of volcanic rock, dust sized or larger, that settle out of the air to form **pyroclastics** (fire fragments), as a special category of extrusive rock (• Fig. 13.10a). When molten rock beneath Earth's surface, that is,

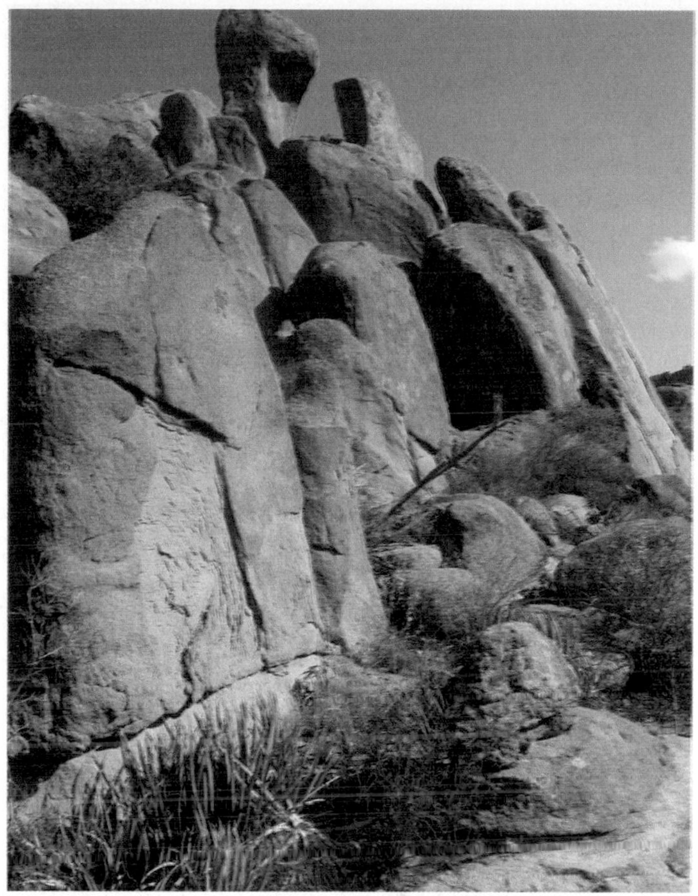
J. Petersen

• FIGURE 13.8
Exposures of solid rock that are exposed (crop out) at the surface are often referred to as outcrops.
What physical characteristics of this rock outcrop caused it to protrude above the general land surface?

• FIGURE 13.9
Basalt, a fine-grained extrusive igneous rock, forms these fresh lava flows on the island of Hawaii.
What visible evidence in this photograph indicates that the lava flow is very recent?

USGS

Copyright and photograph by Dr. Parvinder S. Sethi

(a)

Copyright and photograph by Dr. Parvinder S. Sethi

(b)

• FIGURE 13.10
(a) Pyroclastic rocks are made of fragments ejected during a volcanic eruption. (b) Obsidian is volcanic glass. The molten lava cooled too quickly for crystals to form.

magma, changes to a solid (freezes), it forms **intrusive igneous rock,** also referred to as **plutonic rock** (after Pluto, Roman god of the underworld). Igneous rocks vary in chemical composition, texture, crystalline structure, tendency to fracture, and presence or absence of layering. They are grouped or classified in terms of their crystal size, or texture, as well as their mineral composition.

Rocks composed of small-sized individual minerals not visible to the unaided eye are described as having a fine-grained texture, while those with large minerals that are visible without magnification are referred to as coarse-grained. Extrusive rocks cool very quickly—at Earth surface temperatures—and are fine-grained as a result of the little time available for crystal growth prior to solidification. An extreme example is the extrusive rock *obsidian,* which has cooled so rapidly that it is essentially a glass (Fig. 13.10b). Large masses of intrusive rocks that solidify deep inside Earth cool very slowly because surrounding rock slows the loss of heat from molten magma. Slow cooling allows more time for crystal formation prior to solidification. Thin stringers of intrusive rocks and those that solidified close to the surface may cool rapidly and be fine-grained as a result.

The chemical composition of igneous rocks varies from *felsic,* which is rich in light-colored, lighter weight minerals, especially silicon and aluminum, to *mafic,* which is lower in silica and rich in heavy minerals, such as compounds of magnesium and iron (*ma* = magnesium and *f* = iron). Granite, a felsic, coarse-grained, intrusive rock, has the same chemical and mineral composition as rhyolite, a fine-grained, extrusive rock. Likewise, basalt is the dark-colored, mafic, fine-grained extrusive chemical and mineral equivalent to gabbro, a coarse-grained intrusive rock that cools at depth (• Fig. 13.11).

Igneous rocks also form with an intermediate composition, a rough balance between felsic and mafic minerals. The intrusive rock *diorite* and the extrusive rock *andesite* (named after the Andes where many volcanoes erupt lava of this composition) represent this intermediate composition (see again Fig. 13.11).

Many igneous rocks are broken along fractures that may be spaced or arranged in regular geometric patterns. In the Earth sciences, simple fractures or cracks in bedrock are called **joints.** Although joints are common in all rock types, one way they develop in igneous rocks is by a molten mass shrinking in volume and fracturing as it cools and solidifies. Solidified lava flows in particular tend to have many fractures. Basalt is famous for its distinctive jointing, which commonly forms hexagonal columns of solidified lava. Devil's Postpile in California and Devil's Tower in Wyoming are well-known landforms that consist of hexagonal columns of lava formed by **columnar joints** (• Fig. 13.12). Jointing in rocks of all types is also caused by regional stresses in the crust and can be a major influence in the development of landforms.

Sedimentary Rocks

As their name implies, **sedimentary rocks** are derived from accumulated sediment, that is, unconsolidated mineral materials that have been eroded, transported, and deposited. After the materials have accumulated, often in horizontal layers, pressure from the material above compacts the sediment, expelling water and reducing pore space. Cementation occurs when silica, calcium carbonate, or iron oxide precipitates between particles of sediment. The processes of compaction and cementation transform (lithify) sediments into solid, coherent layers of rock. There are three major categories of sedimentary rocks: clastic, organic, and chemical precipitates.

Broken fragments of solids are called **clasts** (from Latin: *clastus,* broken). In order of increasing size, clasts may range from clay, silt, and sand (see again Fig. 12.7) to gravel, which is a general category for any fragment larger than sand (larger than 2.0 mm) and includes granules, pebbles, cobbles, and boulders. Most sediments consist of fragments of previously existing rocks, shell, or bone that were deposited on a river bed, beach, sand dune, lake bottom, the ocean floor, and other environments where clasts accumulate. Sedimentary rocks that form from fragments of preexisting rocks are called **clastic sedimentary rocks.**

Examples of clastic sedimentary rocks include *conglomerate, sandstone, siltstone,* and *shale* (• Fig. 13.13). Conglomerate is a solid mass of cemented, roughly rounded pebbles, cobbles, and boulders

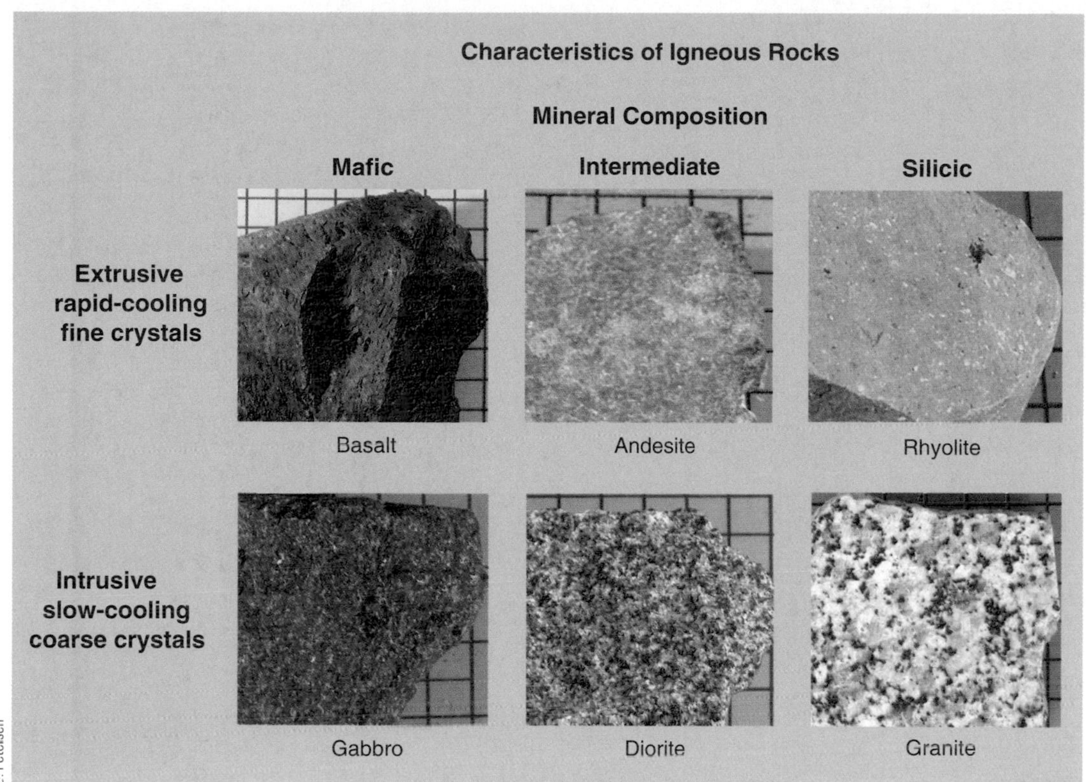

• FIGURE 13.11

Igneous rocks are distinguished on the basis of texture (mineral grain size) and mineral composition. Igneous rocks that cooled rapidly, such as those that cooled at and near the surface, have fine (small) crystals. Rocks that cooled slowly, typically in large masses deep beneath the surface, have a coarse crystalline texture (large mineral grains). Mineral composition can be mafic, silicic, or intermediate.

What is the difference between granite and basalt?

(a)

(b)

• FIGURE 13.12

Large outcrops of fine-grained igneous rock can display hexagonal columnar jointing due to shrinkage as the molten rock matter cooled and solidified. (a) Devil's Postpile National Monument, California. (b) Devil's Tower National Monument, Wyoming.

Why are the cliffs shown in these photographs so steep?

and may have clay, silt, or sand filling in spaces between the larger particles. A somewhat similar rock that has cemented fragments that are angular rather than rounded is called *breccia*. Sandstone is formed of cemented sand-sized particles, most commonly grains of quartz. Sandstone is usually granular (with visible grains), porous, and resistant to weathering, but the cementing material influences its strength and hardness. If cemented by substances other than silica (such as calcium carbonate or iron oxide), sandstone tends to be more easily weathered. Individual grains are not visible in siltstone, which is composed of silt-sized particles. Shale is produced from

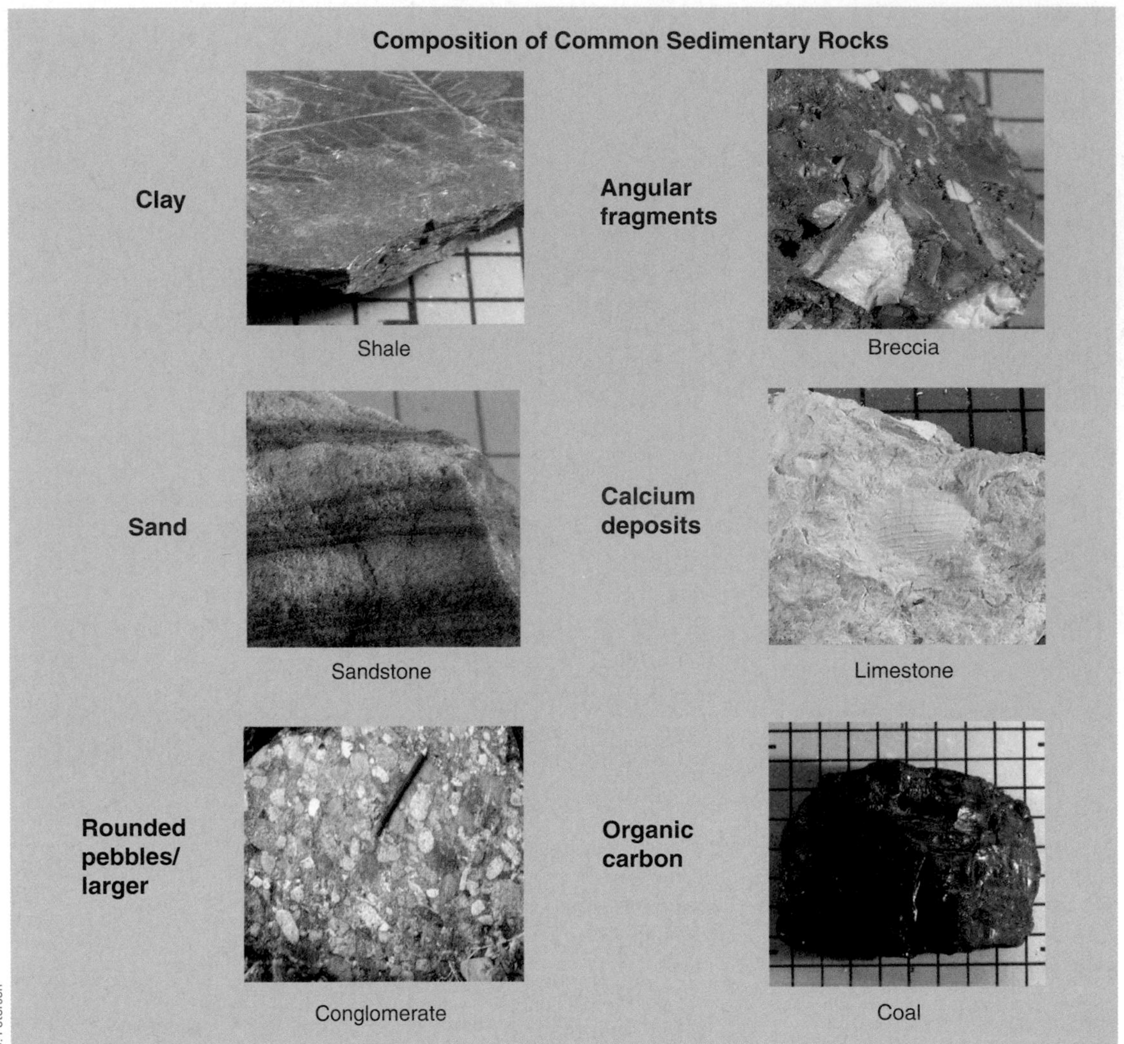

• **FIGURE 13.13**
Clastic sedimentary rocks are classified by the size and/or shape of the sediment particles they contain.
Why do the shapes and sizes of the sediments in sedimentary rocks differ?

the compaction of very fine-grained sediments, especially clays. Shale is finely bedded, smooth-textured, and has low permeability. It is also brittle and easily cracked, broken, or flaked apart.

Sedimentary rocks may be further classified by their origin as either marine or terrestrial (continental). Marine sandstones typically formed in nearshore coastal zones; terrestrial sandstones generally originated in desert or floodplain environments on land. The nature and arrangement of sediments in a sedimentary rock provide a great deal of evidence for the kind of environment in which they were deposited and the processes of deposition, whether on a stream bed, a beach, or the deep-ocean floor.

Organic sedimentary rocks lithified from the remains of organisms, both plants and animals. *Coal,* for example, was created by the accumulation and compaction of partially decayed vegetation in acidic, swampy environments where water-saturated ground prevented oxidation and complete decay of the organic matter. The initial transformation of such organic material produces peat, which, when subjected to deeper burial and further compaction, is lithified to produce coal. Most of the world's greatest coal deposits originated between 300 and 354 million years ago during an interval of geologic time known as the Mississippian and Pennsylvanian Periods.

Other organic sedimentary rocks developed from the remains of organisms in lakes and seas. The remains of shellfish, corals, and microscopic drifting organisms called plankton sank to the bottom of such water bodies where they were cemented and compacted together to form a type of *limestone,* which typically contains fossils such as shells and coral fragments (• Fig. 13.14).

At times in Earth history, dissolved minerals accumulated in ocean and lake water until saturation when it began to precipitate and build up as a deposit on the sea or lake bottom. Many fine-grained limestones formed as **chemical sedimentary rocks** in this manner from precipitates of calcium carbonate ($CaCO_3$). Limestone therefore may vary from a jagged and cemented complex of visible shells or fossil skeletal material to a smooth-textured rock. Where magnesium is a major constituent along with calcium carbonate, the rock is called *dolomite.* Because the calcium carbonate in limestone can slowly dissolve in water, limestone in arid or semiarid climates tends to be resistant, but in humid environments it tends to be weak.

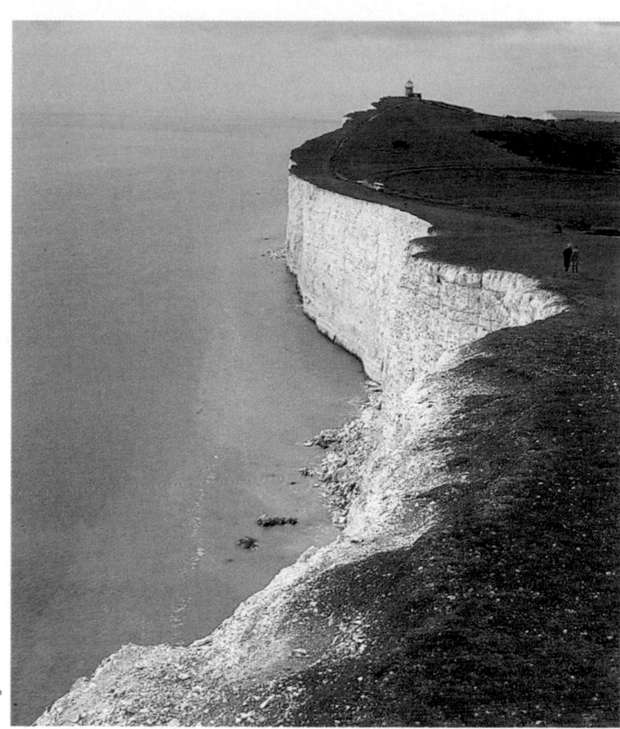

R. Sager

● **FIGURE 13.14**
The White Cliffs of Dover, England. These striking, steep white cliffs along the English Channel are made of chalky limestone from the skeletal remains of microscopic marine organisms.

Mineral salts that have reached saturation in evaporating seas or lakes will precipitate to form a variety of sedimentary deposits that are useful to humans. These include *gypsum* (used in wallboard), *halite* (common salt), and *borates,* which are important in hundreds of products such as fertilizer, fiberglass, detergents, and pharmaceuticals.

Most sedimentary rocks display distinctive layering referred to as **stratification.** The many types of sedimentary deposits produce distinctive **strata** (layers or beds) within the rocks. The **bedding planes,** or boundaries between sedimentary layers, indicate changes in energy in the depositional environment but no real break in the sequence of deposition (● Fig. 13.15). Where a marked mismatch and an irregular, eroded surface occur between beds, the contact between the rocks is called an **unconformity.** This indicates a gap in the section caused by erosion, rather than deposition, of sediment. Within some sedimentary rocks, especially sandstones, thin "microbedding" may occur. A type of microbedding, called **cross bedding,** is characterized by a pattern of thin layers that accumulated at an angle to the main strata, often reflecting shifts of direction by waves along a coast, currents in streams, or winds over sand dunes (● Fig. 13.16). All types of stratification provide evidence about the environment within which the sediments were deposited, and changes from one layer to the next reflect elements of the local geologic history. For example, a layer of sandstone representing an ancient beach may lie directly beneath shale layers that represent an offshore environment, suggesting that first this was a beach that the sea later covered.

Sedimentary rocks become jointed, or fractured, when they are subjected to crustal stresses after they lithify. The impressive "fins" of rock at Arches National Park, Utah, owe their vertical, tabular shape to joints in great beds of sandstone (● Fig. 13.17).

● **FIGURE 13.15**
Bedding planes are the boundaries between differing layers (strata) of sediment that mark some change in the nature of the deposited sediment. Numerous bedding planes are visible in this scene from Grand Canyon National Park, Arizona, many represented by color changes. **Where would the youngest strata in this photo be located?**

National Park Service

● **FIGURE 13.16**
Cross bedding in sandstone at Zion National Park, Utah.
Under what circumstances might sediment be deposited at a substantial angle like this, rather than in a more horizontal layer?

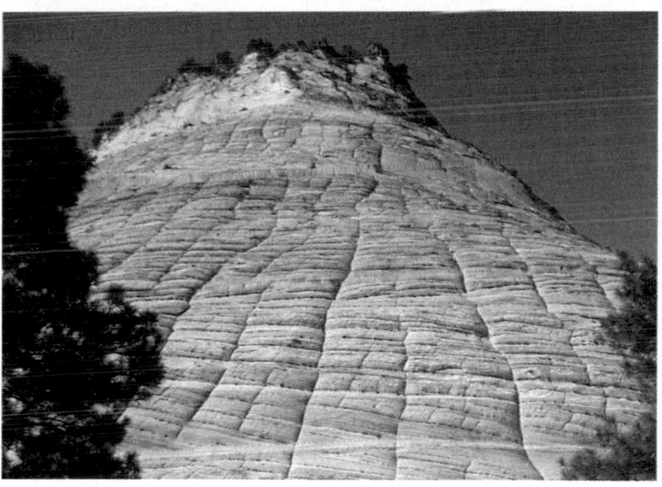

J. Petersen

● **FIGURE 13.17**
Vertical jointing of sandstone in Arches National Park, Utah, is responsible for creation of these vertical rock walls, called fins. Rock has been preferentially eroded along the joints. Only rock that was far from the locations of the joints remains standing.

Copyright and photograph by Dr. Parvinder S. Sethi

Structures such as bedding planes, cross bedding, and joints are important in the development of physical landscapes because these structures are weak points in the rock that weathering and erosion can attack with relative ease. Joints allow water to penetrate deeply into some rock masses, causing them to be removed at a faster rate than the surrounding rock.

Metamorphic Rocks *Metamorphic* means "changed form." Enormous heat and pressure deep in Earth's crust can alter (metamorphose) an existing rock into a new rock type that is completely different from the original by recrystallizing the minerals without creating molten rock matter. Compared to the original rocks, the resulting **metamorphic rocks** are typically harder and more compact, have a reoriented crystalline structure, and are more resistant to weathering. There are two major types of metamorphic rocks, based on the presence (foliated) or absence (nonfoliated) of platy surfaces or wavy alignments of light and dark minerals that formed during metamorphism.

Metamorphism occurs most commonly where crustal rocks are subjected to great pressures by tectonic processes or deep burial, or where rising magma generates heat that modifies the nearby rock. Metamorphism causes minerals to recrystallize and, with enough heat and pressure, to reprecipitate perpendicular to the applied stress, forming platy surfaces (cleavage) or wavy bands known as **foliations** (● Fig. 13.18). Some shales produce a hard metamorphic rock known as *slate,* which exhibits a tendency to break apart, or cleave, along smooth, flat surfaces that are actually extremely thin foliations (● Fig. 13.19a). Where the foliations are moderately thin, individual minerals have a flattened but wavy, "platy" structure, and the rocks tend to flake apart along these bands. A common metamorphic rock with thin foliations is called *schist.*

● FIGURE 13.18
Metamorphic rocks and foliations. Metamorphic foliations develop at right angles to the stress directions (arrows). (a) Layered rocks under moderate pressure. (b) Greater pressure can cause metamorphism and the development of platy foliations in rocks. (c) Stronger metamorphism can cause the foliations to widen into wavy bands of light and dark minerals.
How are foliations different from bedding planes?

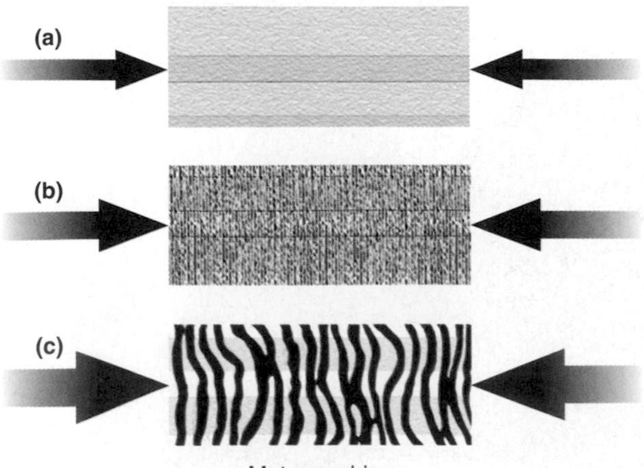

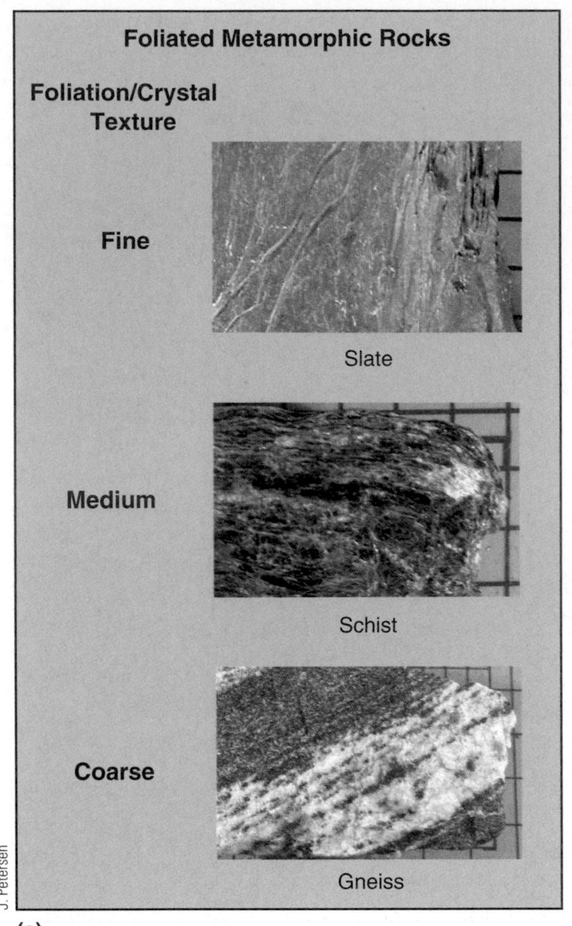

Nonfoliated Metamorphic Rocks

Original Rock Type | Metamorphic Rock

Limestone — Marble

Sandstone — Quartzite

J. Petersen

(b)

● FIGURE 13.19
Metamorphic rocks. (a) Slate, schist, and gneiss illustrate differences in metamorphism and the size of foliations. (b) Marble and quartzite are nonfoliated metamorphic rocks with a harder, recrystallized composition compared to the limestone and sandstone from which they were made.

Where the foliations develop into broad mineral bands, the rock is extremely hard and is known as *gneiss* (pronounced "nice"). Coarse-grained rocks such as granite generally metamorphose into gneiss, whereas finer-grained rocks tend to produce schists.

Rocks that originally were composed of one dominant mineral are not foliated by metamorphism (Fig. 13.19b). Limestone is metamorphosed into much denser *marble,* and impurities in the rock can produce a beautiful variety of colors. Silica-rich sandstones fuse into *quartzite.* Quartzite is brittle, harder than steel, and almost inert chemically. Thus, it is virtually immune to chemical weathering and commonly forms cliffs or rugged mountain peaks after the surrounding, less resistant rocks have been removed by erosion. The physical and chemical characteristics of rocks are important factors in the development of landforms.

The Rock Cycle Like landforms, many rocks do not remain in their original form indefinitely but instead, over a long time, tend to undergo processes of transformation. The **rock cycle** is a conceptual model for understanding processes that generate, alter, transport, and deposit mineral materials to form different kinds of rocks (● Fig. 13.20). The term *cycle* means that existing rocks supply the materials to make new and sometimes very different rocks. Whole existing rocks can be "recycled" to form new rocks. The geologic age of a rock is based on the time when it assumed its current state; metamorphism, or melting, and other rock-forming processes reset the age of origin.

Although a complete cycle is shown in Figure 13.20, many rocks do not go through every step of the rock cycle, as shown by the arrows that cut across the diagram. Igneous rocks form by the cooling and crystallizing of molten lava or magma. Igneous rocks can be remelted and recrystallized to form new igneous rocks; can be changed into metamorphic rocks by heat and/or pressure; or can be weathered into fragments that are eroded, transported, and deposited to form sedimentary rocks. Sedimentary rocks consist of particles and deposits derived from any of the three basic rock types. Metamorphic rocks can be created by means of heat and pressure applied to igneous or sedimentary rock or through further metamorphism into a new rock type. In addition, metamorphic rocks can be heated sufficiently to melt into magma and cool to form igneous rocks.

● FIGURE 13.20

The rock cycle helps explain the formation of igneous, sedimentary, and metamorphic rocks. Note the links that bypass some parts of the cycle.

What conditions are necessary to change igneous rock to metamorphic rock? Can a metamorphic rock be metamorphosed?

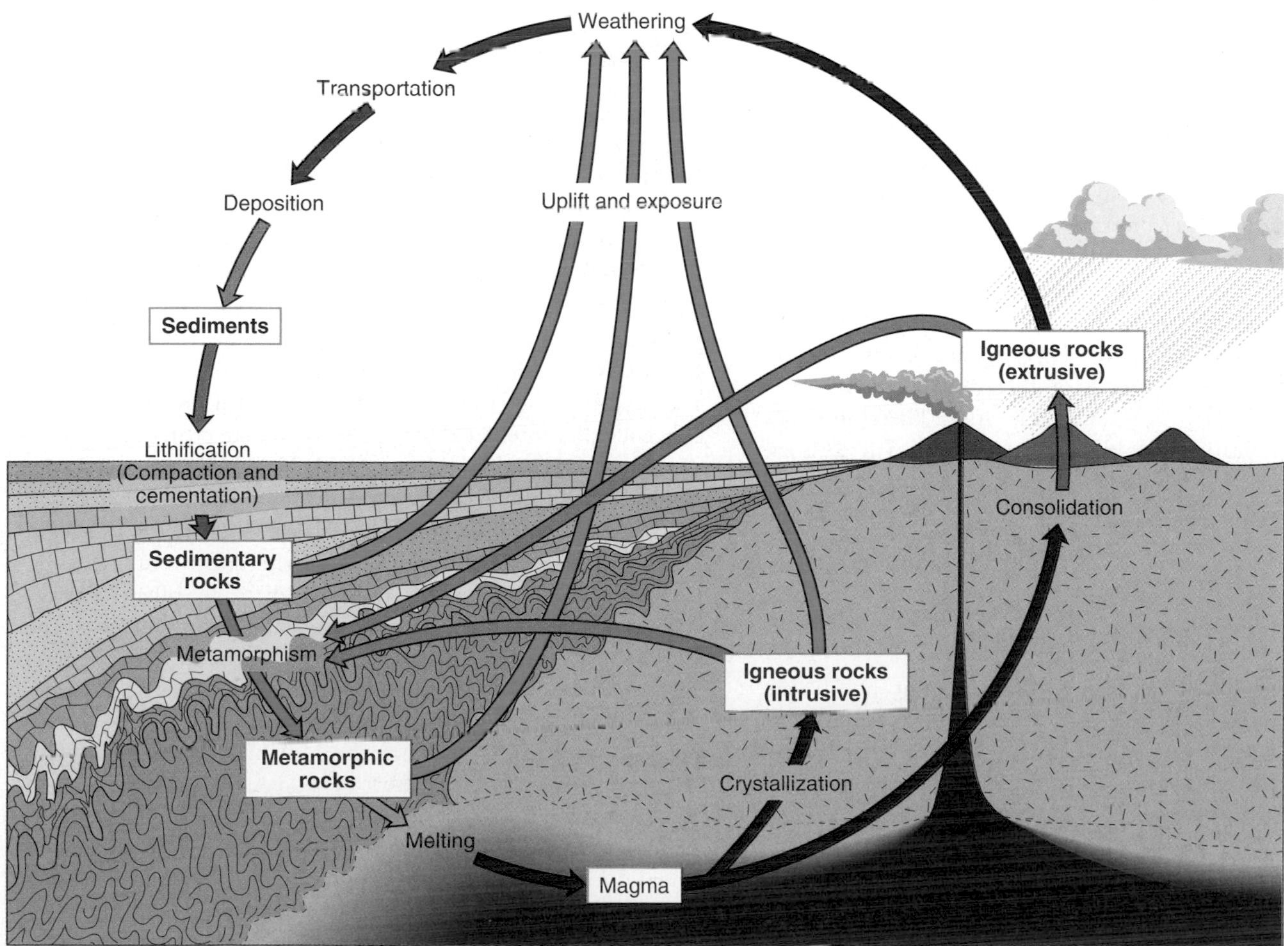

Continents in Motion: The Search for a Unifying Theory

Scientists in all disciplines constantly search for broad explanations that shed light on the detailed facts, recurring patterns, and interrelated processes that they observe and analyze. Is there one broad theory that can help explain how and why Earth's lithospheric processes work? Can it explain such diverse phenomena as the growth of continents, the movement of solid rock beneath Los Angeles, the location of great mountain ranges, differing patterns of temperature in the rocks of the seafloor, and the violent volcanic eruptions on the island of Montserrat in the West Indies? The answer is yes, and the concept is that of the continual movement of landmasses on Earth's surface over millions of years of time.

Sometimes it requires years to develop, test, and refine a scientific theory to the point where it is more fully understood and broadly acceptable. As data and information are gathered and analyzed, new methods and technologies contribute to the process of testing hypotheses via the scientific method, and bit by bit an acceptable explanatory framework emerges. Over the past century, the theoretical framework of *continental drift* has been refined into a well-established theory called *plate tectonics,* which has been tested by collecting a great deal of evidence from the lithosphere. The theory of plate tectonics has revolutionized the Earth sciences and our understanding of Earth's history.

Long ago, some scientists believed that Earth's landscapes were created by great cataclysms. They might have believed, for example, that the Grand Canyon split open one violent day and has remained that way ever since, or that the Rocky Mountains appeared overnight. This theory, called **catastrophism,** has been rejected. For almost two centuries, physical geographers, geologists, and other Earth scientists have accept instead the theory of **uniformitarianism,** which is the idea that internal and external Earth processes operate today in the same manner as they have for millions of years.

Uniformitarianism, however, does not mean that processes have always operated at the same rate or with equal strength everywhere on Earth. In fact, our planet's surface features are the result of variations in the intensity of internal and external processes, influenced by their geographical location. These processes have varied in intensity and location throughout Earth's history. Furthermore, regular or episodic changes in the Earth system that may seem relatively small to us can dramatically alter a landscape after progressing, even on an irregular basis, for millions of years.

Continental Drift

Most of us have probably noted on a world map that the Atlantic coasts of South America and Africa look as if they could fit together. In fact, if a globe were made into a spherical jigsaw puzzle, several widely separated landmasses could fit alongside each other without large gaps or overlaps (• Fig. 13.21). Is there a scientific explanation for this phenomenon?

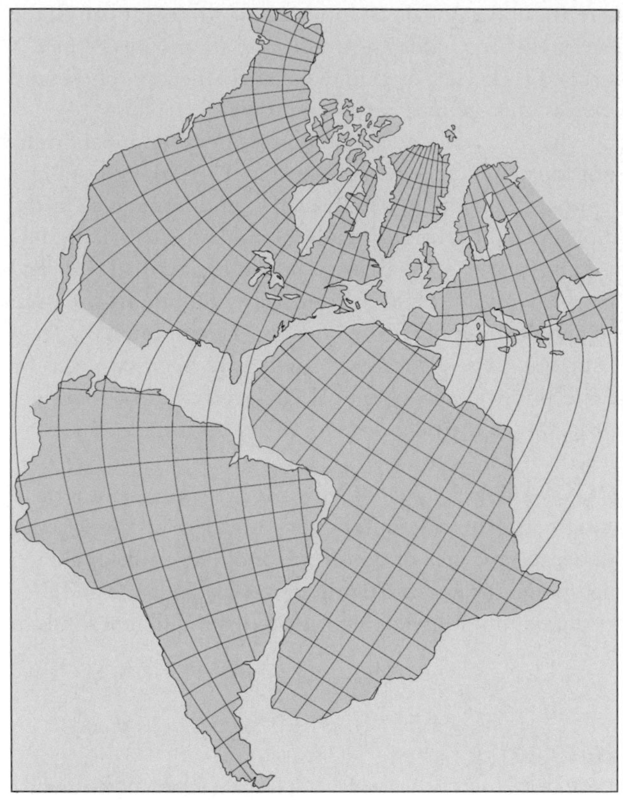

• FIGURE 13.21
The geographic basis for Wegener's continental drift hypothesis. Note the close correlation of the edges of the continents that face one another across the width of the Atlantic Ocean. The actual fit is even closer if the continental slopes are matched.

In the early 1900s, Alfred Wegener, a German climatologist, proposed the theory of **continental drift,** the idea that continents and other landmasses have shifted their positions during Earth history. Wegener's evidence for continental drift included the close fit of continental coastlines on opposite sides of oceans and the trends of mountain ranges on land areas that also match across oceans. He cited comparable geographical patterns of fossils and rock types found on different continents that he felt could not result from chance and did not reflect current climatic conditions. To explain the spatial distributions of these features, he reasoned that the continents must have been previously joined. Wegener also noted evidence of great climate change, such as ancient evidence of glaciation where the Sahara Desert is today and tropical fossils found in Antarctica, that could be explained best by large landmasses moving from one climate zone to another.

Wegener hypothesized that all the continents had once been part of a single supercontinent, which he called **Pangaea,** that later divided into two large landmasses, one in the Southern Hemisphere (Gondwana), and one in the Northern Hemisphere (Laurasia). Later, these two supercontinents also broke apart into sections (the present continents) and drifted to their current positions. Laurasia in the Northern Hemisphere consisted of North America, Europe, and Asia. Gondwana in the Southern Hemisphere was made up of South America, Africa, Australia, Antarctica, and India (• Fig. 13.22). Continued continental movement created the geographical configuration of the landmasses that exist on Earth today.

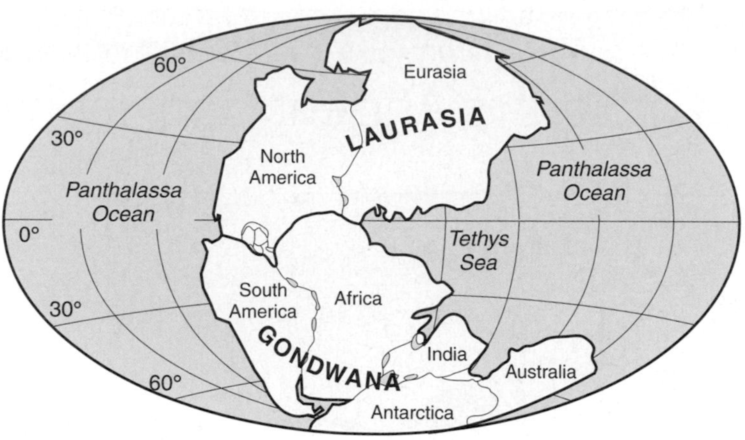

• FIGURE 13.22

The supercontinent of Pangaea included all of today's major landmasses joined together. Pangaea later split to make Laurasia in the Northern Hemisphere and Gondwana in the Southern Hemisphere. Further plate motion has produced the continents as they are today.

How has continental movement affected the climates of landmasses?

The reaction of most of the scientific community to Wegener's proposal ranged from skepticism to ridicule. A major objection to his hypothesis was that neither he nor anyone else could provide an acceptable explanation for the energy needed to break up huge landmasses and slide them over the rigid crust and across vast oceans.

Supporting Evidence for Continental Drift

About a half century later, in the late 1950s and 1960s, Earth scientists began giving serious consideration to Wegener's notion of moving continents. New information appeared from research in oceanography, geophysics, and other Earth sciences, aided by sonar, radioactive dating of rocks, and improvements in equipment for measuring Earth's magnetism. These scientific efforts discovered much new evidence that indicated the movement of portions of the lithosphere, including the continents.

As one example, scientists were originally unable to explain the varied orientations of magnetic fields found in basaltic rocks that had cooled millions of years ago. They knew that iron-bearing minerals in rocks display the magnetic field of Earth as it existed when the rocks solidified, which is a phenomenon known as **paleomagnetism.** Scientists at that time also knew that the exact position of the magnetic poles wandered through time, but they could not account for the confusing range of magnetic field orientations indicated by the basaltic rocks they studied. Magnetic field orientations of rocks of the same age did not point toward a single spot on Earth, and the indicated positions for the magnetic north pole ranged widely, including some that pointed toward the present south magnetic pole. The observed variations were more than could be accounted for by the known magnetic polar wandering.

Scientists eventually used the paleomagnetic data to model where the sampled rocks would have to have been relative to a common magnetic north pole. Successful alignment was only possible if the continents had been in different positions than they are today. Using rocks of different age, they reconstructed locations of the continents during past periods in geologic history (• Fig. 13.23). Paleomagnetic data revealed that the continents were grouped together about 200 million years ago, just as Wegener's hypothesized two supercontinents began to split apart to form the beginnings of the modern Atlantic Ocean. Paleomagnetic data also revealed that the polarity of Earth's magnetic field had reversed many times in the past. A record of these polarity reversals was imprinted within the iron-rich basaltic rocks of the seafloor.

Supporting evidence for crustal movement came from a variety of other sources in the mid-20th century. The widely separated patterns of similar fossil reptiles and plants found in Australia, India, South Africa, South America, and Antarctica, previously noted by Wegener, were mapped in detail. The fossils represented organisms that in each instance were so similar and specialized that they could not have developed without their now-distant locations being either connected or at least much closer together than they are today. When the positions of the continents were reassembled on a paleomap derived from paleomagnetic data and representing the time when the organisms were living, the fossil locations came together spatially. Other types of ancient environmental evidence, such as left by glaciations, could also be fit together in logical geographical patterns on reconstructed paleomaps of the continents or the world (• Fig. 13.24).

How well Earth's landmasses match up when they are brought together on a paleomap was found to be even better

• FIGURE 13.23

Paleomagnetic properties of rocks that formed when the Northern Hemisphere continents were joined point to the location of magnetic north at that time. It requires rejoining the continents to their original positions, as shown on this map, in order for the magnetic orientations to point to a common magnetic pole.

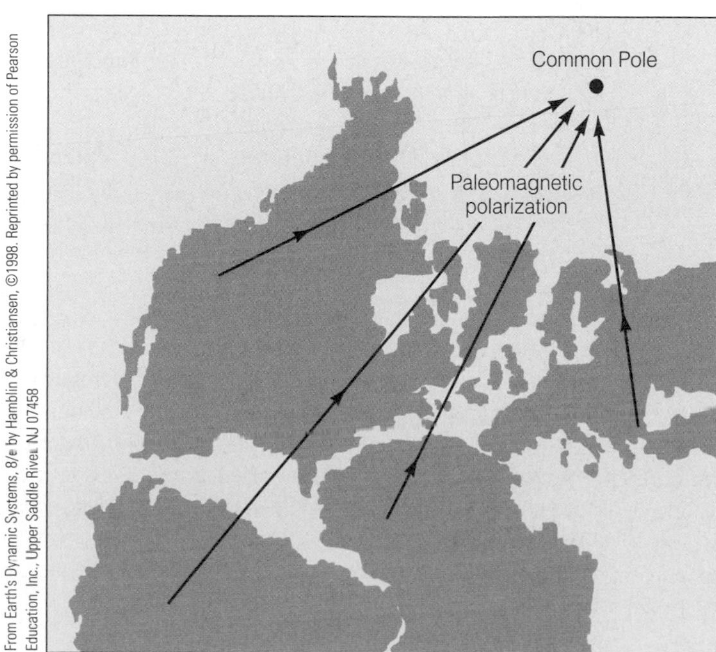

From Earth's Dynamic Systems, 8/e by Hamblin & Christiansen, ©1998. Reprinted by permission of Pearson Education, Inc., Upper Saddle River NJ 07458

GEOGRAPHY'S SPATIAL SCIENCE PERSPECTIVE

Paleomagnetism: Evidence of Earth's Ancient Geography

Earth's magnetic field encircles the globe with field lines that converge at two opposite magnetic poles. The geographic North and South Poles do not coincide with their magnetic counterparts, but outside of the polar regions the magnetic poles are useful for navigation by compass. It is necessary to account for the magnetic declination (see again Figs. 2.26 and 2.27) for directional accuracy.

Earth's magnetic field has changed over geologic time by increasing and decreasing intensity, and the polarity of the magnetic poles has reversed many times. Before the last reversal, about 700,000 years ago (to what we call normal polarity today), a compass would have pointed to the south. *Paleomagnetism* deals with changes in Earth's magnetic field through time. Paleomagnetic studies have yielded much evidence to help us understand plate tectonics and assist in reconstructing the shifting geographical positions of landmasses during Earth history.

By studying orientations of magnetic fields in mineral crystals within rocks of varying ages, we know that magnetic pole reversals have occurred. Knowing the age of the rocks by radiometric dating, we can determine their location when they cooled as well as the nature of the magnetic field at that time. Ancient basaltic rocks, which are iron rich, are most commonly used for this research. When basalt solidifies, iron oxide crystals in the rock become magnetized in a way that records several magnetic properties, which are related to Earth's magnetic field at the time of cooling.

Three important characteristics that these rocks record are polarity (normal, like that of today, or reversed), declination, and inclination, which is measured with a vertically mounted compass needle. Each property provides different evidence about changes in the magnetic field and about how Earth's paleogeography varied as plate tectonics moved the landmasses.

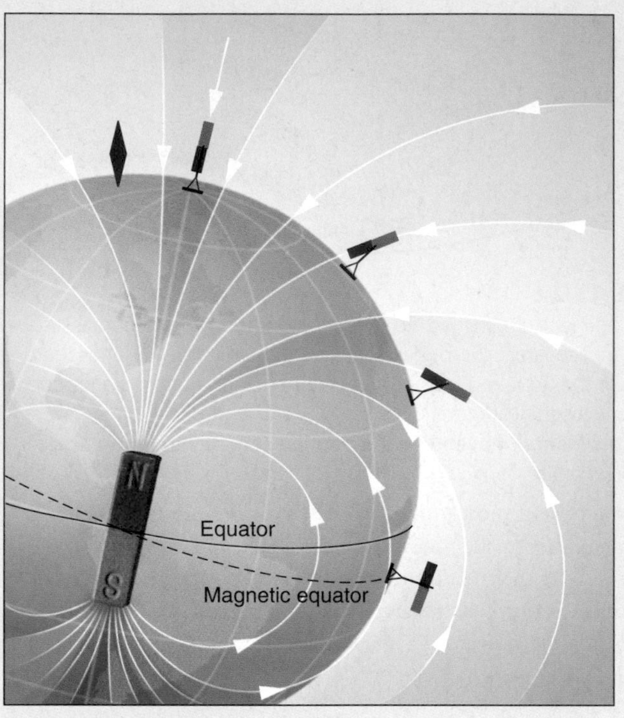

Earth's magnetic field, circling the planet, makes a magnetized dip needle point downward at an angle that equals the latitude of the needle's location. At the equator, the magnetic dip would be 0° (horizontal), and at the magnetic pole the needle would point straight down (90°).

Numerous measurements of these three paleomagnetic qualities worldwide have given scientists a good picture of Earth's continually changing paleogeography throughout the last several hundred million years.

- **Polarity** Seafloor spreading was confirmed by polarity changes discovered in stripelike patterns of basalts that matched on opposite sides of the spreading center where they formed. Going farther away from the Mid-Atlantic Ridge, the rocks were progressively older, and each stripe had a counterpart of the same age and same magnetic polarity on opposite sides of the ridge. The basaltic seafloor had recorded the polarity history of the magnetic field and the widening of the Atlantic Ocean.
- **Declination** Declination shows the direction to the magnetic pole. By studying basalts of the same age but on several continents, it is possible to triangulate directions to the magnetic north pole at the time they formed (see Fig. 13.23). The information provided by these paleodeclinations is the *orientation* of ancient landmasses, in other words, whether or not they rotated relative to north as they drifted.
- **Inclination** The magnetic field surrounding Earth causes not only a magnetic compass needle to point north but also to dip downward in a straight-line direction to north. This is called magnetic dip, and a needle's angle off of horizontal approximates its *latitudinal* location. Paleoinclinations recorded from ancient basalts provide the latitude of their location at the time of cooling.

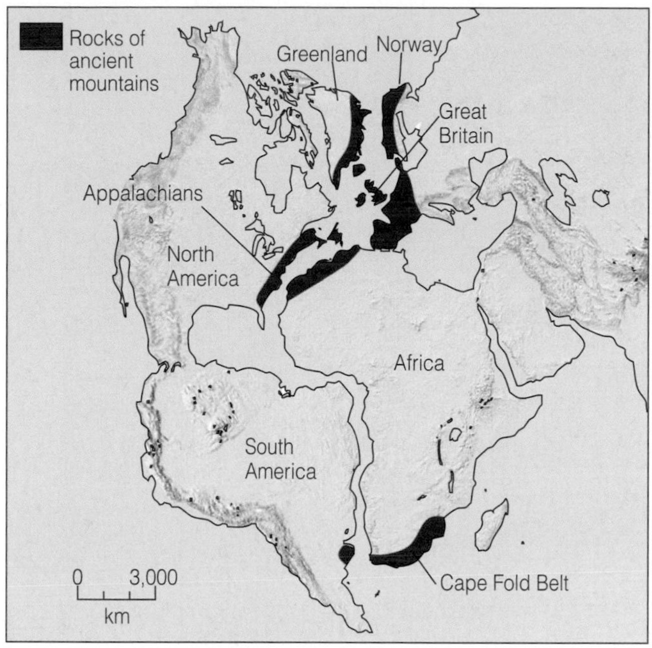

(a)

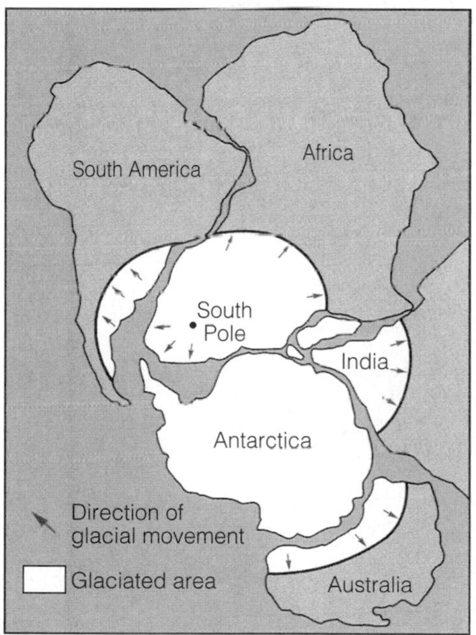

(b)

• FIGURE 13.24
A wide variety of paleogeographical evidence indicates the previous locations and distributions of Earth's landmasses in the geologic past: (a) rocks of ancient mountain ranges; (b) evidence of ancient glaciation.

when using the true continental edges—the continental slopes—which lie a few hundred meters below sea level. In this case, as also had been noted by Wegener, mountain ranges on opposite sides of oceans line up and rock ages and types match where the continents join. Knowledge of the geographical distribution of Earth's environments relative to latitude and climate zones provided additional insight. Evidence that ancient glaciation occurred simultaneously in India and South Africa while tropical forest climates (represented by coal deposits) existed in the northeastern United States and in Great Britain could only be explained by the latitudinal movement of landmasses, and their locations came together well on *paleogeographic* reconstructions.

Plate Tectonics

Plate tectonics, the modern theory to explain the movement of continents, suggests that the rigid and brittle outer shell of Earth, that is, the lithosphere (crust and uppermost mantle), is broken into several separate segments called **lithospheric plates** that rest on, and are carried along with, the flowing plastic asthenosphere (• Fig. 13.25). Tectonics involves large-scale forces originating within Earth that cause parts of the lithosphere to move around. In plate tectonics, the lithospheric plates move as distinct and discrete units. In some places they pull away from each other (diverge), in other places they push together (converge), and elsewhere they slide alongside each other (move laterally). Seven major plates have proportions as large as or larger than continents or ocean basins. Five other plates are of minor size, although they have maintained their own identity and direction of movement for some time. Several additional plates are even smaller and exist in active zones at the boundaries between major plates. All major plates consist of both continental and oceanic crust although the largest, the Pacific plate, is primarily oceanic. To understand how plate tectonics operates and why plates move, we must consider the scientific evidence that was gathered to test this theory. We should also evaluate how well this theory holds up under rigorous examination. The supporting evidence, however, is overwhelming.

Seafloor Spreading and Convection Currents

In the 1960s, several keys to plate tectonics theory were found while studying and mapping the ocean floors. First, detailed undersea mapping was conducted on a system of midoceanic ridges (also called oceanic ridges or rises) that revealed configurations remarkably similar to the continental coastlines. Second, it was discovered in the Atlantic and Pacific Oceans that basaltic seafloor displayed parallel bands of matching patterns of magnetic properties in rocks of the same age but on opposite sides of midoceanic ridges. Third, scientists made the surprising discovery that although some continental rocks are 3.6 billion years old, rocks on the ocean floor are all geologically young, having been in existence less than 250 million years. Fourth, the oldest rocks of the seafloor lie beneath the deepest ocean waters or close to the continents, and rocks become progressively younger toward the midoceanic ridges where the youngest basaltic rocks are found (• Fig. 13.26). Finally, temperatures of rocks on the ocean floor vary significantly, being hottest near the ridges and becoming progressively cooler farther away.

Only one logical explanation emerged to fit all of this evidence. It became apparent that new oceanic crust is being formed at the midoceanic ridges while older oceanic crust is being destroyed along other margins of ocean basins. The emergence of new oceanic crust is associated with the movement of great sections or plates of the lithosphere away from the midoceanic ridges. This phenomenon, which represents a major advance in our understanding of how

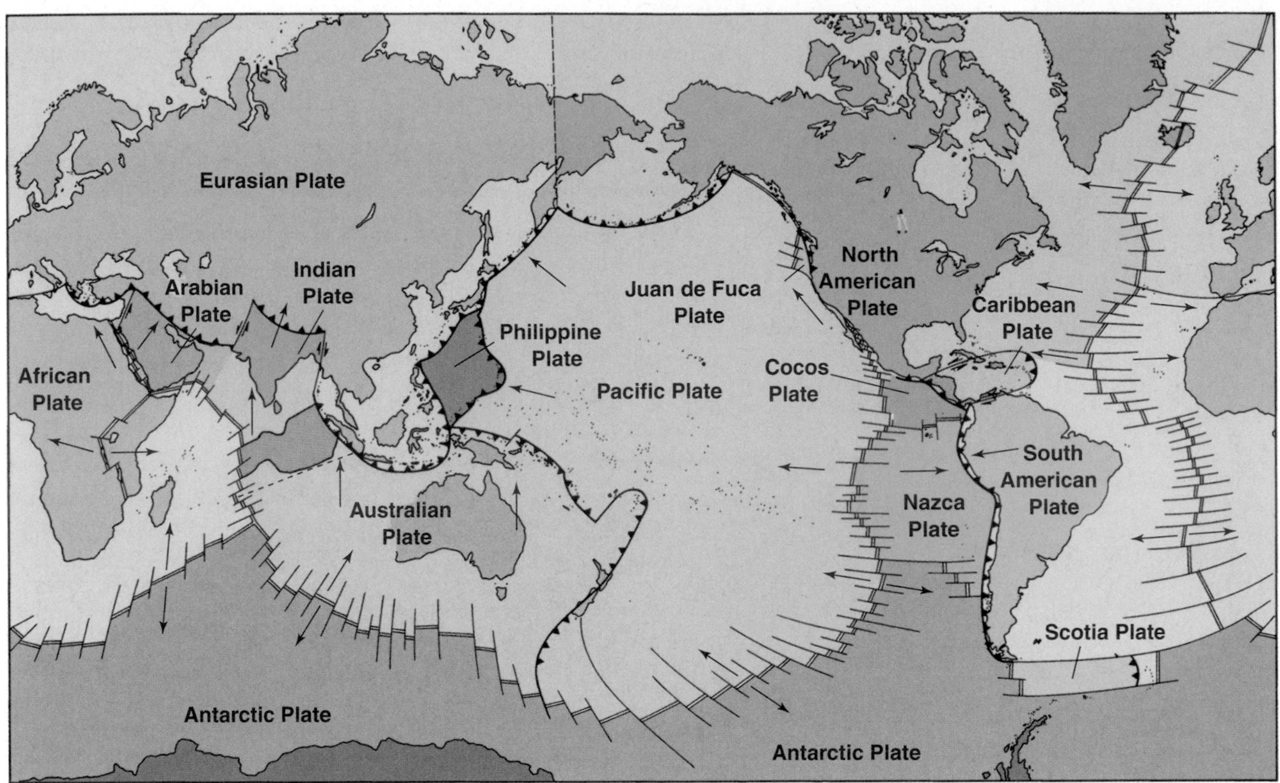

• FIGURE 13.25
Earth's solid exterior (the lithosphere) is broken into giant segments called plates. This map shows Earth's major tectonic plates and their general directions of movement. Most tectonic and volcanic activity occurs along plate boundaries where the large segments separate, collide, or slide past each other. Barbs indicate boundaries where one plate is overriding another, with the barbs on the side of the overriding plate.
Does every lithospheric plate include a continent?

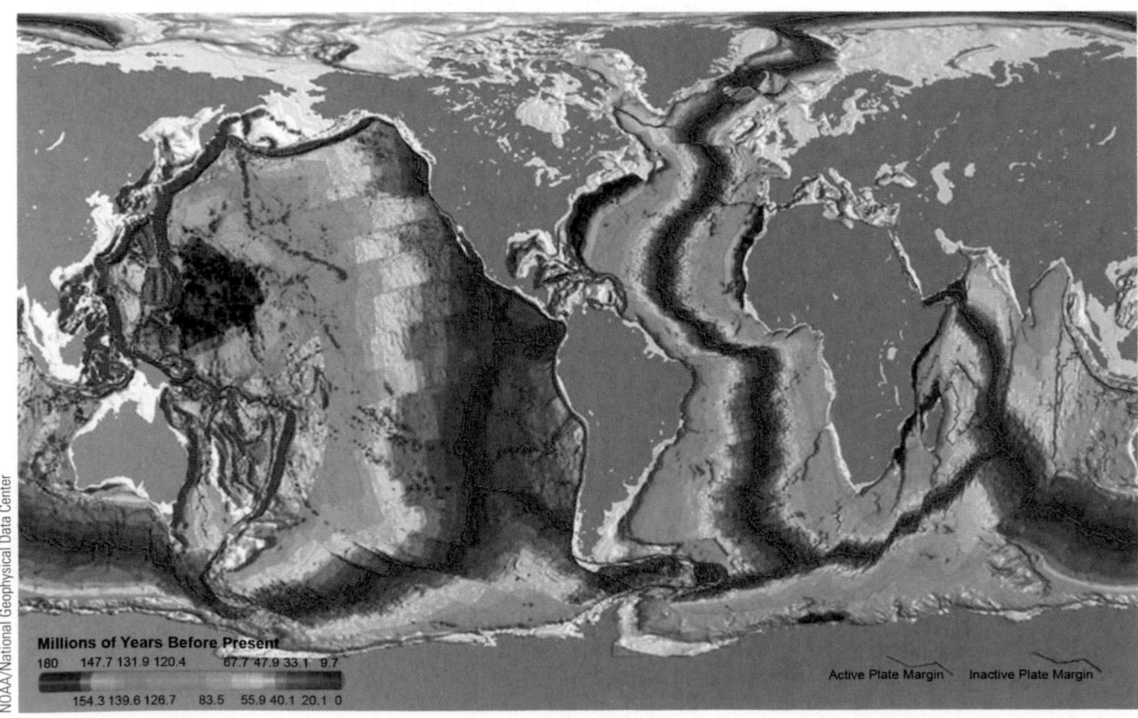

• FIGURE 13.26
The global oceanic ridge system and the age of the seafloor. Red represents the youngest seafloor, and blue the oldest. Detailed mapping and study of the ocean floors yielded much evidence to support the theory of plate tectonics by identifying the process of seafloor spreading.

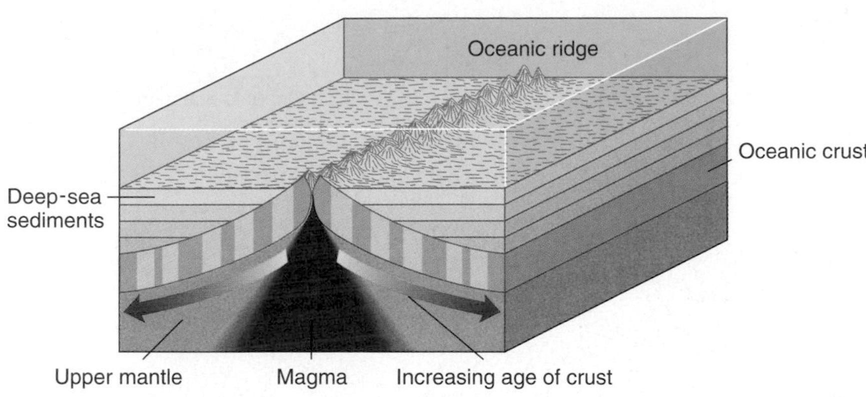

• **FIGURE 13.27**
Seafloor spreading at an oceanic ridge produces new seafloor.

continents move, is called **seafloor spreading** (• Fig. 13.27). The rigid lithospheric plates diverge along the oceanic ridges and separate at an average rate of 2 to 5 centimeters (1–2 in.) per year as they are carried along with the flowing plastic asthenosphere in the mantle. The young age of oceanic crust results from the creation of new basaltic rock at undersea ridges and the movement of the seafloor with lithospheric plates toward ocean basin margins where the older rock is remelted and destroyed. As molten basalt cooled and crystallized in the seafloor, the iron minerals that they contain became magnetized in a manner that replicated the orientation of Earth's magnetic field at that time. The iron-rich basalts of the seafloor have preserved a historical record of Earth's magnetic field, including **polarity reversals** (times when the north pole became south, and vice versa).

Plate tectonics includes a plausible explanation of the mechanism for continental movement, which had eluded Wegener. The mechanism is **convection.** Hot mantle material travels upward toward Earth's surface and cooler material moves downward as part of huge subcrustal convection cells (• Fig. 13.28). Mantle material rises to the asthenosphere where it spreads laterally and flows in opposite directions, dragging the lithospheric plates with it. Pulling apart the brittle lithosphere breaks open a midoceanic ridge. Molten basalt wells up into the fractures, cooling and sealing them to form new seafloor. In this process, the ocean becomes wider by the width of the now-sealed fracture. The convective motion continues as solidified crustal material moves away from the ridges. In a time frame of up to 250 million years, older oceanic crust is consumed in the deep trenches near plate boundaries where sections of the lithosphere meet and are recycled into Earth's interior.

Tectonic Plate Movement

The shifting of tectonic plates relative to one another provides an explanation for many of Earth's surface features. Plate tectonics theory enables physical geographers to better understand not only our planet's ancient geography but also the modern global distributions and spatial relationships among such diverse, but often related, phenomena as earthquakes, volcanic activity, zones of crustal movement, and major landform features (• Fig. 13.29). Let's briefly examine the three ways in which lithospheric plates relate to one another along their boundaries as a result of tectonic movement: by pulling apart, pushing together, or sliding alongside each other.

Plate Divergence The pulling apart of plates, tectonic **plate divergence,** is directly related to seafloor spreading (see again Fig. 13.27). Tectonic forces that act to pull objects apart cause the crust to thin and weaken. Shallow earthquakes are often associated with this crustal stretching, and asthenospheric magma wells up between crustal fractures. This creates new crustal ridges and new ocean floor as the plates move away from each other. The formation of new crust in these spreading centers gives the label *constructive plate margins* to these zones. Occasional "oceanic" volcanoes,

• **FIGURE 13.28**
Convection is the mechanism for plate tectonics. Heat causes convection currents of material in the mantle to rise toward the base of the solid lithosphere where the flow becomes more horizontal. As the asthenosphere undergoes its slow, lateral flow, the overlying lithospheric plates are carried along because of friction at the boundary between the asthenosphere and lithosphere.
Why is plate tectonics a better name than continental drift for the lateral movement of Earth's solid outer shell?

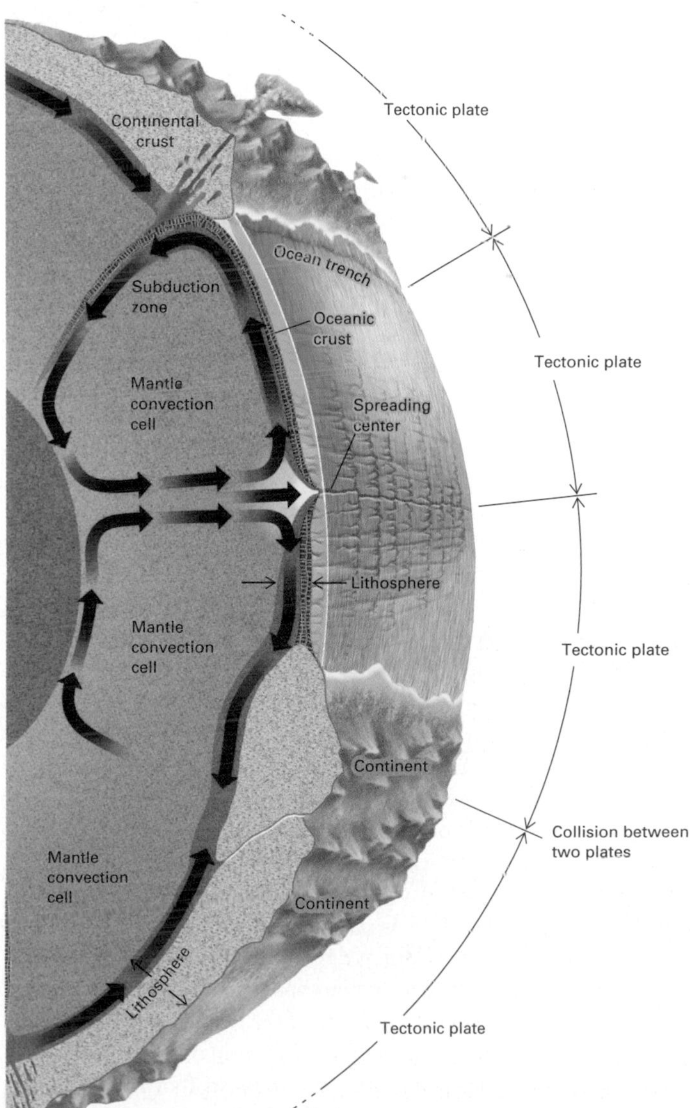

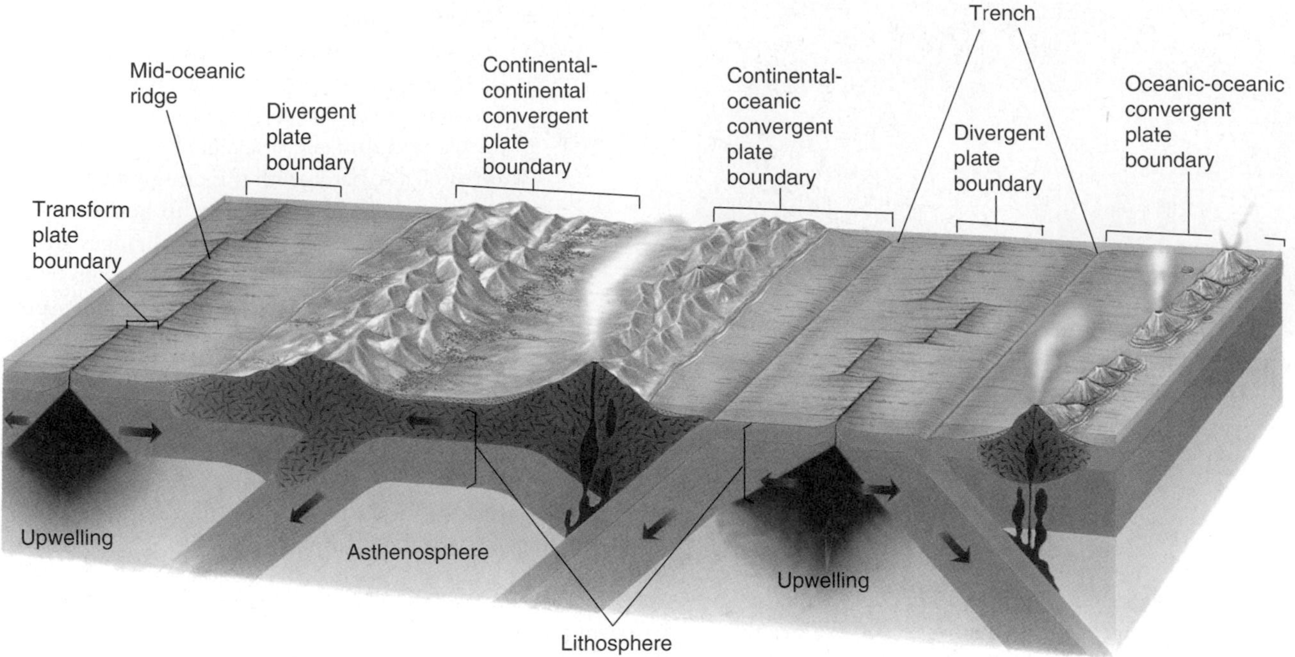

• FIGURE 13.29
Plate tectonic movement. Unlike the other major Earth systems, the plate tectonics system does not obtain its energy from the sun. Instead, movements of the lithosphere result from heat energy derived from Earth's interior. As lithospheric plates move due to heat-driven convection cells in the mantle, they interact with adjoining plates, forming different boundary types, each displaying distinct landform features. This diagram shows three major plate boundary types: spreading centers, subduction zones, and continental collision zones.

Spreading centers (far left and right of middle on diagram) are divergent plate boundaries. These are constructional boundaries at which new crustal material emerges along active rift zones. Over time, newer material pushes older rock progressively away from the active rift zone in both directions. Earth's oceanic divergent plate boundaries form the midoceanic ridge system, which extends through all of the major oceans.

Subduction zones (right side of diagram) occur where two plates converge, with the margin of at least one of them consisting of oceanic crust. This is a destructive type of boundary where crustal material returns to Earth's interior. The denser oceanic plate is forced by gravity and plate movement to subduct beneath the less dense plate, whether that consists of continental crust or oceanic crust. Surface features common to subduction zones are deep ocean trenches and volcanic mountain ranges or island arcs. The best examples of subduction are found around the Pacific Ring of Fire, such as those along Japan, Chile, New Zealand, and the northwest coast of the United States.

Continental collision zones (middle of diagram) are found where two continental plates collide. Massive mountain building occurs as the crust thickens because of compression. Volcanoes tend to be absent in these regions. The world's highest mountains, the Himalayas, were formed when the Indian plate collided with Eurasia. The Alps were formed in a similar manner in a collision between the African and Eurasian plates.

like those of Iceland, the Azores, and Tristan da Cunha, mark such boundaries (● Fig. 13.30).

Most plate divergence occurs along oceanic ridges, but this process can also break apart continental crust, eventually reducing the size of the landmasses involved (● Fig. 13.31a). The Atlantic Ocean floor formed as the continent that included South America and Africa broke up and moved apart 2 to 4 centimeters (1–2 in.) per year over millions of years. The Atlantic Ocean continues to grow today at about the same rate. The best modern example of divergence on a continent is the rift valley system of East Africa, stretching from the Red Sea south to Lake Malawi. Crustal blocks that have moved downward with respect to the land on either side, with lakes occupying many of the depressions, characterize the entire system, including the Sinai Peninsula and the Dead Sea. Measurable widening of the Red Sea suggests that it may be the beginning of a future ocean that is forming between Africa and the Arabian Peninsula, similar to the young Atlantic between Africa and South America about 200 million years ago (Fig. 13.31b).

Plate Convergence A vide variety of crustal activity occurs at areas of tectonic **plate convergence.** Despite the relatively slow rates of plate movement (in terms of human perception), the incredible energy involved in convergence causes the crust to crumple as one plate overrides another. The denser plate is forced deep below the surface in a process called **subduction.** Subduction is most common where dense oceanic crust collides with and descends beneath less dense continental crust (● Fig. 13.32). This is the situation along South America's Pacific coast, where the Nazca plate subducts beneath the South American plate, and in Japan, where the Pacific plate dips under the Eurasian plate. As oceanic crust, and the lithospheric plate of which it forms a part, is subducted, it descends into the asthenosphere to be melted and recycled into Earth's interior.

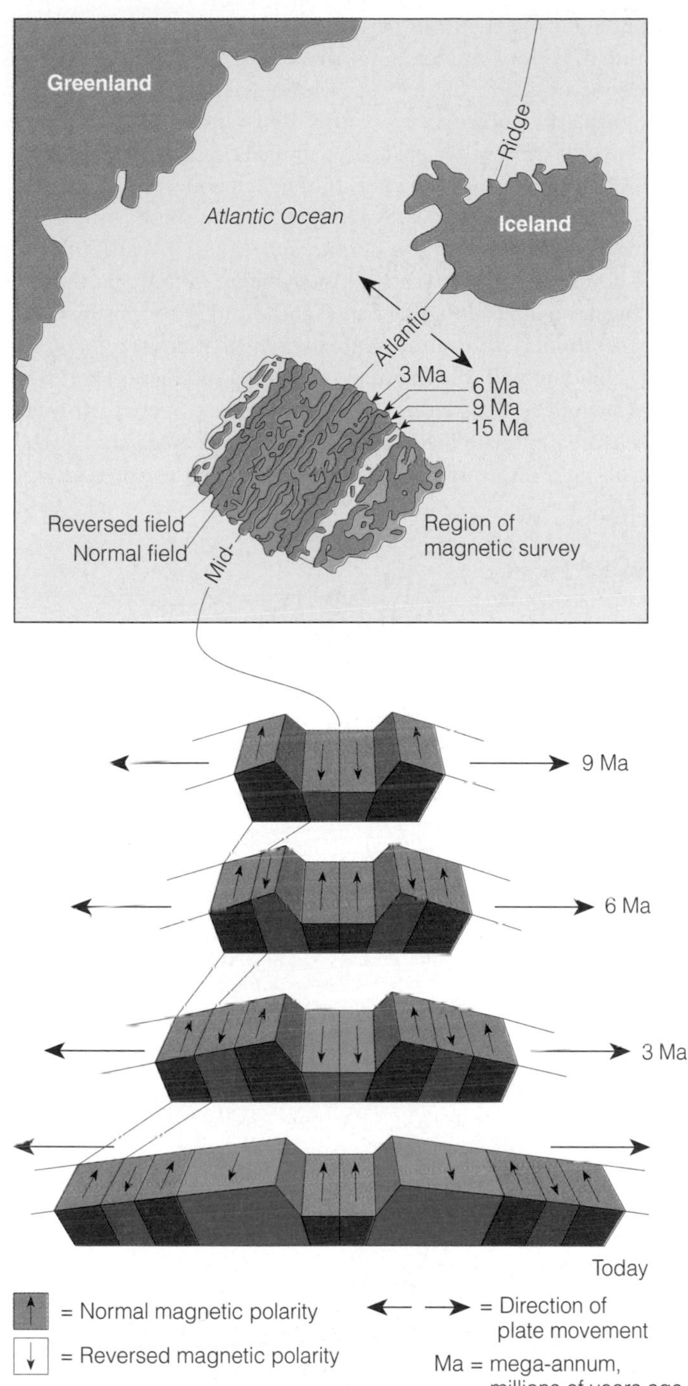

• FIGURE 13.30
Iceland represents part of the Mid-Atlantic Ridge where it stands above sea level to form a volcanic island. The "striped" pattern of polarity reversals documented in the basaltic rocks along the Mid-Atlantic Ridge helped scientists understand the process of seafloor spreading.

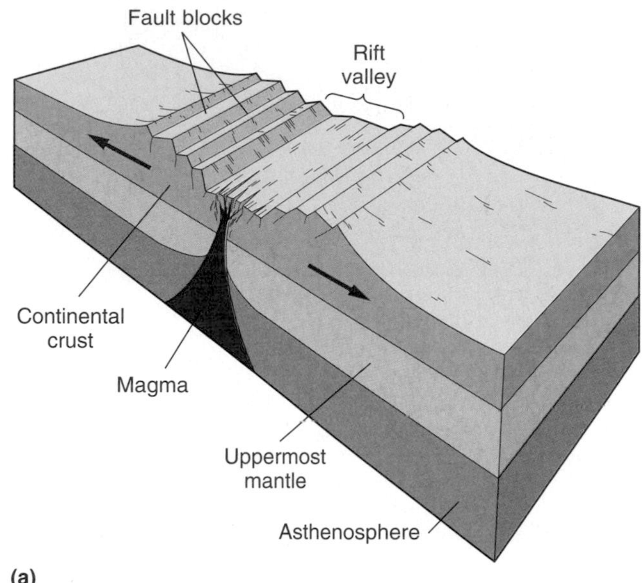

(a)

(b)

• FIGURE 13.31
(a) A continental divergent plate boundary breaks continents into smaller landmasses. (b) The roughly triangular-shaped Sinai Peninsula, flanked by the Red Sea (lower left) to the south, Gulf of Suez (photo center) on the west, and Gulf of Aqaba (lower right) toward the east, illustrates the breakup of a continental landmass. The Red Sea rift and the narrow Gulf of Aqaba are both zones of spreading. The irrigated valley of the Nile River (upper left) in Egypt can be seen heading northward across the desert into the Mediterranean Sea.

Deep ocean trenches, such as the Peru–Chile trench and the Japanese trench, occur where the crust is dragged downward into the mantle. Frequently, hundreds of meters of sediments that are deposited on the seafloor or along continental margins are carried down into these trenches. At such convergent boundaries, rocks can be squeezed and contorted between colliding plates, becoming uplifted and greatly deformed or metamorphosed. These processes have produced many great mountain ranges, such as the Andes, at convergent plate margins. A subducting plate is heated as it plunges downward into the mantle. Its rocks are melted, and the resulting magma migrates upward into the overriding plate. Where molten rock reaches the surface, it forms a series of volcanic peaks, as in the Cascade Range of the northwestern United States. Where two oceanic plates meet, the older, denser one will subduct below the younger, less dense oceanic plate, and volcanoes may develop, creating major **island arcs** on the overriding plate between the continents and the ocean trenches. The Aleutians, the Kuriles, and the Marianas are all examples of island arcs near oceanic trenches that border the Pacific plate.

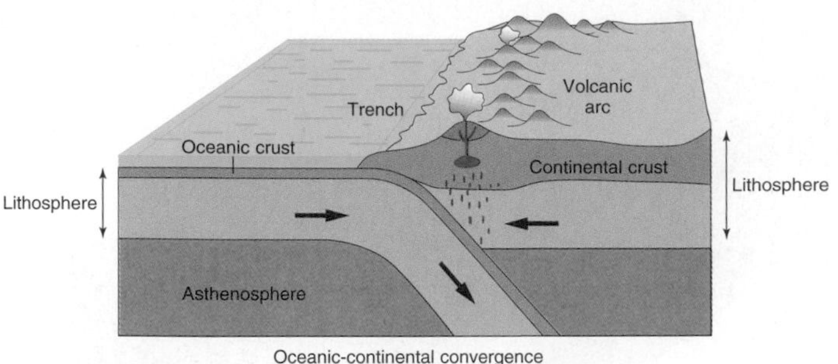

• FIGURE 13.32
An oceanic–continental convergent plate boundary where continent and seafloor collide. The west coast of South America is an excellent example of this kind of plate margin. Collision has contributed to the development of the Andes and a deep ocean trench offshore.

As the subducting plate grinds downward, enormous friction is produced, which explains the occurrence of major earthquakes in these regions. Subduction zones are sometimes referred to as Benioff zones, after the seismologist Hugo Benioff, who first plotted the position of earthquakes extending downward at a steep angle on the leading edge of a subducting plate (see again Fig. 13.29).

Continental collision causes two continents or major landmasses to fuse or join together, creating a new larger landmass (• Fig. 13.33). This process, which closes an ocean basin that once separated the colliding landmasses, has been called *continental suturing*. Where two continental masses collide, the result is massive folding and crustal block movement rather than volcanic activity. This crustal thickening generally produces major mountain ranges at sites of continental collision. The Himalayas, the Tibetan Plateau, and other high Eurasian ranges formed in this way as the plate containing the Indian subcontinent collided with Eurasia some 40 million years ago. India is still pushing into Asia today to produce the highest mountains in the world. In a similar fashion, the Alps were formed as the African plate was thrust against the Eurasian plate.

Zones where plates are converging mark locations of major, and some of the tectonically more active, landforms on our planet: huge mountain ranges, chains of volcanoes, and deep ocean trenches. The distinctive spatial arrangement of these features worldwide can best be understood within the framework of plate tectonics.

Transform Movement Lateral sliding along plate boundaries, called **transform movement,** occurs where plates neither pull apart nor converge but instead slide past each other as they move in opposite directions. Such a boundary exists along the San Andreas Fault zone in California (• Fig. 13.34). Mexico's Baja peninsula and Southern California are west of the fault on the Pacific plate. San Francisco and other parts of California east of the fault zone are on the North American plate. In the fault zone, the Pacific plate is moving laterally northwestward in relation to the North American plate at a rate of about 8 centimeters (3 in.) a year (80 km or about 50 mi per million years). If movement continues at this rate, Los Angeles will lie alongside San Francisco (450 mi northwest) in about 10 million years and eventually pass that city on its way to finally colliding with the Aleutian Islands at a subduction zone.

Another type of lateral plate movement occurs on ocean floors in areas of plate divergence. As plates pull apart, they usually do so along a series of fracture zones that tend to form at right

• FIGURE 13.33
Continental collision along a convergent plate boundary fuses two landmasses together. The Himalayas, the world's highest mountains, were formed when India drifted northward to collide with Asia.

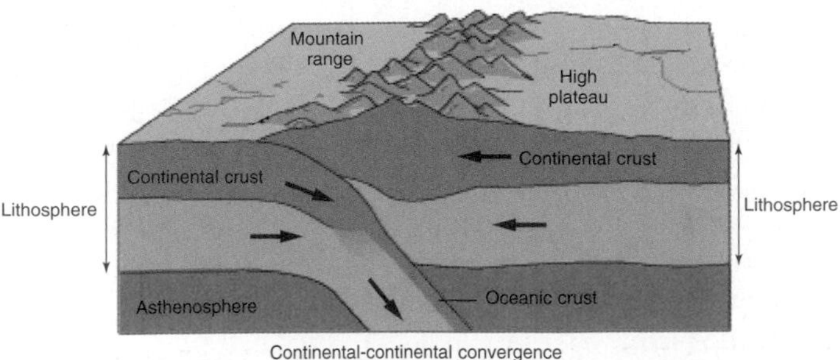

• FIGURE 13.34
Along this lateral plate boundary, marked by the San Andreas Fault in western North America, the Pacific plate moves northwestward relative to the North American plate. Note that north of San Francisco the boundary type changes.
What boundary type is found north of San Francisco and what types of surface features indicate this change?

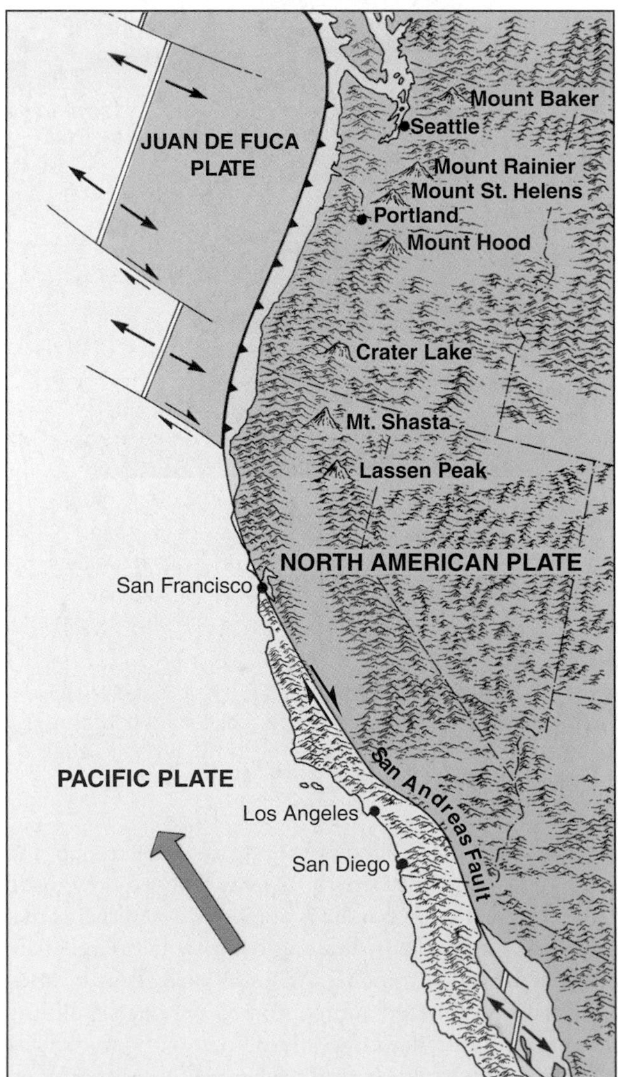

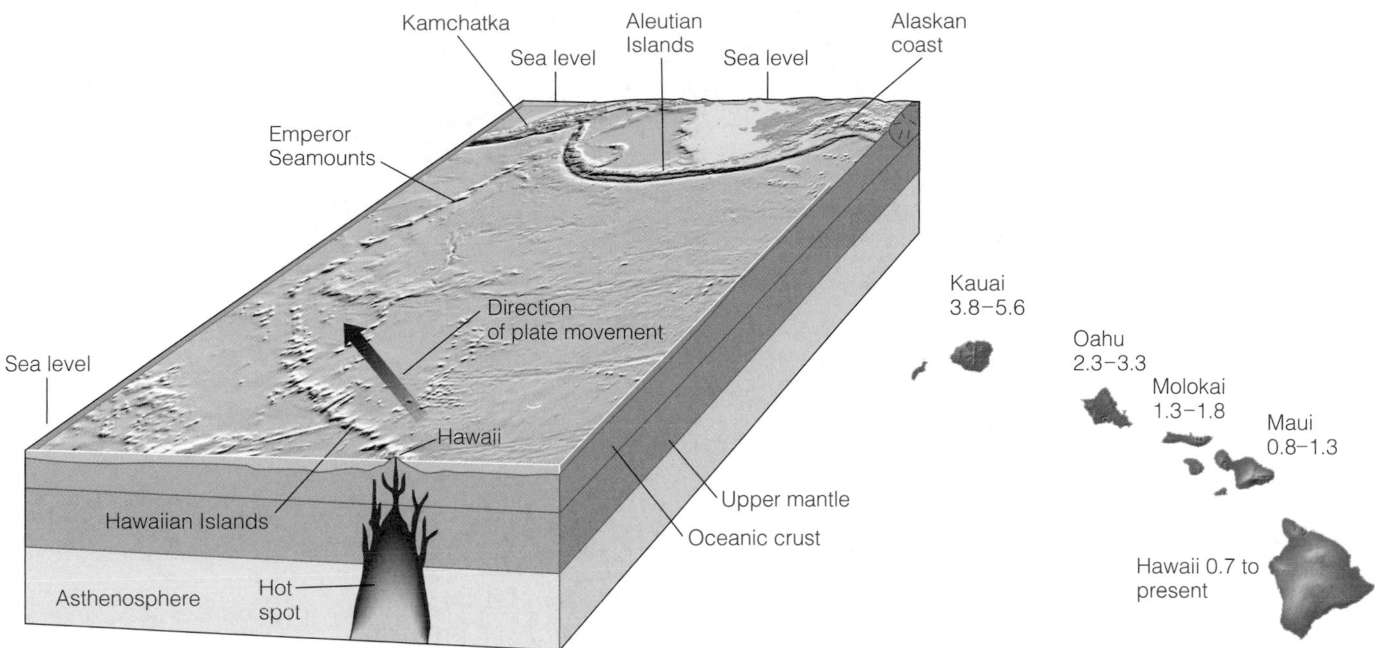

• FIGURE 13.35
The Hawaiian Islands were created by a mantle hot spot. A stationary zone of molten material in the mantle has caused volcanoes to form at the same location in the Pacific Ocean for millions of years. As the Pacific plate has drifted to the northwest, each of the Hawaiian Islands has moved with it, away from the active volcanic zone. The islands are progressively older toward the northwest (ages are in millions of years). The hot spot is currently located at the island of Hawaii. It is about 300 kilometers from the island of Hawaii to Honolulu on the island of Oahu.
How long did it take the Pacific plate to move Oahu to its current position?

angles to the major zone of plate contact. These crosshatched plate boundaries along which lateral movement takes place are called *transform faults.* Transform faults, or fracture zones, are common along midoceanic ridges, but examples can also be seen elsewhere, as on the seafloor offshore from the Pacific Northwest coast between the Pacific and Juan de Fuca plates (see again Fig. 13.34). Transform faults are caused as adjacent plates travel at variable rates, causing lateral movement of one plate relative to the other. The most rapid plate motion is on the East Pacific rise where the rate of movement is more than 17 centimeters (5 in.) per year.

Hot Spots in the Mantle

The Hawaiian Islands, like many major landform features, owe their existence to processes associated with plate tectonics. As the Pacific plate moves toward the northwest near these islands, it passes over a mass of molten rock in the mantle that does not move with the lithospheric plate. Called **hot spots,** these almost stationary molten masses occur in a few other places in both continental and oceanic locations. Melting of the upper mantle and oceanic crust causes undersea eruptions and the outpouring of basaltic lava on the seafloor, eventually constructing a volcanic island. This process is responsible for building the Hawaiian Islands, as well as the chain of islands and undersea volcanoes that extend for thousands of miles northwest of Hawaii. Today the hot spot causes active volcanic eruptions on the island of Hawaii. The other islands in the Hawaiian chain came from a similar origin, having formed over the hot spot as well, but these volcanoes have now drifted along with the Pacific plate away from their magmatic source. Evidence of the plate motion is indicated by the fact that the youngest islands of the Hawaiian chain, Hawaii and Maui, are to the southeast, and the older islands, such as Kauai and Oahu, are located to the northwest (• Fig. 13.35). A newly forming undersea volcano, named Loihi, is now developing southeast of the island of Hawaii and will someday be the next member of the Hawaiian chain.

Growth of Continents

The origin of continents is still being debated. It is clear that the continents tend to have a core area of very old igneous and metamorphic rocks that may represent the deeply eroded roots of ancient mountains. These core regions have been worn down by hundreds of millions of years of erosion to create areas of relatively low relief that are located far from active plate boundaries. As a result, they have a history of tectonic stability over an immense period of time. These ancient crystalline rock areas are called **continental shields** (• Fig. 13.36). The Canadian, Scandinavian, and Siberian shields are outstanding examples. Around the peripheries of the exposed shields, flat-lying, younger sedimentary rocks at the surface indicate the presence of a stable and rigid rock mass below, as in the American Midwest, western Siberia, and much of Africa.

Most Earth scientists consider continents to grow by **accretion,** that is, by adding numerous chunks of crust to the main continent by collision. Western North America grew in this manner over the past 200 million years by adding segments of crust,

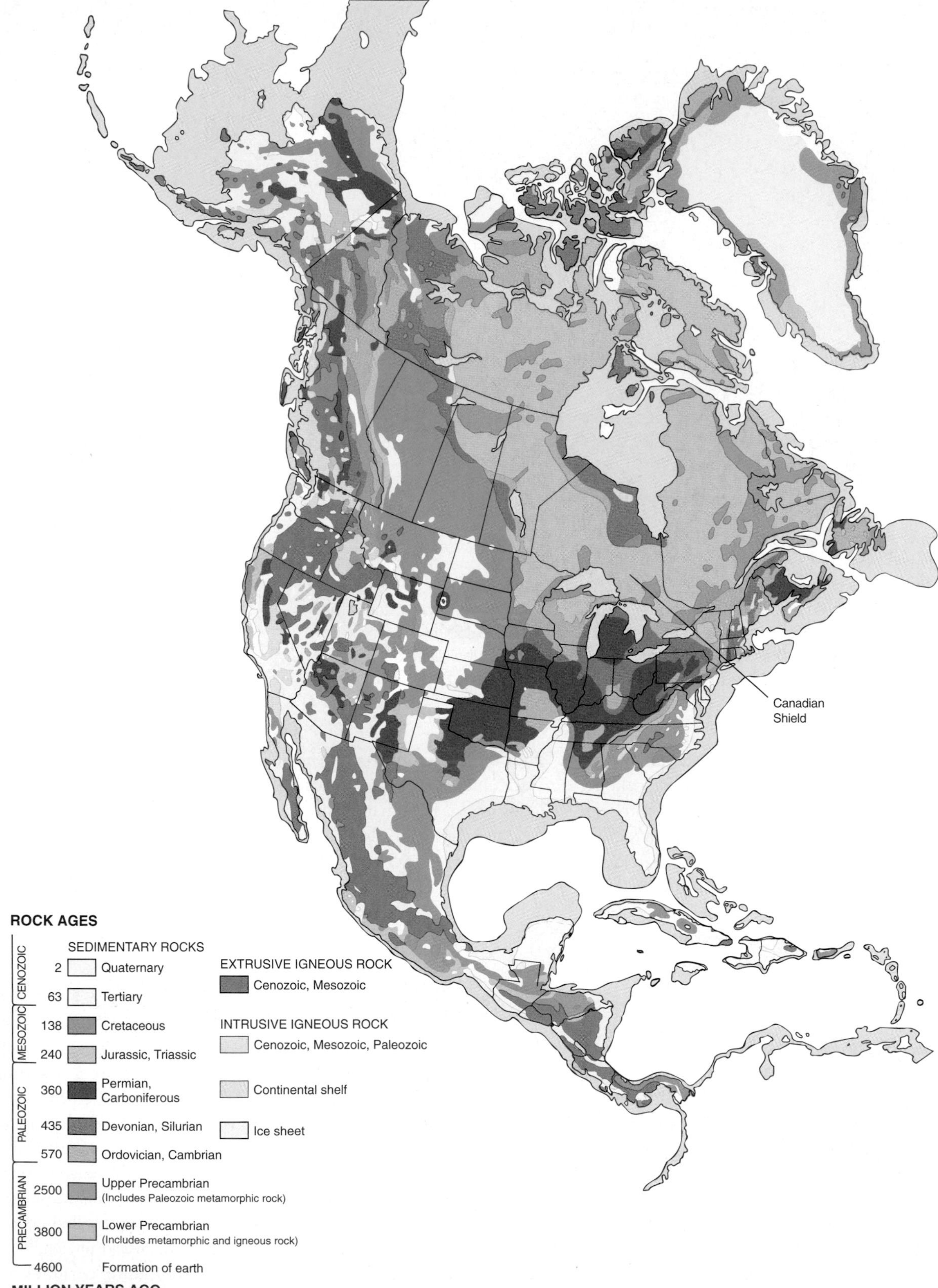

• FIGURE 13.36
Map of North America showing the continental shield and the general ages of rocks.
Going outward from the shield toward the coast, what generally happens to the ages of rocks? What does this suggest about the size of the continent during the time span represented by the rocks along the continental margins?

GEOGRAPHY'S PHYSICAL SCIENCE PERSPECTIVE

Isostasy: Balancing Earth's Lithosphere

Plate tectonics explains that the continents are parts of lithospheric plates that act like rafts, moving along with the slowly flowing asthenosphere. The solid upper mantle, oceanic crust, and continental crust constitute the lithosphere, which lies on top of the flowing asthenosphere. The mantle material in the asthenosphere flows at about 2–5 centimeters (1–2 in.) per year, like a very thick fluid. Gravity does not cause the lithosphere to sink because its material is less dense than that of the asthenosphere.

The principle of buoyancy tells us that an object will sink if its density (mass divided by volume) is greater than that of the fluid. The volume of fluid displaced by a floating object will weigh the same as the object. If the object floats, the proportion floating above the surface equals the percentage of density difference between it and the fluid. An ice cube having 90% of the density of water floats with 10% of the cube extending above the water surface. As long as the weight of a cargo ship and its load is less than the weight of the water they displace, a balance (equilibrium) will be maintained and the ship will float. If the ship and its cargo become heavier than the volume of water they displace, the ship will sink. Ships float higher when empty and lower when full of cargo.

Isostasy is the term for the equalization of hydrostatic pressure (fluid balance) that affects Earth's lithosphere and in turn its topography. One concept of isostasy suggests that material of the lithosphere exists in a density in equilibrium with the material of the asthenosphere. A column of lithosphere (and the overlying hydrosphere) anywhere on Earth weighs about the same as a column of equal diameter from anywhere else regardless of vertical thickness. The lithosphere is thicker (taller and deeper) where it contains a higher percentage of low-density materials. The lithosphere is thinner where it contains a higher percentage of high-density materials. Continental crust has a lower density than oceanic crust, which is why it is the thinner, denser oceanic crust that is subducted along ocean trenches.

If an additional load is placed in an area by a massive accumulation of sediments, lake water, or glacial ice, the lithosphere there will subside to a new equilibrium level. If these materials are later removed, the region will tend to rise in a process called *isostatic rebound.* Neither uplift nor subsidence of the lithosphere will be instantaneous because flow in the asthenosphere is only a few centimeters per year. Imagine a waterbed filled with molasses. If you lie on it, you will sink slowly, because molasses is thicker than water, until you reach a floating equilibrium. When you get out of the bed, the depression that you made will slowly rise back up as the molasses fills in the space from below.

Isostasy suggests that mountains are made of relatively light crustal materials but exist in areas of very thick crust, while regions of low elevation have thin crust. Here the analogy is like that of an iceberg: A tall iceberg requires a massive amount of ice below the surface in order to expose ice so high above sea level, and as ice above the surface melts, ice from below will rise above sea level to replace it until the iceberg has completely melted.

Isostatic balance helps to explain many aspects of Earth's surface, including the following:

- Why most of the continental crust lies above sea level
- Why wide areas of the seafloor are at a uniform depth
- Why many mountain ranges continue to rise even though erosion removes material from them
- Why some regions where rivers are depositing great amounts of sediments are subsiding
- Why the crust subsided in areas that were covered by 2000 to 3000 meters of ice during the last glacial age and now continues to rebound after deglaciation

The density of ice is 90% water, thus icebergs (and ice cubes) float with nine tenths of their volume below the surface and 10% above.

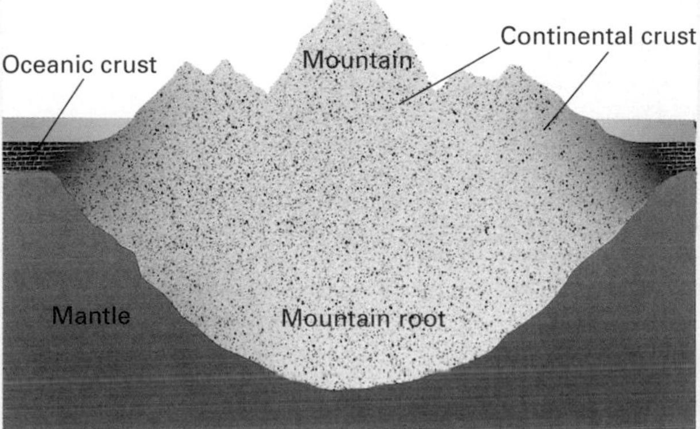

Because continental crust is considerably less dense than the material in the asthenosphere, where continental crust reaches high elevations it also extends far below the surface. Oceanic crust is also less dense than mantle material, but because it is denser than continental crust, it is thinner than continental crust.

known as **microplate terranes** (a term that should not be confused with the term *terrain*), as it moved westward over the Pacific and former oceanic plates. Paleomagnetic data show that parts of western North America from Alaska to California originated south of the equator and moved to join the continent. Terranes, which have their own distinct geology from that of the continent to which they are now joined, may have originally been offshore island arcs, undersea volcanoes, or islands made of continental fragments, such as New Zealand or Madagascar are today.

Paleogeography

The study of past geographical environments is known as **paleogeography.** The goal of paleogeography is to try to reconstruct the past environment of a geographical region based on geologic and climatic evidence. For students of physical geography, it generally seems that the present is complex enough without trying to know what the geography of ancient times was like. However, peering into the past helps us forecast and prepare for changes in the future.

The immensity of geologic time over which major events or processes (such as plate tectonics, ice ages, or the formation and erosion of mountain ranges) have taken place is difficult to picture in a human time frame of days, months, and years. The geologic timescale is a calendar of Earth history (Table 13.2). It is divided into *eras,* which are typically long units of time, such as the Mesozoic Era (which means "middle life"), and eras are divided into *periods,* such as the Cretaceous Period. *Epochs,* as for example the Pleistocene Epoch (recent ice ages), are shorter time units and are used to subdivide the periods of the Cenozoic Era ("recent life"), for which geologic evidence is more abundant. Today we are in the Holocene Epoch (last 10,000 years), of the

TABLE 13.2
Geologic Timescale

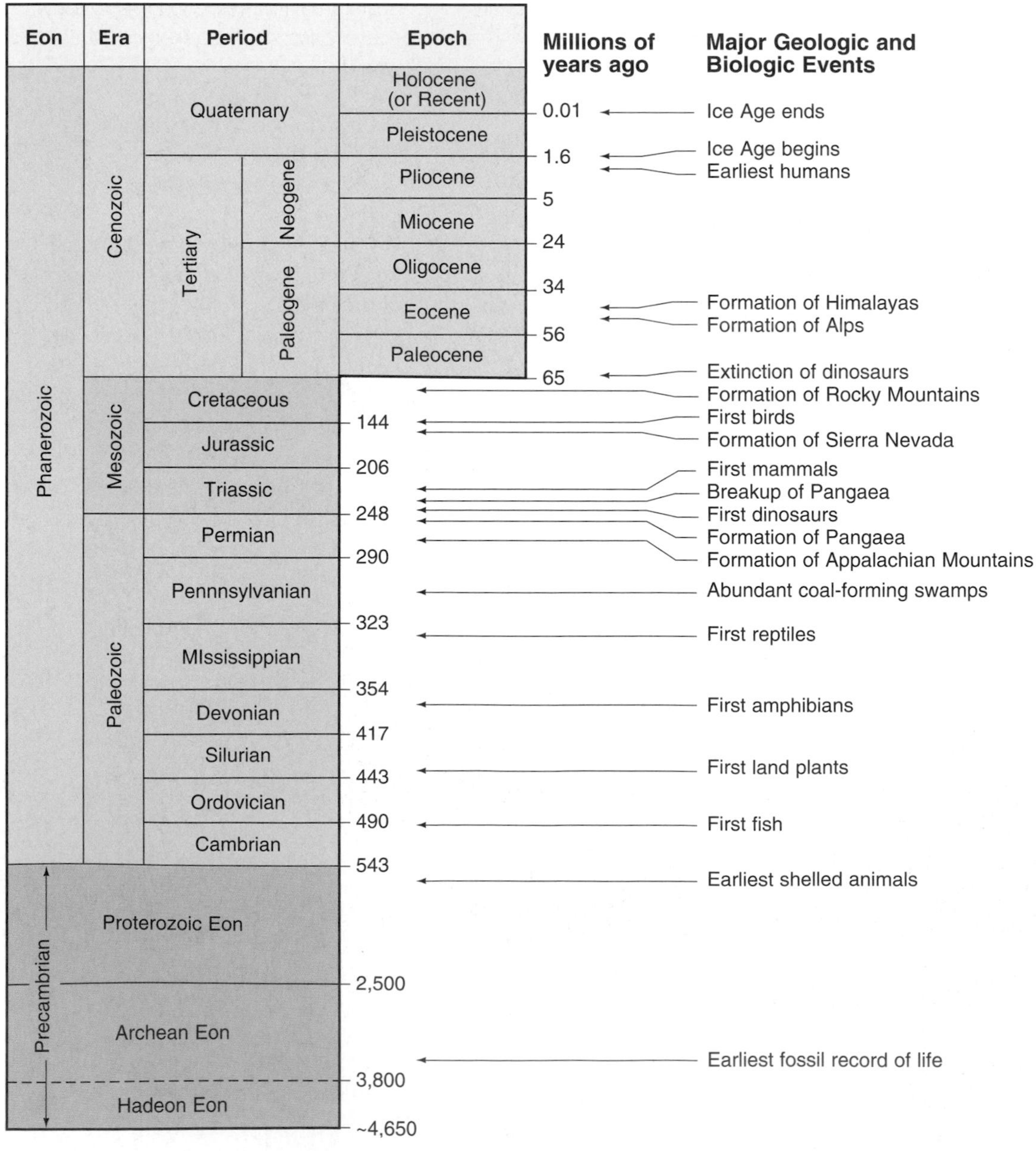

Eon	Era	Period		Epoch	Millions of years ago	Major Geologic and Biologic Events
Phanerozoic	Cenozoic	Quaternary		Holocene (or Recent)	0.01	Ice Age ends
				Pleistocene	1.6	Ice Age begins; Earliest humans
		Tertiary	Neogene	Pliocene	5	
				Miocene	24	
			Paleogene	Oligocene	34	
				Eocene	56	Formation of Himalayas; Formation of Alps
				Paleocene	65	Extinction of dinosaurs
	Mesozoic	Cretaceous			144	Formation of Rocky Mountains; First birds
		Jurassic			206	Formation of Sierra Nevada
		Triassic			248	First mammals; Breakup of Pangaea; First dinosaurs
	Paleozoic	Permian			290	Formation of Pangaea; Formation of Appalachian Mountains
		Pennnsylvanian			323	Abundant coal-forming swamps
		MIssissippian			354	First reptiles
		Devonian			417	First amphibians
		Silurian			443	First land plants
		Ordovician			490	First fish
		Cambrian			543	
Precambrian: Proterozoic Eon					2,500	Earliest shelled animals
Precambrian: Archean Eon					3,800	Earliest fossil record of life
Precambrian: Hadeon Eon					~4,650	

Quaternary Period (last 1.6 million years), of the Cenozoic Era (last 65 million years). In a sense, these divisions are used like we would use days, months, and years to record time.

If we took a 24-hour day to represent the approximately 4.6 billion-year history of Earth, the Precambrian, an era of which we know very little, would consume the first 21 hours. The current period, the Quaternary, which has lasted about 1.6 million years, would take less than 30 seconds, and human beginnings, over about the last 4 million years, about 1 minute.

Each era, period, and epoch in Earth's geologic history had a unique paleogeography with its own distribution of land and sea, climate regions, plants, and animal life. If we look at evidence for the paleogeography of the Mesozoic Era (245 million to 65 million years ago), for instance, we would find a much different physical geography than exists now. This was a time when the supercontinents, Gondwana and Laurasia, each gradually split apart as new ocean floors widened, creating the continents that are familiar to us today. Global and local Mesozoic climates were very different from those of today but were changing as North America drifted to the northwest. During the Cretaceous Period, much of the present United States experienced warmer climates than now. Ferns and conifer forests were common. The Mesozoic was the "age of the dinosaurs," a class of large animals that ruled the land and the sea. Other life also thrived, including marine plants and invertebrates, insects, mammals, and the earliest birds.

The Mesozoic Era ended with an episode of great extinctions, including the end of the dinosaurs. Geologists, paleontologists, and paleogeographers are not in agreement as to what caused these great extinctions. Some of the strongest evidence points to a large meteorite striking Earth 65 million years ago, disrupting global climate and causing global environmental change. Other evidence points to plate tectonic changes in the distribution of oceans and continents or increased volcanic activity, either of which could cause rapid climate changes that might possibly trigger mass extinctions.

Our maps of Earth in early geologic times show only approximate and generalized patterns of mountains, plains, coasts, and oceans, with the addition of some environmental characteristics. These maps portray a general picture of how global geography has changed through geologic time (● Fig. 13.37). Much of

● FIGURE 13.37

Paleomaps showing Earth's tectonic history over the last 250 million years of geologic time. A preponderance of evidence from paleomagnetism, ages and distributions of rocks and fossils, patterns of earthquakes and volcanoes, configurations of landmasses and mountain ranges, and studies of the ocean floor supports plate tectonics. These lines of evidence make it possible to produce a generalized historic sequence of how Earth's global geography has changed over that time frame.

How has the environment at the location where you live changed through geologic time?

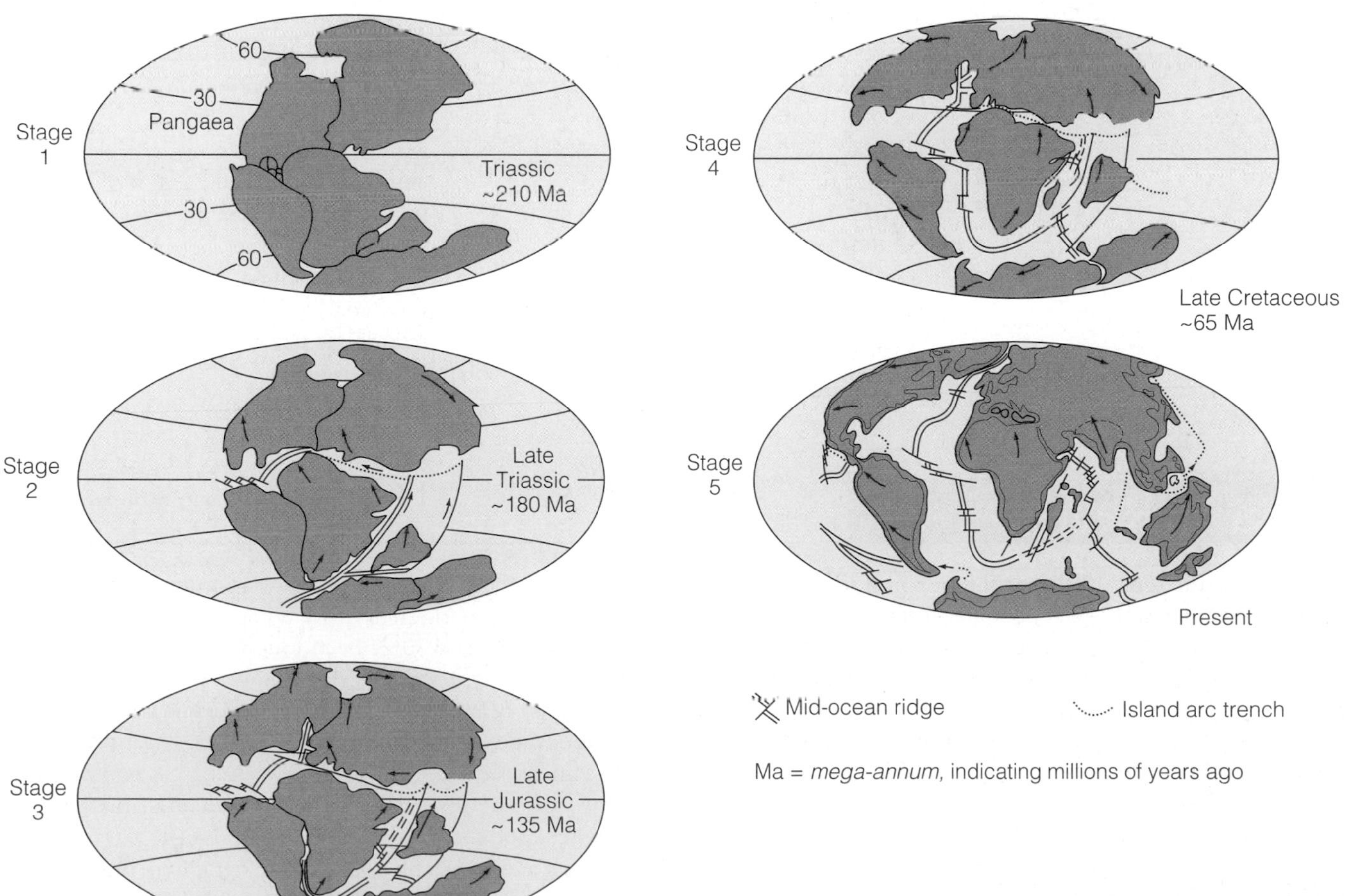

the evidence and the rocks that bear this information have been lost through metamorphism or erosion, buried under younger sediments or lava flows, or recycled into Earth's interior. The further back in time, the sketchier is the paleoenvironmental information presented on the map. Paleomaps, like other maps, are simplified models of the regions and times they represent.

As time passes and additional evidence is collected, paleogeographers may be able to fill in more of the empty spaces on those maps of the past that are so unfamiliar to us. These paleogeographic studies aim not only at understanding the past but also at understanding today's environments and physical landscapes, how they have developed, and how processes act to change them. By applying the concept of uniformitarianism and the theory of plate tectonics to our knowledge of how the Earth system and its subsystems function, we can gain a better understanding of our planet's geologic past, as well as its present, and this will facilitate better forecasts of its potential future.

Chapter 13 Activities

Define & Recall

seismic waves
seismograph
core
inner core
outer core
mantle
plastic solid
elastic solid
lithosphere (as an element of planetary structure)
asthenosphere
tectonic forces
Mohovičić discontinuity (Moho)
crust
oceanic crust
continental crust
mineral
rock
silicate
bedrock
regolith
outcrop
igneous rock
magma
lava
extrusive igneous rock
pyroclastics
intrusive igneous rock
plutonic rock
joint
columnar joint
sedimentary rocks
clasts
clastic sedimentary rock
organic sedimentary rock
chemical sedimentary rock
stratification
strata
bedding planes
unconformity
cross bedding
metamorphic rock
foliations
rock cycle
catastrophism
uniformitarianism
continental drift
Pangaea
paleomagnetism
plate tectonics
lithospheric plate
seafloor spreading
polarity reversal
convection
plate divergence
plate convergence
subduction
island arc
continental collision
transform movement
hot spot
continental shield
accretion
microplate terrane
paleogeography

Discuss & Review

1. Identify the major zones of Earth's interior from the center to the surface. How do these zones differ from one another?
2. Define and distinguish continental crust and oceanic crust. Define and distinguish the lithosphere from the asthenosphere.
3. List the eight most common elements in Earth's crust. What is a mineral? What is a rock?
4. Describe the three major categories of rock and the principal means by which each is formed. Give an example of each.
5. What is the rock cycle?
6. What evidence did Wegener rely on in the formulation of his theory of continental drift? What evidence did he lack? What evidence has since been found to support the theory that landmasses at Earth's surface move around?
7. What type of lithospheric plate boundary is found paralleling the Andes, at the San Andreas Fault, in Iceland, and near the Himalayas?
8. How does the formation of the Hawaiian Islands support plate tectonic theory?
9. Define paleogeography. Why are geographers interested in this topic?

Consider & Respond

1. Explain why the eastern United States has relatively little tectonic activity compared to the western United States.
2. How is the concept of uniformitarianism related to plate tectonics theory and the arrangement of the continents today? How is it related to the geographical distribution of tectonically active regions?
3. List four of the major plate boundary types and describe where an example of each occurs today.

Apply & Learn

1. Two plates are diverging and both are moving at a rate of 3 centimeters per year. How long will it take for them to move 1000 miles apart?
2. An area of oceanic crust has a density of 3.0 grams per cubic centimeter and a thickness of 4 kilometers. An area of continental crust has a density of 2.7 grams per cubic centimeter. How thick in total mass would the continental crust have to be to equal a column of oceanic crust?

Volcanic and Tectonic Processes and Landforms

14

CHAPTER PREVIEW

Earth's surface topography results from the interaction of internal and external processes that act to either increase or decrease relief.

- What are the two major processes that increase relief?
- What are the major processes that work to reduce relief?
- What are some of the relationships that might exist between the internal and external processes where mountains exist?

Intrusive igneous rock masses of various shapes and sizes form when magma cools slowly deep beneath Earth's surface.

- How can such rock masses come to influence Earth's surface topography?
- In what different ways might the various shapes and sizes of intrusive rock masses influence surface topography?

Many natural processes that build and shape Earth's surface operate in a manner that is not predictable, steady, and continuous but rather unpredictable, variable, and episodic.

- In what ways are the various processes that increase the elevation of Earth's surface unpredictable, variable, and episodic?
- How do these processes affect humans?
- What role does scientific understanding of these natural processes play in our coping with them?

The locations of tectonic plate boundaries and the natural hazards associated with these regions have a direct correlation with the spatial distribution of some of Earth's great "hazard zones."

- In what ways are volcanism and faulting involved in this relationship?
- What can humans do to mitigate these processes?

The degree of influence of rock structure on the surface landforms and landscapes varies widely.

- How does the relative interaction between internal and external processes affect this degree of influence?
- How does rock structure differ from topography?

◄ Opposite: Basaltic lava flows from a volcano on the Island of Hawaii. The lava has solidified and turned dark in color at the surface, but the lava still flowing in the channel glows red, yellow, and orange.
USGS/HVO

Our planet's surface **topography,** the distribution of landscape highs and lows, is intriguing and complex. Landscapes may consist of rugged mountains, gently sloping plains, rolling hills and valleys, or elevated plateaus cut by steep canyons. These are just a few examples of the varied types of surface terrain features, referred to as **landforms,** that contribute to the beauty and diversity of the environments on Earth. Landforms comprise one of the most appealing and impressive elements of Earth's surface. Local, state, and national parks attract millions of visitors annually seeking to observe and experience firsthand spectacular examples of landforms and associated environmental features. Landforms are the surface expression of the lithosphere and owe their development to processes and materials that originate within Earth's interior, at its surface, or, most typically, some combination of both.

Understanding landforms and landscapes—how they form, why they vary, and their significance in a local, regional, or global context—is the primary goal of **geomorphology,** a major subfield of physical geography devoted to the scientific study of landforms. Geomorphologists seek explanations for the shape, origin, spatial distribution, and development of terrain features of all kinds, including the processes that modify and destroy them. Variations in landforms result from interactions among the processes that elevate, depress, or disrupt Earth's surface, creating topographic inequalities, and the processes that wear down, fill in, and work to level the landscape.

Landforming **igneous processes** (from Latin: *ignis,* fire), which are related to the eruption and solidification of rock matter, and **tectonic processes** (from Greek: *tekton,* carpenter, builder), which are movements of parts of the crust and upper mantle, constitute the major geomorphic mechanisms that increase the topographic irregularities on Earth's surface. Although igneous and tectonic forces originate in the planet's interior, they contribute significantly to the nature of Earth's exterior topography, affecting the appearance, composition, shape, and size of surface terrain features in many regions. Areas of the crust can be built up by igneous processes, which include the ejection of volcanic rock matter from Earth's interior onto the surface, or can be uplifted or downdropped by tectonic processes. Igneous and tectonic processes build extensive mountain systems, but they also produce a great variety of other landforms. The geographical distribution of these terrain features, moreover, is not random. Volcanic landforms occur most commonly in association with lithospheric plate margins. Likewise, the greatest tectonic forces, largest mountain ranges, and dangerous earthquake zones lie along plate boundaries.

Igneous and tectonic processes have produced many impressively scenic landscapes, but they can also present serious natural hazards to people and their property. This chapter and those that follow focus on understanding how various landforms and landscapes develop and on the potential hazards that are related to those landforms and landscapes. It is extremely important for us to understand how geomorphic processes work to shape Earth's surface landforms because they are active, ongoing, and often powerful processes that can impact human welfare. Landforms are a dynamic, fundamental, beautiful, diverse, and sometimes dangerous aspect of the human habitat.

Landforms and Geomorphology

Landforms and landscapes are often described by their relative amount of **relief,** which is the difference in elevation between the highest and lowest points within a specified area or on a particular surface feature (• Figs. 14.1a and 14.1b). With no variations in relief our planet would be a smooth, featureless sphere and certainly much less interesting. It is hard to imagine Earth without dramatic terrain as seen in the high-relief mountainous regions of the Himalayas, Alps, Andes, Rockies, and Appalachians or in the huge chasm that we call the Grand Canyon. Interspersed with high-relief features, large expanses of low-relief features, like the Great Plains, can be equally impressive and inspiring.

Earth's landforms result from mechanisms that act to increase relief by raising or lowering the land surface and mechanisms that work to reduce relief by removing rock from high places and filling in depressions. In general, geomorphic processes that originate within Earth, called **endogenic processes** (*endo,* within; *genic,* originating), result in an increase in surface relief, while the **exogenic processes** (*exo,* external), those that originate at Earth's surface, tend to decrease relief. Igneous and tectonic processes constitute the endogenic geomorphic processes. Exogenic processes consist of various means of rock breakdown, collectively known as **weathering,** and the removal, movement, and relocation of those weathered rock products in the continuum of processes known as **erosion, transportation,** and **deposition.** Erosion, transportation, and deposition occur through the force of gravity alone, as in the fall of a weathered clast from a cliff to the ground below, or operate with the help of a **geomorphic agent,** a medium that picks up, moves, and eventually lays down broken rock matter. The most common geomorphic agents are flowing water, wind, moving ice, and waves, but people and other organisms can also accomplish some erosion, transportation, and deposition of weathered pieces of Earth material.

Mountains and other landforms of high relief are made predominantly of resistant rocks that undergo slow rates of weathering and erosion, or have formed so recently that there has not been enough time for the exogenic processes to have much of an impact. In some cases, recent uplift and resistant rocks are both involved. High-relief features, including mountains and hills, also exist where endogenic processes operate, or have operated, at faster rates than weathering and erosion (• Fig. 14.2). The exogenic process of deposition predominates in valleys and other areas of low relief that are being filled in with sediment eroded and transported from highlands.

Natural processes proceed at a wide variety of rates, usually expressed in millimeters or centimeters of change per year, century, or millennium, averaged over the time span of available data. However, it is important to remember that these are average rates, and many natural processes operate in an episodic, rather than in a steady and continuous, manner. In episodic processes, a period of relative calm and little change continues until a certain threshold condition is reached, resulting in a short burst of rapid, often intense activity that causes significant environmental modification. The exact timing of these periods of increased activity is difficult or impossible to predict with certainty. Many physical and life processes in the Earth system operate according to this **punctuated equilibrium** (• Fig. 14.3). Earthquakes and volcanic eruptions provide good examples of this type of Earth change. For example, major earthquake activity or volcanic eruption can dramatically alter a landscape that had previously been undergoing only slow change by exogenic processes. Long periods of limited activity are occasionally interrupted by large earthquakes or major eruptions that are short-lived but cause great change in the landscape. After an occurrence of major change, the landscape adjusts to the newly created conditions and stays in relative equilibrium, experiencing comparatively limited modification by exogenic processes until the next major event again disrupts the system (• Fig. 14.4). Exogenic processes may also operate episodically with extended periods of slow or moderate activity interrupted by short periods of rapid change, such as occurs during a stream flood or landslide.

(a) (b)

• FIGURE 14.1
An area displaying (a) low relief in western Utah lies about 100 kilometers (60 mi) east of (b) an area of high relief in Great Basin National Park of eastern Nevada.

• FIGURE 14.2
The Teton range of Wyoming stands high above the valley floor because rates of uplift due to endogenic processes exceed the rates of the exogenic processes of weathering and erosion, which operate to reduce their height.

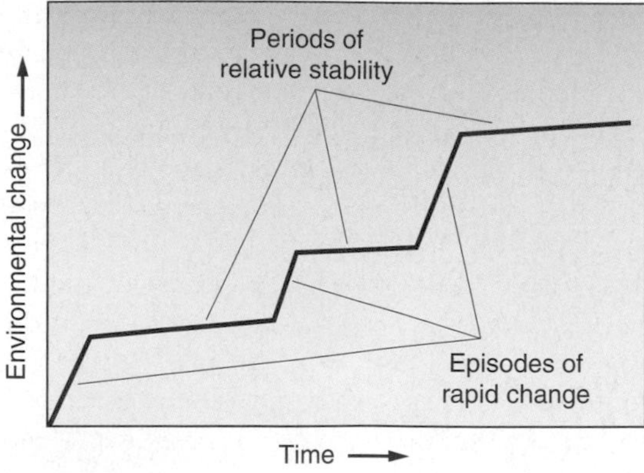

• FIGURE 14.3
Punctuated equilibrium describes Earth processes that operate slowly most of the time but that are occasionally punctuated by events that cause relatively major change in a landscape or natural system. For example, many small volcanic eruptions or earthquakes occur over time, resulting in little landscape change. Eventually a large eruption or earthquake may strike. This punctuates the long period of relative stability with a major landscape change that takes place over a short period of time. A great variety of processes that affect the Earth system operate in this episodic manner.
What Earth processes can you think of that operate this way?

In this chapter, we study the processes related to the buildup of relief through igneous and tectonic activity and examine the landforms and rock structures associated with these endogenic processes. Subsequent chapters focus on the exogenic processes and the landforms and landscapes formed by the various geomorphic agents.

Igneous Processes and Landforms

Landforms resulting from igneous processes may be related to eruptions of extrusive igneous rock material or emplacements of intrusive igneous rock. **Volcanism** refers to the extrusion of rock matter from Earth's subsurface to the exterior and the creation of surface terrain features as a result. **Volcanoes** are mountains or hills that form in this way. **Plutonism** refers to igneous processes that occur below Earth's surface including the cooling of magma to form intrusive igneous rocks and rock masses. Some masses of intrusive igneous rock are eventually exposed at Earth's surface where they comprise landforms of distinctive shapes and properties.

Volcanic Eruptions

Few spectacles in nature are as awesome as a volcanic eruption (• Fig. 14.5). Although large, violent eruptions tend to be infrequent events, they can devastate the surrounding environment and completely change the nearby terrain. Yet volcanic eruptions are natural processes and should not be unexpected by people who live in the vicinity of active volcanoes.

Eruptions can vary greatly in their size and character, and the volcanic landforms that result are extremely diverse. **Explosive eruptions** violently blast pieces of molten and solid rock into the air, whereas molten rock pours less violently onto the surface as flowing streams of lava in **effusive eruptions.** Variations in eruptive style and in the landforms produced by volcanism result mainly from temperature and chemical differences in the magma that feeds the eruption.

• FIGURE 14.4
(a) Mount Vesuvius overlooks the ancient city center of Pompeii, near Naples, Italy. The eruption of Vesuvius in AD 79, which destroyed Pompeii, is an example of an episodic process. It is often difficult for humans to fully comprehend the potential danger from Earth processes that operate with bursts of intense activity, separated by years, decades, centuries, or even millennia of relative quiescence. (b) A plaster cast shows a victim who attempted to cover his face from hot gases and the volcanic ash that buried Pompeii.

J. Petersen
(a)

J. Petersen
(b)

USGS/VHP/B. Chouet

• **FIGURE 14.5**
Few natural events are as spectacular as a volcanic eruption. This eruption of Italy's Stromboli volcano, on an island off Sicily, lit up the night sky.

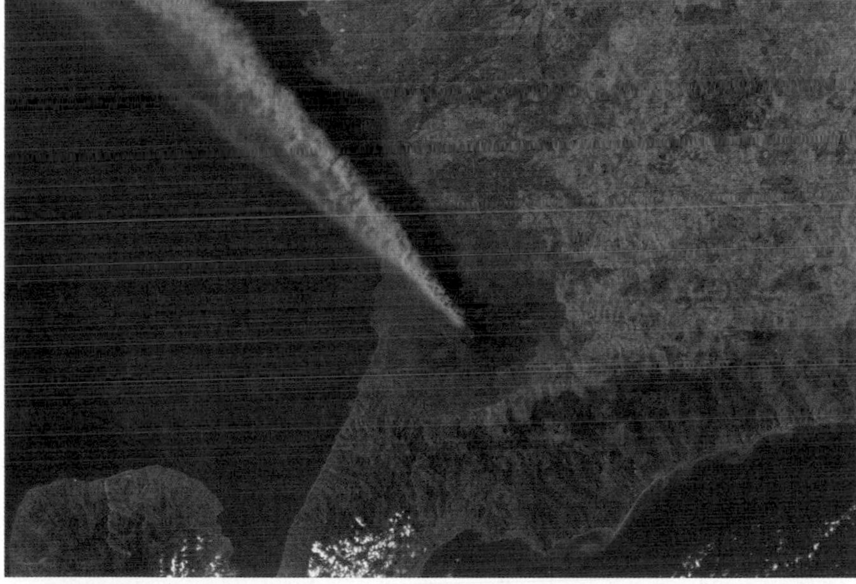

NASA Visible Earth

• **FIGURE 14.6**
Volcanic ash streaming to the southeast from Mount Etna on the Italian island of Sicily was captured on this photograph (south is at the top) taken from the International Space Station in July of 2001. The ash cloud reportedly reached a height of about 5200 meters (17,000 ft) on that day.
What do you think conditions were like at the time of this eruption for settlements located under the ash cloud?

The mineral composition that exists in a magma source is the most important factor determining the nature of a volcanic eruption. Silica-rich *felsic* magmas tend to be relatively cool in temperature while molten and have a viscous (thick, resistant to flowing) consistency. *Mafic* magmas are more likely to be extremely hot and less viscous, and thus flow readily in comparison to silica-rich magmas. Magmas contain large amounts of gases that remain dissolved when under high pressure at great depths. As molten rock rises closer to the surface, the pressure decreases, which tends to release expanding gases. If the gases trapped beneath the surface cannot be readily vented to the atmosphere or do not remain dissolved in the magma, explosive expansion of gases produces a violent, eruptive blast. Highly viscous, silica-rich magmas and lavas (rhyolitic in composition) have the potential to erupt with violent explosions. Mafic magmas and lavas, such as those with a basaltic composition, are hotter and less viscous (more fluid) and therefore tend to vent the gases more readily. When basaltic magma is forced to the surface, the resulting eruptions are usually effusive rather than explosive, and enormous amounts of fluid lava may be produced.

Molten material that is hotter, less viscous, and more mafic tends to erupt in the less violent effusive fashion, with streams of flowing lava. By contrast, the cooler, more viscous, silicic magma can produce explosive eruptions that hurl into the air molten material that solidifies in flight or on the surface or expel solid lava fragments of various sizes. These **pyroclastic materials** (from Greek: *pyros,* fire; *clastus,* broken), also referred to as **tephra,** vary in size from **volcanic ash,** which is sand-sized or smaller, to gravel-sized cinders (2–4 mm), lapilli (4–64 mm), and blocks (>64 mm). They may also include volcanic "bombs," which are large spindle shaped clasts. In the most explosive eruptions, clay and silt-sized volcanic ash may be hurled into the atmosphere to an altitude of 10,000 meters (32,800 ft) or more (• Fig. 14.6). The 1991 eruptions of Mount Pinatubo in the Philippines ejected a volcanic aerosol cloud that circled the globe. The suspended material caused spectacular reddish orange sunsets due to increased scattering and lowered global temperatures slightly for 3 years by increasing reflection of solar energy back to space.

Volcanic Landforms

The landforms that result from volcanic eruptions depend primarily on the explosiveness of the eruptions. We will consider six major kinds of volcanic landforms, beginning with those associated with the most effusive (least explosive) eruptions. Four of the six major landforms are types of volcanoes.

Lava Flows **Lava flows** are layers of erupted rock matter that when molten poured or oozed over the landscape. After they cool and solidify they retain the appearance of having flowed. Lava flows can form from any lava type (see Appendix C), but basalt is by far the most common because its hot eruptive temperature and low viscosity allow gases to escape, greatly reducing the potential for an explosive eruption. Basaltic lava flows may develop vertical fractures, called *joints,* due to shrinking of the lava during cooling. This creates *columnar-jointed* basalt flows (• Fig. 14.7).

Lava flows display variable surface characteristics. Extremely fluid lavas can flow rapidly and for long distances before solidifying. In this case, a thin surface layer of lava in contact with the atmosphere solidifies, while the molten lava beneath continues to move, carrying the thin, hardened crust along and wrinkling it into a ropy surface form called **pahoehoe.** Lavas of slightly greater viscosity flow more slowly, allowing a thicker surface layer to harden while the still-molten interior lava keeps on flowing. This causes the thick layer of hardened crust to break up into sharp-edged, jagged blocks, making a surface known as **aa.** The terms pahoehoe and aa both originated in Hawaii, where effusive eruptions of basalt are common (● Figs. 14.8a and 14.8b).

● FIGURE 14.7
Basalt shrinks when it cools and solidifies. Some basaltic lava flows acquire a network of vertical cracks, called joints, upon cooling in order to accommodate the shrinkage. Often, polygonal joint systems separate vertical columns of basaltic rock creating columnar-jointed basalt as in this basalt flow in west-central Utah.

D. Sack

Lava flows do not have to emanate directly from volcanoes, but can pour out of deep fractures in the crust, called **fissures,** that can be independent of mountains or hills of volcanic origin. In some continental locations, very fluid basaltic lava that erupted from fissures was able to travel up to 150 kilometers (93 mi) before solidifying. These very extensive flows are often called **flood basalts.** In some regions, multiple layers of basalt flows have constructed relatively flat-topped, but elevated, tablelands known as **basalt plateaus.** In the geologic past, huge amounts of basalt have poured out of fissures in some regions, eventually burying existing landscapes under thousands of meters of lava flows. The Columbia Plateau in Washington, Oregon, and Idaho, covering 520,000 square kilometers (200,000 sq mi), is a major example of a basaltic plateau (● Fig. 14.9), as is the Deccan Plateau in India.

Shield Volcanoes

When numerous successive basaltic lava flows occur in a given region they can eventually pile up into the shape of a large mountain, called a **shield volcano,** which resembles a giant knight's shield resting on Earth's

● FIGURE 14.8
Scientists use Hawaiian terminology to refer to the two major surface textures commonly found on lava flows. Although all lava flows have low viscosity, slight variations exist from one flow to another. (a) Very low viscosity lava forms a ropy surface, called pahoehoe. (b) Somewhat more viscous lava leaves a blocky surface texture, called aa.
In which direction relative to the photo did the pahoehoe flow?

D. Sack
(a)

D. Sack
(b)

surface (● Fig. 14.10a). The gently sloping, dome-shaped cones of Hawaii best illustrate this largest type of volcano (● Fig. 14.11). Shield volcanoes erupt extremely hot, mafic lava with temperatures of more than 1090°C (2000°F). Escape of gases and steam may hurl fountains of molten lava a few hundred meters into the air, with some buildup of cinders (fragments or lava clots that congeal in the air), but the major feature is the outpouring of fluid basaltic lava flows (● Fig. 14.12). Compared to other volcano types, these eruptions are not very explosive, although still potentially damaging and dangerous. The extremely hot and fluid basalt can flow long distances before solidifying, and the accumulation of flow layers develops broad, dome-shaped volcanoes with very gentle slopes. On the island of Hawaii, active shield volcanoes also erupt lava from fissures on their flanks so that living on the island's edges, away from the summit craters, does not guarantee safety from volcanic hazards. Neighborhoods in Hawaii have been destroyed or threatened by lava flows. The Hawaiian shield volcanoes form the largest volcanoes on Earth in terms of both their height—beginning at the ocean floor—and diameter.

● FIGURE 14.9
River erosion has cut a deep canyon to expose the uppermost layers of basalt in the Columbia Plateau flood basalts in southwestern Idaho.

©Jef Gnass

Cinder Cones The smallest type of volcano, typically only a couple of hundred meters high, is known as a cinder cone. **Cinder cones** generally consist largely of gravel-sized pyroclastics. Gas-charged eruptions throw molten lava and solid pyroclastic fragments into the air. Falling under the influence of gravity, these particles accumulate around the almost pipelike conduit for the eruption, the *vent,* in a large pile of tephra (Fig. 14.10b). Each eruptive burst ejects more pyroclastics that fall and cascade down the sides to build an internally layered volcanic cone. Cinder cone volcanoes typically have a rhyolitic composition, but can be made of basalt if conditions of temperature and viscosity keep gases from escaping easily. The form of a cinder cone is

● FIGURE 14.10
The four basic types of volcanoes are: (a) shield volcano, (b) cinder cone, (c) composite cone, also known as stratovolcano, and (d) plug dome.
What are the key differences in their shapes? What properties are alike or different in their internal structure?

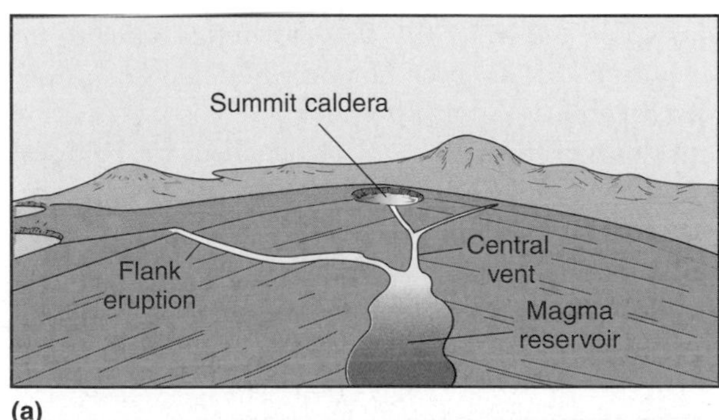

(a)

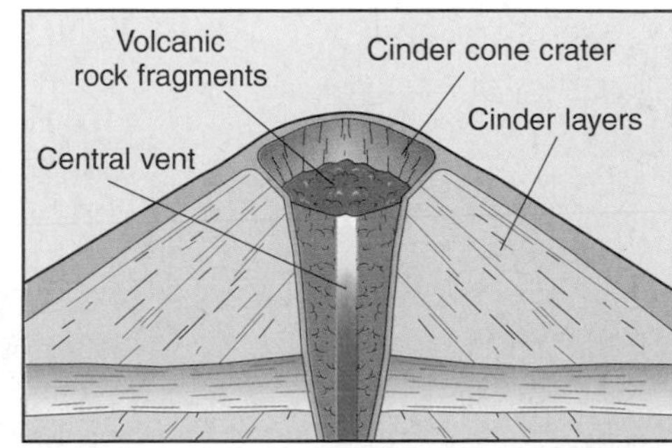

(b)

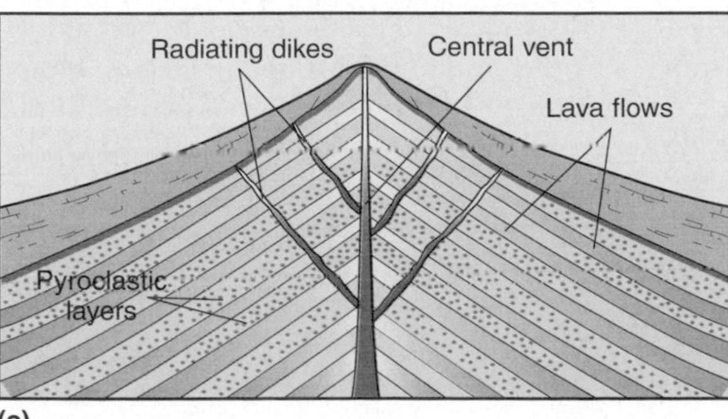

(c)

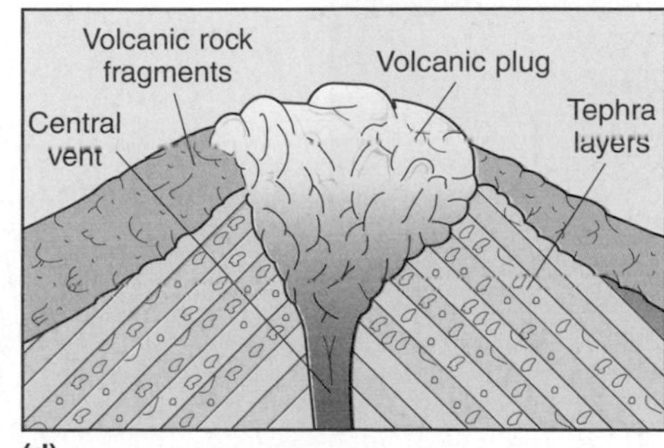

(d)

D. Sack

• **FIGURE 14.11**
Mauna Loa, on the island of Hawaii, is the largest volcano on Earth and clearly displays the dome, or convex, shape of a classic shield volcano. Mauna Loa reaches to 4170 meters (13,681 ft) above sea level, but its base lies far beneath sea level, creating almost 17 kilometers (56,000 ft) of relief from base to summit.
Why do Hawaiian volcanoes erupt less explosively than volcanoes of the Cascades or Andes?

J. D. Griggs/USGS

• **FIGURE 14.12**
This fountain of lava in Hawaii reached a height of 300 meters (1000 ft).

very distinctive, with steep straight sides and a large crater in the center, given the size of the volcano (• Fig. 14.13). Cinder cone examples include several in the Craters of the Moon area in Idaho, Capulin Mountain in New Mexico, and Sunset Crater, Arizona. In 1943 a remarkable cinder cone called Paricutín grew from a fissure in a Mexican cornfield to a height of 92 meters (300 ft) in 5 days and to more than 360 meters (1200 ft) in a year. Eventually, the volcano began erupting basaltic lava flows, which buried a nearby village except for the top of a church steeple.

Composite Cones A third kind of volcano, a **composite cone,** results when formative eruptions are sometimes effusive and sometimes explosive. Composite cones are therefore composed of a combination, that is, they represent a composite of lava flows and pyroclastic materials (Fig. 14.10c). They are also called *stratovolcanoes* because they are constructed of layers (strata) of pyroclastics and lava. The topographic profile of a composite cone represents what many people might consider the classic volcano shape, with concave slopes that are gentle near the base and steep near the top (• Fig. 14.14). Composite volcanoes form from andesite, which is a volcanic rock intermediate in silica content and explosiveness between basalt and rhyolite. Although andesite is only intermediate in these characteristics, composite, cones are dangerous. As a composite cone grows larger, the vent eventually becomes plugged with unerupted andesitic rock. When this happens, the pressure driving an eruption can build to the point where either the plug is explosively forced out or the mountain side is pushed outward until it fails, allowing the great accumulation of pressure to be relieved in a lateral explosion. Such explosive eruptions may be accompanied by **pyroclastic flows,** fast-moving density currents of airborne volcanic ash, hot gases, and steam that flow downslope close to the ground like avalanches. The speed of a pyroclastic flow can reach 100 kilometers per hour (62 mi/hr) or more.

Most of the world's best-known volcanoes are composite cones. Some examples include Fujiyama in Japan, Cotopaxi in Ecuador, Vesuvius and Etna in Italy, Mount Rainier in Washington, and Mount Shasta in California. The highest volcano on Earth, Nevados Ojos del Salado, is an andesitic composite cone that reaches an elevation of 6887 meters (22,595 ft) on the border between Chile and Argentina in the Andes, the mountain range after which andesite was named.

On May 18, 1980, residents of the American Pacific Northwest were stunned by the eruption of Mount St. Helens. Mount St. Helens, a composite cone in southwestern Washington that had been venting steam and ash for several weeks, exploded with incredible force on that day. A menacing bulge had been growing on the side of Mount St. Helens, and Earth scientists warned of a possible major eruption, but no one could forecast the magnitude or the exact timing of the blast. Within minutes, nearly 400 meters (1300 ft) of the mountain's north summit had disappeared by being blasted into the sky and down the

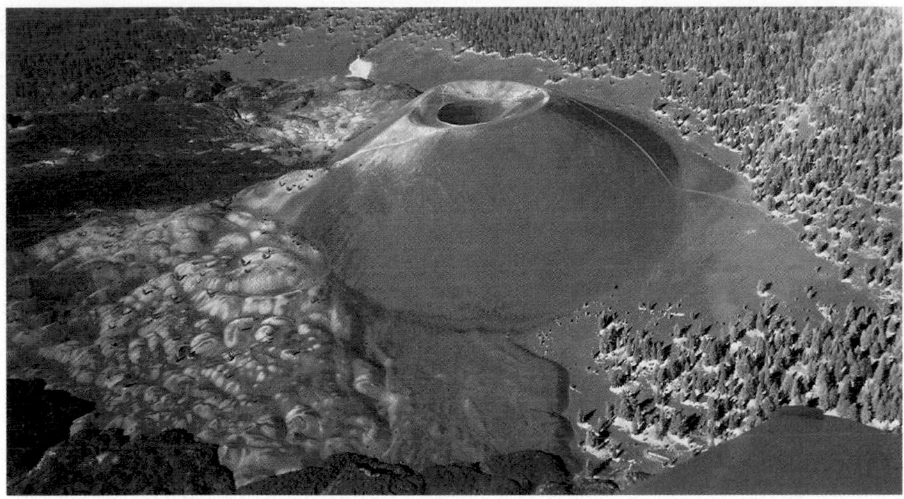
D. R. Crandell/USGS

• **FIGURE 14.13**
Cinder cones grow as volcanic fragments (pyroclastics) ejected during gas-charged eruptions pile up around the eruptive vent. Here, a cinder cone stands among lava flows in Lassen Volcanic National Park, California.
Why is the crater so prominent on this volcano?

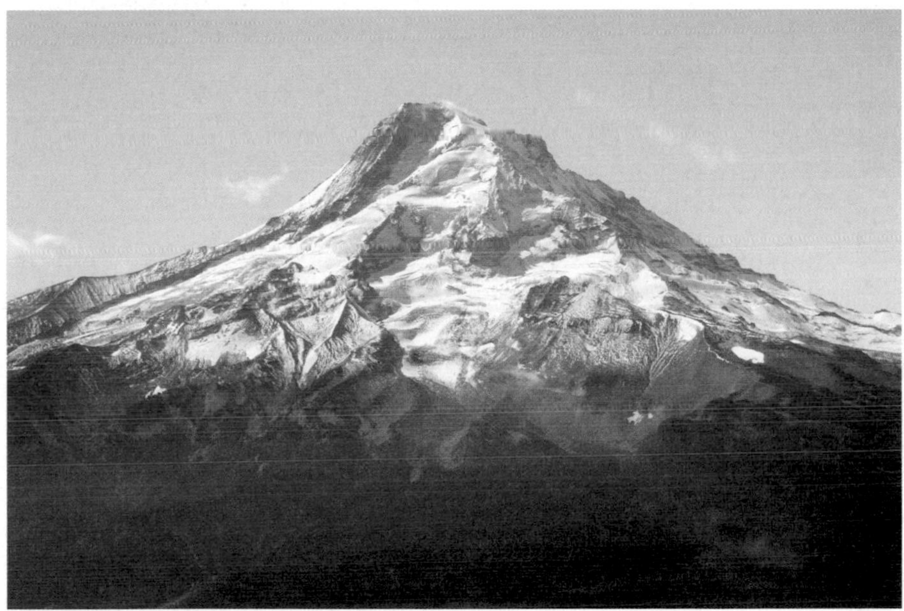
USGS/CVO Oregon

• **FIGURE 14.14**
Composite cones are composed of both lava flows and pyroclastic material and have distinctive concave side slopes. Oregon's Mount Hood is a composite cone in the Cascade Range.
Along what type of lithospheric plate boundary is this volcano located?

mountainside (• Fig. 14.15). Unlike most volcanic eruptions, in which the eruptive force is directed vertically, much of the explosion blew pyroclastic debris laterally outward from the site of the bulge. An eruptive blast composed of an intensely hot cloud of steam, noxious gases, and volcanic ash burst outward at more than 320 kilometers per hour (200 mi/hr), obliterating forests, lakes, streams, and campsites for nearly 32 kilometers (20 mi). Volcanic ash and water from melted snow and ice formed huge mudflows that choked streams, buried valleys, and engulfed everything in their paths. More than 500 square kilometers (193 sq mi) of forests and recreational lands were destroyed. Hundreds of homes were buried or badly damaged. Choking ash several centimeters thick covered nearby cities, untold numbers of wildlife were killed, and more than 60 people lost their lives in the eruption. It was a minor event in Earth's history but a sharp reminder to the region's residents of the awesome power of natural forces.

Some of the worst natural disasters in history have occurred in the shadows of composite cones. Mount Vesuvius, in Italy, killed more than 20,000 people in the cities of Pompeii and Herculaneum in AD 79. Mount Etna, on the Italian island of Sicily, destroyed 14 cities in 1669, killing more than 20,000 people. Today, Mount Etna is active much of the time. The greatest volcanic eruption in recent history was the explosion of Krakatoa in the Dutch East Indies (now Indonesia) in 1883. The explosive eruption killed more than 36,000 people, many as a result of the subsequent *tsunamis,* large sets of ocean waves generated by a sudden offset of the water, that swept the coasts of Java and Sumatra. In 1985 the Andean composite cone Nevado del Ruiz, in the center of Colombia's coffee-growing region, erupted and melted its snowcap, sending torrents of mud and debris down its slopes to bury cities and villages, resulting in a death toll in excess of 23,000. The 1991 eruption of Mount Pinatubo in the Philippines killed more than 300 people and airborne ash caused climatic effects for 3 years following the eruption. In 1997 a series of violent eruptions from the Soufriere Hills volcano destroyed more than half of the Caribbean island of Montserrat with volcanic ash and pyroclastic flows (• Fig. 14.16). In recent years, Mexico City, one of the world's most populous urban areas, has been threatened by continued eruptions of Popocatepetl, a large, active composite cone that is 70 kilometers (45 mi) away. At this distance, ash falls from a major eruption would be the most severe hazard to be expected. Volcanic ash is much like tiny slivers of glass. It can cause breathing problems in people and other organisms. Vehicles stall when ash chokes the air intakes of combustion engines. In addition, the heavy weight of significant ash accumulations on roofs can cause buildings to collapse.

Plug Domes Where extremely viscous silica-rich magma has pushed up into the vent of a volcanic cone without flowing beyond it, it creates a **plug dome** (Fig. 14.10d). Solidified outer parts of the blockage create the dome-shaped summit, and jagged blocks that broke away from the plug or preexisting parts of the cone form the steep, sloping sides of the volcano.

USDA Forest Service

(a)

USGS/J. Rosenbaum

(b)

USGS/Lyn Topinka

(c)

• FIGURE 14.15

Mount St. Helens, Washington, in the Cascade Range of the Pacific Northwest, illustrates the massive change that a composite volcano can undergo in a short period of time. (a) Prior to the 1980 eruption, Mount St. Helens towered majestically over Spirit Lake in the foreground. (b) On May 18, 1980, at 8:32 a.m., Mount St. Helens erupted violently. The massive landslide and blast removed more than 4.2 cubic kilometers (1 cu mi) of material from the mountain's north slope, leaving a crater more than 400 meters (1300 ft) deep. The blast cloud and monstrous mudflows destroyed the surrounding forests and lakes and took 60 human lives. (c) Two years after the 1980 eruption, the volcano continued to spew much smaller amounts of gas, steam, and ash. Mount St. Helens is currently experiencing a phase of eruptive activity that began in fall of 2004. **Could other volcanoes in the Cascade Range, such as Oregon's Mount Hood, erupt with the kind of violence that Mount St. Helens displayed in 1980?**

Great pressures can build up causing more blocks to break off, and creating the potential for extremely violent explosive eruptions, including pyroclastic flows. In 1903 Mount Pelée, a plug dome on the French West Indies island of Martinique, caused the deaths in a single blast of all but one person from a town of 30,000. Lassen Peak in California is a large plug dome that has been active in the last 100 years (• Fig. 14.17). Other plug domes exist in Japan, Guatemala, the Caribbean, and the Aleutian Islands.

• FIGURE 14.16

Beginning in 1995 the Caribbean island of Montserrat was struck by a series of volcanic eruptions, including pyroclastic flows, that devastated much of the island. The town of Plymouth, shown here, has been completely abandoned because of the amount of destruction and threat of future eruptions. Prior to the 1995 disaster, the volcano had not erupted for 400 years.

R.P. Hoblitt/USGS Volcano Hazards Program

• FIGURE 14.17

Plug dome volcanoes extrude stiff silica-rich lava and have steep slopes. Lassen Peak, located in northern California, is a plug dome and the southernmost volcano in the Cascade Range. The lava plugs are the darker areas protruding from the volcanic peak. Lassen was last active between 1914 and 1921.

Why are plug dome volcanoes considered dangerous?

USGS

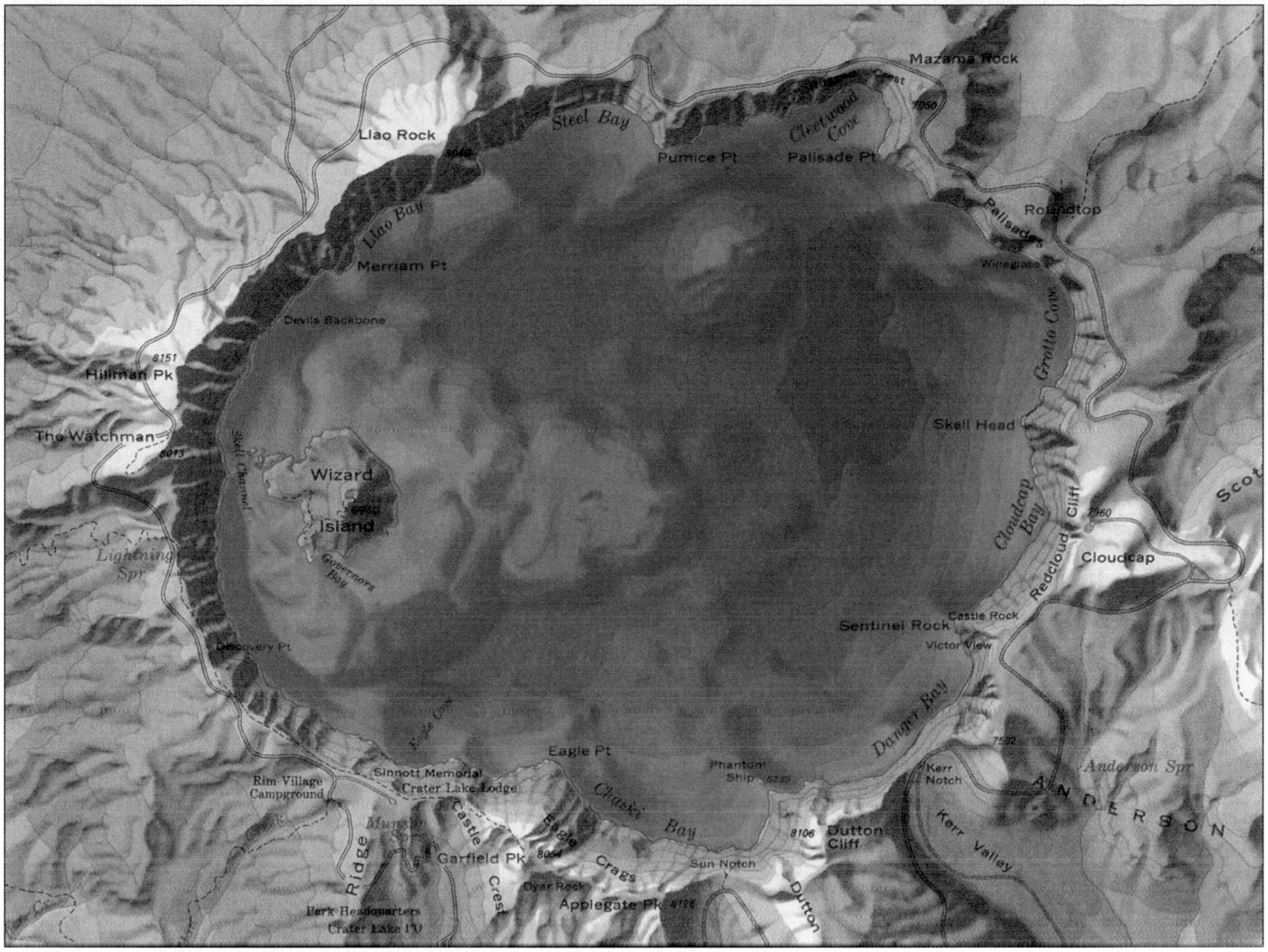

(a)

National Park Service

(b)

• FIGURE 14.18

(a) Crater Lake, Oregon, is the best-known caldera in North America. It developed when a violent eruption of Mount Mazama about 6000 years ago blasted out solid and molten rock matter, leaving behind a deep crater. (b) In the humid climate of south-central Oregon, water has accumulated in the crater, creating the 610-meter-deep (2000 ft) Crater Lake. Wizard Island is a later, secondary volcano that has risen within the caldera.

Could other Cascade volcanoes erupt to the point of destroying the volcano summit and leaving a caldera?

Calderas Occasionally, the eruption of a volcano expels so much material and relieves so much pressure within the magma chamber that only a large and deep depression remains in the area that previously contained the volcano's summit. A large depression made in this way is termed a **caldera.** The best-known caldera in North America is the basin in south-central Oregon that contains Crater Lake, a circular body of water 10 kilometers (6 mi) across and almost 610 meters (2000 ft) deep, surrounded by near-vertical cliffs as much as 610 meters (2000 ft) high. The caldera that contains Crater Lake was formed by the prehistoric eruption and collapse of a composite volcano. A cinder cone, Wizard Island, has built up from the floor of the caldera and rises above the lake's surface (● Fig. 14.18). The area of Yellowstone National Park is the site of three ancient calderas, and the Valles Caldera in New Mexico is another excellent example. Krakatoa in Indonesia and Santorini (Thera) in Greece have left island remnants of their calderas. Calderas are also found in the Philippines, the Azores, Japan, Nicaragua, Tanzania, and Italy, many of them occupied by deep lakes.

Plutonism and Intrusions

Bodies of magma that exist beneath the surface of Earth or masses of intrusive igneous rock that cooled and solidified beneath the surface are called **igneous intrusions,** or *plutons*. A great variety of shapes and sizes of magma bodies can result from intrusive igneous activity, also called *plutonism*. When they are first formed, smaller plutons have little or no effect on the surface terrain. Larger plutons, however, may be associated with uplift of the land surface under which they are intruded.

The many different kinds of intrusions are classified by their size, shape, and relationship to the surrounding rocks (• Fig. 14.19). After millions of years of uplift and erosion of overlying rocks, even small intrusions may be located at the surface to become part of the landscape. Uplifted plutons composed of granite or other intrusive igneous rocks that are eventually exposed at the surface tend to stand higher than the landscape around them because their resistance to weathering and erosion exceeds that of many other kinds of rocks.

When exposed at Earth's surface, a relatively small, irregularly shaped intrusion is called a **stock.** A stock is usually limited in area to less than 100 square kilometers (40 sq mi). The largest intrusions, called **batholiths** when visible at the surface, are larger than 100 square kilometers and are complex masses of solidified magma, usually granite. Batholiths represent large plutons that melted, metamorphosed, or pushed aside other rocks as they developed kilometers beneath Earth's surface. Batholiths vary in size; some are as much as several hundred kilometers across and thousands of meters thick. They form the core of many major mountain ranges primarily because older covering rocks were eroded away, leaving the more resistant intrusive igneous rocks that comprise the batholith. The Sierra Nevada, Idaho, Rocky Mountain, Coast, and Baja California batholiths cover areas of hundreds of thousands of square kilometers of granite landscapes in western North America.

Magma can create other kinds of igneous intrusions by forcing its way into fractures and between rock layers without melting the surrounding rock. A **laccolith** develops when molten magma flows horizontally between rock layers, bulging the overlying layers upward, making a solidified mushroom-shaped structure. Laccoliths have a mushroomlike shape because they are usually connected to a magma source by a pipe or stem. The resulting uplift on Earth's surface is like a giant blister, with magma beneath the overlying layers comparable to the fluid beneath the skin of a blister. Laccoliths are generally much smaller than batholiths, but both can form the core of mountains or hills after erosion has worn away the overlying less resistant rocks. The La Sal, Abajo, and Henry Mountains in southern Utah are composed of exposed laccoliths, as are other mountains in the American West (• Fig. 14.20).

Smaller but no less interesting landforms created by intrusive activity may also be exposed at the surface by erosion of the overlying rocks. Magma can intrude between rock layers without bulging them upward, solidifying into a horizontal sheet of intrusive igneous rock called a **sill.** The Palisades, along New York's Hudson River, provide an example of a sill made of gabbro, the intrusive compositional equivalent of basalt (• Fig. 14.21). Molten rock under pressure may also intrude into a nonhorizontal fracture that cuts into the

• FIGURE 14.19
Igneous intrusions solidify below Earth's surface. Because intrusive igneous rocks tend to be more resistant to erosion than sedimentary rocks, when they are eventually exposed at the surface sills, dikes, laccoliths, stocks, and batholiths generally stand higher than the surrounding rocks. Irregular, pod-shaped plutons less than 100 square kilometers (40 sq mi) in area form stocks when exposed, while larger ones form extensive batholiths.

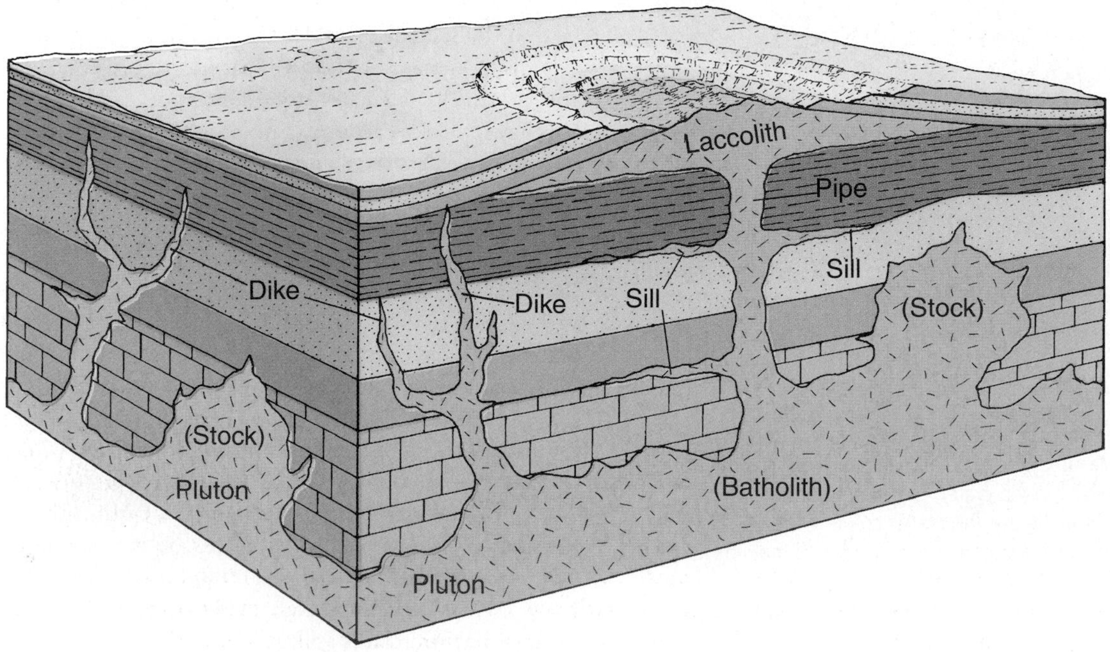

Copyright and photograph by Dr. Parvinder S. Sethi

• FIGURE 14.20
The La Sal Mountains in southern Utah, near Moab, are composed of a laccolith that was exposed at the surface by uplift and subsequent erosion of the overlying sedimentary rocks.
How do laccoliths deform the rocks they are intruded into?

Anthony G. Taranto Jr., Palisades Interstate Park – NJ Section

• FIGURE 14.21
Sills develop where magma intrudes between parallel layers of surrounding rocks. The Palisades of the Hudson River, the impressive cliffs found along the river's western bank in the vicinity of New York City, are made from a thick sill of igneous rock that was intruded between layers of sedimentary rocks.
Why does the sill at the Palisades form a cliff?

surrounding rocks. As it solidifies, the magma forms a wall-like structure of igneous rock known as a **dike.** When exposed by erosion, dikes often appear as vertical or near-vertical walls of resistant rock rising above the surrounding topography (• Fig. 14.22). At Shiprock, in New Mexico, resistant dikes many kilometers long rise vertically to more than 90 meters (300 ft) above the surrounding plateau (• Fig. 14.23). Shiprock is a **volcanic neck,** a tall rock spire made of the exposed (formerly subsurface) pipe that fed a long-extinct volcano situated above it about 30 million years ago. Erosion has removed the volcanic cone, exposing the resistant dikes and neck that were once internal features of the volcano at Shiprock.

Copyright and photograph by Dr. Parvinder S. Sethi

• FIGURE 14.22
The igneous rock of this exposed dike in New Mexico was intruded into a near-vertical fracture in weaker sandstone. Later much of the sandstone was eroded away, leaving the resistant dike exposed.
How does a dike differ from a sill? How are they alike?

Copyright and photograph by Dr. Parvinder S. Sethi

• FIGURE 14.23
Shiprock, New Mexico, is a volcanic neck exposed by erosion of surrounding rock. Volcanic necks are resistant remnants of the intrusive pipe of a volcano.
Why do you think this feature is called Shiprock?

Tectonic Forces, Rock Structure, and Landforms

Tectonic forces, which at the largest scale move the lithospheric plates, also cause bending, warping, folding, and fracturing of Earth's crust at continental, regional, and even local scales. Such deformation is documented by **rock structure,** the nature, orientation, inclination, and arrangement of affected rock

GEOGRAPHY'S SPATIAL SCIENCE PERSPECTIVE

Spatial Relationships between Plate Boundaries, Volcanoes, and Earthquakes

The geographic distributions of volcanism and earthquake activity are quite similar. Both tend to be concentrated in linear patterns along the boundaries of lithospheric plates. Although the locations of volcanic and earthquake activity correlate fairly well, there are exceptions, and their nature and severity differ from place to place. In general, the frequency and severity of volcanic eruptions or earthquakes vary according to their proximity to a specific type of lithospheric plate boundary or specific site in the central part of a plate.

Regardless of whether it breaks a continent or the seafloor, plate divergence creates fractures that provide avenues for molten rock to reach the surface. The divergent midoceanic ridges experience rather mild volcanic eruptions and small to moderate earthquakes that originate at a shallow depth. People are impacted when these volcanic and tectonic activities occur on islands associated with midocean ridges, such as the Azores and Iceland.

Volcanism also arises where continental crust is breaking and diverging. In these regions, earthquakes tend to be small to moderate, but continental crust mixed with mafic magma produces a wider variety of volcanic eruptions, some of which are potentially quite violent. Examples of resulting volcanoes in the East African rift valleys include Mount Kilimanjaro and Mount Kenya.

The potential severity of earthquakes and volcanic eruptions is much greater where plates are converging rather than diverging. Along the oceanic trenches where crustal rock material is subducted, volcanoes typically develop along the edge of the overriding plate. The largest region where this occurs is the "Pacific Ring of Fire," the volcanically active and earthquake-prone margin around the Pacific Ocean. Where oceanic crust subducts beneath continental crust along an oceanic trench, some of it melts into magma that moves upward under the continental crust. Subduction along the Pacific Ocean is associated with extensive volcanoes in the Andes, the Cascades, and the Aleutians; the Kuril Islands and the Kamchatka Peninsula in Russia; and Japan, the Philippines, New Guinea, Tonga, and New Zealand. Many of these volcanoes erupt rock and lava of andesitic composition and can be dangerously explosive. Earthquakes are also common events along the Pacific Rim. Although most are small to moderate, the largest earthquakes ever recorded have been related to subduction in this region. Points where earthquakes originate along an oceanic trench become deeper toward the overriding plate, indicating the subducting plate's progress downward toward where it is recycled into the mantle.

Another volcanic and seismic belt occupies the collision zone between northward-moving Southern Hemisphere lithospheric plates and the Eurasian plate. The volcanoes of the Mediterranean region, Turkey, Iran, and Indonesia are located along this collision zone. Seismic activity is common in that zone and has included some major, deadly earthquakes.

Transform plate boundaries, where lateral sliding occurs, also experience many earthquakes. The potential for major earthquakes mainly exists in places such as along the San Andreas Fault zone in California where thick continental crust is resistant to sliding easily. Volcanic activity along transform plate boundaries ranges from moderate on the seafloor to slight in continental locations.

Areas far from active plate boundaries are not necessarily immune from earthquakes and volcanism. The Hawaiian Islands, the Galapagos Islands, and the Yellowstone National Park area are examples of intraplate "hot spots" located away from plate margins and associated with a plume of magma rising from the mantle. Oceanic crustal areas that lie over hot spots, like the Hawaiian Islands, have strong volcanic activity and moderate earthquake activity. In midcontinental areas large earthquakes occur in suture zones where continents are colliding, such as in the Himalayas, or where broken edges of ancient landmasses shift even though they are today situated in midcontinent and are deeply buried by more recent rocks.

Volcanic and earthquake activities that are located away from active plate margins are intriguing and show that we still have much to learn about Earth's internal processes and their impact on the surface. Still, plate tectonics has contributed greatly to our understanding of the variations in volcanism, earthquake activity, and the landforms associated with these processes.

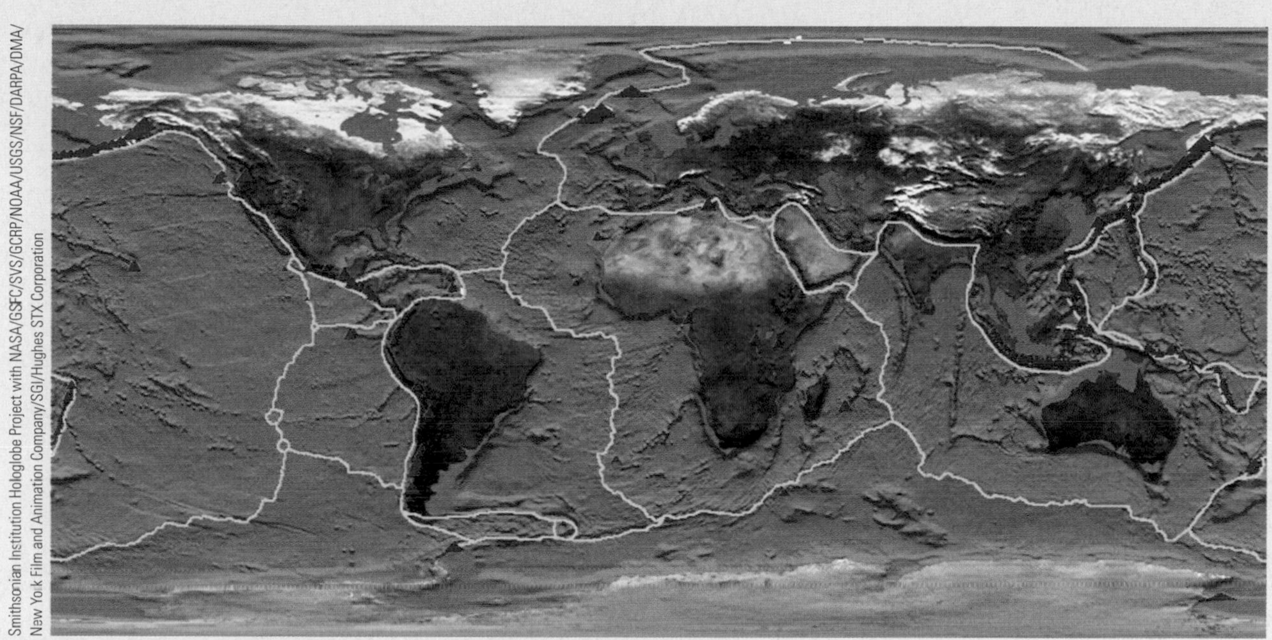
Smithsonian Institution Hologlobe Project with NASA/GSFC/SVS/GCRP/NOAA/USGS/NSF/DARPA/DMA/
New York Film and Animation Company/SGI/Hughes STX Corporation

The spatial correspondence among plate margins, active volcanoes, earthquake activity, and hot spots is not coincidental but is strongly related to lithospheric plate boundaries. This map shows plate boundaries and the global distribution of active volcanoes (1960–1994).

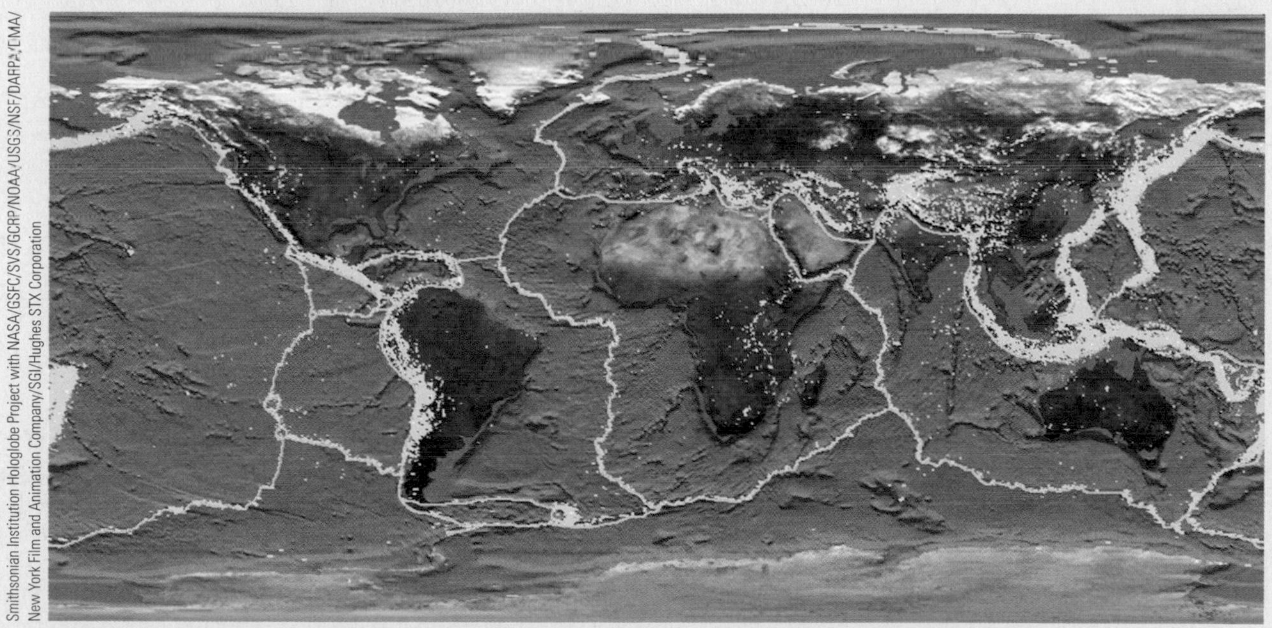
Smithsonian Institution Hologlobe Project with NASA/GSFC/SVS/GCRP/NOAA/USGS/NSF/DARPA/DMA/
New York Film and Animation Company/SGI/Hughes STX Corporation

This map shows plate boundaries and the global distribution of earthquake activity (magnitude 4.5+, 1990–1995).

layers. For example, rock layers that have undergone significant tectonic forces may be tilted, folded, or fractured, or, relative to adjacent rocks masses, offset, uplifted, or downdropped. Sedimentary rocks are particularly useful for identifying tectonic deformation because they are usually horizontal when they are formed, and older rock layers are originally overlain by successively younger rock layers. If strata are bent, fractured, offset, or otherwise out of sequence, some kind of structural deformation has occurred.

Earth scientists describe the orientations of inclined rock layers by measuring their strike and dip. **Strike** is the compass direction of the line that forms at the intersection of a tilted rock layer and a horizontal plane. A rock layer, for example, might strike northeast, which could also be expressed correctly as striking southwest (• Fig. 14.24). The inclination of the rock layer, the **dip,** is always measured at right angles to the strike and in degrees of angle from the horizontal (0° dip = horizontal). The direction toward which the rock dips down is expressed with the general compass direction. For example, a rock layer that strikes northeast and dips 11° from the horizontal down to the southeast would have a dip of 11° to the southeast (see again Fig. 14.24).

Earth's crust has been subjected to tectonic forces throughout its history, although the forces have been greater during some geologic periods than others and have varied widely over Earth's surface. Most of the resulting changes in the crust have occurred over hundreds of thousands or millions of years, but others have been rapid and cataclysmic. The response of crustal rocks to tectonic forces can yield a variety of configurations in rock structure, depending on the nature of the rocks and the nature of the applied forces.

Tectonic forces are divided into three principal types that differ in the direction of the applied forces (• Fig. 14.25). **Compressional tectonic forces** push crustal rocks together. **Tensional tectonic forces** pull parts of the crust away from each other. **Shearing tectonic forces** slide parts of Earth's crust past each other.

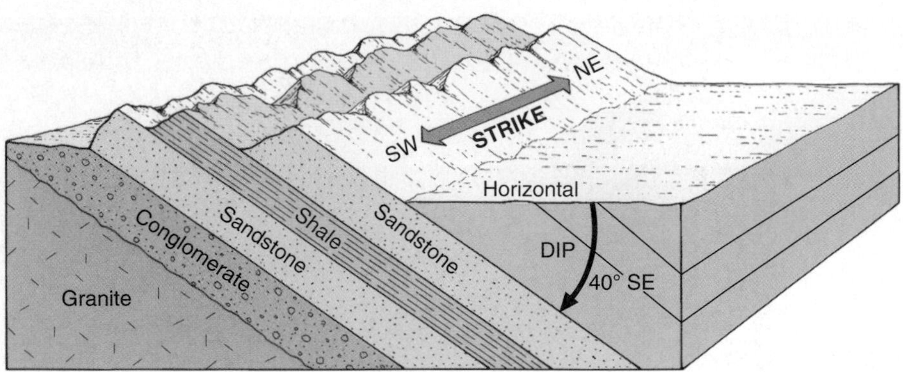

• **FIGURE 14.24**

Geoscientists use the properties of strike and dip to describe the orientation of sedimentary rock layers. Strike is the compass direction of the line created by the intersection of a rock layer with a horizontal plane. Dip is the angle from the horizontal and compass direction toward which the rock layer angles down. Dip direction lies at a 90° angle to the strike.

What are the strike and dip of the upper layer of sandstone in this diagram?

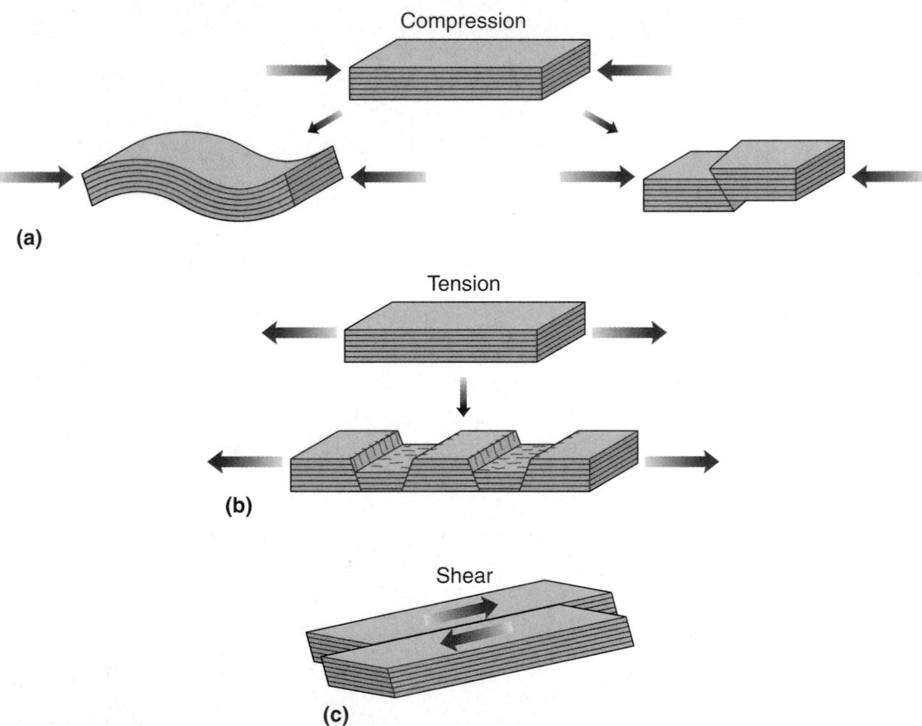

• **FIGURE 14.25**

Three types of tectonic force cause deformation of rock layers. (a) Compressional forces push rocks together. Compressional forces can bend (fold) rocks, or they can cause the rocks to break and slide along the breakage zone, which is called a fault. (b) Tensional forces pull rocks apart and may also lead to the breaking and shifting of rock masses along faults. (c) Shearing forces work to slide rocks past each other horizontally, rather than into or away from each other. If the shearing forces are greater than the resistance of the rocks to them, the rocks will break and slide in opposite directions past each other along the breakage zone (fault).

Compressional Tectonic Forces

Tectonic forces that push two areas of crustal rocks together tend to shorten and thicken the crust. How the affected rocks respond to compressional forces depends on how brittle (breakable) the rocks are and the speed with which the forces are applied.

Folding, which is a bending or wrinkling of rock layers, occurs when compressional forces are applied to rocks that are ductile (bendable), as opposed to brittle. Rocks that lie deep within the crust and that are therefore under high pressure are generally ductile and particularly susceptible to behaving plastically, that is, deforming without breaking. As a result rocks deep within the crust typically fold rather than break in response to compressional forces (• Fig. 14.26). Folding is also more likely than fracturing when the compressional forces are applied slowly. Eventually, however, if the force per unit area, the **stress,** is great enough, the rocks may still break with one section pushed over another.

• FIGURE 14.26
Compressional forces have made complex folds in these layers of sedimentary rock. **How can solid rock be folded without breaking?**

As elements of rock structure, upfolds are called **anticlines,** and downfolds are called **synclines** (• Fig. 14.27). The rock layers that form the flanks of anticlinal crests and synclinal troughs are the *fold limbs*. Folds in some rock layers are very small, covering a few centimeters, while others are enormous with vertical distances between the upfolds and downfolds measured in kilometers. Folds can be tight or broad, symmetrical or asymmetrical. Folds are symmetrical—that is, each limb has about the same dip angle—if they formed by compressional forces that were relatively equal from both sides. If compressional forces were stronger from one side, a fold may be asymmetrical, with the dip of one limb being much steeper than that of the other. Eventually, asymmetrically folded rocks may become overturned and perhaps so compressed that the fold lies horizontally; these are known as **recumbent folds** (see again Fig. 14.27).

Much of the Appalachian Mountain system is an example of folding on a large scale. Spectacular folds exist in the Rocky Mountains of Colorado, Wyoming, and Montana and in the Canadian Rockies. Highly complex folding created the Alps, where folds are overturned, sheared off, and piled on top of one another. Almost all mountain systems exhibit some degree of folding.

Rock layers that are near Earth's surface, and not under high confining pressures, are too rigid to bend into folds when experiencing compressional forces. If the tectonic force is large enough, these rocks will break rather than bend and the rock masses will move

• FIGURE 14.27
Folded rock structures become increasingly complex as the applied compressional forces become more unequal from the two directions.

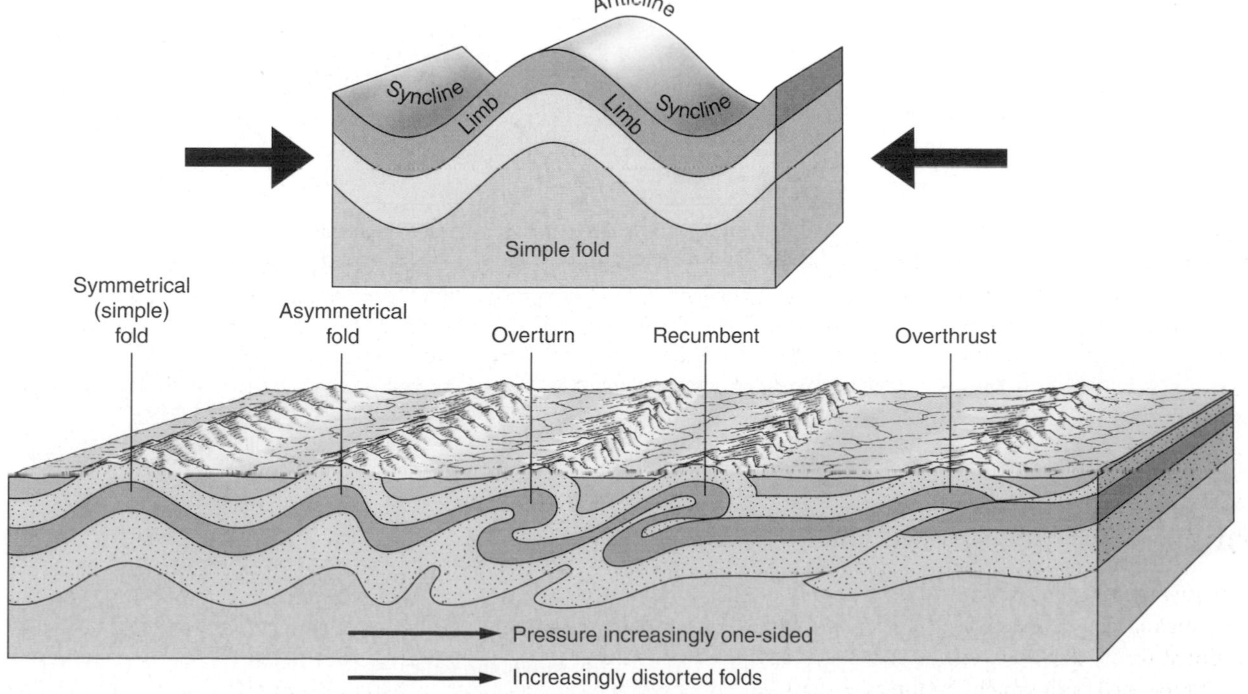

relative to each other along the fracture. **Faulting** is the slippage or displacement of rocks along a fracture surface, and the fracture along which movement has occurred is a **fault.** When compressional forces cause faulting either one mass of rock is pushed up along a steep-angled fault relative to the other or one mass of rock slides along a shallow, low-angle fault over the other. The steep, high-angle fault resulting from compressional forces is termed a **reverse fault** (• Fig. 14.28a). Where compression pushes rocks along a low-angle fault so that they override rocks on the other side of the fault, the fracture surface is called a **thrust fault,** and the shallow displacement is an **overthrust** (Fig. 14.28b). In both reverse and thrust faults, one block of crustal rocks is wedged up relative to the other. Direction of motion along all faults is always given in relative terms because even though it may seem obvious that one block was pushed up along the fault, the other block may have slid down some distance as well, and it is not always possible to determine with certainty if one or both blocks moved. Reverse or thrust faulting can also result from compressional forces that are applied rapidly and in some cases to rocks that have already responded to the force by folding. In the latter case, the upper part of a fold breaks, sliding over the lower rock layers along a thrust fault forming an overthrust. Major overthrusts occur along the northern Rocky Mountains and in the southern Appalachians. Together, recumbent folds and overthrusts are important rock structures that have formed in complex mountain ranges such as the Andes, Alps, and Himalayas.

Tensional Tectonic Forces

Tensional tectonic forces pull in opposite directions in a way that stretches and thins the impacted part of the crust. Rocks, however, typically respond by faulting, rather than bending or stretching plastically, when subjected to tensional forces. Tensional forces commonly cause the crust to be broken into discrete blocks, called **fault blocks,** that are separated from each other by **normal faults** (Fig. 14.28c). In order to accommodate the extension of the crust, one crustal fault block slides downward along the normal fault relative to the adjacent fault block. Notice that the direction of motion along a normal fault is opposite to that along a reverse or thrust fault (see again Fig. 14.28a).

In map view, regional scale tensional forces frequently cause a roughly parallel succession of normal faults to occur, creating a series of alternating downdropped and upthrown fault blocks. Each block that slid downward between two normal faults, or that remained in place while blocks on either side slid upward

• FIGURE 14.28

The major types of faults are illustrated here along with the direction of tectonic forces that cause them (indicated by large arrows). Compressional forces may create reverse (a) or thrust (b) faults. Tensional tectonic forces break rocks along normal faults (c). Shearing forces move rocks horizontally past each other along strike-slip faults (d).

How does motion along a normal fault differ from that along a reverse fault?

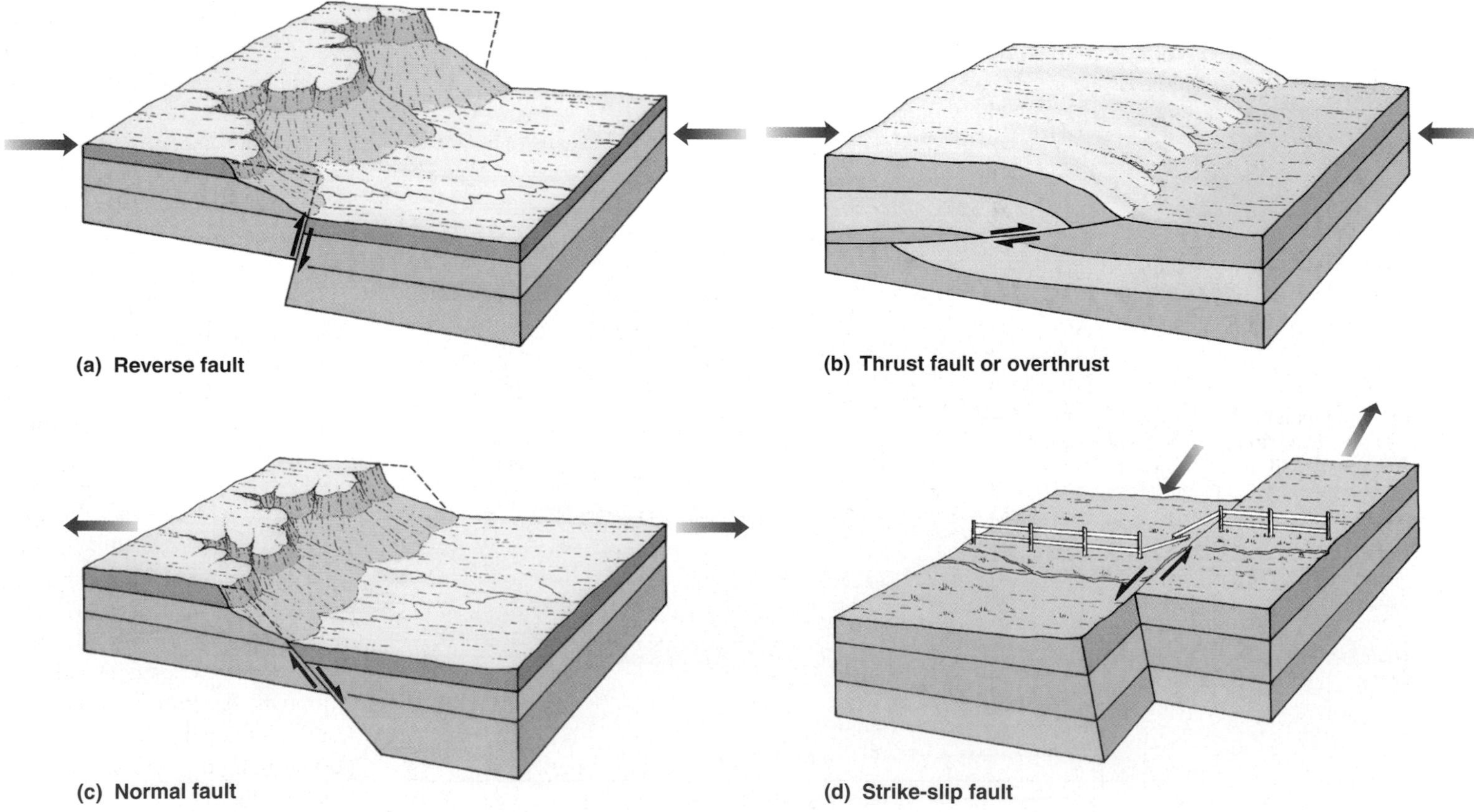

along the faults, is called a **graben** (● Fig. 14.29). A fault block that moved relatively upward between two normal faults—that is, it actually moved up or remained in place while adjacent blocks slid downward—is a **horst.** The great Ruwenzori Range of East Africa is a horst, as is the Sinai Peninsula between the fault troughs in the Gulfs of Suez and Aqaba (see again Fig. 13.31). Horsts and grabens are rock structural features that can be identified by the nature of the offset of rock units along normal faults. Topographically, horsts form mountain ranges and grabens form basins. The Basin and Range region of the western United States that extends eastward from California to Utah and southward from Oregon to New Mexico is an area undergoing tensional tectonic forces that are pulling the region apart to the west and east. A transect from west to east across that region, for example from Reno, Nevada, to Salt Lake City, Utah, encounters an extensive series of alternating downdropped and upthrown fault blocks comprising the basins and ranges for which the region is named. Some of the ranges and basins are simple horsts and grabens, but others are **tilted fault blocks** that result from the uplift of one side of a fault block while the other end of the same block rotates downward (● Fig. 14.30). Death Valley, California, is a classic example of the down-tilted side of a tilted fault block (● Fig. 14.31).

Large-scale tensional tectonic forces can create **rift valleys,** which are composed of relatively narrow but long regions of crust downdropped along normal faults. Examples of rift valleys include the Rio Grande rift of New Mexico and Colorado, the Great Rift Valley of East Africa, and the Dead Sea rift valley where that body of water lies at an elevation some 390 meters (1280 ft) below the Mediterranean Sea, which is only 64 kilometers (40 mi) away. Rift valleys also run along the centers of oceanic ridges.

An **escarpment,** often shortened to **scarp,** is a steep cliff, which may be tall or short. Scarps can form on Earth surface terrain for many reasons and in many different settings. A cliff that results from movement along a fault is specifically a **fault scarp.** Fault scarps are commonly visible in the landscape along normal fault zones, where they may consist of rock faces on fault blocks that have undergone extensive amounts of uplift over long periods of time. **Piedmont fault scarps** offset unconsolidated sediments that have been eroded from uplifted fault blocks and deposited along the base of the fault block (● Fig. 14.32).

● FIGURE 14.29
Horsts and grabens are blocks of Earth material that are bounded by normal faults. A block that has moved upward along a normal fault relative to adjacent blocks is a horst. A block that has slid down along a normal fault relative to adjacent blocks is a graben.
What kind of tectonic force causes these kinds of fault blocks?

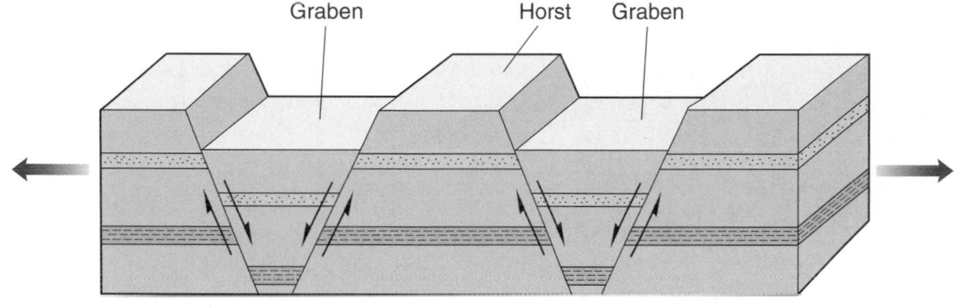

● FIGURE 14.30
This diagram of a tilted fault block indicates its strike and dip. The east facing cliff is an erosion-modified fault scarp. This configuration is a simplified version of the kind of faulting that produced Death Valley, which occupies the downtilted part of a tilted fault block.

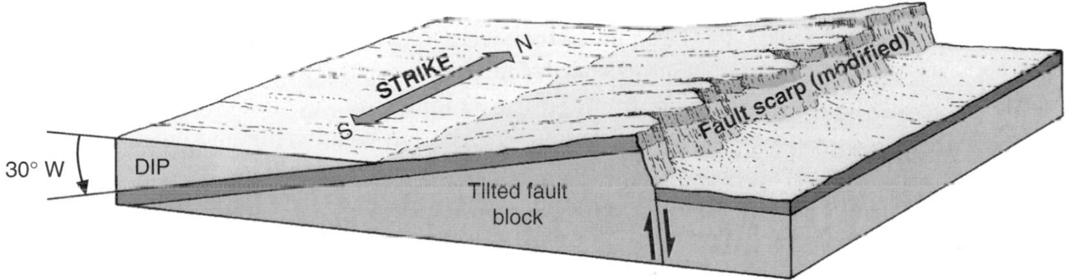

● FIGURE 14.31
Death Valley, California, is a classic example of a topographic basin created by tilted fault blocks. The valley floor is 86 meters (282 ft) below sea level, which is the lowest elevation in North America.

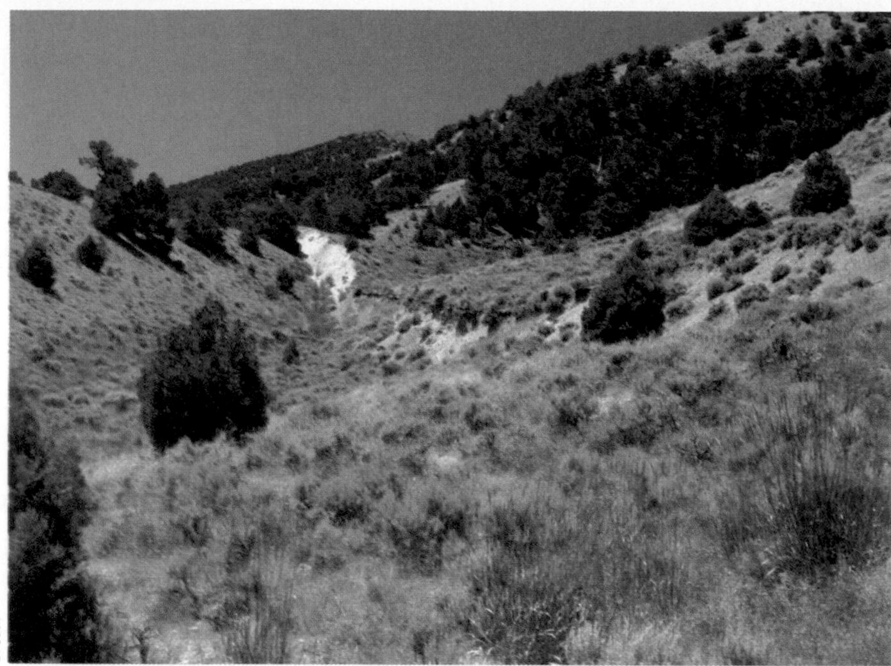

D. Sack

• FIGURE 14.32
This piedmont fault scarp in Nevada is the topographic expression of a normal fault. Movement along the fault that created this scarp occurred about 30 years before the photograph was taken.
On which side of the fault does the horst lie?

Fault scarps can account for spectacular mountain walls, especially in regions like much of the western United States with a history of recent tectonic activity. The east face of the 645-kilometer-long (405 mi) Sierra Nevada Range in California is a classic example of a fault scarp that rises steeply 3350 meters (11,000 ft) above the desert (• Fig. 14.33). In contrast, the west side of the Sierra (the "back slope") descends very gently over a distance of 100 kilometers (60 mi) through rolling foothills. The Sierra Nevada Range is a great tilted fault block where the east side was faulted upward and the west side tilted down (see again Fig. 14.30). The equally dramatic Grand Tetons of Wyoming also rise along a fault scarp facing eastward. In Big Bend National Park, Texas, the fault block that forms the walls of Santa Elena Canyon is an excellent example of a fault scarp. Other than the 500-meter-deep canyon that the Rio Grande has cut, the fault block is modified so little by erosion that it preserves much of its blocklike shape (• Fig. 14.34). In the southwestern United States, the Colorado Plateau steps down to the Great Basin by a series of fault scarps that face westward in southern Utah and northern Arizona.

Major uplift of faulted mountain ranges can have a strong impact on other physical systems, and an excellent example is the Sierra Nevada. As the mountains rose, stream erosion accelerated because of the increase in slope. Precipitation on the windward side of the Sierra increased because of orographic lifting. The steep lee side of the tilted fault block became more arid than before because it was situated in the rain shadow of the Sierra. Increased precipitation and lower temperatures at higher elevations changed the climate of the uplifted range significantly, and climate change influenced the vegetation, soils, and animal life. Soils have also been affected by increased runoff and erosion. The uplift of the Sierra has extended over several million years in an episodic sequence of faulting. The Sierra Nevada Range is continuing to rise rapidly, in a geologic sense—on average about a centimeter per year. Weathering and erosion have attacked the rocks as uplift progressed. The Sierra Nevada, like most high mountain ranges, have been altered and etched by glaciation, stream erosion, and downslope gravitational movement of rock material. These processes have carved and shaped

• FIGURE 14.33
The east front of the Sierra Nevada in California is essentially the steep scarp side of a tilted fault block.

©Terry Husebye/Getty Images

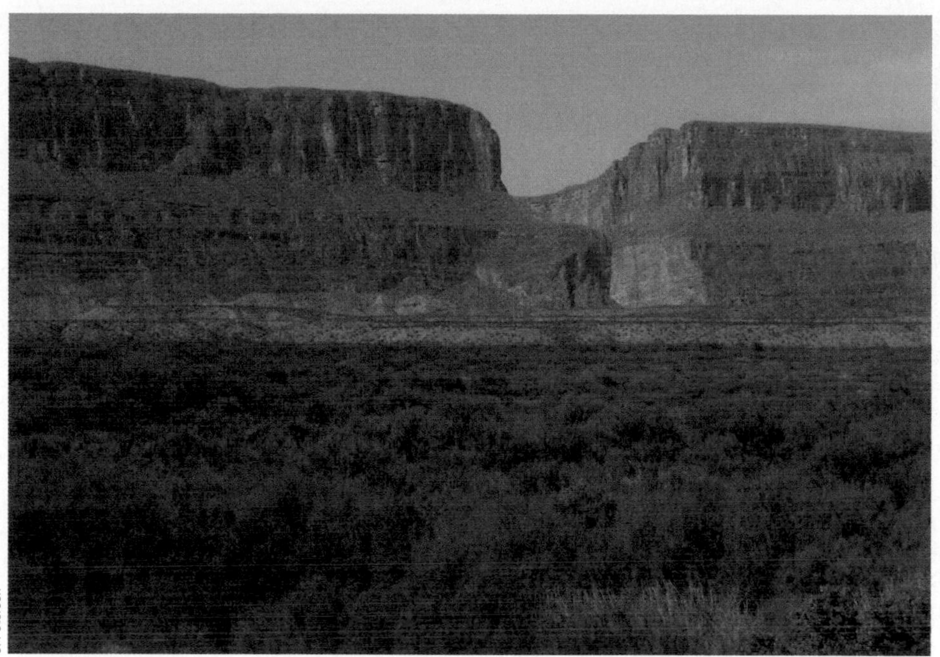

J. Petersen

• FIGURE 14.34
The steep fault scarp at Santa Elena Canyon, along the Texas–Mexico border, has undergone limited modification by weathering and erosion. The Rio Grande has cut a canyon into the uplifted and tilted fault block. In this photo, the wall to the left of the canyon is in Mexico and that to the right is in the United States.

valleys in the Sierran fault block, leaving the spectacular canyons and mountain peaks.

Shearing Tectonic Forces

Vertical displacement along a fault occurs when the rocks on one side move up or drop down in relation to rocks on the other side. Faults with this kind of movement, up or down along the *dip* of the fault plane extending into Earth, are known as **dip-slip faults.** Normal and reverse faults, for example, have dip-slip motion. There also exists, however, a completely different category of fault along which displacement of rock units is horizontal rather than vertical. In this case, the direction of slippage is parallel to the surface trace, or *strike,* of the fault; thus it is called a **strike-slip fault** or, because of the horizontal motion, a **lateral fault** (see again Fig. 14.28d). Offset along strike-slip faults is most easily seen in map view (from above), rather than in cross-sectional view. Active strike-slip faults can cause horizontal displacement of roads, railroad tracks, fences, streambeds, and other features that extend across the fault. The motion along a strike-slip fault is described as left lateral or right lateral, depending on the direction of movement of the blocks. To determine whether motion is left or right lateral, imagine yourself standing on one block and looking across the strike-slip fault to the other block. The relative direction of motion of the block across the fault determines whether it is a left lateral or right lateral fault. The San Andreas Fault, which runs through much of California, has right lateral strike-slip movement. A long and narrow, rather linear valley composed of rocks that have been crushed and weakened by faulting marks the trace of the San Andreas Fault zone (• Fig. 14.35).

The amount that Earth's surface can be offset during instantaneous movement along a fault varies from fractions of a centimeter to several meters. Faulting can move rocks laterally, vertically, or both. The maximum horizontal displacement along the San Andreas Fault in California during the 1906 San Francisco earthquake was more than 6 meters (21 ft). A vertical displacement of more than 10 meters (33 ft) occurred during the Alaskan earthquake of 1964. Over millions of years, the cumulative displacement along a major fault may be tens of kilometers vertically or hundreds of kilometers horizontally, although the majority of faults have offsets that are much smaller.

Relationships between Rock Structure and Topography

Tectonic activity can result in a variety of structural features that range from microscopic fractures to major folds and fault blocks. At the surface, structural features comprise various topographic features (landforms) and are subject to modification by weathering, erosion, transportation, and deposition. It is important to distinguish between structural elements and topographic features because rock structure reflects endogenic factors while landforms reflect the balance between endogenic and exogenic factors. As a result, a specific

• FIGURE 14.35
The San Andreas Fault along the Carrizo Plain in California runs from left to right across the center of this photo. The area west (background) of the fault is moving northwestward, in relation to the area on the east (foreground) side. Valleys of creeks that cross the fault have been offset about 130 meters (427 ft) by numerous episodes of earthquake displacement.
What type of fault is the San Andreas?

USGS/R.E. Wallace

GEOGRAPHY'S ENVIRONMENTAL SCIENCE PERSPECTIVE

Mapping the Distribution of Earthquake Intensity

When an earthquake strikes a populated area, one of the first pieces of scientific information released is the magnitude of the tremor. Magnitude is a numerical expression of an earthquake's size at its focus in terms of energy released. In this sense, earthquakes can be compared to explosions. For example, a magnitude 4.0 earthquake releases energy equivalent to exploding 1000 tons of TNT. Because the scale is logarithmic, a 6.9 magnitude is the equivalent of 22.7 million tons of TNT.

Because of their greater energy, earthquakes of greater magnitude have the potential to cause much more damage and human suffering than those of smaller magnitude, but the reality is much more complex than that. A moderate earthquake in a densely populated area may cause great injury and damage, while a very large earthquake in an isolated region may not affect humans at all. Many factors relating to physical geography can influence an earthquake's impact on people and their built environment. In general, the farther a location is from the earthquake epicenter, the less the effect of shaking, but this generalization does not apply in every case. An earthquake in 1985 caused great damage in Mexico City, including the complete collapse of buildings, even though the epicenter was 385 kilometers (240 mi) away.

The Mercalli Scale of earthquake intensity (I–XII) was devised to measure the impact of a tremor on people, their homes, buildings, bridges, and other elements of human habitation. Although every earthquake has only one magnitude, intensity can vary greatly from place to place, so a range of intensities will typically be encountered for a single tremor. The impact of an earthquake on a region varies spatially, and the patterns of Mercalli intensity can be mapped. Earthquake intensity maps use lines of equal shaking and earthquake damage, called *isoseismals,* expressed in Mercalli intensity levels. Patterns of isoseismals are useful in assessing what local conditions

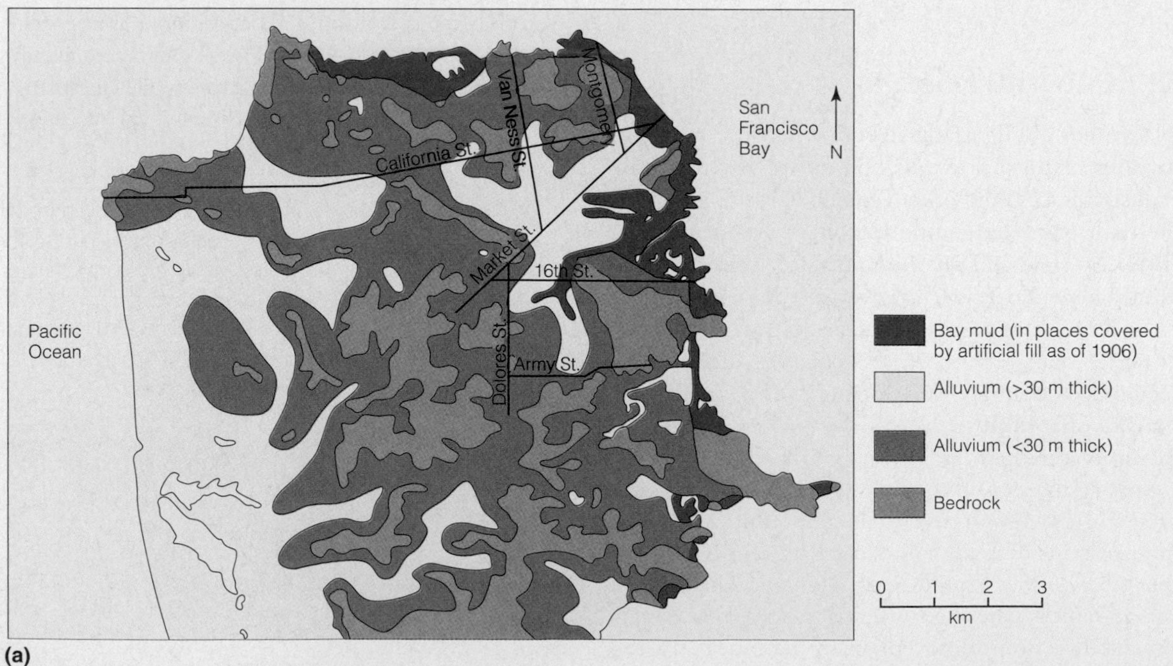

The location of different Earth materials during the 1906 San Francisco earthquake.

contributed to spatial variations in shaking and impact. Earthquake intensity factors vary geographically according to the nature of the substrate, population density, construction type and quality, and topography. Areas with unconsolidated Earth materials, poor construction, or high population densities generally suffer more from shaking and experience greater damage.

The 1906 San Francisco earthquake and ensuing fire caused the destruction of a great many buildings, numerous injuries, and an estimated 3000 deaths. The fire resulted from earthquake-damaged electrical and gas lines. Neither the magnitude nor the intensity scales existed in 1906. Subsequent studies, however, suggest that the earthquake magnitude was about 8.3, and cartographers have prepared maps of the spatial distributions of Mercalli intensity. These show the great variations in ground shaking and damage that the earthquake caused. The geographic patterns of the isoseismals have been analyzed to explain why certain areas suffered more than others did. Areas of bedrock were shown to have experienced lower intensities (less damage) in comparison to areas of unconsolidated Earth materials. Much of the worst damage occurred on artificially filled lands along the bayfront and on areas that had been stream valleys but were covered over with loose Earth materials in order to construct buildings on the land. In some cases, buildings on one side of a street were destroyed, while those on the other side of the street suffered little damage.

Analyzing the nature of sites where intensities in an earthquake were higher or lower than expected helps us understand the reasons for spatial variations in earthquake hazards. The geographic patterns of Mercalli intensity that are generated even by small tremors help in planning for larger earthquakes in the same area. The overall patterns of ground shaking and isoseismals should be similar for a larger earthquake having the same epicenter as a smaller one, but the amount of ground shaking, the level of intensity, and the size of the area affected would be greater for the larger magnitude tremor.

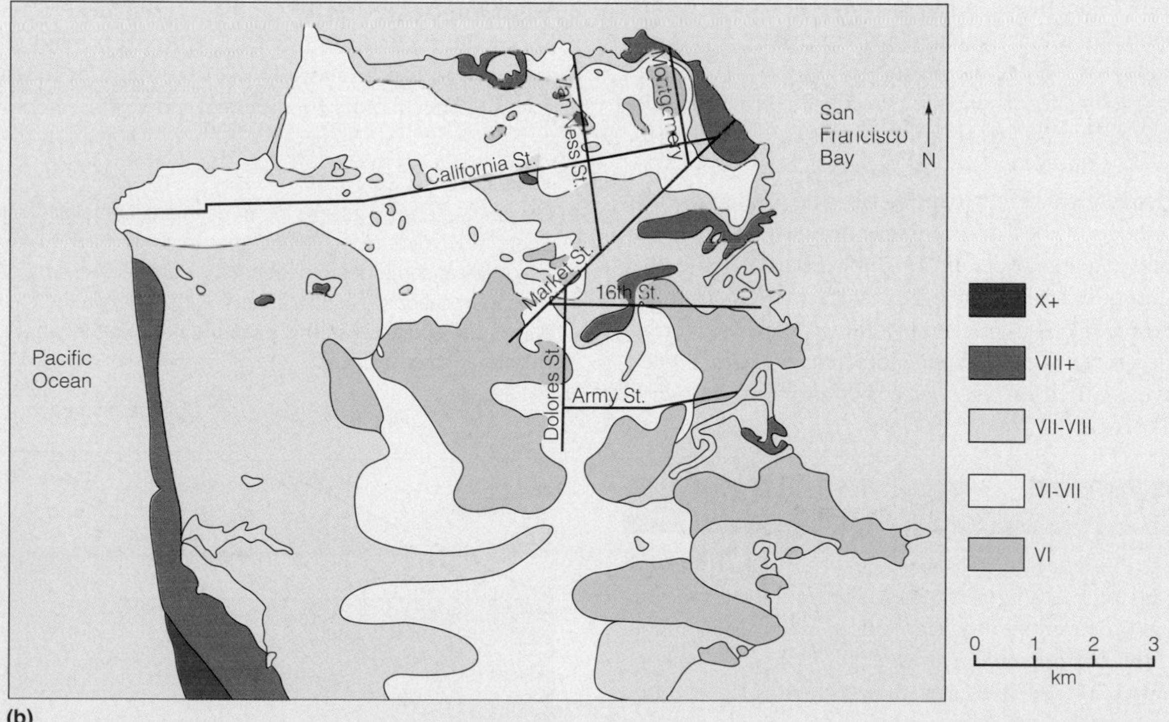

Geographic patterns of Mercalli Intensity caused by the 1906 earthquake. The areas of intensity VIII+ in the northeast quarter of the city are on artificial landfill.

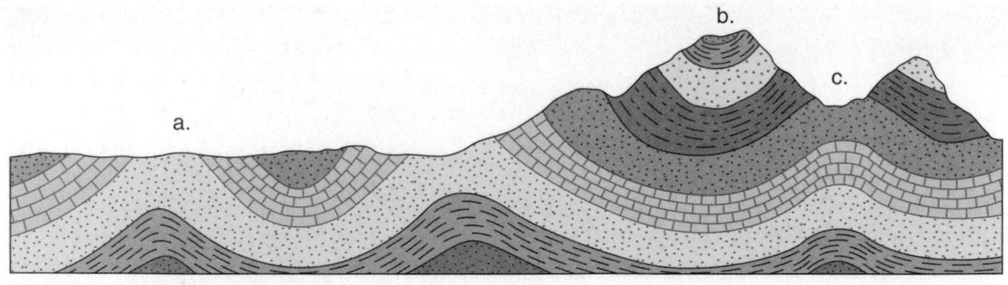

• FIGURE 14.36
This example cross section from a region of folded rocks illustrates the distinction between rock structure and surface topography (landforms). Structure is the rock response to applied tectonic forces. Rock structure may or may not be represented directly in the surface topography, which depends on the nature and rate of exogenic as well as endogenic geomorphic processes. Structural upfolds do not always comprise topographic mountains, nor do all downfolds form valleys. (a) The structure is an anticline, but the surface landform is a plain of low relief. (b) Here, the erosionally resistant center of a downfold (a syncline) supports a mountain peak. (c) A valley has been eroded into the crest of an anticline.
Why is it that not all anticlines form mountains?

type of structural element can assume a variety of topographic expressions (• Fig. 14.36). For instance, an upfolded structural feature is an anticline even though geomorphically it may comprise a ridge, a valley, or a plain, depending on erosion of broken or weak rocks. Nashville, Tennessee, occupies a topographic valley, yet it is sited in the remains of a structural dome (a circular domal anticline). Likewise, even though synclines are structural downfolds, topographically a syncline may contribute to the formation of a valley or a ridge. Some mountain tops in the Alps are the erosional remnants of synclines. Words like *mountain, ridge, valley, basin,* and *fault scarp* are geomorphic terms that describe the surface topography, while *anticline, syncline, horst, graben,* and *normal fault* are structural terms that describe the arrangement of rock layers. Elements of rock structure may or may not be directly reflected in the surface topography. It is important to remember that the topographic variation on Earth's surface results from the interaction of three major factors: endogenic processes that create relief, exogenic processes that shape landforms and reduce relief, and the relative strength or resistance of different rock types to weathering and erosion.

Earthquakes

Earthquakes, evidence of present-day tectonic activity, are ground motions of Earth caused when accumulating tectonic stress is suddenly relieved by displacement of rocks along a fault. The sudden, lurching movement of crustal blocks past one another represents a release of energy that generates these internal earthquake motions, the *seismic waves* that were discussed in Chapter 13 as helpful in understanding Earth's interior. Seismic waves, however, can also have a great impact on Earth's surface. It is primarily when these waves pass along the crustal exterior or emerge at Earth's surface from below that they cause the damage and subsequent loss of life that we associate with major tremors. The subsurface location where the rock displacement and resulting earthquake originated is the earthquake **focus,** which may be located anywhere from near the surface to a depth of 700 kilometers (435 mi). The earthquake **epicenter** is the point on Earth's surface that lies directly above the focus and is where the strongest shock is normally felt (• Fig. 14.37).

The vast majority of earthquakes are so slight that we cannot feel them, and they produce no injuries or damage. Most earthquakes occur at a focus that is deep enough so that no displacement is visible at the surface. Others may cause mild shaking that rattles a few dishes, while a few are strong enough to topple buildings and break power lines, gas mains, and water pipes. Surface offset or ground shaking during an earthquake can also trigger rockfalls, landslides, avalanches, and tsunamis. Aftershocks may follow the main earthquake as crustal adjustments continue to occur. Geophysicists are currently investigating the possibility that foreshocks may alert us to major earthquakes, although evidence is at present inconclusive.

Measuring Earthquake Size

An earthquake's severity can be expressed in two ways: (1) the size of the event as a physical Earth process, and (2) the degree of its impact on humans. These two methods may be related in that, all other factors being equal, powerful earthquakes should

• FIGURE 14.37
The point of energy release for an earthquake, that is, the location where movement along the fault began, is the earthquake focus, which is typically at some depth beneath the surface. The earthquake epicenter is the location on Earth's surface directly above the focus.
Why is the epicenter in this example not located where the fault crosses Earth's surface?

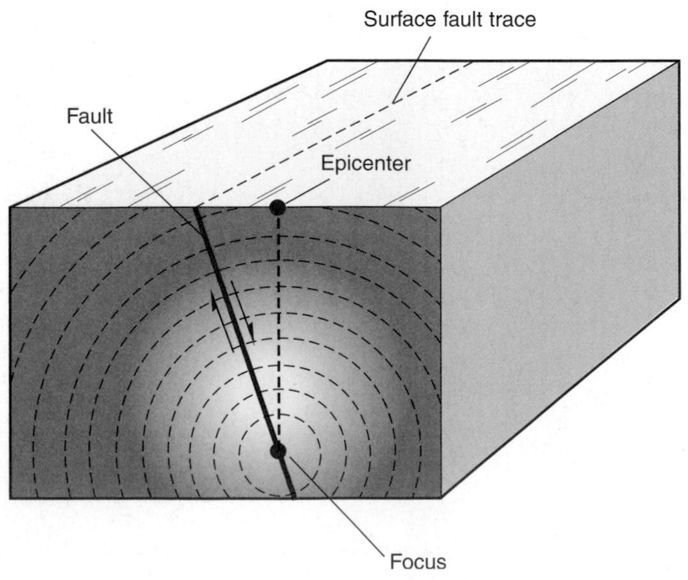

have a greater effect on humans than smaller earthquakes. However, large tremors that strike in sparsely inhabited places will have limited human impact, while small earthquakes that strike densely populated areas can cause considerable damage and suffering. Many factors other than earthquake size affect the damage and loss of human life resulting from a tremor. Measuring the physical size of earthquakes and, separately, their effects on people, help scientists and planners understand local and regional earthquake hazard potential.

The scale of **earthquake magnitude,** originally developed by Charles F. Richter in 1935, is based on the energy released by an earthquake as recorded by seismographs (see again Fig. 13.1). Earthquake magnitude is expressed in a number, generally with one decimal place. Every increase of one whole number in magnitude (for example, from 6.0 to 7.0) means that the earthquake wave motion is 10 times greater, but the actual energy released is about 30 times greater. The extremely destructive 1906 San Francisco earthquake occurred before the magnitude scale was devised but is estimated to have been a magnitude 8.3. The 1989 Loma Prieta earthquake near San Francisco had a Richter magnitude of 7.1, but was still responsible for dozens of fatalities. The strongest earthquake in North America to date, a magnitude 8.6, occurred in Alaska in 1964, and the strongest in the world in recent history was a 9.5 magnitude earthquake that occurred off the coast of Chile in 1960.

In recent years, scientists have modified how earthquake size is measured by using the *moment magnitude*. Moment magnitudes are similar in value to magnitudes determined using Richter's approach—for example, the moment magnitude of the 1989 Loma Prieta earthquake was 6.9—but the moment magnitude is measured in a different way and with greater accuracy. As with Richter magnitude, each earthquake has only one moment magnitude and it represents the earthquake's size in terms of energy released. The tragic December 2004 earthquake off the west coast of northern Sumatra, Indonesia, called the Sumatra–Andaman earthquake, had a moment magnitude of 9.1 and generated the deadliest tsunami on record (• Fig. 14.38).

A very different type of scale is that used to record and understand patterns of **earthquake intensity,** the damage caused by an earthquake and the degree of its impact on people and their property. The **modified Mercalli scale** of earthquake intensity utilizes categories numbered from I to XII (Table 14.1) to describe the effects of an earthquake on humans and the spatial variation of those impacts. The categories are represented with Roman numerals to avoid confusion with earthquake magnitude scales. Although a maximum intensity is determined for each earthquake, one earthquake typically produces a variety of intensity levels, depending on variable local conditions. These conditions include distance from the epicenter, how long shaking lasted, the severity of shaking resulting from the local surface materials affected, population density, and building construction in the affected area. After an earthquake, observers gather information about Mercalli intensity levels by noting the damage and talking to residents about their experiences. The variation in intensity levels is then mapped so that geographic patterns of damage and shaking intensity can be analyzed. Understanding the spatial patterns of damage and ground response helps us plan and prepare so that we might reduce the hazards of future earthquakes.

Earthquake Hazards

Unfortunately, there is plenty of evidence concerning the deadliness of earthquakes. Together with the ensuing tsunami, the 2004 Sumatra–Andaman earthquake resulted in almost 300,000 fatalities. Major earthquake tragedies have also occurred in other parts of the world. For example, 86,000 lives were lost in Pakistan due to a magnitude 7.6 earthquake in 2005. In 2003, 31,000 people perished from a magnitude 6.6 earthquake in southeastern Iran, and at least 40,000 died in western Iran from a magnitude 7.4 earthquake in 1990. More than 5000 people were killed and more than 50,000 buildings were destroyed in Kobe, Japan, when a 7.2 magnitude earthquake struck there in 1995 (• Fig. 14.39). In Armenia (in the former Soviet Union) a 6.9 magnitude earthquake in 1988 destroyed two cities, with a death toll of 25,000. A magnitude 7.8 earthquake in Peru caused 65,000 deaths in 1970.

The impact of an earthquake on humans and their property depends on many factors. Some of the most important pertain to the likelihood of, and potential number of people affected by, structural collapse. Factors influencing structural collapse include the location of an earthquake epicenter relative to population density, construction materials and methods, and the stability of Earth materials on which buildings were erected. A strong tremor of magnitude 7.5 that struck the Mojave Desert of Southern California in

• FIGURE 14.38
The 9.1 magnitude Sumatra–Andaman earthquake, and the tsunami that it generated, devastated the city of Banda Aceh on the Indonesian island of Sumatra. Extensive parts of the city were flattened by the catastrophe. Here, only a lone piece of a building is left standing. The dark scratches above the lowest set of windows were made by debris carried in the tsunami.

USGS/Dr. Guy Gelfenbaum

TABLE 14.1
Modified Mercalli Intensity Scale

I.	Not felt except by a very few under especially favorable conditions.
II.	Felt only by a few persons at rest, especially on upper floors of buildings.
III.	Felt quite noticeably by persons indoors, especially on upper floors of buildings. Many people do not recognize it as an earthquake. Standing motor cars may rock slightly. Vibrations similar to the passing of a truck. Duration estimated.
IV.	Felt indoors by many, outdoors by few during the day. At night, some awakened. Dishes, windows, doors disturbed; walls make cracking sound. Sensation like heavy truck striking building. Standing motor cars rocked noticeably.
V.	Felt by nearly everyone; many awakened. Some dishes, windows broken. Unstable objects overturned. Pendulum clocks may stop.
VI.	Felt by all, many frightened. Some heavy furniture moved; a few instances of fallen plaster. Damage slight.
VII.	Damage negligible in buildings of good design and construction; slight to moderate in well-built ordinary structures; considerable damage in poorly built or badly designed structures; some chimneys broken.
VIII.	Damage slight in specially designed structures; considerable damage in ordinary substantial buildings with partial collapse. Damage great in poorly built structures. Fall of chimneys, factory stacks, columns, monuments, walls. Heavy furniture overturned.
IX.	Damage considerable in specially designed structures; well-designed frame structures thrown out of plumb. Damage great in substantial buildings, with partial collapse. Buildings shifted off foundations.
X.	Some well-built wooden structures destroyed; most masonry and frame structures destroyed with foundations. Rails bent.
XI.	Few, if any (masonry) structures remain standing. Bridges destroyed. Rails bent greatly.
XII.	Damage total. Lines of sight and level are distorted. Objects thrown into the air.

Source: Abridged from *The Severity of an Earthquake: A U.S. Geological Survey General Interest Publication.* U.S. Government Printing Office: 1989, 288–913.

• FIGURE 14.39
The 7.2 magnitude earthquake that struck Kobe, Japan, in 1995 killed more than 5000 people and destroyed more than 50,000 buildings. In terms of the value of damaged property, estimated at $100 billion, this is one of the costliest earthquakes on record.
What human factors were related to the great damage and death totals?

©Robert Patrick/CORBIS SYGMA

1992 caused little loss of life because of sparse settlement. Structures made of brick, unreinforced masonry, or other inflexible materials or without adequate structural support do not withstand ground shaking well. Freeway spans collapsed in the San Francisco Bay area in the Loma Prieta earthquake of 1989 and in the 1994 6.7 magnitude Northridge earthquake in the San Fernando Valley area of Los Angeles (• Fig. 14.40). Wood frame houses generally fare better than brick, adobe, or block structures because the wood frame has greater flexibility during ground shaking. Buildings located on loosely deposited Earth materials (unconsolidated sediment) tend to shake violently compared to areas of solid bedrock. The effects of an 8.1 magnitude earthquake with an epicenter 385 kilometers (240 mi) from Mexico City in 1985 caused a death toll of more than 9000 and extensive damage to many buildings, including high-rise sections of the city. Mexico's densely populated capital city is built on an ancient lake bed made of soft sediments that shook greatly, causing much destruction even though the earthquake was centered in a distant mountain region.

In addition to structural collapse, loss of life in an earthquake is influenced by several other factors including time of day, time of year, weather conditions at the time of the quake, resulting fires from downed power lines and broken gas mains, and whether ground shaking triggers rockfalls, landslides, avalanches,

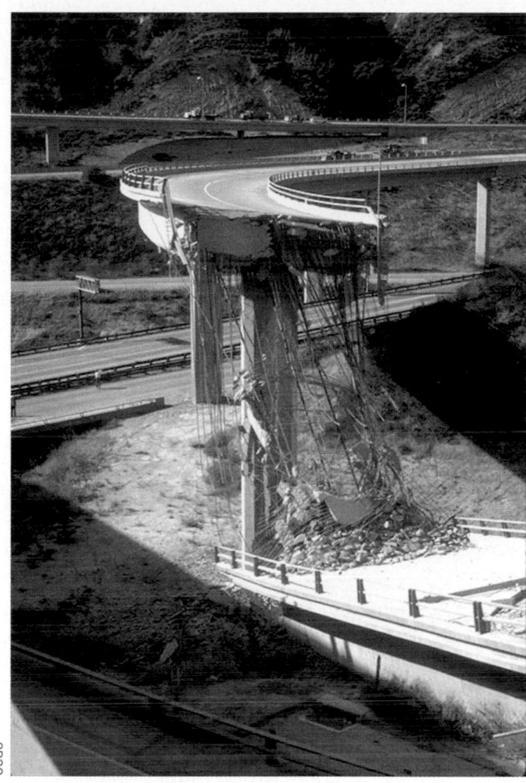

• **FIGURE 14.40**
Spans of elevated freeways collapsed in both the 1994 earthquake epicentered in Northridge in the Los Angeles area, damage from which is shown here, and the Loma Prieta earthquake in the San Francisco Bay area in 1989. Each killed close to 60 people, but the 1994 tremor would have been more deadly had it occurred on a regular workday during rush hour instead of 4:30 a.m. local time on Martin Luther King Jr. Day.

or tsunamis. Much greater loss of life will occur from freeways that collapse in an earthquake that occurs during rush hour than in one that strikes in the middle of the night. Contributing to the large death toll from the 1970 earthquake in Peru were enormous earthquake-induced avalanches that obliterated entire mountain villages. Even small slope failures that block roads can increase the death toll by hampering access for rescue operations.

Although most major earthquakes are related to known faults and have epicenters in or near mountainous regions, the one most widely felt in North America was not. It was one of a series of tremors that occurred during 1811 and 1812, centered near New Madrid, Missouri, and was felt from Canada to the Gulf of Mexico and from the Rocky Mountains to the Atlantic Ocean. Fortunately, the region was not densely settled at that time in history. Because of that huge earthquake, St. Louis is considered to be at risk from a major earthquake, as shown on the seismic-risk map for the conterminous United States (• Fig. 14.41). Although not common in these regions, recent earthquakes have occurred in New England, New York, and the Mississippi Valley. Probably no area on Earth could be called entirely "earthquake safe."

Like volcanic processes, tectonic processes are natural-functioning parts of the Earth system. Many regions of active earthquake or volcanic activity are incredibly scenic and offer attractive environments in which to live, so it is not surprising that some of these hazard-prone areas are densely populated. It is essential, however, that residents and governmental agencies in areas where volcanic, tectonic, or other potentially hazardous natural processes are active make detailed preparations for coping with disasters before they occur.

• **FIGURE 14.41**
Earthquake hazard potential in the conterminous United States. This map shows a very generalized determination of the greatest expectable earthquake intensities for the 48 states shown.
What is the earthquake hazard potential where you live, and what does that level of intensity mean according to the Mercalli scale in Table 14.1?

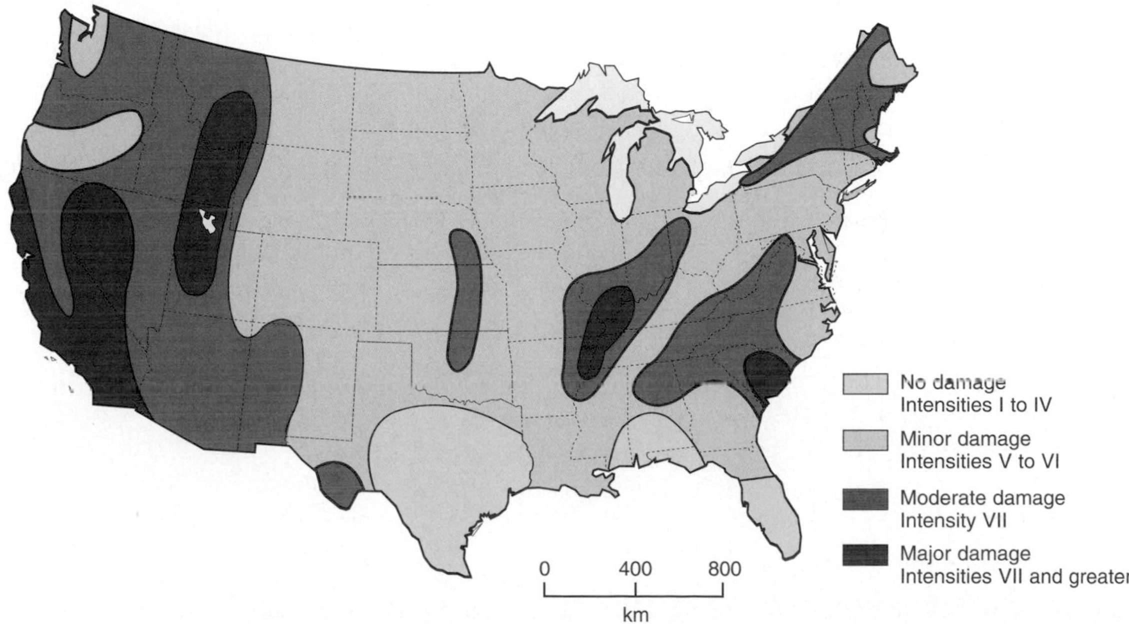

Chapter 14 Activities

Define & Recall

topography
landform
geomorphology
igneous processes
tectonic processes
relief
endogenic processes
exogenic processes
weathering
erosion
transportation
deposition
geomorphic agent
punctuated equilibrium
volcanism
volcano
plutonism
explosive eruption
effusive eruption
pyroclastic material
tephra
volcanic ash
lava flow
pahoehoe
aa
fissure
flood basalt
basalt plateau
shield volcano
cinder cone
composite cone
pyroclastic flow
plug dome
caldera
igneous intrusion (pluton)
stock
batholith
laccolith
sill
dike
volcanic neck
rock structure
strike
dip
compressional tectonic force
tensional tectonic force
shearing tectonic force
folding
stress
anticline
syncline
recumbent fold
faulting
fault
reverse fault
thrust fault
overthrust
fault block
normal fault
graben
horst
tilted fault block
rift valley
escarpment (scarp)
fault scarp
piedmont fault scarp
dip-slip fault
strike-slip fault (lateral fault)
earthquake
focus
epicenter
earthquake magnitude
earthquake intensity
modified Mercalli scale

Discuss & Review

1. How do endogenic processes differ from exogenic processes?
2. What are the four basic types of volcanoes? Give an example of each.
3. Which is the most dangerous type of volcano? Explain why.
4. How do plutonism and volcanism differ from each other? How are they alike?
5. What is a batholith? What is the difference between a sill and a dike?
6. Distinguish among compressional, tensional, and shearing forces.
7. What are the major differences between folding and faulting? What causes these types of crustal activity?
8. Draw a diagram to illustrate folding, showing anticlines and synclines.
9. What are the differences between a reverse and a normal fault?
10. How is a fault different from a joint?
11. What causes an earthquake?
12. What is the relationship between the focus and the epicenter of an earthquake?
13. How do earthquake magnitude and earthquake intensity differ? Why are there two systems for evaluating earthquakes?

Consider & Respond

1. Name the type of tectonic or volcanic activity that is primarily responsible for the following:
   a. Sierra Nevada
   b. Cascades
   c. Basin and Range region
   d. Ridge and Valley section of the Appalachians
2. Name several areas in the United States that are highly susceptible to natural hazards from earthquakes or volcanic activity.
3. List five countries bordering the Pacific Ocean that exhibit evidence of major volcanic activity.
4. Name a few countries not in the Pacific region that face tectonic hazards.
5. Can you recall any recent earthquakes or volcanic eruptions in the news? If so, where did these occur?
6. Assume that you are a regional planner for an urban area in the western United States. What hazards must you plan for if the region has active fault zones and volcanoes?
7. What recommendations for land use and settlement patterns might lessen the danger from earthquake and volcanic hazards?

Apply & Learn

1. If the first seismic wave to arrive at the epicenter of an earthquake traveled 6 kilometers per second and took 4.3 seconds to reach the surface, how deep was the earthquake focus?
2. A 6.5 magnitude earthquake rocked a city on the California coast. In an older community along a bay everyone felt it, and those who were asleep were awakened by it. People were very frightened and stood under doorways for protection. Ultimately, there were no fatalities and only minor injuries. Almost every house experienced at least broken dishes, and there were some reports of fallen plaster, broken windows, and damaged chimneys. Three homes experienced collapse of overhangs that had been added to cover the porch after the homes were built. What intensity was the earthquake for this coastal community?

Locate & Explore

Note: Please read the About Locate & Explore Activities section of the Preface before beginning these exercises.

1. Using Google Earth and the Earthquake Layer from the United States Geological Survey (USGS), describe the distribution of earthquakes globally and in the United States. In what areas do they tend to cluster and why? What was the largest earthquake in the United States within the last 7 days? Was the event large enough to reach the national media?
2. The Hawaiian hot spot is a mantle plume that has persisted for millions of years. As the Pacific plate moved northwest, a progression of volcanoes was created and then died as their source of magma was shut off. As shown in Figure 13.35 of your text, the island of Kauai (22.05°N, 159.54°W) was formed between 3.8 and 5.6 million years before present, while the island of Hawaii (19.60°N, 155.50°W) started to form approximately 700,000 years before present. Using the ruler tool in Google Earth, calculate the distance between the islands and use the distance to calculate the speed of the Pacific Plate in miles/year.
3. Using Google Earth, identify the landforms at the following locations (latitude, longitude) and provide a brief discussion of how the landform developed. Include a brief discussion of why the landform is found in that general area of the United States. Tip: Use the zoom, tilt, rotate, and elevation exaggeration tools to help view and interpret the landform and the area in which it is found.
   a. 37.82°N, 117.64°W
   b. 40.43°N, 77.67°W
   c. 42.94°N, 122.10°W
   d. 26.20°N, 122.19°W
   e. 39.98°N, 105.29°W
   f. 35.27°N, 119.825°W
   g. 19.47°N, 155.59°W
   h. 36.688°N, 108.837°W

Map Interpretation

VOLCANIC LANDFORMS

The Map

The Menan Buttes map area is located on the upper Snake River Plain in eastern Idaho. The Snake River Plain is a region of recent lava flows that extends across southern Idaho. It is part of the vast volcanic Columbia Plateau, which covers more than 520,000 square kilometers (200,000 sq mi) of the northwestern United States. The lava originated from fissure eruptions that spread vast amounts of fluid basaltic lava across the landscape, accumulating to a thickness of a few thousand meters. Most of the Snake River Plain has an elevation between 900 and 1500 meters (3000 and 5000 ft). Rising above the basalt plain are numerous volcanic peaks, including Menan Buttes and Craters of the Moon. The Snake River flows westward across the region.

The Snake River Plain has a semiarid, or steppe, climate. Because of the moderately high elevation of the plain, temperatures are cooler than in nearby lowlands, with an average annual temperature of about 10°C (50°F). Precipitation is between 25 and 50 centimeters (10 and 20 in.). It is unpredictable from year to year, and there are thunderstorms during the summer months. The low rainfall total is mainly a result of a rain-shadow effect. Moisture-producing storms from the Pacific Ocean are blocked from reaching the region by the Cascade Range and the Idaho Batholith section of the northern Rockies. The upper Snake River Plain and the Menan Buttes area are left with dry, adiabatically warmed air descending from the lee slopes of the mountains. The vegetation cover in the area is sparse, mainly characterized by sagebrush and bunch grasses.

Interpreting the Map

1. What type of volcano are the Menan Buttes? On what landform characteristics is your decision based?
2. What is the local relief of the northern Menan Butte? What is the depth of the crater of each butte?
3. Is the overall shape of each butte symmetrical or asymmetrical? What might account for that aspect of their shape?
4. Do you think that these volcanoes are active at present? What evidence from the map and aerial photograph indicates activity or a period of inactivity?
5. Sketch an east–west profile across the center of northern Menan Butte from the railroad tracks to the channel of Henry's Fork (north is at the top). Is this profile typical of a volcanic summit and crater? What is the horizontal distance of the profile?
6. Slope ratio can be calculated by dividing the relief by the horizontal distance. For example, a 1000-meter-high mountain slope with a horizontal distance of 3000 meters would have a slope ratio of 1: 3. What is the slope ratio for the western slope of northern Menan Butte from the crater ridge down to the railroad tracks at the foot of the butte?
7. This is a shaded relief topographic map and differs from most of the other topographic maps in this book. What is the major advantage of this shaded relief mapping technique? Are there any disadvantages, compared with the regular contour topographic maps?
8. Compare the southern Menan Butte on the map with the vertical aerial photograph on this page. Why would it be useful to have both a map and an aerial photograph when studying landforms? What is the chief advantage of each?

USDA/NCRS

Vertical aerial photograph of one of the Menan Buttes, Idaho.

Opposite:
Menan Buttes, Idaho
Scale 1:24,000
Contour interval = 10 ft
U.S. Geological Survey

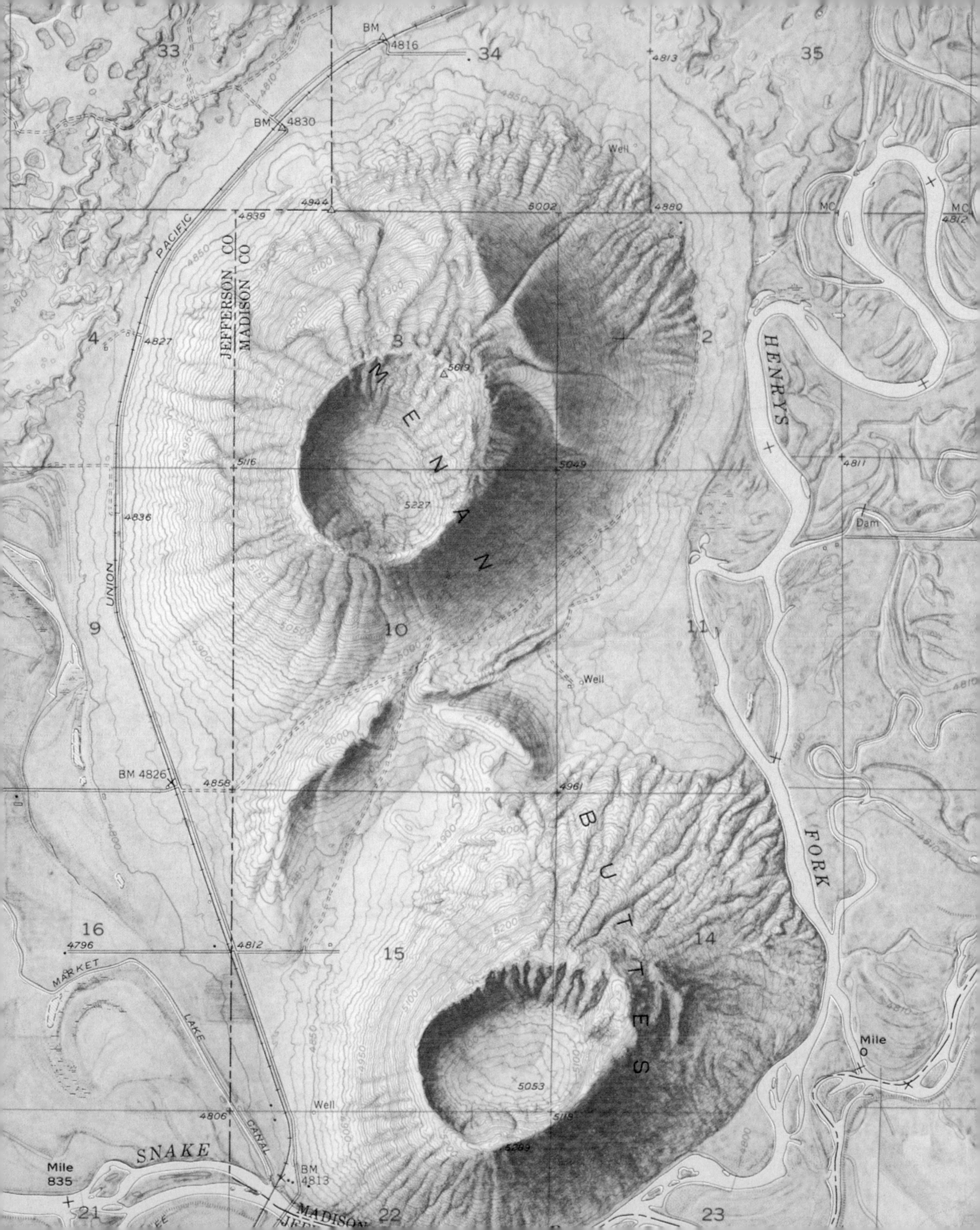

MENAN
BUTTES
HENRYS
FORK
SNAKE
JEFFERSON CO
MADISON CO
UNION
PACIFIC
MARKET
LAKE
CANAL
MADISON
BM 4816
BM 4830
BM 4826
BM 4813
Well
Dam
Mile 835
Mile 0
5619
5227
5053
5283
33
34
35
4
3
2
9
10
11
16
15
14
21
22
23

Weathering and Mass Wasting

15

CHAPTER PREVIEW

Rocks exposed at Earth's surface are subject to disintegration and decomposition by physical and chemical weathering processes.

- What is there about the environment at Earth's surface that makes rocks vulnerable to this breakdown?
- Under what climatic conditions are chemical weathering processes generally more effective than physical weathering processes?

Not all rocks weather at the same rate in the same environmental setting.

- Why are some rocks more vulnerable to weathering than others in a given environment?
- What factors might cause different outcrops of the same rock type to vary in their vulnerability to weathering in the same region?

Weathering and mass wasting are exogenic processes, which means they originate at the surface of Earth and contribute to an overall decrease in relief.

- Why are weathering and mass wasting the initial exogenic processes experienced by rock matter at Earth's surface?
- How are weathering and mass wasting related to erosion, transportation, and deposition?

Various categories of mass wasting are distinguished on the basis of the materials moved and the nature of their motion.

- Why is climate important in determining the type of mass wasting that can occur?
- Why is rock type important in determining the type of mass wasting that can occur?

Weathering and mass wasting are the most commonly occurring exogenic processes.

- How do they affect the appearance of the landscape?
- What makes weathering and mass wasting important to people and society?

◄ Opposite: Rockfall deposits in Glacier National Park, Montana, have accumulated into a talus cone, the slope of angular clasts.
USGS/P. Carrara

The previous two chapters concerned the materials, processes, and structures associated with the construction of topographic relief at the surface of Earth. However, the relief-building *endogenic* geomorphic processes of tectonism and volcanism, which arise from within Earth, are opposed by the relief-reducing *exogenic* geomorphic processes, which originate at the surface. Exogenic processes break down rocks and erode rock materials from higher energy sites and transport them to locations of lower energy. The relocation of rock fragments can be accomplished by the force of gravity alone or with the help of one of the *geomorphic agents*—flowing water, wind, moving ice, or waves. This chapter focuses on the exogenic processes that cause rocks to decay and on the ways that erosion, transportation, and deposition of surficial Earth materials are accomplished when gravity, rather than a geomorphic agent, is the dominant factor in transporting the Earth materials.

Gravity constantly pulls downward on all Earth surface materials. Weakened rock and broken rock fragments are especially susceptible to downslope movement by gravity, which may be slow and barely noticeable, or rapid and catastrophic. Slope instability causes costly damage to buildings, roadways, pipelines, and other types of construction, and is also responsible for injury and loss of life.

Some gravity-induced slope movements are entirely natural in origin, but human actions contribute to the occurrence of others. Understanding the processes involved and circumstances that lead to slope movements can help people avoid these costly and often hazardous events.

Nature of Exogenic Processes

Earth's surface provides a harsh environment for rocks. Most rocks originate under much higher temperatures and pressures and in very different chemical settings than those occurring at Earth's surface. Thus, surface and near-surface conditions of comparatively low temperature, low pressure, and extensive contact with water cause rocks to undergo varying amounts of disintegration and decomposition (• Fig. 15.1). This breakdown of rock material at and near Earth's surface is known as **weathering.** Rocks weakened and broken by weathering become susceptible to the other exogenic processes—erosion, transportation, and deposition. A rock fragment broken (weathered) from a larger mass will be removed from that mass (eroded), moved (transported), and set down (deposited) in a new location. Together, weathering, erosion, transportation, and deposition actually represent a chain or continuum of processes that begins with the breakdown of rock.

Erosion, transportation, and deposition of weathered rock material often occurs with the assistance of a geomorphic agent, such as stream flow, wind, moving ice, and waves. Sometimes, however, the only factor involved is gravity. Gravity-induced downslope movement of rock material that occurs without the assistance of a geomorphic agent, as in the case of a rock falling from a cliff, is known as **mass wasting.** Although gravity also plays a role in the redistribution of rock material by geomorphic agents, the term mass wasting is reserved for movement caused by gravity alone. Whether it is mass wasting or a geomorphic agent doing the work, fragments and ions of weathered rock are removed from high-energy locations and transported to positions of low energy, where they are deposited.

The variations in elevation at Earth's surface as well as the appearance of various landforms reflect the opposing tendencies of endogenic and exogenic processes. The relief created by tectonism and volcanism will decrease over time if endogenic processes cease or operate at a slow rate compared to the exogenic processes (• Fig. 15.2). Rates of the exogenic processes depend on such factors as rock resistance to weathering and erosion, the amount of relief, and climate.

Different weathering processes often impart visually distinctive features to a landform or landscape. Typically, weathering, mass wasting, or one of the various geomorphic agents do not work alone in shaping and developing a landform. More often they work together to modify the landscape, and evidence of multiple processes can be discerned in the appearance of the resulting landform (• Fig. 15.3). For example, both the northern Rockies and the Sierra Nevada were produced by tectonic uplift, but much of the spectacular terrain seen there today is the result of weathering, mass wasting, running water, and glacial activity that have sculpted the mountains and valleys, often in distinctive ways (• Fig. 15.4).

Weathering

Environmental conditions at and near Earth's surface subject rocks to temperatures, pressures, and substances, especially water, that contribute to physical and chemical breakdown of exposed rock. Broken fragments of rock, called *clasts*, that detach from the original rock mass can be large or small. These detached pieces continue to weather into smaller particles. Fragments may accumulate close to their source or be widely dispersed by mass wasting and the geomorphic agents. Many weathered rock fragments become sediments deposited to form landforms such as floodplains, beaches, or sand dunes, while others blanket hillslopes and comprise *regolith*, the inorganic portion of soils. Weathering is the principal source of inorganic soil constituents, without which most vegetation could not grow. Likewise, ions chemically removed from rocks during weathering are transported in surface or subsurface water to close or distant locations. They are a major source of nutrients in terrestrial as well as aquatic ecosystems, including rivers, ponds, lakes, and the oceans.

The kinds of rock weathering fall into two basic categories. **Physical weathering,** also known as **mechanical weathering,** disintegrates rocks, breaking smaller fragments from a larger block or outcrop of rock. **Chemical weathering** decomposes rock through chemical reactions that change the original rock-forming minerals. Many different physical and chemical processes lead to rock weathering, and water plays an important role in almost all of them.

• FIGURE 15.1
This boulder, which was once hard and solid, has been subjected to conditions at Earth's surface leading to its disintegration and decomposition.
Why are some exposed parts of the boulder darker than others?

J. Petersen

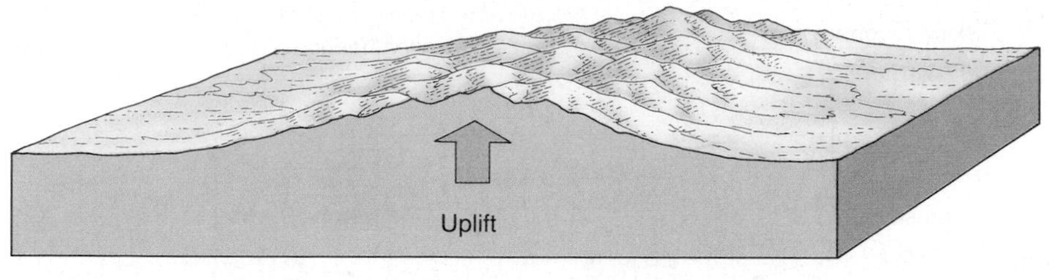

(a) Tectonic uplift

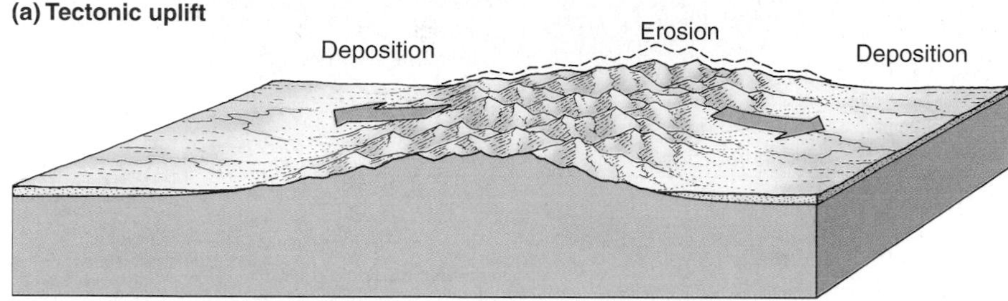

(b) Exogenic processes dominate

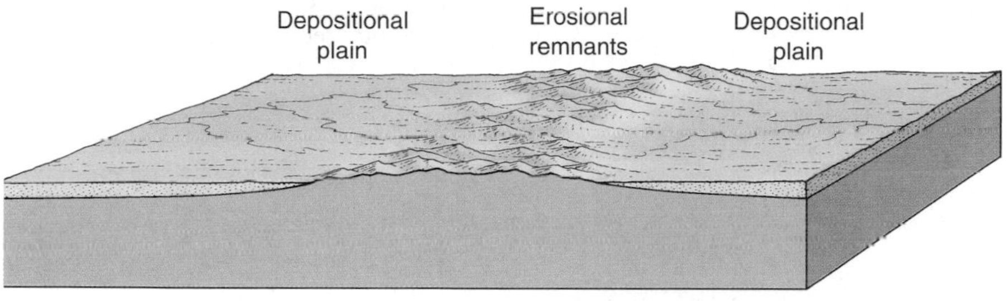

(c) Reduced relief

● **FIGURE 15.2**
(a) Tectonic (endogenic) uplift is opposed by (b) the exogenic processes of weathering, mass wasting, erosion, transportation, and deposition that (c) eventually decrease relief at Earth's surface significantly if no additional uplift occurs.

Physical Weathering

The mechanical disintegration of rocks by physical weathering is especially important to landscape modification in two ways. First, smaller clasts are more easily eroded and transported than larger ones. Second, the breakup of a large rock into smaller ones encourages additional weathering because it increases the surface area exposed to weathering processes. There are several ways by which rocks can be physically weathered. For example, a person breaking a rock with a hammer is carrying out physical weathering. Although people, animals, and plants are responsible for breaking some rocks (● Fig. 15.5), most physical weathering occurs in other ways. Five principal types of physically weathering are discussed here.

Unloading Most rocks form under much higher pressures (weight per unit area) than the 1013.2 millibars (29.92 in. Hg), or 10 Newtons per square centimeter (15 lbs/in^2), of average atmospheric pressure that exists at Earth's surface. Intrusive igneous rocks solidify slowly deep beneath the surface under great pressure from the weight of the overlying rocks. Sedimentary rocks solidify partly or wholly due to compaction from the weight of overlying sediments. Most metamorphic rocks originate when high pressure and temperature substantially change preexisting rocks. Commonly, rocks that originated under conditions of high pressure from deep burial are uplifted through mountain-building tec tonic processes to eventually be exposed at the surface. For example, a large pluton of the intrusive igneous rock granite can be uplifted in a fault block during tectonism. High elevation helps drive erosional stripping of the overlying rocks, and ultimately through this removal of overlying weight, the **unloading** process, the rock is exposed at the surface, where it is subjected to the low pressure of the atmosphere. As a result of the pressure differential between high pressure at depth and low atmospheric pressure, the outer few centimeters to meters of the rock mass expand outward toward the atmosphere (● Fig. 15.6). This expansion causes the rock mass to crack in a roughly concentric form. These expansion cracks are joints, not faults. Concentric sheets of granite or other massive rocks form, broken and separated by these concentric joints. The granite expands and breaks, weathering mechanically like this because of the erosional removal of overlying rock masses. This physical weathering process of unloading is especially common with granite, but can affect other types of rocks as well.

As the outer sheet of an unloaded rock breaks further, segments of it may slide off and slough away or weather away, reducing the load on underlying rock and allowing additional concentric joints to form. The successive removal of these outer rock sheets is known as **exfoliation.** Weathering by unloading leads to exfoliation, and each concentric broken layer of rock is called an **exfoliation sheet.** Unloaded granite rock masses often have a domelike surface form, a topographic feature called an **exfoliation dome.** Well-known examples of granite exfoliation domes are Stone Mountain in Georgia, Half Dome in Yosemite National Park, Sugar Loaf Mountain, which overlooks Rio de Janeiro, Brazil, and Enchanted Rock in central Texas (● Fig. 15.7).

Thermal Expansion and Contraction Many early Earth scientists believed that the extreme diurnal temperature changes common in deserts caused physical weathering of rocks by expansion and contraction as they warmed and cooled. They cited widespread existence of split rocks in arid regions as evidence of the effectiveness of this **thermal expansion and contraction** weathering (● Fig. 15.8). Laboratory studies in the early 20th century seemed to refute this idea, but the notion remained somewhat popular, although controversial. Recent field studies in the American Southwest deserts lend support to the notion that alternating heating and cooling can indeed lead to

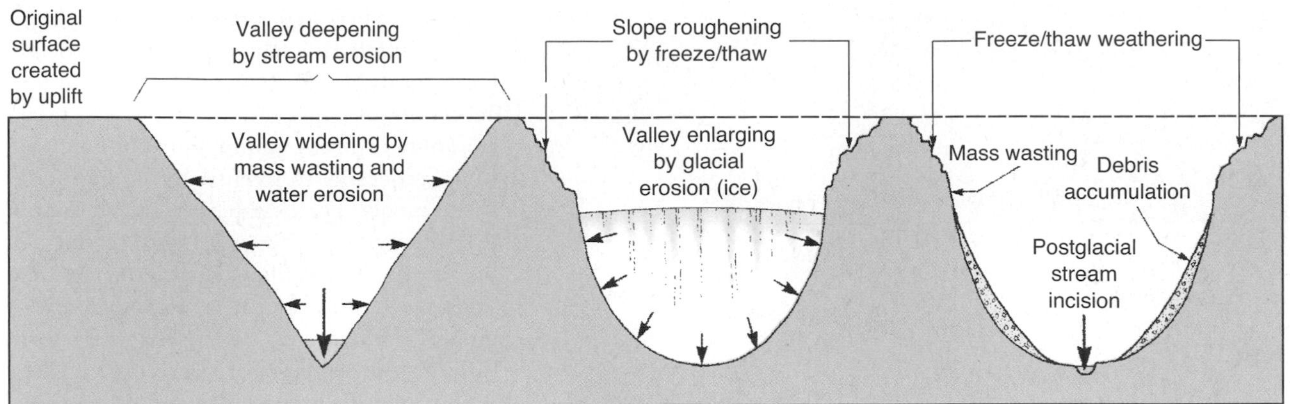

• FIGURE 15.3
The landscape of most high mountain regions has been produced by at least three different phases of landforming processes: stream cutting and valley formation during tectonic uplift; glacial enlargement of former stream valleys, with intense freeze–thaw weathering above the level of the ice; and postglacial weathering and mass wasting with recent stream cutting.
How does the cross-sectional profile of the valley change at each phase?

USGS/Austin Post

• FIGURE 15.4
In the foreground of this aerial view of mountain summits in the northern Rockies are the White Cloud Peaks in the Sawtooth Range of Idaho.
Can you identify evidence of the three phases shown in Figure 15.3?

the mechanical splitting of rocks. Less controversial has been the notion that differential thermal expansion and contraction of individual mineral grains in coarse crystalline rocks contributes to **granular disintegration,** the breaking free of individual mineral grains from a rock (• Fig.15.9). The broken rocks found after a forest or brush fire provide additional evidence for the effectiveness of thermal expansion and contraction weathering, but under much greater extremes of temperature.

J. Petersen

● **FIGURE 15.5**
A growing tree has broken up a sidewalk and retaining wall made of concrete, just as trees and other plants can break up natural rock into smaller fragments, contributing to physical weathering.
How might an animal cause physical weathering?

J. Petersen

● **FIGURE 15.6**
Outward expansion due to weathering by unloading has caused this granite in the Sierra Nevada, California, to break, forming joints and sheets of rock that parallel the surface.

Freeze–Thaw Weathering In areas subject to numerous diurnal cycles of freeze/thaw, water repeatedly freezing in fractures and small cracks in rocks contributes significantly to rock breakage by **freeze–thaw weathering,** sometimes referred to as *frost weathering,* or *ice wedging.* When water freezes, it expands in volume up to 9%, and this can cause large pressures to be exerted on the walls and bottom of the crack, widening it and eventually leading to a piece of rock breaking off (● Fig. 15.10). The damaging effects of freezing water expanding is why vehicles driven in regions that experience freezing temperatures must have antifreeze instead

J. Petersen

● **FIGURE 15.7**
Enchanted Rock in central Texas is a huge granite exfoliation dome.
Why is granite so susceptible to unloading and exfoliation?

J. Petersen

● **FIGURE 15.8**
This split rock may be the result of physical weathering by thermal expansion and contraction.

of water in their radiator systems. Likewise, water pipes to and within buildings will burst if they are not sufficiently insulated to protect the water inside them from freezing. Freeze–thaw weathering is particularly effective in the upper-middle and lower-high latitudes, and results of freeze–thaw weathering are especially noticeable in mountainous regions near tree lines where angular blocks of rock attributed to freeze–thaw weathering are common (● Fig. 15.11). Freeze–thaw weathering is not significant at lower latitudes except in areas of high elevation.

Salt Crystal Growth The development of salt crystals in cracks, fractures, and other void spaces in rocks causes physical disintegration in a way that is similar to freeze–thaw weathering. With **salt crystal growth,** water with dissolved salts accumulates in these

• **FIGURE 15.9**
Weathering of this intrusive igneous boulder has taken the form of granular disintegration, as evidenced by the numerous, individual mineral grains that have collected around the base of the boulder.
What other evidence exists on the boulder to suggest that it has been subjected to considerable weathering?

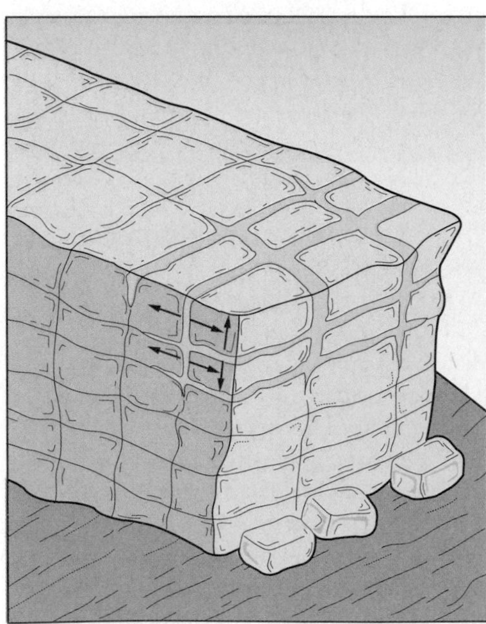

• **FIGURE 15.10**
Water expands when it freezes into ice. The freezing of water within rock fractures creates a force of expansion great enough to cause some rock weathering.
How important is freeze–thaw weathering where you live?

spaces and then evaporates, and the growing salt crystals wedge pieces of rock apart. This physical weathering process is most common in arid regions and in rocky coastal locations where salts are abundant, but people also contribute to salt weathering when they use salt to melt ice from roads and sidewalks in winter. Salt crystal growth can lead to granular disintegration in coarse crystalline (intrusive igneous) rocks or to the removal of clastic particles from sedimentary rocks, especially sandstones. Continued salt weathering, accompanied by removal of the weathered fragments, contributes to the creation of numerous hollows, small caves, and large overhangs in exposed sandstone cliffs (• Fig. 15.12). Large caves and overhangs (called *alcoves*) provided important living spaces for prehistoric Native American communities in the American Southwest, for example, at Mesa Verde, Colorado, and Canyon de Chelly, Arizona.

Hydration In weathering by **hydration,** water molecules attach to the crystalline structure of a mineral without causing a permanent change in that mineral's composition. The water molecules are able to join and leave the "host" mineral during hydration and dehydration, respectively. A mineral will expand when hydrated and shrink when dehydrated. As in freeze–thaw and salt crystal growth weathering, when hydrating materials expand in cracks or voids, pieces of the rock can wedge apart. Clasts, mineral grains, and thin flakes can be broken from a rock mass because of hydration weathering. Salts and *clay minerals,* which are clay-sized materials formed during chemical weathering, commonly occupy cracks and voids in rocks and are subject to hydration and dehydration.

Chemical Weathering

Chemical reactions between substances at Earth's surface and rock-forming minerals also work to break down rocks. In chemical weathering, ions from a rock are either released into water or recombine with other substances to form new materials, such as clay minerals. New materials made by chemical weathering are more stable at Earth's surface than the original rocks. The most important catalysts and reactive agents performing chemical weathering are water, oxygen, and carbon dioxide, all of which are common in soil, precipitation, surface water, groundwater, and air.

Oxidation Water that has regular contact with the atmosphere contains plenty of oxygen. When oxygen in water comes in contact with certain elements in rock-forming minerals, a chemical reaction can occur. In this reaction, the element releases its bond with the mineral, leaving behind a substance with an altered chemical formula, and establishes a new bond with the oxygen. This chemical union of oxygen atoms with another substance to create a new product is called **oxidation**. Metals, particularly iron and aluminum, are commonly oxidized in rock weathering and

D. Sack

• **FIGURE 15.11**
Angular blocks of rock attributed to freeze–thaw weathering, like these near the tree line in Great Basin National Park, Nevada, are common in mountainous areas that experience numerous freeze–thaw cycles every year.
Why are these rocks angular in shape rather than rounded?

• **FIGURE 15.12**
Physical weathering by salt crystal growth helps break apart rocks, especially along bedding planes exposed in cliffs, and on rocky seacoasts. Weathered fragments are removed by gravity, wind, and water, leaving behind hollows.
Once a small hollow is formed, how might it impact further weathering at that site?

Dawn Endico/ www.flickr.com/photos/candiedwomanire/104320384/

form *iron and aluminum oxides* as the new products. Compared to the original rock, these oxides, including Fe_2O_3 and Al_2O_3, are chemically more stable, not as hard, larger in volume, and have a distinctive color. Iron oxides produced in this way often have a red, orange, or yellow color, while oxidized aluminum from rock weathering frequently appears yellow (• Fig. 15.13). Oxidation of iron is very common, and when it affects steel and iron objects we call it rust.

Carbonation and Solution

Water at and near the surface also typically contains a considerable amount of carbon dioxide obtained from the air and from decaying organic matter in the soil. Weathering by **carbonation** occurs when carbon dioxide in water reacts with rock material to produce bicarbonate ions (HCO_3^-) and other ions that vary with the composition of the decomposing rock. Carbonation weathering is most effective on carbonate rocks (those containing CO_3), particularly limestone, which is an abundant chemical precipitate sedimentary rock

D. Sack

• **FIGURE 15.13**
The reddish orange coloration on the surface of this boulder reveals that its constituents include minerals containing iron, and that it has been subjected to chemical weathering by oxidation.
What is a likely chemical formula for the reddish orange substance?

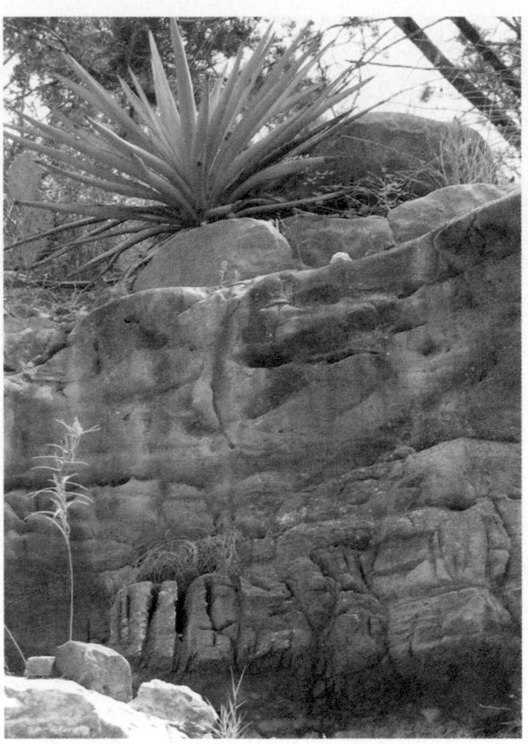

J. Petersen

• **FIGURE 15.14**
This limestone shows the effects of chemical weathering by carbonation. Weathered limestone often takes on a fluted, pitted, or even honeycombed appearance.
Why does the rock near the bottom seem to be more weathered than that at the top of this rock ledge?

composed of calcium carbonate ($CaCO_3$). When water with sufficient carbon dioxide comes into contact with limestone, the chemical reaction creates calcium ions as well as bicarbonate ions. During carbonation, the calcium and carbonate in limestone detach from each other, thereby decomposing the limestone (• Fig. 15.14). The weathering event occurs as follows: $H_2O + CO_2 + CaCO_3 = Ca^{2+} + 2HCO_3^-$. Water carries away the ions of calcium and bicarbonate produced in the reaction. Similar reactions take place when other carbonate rocks undergo carbonation weathering. Because of the role of water in carbonation, limestone in particular, but carbonate rocks overall, tend to weather extensively in humid regions, but they are resistant and often form cliffs in arid climates. Because water can obtain carbon dioxide by moving through the soil, carbonation operates on rocks impacted by soil water or groundwater in the subsurface as well as on rocks exposed at the surface.

Carbonation is a specific chemical reaction that leads to **solution,** the dissolving of rock matter in water. Rock salt, which contains the mineral halite (NaCl), is especially susceptible to solution in water more directly, without involving the carbonation reaction. Most minerals that are insoluble or only slightly soluble in pure water will dissolve more readily if the water is acidic. Carbon dioxide in water creates acidic conditions because the two substances react to make carbonic acid (H_2CO_3). Other acids present in water, derived primarily from decaying organic matter in the soil, can also facilitate the solution of minerals. In some cases this operates on exposed rock, as when lichens and mosses that grow on rock surfaces secrete acidic substances and that mix with water from precipitation (• Fig. 15.15).

G. Nadon/Ohio University

• **FIGURE 15.15**
The green patches of lichens growing on this black and white granitic rock contribute to solution weathering with the acids that they secrete.
How does the fact that lichens retain moisture also contribute to weathering?

Hydrolysis In the weathering process of **hydrolysis,** water molecules alone, rather than oxygen or carbon dioxide in water, react with chemical components of rock-forming minerals to create new compounds, of which the H^+ and OH^- ions of water are a part. Many common minerals are susceptible to hydrolysis, particularly the silicate minerals that comprise igneous rocks. Hydrolysis of silicate minerals often produces clay minerals.

GEOGRAPHY'S PHYSICAL SCIENCE PERSPECTIVE

Expanding and Contracting Soils

Expansion and contraction of rocks, sediments, or of materials in fractures and other void spaces in rocks and sediments is a common theme in weathering and mass wasting. Physical weathering by thermal expansion and contraction, freeze/thaw of water, and hydration/dehydration of salts and clay minerals all involve cycles of expansion and contraction, as does downslope movement of soil by creep.

Soils consist of inorganic and organic materials, soil water, and soil air. The inorganic fraction of soils—rock fragments and minerals—derives from rock weathering. These fragments are of sand, silt, and clay size. As weathering continues, clasts become smaller and smaller. In addition, the chemical weathering process of hydrolysis produces very small substances called clay minerals. Clay minerals are *colloidal* in size, meaning they are smaller than 0.0001 millimeter. In addition to rock fragments and clay minerals, some decaying organic matter (humus) in the soil is colloidal in size.

Having colloidal-sized materials in the soil benefits soil fertility because colloids increase the cation exchange capacity of the soil, which is essentially the ability of the soil to hold plant nutrients. Colloids are so small that their negative electrical charge becomes an important characteristic. Plant nutrients, such as potassium, calcium, magnesium, phosphorous, copper, zinc, and several others, are positively charged ions, or cations. The negatively charged colloids, therefore, attract nutrient cations. When a plant takes up soil water, nutrient cations are drawn up as well. A high cation exchange capacity means that the soil has plenty of colloids and therefore the potential to maintain a large number of nutrient cations in the soil for plants to use. The specific capacity of a colloid to hold cations is a function of its composition and structure. The cation exchange capacity of a clay mineral depends on the type of clay mineral, which, in turn, depends on the kind of parent material and the extent of weathering. Illite and vermiculite are two clay minerals associated with slightly weathered soils; montmorillonite is an important component of moderately weathered soils; and kaolinite is characteristic of highly weathered soils.

Nutrient cations are not the only substances that join and leave some colloids. Water molecules come into and leave the crystal structure of colloidal-sized clay minerals during hydration and dehydration. Clay minerals have layered, "sandwichlike" structures. Water molecules enter and exit their layered crystalline structure and cause the clay mineral to expand when wet and contract when dry. Of the four basic clay minerals, montmorillonite expands and contracts the most upon hydration and dehydration. The expansion and contraction, or swelling and shrinking, of clay minerals applies physical pressure on surrounding soil particles and rock surfaces and contributes to physical weathering. According to some sources, the expansion in clay mineral volume can vary from a very small percentage to more than 100%. In most cases, however, the expansion in volume is less than 50%.

A high content of clay minerals advantageous for soil fertility sometimes interferes with the human-built environment. Soils with a significant amount of expansive clays swell and shrink considerably as they become wet and dry. The expansion and contraction in soil volume can shift and crack roads, sidewalks, and building foundations. In regions with typically fertile, expansive clay-rich soils, this causes an estimated $7 billion damage a year to structures in the United States alone. Special construction techniques can be used to mitigate the negative effects of shrinking and swelling on the built environment.

© Jeff Pope/JPI Home Inspection Service/Courtesy National Association of Certified Home Inspectors

The concrete driveway slab here has been heaved upward by expansive soils, also causing damage to the adjacent railing and retaining wall.

Jeff Griffith, Quality Home Inspectors/Photo courtesy of National Association of Certified Home Inspectors

Foundation cracks due to expansive soils can cause serious damage to buildings that is expensive to repair.

The specific type of clay mineral produced by hydrolysis depends on the composition of the preexisting mineral and other substances that may be present and active in the water. Because water is the weathering agent, hydrolysis is not limited to rocks that are exposed at the surface, but can occur in the subsurface through the action of soil and groundwater; in tropical humid climates, it can decay rock to a depth of 30 meters (100 ft) or more.

The chemical weathering process of hydrolysis differs from the physical weathering process of hydration, in which water molecules join and leave a substance, causing it to swell and shrink without changing its inherent chemical formula. Both weathering processes, however, may involve clay minerals because those clay-sized substances are often the product of hydrolysis, and many clay minerals swell and shrink substantially during hydration and dehydration.

Variability in Weathering

How and why types and rates of weathering vary, at both the regional and local spatial scales, are of particular interest to physical geographers. In addition to climate, the type of rock (lithology) and the nature and amount of fractures or other weaknesses in it are major influences on the effectiveness of the various rock weathering processes. How specific rocks weather in different natural settings is important for explaining the amount and formation of regolith, soil, and relief, but it also impacts the cultural environment because weathering affects building stone as well as rock in its natural setting.

Climate

In almost all environments, physical and chemical weathering processes operate together, but usually one of these categories dominates. Although water plays a role in all but two of the physical weathering processes, it is essential for all types of chemical weathering. Also, chemical weathering increases as more water, comes in contact with rocks. Chemical weathering, then, is particularly effective and rapid in humid climates (● Fig. 15.16). Most arid regions have enough moisture to allow some chemical weathering, but it is much more restricted than in humid climates. Arid regions typically receive sufficient moisture for physical weathering by salt crystal growth and the hydration of salts. Abundant salts, high humidity, and contact with seawater make salt weathering processes very effective in marine coastal locations.

The other principal climatic variable, temperature, also influences dominant types and rates of weathering. Most chemical reactions proceed faster at higher temperatures. Low-latitude regions with humid climates consequently experience the most intense chemical weathering. In the tropical rainforest, savanna, and monsoon climates, chemical weathering is more significant than physical weathering, soils are deep, and landforms appear rounded. Although chemical weathering is somewhat less extreme in the mid-latitude humid climates, its influence is apparent in the moderate soil depth and rounded forms of most landscapes in those regions. In contrast, the landforms and rocks of both arid and cold regions, where physical weathering dominates, tend to be sharper, angular, and jagged, but this depends to some extent on rock type, and rounded features may remain in an arid landscape as relicts from prehistoric times when the climate was wetter (● Fig. 15.17). Comparatively low rates of chemical weathering are reflected in the thin (or absent) soils found in arid, subarctic, and polar climatic regimes. We have already noted the role of daily temperature ranges in weathering processes, including thermal expansion and contraction weathering in arid climates and freeze–thaw weathering in areas with cold winters.

Air pollution that contributes to the acidity of atmospheric moisture accelerates weathering rates. Extensive damage has

● FIGURE 15.16

This diagram of weathering regions summarizes the relationships between climate and weathering processes. Physical weathering is most active where temperature and rainfall are both low. Chemical weathering is most active in regions of high temperature and rainfall. Most world regions experience a combination of both physical and chemical weathering.

In which weathering region would we find a site that has an annual mean temperature of 5°C (41°F) and an annual rainfall of 100 centimeters (40 in.)?

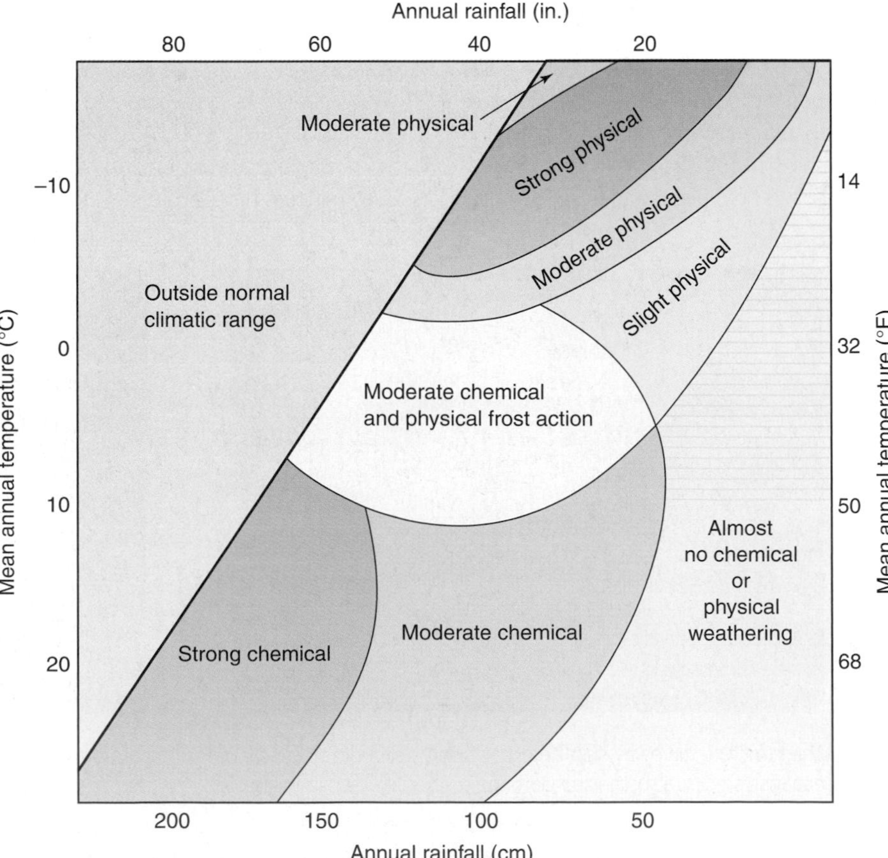

(a)

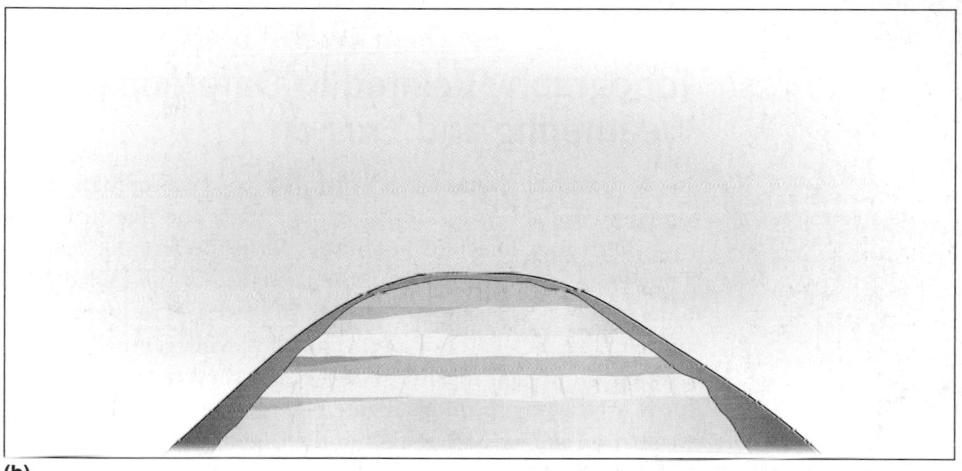

(b)

• FIGURE 15.17
(a) Due to the dominance of slow physical weathering and the sparseness of vegetation, slopes in arid and semiarid environments tend to be bare and angular. Slope angles reflect differences in component rock resistance to weathering and erosion. (b) Because humid region hillsides are affected by the more rapid chemical weathering processes and regolith is held longer on the slope by vegetation, slopes in wetter climates are much more rounded with a deeper weathered mantle than arid region slopes.

already occurred in some regions to historic cultural artifacts made of limestone and marble (metamorphosed limestone), both containing calcium carbonate (● Fig. 15.18). Because many of the world's great monuments and sculptures are made of limestone or marble, there is a growing concern about weathering damage to these treasures. The Parthenon in Greece, the Taj Mahal in India, and the Great Sphinx in Egypt are examples of structures made of rock where pollution-induced chemical solution, and related growth and hydration of salts, are damaging and rotting away monument surfaces.

Rock Type

Wherever many different rock types occupy a landscape, some will be more resistant and others will be less resistant to the weathering processes operating there. Because erosion removes weathered rock fragments more easily than large, intact rock masses, areas of diverse rock types undergo **differential weathering and erosion;** easily eroded rocks exhibit more extensive effects of weathering and erosion than the resistant rocks.

A rock that is strong under certain environmental conditions may be easily weathered and eroded in a different environmental setting. Rocks that are resistant in a climate dominated by chemical weathering may be weak where physical weathering processes dominate, and vice versa. Quartzite is a good example. It is chemically nearly inert and harder than steel, but it is brittle and can be fractured by physical weathering. Shale is chemically inert but mechanically weak. In humid regions, limestone is highly susceptible to carbonation and solution, but under arid conditions, limestone is much more resistant. Granitic outcrops in an arid or semiarid region resist weathering. However, the minerals in granite are susceptible to alteration by oxidation, hydration, and hydrolysis, particularly in regions with warm, humid conditions. Accordingly, granitic areas are often covered by a deeply weathered regolith when they have been exposed to a tropical humid environment.

Structural Weaknesses

In addition to rock type, the relative resistance of a rock to weathering depends on other characteristics, such as the presence of joints, faults, folds, and bedding planes that make rocks susceptible to enhanced weathering. Sandstone is only as strong as its cement, which varies from soluble calcium carbonate to inert and resistant silica. In general, the more massive the rock, that is, the fewer the joints and bedding planes it has, the more resistant it is to weathering.

The processes of volcanism, tectonism, and rock formation produce fractures in rocks that can be exploited by exogenic processes, including weathering. Joints can be found in any solid rock that has been subjected to crustal stresses, and some rocks are intensely jointed (● Fig. 15.19). Joints and other fractures that commonly develop in igneous, sedimentary, and metamorphic rocks represent zones of weakness that expose more surface area of rock, provide space for flow or accumulation of water, collect salts and clay minerals, and offer a foothold for plants (● Fig.15.20). Rock surfaces along fractures tend to experience pronounced weathering. Chemical and physical weathering both proceed faster along any kind of gap, crack, or fracture than in places without such voids.

Because joints are sites of concentrated weathering, the spatial pattern of joints strongly influences the landforms and the appearance of the landscapes that develop. Multiple joints that parallel each other form a **joint set,** and two sets, each composed of multiple parallel joints, will cross each other at an angle (• Fig. 15.21). Joints divide rocks into many different configurations, most commonly resembling blocks or columns, which are often visible in the topography. Over time, preferential weathering and erosion in crossing joint sets leave rock in the central area between the fractures only slightly weathered while the rock near the fractures acquires a more rounded appearance. This distinctive, rounded weathered form, known as **spheroidal weathering,** develops especially well on jointed crystalline rocks, such as granite (• Fig. 15.22). Spheroidal weathering is a *result* that can occur from the interaction of many weathering processes; it does not refer to a specific weathering process. Once a rock becomes rounded, weathering rates of spheroidal outcrops and boulders decrease because there are no more sharp, narrow corners or edges for weathering to attack, and a sphere exposes the least amount of surface area for a given volume of rock.

• FIGURE 15.18
These limestone tombstones have undergone extensive chemical weathering, accelerated by air pollution.
What kind of chemical weathering has impacted the iron fence?

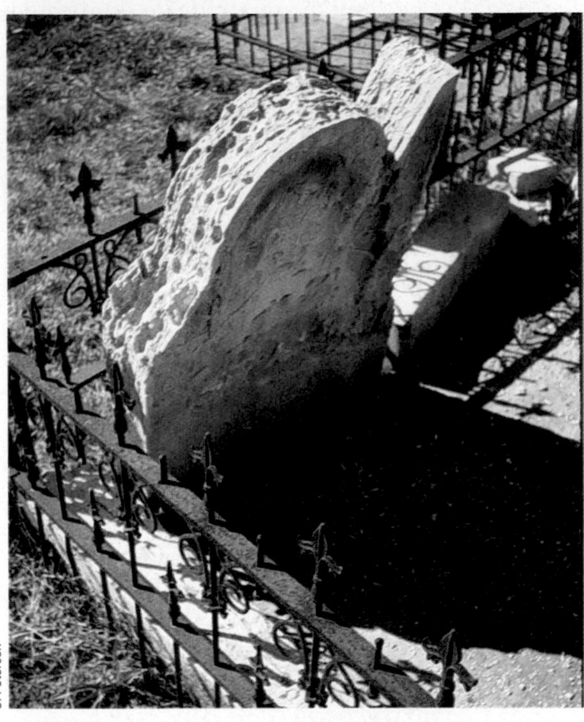

J. Petersen

Topography Related to Differential Weathering and Erosion

In the previous chapter, we learned that structural upfolds (anticlines) do not always form topographic ridges and that structural downfolds (synclines) do not always form topographic valleys. The reason for this stems from the resistance of the folded rock units to the weathering and erosion processes. If resistant rocks are found in the center of a syncline, they will eventually create a topographic high regardless of the structural downfold. Weak rocks, even when forming an anticline, are too easily attacked by weathering and erosion to exist in the landscape as a topographic high for very long. Variation in rock resistance to weathering

• FIGURE 15.19
(a) Massive jointing influences the location and shape of canyons in and around Zion National Park, Utah. To get an idea of the scale of this satellite image, note the smoke from the wildfire at upper left. (b) Farther north, in Bryce Canyon National Park, Utah, numerous closely spaced vertical joints in sedimentary rocks are sites of preferential weathering and erosion, leaving narrow rock spires between joints.

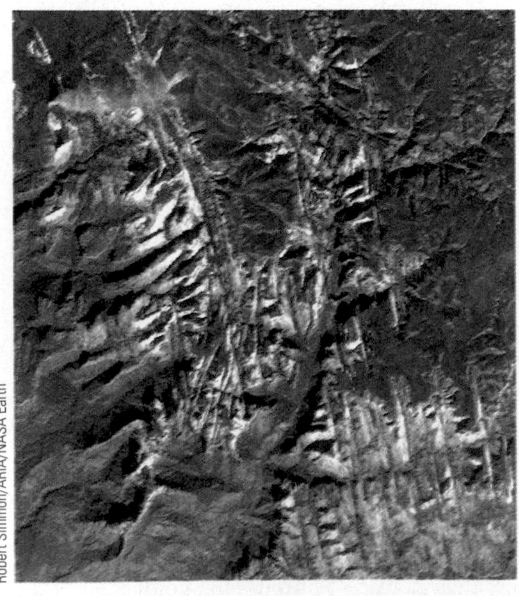

Robert Simmon/ARIA/NASA Earth

(a)

J. Petersen

(b)

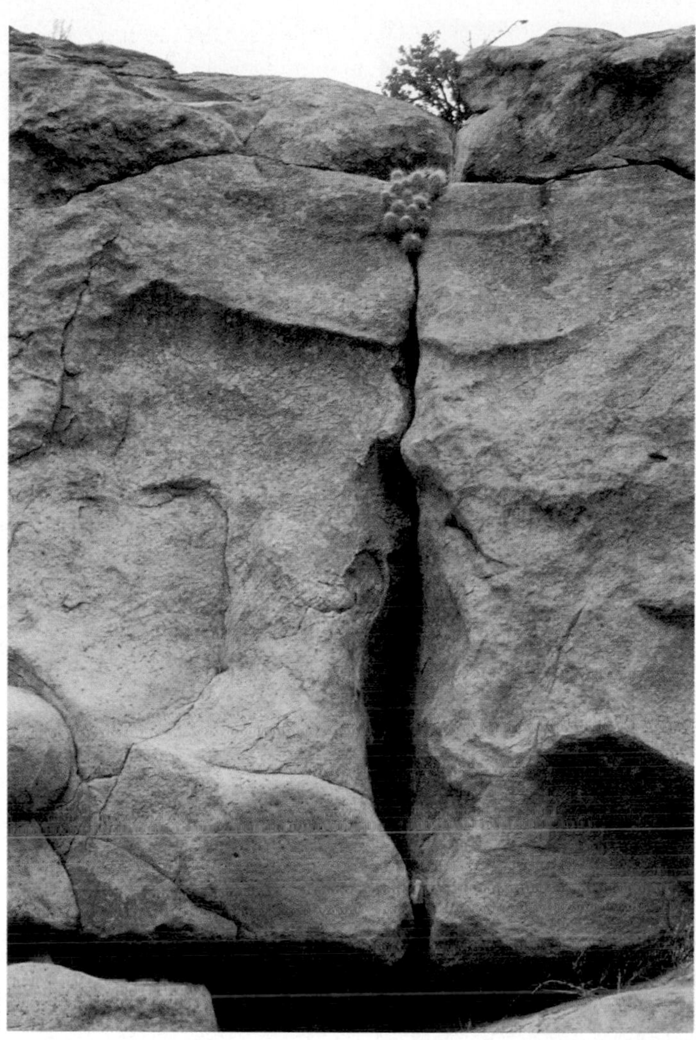
D. Sack

• **FIGURE 15.20**
Vertical joints in rocks are sites of concentrated weathering and erosion as water preferentially accumulates in and moves along them. Higher up at a narrow spot, this joint has accumulated enough soil for the cactus to grow, while enhanced weathering and erosion lower down has noticeably widened the joint.

exerts a strong and often highly visible influence on the appearance of landforms and landscapes. Given sufficient time, rocks that are resistant to weathering and erosion tend to stand higher than less resistant rocks. Resistant rocks stand out in the topography as cliffs, ridges, or mountains, while weaker rocks undergo greater weathering and erosion to create gentler slopes, valleys, and subdued hills.

An outstanding example of how differential weathering and erosion can expose rock structure and enhance its expression in the landscape is the scenery at Arizona's Grand Canyon (• Fig. 15.23). In the arid climate of that region, limestone is resistant, as are sandstones and conglomerates, but shale is relatively weak. Strong and resistant rocks are necessary to maintain steep or vertical cliffs. Thus, the stair-stepped walls of the Grand Canyon have cliffs composed of limestone, sandstone, or conglomerate, separated by gentler slopes of shale. At the canyon base, ancient resistant metamorphic rocks have produced a steep-walled inner gorge. The topographic effects of differential weathering and erosion tend to be more prominent and obvious in landscapes of arid and semiarid climates. In these dry environments, chemical weathering is minimal, so slopes and varying rock units are not generally covered under a significant mantle of soil or weathered rocks. In addition, vegetation typically does not mask the topography in arid regions as it does in humid regions.

Another example of differential weathering and erosion can be seen in the Appalachian Ridge and Valley region of the eastern United States (• Fig. 15.24). The rock structure here

• **FIGURE 15.21**
Multiple cross-cutting joint sets are visible in this aerial view of part of the Colorado Plateau.
With north at the top of this photo, what directions do the two most apparent joint sets trend?

D. Sack

• **FIGURE 15.22**
Cross-jointed granite east of the Sierra Nevada in California forms this hill of spheroidally weathering blocks of rock.

J. Petersen

National Park Service/Mark Lellouch

• FIGURE 15.23
The Grand Canyon of the Colorado River in Arizona is a classic example of differential weathering and erosion in an arid climate. Weathering, mass wasting, and erosion work together to make differences in rock structure and strength visible in the landscape. Rock layers of varying thickness and resistance result in a distinctive array of cliffs formed by strong rocks and slopes formed by less resistant rocks.

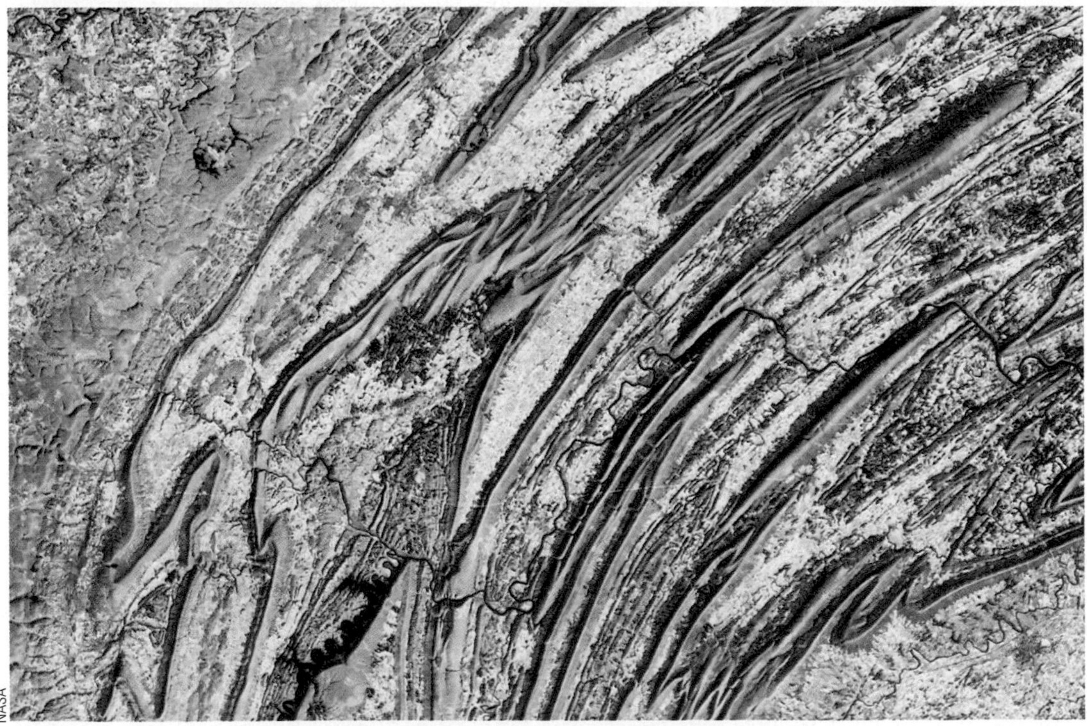

NASA

• FIGURE 15.24
A satellite image of Pennsylvania's Ridge and Valley section of the Appalachians clearly shows the effects of weathering and erosion on folded rock layers of different resistance. Resistant rocks form ridges, and weaker rocks form valleys.
Can you see how the topography of the Ridge and Valley section influences human settlement patterns?

consists of sandstone, conglomerate, shale, and limestone folded into anticlines and synclines. These folds have been eroded so that the edges of steeply dipping rock layers are exposed as prominent ridges. In this humid climate region, forested ridges composed of resistant sandstones and conglomerates stand up to 700 meters (2000 ft) above agricultural lowlands that have been excavated by weathering and erosion out of weaker shales and soluble limestones.

Mass Wasting

Mass wasting, also called *mass movement,* is a collective term for the downslope transport of surface materials in direct response to gravity. Everywhere on the planet's surface, gravity pulls objects toward Earth's center. This gravitational force is represented by the weight of each object. Heavier objects have a greater downward pull from gravity than lighter objects. The force of gravity encourages rock, sediment, and soil to move downhill on sloping surfaces.

Mass wasting operates in a wide variety of ways and at many scales. A single rock rolling and tumbling downhill is a form of this gravitational transfer of materials (● Fig. 15.25), as is an entire hillside sliding hundreds or thousands of meters downslope, burying homes, cars, and trees. Some mass movements act so slowly that they are imperceptible by direct observation and their effects appear gradually over long periods of time. In these cases, tilted telephone poles, gravestones, fence posts, retaining walls, trees, or cracks in buildings can reveal how mass wasting processes are affecting the ground beneath those objects. Other types of mass wasting are disastrous in scale and produce instantaneous violence.

The cumulative impact of all forms of mass wasting rivals the work of running water as a modifier of physical landscapes because gravitational force is always present. Wherever there is loose rock, regolith, or soil on a slope, gravity will cause some movement downslope. Friction and rock strength are factors that resist this downslope movement of materials. Friction increases with the roughness and angularity of a rock fragment and the roughness of the surface on which it rests. Rock strength depends on physical and chemical properties of the rock and is especially decreased by any kind of break or gap in the rock. Fractures, joints, faults, bedding planes, and spaces between mineral grains or clasts all weaken the rock. Furthermore, because all of these gaps invite the accumulation of water, their bonds to the outcrop continue to weaken further over time through weathering.

Slope angle also helps determine whether or not mass wasting will occur. Gravitational forces act to pull objects straight downward, toward the center of Earth. The closer a slope is to being parallel to that downward direction, in other words, the steeper the slope angle, the easier it is for the gravitational forces to overcome the resistance of friction and rock strength. Gravity is more effective at pulling rock materials downslope on steep hillsides and cliffs than on gently sloping or level surfaces. The steeper the slope, the stronger the friction or rock strength must be to resist downslope motion (● Fig. 15.26). Any surface materials on a slope that do not have the strength or stability to resist the force of gravity will respond by *creeping, falling, sliding,* or *flowing* downslope until stopping at the bottom of the slope or wherever there is enough friction to resist further movement. As a result, soil and regolith are thinner on steep slopes and thicker on gentle slopes, and intense mass wasting is one reason why bedrock tends to be exposed in areas of steep terrain.

Gravity is the principal force responsible for mass wasting, but water is often a contributing factor. Water (1) contributes to weathering, which prepares rock material for mass movement, (2) adds weight to porous materials on a slope, (3) decreases the strength of unconsolidated slope material, and (4) can increase the slope angle. We have seen that water is involved in many weathering processes that break and weaken rocks, making them more susceptible to mass movement. Unconsolidated soil and regolith have a considerable volume of voids, or pore spaces, between particles. Usually some of these voids contain air and some contain water, but storms, wet seasons, broken pipelines, irrigation, and other situations can cause the voids to fill with water. Saturated conditions encourage mass wasting because water adds weight

● FIGURE 15.25

A massive boulder, which was loosened by heavy rains and pulled downhill by the force of gravity, blocks a road in Southern California. The boulder weighed an estimated 270,000 kilograms (300 tons) and had to be dynamited to clear the highway.

What other kinds of problems on roads are related to mass wasting?

©AP/Wide World Photos

● FIGURE 15.26

The occurrence of mass wasting is strongly related to the slope angle and the strength of materials that make up a slope. Other factors, such as amount and type of vegetation cover and amount of soil moisture or groundwater, also influence downslope movement of materials by mass wasting. G = total force of gravity (weight of the block); F = component of the block's weight resisting motion; f = frictional forces resisting motion; D = downslope component of gravity.

How might vegetative cover or moisture content affect the potential for downslope movement of soil?

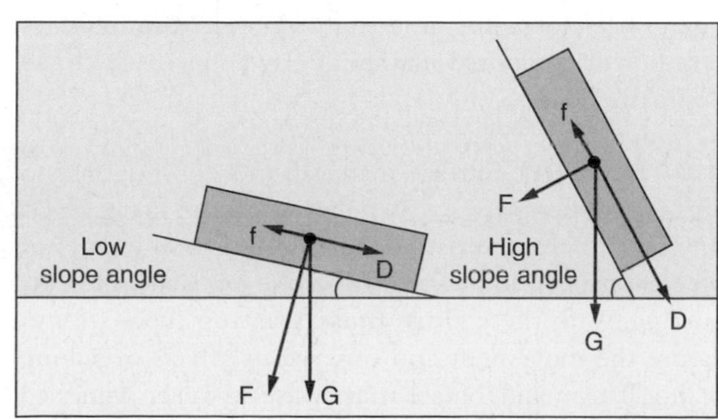

to the sediments, and an object's weight represents the amount of gravitational force pulling on it. As water replaces air in the voids, unconsolidated sediments and soil also experience decreasing strength as the rock fragments come into greater contact with the liquid, which tends to flow downslope. Finally, streams flowing along the base of a slope can undercut it by erosion. This increases the slope angle, thereby facilitating mass movement.

Undercutting and steepening of slopes also occurs by waves in coastal locations, and through the actions of people, especially when removing rock material from the base of a slope for construction projects. The strength or stability of slope material is often reduced by a triggering event, such as an earthquake. But, the vibrations produced by explosions and even the movements of heavy trucks or trains can be enough to shake material loose from a slope. Shaking reduces the support and friction between particles, so it can trigger mass movement.

Understanding the conditions and processes that affect mass wasting is important because gravity-induced movements of Earth materials are common and frequently impact people and the built environments in which they live. Although this natural hazard cannot be eradicated, people can avoid actions that aggravate the hazard potential and pay close attention to evidence of impending failure in susceptible terrain.

Classification of Mass Wasting

Physical geographers classify mass wasting events according to the kinds of Earth materials involved and the ways in which they move. Mass wasting events are categorized by using a descriptive name, for example *rockfall,* that summarizes the type of material and the type of motion. The various kinds of gravity-induced motions are separated into two general groups according to the speed with which they occur. After an overview of material categories and speed, types of motion are described.

Types of Earth Material Anything on Earth's surface that exists in or on an unstable, or potentially unstable, land surface is susceptible to gravity-induced movement and can therefore be transported downslope as a result of mass wasting. Mass wasting involves almost all kinds of surface materials. Rock, snow and ice, soil, earth, debris, and mud commonly experience downslope movements. In a mass wasting sense, **soil** refers to a relatively thin unit of predominantly fine-grained, unconsolidated surface material. A thicker unit of the same type of material is referred to as earth. **Debris** specifies sediment with a wide range of grain sizes, including at least 20% gravel. **Mud** indicates saturated sediment composed mainly of clay and silt, which are the smallest particle sizes.

Speed of Motion Surface materials move in response to gravity in many different ways, depending on the material, its water content, and characteristics of the setting. Some types of mass movement happen so slowly that no one can watch the motion occurring. With these **slow mass wasting** types we can only measure the movement and observe its effects over long periods of time. The motion of **fast mass wasting** can be witnessed by people. The speed of downslope movement of material varies greatly according to details of the slope, the material, and if a triggering factor is involved. In addition, slow and fast mass movements often work in combination. Mass movement that is initially slow may be a precursor to more destructive rapid motion, and most materials that have undergone rapid mass wasting continue to shift with slow movements. Specific types of motion, and common Earth materials associated with them, are discussed in the following sections on slow and fast mass wasting. These are summarized in Table 15.1.

TABLE 15.1
Different Kinds of Mass Wasting Processes

Motion	Common Material	Typical Speed	Effect
Creep	Soil	Slow	
Solifluction	Soil	Slow	
Fall	Rock	Fast	
Avalanche	Ice and snow or debris or rock	Fast	
Slump (rotational slide)	Earth	Fast	
Slide (linear)	Rock or debris	Fast	
Flow	Debris or mud	Fast	

Slow Mass Wasting

Slow mass wasting has a significant, cumulative effect on Earth's surface. In general, slow mass wasting produces rounded hillcrests and a landscape free of sharp angular features.

Creep Most hillslopes covered with weathered rock material or soils undergo **creep,** the slow migration of particles to successively lower elevations. This gradual downslope motion, often occurs as *soil creep,* primarily affecting a relatively thin layer of weathered rock material. Creep is so gradual that it is visually imperceptible; the rate of movement is usually less than a few centimeters per year. Yet creep is the most widespread and persistent form of mass wasting because it affects nearly all slopes where there are weathered materials at the surface.

Creep typically results from some kind of **heaving** process, which causes individual soil particles or rock fragments to be first pushed upward perpendicular to the slope, and then eventually fall straight downward due to gravity. The freezing and thawing of soil water, as well as the wetting and drying of soils or clays, can lead to soil heaving. For example, when soil water freezes, it expands, pushing overlying soil particles upward relative to the surface of the slope. When that ice thaws, the soil particles move back, but not to their original position because the force of gravity pulls downward on them, as shown in ● Figure 15.27. Soil creep results from repeated cycles of expansion and contraction related to freezing and thawing, or wetting and drying, which cause lifting followed by the downslope movement of soil and rock particles.

Organisms can also contribute to soil creep and other kinds of mass wasting. Small burrowing animals, including ground squirrels and chipmunks are effective soil movers. When they dig their tunnels on a sloping surface, the excavated material tends to fall downslope. Plant roots can also move soils outward and in a downward direction on a slope. Even the traversing of slopes by people and animals tends to push surface material downhill. Every step up, down, or across a steep slope shifts some surface material downhill to a slightly lower position.

Rates of soil creep are greater near the surface of a slope and diminish into the subsurface, because the factors instigating it are more frequent near the surface, and because frictional resistance to movement increases with depth. As a result, telephone poles, fence posts, other human structures, and even trees (● Fig. 15.28)—all of which are anchored at a level below the surface—exhibit a downslope tilt when affected by the downward movement of creep.

For the most part, creep does not produce discrete distinctive landforms, but it contributes to rounded, rather than angular, topography in hilly terrain. Once deposited at the base of the slope by gravity, these surface sediments can subsequently be carried away by one of the geomorphic agents, usually running water.

● FIGURE 15.27

Repeated cycles of expansion and contraction cause soil particles to be lifted at right angles to the surface slope but to fall straight downward by the force of gravity, resulting in soil creep.

Are there places near where you live that show evidence of soil creep?

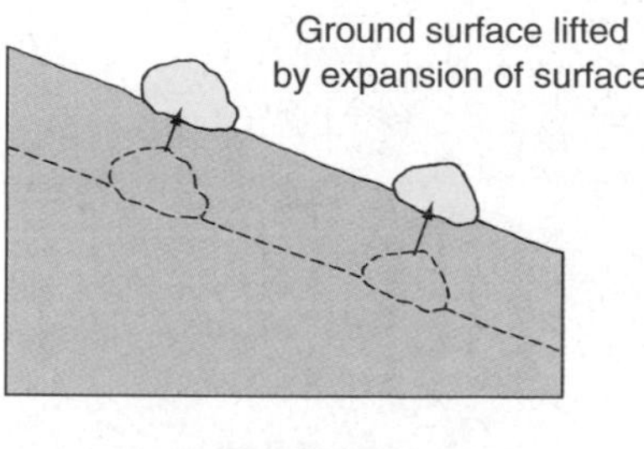

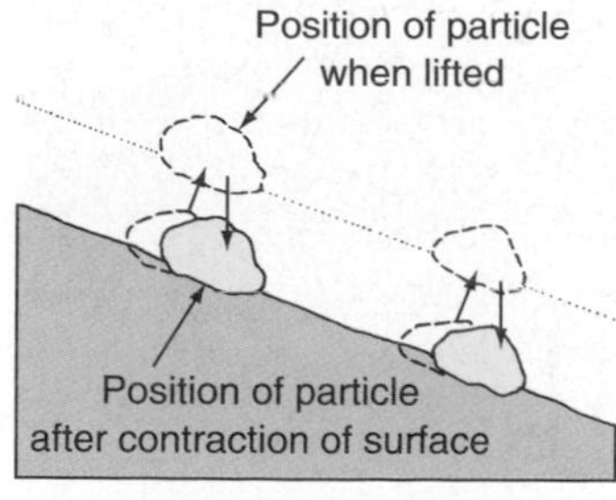

● FIGURE 15.28

(a) Visible landscape effects of soil creep are common on natural and cultural features. (b) Trees attempt to grow vertically, but their trunks become bent if surface creep is occurring.

What other constructed features might be damaged by creep?

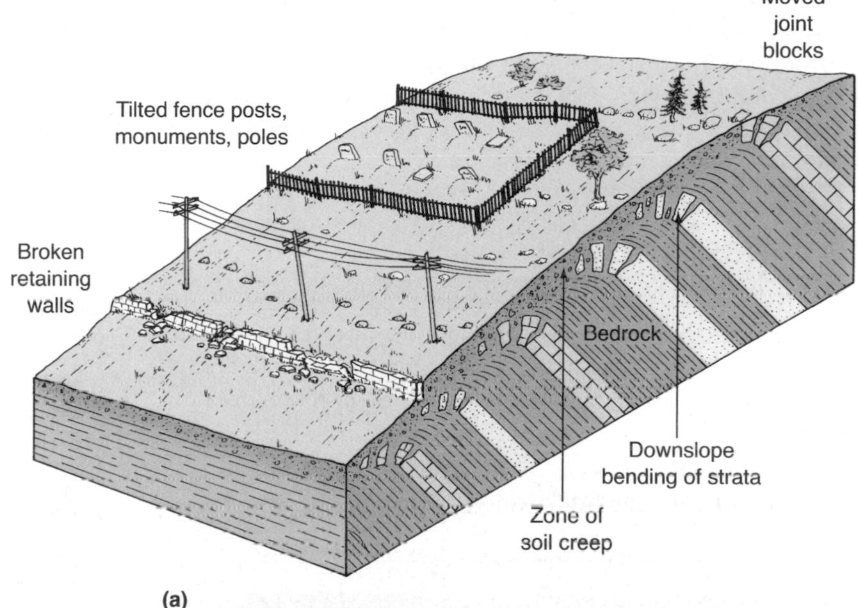

(a)

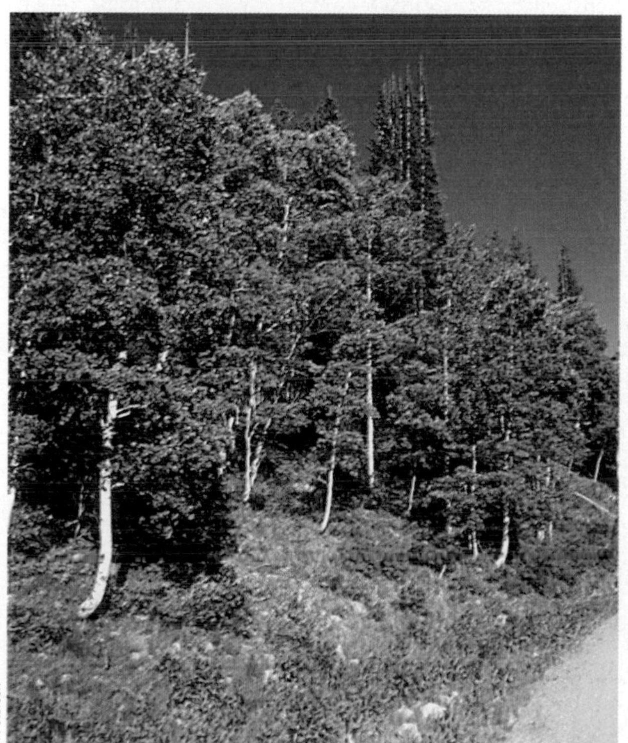

(b)

Solifluction The word **solifluction,** which literally means "soil flow," refers to the relatively slow downslope movement of water-saturated soil and/or regolith. Solifluction is most common in high-latitude or high-elevation tundra regions that have *permafrost,* a subsurface layer of permanently frozen ground. Above the permafrost layer lies the **active layer,** which freezes during winter and thaws during summer. During the summer thaw of the centimeters- to meters-thick active layer, the permanently frozen ground beneath it prevents downward percolation of melted soil water. As a result, the active layer becomes a heavy, water-saturated soil mass that, even on a gentle incline, sags slowly downslope by the force of gravity until the next surface freeze arrives. Movement rates are typically only several centimeters per year.

Evidence of solifluction exists in many tundra landscapes. It consists of irregular lobes of soil that produce hummocky terrain or mounds (● Fig. 15.29). Slopes affected by solifluction typically exhibit cracks and tongue-shaped lobes formed during downslope movement.

Fast Mass Wasting

Four major kinds of mass wasting usually occur so quickly—from seconds to days—that people can watch the material move. The speed of movement varies with the situation and depends on the quantity and composition of the material, the steepness of slope, the amount of water involved, the vegetative cover, and the triggering factor. The effects of fast mass wasting events on the land surface tend to be more dramatic than those of slow mass wasting. Rapid mass movements usually leave a visible upslope scar on the landscape, revealing where material has been removed, and a definite deposit where transported Earth material has come to rest at a lower elevation.

● **FIGURE 15.29**
Solifluction has formed these tongue-shaped masses of soil on a slope near Suslositna, Alaska.
How does solifluction differ from soil creep?

B. Bradley/NOAA, National Geophysical Data Center

Falls Mass wasting events that consist of Earth materials plummeting downward freely through the air are called **falls. Rockfalls** are probably the most common type of fall. Rocks fall from steep bedrock cliffs, either (1) one by one as weathering weakens the bonds between individual clasts and the rest of the cliff, or (2) as large rock masses that fall from a cliff face or an overhanging ledge (● Fig. 15.30). Large slabs that fall typically

● **FIGURE 15.30**
Eventually this overhanging sandstone ledge will fail in a rockfall. Rockfall beneath the ledge has already occurred, creating a zone with enhanced weathering because it stays wet longer in the shade. As more rocks fall from beneath, increasing the size of the overhang, the force of gravity will increase until it exceeds the strength of the bonds holding the ledge in place.
What weathering processes might be acting on the sandstone beneath the overhang when it becomes wet?

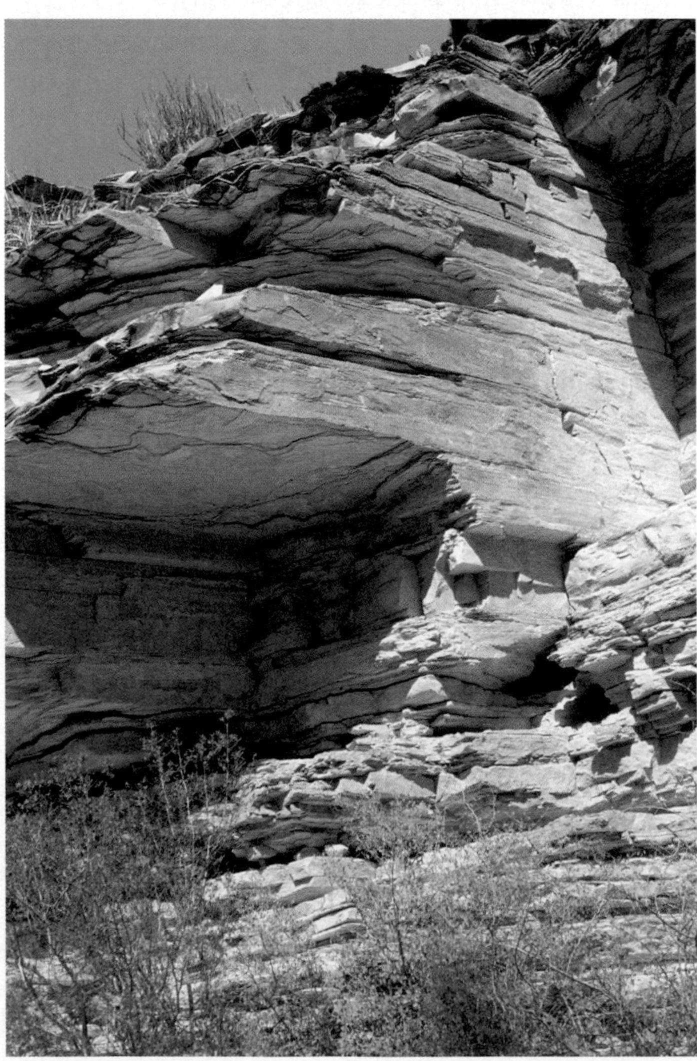

D. Sack

break into angular individual clasts when they hit the ground at the base of the cliff.

In steep mountainous areas, rockfalls are particularly common during the spring when snowmelt, rains, and alternating freezing and thawing can disturb precariously balanced rock masses, loosening them from their previously secure positions. Ground shaking caused by earthquakes is another common trigger for rockfalls.

Over time, a sloping accumulation of angular, broken clasts piles up at the base of a cliff that is subject to rockfall. This slope is known as a **talus, talus slope,** or, where cone-shaped, a **talus cone** (● Fig. 15.31). The presence of talus is good evidence that the cliff above is undergoing rockfall. The steepest slope angle that a pile of loose sediment can stand at without having particles tumble or slide downslope is called the **angle of repose.** Like other accumulations of unconsolidated sediment, such as the slopes of cinder cones or piles of gravel in a gravel pit, talus slopes typically lie at or near their angle of repose (● Fig. 15.32). The angle of repose for these different features varies somewhat depending on the size and angularity of the particles, but commonly ranges between about 30° and 34°. Large angular clasts have a steeper angle of repose than small, more rounded rock fragments.

Falling rocks create hazardous conditions in mountainous regions and wherever steep roadcuts expose bedrock cliffs. Rockfall hazard mitigation is a high priority along mountainous stretches of Interstate 70 and other highways in Colorado. In Yosemite Valley, California, massive rockfalls have originated from the area's towering and steep granite cliffs. In July 1996, one hiker was killed and numerous others injured by such an event. A huge 180,000-kilogram (200-ton) mass of rock broke away from a cliff, slid 200 meters (650 ft) down a steep slope, and then fell airborne for another 550 meters (1800 ft) before hitting the ground with great force. The rockfall was estimated to have moved downslope at more than 250 kilometers per hour (160 mph). The huge mass of moving rock also generated a destructive blast of compressed air that destroyed trees hundreds of meters from the cliff as the rock crashed to the valley floor (● Fig. 15.33).

Avalanches

An **avalanche** is a type of mass movement in which much of the involved material is pulverized, that is, broken into small, powdery fragments, which then flow rapidly as a density current along Earth's surface. Although the word *avalanche* may bring to mind billowing torrents of snow and ice roaring down a steep mountainside, *snow avalanches* are not the only kind. Avalanches of pulverized bedrock, called *rock avalanches,* and those of a very poorly sorted mixture of gravel, sands, silts, and clays, called *debris avalanches,* are also common and have caused considerable loss of life and destruction to mountain communities. Many avalanches are triggered by falls of snow and ice, rock, or debris that pulverize when they impact a lower surface. Snow avalanches are the best-known type of avalanche to the public, and they present serious hazards to skiers, mountaineers, and people who live in steep mountain communities that experience snowy winters. Regardless of the specific type of Earth material involved, avalanches can be very dangerous and powerful, traveling up to 100 kilometers per hour (60 mph). They easily knock down trees and demolish buildings (● Fig. 15.34), and have even destroyed entire towns.

● FIGURE 15.31
A steep slope of large, angular clasts has accumulated by rockfall in the talus built at the base of a limestone cliff in southern Idaho.

D. Sack

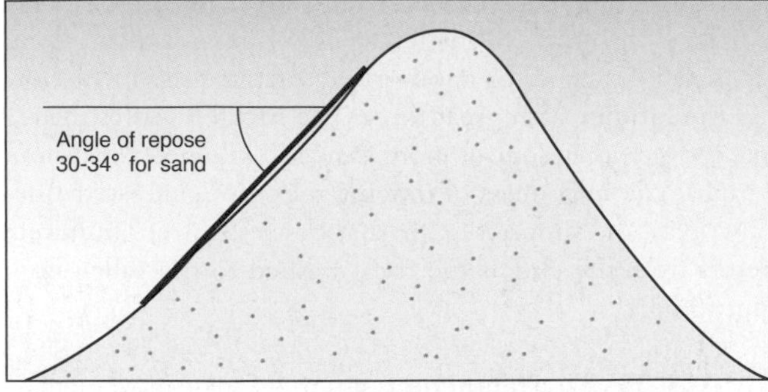

(a)

(b)

• FIGURE 15.32
(a) The angle of repose is the steepest natural slope angle that loose material can maintain. Particles are held at this angle by friction between clasts. (b) The same size and shape of particles and the same-sized dump truck loads formed these nearly identical piles, all with the same height and slope angle.
Would angular particles form steeper or gentler slopes in comparison to rounded particles if dumped in this manner?

• FIGURE 15.33
The part of this cliff that is lighter in tone marks the point of origin for the 1996 Happy Isles rockfall in Yosemite Valley, California. The rockfall killed one hiker and injured many others. The fall was so large and rapid, traveling at an estimated 250 kilometers per hour (160 mph), that it created an air blast which set off a giant dust cloud from the valley floor.

Slides In **slides,** a cohesive or semicohesive unit of Earth material slips downslope in continuous contact with the land surface. Water plays a somewhat greater role in most slides than it does in falls. Slides of all kinds threaten the lives and property of people who live in regions with considerable slope as well as such characteristics as tilted layers of alternating strong and weak rocks.

Slides of large units of bedrock, called *rockslides,* are frequent in mountainous terrain where originally horizontal sedimentary rock layers have been tilted by tectonism. The importance of water in reducing the resisting forces of rock strength and friction is seen in the fact that rockslides are most common in wet years, or after a rainstorm or snowmelt. As weathering and erosion by water weaken contacts between successive rock layers, the force of gravity can exceed the strength of the bonds between two rock layers. When this happens, a unit of rock, often of massive size, detaches and slides along the tilted planar surface of the contact (• Fig. 15.35). Rockslides sometimes end as rockfalls if the topography is such that free fall is needed to transport the rock to a stable position on more level ground. Deposits from rockslides tend to consist of larger blocks of rock than comprise rockfall deposits.

Some rockslides are enormous, with volumes measured in cubic kilometers. Anything in their path is obliterated. In addition, rockslides may form dams across river valleys, which soon become filled with lakes. When the lakes become deep enough, they may wash out the rockslide dams, producing sudden and disastrous downstream floods. Thus, immediately after this kind of a major rockslide, workers do what is necessary to stabilize the resulting dam and control the overflow outlet of the water trapped in the newly formed lake. This was done successfully

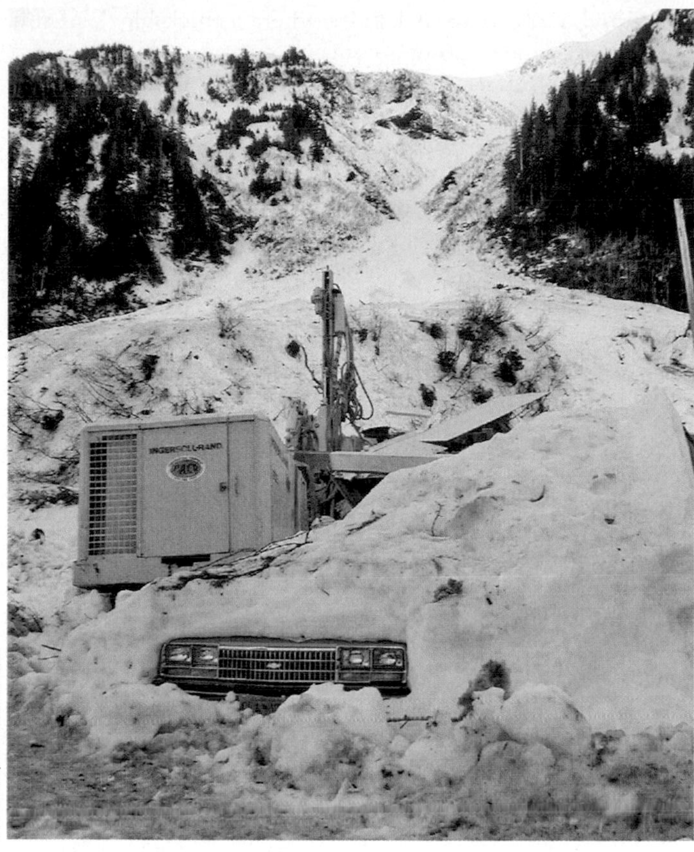
Dave Saville/FEMA News Photos

• **FIGURE 15.34**
Snow avalanches like this one in Alaska can block roads, knock down trees, carry rocks and tree trunks downslope, and damage structures. Note the heavy-packed nature of the snow, partly a result of pressure at impact. Many people erroneously believe that snow avalanche deposits are light and powdery.

after the Hebgen Lake slide in southwestern Montana in 1959 (• Fig. 15.36). Triggered by an earthquake, this rockslide, one of the largest in North American history, killed 28 people camped along the Madison River.

Huge rockslides have also resulted from instability related to rock structure and to the undercutting of slopes by streams, glaciers, or waves. Today, there are many locations in mountain

©Lloyd Cluff/CORBIS

• **FIGURE 15.36**
The 1959 earthquake-induced rockslide on the Madison River, Montana, completely blocked the river valley and created a new body of water, Earthquake Lake, seen in the background. The massive slide killed 28 people in a valley campground.
Why can earthquakes trigger landslides?

• **FIGURE 15.35**
Rock units that dip in the same direction as the topographic slope of the land are especially susceptible to rock-slide, as recognized along this stretch of highway in Wyoming.

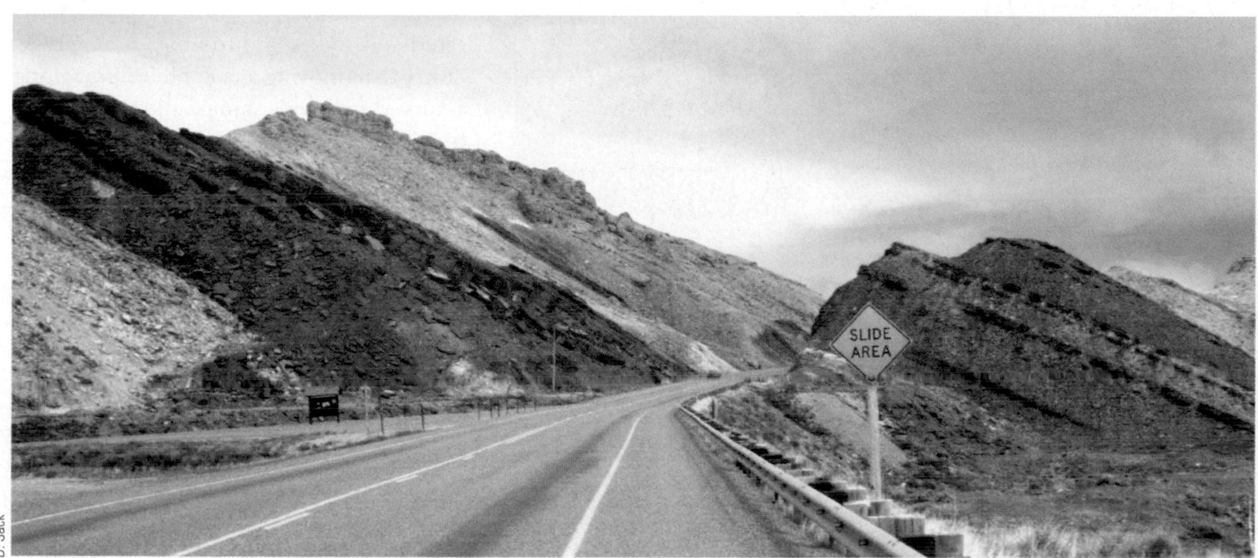

D. Sack

regions where enormous slabs of rock supported by weak materials are poised on the brink of detachment, waiting only for an unusually wet year or a jarring earthquake to set them in motion.

Rock is not the only Earth material prone to mass wasting by sliding. *Debris slides,* which contain a poorly sorted mixture of gravel and fines, and *mudslides,* which are dominated by wet silts and clays, are also common.

• FIGURE 15.37
Slump is the common name for a rotational earthslide. Many slumps transition into the more fluid motion of an earthflow in their lower reaches.
How does the earthflow component differ from the slump component?

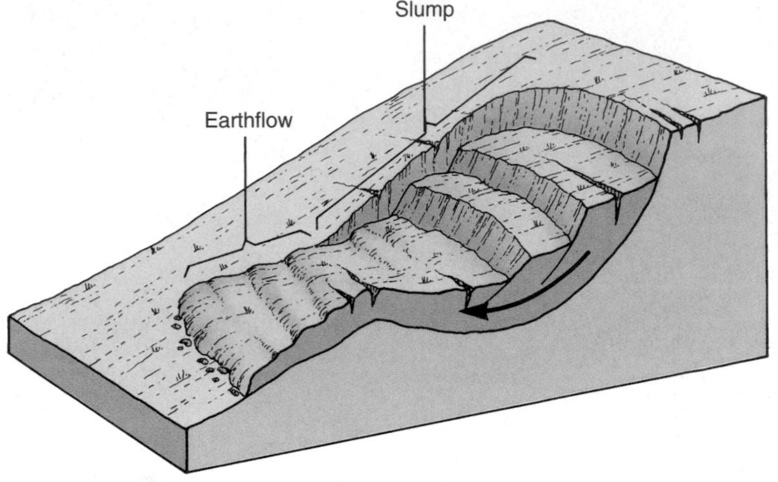

Slumps are rotational slides where a thick block of soil, called earth, moves along a concave, curved surface. Because of this curved surface of failure, slump blocks undergo a backward rotation as they slide (• Fig. 15.37), causing what used to be the ground surface at the top of the slump to tilt backward. Slumps are most common in wet years and during wet seasons in many regions with substantial relief, including the Appalachians, New England, and mountainous parts of the western United States. During exceptionally wet winters in Mediterranean climate regions, like California, slumps frequently damage hillside homes. Like rockslides, slumps can be triggered by earthquakes, which greatly reduce friction and material strength. People can also contribute to slumping by purposefully or accidentally adding water to hillslope sediments and by increasing slope angles by excavating for construction purposes.

Landslide has become a general term popularly used to refer to any form of rapid mass movement. However, sometimes large, rapid mass wasting events are difficult to classify because they contain elements of more than one category of motion or because multiple types of materials—rock, debris, earth, soil, and mud—are involved in a single massive slide. In some cases, Earth scientists call these large failures landslides. Such large slides are relatively rare but are often newsworthy because of their destructive qualities (• Fig. 15.38).

• FIGURE 15.38
Very large slides that move a variety of materials are referred to as landslides. Landslides modify the landscape but can also cause much destruction when buildings are constructed in areas susceptible to these large mass movements.

©AFP/Getty Images

Flows

Mass wasting **flows** are masses of water-saturated unconsolidated sediments that move downslope by the force of gravity. Flows carry water in moving sediments whereas rivers carry sediments in moving water. Compared to slides, which tend to move as cohesive units, flows involve considerable churning and mixing of the materials as they move.

When a relatively thick unit of predominantly fine-grained, unconsolidated hillside sediment or shale becomes saturated and mixes and tumbles as it moves, rather than moving as a cohesive unit along a curved surface (a slump), the mass movement is an *earthflow.* Earthflows occur as independent gravity-induced events or in association with slumps in a compound feature called a *slump-earthflow* (see again Fig. 15.38). A slump-earthflow moves as a cohesive unit along a concave surface in the middle and upper reaches of the failure. In the downslope reach, however, the sediments flow beyond the failure plane, and the mass flows in a less cohesive manner as an earthflow.

Debris flows and **mudflows** differ from each other primarily grain size and sediment attributes. Both flow faster

GEOGRAPHY'S ENVIRONMENTAL SCIENCE PERSPECTIVE

The Frank Slide

Many mass wasting events are compound, having elements of more than one type of motion. Avalanches and flows often begin as falls or slides. Slumps commonly grade into earthflows in their lower reaches. The massive and deadly slope failure that occurred over a century ago at Turtle Mountain in the Canadian Rockies appears to have comprised two of the most catastrophic types of mass movement, rockfall and rock avalanche. This 1903 Turtle Mountain failure is known as the Frank Slide after the town of Frank, Alberta, a portion of which was obliterated by the very rapidly moving 30 million cubic meters (82 million tons) of rock, resulting in the loss of an estimated 70 lives.

Frank was situated at the base of Turtle Mountain along the Canadian Pacific Railroad line. Many of the town's 600 citizens worked as underground coal miners within the steep mountain. Most of the townsfolk didn't know what hit them as they lay sleeping at 4:10 a.m. when the mountain gave way. Others were at work in the mine inside the mountain when the force of gravity overcame the strength of the 1 kilometer (3280 ft) wide, 425 meter (1395 ft) high, and 150 meter (490 ft) thick mass of limestone. The resistance of the rock mass was weakened by underground mining activities, including blasting, and weathering and erosion along fractures near the mountain summit. Severe weather conditions may have also played a role. In less than 2 minutes, the rockfall that became an avalanche on impact destroyed homes, buildings, roads, and the railroad line in its path, and left a huge expanse of broken rock that extends across to the far side of the valley. Amazingly, 17 miners survived and dug their way out of the mountain through the rubble.

Despite having occurred over a century ago, tremendous evidence of the huge rock failure still exists in the landscape today. The scar on the flank of Turtle Mountain and the rock rubble strewn over more than 3 square kilometers (1.2 sq mi) across the valley floor serve as a reminder of the incredible and deadly power that can be unleashed by the force of gravity.

D. Sack

The massive 1903 rockfall-avalanche known as the Frank Slide left a huge scar on Turtle Mountain, Alberta, that remains very obvious in the landscape more than a century later.

G. Nadon/Ohio University

Rubble from the rapidly moving rockfall-avalanche was strewn across the valley floor, well beyond the partly buried town of Frank, in Alberta, Canada.

than earthflows, often move down gullies or canyon stream channels for at least part of their travel, create raised channel rims called *flow levees,* and leave lobate (tongue-shaped) deposits where they spill out of the channel. They result from torrential rainfall or rapid snowmelt on steep, poorly vegetated slopes, and are the most fluid of all mass movements. Debris flows transport more coarse-grained sediment than mudflows do.

Debris flows often originate on steep slopes, especially in arid or seasonally dry regions. They also occur on steep slopes in humid regions that have been deforested by human activity or wildfire. The rain or meltwater flush weathered rock material into canyons where it acquires additional water from surface runoff. The result is a chaotic, saturated mixture of fine and coarse sediment, ranging in size from tiny clays to large boulders. As it flows down stream channels, some of the debris is piled along the sides as levees. Where a flow spills out of the channel, the unconfined mass spreads out and velocity decreases, resulting in deposition of a lobe of sediment (• Fig. 15.39). Debris flows are powerful mass wasting events that can destroy bridges, buildings, and roads (• Fig. 15.40). With dry summers and wet winters, Mediterranean climate regions are susceptible to mudflows and debris flows, particularly in rainy seasons that follow dry seasons during which devastating wildfires destroyed hillside vegetation.

Serious mudflow hazards exist in many active volcanic regions. Here, steep slopes may be covered with hundreds of meters of volcanic ash. During eruptions, emitted steam, cooling and falling as rain, saturates the ash, sending down dangerous and fast-moving volcanic mudflows, known as **lahars.** Of particular concern are high volcanic peaks capped with glaciers and snowfields. Should an eruption melt the ice and snow, rapid and catastrophic lahars rush down the mountains with little warning and bury entire valleys and towns. In the United States, there is concern that some of the high Cascade volcanoes in the Pacific Northwest may pose a risk of eruptions and resulting lahars. Lahars accompanied the 1980 eruption of Mount St. Helens, and Mounts Rainier, Baker, Hood, and Shasta all have the conditions in place, including nearby populated areas, for potentially disastrous mudflows to occur (• Fig. 15.41).

• FIGURE 15.39
Although small, this recent debris flow in western Utah left well-developed levees on either side of the fresh channel and deposited a tongue-shaped mass (lobe) of sediment where the flow spread out at the base of the slope.
What evidence is there to indicate this is a site of repeated debris flows?

D. Sack

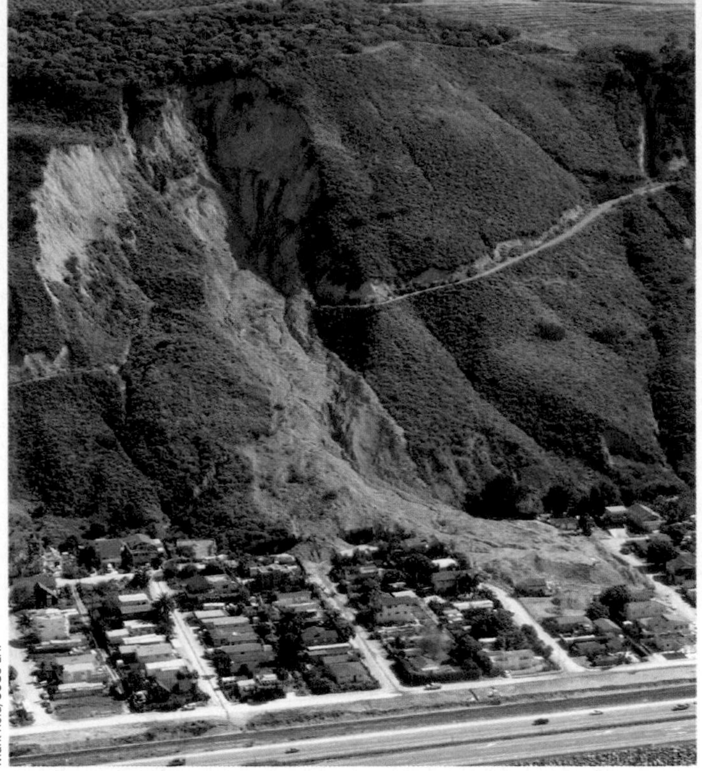

Mark Reid/USGS LHP

• FIGURE 15.40
A 1995 debris flow in La Conchita, California, destroyed several homes. Steep slopes consisting of weak unstable sediments failed during a period of heavy rainfall. A similar precipitation event triggered massive movement there again 10 years later, damaging 36 homes and killing ten people.
Why might a specific site experience repeated slope failures over time?

Weathering, Mass Wasting, and the Landscape

In this chapter we have concentrated on the exogenic processes of weathering and mass movement. Although neither weathering nor the slower forms of mass movement usually attract much attention from the general public, they are critical to soil formation and, like faster forms of gravity-induced motion, they are significant processes in shaping the landscape. While weathering

Lyn Topinka/USGS CVO

• FIGURE 15.41
The violent 1980 eruption of Mount St. Helens in Washington generated lahars—mudflows consisting of volcanic ash. This house was half buried in lahar deposits associated with that volcanic eruption.

and mass wasting processes shape the landscape and create new landforms, they also have impacts, at times catastrophic and deadly, on people and the built environment. The impact is reciprocal, however; our construction and recreational activities can influence the breakdown of rocks and induce the occurrence of mass movements.

Every slope reflects the local weathering and mass wasting processes that have occurred. These, in turn, are largely determined by the properties of the rocks and the local climatic factors. Slow weathering of resistant rocks leaves steep hillslopes, while rapid weathering of weak rocks produces gentle hillslopes that are typically blanketed by a thick mantle of soil or regolith. Differential weathering and erosion in areas of multiple rock types or variations in structural weakness produce complex landscapes of variable slopes.

Chemical weathering is most intense in warm humid climates where moisture and warmth help to accelerate the chemical reactions that cause minerals and rocks to decompose. In contrast, rocks in arid, semiarid, and cold climates tend to weather slowly, mainly by physical processes. Physical weathering, particularly those processes related to the freezing of water, is especially intense in climates that experience many cycles of freezing and thawing in a year, such as high-latitude and high-elevation locations. Because arid and semiarid regions often lack a weathered mantle and have a sparse vegetative cover, these regions typically exhibit barren, angular slopes of exposed bedrock that reflect a tendency toward fast mass wasting. Loose, weathered rock fragments in arid lands are also easily mobilized during intense precipitation events. In the following chapters, it will be important to remember that rock weathering and mass wasting are key parts of the geomorphic system of processes that interact to shape the landforms and topography of Earth.

Chapter 15 Activities

Define & Recall

weathering
mass wasting
physical (mechanical) weathering
chemical weathering
unloading
exfoliation
exfoliation sheet
exfoliation dome
thermal expansion and contraction
granular disintegration
freeze–thaw weathering
salt crystal growth
hydration
oxidation
carbonation
solution
hydrolysis
differential weathering and erosion
joint set
spheroidal weathering
soil (as a mass wasting material)
earth (as a mass wasting material)
debris
mud
slow mass wasting
fast mass wasting
creep
heaving
solifluction
active layer

fall
rockfall
talus (talus slope, talus cone)
angle of repose
avalanche
slide
slump
landslide
flow
debris flow
mudflow
lahar

Discuss & Review

1. In what ways is mass wasting similar to, yet different from, the action of the geomorphic agents?
2. How does physical weathering encourage chemical weathering in rock?
3. How are joints, fractures, and other voids in a rock related to the rate at which weathering takes place?
4. What are several ways in which expansion and contraction can affect the weathering of rock?
5. Why is chemical weathering more rapid in humid climates than in more arid climates?
6. Distinguish between hydration and hydrolysis.
7. What are the impacts of differential weathering and erosion on shaping landforms?
8. Distinguish among the principal types of Earth materials moved in mass wasting.
9. What factors facilitate creep?
10. Describe the principal differences between a rockslide and a mudflow.
11. What role does climate play in mass wasting?

Consider & Respond

1. Based on Figure 15.16, in what weathering regions are the following sites located?
   a. Your local area
   b. Brazil's Amazon Basin
   c. The North Slope of Alaska
   d. The summit of Pike's Peak, Colorado
   e. The Mojave Desert of Southern California
   f. The Appalachian Mountains of Pennsylvania
2. What would you recommend as a solution to prevent the loss of valuable historical monuments to weathering processes?
3. If you were an urban planner in a city with numerous steep slopes, what major hazards would you have to plan for? What recommendations would you make to lessen these dangers to the community?

Apply & Learn

1. Over a period of 2 hours, a thick, wet mass of unconsolidated sediment traveled down a mountain canyon stream channel before reaching and spreading out onto a desert plain in the arid western United States. Analyses determined that the sediment contained 18% clay, 29% silt, 27% sand, and 26% gravel, consisting of some very large blocks of rock. Based on this information, what specific type of event was it? What landform features would you look for to support your answer?
2. Find the climate data for the place where you live, or for the nearest climate data station with a similar climate to that of where you live. Using that data, determine which theoretical weathering region you live in, from the graph in Figure 15.16. Based on your observational evidence of weathering in the local environment (perhaps of natural rocks, tombstones, or building materials like stone, asphalt, and so on), write a short explanation and cite examples of why or why not the rock weathering in your area fits the theoretical weathering region from Figure 15.16.

Locate & Explore

Note: Please read the About Locate & Explore Activities section of the Preface before beginning these exercises.

1. Using Google Earth, identify the landforms at the following locations (latitude, longitude) and provide a brief discussion of how the landform developed. Include a brief discussion of why the landform is found in that general area of North America.
   Tip: Use the zoom, tilt, rotate, and elevation exaggeration tools to help view and interpret the landform and the area in which it is found.
   a. 33.805°N, 84.145°W
   b. 51.56°N, 116.36°W
   c. 37.75°N, 119.53°W
   d. 22.19°N, 159.61°W
   e. 49.305°N, 121.241°W
   f. 49.59°N, 114.40°W
2. Using Google Earth and the Landslide Hazard Layer, provide an explanation for why the landslide hazard is high in some areas of the United States and low in others. Make sure that you consider geology, topography, precipitation, soils, and vegetation cover.
   Tip: Use the zoom, tilt, rotate and elevation exaggeration tools to help view and interpret the landscape where there is a high landslide hazard.

Underground Water and Karst Landforms

16

CHAPTER PREVIEW

Like surface water, subsurface water moves, carries other substances, and is a principal component of the hydrologic cycle.

- What are possible flow pathways for water that infiltrates into the subsurface?
- How might subsurface water return to Earth's surface?

The subsurface part of the hydrologic cycle includes groundwater, a major source of fresh water for human use.

- How do people access groundwater?
- What are some problems associated with using groundwater?

The amount and distribution of subsurface water can vary substantially over time and space.

- How do climate and climate change impact the amount and distribution of subsurface water?
- What role does surface topography play in helping to determine properties of groundwater and the water table?

Karst landscapes are found primarily in regions where limestone bedrock lies at or near the surface.

- Why is this so?
- What other environmental characteristics encourage the formation of karst topography?

Landforms made by solution of bedrock are often distinctive and exist in many parts of the world.

- What are some principal surface and subsurface karst features?
- How does rock structure impact the formation of surface and subsurface karst forms?

◄ **Opposite: Solution of limestone by underground water can produce sizable caves like those at Natural Bridge Caverns, Texas. The impressive cave features seen here developed slowly, after the caverns were formed, by deposition of limestone.**
Courtesy of Natural Bridge Caverns, Texas

As part of our understanding of the distribution and effects of flowing water on Earth, we must consider the part of the hydrologic cycle that operates beneath the surface as well as surface flow. Like water flowing at the surface, water beneath Earth's surface moves, carries other substances, influences the form and appearance of the landscape, and represents an important source of fresh water for human use.

Fresh water is a precious, but limited, natural resource. There is much concern today throughout the world about the quantity and quality of our freshwater resources. Many populated regions have limited supplies of fresh water, while ironically in some sparsely populated or uninhabited areas, such as tundra and tropical rainforest regions, potable water is plentiful. Ice and snow in the polar regions represent 70% of the fresh water on Earth, but it is generally unavailable for human use. When most people think of fresh water, they typically envision rivers and lakes, both of which are important sources, but together they represent less than 1% of this resource. The remaining fresh water, nearly 30%, lies close to, but beneath, Earth's surface. These underground resources make up an impressive 90% of the fresh water that is readily available for human use (● Fig. 16.1).

In the last 100 years, worldwide usage of fresh water has grown twice as fast as the population. At the same

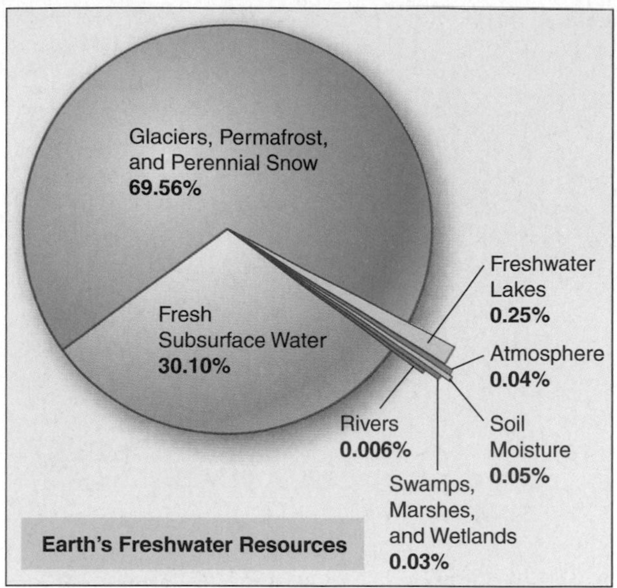

• FIGURE 16.1
With 97% of Earth's water existing as salt water in the oceans, Earth's freshwater resources are very limited. This global distribution of just the freshwater resources shows the dominance of ice and subsurface water. Most of Earth's fresh water (almost 70%) is stored as glacier ice mainly in remote polar regions. Much of the rest (about 30%) lies underground.
How can we work to conserve our freshwater resources?

time, the quality of freshwater resources has been declining as both surface and underground sources are subject to pollution. Because of the severity of the global freshwater problem, and the crucial role of water for life on Earth, the United Nations has declared 2005–2015 the *Decade for Action: Water for Life* to promote development, conservation, and wise use of water resources. Understanding the nature and distribution of the largest source of our freshwater resources—underground water—is critical to maintaining enough water of suitable quality for domestic, agricultural, and industrial purposes, and to maintaining environmental quality in general. Thus, in this chapter we investigate the nature and distribution of underground water as well as its impact on the landscape.

The Nature of Underground Water

Subsurface water is a general term encompassing all water that lies beneath Earth's surface. It includes water contained within soil, sediments, and rock. By far, the largest proportion of potable subsurface water is originally derived from the atmosphere—as precipitation. Water from precipitation, or meltwater from frozen precipitation, that soaks into the ground does so by the process of **infiltration.** During infiltration, water moves from the ground surface into void spaces in soil and loose sediments, and into cracks, joints, and other fractures in rock. Infiltration rejuvenates, that is, **recharges** or adds to, the amount of water in subsurface locations. Water from underground sources, in turn, reaches the surface in seeps, springs, and wells, and it contributes substantially to water in streams and in standing water bodies, such as lakes and ponds.

Some underground water resources tapped and used by people today are irreplaceable because they accumulated during previous wetter times in geologic history. A small portion of subsurface water is so deep beneath Earth's surface that it may never have been part of the hydrologic cycle. Elsewhere, because of changes in Earth's surface, subsurface water has been cut off from the hydrologic cycle for a long period of time. This water is contained deep within sediment layers that were deposited by ancient rivers or seas. Future changes in the lithosphere could release these trapped waters and return them to the hydrologic cycle. Volcanic activity, for example, can release some of this water in the form of steam during eruptions and as steam or hot water in geysers and hot springs.

Subsurface Water Zones and the Water Table

Organized by their depth and water content, three distinct subsurface water zones exist in humid regions (• Fig. 16.2). Under conditions of moderate precipitation and good drainage, water infiltrating into the ground first passes through a layer called the **zone of aeration** in which pore spaces in the soil and rocks almost always contain both air and water. This uppermost zone only rarely becomes saturated. If all the pore spaces do become filled with water because of a large rainfall or snowmelt event, it is very temporary. Water will soon drain downward by gravity beyond the zone of aeration to lower levels by the process of **percolation.** Water in the zone of aeration is known as **soil water.**

In the lowest of the three layers of underground water, the openings in sediments and rocks are completely filled with water. The water in this **zone of saturation** is called **groundwater.** The **water table** is a surface that marks the upper limit of the zone of saturation. The water table does not remain at a fixed depth below the land surface, but in any given area it fluctuates with the quantity of recent precipitation, loss by outflow to the surface, and the amount of removal by pumping. After an unusually wet period, the water table will rise. Because the depth to the water table generally reflects the precipitation amount for a given location (minus evaporation and other losses), it typically lies closer to the surface in humid regions and deeper underground in arid regions.

Between the zones of aeration and saturation is an **intermediate zone** that is saturated during periods of ample precipitation (and infiltration), but not saturated during intervals of low precipitation. The water table fluctuates through this middle layer that alternates between unsaturated and saturated conditions. A well or spring originating within the zone of saturation will always bear water, but one originating in the intermediate zone of fluctuation will run dry when the water table falls below it.

In some desert regions there is no saturated zone at all because water at or just below the surface evaporates soon after rainstorms. In many arid and semiarid landscapes, if considerable groundwater is present, it may be very old, having accumulated during a

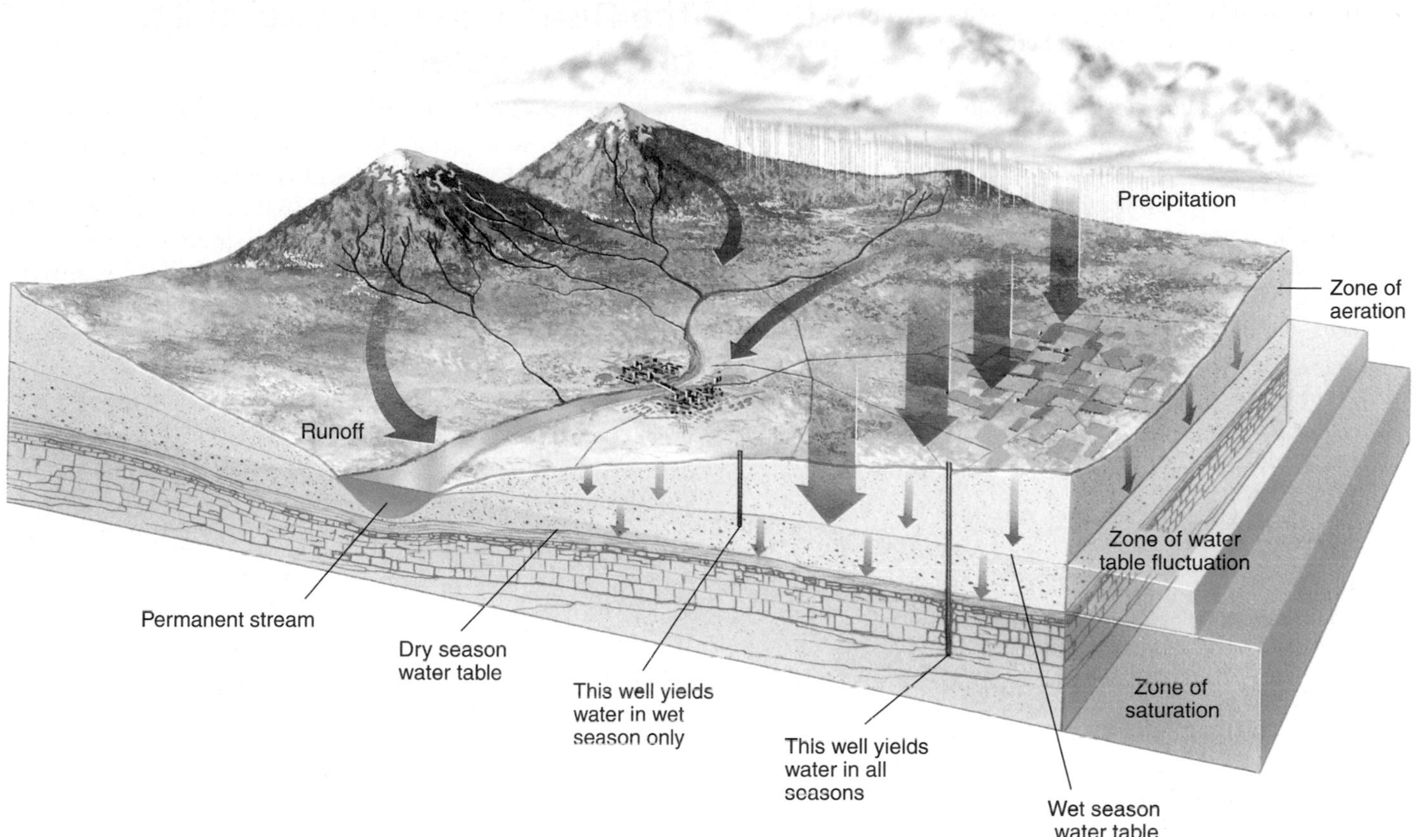

• FIGURE 16.2
Environmental Systems: Underground Water. The primary inputs to the underground subsystem of the hydrologic cycle are precipitation and snowmelt that infiltrate into the ground. The major output (outflow) of subsurface water occurs through dug wells or natural springs. In humid regions, water seeps from underground sources to streams and lakes, but in arid regions water seeps from these surface sources into the subsurface.

From the surface down, the subsurface water system consists of three layers: the zone of aeration, the intermediate zone, and the zone of saturation. Air almost always occupies many of the void spaces in the zone of aeration; water occupies all of the void spaces in the zone of saturation, the top of which is the water table. Infiltrated water percolates down beyond the zone of aeration to the transitional intermediate zone and eventually into the zone of saturation, that is, the groundwater zone. Water table depth responds to changes in infiltration and outflow, falling during dry seasons or years and rising during wet seasons or years.

Over half of the U.S. population receives their drinking water from groundwater. In the arid western United States, groundwater is also the major source of irrigation water, and overuse has caused the water table to fall drastically in some western areas. Like surface water, subsurface water is subject to pollution. Major sources of groundwater pollution include leaking septic systems, animal feed lots, leaking underground storage tanks, and seepage related to disposal of toxic materials.

past period of greater precipitation. Groundwater extracted from wells in these regions is not replaced from the atmosphere under current conditions of aridity, and as pumping continues, the water table will fall lower and lower. Pumping this ancient water from the subsurface faster than it is replenished through recharge is called **water mining** to emphasize that these groundwater resources are of limited supply and will not last indefinitely.

Despite its name, the water table is typically not level but tends to vary with the general contours of the land surface; it lies at a higher elevation under hills or other elevated topography and at a lower elevation under valleys or other depressions. Affected by gravitational force, the groundwater under higher land surfaces generally flows downslope toward lower elevations, as would a stream on the surface. Thus, in an area with hilly terrain, the water table is usually closer to the ground surface under low places than under high places.

In humid regions of low relief, the water table may be so high that it intersects the ground surface, producing lakes, ponds, or marshes, such as those common in New England and along the Gulf Coast from Louisiana to Florida. Where the landscape is one of hills and narrow valleys, the lowest points of the water table are often indicated by the location of stream channels on valley floors. The bottom of most humid region stream channels lies below the elevation of the water table. In these cases, groundwater flows underground downslope from the adjacent hills to the stream channel in the valley, where it seeps into the channel.

This **effluent** condition, where groundwater is seeping into a stream, helps keep the stream flowing between rains or during dry seasons (● Fig. 16.3a).

Many streams in semiarid and arid regions flow only seasonally or immediately following a significant rainfall. In semiarid regions, the water table typically lies beneath the streambed during dry periods and rises to intersect the streambed during wet periods. These streams are fed by groundwater only during wet seasons; during dry, **influent** periods they lose water by seepage into the ground (Fig. 16.3b). In most truly arid regions, surface streams flow only during and immediately after rains. Under these conditions, only influent flow occurs because the water table is so deep beneath the surface that no groundwater is available to feed the stream. Much of the surface flow is lost by seepage into the ground beneath the bed of the stream channel (Fig. 16.3c).

● FIGURE 16.3
Water may exit or enter the subsurface system at surface streams. (a) In humid regions, groundwater seeps out (effluent condition) into major stream channels all year, providing them with a source of continuous flow. (b) In semiarid or seasonally dry regions, the water table may fall below the stream bed causing the stream to dry up until the next wet period (seasonally influent condition). (c) In arid regions, significant depth to the water table means that streams flow only immediately after a rainfall event, and water seeps from the streambed into the subsurface (influent condition).

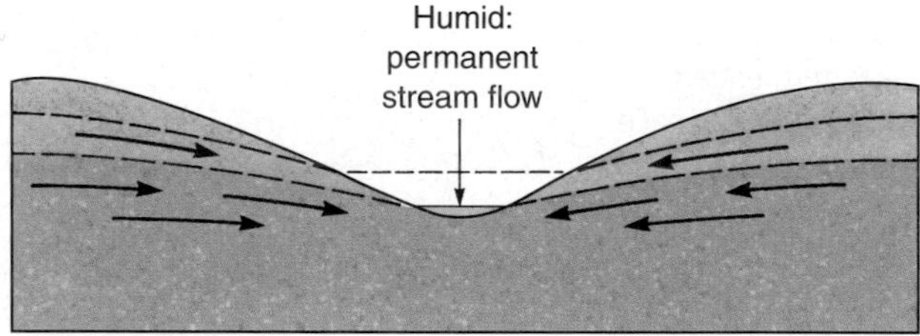

(a) Effluent condition

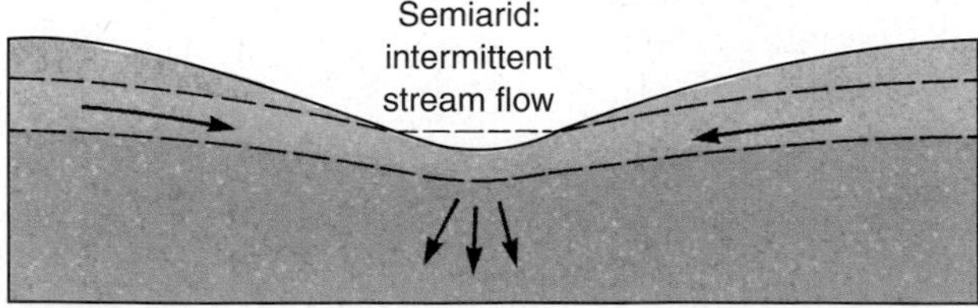

(b) Seasonally influent condition

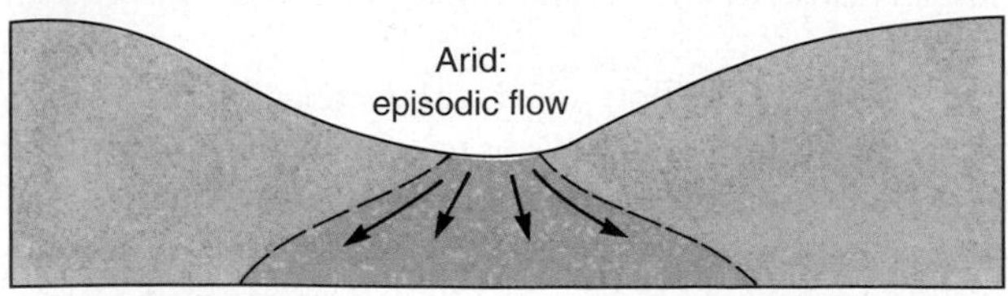

(c) Influent condition

No water table except by seepage from stream

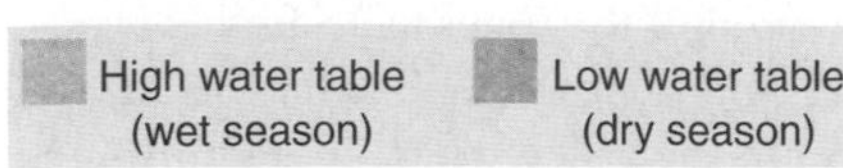

The Distribution and Availability of Groundwater

The quantity, quality, and availability of groundwater in an area depend on a variety of factors. Most fundamental is the amount of precipitation that falls in a given location and in the areas that drain into it. Second is the rate of evaporation. Third is the ability of the ground surface to allow water to infiltrate into the underground water system. A fourth factor is the amount and type of vegetation cover. Although dense vegetation transpires great amounts of moisture back to the atmosphere, it also inhibits rapid runoff of rainfall, encourages infiltration of water into the ground, and lowers evaporation rates by providing shade. Thus, the overall effect of vegetation in humid regions is to increase the supply of groundwater.

Two additional factors that affect the amount and availability of groundwater are the porosity and permeability of the sediments and rocks (● Fig. 16.4). **Porosity** refers to the

● FIGURE 16.4
The relationship between porosity and permeability is illustrated in three different kinds of rocks. (a) Sandy sediment and unjointed sandstones tend to be low in porosity and high in permeability, but the presence of the fracture in this jointed sandstone increases both, making it porous and highly permeable. (b) Unjointed shale (clay-sized clasts) may have a high porosity but the large number of small pores are very poorly connected, making for a low permeability. (c) Jointed rock that is nonporous gains porosity and permeability through its fractures.
Which of these three rock types would make the best aquifer?

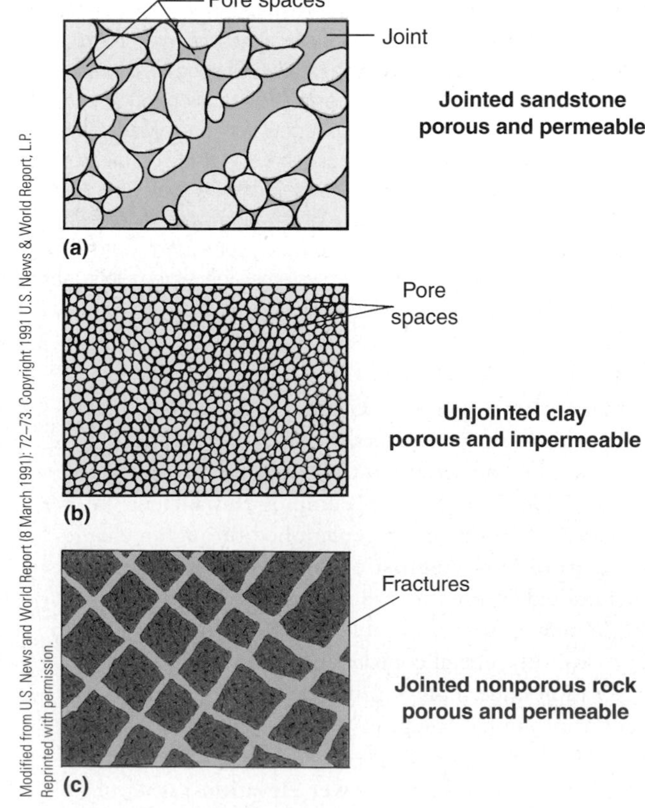

Modified from U.S. News and World Report (8 March 1991): 72–73. Copyright 1991 U.S. News & World Report, L.P. Reprinted with permission.

amount of space between the particles that make up sediments or rocks, expressed as the volume of voids compared to the total volume of the material (including voids). Sediments and rocks consisting of clay-sized clasts, perhaps surprisingly, have a relatively high porosity, and therefore can contain considerable amounts of water in the large number of very tiny pores. Sand and gravel have comparatively low porosity, but the actual value depends on the packing of grains and uniformity of the grain sizes present. **Permeability** expresses the relative ease with which water flows through void spaces in Earth material. Despite the high porosity of clay sediments and shale, the pore spaces are poorly interconnected, giving them a low permeability. It is therefore very difficult to obtain the water held within unjointed clays and shales. Permeability will increase significantly if these materials have joints or fractures that provide interconnections to facilitate flow. The inherently high permeability of sands and gravels often offsets their low porosity to make them good sources from which to obtain water.

Rocks that are composed of interlocking crystals (rocks such as granite) have virtually no pore space and can hold little water within the rock itself. These crystalline rocks, however, may contain water within joints, which allow the passage of groundwater rather freely. Thus, jointed granite can be described as permeable, even though the rock itself is not porous. Basaltic lavas may contain much pore space formed by gas bubbles frozen into the rock, but typically these holes are not interconnected, so the rock itself is porous but not permeable. In volcanic areas, groundwater permeability is usually provided by numerous fractures.

It is important to remember that porosity and permeability are not synonymous qualities. Porosity affects the potential amount of groundwater storage by providing available spaces for the water. Permeability affects the rates and volumes of groundwater movement and is facilitated by large pore spaces, bedding planes, joints, faults, and even caverns. Both of these factors affect the availability of groundwater for wells and springs.

An **aquifer** (from Latin: *aqua,* water; *ferre,* to carry) is a sequence of porous and permeable layers of sediments or rock that acts as a storage medium and transmitter of water (• Fig. 16.5). Although any rock material that is sufficiently porous and permeable can serve as an aquifer, most aquifers that supply water for human use are sandstones, limestones, or deposits of loose, coarse sediments (sand and gravel). A rock layer that is relatively impermeable, such as slate or shale, restricts the passage of water and therefore is called an **aquiclude** (from Latin: *aqua,* water; *claudere,* to close off).

Sometimes an aquifer will exist between two aquicludes. In this case, water flows in the aquifer much as it would in a water pipe or hose. Water can pass through the aquifer but does not escape through the aquicludes. Furthermore, soil water percolating downward may be prevented from reaching the zone of saturation by an aquiclude. An accumulation of groundwater above an

• FIGURE 16.5

An aquifer is a natural underground storage medium for groundwater. A perched water table can develop where impermeable rock exists beneath an aquifer. In this example, the perched water table is also underlain by a dry zone in between impermeable beds of shale and clay. Below the dry zone is the regional water table, the surface of the zone of saturation through which water flows toward the nearby river.

Is a perched water table a reliable source of groundwater?

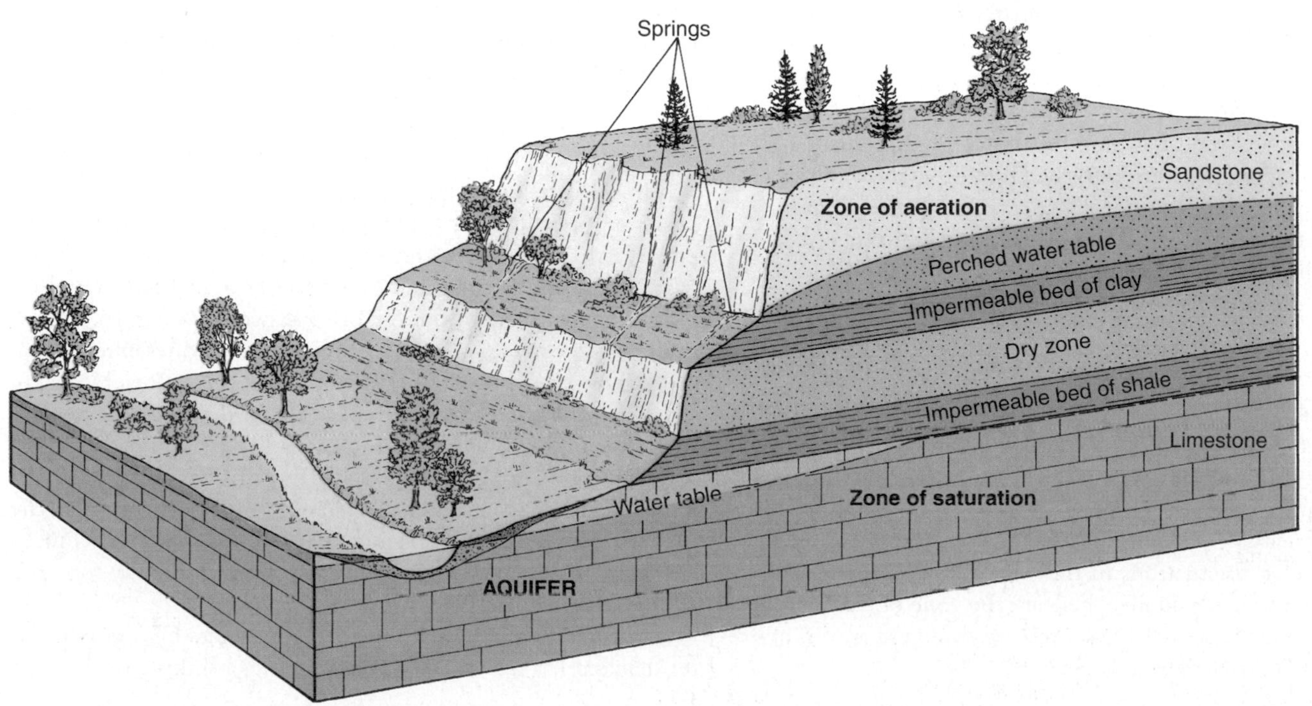

aquiclude is called a **perched water table** (see again Fig. 16.5). Careless drilling can puncture the aquiclude supporting a perched water table so that the water drains farther down into the subsurface. A well originating in the perceived water table must then be deepened to reach the true water table.

Springs are natural outflows of groundwater to the surface. They are related to many causes—landform configuration, bedrock structure, level of the water table, and the relative position of various types of aquicludes. Springs may occur along a valley wall where a stream or river has cut through the land to a level lower than a perched water table. The impermeable aquiclude below the perched water table prevents further downward percolation of water, forcing the water above to move horizontally until it reaches an outlet on the land surface. A spring flows continuously if the water table always remains at a level above the spring's outlet; otherwise the spring flow is intermittent, flowing only when the water table is at a high enough level to feed water to the outlet.

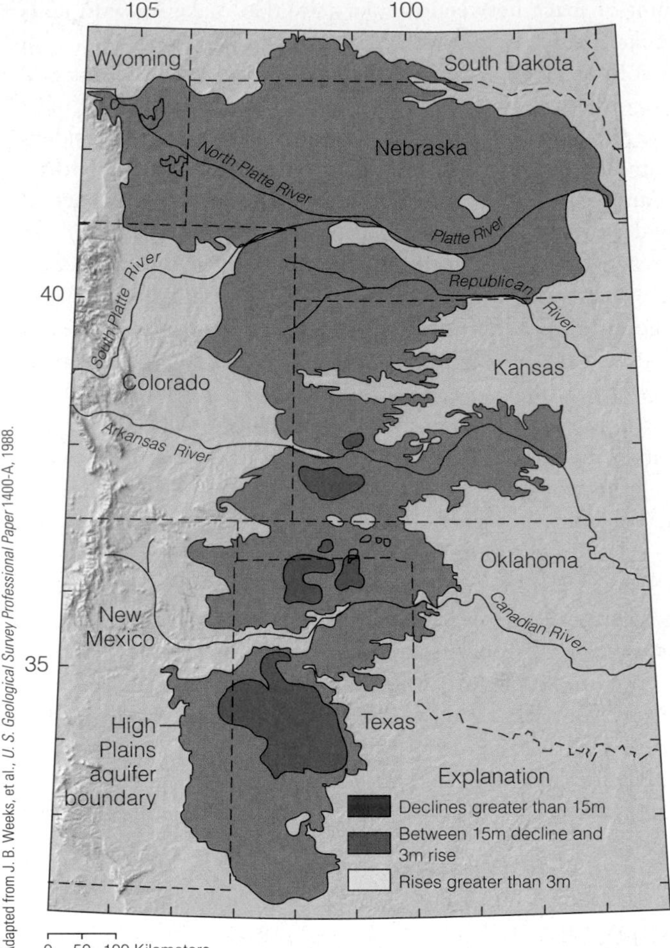

Adapted from J. B. Weeks, et al., *U. S. Geological Survey Professional Paper* 1400-A, 1988.

• FIGURE 16.6
The Ogallala Aquifer supplies water to a large, semiarid area of the High Plains. Said to be the largest freshwater aquifer in the world, much of the water in the Ogallala accumulated during wetter times thousands of years ago.
Why do you think the drop in water supply has been greatest in the southern part of the aquifer?

Groundwater Utilization

Groundwater is a vital resource to most of the world. Half of the U.S. population derives its drinking water from groundwater, and some states, such as Florida, draw almost all of their drinking water from this source. Irrigation, however, consumes the bulk of the groundwater—over two thirds—used in the United States today. One of the largest aquifers supplying groundwater for irrigation is the Ogallala Aquifer, which underlies the Great Plains from west Texas northward to South Dakota. The Ogallala Aquifer alone supplies more than 30% of the groundwater used for irrigation in the United States (• Fig. 16.6). Considerable concern exists about the future of this aquifer. Much of the water withdrawn from it accumulated thousands of years ago, and with the now-semiarid climate of the region, recharge is limited.

Groundwater also plays a major role in supporting many wetlands and in forming shallow lakes and ponds, all of which are ecologically invaluable resources. Water bodies and wetlands, fed by groundwater, are critical habitats for thousands of resident and migratory birds. Adequate groundwater flow is vital for the survival of the Everglades in southern Florida. This "river of grass," its great variety of birds, and many other animals are totally dependent on the continued southward movement of groundwater flow through the region.

Wells

Wells are artificial openings dug or drilled below the water table to extract water. Water is drawn from wells by lifting devices ranging from simple rope-drawn water buckets to pumps powered by gasoline, electricity, or wind. In shallow wells, the supply of water often depends on fluctuations in the water table. Deeper wells that penetrate into lower aquifers beneath the zone of water table fluctuation provide more reliable sources of water and are less affected by seasonal periods of drought.

In areas where there are many wells or a limited groundwater supply, the rate of groundwater removal may exceed the rate of its natural replenishment through groundwater recharge. Many areas that are irrigated from wells have had their water table fall below the depth of the original wells (• Fig. 16.7). Progressively deeper wells must be dug, or the old ones extended, in order to reach the supply of water. In the Ganges River Valley of northern India, the excavation of deep, modern wells to replace shallow hand-dug wells has increased the amount of groundwater being brought to the surface, but this has also resulted in a significantly lower water table. As previously mentioned, the Ogallala Aquifer is experiencing alarming water level declines, a condition that can be attributed to wells used mainly for agricultural irrigation.

In certain environments, particularly where high groundwater demand has led to extensive pumping, a sinking of the land, called *subsidence,* can occur due to compaction related to the water withdrawal. Mexico City, Venice (Italy), and the Central Valley of California, among many other places, have subsidence problems. In parts of Southern California, groundwater has been artificially replaced by diverting rivers so that they flow over permeable deposits. This process is known as **artificial recharge.**

Artesian Systems

In some cases, groundwater exists in **artesian** conditions, meaning the water is under so much pressure that if it finds an outlet it will flow upward to a level above the local water table. Where this water achieves outflow to the surface, it creates an **artesian spring** if natural; where the outlet to the surface is artificial, the result is a **flowing artesian well.** In flowing artesian wells, water tapped by the well rises to the surface and flows out under its own pressure, without pumping. In fact, the word artesian is derived from the Artois region of France, where the first known free-flowing well of this kind was dug in the Middle Ages. If water in a well rises above the local water table, but not to the point of flowing out of the well, it is a **nonflowing artesian well.**

Certain conditions are required for artesian water flow (● Fig. 16.8). First, a permeable aquifer, such as sandstone or limestone, must be exposed at the surface in an area of high recharge by precipitation or infiltration. This aquifer must receive water from the surface, incline downward often hundreds of meters below the surface, and be confined between impermeable layers that prevent escape of the water, except to the artesian springs or wells. These conditions cause the aquifer to act as a pipe that conducts water through the subsurface. The water in the "pipe" is under pressure from the water above it, which lies closer to the area of intake at the surface. As a consequence of this pressure, water will flow toward any available outlet. If that outlet happens to be a well drilled through the impervious layer and into the aquifer, the water will rise in the well, gushing out at the surface in the case of a flowing artesian well. The height to which the water rises in a well depends on the amount of pressure exerted on the water. Pressure in turn depends on the quantity of water in the aquifer, the angle of incline, and the number of other outlets, usually wells, available to the water. The pressure and water height increase with greater amounts of water, steeper inclines, and fewer outlets.

Sandstone exposed at the surface in Colorado and South Dakota transmits artesian water eastward to wells as far as

● FIGURE 16.7
Cones of depression may develop in the water table due to the pumping of water from wells. In areas with many wells, adjacent cones of depression intersect, lowering the regional water table and causing shallow wells to go dry.
What impact might this scenario have on some of the local natural vegetation?

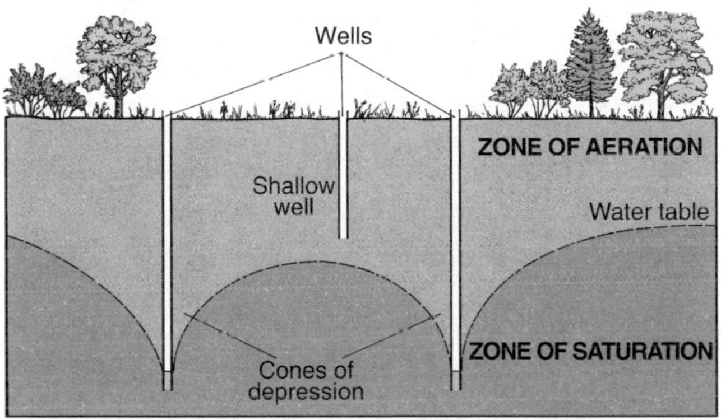

● FIGURE 16.8
Special conditions produce an artesian system. The Dakota Sandstone, an aquifer that averages 30 meters (100 ft) in thickness, transmits water from the Black Hills to locations more than 320 kilometers (200 mi) eastward beneath South Dakota.
What is unique about artesian wells?

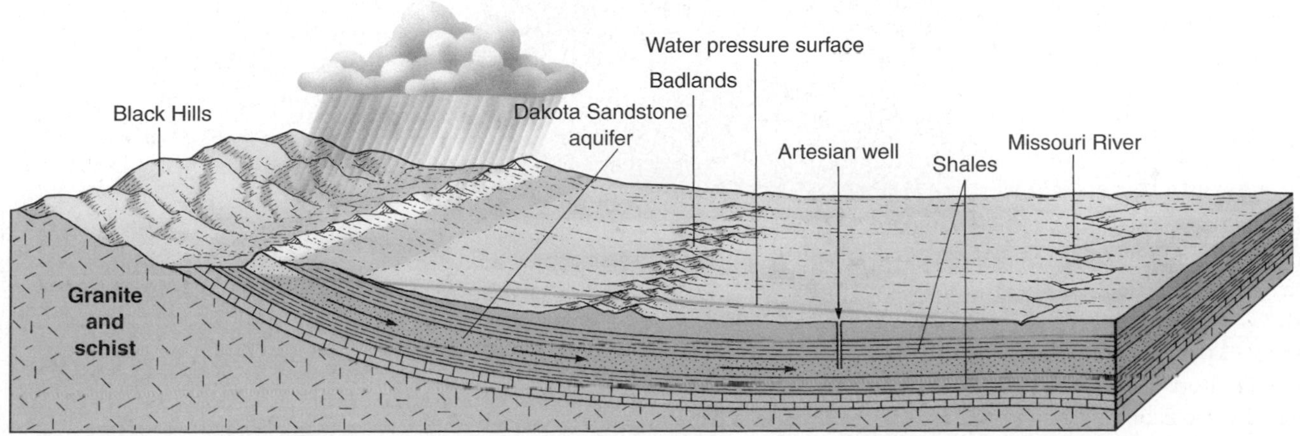

GEOGRAPHY'S PHYSICAL SCIENCE PERSPECTIVE

Acid Mine Drainage

Acid mine drainage (AMD) typically occurs when subsurface water flowing through mines or mine tailings undergoes chemical reactions that leave the water highly acidic (low pH) and metal-rich. AMD is a serious environmental concern in some coal- and metal-mining regions, including much of the coal-mining area of the eastern United States, parts of Australia, South America, South Africa, and elsewhere. When sufficient quantities of the affected water flow onto the surface as springs or seeps and into streams, lakes, or ponds, the mineral content and low pH endanger aquatic organisms and make the water unsuitable for human consumption. Fish have completely disappeared from some streams with very low pH and high metal content due to AMD.

The chemical reactions involved in the formation of AMD occasionally happen under natural conditions, but at much slower rates and in much smaller amounts than on disturbed lands. Mining greatly increases the permeability of susceptible rocks, allowing much more water to undergo the chemical reactions. Susceptible rocks contain pyrite (FeS_2), a common substance in the extensive coals of Pennsylvanian age in the eastern United States. When water containing oxygen comes into contact with pyrite in an abandoned underground coal mine, the pyrite oxidizes readily. Iron, sulfate (SO_4), and hydrogen ions are released into the water as a result of the chemical weathering; it is the presence of hydrogen ions that increases the water's acidity. Additional chemical reactions lead to hydrolysis of the iron ions in the water, which releases more hydrogen ions, and a further increase in acidity. Rate of hydrogen production is greatly accelerated if certain microorganisms that thrive in conditions of low pH are present.

Mines that lie in the zone of the fluctuating water table between the zone of aeration and the zone of saturation, like many mines in the Appalachian coal fields, are particularly susceptible to AMD because of the frequent introduction of new, moving, oxygenated water. Pumping removes water from these mines while they are operational. When they are abandoned, however, flowing underground water returns to the now highly permeable and chemically reactive environment producing AMD.

Controlling the flow of underground water represents a principal way to limit production of AMD. Approaches include locking susceptible mines out of subsurface water circulation by sealing water out, or flooding them with very slow moving, stagnant groundwater that has little opportunity to obtain new oxygen. AMD prevention and reduction remains an active field of research, and understanding the chemistry and circulation of underground water is critical to the ongoing investigation.

D. Sack

Acid mine drainage related to Appalachian coal mining drastically lowers the pH of local streams, and in some cases leads to precipitation of orange-colored oxidized iron compounds in the water and on the rocks in the channel.

320 kilometers (200 mi) away (see again Fig. 16.8). Other well-known artesian systems are found in Olympia, Washington; the western Sahara; and eastern Australia's Great Artesian Basin, which is the largest artesian system in the world.

Groundwater Quality

Because most subsurface water percolates down through a considerable amount of soil and rock, by the time it reaches the zone of saturation it is mostly free of clastic sediment. However, it often carries a large amount of minerals and ions dissolved from the materials through which it passed. As a result of this large mineral content, groundwater is often described as "hard water," in comparison with "softer" (less mineralized) rainwater. Moreover, just as increases in population, urbanization, and industrialization have resulted in the pollution of some of our surface waters, they have also resulted in the pollution of some of our groundwater supplies. For example, subsurface water moving through underground mines in certain types of coal deposits that are widespread in the eastern United States becomes highly acidic. If this **acid mine drainage** reaches the surface, it can have very detrimental effects on the local aquatic organisms.

Other dangers to groundwater quality stem from the percolation of toxic substances into the zone of saturation. Excessive applications of pesticides and incompletely sealed surface or subsurface storage facilities for toxic substances, including gasoline and oil, are situations that can lead to groundwater pollution.

In coastal regions with excessive groundwater pumping, denser salt water from the ocean seeps in to fill the voids. This problem with saltwater replacement has occurred in many localities, notably in southern Florida, New York's Long Island, and Israel.

Landform Development by Subsurface Water

In areas where the bedrock is soluble in water, subsurface water is an important agent in shaping landform features at the surface as well as underground. Underground water is a vital ingredient in subsurface chemical weathering processes, and, like surface water, subsurface water dissolves, removes, transports, and deposits materials.

The principal mechanical role of subsurface water in landform development is to encourage mass movement by adding weight and reducing the strength of soil and sediments, thereby contributing to slumps, debris flows, mudflows, and landslides. Through chemical activity, subsurface water contributes to many other and sometimes quite distinctive processes of landform development. Through the chemical removal of rock materials by carbonation and other forms of solution, and the deposition of those materials elsewhere, underground water is an effective land-shaping agent, especially in areas where limestone is present. Subsurface as well as surface water can dissolve limestone through carbonation or simple solution in acidic water. In many of these areas, surface outcrops of limestone are pitted and pockmarked by chemical solution, especially along joints, sometimes forming large, flat, furrowed limestone platforms (• Fig. 16.9). Wherever water can act on any rock type that is significantly soluble in water, a distinctive landscape will develop.

Karst Landscapes and Landforms

Overwhelmingly the most common soluble rock is limestone, the chemical precipitate sedimentary rock composed of calcium carbonate ($CaCO_3$). Landform features created by the solution and reprecipitation (redeposition) of calcium carbonate by surface or subsurface water are found in many parts of the world. The eastern Mediterranean region in particular exhibits large-scale limestone solution features. These are most clearly developed on the Karst Plateau along Croatia's scenic Dalmatian Coast. Landforms developed by solution are therefore called **karst** landforms after this classic locality. Other extensive karst regions are located in Mexico's Yucatán Peninsula, the larger Caribbean islands, central France, southern China, Laos, and many areas of the United States (• Fig. 16.10).

The development of a classic karst landscape, in which solution has been the dominant landforming process, requires several special circumstances. A warm, humid climate with ample precipitation is most conducive to karst development. In arid climates, karst features are typically absent or are not well developed. However, some arid regions have karst features that originated during previous periods of geologic time when the climate was much wetter than it is today. Compared to colder

• FIGURE 16.9
Solution of limestone is most intense in fracture zones where the dissolved minerals from the rock are removed by surface water infiltrating to the subsurface. This landscape shows a limestone "platform" where intersecting joints have been widened by solution.

J. Petersen

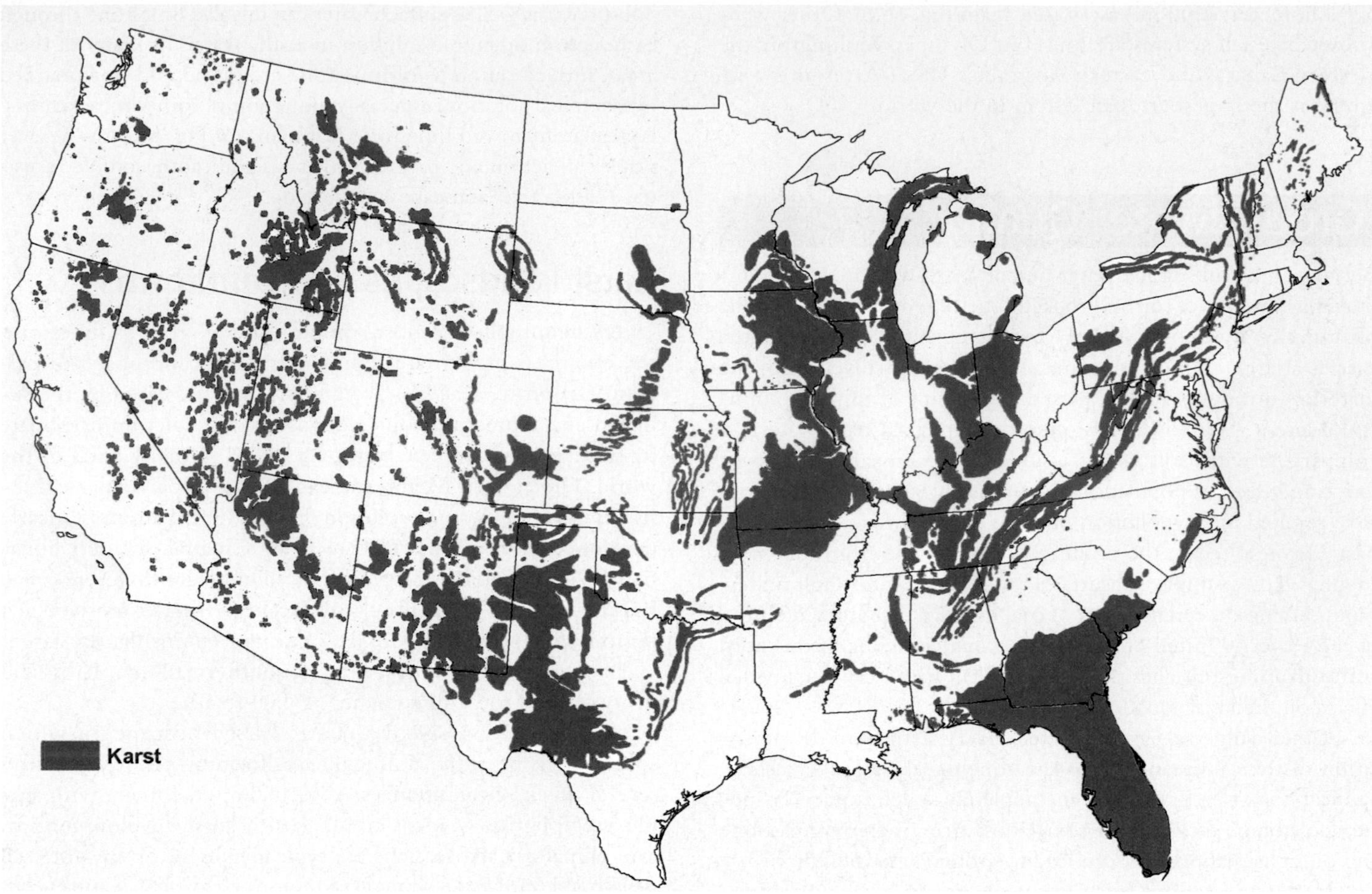

• FIGURE 16.10
The distribution of limestone in the conterminous United States indicates where varying degrees of karst landform development exists, depending on climate and local bedrock conditions.
Where is the nearest karst area to where you live?

humid climates, warmer humid climates have greater amounts of vegetation, which supplies carbon dioxide to subsurface water. Carbon dioxide is necessary for carbonation of limestone and it increases the acidity of the water, which encourages solution in general.

Another important factor in the development of karst landforms is the active movement of subsurface water. This allows water that has become saturated with dissolved calcium carbonate to flow away and become quickly replaced by water unsaturated with dissolved calcium carbonate. Groundwater undergoes vigorous movement when an outlet is available at a low level, such as by a deeply cut stream valley or a tectonic depression. In addition, the greater the permeability of the Earth material, the faster groundwater will flow.

Because infiltration of water into the subsurface tends to be concentrated where cross-cutting sets of joints intersect, these intersections are subject to accelerated solution. Such concentrated solution can cause the development of roughly circular surface depressions, called **sinkholes** or **dolines,** that are prominent features of many karst landscapes (• Figs. 16.11a and b). There are two dominant types of sinkholes identified by their differing formation processes (• Fig. 16.12). If the depressions are due primarily to the solution process at or near the surface and the removal of dissolved rock by water infiltrating downward into the subsurface, the depressions are called **solution sinkholes** (• Fig. 16.13). If the depressions are caused by the caving-in of the land surface above voids created by subsurface solution in bedrock below, they are referred to as **collapse sinkholes** (• Fig. 16.14).

The two processes, solution and collapse, cooperate to create most sinkholes in soluble rocks. Whether the depressions are termed solution or collapse sinkholes depends on which of these two processes was dominant in their formation. Collapse and solution sinkholes often occur together in a region, and they may be difficult to distinguish from one another based on their form. There is a tendency for solution sinkholes to be funnel-shaped, and collapse sinkholes to be steep-walled, but these shapes vary greatly.

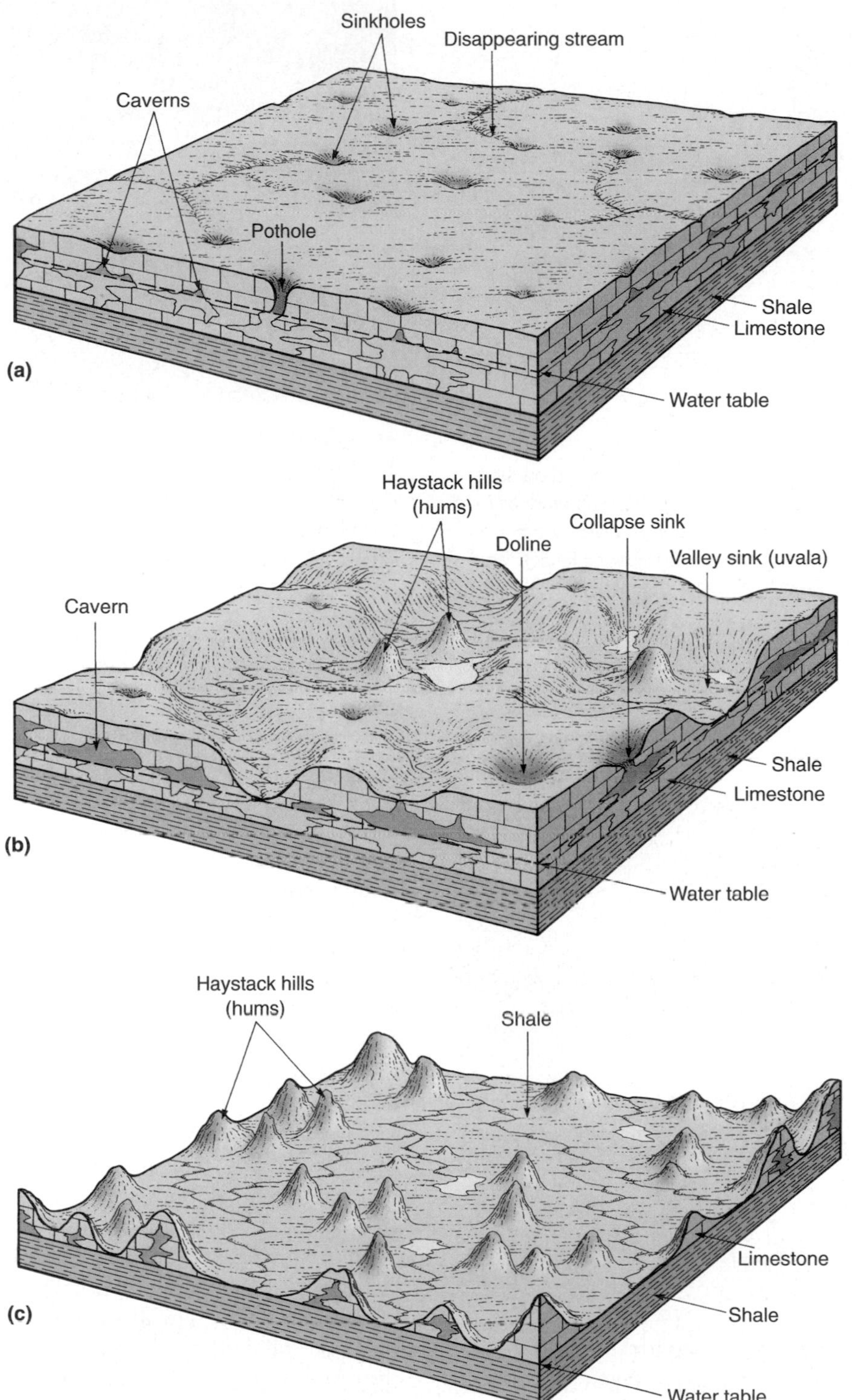

• FIGURE 16.11
Karst landscapes can be quite varied. (a) Solution at joint intersections in limestone encourages sinkhole development. Caverns form by groundwater solution along fracture patterns and between bedding planes in the limestone. Cavern ceilings may collapse, causing larger and deeper sinkholes. Surface streams may disappear into sinkholes to join the groundwater flow. (b) Some limestone landscapes have more relief, such as this irregular terrain with merged sinkholes that make karst valleys, called uvalas, and some conical haystack hills. (c) Areas that have experienced more intense solution may be dominated by limestone remnants, with the haystack hills (hums), isolated above an exposed surface of insoluble shale.
Why are there no major depositional landforms created at the surface in areas of karst terrain?

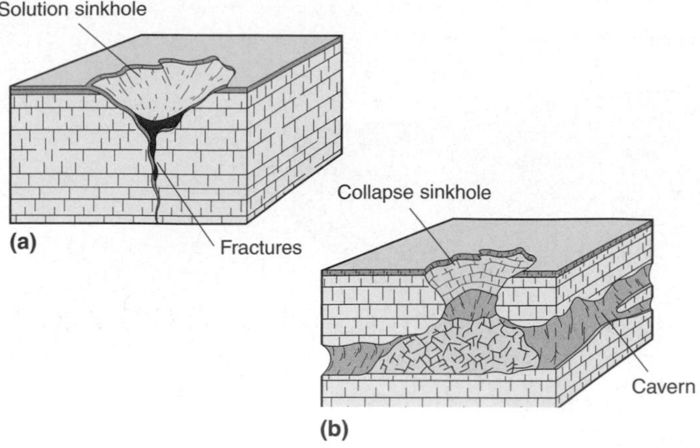

• FIGURE 16.12
Sinkholes (dolines) are divided into two major categories based on their principal mode of formation. (a) Solution sinkholes develop gradually where surface water funneling into the subsurface dissolves bedrock to create a closed depression in the landscape. (b) Collapse sinkholes form when either the bedrock or the regolith above a large subsurface void fails, falling into the void.

• FIGURE 16.14
This large collapse sinkhole developed in Winter Park, Florida, when a falling water table caused the surface to fall into an underground cavern system during a time of severe drought.
What human activities might contribute to such hazards?

• FIGURE 16.13
Slowly forming solution sinkholes form closed depressions in the landscape. This scene is in southern Indiana.

Sudden collapse of sinkholes is a significant natural hazard that every year causes severe property damage and human injury. Rapidly forming sinkholes may be caused by excessive groundwater withdrawal for human use, or they may occur during drought periods. Either of these conditions will lower the water table, causing a loss of buoyant support for the ground above, followed by collapse. Rapid sinkhole collapse has damaged roads and railroads and has even swallowed buildings.

Despite their humid climates, many karst regions have few continually flowing surface streams. Surface water seeping into fractures in limestone widens the fractures by solution. These widened avenues for water flow increase the downward permeability, accelerate infiltration, and direct water rapidly toward the zone of saturation.

Groundwater flowing along joints and bedding planes below the surface can also dissolve limestone, sometimes creating a system of connected passageways within the soluble bedrock (see again Figs. 16.11a and b). If the water table falls leaving these passageways above the zone of saturation, they are called **caverns** or **caves.**

In some cases, surface streams flowing on less permeable bedrock upstream will encounter a highly permeable rock downstream, where it rapidly loses its surface flow by infiltration (see again Fig. 16.9). These are called **disappearing streams** because they "vanish" from the surface as the water flows into the subsurface (● Fig. 16.15a).

In many well-developed karst landscapes, including the Mammoth Cave area of Kentucky, a complex underground drainage system all but replaces surface patterns of water flow. In these cases, the landscape consists of many large valleys that contain no streams. Surface streams originally existed and excavated the valleys, but eventually broke through into a cavern system below. This diverted the surface stream water to underground conduits. The site where a stream disappears into the cavern system is called a **swallow hole** (Fig. 16.15b). In some cases these underground-flowing "lost rivers" reemerge at the surface as springs where they encounter impervious beds below the limestone.

Sinkholes (or dolines), which are also known as *cenotes* in Mexico, may enlarge and merge over time to form larger karst depressions. As sinkholes coalesce, the larger depressions that develop, called **uvalas,** or **valley sinks,** are often linearly arranged along former underground water courses (see again Fig. 16.11b). The terms *doline* and *uvala* are derived from Slavic languages used in the former Yugoslavia.

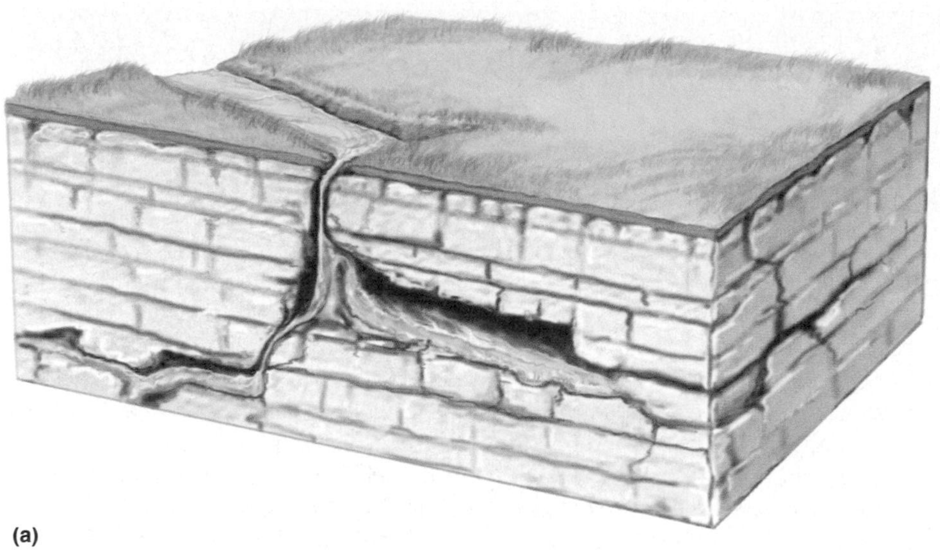

(a)

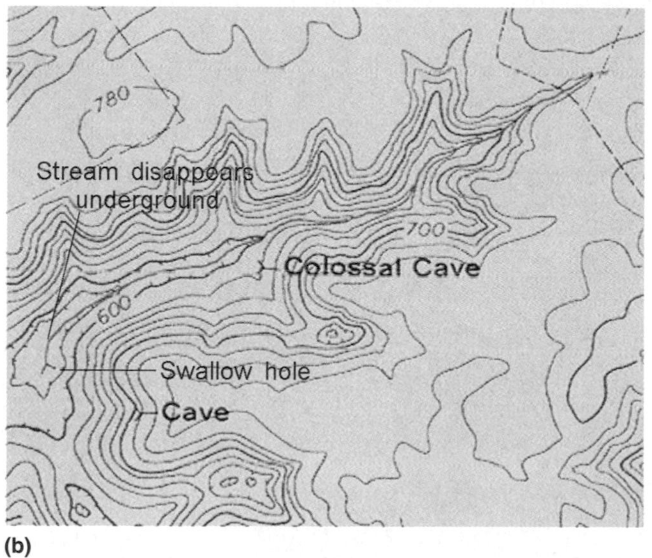

(b)

• FIGURE 16.15
Disappearing streams and swallow holes are common in some karst areas. (a) A surface stream flows into a solution-widened hole in the ground, the swallow hole, and disappears from the surface, flowing instead through a subsurface cavern system. Flow may emerge back onto the surface at a major spring. (b) A topographic map shows a disappearing stream and a swallow hole. Note the hachured contours that indicate the closed depression into which the stream is flowing.

After intense and long-term karst development, especially in wet tropical conditions, only limestone remnants are left standing above insoluble rock below. These remnants usually take the form of small, steep-sided, and cave-riddled karst hills called **haystack hills, conical hills,** or **hums** (see again Figs. 16.11b and c). Examples of this landscape are found in Puerto Rico, Cuba, and Jamaica. They have been described as "egg box" landscapes because an aerial view of the numerous sinkholes and hums resembles the shape of an egg carton (• Fig. 16.16). If the limestone hills are particularly high and steep-sided, the landscape is called **tower karst.** Spectacular examples of tower karst landscapes are found in southern China and Southeast Asia (• Fig. 16.17).

Limestone Caverns and Cave Features

Caverns created by the solution of limestone are the most spectacular and best-known karst landforms. Groundwater, sometimes flowing much like an underground stream, dissolves rock, leaving networks of passageways. Should the water table drop to the floor of the cavern or lower, typically resulting from climatic change or tectonic uplift, the caves will become filled with air. Interaction between the cave air and mineral-saturated water percolating down to the cavern from above will then begin to precipitate minerals, especially calcium carbonate, on the cave ceiling, walls, and floor, decorating them with often-intricate depositional forms.

Examples of limestone caverns in the United States are Carlsbad Caverns in New Mexico, Mammoth and Colossal Caves in Kentucky, and Luray and Shenandoah Caverns in Virginia. In fact, 34 states have caverns that are open to the public. Some are quite extensive with rooms more than 30 meters (100 ft) high and with kilometers of connecting passageways. Every year, millions of visitors marvel at the intriguing variety of forms, colors, and passageways that they see and experience on tours of limestone caverns. The vast majority of these features in limestone caverns are related to solution and deposition by groundwater.

• FIGURE 16.16
Intense solution in wet tropical environments can form a karst landscape consisting of a maze of hums and intergrown sinkholes.

Courtesy Parris Lyew-Ayee

GEOGRAPHY'S ENVIRONMENTAL SCIENCE PERSPECTIVE

Sudden Sinkhole Formation

Although many sinkholes develop slowly through solution and infiltration of water, regolith, and soil along joint intersections in soluble rocks, others appear almost instantaneously as the ground unexpectedly collapses into subsurface voids, including caverns. The sudden collapse of the ground caused by the creation of a sinkhole is a problem in many areas of the United States that have soluble rocks at or near the surface. An example is shown in the accompanying photographs of a road collapse near Bowling Green, Kentucky.

Earth scientists recognize two types of collapse processes that form sinkholes: regolith collapse (or cover collapse) and bedrock collapse. In both cases, subsurface solution begins long before the surface sinkhole appears.

Sinkhole formation by sudden regolith collapse occurs where vertical gaps in the subsurface rock exist due to solution along joints or other forms of weakness in the rock. The gaps often open to larger passageways at depth. Regolith and soil lie on top of the bedrock and also cross over the gaps as regolith bridges or arches. Initially, the water table lies above the rock within the regolith layer. Sudden regolith collapse is most common after the water table has fallen below the subsurface gaps, such as during drought periods or by excessive pumping from wells. The drop in the water table leaves the regolith in a zone in which some pore spaces fill with air and through which water percolates downward toward the air-filled gap or cavern below. If the force of gravity pulling downward on the soil and regolith exceeds the frictional and cohesive strength holding up the regolith bridge, the regolith collapses into the gap, creating a sinkhole at the ground surface. Adding additional weight to the regolith and soil layer by constructing roads and buildings or by impounding of water at the surface increases the force of gravity and therefore the likelihood that the regolith bridges will suddenly collapse.

The formation of a sinkhole by bedrock collapse is associated with a growing underground cavern relatively near the ground surface. Caverns can grow vertically as well as horizontally. If continued solution makes the rock roof of the cavern thinner and thinner, eventually the rock can become so thin that it fails, crashing into the cavern void below. A sinkhole is formed when collapse of the bedrock cavern roof causes overlying Earth material up to the ground surface to lose its support and also slump into the hole.

Physical geographers who have examined the Dishman Lane sinkhole in Bowling Green report evidence of cave roof collapse in this case. It is likely that much of the roof collapsed thousands of years ago, but the fact that limestone blocks were found in the rubble indicated that part of the cave roof collapsed in this event. The resulting sinkhole affected over 0.4 hectares (1 acre) of surface area and caused considerable property damage.

Courtesy of the Center for Cave and Karst Studies at Western Kentucky University

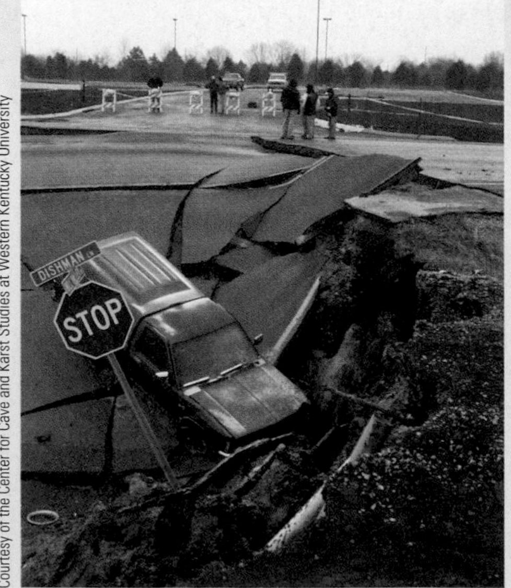

Courtesy of the Center for Cave and Karst Studies at Western Kentucky University

Sinkhole collapse near Bowling Green, Kentucky, caused severe road damage in an area larger than a football field. Luckily, no one was hurt during the rush-hour collapse, although several vehicles were damaged. Fixing the problem required completely filling in the sinkhole with rock and repaving the road at a cost of more than $1 million.

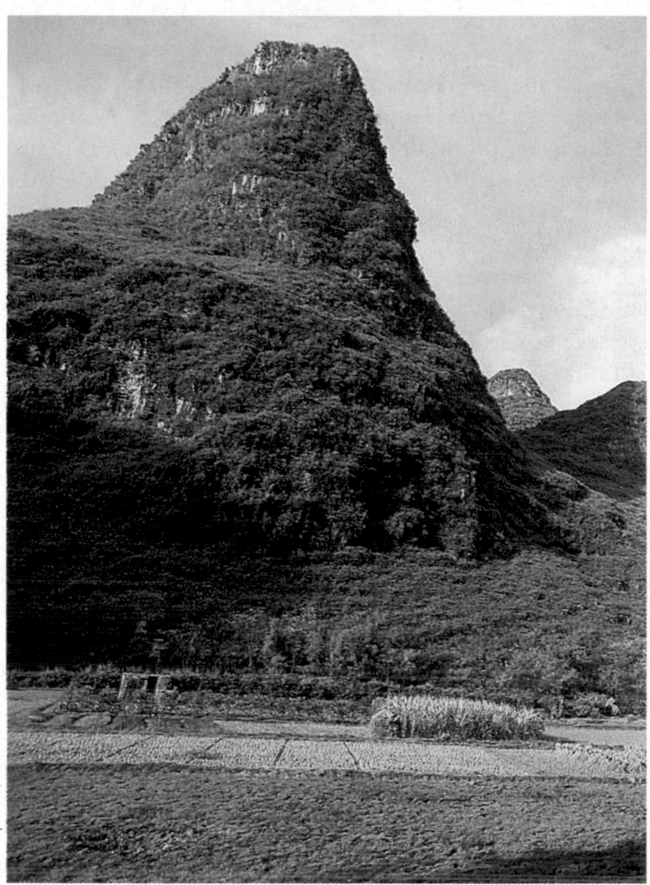

• **FIGURE 16.17**
Guilin, a limestone region in southern China, is famous for its beautiful karst towers (hums).
At one time in the geologic past, could this region have looked much like the landscape in Figure 16.16?

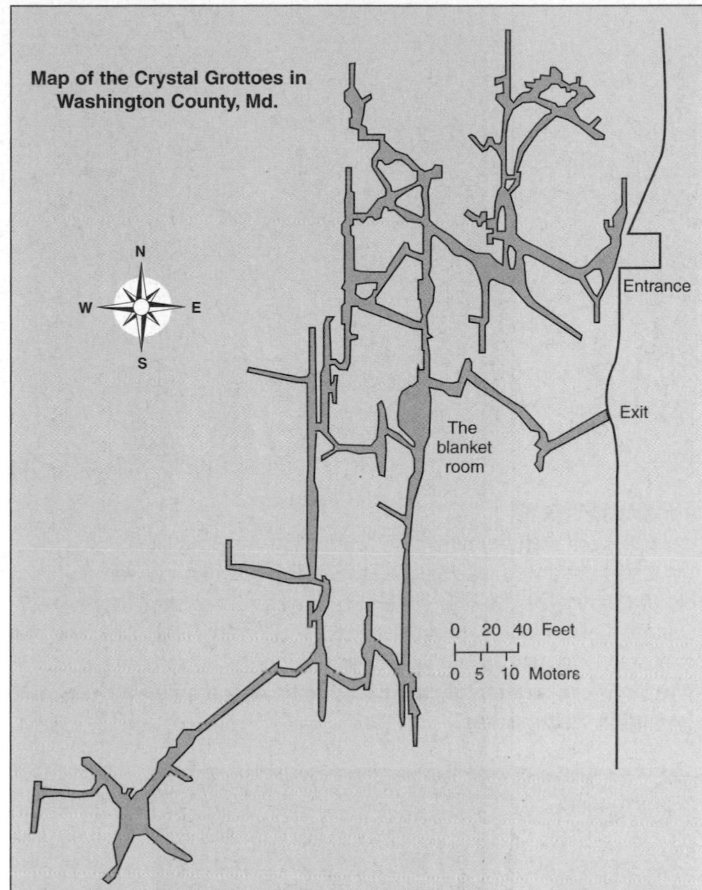

• **FIGURE 16.18**
This map of the Crystal Grottoes in Washington County, Maryland, illustrates the influence of fractures on the growth of the cavern system, as evidenced by the geometric arrangement and spacing of passageways. Caverns develop along zones of weakness, and groundwater flow widens fractures to develop a cave system.
Why are limestone fractures so susceptible to widening by dissolution?

The nature of fracturing that exists in soluble bedrock exerts a strong influence on cavern development in karst regions. Groundwater solution widens the space between opposing surfaces of joints, faults, and bedding planes to produce passageways. The relationship between caverns and fracture distributions is evident on cave maps that show linear and parallel patterns of cave passageways (• Fig. 16.18). Not all caves contain actively flowing water, but all caves formed by solution show some evidence of previous water flow, such as deposits of clay and silt on the cavern floor.

Speleothem is the generic term for any chemical precipitate feature deposited in caves. Speleothems develop in a great variety of textures and shapes, and many are both delicate and ornate. Speleothems develop when previously dissolved substances, particularly calcium carbonate ($CaCO_3$), precipitate out of subsurface water to produce some of the most beautiful and intricate forms found in nature. The dripping water leaves behind a deposit of calcium carbonate called travertine, or dripstone. As these travertine deposits grow downward, they form icicle-like spikes called **stalactites** that hang from the ceiling (• Fig. 16.19). Water saturated with calcium carbonate dripping onto the floor of a cavern builds up similar but more massive structures called stalagmites. Stalactites and **stalagmites** often meet and continue growing to form **columns** or pillars (• Fig. 16.20).

Most people believe that evaporation, leaving deposits of calcium carbonate behind, is the dominant process that produces the spectacular speleothems seen in caves, but that is not the case. Caves that foster active development of speleothems typically have air that is fully saturated with water, having a relative humidity near 100%, so evaporation is minimal. Instead, much of the water that percolates into the cave from above picked up carbon dioxide from the soil and dissolved calcium carbonate by carbonation on its way to the cave. Cave air, in contrast, contains a comparatively low amount of carbon dioxide. When the dripping water contacts the cave air, it therefore releases carbon dioxide gas to the air. This degassing of carbon dioxide from the water essentially reverses the carbonation process, causing the water to precipitate calcium carbonate. Eventually, enough calcium carbonate is

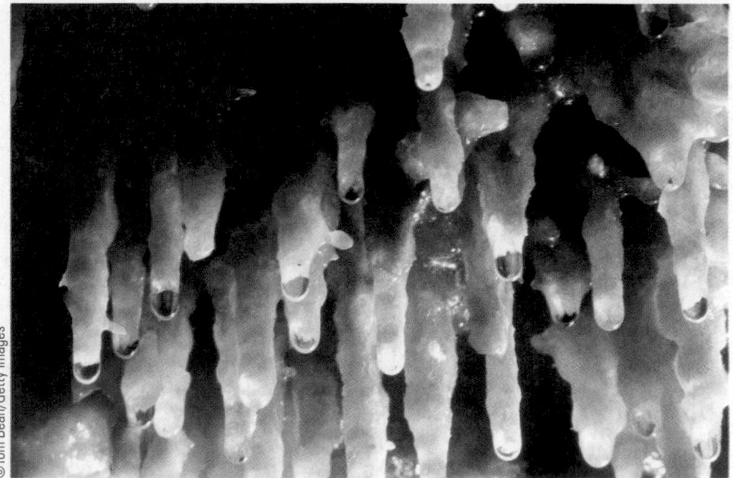
©Tom Bean/Getty Images

• **FIGURE 16.19**
Stalactites form where mineralized water that has percolated down to the cave from above reaches the cave ceiling and deposits some of its dissolved minerals, typically calcium carbonate. These stalactites in the Lehman Caves of Great Basin National Park, Nevada, are hollow and may grow into long, narrow, and very delicate tubes called soda straws.
Why does evaporation of water tend to be only a minor process in the formation of stalactites?

National Park Service–Carlsbad Caverns

• **FIGURE 16.20**
Carlsbad Caverns National Park in New Mexico contains some large speleothems (cavern deposits of dripstone).
How do you explain the presence of this huge cavern in a desert climate?

deposited in this way to form a stalactite, stalagmite, or other depositional cave feature.

Limestone caverns vary greatly in size, shape, and interior character. Some are well decorated with speleothems, and others are not. Many have several levels and are almost spongelike in the pattern of their passageways, whereas others are linear in pattern. The variations in cavern size and form may indicate differences in mode of origin. Some small caves may have formed above the water table by water percolating downward through the zone of aeration. Most, however, developed just below the water table, where the rate of solution is most rapid. A subsequent decline in the water table level, caused by the incision of surface streams, climatic change, or tectonic uplift, fills the cavern with air allowing speleothem formation to begin. Subterranean flow of water deepens some caverns, and collapse of their ceilings enlarges them upward.

Cavern development is a complex process, involving such variables as rock structure, groundwater chemistry, and hydrology, as well as the regional tectonic and erosional history. As a result, the scientific study of caverns, **speleology,** is particularly challenging. Adding to the challenge are the field conditions. Much of our knowledge of cavern systems has come from explorations as deep as hundreds of meters underground by individuals making scientific observations while crawling through mud, water, and even bat droppings in dark, narrow passages (• Fig. 16.21). Recent exploration and mapping of the water-filled caves below the surface in Florida and other karst regions have involved scuba diving. Cave diving in fully submerged, totally dark, and confined passageways, which may contain dangerous currents, is an extremely risky operation.

• **FIGURE 16.21**
Cave exploration is called caving, or spelunking. Here, a spelunker explores Spider Cave in Carlsbad Caverns, National Park. Caving can be exciting but also quite hazardous. It requires knowledge, skill, good equipment, and proper judgment of conditions.
What are some of the potential hazards of caving?

National Park Service–Carlsbad Caverns

Geothermal Water

Water heated by contact with hot rocks in the subsurface is referred to as **geothermal water.** Geothermal waters that flow out onto the surface fairly continuously form **hot springs.** When geothermal water flow is intermittent and somewhat eruptive it produces a **geyser,** a phenomenon that impresses with its sporadic bursts of expelled water and steam. The Old Faithful geyser in Yellowstone National Park is a well-known example (● Fig. 16.22). The word *geyser* is an Icelandic term for the steam eruptions that are so common on that volcanic island. Geysers erupt when temperature and pressure of the water at depth reach a critical level, forcing a column of superheated water and steam out of the fissure in an explosive manner.

Hot springs and geysers are both intriguing groundwater-related phenomena. Most hot springs and geysers contain significant amounts of minerals in solution, which become deposited at Earth's surface in various forms, often as terraces or cones around the vent or spring. These calcareous travertine and siliceous geyserite deposits often accumulate in colorful and impressive forms (● Fig. 16.23).

Geothermal activity is usually associated with areas of tectonic and volcanic activity, especially along plate boundaries and over hot spots. Geothermal energy has been used to produce electricity in areas such as California, Mexico, New Zealand, Italy, and Iceland. The best geothermal water for harnessing energy is not only very hot for generating steam but also "clean," that is, relatively free of dissolved minerals that can clog pipes and generating equipment.

● FIGURE 16.22
One of the world's most famous geysers is Old Faithful in Yellowstone National Park, Wyoming.
How do geysers differ from hot springs?

National Park Service–Yellowstone

● FIGURE 16.23
Because of the water's high mineral content, this hot spring in Yellowstone National Park, Wyoming, has left extensive calcareous deposits, known as travertine.

R. Gabler

Chapter 16 Activities

Define & Recall

subsurface water
infiltration
recharge
zone of aeration
percolation
soil water
zone of saturation
groundwater
water table
intermediate zone
water mining
effluent
influent
porosity
permeability
aquifer
aquiclude
perched water table
spring
well
artificial recharge
artesian
artesian spring
flowing artesian well
nonflowing artesian well
acid mine drainage
karst
sinkhole (doline)
solution sinkhole
collapse sinkhole
cavern (cave)
disappearing stream
swallow hole
uvala (valley sink)
haystack hill (conical hill or hum)
tower karst
stalactite
stalagmite
column
speleology
geothermal water
hot spring
geyser

Discuss & Review

1. Why is it important to understand the basic nature and properties of underground water?
2. What is the difference between infiltration and percolation?
3. When water soaks into the subsurface, what two zones does it pass through on its way to the zone of saturation? How is the zone of saturation related to groundwater and the water table?
4. Define porosity and permeability. How are these two properties related to groundwater?
5. What is the difference between a spring and an artesian spring? What is the difference between a well and a flowing artesian well?
6. What is an aquifer? Describe the conditions necessary for an aquifer.
7. Define karst. What conditions encourage the development of karst landscapes?
8. Describe a sinkhole. How are sinkholes formed?
9. Describe a cavern. How do caverns form?
10. Explain how calcium carbonate is deposited by subsurface water to form a stalactite in a humid cave.

Consider & Respond

1. If you were a water resources geographer and had to plan for the development of groundwater resources for a community, what would be your major considerations?
2. Describe the major landform features in a region of karst topography. Over time, what changes might be expected in the landscape?
3. What are some human impacts and environmental problems related to groundwater resources?

Apply & Learn

1. Drilling at seven sites spaced every 5 kilometers along a straight-line transect heading from west to east has provided the following data on depth to the water table and rock type just below the water table: (a) 15 meters in sandstone, (b) 14.5 meters in sandstone, (c) 14 meters in sandstone, (d) 5 meters in limestone, (e) 13 meters in sandstone, (f) 12.5 meters in sandstone, (g) 12 meters in sandstone. What is the regional slope of the water table? Provide a reasonable explanation for the data obtained from site d.
2. Scientists studying an air-filled cavern located in an arid region determined that the speleothems present began forming 20,000 years ago and stopped growing 12,000 years ago. Suggest a reasonable history of climatic and hydrologic events related to the formation of the cavern evidence.

Locate & Explore

Note: Please read the About Locate & Explore Activities section of the Preface before beginning these exercises.

1. Using Google Earth and Figure 16.10 from your text, examine the spatial distribution of karst regions in the United States. In general, the development of karst landscapes depends on geology, precipitation, topography, tectonic activity, vegetation cover, and climate change history. Provide an explanation for karst development in Kentucky, Texas, and Florida.
2. Using Google Earth, identify the landforms at the following locations (latitude, longitude) and provide a brief discussion of how the landform developed. Include a brief discussion of why the landform is found in that general area of the United States or the world.

   Tip: Use the zoom, tilt, rotate, and elevation exaggeration tools to help view and interpret the landform and the area in which it is found.

   a. 18.40°N, 66.53°W
   b. 25.05°N, 110.37°E
   c. 29.66°N, 81.87°W
   d. 38.207°N, 90.175°W
   e. 20.87°N, 107.19°E
   f. 32.12°S, 125.29°E

Map Interpretation

KARST TOPOGRAPHY

The Map

The Interlachen area is in northern central Florida. Florida's peninsula is the emerged portion of a gentle anticline called the Peninsular Arch. The region is underlain by thousands of meters of marine limestones and shales. This great thickness of marine sediments originated in the Mesozoic Era when Florida was a marine basin. As the arch rose, Florida became a shallow shelf and was eventually elevated above sea level.

Although outsiders think of Florida mainly as a state with magnificent beaches and warm winter weather due to its humid subtropical climate, it is also a state with hundreds of lakes dotting its center.

The lake region is formed on the Ocala Uplift, a gentle arch of limestone that reaches to 46 meters (150 ft) above sea level. Lake Okeechobee is the largest of these lakes and has an average depth of less than 4.5 meters (15 ft). Most of the lakes, such as those in the Interlachen map area, are much smaller.

Florida's central lake region is an ideal area for studying karst topography. Both the surface and the subsurface features express the geomorphic effects of subsurface water. Extensive cavern systems exist beneath the surface. Much of the state's runoff is channeled through huge aquifers, and springs are quite common.

Interpreting the Map

1. This area would be classified as what major type of landform (for example, mountain)? What is its average elevation?
2. What contour interval is used on this map? Why do you think the cartographers chose this interval?
3. On what type of bedrock is the map area situated? Do you think the climate has any influence on the landforms in the area? Explain.
4. What landform features on the map indicate that this is a karst region?
5. What are the round, steep depressions called? Why do lakes occupy some of the depressions?
6. Locate Clubhouse Lake on the full-page map (scale 1: 62,500) and the smaller map (1: 24,000). What is the elevation of Clubhouse Lake? What is its maximum width?
7. What type of feature is the area north of Lake Grandin?
8. What is the approximate elevation of the water table? (Note: You can determine this from the elevation of the lakes' water surface.)
9. Underground, the water flows through an aquifer. Define an aquifer and list the characteristics an aquifer must have. What is the general direction of groundwater flow in the aquifer underlying the Interlachen area?
10. Because much of central Florida is rapidly urbanizing, what problems and hazards do you anticipate in this karst area?

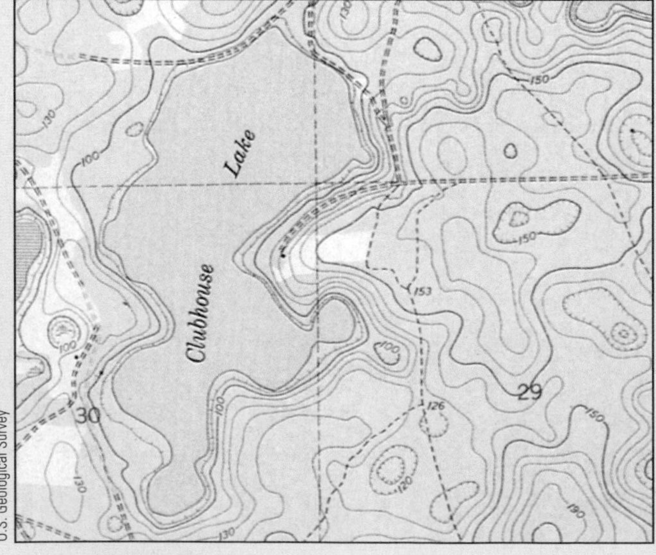

Putnam Hall, Florida
Scale 1: 24,000
Contour interval = 10 ft

Opposite:
Interlachen, Florida
Scale 1:62,500
Contour interval = 10 ft
U.S. Geological Survey

Clubhouse Lake
Brantley Lake
Boyds Lake
Long Pond
Lake Grandin
Cue Lake
Mason Lake
Trotting Pond
Clearwater Lake
Junior Lake
Church Lake
Lake Loyal
Big Pond
Hubbard Pond
Mariner Lake
Jewel Lake
Sand Hill Pond
Grassy Lake
Menneville Cem
Mirror Lake
Chipco Lake
Cooper Lake
East Lake
West Lake
Lake Galilee
Clear Lake
Ladyslipper Lake
Interlachen
Water Tank
Lake Lagonda
Morris Lake
Cambo Lake
Mt Bethel Ch
Oak Grove Sch
Keuka
Lake Keuka
Pipe Conveyor
Sub Sta
Edgar
Sewage Disposal
Water
ATLANTIC
COAST
LINE
Gum Creek
Johnson Park Sch
Woodlawn Ch
Johnson
Darkwater Lake
Cem
Cabbage Creek
Higginbotham Lake
Lake Fanny

Fluvial Processes and Landforms

17

CHAPTER PREVIEW

Running water, as sheet wash and stream flow, is the most important geomorphic agent, doing more landforming work than any of the other agents—wind, glaciers, or waves.

- What does this indicate about Earth's landforms?
- How is this significant to people?

Perennial streams flow every day of the year even if it has been several weeks since the last rainstorm.

- What does this indicate about the source of water in perennial streams?
- How will stream flow differ near the end of a large storm compared to after several weeks without rain?

Moderate increases in the volume, velocity, and turbulence of stream flow cause major increases in the sediment-carrying capacity of a stream.

- What does this imply about stream erosion and deposition?
- What does it suggest about a stream at flood stage?

The level of the average stream rises to the brink of flooding roughly once every 1.5 years.

- What is the geomorphic significance of this to the floodplain, the gently sloping, low-relief land commonly found adjacent to streams?
- What are the pros and cons of people artificially adding to the height of the natural levees that lie adjacent to many stream channels?

Streams are natural systems that convey inputs, throughputs, and outputs of water, sediment, and energy from source to mouth.

- How might a stream adjust to a lowering of sea level at its mouth?
- What changes might a stream channel undergo if a large fire stripped the vegetation off of a large portion of its drainage basin?

◄ Opposite: Streams are the dominant geomorphic agent on Earth's surface. This stream flows in rapids over the rocks at Great Falls Park, near Washington, D.C., in Virginia.

© Virginia Natural Heritage Program/Gary P. Fleming

Flowing water is more influential in shaping the surface form of our planet than any other exogenic geomorphic process, primarily because of the sheer number of streams on Earth. Through both erosion and deposition, water flowing downslope over the land surface, particularly when concentrated in channels, modifies existing landforms and creates others. Nearly every region of Earth's land surface exhibits at least some topography that has been shaped by the power of flowing water, and many regions exhibit extensive evidence of stream action. Flowing water is the main geomorphic agent in arid as well as humid environments. Polar landscapes buried under thick accumulations of perennial ice are the major exception to Earth's extensive areas of stream-dominated topography.

The study of flowing water as a land-shaping process, together with the study of the resulting landforms, is termed **fluvial geomorphology** (from Latin: *fluvius,* river). Fluvial geomorphology includes the action of both channelized and unchannelized flow moving downslope by the force of gravity.

Stream is the general term for natural, channelized flow. In the Earth sciences, the term *stream* pertains to water flowing in a channel of any size, although in general usage we describe large streams as rivers and use local terms, such as creek, brook, run, draw, and bayou, for smaller streams.

The land between adjacent channels in a stream-dominated landscape is referred to as the **interfluve** (from Latin: *inter,* between; *fluvius,* river).

Because of the common and widespread occurrence of stream systems and their key role in providing fresh water for people and our agricultural, industrial, and commercial activities, a substantial portion of the world's population lives in close proximity to streams. This makes understanding stream processes, landforms, and hazards fundamental for maintaining human safety and quality of life.

Most streams occasionally expand out of the confines of their channel. Although these **floods** typically last only a few days at most, they reveal the tremendous—and often very dangerous—geomorphic power potential of flowing water. The long-term effects of stream flow, whether dominated by erosion or deposition, are sometimes also quite dramatic. Two prime examples in the United States that illustrate the effectiveness of flowing water in creating landforms are the Black Canyon of the Gunnison River (• Fig. 17.1a), carved by long-term river erosion into the Rocky Mountains, and the Mississippi River delta (Fig. 17.1b) where fluvial deposition is building new land into the Gulf of Mexico.

Surface Runoff

Liquid water flowing over the surface of Earth, that is, **surface runoff,** can originate as ice and snow melt or as outflow from springs, but most runoff originates from direct precipitation. When precipitation strikes the ground, several factors interact to determine whether surface runoff will occur. Basically, runoff is generated when the amount, duration, and/or rate of precipitation exceed the ability of the ground to soak up the moisture. Because the process of water soaking into the ground is *infiltration,* the amount of water the soil and surface sediments can hold is known as the **infiltration capacity.** A portion of infiltrated water will seep (percolate) down to lower positions and reach the zone of saturation beneath the water table, while much of the rest will eventually return to the atmosphere by evaporation from the soil or by transpiration from plants. When more precipitation falls than can be infiltrated into the ground, the excess water flows downslope by the force of gravity as surface runoff.

Various factors act individually or together to either enhance or inhibit the generation of surface runoff. Greater infiltration to the subsurface, and therefore less runoff, occurs under conditions

• **FIGURE 17.1**
(a) Long-term fluvial erosion by the Gunnison River, with the aid of weathering and mass wasting, has carved the Black Canyon of the Gunnison through a part of the Rocky Mountains in Colorado. The canyon is 829 meters (2722 ft) deep, 80 kilometers (50 mi) long, and very narrow in width, with resistant rocks comprising the steep walls. The Gunnison River flows beyond the canyon westward into the Colorado River toward the Gulf of California. (b) Through fluvial deposition, the Mississippi River has been building its delta outward into the Gulf of Mexico for thousands of years. Where the river enters the Gulf, the slowing current deposits large amounts of muddy sediment that came from the river's drainage basin. This is a false-color composite, digital image. Muddy water appears light blue, and clearer, deeper water is dark blue.

L. Lynch/National Park Service
(a)

© NASA Earth Observatory/ Image courtesy NASA/GSFC/METI/ERSDAC/JAROS, and U.S./Japan ASTER Science Team
(b)

of permeable surface materials, deeply weathered sediments and soils, gentle slopes, dry initial soil conditions, and a dense cover of vegetation. **Interception** of precipitation by vegetation allows greater infiltration by slowing down the rate of delivery of precipitation to the ground. Vegetation also enhances infiltration when it takes up soil water and returns it to the atmosphere through transpiration. Given the same precipitation event, surface materials of low permeability and limited weathering, thin soils, steep slopes, preexisting soil moisture, and sparse vegetation each contribute to increased runoff by decreasing infiltration (• Fig. 17.2). Human activities can impact many of these variables, and in some places the generation of runoff has been greatly modified by urbanization, mining, logging, or agriculture.

Once surface runoff forms, it first starts to flow downslope as a thin sheet of unchannelized water, known as **sheet wash,** or unconcentrated flow. Because of gravity, after a short distance the sheet wash will begin to move preferentially into any preexisting swales or depressions in the terrain. This concentration of flow leads to the formation of tiny channels, called **rills,** or somewhat larger channels, called **gullies.** Rills are on the order of a couple of inches deep and a couple of inches wide (• Fig. 17.3), whereas gully depth and width may approach as much as a couple of feet.

Water does not flow in rills and gullies all the time but only during and shortly after a precipitation (or snowmelt) event. Channels that are empty of water much of the time like this are described as having **ephemeral flow.** As these small, ephemeral channels continue downslope, rills join to form slightly larger rills, which may join to make gullies. In humid climates, following these successively larger ephemeral channels downslope will

• FIGURE 17.3
Rills are the smallest type of channel. They are commonly visible on devegetated surfaces, like these mine tailings.

eventually lead us to a point where we first encounter **perennial flow.** Perennial streams flow all year, but not always with the same volume or at the same velocity. Most arid region streams flow on an ephemeral basis although some may have **intermittent flow,** which lasts for a couple of months in response to an annual rainy season or spring snowmelt. Because of this contrast in flow duration, and other differences between arid and humid region streams, a full discussion of arid region stream systems appears in the separate chapter on arid region landforms, Chapter 18.

Perennial streams flow throughout the year even if it has been several weeks since a precipitation event. In most cases, this is possible only because the perennial streams continue to receive direct inflow of groundwater (Chapter 16) regardless of the date of most recent precipitation. Slow-moving groundwater seeps directly into the stream through the channel bottom and sides at and below the level of the water surface as **base flow.** Except in rare instances, it takes a humid climate to generate sufficient base flow to maintain a perennial stream between rainstorms.

• FIGURE 17.2
The amount of runoff that occurs is a function of several factors, including the intensity and duration of a rainstorm. Surface features, however, are also important. Any factor that increases infiltration and evapotranspiration will reduce the amount of water available to run off, and vice versa. Deep soil, dense vegetation, fractured bedrock, and gentle slopes tend to reduce runoff. Thin or absent soils, thin vegetation, and steep slopes tend to increase runoff.

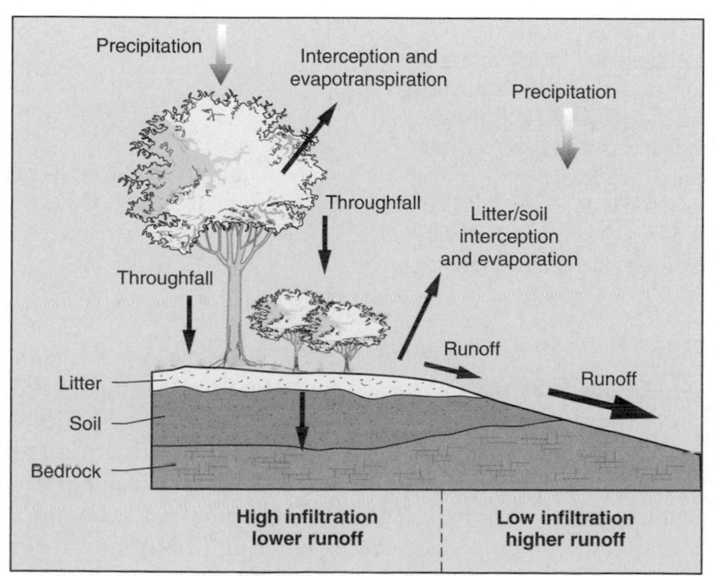

The Stream System

Most flowing water becomes quickly channelized into streams as it is pulled downhill by the force of gravity. Continuing downslope, streams form organized channel systems in which small perennial channels join to make larger perennial channels, and larger perennial channels join to create even bigger streams. Smaller streams that contribute their water and sediment to a larger one in this way are **tributaries** of the larger channel, which is called the **trunk stream** (• Fig. 17.4).

Drainage Basins

Each individual stream occupies its own **drainage basin** (also known as **watershed,** or **catchment area**), the expanse of land from which it receives runoff. **Drainage area** refers to the measured extent of a drainage basin, and is typically expressed

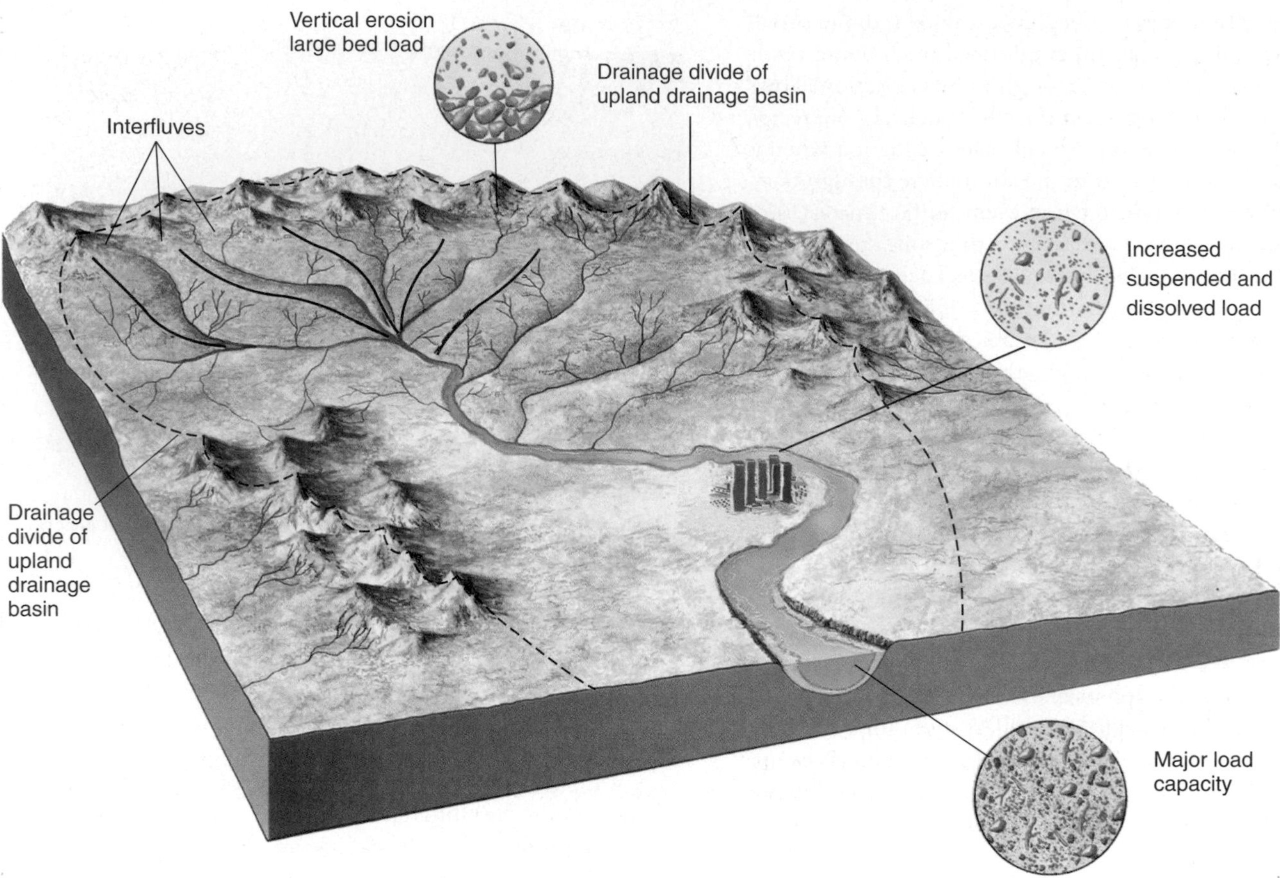

• FIGURE 17.4

Environmental Systems: The Hydrologic System—Streams. The stream, or surface runoff, system is a subsystem of the hydrologic cycle. Its major water input is from precipitation. However, groundwater may also contribute to the stream system, particularly in regions of humid climate. The major water output for most stream systems returns water into the ocean. Output or loss of water also occurs by evaporation back to the atmosphere and by infiltration into the groundwater system. Stream systems are divided into natural regions known as drainage basins (also called watersheds), separated from each other by divides; interfluves are the land areas between stream channels in the same drainage basin. Drainage basins are fundamental natural regions of critical importance to life in both humid and arid regions.

Streams are complex systems of moving water that involve energy transfers and the transport of a variety of surface materials. Energy enters the system with precipitation. The runoff flows downslope, increasing the amount of energy available to the stream for cutting and eroding channels.

Materials transported by streams, known as load, enter the stream system by erosion and mass movement, particularly in the headwaters of a drainage basin. Much of the surface material eroded there consists of large particles, including boulders. Coarse material is carried along the channel bottom as bed load. As the number and size of tributaries increase downstream, the amount of load carried by the stream generally increases dramatically. This is especially true for finer materials suspended in water (suspended load) and dissolved minerals (dissolved load). Load leaves the stream system when carried and deposited in the sea at the river mouth. Streams also deposit sediment adjacent to their channels as they overflow their banks during floods. Human activities can change the amount of load available for stream systems by building dams, altering land with construction projects, overgrazing, and clearing forests. These activities may also affect water quality downstream, where communities may depend on the stream system for their water supply.

in square kilometers or square miles. Because the runoff from a tributary's drainage basin is delivered by the tributary to the trunk stream, the tributary's drainage basin also constitutes part of the drainage basin of the trunk stream. In this way, small tributary basins are nested within, or are *subbasins* of, a succession of larger and larger trunk stream drainage basins. Large river systems drain extensive watersheds that consist of numerous inset subbasins.

Drainage basins are open systems that involve inputs and outputs of water, sediment, and energy. Knowing the boundaries of a drainage basin and its component subbasins is critical to properly managing the water resources of a watershed. For example, pollution discovered in a river almost always comes from a source within its drainage basin, entering the stream system either at the point where the pollutant was first detected or at a location upstream from that site. This knowledge helps us track, detect, and correct sources of pollution.

The **drainage divide** represents the outside perimeter of a drainage basin and thus also the boundary between it and adjacent basins (● Fig. 17.5). The drainage divide follows the crest of the interfluve between two adjacent drainage basins. In some places, this crest is a definite ridge, but the higher land that constitutes the divide is not always ridge-shaped, nor is it necessarily much higher than the rest of the interfluve. Surface runoff generated on one side of the divide flows toward the channel in one drainage basin, while runoff on the other side travels in a very different direction toward the channel in the adjacent drainage basin. The *Continental Divide* separates North America into a western region where most runoff flows to the Pacific Ocean and an eastern region where runoff flows to the Atlantic Ocean. The Continental Divide generally follows the crest line of high ridges in the Rocky Mountains, but in some locations the highest point between the two huge basins lies along the crest of gently sloping high plains.

A stream, like the Mississippi River, that has a very large number of tributaries and encompasses several levels of nested sub-basins, will have some major differences from a small creek that has no perennial tributaries and lies high up in the drainage basin near the divide. Knowing where each stream lies in the hierarchical order of tributaries helps Earth scientists make more meaningful comparisons among different streams. The importance of quantitatively describing a stream or drainage basin's position in the hierarchy was realized long ago. In the 1940s, a hydrologist, Robert Horton, first proposed a system for determining this **stream order.** The stream-ordering system in most common use today is a modified version of what Horton suggested. In this system, *first-order streams* have no perennial tributaries. Even though they are the smallest perennial channels in the drainage basin, first-order channels can be mapped on large-scale topographic maps. Most first-order streams lie high up in the drainage basin near the drainage divide, the **source** area of the stream system. Two first-order streams must meet in order to form a *second-order stream,* which is larger than each of the first-order streams. It takes the intersection of two second-order channels to make a *third-order stream* regardless of how many first-order streams might independently join the second-order channels. The ordering system continues in this way, requiring two streams of a given order to combine to create a stream of the next higher order. The order of a drainage basin derives from the largest stream order found within it (● Fig. 17.6). For example, the Mississippi is a tenth-order drainage basin because the Mississippi River is a tenth-order stream. Stream ordering allows us to compare various attributes of streams quantitatively by relative size, which helps us better understand how stream systems work. Among other things, comparing streams on the basis of order has shown that as stream order increases, basin area, channel length, channel size, and amount of flow also increase.

Water moves through a stream system via channels of ever-increasing order as gravity pulls it downslope toward the downstream end, or **mouth** of the stream. The mouth of most humid region, perennial stream systems lies at sea level where the channel system finally ends and the stream water is delivered to the ocean. Drainage basins with channel systems that convey water to the ocean have **exterior drainage.** Many arid region streams have **interior drainage** because they do not have enough flow to reach the ocean but terminate instead in local or regional areas of low elevation.

● FIGURE 17.5
An aerial photograph of a small stream channel network, basin, and divide. Every arm of the stream occupies its own subbasin (separated by divides) that combines with others to form the basin of the main stream channel.
Can you mentally trace the outline of the main drainage basin shown here?

©Texas Natural Resources Information System

● FIGURE 17.6
The hierarchy of stream ordering is illustrated by the channels of this fourth-order watershed. Note that when two streams of a given order meet, they join to form a segment of the next higher order. Order does not change unless two streams of the same order flow together. When the stream order changes, the channel is considered a new segment.
What is the highest-order stream in a selected drainage basin called?

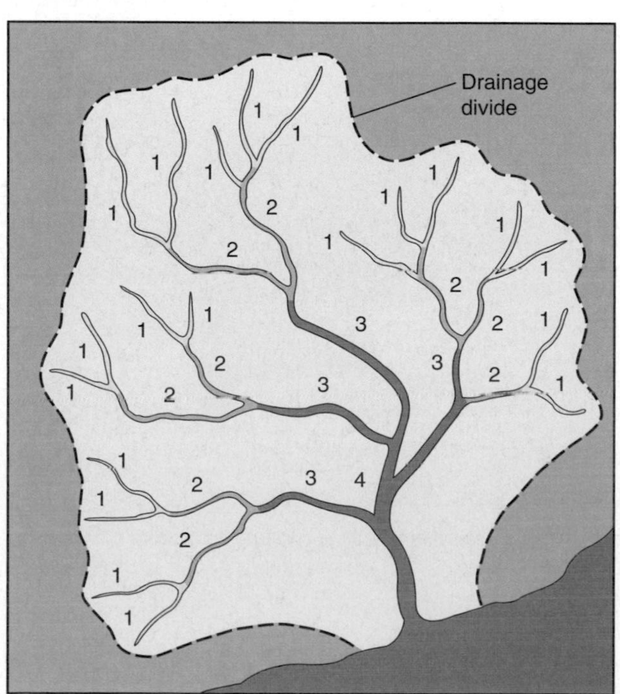

GEOGRAPHY'S SPATIAL SCIENCE PERSPECTIVE

Watersheds as Critical Natural Regions

Perhaps the most environmentally logical way to divide Earth's land surface into regions is by watersheds, the drainage basins for stream systems. Virtually all of Earth's land surface comprises part of a watershed. Like the stream systems that they contain, watersheds are hierarchical and are conveniently subdivided into smaller subbasins for local studies and management. At the same time, higher-order watersheds of large river systems are subject to broad, integrated, regional-scale analyses.

The environmental processes and physical components within a watershed are strongly interrelated. Problems in one part of the system are likely to cause problems in other parts of the system. Along with the channel network that occupies each basin, watersheds represent well-integrated natural systems that include water, soil, rock, terrain, vegetation, people, wildlife, and domesticated animals. Complex natural biotic habitats exist within watersheds, and they are greatly affected by the environmental quality of the basin. Monitoring and managing these complex watershed systems requires an interdisciplinary effort that considers aspects of all four of the world's major spheres.

Surface water from watershed sources provides much of the potable water resources for the world's population. Increasing human populations and land-use intensities in previously natural drainage basins place pressures on these habitats and their water quality. Maintaining water quality within local subbasins contributes to maintaining water quality of larger river systems at the regional scale. Because streams comprise a network of flow in one direction (downstream), identifying sources of pollution involves upstream tracing of pollutants to the source and downstream spreading of pollution problems. Humans have a vital responsibility to monitor, maintain, and protect the quality of freshwater resources, and doing so at the local watershed scale has far-reaching as well as local benefits.

In recent years, many governmental agencies have designated watersheds as critical regions for environmental management. Some of these efforts involve the establishment of locally administered river conservation districts representing relatively small-order drainage basins in which community involvement plays a vital role. There are several sound reasons, including the connectivity and hierarchical nature of watersheds, for this spatial strategy.

Watersheds are clearly defined, well-integrated, natural regions of critical importance to life on Earth, and they make a logical spatial division for environmental management. However, a stream system may flow through many cities, counties, states, and countries, and management problems can arise where regional political and administrative boundaries do not coincide with the divide that defines the edge of a watershed. Each of these political jurisdictions may have very different needs, goals, and strategies for using and managing their part of a watershed and often a variety of strategies results in conflict. Still, most administrative units recognize that cooperative management of the watershed system as a whole is the best approach. Many cooperative river basin authorities have been established to encourage a united effort to protect a shared watershed. This watershed-oriented management strategy, based on a fundamental natural region, is an important step in protecting our freshwater resources.

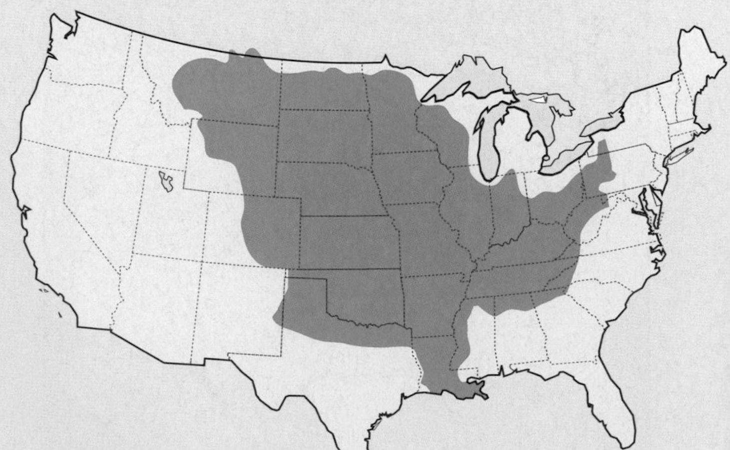

The Mississippi River drainage area (watershed) covers an extensive region of the United States.

As a tributary of the Ohio River, the fourth-order Hocking River watershed in southeastern Ohio is a subbasin within the Mississippi River system. Local involvement in improving and maintaining the health of a watershed has regional as well as local benefits.

Every stream has a **base level,** the elevation below which it cannot flow. Sea level is the *ultimate base level* for virtually all stream action. Streams with exterior drainage reach ultimate base level; the low point of flow for a stream with interior drainage is referred to as a *regional base level*. In some drainage basins, a very resistant rock layer located somewhere upstream from the river mouth can act as *temporary base level,* temporarily controlling the lowest elevation of the flow upstream from it until the stream is finally able to cut down through it (• Fig. 17.7).

Drainage Density and Drainage Patterns

Each organized system of stream tributaries exhibits spatial characteristics that provide important information about the nature of the drainage basin. The extent of channelization can be represented by measuring **drainage density** (D_d), where $D_d = L/A_d$, the total length of all channels (L) divided by the area of the drainage basin (A_d). Drainage density indicates how dissected the landscape is by channels, thus it reflects both the tendency of the drainage basin to generate surface runoff and the erodibility of the surface materials (• Fig. 17.8). Regions with high drainage densities will have limited infiltration, promote considerable runoff, and have at least moderately erodible surface materials.

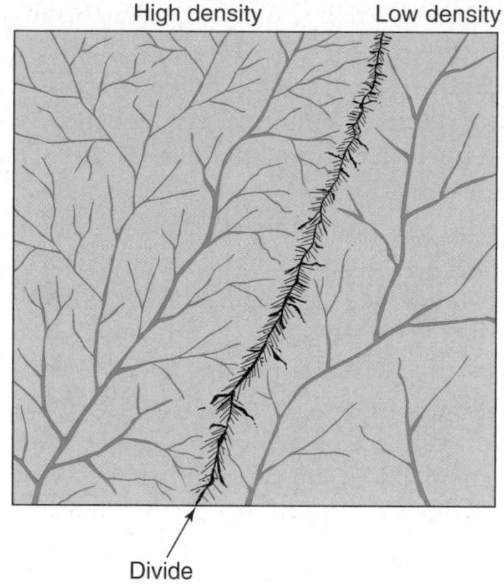

• FIGURE 17.8
Drainage density—the length of channels per unit area—varies according to several environmental factors. For example, everything else being equal, highly erodible and impermeable rocks tend to have higher drainage density than areas dominated by resistant or permeable rocks. Slope and vegetation cover can also affect drainage density.
What kind of drainage density would you expect in an area of steep slopes and sparse vegetation cover?

• FIGURE 17.7
The lowest point to which a stream can flow is its base level. Stream water travels downslope by the force of gravity until it can flow no lower due to factors of topography, climate, or both. Most humid region streams have sufficient flow to make it all the way to an ocean basin. Thus, sea level represents ultimate base level for all of Earth's streams. In arid climates, many streams lose so much water by evaporation to the atmosphere and infiltration into the channel bed that they cannot flow to the sea. The lowest point they can reach is instead a regional base level, a topographic basin on the continent. A temporary base level is formed when a rock unit lying in the pathway of a stream is significantly more resistant than the rock upstream from it. The stream will not be able to cut into the less resistant rock any faster than it can cut into the resistant rock of the temporary base level.

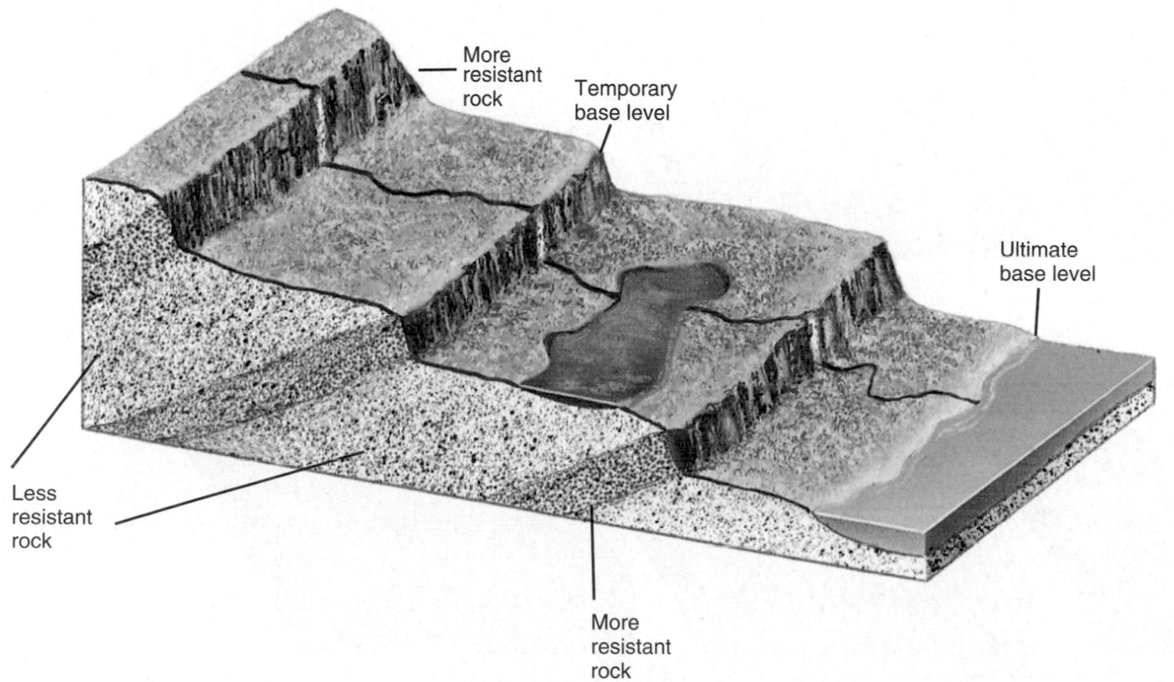

In addition to the factors noted previously that reduce infiltration and promote runoff—impermeable sediments, thin soils, steep slopes, sparse vegetation—the ideal climate for high drainage densities is one that is semiarid. Humid climates encourage extensive vegetation cover, which promotes infiltration through interception and reduces channel formation by holding soils and surface sediment in place. In arid climates, although the vegetation cover is sparse, there is insufficient precipitation to create enough runoff to carve many channels. Semiarid climates have enough precipitation input to produce overland flow, but not enough to support an extensive vegetative cover. The easily eroded Dakota Badlands, which are located in a steppe climate, have an extremely high drainage density of over 125 kilometers per square kilometer (125 km of channel per 1 sq km of land), whereas very resistant granite hills in a humid climate may have a drainage density of only 5 kilometers per square kilometer. Another way to understand the concept of drainage density is to think about what would happen on a hillside in a humid climate if the natural vegetation were burned off in a fire. Erosion would rapidly cut gullies, creating more channels than previously existed there. In other words, the drainage density would increase. We could use the quantitative measure of $D_d = L/A_d$ to determine the change in channelization precisely and to monitor it over time.

When viewed by looking down at Earth's surface from the air or on maps, the tributaries of various stream systems may form distinct **drainage patterns** (also called **stream patterns**). Two primary factors that influence drainage pattern are bedrock structure and surface topography. A *dendritic* (from Greek: *dendros,* tree) stream pattern (● Fig. 17.9a) is an irregular branching arrangement with tributaries joining larger streams at acute angles (less than 90°). A dendritic stream pattern is the most common type, in part because water flow in this spatial arrangement is highly efficient. Dendritic patterns form where the underlying rock structure does not strongly control the position of stream channels. Hence, dendritic patterns tend to develop in areas where the rocks have a roughly equal resistance to weathering and erosion and are not intensely jointed. In contrast, a *trellis* drainage pattern consists of long, parallel streams linked by short, right-angled segments (Fig. 17.9b). Trellis drainage is usually evidence of folding where parallel outcrops of erodible rocks form valleys between more resistant ridges, as in the Ridge and Valley region of the Appalachians. A *radial* pattern develops where streams flow away from a common high point on cone- or dome-shaped geologic structures, such as volcanoes and domal uplifts (Fig. 17.9c). The opposite pattern is *centripetal,* with the streams converging on a central area as in an arid region basin of interior drainage (Fig. 17.9d). *Rectangular* patterns occur where

● FIGURE 17.9

Drainage patterns often reflect bedrock structure. (a) The dendritic pattern is found where rocks have uniform resistance to weathering and erosion. (b) A trellis pattern indicates parallel valleys of weak rock between ridges of resistant rock. (c) Multiple channels trending away from the top of a domed upland or volcano form the radial pattern. (d) The centripetal pattern shows multiple channels flowing inward toward the center of a structural lowland or basin. (e) Rectangular patterns indicate linear joint patterns in the bedrock structure. (f) A deranged pattern typically results following the retreat of continental ice sheets; it is characterized by a chaotic arrangement of channels connecting small lakes and marshes.

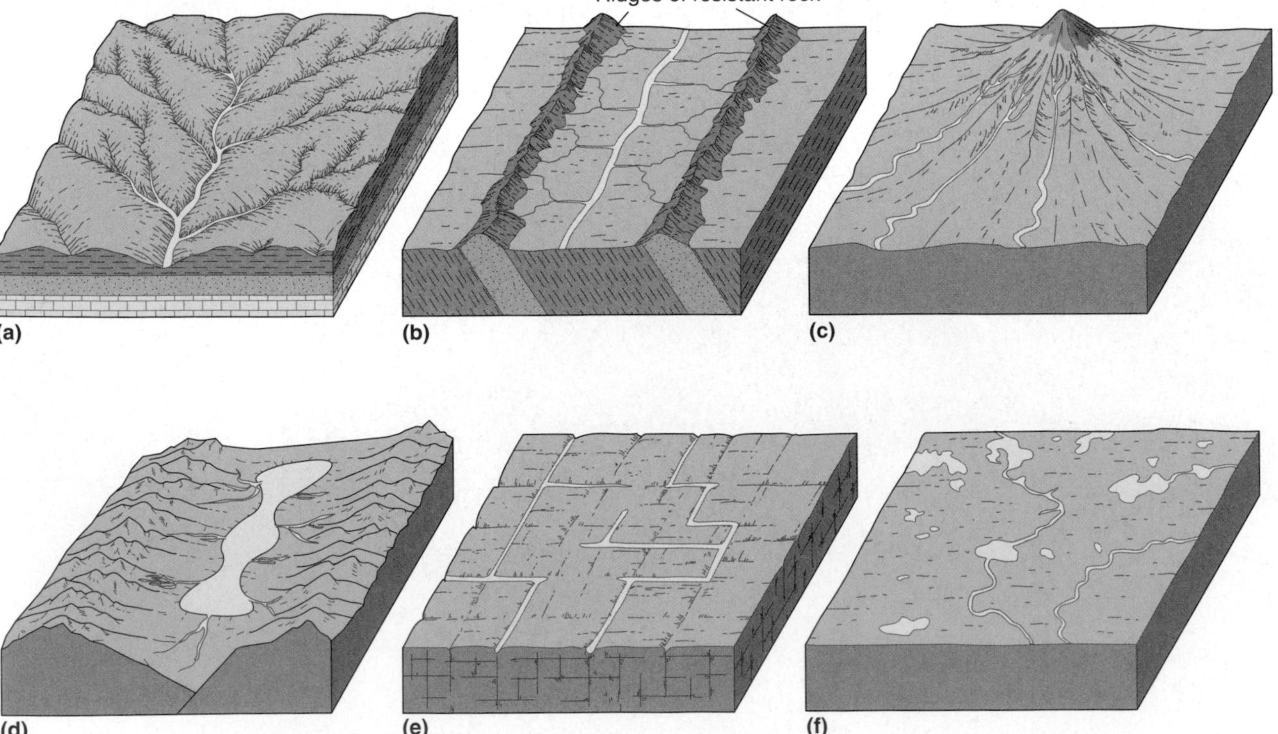

streams follow sets of intersecting fractures to produce a blocky network of straight channels with right-angle bends (Fig. 17.9e). In some regions that were recently covered by extensive glacier ice, streams flow on low-gradient terrain left by the receding glaciers, wandering between marshes and small lakes in a chaotic pattern called *deranged* drainage (Fig. 17.9f).

Many streams follow the "grain" of the topography or the bedrock structure, eroding valleys in weaker rocks and flowing away from divides formed on resistant rocks. Many examples also exist of streams that flow across, that is, transverse to, the structure, cutting a gorge or canyon through mountains or ridges. These **transverse streams** can be puzzling, giving rise to questions as to how a stream can cut a gorge through a mountain range or how a stream can move from one side of a mountain range to the other. Such streams are probably either *antecedent* or *superimposed.*

Antecedent streams existed before the formation of the mountains that they flow across, maintaining their courses by cutting an ever-deepening canyon as gradual mountain building took place across their paths. The Columbia River Gorge through the Cascade Range in Washington and Oregon and many of the great canyons in the Rocky Mountain region, such as Royal Gorge and the Black Canyon of the Gunnison, both in Colorado, probably originated as antecedent streams (● Fig. 17.10a). Other rivers, including some in the central Appalachians, have cut gaps through mountains in a very different manner. These streams originated on earlier rock strata, since stripped away by erosion, so that the streams have been superimposed onto the rocks beneath (Fig. 17.10b). This sequence would explain why, in many instances, the rivers flow across folded rock structure, creating water gaps through mountain ridges. Examples include the Cumberland Gap and the gap formed by the Susquehanna River in Pennsylvania, both of which were important travel routes for the first European settlers crossing the Appalachians.

Stream Discharge

The amount of water flowing in a stream depends not only on the impact of recent weather patterns but also on such drainage basin factors as its size, relief, climate, vegetation, rock types, and land-use history. Stream flow varies considerably from time to time and place to place. Most streams experience occasional brief periods when the amount of flow exceeds the ability of the channel to contain it, resulting in the flooding of channel-adjacent land areas.

Just as it was important to develop the technique of stream ordering to indicate quantitatively a channel's place in the hierarchy of tributaries, it is also crucial to be able to describe quan-

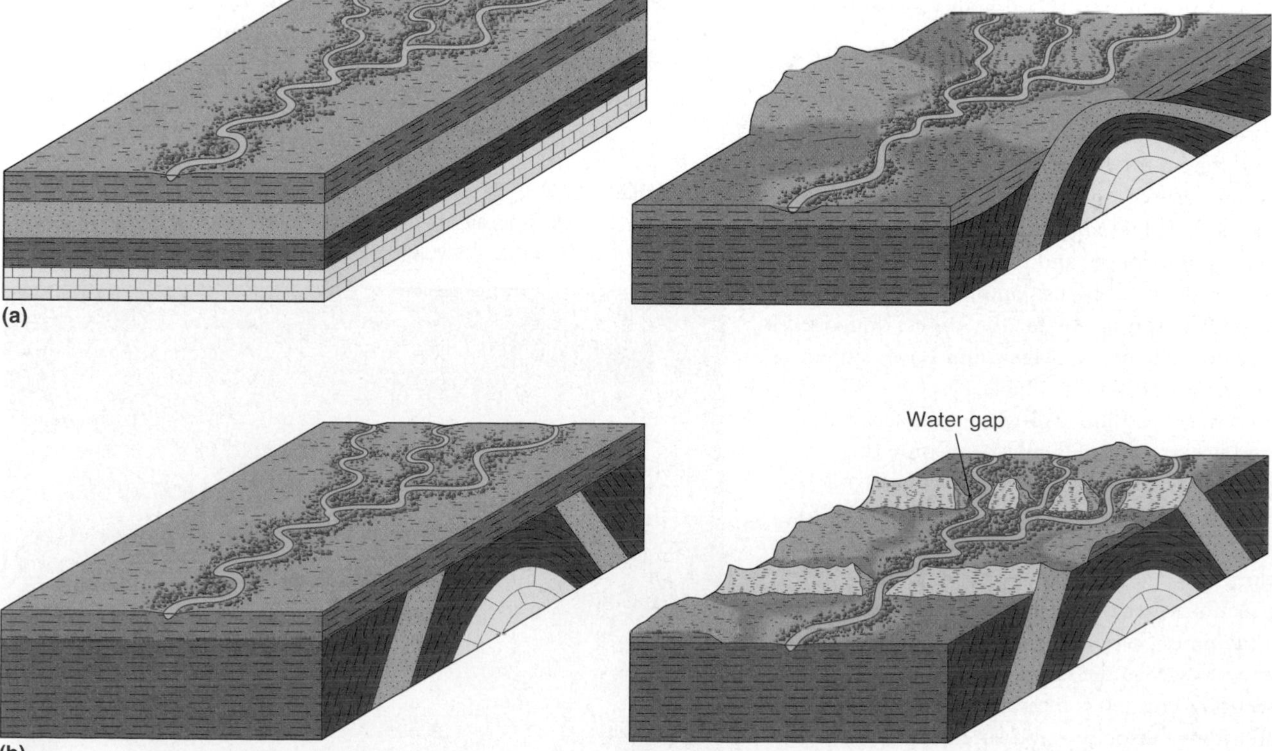

● FIGURE 17.10
Transverse streams form valleys and canyons across mountains. (a) An antecedent stream maintains its course by cutting a valley through a mountain as it is uplifted gradually over much time. (b) A superimposed stream has uncovered and excavated ancient structures that were buried beneath the surface. As the stream erodes the landscape downward, it cuts across and through the ancient structures.

TABLE 17.1
Ten Largest Rivers of the World

	Length		Area of Drainage Basin		Discharge	
	km	**mi**	**sq km (×1000)**	**sq mi**	**1000 m^3/s**	**1000 cfs**
Amazon	6276	3900	6133	2368	112–140	4000–5000
Congo (Zaire)	4666	2900	4014	1550	39.2	1400
Chang Jiang (Yangtze)	5793	3600	1942	750	21.5	770
Mississippi–Missouri	6260	3890	3222	1244	17.4	620
Yenisei	4506	2800	2590	1000	17.2	615
Lena	4280	2660	2424	936	15.3	547
Paraná	2414	1500	2305	890	14.7	526
Ob	5150	3200	2484	959	12.3	441
Amur	4666	2900	1844	712	9.5	338
Nile	6695	4160	2978	1150	2.8	100

Source: Adapted from Morisawa, *Streams: Their Dynamics and Morphology.* New York: McGraw-Hill Book Company, 1968.

titatively the amount of flow being conveyed through a stream channel. **Stream discharge** (Q) is the volume of water (V) flowing past a given cross section of the channel per unit time (t): $Q = V/t$. Discharge is most commonly expressed in units of cubic meters per second (m^3/s) or cubic feet per second (ft^3/s). A cross section is essentially a thin slice extending from one stream bank straight across the channel to the other stream bank and oriented perpendicular to the channel. If a drainage basin experiences a rainfall event that produces significant runoff, the volume of water (V) reaching the channel will increase. Notice from the discharge equation that this increase in volume (V) will cause an increase in stream discharge (Q).

It is important to collect and analyze discharge data for several reasons. For example, it can be used to compare the amount of flow carried in different streams, at different sites along a single stream, or at different times at a single cross section. Discharge data indicate the size of a stream and, in times of excessive flow, provide an index of flood severity. In general, streams in larger drainage basins are longer and have greater discharge than streams in smaller drainage basins. Among the major rivers of the world, the Amazon has by far the largest drainage basin and the greatest discharge; the Mississippi River system is ranked fourth in terms of discharge (Table 17.1).

The volume of water rushing through a cross section of a stream per second is extremely difficult to measure directly. In reality, discharge is determined not by measuring $Q = V/t$ directly but by using the fact that discharge (Q) is also equal to the area of the cross section (A) times the average stream velocity (v). This equation, $Q = Av$, can also be expressed as $Q = wdv$ because the cross-sectional area (A) is approximately equivalent to channel width (w) times channel depth (d), two factors that are relatively easy to measure in the field (• Fig. 17.11). Notice that, since cross-sectional area (A) is measured in square meters or square feet and average velocity is measured in meters per second or feet per second, solving the equation $Q = Av$ yields discharge values in units of volume per unit time (cubic meters per second or cubic feet per second). This analysis of the measurement units should convince you that volume per unit time is indeed equivalent to cross-sectional area times average velocity—that stream discharge is both volume per unit time and cross-sectional area times average velocity, $Q = V/t = Av$.

As is true for any equation, a change on one side of the discharge equation must be accompanied by a change on the other. If discharge increases because a rainstorm delivers a large volume of water to the stream, that increase in volume (V) will occupy a larger cross-sectional area (A) or flow through the cross section at a faster rate (v), and it usually does some of both. In other words, a large rainstorm will cause the level of a stream to rise and the water to flow faster. As the water level (flow depth, d) rises,

• FIGURE 17.11
These university students are measuring a stream's velocity, depth, and width in order to find discharge, the volume flowing in a stream per unit time.

J. Petersen

streams also experience an increase in width (w) because most channels flare out a bit as they rise up toward their banks. Stream channels continually adjust their cross-sectional area and/or flow velocity in response to changes in flow volume. Understanding the relationships among the factors involved in discharge is very important for understanding how streams work.

Stream Energy

When people have more energy, we can accomplish more, or do more work, than when we have less energy. The same is true for streams and the other geomorphic agents. The ability of a stream to erode and transport sediment, that is, to perform geomorphic work, depends on its available energy. Picking up pieces of rock and moving them require the stream to have **kinetic energy,** the energy of motion. When a stream has more kinetic energy, it can pick up and move more clasts (rock particles) and heavier clasts than when it has less energy. The kinetic energy (E_k) equation, $E_k = \frac{1}{2}mv^2$, shows that the amount of energy a stream has depends on its mass (m), but especially on its *velocity* of flow (v) because kinetic energy varies with velocity squared (v^2). Thus, stream velocity is a critical factor in determining the amount of geomorphic work accomplished by a stream. As we saw in the previous section, when stream discharge (Q) varies due to changes in runoff, so does stream velocity (v) because $Q = V/t = wdv$.

In addition to variations in flow discharge, another major way to alter stream velocity, and thereby stream energy, is through a change in the slope, or gradient, of the stream. **Stream gradient** is the drop in stream elevation over a given downstream distance and is typically expressed in meters per kilometer (m/km) or feet per mile (ft/mi). Everything else being equal, water flows faster down channels with steeper slopes and slows down over gentler slopes. Steeper channels typically occupy locations that are farther upstream and higher in the drainage basin, as well as places where a stream flows over rock types that are more resistant to erosion. Gentler stream gradients tend to occur closer to the mouth of a stream and where the stream crosses easily eroded rock types. As water moves through a channel system, its ability to erode and transport sediment—that is, its energy—is continuously changing as variations in stream gradient and discharge cause changes in flow velocity.

Different factors work to decrease the energy of a stream. Friction along the bottom and sides of a channel and even between the stream surface and the atmosphere slows down the stream velocity and therefore contributes to a decrease in stream energy. A channel with numerous substantial irregularities, such as those due to large rocks and vegetation sticking up into the flow, has a great amount of **channel roughness,** which causes a considerable decrease in stream energy due to the resulting frictional effects. Smooth, bedrock channels without these irregularities have lower channel roughness and a smaller loss in stream energy due to frictional effects. In all cases, however, friction along the bottom and sides of a channel slows down the flow and slows it down the most at the channel boundaries so that the maximum velocity usually occurs somewhat below the stream surface at the deepest part of the cross section. In addition to the external friction at the channel boundary, streams lose energy because of internal friction in the flow related to eddies, currents, and the interaction among water molecules. About 95% of a stream's energy is consumed in overcoming all types of frictional effects. Only the remaining energy, probably less than 5%, is available for eroding and transporting sediment.

Stream gradients are usually steepest at the headwaters and in new tributaries and diminish in the downstream direction (● Fig. 17.12). Discharge, on the other hand, increases downstream in humid regions as the size of the contributing drainage basin increases. As the water originally conveyed in numerous small channels high in the drainage basin is collected in the downstream direction into a shrinking number of larger and larger channels, flow efficiency improves and frictional resistance decreases. As a result, flow velocity tends to be higher in downstream parts of a channel system than over the steep gradients in the headwaters. The ability of the stream to carry sediment, therefore, may be equivalent or even greater downstream than upstream.

The sediment being transported by a stream is called the **stream load.** Carrying sediment is a major part of the geomorphic work accomplished by a stream. In order from smallest to largest, the size of sediment that a stream may transport includes clay, silt, sand, granules, pebbles, cobbles, and boulders. Sand marks the boundary between fine-grained (small) clasts and coarse-grained (large) clasts. Gravel is a general term for any sediment larger in size than sand. The maximum size of rock particles that a stream is able to transport, referred to as the **stream competence** (measured as particle diameter in centimeters or inches), and the total amount of load being moved by a stream, termed **stream capacity** (measured as weight per unit time), depend on available stream energy and thus on the flow velocity. As a stream flows along, its velocity is always changing, reflecting the constant variations in stream discharge, stream gradient,

● FIGURE 17.12
Stream gradient decreases from source to mouth for the ideal stream.

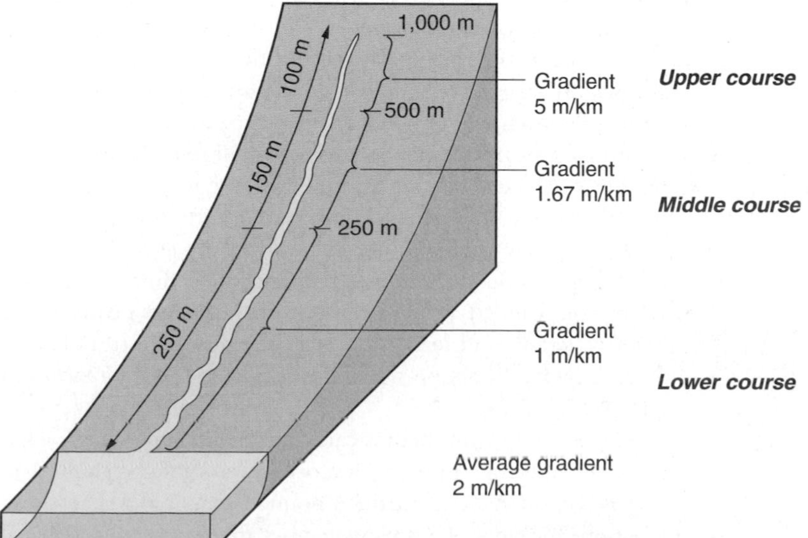

and frictional resistance factors. As a result, the size and amount of load that the stream can carry is also changing constantly. If the material is available when flow velocity increases, the stream will pick up from its own channel bed larger clasts and more load. When its velocity decreases and the stream can no longer transport such large clasts and so much material, it drops the larger clasts onto its bed, depositing them there until a new increase in velocity provides the energy to entrain and transport them. Thus, the particular geomorphic process being carried out by a stream at any moment, whether it is eroding or depositing, is determined by the complex of factors that control its energy.

Both stream capacity and stream competence increase in response to a relatively small increase in velocity. A stream that doubles its velocity during a flood may increase its amount of sediment load six to eight times. The boulders seen in many mountain streams arrived there during some past flood that greatly increased stream competence; they will be moved again when a flow of similar magnitude occurs. In fact, rivers do most of their heavy earth-moving work during short periods of flood (● Fig. 17.13).

Streams have no influence over the amount of water entering the channel system, nor can they change the type of rocks over which they flow. Streams do, however, have some control over channel size, shape, and gradient. For example, when a stream undergoes a decrease in energy so that it deposits some of its sediment, the deposit raises the channel bottom and locally creates a steeper gradient downchannel from the deposit. With continued deposition, the location will eventually attain a slope steep enough to cause a sufficient increase in velocity so that the flow will again entrain and carry that deposited sediment. Streams tend to adjust their channel properties so they can move the sediment supplied to them. *Dynamic equilibrium* is maintained by adjustments among channel slope, shape, and roughness; amount of load eroded, carried, or deposited; and the velocity and discharge of stream flow. A **graded stream** has just the velocity and discharge necessary to transport the load eroded from the drainage basin.

● FIGURE 17.13
It is not just the power of flowing water that causes damage in a flood. A flooding stream typically has a high suspended load, can transport large, heavy materials, and may carry floating debris that can damage homes and other structures, as seen here in the village of Orosi de Cartago, Costa Rica, 35 kilometers (22 mi) southeast of the capital, San Jose.

© Reuters New Media, Inc./CORBIS

Fluvial Processes

Stream Erosion

Fluvial erosion is the removal of rock material by flowing water. Fluvial erosion may take the form of the chemical removal of ions from rocks or the physical removal of rock fragments (clasts). Physical removal of rock fragments includes breaking off new pieces of bedrock from the channel bed or sides and moving them as well as picking up and removing preexisting clasts that were temporarily resting on the channel bottom. Breaking off new pieces of bedrock proceeds very slowly where highly resistant rock types are found.

Erosion is simply the removal of rock material; erosion of sediments from the bottom of a stream channel does not necessarily mean that the channel will occupy a lower position in the landscape. If the eroded rock fragments are replaced by the deposition of other fragments transported in from upstream, there will be no lowering of the channel bottom. Such lowering, or channel incision, occurs only when there is *net erosion* compared to deposition. Net erosion results in the lowering of the affected part of the landscape and is termed **degradation.** Net deposition of sediments results in a building up, or **aggradation,** of the landscape.

One way that streams erode occurs when stream water chemically dissolves rock material and then transports the ions away in the flow. This fluvial erosion process, called **corrosion** (or *solution,* or *dissolution*), has a limited effect on many rocks but can be significant in certain rock types, such as limestone.

Hydraulic action refers to the physical, as opposed to chemical, process of stream water alone removing pieces of rock. As stream water flows downslope by the force of gravity, it exerts stress on the streambed. Whether this stress results in entrainment and removal of a preexisting clast currently resting on the channel bottom, or even the breaking off of a new piece of bedrock from the channel, depends on several factors including the volume of water, flow velocity, flow depth, stream gradient, friction with the streambed, the strength and size of the rocks over which the stream flows, and the degree of stream turbulence. **Turbulence** is chaotic flow that mixes and churns the water, often with a significant upward component, that greatly increases the rate of erosion as well as the load-carrying capacity of the stream. Turbulence is controlled by channel roughness and the gradient over which the stream is flowing. A rough channel bottom increases the intensity of turbulent flow. Likewise, even a small increase in velocity caused by a steeper gradient can result in a significant increase in turbulence. Turbulent currents contribute to erosion by hydraulic action when they wedge under or pound away at rock slabs and loose fragments on the channel bed and sides, dislodging clasts that are then carried away in the current. **Plunge pools** at the base of waterfalls and in rapids reveal the power of turbulence-enhanced hydraulic action where it is directed toward a localized point (● Fig. 17.14).

As soon as a stream begins carrying rock fragments as load, it can start to erode by **abrasion,** a process even more powerful than hydraulic action. As rock particles bounce, scrape, and drag along the bottom and sides of a stream channel, they break off additional rock fragments. Because solid rock particles are denser than water,

National Park Service—Nay Aug Park

• **FIGURE 17.14**
A plunge pool has been eroded at the base of Nay Aug Falls by Roaring Brook in Scranton, Pennsylvania.
Why do deep plunge pools form at the base of most waterfalls?

Courtesy Sheila Brazier

• **FIGURE 17.15**
These potholes were cut into a streambed of solid rock during times of high flow. Potholes result from stream abrasion and the swirling currents in a fast-flowing river.

the impact of having clastic load thrown against the channel bottom and sides by the current is much more effective than the impact of water alone. Under certain conditions, stream abrasion makes distinctive round depressions called **potholes** in the rock of a bedrock streambed (• Fig. 17.15). Potholes generally originate in special circumstances, such as below waterfalls or swirling rapids, or at points of structural weakness, which include joint intersections in the streambed. Potholes range in diameter and depth from a few centimeters to many meters. If you peer into a pothole, you can often see one or more round stones at the bottom. These are the abraders, or *grinders*. Swirling whirlpool movements of the stream water cause such stones to grind the bedrock and enlarge the pothole by abrasion while finer sediments are carried away in the current.

In a process related to abrasion, as rock fragments moving as load are transported downstream, they are gradually reduced in size, and their shape changes from angular to rounded. This wear and tear experienced by sediments as they tumble and bounce against one another and against the stream channel is called **attrition.** Attrition explains why gravels found in streambeds are rounded and why the load carried in the lower reaches of most large rivers is composed primarily of fine-grained sediments and dissolved minerals.

Stream erosion widens and lengthens stream channels and the valleys they occupy. Lengthening occurs primarily at the source through **headward erosion,** accomplished partly by surface runoff flowing into a stream and partly by springs undermining the slope. The lengthening of a river's course in an upstream direction is particularly important where erosional gullies are rapidly dissecting agricultural land. Such gullying may be counteracted by soil conservation practices to reduce erosional soil loss. Channel lengthening, which results in a decrease in stream gradient, also occurs if the path of the stream channel becomes more winding, or *sinuous*.

Stream Transportation

A stream directly erodes some of the sediment that it transports, and most chemical sediments are delivered to the channel in base flow, but a far greater proportion of its load is delivered to the stream channel by surface runoff and mass movement. Regardless of the sediment source, streams transport their load in several ways (• Fig. 17.16). Some minerals are dissolved in the water and are thus carried in the transportation processes of **solution.** The finest solid particles are carried in **suspension,** buoyed by vertical turbulence. Such small grains can remain suspended in the water column for long periods, as long as the force of upward turbulence is stronger than the downward settling tendency of the particles. Some grains too large and heavy to be carried in suspension bounce along the channel bottom in a process known as **saltation** (from French: *sauter,* to jump). Particles that are too large and heavy to move by saltation may slide and roll along the channel bottom in the transportation process of **traction.**

There are three main types of stream load. Ions of rock material held in solution constitute the **dissolved load. Suspended load** consists of the small clastic particles being moved in suspension. Larger particles that saltate or move in traction along the streambed comprise the **bed load.** The total amount of load that a stream carries is expressed in terms of the weight of the transported material per unit time.

The relative proportion of each load type present in a given stream varies with such drainage basin characteristics as climate, vegetative cover, slope, rock type, and the infiltration capacities and permeabilities of the rock and soil types. Dissolved loads will be larger than average in basins with high amounts of infiltration and base flow, and therefore limited surface runoff, because slow-moving groundwater that feeds the base flow acquires ions from the rocks through which it moves. Humid regions experience considerable weathering, which produces much fine-grained sediment, and thus humid region streams tend to have a large amount of suspended load. Rivers that are carrying a high suspended load look characteristically muddy (• Fig. 17.17). The Huang He in northern China, known as the "Yellow River" because of the color of its silty suspended load, carries a huge amount of sediment in

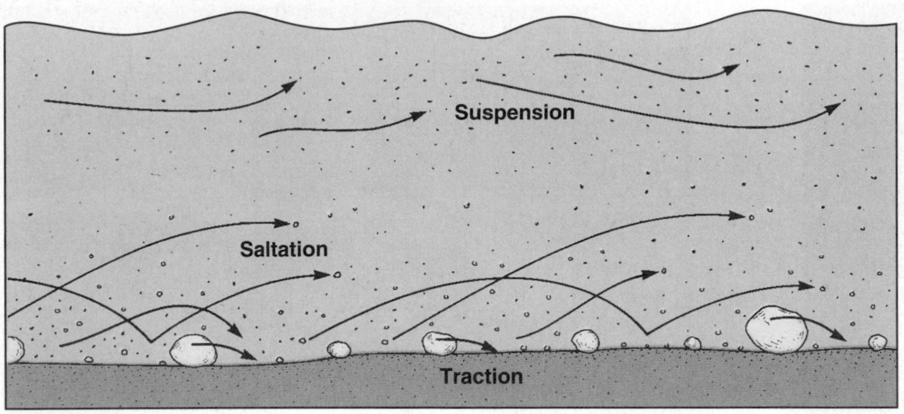

• **FIGURE 17.16**
Transport of solid load in a stream. Clay and silt particles are carried in suspension. Sand typically travels by suspension and saltation. The largest (heaviest) particles move by traction.
What is the difference between traction and saltation?

© Ron Sherman/Getty Images

• **FIGURE 17.17**
Some rivers carry a tremendous load of sediment in suspension and show it in their muddy appearance, as in this aerial view of the Mississippi River in Louisiana. Much of the load carried by the Mississippi River is brought to it by its major tributaries.
What are some of the major tributaries that enter the Mississippi River?

suspension, with more than 1 million tons of suspended load per year. Compared to the "muddy" Mississippi River, the Huang He transports five times the suspended sediment load with only one fifth the discharge. Streams dominated by bed load tend to occur in arid regions because of the limited weathering rate in arid climates. Limited weathering leaves considerable coarse-grained sediment in the landscape available for transportation by the stream system.

Stream Deposition

Because the capacity and competence of a stream to carry material depend on flow velocity, a decrease in velocity will cause a stream to reduce its load through deposition. Velocity decreases over time when flow subsides—for example, after the impact of a storm—but it also varies from place to place along the stream. Shallow parts of a channel that in cross section lie far from the deepest and fastest flow typically experience low flow velocity and become sites of recurring deposition. The resulting accumulation of sediment, like what forms on the inside of a channel bend, is referred to as a **bar.** Sediment also collects in locations where velocity falls due to a reduction in stream gradient, where the river current meets the standing body of water at its mouth, and on the land adjacent to the stream channel during floods.

Alluvium is the general name given to fluvial deposits, regardless of the type or size of material. Alluvium is recognized by the characteristic sorting and/or rounding of sediments that streams perform. A stream sorts particles by size, transporting the sizes that it can and depositing larger ones. As velocity fluctuates due to changes in discharge, channel gradient, and roughness, particle sizes that can be picked up, transported, and deposited vary accordingly (• Fig. 17.18). The alluvium deposited by a stream with fluctuating velocity will exhibit alternating layers of coarser and finer sediment.

When streams leave the confines of their channels during floods, the channel cross-sectional width is suddenly enlarged so much that the velocity of flow must slow down to counterbalance it ($Q = wdv$). The resulting decrease in stream competence and capacity cause deposition of sediment on the flooded land adjacent to the channel. This sedimentation is greatest right next to the channel where aggradation constructs channel-bounding ridges known as **natural levees,** but some alluvium will be left behind wherever load settled out of the receding flood waters.

Floodplains constitute the often extensive, low-gradient land areas composed of alluvium that lie adjacent to many stream channels (• Fig. 17.19). Floodplains are aptly named because they are inundated during floods and because they are at least partially composed of *vertical accretion deposits,* the sediment that settles out of slowing and standing floodwater. Most floodplains also contain *lateral accretion deposits.* These are generally channel bar deposits that get left behind as a channel gradually shifts its position in a sideways fashion (laterally) across the floodplain (• Fig. 17.20).

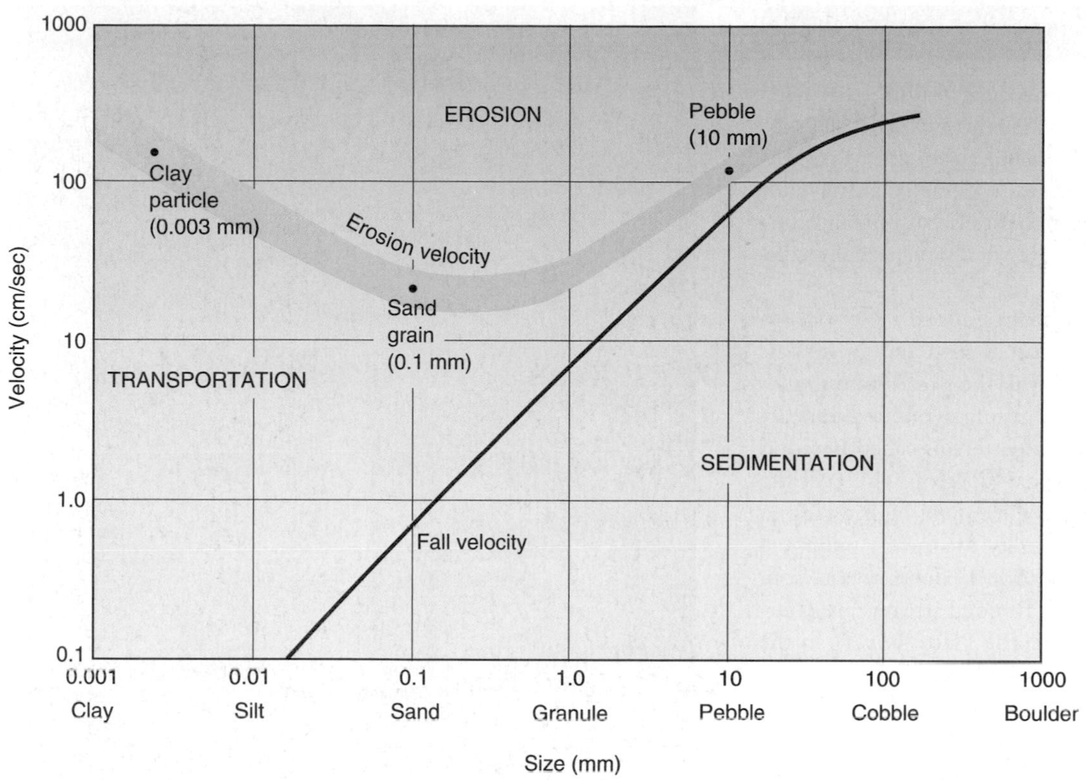

• FIGURE 17.18
This graph shows the relationship between stream flow velocity and the ability to erode or transport material of varying sizes (inability to erode or transport particles of a particular size or larger will result in deposition). Note that small pebbles (particles with a diameter of 10 mm, for example) need a high stream flow velocity to be moved because of their size and weight. The fine silts and clays (smaller than 0.05 mm) also need high velocities for erosion because they stick together cohesively. Sand-sized particles (between 0.05 and 2.0 mm) are relatively easily eroded and transported, compared to clays or gravel (particles larger than 2.0 mm).

Nick Dekker, Missouri Department of Natural Resources

• FIGURE 17.19
During floods, low areas adjacent to the river are inundated with sediment-laden water that flows over the banks to deposit alluvium, mainly silts and clays, on the floodplain. This is the Missouri River floodplain at Jefferson City, Missouri, during the 1993 Midwest flood.
What would the river floodwaters leave behind in flooded homes after the water recedes?

D. Sack

• FIGURE 17.20
Sediment deposited in a bar on the inside of a channel bend becomes part of the floodplain alluvium if the stream migrates away leaving the bar deposits behind, as occurred with these lateral accretion deposits. In this photo, remnants of winter ice fill swales between ridges that mark successive crests of the laterally accreted bar deposits.

Channel Patterns

Three principal types of stream channel have traditionally been recognized when considering the form of a given channel segment in map view. Although *straight channels* may exist for short distances under natural circumstances, especially along fault zones, joints, or steep gradients, most channels with parallel, linear banks are artificial features that were totally or partially constructed by people.

If a stream has a high proportion of bed load in relation to its discharge, it deposits much of its load as sand and gravel bars in the streambed. These obstructions break the stream into strands that interweave, separate, and rejoin, giving a braided appearance to the channel, and indeed such a pattern is called **braided** (• Fig. 17.21). This channel pattern may develop wherever the coarse-sediment input into a stream is extremely high owing to banks of loose sand and gravel or proximity to a nearly unlimited supply of coarse bed load, such as that found downstream from glaciers and also in many desert areas. Braided streams are common on the Great Plains (for example, the Platte River), in the desert Southwest, in Alaska, and in Canada's Yukon.

The most common channel pattern in humid climates displays broad, sweeping bends in map view. Over time, these sinuous, **meandering channels** also wander from side to side across their low-gradient floodplains widening the valley by lateral erosion on the outside of meander bends and leaving behind lateral accretion (bar) deposits on the inside of meander bends (• Fig. 17.22). These streams and their floodplains have a higher proportion of fine-grained sediment and thus greater bank cohesion than the typical braided stream.

• FIGURE 17.21
The braided stream channel of the Brahmaputra River in Tibet, viewed from the International Space Station. Stream braiding results from an abundant bed load of coarse sediment that obstructs flow and separates the main stream into numerous strands. Braided channels typically occur when a stream has low discharge compared to the amount of bed load.
What does the common occurrence of braided channels just downstream from glaciers tell you about the sediment transported and deposited by moving ice?

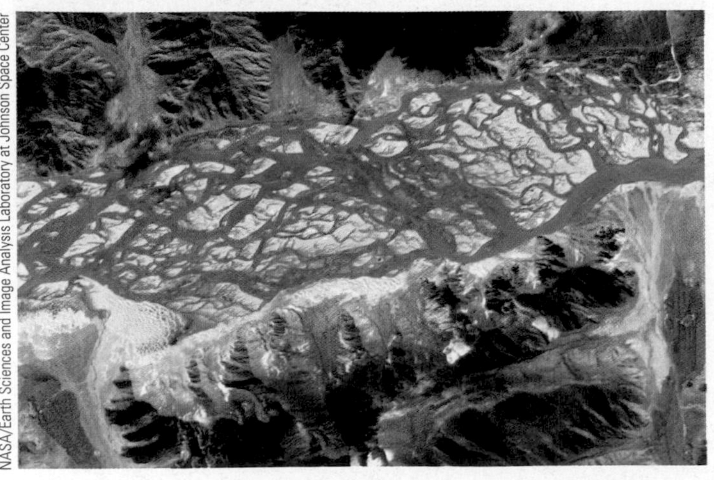

NASA/Earth Sciences and Image Analysis Laboratory at Johnson Space Center

NASA/JPL TOPSAR; image generation performed at Washington University, St. Louis

• FIGURE 17.22
Over time, a meandering (sinuous) stream channel, like the Missouri River shown on this colorized radar image, may swing back and forth across its valley. Where the outside of a meander bend on the floodplain (purple and blue tones) impinges on the edge of higher terrain (orange areas), stream erosion can undercut the valley side wall and, with the assistance of mass wasting, contribute to floodplain widening.

Land Sculpture by Streams

One way to understand the variety of landform features resulting from fluvial processes is to examine the course of an idealized river as it flows from headwatersw in the mountains to its mouth at the ocean. The gradient of this river would diminish continually downstream as it flows from its source toward base level. In nature, exceptions exist to this idealized profile because some streams flow entirely over a low gradient (● Fig. 17.23), while other streams, particularly small ones on mountainous coasts, flow down a steep gradient all the way to the standing body of water at their mouth. Rather than having a smoothly decreasing slope from headwaters to mouth, we would expect real streams to have some irregularities in this **longitudinal profile,** the stream gradient from source to mouth (see again Fig. 17.12).

The following discussion subdivides the ideal stream course into upper, middle, and lower river sections flowing over steep, moderate, and low gradients, respectively. Fluvial erosion processes dominate the steep upper course, whereas deposition predominates in the lower course. The middle course displays important elements of both fluvial erosion and deposition.

Features of the Upper Course

At the headwaters in the upper course of a river, the stream primarily flows in contact with bedrock. Over the steep gradient high above its base level, the stream works to erode vertically downward by hydraulic action and abrasion. Erosion in the upper course creates a steep-sided valley, gorge, or ravine as the stream channel in the bottom of the valley cuts deeply into the land. Little if any floodplain is present, and the valley walls typically slope directly to the edge of the stream channel. Steep valley sides encourage mass movement of rock material directly into the flowing stream. Valleys of this type, dominated by the downcutting activity of the stream, are often called **V-shaped valleys** because with their steep slopes they attain the form of the letter *V* (● Fig. 17.24).

The effects of *differential erosion* can be significant in the upper course where rivers cut through rock layers of varying resistance. Rivers flowing over resistant rock have a steeper gradient than where they encounter weaker rock. A steep gradient gives the stream flow more energy, which the stream needs to erode the resistant rock. Rapids and waterfalls may mark the location of resistant materials in a stream's upper course. Where rocks are particularly resistant to weathering and erosion, valleys will be narrow, steep-sided gorges or canyons; where rocks are less resistant, valleys tend to be more spacious.

Many streams spill from lake to lake in their upper courses, either over open land (like the Niagara River at Niagara Falls, between Lake Erie and Lake Ontario) or through gorges. In either case, the lakes will eventually be eliminated if stream erosion lowers their outlets enough or if fluvial sediment deposited at the inflow points fills the lakes.

● **FIGURE 17.23**
Not all streams fit the generalized pattern of characteristics for upper, middle, and lower stream segments. The Mississippi River, here a relatively small stream, meanders on a low gradient near its headwaters in Minnesota, far upstream from its mouth.

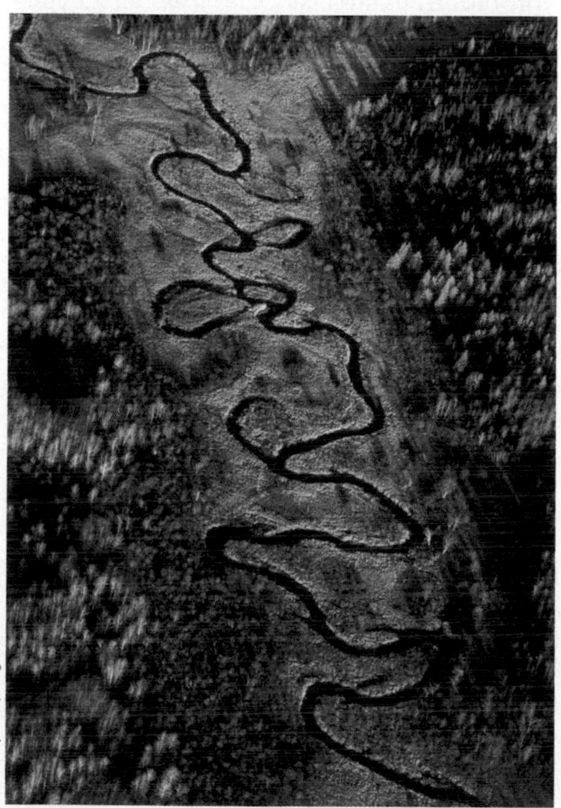

© Jake Rajs/Getty Images

● **FIGURE 17.24**
Where the upper course of a stream lies in a mountainous region, its valley typically has a characteristic "V" shape near the headwaters. Such a stream flows in a steep-walled valley, with rapids and waterfalls, as shown here in Yellowstone Canyon, Wyoming.
How does the gradient of the Yellowstone River compare with that of the stretch of the Mississippi River shown in Figure 17.23?

R. Sager

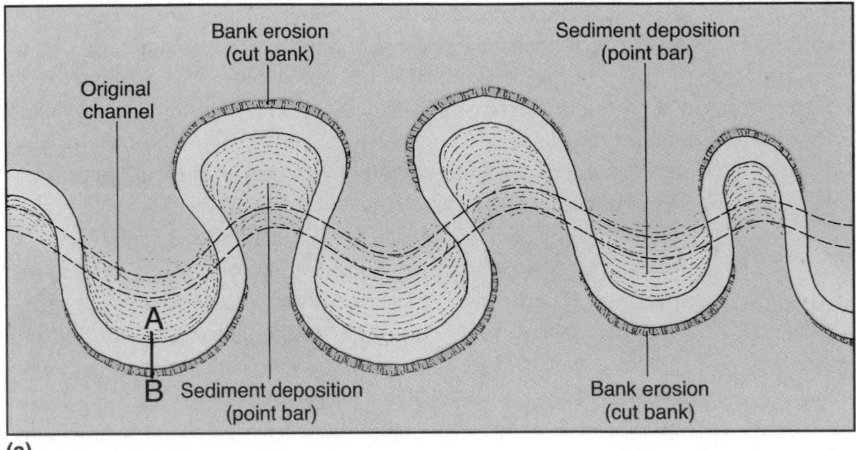

(a)

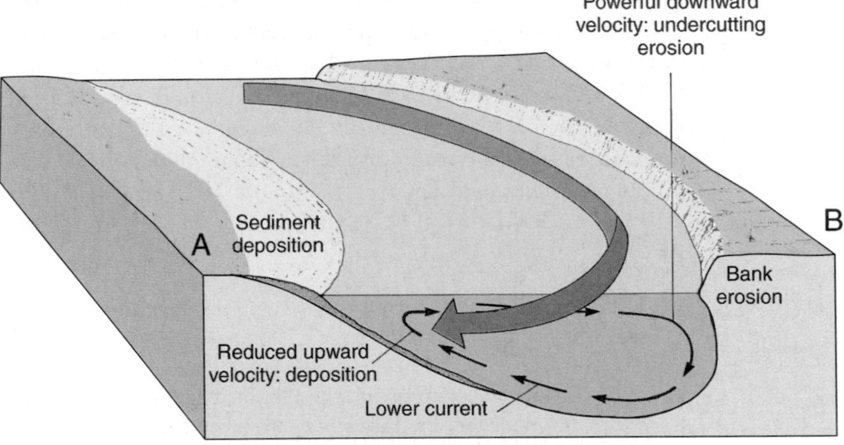

(b)

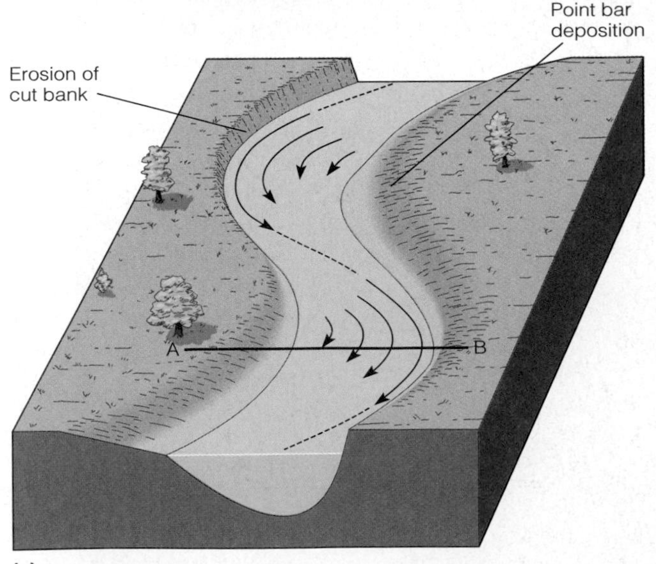

(c)

• **FIGURE 17.25**
Characteristics of a meandering river channel. Note that water flowing in a channel has a tendency to flow downstream in a helical, or "corkscrew," fashion, which moves water against one side of the channel and then to the opposite side. The up-and-down motion of the water contributes to the processes of erosion, transportation and deposition.

Features of the Middle Course

In the middle section of the ideal longitudinal profile, the stream flows over a moderate gradient and on a moderately smooth channel bed. Here the river valley includes a floodplain, but remaining ridges beyond the floodplain still form definite valley walls. The stream lies closer to its base level, flows over a gentler gradient, and thus directs less energy toward vertical erosion than in its upper course. The stream still has considerable energy, however, due to the downstream increase in flow volume and reduction in bed friction. The river now uses much of its available energy for transporting the considerable load that it has accumulated and toward lateral erosion of the channel sides. The stream displays a definite meandering channel pattern with its sinuous bends that wander over time across the valley floor. The stream erodes a **cut bank** on the outside of meander loops, where the channel is deep and centrifugal force accelerates stream velocity. The cut bank is a steep slope, and slumping may occur there particularly when there is a rapid fall in water level. Slumping on the outside of meander bends contributes to the effect of lateral erosion by the stream and adds load to the stream. In the low velocity and shallow flow on the inside of the meander bends, the stream deposits a **point bar** (• Fig. 17.25). Erosion on the outside and deposition on the inside of river meander bends result in the sideways displacement, or **lateral migration,** of meanders. This helps increase the area of the gently sloping floodplain when cut banks impact the confining valley walls. Tributaries flowing into a larger stream also aid in widening the valley through which the trunk stream flows. Although flooding of the valley floor is always a potential hazard, the richness of floodplain soils offers an irresistible lure for farmers.

Features of the Lower Course

The minimal gradient and close proximity to base level along the ideal lower river course make downcutting virtually impossible. Stream energy, now derived almost exclusively from the higher discharge rather than the downslope pull of gravity, leads to considerable lateral shifting of the river channel. The river meanders around helping to create a large depositional plain (see Map Interpretation: Fluvial Landforms). The lower floodplain of a major river is much wider than the width of its meander belt and shows evidence of many changes in course (• Fig. 17.26). The stream migrates laterally through its own previously deposited sediment in a channel composed exclusively of alluvium. During floods, these extensive floodplains, or **alluvial plains,** become inundated with sediment-laden water that contributes vertical accretion deposits to the large natural levees and to the already thick alluvial valley fill of the floodplain in general. Natural levees along the Mississippi River rise up to 5 meters (16 ft) above the rest of the floodplain.

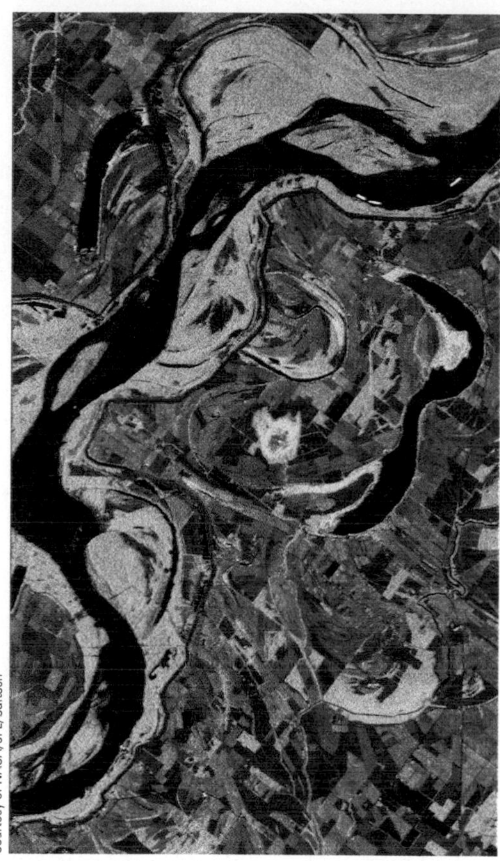

• FIGURE 17.26
This colorized radar image shows part of the Mississippi River floodplain along the Arkansas–Louisiana–Mississippi state lines. Images like this help us assess flood potential and learn much about the geomorphic history of the river and its floodplain. The colors are used to enhance landscape features such as water bodies (dark), field patterns, and forested areas (green). Note how the river has changed its channel position many times, leaving oxbow lakes and meander scars on the floodplain.

A common landform in this deposition-dominated environment provides evidence of the meandering of a river over time. Especially during floods, **meander cut-offs** occur when a stream seeks a shorter, steeper, and straighter path; breaches through the levees; and leaves a former meander bend isolated from the new channel position. If the cut-off meander remains filled with water, which is common, it forms an **oxbow lake** (• Fig. 17.27).

Sometimes people attempt to control streams by building up levees artificially in order to keep the river in its channel. During times of reduced discharge, however, when a river has less energy, deposition occurs in the channel. Thus, in an artificially constrained channel, a river may raise the level of its channel bed. In some instances, as in China's Huang He and the Yuba River in northern California, deposition has raised the streambed above the surrounding floodplains. Flooding presents a very serious danger in this situation with much of the floodplain lying below the level of the river. Unfortunately, when floodwaters eventually overtop or breach the levees, they can be even more extensive and destructive than they would have been in the natural case.

The presence of levees—both natural and artificial—can prevent tributaries in the lower course from joining the main stream. Smaller streams are forced to flow parallel to the main river until a convenient junction is found. These parallel tributaries are called **yazoo streams,** named after the Yazoo River, which parallels the Mississippi River for more than 160 kilometers (100 mi) until it finally joins the larger river near Vicksburg, Mississippi.

Deltas

Where a stream flows into a standing body of water, such as a lake or the ocean, the flow is no longer confined in a channel. The current expands in width, causing a reduction in flow velocity and thus a decrease in load competence and capacity. If the stream is carrying much load, the sediment will begin to settle out, with larger particles deposited first, closer to the river mouth, and smaller particles deposited farther out in the water body. With continued aggradation, a distinctive landform, called a **delta** because the map view shapes of some resemble the Greek letter delta (Δ), may be constructed (• Fig. 17.28a).

Deltas form at the interface between fluvial systems and coastal environments of lakes or the ocean and therefore originate in part from fluvial and in part from coastal processes. Deltas have a subaqueous (underwater) coastal component, called the **prodelta,** and a fluvial part, the **delta plain,** that exists at, to slightly above, the lake level or

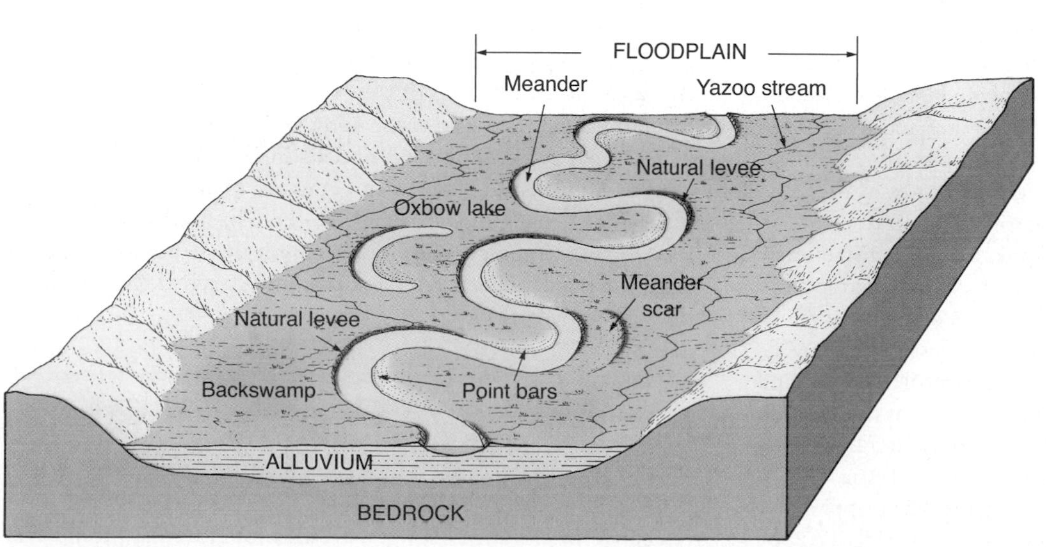

• FIGURE 17.27
Features of a large floodplain common in the lower courses of major rivers. Low marshy or swampy parts of the floodplain, generally at the water table, are called backswamps.
What is the origin of an oxbow lake?

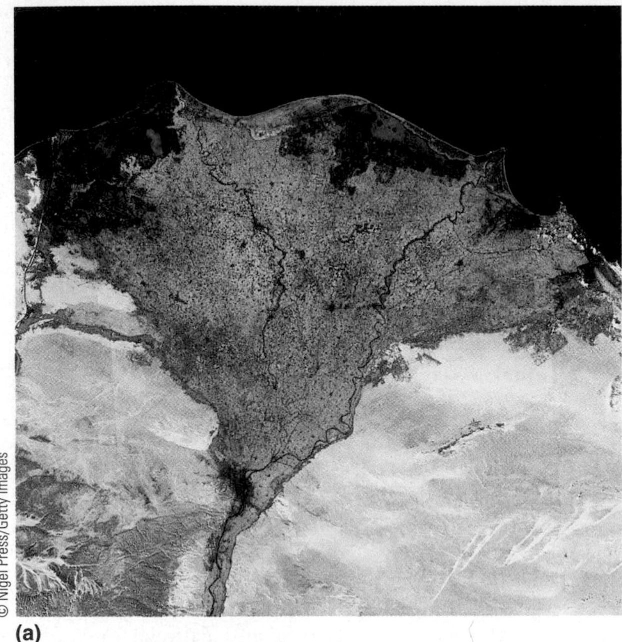
© Nigel Press/Getty Images
(a)

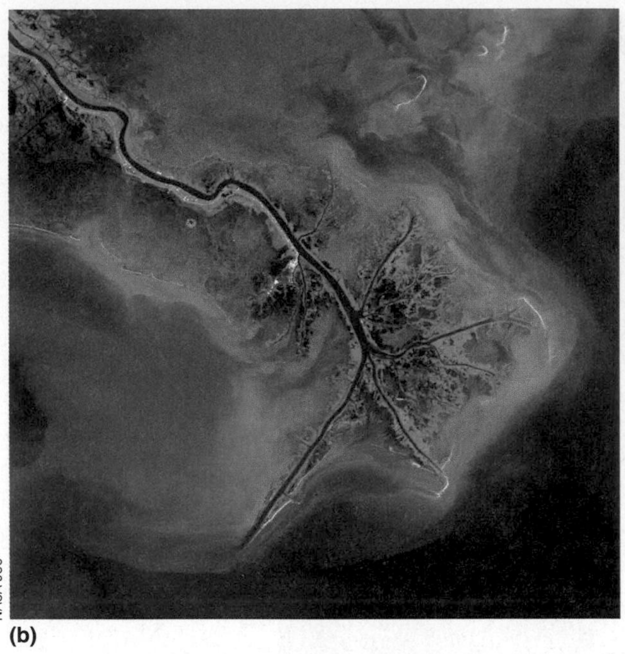
NASA JSC
(b)

• FIGURE 17.28
Satellite views of two different types of deltas. (a) The Nile River delta at the edge of the Mediterranean Sea is an arcuate delta, displaying the classic triangular arc shape. Waves and currents smooth out irregularities along the seaward edge of the delta. (b) The unusual shape of the Mississippi River delta resembles a bird's foot. Waves, currents, and tides in the Gulf of Mexico do little to change the visible shape of this type of delta.
Why are the shapes of some deltas controlled more by fluvial processes whereas the shapes of others are strongly influenced by coastal processes?

sea level. Deltas can form only at those river mouths where the fluvial sediment supply is high, where the underwater topography does not drop too sharply, and where waves, currents, and tides cannot transport away all the sediments delivered by the river. Although these circumstances exist at the mouths of many rivers, not all rivers have deltas.

Delta construction is a slow, ongoing process. A river channel that approaches its base level at a large standing body of water typically has a very low gradient. Lacking the ability to incise its channel below base level, the stream may divide into two channels, and may do so multiple times, to convey its water and load to the lake or ocean. These multiple channels flowing out from the main stream are called **distributaries,** are typical features of the delta plain, and help direct flow and sediment toward the lake or ocean. Natural levees accumulate along the banks of these distributary channels. Continued deposition and delta formation extend the delta plain and create new land far out from the original shoreline. Rich alluvial deposits and the abundance of moisture allow vegetation to quickly become established on these fertile deposits and further secure the delta's position. Delta plains, such as those of the Mekong, Indus, and Ganges Rivers, form important agricultural areas that feed the dense populations of many parts of Asia.

Where it flows into the Gulf of Mexico, the Mississippi River has constructed a type of delta called a *bird's-foot delta* (Fig. 17.28b). Bird's-foot deltas form in settings where the influence of the fluvial system far exceeds the ability of waves, currents, and tides of the standing water body to rework the deltaic sediment into coastal landforms or to transport it away. Natural levee crests along numerous distributaries remain intact slightly above sea level and extend far out into the receiving water body. Occasional changes in the distributary channel system occur when a major new distributary is cut that siphons flow away from a previous one, causing the center of deposition to switch to a new location far from its previous center. The appearance in map view of the natural levees extending toward the present and former depositional centers leaves the delta resembling a bird's foot.

Different types of deltas are found in other kinds of settings. An *arcuate delta,* like that of the Nile River, projects to a limited extent into the receiving water body, but the smoother, more regular seaward edge of this kind of delta shows greater reworking of the fluvial deposits by waves and currents than in the case of the bird's-foot delta. *Cuspate deltas,* like the São Francisco in Brazil, form where strong coastal processes push the sediments back toward the mainland and rework them into beach ridges on either side of the river mouth.

Base-Level Changes and Tectonism

A change in elevation along a stream's longitudinal profile will cause an increase or decrease in the stream's gradient, which in turn impacts the amount of geomorphic work the stream is capable of accomplishing. Elevation changes can occur within the drainage basin due to tectonic uplift or depression. Changes of base level for basins of exterior drainage result principally from climate change. Sea level drops in response to large-scale growth of glaciers and rises with substantial glacier shrinking. Tectonic uplift or a drop in base level give the stream a steeper gradient and increased energy for erosion and transportation. The landscape and its stream are then said to be **rejuvenated** because the stream uses its renewed energy to incise its channel to the new base level. Waterfalls and rapids may develop as rejuvenated channels are deepened by erosion. Tectonic depression of the drainage basin or a rise in sea level reduces the stream's gradient and energy, enhancing deposition.

If new uplift occurs gradually in an area where stream meanders have formed, these meanders may become **entrenched** as the stream deepens its valley (• Fig. 17.29). Now, instead of eroding the land laterally, with meanders migrating across an alluvial plain, the rejuvenated stream's primary activity is vertical incision.

It is important to note that virtually all rivers reaching the sea incised their valleys during the Pleistocene in response to the lowering of sea level associated with continental glaciation. The accumulation of water on the land in the form of glacial ice caused sea level to drop as much as 120 meters (400 ft) and lowered the base level for streams of exterior drainage. Consequently, near their mouths, the streams eroded deep valleys for their channels. Subsequent melting of the glaciers again elevated base level. This base-level rise caused the streams to deposit their sediment loads, filling valleys with sediment, as the streams adjusted their channels to a second new base level. These consecutive changes in base level produced broad, flat, alluvial floodplains above buried valleys cut far below today's sea level.

While a drop in base level causes downcutting and a rise causes deposition, an upset of stream equilibrium resulting from sizable increases or decreases in discharge or load can have similar effects on the landscape. Research has shown that variations in base level, tectonic movements, and changes in stream equilibrium can each cause downcutting in stream valleys, so the valley is slightly deepened and remnants of the older, higher valley floor are preserved in "stair-stepped" banks along the walls of the valley. These remnants of previous valley floors are **stream terraces.** Multiple terraces are a consequence of successive periods of downcutting and deposition (• Fig. 17.30). Stream terraces provide a great deal of evidence about the geomorphic history of the river and its surrounding region.

• FIGURE 17.29
A spectacular example of an entrenched meander from the San Juan River in the Colorado Plateau region of southern Utah.

© Rainer Duttmann

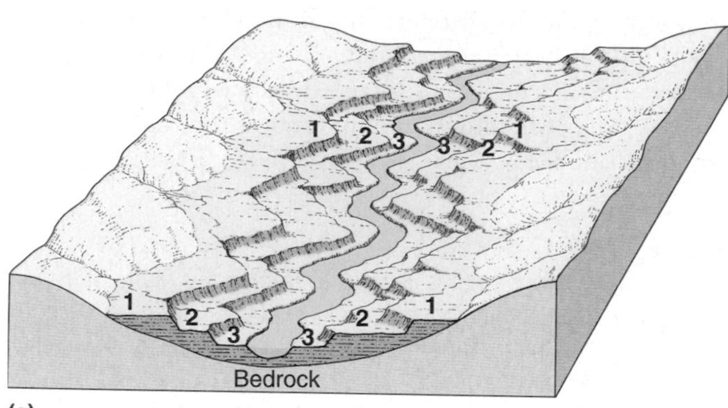

(a)

Courtesy Sheila Brazier

(b)

• FIGURE 17.30
(a) Diagram of a stream valley displaying two sets of alluvial stream terraces (labeled 1 and 2) and the present floodplain (labeled 3). The stream terraces are remnants of previous positions of the valley floor, with the higher (1) being the older. Stream incision into and abandonment of each of the two former floodplains (1 and 2) may have occurred by the stream experiencing rejuvenation due to a lowering of its base level or uplift in the section pictured or upstream from it. Major changes in the equilibrium of streams (from erosion-dominated to deposition-dominated, or vice versa) can also form terraces. (b) River terraces in the Tien Shan Mountains of China.
How many terraces can you identify in this photo?

Stream Hazards

Although there are many benefits to living near streams, settlement along a river has its risks, particularly in the form of floods. Variability of stream flow constitutes the greatest problem for life along rivers and is also an impediment to their use. Stream channels usually contain the maximum flows that are estimated to occur once every year or two. The maximum flows that are probable over longer periods of 5, 10, 100, or 1000 years overflow the channel and inundate the surrounding land, sometimes with disastrous results (• Fig. 17.31). Similarly, exceptionally low flows may produce crises in water supply and bring river transportation to a halt.

The U.S. Geological Survey maintains more than 6000 gaging stations for the measurement of stream discharge in the United States (• Fig. 17.32). Many of these gaging stations operate on solar energy, measure stream discharge in an automated fashion, and beam their data to a satellite that relays them to a receiving station. With this system, stream flow changes can be monitored at the time they occur, which is very beneficial in issuing flood warnings as well as in understanding how streams change in response to variations in discharge.

• FIGURE 17.31
Living on an active floodplain has its risks, as seen in this photo of the Coast Guard rescuing a man stranded on a rooftop in Olivehurst, California. A levee along the Feather River ruptured, sending acres of water into the Sutter County community.
What can be done to prepare for or to avoid flood problems?

© AP/Wide World Photos

The record of changes in discharge in a stream over time is a **stream hydrograph** (• Fig. 17.33). Hydrographs cover a day, a few days, a month, or even a year depending on a scientist's purpose. Because of the relationship of discharge to water depth and velocity ($Q = wdv$), hydrographs are often used to indicate how high and fast the water level rises in response to a precipitation event.

During and just after a rainfall that produces runoff, the level of the stream will rise in response to the increased discharge. After the water level peaks at the time of maximum discharge, it will then fall as the river eventually returns to a more average level of flow. The discharge of a stream is recorded by a gaging station in the form of a hydrograph curve plotted on a graph that represents this rise, peak, and fall of stream level. The shape of the curve can be used to understand a great deal about how a watershed and a stream channel respond to an increase in runoff, particularly during and after floods. The higher the peak, the greater the discharge when a flood is at its greatest level, called the *flood crest*. How high a river rises and how fast it reaches peak flow in response to a certain amount of precipitation are both important when scientists make preparations for future floods.

The hydrograph curve recorded during a flood event on a stream consists of three major parts: (1) the *rising limb,* (2) the *peak flow* or *flood crest,* and (3) the *receding limb*. Any conditions in the watershed that contribute to high rates of runoff will make the stream rise faster; the rising limb will be steep, representing a rapid rise in discharge to the peak flow in a short amount of time.

• FIGURE 17.32
(a) The U.S. Geological Survey (USGS) monitors the flow of streams and rivers using gaging stations like the one illustrated here. When river level rises or falls, so does the water in the lower part of the station, connected to the channel by intake pipes. A gaging float moves up and down with the water level, and this motion is measured and recorded. Stream gaging stations electronically beam flow data to a satellite that transfers the information to a receiving station. This allows for data to be obtained from many stations at once and for real-time monitoring of how a stream is responding to a storm. (b) USGS stream gaging station. These are commonly located where highway bridges cross rivers. Note the antenna for sending stream flow data through a satellite link to the USGS receiving station.

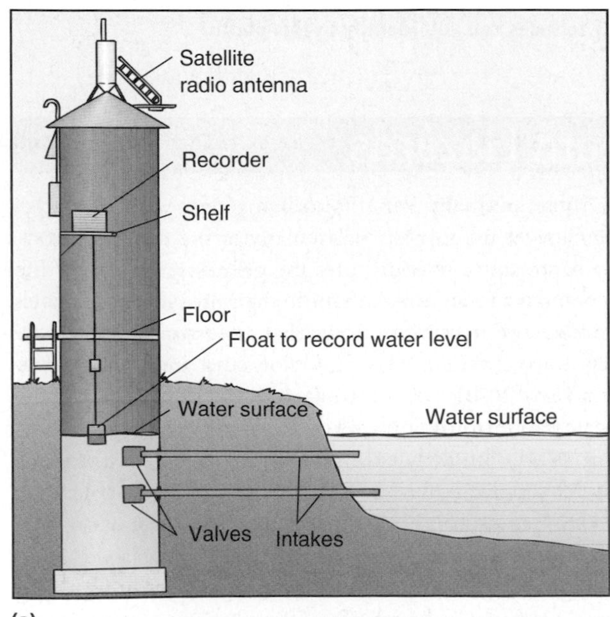

(a)

J. Petersen
(b)

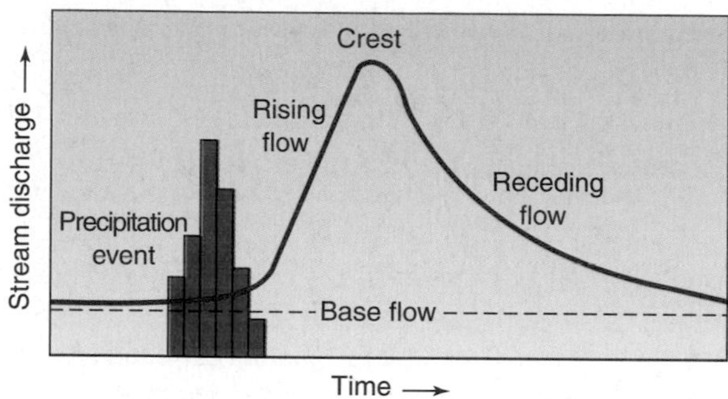

• **FIGURE 17.33**
A stream hydrograph shows changes in discharge, which implies changes in flow depth, recorded over a period of time by a gaging station. This hydrograph example shows the rise in a river in response to flood runoff, the flood peak, and the recession of flow waters following the end of a storm. Note the lag between the time that precipitation starts and the rise in the river.
Why would such a time lag occur between the rainfall and rise in the river?

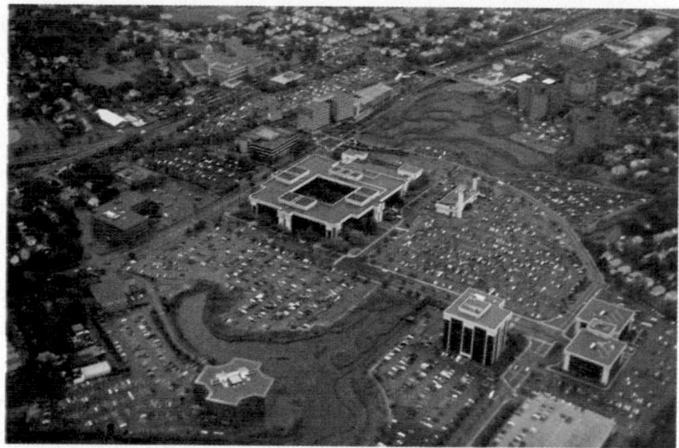

D. Sack

• **FIGURE 17.34**
Most aspects of urbanization and suburbanization, as in this area near Boston, increase the extent of impermeable cover within the drainage basin. As a result, the amount and rate of runoff from urbanized areas increase compared to their presettlement state.
What features of the urbanized landscape shown here enhance runoff?

Rising limbs that are represented by very steep curves indicate flash flood conditions. After the flood crest passes, the stream discharge will decline, but typically the return to a more average flow takes longer (a more gentle curve on the receding limb) because water continues to seep into the channel from the precipitation-saturated ground of the watershed. Studies have shown that urbanization of a watershed (particularly small drainage basins) tends to make the flood peak rise higher and faster than was the case before human development in the drainage basin. Reduction of vegetation and its replacement with impermeable surfaces, like roads, roofs, and parking lots, contribute to higher rates of runoff and increased runoff, two conditions that directly affect the flow of the stream (• Fig. 17.34). When a drainage basin becomes more highly developed and populated, the potential flood size, the rapidness of flood onset, and the flood hazard all tend to become greater.

Obviously, it is very important to be able to determine how often we might expect flows of a given magnitude to occur in a particular river, especially for those discharges that cause flooding. To accomplish this, Earth scientists assess the river's previous flow history, typically using maximum annual discharge data. These data are used to calculate different **recurrence intervals,** the average number of years between past flow events that equaled or exceeded various discharges. For example, if flooding along a river resulted from a discharge of 7000 cubic meters per second (250,000 cu ft/sec) and larger in 2 out of 100 years of data, that discharge has a recurrence interval of 50 years—it is the 50-year flood—even if those flows occurred in two consecutive years. Because that discharge (and possibly greater) happened twice in the past 100 years, there is an estimated 2 out of 100, or 2%, probability that it will happen in any given year. Because recurrence intervals are averages determined from historic data of limited duration, the size of a 10-, 50-, 100-year, or other flood is not absolute, but rather an estimate subject to change, and the length of time between 50-year floods, for example, will not necessarily be 50 years.

The Importance of Surface Waters

Streams

Historically, people have used rivers and smaller streams for a variety of purposes. The settlement and growth of the United States would have been very different without the Mississippi River and its far-reaching tributary system that drains most of the country between the Appalachians and the Rockies. The Mississippi, like many other rivers, has been used for exploration, migration, and settlement, and the number of major cities along it—Minneapolis/St. Paul, St. Louis, Memphis, and New Orleans, to name a few—illustrates people's tendency to settle along rivers. The Mississippi also provides inexpensive transportation for bulk cargo. Today, our navigable rivers still compete successfully with railroads and trucks as carriers of grain, lumber, and mineral fuels.

Streams have long generated power for mills and, more recently, for hydroelectricity. They supply irrigation water, and the alluvial soils of floodplains and river banks are often productive agricultural lands. We also use streams as sources of food and water, for many types of recreation, and as a depository for waste.

Because stream flow can be so variable and in some cases unreliable, humans today regulate most rivers in some way. Many river systems now consist of a series of **reservoirs**—artificial lakes impounded behind dams. The dams hold back potentially devastating floodwaters and store the discharge of wet periods to make the water available during dry seasons or drought years (• Fig. 17.35). An example of this kind of river-basin management comes from the Tennessee River Valley, which was "tamed" during the 1930s. A look at a hydrographic map will show that most of our great rivers, such as the Missouri, Columbia, and Colorado, have been transformed by dam construction into a string of reservoirs. Unfortunately, the life of these reservoirs, like that of

GEOGRAPHY'S ENVIRONMENTAL SCIENCE PERSPECTIVE

Dams

Dams of many types and sizes have been built across America's streams for various reasons, including flood control, maintaining water supplies through drought periods, generating hydroelectric power, improving river navigation, providing water for irrigation, making drinking ponds for stock, and supplying recreational facilities. Some dams in the United States are very old, but most in existence today date from the mid-20th-century period of extensive dam building. Researchers estimate that by the end of that century, U.S. streams were crossed by more than 80,000 dams. Although the impact of dams was occasionally questioned by individuals at various times throughout history, aging of the large number of mid-20th-century dams, and the current trend toward restoring streams to more natural conditions, has recently encouraged this questioning along many broad scientific fronts, including within physical geography. Because of our knowledge of the many and varied aspects of Earth surface systems, physical geographers are well placed to contribute to this discussion.

There are many different types and sizes of dams, some privately owned, but constructing an artificial flow-retarding barrier across any stream changes the fluvial dynamics. Reservoirs behind dams will eventually fill with sediment that the water naturally would have transported downstream and reduced in size by attrition. Underloaded flows released from reservoirs may scour the channel downstream of biologically important sediment sizes. Dams interrupt the natural behavior of streams for some distance upstream and downstream of the site. Assessing the extent of the environmental disruption versus the benefits derived by a dam is typically a complex issue involving economic, social, and cultural factors as well as knowledge of the physical, chemical, and biological systems. Furthermore, different stakeholders will have different goals and perspectives.

With many aging dams in the United States needing expensive repair, the notion of dam removal is now being considered, and acted upon, more frequently than ever before. Earth scientists are also currently working to gain a better understanding of the impact of dam removal on the living and nonliving aspects of the stream system. Sudden release of large amounts of reservoir-stored sediment into the downstream channel will influence the behavior of the flow and likely have ecological impacts. Understanding how streams erode, transport, and deposit sediment, and the role of stream energy, velocity, discharge, and gradient on those processes, is an important component of assessing the environmental costs and benefits of dam removal.

Lynn Betts/NRCS

Hoover Dam and Lake Mead straddle the Nevada-Arizona state line. The road across the top of the dam serves as a bridge between the two states.

almost any lake, will be only a few centuries at most because they will gradually fill with sediment carried by the inflowing streams. To protect the natural flow and environments of the few remaining undeveloped streams and rivers in the United States, Congress enacted the Wild and Scenic Rivers Act in 1968. In recent years, increasing efforts have been directed toward assessing the impacts of dams and dam removal on stream geomorphology and ecology. This, in fact, is just one part of a larger trend in the natural sciences aimed at studying and mitigating the effects of human disturbance on natural stream and other systems.

U.S. Army Corps of Engineers, Portland District

• **FIGURE 17.35**
The multipurpose Lookout Point Dam on the Middle Fork of the Willamette River, Oregon.
What are some of the functions that multipurpose dams serve?

USGS

• **FIGURE 17.36**
Many of the world's lakes are of glacial origin, like this one in Alaska.
Why might this lake be considered a temporary feature?

Lakes

Lakes are standing bodies of inland water. Most of the world's lakes, like Lake Superior, Lake Tahoe, and Lake Victoria, hold freshwater surface runoff in temporary storage along stream systems. However, some lakes, such as the Caspian Sea, Dead Sea, and Great Salt Lake in Utah, contain salty water because they exist in closed basins of interior drainage with no outflowing streams. Evaporation of water to the atmosphere from these lakes leaves behind the dissolved minerals that comprised the dissolved load of inflowing water.

Natural lakes form wherever the water supply is adequate and geomorphic or topographic processes have created depressions on the land surface. The majority of the world's lake basins, such as those of North America's five Great Lakes and Minnesota's "10,000 lakes," are products of glaciation. Rivers, groundwater, tectonic activity, volcanism, and human activities also produce lakes. With a depth of more than 1525 meters (5000 ft), Lake Baikal in Russia, occupying a tectonic depression, is the world's deepest lake. Crater Lake in Oregon, North America's deepest lake, lies in a caldera formed by the collapse of a composite volcano.

Most lakes and ponds (small, shallow lakes) are temporary features on Earth's surface. Few have been in existence for more than 10,000 years, and the majority are very recent in terms of Earth history (• Fig. 17.36). Sedimentation, biological activity, or downcutting of an outlet by a stream eventually lead to the destruction of most lakes.

Lakes are important to humans for more than their scenic appeal and their value for fishing or recreational activities. Like oceans, lakes affect the nearby climates, particularly by reducing daily and seasonal temperature ranges and by increasing humidity. Major fruit-producing areas exist near lakes in Florida, New York, Michigan, and Wisconsin because of the moderating temperature effects of lakes. Lakes can also cause a downwind increase in precipitation—snow or rain generated by the *lake effect,* in which storms either pick up more moisture from the lake or undergo uplift as they move over a relatively warm body of water.

The benefits of lakes are such that humans have produced tens of thousands of artificial lakes through the construction of dams. Reservoirs represent some of the most ambitious construction projects that people have undertaken. However, because lake water tends to stratify by temperature, it does not mix as well as rivers or the oceans. Poor circulation makes lakes easily susceptible to destruction by the chemical, thermal, and biological pollution often resulting from human activities. The Great Lakes, and especially Lake Erie, provide instructive examples of the damage that can be done to a large, complex natural system by human misuse over a short period of time.

Quantitative Fluvial Geomorphology

Quantitative methods are important in studying virtually all aspects of the Earth system, and they are used by climatologists, meteorologists, biogeographers, soil scientists, and hydrologists as well as geomorphologists. The importance of quantitative methods to the objective analysis of fluvial systems in particular may be discerned, at least in part, from the material presented in this chapter. Streams are complicated, dynamic systems with input, throughput, and output of energy and matter that depend on numerous, often-interrelated variables. Geomorphologists routinely measure stream channel, drainage basin, and flow properties and analyze them using statistical methods and the principles of fluid mechanics so that they can describe, compare, monitor, predict, and learn more about streams and the geomorphic work that they perform. Drainage area, stream order, drainage density, stream discharge, stream velocity, channel width, and channel depth are just some of the numeric data that are collected in the field, on topographic maps, from digital elevation models, or from remotely sensed imagery to facilitate the study of stream systems. Extensive efforts are made to gather and analyze numeric stream data because of the widespread occurrence of streams and their great importance to human existence. Future quantitative studies will not only help us better understand the origins and formational processes of landforms and landscapes, but they will also help us better predict water supplies and flood hazards, estimate soil erosion, and trace sources of pollution.

Chapter 17 Activities

Define & Recall

fluvial geomorphology
stream
interfluve
flood
surface runoff
infiltration capacity
interception
sheet wash
rill
gully
ephemeral flow
perennial flow
intermittent flow
base flow
tributary
trunk stream
drainage basin (watershed, catchment area)
drainage area
drainage divide
stream order
source
mouth
exterior drainage
interior drainage
base level
drainage density
drainage pattern (stream pattern)
transverse stream
stream discharge
kinetic energy
stream gradient
channel roughness
stream load
stream competence
stream capacity
graded stream
degradation
aggradation
corrosion
hydraulic action
turbulence
plunge pool
abrasion
pothole
attrition
headward erosion
solution
suspension
saltation
traction
dissolved load
suspended load
bed load
bar
alluvium
natural levee
floodplain
braided channel
meandering channel
longitudinal profile
V-shaped valley
cut bank
point bar
lateral migration
alluvial plain
meander cut-off
oxbow lake
yazoo stream
delta
prodelta
delta plain
distributary
rejuvenated stream
entrenched stream
stream terrace
stream hydrograph
recurrence interval
reservoir

Discuss & Review

1. On a worldwide basis, what is the most important geomorphic agent operating to shape the landscape of the continents? Why is that agent so important?
2. What is the relationship between infiltration and surface runoff? What are some of the factors that enhance infiltration? What are some of the factors that enhance surface runoff?
3. Consider a fifth-order drainage basin. What difference would you expect between the first- and fifth-order channels in terms of overall number of channels? How would a typical first-order channel differ from the fifth-order channel in terms of length, discharge, velocity, and gradient?
4. What factors affect the discharge of a stream? What are the two very different equations that can be used to represent stream discharge?
5. Why do drainage basins in semiarid climates tend to have greater drainage density than those in humid climates?
6. Which is the most effective fluvial erosion process? Why is it so effective?
7. What are the differences among the fluvial transportation processes? Which moves the largest particles?
8. How does a stream sort alluvium? Explain the relationship of this sorting to velocity changes in a stream.
9. Describe the development of natural levees.
10. What are stream terraces? How are they formed? How did changes in sea level during the Pleistocene cause stream terraces to form on land?
11. Why is it important to monitor stream gaging stations and the stream flow data that they provide?
12. What is the relationship between rivers and lakes, and how do the two types of features interact?

Consider & Respond

1. How do streams represent the concept of dynamic equilibrium? What are some examples of negative feedback in a stream system?
2. In the study of a drainage basin, what types of geographic observations and quantitative data might prove useful in planning for flood control and water supply?
3. How does urbanization affect runoff, discharge, and flood potential in a drainage basin?

Apply & Learn

1. A measured cross section of a stream has a width of 6.5 meters, a depth of 2.5 meters, and a flow velocity of 2.0 meters per second. What is the area of the cross section? What is the stream discharge at the cross section?
2. A stream channel lies at an elevation of 1200 meters above sea level at one location, and at 1075 meters above sea level at a site 11 kilometers downstream. What is the channel gradient between those two locations?
3. One hundred years of maximum annual flow data show that the stream under study had a maximum annual discharge of 3700 cubic meters per second or greater five times. What is the recurrence interval of this flow? What is the probability that a discharge of 3700 cubic meters per second will occur this year?

Locate & Explore

Note: Please read the About Locate & Explore section of the Preface before beginning these exercises.

1. Using Google Earth, identify the landforms at the following locations (latitude, longitude) and provide a brief discussion of how the landform developed. Include a brief discussion of why the landform is found in that general area of the United States or the world.

   Tip: Use the zoom, tilt, rotate, and elevation exaggeration tools to help view and interpret the landform and the area in which it is found.

   a. 43.73°S, 172.02°E
   b. 52.12°N, 28.89°E
   c. 30.40°N, 85.03°W
   d. 45.13°N, 111.66°W
   e. 37.20°N, 110.00°W
   f. 45.36°N, 111.71°W
   g. 39.27°N, 119.55°W
   h. 30.57°N, 87.19°W
2. Using Google Earth and the Streamflow Layer from the USGS, track the daily changes in stream flow for the rivers listed below. For each river, make a table that shows the historic minimum, maximum, and mean discharge. Provide an explanation of the variation based on the location of the river within its drainage basin (upper, middle, or lower) and with respect to the Köppen Climate Classification system shown in Figure 8.6 of your text.
   a. Mississippi River at Baton Rouge, Louisiana (30.443801°N, 91.194420°W)
   b. Mississippi River at St. Louis, Missouri (38.629097°N, 90.179704°W)
   c. Verde River near Scottsdale, Arizona (33.546048°N, 111.687700°W)
   d. Columbia River below Priest Rapids Dam, Washington (46.618490°N, 119.86048°W)
   e. Stikine River near Wrangell, Alaska (56.705416°N, 132.132541°W)
   f. Red River of the North at Drayton, North Dakota (48.572275° N, 97.142759°W)
   g. Hudson River at Fort Edwards, New York (43.259648°N, 73.594598°W)
   h. Savannah River at Augusta, Georgia (32,920327°N, 83.337578°W)
   i. Manatee River, near Myakka Head, Florida (27.289716°N, 82.538898°W)
   j. Brazos River at Hempstead, Texas (29.985725°N, 96.474436°W)
3. Using Google Earth and the Meandering River Layer, identify the marked features (A–E). Rank the features from youngest to oldest with an explanation for your order.

   Tip: Use the zoom, tilt, rotate, and elevation exaggeration tools to help view and interpret the landform and the area in which it is found.

   a. 34.543999°N, 90.568936°W
   b. 34.659545°N, 90.480867°W
   c. 34.930455°N, 90.333536°W
   d. 34.642015°N, 90.411898°W
   e. 34.520741°N, 90.254868°W

Map Interpretation

FLUVIAL LANDFORMS

The Map

Campti is in northwestern Louisiana on the Gulf Coastal Plain. The coastal plain region stretches westward from northern Florida to the Texas–Mexico border, and extends inland in some places for more than 320 kilometers (200 mi). Elevations on the Gulf Coastal Plain gradually increase from sea level at the shoreline to a few hundred meters far inland. The region is underlain by gently dipping sedimentary rock layers. The surface material includes marine sediments and alluvial deposits from rivers that cross the coastal plain, especially those of the Mississippi drainage system. This is a landscape of meanders, natural levees, and bayous.

The Red River's headwaters are located in the semiarid plains of the Texas Panhandle, but it flows eastward toward an increasingly more humid climate. About midcourse, the Red River enters a humid subtropical climate region, which supports rich farmland and dense forests. The river flows into the Mississippi in southern Louisiana, about 160 kilometers (100 mi) downstream of the map area. The Red River is the southernmost major tributary of the Mississippi.

Louisiana has mild winters with hot, humid summers. Annual rainfall totals for Campti average about 127 centimeters (50 in.). The warm waters of the Gulf of Mexico supply vast amounts of atmospheric energy and moisture, producing a high frequency of thunderstorms, tornadoes, and, on occasion, hurricanes that strike the Gulf Coast.

Interpreting the Map

1. How would you describe the general topography of the Campti map area? What is the local relief? What is the elevation of the banks of the Red River at the town of Campti?
2. What kind of landform is the low-relief surface that the river is flowing on? Is this mainly an erosional or depositional landform?
3. In what general direction does the Red River flow? Is the direction of flow easy or hard to determine just from the map area? Why?
4. Does the Red River have a gentle or steep gradient? Why is it difficult to determine the river gradient from the map area?
5. What is the origin of Smith Island? Adjacent to Smith Island is Old River; what is this type of feature called?
6. Explain the stippled brown areas in the meanders south of Campti. Are these areas on the inside or the outside of the meander bend? Explain why.
7. How would you describe the features labeled as "bayou"?
8. Is this map area more typical of the upper, middle, or lower course of a river?
9. Although this is not a tectonically active region, how would the river change if it experienced tectonic uplift?

U.S. Department of Agriculture

Healthy green vegetation appears red on color infrared images, here showing the Red River and Smith Island, Louisiana.

Opposite:
Campti, Louisiana
Scale 1:62,500
Contour interval = 20 ft
U.S. Geological Survey

Bright Morning Star Ch
Lake Poule d'eau
RED RIVER
SOUTHERN
Campti
Powhatan
Smith Island
Old River
Passon Lake
PACIFIC
PIERRE
North Star Ch
New Zion Ch
Flat River
Three League Bayou
Red Bayou
St Marys Bayou
Hyams Bayou
Boggy Bayou
Johnsons Chute
BAYOU PIERRE
Berry Bridge
St Lukes Cem
St Lukes Ch
Calvary Ch
Grand Ecore

Arid Region Landforms and Eolian Processes

18

CHAPTER PREVIEW

In most deserts, running water operates only occasionally, so it may seem odd that water rather than wind is the chief landforming agent in dry regions.

- How can this be?
- What features in arid regions are produced predominantly by the action of flowing water?

Stream channels in desert areas are often quite different from those in humid regions.

- What are some of those differences?
- What combination of factors causes these differences?

A variety of landforms in many deserts today could not have been created under present conditions of aridity; they are evidence of landscapes formed under earlier, wetter climatic conditions.

- What is some landform evidence of climatic change in deserts?
- How might the most recent cooler, wetter period experienced by midlatitude deserts be associated with glaciation?

Over most land surfaces the wind exerts little influence on landform development, but under certain circumstances it can be an important geomorphic agent.

- Why is wind so limited as a major agent in landform development?
- Under what conditions and in what environments is the wind an important agent in shaping the land?

Sand dunes, and the landscapes that they exist in, are often fragile environments that require protection from certain human activities.

- Why are these environments fragile, and why is it important to protect them?
- What measures can be taken to minimize erosion, soil loss, and environmental degradation in sand dune areas?

◀ Opposite: Joshua Tree National Park, in the arid climate of California's Mojave Desert, features a distinctive landscape of fractured granite hills and boulders.
J. Petersen

Because of their low amounts of precipitation, arid region landscapes look quite different from those of other climatic environments in many ways. The limited water supply restricts rock weathering as well as the amount of vegetation present. Without extensive vegetation to hold weathered rock matter (regolith) in place, the weathered rock particles that are produced are often stripped away when storms do occur. As a result, while hillslopes in humid regions tend to be rounded and mantled by soil, mountains and hill slopes in arid regions are generally angular, with extensive, barren exposures of bedrock. Desert lowlands may be filled in with sediments eroded from uplands, or they may consist of just a thin cover of sediments overlying rock strata.

Many desert landscapes have a majestic beauty—a fact underscored by the frequent use of desert locations by filmmakers throughout the history of movie making. One reason desert scenery is so striking is that desert landscapes often display, in stark beauty, the colors, characteristics, and structure of the rocks that make up the area. The desert's barrenness reveals considerable evidence about landforms and geomorphic processes that is much more difficult to observe in humid environments, with their extensive cover of soil and vegetation. Much of our understanding of how landscapes and landforms develop in a wide variety of environments has come from important studies and scientific explorations conducted in desert regions.

Although the wind plays an important role in arid region geomorphology, you may be surprised to learn that, overall, running water does more geomorphic work than the wind does in arid regions. Wind erosion is mainly confined to picking up fine, dust-sized (silt and clay) particles from desert regions and to dislodging loose rock fragments of sand-sized materials. Still, we tend to associate arid environments with eolian (wind) geomorphic processes because of the notable accumulations of wind-deposited sediment displayed in some desert areas, usually in the form of sand dunes. With its sparse vegetation and other environmental characteristics, eolian geomorphology reaches its optimum in arid environments. However, because the air has a much lower density than water, even in deserts the geomorphic work of the wind is outmatched by fluvial geomorphic processes. We should understand, too, that eolian processes and landforms are not confined to arid regions; they are also conspicuous in many coastal areas and in any area where loose sediments are frequently exposed to winds strong enough to move them.

© Rainer Duttmann

• **FIGURE 18.1**
In almost all deserts, even in the most arid locations, the effects of erosion and deposition by running water are prominent in the landscape. This is Death Valley, California.
Why do you think the drainage density is so high here?

Surface Runoff in the Desert

Landforms, rather than vegetation, typically dominate desert scenery. The precipitation and evaporation regimes of an arid climate result in a sparse cover of vegetation and in addition, because many weathering processes require water, relatively low rates of weathering. With low weathering rates, insufficient vegetation to break the force of raindrop impacts, and a lack of extensive plant root networks to help hold rock fragments in place, a blanket of moisture-retentive soil cannot accumulate on slopes. Soils tend to be thin, rocky, and discontinuous. This absence of a continuous vegetative and soil cover gives desert landforms their unique character. Under these surface conditions of very limited interception and low permeability, much of the rain that falls in the desert quickly becomes surface runoff available to perform fluvial geomorphic work. With little to hold them in place, any grains of rock that have been loosened by weathering may be swept away in surface runoff produced by the next storm. Ironically, although desert landscapes strongly reflect a deficiency of water, the effects of running water are widely evident on slopes as well as in valley bottoms (• Fig. 18.1). Where vegetation is sparse, running water, when it is available, is extremely effective in shaping the land.

Desert climates characteristically receive small amounts of precipitation and are subjected to high rates of potential evapotranspiration. In exceptional circumstances, years may pass without any rain in certain desert areas. Most desert locations, however, receive some precipitation each year, although the frequency and amount are highly unpredictable. Rains that do fall are often brief and limited in their spatial coverage, but they can also be quite intense. While times of rainfall are short, unreliable, and difficult to predict, potential evapotranspiration remains high throughout the year in most arid regions. The most important impact of rain on landform development in deserts is that when rainfall does occur, much of it falls on impermeable surfaces, producing intense runoff, generating flash floods, and operating as a powerful agent of erosion.

The visible evidence of water as a geomorphic agent in arid regions stems not only from the climate of those areas today, but also from past climates. Paleogeographic studies reveal that most deserts have not always had the arid climates that exist today. Geomorphologists studying arid regions have found certain landforms that are incompatible with the present climate, and they attribute these features to the work of water under earlier, wetter climates. A great majority of desert areas were wetter at intervals in the past. For midlatitude and subtropical deserts, the most recent major wet period was during the Pleistocene Epoch. While glaciers were advancing in high latitudes and in high-elevation mountain regions during the Pleistocene, precipitation was also greater than it is today in those basins, valleys, and plains of the middle and subtropical latitudes that are now desert areas. At the same time, cooler temperatures for these regions meant that they also experienced lower evaporation rates. In many of today's deserts, evidence of past wet periods includes deposits and wave-cut shorelines of now-extinct lakes (• Fig. 18.2) and immense canyons occupied by streams that are now too small to have eroded such large valleys.

Running water is a highly effective agent of landform development in deserts even though it operates only occasionally. In most desert regions, running water is active just during and shortly after rainstorms. Desert streams, therefore, are typically *ephemeral channels,* containing water only for brief intervals. Ephemeral stream channels are exposed and dry the rest of the time. In contrast to *perennial channels,* which flow all year and are typical of humid environments, ephemeral streams do not receive seepage from groundwater to sustain them between episodes of surface runoff. Ephemeral streams instead generally lose water to the groundwater system through infiltration into the channel bed. Because of the low weathering rates in desert environments, most arid region

D. Sack

• **FIGURE 18.2**

Many desert basins in Nevada, Utah, and California have remnant shorelines that were created by wave action from lakes they contained during the Pleistocene. The linear feature extending across this hillslope in Utah is a shoreline from one of these ancient lakes.

What can we learn about climate change from studying these relict lake features?

D. Sack

• **FIGURE 18.3**

This braided stream in Canyon de Chelly National Monument near Chinle, Arizona, splits and rejoins multiple times as it works to carry its extensive bed load of coarse sand.

Why do you think the number and position of the multiple channels can change rapidly?

streams receive an abundance of coarse sediment that they must transport as bed load. As a result, *braided channels,* in which multiple threads of flow split and rejoin around temporary deposits of coarse-grained sediments, are common in deserts (● Fig. 18.3).

Unlike the typical situation for humid region streams, many desert streams undergo a downstream decrease, rather than increase, in discharge. A discharge decrease occurs for two major reasons: (1) water losses by infiltration into the gravelly stream channel accumulate downstream, and (2) evaporation losses increase downstream due to warmer temperatures at lower elevations. As a result of the diminishing discharge, many desert streams terminate before reaching the ocean. The same mountains that contribute to aridity through the rain-shadow effect can effectively block desert streams from flowing to the sea. Without sufficient discharge to reach ultimate base level, desert streams terminate in depressions in the continental interior where they commonly form shallow, ephemeral lakes. Ephemeral lakes evaporate and disappear and then reappear when rain provides another episode of adequate inflow. During the cooler and wetter times of the Pleistocene Epoch, many closed basins in now-arid regions were filled with considerable amounts of water that in some cases formed large perennial freshwater lakes instead of the shallow ephemeral lakes that they contain today.

Where surface runoff drains into closed desert basins, sea level does not govern erosional base level as it does for streams that flow into the ocean and thereby attain *exterior drainage*. Desert drainage basins characterized by streams that terminate in interior depressions are known as basins of *interior drainage* (● Fig. 18.4); such streams are controlled by a **regional base level** instead of ultimate base level. When sedimentation raises the elevation of the desert basin floor located at the stream's terminus, the stream's base level rises, which causes a decrease in the stream's slope, velocity, and energy. If tectonic activity lowers the basin floor, the regional base level is depressed, which may lead to rejuvenation of the desert stream. Tectonism has even created some desert basins of interior drainage with floors below sea level, as in Death Valley, California, the Dead Sea Basin in the Middle East, the Turfan Basin in western China, and Australia's Lake Eyre (see Map Interpretation: Desert Landforms).

Many streams found in deserts originate in nearby humid regions or in cooler, wetter mountain areas adjacent to the desert. Even these, however, rarely have sufficient discharge to sustain flow across a large desert (● Fig. 18.5). With few tributaries and virtually no inflow from groundwater, stream water losses to evaporation and underground seepage are not replenished. In most cases, the flow dwindles and finally disappears. The Humboldt River in Nevada is an outstanding example; after rising in the mountains of central Nevada and flowing 465 kilometers (290 mi), the river

● FIGURE 18.4
The Sierra Nevada (across entire upper part of photo) poses a topographic barrier to streams on its rain-shadow side (lower part of photo) and did so even during the wetter times of the Pleistocene so that few flowed to the sea. The other streams filled depressions to form lakes, most of which are completely dry today. This image, oriented with north to the right, shows the bed of Owens Lake (large white area in photo center), which shrank because of climate change and then desiccated when its waters were diverted to urban areas in Southern California. A small amount of moisture (elongated dark zones) occupied part of the dry lake bed when the photo was taken.

NASA/Earth Observations Lab/Johnson Space Center

J. Petersen

• **FIGURE 18.5**
A stream flows through a deep gorge in the Atlas Mountains of Morocco. This is the arid, rain-shadow side of the mountains, facing the Sahara. This stream loses water by infiltration and evaporation and disappears into the Sahara. Note the steeply dipping, folded rocks of the Atlas and the thin line of vegetation along the stream channel.
Was the gorge eroded by the stream with this amount of flow? If not, what factors might have produced more discharge to erode the deep canyon?

disappears into the Humboldt Basin, a closed depression. Only a few large rivers that originate in humid uplands have sufficient volume to survive the long journey across hundreds of kilometers of desert to the sea (• Fig. 18.6). Called **exotic streams,** the rivers that successfully traverse the desert erode toward a base level governed by sea level and provide drainage that is external to the arid region. Classic examples of exotic streams include the Nile (Egypt and Sudan), Tigris-Euphrates (Iraq), Indus (Pakistan), and Murray (Australia) Rivers. Under natural conditions, the Colorado River of the United States and Mexico would reach ultimate base level at the head of the Gulf of California most of the time, but because of huge water withdrawals from the river by people, in actuality, it rarely flows all the way to the gulf.

NASA

• **FIGURE 18.6**
A false-color satellite image of the Nile River meandering across the Sahara in Egypt. The irrigated croplands that appear dark red contrast with the lighter tones of the barren desert terrain. The Nile is an exotic stream. Its headwaters are in the wetter climates of the Ethiopian Highlands and lakes in the East African Rift Zone, which support its northward flow across the Sahara to the Mediterranean Sea.

Water as a Geomorphic Agent in Arid Lands

When rain falls in the desert, sheets of water run down unprotected slopes, picking up and moving sediment. Dry channels quickly change to flooding streams. The material removed by runoff and surface streams is transported, just as in humid lands, until flow velocity decreases sufficiently for deposition to occur. Eventually these streams disappear when seepage and evaporation losses exceed their discharge. Huge amounts of sediment can be deposited along the way as a stream loses volume and velocity. The processes of erosion, transportation, and deposition by running water are essentially the same in both arid and humid lands. However, the resulting landforms differ because of the sporadic nature of desert runoff, the lack of vegetation to protect surface materials against rapid erosion, and the common occurrence of streams that do not reach the sea.

Arid Region Landforms of Fluvial Erosion

Among the most common desert landforms created by surface runoff and erosion are the channels of ephemeral streams. Known as **washes** or **arroyos** in the southwestern United States, **barrancas** in Latin America, and **wadis** in North Africa and Southwest Asia, these channels usually form where rushing surface waters cut into unconsolidated alluvium (• Fig. 18.7). These typically gravelly, braided channels are prone to flash floods, which makes them potentially very dangerous sites. Though it may sound strange, many people have drowned in the desert—during flash floods.

In areas of weak, easily eroded clays or shales, rapid erosion from surface runoff can produce a dense network of barren

D. Sack

• **FIGURE 18.7**
This dry streambed, or wash, has a channel bed of coarse alluvium and conveys water only during and slightly after a rainstorm.
Why would this desert stream channel have a high risk for flash floods?

slopes and ridges dissected by a maze of steep, dry gullies and ravines. Early fur trappers in the Dakotas called such areas "bad lands to cross" (• Fig. 18.8). The phrase stuck, and those regions are still called the Badlands, while that type of rugged, barren, and highly dissected terrain is termed **badlands** topography. Badlands topography has an extremely high *drainage density,* defined as the length of stream channels per unit area of the drainage basin. Besides the Dakotas, extensive badlands can be seen in Death Valley National Park in California, Big Bend National Park in Texas, and southern Alberta, Canada. Badlands generally do not form naturally in humid climates because the vegetation there slows runoff and erosion, leading to lower drainage densities. Removing the vegetation from clay or shale areas by overgrazing, mining, or logging, however, can cause badlands topography to develop even in humid environments.

A **plateau** is an extensive, elevated region with a fairly flat top surface. Plateaus

Copyright and photograph by Dr. Parvinder S. Sethi

• **FIGURE 18.8**
In badlands, such as these in South Dakota, impermeable clays that lack a soil cover produce rapid runoff, leading to intensive gully erosion and a high drainage density.
Was this rugged terrain named appropriately?

are generally dominated by a structure consisting of horizontal rock layers. Many striking plateaus exist in the deserts and semiarid regions of the world. An excellent example in the United States is the Colorado Plateau, centered on the Four Corners area of Arizona, Colorado, New Mexico, and Utah. In tectonically uplifted desert plateau regions such as this, streams and their tributaries respond to uplift by cutting narrow, steep-sided canyons. Where the canyon walls consist of horizontal layers of alternating resistant and erodible rocks, differential weathering and erosion exert a strong influence on the canyon walls. Canyons in these areas tend to have stair-stepped walls, with near-vertical cliffs marking the resistant layers (ordinarily sandstone, limestone, or basalt) and weaker rocks (often shales) forming the slopes. The distinctive walls of the Grand Canyon have this appearance, which exposes the structure of horizontal rock layers of varying thickness and resistance (• Fig. 18.9). The rim of the Grand Canyon is a flat-topped cliff made of a **caprock,** a term that refers to a resistant horizontal layer that forms (caps) the top of a landform.

• FIGURE 18.9
The earliest European American explorers of the Grand Canyon took along an artist to record the geomorphology of the canyon, here beautifully shown in great detail. Aridity creates an environment where the bare rocks are exposed to our view; differential weathering and erosion give the stair-stepped quality to the walls of the Grand Canyon.

USGS

Caprocks top plateaus and constitute canyon rims, but they also form the summits of other, smaller kinds of flat-topped landforms that, although they are found in many climate regions, are most characteristic of deserts. Weathering and erosion will eventually reduce the extent of a caprock until only flat-topped, steep-sided **mesas** remain (*mesa* means "table" in Spanish). A mesa has a smaller surface area than a plateau and is roughly as broad across as it is tall. Mesas are relatively common landscape features in the Colorado Plateau region. Through additional erosion of the caprock from all sides, a mesa may be reduced to a **butte,** which is a similar, flat-topped erosional remnant but with a smaller surface area than a mesa (• Fig. 18.10). Mesas and buttes in a landscape are generally evidence that uplift occurred in the past and that weathering and erosion have been extensive since that time. Variations in the form of the slope extending down the sides of buttes, mesas, and plateaus are related to the height of the cliff at the top, which is controlled by the thickness of the caprock in comparison to the size of landform feature. Monument Valley, in the Navajo Tribal Reservation on the Utah–Arizona state line, is an exquisite example of such a landscape formed with a caprock that is particularly thick, contributing to the distinctive scenery (• Fig. 18.11). Many famous western movies have been filmed in Monument Valley and in nearby areas of the Colorado Plateau because of the striking, colorful, and photogenic desert landscape.

Sheet wash and gully development generally accomplish extensive erosion of mountainsides and hillslopes fringing a desert basin or plain. Particularly in desert regions with exterior drainage or a sizable trunk stream on the basin or plain, this fluvial action, aided by weathering, can lead to the gradual erosional retreat of bedrock slopes. This retreat of the steep mountain front can leave behind a more gently sloping surface of eroded bedrock, called a **pediment** (• Fig. 18.12). Characteristically in desert areas, there tends to be a sharp break in slope between the base of steep hills or mountains, which rise at angles of 20 to 30° or steeper, and the gentle pediment, whose slope is usually only 2 to 7°. Resistant knobs of the bedrock comprising the pediment may project up above the surface on some pediments. These resistant knobs are referred to as **inselbergs** (from German: *insel,* island; *berg,* mountain).

Geomorphologists do not agree on exactly how pediments are formed, perhaps because different processes may be responsible for their formation in different regions. However, there is general agreement that most pediments are erosional surfaces created or partially created by the action of running water. In some areas, weathering, perhaps when the climate was wetter in the past, may also have played a strong role in the development of pediments.

• FIGURE 18.10
Plateaus, mesas, and buttes are developed through weathering and erosion in areas of horizontal rock layers with a resistant caprock. An excellent example of this terrain is the Colorado Plateau of Arizona, Colorado, New Mexico, and Utah.

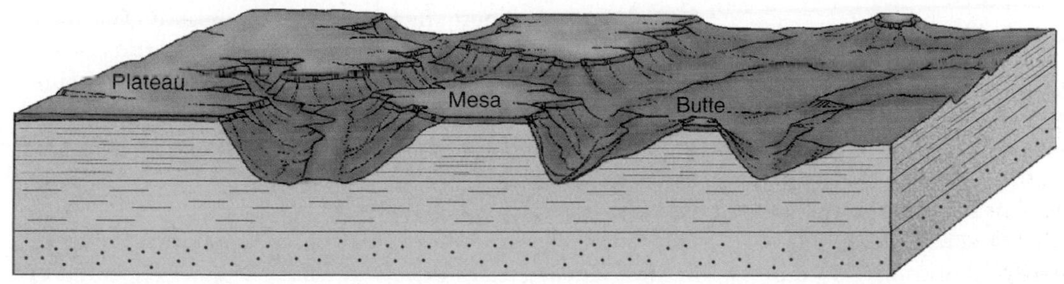

Copyright and photograph by Dr. Parvinder S. Sethi

• **FIGURE 18.11**
The caprock in Monument Valley, Arizona, is particularly thick and represents a rock layer that once covered the entire region. The prominent buttes are erosional remnants of that layer.
Compare this photo to the diagram in Figure 18.10. How were these landforms produced?

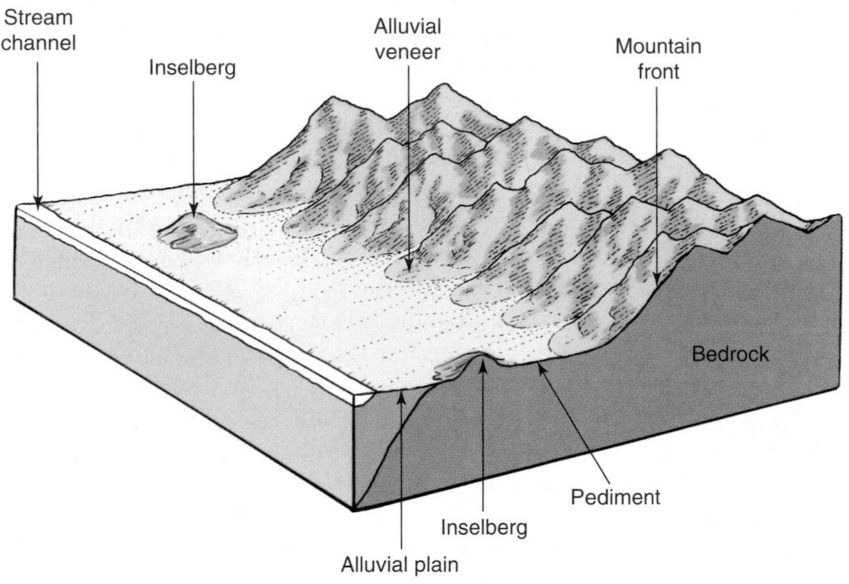

• **FIGURE 18.12**
Pediments are erosion surfaces cut into bedrock beyond the present mountain front in arid regions. Pediments most commonly occur in desert areas with exterior drainage that can remove a portion of the erosion products. At many locations, pediment surfaces are covered with a thin veneer of alluvium, and their gently sloping surface may be interrupted by resistant knobs of bedrock, called inselbergs, that stick up above the pediment surface.

Arid Region Landforms of Fluvial Deposition

Deposition is as important as erosion in creating landform features in arid regions, and in many areas sedimentation by water does as much to level the land as does erosion. Many desert areas have wide expanses of alluvium deposited either in closed basins or at the base of mountains by streams as they lose water in the arid environment. As the flow of a stream diminishes, so does its *capacity*—the amount of load it can transport. Most landforms developed by fluvial deposition in arid lands are not exclusive to desert regions but are particularly common and visible in dry environments due to their thin soil and sparse vegetative cover.

Alluvial Fans Where streams, particularly ephemeral (sporadic) or *intermittent* (seasonal) ones, flow out of uplands through narrow canyons and onto open plains, their channels may flare out to become wide and shallow. Because stream discharge (Q) equals channel cross-sectional width (w) and depth (d) multiplied by flow velocity (v) ($Q = wdv$), the increase in width causes a decrease in flow velocity, reducing stream competence (maximum size of load) and stream capacity (maximum amount of load). Discharge also decreases as water seeps from the channel into coarse alluvium below. As a result, most of the sediment load carried by such streams is deposited along the base of the highlands.

Upstream of the mouth of the canyon, the channel is constrained by bedrock valley walls, but as it flows out of the canyon, the channel is free to not only widen but also shift its position laterally. Sediments are initially deposited just beyond the canyon mouth, building up the lowland area near the mountain front. Eventually this aggradation causes the channel to shift laterally where it begins to deposit and build up another zone close to the mountain front and adjacent to the first aggraded area. The canyon mouth serves as a pivot point anchoring the channel as it swings back and forth over the lowlands near the mountain front, leaving alluvium behind. This creates a fan-shaped depositional landform, called an **alluvial fan,** in which deposition takes place radially away from that pivot point, or **fan apex** (• Fig. 18.13).

An important characteristic of an alluvial fan is the sorting of sediment that typically occurs on its surface. Coarse sediments, like boulders and cobbles, are deposited near the fan apex where the stream first undergoes a decrease in competence and capacity as it emerges from the confinement of the canyon. In part because of the large size of the clasts deposited there, the slope of an alluvial fan is steepest at its apex and gradually diminishes, along with grain size, with increasing distance downstream from the canyon mouth. In areas where the uplands generate debris flows rather than stream flows, **debris flow fans** or even mixed debris and alluvial fans are constructed instead of purely alluvial fans. Debris flow fans tend to be steeper than alluvial fans and do not display the same degree of downslope sorting shown by the fluvial counterpart.

Although they can be found in mountainous areas of almost any climate, alluvial fans are particularly common where ephemeral or intermittent streams laden with coarse sediment flow out of a mountainous region onto desert plains or into arid interior basins.

Copyright and photograph by Dr. Parvinder S. Sethi

• **FIGURE 18.13**
Alluvial fans are constructed at the base of a mountain range. Where streams come out of confined canyons, they are free to widen, which causes a decrease in stream velocity, capacity, and competence. The apex of a fan lies at the mouth of the canyon or wash. Fans, such as this one in Death Valley, California, are particularly common landforms in the arid Basin and Range region of the western United States.
How do alluvial fans differ from pediments?

In the western United States, alluvial fans are a major landform feature in landscapes consisting of fault-block mountains and basins, as in the Great Basin of California, Nevada, and Utah (• Fig. 18.14). Here streams laden with sediment periodically rush from canyons cut into uplifted fault-block mountains and deposit their load in the adjacent desert basins. Everything else being equal, fans associated with larger drainage basins within the uplifted fault-block mountains tend to have greater area and be less steep than fans developed from streams draining smaller upland drainage basins.

Large, conspicuous alluvial fans develop in environmental settings like the Great Basin for several reasons. First, highland areas in desert regions are subject to intense erosion, primarily because of the low density of vegetative cover, steep slopes, and the orographically intensified downpours that can occur over mountains. In addition, streams in arid regions typically carry a greater concentration of sediment load (in comparison to the discharge) than comparable streams in more humid regions. As the streams flow from confined mountain canyons into desert basins, they deposit most of their coarse sediment near the canyon mouth. Flowing into the desert basin, their width increases, their depth and velocity decrease, and their volumes are significantly reduced through infiltration into the alluvial channel bed. Not far from the canyon, the stream itself may disappear, or it may occasionally reach the desert basin floor where it deposits its remaining load, the silts and clays. Extensive alluvial fans are not as common in humid as in arid regions because most highland streams in moist climates are perennial and have sufficient flow to continue across adjacent lowlands.

Along the bases of mountains in arid regions, adjacent alluvial fans may become so large that they join together along their sides to form a continuous ramplike slope of alluvium called a **bajada** (• Fig. 18.15). A bajada consists of adjacent alluvial fans that have coalesced to form an "apron" of alluvium along the mountain base. Where extensive fans coalesce over very wide areas, they form a **piedmont alluvial plain,** like the area surrounding Phoenix, Arizona (• Fig. 18.16).

Piedmont alluvial plains generally have rich soils and the potential to be transformed into productive agricultural lands. The major obstacle is inadequate water supply to grow crops in an arid environment. In many world regions, arid alluvial plains are irrigated with water diverted from mountain areas or obtained from reservoirs on exotic streams. The alluvial farmlands near Phoenix are a good example, producing citrus fruits, dates, cotton, alfalfa, and vegetables.

Where a veneer of alluvium has been deposited on a pediment, the land surface may closely resemble a water-deposited alluvial fan. In some situations, it may not be possible to determine the existence of an underlying pediment without either excavating through the surface alluvium or finding the erosional pediment surface exposed in the walls of washes or gullies. In locations where no extensive pediment exists, alluvial deposits beneath the fans can be tens or even hundreds of meters thick. In contrast, the layer of alluvium overlying a pediment is only a relatively thin layer, no more than a few meters deep and sometimes much less.

Playas Desert basins of interior drainage surrounded by mountains are sometimes called **bolsons.** Most bolsons were formed by faulting that created basins between uplifted mountains. The lowest part of most bolsons is occupied by a landform called a **playa** (in Spanish: *playa,* beach or shore), which is the fine-grained bed of an ephemeral lake. Occasionally, large rainfall (or snowmelt) events or wet seasons cover the playa with a very shallow body of water, called a **playa lake.** Direct precipitation onto the playa, inflow from surface runoff, or discharge from the groundwater zone can contribute water to the playa lake. The playa lake may persist for a day or two or for several weeks (• Fig. 18.17). Wind blowing over the playa lake moves the shallow water, along with its suspended and dissolved load, around on the playa surface. This helps to fill in any low spots on the playa and contributes to making playas one of the flattest of all landforms on Earth. Playa lakes lose most of their water by evaporation to the desert air.

Although playas are very flat, considerable variation exists in the nature of playa surfaces. Playas that receive most of their water from surface runoff typically have a smooth clay surface, sometimes called a **clay pan,** baked hard by the desert sun when it is dry but extremely gooey and slippery when wet. In contrast, **salt-crust playas,** also known as **salt flats** or **salinas,** receive much of their water from groundwater, are damp most of the time, and are encrusted with salt mineral deposits crystallizing out of the evaporating groundwater (• Fig. 18.18). Some salt-crust playas composed of an orderly sequence toward the lowest part of the playa of carbonates, sulfates, and chlorides (for example, calcite, gypsum, and halite, respectively) are the floors of desiccated, ancient lakes that occupied now-desert basins during the Pleistocene Epoch.

Playas are useful in several ways. For one, companies mine the rich deposits of evaporite minerals, including such important

• **FIGURE 18.14**
Map of the Great Basin of the western United States showing major lakes that existed during glacial times. More than 100 lakes formed in the fault-block basins of this region during the Pleistocene Epoch.
Why are there only a few remnant lakes in this region today?

industrial chemicals as potassium chloride, sodium chloride, sodium nitrate, and borates, that have been deposited in some playa beds. Also, the extensive, flat surfaces of some playas make them suitable as racetracks and airstrips. Utah's famous Bonneville Salt Flats mark the bed of an extinct Pleistocene lake. The western portion of the Salt Flats, where world land-speed records are set, still floods to a depth of 30 to 60 centimeters (1–2 ft) in the cool, wet, winter season. The hard, flat playa surface at Edwards Air Force Base, in California's Mojave Desert, has served for many decades as a landing site for military aircraft and, in recent decades, for the space shuttle. These landings have occasionally been disrupted due to flooding of the playa.

Copyright and photograph by Dr. Parvinder S. Sethi

• **FIGURE 18.15**
A bajada is formed when a series of alluvial fans coalesce, forming a continuous alluvial slope along the front of an eroding mountain range. This example is from Death Valley, California. **Why would a series of alluvial fans have a tendency to eventually join to form a bajada?**

NASA/Goddard Space Flight Center Scientific Visualization Studio

• **FIGURE 18.16**
A large desert alluvial plain in Arizona that is extensively urbanized.

Wind as a Geomorphic Agent

On a worldwide basis, wind is less effective than running water, waves, groundwater, moving ice, or mass movement in accomplishing geomorphic work. Under certain circumstances, however, wind can be a significant agent in the modification of topography. Landforms—whether in the desert or elsewhere—that are created by wind are called **eolian** (or **aeolian**) landforms, after Aeolus, the god of winds in classical Greek mythology (• Fig. 18.19). The three principal conditions necessary for wind to become an effective geomorphic agent are a sparse vegetative cover, the presence of dry, loose materials at the surface, and a wind velocity that is high enough to pick up and move those surface materials. These three conditions occur most widely in arid regions and on beaches, although they are also found on or adjacent to exposed lake beds, areas of recent alluvial or glacial deposition, newly plowed fields, and overgrazed lands.

A dense vegetative cover reduces wind velocity near the surface by providing frictional resistance. It also prevents wind from being directed against the land surface and holds materials in place with its root network. Without such a protective cover, fine-grained and sufficiently dry surface materials are subject to removal by strong gusts of wind. If surface particles are damp, they tend to adhere together in wind-resistant aggregates due to increased cohesion provided by the water. The arid conditions of deserts therefore make those regions most susceptible to wind erosion.

Eolian processes have many things in common with fluvial processes because air and water are both fluids. Some important contrasts also exist, however, due to fundamental differences between gases and liquids. For example, rock-forming materials cannot dissolve in air, as some can in water; thus air does not erode by corrosion or move load in solution. Otherwise, the wind detaches and transports rock fragments in ways comparable to flowing water, but it does so with less overall effectiveness because air has a much lower density than water. Another difference is that, compared to streams, the wind has fewer lateral or vertical limitations on movement. As a result, the dissemination of material by the wind can be more widespread and unpredictable than that by streams.

A principal similarity between the geomorphic properties of wind and running water is that flow velocity controls their competence—that is, the size of particles each can pick up and carry. However, because of its low density, the competence of moving air is generally limited to rock fragments that are sand sized or smaller. Wind erosion selectively entrains small particles, leaving behind the coarser and heavier particles that it is not able to lift. Like water-laid sediments, wind deposits are stratified according to changes in its velocity although within a much narrower range of grain sizes than occurs with most alluvium.

Wind Erosion and Transportation

Strong winds blow frequently in arid regions, whipping up loose surface materials and transporting them within turbulent air currents. Wind erodes surface materials by two main processes. The first of these is **deflation,** which is similar to the hydraulic force

Courtesy Sheila Brazier

• **FIGURE 18.17**
Badwater, in Death Valley, California. The playa surface, seen just behind the edge of the playa lake water, is the lowest elevation in North America at 86 meters (282 ft) below sea level. Badwater, the small lake, is fed by groundwater that flows down from the surrounding mountains into alluvial deposits under the surface of the basin. The water quickly evaporates in this extremely arid environment.
Why was this small lake called Badwater?

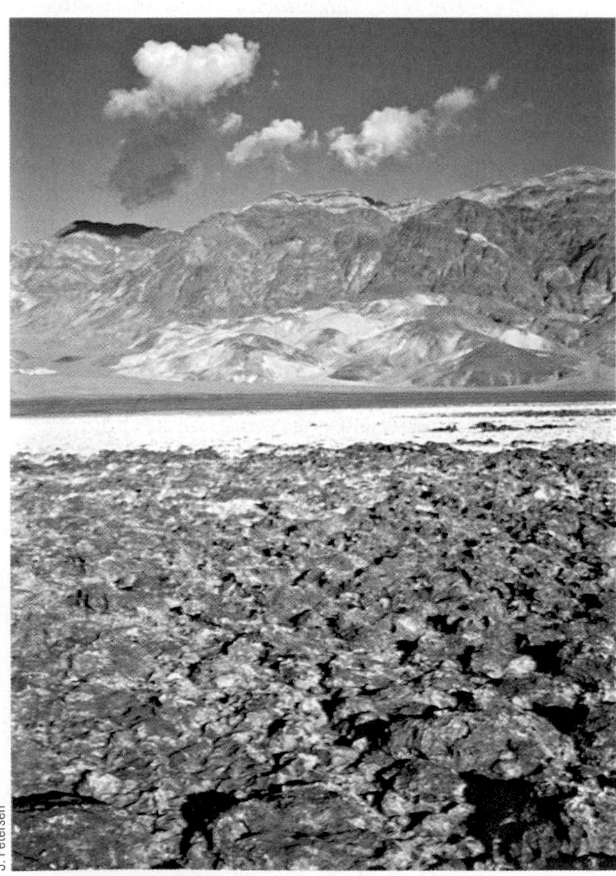
J. Petersen

• **FIGURE 18.18**
This view of Death Valley shows deposits of salt left behind by evaporation primarily of groundwater derived from the surrounding mountains. Evaporation is accumulating salts in the mud in the foreground, creating the playa microtopography known as puffy ground.

of running water. Deflation occurs when wind blowing fast enough or with enough turbulence over an area of loose sediment is able to pick up and remove small fragments of rock. The finest particles transported by winds, clays and silts, are moved in *suspension,* buoyed by vertical currents (• Fig. 18.20). Such particles essentially comprise a fine dust that will remain in suspension as long as the strength of upward air currents exceeds the tendency of the particles to settle out to the ground due to the force of gravity. The sediments carried in suspension by the wind make up its suspended load. If the wind velocity surpasses 16 kilometers (10 mi) per hour, surface sand grains can be put into motion. As with fluvial transportation, particles that are too large to be carried in eolian suspension are bounced along the ground as part of the bed load in the transportation process of *saltation.* When particles moving in eolian saltation, which are typically sand sized, bounce on the ground, they generally dislodge other particles that are then added to the wind's suspended or saltating load. Even larger sand grains too heavy to be lifted into the air are pushed forward along the ground surface, in a process called **surface creep.** Grains with low saltation trajectories become organized into small wave forms termed **ripples.** The velocity required for the wind to pick up and start to move a grain is greater than that required to keep it moving, and this is primarily due to such surface factors as roughness and cohesion.

The second way that the wind erodes is by *abrasion.* Once the wind obtains some load, the impact of those wind-driven solid particles is more effective than the wind alone in dislodging and entraining other grains or for breaking off new fragments of rock. This process is analogous to abrasion that occurs with stream flow and the other geomorphic agents, but eolian abrasion operates on a much more limited scale. Most eolian abrasion is quite literally sandblasting, and quartz sand, which is common in many desert areas, can be a very effective abrasive agent in eolian processes. Yet sand grains are typically the largest size of clast that the wind can move, and they rarely are lifted higher than 1 meter (3 ft) above the surface. Consequently, the effect of this natural sandblast is limited to a zone close to ground level.

Where loose dust particles exist on the land surface, they will be picked up and carried in suspension by strong winds. The result is a thick, dark, swiftly moving dust cloud swirling over the land. **Dust storms** (• Fig. 18.21) can be so severe that visibility drops to nearly zero and almost all sunlight is blocked. They can also be highly destructive, removing layers of surface materials and depositing them elsewhere, sometimes in thick, choking new layers, all within a matter of a few hours. The infamous Dust Bowl era of the 1930s particularly impacted the southern Great Plains of the United States in this way when devastating dust storms were brought about by years of drought and poor agricultural practices. **Sandstorms** may occur in areas where sand is abundant at the surface. As previously noted, sand grains are larger and heavier than dust particles, thus most sandstorms are confined to a low level near the surface. Evidence of the restricted

Copyright and photograph by Dr. Parvinder S. Sethi

• FIGURE 18.19
This view of sand dunes, again in Death Valley, shows how eolian processes can create a stunningly beautiful landscape.
What makes the processes that formed these dunes so different from those that form alluvial fans?

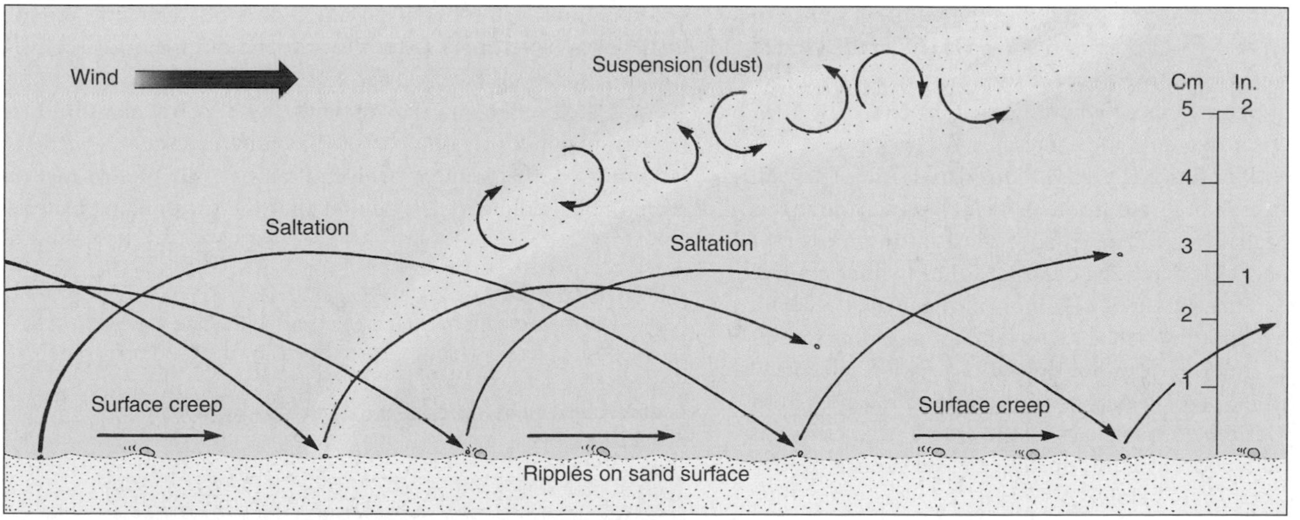

• FIGURE 18.20
Wind moves sediment in the transportation processes of suspension, saltation, and surface creep. Some of the bed load forms ripples, which can be seen moving forward when the wind is strong.
Why are grains larger than sand not generally moved by the wind?

height of desert sandstorms can be seen on vehicles that have traveled through the desert, as well as on fence posts, utility poles, and other structures. The pitting, gouging, and abrading effects of natural sandblast are more damaging to the lower portions of vehicles and other objects that were subjected to the abrasion.

Erosion by deflation can produce shallow depressions in a barren surface of unconsolidated materials. These depressions,

NOAA/George E. Marsh Album

• **FIGURE 18.21**
Dust storms occur when strong winds mobilize, erode, and suspend silt particles, picked up in areas of barren soil or alluvium. This is a dust storm in Texas during the "Dust Bowl" era of the 1930s.
Can you suggest a continent that might be a source of major dust storms today?

which can vary in diameter from a few centimeters to a few kilometers, are called **deflation hollows.** Deflation hollows are particularly common in nonmountainous arid regions. They tend to collect rainwater and may hold water for a time, depending on permeability and evaporation rates. Thus deflation hollows, like bolsons, frequently contain playas. Many thousands of deflation hollows that contain playas occur in the semiarid southern High Plains of West Texas and eastern New Mexico in the United States. Often deflation hollows form at sites that were already exhibiting a slight depression or where vegetation cover has been breached by overgrazing, fire, and other means.

Deflation has traditionally been considered one of several possible factors that help to produce **desert pavement** (**reg** in North Africa, **gibber** in Australia), a close-fitting mosaic of gravel-sized stones that overlies a deposit of mostly finer grained sediments. Desert pavement is common in many arid regions, particularly in parts of the Sahara, interior Australia, the Gobi in central Asia, and the American Southwest. If deflation selectively removes the smaller particles (clays, silts, and possibly sands) from a desert surface of mixed particle sizes, the gravel-sized clasts left behind can form a concentration of stones at the surface overlying the mixed grain sizes below (• Fig. 18.22). Sheet wash (unchannelized running water) may contribute to the formation of desert pavement by selectively eroding only the fine-grained clasts from an area of mixed grain sizes. Research has also shown that pavements can form by eolian deposition, rather than erosion, of the fine-grained sediments found beneath the stony surface layer. Regardless of its origin, desert pavement is important for the protection it affords the finer material below the surface layer of gravel. Pavement formation stabilizes desert surfaces by preventing continuous erosion. Unfortunately, off-road recreational vehicles can disrupt the surface stones and disturb this stability, thus damaging desert ecological systems.

Like deflation, eolian abrasion is also responsible for creating interesting desert landform features. Where the land surface is exposed bedrock, wind abrasion can polish, groove, or pit the rock surface and in some cases produces **ventifacts,** which are individual wind-fashioned rocks. A ventifact is a rock that has been trimmed back to a smooth slope on one or more sides by sandblast. Because of frictional effects at the surface, the ability of the wind to erode by abrasion increases with increasing distance from the ground surface, at least up to a certain height. Thus, abrasion carves the windward side of a rock into a smooth, sloping surface, or face. Ventifacts subjected to multiple sand-transporting wind directions have multiple faces, called facets, which meet along sharp edges (• Fig. 18.23). Although not extremely common, ventifacts are plentiful in local areas where wind and surface rock conditions are ideal for their formation.

Another feature often attributed to wind abrasion is the pedestaled, or balanced, rock—commonly and incorrectly thought to form where eolian abrasion attacks the base of an individual rock so that the larger top part appears balanced on a thinner pedestal below. Actually, such forms result from various physical and chemical weathering processes in the damper environment at the base of an exposed rock and are not typically related to eolian abrasion (• Fig. 18.24).

Rates of eolian erosion in arid regions often reflect the strength and kinds of the exposed surface materials. Where eolian abrasion affects rocks of varying resistance, differential erosion etches away softer rocks faster than the more resistant rocks. Even in desert locations of extensive soft rock, such as shale, or semiconsolidated sediments, like ancient lake deposits, abrasion may not act in a uniform fashion over the entire exposure. A **yardang** is a wind-sculpted remnant ridge, often of easily eroded rock or semilithified sediments, left behind after the surrounding material

• **FIGURE 18.22**
Some desert pavement may be created when the fine-grained fraction of a deposit of mixed grain sizes is removed by the wind or by sheet wash, leaving a lag of stones at the surface.
Is desert pavement a surface indestructible by human activities? Why?

J. Petersen

NASA/JPL

• **FIGURE 18.23**
This nearly true-color image of a ventifact on Mars was taken by the Mars Exploration Rover *Spirit*. With high wind speeds, plenty of loose surface materials, and lack of vegetation cover, the rocks on the surface of Mars display strong evidence of abrasion by the wind.

Michael Vanden Berg/Utah Geological Survey

• **FIGURE 18.24**
A pedestal rock in Utah's Goblin Valley State Park. Rocks with this shape are also called rock mushrooms, and they form by desert weathering and erosion processes. Abrasion by wind-blown sand can contribute to the narrowing at the base.
What other processes or rock factors could account for such an unusual shape?

has been eroded by abrasion (• Fig. 18.25), perhaps with deflation assisting in removal of fine-grained fragments. Everything else being equal, abrasion and deflation are most effective where rocks are soft or weak.

Wind Deposition

All materials transported by the wind are deposited somewhere, generally in a distinctive manner that is related to characteristics of the wind as well as the nature and grain size of the deposits. Coarser, sand-sized material is often deposited in drifts in the shape of hills, mounds, or ridges, called **sand dunes.** Fine-grained sediment, such as silt, can be transported in suspension long distances from its source area before blanketing and sometimes modifying the existing topography as a deposit called **loess** (• Fig. 18.26).

Courtesy of Marion I. Whitney

• **FIGURE 18.25**
Eolian erosion can leave behind yardangs, aerodynamically shaped ridges like this one in the Kharga Depression of Egypt.
Which is the upwind side of the yardang?

J. Petersen

• **FIGURE 18.26**
Loess, seen here near Ogden, Utah, is wind-deposited silt.

Sand Dunes To many people, the word *desert* evokes the image of endless sand dunes, blinding sandstorms, a blazing sun, mirages, and an occasional palm oasis. Although these features do exist, particularly in Arabia and North Africa, most of the areas of the world's deserts have rocky or gravelly surfaces, scrubby vegetation, and few or no sand dunes. Nevertheless, sand dunes are certainly the most spectacular features of wind deposition, whether they occur as seemingly endless dune regions, called **sand seas** (or **ergs**), as small dune fields, or as sandy ridges inland from a beach (• Fig. 18.27).

Dune topography is highly variable. For instance, dunes in the great sand seas of the Sahara and Arabia look like rolling ocean waves. Others have aerodynamic crescent forms. Eolian sand deposits can also form **sand sheets,** with no dune formation at all. Research has shown that the specific type of dune that forms depends on the amount of sand available, the strength and direction of sand-transporting winds, and the amount of vegetative cover. As wind that is carrying sand encounters surface obstacles or topographic obstructions that decrease its velocity, the sand is deposited and piles up in drifts. These sand piles also interfere with wind velocity and the sand-transporting capabilities of the wind, so the dunes grow larger until equilibrium is reached between dune size and the ability of winds to feed sand to the dune.

Sand dunes may be classified as either *active* or *stabilized* (• Fig. 18.28). Active dunes change their shape or advance downwind as a result of wind action. Dunes may change their shape with variations in wind direction and/or wind strength. Dunes travel forward as the wind erodes sand from their upwind (windward) slope, depositing it on their downwind (leeward) side. Sand entrained on the upwind slope moves by saltation and surface creep up to the dune crest and over and onto the steep leeward slope, which is the **slip face.** The slip face of a dune lies at the *angle of repose* for dry, loose sand. The angle of repose (about 35° for sand) is the steepest slope that a pile of dry, loose material can maintain without experiencing slipping or sliding down the slope.

• FIGURE 18.27
Grass has been planted on this dune ridge along the Maryland coast to help stabilize the sand.
Why are coastlines such good locations for dune formation?

D. Sack

D. Sack
(a)

D. Sack
(b)

• FIGURE 18.28
Active and stabilized dunes. (a) Sand moves freely in active dunes. The gentle back slopes face upwind, while the downwind advancing slip faces are steep. Sediments transported across the dune, some of which may have been eroded from the back slope, are deposited on the slip face. The slip face is at the angle of repose for dry sand, as shown by these active dunes. (b) If plants can establish themselves in a dune area, they create friction that reduces wind velocity and helps increase sediment cohesion with roots, decaying organic matter, and moisture. All of these factors help stabilize the dune movement. Stabilized dunes tend to have more rounded forms than active dunes.
These stabilized dunes are crossed by vehicle trails. How might these trails affect the stabilized dunes?

When wind direction and velocity are relatively constant, a dune can move forward while maintaining its general form by this downwind transfer of sediment (• Fig. 18.29). The speed at which active dunes move downwind varies greatly, but as with many processes, the movement is episodic; the dune advances only when the wind is strong enough to move sand from the upwind to the downwind side. Because of their greater height and especially their greater amount of sand, large dunes travel more slowly than smaller dunes, which may migrate up to 40 meters (130 ft) a year. During sandstorms, a dune may migrate more than 1 meter (3 ft) in a single day. Some dunes are affected by seasonal wind reversals so that they do not advance, but the crest at the top moves back and forth annually under the influence of seasonally opposing winds.

Stabilized dunes maintain their shape and position over time. Vegetative cover normally stabilizes dunes. If the vegetation cover becomes breached on a stabilized dune, perhaps due to the effects of range animals or off-road vehicles, the wind can then remove some of the sand, creating a **blowout.** In places where plants, including trees, lie in the path of an advancing dune, the sand cover may move over and smother the vegetation. Where invading dunes and blowing sands are a problem, attempts are frequently made to plant grasses or other vegetation to stabilize the dunes, halting their advance. Vegetation can stabilize a sand dune if plants can gain a foothold and send roots down to moisture deep within the dune (● Fig. 18.30). This task is difficult for most plants because the sand itself offers little in the way of nutrients and because of its high permeability and limited moisture.

One extensive area of stabilized dunes in North America is the Sand Hills of Nebraska. This region features large dunes that formed during a drier period between glacial advances in the Pleistocene Epoch. These impressive dunes are now stabilized by a cover of grasses (● Fig. 18.31). Similar stabilized dunes are found along the southern edge of the Sahara, where sand dunes that were active during drier conditions in the recent geologic past are now stabilized by vegetation. Both locations involve changes in climate that affected sand supply, wind patterns, and moisture availability (see Map Interpretation: Eolian Landforms).

● FIGURE 18.29
Active dunes move downwind. The wind transports sand from the dune's upwind side, up its back slope toward the sharp dune crest. The sand then slides down the steeper slip face on the downwind (leeward) side of the dune, causing the dune to advance, or migrate. Sand supply, dune size, wind speed, and duration of the wind are major factors controlling the speed of dune advance.
Why does the inside of the migrating dune consist of former slip faces?

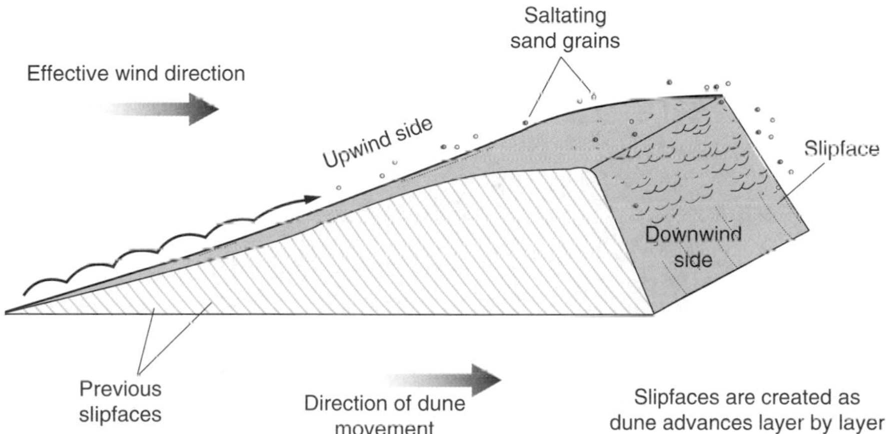

Types of Sand Dunes

Sand dunes are classified according to their shape and their relationship to the wind direction. The different types are also related to the amount of available sand, which affects not only the size but also the shape of sand dunes.

Barchans are one kind of crescent-shaped dune (● Fig. 18.32a). The two arms of the crescent, called the dune's horns, point downwind (● Fig. 18.33). The main body of the crescent lies on the upwind side of the dune. From the desert floor at its upwind edge, the dune rises as a gentle slope up which the sand moves until it reaches the highest point, or crest, of the dune and, just beyond that, the slip face at the angle of repose. The slip face is oriented perpendicular to the barchan's arms. The arms extend downwind beyond the location of the slip face. Barchans form in areas of minimal sand supply where winds are strong enough to move

● FIGURE 18.30
A sand dune advances into a vegetated area on the Oregon coast. In coastal dune regions, landscape change is common as dunes move inland and are subsequently invaded by plants.
Explain how plants can stabilize dunes.

D. Sack

● FIGURE 18.31
The rolling grazing lands of the Sand Hills in central Nebraska were once a major region of active sand dunes.
Why are these dunes no longer active today?

© David B. Loope/ University of Nebraska

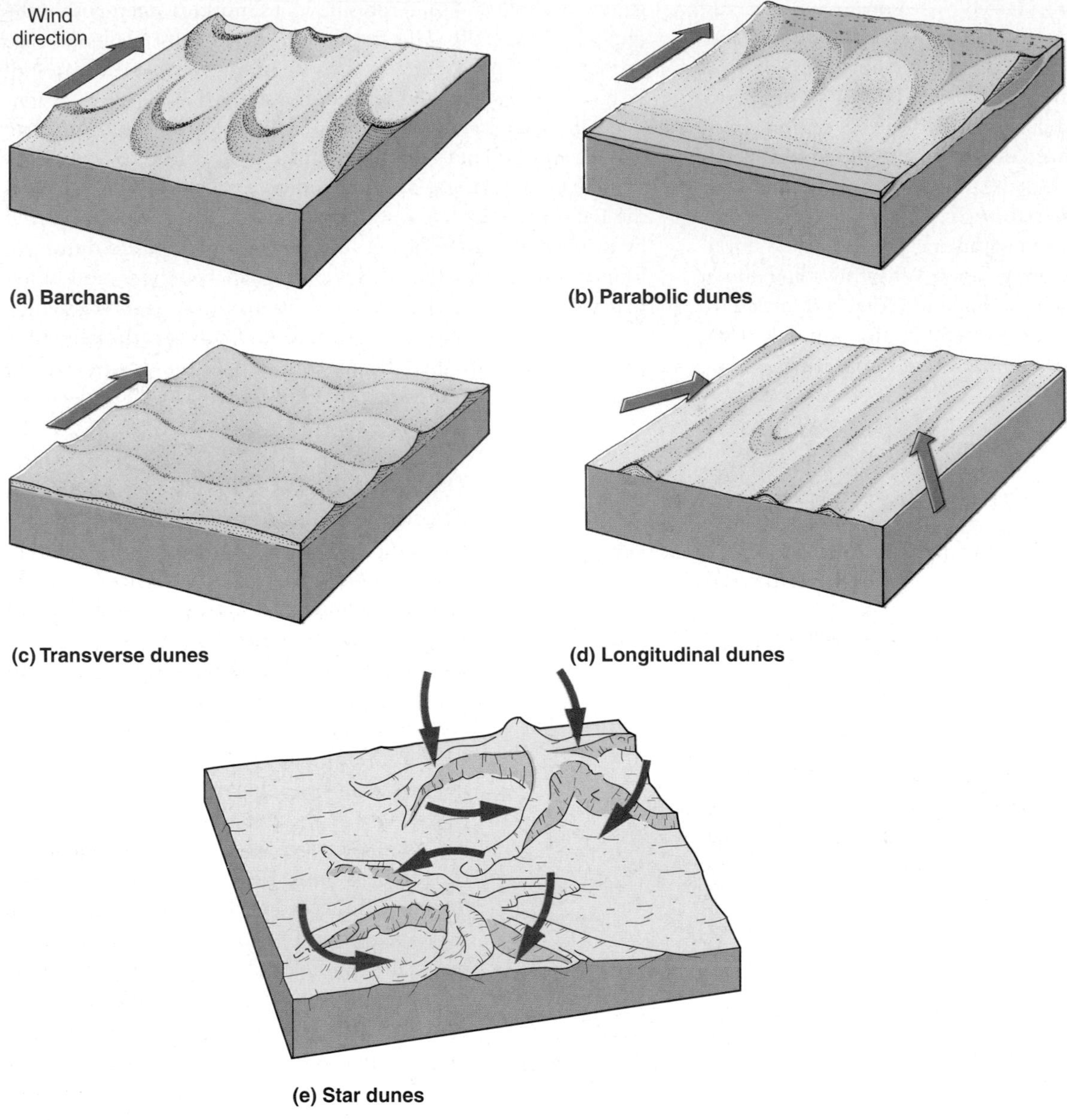

• FIGURE 18.32
Five principal sand dune types are: (a) barchan, (b) parabolic, (c) transverse, (d) longitudinal, and (e) star. The wind direction is shown in all figures, as indicated by arrows.
What factors play a role in which type of dune will be found in a region?

sand downwind in a single prevailing direction. They may be most common in smaller desert basins surrounded by highlands where they tend to form near the downslope, sandy edge (toe) of alluvial fans, and adjacent to small playas. Although they may form as isolated dunes, barchans often appear in small groups, called barchan fields.

Parabolic dunes are similar to barchans in that they are also crescent-shaped dunes, but their orientation is reversed from that of barchans (Fig. 18.32b). Here, the arms of the crescent tend to be stabilized by vegetation, long, and pointing upwind, trailing behind the unvegetated main body and crest of the dune, rather than extending downwind from it. The main body of a parabolic dune points downwind, and the slip face along its downwind edge has a convex shape when viewed from above. Parabolic dunes commonly occur just inland of beaches and along the margin of active dune areas in deserts.

Transverse dunes are created where sand-transporting winds blow from a constant direction and the supply of sand is abundant (Fig. 18.32c). The upwind slope of a transverse dune ridge is gentle like that of barchans, while the steeper downwind slip face is at the angle of repose. In the downwind direction, transverse dunes form ridge after ridge separated from each other by low swales in a repeating wavelike fashion. Each dune ridge is laterally extensive perpendicular to the sand-transporting wind direction. The dune ridges, slip faces, and interdune swales trend perpendicular to the direction of prevailing winds, hence the name *transverse*. Abundant sand supply may derive from such sources as widespread exposure of easily eroded sandstone bedrock, sandy

D. Sack

• **FIGURE 18.33**
A small barchan in southern Utah.
On this photo, which side of this dune faces the wind, right or left?

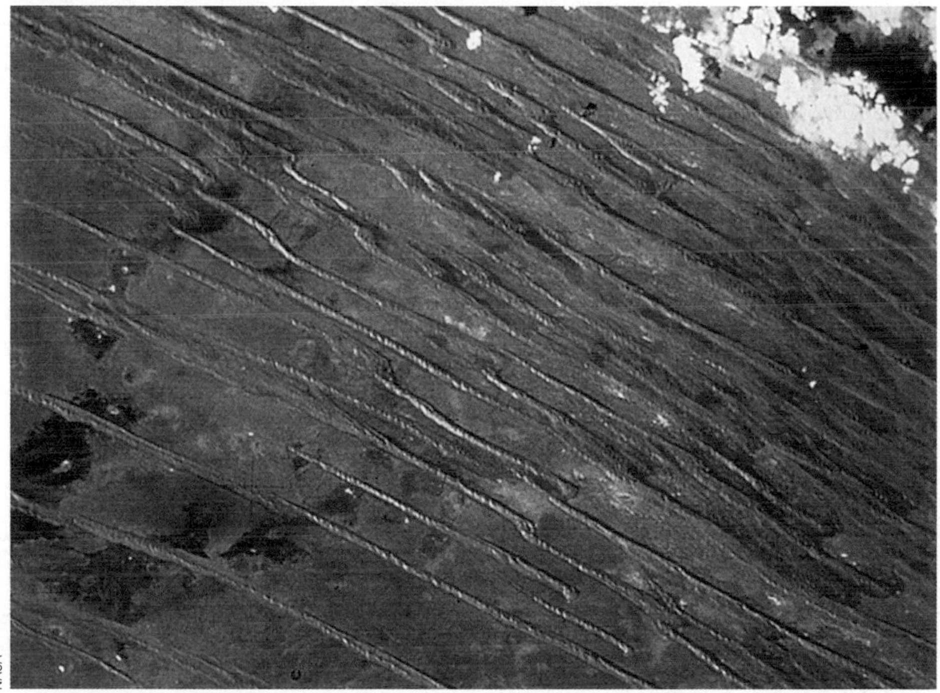

NASA

• **FIGURE 18.34**
A satellite view of longitudinal dunes in the Sahara. The width of this image represents approximately 160 kilometers (100 mi) of terrain, from left to right.
Estimate the ground length of the dunes in this satellite image.

alluvium deposited by exotic streams or during wetter climates in the Pleistocene Epoch, or from sandy deltas and beaches left in the landscape after the desiccation of ancient lakes.

Longitudinal dunes are long dunes aligned parallel to the average wind direction (Fig. 18.32d). There is no consistent distinction between the back slopes and slip faces of these dunes, and their summits may be either rounded or sharp. Strong winds are important to the formation of most longitudinal dunes, which do not migrate but instead elongate in the downwind direction. Small longitudinal dunes, such as those found in North America, may simply represent the long trailing ridges of breached parabolic dunes or a sand streak extending from a somewhat isolated source of sand. Much higher and considerably longer than these are the impressive longitudinal dunes that cross vast areas of the flatter, more open desert topography of North Africa, the Kalahari, the Arabian peninsula, and interior Australia (• Fig. 18.34). These dunes develop under bidirectional wind regimes, where the two major sand-transporting wind directions come from the same quadrant, such as a northwesterly and southwesterly wind. A type of large longitudinal dune called a **seif** (pronounced *safe;* from the Arabic for "sword") is found in the deserts of Arabia and North Africa. Seif dunes are huge, sinuous rather than linear, sharp-crested dunes, sometimes hundreds of kilometers long, whose troughs are almost free of sand. They may reach 180 meters (600 ft) in height.

Star dunes are large, widely spaced, pyramid-shaped dunes in which ridges of sand radiate out from a peaklike center to resemble a star in map view (Fig. 18.32e). These dunes are most common in areas where there is a great quantity of sand, changing wind directions, and an extremely hot and dry climate. Star dunes are stationary, but ridges and slip faces shift orientation with wind variations.

Dune Protection To many who visit the desert or beaches, dunes are one of nature's most beautiful landforms. However, dune areas are also very attractive sites for recreation, and they are particularly inviting for drivers of off-road vehicles. Although dunes appear to be indestructible and rapidly changing environments that do not damage easily, this is far from the truth. Dunes are fragile environments with easily impacted ecologies. Because dune regions are the result of an environmental balance between moving dunes and the plants trying to stabilize them, the environmental equilibrium is easily upset. Many of the most spectacular dune areas in the United States have special protection in national parks, national monuments, or national seashores, such as White Sands, New Mexico; Great Sand Dunes, Colorado;

Indiana Dunes, Indiana; and Cape Cod, Massachusetts. Many dune areas, however, do not have special protection, and environmental degradation is a constant threat.

There are many practical reasons for dune preservation. In coastal zones, dunes play an important role in coastal protection and are sometimes the last defense of coastal communities from storm waves (• Fig. 18.35). They are particularly important along the low-lying Gulf and Atlantic coasts of the United States where occasional hurricanes or "nor'easters" batter coastlines and erode beaches in front of the dunes. In nations such as the Netherlands, coastal dunes are extremely important because the land behind them is below sea level; thus a breach through the dunes could mean disaster. Coastal dune regions also are critical wildlife habitats, especially for many bird species.

Loess Deposits The wind can carry in suspension dust-sized particles of clay and silt, removed by deflation, for hundreds or thousands of kilometers before depositing them. Eventually these particles settle out to form a tan or grayish blanket of loess that may cover or bury the existing topography over widespread areas. These deposits vary in thickness from a few centimeters or less to more than 100 meters (330 ft). In northern China downwind from the Gobi Desert, the loess is 30 to 90 meters (100–300 ft) thick (• Fig. 18.36). Loess may originate from deserts, other sparsely vegetated surfaces, or river floodplains. The widespread loess deposits of the American Midwest and Europe were derived from extensive glacial and meltwater deposits of retreating glaciers during ice ages of the recent geologic past. As winds blew across the barren glaciated regions or glacial meltwater areas, they picked up a large load of fine sediment and deposited it as loess in downwind regions.

Certain interesting characteristics of loess affect the landscape where it forms the surface material. For example, though fine and dusty to the touch, because of its high cohesion, particularly when damp, loess maintains vertical walls when cut through naturally by a stream or artificially by a road. Sometimes slumping will occur on these steep faces. Slumping gives a steplike profile to many loess bluffs. In addition, loess is easily eroded by either the wind or running water because of its fine texture and unconsolidated character. As a result, loess-covered regions that are unprotected by vegetation often become gullied. Where loess covers hills, gully erosion and slumping are conspicuous processes. A particularly severe erosion problem is responsible for the recent collapse of high loess bluffs along the Mississippi River at Vicksburg, Mississippi (• Fig. 18.37).

• FIGURE 18.35
A sign directs visitors on how to help protect this dune area along the South Carolina coast.
Why should some dune areas be protected from human activities such as driving dune buggies and other recreational vehicles?

• FIGURE 18.36
A steep gully eroded into the thick loess deposits of northern China.
Where is the origin of these loess deposits?

GEOGRAPHY'S ENVIRONMENTAL SCIENCE PERSPECTIVE

Off-Road Vehicle Impacts on Desert Landscapes

Whether you prefer to call them off-road vehicles (ORVs) or all-terrain vehicles (ATVs), driving motorized vehicles of any kind over the desert landscape off established roadways has tremendous negative impacts on desert flora and fauna, the habitat of those organisms, the stability of landforms and landforming processes, and the aesthetic beauty of the natural desert environment. Deserts are particularly fragile environments primarily because of the low amount and high variability of precipitation. The climatic regime leads to slow rates of weathering, slow rates of soil formation, low density of vegetation, unusual species of plants and animals specially adapted to the hostile conditions of severe moisture stress, and rapid generation of surface runoff when precipitation does occur.

ORVs damage desert biota in several ways. ORVs kill and injure plants, animals (for example, birds, badgers, foxes, snakes, lizards, and tortoises, to name just a few), and insects when they hit or ride over top of them. Even careful ORV riders who drive slowly cannot avoid crushing small animals and insects that lie hidden under a loose cover of sand, or shallow roots that extend out far from a plant. Driving ORVs at night is particularly deadly for desert animals, which tend to be nocturnal. ORVs do not have to hit or ride over organisms to cause populations to decrease. Hearing loss experienced by animals in areas frequented by ORVs puts them at a major disadvantage for feeding, defense, and mating. ORVs crush and destroy burrows and other animal and insect homes and nesting sites, including plants. Oil and gas leaking from poorly maintained vehicles are another danger for desert dwellers, as are grass and range fires inadvertently started from such vehicles. No desert subenvironment, not even sand dunes, are immune from these negative impacts.

In addition to the direct impacts on organisms, physical properties of the landscape are also negatively affected by ORVs. One major problem is that ORVs compact the desert surface sediments and soil. Compaction causes a decrease in permeability, which greatly restricts the ability of water to infiltrate into the subsurface. More surface runoff translates to greater erosion of the already thin desert soils. In addition, even long after ORV use ceases in a disturbed area, soil compaction makes it very difficult for desert plants, animals, and insects to become reestablished there. Denuded, eroded swaths made by ORVs often remain visible as ugly scars in the landscape for decades, and act as sources of sediment, adding to the severity of dust storms. Where ORV trails traverse steep slopes, they have contributed to the occurrence of debris flows, mudflows, and other forms of mass wasting.

The attributes of desert environments that give them such stark beauty also make them very vulnerable to disturbance. With limited, but often torrential, precipitation, sparse and slow-growing vegetation, and slow rates of weathering and soil formation, deserts require long periods of time to recover from environmental disturbance. Recovery, moreover, will likely not return the landscape to the condition it would have been in if the disturbance had not occurred.

D. Sack

Motorcycles ridden on this desert hillslope in Utah have stripped the vegetation, compacted the soil, instigated accelerated erosion, and seriously marred the aesthetic beauty of the landscape, which is referred to as the viewshed.

D. Sack

Riding ORVs on active sand dunes harms the sensitive organisms that live there and interferes with the natural dune processes by compacting the sand.

© Vicksburg Convention and Visitors Bureau

• **FIGURE 18.37**
The steep and unstable loess bluffs of Vicksburg, Mississippi.
Why might the instability of loess cliffs be a problem?

Because of its high calcium carbonate content and unleached characteristics, loess is the parent material for many of Earth's most fertile agricultural soils. Extensive loess deposits are found in northern China, the Pampas of Argentina, the North European Plain, Ukraine, and Kazakhstan. In the United States, the midwestern plains, the Mississippi Valley, and the Palouse region of eastern Washington are underlain by rich loess soils (• Fig. 18.38). All these areas are extremely productive grain-farming regions.

Landscape Development in Deserts

Geomorphic landscape development in arid climates is comparable in many ways to that in humid climates, but not in all ways. Weathering and mass movement processes operate, and fluvial processes predominate, in both environments but at different rates and with the tendency to produce some different landform and landscape features in the two contrasting environments. Arid environments experience the added regionally or locally important effect of eolian processes, which are not very common in humid settings beyond the coastal zone. The major differences in the results of geomorphic work in arid climates, as compared to humid climates, are caused by the great expanses of exposed bedrock, a lack of continuous water flow, and a more active role of the wind in arid regions.

Some desert landscapes, such as most of those in the American Southwest, are found in regions of considerable topographic relief, whereas others, like much of the Sahara and interior Australia, occupy huge expanses of open terrain with few mountains. An excellent example of a typical desert landscape in a region of considerable structural relief is found in

• **FIGURE 18.38**
Map of the major loess regions of the world (shown in yellow). Most loess deposits are peripheral to deserts and recently glaciated regions.

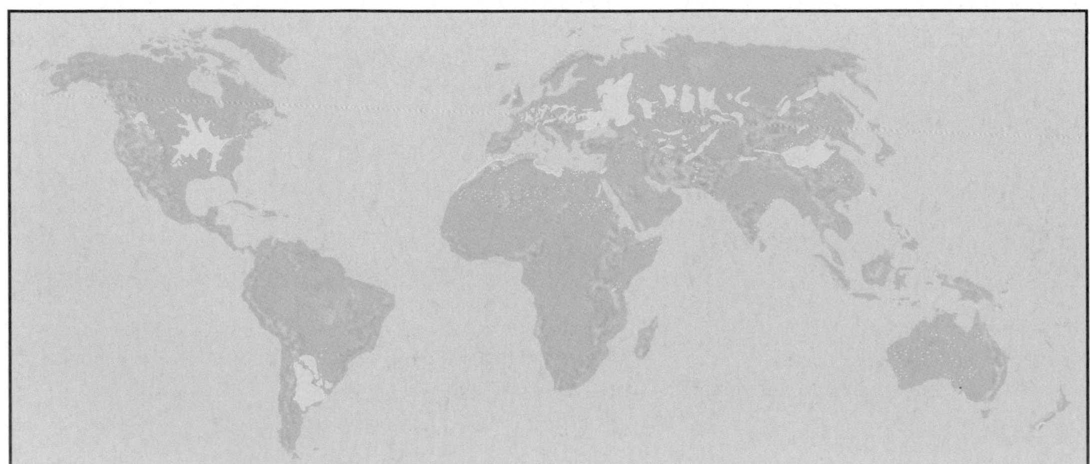

the Basin and Range region of western North America. The region extends from west Texas and northwestern Mexico to eastern Oregon. It includes all of Nevada and large portions of New Mexico, Arizona, Utah, and eastern California. Here, more than 200 mountain ranges, with basins between them, dominate the topography. The Great Basin—a large subregion characterized by interior drainage, numerous alternating mountain ranges and basins, active tectonism, and centered on Nevada—occupies much of the central and northern part of the Basin and Range.

Fault-block mountains in the Basin and Range region rise thousands of meters above the desert basins, and many form continuous ranges (• Fig. 18.39). These high ranges, such as the Guadalupe Mountains, Sandia Mountains, Warner Mountains, and the Panamint Range, to name just a few, encourage orographic rainfall. Fluvial erosion dissects the mountain blocks to carve canyons between peaks and cut washes between interfluves. Where active tectonism continues so that the uplift of mountain ranges matches or exceeds the rate of their erosion, as in the Great Basin, fluvial deposition constructs alluvial fans extending from canyon mouths outward toward the basin floor. In many basins, the alluvial fans have coalesced to create a bajada. In these tectonically active basins with interior drainage, playas often occupy the lowest part of the basin, beyond the toe of the alluvial fans (• Fig. 18.40a). Although the mountain ranges create considerable roughness, the wind can remove sandy alluvium from the toe areas of alluvial fans, sand-sized aggregates of playa sediments,

• FIGURE 18.39
A false-color satellite image of Death Valley, California, which is a desert bolson in the Basin and Range region. Death Valley lies between the Amargosa Range to the northeast and the Panamint Range to the southwest. Alluvial fans, which constitute much of the red areas of this image, dominate the edges of the valley.
What do you think the white areas are in the center of the valley?

Courtesy NASA GSFC, MIT, ERSDAC, JAROS, and U.S./Japan ASTER Science Team

• FIGURE 18.40
(a) Alluvial fans and playas tend to occur in mountain and basin deserts of active tectonism, which typically have interior drainage. (b) Pediments and inselbergs are common in desert areas that have been tectonically stable since a distant period of mountain formation. Development of exterior drainage helps transport sediment eroded from the mountains out of the basin.

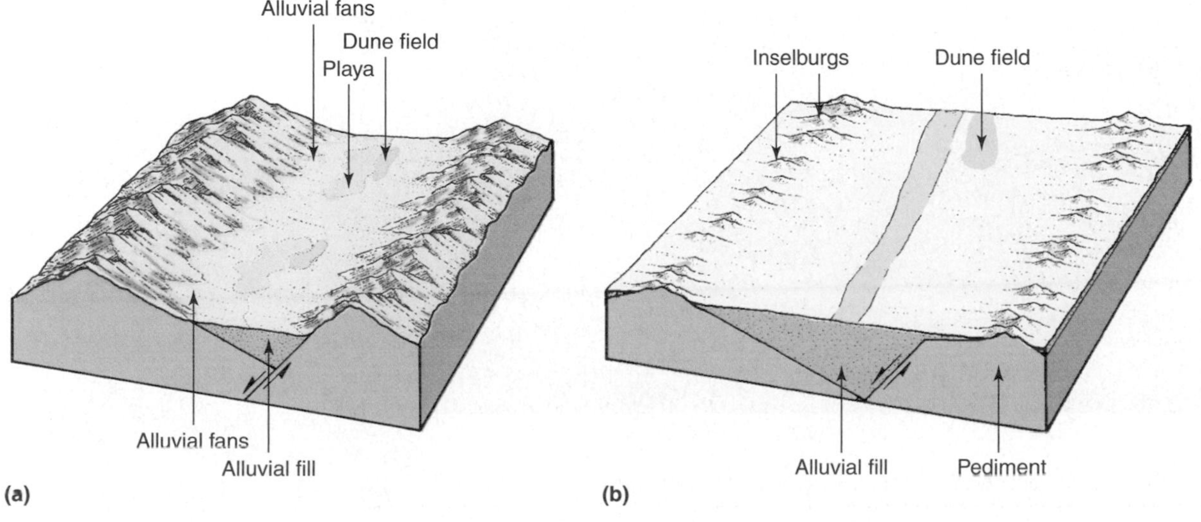

and, in some cases, sandy sediment from beaches left behind by ancient perennial lakes. These sediments may be transported relatively short distances before being deposited in local dune fields.

Pediments lie along the base of some mountains, particularly in the tectonically less active areas south of the Great Basin, where much of the terrain is open to exterior drainage. In these areas, mountains are being lowered and the pediments extended. Resistant inselbergs remain on some of the pediments. Sandy alluvium along streams and sandy deposits left by ancient rivers and deltas provide sediment to be reworked by the wind into occasional dune fields. The landscapes of the Mojave Desert in California and parts of the Sonoran Desert in Arizona have localities where uplift along faults has been inactive long enough for the landscape to be dominated by extensive desert plains interrupted by a few isolated inselbergs as reminders of earlier, tectonically active, mountainous landscapes (Fig. 18.40b).

The geologic structures and geomorphic processes found in desert areas are, for the most part, the same as those found in humid regions. It is variations in the effects and rates of these processes that make the desert landscape distinctive. Although fault-block mountains and fault-block basins dominate the

© David Ball

• **FIGURE 18.41**
Uluru (Ayers Rock) is a striking sandstone inselberg, an erosional remnant, rising above the arid, flat interior of the Australian Outback. Kata Tjuta (The Olgas), in the background, formed as another erosional remnant divided into many bedrock hills by the widening of joints.
Explain how inselbergs form.

National Park Service - Canyonlands

• **FIGURE 18.42**
This view of the Green River Overlook in Canyonlands National Park, Utah, illustrates the dramatic beauty of desert landscapes.
What aspects of this environment make this an attractive landscape?

geologic structure of immense regions of the American West and other arid locations around the world, it is important to know that desert landscapes are as varied as those of other climatic environments. Deserts exist at localities where the landscape has developed in nearly every imaginable geologic setting, including volcanoes, ash deposits and lava flows, folded rocks forming ridges and valleys, horizontal strata, and exposures of massive intrusive rocks. Where arid climates occur in expansive regions of largely open, low-relief terrain, as in the ancient and geologically stable deserts of Australia and the Sahara, common landforms include inselbergs surrounded by extensive desert plains (● Fig. 18.41), deflation hollows, playas, washes, the channels of ancient streams, and the beds of shallow, ancient lakes. With limited terrain roughness, areas of large longitudinal dunes can develop in these settings.

Arid landscapes and eolian landscapes in any climatic environment can be beautiful and stark, but those who are unfamiliar with these environments too often misunderstand them. Deserts are not wastelands, and dunes are not merely piles of sand. It is true that most desert and eolian areas have little to offer in terms of directly exploitable economic value, but their unique characteristics and scenic beauty fully qualify them for preservation and protection. The austere, angular character of their landforms, the fragility of their environments, the special nature of their biota, and the opportunities they provide for learning about how certain Earth systems operate continue to attract those who seek to understand and experience arid and eolian environments (● Fig. 18.42). One only needs to count the number of national parks, national monuments, and other scenic attractions in the arid southwestern United States and in regions of sand dunes to find ample support for their survival. Deserts and dune localities are places with a beauty all their own and are areas worthy of appreciation and appropriate environmental protection.

Chapter 18 Activities

Define & Recall

regional base level
exotic stream
wash (arroyo, barranca, wadi)
badlands
plateau
caprock
mesa
butte
pediment
inselberg
alluvial fan
fan apex
debris flow fan
bajada
piedmont alluvial plain
bolson
playa
playa lake
clay pan
salt-crust playa (salt flat, salina)
eolian (aeolian)
deflation
surface creep
ripples
dust storm
sandstorm
deflation hollow
desert pavement (reg, gibber)
ventifact
yardang
sand dune
loess
sand sea (erg)
sand sheet
slip face
blowout
barchan
parabolic dune
transverse dune
longitudinal dune
seif
star dune

Discuss & Review

1. What are eolian landforms? What geomorphic agent most significantly affects the desert landscape?
2. How do climate and vegetation affect landforms in the desert?
3. How do basins of interior drainage differ from basins of exterior drainage?
4. How does an exotic stream differ from an ephemeral stream? Give three examples of exotic streams.
5. How does the formation of a mesa or butte differ from the formation of a pediment?
6. Describe the main characteristics of an alluvial fan.
7. Why do temporary lakes form in bolsons? How are such lakes related to playas?
8. What are the major differences between deflation and eolian abrasion?

9. Why are most sandstorms confined to a low height? How do they differ from dust storms?
10. Describe how a ventifact is formed.
11. What is the difference between a barchan and a transverse dune?
12. What is loess? Where are some major regions in the world where loess deposits are located? What is an important economic activity related to loess in many regions?

Consider & Respond

1. How are eolian erosion and transportation processes similar to fluvial erosion and transportation processes, and how do they differ? What are the main reasons for these similarities and differences? Is eolian deposition similar to fluvial deposition? If so, what are some of the similarities?
2. What physical geographic factors contribute to create so many basins of interior drainage in arid regions? What are some ways in which base-level changes can occur in basins of interior drainage?
3. How does stream action in arid regions differ from stream action in humid regions? What are the most common erosional and depositional landforms associated with stream processes in arid regions?
4. Name several types of arid region landforms commonly found in tectonically active desert regions, such as the American Great Basin. Name several types of arid region landforms commonly found in deserts, like interior Australia, that have been tectonically stable for a very long time.
5. Deserts are considered fragile environments. What are some ways in which desert landforms fit or contribute to this designation?
6. Why are there so many national parks and monuments in the arid parts of the American Southwest? What landform characteristics and other factors of desert landscapes draw people to these locations?

Apply & Learn

1. At the fan apex, stream flow issuing from the mouth of a canyon onto an alluvial fan has a discharge of 12 cubic meters per second, a flow depth of 2 meters, and flow width of 3 meters. What is the velocity of flow at the fan apex? If the velocity of flow is decreasing downfan by 0.1 meter per second, at what distance from the apex will it reach a velocity of zero?
2. A barchan in southeastern California has migrated the following amounts in recent years: 9 meters in 1995, 12 meters in 1996, 14 meters in 1997, 15 meters in 1998, 15 meters in 1999, 19 meters in 2000, 22 meters in 2001, 23 meters in 2002, 26 meters in 2003, 28 meters in 2004, 29 meters in 2005, 29 meters in 2006, 30 meters in 2007. Over this 13-year period, what was the average annual migration rate of this dune? What was the total distance that the dune migrated in the whole 13-year period? How would you describe the overall trend in dune migration for this period? What factor or factors might be responsible for the observed trend?

Locate & Explore

Note: Please read the About Locate & Explore Activities section of the Preface before beginning these exercises.

1. Using Google Earth, identify the landforms at the following locations (latitude, longitude) and provide a brief discussion of how the landform developed. Include a brief discussion of why the landform is found in that general area of the United States or the world.

   Tip: Use the zoom, tilt, rotate, and elevation exaggeration tools to help view and interpret the landform and the area in which it is found.

   a. 36.415°N, 116.810°W
   b. 37.66°N, 117.625°W
   c. 36.25°N, 116.82°W
   d. 23.42°S, 14.77°E
   e. 36.96°N, 110.11°W
   f. 25.35°S, 131.03°E
   g. 28.07°N, 114.06°W
   h. 36.25°N, 116.82°W
   i. 36.32°N, 116.94°W
   j. 32.77°N, 106.21°W

2. Following are latitude and longitude points for the center of major world deserts. Using Google Earth, provide an explanation of why these deserts are located in these areas and why. Make sure that you consider the geology of the area (location of the mountains), continentality, and the location of the desert relative to the global circulation patterns shown in Figure 5.12 of your text.

   a. Sahara Desert (22.4°N, 12.6°E)
   b. Atacama Desert (23.7°S, 68.8°W)
   c. Simpson Desert (25.3°S, 133.8°E)
   d. Mojave Desert (35.7°N, 115.1°W)
   e. Gobi Desert (38.9°N, 82.6°E)
   f. Namib Desert (24.3°S, 15.1°E)

Map Interpretation

DESERT BASIN LANDFORMS

The Map

The map shows a section of Death Valley National Park, California. This is part of the Basin and Range region, characterized by fault-block mountains (ranges) separated by down-faulted valleys (basins). As the rugged ranges erode, the basins fill with sediment carried by infrequent flash floods.

The mountain block that forms the western slope of Death Valley (in the map area) is the Panamint Range. The highest summit in that range, Telescope Peak, reaches an elevation of 3368 meters (11,049 ft). The Amargosa Range forms the eastern boundary. Death Valley's lowest elevation is 286 meters below mean sea level (2282 ft).

The present climate of the Basin and Range is arid, except for high-elevation mountains that receive more precipitation. With an average annual rainfall of 3.5 centimeters (1.7 in.) and a potential evapotranspiration that may exceed 380 centimeters (150 in.), Death Valley has the most extreme aridity in the region. Summer temperatures commonly exceed 40°C (104°F), and the record maximum is 57°C (134°F). In winter, much of the Basin and Range region has freezing nighttime temperatures.

During the Pleistocene, deep lakes filled numerous basins in the region. Death Valley was occupied by Lake Manly, and traces of this 183-meter (550-ft) deep lake's shoreline can be seen from the valley floor. This is a classic site for studying desert landforms.

Interpreting the Map

Note: In answering these questions it may be helpful to refer to both the figures and text on pages 510–514.

1. Based on the general location of Death Valley, why is the area so arid?
2. Describe the general topography of the map area.
3. What is the lowest elevation on the map? What is significant about the elevation of Death Valley?
4. What specific type of arid landform is the blue-striped feature located in the depression? Why are the edges of this feature shown with blue-dashed lines?
5. What types of surface materials compose the feature in Question 4? Explain how this feature was formed.
6. What specific type of landform is indicated on the map by the curved contours at the base of the mountains? What do these blue lines —— . . . —— . . . —— crossing the contours represent?
7. What types of surface materials compose the feature in Question 6? Explain how this feature was formed.
8. Note that the large, curved landforms at the mountain base coalesce. What is the specific landform name for such broad alluvial features?
9. What evidence from the map indicates that this is an interior drainage basin?
10. Sketch a general east–west profile from the benchmark (elevation –270 ft) at Devil's Speedway to the 2389-foot benchmark on the mountain straight to the west. Label the following landforms: mountain front, mountain peak, alluvial fan, and basin.

NASA/JPL/CACR

A digital terrain model shows the topography of Death Valley. Figure 18.39 presents an alternate view with a topographic profile (black foreground) at about 3× vertical exaggeration.

Opposite:
Furnace Creek, California
Scale 1:62,500
Contour interval = 80 ft
U.S. Geological Survey

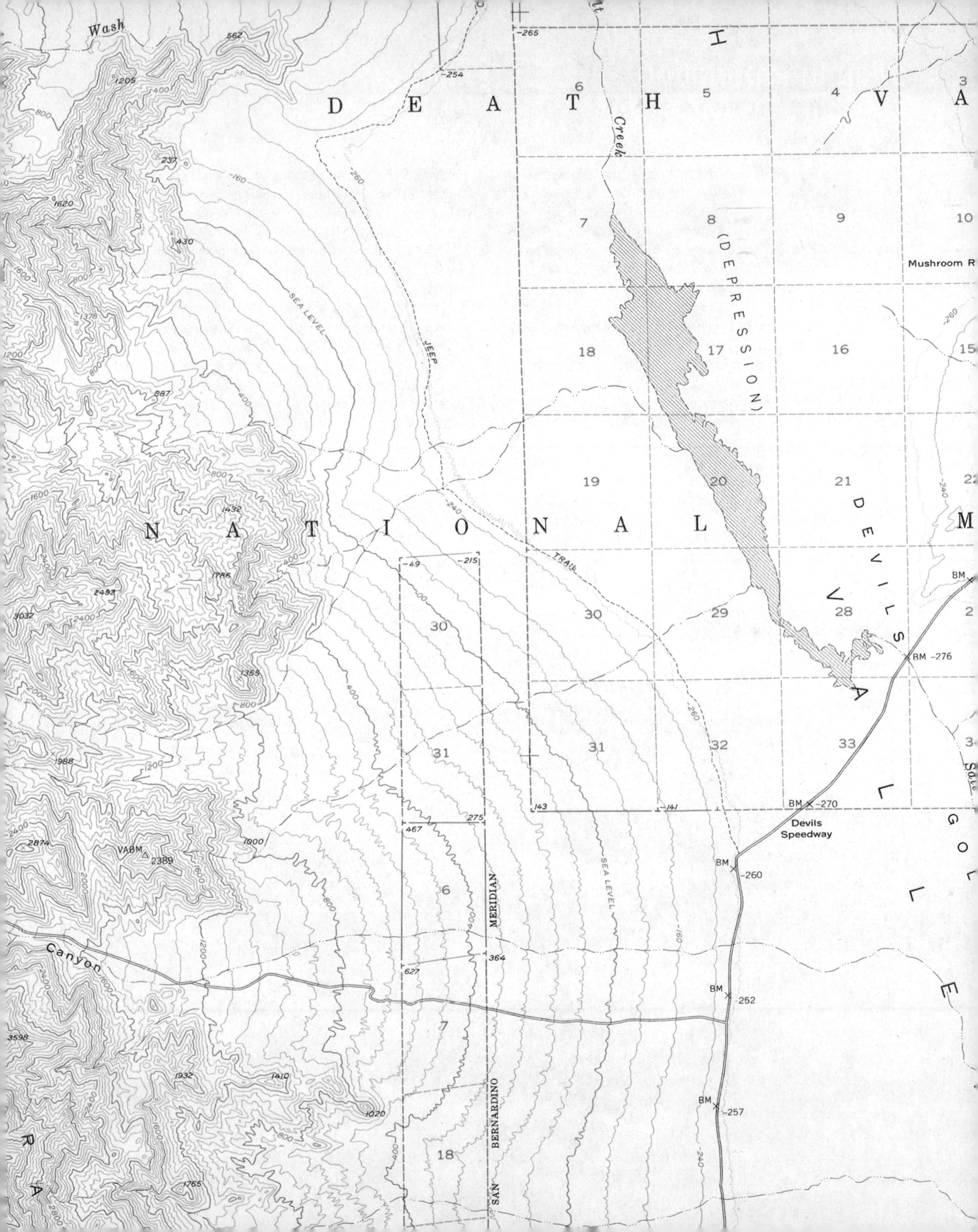

D E A T H
V A
N A T I O N A L
(DEPRESSION)
DEVILS
GOLF
Mushroom R
Devils
Speedway
Wash
Creek
JEEP
TRAIL
SEA LEVEL
MERIDIAN
SAN BERNARDINO
Canyon
VABM
2389
BM -276
BM -270
BM -260
BM 252
BM -257

Map Interpretation

EOLIAN LANDFORMS

The Map

Eolian processes formed the Sand Hills region, the largest expanse of sand in North America. The region covers over 52,000 square kilometers (20,000 sq mi) of central and western Nebraska.

The Sand Hills region was part of an extensive North American desert some 5000 years ago. The sand dunes here reached more than 120 meters (400 ft) high and inundated postglacial peat bogs and rivers. As the climate became wetter, vegetation growth invaded the dunes, greatly reducing eolian erosion and transportation. The vegetation anchored the sand, and the stabilized dunes developed a more rounded form. Underlying the Sand Hills is the Ogallala aquifer. The high water table of the aquifer supports the many lakes nestled between the dunes.

The Sand Hills area has a middle-latitude steppe climate (*BSk*) and receives about 50 centimeters (20 in.) of precipitation annually. Temperatures have a great annual range, from freezing winters to very hot summers. During summer months, the region is often pelted by thunderstorms and hail; during the winter months, the area is subject to blinding blizzards.

Vegetation is mainly bunch grasses that can survive on the dry, sandy, and hilly slopes. Some species of bunch grasses have extensive root systems that may extend more than 1 meter (3 ft) into the sandy soil. The lakes and marshes in the interdunal valleys support a marsh plant community that in turn supports thousands of migratory and local birds. Currently, the main land use in the region is cattle grazing. Some scientists are predicting that the Sand Hills area will lose its protective grass cover if global warming continues and will revert to an active area of migrating desert sand dunes.

Interpreting the Map

1. What is the approximate relief between the dune crests and the interdunal valleys?
2. What is the general linear direction of the dunes and valleys?
3. To which direction do the steepest sides of the dunes generally face?
4. If the slip face is the steepest slope of the dunes, what was the prevailing wind direction when the dunes were active?
5. Based on your answers to the previous three questions, determine what type of sand dune formed the Sand Hills.
6. What is the general direction of groundwater flow in the aquifer beneath the Sand Hills? (*Note:* Use the elevation of the lakes to determine the water table elevation.)
7. Sketch a north–south profile across the middle of the map from School No. 94 to the eastern end of School Section Lake (in section 16). Label the following landform features: dune crests, dune slip faces, interdunal valleys, and lakes.
8. What cultural features on the map indicate the dominant land use for the region?

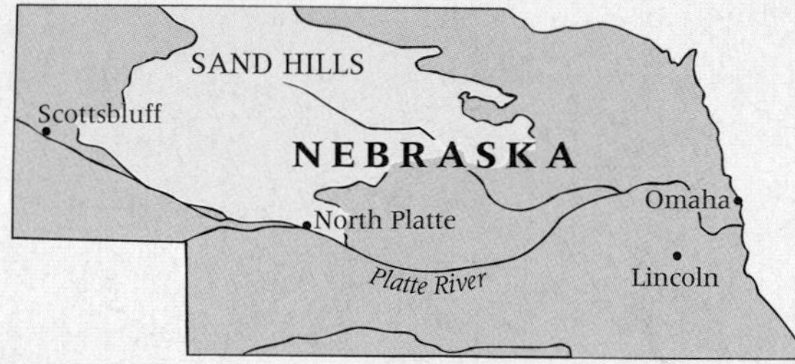

A location map of the Nebraska Sand Hills, which cover almost one third of the state.

Opposite:
Steverson Lake, Nebraska
Scale 1:62,500
Contour interval = 20 ft
U.S. Geological Survey

Home Valley Lake
3568
Pelican Lake
3546
Round Lake
3492
Carver
Cottonwood Lake
3602
Flowing Well
Windmill
Lake
3602
F Marshall Ranch
Silver Lake
3527
B U
Piester Lake
School No 94
BM
Spring Lake
3592
Downing Ranch
Rat Lake
3558
Marshall & Son Ranch
Sweeney Ranch
W Marshall Ranch
SURVEY
SURVEY VALLEY
VALL
Cem
Windmill
L Hoyt Ranch
H Hoyt Ranch
School Section Lake
F Henderson Ranch
Henderson Ranch
L A C K E Y
Windmill

Glacial Systems and Landforms

19

CHAPTER PREVIEW

A glacier is an excellent example of an open system.

- What materials enter and leave the glacial system?
- How does energy move through the system?

The two principal categories of glaciers are similar in many respects, yet they leave their own distinct impacts on the landscape and originate in different environmental settings.

- In what environmental setting does each category of glacier originate?
- Why does their impact on the landscape differ?

The ice budget of a glacier is determined by the annual accumulation of frozen water and the annual loss of frozen water from the glacier.

- What happens to the glacier when annual addition exceeds annual loss of frozen water?
- What happens when dynamic equilibrium exists in a glacier?

A glacier flows only after a sufficient thickness of ice has accumulated.

- Why must the frozen mass be deep before it can flow?
- How does the thickness of flowing ice affect glacial landforms?

The most visible effects of continental glaciation in North America are from the Wisconsinan, the last stage of Pleistocene glaciation.

- How did Pleistocene glaciation change the physical appearance of North America?
- Where is continental glaciation an ongoing process today?

◀ **Opposite: Mountain tops manage to project up above the Antarctic ice sheet near its juncture with the Larsen ice shelf (lower left) at the continent's coast.**
NASA/Dryden Flight Research Center Photo Collection

Glaciers, large masses of flowing ice, play several important roles in the Earth system. They are excellent climate indicators because certain environmental conditions are required for glaciers to exist and glaciers respond visibly to climate variation. Glaciers become established, expand, contract, and disappear in response to changes in climate. Their long-term storage of fresh water as ice has a tremendous impact on the hydrologic cycle and the oceans, and the accumulation of ice by glaciers provides a record of past climates that can be studied in ice cores. Where glaciers once existed or where they were once much larger than they are today, much evidence can be found concerning past climatic conditions.

The processes of erosion, transportation, and deposition by glaciers, whether ongoing or in the past, leave a distinctive stamp on a landscape. Some of the most beautiful and rugged terrain in the world exists in mountainous and other highland regions that have been sculpted by glaciers. Virtually every high-mountain region in the world displays glacial landscapes, including the Alps, the Rocky Mountains, the Himalayas, and the Andes. Glaciers have also carved impressive steep-sided coastal valleys in Norway, Chile, New Zealand, and Alaska. Rugged mountain peaks rising high above lake-filled valleys or narrow and deep-sea lanes create the ultimate in scenic appeal for many people.

Masses of moving ice have transformed the appearance of high mountains, as well as large portions of continental plains, into distinctive glacial landscapes. The flowing ice of glaciers is an effective and spectacular geomorphic agent on major portions of Earth's surface.

Glacier Formation and the Hydrologic Cycle

Glaciers are masses of flowing ice that have accumulated on land in areas where the annual input of frozen precipitation, especially snowfall, has exceeded its yearly loss by melting and other processes. Snow falls as hexagonal ice crystals that form flakes of intricate beauty and variety. Snowflakes have a low density (mass per unit volume) of about 0.1 grams per cubic centimeter (0.06 oz/in.^{3}). Once snow accumulates on the land, it becomes transformed by compaction along with melting and refreezing at pressure points into a mass of smaller, rounded grains (● Fig. 19.1). Density increases as the air space around this more granular snow continues to decrease by compaction and melting and refreezing. Through melting, refreezing, and pressure caused by the increasing weight from burial under newer snowfalls, the snow compacts further into a crystalline granular stage known as **firn,** which has a density of about 0.5 grams per cubic centimeter (0.29 oz/in.^{3}). After additional time, small firn granules become larger by recrystallizing into interlocked ice crystals through pressure, partial melting, and refreezing. When the ice is deep enough and has a density up to about 0.9 grams per cubic centimeter (0.52 oz/in.^{3}), it becomes glacial ice. Pressure from burial under many layers of snow, firn, and ice causes the glacial ice below to become plastic and flow outward or downward away from the area of greatest snow and ice accumulation.

● FIGURE 19.1
The transformation of frozen water from snow to glacial ice changes the size and shape of the crystals and greatly increases the material density. Compaction and pressure melting, followed by refreezing, accomplish this alteration.

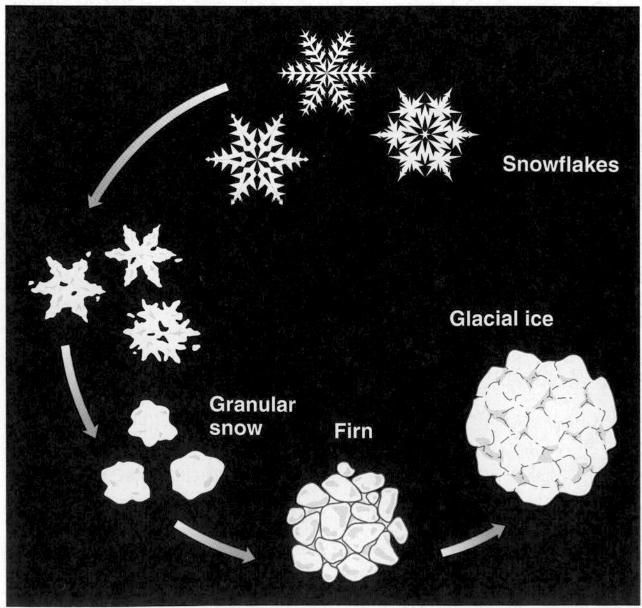

Glaciers are open systems with input, storage, and output of material. Any addition of frozen water to a glacier is termed **accumulation.** Most accumulation occurs in winter and consists of snowfall onto a glacier, but there are many ways that frozen water can accumulate, including other forms of precipitation onto the ice, atmospheric water vapor freezing directly onto the ice (gas to solid), and others. Some glaciers originate not by snowfall on the glacier itself but by the accumulation of snow blowing, drifting, and avalanching onto the glacier's surface. The Colorado Rockies and the Ural Mountains in Russia provide some good examples of highland glaciers formed by accumulations of this *snowdrift.*

Ablation is the opposite of accumulation, representing any removal of frozen water mass from a glacier. Although most ablation is accomplished in summer through melting, direct change from ice to water vapor (solid to gas) and other processes may also be at work. Interestingly, the term **sublimation** refers to the direct change from solid to gas as well as the direct change from gas to solid; therefore it can contribute to ablation or accumulation. **Calving,** another form of glacial ice loss, occurs when a glacier loses large chunks of ice that break away as **icebergs** that float in the ocean or a large lake (● Fig. 19.2).

A glacial system is controlled by two basic climatic conditions: precipitation in the form of snow, and freezing temperatures. First, to establish a glacier, there must be sufficient accumulation (input) to exceed the annual loss through ablation. Most of the accumulation is a result of snowfall. Mountains along midlatitude coastlines and high mountains near the equator can support glaciers because of heavy orographic snowfalls, despite intense sunshine and warm climates in the surrounding lowlands. Yet some very cold polar regions in subarctic Alaska and Siberia and a few valleys in Antarctica have no glaciers because the climate is too dry.

The second climatic condition that affects a glacier is temperature. Summer temperatures must not be high for too long, or all of the accumulation (primarily snowfall) from the previous winter will be lost through ablation (primarily melting). Surplus snowfall is essential for glacial formation because it allows the pressure from years of accumulated snow layers to transform older buried snow into firn and glacial ice. When the ice reaches a depth of about 30 meters (100 ft), a pressure threshold is reached that enables the solid ice to flow, thereby creating a glacier.

Glaciers are an important part of Earth's hydrologic cycle and are second only to the ocean in the amount of water they contain. Approximately 2.25% of Earth's total water is currently frozen in glaciers. This frozen water, however, makes up about 70% of the world's *fresh* water, with the vast majority stored on Greenland and Antarctica. The total amount of ice is even more impressive if we estimate the water that would be released if all of the world's glaciers were to melt. Sea level would rise about 65 meters (215 ft). This would change the geography of the planet considerably. In contrast, if another ice age were to occur, sea level would drop drastically. During the last major glacial advance, sea level fell about 120 meters (400 ft).

Unlike the water in a stream system, much of which returns rapidly to the sea or atmosphere, the snow that becomes glacier ice is stored for a long time in a much more slowly flowing system.

NASA/GSFC/MITI/ERSDAC/JAROS, and U.S./Japan ASTER Science Team

(a)

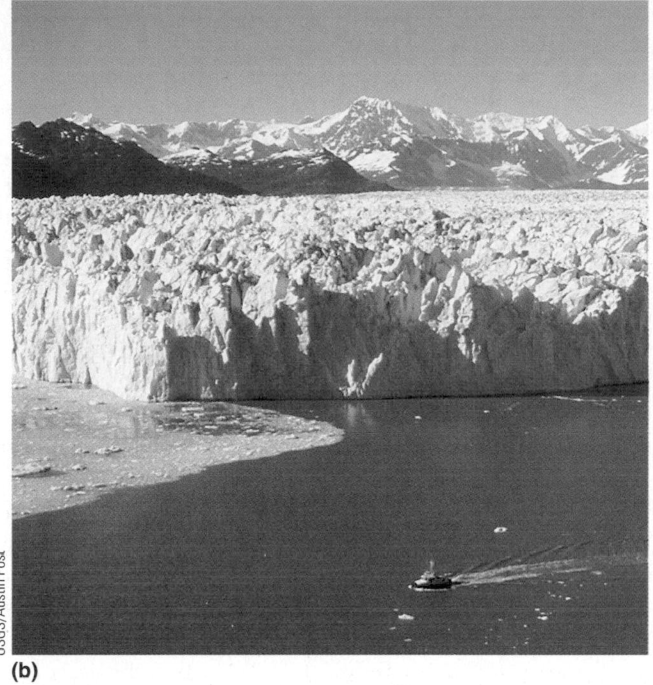
USGS/Austin Post

(b)

• **FIGURE 19.2**

Glaciers that flow into the sea or a deep lake undergo calving, the process by which large chunks of ice break off and become icebergs in the water. (a) Large icebergs derived from the toe of a glacier on Ellesmere Island, Canada, are visible on this satellite image of Greely Fjord taken in August of 2003. (b) The edge of the Columbia Glacier in Prince William Sound, Alaska, is a source of icebergs in that region.

Glaciers may store water as ice for hundreds or even hundreds of thousands of years before it is released as meltwater into the liquid part of the hydrologic system. Yet, glacial ice is not stagnant. It moves slowly but with tremendous energy across the land. Glaciers reshape the landscape by engulfing, eroding, pushing, dragging, carrying, and finally depositing rock debris, often in places far from its original location. Long after glaciers recede from a landscape, glacial landforms remain as a reminder of the energy of the glacial system and as evidence of past climates (• Fig. 19.3).

Glaciers have not existed on the planet during most of Earth history. However, when a period occurs during which significant areas of the middle latitudes are covered by glaciers, we call it an *ice age*. Today, glaciers cover about 10% of Earth's land surface. Present-day glaciers are found on Greenland, on Antarctica, and at high latitudes and high elevations on all continents except Australia. During recent Earth history, from about 2.4 million years ago to about 10,000 years before the present, during the Pleistocene Epoch, glaciers periodically covered nearly a third of Earth's land area. Other ice ages occurred in the much more distant geologic past.

D. Sack

• **FIGURE 19.3**

Glaciers are powerful geomorphic agents that leave distinctive landform evidence of their previous, widespread existence. This system of trough-shaped valleys, narrow bedrock ridges, and lakes in bowl-shaped depressions lies in the previously glaciated Ruby Mountains of northeastern Nevada.

Types of Glaciers

The two major categories of glaciers are alpine and continental. **Alpine glaciers** exist where the precipitation and temperature conditions required for glacier formation result from high elevation. Alpine glaciers are fed by ice and snow in mountain areas and usually occupy preexisting valleys created by a previous period of stream erosion. The ice masses flow downslope because of their own weight, that is, due to the force of gravity. Alpine glaciers that are confined by the rock walls of the valley they occupy are **valley glaciers** (• Fig. 19.4). They are known as **piedmont glaciers** when the ice flows beyond the confines of the valley, spreading out over flatter land. Some alpine glaciers, however, do not reach the valleys below the zone of high peaks. Instead, they occupy distinctive steep-sided, amphitheater-like

GEOGRAPHY'S PHYSICAL SCIENCE PERSPECTIVE

Glacial Ice Is Blue!

When we make ice in our freezers, clear colorless water turns to relatively clear ice cubes. The ice cubes may contain some white crystalline forms and air bubbles, but in general the ice is clear. In nature, the process of making ice is very different from that of an ice maker in a refrigerator-freezer. As snow falls at colder, higher latitudes and elevations, it forms a layer of snow on the surface. Each successive snowfall makes another layer as it piles onto the previous snowpack. The weight of the successive layers of snow creates pressure that compresses the older layers beneath. Through time, the layers of low-density snow become layers of much denser solid ice. Some of this change is due to compaction, but pressure also causes some melting and refreezing of the ice. The temperature at which ice melts is 0°C (32°F) at atmospheric pressure, but ice can melt at lower temperatures if it is under enough pressure. Both compaction and pressure melting and refreezing work to reduce the amount of air in the frozen mass and thereby increase the material's density.

Objects that appear white to the human eye reflect all wavelengths of light with equal intensity, and this is what the hexagonal crystalline structure of snowflakes does. As the snow strata under great pressure in a glacier become compacted over the years (sometimes hundreds or thousands of years) the ice becomes denser. Basically, under this pressure, more ice crystals are squeezed into the same volume. As the density of ice increases, it reflects increasing amounts of shorter wavelengths of light, which is the blue part of the spectrum. The denser the ice, the bluer it appears. Ice density can be influenced by factors other than time, though, so we must be careful not to assume that deeper blue layers in a glacier are necessarily the older layers. For example, the packing of higher density wetter snow as opposed to lower density drier snow can affect the density of specific layers. Nevertheless, what is certain when looking at massive glacial ice accumulations in nature, such as in ice caps and ice sheets, is that the ice will appear as shades of blue.

NASA/Jim Rossi

This iceberg in Antarctica displays very old layers of glacial ice.

depressions called **cirques** that are eroded by ice flow at the heads of valleys, and are thus termed **cirque glaciers** (● Fig. 19.5). Cirque glaciers are the smallest type of glacier. Alpine glaciers begin as cirque glaciers at the start of an ice age, expanding into valley glaciers, and perhaps eventually piedmont glaciers as the ice age intensifies. Most cirque glaciers today represent small remnants of previously larger alpine glaciers.

Alpine glaciers created the characteristic rugged scenery of much of the world's high-mountain regions. Today alpine glaciers are found in the Rockies, the Sierra Nevada, the Cascades, the Olympic Mountains, the Coast Ranges, and numerous Alaskan ranges of North America. They also exist in the Andes, the Alps, the Southern Alps of New Zealand, the Himalayas, the Pamirs, and other high Asian mountain ranges. Small alpine glaciers are even found at high elevations on tropical mountains in New Guinea and in East Africa on Mounts Kenya and Kilimanjaro. The largest alpine glaciers currently in existence are located in Alaska and the Himalayas, where some reach lengths of more than 100 kilometers (62 mi).

The second category of glacier, **continental glaciers,** includes glaciers that are much larger and thicker than the alpine types, and that exist where the appropriate conditions for ice formation occur because of high latitude (● Fig. 19.6). At one time, continental glaciers covered as much as 30% of Earth's land area. Continental glaciers are subdivided by size into ice sheets and ice caps, with ice sheets exceeding 50,000 square kilometers (19,000 sq mi) in extent. Earth's two polar **ice sheets,** the largest type of glacier, still blanket Greenland and Antarctica in the high

USGS/Austin Post

• FIGURE 19.4
Separate valley glaciers join together to form the larger Susitna Glacier in Alaska. Mount Hayes, the highest mountain in the Alaska Range, is the peak in the background.
How are valley glaciers similar to rivers?

USGS/Austin Post

• FIGURE 19.5
This cirque glacier in Alaska occupies a classic bowl-shaped basin, the cirque.
Why is only part of the ice a bright white color?

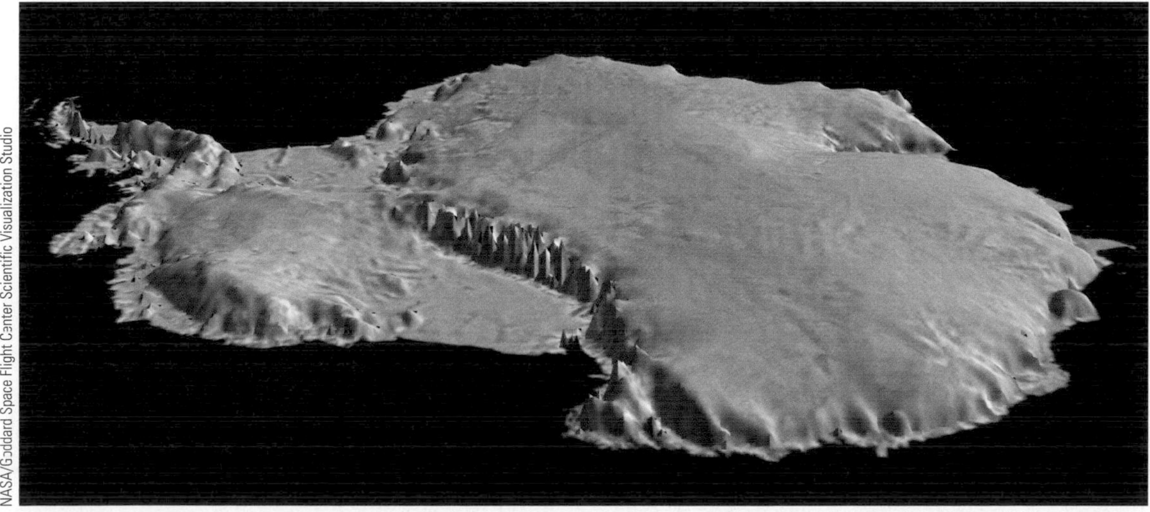
NASA/Goddard Space Flight Center Scientific Visualization Studio

• FIGURE 19.6
With areas exceeding 50,000 square kilometers (19,000 sq mi), ice sheets are the largest type of continental glacier. Here, a terrain model of Antarctica shows the domelike form of its ice sheet. Ice flows away from the center of the continent, near the South Pole, because the ice there is thickest and the resulting pressure is greatest. This pattern of continental ice flow, from a thicker center to the thinner outer edges, is described as being radial.
How is radial ice flow both similar to and different from the radial drainage pattern observed for some stream systems?

latitudes to a depth of at least 3 kilometers (2 mi). Somewhat smaller, but still large, **ice caps** are present on some Arctic islands and on Iceland. In contrast to alpine glaciers, ice sheets and ice caps are not confined by the underlying topography, but more or less drown the underlying topography in ice. Direction of flow within ice sheets and ice caps is from thicker to thinner ice, which is radially outward in all directions from a central source area of maximum ice thickness.

How Do Glaciers Flow?

Like the slow forms of mass wasting, we normally cannot view glacier movement directly. Nevertheless, flowing ice has a tremendous geomorphic impact on the landscape. Most glaciers move through a combination of processes, but *internal plastic deformation* is the dominant process; virtually all moving glaciers experience this type of flow. Glaciers move in this way when the weight of overlying ice, firn, and snow causes ice crystals at depth to arrange themselves in parallel layers that glide over each other, much like spreading a deck of cards (● Fig. 19.7). This internal plastic deformation happens when a threshold pressure (weight per unit area) from the overlying mass is exceeded. The threshold pressure is achieved at an ice thickness of about 30 meters (100 ft), and the zone experiencing plastic flow extends within (is internal to) the ice mass from that depth to the base of the glacier. The speed with which the ice flows increases as pressure from overlying material increases and with steeper slopes. Pressure is greater under thicker accumulations of ice and on the upflow side of obstacles at the base of a glacier. Internal plastic deformation causes continental glaciers to flow radially outward from central areas of thicker ice (higher pressure) to marginal areas of thinner ice (lower pressure).

In addition to internal ice movement through plastic deformation, many glaciers move by processes concentrated at the base of the ice mass. Many temperate glaciers—those with temperatures at and near the melting point—move by the process of *basal sliding* (see again Fig. 19.7). In this case, meltwater at the base of the glacier reduces friction between the ice and ground through lubrication and hydrostatic pressure. As a result, when a location at the glacier base has insufficient frictional resistance to oppose the downslope pull of gravity on the ice mass, the affected part of the glacier jerks forward. Steep slopes contribute to the tendency toward basal sliding. This type of motion is most important in midlatitude glaciers on steep slopes, particularly during summer when much of the glacier is near the melting point and meltwater is available. Little if any basal sliding occurs during winter and in the colder, polar glaciers with little available meltwater. Another type of basal ice flow involves local melting at the base of the glacier, downslope flow of the meltwater, then its refreezing onto the glacier base.

The upper surface of a glacier consists of brittle ice that does not experience plastic deformation. It moves instead by being carried along as ice flows in the underlying zone of plastic flow. As it moves, the ice in the brittle zone fractures and cracks. These ice cracks, called **crevasses,** are common wherever a glacier becomes stretched, experiencing tensional stress, particularly where it flows over a break in slope (● Fig. 19.8). Where a glacier locally flows over a steep descent, such as over a subglacial cliff, an **icefall** develops in the brittle upper ice (● Fig. 19.9). Here, intersecting crevasses break the ice into a morass of unstable ice blocks that ride on rapid flowing ice below. Ice falls and crevasses are extremely dangerous areas for mountain climbers and scientists who venture onto the ice.

Glacial flow rates vary from imperceptible fractions of a centimeter per day to as much as 30 meters (100 ft) per day. Glaciers flow more rapidly where the slopes are steep, where the ice is thickest, and where temperatures are warmest. For example, the Nisqually Glacier, on the steep slopes of Washington's Mount Rainier, flows 38 centimeters (16 in.) per day in summer. As a general rule, temperate alpine glaciers flow much faster than the cold polar continental glaciers.

The flow of an individual glacier varies from time to time with changes in the dynamic equilibrium and from place to place because of variations in the gradient over which it flows or differences in the friction encountered with adjacent rock. Within an alpine glacier, the rate of movement is greatest on the glacier surface toward the middle of the ice because this location experiences accumulated movement from layers

● FIGURE 19.7
Most glacier movement is by internal plastic deformation–flow as a plastic solid–resulting from the weight of the overlying ice, firn, and snow. Glaciers with meltwater at their base also move by basal slippage when the resulting reduction in friction allows the ice to slip over the ground. The entire vertical column of the glacier moves the same amount, B-B', due to basal sliding. Although pressure is greater lower in the ice, plastic flow is cumulative upward so that flow deeper in the ice column carries along overlying ice layers. As a result, ice flow velocity is greatest at the glacier's surface, A-A'.
Does the ice on the top of the glacier flow or just ride along on the ice below?

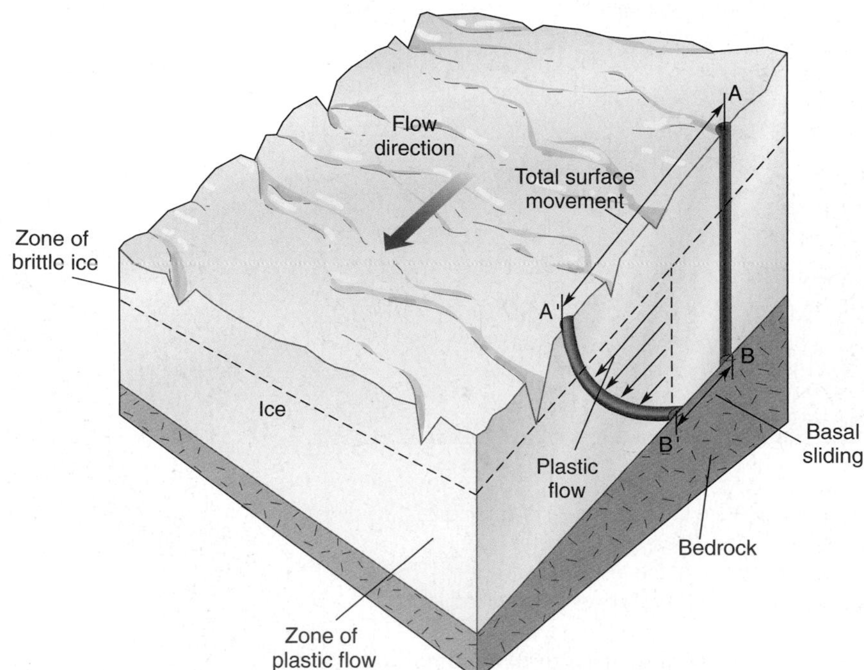

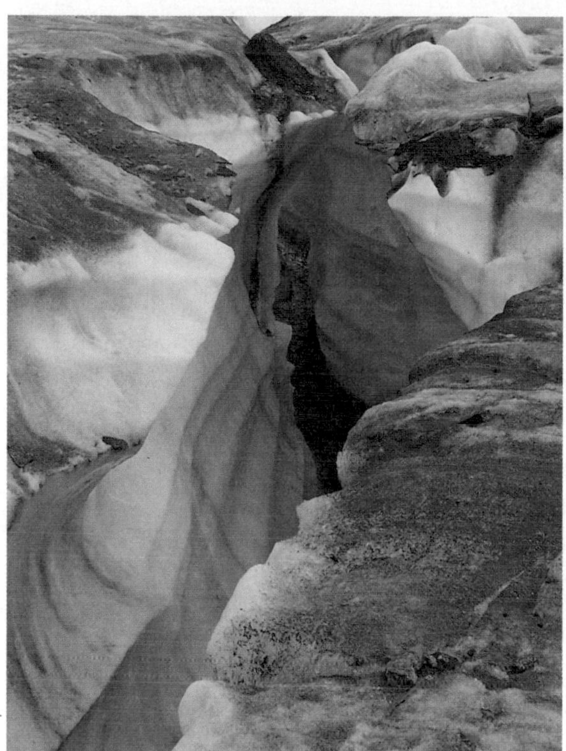
M. Trapasso

• **FIGURE 19.8**
A large crevasse on the Yanert glacier in Denali National Park, Alaska.
Why does the surface of a glacier break into crevasses?

© Matt Ebiner

• **FIGURE 19.9**
Icefalls are the glacial equivalent of rapids or waterfalls in a river and are riddled with crevasses that break the ice into huge, unstable blocks. Although glacial ice flows much more slowly than running water, icefalls are the most rapidly moving and changing part of a glacier. They are one of the most treacherous parts of a glacier to cross because the huge ice blocks can shift at any time.

of plastic flow below and is farthest from frictional resistance with the valley sides.

Sometimes a glacier's velocity will increase by many times its normal rate, causing the glacier to advance hundreds of meters per year. The reasons for such enormous *glacial surges* are not completely clear, although lubrication of a glacier's bed by pockets of meltwater explains some of them.

Glaciers as Geomorphic Agents

Because a deep, and therefore heavy, accumulation of ice is required for glaciers to flow, even the smaller alpine glaciers are particularly powerful geomorphic agents able to perform great amounts of geomorphic work. Whether it is an alpine glacier carving out a trough-shaped valley or a continental glacier gouging out the basins of the North American Great Lakes, the work done by glaciers is impressive.

Glaciers remove and entrain rock particles by two erosional processes. Glacial **plucking** is the process by which moving ice freezes onto loosened rocks and sediments, incorporating them into the flow. Weathering, particularly the freezing of water in bedrock joints and fractures, breaks rock fragments loose, encouraging plucking. Once load is entrained at the base and sides of the ice, moving glaciers are armed with clastic particles that are very effective tools for scraping and gouging out more rock material by the erosional process of *abrasion* (• Fig. 19.10). Bedrock obstructions subjected to intense glacial abrasion are typically smoother and more rounded than those produced by plucking.

• **FIGURE 19.10**
The sediment load transported by glaciers is a poorly sorted mix of grain sizes. As soon as a glacier obtains some load, those clastic particles are used as tools to help erode more rock by abrasion. Here, a cobble from a glacial deposit shows scrapes and scratches obtained by grinding against bedrock and other particles as it was carried in the ice.
How does sediment load of a glacier differ from sediment load of a stream?

D.Sack

Unlike the situation with liquid water in streams, volume and velocity of flow do not directly determine the particle sizes that plastic-flowing solid ice can erode and transport. Plucking and abrasion provide the bottom and sides of glaciers with a chaotic load of rock fragments of all sizes, from clay-sized crushed rock, called *rock flour,* to giant boulders. Mass wasting along steep mountain slopes, especially above alpine glaciers, contributes sediment, also of a variety of grain sizes, to the ice surface and sides. Also in contrast to streams, little sorting of sediment by size is accomplished by glaciers during transportation and deposition. This lack of sorting by size makes glacial deposits look very different from accumulations of stream deposits. Because of this contrast in the two types of sediment, it is logical that they are referred to by two different terms. Whereas stream-deposited sediment is called *alluvium,* sediment deposited directly by moving ice is **till.**

Alpine Glaciers

From a mass balance perspective, alpine glaciers consist of two functional parts, or zones (• Fig. 19.11). The colder, snowier upslope portion of a glacier, where annual accumulation (input) exceeds annual ablation (output), is the **zone of accumulation.** In contrast, in the warmer downslope portion of an alpine glacier, the

• FIGURE 19.11

Environmental Systems: The Hydrologic System—Glaciers. Glacial systems are controlled by the input of frozen water (accumulation), primarily from winter snowfall, and the loss of frozen water (ablation), primarily from summer melting. In the higher-elevation, colder part of the glacier, called the zone of accumulation, annual input exceeds annual loss. In the lower-elevation, warmer zone of ablation, annual ice loss exceeds addition of frozen water. The equilibrium line marks the elevation where annual accumulation equals annual ablation.

Over the years snowfall on the zone of accumulation, buried by successive snow layers, changes into denser firn and later into even denser glacial ice. When the ice reaches a threshold depth, the deeply buried ice deforms plastically in response to the overlying weight and flows downslope. The ice flows faster (1) as the overlying weight increases due to greater input, and (2) as the ice flows over steeper slopes. This continuous, internal plastic flow transports ice from the zone of accumulation toward the toe of the glacier in the zone of ablation.

If the glacier as a whole experiences more accumulation than ablation in a year, the increase in mass will cause some thickening of the ice, but it will also cause the toe of the glacier to advance farther downvalley. If ablation exceeds accumulation for the year, the glacier will lose mass and retreat. Equilibrium exists if annual accumulation equals annual ablation.

At present, most of the world's glaciers are receding, some quite rapidly. Scientists are concerned about the potential effects of global warming on glacial systems, especially on the continental ice sheets of Antarctica and Greenland. Increasing world temperatures can shift glacial systems toward a new equilibrium, at a smaller size and mass, with more rapid ice loss through melting, calving into water bodies, and other means. A significant decrease in the amount of glacial ice on Earth will result in a global rise in sea level that would affect low-lying coastal areas.

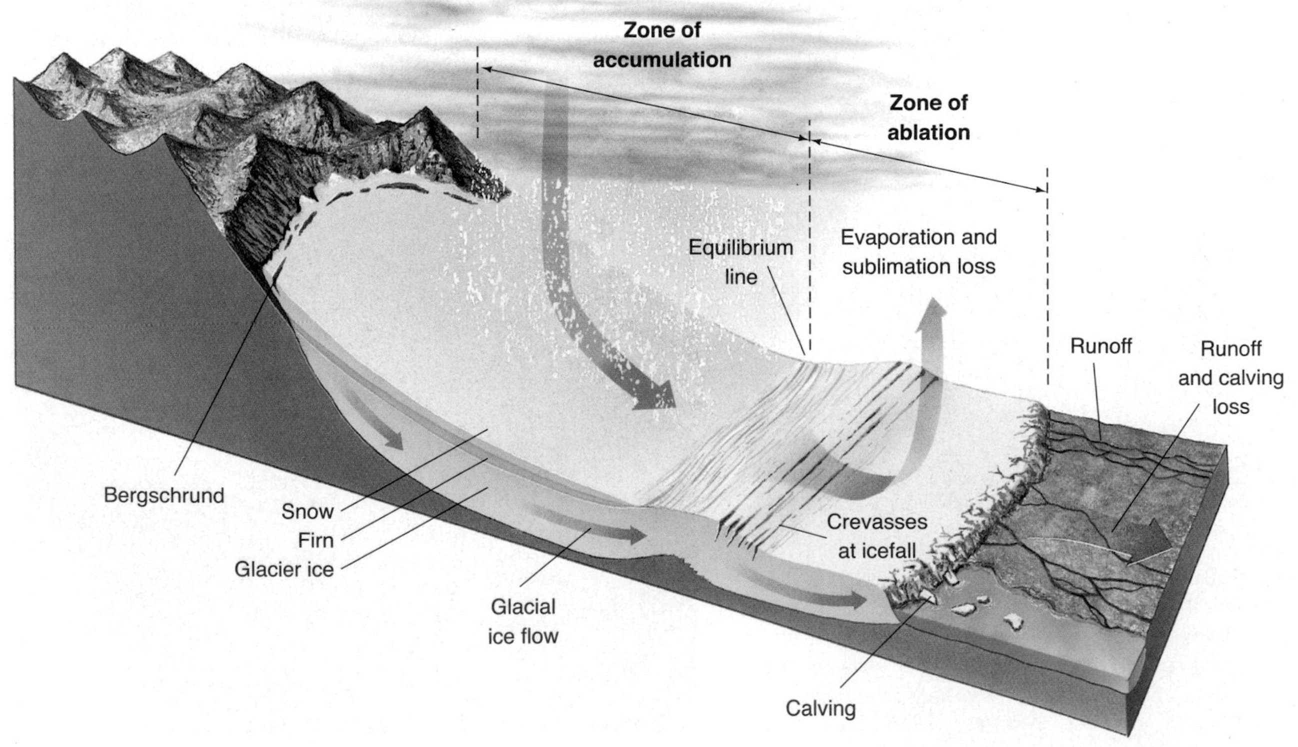

zone of ablation, annual ablation exceeds annual accumulation. Winter is the dominant accumulation season and summer is the dominant ablation season. Alpine glaciers change size, sometimes quite dramatically, over the course of a year. The toe of a glacier lies farthest downvalley near the end of winter and farthest upvalley at the end of the ablation season near the end of summer.

The **equilibrium line** marks the boundary between the zones of accumulation and ablation on an alpine glacier. It indicates where annual accumulation equals annual ablation for the glacier (● Fig. 19.12). The equilibrium line differs from the *snow line,* which is the elevation where snow cover begins on a landscape. The snow line regularly changes position through the seasons and in response to the weather, including after every snowfall. The equilibrium line, in contrast, represents the elevation at which the ground is covered by snow all year long.

● FIGURE 19.12
A valley glacier on Alaska's Kenai Peninsula displays a blue zone undergoing ablation at lower elevations contrasted with a white accumulation zone at the glacier's higher elevations. The visible contact between the two zones on the photograph, however, is not necessarily *the* equilibrium line because it may not represent the elevation at which accumulation equaled ablation for the year.
What additional information would be needed to assess if the boundary between the white and blue zones on this photo is the glacier's annual equilibrium line?

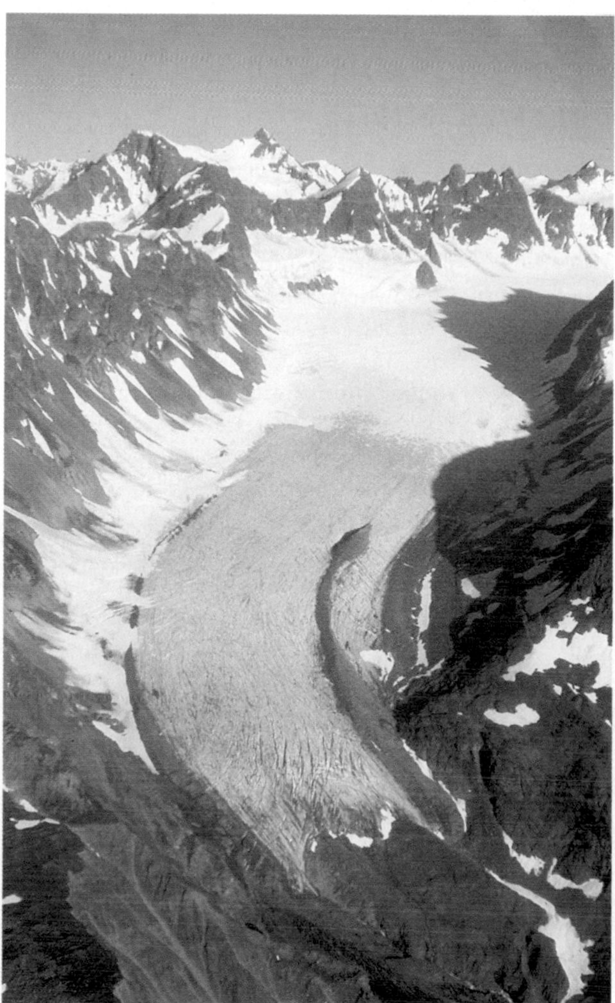

USGS/Austin Post

Several factors influence the location of the equilibrium line. The interaction between latitude and elevation, both of which affect temperature, is an important factor. On mountains near the equator, the equilibrium line lies at very high elevations. Elevation of the equilibrium line decreases with increasing latitude until it coincides with sea level in the polar regions. Equally important to temperature in determining the position of the equilibrium line is the amount of snowfall received during winter. With colder temperatures and greater snowfall, the equilibrium line will decrease in elevation; it retreats to higher elevations if the climate warms. Other attributes causing variations in the equilibrium-line elevation include the amount of insolation. A shady mountain slope will have a lower equilibrium line than one that receives more insolation. Wind is another factor because it produces snowdrifts on the leeward side of mountain ranges. In the middle latitudes of the Northern Hemisphere, the equilibrium line is lower on the north (shaded) and east (leeward) slopes of mountains. Consequently, the most significant glacier development in this region is on north-facing and east-facing slopes (● Fig. 19.13).

● FIGURE 19.13
This topographic map of part of Rocky Mountain National Park in Colorado illustrates the impact of slope aspect (compass direction) on glaciation. In the Northern Hemisphere at middle to high latitudes, slopes facing north or northeast tend to be shaded from the sun, allowing greater amounts of snow and ice to accumulate there. This encourages glaciation on north- and east-facing slopes.

USGS

At the farthest upslope edge of an alpine glacier, the upslope end of the zone of accumulation, lies the **glacier's head.** The head of the glacier abuts the steep bedrock cliff that comprises the *cirque headwall.* Ice within an alpine glacier flows downslope from the zone of accumulation to the zone of ablation. This downslope movement is sometimes evidenced by a large crevasse, known as a *bergschrund,* that may develop between the head of the glacier and the cirque headwall (see again Fig. 19.11). Presence of a bergschrund shows that the ice mass is pulling away from the confining rock walls of the cirque. The downslope end of a glacier is called its **terminus,** or **glacier toe.**

Equilibrium and the Glacial Budget

Because the location of the toe of a glacier changes throughout the year with accumulation and ablation, to determine if an alpine glacier is growing or shrinking over a period of years requires annually noting the location of its terminus at the same time of year (• Fig. 19.14). Typically this is done at the end of the ablation season when the glacier is at its minimum size for the year. If a glacier received more input of frozen water (accumulation) during a year than was removed from it (ablation) that year, it experienced *net accumulation.* The result of net accumulation is a larger glacier, and as alpine glaciers grow they **advance;** that is, their toes extend farther downvalley. A glacier that undergoes *net ablation,* more removal than addition of frozen water for the year, shrinks in size causing the toe to **retreat** upvalley.

If the annually measured toe of a glacier neither advances nor retreats over a period of years, the glacier is in a state of equilibrium in which a balance has been achieved between accumulation and ablation of ice and snow. As long as equilibrium is maintained, the location of the glacier's toe at the end of the ablation season will remain constant.

It is crucial to understand that whether an alpine glacier is advancing, retreating, or in a state of equilibrium, the ice comprising it continues to flow downslope. Even a glacier that is retreating over the long term will receive winter snow, with more at its higher than at its lower elevations. The weight per unit area of this frozen water drives glacier movement through internal plastic deformation. Winter snow buried by additional snow eventually turns to firn and glacial ice, and over a period of decades makes its way along the glacier to its terminus. Most ice at the terminus of a glacier today made its way there slowly from the zone of accumulation. Downslope movement stops only if net ablation proceeds so far that the ice becomes too thin to maintain plastic flow. In fact, if ice flow ceases, the mass is no longer considered an active glacier.

From about 1890 to the present, most Northern Hemisphere glaciers have been receding. This overall retreat may be an indication of global warming, yet each individual glacier has its own balance. For example, in 1986, Alaska's Hubbard Glacier advanced so rapidly that it cut off Russell Fjord from the ocean, trapping many seals and porpoises. Yet just a few hundred kilometers away, the giant Columbia Glacier is rapidly receding, and calving has increased the numbers of icebergs in Prince William Sound. Scientists and the Coast Guard are concerned because of the increased hazard from icebergs in oil tanker lanes from the Trans-Alaska Pipeline.

• FIGURE 19.14
The Jacobshavn Glacier, Greenland's largest outlet glacier, has been generally retreating since the beginning of measurement and monitoring by scientists in 1850. The colored lines mark the former position of the glacier's terminus.
Why is the rapid receding of this glacier of particular concern to scientists?

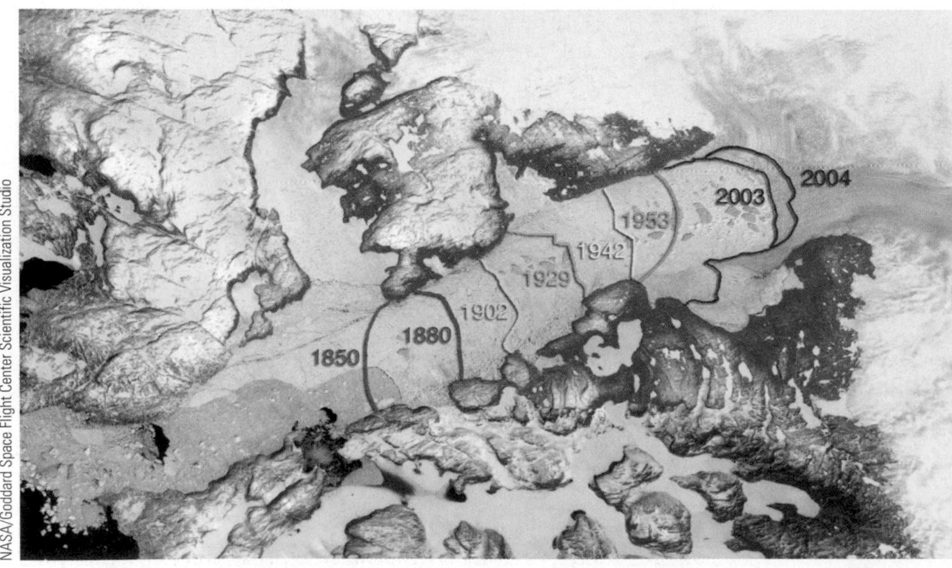

NASA/Goddard Space Flight Center Scientific Visualization Studio

Erosional Landforms of Alpine Glaciation

Glacial abrasion leaves **striations**—linear scratches, grooves, and gouges—where sharp-edged rocks scraped across bedrock (• Fig. 19.15). In areas devoid of glaciation today, striations indicate direction of ice flow long after the ice disappeared from the landscape. Abrasion and plucking at the base of a glacier work together to form **roches moutonnées,** asymmetric bedrock hills or knobs that are smoothly rounded on the up-ice side by abrasion, with plucking evident on the abrupt down-ice side (• Fig. 19.16).

When an alpine glacier first develops in a hollow high in the mountains, its small size results in almost rotational flow lines for the moving ice. In the zone of accumulation at the glacier's head, ice movement has a large downward component. For ice to reach the toe in the zone of ablation a short distance away may require an upward component to the flow there. Ice movement, accompanied by weathering and mass wasting, steepens the bedrock wall at the head of the small glacier while it deepens the hollow into the semispherical or amphitheater-shaped depression that is the *cirque*. In fact, in many mountain areas, cirques are formally or informally referred to as "bowls" because of their distinctive shape. Cirque glaciers often deposit a ridge of till at the down-ice edge of a cirque, and this accentuates the bowl-shaped appearance. When the ice disappears due to climate change, the erosional cirque is left behind, often forming a natural basin in which water accumulates. Lakes that form in cirques are called **tarns,** or they may be simply referred to as *cirque lakes* (● Fig. 19.17a).

Often two or more cirques develop in neighboring hollows on a mountainside. As the cirques or valleys of two adjacent glaciers enlarge, the bedrock ridge between them will be shaped into a jagged, sawtooth-shaped spine of rock, called an **arête** (Fig. 19.17b). Where three or more cirques meet at a mountain summit, headward erosion carves the high ground between them into a characteristic pyramid-like peak called a **horn** (Fig. 19.17c). The Matterhorn in the Swiss Alps is the world's prime example. A **col** is a pass formed by the headward erosion of two cirques that have intersected to produce a low saddle in a high-mountain ridge, or arête.

As they expand downslope out of their cirques, alpine glaciers take over the downslope pathways established by streams before the ice accumulated. Steep mountain streams carve valleys that in cross section resemble the letter *V*. Because glaciers are much thicker than streams are deep, they erode the sides as well as the bottom of these valleys, bowing the cross-sectional shape out to a U-shaped **glacial trough** (Fig. 19.17d). In addition, a glacier's tendency to flow straight ahead rather than to meander causes it to straighten out the preexisting valley that it occupies.

By preferentially eroding weaker rocks on the valley floor, some alpine glaciers create a sequence of rock steps and excavated basins. While the ice is still present, crevasses or even icefalls develop in its surface where it flows over the steps. When the ice recedes, rockbound lakes sometimes fill the basins, often looking like beads connected by a glacial stream flowing down the glacial trough. Such lake chains are **paternoster lakes.**

At higher latitudes, many glacial troughs extend down to the ocean. Today, glaciers flow into the sea where they calve icebergs along the coasts of British Columbia, southern Alaska, Chile, Greenland, and Antarctica. The toes of these glaciers are dangerous areas because large waves can be created as huge blocks of ice calve off and topple into the water. As a coastal glacier retreats landward through net ablation, the ocean invades the abandoned glacial trough, creating a deep, narrow ocean inlet called a **fjord**

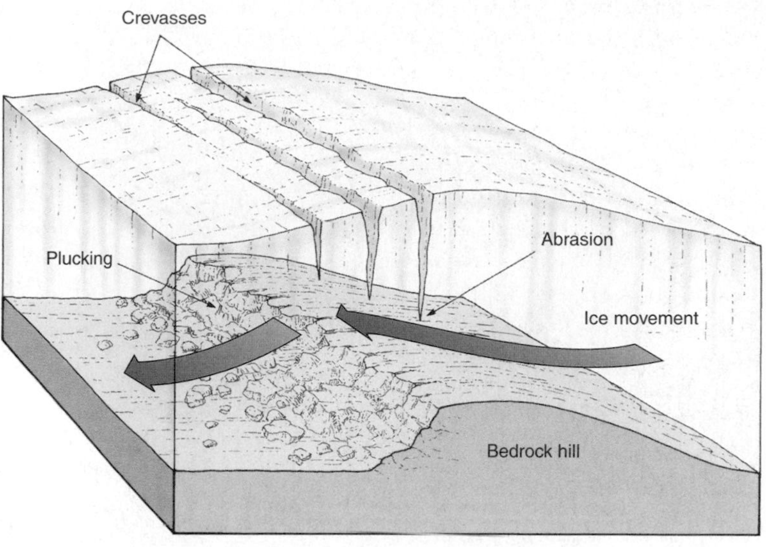

● FIGURE 19.16
(a) Glacial erosion of bedrock hills typically produces asymmetric landforms, known as roches moutonnées (sheep-back rocks), caused by abrasion upstream and plucking downstream. Arrows indicate the direction of ice flow. (b) This example of a roche moutonnée is located in Yosemite National Park, California.
Why would crevasses form in ice flowing above this feature?

● FIGURE 19.15
Glacial abrasion produces smooth rock surfaces that are cut by striations (scratches and grooves) oriented parallel to the direction of ice movement.
Can the direction of ice flow be determined with certainty from the evidence in this photograph?

● **FIGURE 19.17**
Alpine glaciation produced each of these erosional landforms. (a) This glacially carved cirque in the Sierra Nevada contains a tarn (lake in a cirque). (b) Jagged, narrow ridges of rock, such as this one in the Sierra Nevada, are known as arêtes. (c) The Matterhorn in the Swiss Alps is a classic example of a horn. Horns are formed when several glaciers cut headward into a mountain peak. (d) Glaciers carve steep-sided U-shaped valleys called glacial troughs. Little Cottonwood Canyon, east of Salt Lake City in the Wasatch Range of the Rocky Mountains, is an excellent example of a glacial trough.
Can you identify on the photo in (d) the height to which ice filled the valley?

(● Fig. 19.18). The fjords of Scotland, Norway, Iceland, and New Zealand show that glaciers in those regions reached the sea during the Pleistocene. Most of the deep, narrow channels of Washington's Puget Sound were carved into bedrock by glacial erosion and later invaded by the sea. Unlike streams, which erode only to base level, glaciers can erode somewhat below sea level, but most fjords were carved during times of extensive Pleistocene glaciation when sea level was lower than it is today, and they were later submerged as sea level rose with melting of the glaciers.

Most large valley glaciers have *tributary glaciers*. These tributary glaciers, like the main ice stream, also carve U-shaped channels (see again Fig. 19.3). However, because these tributaries have less ice volume than the main glacier, they also have lower rates of erosion and less ability to erode their channels. As a result, their troughs are smaller and not as deep as those of the main glacier. Nevertheless, during peak glacial phases, the ice surface of smaller tributary glaciers flows in at the surface level of the larger glacier. Not until the glaciers begin to wane does the difference in

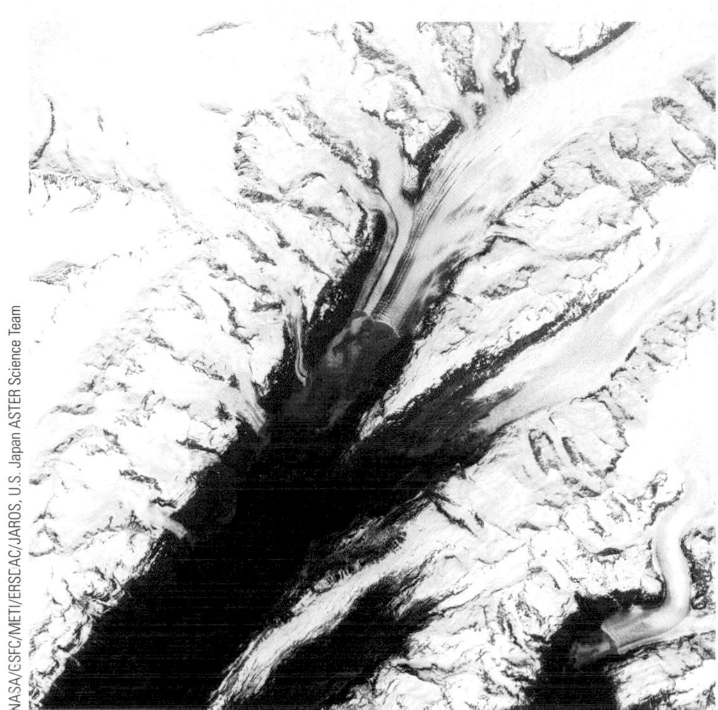

• **FIGURE 19.18**
Glaciers carved the deep College Fjord, seen here in a satellite image, in northwestern Alaska.
How many glaciers can you see on this image?

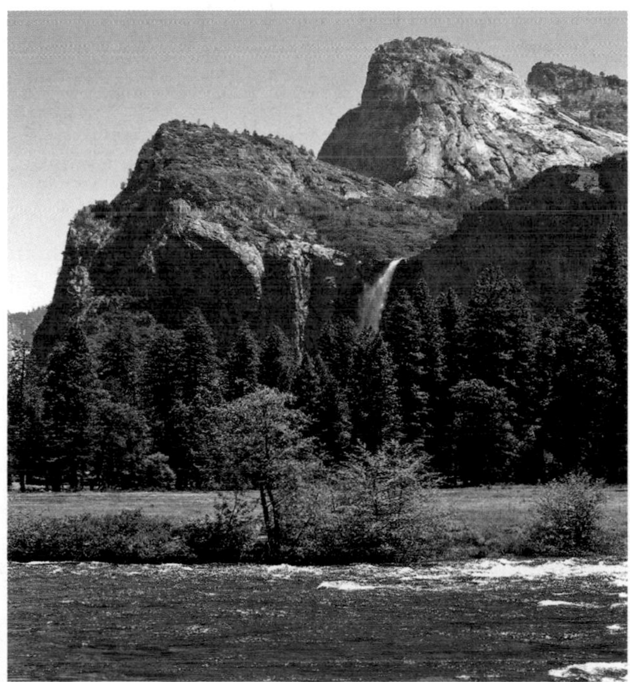

• **FIGURE 19.19**
At the top of the cliff, the water of Bridalveil Falls in Yosemite National Park cascades off the floor of a U-shaped hanging valley and plunges toward Yosemite Valley below. The glacier that occupied the hanging valley was tributary to that in the main valley below, and the hanging valley's elevation indicates the approximate height of the ice mass in Yosemite Valley.

height between their trough floors become apparent. Once the troughs are no longer occupied by glaciers, the tributary troughs form **hanging valleys** (• Fig. 19.19). A stream that flows down such a channel will drop down to the lower glacial valley by a high waterfall or a series of steep rapids. Yosemite Falls and Bridalveil Falls in Yosemite National Park are excellent examples of waterfalls cascading out of hanging valleys. Yosemite Valley itself is a beautiful example of a glacial trough. A possible scenario in the development of alpine glacial erosional topography is illustrated in • Figure 19.20.

Landscapes eroded by alpine glaciers show a sharp contrast between the deep, U-shaped glacial troughs scoured smooth by ice flow and the jagged peaks above the former ice levels. The rugged quality of these upper surfaces is caused primarily by mechanical weathering above the ice surface and by glacial undercutting to create horns and arêtes (• Fig. 19.21). In North America, spectacular rugged alpine glacial terrains are found in the mountains of Alaska, in California's Sierra Nevada, and in the Rockies (see Map Interpretation: Alpine Glaciation).

Depositional Landforms of Alpine Glaciation

Like glacial load, glacial deposits include clastic sediments of a wide range of sizes, frequently mixed with layers of pollen, other plant matter, and soil. In addition to the poorly sorted till deposited directly by glacial ice, meltwater streams, lakes, and wind occurring in association with glaciers contribute to the deposition of sediments and creation of landforms in glacial terrain. *Glaciofluvial* is used to specify the better sorted and stratified fluvial deposits related to glacial meltwater. All deposits of glacial ice or its meltwater, and therefore including till and glaciofluvial deposits, are included within the general term **drift** (• Fig. 19.22).

Active alpine glaciers deposit load primarily along the sides and toe of the ice. Landforms constructed from glacial deposits, typically ridges of till along these margins of glaciers, are **moraines.** Till deposited as ridges paralleling the side margins of a glacier are **lateral moraines** (• Fig. 19.23a). Where two tributary valley glaciers join together, their lateral moraines merge downflow creating a **medial moraine** in the center of the trunk glacier. Medial moraines cause the characteristic dark stripes seen on the surface of many alpine glaciers (Fig. 19.23b). At the toe of a glacier, sediment carried forward by the "conveyor belt" of ice or pushed ahead of the glacier is deposited in a jumbled heap of material of all grain sizes, forming a curved depositional ridge called an **end moraine** (Fig. 19.23c). End moraines that mark the farthest advance of a glacier are **terminal moraines.** End moraines deposited as a consequence of a temporary pause by a retreating glacier, followed by a stabilization of the ice front prior to further recession, are called **recessional moraines.** A retreating glacier also deposits a great deal of till on the floor of the glacial trough as the ice melts away and leaves its load behind. The hummocky landscape created by these glacial deposits is called **ground moraine.**

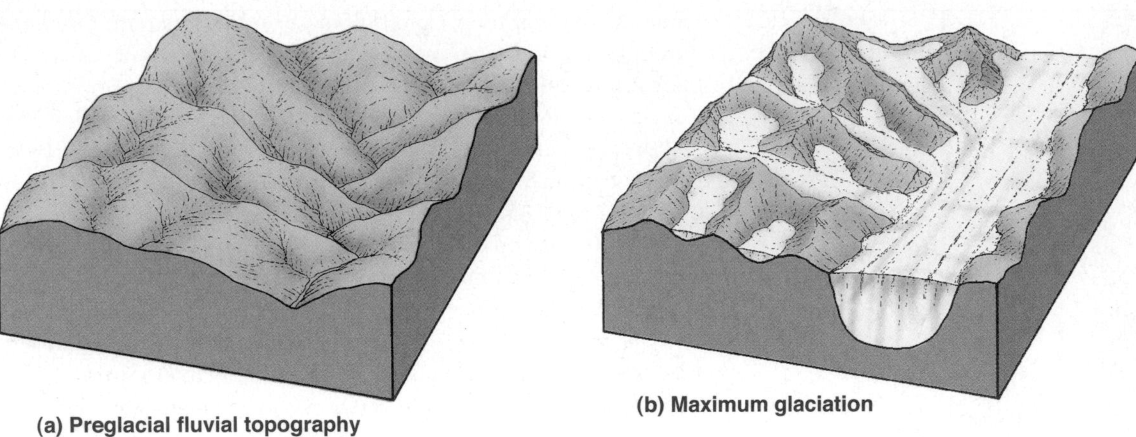

(a) Preglacial fluvial topography

(b) Maximum glaciation

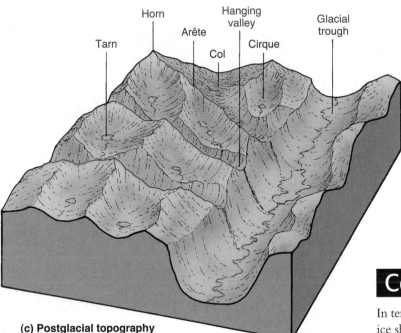

(c) Postglacial topography

• FIGURE 19.20
(a) Preexisting mountain stream valleys provide the path of least resistance toward lower elevations, and (b) are therefore preferred locations for advancing alpine glaciers. (c) After the ice disappears, the tremendous geomorphic work accomplished by the alpine glaciers is evident in the distinctive erosional landforms created by the moving ice.
How do the valley profiles change from preglacial to postglacial times?

Braided meltwater streams laden with sediment commonly issue from the glacier terminus (• Fig. 19.24). The sediment, called **glacial outwash,** is deposited beyond the terminal moraine, with larger rocks deposited first, followed downstream by progressively finer particles. Often resembling an alluvial fan confined by valley walls, this depositional form left by braided streams is called a **valley train.** Valleys in glaciated regions may be filled to depths of a few hundred meters by outwash or by deposits from moraine-dammed lakes.

Continental Glaciers

In terms of their size and shape, continental glaciers, which consist of ice sheets or the somewhat smaller ice caps, are very different from alpine glaciers. However, all glaciers share certain characteristics and processes, and much of what we have discussed about alpine glaciers also applies to continental glaciers. The geomorphic work of the two categories of glaciers differs primarily in scale, attributable to the enormous disparity in size between continental and alpine glaciers.

Ice sheets and ice caps are shaped somewhat like a convex lens in cross section, thicker in the center and thinning toward the edges. They flow radially outward in all directions from where the pressure is greatest, in the thick, central zone of accumulation, to the surrounding zone of ablation (• Fig. 19.25). Like all glaciers, ice sheets and ice caps advance and retreat by responding to changes in temperature and snowfall. As with alpine glaciers flowing down preexisting stream valleys, movement of advancing continental glaciers takes advantage of paths of least resistance found in preexisting valleys and belts of softer rock.

Existing Continental Glaciers

Glaciers of all categories currently cover about 10% of Earth's land area. In area and ice mass, alpine glaciers are almost insignificant compared to the huge ice sheets of Greenland and Antarctica,

USGS/Bruce F. Molnia

• **FIGURE 19.21**
The alpine glacial topography of the north-central Chugach Mountains, Alaska, includes numerous cirques, arêtes, and horns as well as deposits of alpine glaciation.

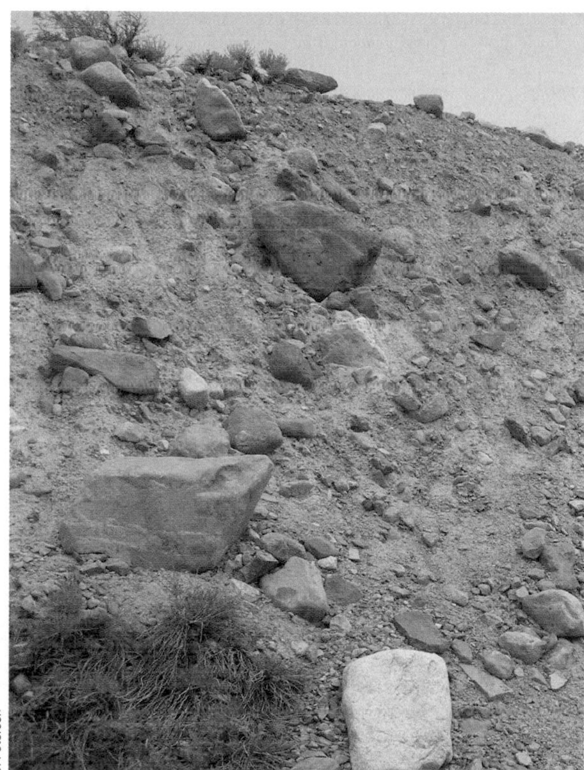

J. Petersen

• **FIGURE 19.22**
Glacial till, here deposited by an alpine glacier in the Sierra Nevada, consists of an unsorted, unstratified, rather jumbled mass of gravel, sand, silt, and clay.
Why does till have these disorganized characteristics?

which account for 96% of the area occupied by glaciers today. Ice caps currently exist in Iceland, on the arctic islands of Canada and Russia, in Alaska, and in the Canadian Rockies.

The Greenland ice sheet covers the world's largest island with a glacier that is more than 3 kilometers (2 mi) thick in the center.

USGS

(a)

USGS/Austin Post

(b)

© Matt Ebiner

(c)

• **FIGURE 19.23**
(a) Lateral moraines on the Kenai Peninsula of Alaska clearly mark the former position of the side margins of a glacier, thereby indicating its width as well. (b) Where two valley glaciers flow together, their adjoining lateral moraines merge into a medial moraine of the larger glacier that they create. Medial moraines comprise the characteristic stripes that appear on these glaciers in the Alaska Range. (c) The glacial deposits of an end moraine form a ridge that rims the position of the terminus of this glacier in Pakistan.
What can we learn from studying moraines?

USGS

• **FIGURE 19.24**
These braided stream channels carry meltwater and large amounts of outwash derived from glaciers in Denali National Park, Alaska.
Why are braided channels often associated with the deposition of glacial outwash?

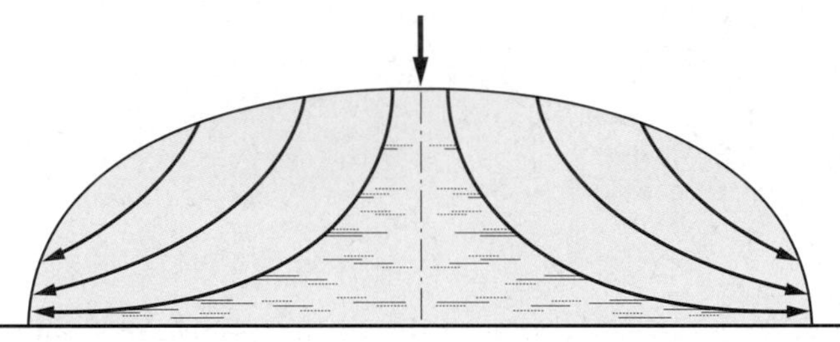

• **FIGURE 19.25**
This diagram shows how ice in an ice sheet flows outward and downward from the center where the glacier is thickest.
How is this manner of ice flow different from and similar to that of a valley glacier?

The only land exposed in Greenland is a narrow, mountainous coastal strip (• Fig. 19.26). Where the ice reaches the ocean, it usually does so through fjords. These flows of ice to the sea resemble alpine glaciers and are called **outlet glaciers.** The action of waves and tides on the outlet glaciers breaks off huge ice masses that float away. The resulting icebergs are a hazard to vessels in the North Atlantic shipping lanes south of Greenland. Tragic maritime disasters, such as the sinking of the *Titanic,* have been caused by collisions with these large irregular chunks of ice, which are nine-tenths submerged and thus mostly invisible to ships. Today, with iceberg tracking by radar, satellites, and the ships and aircraft of the International Ice Patrol, these sea disasters are minimized.

The Antarctic ice sheet covers some 13 million square kilometers (5 million sq mi), almost 7.5 times the area of the Greenland ice sheet (• Fig. 19.27). As in Greenland, little land is exposed in Antarctica, and the weight of the 4.5-kilometer- (nearly 3 mi) thick ice in some interior areas has depressed the land well below sea level. Where the ice reaches the sea, it contributes to the **ice shelves,** enormous flat-topped plates of ice attached to land along at least one side (• Fig. 19.28). These ice shelves are the source of icebergs in Antarctic waters, which do not have the irregular surface form of Greenland's icebergs. Because they do not float into heavily used shipping lanes, these extensive tabular-shaped Antarctic icebergs are not as much of a hazard to navigation (• Fig. 19.29). They do, however, add to the problem of access to Antarctica for scientists. The huge wall of the ice shelf itself, the massive, broken-up sea ice, and the extreme climate combine to make Antarctica inaccessible to all but the hardiest individuals and equipment. This icy continent serves, however, as a natural laboratory for scientists from many countries to study Antarctic glaciology, climatology, and ecology.

Pleistocene Glaciation

The Pleistocene Epoch of geologic time was an interval of great climate change that began about 2.4 million years ago and ended around 10,000 years before the present. There were a great number of glacial fluctuations during the Pleistocene, marked by numerous major advances and retreats of ice over large portions of the world's landmasses. It is a common misconception that the continental ice sheets originated, and remain thickest, at the poles. When the Pleistocene glaciers advanced, ice spread outward from centers in Canada, Scandinavia, and eastern Siberia, as well as Greenland and Antarctica, while alpine glaciers spread to lower elevations. At their maximum extent in the Pleistocene, glaciers covered nearly a third of Earth's land surface (• Fig. 19.30). At the same time, the extent of *sea ice* (frozen sea water) expanded equatorward. In the Northern Hemisphere, sea ice was present along coasts as far south as Delaware in North America and Spain in Europe. Between each glacial advance, a warmer time called an *interglacial* occurred, during which the enormous continental ice sheets, ice caps, and sea ice retreated and almost completely disappeared. Studies of glacial deposits and landforms have revealed that within each major glacial advance, many minor retreats and advances occurred, reflecting smaller changes in global temperature and precipitation. During each advance of the ice sheets, alpine glaciers were much more numerous, extensive, and massive in highland areas than they are today, but their total extent was still dwarfed by that of the continental glaciers.

North America and Eurasia experienced major glacial expansion during the Pleistocene. In North America, ice sheets extended as far south as the Missouri and Ohio Rivers and covered nearly all of Canada and much of the northern Great Plains, the Midwest, and the northeastern United States. In New England, the ice was thick enough to overrun the highest mountains, including Mount Washington, which has an elevation of 2063 meters (6288 ft). The ice was more than 2000 meters (6500 ft) thick in the Great Lakes region. In Europe, glaciers spread over most of what is now Great Britain, Ireland, Scandinavia, northern Germany, Poland, and western Russia. The weight of the ice depressed the land surface several hundred meters. As the ice receded, the land rose by isostatic rebound. Measurable isostatic rebound still raises elevations in parts of Sweden, Canada, and eastern Siberia by up to 2 centimeters (1 in.) per year and it may cause Hudson Bay and the Baltic Sea to emerge someday above sea level. Should Greenland and Antarctica lose their ice sheets, their depressed central land areas would also rise to reach isostatic balance.

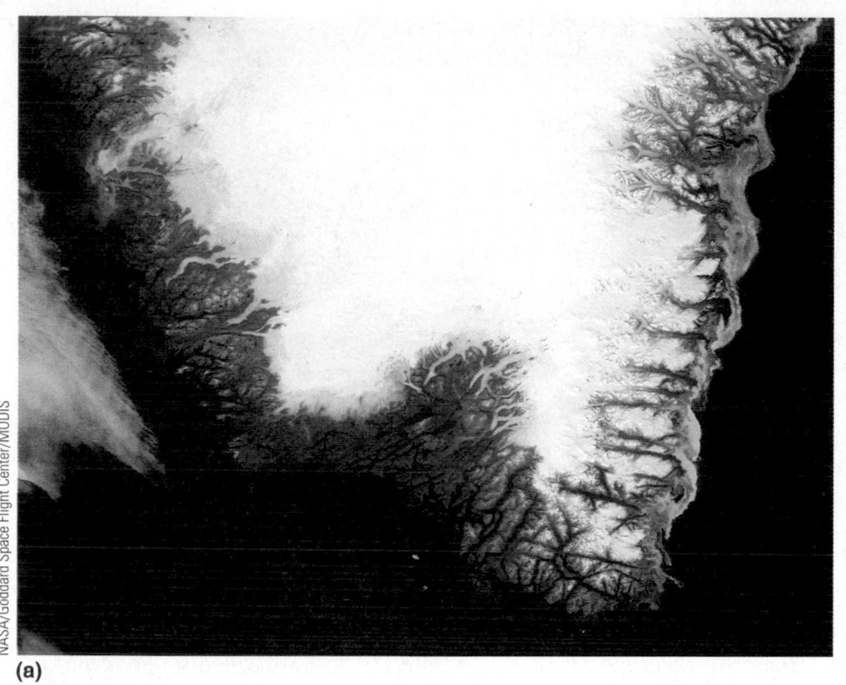

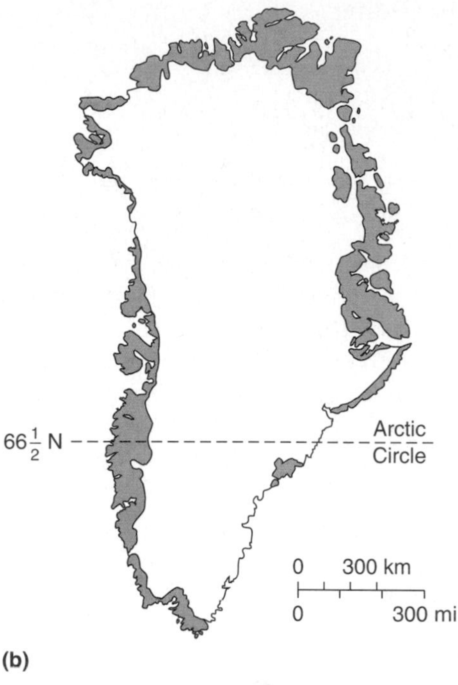

• **FIGURE 19.26**
The Greenland ice sheet. (a) Except for the mountainous edges, the Greenland ice sheet almost completely covers the world's largest island. Ice thickness is more than 3000 meters (10,000 ft) and depresses the bedrock below sea level. In this satellite view, several outlet glaciers flow seaward from the ice sheet to the east coast of the island. (b) The extent of the Greenland ice sheet.

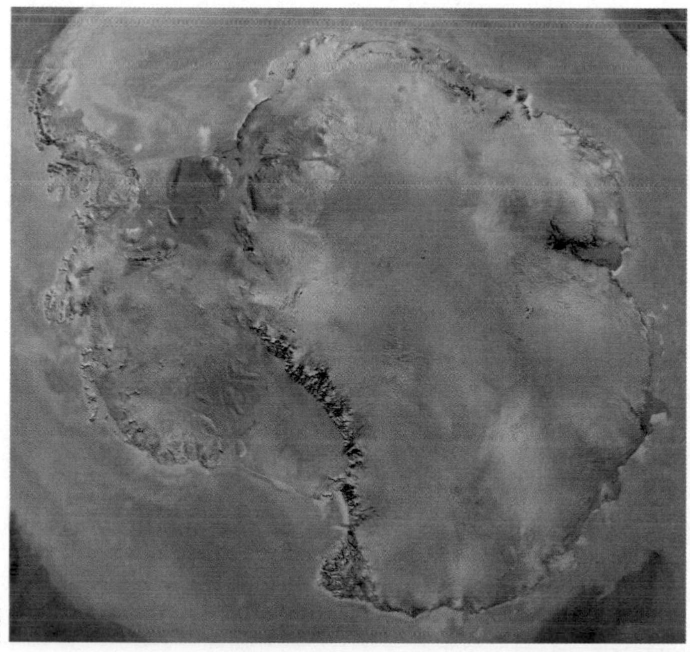

• **FIGURE 19.27**
The Antarctic ice sheet. The world's largest ice sheet covers an area larger than the United States and Mexico combined, and has a thickness of more than 4500 meters (14,000 ft). This satellite mosaic image covers the whole south polar continent. Notice that most of Antarctica is ice-covered (white and blue on the image). The only rocky areas (darker areas on the image) are the Antarctic Peninsula and the Transantarctic Mountains. Large ice shelves flow to the coastline, the largest being the Ross Ice Shelf. The image is oriented with the Greenwich Meridian at the top.

The geomorphic effects of the last major glacial advance, known in North America as the Wisconsinan stage, are the most visible in the landscape today. The glacial landforms created during the Wisconsinan stage, which ended about 10,000 years ago, are relatively recent and have not been destroyed to any great extent by subsequent geomorphic processes. Consequently, we can derive a fairly clear picture of the extent and actions of the ice sheets, as well as alpine glaciers, at that time.

Where did the water locked up in all the ice and snow come from? Its original source was the oceans. During the periods of glacial advance, there was a general lowering of sea level, exposing large portions of the continental shelf and forming land bridges across the present-day North, Bering, and Java Seas. The most recent Pleistocene melting and glacial retreat raised sea level a similar amount—about 120 meters (400 ft). Evidence for this rise in sea level can be seen along many coastlines around the world.

Continental Glaciers and Erosional Landforms

Ice sheets and ice caps erode the land through plucking and abrasion, carving landscape features that have many similarities to those of alpine glaciers, but on a much larger scale. Erosional landforms created by ice sheets are far more extensive than those formed by alpine glaciation, stretching over millions of square kilometers of North America, Scandinavia, and Russia. As ice sheets flowed over the land, they gouged Earth's surface with striations, enlarged valleys that already existed, scoured out rock basins, and

GEOGRAPHY'S SPATIAL SCIENCE PERSPECTIVE

The Driftless Area—A Natural Region

The Driftless Area, a region in the upper midwestern United States, mainly covers parts of Wisconsin, but also sections of the adjacent states of Iowa, Minnesota, and Illinois. This region was not glaciated by the last Pleistocene ice sheets that extended over the rest of the northern Midwest, so it is free of glacial deposits from that time. Drift is a general term for all deposits of glacial ice and its meltwater, thus the regional name is linked to its geomorphic history. There is some debate about whether earlier Pleistocene glaciations covered the region, but the most recent glacial advance did not override this locality.

The landscape of the Driftless Area is very different from the muted terrain and low rounded hills of adjoining glaciated terrain. Narrow stream valleys, steep bluffs, caves, sinkholes, and loess-covered hills produce a scenic landscape with many landforms that would not have survived erosion by a massive glacier.

In some parts of the United States, lobes of the ice sheet extended as much as 485 kilometers (300 mi) farther south than the latitude of the Driftless Area. Why this region remained glacier-free is a result of the topography directly to the north. A highland of resistant rock, called the Superior Upland, caused the front of the glacier to split into two masses, called *lobes*, diverting the southward-flowing ice around this topographic obstruction.

The diverging lobes flowed into two valleys that today hold Lake Superior and Lake Michigan. These two lowlands were oriented in directions that channeled the glacial lobes away from the Driftless Area. Although not overridden by these glaciers, the Driftless Area did receive some outwash deposits and a cover of wind-deposited loess, both derived from the surrounding glaciers and their sediments.

The Driftless Area has a unique landscape because of its isolation, an unglaciated "island" or "peninsula" almost completely surrounded by extensive glaciated regions. The rocks, soils, terrain, relief, vegetation, and habitats contrast strongly with the nearby drift-covered terrains and provide a glimpse of what the preglacial landscapes of this midwestern region may have been like. The Driftless Area also provides distinctive habitats for flora and fauna that do not exist in the adjacent glaciated terrain. Unique and unusual aspects of its topography and ecology attract ecotourists and offer countless opportunities for scientific study. For these reasons, there are many protected natural lands in the Driftless Area, an excellent example of a natural region defined by its physical geographic characteristics.

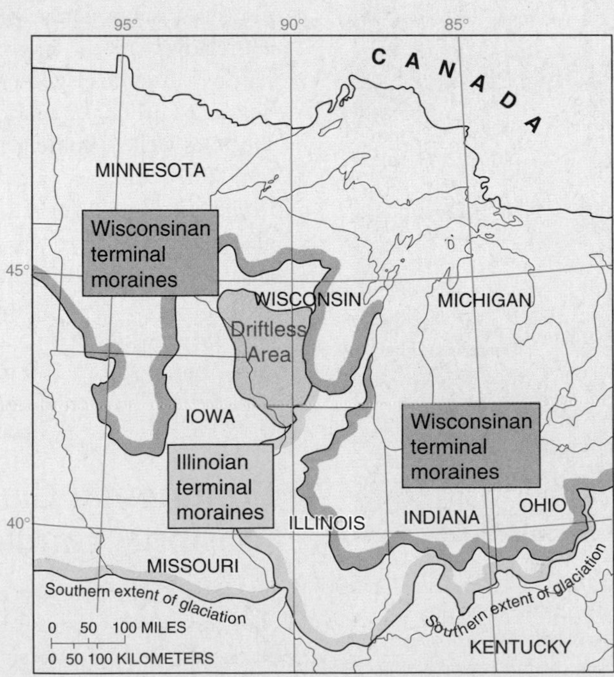

A map of the Midwest shows that the Driftless Area was virtually surrounded by the advancing Pleistocene ice sheets yet was not glaciated.

• FIGURE 19.28
(a) Forming a coastal buffer of ice between the open ocean and the ice-covered continent, Antarctica's flat-topped ice shelves are impressively beautiful. (b) In recent years, the ice shelves have experienced significant destruction, in some cases generating huge icebergs, as with the breakup of the Ross Ice Shelf in 2000. This image records the subsequent collapse of Antarctica's Larsen B Ice Shelf, which occurred in 2002.

• FIGURE 19.29
Antarctica's flat-topped (tabular) icebergs, here with penguin passengers, look quite different from the irregularly shaped icebergs of the Northern Hemisphere.
What portion of an iceberg is hidden below the ocean surface?

• FIGURE 19.30
Glacial ice coverage in the Northern Hemisphere was extensive during the Pleistocene. Glaciers up to several thousand meters thick covered much of North America and Eurasia.
What might be a reason for some areas that were very cold during this time, such as portions of interior Alaska and Siberia, being ice-free?

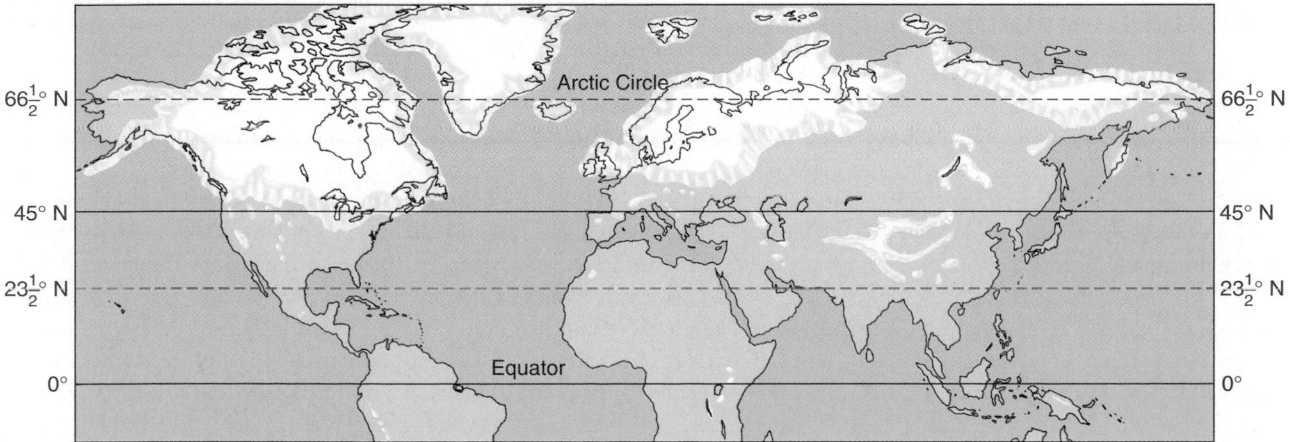

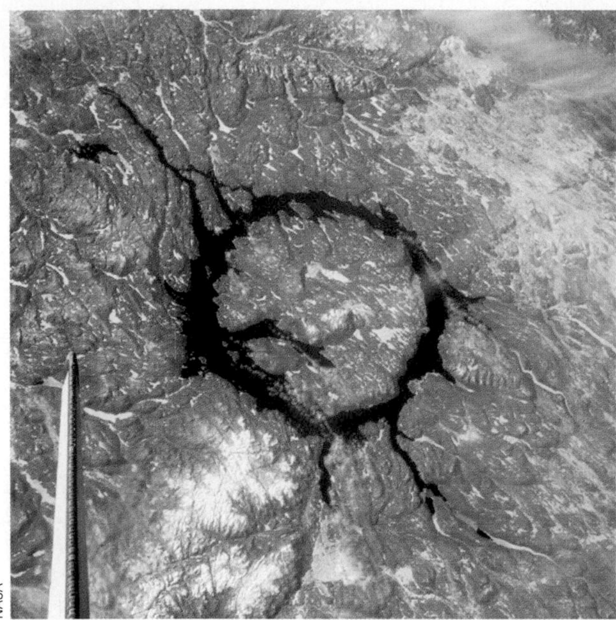

• **FIGURE 19.31**
Seen from the space shuttle, Lake Manicouagan on the Canadian Shield in Quebec occupies a circular depression that continental glaciers erosionally enhanced in rocks surrounding an asteroid impact structure. Although the overall relief in the region is low, note how erosion has exploited fractures to make linear valleys and lakes. The annular (ring-shaped) lake has a diameter of approximately 70 kilometers (43 mi).

smoothed off existing hills. The ice sheets removed most of the soil and then eroded the bedrock below. Today, these **ice-scoured plains** are areas of low, rounded hills, lake-filled depressions, and wide exposures of bedrock (• Fig. 19.31).

When ice sheets expand, they cover and totally disrupt the former stream patterns. Because the last glaciation was so recent in terms of landscape development, new drainage systems have not had time to form a well-integrated system of stream channels. In addition to large expanses of exposed gouged bedrock, ice-scoured plains are characterized by extensive areas of standing water, including lakes, marshes, and muskeg (poorly drained areas grown over with vegetation that form in cold climates).

Continental Glaciers and Depositional Landforms

The sheer disparity in scale causes depositional landforms of ice sheets and ice caps to differ from those of alpine glaciers. Although terminal and recessional moraines, ground moraines, and glaciofluvial deposits are produced by both categories of glaciers, retreating continental glaciers leave significantly more extensive versions of these features than alpine glaciers do (• Fig. 19.32 and Map Interpretation: Continental Glaciation).

End Moraines Terminal and recessional end moraines deposited by Pleistocene ice sheets comprise substantial belts of low hills and ridges on the land in areas affected by continental glaciation. These landforms generally range up to about 60 meters (200 ft) in height (• Fig. 19.33). The last major Pleistocene glacial advance through New England left its terminal moraine running the length of New York's Long Island and created the offshore islands of Martha's Vineyard and Nantucket, Massachusetts. Glacial retreat left recessional moraines forming both Cape Cod and the rounded southern end of Lake Michigan. Both types of end moraine are usually arc-shaped and convex toward the direction of ice flow. Their pattern and placement indicate that the ice sheets did not maintain an even front but spread out in tongue-shaped lobes channeled by the underlying terrain (• Fig. 19.34). The positions of terminal and recessional moraines provide more evidence than simply the direction of ice flow. Examining the characteristics of deposited materials helps us detect the sequence of advances and retreats of each successive ice sheet.

Till Plains In the zone of ice sheet deposition, massive amounts of poorly sorted glacial till accumulated, often to depths of 30 meters (100 ft) or more to form **till plains.** Because of the uneven nature of deposition from the wasting ice, the topographic configuration of land covered by till varies from place to place. In some areas, the till is too thin to hide the original contours of the land, while in other regions, thick deposits of till make broad, rolling plains of low relief. Small hills and slight depressions, some filled with water, characterize most till plains, reflecting the uneven glacial deposition. Some of the best agricultural land of the United States is found on the gently rolling till plains of Illinois and Iowa. The young, dark-colored, grassland soils (mostly mollisols) that developed on the till are extremely fertile.

Outwash Plains Beyond the belts of hills that represent terminal and recessional moraines lie **outwash plains** composed of meltwater deposits. These extensive areas of relatively low relief consist of glaciofluvial deposits that were sorted as they were transported by meltwater from the ice sheets. Outwash plains, which may cover hundreds of square kilometers, are analogous to the valley trains of alpine glaciers.

Small depressions or pits, called **kettles,** mark some outwash plains, till plains, and moraines. Kettles represent places where blocks of ice were originally buried in glacial deposits. When the blocks of ice eventually melted, they left surface depressions, and many kettles now contain **kettle lakes** (• Fig. 19.35). For example, most of Minnesota's famous 10,000 lakes are kettle lakes. Some kettles occur in association with alpine glacial deposits, but the vast majority are found in landscapes that were occupied by ice sheets or ice caps.

Drumlins A **drumlin** is a streamlined hill, often about 0.5 kilometer (0.3 mi) in length and less than 50 meters (160 ft) high, molded in glacial drift on till plains (• Fig. 19.36a). The most conspicuous feature of drumlins is the elongated, streamlined shape that resembles half an egg or the convex side of a teaspoon. The broad, steep end faces in the up-ice direction, while the gently sloping tapered end points in the direction that the ice flowed; thus the geometry of a drumlin is the reverse of that of roches moutonnées. Drumlins are usually found in swarms, called *drumlin fields,* with as many as a hundred or more clustered together.

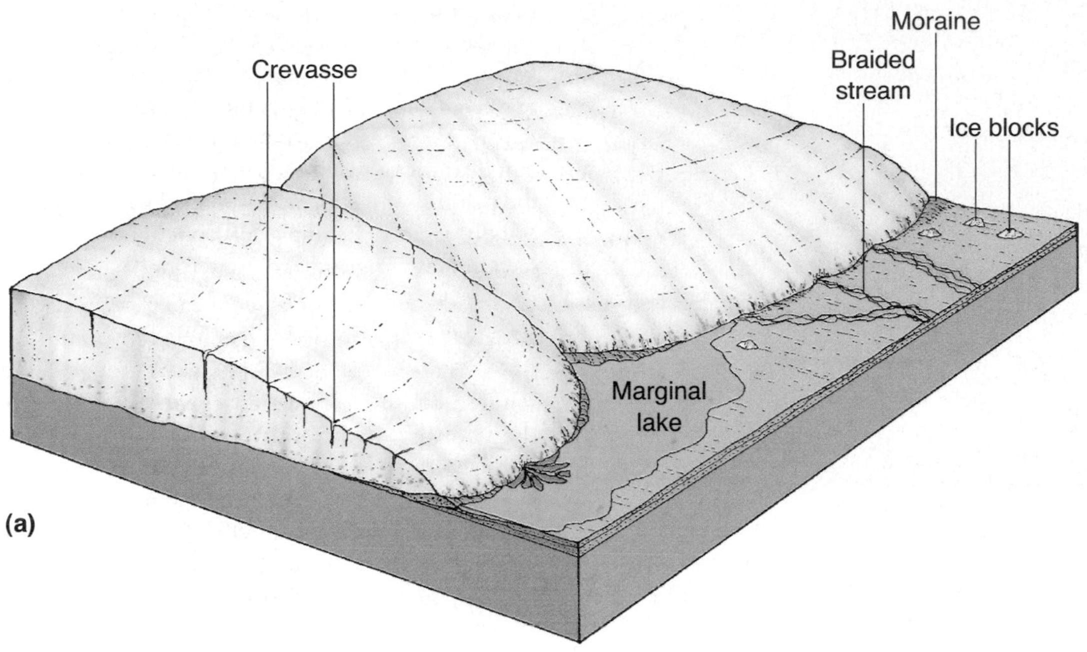

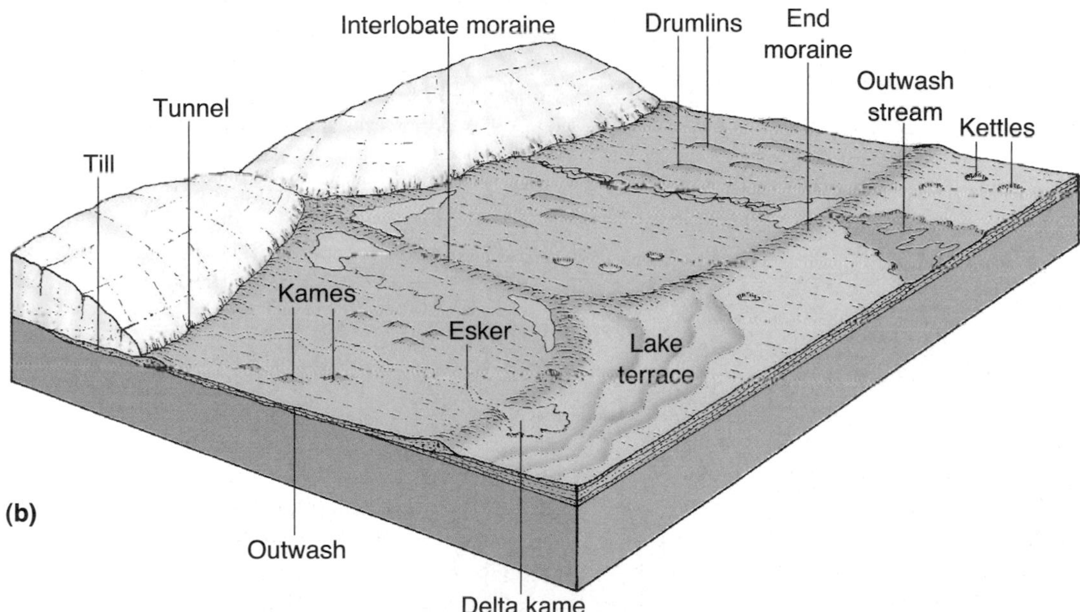

• FIGURE 19.32
Landscape alteration by continental glacier sedimentation creates (a) depositional features associated with ice stagnation at the edge of the glacier, and (b) landforms resulting from further modification of the ice-marginal terrain as the glacier retreats.
How important is liquid water in creating the landforms shown here?

Opinions among scientists differ regarding the origin of drumlins, particularly with respect to the relative importance of ice versus meltwater processes in their formation. Drumlins are well developed in Canada, in Ireland, and in the states of New York and Wisconsin. Boston's Bunker Hill, a drumlin, is one of America's best-known historical sites.

Eskers An **esker** is a narrow and typically winding ridge composed of glaciofluvial sands and gravels (Fig. 19.36b). Some eskers are as long as 200 kilometers (130 mi), although several kilometers is more typical of esker length. Most eskers probably formed by meltwater streams flowing in ice tunnels at the base of ice sheets. Eskers are prime sources of gravel and sand for the construction industry. Being natural embankments, they are frequently used in marshy, glaciated landscapes as highway and railroad beds. Eskers are especially well developed in Finland, Sweden, and Russia.

Kames Roughly conical hills composed of sorted glaciofluvial deposits are known as **kames.** Kames may develop from sediments

© John S. Shelton

• **FIGURE 19.33**
The hilly topography of an end moraine deposited by a continental ice sheet. This aerial photo is of an end moraine on the Waterville Plateau in eastern Washington.
What makes the terrain at the left of the photo appear bumpier compared to the smoother surface of the plain at the right?

that accumulated in glacial ice pits, in crevasses, and among jumbles of detached ice blocks. Like eskers, kames are excellent sources for mining sand and gravel and are especially common in New England. **Kame terraces** are landforms resulting from accumulations of glaciofluvial sand and gravel along the margins of ice lobes that melted away in valleys of hilly regions. Examples of kame terraces can be seen in New England and New York.

Erratics Large boulders scattered in and on the surface of glacial deposits or on glacially scoured bedrock are called **erratics** if the rock they consist of differs from the local bedrock (• Fig. 19.37). Moving ice is capable of transporting large rocks very far from their source. The source regions of erratics can be identified by rock types, which provide evidence of the direction of ice flow. Erratics are known to be of glacial origin because

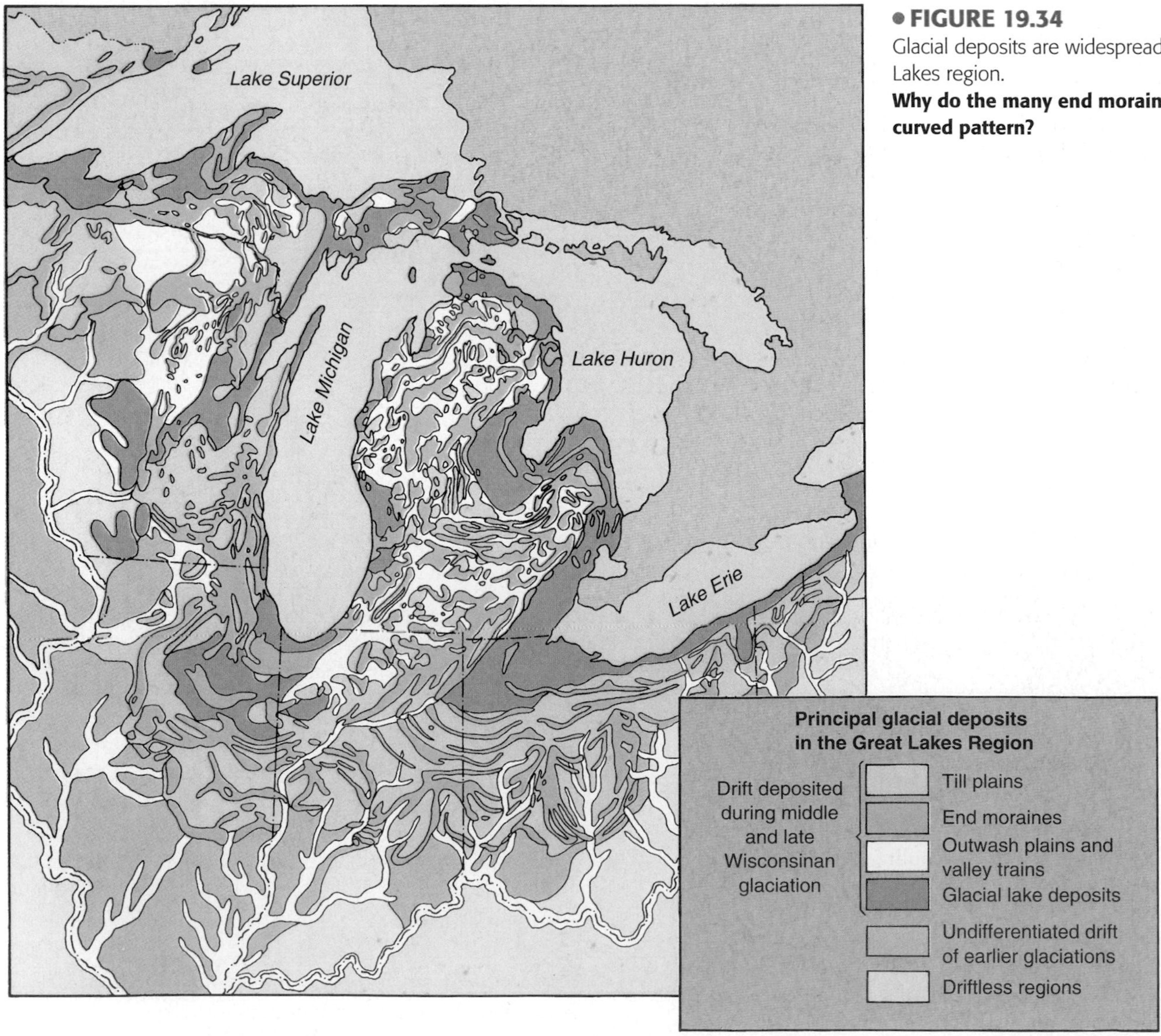

Lake Superior
Lake Michigan
Lake Huron
Lake Erie

Principal glacial deposits in the Great Lakes Region

Drift deposited during middle and late Wisconsinan glaciation:
- Till plains
- End moraines
- Outwash plains and valley trains
- Glacial lake deposits

- Undifferentiated drift of earlier glaciations
- Driftless regions

• **FIGURE 19.34**
Glacial deposits are widespread in the Great Lakes region.
Why do the many end moraines have such a curved pattern?

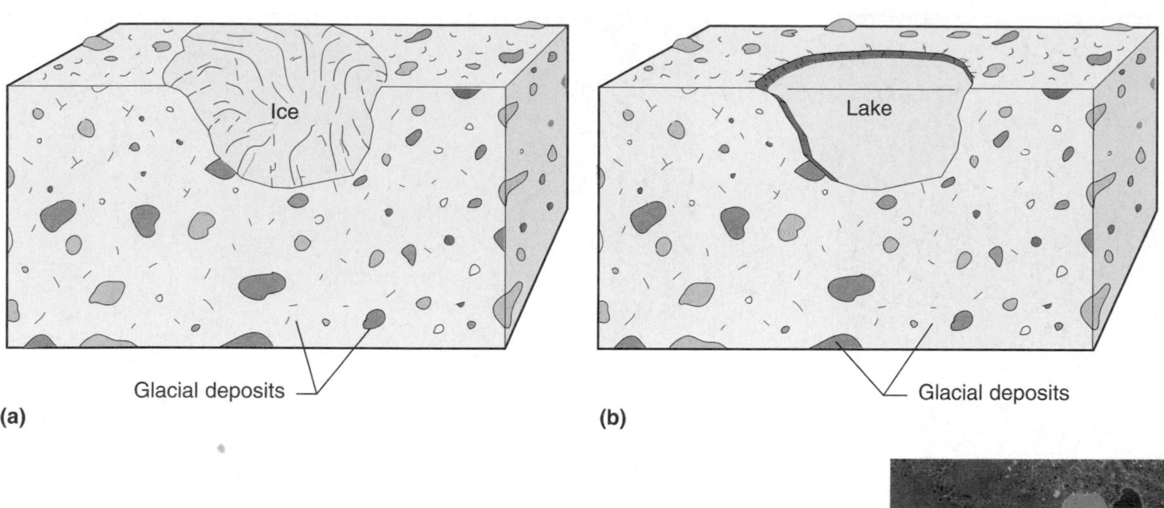

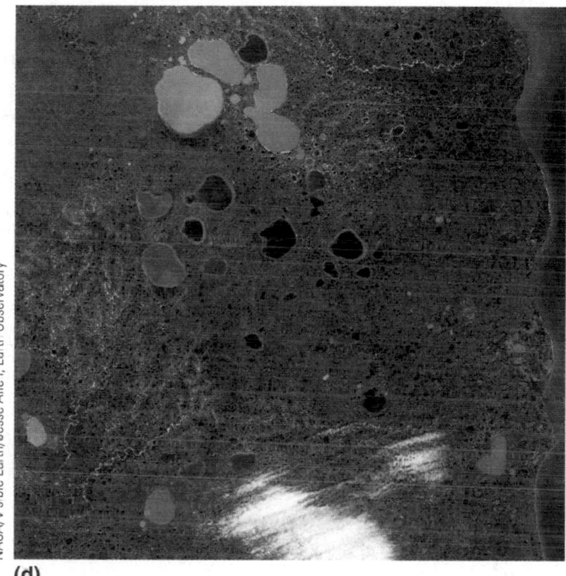

• **FIGURE 19.35**

Kettle lakes form (a) where large, discrete blocks of ice buried or partly buried in glacial deposits (b) melt away leaving a depression. (c) This kettle, near the headwaters of the Thelon River in the Northwest Territories, Canada, formed in an area of ground moraine deposits, as can be seen by the hummocky topography. (d) Numerous large kettle lakes dot part of Siberia due to Pleistocene continental glaciation.

• **FIGURE 19.36**

Drumlins and eskers can be identified by their distinctive shapes. (a) Drumlins, such as this one in Montana, are streamlined hills elongated in the direction of ice flow. (b) An esker near Albert Lea, Minnesota, illustrates the form of these ridges of meltwater sediments deposited in a tunnel under the ice.

What economic importance do eskers have?

© Robert B. Jorstad

• **FIGURE 19.37**
This huge boulder, being examined by a class on a field trip to Yellowstone National Park, is a glacial erratic, transported and deposited by a glacier. **What does this erratic illustrate about the ability of flowing ice to modify the terrain?**

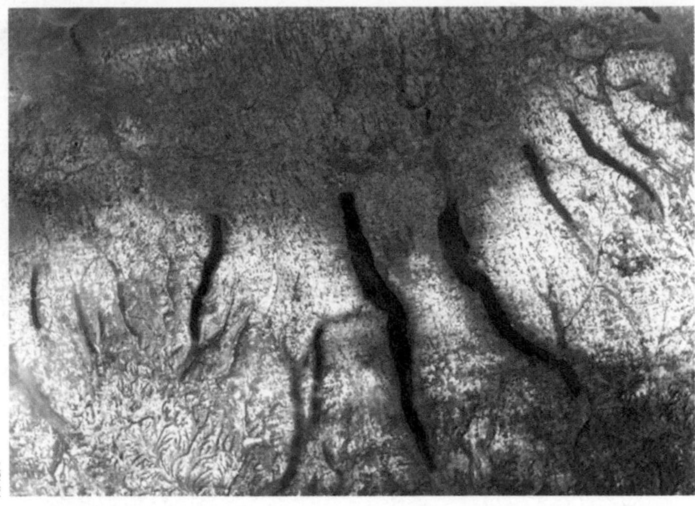

NASA

• **FIGURE 19.38**
A satellite image shows the Finger Lakes of New York that occupy linear, glacially eroded basins excavated during the Pleistocene. A Pleistocene ice sheet glaciated the entire region shown here.
What characteristics of the bedrock caused ice to form these narrow lake basins?

they are marked by glacial striations, and are found only in glaciated terrain. Erratics can occur in association with alpine glaciers, but they are best known and more impressive when deposited by ice sheets, which have moved boulders weighing hundreds of tons over hundreds of kilometers. In Illinois, for example, glacially deposited erratics have come from source regions as far away as Canada.

Glacial Lakes

A recurring theme in this chapter has been the association of lakes with glaciated terrain. Many, many thousands of glacially created lakes exist in depressions within deposits of the continental glaciers that once covered much of North America and Eurasia. The Pleistocene ice sheets created numerous other lake basins by erosion, scooping out deep elongated basins along zones of weak rock or along former stream valleys. New York's beautiful Finger Lakes are excellent examples of lakes in elongated, ice-deepened basins (• Fig. 19.38). Lakes are also common in areas that were impacted by erosion and deposition from alpine glaciers. Lakes in ice-free cirques and glacial troughs are commonly contained on one side by end moraines, as in the case of Washington's Lake Chelan and Lakes Maggiore, Como, and Garda in the Italian Alps (• Fig. 19.39). There is evidence for many other glacial lakes that no longer exist. The *glaciolacustrine* (from *glacial,* ice; *lacustrine,* lake) deposits of these ancient lakes prove their former existence and size.

Some lakes developed while the Pleistocene ice was present where glacial deposition disrupted the surface drainage, or a glacier prevented depressions from being drained of meltwater. These lakes usually accumulated where water became trapped between a large end moraine and the ice front, or where the land sloped toward, instead of away from, the ice front. In both situations, **ice-marginal lakes** filled with meltwater. They drained and ceased to exist when the retreat of the ice front uncovered an outlet route for the water body.

During their existence, fine-grained sediment accumulated on the floors of these ice-marginal lakes, filling in topographic irregularities. As a result of this sedimentation, extremely flat surfaces characterize glacial plains where they consist of glaciolacustrine deposits. An outstanding example of such a plain is the valley of the Red River in North Dakota, Minnesota, and Manitoba. This plain, one of the flattest landscapes in the world, is of great agricultural significance because it is well suited to growing wheat. The plain was created by deposition in a vast Pleistocene lake held between the front of the receding continental ice sheet on the north and moraine dams and higher topography to the south. This ancient body of water is named Lake Agassiz for the Swiss scientist who early on championed the theory of an ice age. The Red River flows northward eventually into the last remnant of Lake Agassiz, Lake Winnipeg, which occupies the deepest part of an ice-scoured and sediment-filled lowland.

Another ice-marginal lake in North America produced much more spectacular landscape features, but not in the area of

From Mortara G., Mercalli L., 2002–Il lago epiglaciale "Effimero" sul ghiacciaio del Belvedere, Macugnaga, Monte Rosa. Nimbus, n. 23–24, p. 10–17.

• **FIGURE 19.39**
After withdrawal of the ice, end moraines often contribute to the formation of closed depressions in which glacial lakes accumulate.

© Jeff Schmaltz, MODIS Rapid Response Team, NASA/GSFC

• FIGURE 19.40
Much of eastern Washington State (tan colored) is called the channeled scablands because of the gigantic, largely abandoned river channels that cross the region (dark gray). These channels represent huge outpourings of water that occurred when glacial ice damming large lakes in the Pleistocene failed, releasing an amount of water perhaps ten times the flow of all rivers of the world today.

the lake itself. In northern Idaho, a glacial lobe moving southward from Canada blocked the valley of a major tributary of the Columbia River, creating an enormous ice-dammed lake known as Lake Missoula. This lake covered almost 7800 square kilometers (3000 sq mi) and was 610 meters (2000 ft) deep at the ice dam. On occasions when the ice dam failed, Lake Missoula emptied in tremendous floods that engulfed much of eastern Washington. The racing floodwaters scoured the basaltic terrain, producing Washington's channeled scablands consisting of intertwining steep-sided troughs (*coulees*), dry waterfalls, scoured-out basins, and other features quite unlike those associated with normal stream erosion, particularly because of their gigantic size (● Fig. 19.40).

The Great Lakes of the eastern United States and Canada make up the world's largest lake system. Lakes Superior, Michigan, Huron, Erie, and Ontario occupy former river valleys that were vastly enlarged and deepened by glacial erosion. All the lake basins except that of Lake Erie have been gouged out to depths below sea level and have irregular bedrock floors lying beneath thick blankets of glacial till. The history of the Great Lakes is exceedingly complex, resulting from the back-and-forth movement of the ice front that produced many changes of lake levels and overflow in varying directions at different times (● Fig. 19.41).

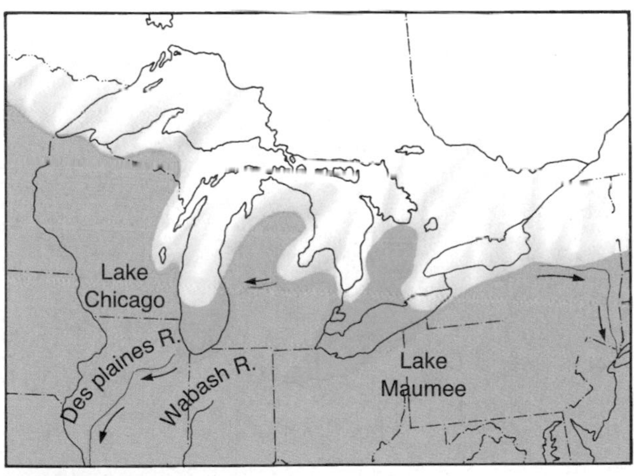

(a) Glacial retreat

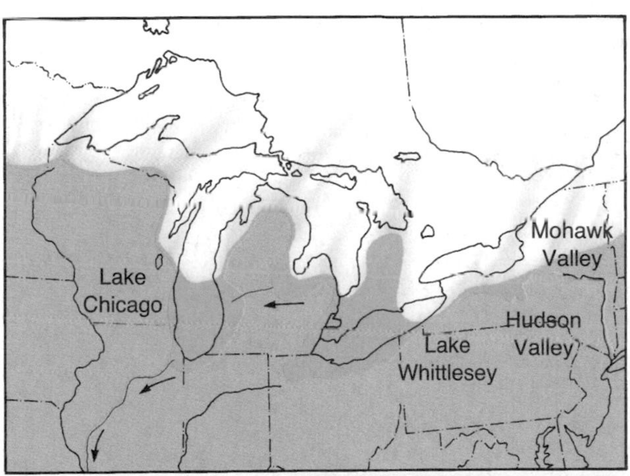

(b) Port Huron glacial advance

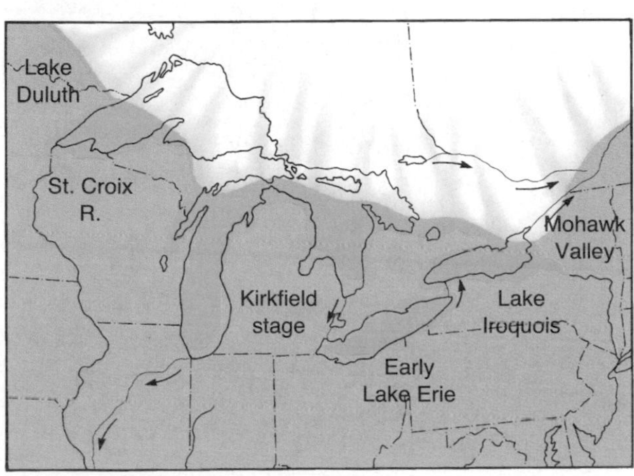

(c) Glacial retreat (post-Valderan)

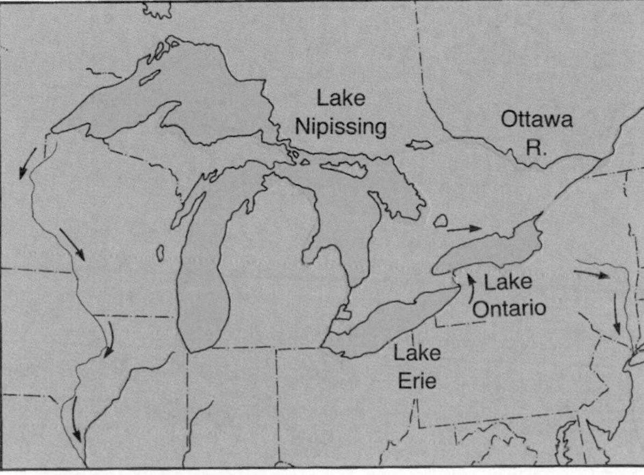

(d) Postglacial Great Lakes

• FIGURE 19.41
The Great Lakes of North America formed as the ice sheet receded at the close of the Pleistocene Epoch.
Name and locate the five Great Lakes.

Periglacial Landscapes

Not all cold regions have sufficient precipitation to lead to permanent accumulation of thick masses of ice. In much of Siberia and interior Alaska, it was too cold and dry during the Pleistocene to generate the massive snow and ice accumulations that occurred elsewhere in North America and northwestern Europe. Instead of glacial processes, these **periglacial** environments (*peri,* near) lacking year-round snow or ice cover undergo intense frost action and are frequently regions of permafrost (permanently frozen ground). Large areas of periglacial terrain exist today in Alaska, Canada, Russia, and some areas of high elevation, including in mountain and plateau regions of China. During the Pleistocene ice advances, periglacial environments also migrated to lower latitudes and lower elevations, leaving relict periglacial features in places, including parts of the Appalachians, where it is no longer actively forming today.

The intense frost action of periglacial landscapes includes freezing of soil moisture, and produces angular, shattered rocks. Frost action also causes heaving, thrusting, and size-sorting of stones in the soil that lead to formation of fascinating repeating designs in **patterned ground** features (● Fig. 19.42).

Where mean annual temperatures are cold enough, ice-free landscapes develop permafrost (● Fig. 19.43). As we saw in Chapter 15, extensive areas of permafrost result in considerable *solifluction* (slowly flowing soil) on slopes when the active (upper) layer of the permanently frozen ground thaws in the summer and becomes saturated. Permafrost areas are also prone to the formation of fissures that accumulate ice, leading to the formation of large **ice wedges** in the ground.

Understanding permafrost is important for human activity in periglacial regions. Unless proper construction techniques are used, erecting buildings, roads, pipelines, and other structures on permafrost disrupts the natural thermal environment, often leading to permafrost melting. Saturated ground cannot support the weight of structures resting on top of them. As the ground deforms and slowly flows, the structures are destroyed. To avoid these problems, construction techniques have been developed that keep the permafrost frozen or allow it to experience its natural temperature fluctuations. The latter is generally accomplished by building above the surface so that air can circulate to the permafrost.

● FIGURE 19.42
Intensive frost action in periglacial regions leads to the formation of intricate, repetitive patterned ground features.

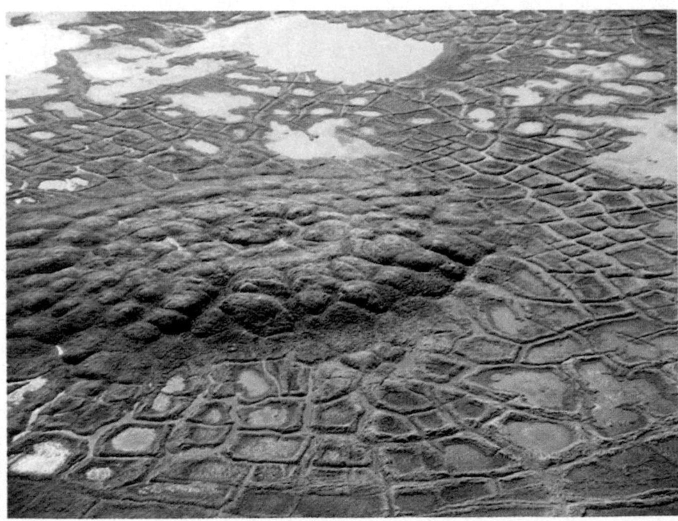

© Emma Pike

● FIGURE 19.43
Many periglacial regions contain permafrost. This map shows the distribution of permafrost in the Northern Hemisphere.

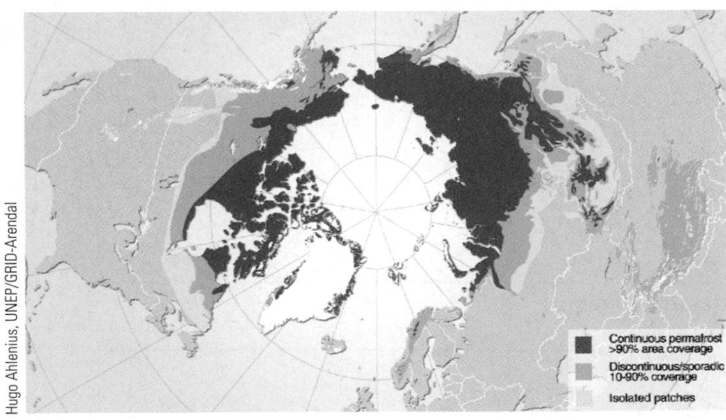

Hugo Ahlenius, UNEP/GRID-Arendal

Chapter 19 Activities

Define & Recall

glacier
firn
accumulation
ablation
sublimation
calving
iceberg
alpine glacier
valley glacier
piedmont glacier
cirque
cirque glacier
continental glacier
ice sheet
ice cap
crevasse
icefall
plucking
till
zone of accumulation
zone of ablation
equilibrium line
glacier head
glacier terminus (toe)
glacier advance
glacier retreat
striation
roche moutonnée
tarn
arête
horn
col
glacial trough
paternoster lakes
fjord
hanging valley
drift
moraine
lateral moraine
medial moraine
end moraine
terminal moraine
recessional moraine
ground moraine
glacial outwash
valley train
outlet glacier
ice shelf
ice-scoured plain
till plain
outwash plain
kettle
kettle lake
drumlin
esker
kame
kame terrace
erratic
ice-marginal lake
periglacial
patterned ground
ice wedge

Discuss & Review

1. How does glacial ice differ from snow?
2. Explain how glaciers move.
3. What are the three main types of alpine glacier and what distinguishes them from each other?
4. What are glacial advance and retreat, and what is their relationship to a glacial state of equilibrium?
5. Diagram and label the characteristic parts of an alpine glacier.
6. How do glaciers accumulate load? Provide some examples of evidence of glacial erosion and movement.
7. What landscape features would you look for to determine if a high-mountain region currently without glaciers had experienced a previous period of glaciation?
8. What are some major similarities and some major differences between continental and alpine glaciers?
9. Distinguish till from outwash and drift.
10. Where are the two major existing ice sheets located? How do they compare in area and thickness?
11. How does the present extent of continental ice-sheet coverage compare with the maximum extent of the Pleistocene ice-sheet coverage?
12. How has ice-sheet erosion altered the landscape? What kinds of landscape features are produced by continental glacier deposition and recession?
13. What landscape features would indicate that a region is undergoing periglacial processes?

Consider & Respond

1. How do alpine glaciers differ from streams in terms of flow processes, erosion processes, load characteristics, valley shaping, and nature of tributary valleys?
2. Name four coastal areas of the world where fjords can be found. What is the dominant climate type in coastal areas with fjords? Why are fjords associated with that climate? In what way are fjords related to both alpine and continental glaciers?
3. Describe the ice budget of a glacial system. What two major factors control this budget? How is the ice budget related to the movement of glaciers? Explain how a glacier that is calving can also be undergoing advance.
4. Glaciation is truly an interdisciplinary topic. In addition to physical geographers, what other scientists do you think are involved in the study of glaciers and why?

Apply & Learn

1. The position of the toe of a valley glacier in the Northern Hemisphere was measured annually during the 1970s, then again during the 1990s. Each time, the position was measured at the ice minimum, on September 30, as elevation above sea level. The following tables contain those data.
   a. What was the average elevation of the glacier toe during the decade of the 1970s, and during the decade of the 1990s?
   b. Describe the glacier's relative mass balance over each of the two decades.
   c. What was the average annual rate of change in the glacier's position in each of the two decades?
   d. Summarize how the behavior of the glacier has changed in the 1990s compared to the 1970s.

Year	1970	1971	1972	1973	1974	1975	1976	1977	1978	1979
Elevation (meters)	2634	2631	2632	2629	2630	2629	2627	2625	2624	2623

Year	1990	1991	1992	1993	1994	1995	1996	1997	1998	1999
Elevation (meters)	2642	2648	2654	2661	2668	2676	2684	2692	2701	2710

2. Some diamonds were recently found at three separate sites in Canada in till deposited by the Pleistocene ice sheet. At Site A, nearby ice-scoured troughs now occupied by lakes are elongated from NNE to SSW. Site B lies 60 kilometers (37 mi) west of Site A, and striations in adjacent bedrock show a NE to SW orientation. Lying 60 kilometers (37 mi) east of Site A, Site C has nearby drumlins with their tail end pointed SSE and their blunt end NNW. Using this information, how might you proceed in narrowing down a search area for the bedrock from which the diamonds originated?

Locate & Explore

Note: Please read the About Locate & Explore Activities section of the Preface before beginning these exercises.

1. Using Google Earth and the Glacial Layer at a glacier in Alaska, identify the marked features (A–E). Rank the features from oldest to youngest with an explanation for your order.

Tip: Use the zoom, tilt, rotate, and elevation exaggeration tools to help view and interpret the landform and the area in which it is found.

   a. 56.973822°N, 131.763523°W
   b. 56.971071°N, 131.796291°W
   c. 56.969477°N 131.832572°W
   d. 56.977346°N, 131.812205°W
   e. 56.986124°N, 131.823604°W
   f. 56.983097°N, 131.831811°W
   g. 56.993715°N, 131.856154°W
   h. 56.982203°N, 131.840695°W

2. Using Google Earth and the Topographic Map Layer for Gilkey Glacier from the USGS, look at the change in the glacier (58.76°N, 134.57°W) between 1948 and today and answer the following questions:
   a. How far has the glacier retreated? What feature remains in place of the glacier? Tip: Use the measuring tool.
   b. What is the average rate of retreat (feet/year) between 1948 and today?
   c. Battle Glacier used to be joined to Gilkey Glacier, but it has also retreated. What river now occupies the valley formerly filled by Battle and Three Glaciers? Why is this type of river commonly associated with glaciers?
   d. What glacial landform is found along the edge of the valley of Battle Glacier (58.743911°N, 134.582156°W)?

Tip: Use the zoom, tilt, rotate, and elevation exaggeration tools to help view and interpret the landform and the area in which it is found.

3. Using Google Earth, identify the landforms at the following locations (latitude, longitude) and provide a brief discussion of how the landform developed. Include a brief discussion of why the landform is found in that general area of the United States or the world. Based on the orientation and shape of the landform, which direction was the glacier moving in that area?

Tip: Use the zoom, tilt, rotate, and elevation exaggeration tools to help view and interpret the landform and the area in which it is found.

   a. 59.75°N, 140.59°W
   b. 42.73°N, 76.73°W
   c. 37.72°N, 119.66°W
   d. 54.06°N, 98.71°W
   e. 61.34°N, 25.47°E
   f. 44.52°S, 168.84°E
   g. 42.35°N, 71.92°W
   h. 58.44°N, 107.58°W

Map Interpretation

ALPINE GLACIATION

The Map

The Chief Mountain, Montana, map area is located in Glacier National Park, which adjoins Canada's Waterton Lakes National Park located across the international border. Alpine glaciation has created much of the spectacular scenery of this region of the northern Rocky Mountains.

Most of the glaciated landscape in this region was produced during the Pleistocene. Today, glaciers still exist in Glacier National Park, but because of rapid melting in recent decades, it is estimated that they may be gone within 70 years. Prior to glaciation, this region was severely faulted and folded during the formation of the northern Rocky Mountains.

The photograph and map clearly show the rugged nature of the terrain of this map area. Steep slopes, horns, U-shaped valleys, lakes, arêtes, and glaciers are obvious landform characteristics of alpine glaciation. Temperature is a primary control of the highland climate of this region. Elevation, in turn, influences temperature and precipitation amounts.

As you would expect, the rapid decrease in temperature with increasing elevation results in a variety of microclimates within alpine regions. Exposure is also an important highland climate control. West-facing slopes receive the warm afternoon sun, whereas east-facing slopes are sunlit only in the cool of the early morning.

Interpreting the Map

1. What is the approximate amount of local relief depicted on this map? How does the scale of this topographic map compare to those previously shown in the book (is this map of larger or smaller scale)?
2. Examining the topographic map and photograph of this mountain region, do you think most of the glaciated landscape was produced by erosion or deposition?
3. Locate Grinnell, Swiftcurrent, and Sperry Glaciers. What specific type of glaciers are they?
4. What evidence indicates that the glaciers were once larger and extended farther down the valleys?
5. Note that most of the existing glaciers are located to the northeast of the mountain summits. Explain this orientation. At what elevation are the glaciers found?
6. What types of glacial landform are the following features?
   a. The features occupied by Kennedy, Iceberg, and Ipasha Lakes
   b. The feature occupied by McDonald Lake (in the southwest corner of the map) and Lake Josephine
   c. Mount Gould and Mount Wilbur
   d. The series of lakes that occupy the valley of Swiftcurrent Creek
7. Along the high ridges runs a dashed line labeled "Continental Divide." What is its significance?
8. If you were to hike southeast from Auto Camp on McDonald Creek up Avalanche Creek to the base of Sperry Glacier, how far would you travel, and how much elevation change would you encounter?

D. Sack

Glaciated terrain in Glacier National Park.

Opposite:
Chief Mountain, Montana
Scale 1:125,000
Contour interval = 100 ft
U.S. Geological Survey

Mt Kipp
Fifty Mtn Camp
Kootenai Pass
West Flattop Mtn.
GLACIER
FLATHEAD
FLATTOP MOUNTAIN
Ipasha Lake
Ipasha Pk
Ipasha Glacier
Ahern Glacier
Helen Lake
Ptarmigan Wall
Ptarmigan Lake
Hedgap Pass
Kennedy
Pota Lake
Appekunny Mtn
Kennedy Lake
Ahern Pk
Ahern Pass
Ahern Cr
Cattle Queen Cr
Ptarmigan Falls
Iceberg Cr
Mt Henkel
Natahki Lake
Altyn Pk
Appekunny Falls
Iceberg Lake
Mt Wilbur
Many Glacier
Chalet
Swiftcurrent Lake
Lake Sherburne
North Swiftcurrent Glacier
Redrock Falls
Bullhead Lake
Grinnell Pt
Swiftcurrent Mtn
Swiftcurrent Pass
Mt Grinnell
Lake Josephine
Swiftcurrent Glacier
Grinnell Lake
Grinnell Falls
Grinnell Glacier
Snow Moon Lake
Wynn Mtn
Morning Eagle Falls
Allen Mtn
Feather Plume Falls
Granite Park
Chalet
Mineral Creek
Flattop Cr
McDonald Creek
Longfellow Cr
Lake Evangeline
Longfellow Peak
Camas Lake
RANGE
The Garden Wall
Mt Gould
Cataract Creek
Cracker Lake
Siyeh Glacier
MT SIYEH
Haystack Butte
Logan Cr
Cataract Mtn
Piegan Pass
Siyeh Pass
Preston Park
Otokomi Lake
Pollock Mtn
Piegan Mtn
Piegan Glacier
Glacier Wall
Heavens Pk.
Arrow L
Mt Oberlin
Oberlin Falls
Matahpi Peak
Sexton Glacier
Logan Pass
HIGHWAY
Going-to-the-Sun Mtn
McPartland Mtn
Mt Cannon
Clements Mtn
Hanging Gardens
Reynolds Cr
Baring Falls
Mt Vaught
Auto Camp
Hidden Cr
Hidden Lake
Heavy Runner Mtn
Trout Lake
Avalanche Cr
Bearhat Mtn
Reynolds Mtn
Mirror Pond
Stanton Mtn
Avalanche Lake
St.Mary Falls
Virginia Falls
Avalanche Basin
Twin Lakes
Florence Falls
St. Mary River
Ranger Sta.
McDonald Falls
Mt.Brown
Fusillade Mtn
Citadel Mtn
Kellys Camp
Snyder Lake
Edwards Mtn
Sperry Glacier
Gunsight Lake
Almost-a-Dog Mtn
Lake McDonald
Snyder Cr
Akaiyan Lake
Gunsight Mtn
Gunsight Pass
Sperry Chalet
Glacier Basin
Lake Ellen Wilson
MT JACKSON
Blackfoot Glacier
Logan Glacier
Mt. Logan
Sprague Creek
Fish Lake
Lincoln Pk
Lincoln Lake
Lincoln Falls
Harrison Glacier
Red Eagle Glacier
Blackfoot Mtn
Pumpelly
Walton Mtn
CONTINENTAL DIVIDE
GLACIER NATIONAL PARK
LOOKOUT

Map Interpretation

CONTINENTAL GLACIATION

The Map

The Jackson, Michigan, map area is located in the Great Lakes section of the Central Lowlands. Massive continental ice sheets covered this region during the Pleistocene. As the glaciers melted and the ice sheets retreated, a totally new terrain, vastly different from that of preglacial times, was exposed. Today, evidence of glaciation is found throughout the region. The most obvious glacially produced landforms are the thousands of lakes (including the Great Lakes), the knobby terrain, and moraine ridges. Moraines left by advancing and retreating tongue-shaped ice lobes that extended generally southward from the main continental ice sheet also influenced the shapes of the Great Lakes.

The ice sheet, its deposits of sediment, and its meltwaters also created other, smaller-scale features. The glaciers left a jumbled mosaic of deposits from boulders through smaller gravel to sand, silt, and clay that has produced a hilly and hummocky terrain. The overall relief is low, in part because of glacial erosion. Many landform features of continental glaciation are well illustrated on the Jackson, Michigan, 7.5′ quadrangle map.

This region has a humid continental climate. The summers are mild and pleasant. Excessively warm and humid air seldom reaches this area for more than a few days at a time. Instead, cool but pleasant evening temperatures tend to be the rule in summer. Winters, however, are long and often harsh. Snow can be abundant and on the ground continuously for many weeks or even months at a time. The annual temperature range is quite large; precipitation occurs year-round provided primarily by midlatitude cyclonic storms.

Interpreting the Map

1. Describe the general topography of the map area.
2. What is the local relief? Why is it so difficult to find the exact highest and lowest points on this map?
3. Does the topography of this region indicate glacial erosion or glacial deposition?
4. Does the area appear to be well drained? What are the three main hydrographic features that indicate the drainage conditions?
5. What type of glacial landform is Blue Ridge? What are its dimensions (length and average height)?
6. How is a feature such as Blue Ridge formed? What economic value might it have?
7. What is the majority of the surface material in the map area? What is the term for this type of surface cover?
8. What probably caused the many small depressions and rounded lakes? What are these depressions called?
9. Describe how the topography of the Jackson, Michigan, area appears on the satellite image.
10. What is the advantage of having both satellite image and topographic map coverage of an area of study?

Michigan Center for Geographical Information

Digital aerial image of the region near Jackson, Michigan.

Opposite:
Jackson, Michigan
Scale 1:62,500
Contour interval = 10 ft
U.S. Geological Survey

Williams Lake
Brown Lake
Vandercook Lake
Hutchins Sch
Cullum Sch
Grand
Peter White Lake
Greens Lake
Gravel Sch
Ackerson Lake
Blue Ridge
Cranberry Lake
School House Lake
Eldred Sch
N A P
Spring Brook Sch
Prospect Hill
River
Crouch Sch
NORTHERN
CINCINNATI
MERIDIAN
Kerr Sch
Skiff Lake
East Liberty Sch
Clarkdale
Clark
Mud Lake
L I B E R T Y
Crispell Lake
River
Liberty
Grand
Infant Sch
Putney Mill Pond
MICHIGAN
TRANSMISSION LINE
Sutlin Sch
Grand Lakes
JACKSON CO
HILLSDALE CO
Brazee Lake
Crystal Lake
CENTRAL
Lime Lake
Perch Lake
Goose Lake
Bunday Hill

Coastal Processes and Landforms

20

CHAPTER PREVIEW

The ever-changing coastal regions of Earth are greatly affected by human activities.

- Why are coasts such dynamic environments?
- What are some principal ways in which people interfere with natural processes acting in the coastal zone?

Waves are the dominant geomorphic agent acting in the coastal zone.

- What are the three major types of water waves?
- How do they differ in their effects on the coastline?

Waves are powerful agents of erosion that create spectacular scenery.

- How is wave erosion similar to erosion by streams?
- Besides those related to waves, what other processes contribute to coastal erosion?

Beaches represent a balance between input and removal of sediment by waves and currents.

- Of what types of material are beaches made?
- How do coastal currents affect the formation of beaches?

Coasts are dynamic and complex systems that are hard to classify because of their great variety and changeable nature.

- Based on global tectonics, what are the two major types of coastline?
- What are the differences between coastlines of emergence and coastlines of submergence?

◀ **Opposite: The shorelines of the world are extensive, complex, and often spectacular environments.**

© Richard Price/ Getty Images

The world ocean covers 71% of Earth's surface, and the boundary between the ocean and dry land is of enormous length. A large percentage of the world's population lives near the coast, with cities and towns established there for reasons of transportation, resources, industry, commerce, defense, and in some cases, tourism and spectacular scenery. Earth's coastlines have tremendous resources and are biologically and geomorphically diverse. They draw more tourists than any other natural environment and continue to attract new residents.

Coastal zones are popular, but they are also subject to an array of natural hazards and human-induced environmental problems. Coastal communities must cope with powerful storms, the influence of tides, waves, currents, and moving sediment. Low-lying coasts are subject to flooding, storm surges, and, in some places, tsunamis. Coasts of high relief are susceptible to rockfalls, landslides, and other forms of mass wasting. Environmental problems stem from rapid urban development, high population densities, and economic and industrial activities ranging from tourism to port operations, offshore oil production, and agricultural runoff. Some of our most polluted waters are found in coastal locations. If global warming reduces the extent of the continental ice sheets, the ensuing rise in sea level will have a profound impact on the human-built infrastructure in coastal regions as well as on coastal geography and geomorphology. Understanding the natural processes that operate in the coastal zone is fundamental to solving the present and future problems in this dynamic part of Earth's landscape.

The Coastal Zone

Most of the processes and landforms of the marine coastal zone are also found along the coastlines of large lakes. All are considered *standing bodies of water* because the water in each occupies a basin and has an approximately uniform still-water level around the basin. This contrasts with the sloping, channelized flow toward lower elevations that constitute streams.

The **shoreline** of a standing body of water is the exact and constantly changing contact between the ocean or lake surface and dry land. The position of this boundary fluctuates with incoming waves, with storms, and, in the case of the ocean, with the tides. Over the long term, the position of the shoreline is also affected by tectonic movements and by the amount of water held in the ocean or lake basin. **Sea level** is a complexly determined average position of the ocean shoreline, and the vertical position (the reference, or *datum*) above and below which other elevations are measured. The **coastal zone** consists of the general region of interaction between the land and the ocean or lake. It ranges from the inland limit of coastal influence through the present shoreline to the lowest submerged elevation to which the shoreline fluctuates.

As waves approach the mainland from the open body of water, they eventually become unstable and break, sending a rush of water toward land. The **nearshore zone** extends from the seaward or lakeward edge of breakers to the landward limit reached by the broken wave water (• Fig. 20.1). The nearshore zone contains the **breaker zone** where waves break, the **surf zone** through which a bore of broken wave water moves, and, most landward, the **swash zone** over which a thin sheet of water rushes up to the inland limit of water and then back toward the surf zone. This thin sheet of water rushing toward the shoreline is known as **swash,** and the return flow is **backwash.** The **offshore zone** accounts for the remainder of the standing body of water, that part lying seaward or lakeward of the outer edge of the breaker zone.

• FIGURE 20.1
Principal divisions of the coastal zone.

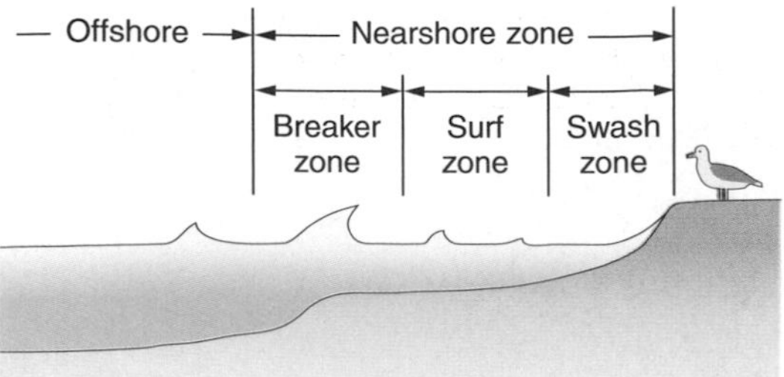

Origin and Nature of Waves

Waves are traveling, repeating forms that consist of alternating highs and lows, called **wave crests** and **wave troughs,** respectively (• Fig. 20.2). The vertical distance between a trough and the adjacent crest is **wave height. Wavelength** is the horizontal distance between successive wave crests. Other important attributes are **wave steepness,** or the ratio of wave height to wavelength, and **wave period,** the time it takes for one wavelength to pass a fixed point.

Waves that have traveled across the surface of a water body are the principal geomorphic agent responsible for coastal landforms. Like streams, glaciers, and the wind, waves erode, transport, and deposit Earth materials, continually reworking the narrow strip of coastal land with which they come in contact. Most of the waves that impact the coastal zone originate in one of three ways. The **tides** consist of two very long wavelength waves caused by interactions between Earth and the moon and sun. **Tsunamis** result from the sudden displacement of water by movement along faults, landslides, volcanic eruptions, or other impulsive events. Most of the waves that impact the coastal zone, however, are **wind waves,** created when air currents push along the water surface.

Tides

The two long-wavelength waves that comprise the tides always exist on Earth. There are two wave crests (high tides), each followed by a trough (low tide). As a crest then a trough move slowly through an area of the ocean, they cause a gradual rising and subsequent falling of the ocean surface. Along the marine coastline, the change in water level caused by the tides brings the influence of coastal processes to a range of elevations. Tides are so small on lakes that they have virtually no effect on coastal processes, even in large lakes.

The gravitational pull of the moon, and to a lesser extent the sun, and the force produced by motion of the combined Earth–moon system are the major causes of the tides (• Fig. 20.3). The moon is much smaller than the sun, but because it is significantly closer to Earth, its gravitational influence on Earth exceeds that of the sun. The moon completes one revolution around Earth every 29.5 days, but it does not revolve around the center of Earth. Instead, the moon and Earth are a combined system that as a unit moves around the system's center of gravity. Because of the large mass of Earth compared to the moon, the system's center of gravity occupies a point within Earth on the side that is facing the moon. Being closest to the moon and more easily deformed than land, ocean at Earth's surface above the center of gravity is pulled toward the moon, making the first tidal bulge (high tide). At the same time, ocean water on the opposite side of Earth experiences the outward-flying, or *centrifugal,* force of inertia and forms the other tidal bulge. Troughs (low tides) occupy the sides of Earth midway between the two tidal bulges. As Earth rotates on its axis each day, these bulges and troughs sweep across Earth's surface.

The sun has a secondary tidal influence on Earth's ocean waters, but because it is so much farther away, its tidal effect is less than half that of the moon. When the sun, moon, and Earth are aligned, as they are during new and full moons, the added influence of the sun on ocean waters causes higher than average high tides and lower than average low tides. The difference in sea level between high tide and low tide is called the **tidal range.** The increased tidal interval due to the alignment of Earth, the moon, and the sun, known as **spring tide,** occurs every 2 weeks. A week after a spring tide, when the moon has revolved a quarter of the way around Earth, its gravitational pull on Earth is exerted at a 90° angle to that of the sun. In this position, the forces of the sun and moon detract from one another. At the time of the first-quarter and last-quarter moon, the counteracting force of the sun's gravitational pull diminishes the moon's attraction. Consequently, the high tides are not as high, and the low tides are not as low at those times. This moderated situation, which like spring tides occurs every 2 weeks, is **neap tide** (● Fig. 20.4).

The moon completes its 360° orbit around Earth in a month, traveling about 12° per day in the same direction that Earth rotates daily around its axis. Thus, by the time Earth completes one rotation in 24 hours, the moon has moved 12° in its orbit around Earth (see again Fig. 20.3). To return to a given position with respect to the moon takes the Earth an additional 50 minutes. As a result, the moon rises 50 minutes later every day at any given spot on Earth, the tidal day is 24 hours and 50 minutes, and two successive high tides are ideally 12 hours and 25 minutes apart.

The most common tidal pattern approaches the ideal of two high tides and two lows in a tidal day. This *semidiurnal* tidal regime is characteristic along the Atlantic coast of the United States, for example, but it does not occur everywhere, as seen in ● Figure 20.5. In a few seas that have restricted access to the open ocean, such as the Gulf of Mexico, tidal patterns of only one high and one low tide occur during a tidal day. This type of tide, called *diurnal,* is not very common. A third type of tidal pattern consists of two high tides of unequal height and two low tides, one lower than the

● FIGURE 20.2
Important dimensions of waves.

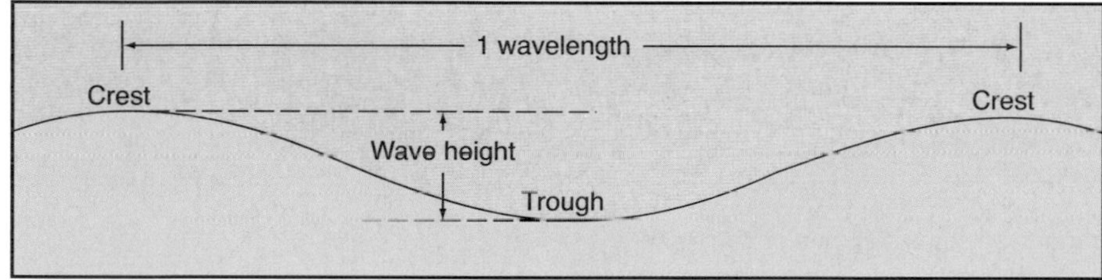

● FIGURE 20.3
The tides are a response to the moon's gravitational attraction (periodically reinforced or opposed by the sun), which pulls a bulge of water toward it while the centrifugal force of rotation of the Earth–moon system forces an opposing mass of water to be flung outward on the opposite side of Earth. Earth rotates through these two bulges each day. A "tidal day," however, is 24 hours and 50 minutes long because the moon continues in its orbit around Earth while Earth is rotating.
How many high tides and low tides are there during each tidal day?

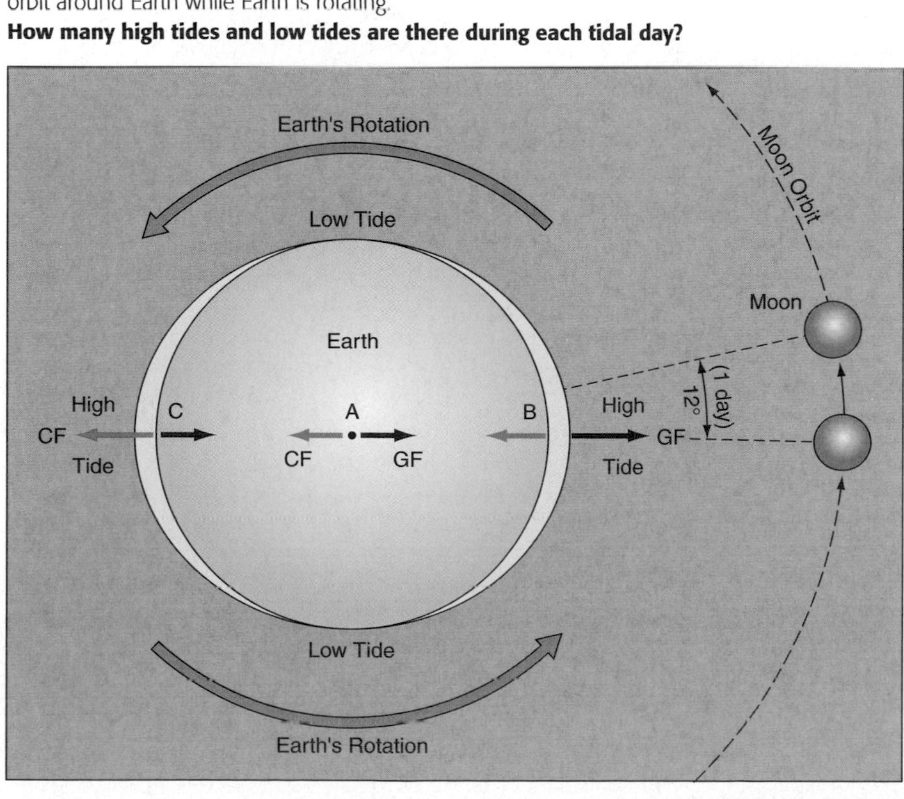

A. Gravitational force (GF) and centrifugal force (CF) are equal. Thus separation between Earth and moon remains constant.

B. Gravitational force exceeds centrifugal force, causing ocean water to be pulled toward moon.

C. Centrifugal force exceeds gravitational force, causing ocean water to be forced outward away from moon.

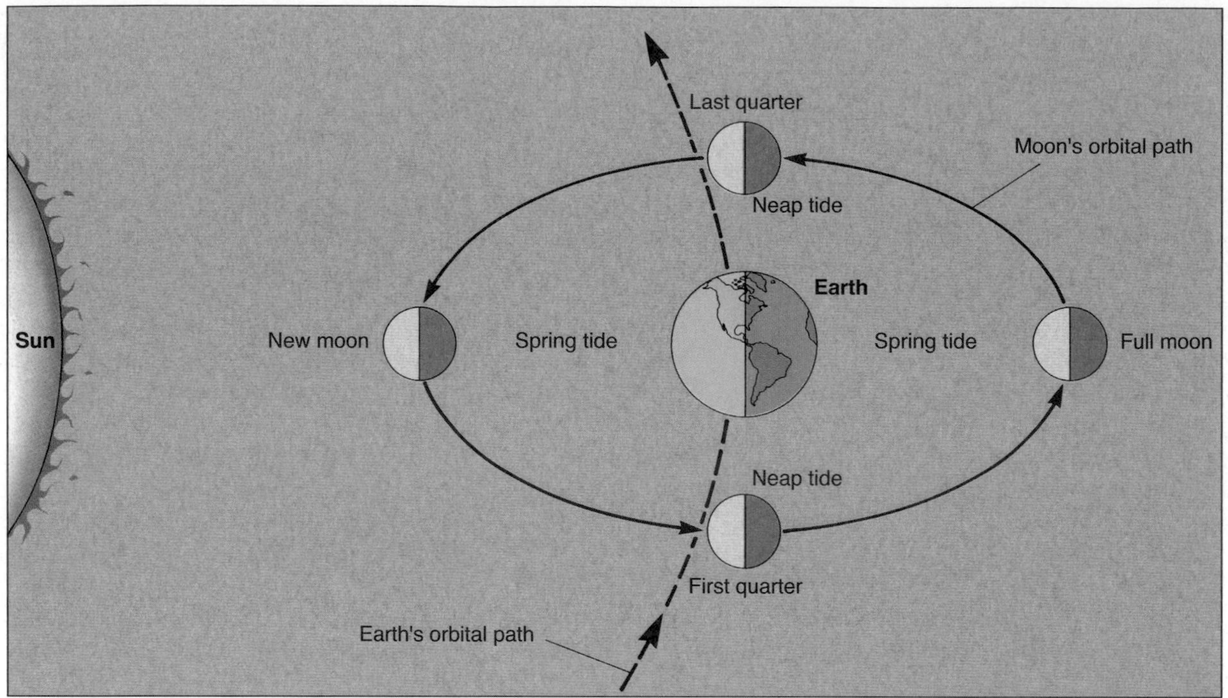

• FIGURE 20.4

The maximum tidal ranges of spring tides occur when the moon and sun are aligned on the same side of Earth or on opposite sides of Earth, which are new moon and full moon, respectively. The minimum tidal ranges of neap tides occur when the gravitational forces of the moon and sun are acting at right angles to each other, at first- and last-quarter moons.

How many spring tides and neap tides occur each month?

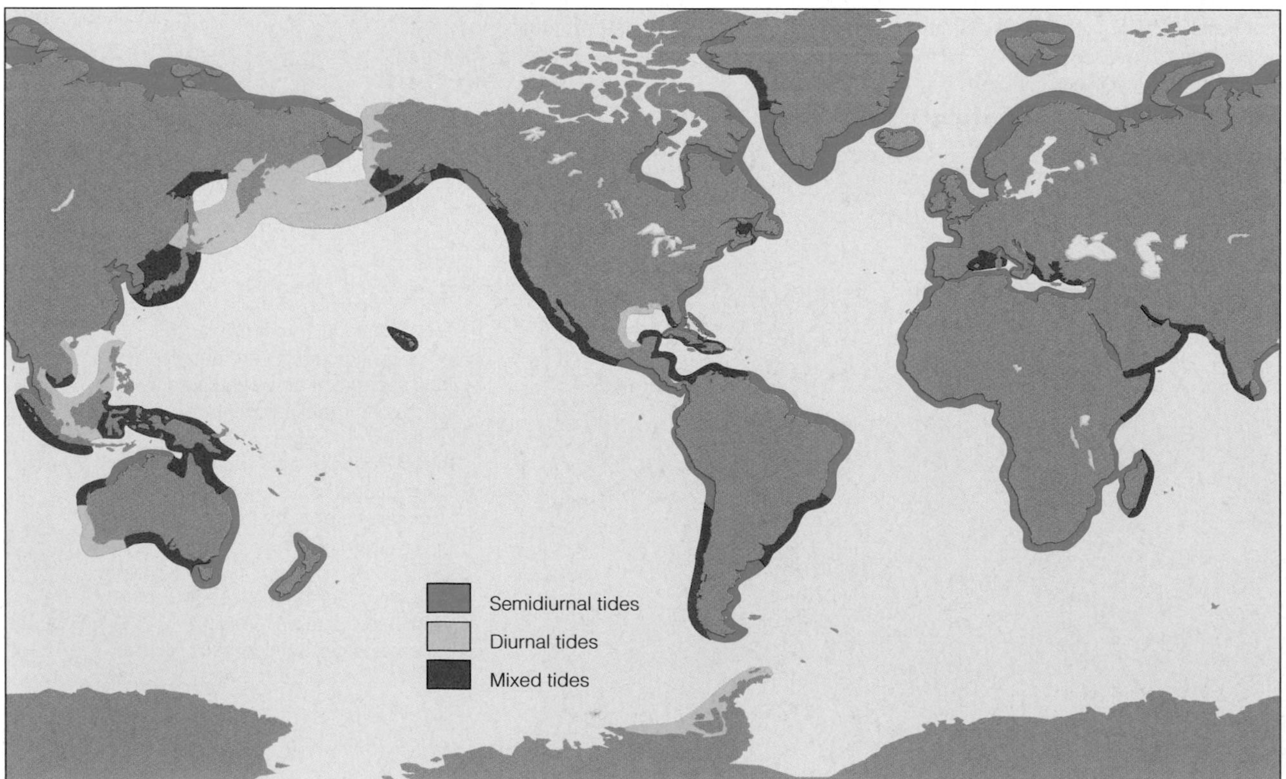

• FIGURE 20.5

This map of world tidal patterns shows the geographic distribution of diurnal, semidiurnal, and mixed tides.

What is the tidal pattern on the coastal area nearest where you live?

other. The waters of the Pacific coast of the United States exhibit this *mixed tide* pattern.

Tidal range varies from place to place in response to the shape of the coastline, water depth, access to the open ocean, submarine topography, and other factors. The tidal range along open-ocean coastlines, like the Pacific coast of the United States, averages between 2 and 5 meters (6–15 ft). In restricted or partially enclosed seas, like the Baltic or Mediterranean Sea, the tidal range is usually 0.7 meters (2 ft) or less. Funnel-shaped bays off major oceans, especially the Bay of Fundy on Canada's east coast, produce extremely high tidal ranges. The Bay of Fundy is famous for its enormous tidal range, which averages 15 meters (50 ft) and may have reached a maximum of 21 meters (70 ft) (• Fig. 20.6). Other narrow, elongated coastal inlets that exhibit great tidal ranges are Cook Inlet in Alaska, Washington's Puget Sound, and the Gulf of California in Mexico.

Tsunamis

Tsunamis are long-wavelength waves that form when a large mass of water displaced upward or downward by an earthquake, volcanic eruption, landslide, or other sudden event works to regain its equilibrium condition. Resulting oscillations of the water surface travel outward from the origin as one wave or a series of waves. In deep water, the displacement may cause wave heights of a meter or more that can travel at speeds of up to 725 kilometers (450 mi) per hour and yet pass beneath a ship unnoticed. As the long-wavelength waves approach the shallow water of a coastline, their height can grow substantially. As these very large and extremely dangerous waves surge into low-lying land areas, they acquire huge amounts of debris and can cause tremendous damage, injury, and loss of life, as well as erosion, transportation, and deposition of Earth materials.

In 1946 an earthquake in Alaska caused a tsunami that reached Hilo, Hawaii, where it attained a maximum height greater than 10 meters (33 ft) and killed more than 150 people. When the Krakatoa volcano erupted in 1883, it generated a powerful 40-meter- (130 ft) high tsunami that killed more than 37,000 people in the nearby Indonesian islands. In December 2004, the devastating earthquake-generated Indian Ocean tsunami struck the shorelines of Indonesia, Thailand, Myanmar, Sri Lanka, India, Somalia, and other countries causing approximately 230,000 fatalities from the tsunami alone. This tragic event reinforced the importance of tsunami early-warning systems. Although an early-warning system had been established for the Pacific Ocean, none was operating in the Indian Ocean in 2004. The United Nations Educational, Scientific, and Cultural Organization (UNESCO) has since been helping member nations in the region develop a more comprehensive tsunami warning system.

Wind Waves

Most waves that we see on the surface of standing bodies of water are created by the wind. Where wind blows across the water, frictional drag and pressure differences cause irregularities in the water surface. The wind then pushes on water slopes that face into the wind, transferring energy to the water and building the slopes into larger waves.

If most waves are caused by the wind, why do we see waves at the beach even during calm days? The answer lies in the fact that waves can, and often do, travel very long distances from the storms that created them with limited loss of energy. Waves arriving at a beach on a calm day may have traveled thousands of kilometers to finally expend their energy when they break along the coast.

When a storm develops on the open ocean, gentle breezes first fashion small ripples on the water surface. If the wind increases, it transforms the ripples into larger waves. While under the influence of the storm, waves are steep, choppy, and chaotic,

• FIGURE 20.6
The extreme tidal range of the Bay of Fundy in Nova Scotia, Canada, can be seen in the difference between (a) high tide and (b) low tide at the same point along the coast.
Why does the Bay of Fundy have such a great tidal range?

©2003, Province of New Brunswick, all rights reserved.
(a)

Robert. D. H. Warren / Communications New Brunswick. All rights reserved.
(b)

GEOGRAPHY'S PHYSICAL SCIENCE PERSPECTIVE

Tsunami Forecasts and Warnings

Along with tides and wind waves, tsunamis are one of the three principal types of waves that impact coastal areas; they are by far the most dangerous. Many decades ago, this type of wave was known as a "tidal wave," but that term was abandoned because tsunamis are caused by major, abrupt displacements of water and are not related to the tides. The term "seismic sea wave" replaced it for a time but is also misleading as a general term for this category of wave. Although most tsunamis originate when faulting causes a sudden, major change in the topography of the ocean floor, not all tsunamis are caused by earthquakes. Tsunami, a Japanese term meaning "harbor wave" (*tsu,* harbor; *nami,* wave), was eventually adopted. Submarine landslides, collapse of submarine volcanic structures, and eruption of underwater volcanoes are other causes of tsunamis. Tsunamis caused by coastal landslides and meteor impacts tend to dissipate quickly and rarely affect distant coastlines.

Tsunamis differ from wind-generated waves in their origin, speed, and size. On the U.S. West Coast, wind waves spawned by a storm in the Pacific Ocean might arrive at the coast one after another with a *period* (time interval between successive waves) of about 10 seconds and a *wavelength* (the distance between two successive waves) of 150 meters (500 ft). A tsunami may have a period of about an hour and a wavelength of more than 100 kilometers (60 mi).

The speed at which a tsunami travels across the open ocean is related to the acceleration due to gravity (g), 9.8 meters/second/second, multiplied by ocean depth (d). In the Pacific Ocean, with average depth of 4000 meters (13,000 ft), tsunamis often travel over 700 kilometers (435 mi) per hour. Tsunamis not only move at high speed but also travel great distances. In 1960 a tsunami originating off the coast of Chile traveled more than 17,000 kilometers (10,600 mi) to Japan, where it killed 200 people.

The energy of a tsunami depends on the balance between its wave speed and its wave height. As it moves into shallow coastal water, the speed of a tsunami decreases, but to maintain conservation of energy, the wave height increases. Thus, a tsunami that is 1 meter (3 ft) high in the open ocean can grow to 30 meters (100 ft) high when it reaches the coast.

A tsunami consists of a series of waves, and the first wave is often not the largest. The danger from a tsunami can last for several hours after the arrival of the first wave.

Detecting a tsunami, determining the direction and speed at which it travels, and tracking its progress across the ocean are critical for saving lives through tsunami warning systems. In this effort, the U.S. National Oceanic and Atmospheric Administration (NOAA) established an array of instruments (*tsunameters*) to monitor pressure and temperature on the ocean floor

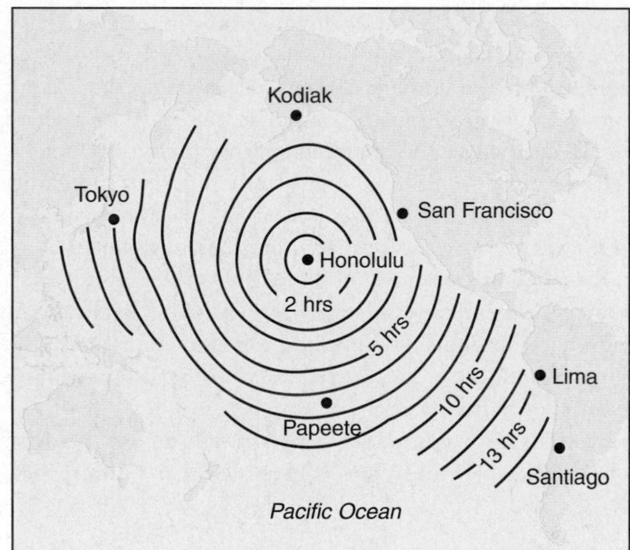

The PTWC locates earthquake epicenters and estimates times of arrival for potential tsunamis in the Pacific region.

and referred to as **sea.** When the waves travel out of the storm area or the wind dies down, the waves become more orderly as they sort themselves into groups of similar speed and length. These gentler, more orderly waves that have traveled beyond the zone of generation are **swell.** It is swell that arrives at coastlines even in the absence of coastal winds.

The energy in a wave is potential energy represented by the wave height. As waves travel they lose a little height, and thus energy, due to friction and to spreading of the wave crest because of the curvature of Earth, but overall they are very efficient means for transporting energy. Three factors that determine the height of wind waves as they form in deep, open bodies of water are (1) wind velocity, (2) duration of the wind, and (3) the area over which the wind blows, the **fetch.** Fetch is the expanse of open water across which the wind can blow without interruption. An increase in any of these three factors produces waves of greater height and greater energy.

When swell that, for example, originated in a storm in the South Pacific arrives at the coast of Southern California, it is local water, not water from the South Pacific, that arrives at the shore in the wave. Recall that waves are traveling *forms.* They do not transport water horizontally from one place to another

and convert these data to height of the water column. The array is maintained by NOAA's Data Buoy Center and constitutes an important part of a growing international tsunami monitoring network. These subsurface sensors enable the Pacific Tsunami Warning Center (PTWC) in Hawaii and its 26-nation group to share warnings throughout the Pacific Basin, and the Alaska Tsunami Warning Center (ATWC) to issue appropriate warnings for the west coast of North America.

When a warning is issued, ships leave the shallow harbors and go out to sea where the tsunamis are not noticeable in deep water. Coastal residents are warned to evacuate the area and move quickly to higher ground. Tsunamis can come with little or no lead time, and when warnings are issued, they need to be taken seriously.

The devastating tsunami that struck coastal areas of the Indian Ocean in December of 2004 caused tremendous death, destruction, and human suffering, in part because no sensor-based warning system was in place in that region. About 230,000 people died, and 1.2 million people were left homeless, according to United Nations estimates, when the ocean surged onshore in some places with waves as high as 15 meters (50 ft).

Australian Agency for International Development / Robin Davies

Tsunami destruction on the west coast of Aceh province, Indonesia, in December 2004.

except where they break along a coastline. The movement of waves in the open water body may be considered similar to the movement of stalks of wheat as wind blows across a wheat field, causing wavelike ripples to roll across its surface. The wheat returns to its original position after the passage of each wave. Water particles likewise return to approximately their original position after transmitting a wave.

Deep-water waves are those traveling through water depth (d) greater than or equal to half the wavelength (L), $d \geq L/2$. Traveling waves have no impact on what is below that depth. For this reason, the depth $L/2$ is sometimes referred to as **wave base.** ● Figure 20.7 illustrates what happens to surface water during the passage of a wave in deep water. There is little if any net forward motion of water molecules during the passage of the wave. As the crest and trough pass through, the water molecules complete an orbital motion. With increasing depth beneath the water surface, the size of the orbits decreases. By a depth of half the wavelength, the orbits are too small to do any significant work. It is only when the wave enters water of $d < L/2$ that it starts to interact with, or "feel," bottom and become affected by friction with the bed (● Fig. 20.8).

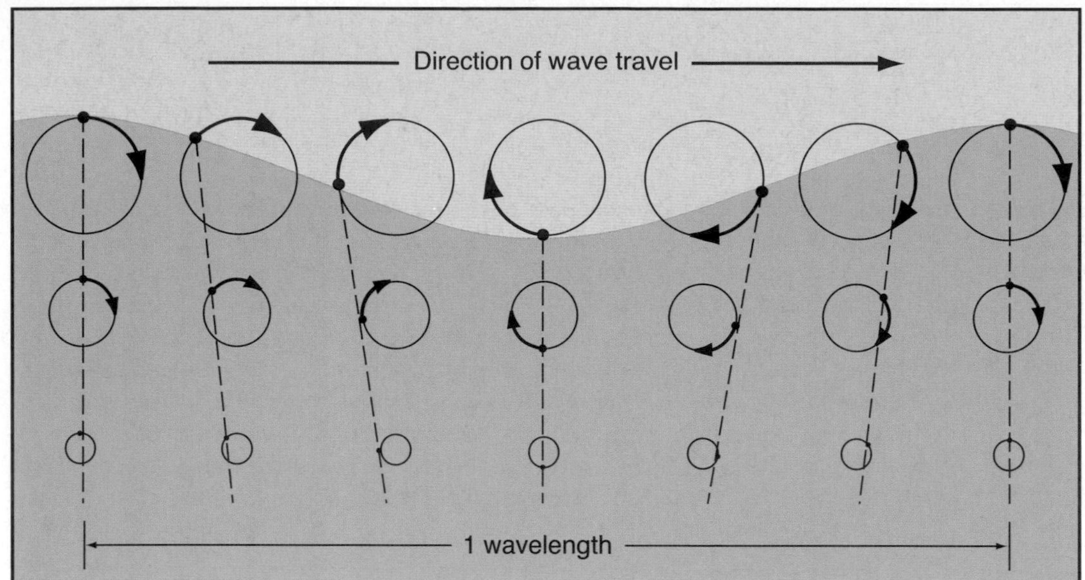

• FIGURE 20.7
Orbital paths of water particles cause oscillatory wave motion in deep water. The diameter of the surface orbit equals wave height, which is the vertical distance from trough to crest.

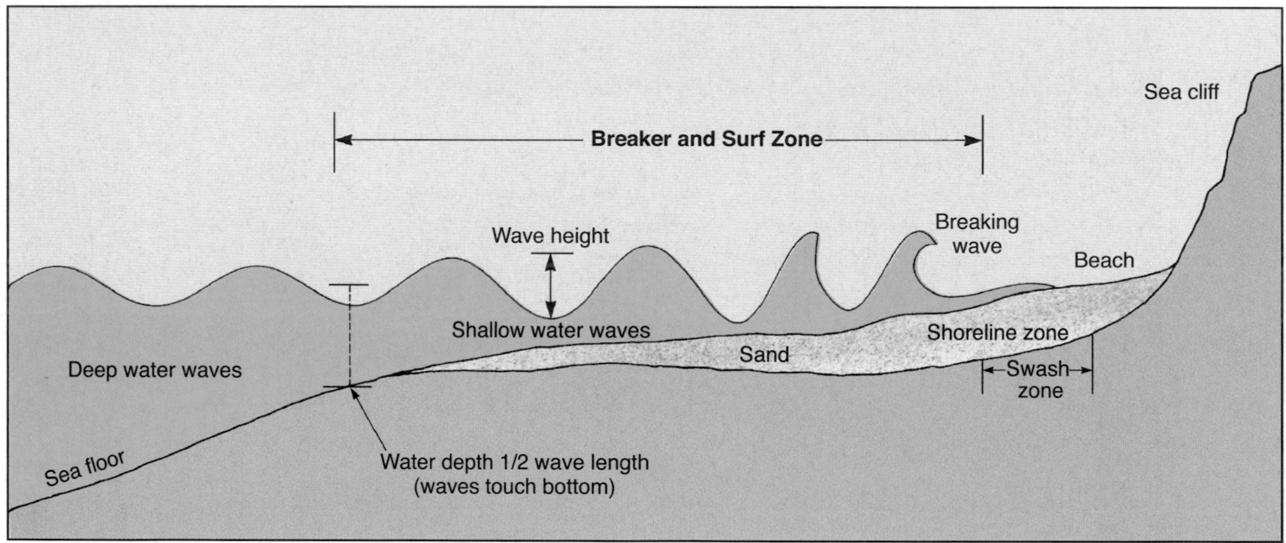

• FIGURE 20.8
Waves begin to "feel bottom" when the water depth becomes half the distance between wave crests. Then the wave velocity and wavelength decrease while the wave height and steepness increase until breaking occurs.
Why don't the waves break in deeper water?

Breaking Waves

As long as they are in deep water relative to their wavelength, $d \geq L/2$, waves roll along without disturbing the bottom and with little loss of energy. As they approach the coast and enter shallower water, $d < L/2$, friction with the bed causes the waves to undergo a decrease in both velocity and wavelength. The wave bunches up, experiencing an increase in wave height (H). As wave height increases and wavelength decreases, wave steepness ($S = H/L$) increases rapidly to the maximum value of 1/7. At this steepness, the wave will become unstable and break, finally expending the energy it had originally obtained in a storm often hundreds or even thousands of kilometers away. Some breaking waves appear to curl over and crash as though trying to complete one last wave form, but with insufficient water available to draw up into that final wave. Once the wave has broken, turbulent surf advances landward, thinning to swash at the water's edge and returning to the surf zone as backwash.

Rip currents (• Fig. 20.9) are relatively narrow zones of strong, offshore-flowing water that occur along some coastal areas. Rip currents are a means for returning broken wave water from the nearshore zone back to deeper water. Rip currents are dangerous. Swimmers who get caught in them often try to swim back to shore against the strong current to keep from being pulled out to deeper water, not always successfully. Rip currents are frequently visible as streaks of foamy, turbid water flowing perpendicular to the shore.

City of Miami Beach, Florida, Public Safety Division

● **FIGURE 20.9**
Rip currents move water seaward from a beach. Here, the current can be seen moving offshore, opposite to the wave direction.
Why are these currents a hazard to swimmers?

Wave Refraction and Littoral Drifting

In map view, or as looking down from an airplane, we often see parallel, linear wave crests steadily approaching the coastal zone at regular intervals from a uniform direction, probably having originated in the same distant storm. They may approach from directly offshore or at an angle to the trend of the coastline. Oftentimes successive wave crests each change orientation relative to the coastline as they move through shallower water. **Wave refraction** is this bending of a wave in map view as it approaches a shoreline.

Wave refraction occurs when part of a wave encounters shallow water before other parts. To understand how this happens, imagine an irregular coast of embayments and headlands (● Fig. 20.10). While in deep water, a wave traveling toward the coast from directly offshore has a straight crest in map view. The wave will feel bottom first in the shallower water off the headlands, while off the embayments it is still traveling in the deeper water. This slows the advance of the wave crest toward the headlands while it continues to speed on toward the embayments. This difference in velocity converts the map view trend of the wave crest from a straight line to a curve that increasingly resembles the shape of the shoreline as it gets closer to land. Wave energy is expended perpendicular to the orientation of the crestline. Thus, when the wave breaks, its energy is focused on the headlands and spread out along the embayments. Over time, the headlands are eroded back toward the mainland, while deposition in the low-energy embayments builds those areas toward the water body. Because of wave refraction, coastlines tend to straighten over time (● Fig. 20.11).

● **FIGURE 20.10**
Wave refraction causes wave energy to be concentrated on headlands, eroding them back, while in embayments, deposition causes beaches to grow seaward.
How will this coastline change over a long period of time?

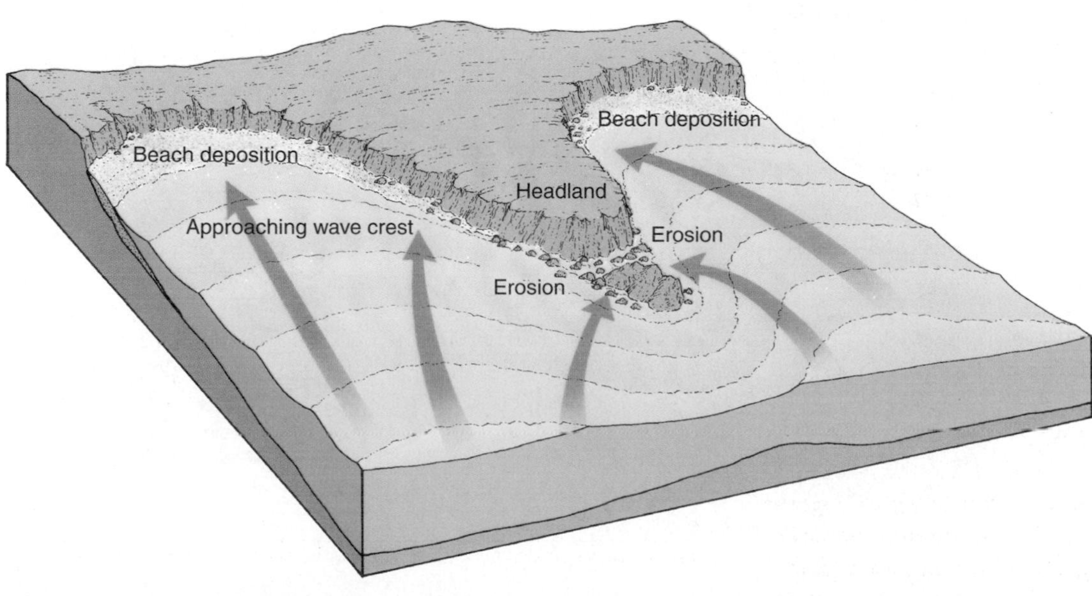

● **FIGURE 20.11**
Embayments along this coastline have been building seaward by filling with sediment, while wave energy focused on the headlands has been eroding them landward.
What happens to sediment eroded from the headlands?

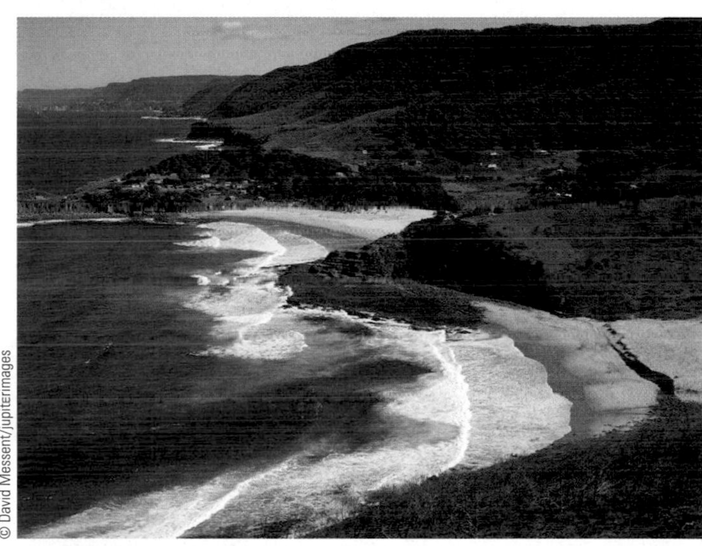
© David Messent/jupiterimages

USGS

• **FIGURE 20.12**
Waves that approach a beach at an angle do not always refract completely, breaking at an angle to the trend of the shoreline.
What factors might keep a wave from refracting completely?

Not all waves refract completely before they break (• Fig. 20.12). Crestlines of incompletely refracted waves do not fully conform to the orientation of the shoreline when they break. Incomplete refraction gives a spatial component to sediment transportation within the *littoral* (coastal) zone. This sediment transportation in the coastal zone, called **littoral drifting,** is accomplished in two ways. Both ways are well demonstrated using the example of a sandy beach along a straight coastline that has smooth underwater topography sloping gently into deeper water.

When in map view a wave crest approaches the straight, gently sloping shoreline at a large angle to the coast (obliquely), it interacts with the bottom and starts to slow down first where it is closest to shore (• Fig. 20.13). This velocity decrease spreads progressively along the crestline as more of the wave enters shallower water. With insufficient time for complete refraction before breaking begins, the crest lies at an angle to the beach, not parallel to it, when it breaks. As a result, the broken wave water, and sediment it has entrained, rushes up the beach face diagonally to the shoreline, rather than directly up its slope. Backwash, however, which also moves sediment, flows straight back down the beach face toward the water body by the force of gravity. In this way, as one incompletely refracted wave after another breaks, sediment zigzags along the beach in the swash zone. **Beach drifting** is this zigzaglike transportation of sediment in the swash zone due to incomplete wave refraction. Over time, beach drifting causes the mass transport of tons of sediment along the shore.

• **FIGURE 20.13**
A wave approaching a straight coastline at a large angle will feel bottom progressively along the wave. The resulting progressive decrease in velocity causes the wave to swing around, but it may not have enough time to conform fully to the shape of the shoreline before breaking, leading to littoral drifting.

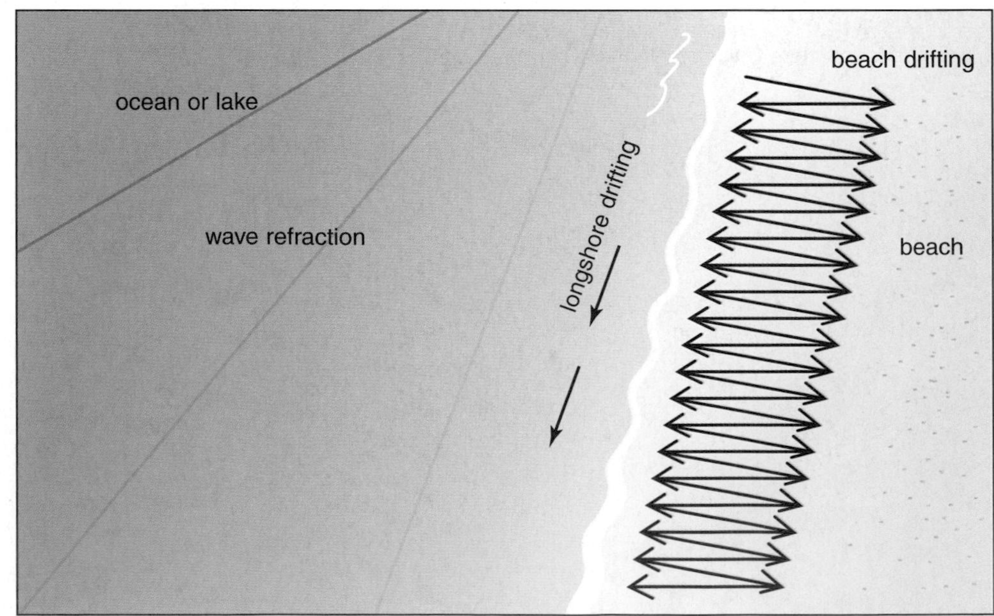

Another outcome of an incompletely refracted oblique wave is that when the crest arrives at the break point at one location, farther along the beach in the direction the waves are traveling, that same position is occupied by a trough. This difference in water level initiates a current of water, called the **longshore current,** flowing parallel to the shoreline near the breaker zone. Considerable amounts of sediment suspended when incompletely refracted waves break are transported along the shore in this process of **longshore drifting.**

Coastal Erosion

Because waves and streams both consist of liquid water, similarities exist in how these two geomorphic agents erode rock matter. Like water in streams, water that has accumulated in basins erodes some rock material chemically through *corrosion*. Corrosion is the removal of the ions that have been separated from rock-forming minerals by solution and other chemical weathering processes. Likewise, the power of *hydraulic action* from the sheer physical force of the water alone pounds against and removes coastal rock material, sometimes compressing air or water into cracks to help in the process. The power of storm waves, combined with the buoyancy of water, enables them at times to dislodge and move even large boulders. Once clastic particles are in motion, waves have solid tools to use to perform even more work through the grinding erosive process of *abrasion*. Abrasion is the most effective form of erosion by each of the geomorphic agents, including waves.

Weathering is an important factor in the breakdown of rocks in the coastal zone, as in other environments, preparing pieces for removal by wave erosion. Water is a key element in most weathering processes, and in addition to normal precipitation, rocks near the shoreline are subjected to spray from breaking waves as well as high relative humidities and condensation. Salt weathering is particularly significant in preparing rocks for removal through chemical and physical weathering along the marine coast and coasts of salt lakes.

Coastal Erosional Landforms

Coasts of high relief are dominated by erosion (● Fig. 20.14). **Sea cliffs** (or **lake cliffs**) are carved where waves pound directly against steep land (● Fig. 20.15a). If a steep coastal slope continues deep beneath the water, it may reflect much of the incoming wave energy until corrosion and hydraulic action eventually take their toll on the rock. The tides present along marine coasts allow these processes to attack a range of shoreline elevations. Once a recess, or **notch,** has been carved out along the base of a cliff (Fig. 20.15b), weathering and rockfall within the shaded overhang supply clasts that can collect on the notch floor and be used by the water as tools for more efficient erosion by abrasion. Abrasion extending the notch landward leaves the cliff above subject to rockfall and other forms of mass wasting. Stones used as tools in abrasion quickly become rounded and may accumulate at the base of the cliff as a **cobble beach.** Where the cliffs are well jointed but cohesive, wave erosion can create **sea caves** along the lines of weakness (Fig. 20.15c). **Sea arches** result where two caves meet from each side of a headland (Fig. 20.15d). When the top of an arch collapses or a sea cliff retreats and a resistant pillar is left standing, the remnant is called a **sea stack** (Fig. 20.15e).

Landward recession of a sea cliff leaves behind a wave-cut bench of rock, an **abrasion platform,** that is sometimes visible at lower water levels, such as at low tide (● Fig. 20.16). Abrasion platforms record the amount of cliff recession. In some cases deposits accumulate as wave-built terraces just seaward of an abrasion platform. If tectonic activity uplifts these wave-cut benches and wave-built terraces above sea level out of the reach of wave action, they become **marine terraces** (● Fig. 20.17). Successive periods of uplift can create a coastal topography of marine terraces that resembles a series of steps. Each step represents a period of time that a terrace was at sea level. The Palos Verdes peninsula just south of Los Angeles has perhaps as many as ten marine terraces, each representing a period of platform formation separated by episodes of uplift.

Rates of coastal erosion are controlled by the interaction between wave energy and rock type. Coastal erosion is greatly accelerated during high-energy events, such as severe storms and tsunamis. Human actions can also accelerate coastal erosion rates. We commonly do so by interfering with coastal sediment and vegetation systems that would naturally protect some coastal segments from excessive erosion rates. We explore the nature of coastal depositional systems next.

Coastal Deposition

Significant amounts of sediment accumulate along coasts where wave energy is low relative to the amount or size of sediment supplied. Embayments and settings where waves break at a distance from the shoreline, such as areas with gently shelving underwater topography, tend to sap wave energy and encourage deposition. Amount and size of sediment supplied to the coastal zone vary with rock type, weathering rates, and other elements of the climatic, biological, and geomorphic environment.

Sediment within coastal deposits comes from three principal sources. Most of it is delivered to the standing body of water by streams. At its mouth, the load of a stream may be deposited for the long term in a delta or within an **estuary,** a biologically very productive embayment that forms at some river mouths where salt and fresh water meet. Elsewhere, stream load may instead be delivered to the ocean or lake for continued transportation. Once in the standing body of water, fine-grained sediments that stay in suspension for long periods may be carried out to deep water where they eventually settle out onto the basin floor. Other clasts are transported by waves and currents in the coastal zone, being deposited when energy decreases and, if accessible, reentrained when wave energy increases. The same is true of the second major source of coastal sediment, coastal cliff erosion. Of less importance is sediment brought to the coast from offshore sources. Although we may tend to think of sand-sized sediment when we think of coastal deposits, coastal depositional landforms may be composed of silt, sand, or any size classes of gravel, from granules and pebbles through cobbles and boulders.

● FIGURE 20.14
This diagram illustrates the major coastal erosional landforms associated with wave activity.

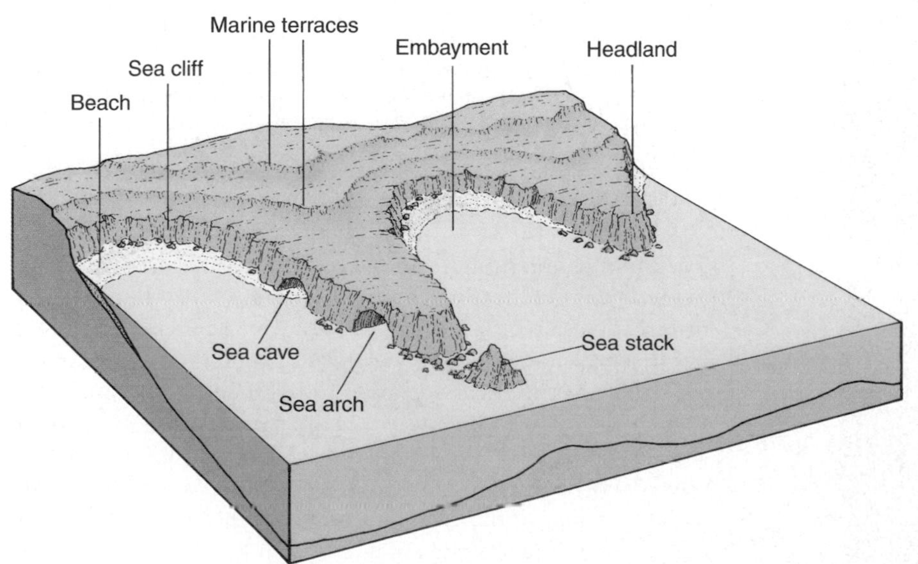

Washington State Department of Ecology/David Byers

(a)

D. Sack

(b)

J. Petersen

(c)

NOAA Image Library

(d)

USGS

(e)

• FIGURE 20.15

Examples of the major landforms associated with erosional coasts. (a) These rugged sea cliffs lie along the uplifted and wave-eroded Washington coastline. (b) Notice the notch carved near the base of this basalt cliff on the island of Hawaii. (c) Sea caves are found along the steep limestone sea cliffs of Italy's Amalfi coast on the Mediterranean Sea. (d) Sea arches, such as this one in Alaska, develop as sea cliffs on opposite sides of a headland are eroded completely through. (e) A sea stack, such as this one off the Oregon coast, forms when sea cliffs retreat, leaving a resistant pillar of rock standing above the waves.

Coastal Depositional Landforms

The most common landform of coastal deposition is the **beach,** a wave-deposited feature that is contiguous with the mainland throughout its length (• Fig. 20.18). Many beaches are sandy, but beaches of other grain sizes are also common, as for example the cobble beach discussed earlier in the section on coastal erosion. In settings with high wave energy, particles tend to be larger and beaches steeper than where only fine material is present and wave energy is low. Beach sediments come in a variety of colors depending on the rock and mineral types represented. Tan quartz, black basalt, white coral, and even green olivine beaches exist on Earth.

D. Sack

• **FIGURE 20.16**
An abrasion platform of strongly dipping sedimentary rocks is exposed at low tide along the erosion-dominated coast of central California.
How was this abrasion platform made?

© Robert Cameron/Getty Images

• **FIGURE 20.17**
This exposed surface along the California coast represents a marine terrace. The former sea cliff lies inland, just beyond the highway.
What does the presence of this marine terrace tell you about the relationship between land and water at this site? What other coastal erosional landforms do you see in this photo?

J. Petersen
(a)

U.S. Coast Guard
(c)

NOAA/Captain Albert E. Theberge
(b)

NOAA/J. Schabel
(d)

• **FIGURE 20.18**
Beaches are the most common evidence of wave deposition and may be made of any material deposited by waves. (a) Drake's Beach, north of San Francisco, California, in Point Reyes National Seashore, is a sandy beach. (b) Beaches can even be made out of boulders, as illustrated by this beach on Mount Desert Island, Acadia National Park, Maine (note the person for scale). (c) White sand beaches are common on tropical islands with coral reefs. This is Palmyra Atoll in the South Pacific. (d) This black-sand beach on the island of Tahiti is composed of volcanic rock material.

GEOGRAPHY'S ENVIRONMENTAL SCIENCE PERSPECTIVE

Beach Protection

Beaches act as buffers to help protect the land behind them from wave erosion. Under natural conditions, sediment eroded from the beach in a storm will eventually be replaced through the action of gentle waves and by littoral drifting bringing in new sediment. People interfere with the natural sediment budget when they dam rivers and build on the normally shifting sediment of beach and dune systems in the coastal zone. Intense human development, like that found at many popular tourist beaches, typically interrupts the supply of replacement sediment and leads to more permanent erosion of beaches. Beach erosion leaves the coastal buildings and infrastructure susceptible to damage from storm waves while eliminating the primary attraction, the beach itself.

Beach protection and restoration strategies include the construction of *groins* built perpendicular to the trend of the coastline to slow down the rate of littoral drifting of sediment out of the area. *Jetties* are also built perpendicular to the coastline, but always in pairs, and act to keep sediment from blocking an inlet, such as a river mouth or a channel for boats. *Breakwaters* are walls built parallel to the shoreline in the breaker zone. Large waves expend the bulk of their energy breaking on the structure, thus limiting the amount of beach erosion. Another strategy has been to add sediment to the beach artificially by dredging harbors or reservoirs and trucking or pumping that sediment through pipes to the beach. Unless the factors that are limiting the natural sediment supply to the beach are addressed, this artificial form of *beach nourishment* will likely have to continue, at least periodically.

Miami Beach, Florida, has long been a popular destination for millions of annual visitors, drawn to the magnificent beaches. After decades of development, by 1970 the beaches had mostly disappeared due to erosion. With the beaches gone, shoreline condominiums and hotels were threatened with serious damage from storm waves, and the hotels were half empty as tourists traveled to other destinations.

To help solve the problem, the U.S. Army Corps of Engineers (the federal agency responsible for the development of inland and coastal waterways) chose Miami Beach to experiment with the *beach building* method of shoreline protection. The technology involves transferring millions of cubic meters of sand from

© AP/Wide World Photos

Visitors and residents of Miami Beach benefit from the artificial replacement of sand to the beach.

offshore to replenish existing beaches or build new ones where beaches have eroded away. Huge barge-mounted dredges dig sand from the sea bottom near the shore, and the sand is pumped as a slurry through massive movable tubes to be deposited on a beach.

The initial project cost $72 million, but it was so successful that within 2 years Miami Beach again had a sandy beach 90 meters (300 ft) wide and 16 kilometers (10 mi) long. With the return of the beach, the number of visitors soon grew to three times the number prior to beach building.

Today, beach building has become the accepted way to respond to beach erosion. Nearly $1.7 billion was spent on beach restoration during the last decade of the 20th century. According to the Corps of Engineers, these projects are meant primarily to protect buildings on or near the beach, rather than to provide beaches for recreational purposes. In the Corps's opinion, the true value of beach building can best be measured in savings from storm damage. For example, it has been estimated that beach replenishment along the coast of Miami-Dade County prevented property losses of $24 million during Hurricane Andrew in 1992.

When beaches are rebuilt or widened to protect against storms, continued erosion is inevitable. In deciding how much sand must be pumped in, it is important to determine the beach width necessary to protect shoreline property from storms. This is usually about 30 meters (100 ft). Then an amount of additional "sacrificial" sand is added to produce a beach twice the width of the desired permanent beach. In about 7 years, the sacrificial sand will be lost to wave erosion. If beach rebuilding is repeated each time the sacrificial sand has been removed, the permanent beach will remain in place and the coastal zone will be protected. An estimated $5.5 billion has already been committed to the continuous rebuilding of existing projects over the next 50 years.

Is the battle with the environment at the beachfront worth the price we are paying? Many people are not so sure, but most environmental scientists respond to the question with a resounding no, for they are concerned with a price that is not measured solely in appropriated dollars. This price is paid in destroyed natural beach environments, reduced offshore water quality, eliminated or displaced species, and repeated damage to food chains for coastal wildlife each time a beach is rebuilt. Clearly, beach building is a mixed blessing.

J. Petersen

Seagulls enjoy the swash on this sand-replenished beach in Alameda, California, on San Francisco Bay.

(a)

(b)

• FIGURE 20.19
Because of seasonal variations in wave energy, the differences in a beach from summer to winter can be striking, particularly in the midlatitudes. (a) Waves in summer are generally mild and deposit sand on the beach. (b) Winter waves from storms nearer to and at the beach remove the sand, leaving boulders and bare bedrock in the beach area.
What attribute of waves represents the amount of energy they have?

Any given stretch of beach may be a permanent feature, but much of the visible sediment deposited in it is not. Individual grains come and go with swash and backwash, wear away through abrasion, are washed offshore in storms, or move into, along, and out of the stretch of beach by littoral drifting. Because waves generated by closer storms tend to be higher than waves generated in distant storms, some beaches undergo seasonal changes in the amount and size of sediment present.

In the middle latitudes, beaches are generally narrower, steeper, and composed of coarser material in winter than they are in summer. The larger winter storm waves are more erosive and destructive, while the smaller summer waves, which often travel from the other hemisphere, are depositional and constructive. On the Pacific coast of the United States, summer beaches are generally temporary accumulations of sand deposited over coarser winter beach materials (• Fig. 20.19). Sand-sized sediment eroded from the beach in winter forms a deposit called a **longshore bar** that lies submerged parallel to shore and returns to the beach in summer. On the Atlantic and Gulf coasts of the United States, the late summer to early fall hurricane season is also a time when beach erosion can be severe.

Whereas beaches are attached to the mainland along their entire length, **spits** are coastal depositional landforms connected to the mainland at just one end (• Fig. 20.20). Spits project out

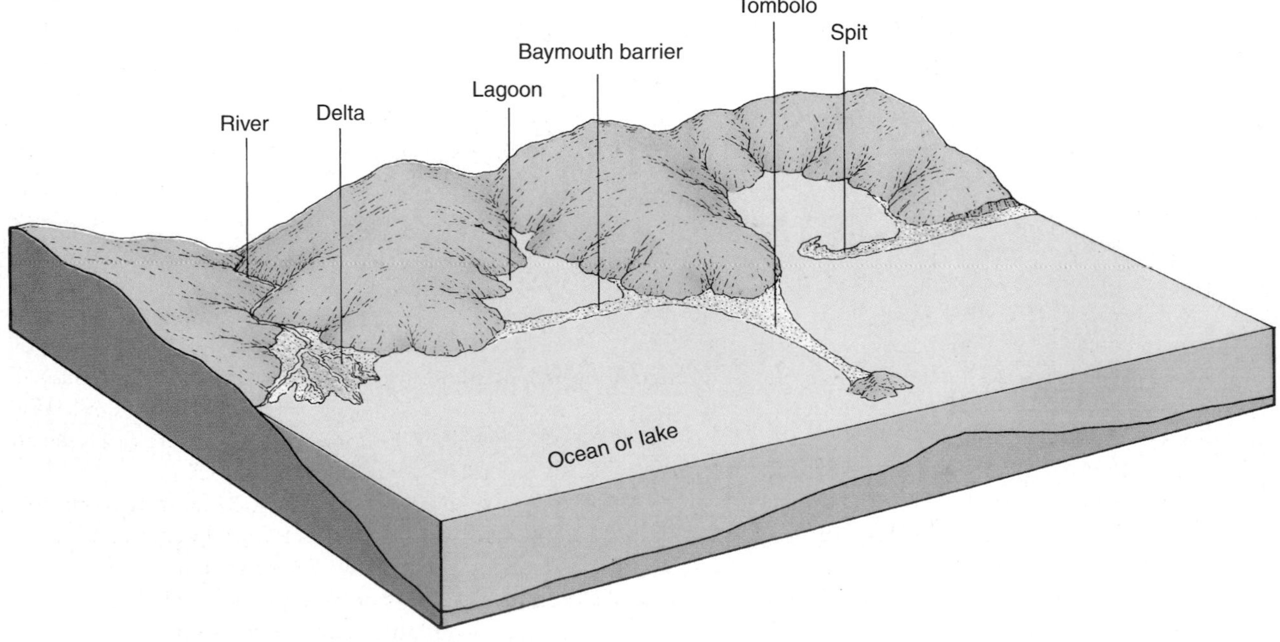

• FIGURE 20.20
This diagram illustrates some of the major landforms found along deposition-dominated coastlines.

USGS Coastal & Marine Geology Program
(a)

USGS Coastal & Marine Geology Program
(c)

NOAA/Captain Albert E. Theberge
(b)

• **FIGURE 20.21**
(a) A spit connects to land at one end, as illustrated by this example on the Oregon coast. (b) A tombolo forms when wave-deposited material connects a nearby island with the mainland, shown here at Point Sur on the California coast. (c) A baymouth barrier, like this one at Big Sur, California, crosses the mouth of an embayment connecting to land at each end.

into the water like peninsulas of sediment. They form where the mainland curves significantly inland while the trend of the longshore current remains at the original orientation. Sediments accumulate into a spit in the direction of the longshore current (• Fig. 20.21a). Where similar processes form a strip of sediment connecting the mainland to an island, the landform is a **tombolo** (Fig. 20.21b).

Another category of coastal landforms are **barrier beaches,** elongate depositional features constructed parallel to the mainland. Barrier beaches act to protect the mainland from direct wave attack. All barrier beaches have restricted waterways, called **lagoons,** that lie between them and the mainland. Salinity in the lagoon varies from that of the open water body, depending on freshwater inflow and evaporation, and affects organisms living in the lagoon. Like beaches and spits, barrier beaches have a submerged part and a portion that is always above water, except in extreme storm conditions or extremely high tide. This contrasts with **bars,** like the longshore bars discussed above, which are submerged except in extreme conditions.

There are three kinds of barrier beaches. A **barrier spit** originated as a spit and thus is attached to the mainland at one end, but has extended almost completely across the mouth of an embayment to restrict the circulation of water between it and the ocean or lake. If the barrier spit crosses the mouth of the embayment to connect with the mainland at both ends, it becomes a **baymouth barrier** (Fig. 20.21c). Limited connection is maintained between the lagoon and the main water body through a breach or *inlet* cut across the barrier somewhere along its length. The position of inlets can change during storms. **Barrier islands** are likewise elongated parallel to the mainland and separate lagoons and the mainland from the open water body, but they are not attached to the mainland at all (• Fig. 20.22).

Barrier islands are common features of low-relief coastlines. They dominate the Atlantic and Gulf coasts of the United States from New York to Texas. Some excellent examples of long barrier islands are Fire Island (New York), Cape Hatteras (North Carolina), Cape Canaveral and Miami Beach (Florida), and Padre Island (Texas).

Rising sea level since the Pleistocene appears to have played a major role in the formation of barrier islands. They migrate landward over long periods of time and may change drastically during severe storms, especially hurricanes (• Fig. 20.23).

Beach systems are in equilibrium when input and output of sediment are in balance. People build artificial obstructions to the longshore current to increase the size of some beaches. A **groin** is an obstruction, usually a concrete or rock wall, built perpendicular to a beach to inhibit sediment removal while sediment input remains the same. This obstruction, however, starves the

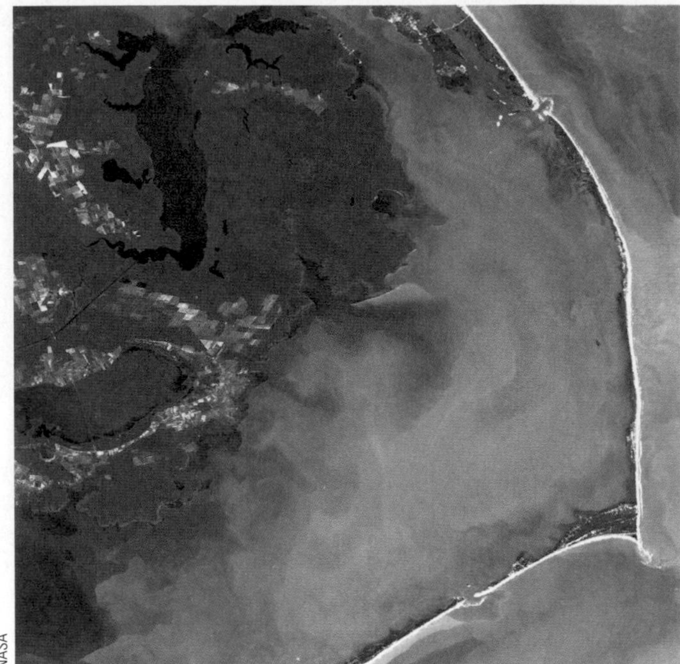
NASA

• **FIGURE 20.22**
Barrier islands are not attached to the mainland, but lie parallel to it. They occur along coasts with gentle slopes and an adequate supply of sediment. This barrier island is located near Pamlico Sound on the North Carolina coast.
What feature separates a barrier island from the mainland?

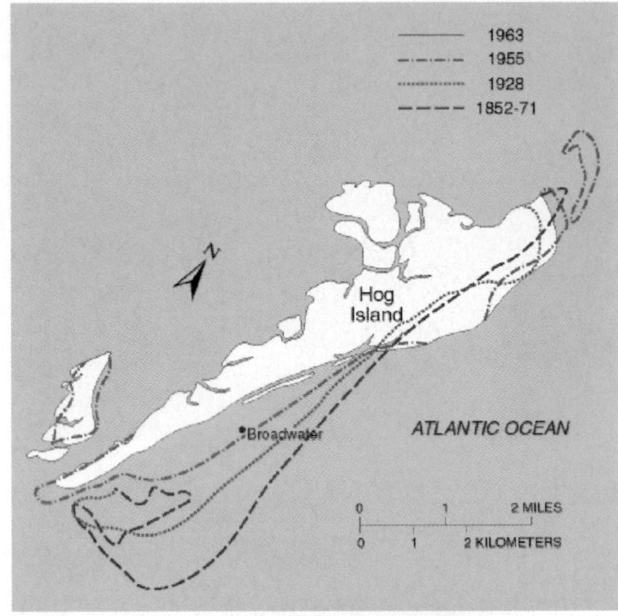

(a)

USGS Coastal & Marine Geology Program
(b)

USGS Coastal & Marine Geology Program
(c)

• **FIGURE 20.23**
(a) Historic maps show us that barrier islands shift their shapes and positions over time, with major changes coming during storm events. (b) and (c) This pair of photos shows hurricane-generated damage to homes built along the shore on a barrier island. Compare the before and after photos.
How can this type of damage be prevented in the future?

adjacent downcurrent beach area of material input from upcurrent while it still has the usual rate of sediment removal (• Fig. 20.24). Beach deposition is also often engineered to keep harbors free of sediment or to encourage growth of recreational beaches. When human actions deplete the natural sediment supply by damming rivers, beaches become narrow and lose some of their ability to protect the coastal region against storms. In Florida, New Jersey, and California, hundreds of millions of dollars have been spent to replenish sandy beaches. The beaches not only serve obvious recreational needs but also help protect coastal settlements from erosion and flooding by storm waves.

Types of Coasts

Coasts are spectacular, dynamic, and complex systems that are influenced by tectonics, global sea-level change, storms, and marine and continental geomorphic processes. Because of this complexity, there is no single, universally accepted classification system for coasts. Coastal classification systems, however, aid our understanding of these natural, complex systems.

On a global scale, coastal classification is based on plate tectonic relationships. This system has two major coastal types: passive-margin coasts and active-margin coasts. **Passive-margin coasts** are well represented by the coastal regions of continents along the Atlantic Ocean (• Fig. 20.25). Most major tectonic activity

USGS

• **FIGURE 20.24**
A series of groins on the Atlantic shoreline at Norfolk, Virginia, captures sand to maintain the beach along one stretch of the coastline.
How do you think the stretch of coast beyond the last groin would be impacted by these structures?

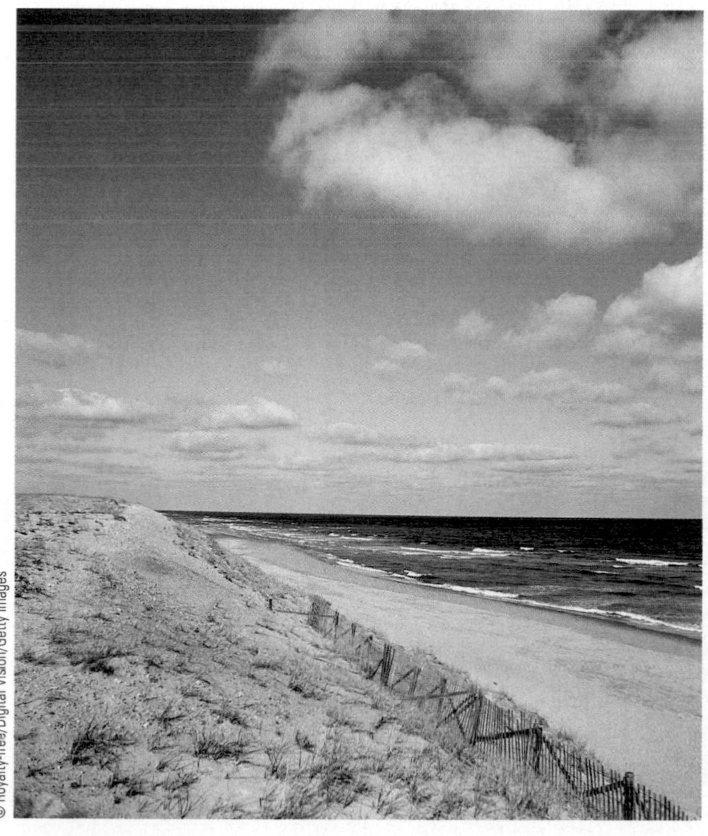
© Royalty-free/Digital Vision/Getty Images

• **FIGURE 20.25**
A sandy passive-margin coast on the Atlantic Ocean at Marconi Beach, Cape Cod National Seashore, Massachusetts.

within the Atlantic occurs in the center of the ocean along the Mid-Atlantic Ridge, whereas the coastal areas are tectonically passive, with little mountain building or volcanic activity. Passive-margin coasts generally have low relief with broad coastal plains and wide, submerged edges of the continent, called **continental shelves** (• Fig. 20.26). Passive-margin coasts that are relatively young, such as those of the Red Sea and Gulf of California, may have somewhat greater relief. Most passive-margin coasts have been modified by marine deposition and some subsidence. The East Coast of the United States is a good example of a low-relief passive-margin coast.

Active-margin coasts are best represented by coastal regions along the Pacific Ocean (• Fig. 20.27). There, most tectonic activity occurs around the ocean margins because of active subduction and transform plate boundaries along the "Pacific Ring of Fire." Active-margin coasts are usually characterized by high relief with narrow coastal plains, narrow continental shelves, earthquake activity, and volcanism. These coasts tend to be erosional, having less time within Earth history for the development of marine or continental depositional features. The West Coast of the United States is an excellent example of an active-margin coast.

On a regional scale, coasts may be classified as coastlines of emergence or coastlines of submergence. **Coastlines of emergence** occur where the water level has fallen or the land has risen in the coastal zone. In either case, land that was once below sea level has emerged above the water. Evidence for emergence includes marine terraces and relict sea cliffs, sea stacks, and beaches found above the reach of present wave action. Coastlines

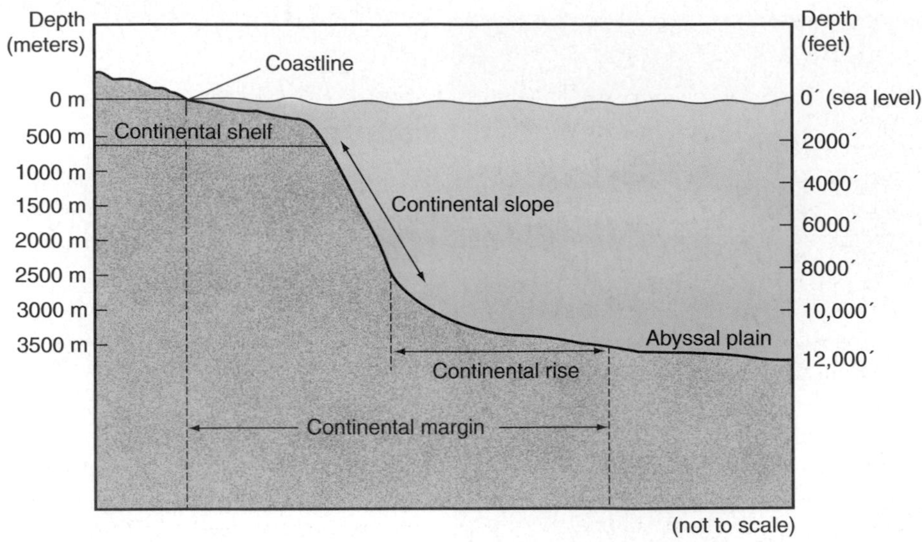

• FIGURE 20.26
This profile shows the general nature of the boundary between the continents (continental crust) and ocean basins (oceanic crust) and their relations to sea level. The gently sloping part of the submerged edge of the continents—the continental shelves—vary in width along different coasts depending on proximity to plate margins and plate tectonic history.

• FIGURE 20.27
The rugged coast of Point Lobos, California, exemplifies an active-margin coastline, having experienced much tectonic unrest that includes general uplift in addition to great lateral movement along the San Andreas Fault.

of emergence were probably common during the glacial phases of the Pleistocene, prior to 12,000 years ago, when sea level fell about 120 meters (400 ft). Features of emergence are best developed along active-margin coasts like those of California, Oregon, and Washington where marine terraces are found as much as 370 meters (1200 ft) above sea level (• Fig. 20.28). Other emergent coastlines, such as around the Baltic Sea and Hudson Bay, are located where isostatic rebound has elevated the land following the retreat of the continental ice sheets.

Along **coastlines of submergence** many features of the former shore lie underwater and the present shoreline crosses land areas that are not fully adjusted to coastal processes. Coastlines of submergence were created as global sea level rose in response to the retreat of the Pleistocene ice sheets. Coastlines of submergence also occur where tectonic forces have lowered the level of the land, as in San Francisco Bay. Great thickness of river deposits and compaction of alluvial sediments, as along the Louisiana coast, may also cause coastal submergence. The features of a new coastline of submergence are related to the character of coastal lands prior to submergence. Plains, for instance, will produce a far more regular shoreline than will a mountainous region or an area of hills and valleys. When areas of low relief with soft sedimentary rocks are submerged, barrier islands form with shallow bays and lagoons behind them. The classic examples of this type of submerging coastline are the Atlantic and Gulf coasts of the United States.

Two special types of submerged coastlines are ria and fjord coasts. **Rias** are created where river valleys are "drowned" by a rise in sea level or a sinking of the coastal area (• Fig. 20.29). These irregular coastlines result when valleys become narrow bays and the ridges form peninsulas. The Aegean coast of Greece and Turkey is an outstanding example of a ria coastline. *Fjords,* which are drowned glacial valleys, form scenically spectacular shorelines (• Fig. 20.30). A fjord shoreline is highly irregular, with deep, steep-sided arms of the sea penetrating far inland in troughs originally deepened by glaciers. Tributary streams cascade down fjord walls that may be a few thousand meters high. Fjord coastlines are found in Norway (where the term originated), Chile, New Zealand, Greenland, and Alaska. Canada, however, has more fjords than any other nation. In many fjords, the glaciers have retreated far inland, but some, especially in Greenland and Alaska, have "tidewater" glaciers that calve icebergs into the cold fjord waters.

Some coastlines, such as those composed of coral reefs and river deltas, cannot be classified as either submerging or emerging. Actually, most shorelines show evidence of more than one type of development largely because the land elevation and the level of the ocean have changed many times during geologic history. For this reason, features of both submerged and emerged shorelines characterize many coastlines.

Because both continental and marine geomorphic processes shape coastlines, another regional classification system recognizes two types of coasts: primary and secondary coastlines. Erosion and depositional processes of the land dominate **primary coastlines.** Primary coastlines result from such rapid changes in the position of the shoreline that coastal processes are not able to have a significant landforming effect. The major types of primary coastlines

USGS

• FIGURE 20.28
Cape Blanco, Oregon, represents an emergent coastline. The flat surface on which the lighthouse is built is a marine terrace formed by wave erosion and deposition prior to tectonic uplift. The elevation of the marine terrace is about 61 meters (200 ft) above sea level.

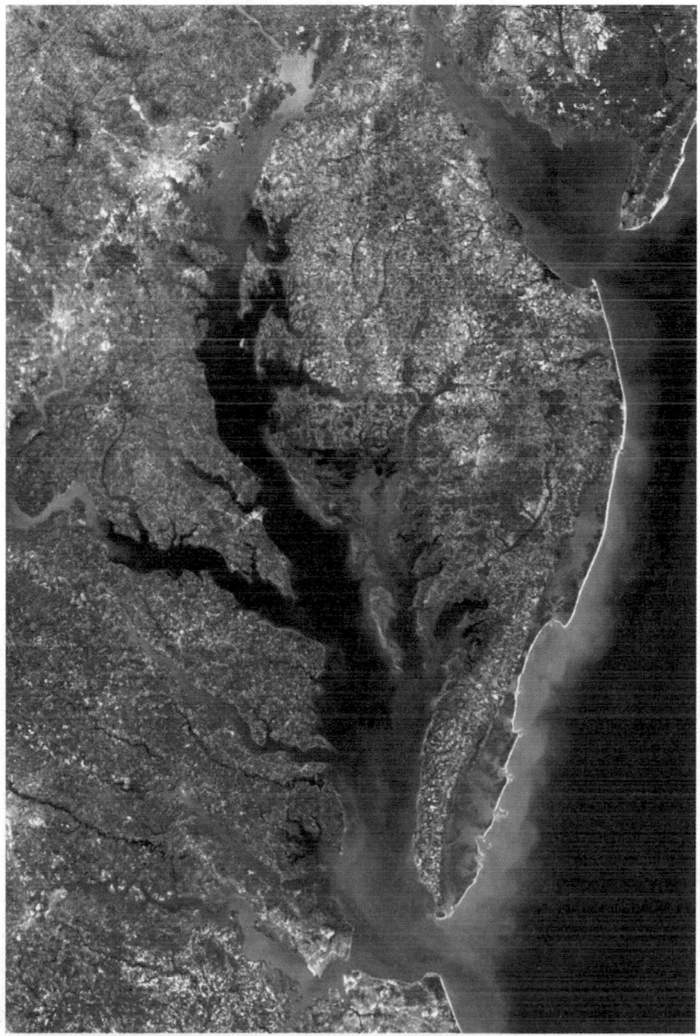

NASA

• FIGURE 20.29
Submergent coasts like the Chesapeake Bay region are characterized by drowned river valleys, known as rias, that developed as sea level rose at the end of the Pleistocene. The white edges on the barrier islands are beaches.

© Matt Ebiner

• **FIGURE 20.30**
Fjords, like this one in Greenland, are glaciated valleys that were drowned by the sea following the Pleistocene Epoch as the glaciers receded and sea level rose.

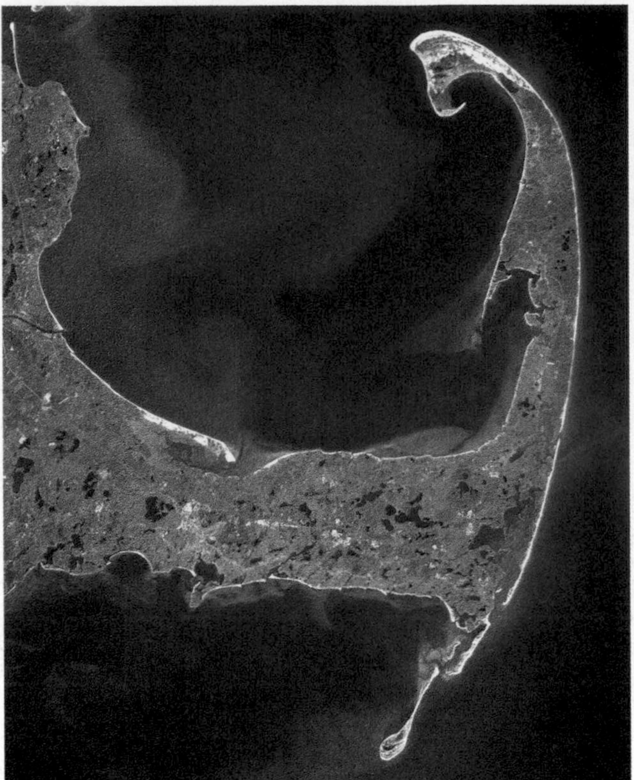
NASA

• **FIGURE 20.31**
Cape Cod is an example of a primary shoreline of glacial deposits being modified by waves and currents.
How has wave action modified the moraine that originally formed Cape Cod?

(with examples) are drowned river valleys (Delaware and Chesapeake Bays), glacial erosion coasts (southeastern Alaska, British Columbia, and Puget Sound in Washington, and from Maine to Newfoundland), glacial deposition coasts (Cape Cod, Massachusetts [• Fig. 20.31], and the north shore of New York's Long Island), river deltas (Mississippi Delta, Louisiana), volcanic coasts (Hawaii), and faulted coasts (California).

Secondary coastlines are those formed mainly by coastal geomorphic agents, especially waves, and by aquatic organisms. Sea cliffs, arches, sea stacks, and sea caves generally dominate erosional secondary coasts, like that of Oregon. Depositional secondary coasts typically display barrier beaches and spits, such as the coastline of North Carolina. An example of coasts built by aquatic organisms is the coral reef (• Fig. 20.32). The Florida Keys were constructed by coral growth. Mangrove trees and saltmarsh grasses also trap sediment to build new land areas in shallow coastal waters.

Islands and Coral Reefs

The perimeters of islands are subject to the same coastal processes as the boundary between a standing body of water and the mainland. Within the ocean there exist three basic types of islands: continental, oceanic, and atolls. **Continental islands** are usually found on the part of a continent submerged by the ocean, the continental shelf. Continental islands are geologically part of the continent but became separated from it because of global sea-level change or regional tectonic activity. The world's largest islands—Greenland, New Guinea, Borneo, and Great Britain—are continental. Smaller continental islands include New York's Long Island, California's Channel Islands, and Vancouver Island off the west coast of Canada. The barrier islands along the Gulf and Atlantic coasts of the United States are also continental. A few large continental islands, such as New Zealand and Madagascar, are isolated "continental fragments" that separated from continents millions of years ago.

Oceanic islands are volcanoes that rise from the deep-ocean floor and are geologically related to oceanic crust, not the continents. Most oceanic islands, such as the Aleutians, Tonga, and the Marianas, occur in island arcs along the edges of the trenches. Others, like Iceland and the Azores, are peaks of oceanic ridges rising above sea level. Many oceanic islands occur in chains, such as the Hawaiian Islands. The oceanic crust sliding over a stationary "hot spot" in the mantle causes these island chains. In the future, the Hawaiian Islands will move northwestward with the Pacific plate and slowly submerge. A new volcanic island, named Loihi, will form to the southeast (• Fig. 20.33). Evidence of the plate motion is indicated by the fact that the youngest islands of the Hawaiian chain, Hawaii and Maui, are to the southeast, while the older islands, such as Kauai and Midway, are located to the northwest.

An **atoll** is an island consisting of a ring of coral reefs that have grown up from a subsiding volcanic island and that encircle a central lagoon (• Fig. 20.34a). In order to understand atolls, we must first consider coral reefs.

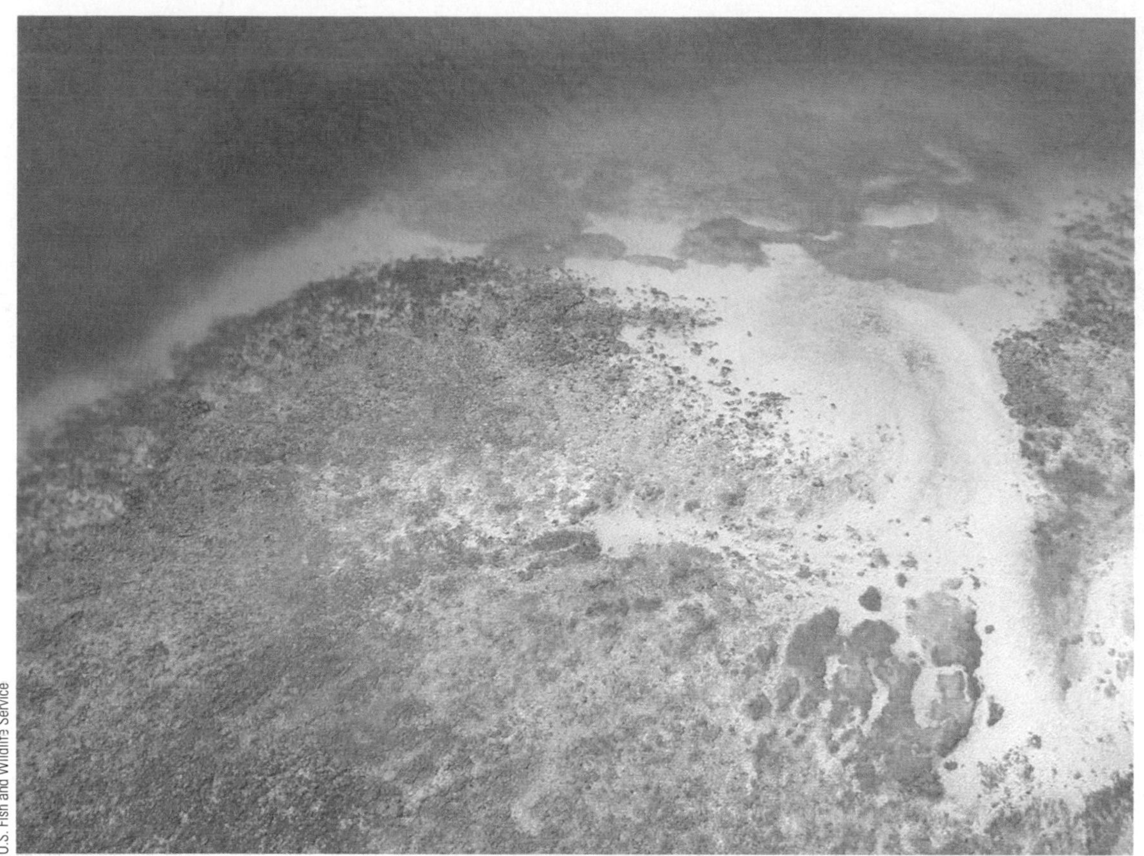

• FIGURE 20.32
Corals building the coastline at Vieques National Wildlife Refuge, Puerto Rico, provide an example of a secondary shoreline.

• FIGURE 20.33
The oceanic island of Hawaii, the largest and youngest Hawaiian island, was formed over a "hot spot" like the other Hawaiian Islands before it. Two large shield volcanoes, Mauna Loa and Mauna Kea, dominate the island. Mauna Loa and Kilauea on its eastern slope are still very active. Loihi, an active submarine volcano known as a *seamount,* is growing to the south and may be above sea level in 50,000 years.

© Tahiti Tourism Board

(a)

© David Hiser/Getty Images

(b)

© Paul Chelsey/Getty Images

(c)

• FIGURE 20.34

The major types of coral reefs are evident in the Society Island chain of French Polynesia. (a) Atolls, like Tetiaroa Atoll, are islands consisting of a ring of coral with no surface evidence remaining of its former volcanic core. (b) Fringing reefs are attached to mainland or island coasts. Moorea is a rugged young oceanic island with a fringing reef. (c) Like coastal barriers in general, barrier reefs are separated from dry land by a lagoon. Bora Bora is a subsiding island with a barrier reef around it.

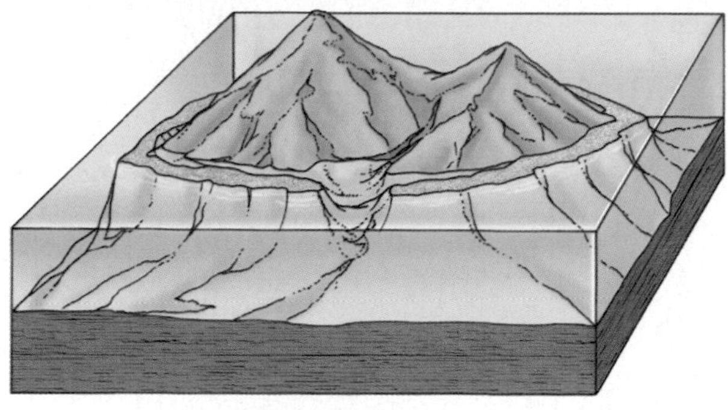

(a)

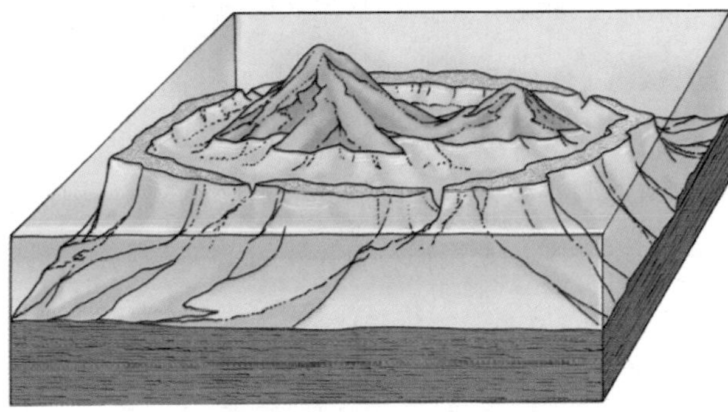

(b)

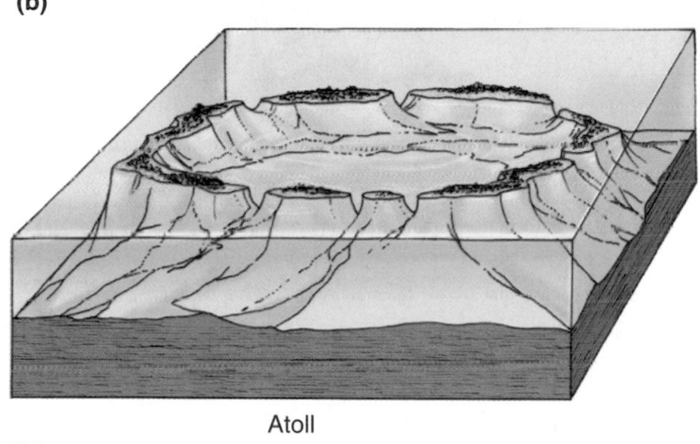

(c)

• FIGURE 20.35
The theory of coral reef development around oceanic islands was proposed by Charles Darwin. Three successive types of reef develop due to island subsidence and coral reef building. (a) First, a fringing reef grows along the shore. (b) Later, as the island erodes and subsides, a barrier reef develops. (c) Further subsidence causes the coral to build upward while the volcanic core of the island is completely submerged below a central lagoon, forming an atoll.
Explain why the island subsides while the coral grows upward.

Coral reefs are shallow, wave-resistant structures made by the accumulation of remains of tiny sea animals that secrete a skeleton of calcium carbonate. Many other organisms, including algae, sponges, and mollusks, add material to the reef structure. Reef corals need special conditions to grow—clear and well-aerated water, water temperatures above 20°C (68°F), plenty of sunlight, and normal marine salinity. These conditions are found in the shallow waters of tropical regions, including Hawaii, the West Indies, Indonesia, the Red Sea, and the coast of Queensland in Australia. Today, coastal water pollution, dredging, souvenir coral collecting, and possibly global warming threaten the survival of many coral reefs.

A **fringing reef** is a coral reef attached to the coast (Fig. 20.34b). Fringing reefs tend to be wider where there is more wave action that brings a continuous supply of well-aerated water and additional nutrients for increased coral growth. They are usually absent near river mouths because the coral cannot grow where the waters are laden with sediment or where river water lowers the salinity of the marine environment.

Sometimes coral forms a **barrier reef,** which lies offshore, separated from the land by a shallow lagoon (Fig. 20.34c). Most barrier reefs occur in association with slowly subsiding oceanic islands, growing at a pace that keeps them above sea level. Other barrier reefs, including Australia's Great Barrier Reef, the Florida Keys, and the Bahamas, were formed on continental shelves and grew upward as sea level rose after the Pleistocene ice age waned. The world's largest organic structure, the Great Barrier Reef of Australia, is more than 1930 kilometers (120 mi) long.

• Figure 20.35 illustrates the manner in which atolls develop. As a volcanic island subsides the fringing reef grows upward, keeping pace with the seafloor subsidence, becoming a barrier reef, and finally an atoll. Charles Darwin proposed this explanation of atoll formation in the 1830s. Drilling evidence indicates that there has been as much as 1200 meters (4000 ft) of subsidence and an equal amount of reef development in the past 60 million years.

Atolls pose severe challenges as environments for human habitation. First, they have a low elevation above sea level and provide no defense against huge storm waves and tsunamis that can inundate the entire atoll, drowning all its inhabitants. Possible future sea-level rise from global warming would also threaten these low islands. Second, there is little fresh water available on the porous coral limestone surface. Third, little vegetation can survive in the lime-rich rock and soil of the atoll islands. The coconut palm is an exception, and coconuts were vital to the survival of early inhabitants of the atoll islands in Polynesia and Micronesia.

Chapter 20 Activities

Define & Recall

shoreline
sea level
coastal zone
nearshore zone
breaker zone
surf zone
swash zone
swash
backwash
offshore zone
wave crest
wave trough
wave height
wavelength
wave steepness
wave period
tide
tsunami
wind wave
tidal range
spring tide
neap tide
sea
swell
fetch
deep-water wave
wave base
rip current
wave refraction
littoral drifting
beach drifting
longshore current
longshore drifting
sea cliff (lake cliff)
notch
cobble beach
sea cave
sea arch
sea stack
abrasion platform
marine terrace
estuary
beach
longshore bar
spit
tombolo
barrier beach
lagoon
bar
barrier spit
baymouth barrier
barrier island
groin
passive-margin coast
continental shelf
active-margin coast
coastlines of emergence
coastlines of submergence
ria
primary coastline
secondary coastline
continental island
oceanic island
atoll
coral reef
fringing reef
barrier reef

Discuss & Review

1. What is the difference between a shoreline and the coastal zone?
2. Describe the major factors that produce the tides. What are some variations in tidal patterns?
3. How do waves change when they enter shallow coastal waters? What is the main factor causing this change in the waves?
4. What is wave refraction? How is wave refraction related to the shape of the coastline?
5. Describe how sea cliffs form. Name three other coastal erosional landforms typically found in areas with sea cliffs.
6. What are the differences between beach drifting and littoral drifting? What causes both?
7. Describe the similarities and differences between beaches and barrier beaches.
8. How does a bar differ from a barrier?
9. What are the major differences between active-margin coastlines and passive-margin coastlines?
10. What is a coral reef and how does it form?

Consider & Respond

1. What major changes would you expect in the world's coastal zones if sea level were to rise due to global warming? How would the landforms change?
2. Explain some of the various ways that people can influence coastal geomorphology.
3. How are systems of coastal classification useful in studying coastal environments?
4. If you were asked to plan for construction of a major tourist resort on a beautiful tropical atoll, what limitations and concerns would you have to evaluate before construction?

Apply & Learn

1. A wind wave with a wavelength of 75 meters (245 ft) and a height of 0.8 meter (2.6 ft) takes 7 seconds to travel past a given point. What is the wave's steepness? What is its period?
2. If a 60-meter- (200 ft) long wave travels through water that is 25 meters (82 ft) deep, is the wave feeling bottom? As the wave comes in toward land, how high can it get before it breaks?

Locate & Explore

Note: Please read the About Locate & Explore Activities section of the Preface before beginning these exercises.

1. Using Google Earth and the Tide Layer, examine the hourly change in tidal elevation over the last week for the coastal sites listed below. Identify the tides as either diurnal, semidiurnal, or mixed and rank the sites by the tidal range. Compare the predicted and observed tidal variations. What, other than tide, causes the water level to vary?
   a. Portland, Maine
   b. Fort Pulaski, Georgia
   c. Key West, Florida
   d. Dauphin Island, Alabama
   e. Rockport, Texas
   f. Los Angeles, California
   g. South Beach, Oregon
   h. Seattle, Washington
2. Using Google Earth, examine the barrier islands north and south on the Maryland coast (38.325°N, 75.09°W). What is the main direction of sediment transport alongshore and what evidence do you see to support your answer? Why do you think the island to the south is displaced landward of the more developed island to the north?
3. Using Google Earth and Hurricane Ivan LIDAR Layer for Santa Rosa Island in northwest Florida, view the topographic data from before and after Hurricane Ivan (September 2004). The topography of the island was created using Light Detection and Ranging (LIDAR), an increasingly used mapping technology that provides spatially dense and accurate topographic data. The data are collected with aircraft-mounted lasers capable of recording elevation measurements at a rate of 2000 to 5000 pulses per second and with a vertical precision of 0.15 meter. In the prestorm data you will see that the dunes along the beach are highly variable in height and extent, while in the poststorm data you will see that the dunes are gone and replaced by overwash fans of different sizes.
   a. Change the transparency of the pre-Ivan LIDAR image and the ruler tool to measure the change in the position of the shoreline as a result of Hurricane Ivan. Now make the post-Ivan image transparent and measure the change in shoreline after Hurricane Ivan.
   b. Using both LIDAR images, describe how the overwash fan development is related to prestorm dune height. Can you develop a simple model to predict the impact of a hurricane on a barrier island using the height of the dunes relative to the elevation of the storm surge?

Map Interpretation

ACTIVE-MARGIN COASTLINES

Point Reyes National Seashore lies north of San Francisco on the rugged California coast. Point Reyes consists of resistant bedrock that is being eroded by the forces of the sea. Marine life is abundant, including many birds and marine mammals (note Sea Lion Cove below Point Reyes). A hilly area, Punta de Los Reyes, separates Drake's Bay and Point Reyes from Tomales Bay (visible on the aerial photograph as a ribbon of water across the top, but barely showing in the northeastern corner of the map).

The oblique aerial photo of Point Reyes was taken from a high-altitude NASA aircraft. Color infrared film makes vegetation appear red. The view is looking northeast. Trending across the upper portion of the photograph is the San Andreas Fault, which forms the linear Tomales Bay. The San Andreas Fault also separates the Pacific plate, on which Point Reyes is located, from the North American plate, which underlies the area east of the fault. Point Reyes is moving northwest along this fault. Emergent coastal features and recent tectonic activity characterize active-margin coastlines.

Point Reyes has a Mediterranean climate, influenced by the cool offshore California current. Here, the sea is not hospitable for swimming because of the uncomfortably cool seawater, high surf, and dangerous rip currents. The coastal location and onshore westerly winds create a truly temperate climate. Although very hot and very cold temperatures are rarely experienced, this is one of the windiest and foggiest coastlines in the United States. It is an area of rugged natural beauty with wind-sculpted trees, grasslands, rocky sea cliffs, and long sandy beaches.

Interpreting the Map

1. Which area of the coast is most exposed to wave erosion? What features indicate this type of high-energy activity?
2. Which area of the map is under the influence of strong longshore currents? What is the general direction of flow, and what coastal feature would indicate this flow?
3. Looking into the future, what may happen to Drakes Estero? What would Limantour Spit become?
4. Which area of the map appears to have strong wind activity? What would indicate this? What do you think the prevailing wind direction is?
5. Locate examples of the following coastal erosional landforms:
   a. Headland
   b. Sea stack
   c. Sea cliffs
6. Note that the bays and *esteros* (Spanish for "estuary") have mud bottoms. Because the creeks and streams are very small in the region, what is the probable source and cause for movement of the mud?
7. Note the offshore bathymetric contours (blue isolines). Which has a steeper gradient, the Pacific coast (west side) or Drakes Bay (east side)? Why do you think there is such a great difference?

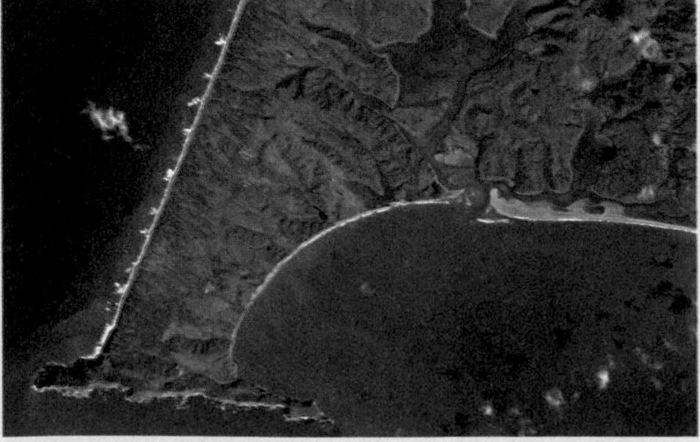

NASA

High-altitude oblique aerial photo of Point Reyes, California.

Opposite:
Point Reyes, California
Scale: 1:62,500
Contour interval = 80 ft
U.S. Geological Survey

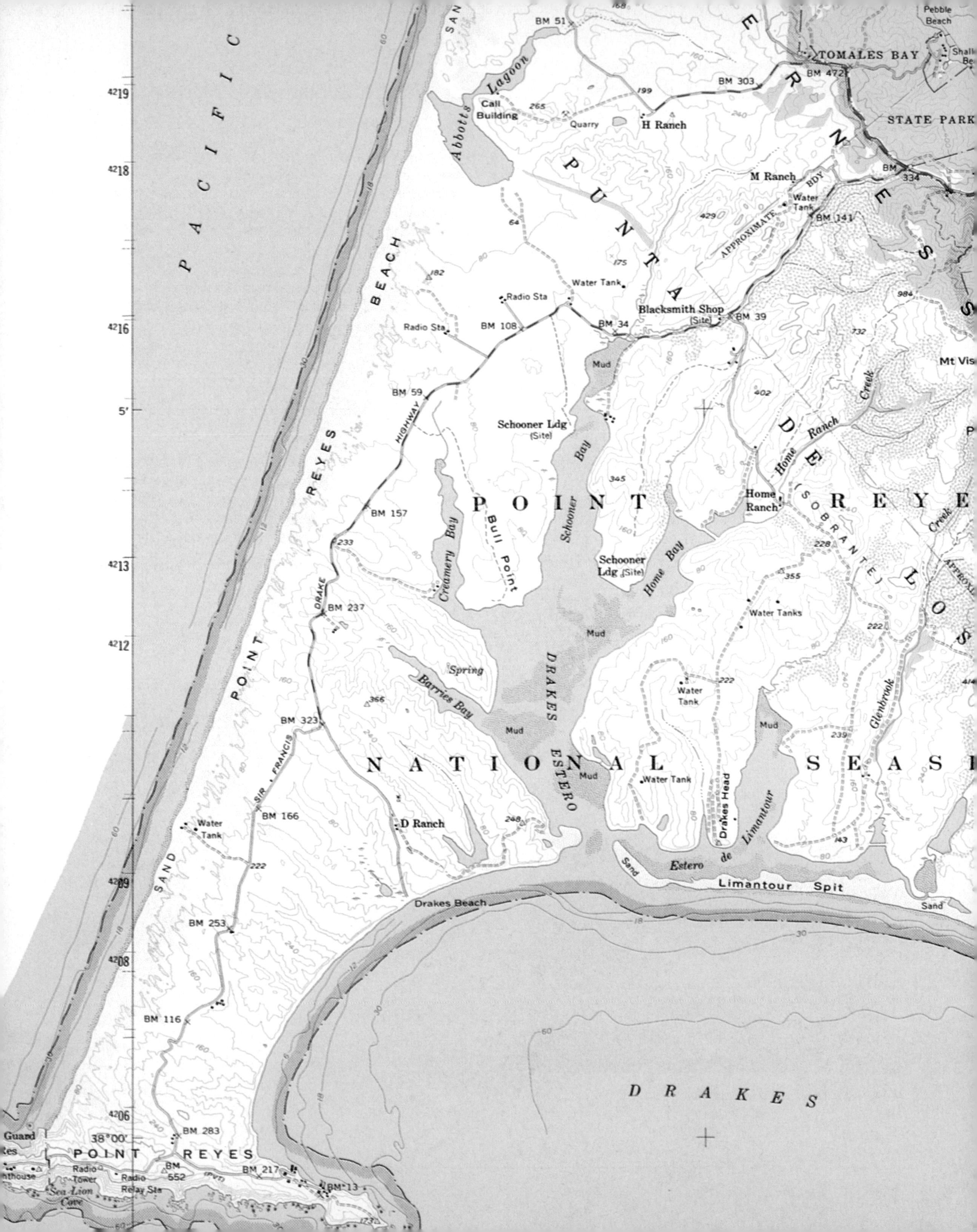
PACIFIC
POINT REYES BEACH
POINT REYES NATIONAL SEASHORE
PUNTA DE LOS REYES
TOMALES BAY
STATE PARK
Pebble Beach
Abbotts Lagoon
Call Building
Quarry
H Ranch
M Ranch
Water Tank
Radio Sta
Blacksmith Shop (Site)
Mud
Schooner Ldg (Site)
Schooner Bay
Home Ranch
Home Ranch Creek
Creamery Bay
Bull Point
Home Bay
Water Tanks
Barries Bay
Spring
DRAKES ESTERO
D Ranch
Drakes Head
Estero de Limantour
Limantour Spit
Sand
Drakes Beach
Glenbrook
Mt Vis
HIGHWAY
SIR FRANCIS DRAKE
APPROXIMATE BDY
(SOBRANTE)
POINT REYES
Radio Tower
Radio Relay Sta
Sea Lion Cove
38°00'
DRAKES

Map Interpretation

PASSIVE-MARGIN COASTLINES

Eastport, New York, is located on the south shore of Long Island, 115 kilometers (70 mi) east of New York City. Long Island is part of the Atlantic Coastal Plain, which extends from Cape Cod, Massachusetts, to Florida. Water bodies such as Delaware Bay, Chesapeake Bay, and Long Island Sound embay much of the Atlantic Coastal Plain in the eastern United States. The south shore of Long Island has low relief, and its coastal location moderates its humid continental climate.

Although this is a coastal region, its recent glacial history influences the landforms that exist today. Two east–west trending glacial terminal moraines deposited during the Pleistocene glacial advances form Long Island. Between the coast and the south moraine is a sandy glacial outwash plain that forms the higher elevations at the northern part of the map area. As the glaciers melted, sea level rose, submerging the lowland now occupied by Long Island Sound. This water body separates Long Island from the mainland of the Atlantic Coastal Plain.

The Atlantic and Gulf coasts of the United States have nearly 300 barrier islands with a combined length of over 2500 kilometers (1600 mi). New York, especially the coastal zone of Long Island's south shore, has 15 barrier islands with a total length of over 240 kilometers (150 mi). Coastal barrier islands protect the mainland from storm waves, and they contribute to the formation of coastal wetlands on their landward side, which are a critical habitat for fish, shellfish, and birds.

Interpreting the Map

1. What type of landform is the entire long, narrrow feature on which lies the straight shoreline labeled Westhampton Beach?
2. What is the highest elevation on the linear coastal feature? What do you think comprises the highest portions of this feature?
3. How much evidence of human use exists on the landform in Question 1? What problems might these cultural features be subjected to?
4. Behind the linear feature is a water body. Is it deep or shallow? Is it a high-energy or low-energy environment? What is a water body such as this called?
5. Is the coastline shown on the Eastport map predominantly erosional or depositional? What features support your answer? Is this a coastline of submergence or emergence?
6. Note Beaverdam Creek in the upper middle of the map area. Does it have a steep or gentle gradient?
7. Based on its gradient, does Beaverdam Creek always flow seaward? If not, what might influence its flow?
8. What kind of topographic feature exists as indicated by the map symbols in the area surrounding Oneck?
9. How do you think geomorphic processes will likely modify this map area in the future? Which area will probably be modified the most? Explain.
10. What type of natural hazard is this coastal area of Long Island most susceptible to? Why?

NASA/GSFC/SVS

Satellite image of Long Island, New York.

Opposite:
Eastport, New York
Scale: 1:24,000
Contour interval = 10 ft
U.S. Geological Survey

Beaverdam Pond
Westhampton
Westhampton Ch
Beaverdam Creek
High Sch
Westhampton Country Club
Westhampton Beach
Tanner Neck
Apacuck Point
Oneck
Moneyboque Bay
Reedy Island
Project Depth 4 Feet
Potunk Pt
Picket Point
Pond Point
Gunning Point
Westhampton Beach
MONTAUK HY
SOUTH COUNTRY ROAD
DUNE ROAD
MAIN ST
Airstrip
Oneck Drain
C E A N

Appendix A

International System of Units (SI), Abbreviations, and Conversions

Symbol	Multiply	By	To Find	Symbol
Area				
in.^{2}	square inches	645.2	square millimeters	mm^2
ft^2	square feet	0.093	square meters	m^2
yd^2	square yards	0.836	square meters	m^2
ac	acres	0.405	hectares	ha
mi^2	square miles	2.59	square kilometers	km
ac	acres	43,560	square feet	ft^2
mm^2	square millimeters	0.0016	square inches	in.^{2}
m^2	square meters	10.764	square feet	ft^2
m^2	square meters	1.195	square yards	yd^2
ha	hectares	2.47	acres	ac
km^2	square kilometers	0.386	square miles	mi^2
Mass				
oz	ounces	28.35	grams	g
lb	pounds	0.454	kilograms	kg
g	grams	0.035	ounces	oz
kg	kilograms	2.202	pounds	lb
Length				
in.	inches	25.4	millimeters	mm
ft	feet	0.305	meters	m
yd	yards	0.914	meters	m
mi	miles	1.61	kilometers	km
ft	feet	63,360	mile	mi
mm	millimeters	0.039	inches	in.
m	meters	3.28	feet	ft
m	meters	1.09	yards	yd
km	kilometers	0.62	miles	mi
Volume				
gal	U.S. gallons	3.785	liters	l
ft^3	cubic feet	0.028	cubic meters	m^3
yd^3	cubic yards	0.765	cubic meters	m^3
l	liters	0.264	U.S. gallons	gal
m^3	cubic meters	35.30	cubic feet	ft^3
m^3	cubic meters	1.307	cubic yards	yd^3
Velocity				
mph	miles/hour	1.61	kilometers/hour	km/h
knot	nautical miles/hour	1.85	kilometers/hour	km/h
km/h	kilometers/hour	0.62	miles/hour	mph
km/h	kilometers/hour	0.54	nautical miles/hour	knots

International System of Units (SI), Abbreviations, and Conversions *(Continued)*

Symbol	Multiply	By	To Find	Symbol
Pressure or Stress				
mb	millibars	0.75	millimeters of mercury	mm Hg
mb	millibars	0.02953	inches of mercury	in Hg
mb	millibars	0.01450	pounds per square inch	($lb/in.^2$ or psi)
$lbs/in.^2$	pounds per square inch	6.89	kilopascals	kPa
in Hg	inch of mercury	33.865	millibars	mb
kPa	kilopascals	0.145	pounds per square inch	($lb/in.^2$ or psi)

Standard Sea-Level Pressure
29.92 in Hg
14.7 $lb/in.^2$
1013.2 mb
760 mm Hg

Temperature				
°F	Fahrenheit	(°F – 232)/1.8	Celsius	°C
°C	Celsius	1.8°C + 32	Fahrenheit	°F
K	Kelvins	K = °C + 273	Celsius	°C

Powers of Ten			
nano	one billionth	$= 10^{-9}$	= 0.000000001
micro	one millionth	$= 10^{-6}$	= 0.000001
milli	one thousandth	$= 10^{-3}$	= 0.001
centi	one hundredth	$= 10^{-2}$	= 0.01
deci	one tenth	$= 10^{-1}$	= 0.1
hecto	one hundred	$= 10^{2}$	= 100
kilo	one thousand	$= 10^{3}$	= 1000
mega	one million	$= 10^{6}$	= 1,000,000
giga	one billion	$= 10^{9}$	= 1,000,000,000

Appendix B
Topographic Maps

Mapping has changed considerably in recent years, and with the ever-increasing capabilities of computers to store, retrieve, and display graphics, this trend will continue well into the future. In the United States the U.S. Geological Survey produces the vast majority of topographic maps available. These maps have long been the tried and true tools of geographers and scientists in many other disciplines who study various aspects of the environment. Today virtually all USGS topographic maps are accessible in digital format, for downloading and printing, on computer disks, or for examining on a computer screen. Computers and the Internet have made maps much more available and accessible than they were just a few years ago. This availability makes it easy to print maps or map segments at home, school, or work. A logical question then would be, will paper maps become obsolete, given computer displays?

There are several reasons why paper maps will still be popular and useful, whether they are purchased from the source or downloaded and printed. Topographic maps are particularly important in fieldwork. Maps are highly portable, require no batteries or electrical power that could fail, and do not suffer from technology glitches. They are reliable and easy to use and they also provide a good base for making field notes, and marking routes.

Today the USGS is working to make and maintain a seamless database of the United States, with maps, imagery, and spatial data in digital form, so areas that once were split on adjacent maps can be printed on a single sheet. This is a great change from dividing the country into quadrangles (the roughly rectangular area a map displays). Topographic quadrangles (quads), however, will continue to be in use for a long time. There are several standard quadrangles, each with a specific scale, and many other special-purpose topographic maps.

7.5-minute quads—these are printed at a scale of 1:24,000 and cover 7.5 minutes of longitude and 7.5 minutes of latitude.

15-minute quads—these are printed at a scale of 1:62,500 and cover 15 minutes of longitude and 15 minutes of latitude.

1° × 2° quads—these are printed at a scale of 1:250,000 and cover 1 degree of latitude and 2 degrees of longitude.

1:100,000 metric quads—these are printed at a scale of 1:100,000, cover 30 minutes of longitude and 60 minutes of latitude, and use metric measurements for distance and elevation.

Brown topographic contours are used to show elevation differences and the terrain. Some rules for interpreting contours are given in Chapter 2.

Determining Distances on a Map, Distances from a Map, or an RF Scale It is important to understand representative fractions, like 1:24,000, which means that any measurement on the map will represent 24,000 of the same measurements on the ground. This knowledge is particularly significant because reproduced maps may not be printed at the original size. On a map that is reduced or enlarged the bar scale will still be accurate, but the printed RF scale will not. Enlarging or reducing a map changes the RF scale from the original.

How to Find the RF of a Map (or Air Photo, or Satellite Image) of Unknown Scale Here is the formula:

$$1/\text{RFD} = \text{MD}/\text{GD}$$

The numerator in an RF scale is always the number 1. The RFD is the denominator of the RF (such as 24,000). MD is map distance measured in any particular units on the map (cm, in.). GD is the true ground distance that the map distance represents (expressed in the same units used to measure the map distance). Never mix units in this calculation, but convert values into desired units afterward.

Example: How long in inches is a mile on a 1:24,000 scale map?

Important information: There are 63,360 inches in a mile.
To find MD, for a known distance (mile) on a map of known scale, use this formula:

$$1/24{,}000 = \text{MD}/63{,}360 \text{ in.}$$

$$1 \text{ mile} = 2.64 \text{ inches at } 1{:}24{,}000$$

This statement has now been converted into a **stated scale** so units may be mixed.

1:24,000 Bar Scale Here is a bar scale that can be used to make distance measurements directly from the 1:24,000 maps printed in the Map Interpretation sections. Note: Check the listed RF, because not all are at a 1:24,000 scale.

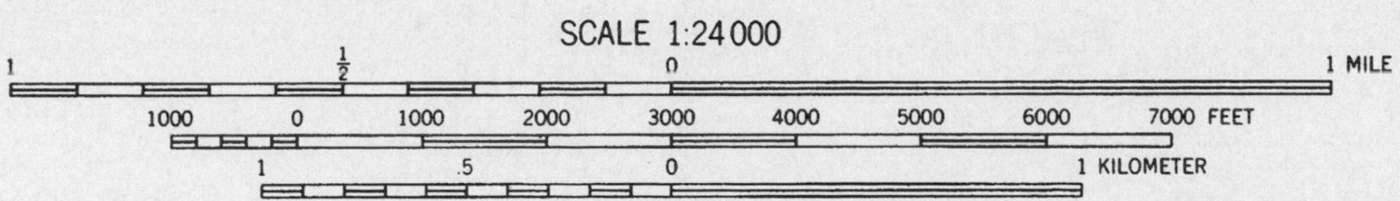

BOUNDARIES
National
State or territorial
County or equivalent
Civil township or equivalent
Incorporated-city or equivalent
Park, reservation, or monument
Small park

LAND SURVEY SYSTEMS
U.S. Public Land Survey System:
Township or range line
Location doubtful
Section line
Location doubtful
Found section corner; found closing corner
Witness corner; meander corner

Other land surveys:
Township or range line
Section line
Land grant or mining claim; monument
Fence line

ROADS AND RELATED FEATURES
Primary highway
Secondary highway
Light duty road
Unimproved road
Trail
Dual highway
Dual highway with median strip
Road under construction
Underpass; overpass
Bridge
Drawbridge
Tunnel

BUILDINGS AND RELATED FEATURES
Dwelling or place of employment: small; large
School; church
Barn, warehouse, etc.: small; large
House omission tint
Racetrack
Airport
Landing strip
Well (other than water); windmill
Water tank: small; large
Other tank: small; large
Covered reservoir
Gaging station
Landmark object
Campground; picnic area
Cemetery: small; large

RAILROADS AND RELATED FEATURES
Standard gauge single track; station
Standard gauge multiple track
Abandoned
Under construction
Narrow gauge single track
Narrow gauge multiple track
Railroad in street
Juxtaposition
Roundhouse and turntable

TRANSMISSION LINES AND PIPELINES
Power transmission line; pole; tower
Telephone or telegraph line
Aboveground oil or gas pipeline
Underground oil or gas pipeline

CONTOURS
Topographic:
Intermediate
Index
Supplementary
Depression
Cut; fill

Bathymetric:
Intermediate
Index
Primary
Index Primary
Supplementary

MINES AND CAVES
Quarry or open pit mine
Gravel, sand, clay, or borrow pit
Mine tunnel or cave entrance
Prospect; mine shaft
Mine dump
Tailings

SURFACE FEATURES
Levee
Sand or mud area, dunes, or shifting sand
Intricate surface area
Gravel beach or glacial moraine
Tailings pond

VEGETATION
Woods
Scrub
Orchard
Vineyard
Mangrove

COASTAL FEATURES
Foreshore flat
Rock or coral reef
Rock bare or awash
Group of rocks bare or awash
Exposed wreck
Depth curve; sounding
Breakwater, pier, jetty, or wharf
Seawall

BATHYMETRIC FEATURES
Area exposed at mean low tide; sounding datum
Channel
Offshore oil or gas: well; platform
Sunken rock

RIVERS, LAKES, AND CANALS
Intermittent stream
Intermittent river
Disappearing stream
Perennial stream
Perennial river
Small falls; small rapids
Large falls; large rapids

Masonry dam

Dam with lock

Dam carrying road

Intermittent lake or pond
Dry lake
Narrow wash
Wide wash
Canal, flume, or aqueduct with lock
Elevated aqueduct, flume, or conduit
Aqueduct tunnel
Water well; spring or seep

GLACIERS AND PERMANENT SNOWFIELDS
Contours and limits
Form lines

SUBMERGED AREAS AND BOGS
Marsh or swamp
Submerged marsh or swamp
Wooded marsh or swamp
Submerged wooded marsh or swamp
Rice field
Land subject to inundation

Appendix C
Understanding and Recognizing Some Common Rocks

Rocks are aggregates of minerals, and although there are thousands of kinds of rocks on our planet, they can be classified into three fundamental kinds based on their origin: igneous, sedimentary, and metamorphic. The formation of rocks was outlined in Chapter 13. Having a solid knowledge of how the Rock Cycle operates (Fig. 13.5) as well as its components and its processes is essential to understanding the solid Earth. Specific rock types are mentioned in several chapters of this book. Although making a positive identification of a rock type requires examining several physical properties, having a mental image of what different rocks look like will be an aid to understanding Earth processes and landforms. The following is an illustrated guide to a few common rocks to help in their identification. Intrusive igneous rocks form from a molten state by cooling and crystallizing underground; they generally cool very slowly, which allows coarse crystals to form that are easy to see with the naked eye.

Igneous Rocks

Igneous rocks are subdivided into intrusive and extrusive, depending on whether they cooled within Earth or on its exterior.

Intrusive Igneous

Copyright and photograph by Dr. Parvinder S. Sethi

Granite
Granite forms deep in the crust, and has easily visible intergrown crystals of light and dark minerals, but is dominated by light-colored silicate minerals. Granitic rocks are typically gray or pink, and their mineral composition is similar to that of the continental crust.

Copyright and photograph by Dr. Parvinder S. Sethi

Diorite
Diorite is an intermediate intrusive rock, meaning that it has a roughly equal mix of easily visible light and dark minerals, which gives it a spotted appearance. Generally diorite is dark gray in color.

Copyright and photograph by Dr. Parvinder S. Sethi

Gabbro
Gabbro is a dark intrusive rock dominated by heavy, iron-rich silicate minerals. The crystals are coarse enough to be easily visible, but because of the overall dark tone, they tend to blend together. Gabbro is black and may contain some very dark green minerals.

Extrusive Igneous

Extrusive igneous rocks cool at or near the surface, and include lavas, as well as rocks made of tephra (pyroclastics), fragments blown out of a volcano. Extrusive rocks sometimes preserve gas bubble holes, and may contain visible crystals, but typically the grains are small. Cooling relatively rapidly at the surface produces fine-grained lavas. Clastic means made up of cemented rock fragments, such as clay, silt, sand, pebbles, cobbles, or even boulders. The sizes and shapes of clasts within a sedimentary rock provide clues about the environments under which the fragments were deposited (fluvial, eolian, glacial, coastal).

Copyright and photograph by Dr. Parvinder S. Sethi

Rhyolite
Rhyolite is a very thick lava when molten (much like melted glass), light in color and high in silica. Colors of rhyolite vary widely, but light gray, very light brown, and pink or reddish are common. Rhyolite is the extrusive equivalent of granite in terms of mineral composition.

Copyright and photograph by Dr. Parvinder S. Sethi

Andesite
Andesite, named for the Andes, is an intermediate lava in terms of both mineral content and color. Associated with composite cone volcanoes, it is relatively thick when molten. Often mineral crystals are visible in a matrix of finer grains, and the color is gray to brown. Andesite is the extrusive equivalent of diorite.

Copyright and photograph by Dr. Parvinder S. Sethi

Basalt
Basalt is dark, typically black, and heavier than other lavas. Associated with fissure flows and shield volcanoes, basalt is relatively low in silica, so it has a lower viscosity than other lavas. Basalt tends to be hotter than other lavas, and is relatively thin when molten, thus it can flow for many miles before cooling enough to stop. Basalt is the rock of the oceanic crust and is the extrusive equivalent of gabbro.

Copyright and photograph by Dr. Parvinder S. Sethi

Tuff
Tuff is a rock made of fine tephra—volcanic ash—that was blown into the atmosphere by a volcanic eruption, and settled out in layers that blanketed the surface. Tephra (loose fragments) was converted into tuff by burial and compaction, or by being welded together from intense heat. Tuff is gray to tan.

Sedimentary Rocks

Sedimentary rocks can also be divided into two major categories, clastic and nonclastic.

Clastic Sedimentary

Clastic means made up of cemented rock fragments, such as clay, silt, sand, pebbles, cobbles, or even boulders. The sizes and shapes of clasts within a sedimentary rock provide clues about the environments under which the fragments were deposited (fluvial, eolian, glacial, coastal).

Copyright and photograph by Dr. Parvinder S. Sethi

Shale

Shale is a fine-grained clastic rock that contains lithified clays, generally deposited in very thin layers. Shales represent a calm water environment, such as a sea or lake bottom. Shales vary widely in color, but most are gray or black, and they break up into smooth, flat surfaces.

Copyright and photograph by Dr. Parvinder S. Sethi

Sandstone

Sandstone is made of cemented fragments of sand size, typically made of quartz or other relatively hard mineral. Sandstones can be virtually any color, feel gritty, and may be banded or layered. Sandstone may represent ancient beaches, dunes, or fluvial deposits.

Copyright and photograph by Dr. Parvinder S. Sethi

Conglomerate

Conglomerate contains rounded pebbles cemented together by finer sediments. Conglomerate may represent deposits from a river's bed load, or a pebble beach.

Copyright and photograph by Dr. Parvinder S. Sethi

Breccia

Breccia is similar to conglomerate, but the cemented fragments are angular. Breccia is associated with mudflows, pyroclastic flows, and fragments deposited by mass wasting.

Nonclastic Sedimentary

Nonclastic sedimentary rocks consist of materials that are not rock fragments. Examples include chemical precipitates, such as limestone, evaporites such as rock salt, and deposits of organic materials, for example coal, or limestones made of shells and coral fragments. Nonclastic rocks also represent the ancient environment under which they were deposited.

Copyright and photograph by Dr. Parvinder S. Sethi

Rock Salt
Rock salt consists of sodium chloride, table salt, often with a mix of other salts. Rock salt represents the deposits left behind by the evaporation of a saline inland lake, or an arm of the sea that was cut off from the ocean by sea-level change or tectonic activity. Common color is white.

Copyright and photograph by Dr. Parvinder S. Sethi

Limestone
Limestone is made of calcium carbonate deposits (lime, $CaCO_3$). Limestone can represent a variety of environments and varies widely in color and appearance. Typical colors are white or gray, and the most common depositional environment was in shallow tropical seas, which were rich in lime. Many cave and spring deposits are also varieties of limestone.

Copyright and photograph by Dr. Parvinder S. Sethi

Coal
Coal is a rock made of the carbonized remains of ancient plants. Coal deposits typically represent a swampy lowland environment that was invaded by sea-level rise, which killed and buried dense vegetation.

Metamorphic Rocks

Metamorphic rocks can also be divided into two general categories, foliated rocks and nonfoliated rocks.

Foliated

Foliated metamorphic rocks have either wavy, roughly parallel plates, or bands of light and dark minerals that formed under intense heat and pressure. The appearance of these foliations indicates the degree of metamorphism or change from the rock's original state.

Copyright and photograph by Dr. Parvinder S. Sethi

Slate
Slate is metamorphosed shale, and looks much like shale, except it is harder and has very thin platy foliations. The most common color is black.

Copyright and photograph by Dr. Parvinder S. Sethi

Schist
Schist has very prominent, wavy, and platy foliations generally covered with mineral crystals that formed during metamorphism. Schist represents a high degree of metamorphism and could originally have been any of a wide variety of rocks, so colors and appearance vary greatly.

Copyright and photograph by Dr. Parvinder S. Sethi

Gneiss
Gneiss (pronounced "nice") is a banded metamorphic rock, with alternating bands of light and dark minerals. Gneiss represents extreme heat and pressure during metamorphism and also may originally have been any of a variety of rocks. Metamorphism of granites commonly produces gneiss.

Nonfoliated

Nonfoliated metamorphic rocks do not display regular patterns of banding or platy foliations. In general, nonfoliated metamorphics represent a rock that has been changed by the fusing and recrystallization of minerals in the original, often identifiable, rock.

Copyright and photograph by Dr. Parvinder S. Sethi

Quartzite
Quartzite is metamorphosed quartz sandstone in which the former sand grains have fused together to produce an extremely hard, resistant rock.

Copyright and photograph by Dr. Parvinder S. Sethi

Marble
Marble is metamorphosed limestone that has been recrystallized. Many colors and patterns of marble exist, and its relative softness compared to other rocks makes it easy to cut and polish.

Glossary

aa a blocky, angular surface of a lava flow.

abiotic natural, nonliving component of an ecosystem.

ablation any removal of frozen water from the mass of a glacier.

abrasion (corrasion) erosion process in which particles already being carried by a geomorphic agent are used as tools to aid in eroding more Earth material.

abrasion platform wave-cut bench of rock just below the water level; indicates the landward extent of coastal cliff erosion.

absolute humidity mass of water vapor present per unit volume of air, expressed as grams per cubic meter, or grains per cubic foot.

absolute location location of an object on the basis of mathematical coordinates on an Earth grid.

accretion growth of a continent by adding large pieces of crust along its border by plate tectonic collision.

accumulation any addition of frozen water to the mass of a glacier.

acid mine drainage seepage to the surface of subsurface water that has become highly acidic by flowing through underground coal mines.

acid rain rain with a pH value of less than 5.6, the pH of natural rain; often linked to the pollution associated with the burning of fossil fuels.

active layer the upper soil zone that thaws in summer in regions underlain by permafrost.

active-margin coast coastal region characterized by active volcanic and tectonism.

actual evapotranspiration the actual amount of moisture loss through evapotranspiration measured from a surface.

adiabatic heating and cooling change of temperature *within* a gas because of compression (resulting in heating) or expansion (resulting in cooling); no heat is added or subtracted from outside.

advection horizontal heat transfer within the atmosphere; air masses moved horizontally, usually by wind.

advection fog fog produced by the movement of warm, moist air across a cold sea or land surface.

aerial photograph photographs of the terrain taken with a camera and film from an aircraft.

aggradation building up of an area or landscape that results from more deposition than erosion over time.

air mass large portion of the atmosphere, sometimes subcontinental in size, that may move over Earth's surface as a distinct, relatively homogeneous entity.

air mass analysis explanation of weather phenomena by a study of the actions and interactions of major portions of the atmosphere.

albedo proportion of solar radiation reflected back from a surface, expressed as a percentage of radiation received on that surface.

Aleutian low center of low atmospheric pressure in the area of the Aleutian Islands, especially persistent in winter.

alfisol soil that has a subsurface clay horizon, is medium to high in bases, and is light colored.

alluvial fan fan-shaped depositional landform, particularly common in arid regions, occurring where a stream emerges from a mountain canyon and deposits sediment on a plain.

alluvial plain extensive floodplain of very low relief.

alluvium general term for clastic particles deposited by a stream.

alpine glacier a mass of flowing ice that exists in a mountainous region due to climatic conditions resulting from high elevation.

altithermal an interval of time about 7000 years ago when the climate was hotter than it is today.

altitude heights of points above Earth's surface.

alto signifies a middle-level cloud (i.e., from 2000 to 6000 m in elevation).

analemma a diagram that shows the declination of the sun throughout the year.

andisol soil that develops on volcanic parent material.

angle of inclination tilt of Earth's polar axis at an angle of 23½° from the vertical to the plane of the ecliptic.

angle of repose maximum angle at which a slope of loose sediment can stand without particles tumbling or sliding downslope.

annual lag of temperature time lag between the maximum (minimum) of insolation during the solstices and the warmest (coldest) temperatures of the year.

annual march of temperatures changes in monthly temperatures throughout the months of the year.

annual temperature range difference between the mean daily temperatures for the warmest and coolest months of the year.

Antarctic circle parallel of latitude at 66½°S; the northern limit of the zone in the Southern Hemisphere that experiences a 24-hour period of sunlight and a 24-hour period of darkness at least once a year.

anticline the upfolded element of folded rock structure.

anticyclone an area of high atmospheric pressure, also known as a *high*.

aphelion position of Earth's orbit at farthest distance from the sun during each Earth revolution.

aquiclude rock layer that restricts flow and storage of ground water; it is impermeable and nonporous.

aquifer rock layer that is a container and transmitter of ground water; it is both porous and permeable.

Arctic circle parallel of latitude at 66½°N; the southern limit of the zone in the Northern Hemisphere that experiences a 24-hour period of sunlight and a 24-hour period of darkness at least once a year.

arête jagged, sawtooth spine or wall of rock separating two expanding cirque basins.

arid climates climate regions or conditions where annual potential evapotranspiration greatly exceeds annual precipitation.
aridisol soil soil that develops in deserts where precipitation is less than half the potential evaporation.
artesian spring natural flow of groundwater to the surface from below due to pressure.
artesian well groundwater that flows toward the surface under its own pressure.
artificial recharge diverting surface water to permeable terrain for the purpose of replenishing groundwater supplies.
asteroid sometimes called a minor planet; any solar system body composed of rock and/or metal not exceeding 500 miles in diameter.
asthenosphere thick, plastic layer within Earth's mantle that flows in response to convection, instigating plate tectonic motion.
atmosphere blanket of air, composed of various gases, that envelops Earth.
atmospheric air pressure (barometric pressure) force per unit area that the atmosphere exerts on any surface at a particular elevation.
atmospheric controls geographic features that affect climate and weather patterns; for example, distance from the ocean, wind direction, altitude.
atmospheric disturbance refers to variation in the secondary circulation of the atmosphere that cannot correctly be classified as a storm; for example, front, air mass.
atmospheric effect the absorption of longwave Earth radiation by water vapor, carbon dioxide, and dust in the atmosphere so that Earth temperatures are moderated.
atoll ring of coral reefs and islands encircling a lagoon, with no inner island.
attrition the reduction of size in sediment as it is transported downstream.
auroras colorful interaction of solar wind with ions in Earth's upper atmosphere; more commonly seen in higher latitudes. Called aurora borealis in the Northern Hemisphere (also known as the northern lights), and the aurora australis (southern lights) in the Southern Hemisphere.
autotroph organism that, because it is capable of photosynthesis, is at the foundation of a food web and is considered a basic producer.
avalanche density current of pulverized (powdered) Earth material traveling rapidly downslope by the pull of gravity.
axis an imaginary line between the geographic North Pole and South Pole, around which the planet rotates.
azimuth an angular direction of a line, point, or route measured clockwise from north 0–360°.
azimuthal map a projection that preserves the true direction from the map center to any other point on the map.
Azores high *see* Bermuda high.

backing wind shift change in wind direction counterclockwise around the compass; for example, from east to northeast, to north, to northwest.
backwash thin sheet of broken wave water that rushes back down the beach face in the swash zone.
badlands barren region of soft rock material intensely eroded into ridges and ravines by numerous gullies and washes.
bajada an extensive intermediate slope of adjacent, coalescing alluvial fans connecting a steep mountain front with a basin or plain.
bar (coastal) shallow, submerged accumulation of sediments located close to shore.
bar (fluvial) a mounded accumulation of sediment in a stream.
barchan crescent-shaped sand dune with arms pointing downwind.
barometer instrument for measuring atmospheric pressure.
barrier beach a category of elongate coastal landforms that lie parallel to the mainland but separated from it by a lagoon.
barrier island a barrier beach constructed parallel to the mainland, but not attached to it; separated from the mainland by a lagoon.
barrier reef coral reef parallel to the coast and separated from it by a lagoon.
barrier spit a barrier beach attached to the mainland at one end like a spit, but enclosing a lagoon.
basalt a dark-colored, fine-grained extrusive igneous rock generally associated with the oceanic crust and oceanic volcanoes.
basalt plateau elevated area of low surface relief consisting of horizontal layers of basaltic lava.
base flow groundwater that seeps into stream channels below the water surface; sustains perennial streams between storms.
base line east–west lines of division in the U.S. Public Lands Survey System.
base level elevation below which a stream cannot flow; most humid region streams flow to sea level (ultimate base level).
batholith a large, irregular mass of intrusive igneous rock (pluton).
baymouth barrier a barrier beach attached to the mainland at either end of the mouth of an embayment to form a lagoon.
beach coastal landform of wave-deposited sediment attached to dry land along its entire length.
beach drifting littoral drifting in which waves breaking at an angle to the shoreline move sediment along the beach in a zigzag fashion in the swash zone.
bearing an angular direction of a line, point, or route measured from north or from a current location to a desired location (often in 90° compass quadrants).
bed load solid particles moved by wind or water by bouncing, rolling, or sliding along the ground or streambed.
bedding plane boundary between different sedimentary layers.
bedrock solid rock of Earth's crust that underlies soil and other unconsolidated materials.
Bergeron (ice crystal) process rain-forming process where cloud droplets begin as ice crystals and melt into rain as they fall toward the surface.
bergschrund the large crevasse at the head of a valley glacier, beneath the cirque headwall.
Bermuda high persistent, high atmospheric pressure center located in the subtropics of the north Atlantic Ocean.
biomass amount of living material or standing crop in an ecosystem or at a particular trophic level within an ecosystem.
biome one of Earth's major terrestrial ecosystems, classified by the vegetation types that dominate the plant communities within the ecosystem.
biosphere the life forms, human, animal, or plant, of Earth that form one of the major Earth subsystems.
blizzard heavy snowstorm accompanied by strong winds (35 mph or greater) and that reduces visibility to less than one-quarter mile.
blowout local, wind-eroded surface depression in an area dominated by wind-deposited sand.

bolson desert basin, surrounded by mountains, with no drainage outlet.

boreal forest (taiga) coniferous forest dominated by spruce, fir, and pine found growing in subarctic conditions around the world north of the 50th parallel of north latitude.

boulder a rock fragment greater than 256 millimeters in diameter.

braided channel stream channel composed of multiple subchannels of simultaneous flow that split and rejoin and frequently shift position.

braided stream stream channel with multiple subchannels that form a braided pattern flowing through alluvial deposits.

breaker zone the part of the nearshore area in which waves break.

butte isolated erosional remnant of a tableland with a flat summit, often bordered by steep-sided escarpments. Buttes are usually found in arid regions of flat-lying sediments and are smaller than mesas.

calcification soil-forming process of subhumid and semiarid climates. Soil types in the mollisol order, the typical end products of the process, are characterized by little leaching or eluviation and by the accumulation of both humus and mineral bases (especially calcium carbonate, $CaCO^{3}$).

caldera a large depression formed by a volcanic eruption.

caliche hardened layers of lime ($CaCO^{3}$) deposited at the surface of a soil by evaporating capillary water.

calorie amount of heat necessary to raise the temperature of 1 gram of water 1°C.

calving breaking off a mass of ice from the toe of a glacier at its junction with the ocean or a lake.

campos region of characteristic tropical savanna vegetation in Brazil, located primarily in the Amazon Basin bordering the tropical rainforest.

Canadian high high atmospheric pressure area that tends to develop over the central North American continent in winter.

capacity the maximum amount of water vapor that can be contained in a given quantity of air at a given temperature.

capillary action the upward movement of water through tiny cracks and pore spaces.

capillary water soil water that clings to soil peds and individual soil particles as a result of surface tension. Capillary water moves in all directions through the soil from areas of surplus water to areas of deficit.

caprock a resistant horizontal layer of rock that forms the flat top of a landform, such as a butte or a mesa.

carbonate a mineral group characterized by carbon's ability to form complex compounds of organic and inorganic origins.

carbonation carbon dioxide in water chemically combining with other substances to create new compounds.

carnivore animal that eats only other animals.

cartography the science of mapmaking.

catastrophism once-popular theory that Earth's landscapes developed in a relatively short time by cataclysmic events.

cavern (cave) natural void in rock created by solution that is large enough for people to enter.

Celsius (or centigrade) scale temperature scale in which 0° is the freezing point of water and 100° its boiling point at standard sea-level pressure.

centrifugal force force that pulls a rotating object away from the center of rotation.

channel roughness an expression of the frictional resistance to stream flow due to irregularities in a stream channel bed and sides.

chaparral sclerophyllous woodland vegetation found growing in the Mediterranean climate of the western United States; these seasonal, drought-resistant plants are low-growing, with small, hard-surfaced leaves and deep, water-probing roots.

chemical sedimentary rocks rocks created from dissolved minerals that have precipitated out of water.

chemical weathering breakdown of rock material by chemical reactions that change the rock's mineral composition (decomposition).

Chinook dry warm wind on the eastern slopes of the Rocky Mountains (*see* foehn wind).

cinder cone hill composed of fragments of volcanic rock (pyroclastics) erupted from a central vent.

circle of illumination line dividing the sunlit (day) hemisphere from the shaded (night) hemisphere; experienced by individuals on Earth's surface as sunrise and/or sunset.

cirque deep, sometimes steep-sided amphitheater formed at the head of an alpine valley by glacial ice erosion.

cirque glacier a generally small alpine glacier restricted to a high-elevation basin (cirque).

cirro signifies a high-level cloud (i.e., above 6000 m in elevation).

cirrus high, detached clouds consisting of ice particles. Cirrus clouds are white and feathery or fibrous in appearance.

Cl, O, R, P, T Hans Jenny's description of the factors of soil formation: Climate, Organics, Relief, Parent material, and Time.

classification process of systematically arranging phenomena into groups, classes, or categories based on some established criteria.

clastic rock sedimentary rock formed by the compaction and cementation of preexisting rock debris.

clasts solid broken pieces of rock, bone, or shell.

clay (clayey) a very fine grained mineral particle with a size less than 0.004 millimeter, often the product of weathering.

climate accumulated and averaged weather patterns of a locality or region; the full description is based on long-term statistics and includes extremes or deviations from the norm.

climatology scientific study of climates of Earth and their distribution.

climax community the final step in the succession of plant communities that occupy a specific location.

climograph graphic means of giving information on mean monthly temperature and rainfall for a select location or station.

closed system system in which no substantial amount of materials can cross its boundaries.

cloud mass of suspended water droplets (or at high altitudes, ice particles) in air above ground level.

coastal zone general region of interaction between the land and a lake or the ocean.

coastline of emergence coast with formerly submerged land that is now above water, due either to uplift of the land or a drop in sea level.

coastline of submergence a coastal area that has undergone sinking or subsidence relative to sea level.

cloud forest rainforest that is produced by nearly constant light rain on the windward slopes of mountains.

cobble rock fragments ranging in diameter from 64 to 256 millimeters.

cobble beach cobble-sized sediment deposited by waves along the shoreline, often found along the base of sea cliffs or lake cliffs.

col a glacially eroded pass between two mountain valleys.

cold front leading edge of a relatively cooler, denser air mass that advances upon a warmer, less dense air mass.

collapse sinkhole topographic depression formed mainly by the cave-in of the land above a cavern.

collision–coalescence process rain-forming process where raindrops form by collision between cloud droplets.

color composite a digital image that combines several wavelength bands—a common color composite that is used resembles a near-infrared color photo.

column pillar-shaped speleothem resulting from the joining of a stalactite and a stalagmite.

columnar joints vertical, polygonal fractures caused by lava shrinking as it cools.

comet a small body of icy and dusty matter that revolves about the sun. When a comet comes near the sun, some of its material vaporizes, forming a large head and often a tail.

composite cone (stratovolcano) volcano formed from alternating layers of lava and pyroclastic materials; generally known for violent eruptions.

compressional tectonic force force originating within Earth that acts to push two adjacent areas of rock toward each other (convergence).

compromise projection maps that compromise true shape and true area in order to display both fairly well.

conceptual model image in the mind of an Earth feature or landscape as derived from personal experiences.

condensation process by which a vapor is converted to a liquid during which energy is released in the form of latent heat.

condensation nuclei minute particles in the atmosphere (e.g., dust, smoke, pollen, sea salt) on which condensation can take place.

conduction transfer of heat within a body or between adjacent matter by means of internal molecular movement.

conformal map projection a map projection that maintains the true shape of small areas on Earth's surface.

connate water groundwater trapped in the pore spaces of sedimentary rock at the time it was first deposited; water locked out of the hydrologic cycle in sedimentary rocks.

Continental Arctic (*cA*) very cold, very dry air mass originating from the arctic region.

continental collision the fusing together of landmasses as tectonic plates converge.

continental crust the less dense (avg. 2.7 g/cm^3), thicker portion of Earth's crust; underlies the continents.

Continental Divide line of separation dividing runoff between the Pacific and Atlantic Oceans. In North America it generally follows the crest of the Rocky Mountains.

continental drift theory proposed by Alfred Wegener stating that the continents joined, broke apart, and moved on Earth's surface; it was later replaced by the theory of plate tectonics.

continental glacier a very large and thick mass of flowing ice that exists due to climatic conditions resulting from high latitude.

continental ice sheet thick ice mass that covers a major portion of a continent and buries all but the highest mountain peaks; it usually flows from one or more areas of accumulation outward in all directions.

continental islands islands that are geologically part of a continent and are usually located on the continental shelf.

Continental Polar (c*P*) cold, dry air mass originating from landmasses approximately 40° to 60° N or S latitude.

continental shelf the gently sloping margin of a continent overlain by ocean water.

continental shield the ancient part of a continent that consists of crystalline rock.

Continental Tropical (c*T*) warm, dry air mass originating from subtropical landmasses.

continentality the distance a particular place is located in respect to a large body of water; the greater the distance, the greater the continentality.

continuous data numerical or locational representations of phenomena that are present everywhere—such as air pressure, temperature, elevation.

contour interval vertical distance represented by two adjacent contour lines on a topographic map.

contour map (topographic map) map that uses contour lines to show differences in elevation (topography).

control (temperature) factors that control temperatures around the world, such as latitude, proximity to water bodies, ocean currents, altitude, and landform barriers and human activities.

convection process by which a circulation is produced within an air mass or fluid body (heated material rises, cooled material sinks); also, in tectonic plate theory, the method whereby heat is transferred to Earth's surface from deep within the mantle.

convectional precipitation precipitation resulting from condensation of water vapor in an air mass that is rising convectionally as it is heated from below.

convective thunderstorm a thunderstorm produced by the convective uplift mechanism.

convergent wind circulation pressure-and-wind system where the airflow is inward toward the center, where pressure is lowest.

coordinate system a precise system of grid lines used to describe locations.

coral reef shallow, wave-resistant structure made by the accumulation of skeletal remains of tiny sea animals.

core extremely hot and dense, innermost portion of Earth's interior; the molten outer core is 2400 kilometers (1500 mi) thick; the solid inner core is 1120 kilometers (700 mi) thick.

Coriolis effect apparent effect of Earth's rotation on horizontally moving bodies, such as wind and ocean currents; such bodies tend to be deflected to the right in the Northern Hemisphere and to the left in the Southern Hemisphere.

corrosion chemical erosion of rock matter by water; the removal of ions from rock-forming minerals in water.

coulee snaking, steep-sided channel cut through lava formations by glacial meltwater.

creep slow downslope movement of Earth material involving the lifting and falling action of sediment particles.

crevasse stress crack commonly found along the margins and at the toe of a glacier.

cross bedding thin layers within sedimentary rocks that were deposited at an angle to the dominant rock layering.

crust relatively thin, approximately 8–64 kilometers (5–40 mi) deep, low-density surface layer of Earth.

crustal warping gentle bending and folding of crustal rocks.

cumulus globular clouds, usually with a horizontal base and strong vertical development.

cut bank the steep slope found on the outside of a bend in a meandering stream channel.

cyclone center of low atmospheric pressure, also known as a *low*.
cyclonic precipitation precipitation formed by cyclonic uplift.

daily march of temperature changes in daily temperatures as we go from an overnight low to a daytime high and back to the overnight low.
debris unconsolidated slope material with a wide range of grain sizes including at least 20% gravel (>2 mm)
debris flow rapid, gravity-induced downslope movement of wet, poorly sorted Earth material.
debris flow fan fan-shaped depositional landform, particularly common in arid regions, created where debris flows emerge onto a plain from a mountain canyon.
declination the latitude on Earth at which the noon sun is directly overhead.
decomposer organism that promotes decay by feeding on dead plant and animal material and returns mineral nutrients to the soil or water in a form that plants can utilize.
decomposition a term that refers to the processes of chemical weathering.
deep-water wave wave traveling in depth of water greater than or equal to half the wavelength.
deflation entrainment and removal of loose surface sediment by the wind.
deflation hollow a wind-eroded depression in an area not dominated by wind-deposited sand.
degradation landscape lowering that results from more erosion than deposition over time.
delta depositional landform constructed where a stream flows into a standing body of water (a lake or the ocean).
delta plain the portion of a delta that lies above the level of the lake or ocean.
dendritic term used to describe a drainage pattern that is tree-like with tributaries joining the main stream at acute angles.
dendrochronology method of determining past climatic conditions using tree rings.
deposition accumulation of Earth materials at a new site after being moved by gravity, water, wind, or glacial ice.
desert climate climate where the amount of precipitation received is less than one half of the potential ET.
desert pavement (reg, gibber) desert surface mosaic of close-fitting stones that overlies a deposit of mostly fine-grained sediment.
detritivore animal that feeds on dead plant and animal material.
dew tiny droplets of water on ground surfaces, grass blades, or solid objects. Dew is formed by condensation when air at the surface reaches the dew point.
dew point the temperature at which an air mass becomes saturated; any further cooling will cause condensation of water vapor in the air.
differential weathering and erosion rock types vary in resistance to weathering and erosion, causing the processes to occur at different rates, often producing distinctive landform features.
digital elevation model (DEM) three-dimensional views of topography.
digital image an image made from computer data displayed like a mosaic of tiny squares, called pixels.
digital mapping mapmaking that employs computer techniques.
digital terrain model a computer-generated graphic representation of topography.
dike igneous intrusion with a wall-like shape.
dip inclination of a rock layer from the horizontal; always measured at right angles to the strike.
dip-slip fault a vertical fault where the movement is up and down the dip of the fault surface.
disappearing stream stream that has its flow diverted entirely to the subsurface.
discharge (stream discharge) rate of stream flow; measured as the volume of water flowing past a cross section of a stream per unit of time (cubic meters or cubic feet per second).
discrete data numerical or locational representations of phenomena that are present only at certain locations—such as earthquake epicenters, sinkholes, tornado paths.
dissolved load soluble minerals or other chemical constituents carried in water as a solution.
distributary a smaller stream that conducts flow away from the larger main channel, especially on deltas; the opposite of a tributary.
diurnal (daily) temperature range difference between the highest and lowest temperatures of the day (usually recorded hourly).
divergent wind circulation pressure-and-wind system where the airflow is outward away from the center, where pressure is highest.
divide line of separation between drainage basins; generally follows high ground or ridge lines.
doldrums zone of low pressure and calms along the equator.
doline *see* sinkhole.
Doppler radar advanced type of radar that can detect motion in storms, specifically motion toward and away from the radar signal.
drainage basin (watershed, catchment) the region that provides runoff to a stream.
drainage density the summed length of all stream channels per unit area in a drainage basin.
drainage divide the outer boundary of a drainage basin.
drainage (stream) pattern the form of the arrangement of channels in a stream system in map view.
drift sediment deposited in association with glacial ice or its meltwater.
drizzle fine mist or haze of very small water droplets with a barely perceptible falling motion.
drumlin streamlined, elongated hill composed of glacial drift with a tapered end indicating direction of continental ice flow.
dry adiabatic lapse rate rate at which a rising mass of air is cooled by expansion when no condensation is occurring (10°C/1000 m or 5.6°F/1000 ft).
dust storm a moving cloud of wind-blown dust (typically silt).
dynamic equilibrium constantly changing relationship among the variables of a system, which produces a balance between the amounts of energy and/or materials that enter a system and the amounts that leave.

earth (as a mass wasting material) thick unit of unconsolidated, predominantly fine-grained slope material.
Earth system set of interrelated components or variables (e.g., atmosphere, lithosphere, biosphere, hydrosphere), which interact and function together to make up Earth as it is currently constituted.
earthquake series of vibrations or shock waves set in motion by sudden movement along a fault.

earthquake intensity a measure of the impact of an earthquake on humans and their built environment.

earthquake magnitude measurement representing an earthquake's size in terms of energy released.

easterly wave trough-shaped, weak, low pressure cell that progresses slowly from east to west in the trade wind belt of the tropics; this type of disturbance sometimes develops into a tropical hurricane.

eccentricity cycle the change in Earth's orbit from slightly elliptical to more circular, and back to its earlier shape every 100,000 years.

ecological niche combination of role and habitat as represented by a particular species in an ecosystem.

ecology science that studies the interactions between organisms and their environment.

ecosystem community of organisms functioning together in an interdependent relationship with the environment that they occupy.

ecotone transition zone of varied natural vegetation occupying the boundary between two adjacent and differing plant communities.

effective precipitation actual precipitation available to supply plants and soil with usable moisture; does not take into consideration storm runoff or evaporation.

effluent the condition of groundwater seeping into a stream.

effusive eruption streaming flows of molten rock matter pouring onto Earth's surface from the subsurface.

El Niño warm countercurrent that influences the central and eastern Pacific.

elastic solid a solid that withstands stress with little deformation until a maximum value is reached, whereupon it breaks.

electromagnetic energy all forms of energy that share the property of moving through space (or any medium) in a wave-like pattern of electric and magnetic fields; also called radiation.

elements (weather and climate) the major elements include solar energy, temperature, pressure, winds, and precipitation.

elevation vertical distance from mean sea level to a point or object on Earth's surface.

eluviation downward removal of soil components by water.

empirical classification classification process based on statistical, physical, or observable characteristics of phenomena; it ignores the causes or theory behind their occurrence.

end moraine a ridge of till deposited at the toe or terminus of a glacier.

endogenic processes landforming processes originating within Earth.

Enhanced Fujita Scale enhancement made to the tornado intensity scale that concentrates more specifically on the types of damages that occur.

entisol soil with little or no development.

entrenched stream a stream that has eroded downward so that it flows in a relatively deep and steep-sided (trenchlike) valley or canyon.

environment surroundings, whether of man or of any other living organism; includes physical, social, and cultural conditions that affect the development of that organism.

eolian (aeolian) pertaining to the landforming work of the wind.

ephemeral flow describes streams that conduct flow only occasionally, during to shortly after precipitation events, or due to ice or snowmelt.

ephemeral stream a stream that flows only at certain times, when adequate discharge is supplied by precipitation events, ice or snowmelt, or irregular spring flow.

epicenter point on Earth's surface directly above the focus of an earthquake.

epipedon surface soil layer that possesses specific characteristics essential to the identification of soils in the National Resources Conservation Service System. (Examples of epipedons may be found in Table 12.1.)

equal-area map projection a map projection on which any given areas of Earth's surface are shown in correct proportional sizes on the map.

equator great circle of Earth midway between the poles; the zero degree parallel of latitude that divides Earth into the Northern and Southern Hemispheres.

equatorial low (equatorial trough) zone of low atmospheric pressure centered more or less over the equator where heated air is rising; *see also* doldrums.

equidistance a property of some maps that depicts distances equally without scale variation.

equilibrium state of balance between the interconnected components of an organized whole.

equilibrium line balance position on a glacier that separates the zone of accumulation from the zone of ablation.

equinox one of two times each year (approximately March 21 and September 22) when the position of the noon sun is overhead (and its vertical rays strike) at the equator; all over Earth, day and night are of equal length.

erg desert region of active sand dunes, most common in the Sahara.

erosion removal of Earth materials from a site by gravity, water, wind, or glacial ice.

erratic large glacially transported boulder deposited on top of bedrock of different composition.

esker narrow, winding ridge of coarse sediment probably deposited in association with a meltwater tunnel at the base of a continental glacier.

estuary coastal waters where salt and fresh water mix.

evaporation process by which a liquid is converted to the gaseous (or vapor) state by the addition of latent heat.

evaporite mineral salts that are soluble in water and accumulate when water evaporates.

evapotranspiration combined water loss to the atmosphere from ground and water surfaces by evaporation and, from plants, by transpiration.

exfoliation successive removal of outer rock sheets or slabs broken from the main rock mass by weathering.

exfoliation dome large, smooth, convex (dome-shaped) mass of exposed rock undergoing exfoliation due to weathering by unloading.

exfoliation sheet relatively thin, outer layer of rock broken from the main rock mass by weathering.

exogenic processes landforming processes originating at or very near Earth's surface.

exotic stream (or river) stream that originates in a humid region and has sufficient water volume to flow across a desert region.

explosive eruption violent blast of molten and solid rock matter into the air.

exposure direction of mountain slopes with respect to prevailing wind direction.

exterior drainage streams and stream systems that flow to the ocean.

extratropical disturbance *see* middle-latitude disturbance.

extrusive igneous rock rock solidified at Earth's surface from lava; also called volcanic rock.

extrusive rock igneous rock that was erupted and solidified on Earth's surface.

Fahrenheit scale temperature scale in which 32° is the freezing point of water, and 212° its boiling point, at standard sea-level pressure.

fall type of fast mass wasting characterized by Earth material plummeting downward freely through air.

fan apex the most upflow point on an alluvial fan; where the fan-forming stream emerges from the mountain canyon.

fast mass wasting gravity-induced downslope movement of Earth material that people can witness directly.

fault breakage zone along which rock masses have slid past each other.

fault block discrete blocklike region of crustal rocks bordered on two opposite sides by normal faults.

fault scarp (escarpment) the steep cliff or exposed face of a fault where one crustal block has been displaced vertically relative to another.

faulting the movement of rock masses past each other along either side of a fault.

feedback sequence of changes in the elements of a system, which ultimately affects the element that was initially altered to begin the sequence.

feedback loop path of change as its effects move through the variables of a system until the effects impact the variable originally experiencing change.

fertilization adding additional nutrients to the soil.

fetch distance over open water that winds blow without interruption.

firn compact granular snow formed by partial melting and refreezing due to overlying layers of snow.

firn line (equilibrium line) boundary between the zones of ablation and accumulation on a glacier, representing the equilibrium point between net snowfall and ablation.

fissure an extensive crack or break in rocks along which lava may be extruded.

fissure flows lava flows that emanated from a crack (fissure) in the surface rather than from a volcano.

fjord deep, glacial trough along the coast invaded by the sea after the removal of the glacier.

flood stream water exceeding the amount that can be contained within its channel.

flood basalts massive outpourings of basaltic lava.

floodplain the low-gradient area adjacent to many stream channels that is subject to flooding and primarily composed of alluvium.

flow rapid downslope movement of wet unconsolidated Earth material that experiences considerable mixing.

flowing artesian well artificial opening that allows groundwater from below to reach the surface and flow out under its own pressure.

fluvial term used to describe landform processes associated with the work of streams and rivers.

fluvial geomorphology the study of streams as landforming agents.

focus point within Earth's crust or upper mantle where an earthquake originated.

foehn wind warm, dry, downslope wind on lee of mountain range, caused by adiabatic heating of descending air.

fog mass of suspended water droplets within the atmosphere that is in contact with the ground.

folding the bending or wrinkling of Earth's crust due to compressional tectonic forces.

foliation the occurrence of banding or platy structure in metamorphic rocks.

food chain sequence of levels in the feeding pattern of an ecosystem.

food web feeding mosaic formed by the interrelated and overlapping food chains of an ecosystem.

freeze–thaw weathering (frost wedging) breaking apart of rock by the expansive force of water freezing in cracks.

freezing rain rainfall that freezes into ice upon coming in contact with a surface or object that is colder than 0°C (32°F).

friction force that acts opposite to the direction of movement or flow; for example, turbulent resistance of Earth's surface on the flow of the atmosphere.

fringing reef coral reef attached to the coast.

front sloping boundary or contact surface between air masses with different properties of temperature, moisture content, density, and atmospheric pressure.

frontal lifting lifting or rising of warmer, lighter air above cooler, denser air along a frontal boundary.

frontal precipitation precipitation resulting from condensation of water vapor in an air mass that is rising over another mass along a front.

frontal thunderstorm a thunderstorm produced by the frontal uplift mechanism.

frost frozen condensation that occurs when air at ground level is cooled to a dew point of 0°C (32°F) or below; also any temperature near or below freezing that threatens sensitive plants.

fusion (thermonuclear reaction) the fusing together of two hydrogen atoms to create one helium atom. This process releases tremendous amounts of energy.

galactic movement movement of the solar system within the Milky Way Galaxy.

galaxy a large assemblage of stars; a typical galaxy contains millions to hundreds of billions of stars.

galeria forest junglelike vegetation extending along and over streams in tropical forest regions.

gap an area within the territory occupied by a plant community when the climax vegetation has been destroyed or damaged by some natural process, such as a hurricane, forest fire, or landslide.

gelisol soil that experiences frequent freezing and thawing.

General Circulation Model (GCM) complex computer simulations based on the relationships of selected variables within the Earth system that are used in attempts to predict future climates.

generalist species that can survive on a wide range of food supplies.

genetic classification classification process based on the causes, theory, or origins of phenomena; generally ignoring their statistical, physical, or observable characteristics.

geocoding the process and reference system used to tie map locations to field locations using a grid system.

geographic grid lines of latitude and longitude form the geographic grid.

geographic information system (GIS) versatile computer software that combines the features of automated (computer) cartography and database management to produce new maps of data for solving spatial problems.

geography study of Earth phenomena; includes an analysis of distributional patterns and interrelationships among these phenomena.

geomorphic agent a medium that erodes, transports, and deposits Earth materials; includes water, wind, and glacial ice.

geomorphology the study of the origin and development of landforms.

geostationary orbit an orbit that synchronizes a satellite's position and speed with Earth rotation so that it continually images the same location.

geostrophic winds upper-level winds in which the Coriolis effect and pressure gradient are balanced, resulting in a wind flowing parallel to the isobars.

geothermal water water heated by contact with hot rocks in the subsurface.

geyser natural eruptive outflow of water that alternates between hot water and steam.

giant planets the four largest planets—Jupiter, Saturn, Uranus, and Neptune.

gibber Australian term for an extensive desert plain covered with pebble or cobble-sized rocks.

glacial outwash the fluvial deposits derived from glacial meltwater streams.

glacial plucking erosive pulling away of rock material underneath a glacier by glacial ice flowing away from a bedrock obstruction.

glacial trough a U-shaped valley carved by glacial erosion.

glacier a large mass of ice that flows as a plastic solid.

glacier advance expansion of the toe or terminus of a glacier to a lower elevation or lower latitude due to an increase in size.

glacier head farthest upslope part of an alpine glacier.

glacier retreat withdrawal of the toe or terminus of a glacier to a higher elevation or higher latitude due to a decrease in size.

glacier terminus (toe) farthest downslope part of an alpine glacier.

glaciofluvial deposit sorted glacial drift deposited by meltwater.

glaciolacustrine deposit sorted glacial drift deposited by meltwater in lakes associated with the margins of glaciers.

glaze (freezing rain) translucent coating of ice that develops when rain strikes a freezing surface.

gleization soil-forming process of poorly drained areas in cold, wet climates. The resulting soils have a heavy surface layer of humus with a water-saturated clay horizon directly beneath.

Global Positioning System (GPS) GPS uses satellites and computers to compute positions and travel routes anywhere on Earth.

global warming climate change that would cause Earth's temperatures to rise.

gnomonic projection planar projection with greatly distorted land and water areas; valuable for navigation because all great circles on the projection appear as straight lines.

graben block of crustal rocks between two parallel normal faults that has slid downward relative to adjacent blocks.

gradational processes processes that derive their energy indirectly from the sun and directly from Earth gravitation and serve to wear down, fill in, and level off Earth's surface.

graded stream stream where slope and channel size provide velocity just sufficient to transport the load supplied by the drainage basin; a theoretical balanced state averaged over a period of many years.

gradient a term for slope often used to describe the angle of a streambed.

granite a coarse-grained intrusive igneous rock generally associated with continental crust.

granular disintegration weathering feature of coarse crystalline rocks in which visible individual mineral grains fall away from the main rock mass.

graphic (bar) scale a rulerlike device placed on maps for making direct measurements in ground distances.

gravel a general term for sediment sizes larger than sand.

gravitation the attractive force one body has for another. The force increases as the mass of the bodies increases, and the distance between them decreases.

gravitational water meteoric water that passes through the soil under the influence of gravitation.

gravity the mutual attraction of bodies or particles.

great circle any circle formed by a full circumference of the globe; the plane of a great circle passes through the center of the globe.

greenhouse effect warming of the atmosphere that occurs because shortwave solar radiation heats the planet's surface, but the loss of longwave heat radiation is hindered by the release of gases associated with human activity (e.g., CO^2).

greenhouse gases atmospheric gases that hinder the escape of Earth's heat energy.

Greenwich mean time (GMT) time at zero degrees longitude used as the base time for Earth's 24 time zones; also called Universal Time or Zulu Time.

groin artificial structure extending out into the water from a beach built to inhibit loss of beach sediment.

ground moraine irregular, hummocky landscape of till deposited on Earth's surface by a wasting glacier.

ground-inversion fog *see* radiation fog.

groundwater water in the saturated zone below the water table.

gullies steep-sided stream channels somewhat larger than rills that even in humid climates flow only in direct response to precipitation events.

gyre broad circular patterns of major surface ocean currents produced by large subtropical high pressure systems.

habitat location within an ecosystem occupied by a particular organism.

hail form of precipitation consisting of pellets or balls of ice with a concentric layered structure usually associated with the strong convection of cumulonimbus clouds.

hanging valley tributary glacial trough that enters a main glaciated valley at a level high above the valley floor.

hardpan dense, compacted, clay-rich layer occasionally found in the subsoil (*B* horizon) that is an end product of excessive illuviation.

haystack hill (conical hill or **hum)** remnant hills of soluble rock remaining after adjacent rock has been dissolved away in karst areas.

headward erosion gullying and valley cutting that extends a stream channel in an upstream direction.

heat the total kinetic energy of all the atoms that make up a substance.

heat energy budget relationship between solar energy input, storage, and output within the Earth system.

heat island mass of warmer air overlying urban areas.

heaving various means by which particles are lifted perpendicular to a sloping surface, then fall straight down by gravity.

hemisphere half of a sphere; for example, the northern or southern half of Earth divided by the equator or the eastern and western half divided by two meridians, the 0° and 180° meridians.

herbivore an animal that eats only living plant material.

heterosphere layer of the atmosphere that lies between 80 kilometers (50 mi) and the outer limits of the atmosphere; here, atmospheric gases separate into individual ionized gases.

heterotroph organism that is incapable of producing its own food and that must survive by consuming other organisms.

high *see* anticyclone.

highland climates a general climate classification for regions of high, yet varying, elevations.

histosol soil that develops in poorly drained areas.

holistic approach considering and examining all phenomena relevant to a problem.

Holocene the most recent time interval of warm, relatively stable climate that began with the retreat of major glaciers about 10,000 years ago.

homosphere layer of the atmosphere where all the atmospheric gases are mixed together in the same proportions; this layer lies between Earth's surface and 80 kilometers (50 mi) aloft.

horizon the visual boundary between Earth and sky.

horn pyramid-like peak created where three or more expanding cirques meet at a mountain summit.

horst crustal block between two parallel normal faults that has slid upward relative to adjacent blocks.

hot spot a mass of hot molten rock material at a fixed location beneath a lithospheric plate.

hot spring natural outflow of geothermal groundwater to the surface.

human geography specialization in the systematic study of geography that focuses on the location, distribution, and spatial interaction of human (cultural) phenomena.

humid continental, hot-summer climate climate type characterized by hot, humid summers and mild, moist winters.

humid continental, mild-summer climate climate type characterized by mild, humid summers and cold, moist winters.

humidity amount of water vapor in an air mass at a given time.

humus organic matter found in the surface soil layers that is in various stages of decomposition as a result of bacterial action.

hurricane severe tropical cyclone of great size with nearly concentric isobars. Its torrential rains and high-velocity winds create unusually high seas and extensive coastal flooding; also called willy willies, tropical cyclones, baguios, and typhoons.

hydration rock weathering due to substances in cracks swelling and shrinking with the addition and removal of water molecules.

hydraulic action erosion resulting from the force of moving water.

hydrologic cycle circulation of water within the Earth system, from evaporation to condensation, precipitation, runoff, storage, and reevaporation back into the atmosphere.

hydrolysis water molecules chemically recombining with other substances to form new compounds.

hygroscopic water water in the soil that adheres to mineral particles.

hydrosphere major Earth subsystem consisting of the waters of Earth, including oceans, ice, freshwater bodies, groundwater, and water within the atmosphere and biomass.

ice age period of Earth history when much of Earth's surface was covered with massive continental glaciers. The most recent ice age is referred to as the Pleistocene Epoch.

ice cap a continental glacier of regional size, less than 50,000 square kilometers.

ice fall portion of a glacier moving over and down a steep slope, creating a rigid white cascade, criss-crossed with deep crevasses.

ice sheet the largest type of glacier; a continental glacier larger than 50,000 square kilometers.

ice shelf large flat-topped plate of ice overlying the ocean but attached to land; a source of icebergs.

iceberg free-floating mass of ice broken off from a glacier where it flows into the ocean or a lake.

Icelandic Low center of low atmospheric pressure located in the north Atlantic, especially persistent in winter.

ice-marginal lake temporary lake formed by the disruption of meltwater drainage by deposition along a glacial margin, usually in the area of an end moraine.

ice-scoured plain a broad area of low relief and bedrock exposures eroded by a continental glacier.

ice-sheet climate climate type where the average temperature of every month of the year is below freezing.

igneous intrusion (pluton or intrusion) a mass of igneous rock that cooled and solidified beneath Earth's surface.

igneous processes processes related to the solidification and eruption of molten rock matter.

igneous rock one of the three major categories of rock; formed from the cooling and solidification of molten rock matter.

illuviation deposition of fine soil components in the subsoil (*B* horizon) by gravitational water.

imaging radar radar systems designed to sense the ground and convert reflections into a maplike image.

inceptisol young soil with weak horizon development.

infiltration water seeping downward into the soil or other surface materials.

infiltration capacity the greatest amount of infiltrated water that a surface material can hold.

influent the condition of stream water seeping into the channel bed and adding to groundwater.

inner core the solid, innermost portion of Earth's core, probably of iron and nickel, that forms the center of Earth.

inputs energy and material entering an Earth system.

inselberg remnant bedrock hill rising above a stream-eroded plain or pediment in an arid or semiarid region.

insolation incoming solar radiation, that is, energy received from the sun.

instability condition of air when it is warmer than the surrounding atmosphere and is buoyant with a tendency to rise; the lapse rate of the surrounding atmosphere is greater than that of *unstable* air.

interception the delay in arriving at the ground surface experienced by precipitation that strikes vegetation.

interfluve the land between two stream channels.

interglacial warmer period between glacial advances, during which continental ice sheets and many valley glaciers retreat and disappear or are greatly reduced in size.

interior drainage streams and stream systems that flow within a closed basin and thus do not reach the ocean.
intermediate zone subsurface water layer between the zone of aeration above and the zone of saturation below; saturated only during times of ample precipitation.
intermittent flow describes streams that conduct flow seasonally.
intermittent stream stream that flows part of the time, usually only during, and shortly after, a rainy period.
International Date Line line roughly along the 180° meridian, where each day begins and ends; it is always a day later west of the line than east of the line.
Intertropical Convergence Zone (ITCZ) zone of low pressure and calms along the equator, where air carried by the trade winds from both sides of the equator converges and is forced to rise.
intrusive igneous rock rock that solidified within Earth from magma; also called plutonic rock.
inversion *see* temperature inversion.
ionosphere layer of ionized gasses concentrated between 80 kilometers (50 mi) and outer limits of the atmosphere.
isarithm line on a map that connects all points of the same numerical value, such as isotherms, isobars, and isobaths.
island arc a chain of volcanic islands along a deep oceanic trench; found near tectonic plate boundaries where subduction is occurring.
isobar line drawn on a map to connect all points with the same atmospheric pressure.
isoline a line on a map that represents equal values of some numerical measurement such as lines of equal temperature or elevation contours.
isostasy theory that holds that Earth's crust *floats* in hydrostatic equilibrium in the denser plastic layer of the mantle.
isotherm line drawn on a map connecting points of equal temperature.

jet stream high-velocity upper-air current with speeds of 120–640 kilometers per hour (75–250 mph).
jetty artificial structure extending into a body of water; built to protect a harbor, inlet, or beach by modifying action of waves or currents.
joint fracture or crack in rock.
joint set system of multiple parallel cracks (joints) in rock.
jungle dense tangle of trees and vines in areas where sunlight reaches the ground surface (not a true rainforest).

kame conical hill composed of sorted glaciofluvial deposits; presumed to have formed in contact with glacial ice when sediment accumulated in ice pits, crevasses, and among jumbles of detached ice blocks.
kame terraces landform resulting from accumulation of glaciofluvial sand and gravel along the margin of a glacier occupying a valley in an area of hilly relief.
karst unique landforms and landscapes derived by the solution of soluble rocks, particularly limestone.
katabatic wind downslope flow of cold, dense air that has accumulated in a high mountain valley or over an elevated plateau or ice cap.
Kelvin scale temperature scale developed by Lord Kelvin, equal to Celsius scale plus 273; no temperature can drop below absolute zero, or 0 degrees Kelvin.
kettle depression formed by the melting of an ice block buried in glacial deposits left by a retreating glacier.
kettle hole water-filled pit formed by the melting of a remnant ice block left buried in drift after the retreat of a glacier.
kettle lake a small lake or pond occupying a kettle.
kinetic energy energy of motion; one half the mass (m) times velocity (v) squared, $E_k = \frac{1}{2}mv^2$.
Köppen system climate classification based on monthly and annual averages of temperature and precipitation; boundaries between climate classes are designed so that climate types coincide with vegetation regions.

La Niña cold sea-surface temperature anomaly in the Equatorial Pacific (opposite of El Niño).
laccolith massive igneous rock intrusion that bows overlying rock layers upwards in a domal fashion.
lahar rapid, gravity-driven downslope movement of wet, fine-grained volcanic sediment.
lamination planes very thin layers in rock.
land breeze air flow at night from the land toward the sea, caused by the movement of air from a zone of higher pressure associated with cooler nighttime temperatures over the land.
landform a terrain feature, such as a mountain, valley, plateau, and so on.
Landsat a family of U.S. satellites that have been returning digital images since the 1970s.
landslide layperson's term for any fast mass wasting; used by some earth scientists for massive slides that involve a variety of Earth materials.
lapse rate *see* normal lapse rate.
latent heat of condensation energy release in the form of heat, as water is converted from the gaseous (vapor) to the liquid state.
latent heat of evaporation amount of heat absorbed by water to evaporate from a surface (i.e., 590 calories/g of water).
latent heat of fusion amount of heat transferred when liquid turns to ice and vice versa; this amounts to 80 calories/gram.
latent heat of sublimation amount of heat that is leased when ice turns to vapor without first going through the liquid phase; this amounts to 670 calories/gram.
lateral migration the sideways shift in the position of a stream channel over time.
lateral moraine a ridge of till deposited along the side margin of a glacier.
laterite iron, aluminum, and manganese rich layer in the subsoil (*B* horizon) that can be an end product of laterization in the wet-dry tropics (tropical savanna climate).
laterization soil-forming process of hot, wet climates. Oxisols, the typical end product of the process, are characterized by the presence of little or no humus, the removal of soluble and most fine soil components, and the heavy accumulation of iron and aluminum compounds.
latitude angular distance (distance measured in degrees) north or south of the equator.
lava molten (melted) rock matter erupted onto Earth's surface; solidifies into extrusive igneous (volcanic) rocks.
lava flow erupted molten rock matter that oozed over the landscape and solidified.
leaching removal by gravitational water of soluble inorganic soil components from the surface layers of the soil.
leeward located on the side facing away from the wind.

legend key to symbols used on a map.

levee natural raised alluvial bank along margins of a river on a floodplain; artificial levees may be constructed along river banks for flood control.

liana woody vine found in tropical forests that roots in the forest floor but uses trees for support as it grows upward toward available sunshine.

life-support system interacting and interdependent units (e.g., oxygen cycle, nitrogen cycle) that together provide an environment within which life can exist.

light year the distance light travels in 1 year—6 trillion miles.

lightning visible electrical discharge produced within a thunderstorm.

lithification the combined processes of compaction and cementation that transform clastic sediments into sedimentary rocks.

lithosphere (planetary structure) rigid and brittle outer layer of Earth consisting of the crust and uppermost mantle.

lithospheric plates Earth's exterior is broken into these several large regions of rigid and brittle crust and upper mantle (lithosphere).

Little Ice Age an especially cold interval of time during the early 14th century that had major impacts on civilizations in the Northern Hemisphere.

littoral drifting general term for sediment transport parallel to shore in the nearshore zone due to incomplete wave refraction.

llanos region of characteristic tropical savanna vegetation in Venezuela, located primarily in the plains of the Orinoco River.

loam soil soil with a texture in which none of the three soil grades (sand, silt, or clay) predominate over the others.

loess wind-deposited silt; usually transported in dust storms and derived from arid or glaciated regions.

longitude angular distance (distance measured in degrees) east or west of the prime meridian.

longitudinal dune a linear ridgelike sand dune that is oriented parallel to the prevailing wind direction.

longitudinal profile the change in stream channel elevation with distance downstream from source to mouth.

longshore bar submerged feature of wave- and current-deposited sediment lying close to and parallel with the shore.

longshore current flow of water parallel to the shoreline just inside the breaker zone; caused by incomplete wave refraction.

longshore drifting transport of sediment parallel to shore by the longshore current.

longwave radiation electromagnetic radiation emitted by Earth in the form of waves more than 4.0 micrometers in amplitude, which includes heat reradiated by Earth's surface.

low *see* cyclone.

magma molten (melted) rock matter located beneath Earth's surface from which intrusive igneous rocks are formed.

magnetic declination horizontal angle between geographic north and magnetic north.

mantle moderately dense, relatively thick (2885 km/1800 mi) middle layer of Earth's interior that separates the crust from the outer core.

map projection any presentation of the spherical Earth on a flat surface.

maquis sclerophyllous woodland and plant community, similar to North American chaparral; can be found growing throughout the Mediterranean region.

marine terrace abrasion platform that has been elevated above sea level and thus abandoned from wave action.

maritime relating to weather, climate, or atmospheric conditions in coastal or oceanic areas.

Maritime Equatorial (*mE*) hot and humid air mass that originates from the ocean region straddling the equator.

Maritime Polar (*mP*) cold, moist air mass originating from the oceans around 40° to 60° N or S latitude.

Maritime Tropical (*mT*) warm, moist air mass that originates from the tropical ocean regions.

mass a measure of the total amount of matter in a body.

mass wasting gravity-induced downslope movement of Earth material.

mathematical/statistical model computer-generated representation of an area or Earth system using statistical data.

matrix the dominant area of a mosaic (ecosystem supporting a particular plant community) where the major plant in the community is concentrated.

meander a broad, sweeping bend in a river or stream.

meander cut-off bend of a meandering stream that has become isolated from the active channel.

meandering channel stream channel with broadly sinuous banks that curve back and forth in sweeping bends.

medial moraine a central moraine in a large valley glacier formed where the interior lateral moraines of two tributary glaciers merge.

Mediterranean climate climate type characterized by warm, dry summers and cool, moist winters.

mental map conceptual model of special significance in geography because it consists of spatial information.

Mercalli Scale, modified an earthquake intensity scale with Roman numerals from I to XII used to assess spatial variations in the degree of impact that a tremor generates.

Mercator projection mathematically produced, conformal map projection showing true compass bearings as straight lines.

mercury barometer instrument measuring atmospheric pressure by balancing it against a column of mercury.

meridian one half of a great circle on the globe connecting all points of equal longitude; all meridians connect the North and South Poles.

mesa flat-topped, steep-sided erosional remnant of a tableland, roughly as broad as tall, characteristic of arid regions with flat-lying sedimentary rocks.

mesopause upper limit of mesosphere, separating it from the thermosphere.

mesosphere layer of atmosphere above the stratosphere; characterized by temperatures that decrease regularly with altitude.

mesothermal climates climate regions or conditions with hot, warm, or mild summers that do not have any months that average below freezing.

metamorphic rock one of the three major categories of rock; formed by heat and pressure changing a preexisting rock.

meteor the luminous phenomenon observed when a small piece of solid matter enters Earth's atmosphere and burns up.

meteorite any fragment of a meteor that reaches Earth's surface.

meteoroid stone or iron mass that enters our atmosphere from outer space becoming a meteor as it burns up in the atmosphere.

meteorology study of the patterns and causes associated with short-term changes in the elements of the atmosphere.

microclimate climate associated with a small area at or near Earth's surface; the area may range from a few inches to 1 mile in size.

microplate terrane segment of crust of distinct geology added to a continent during tectonic plate collision.

microthermal climates climate regions or conditions with warm or mild summers that have winter months with temperatures averaging below freezing.

middle-latitude disturbance convergence of cold polar and warm subtropical air masses over the middle latitudes.

millibar unit of measurement for atmospheric pressure; 1 millibar equals a force of 1000 dynes per square centimeter; 1013.2 millibars is standard sea-level pressure.

mineral naturally occurring inorganic substance with a specific chemical composition and crystalline structure.

mistral cold downslope wind in southern France (*see* katabatic wind).

model a useful simplification of a more complex reality that permits prediction.

Mohorovičić discontinuity (Moho) zone marking the transition between Earth's crust and the denser mantle.

mollisol soil that develops in grassland regions.

monadnock erosional remnant of more resistant rock on a plain of old age; associated with a theoretical cycle of erosion in humid lands.

monsoon seasonal wind that reverses direction during the year in response to a reversal of pressure over a large landmass. The classic monsoons of Southeast Asia blow onshore in response to low pressure over Eurasia in summer and offshore in response to high pressure in winter.

moraines various glacial landforms, mostly ridges, deposited beneath and along the margins of an ice mass.

mosaic a plant community and the ecosystem on which it is based, viewed as a landscape of interlocking parts by ecologists.

mountain breeze air flow downslope from mountains toward valleys during the night.

mouth downflow terminus of a stream.

mud wet, fine-grained sediment, particularly clay and silt sizes.

mudflow rapid mass wasting of wet, fine-grained sediment; may deposit levees and lobate (tongue-shaped) masses.

multispectral remote sensing using and comparing more than one type.

multispectral scanning using a number of energy wavelength bands to create images.

muskeg poorly drained vegetation-rich marshes or swamps usually overlying permafrost areas of polar climatic regions.

natural levees banks of a stream channel (or margins of a mass wasting flow channel) raised by deposition from flood (or flow) deposits; artificial levees are sometimes built along stream banks for flood control.

natural resource any element, material, or organism existing in nature that may be useful to humans.

natural vegetation vegetation that has been allowed to develop naturally without obvious interference from or modification by humans.

navigation the science of location and finding one's way, position, or direction.

neap tide the smaller than average tidal range that occurs during the first and third quarter moon.

Near Earth Objects (NEOs) large celestial bodies like comets, and asteroids, which may come close enough to collide with Earth.

near-infrared (NIR) film photographic film that makes pictures using near-infrared light that is not visible to the human eye.

nearshore zone area from the seaward or lakeward edge of breaking waves to the landward limit of broken wave water.

negative feedback reaction to initial change in a system that counteracts the initial change and leads to dynamic equilibrium in the system.

nekton classification of marine organisms that swim in the oceans.

nimbo a prefix for cloud types that means rain-producing.

nimbus term used in cloud description to indicate precipitation; thus cumulonimbus is a cumulus cloud from which rain is falling.

nonflowing artesian well artificial opening that allows groundwater under pressure from below to be accessible at the surface as a pool in the opening.

normal fault breakage zone with rocks on one side sliding down relative to rocks on the other side because of tensional forces; footwall up, hanging wall down.

normal lapse rate decrease in temperature with altitude under normal atmospheric conditions; approximately 6.5°C/1000 meters (3.6°F/1000 ft).

North Atlantic Oscillation oscillating (see-saw) pressure tendencies between the Azores High and the Icelandic Low.

North Pole maximum north latitude (90°N), at the point marking the axis of rotation.

northeast trades *see* trade winds.

notch a recess, relatively small in height, eroded by wave action along the base of a coastal cliff.

oblate spheroid Earth's shape—a slightly flattened sphere.

obliquity cycle the change in the tilt of the Earth's axis relative to the plane of the ecliptic over a 41,000-year period.

occluded front boundary between a rapidly advancing cold air mass and an uplifted warm air mass cut off from Earth's surface; denotes the last stage of a middle-latitude cyclone.

ocean current horizontal movement of ocean water, usually in response to major patterns of atmospheric circulation.

oceanic crust the denser (avg. 3.0 g/cm^3), thinner, basaltic portion of Earth's crust; underlies the ocean basins.

oceanic islands volcanic islands that rise from the deep ocean floor.

oceanic ridge (midocean ridge) linear seismic mountain range that interconnects through all the major oceans; it is where new molten crustal material rises through the oceanic crust.

oceanic trench (trench) long, narrow depression on the seafloor usually associated with an island arc. Trenches mark the deepest portions of the oceans and are associated with subduction of oceanic crust.

offshore zone the expanse of open water lying seaward or lakeward of the breaker zone.

omnivore animal that can feed on both plants and other animals.

open system system in which energy and/or materials can freely cross its boundaries.

organic sedimentary rocks rocks created from deposits of organic material, such as carbon from plants (coal).

orographic precipitation precipitation resulting from condensation of water vapor in an air mass that is forced to rise over a mountain range or other raised landform.

orographic thunderstorm a thunderstorm produced by the orographic uplift mechanism.

outcrop bedrock exposed at Earth's surface with no overlying regolith or soil.

outer core the upper portion of the Earth's core; considered to be composed of molten iron liquefied by the Earth's internal heat.

outlet glacier a valley glacier that flows outward from the main mass of a continental glacier.

outputs energy and material leaving an Earth system.

outwash glacial drift deposited beyond an end moraine by glacial meltwater.

outwash plain extensive, relatively smooth plain covered with sorted deposits carried forward by the meltwater from an ice sheet.

overthrust low-angle fault with rocks on one side pushed a considerable distance over those of the opposite side by compressional forces; the wedge of rocks that have overridden others in this way.

oxbow lake a lake or pond found in a meander cut-off on a floodplain.

oxidation union of oxygen with other elements to form new chemical compounds.

oxide a mineral group composed of oxygen combining with other Earth elements, especially metallics.

oxisol soil that develops over a long period of time in tropical regions with high temperatures and heavy annual rainfall.

oxygen-isotope analysis a dating method used to reconstruct climate history; it is based on the varying evaporation rates of different oxygen isotopes and the changing ratio between the isotopes revealed in foraminifera fossils.

ozone gas with a molecule consisting of three atoms of oxygen (O^3); forms a layer in the upper atmosphere that serves to screen out ultraviolet radiation harmful at Earth's surface.

ozonosphere also known as the ozone layer; this is a concentration of ozone gas in a layer between 20 and 50 kilometers (13–50 mi) above Earth's surface.

Pacific high persistent cell of high atmospheric pressure located in the subtropics of the North Pacific Ocean.

pahoehoe a smooth, ropy surface on a lava flow.

paleogeography study of the past geographical distribution of environments.

paleomagnetism the historic record of changes in Earth's magnetic field.

palynology method of determining past climatic conditions using pollen analysis.

Pangaea ancient continent that consisted of all of today's continental landmasses.

parabolic dune crescent-shaped sand dune with arms pointing upwind.

parallel circle on the globe connecting all points of equal latitude.

parallelism tendency of Earth's polar axis to remain parallel to itself at all positions in its orbit around the sun.

parent material residual (derived from bedrock directly beneath) or transported (by water, wind, or ice) mineral matter from which soil is formed.

passive-margin coast coastal region that is far removed from the volcanism and tectonism associated with plate boundaries.

patch a gap or area within a matrix (territory occupied by a dominant plant community) where the dominant vegetation is not supported due to natural causes.

paternoster lakes chain of lakes connected by a postglacial stream occupying the trough of a glaciated mountain valley.

patterned ground natural, repeating, often-polygonal designs of sorted sediment on the surface of periglacial environments.

ped naturally forming soil aggregate or clump with a distinctive shape that characterizes a soil's structure.

pediment gently sloping surface of eroded bedrock, thinly covered with fluvial sediments, found at the base of an arid-region mountain.

pediplain desert plain of pediments and alluvial fans; the presumed final erosion stage in an arid region.

peneplain theoretical plain of extreme old age; the last stage in a cycle of erosion, reached when a landmass has been reduced to near base level by stream erosion in a humid region.

perched water table a minor zone of saturation overlying an aquiclude that exists above the regional water table.

percolation subsurface water moving downward to lower zones by the pull of gravity.

perennial flow describes streams that conduct flow continuously all year.

perennial stream a stream with regular and adequate discharge to flow all year.

periglacial pertaining to cold-region landscapes that are impacted by intense frost action but not covered by year-round snow or ice.

perihelion position of Earth at closest distance to sun during each Earth revolution.

permafrost permanently frozen subsoil and underlying rock found in climates where summer thaw penetrates only the surface soil layer.

permeability characteristic of soil or bedrock that determines the ease with which water moves through Earth material.

pH scale scale from 0 to 14 that describes the acidity or alkalinity of a substance and that is based on a measurement of hydrogen ions; pH values below 7 indicate acidic conditions; pH values above 7 indicate alkaline conditions.

photosynthesis the process by which carbohydrates (sugars and starches) are manufactured in plant cells; requires carbon dioxide, water, light, and chlorophyll (the green color in plants).

physical geography specialization in the systematic study of geography that focuses on the location, distribution, and spatial interaction of physical (environmental) phenomena.

physical model three-dimensional representation of all or a portion of Earth's surface.

physical (mechanical) weathering breakdown of rocks into smaller fragments without chemical change by physical forces (disintegration).

phytoplankton tiny plants, algae and bacteria, that float and drift with currents in water bodies.

pictorial/graphic model representation of a portion of Earth's surface by means of maps, photographs, graphs, or diagrams.

piedmont alluvial plain a plain created by stream deposits at the base of an upland, such as a mountain, a hilly region, or a plateau.

piedmont fault scarp steep cliff due to movement along a fault that has offset unconsolidated sediment.

piedmont glacier an alpine glacier that extends beyond a mountain valley spreading out onto lower flatter terrain.

pixel the smallest area that can be resolved in a digital image. Pixels, short for "picture element," are much like pieces in a mosaic, fitted together in a grid to make an image.

plane of the ecliptic plane of Earth's orbit about the sun and the apparent annual path of the sun along the stars.

planet any of the nine largest bodies revolving about the sun, or any similar bodies that may orbit other stars.

plankton passively drifting or weakly swimming marine organisms, including both phytoplankton (plants) and zooplankton (animals).

plant community variety of individual plants living in harmony with each other and the surrounding physical environment.

plastic solid any solid material that changes its shape under stress, and retains that deformed shape after the stress is relieved.

plate convergence movement of lithospheric plates toward each other.

plate divergence movement of lithospheric plates away from each other.

plate tectonics theory that superseded continental drift and is based on the idea that the lithosphere is composed of a number of segments or *plates* that move independently of one another, at varying speeds, over Earth's surface.

plateau an extensive, flat-topped landform or region characterized by relatively high elevation, but low relief.

playa dry lake bed in a desert basin; typically fine-grained clastic (clay pan) or saline (salt crust).

playa lake a temporary lake that forms on a playa from runoff after a rainstorm or during a wet season.

Pleistocene the name given to the most recent "ice age" or period of Earth history experiencing cycles of continental glaciation; it commenced approximately 2.4 million years ago.

plucking erosion process by which a glacier pulls rocks and sediment from the ground along its bed and into the flowing ice.

plug dome a steep-sided, explosive type of volcano with its central vent or vents plugged by the rapid congealing of its highly acidic lava.

plunge pool a depression at the base of a waterfall formed by the impact of cascading water.

pluton *see* igneous intrusion.

plutonic rock *see* intrusive igneous rock.

plutonism the processes associated with the formation of rocks from magma cooling deep beneath Earth's surface.

pluvial rainy time period, usually pertaining to glacial periods when deserts were wetter than at present.

podzolization soil-forming process of humid climates with long cold winter seasons. Spodosols, the typical end product of the process, are characterized by the surface accumulation of raw humus, strong acidity, and the leaching or eluviation of soluble bases and iron and aluminum compounds.

point bar deposit of alluvium found on the inside of a bend in a meandering stream channel.

polar referring to the North or South Polar regions

polar climates climate regions that do not have a warm season and are frozen much or all of the year.

polar easterlies easterly surface winds that move out from the polar highs toward the subpolar lows.

polar front jet stream shifting boundary between cold polar air and warm subtropical air, located within the middle latitudes and strongly influenced by the polar jet stream.

polar highs high pressure systems located near the poles where air is settling and diverging.

polar jet stream high-velocity air current within the upper air westerlies.

polarity reversals times in geologic history when the south magnetic pole became the north magnetic pole and vice versa.

pollution alteration of the physical, chemical, or biological balance of the environment that has adverse effects on the normal functioning of all life forms, including humans.

porosity characteristic of soil or bedrock that relates to the amount of pore space between individual peds or soil and rock particles and that determines the water storage capacity of Earth material.

positive feedback reaction to initial change in a system that reinforces the initial change and leads to imbalance in the system.

potential evapotranspiration hypothetical rate of evapotranspiration if at all times there is a more than adequate amount of soil water for growing plants.

pothole bedrock depression in a streambed drilled by the spinning of abrasive rocks in swirling flow.

prairie grassland regions of the middle latitudes. Tall-grass prairie varied from 2 to 10 feet in height and was native to areas of moderate rainfall; short-grass prairie of lesser height remains common in subhumid and semiarid (steppe) environments.

precession cycle changes in the time (date) of the year that perihelion occurs; the date is determined on the basis of a major period 21,000 years in length and a secondary period 19,000 years in length.

precipitation water in liquid or solid form that falls from the atmosphere and reaches Earth's surface.

pressure belts zones of high or low pressure that tend to circle Earth parallel to the equator in a theoretical model of world atmospheric pressure.

pressure gradient rate of change of atmospheric pressure horizontally with distance, measured along a line perpendicular to the isobars on a map of pressure distribution.

prevailing wind direction from which the wind for a particular location blows during the greatest proportion of the time.

primary coastline coast that has developed its present form primarily from land-based processes, especially fluvial and glacial processes.

primary productivity *see* autotrophs, and productivity.

prime meridian (Greenwich meridian) half of a great circle that connects the North and South Poles and marks zero degrees longitude. By international agreement the meridian passes through the Royal Observatory at Greenwich, England.

prodelta the portion of a delta that lies submerged in the standing body of water.

productivity rate at which new organic material is created at a particular trophic level. Primary productivity through photosynthesis by autotrophs is at the first trophic level; secondary productivity is by heterotrophs at subsequent trophic levels.

profile a graph of changes in height over a linear distance, such as a topographic profile.

punctuated equilibrium concept that periods of relative stability in many Earth systems are interrupted by short bursts of intense action causing major change.

pyroclastic flow airborne density current of hot gases and rock fragments unleashed by an explosive volcanic eruption.

pyroclastic material (tephra) pieces of volcanic rock, including cinders and ash, solidified from molten material erupted into the air.

RADAR *RA*dio *D*etection *A*nd *R*anging.

radiation emission of waves that transmit energy through space; *see also* shortwave radiation and longwave radiation.

radiation fog fog produced by cooling of air in contact with a cold ground surface.

rain falling droplets of liquid water.
rain shadow dry, leeward side of a mountain range, resulting from the adiabatic warming of descending air.
recessional moraine end moraine deposited behind the terminal moraine marking a pause in the overall retreat of a glacier.
recharge replenishing the amount of stored water, particularly in the subsurface.
recumbent fold a fold in rock pushed over onto one side by asymmetric compressional forces; the axial plane of the fold is horizontal rather than vertical.
recurrence interval average length of time between events, such as floods, equal to or exceeding a given magnitude.
regional base level the lowest level to which a stream system in a basin of interior drainage can flow.
regional geography specialization in the systematic study of geography that focuses on the location, distribution, and spatial interaction of phenomena organized within arbitrary areas of Earth space designated as regions.
regions areas identified by certain characteristics they contain that make them distinctive and separates them from surrounding areas.
regolith weathered rock material; usually covers bedrock.
rejuvenated stream a stream that has deepened its channel by erosion because of tectonic uplift in the drainage basin or lowering of its base level.
relative humidity ratio between the amount of water vapor in air of a given temperature and the maximum amount of water vapor that the air could hold at that temperature, if saturated; usually expressed as a percentage.
relative location location of an object in respect to its position relative to some other object or feature.
relief a measurement or expression of the difference between the highest and lowest location in a specified area.
remote sensing mechanical collection of information about the environment from a distance, usually from aircraft or spacecraft, for example, photography, radar, infrared.
remote sensing devices variety of techniques by which information about Earth can be gathered from great heights, typically from very high-flying aircraft or spacecraft.
representative fraction (RF) scale a map scale presented as a fraction or ratio between the size of a unit on the map to the size of the same unit on the ground, as in 1/24,000 or 1:24,000.
reservoir an artificial lake impounded behind a dam.
residual parent material rock fragments that form a soil and have accumulated in place through weathering.
resolution (spatial resolution) size of an area on Earth that is represented by a single pixel.
reverse fault high-angle break with rocks on one side pushed up relative to those on the other side by compressional forces; hanging wall up, footwall down.
revolution (Earth) motion of Earth along a path, or orbit, around the sun. One complete revolution requires approximately 365¼ days and determines an Earth year.
rhumb line line of true compass bearing (heading).
ria coastline with many narrow bays mainly due to submerged river valleys.
ribbon falls high, narrow waterfalls dropping from a hanging glacial valley.
rift valley major lowland consisting of one or more crustal blocks downfaulted as a result of tensional tectonic forces.
rills tiny stream channels that even in a humid climate conduct flow only during precipitation events.
rime ice crystals formed along the windward side of tree branches, airplane wings, and the like, under conditions of supercooling.
rip current strong, narrow surface current flowing away from shore. It is produced by the return flow of water piled up near shore by incoming waves.
ripples small (centimeter-scale) wave forms in water or sediment.
roche moutonnée bedrock hill subjected to intense glacial abrasion on its up-ice side, with some plucking evident on the down-ice side.
rock a solid, natural aggregate of one or more minerals or particles of other rocks.
rock cycle a representation of the processes and pathways by which Earth material becomes different types of rocks.
rock flour rock fragments finely ground between the base of a glacier and the underlying bedrock surface.
rock structure the orientation, inclination, and arrangement of rock layers in Earth's crust.
rockfall nearly vertical drop of individual rocks or a rock mass through air pulled downward by the force of gravity.
rockslide rock unit moving rapidly downslope by gravity in continuous contact with the surface below.
Rossby waves horizontal undulations in the flow of the upper air winds of the middle and upper latitudes.
rotation (Earth) turning of Earth on its polar axis; one complete rotation requires 24 hours and determines one Earth day.
runoff flow of water from the land surface, generally in the form of streams and rivers.

salinas *see* salt flat.
salinization soil-forming process of low-lying areas in desert regions; the resulting soils are characterized by a high concentration of soluble salts as a result of the evaporation of surface water.
salt crystal growth weathering by the expansive force of salts growing in cracks in rocks; common in arid and coastal regions.
salt flat a low-relief deposit of saline minerals, typically in desert regions.
saltation the transportation by running water or wind of particles too large to be carried in suspension; the particles are bounced along on the surface or streambed by repeated lifting and deposition.
sand (sandy) sediment particles ranging in size from about 0.05 millimeter to 2.0 millimeters.
sand dune mound or hill of sand-sized sediment deposited and shaped by the wind.
sand sea an extensive area covered by sand dunes.
sandstorm strong winds blowing sand along the ground surface.
Santa Ana very dry foehn wind occurring in Southern California; *see also* foehn wind.
satellite any body that orbits a larger primary body, for example, the moon orbiting Earth.
saturation (saturated air) point at which sufficient cooling has occurred so that an air mass contains the maximum amount of water vapor it can hold. Further cooling produces condensation of excess water vapor.
savanna tropical vegetation consisting primarily of coarse grasses, often associated with scattered low-growing trees or patches of bare ground.
scale ratio between distance as measured on Earth and the same distance as measured on a map, globe, or other representation of Earth.

sclerophyllous vegetation type commonly associated with the Mediterranean climate; characterized by tough surfaces, deep roots, and thick, shiny leaves that resist moisture loss.

sea steep, choppy, chaotic waves still forming under the influence of a storm.

sea arch span of rock extending from a coastal cliff under which the ocean or lake water freely moves.

sea breeze air flow by day from the sea toward the land; caused by the movement of air toward a zone of lower pressure associated with higher daytime temperatures over the land.

sea cave large wave-eroded opening formed near the water level in coastal cliffs.

sea cliff (lake cliff) steep slope of land eroded at its base by wave action.

sea level average position of the ocean shoreline.

sea stack resistant pillar of rock projecting above water close to shore along an erosion-dominated coast.

seafloor spreading movement of oceanic crust in opposite directions away from the midocean ridges, associated with the formation of new crust at the ridges and subduction of old crust at ocean margins.

secondary coastline coast that has developed its present form primarily through the action of coastal processes (waves, currents, and/or coral reefs).

secondary productivity the formation of new organic matter by heterotrophs, consumers of other life forms; *see* productivity.

section a square parcel of land with an area of 1 square mile as defined by the U.S. Public Lands Survey System.

sedimentary rock one of three major rock categories; formed by compaction and cementation of rock fragments, organic remains, or chemical precipitates.

seif a large, long, somewhat sinuous sand dune elongated parallel to the prevailing wind direction

seismic wave traveling wave of energy released during an earthquake or other shock.

seismograph instrument used to measure amplitude of passing seismic waves.

selva characteristic tropical rainforest comprising multistoried, broad-leaf evergreen trees with significant development of lianas and relatively little undergrowth.

sextant navigation instrument used to determine latitude by star and sun positions.

shearing tectonic force force originating within Earth that moves two adjacent areas of rock alongside each other in opposite directions.

sheet wash thin sheet of unchannelized water flowing over land.

shield volcano dome-shaped accumulation of multiple successive lava flows extruded from one or more vents or fissures.

shoreline exact contact between the edge of a standing body of water and dry land.

short-grass prairie environment where the dominant vegetation type is short grasses.

shortwave radiation radiation energy emitted by the sun in the form of waves of less than 4.0 micrometers (1 micrometer equals one ten thousandth of a centimeter); includes X-rays, gamma rays, ultraviolet rays, and visible light waves.

Siberian high intensively developed center of high atmospheric pressure located in northern central Asia in winter.

side-looking airborne radar (SLAR) a radar system that is used for making maps of terrain features.

silicate the largest mineral group, composed of oxygen and silica and forming most of the Earth's crust.

sill a horizontal sheet of igneous rock intruded and solidified between other rock layers.

silt (silty) sediment particles with a grain size between 0.002 millimeter and 0.05 millimeter.

sinkhole (doline) roughly circular surface depression related to the solution of rock in karst areas.

slash-and-burn (shifting) cultivation also called swidden or shifting cultivation; typical subsistence agriculture of primitive societies in the tropical rainforest. Trees are cut, the smaller residue is burned, and crops are planted between the larger trees or stumps before rapid deterioration of the soil forces a move to a new area.

sleet form of precipitation produced when raindrops freeze as they fall through a layer of cold air; may also, locally, refer to a mixture of rain and snow.

slide fast mass wasting in which Earth material moves downslope in continuous contact with a discrete surface below.

slip face the steep, downwind side of a sand dune.

slope aspect direction a mountain slope faces in respect to the sun's rays.

slow mass wasting gravity-induced downslope movement of Earth material occurring so slowly that people cannot observe it directly.

slump thick unit of unconsolidated fine-grained material sliding downslope on a concave, curved slip plane.

small circle any circle that is not a full circumference of the globe. The plane of a small circle does not pass through the center of the globe.

smog combination of chemical pollutants and particulate matter in the lower atmosphere, typically over urban industrial areas.

snow precipitation in the form of ice crystals.

snow line elevation in mountain regions above which summer melting is insufficient to prevent the accumulation of permanent snow or ice.

snowstorm storm situation where precipitation falls in the form of snow.

soil a dynamic, natural layer on Earth's surface that is a complex mixture of inorganic minerals, organic materials, microorganisms, water, and air.

soil (as a mass wasting material) relatively thin unit of unconsolidated fine-grained slope material.

soil fertilization adding nutrients to the soil to meet the conditions that certain plants require.

soil grade classification of soil texture by particle size: clay (less than 0.002 mm), silt (0.002–0.05 mm), and sand (0.05–2.0 mm) are soil grades.

soil horizon distinct soil layer characteristic of vertical zonation in soils; horizons are distinguished by their general appearances and their specific chemical and physical properties.

soil order largest classification of soils based on development and composition of soil horizons.

soil ped *see* ped.

soil profile vertical cross section of a soil that displays the various horizons or soil layers that characterize it; used for classification.

soil survey a publication of the United States Soil Survey Division of the Natural Resources Conservation Service that includes maps showing the distribution of soil within a given area, usually a county.

soil taxonomy the classification and naming of soils.
soil texture the distribution of particle sizes in a soil that give it a distinctive "feel."
soil water water in the zone of aeration, the uppermost subsurface water layer.
soil-forming regime processes that create soils.
solar constant rate at which insolation is received just outside Earth's atmosphere on a surface at right angles to the incoming radiation.
solar energy *see* insolation.
solar noon the time of day when the sun angle is at a maximum above the horizon (zenith).
solar system the system of the sun and the planets, their satellites, comets, meteoroids, and other objects revolving around the sun.
solar wind streams of hot ions (protons and electrons) traveling outward from the sun.
solid tectonic processes those processes that distort the solid Earth crust by bending, folding, warping, or fracturing (faulting).
solifluction slow movement of saturated soil downslope by the pull of gravity; especially common in permafrost areas.
solstice one of two times each year when the position of the noon sun is overhead at its farthest distance from the equator; this occurs when the sun is overhead at the Tropic of Cancer (about June 21) and the Tropic of Capricorn (about December 21).
solution dissolving material in a fluid, such as water, or the liquid containing dissolved material; water transports dissolved load in solution.
solution sinkhole topographic depression formed mainly by the solution and removal of soluble rock at the surface.
sonar a system that uses sound waves for location and mapping underwater.
source location high in the drainage basin, near the drainage divide, where a stream system's flow begins.
source region nearly homogeneous surface of land or ocean over which an air mass acquires its temperature and humidity characteristics.
South Pole maximum south latitude (90°S), at the point marking the axis of rotation.
southeast trades *see* trade winds.
Southern Oscillation the systematic variation in atmospheric pressure between the eastern and western Pacific Ocean.
spatial distribution location and extent of an area or areas where a feature exists.
spatial interaction process whereby different phenomena are linked or interconnected, and, as a result, impact one another through Earth space.
spatial pattern arrangement of a feature as it is distributed through Earth space.
spatial science term used when defining geography as the science that examines phenomena as it is located, distributed, and interacts with other phenomena throughout Earth space.
specific humidity mass of water vapor present per unit mass of air, expressed as grams per kilogram of moist air.
speleology the scientific study of caverns.
speleothem general term for any cavern feature made by secondary (later) precipitation of minerals from subsurface water.
spheroidal weathering rounded shape of rocks often caused by preferential weathering along joints of cross-jointed rocks.
spit coastal landform of wave- and current-deposited sediment attached to dry land at one end.
spodosol soil that develops in porous substrates such as glacial drift or beach sand.
spring natural outflow of groundwater to the surface.
spring tide the larger than average tidal range that occurs during new and full moon.
squall line narrow line of rapidly advancing storm clouds, strong winds, and heavy precipitation; usually develops in front of a fast-moving cold front.
stability condition of air when it is cooler than the surrounding atmosphere and resists the tendency to rise; the lapse rate of the surrounding atmosphere is less than that of *stable* air.
stalactite spire-shaped speleothem that hangs from the ceiling of a cavern.
stalagmite spire-shaped speleothem that rises up from a cavern floor.
star dune a large pyramid-shaped sand dune with multiple slip faces due to changes in wind direction.
stationary front frontal system between air masses of nearly equal strength; produces stagnation over one location for an extended period of time.
steppe climate characterized by middle-latitude semiarid vegetation, treeless and dominated by short bunch grasses.
stock an irregular mass of intrusive igneous rock (pluton) smaller than a batholith.
storm local atmospheric disturbance often associated with rain, hail, snow, sleet, lightning, or strong winds.
storm surge rise in sea level due to wind and reduced air pressure during a hurricane or other severe storm.
storm track path frequently traveled by a cyclonic storm as it moves in a generally eastward direction from its point of origin.
strata (stratification) distinct layers or beds of sedimentary rock.
strato signifies a low-level cloud (i.e., from the surface to 2000 m in elevation).
stratopause upper limit of stratosphere, separating it from the mesosphere.
stratosphere layer of atmosphere lying above the troposphere and below the mesosphere, characterized by fairly constant temperatures and ozone concentration.
stratovolcano *see* composite cone.
stratus uniform layer of low sheetlike clouds, frequently grayish in appearance.
stream general term for any natural, channelized flow of water regardless of size.
stream capacity the maximum amount of load that a stream can carry; varies with the stream's velocity.
stream competence the largest particle size that a stream can carry; varies with a stream's velocity.
stream discharge volume of water flowing past a point in a stream channel in a given unit of time.
stream gradient vertical drop in a streambed over a given horizontal distance, generally given in meters per kilometer or feet per mile.
stream hydrograph plot showing changes in the amount of stream flow over time.
stream load amount of material transported by a stream at a given instant; includes bed load, suspended load, and dissolved load.
stream order numerical index expressing the position of a stream channel within the hierarchy of a stream system.
stream terrace former floor of a stream valley now abandoned and perched above the present valley floor and stream channel.

stress (pressure) force per unit area.

striations gouges, grooves, and scratches carved in bedrock by abrading rock particles imbedded in a glacier.

strike compass direction of the line formed at the intersection of a tilted rock layer and a horizontal plane.

strike-slip fault a fault with horizontal motion, where movement takes place along the strike of the fault.

structure the descriptive physical characteristics and arrangement of bedrock, such as folded, faulted, layered, fractured, massive.

subarctic climate climate type that produces a tundra landscape.

subduction process associated with plate tectonic theory whereby an oceanic crustal plate is forced downward into the mantle beneath a lighter continental plate when the two converge.

sublimation direct change of state of a material, such as water, from solid to gas or gas to solid.

subpolar lows east–west trending belts or cells of low atmospheric pressure located in the upper middle latitudes.

subsurface horizon buried soil layer that possesses specific characteristics essential to the identification of soils in the National Resources Conservation Service System.

subsurface water general term for all water that lies beneath Earth's surface, including soil water and groundwater.

subsystem separate system operating within the boundaries of a larger Earth system.

subtropical highs cells of high atmospheric pressure centered over the eastern portions of the oceans in the vicinity of 30°N and 30°S latitude; source of the westerlies poleward and the trades equatorward.

subtropical jet stream high-velocity air current flowing above the sinking air of the subtropical high pressure cells; most prominent in the winter season.

succession progression of natural vegetation from one plant community to the next until a final stage of equilibrium has been reached with the natural environment.

sunspots visible dark (cooler) spots on the surface of the sun; their numbers seem to follow an approximate 11-year cycle.

supercooled water liquid water that exists below the freezing point of 0°C or 32°F.

surf zone the part of the nearshore area that consists of a turbulent bore of broken wave water.

surface creep wind-generated transportation consisting of pushing and rolling sediment downwind in continuous contact with the surface.

surface of discontinuity three-dimensional surface with length, width, and height separating two different air masses; also referred to as a *front*.

surface runoff liquid water flowing over Earth's land surface.

surge (glacial) sudden shift downslope of glacial ice, possibly caused by a reduction of basal friction with underlying bedrock.

suspended load solid particles that are small enough to be transported considerable distances while remaining buoyed up in a moving air or water column.

suspension transportation process that moves small solids, often considerable distances, while buoyed up by turbulence in the moving air or water.

swallow hole the site where a surface stream is diverted to the subsurface, such as into a cavern system.

swash thin sheet of broken wave water that rushes up the beach face in the swash zone.

swash zone the most landward part of the nearshore zone; where a thin sheet of water rushes up, then back down, the beach face.

swell orderly lake or ocean waves of rounded form that have traveled beyond the storm zone of wave generation.

symbiotic relationship relationship between two organisms that benefits both organisms.

syncline the downfolded element of folded rock structure.

system group of interacting and interdependent units that together form an organized whole.

taiga term used to describe the northern coniferous forest of subarctic regions on the Eurasian landmass.

taku cold downslope wind in Alaska; *see also* katabatic wind.

tall-grass prairie environment where the dominant vegetation type is tall grasses, with a few scattered trees.

talus (talus slope, talus cone) slope (sometimes cone-shaped) of angular, broken rocks at the base of a cliff deposited by rockfall.

tarn mountain lake in a glacial cirque.

tectonic forces forces originating within Earth that break and deform Earth's crust.

tectonic processes processes that derive their energy from within Earth's interior and serve to create landforms by elevating, disrupting, and roughening Earth's surface.

temperature degree of heat or cold and its measurement.

temperature gradient rate of change of temperature with distance in any direction from a given point; refers to rate of change horizontally; a vertical temperature gradient is referred to as the *lapse rate.*

temperature inversion reverse of the normal pattern of vertical distribution of air temperature; in the case of inversion, temperature *increases* rather than decreases with increasing altitude.

tensional tectonic force force originating within Earth that acts to pull two adjacent areas of rock away from each other (divergence).

tephra *see* pyroclastic material.

terminal moraine end moraine that marks the farthest advance of an alpine or continental glacier.

terminus (snout) the lower end of a glacier.

terra rossa characteristic calcium-rich (developed over limestone bedrock) red-brown soils of the climate regions surrounding the Mediterranean Sea.

terrestrial planets the four closest planets to the sun—Mercury, Venus, Earth, and Mars.

thematic map a map designed to present information or data about a specific theme, as in a population distribution map, a map of climate or vegetation.

thematic mapper (TM) a family of imaging systems that return images of Earth from Landsat satellites.

thermal expansion and contraction notion that rocks can weather due to expansion and contraction effects of alternating heating and cooling.

thermal infrared (TIR) scanning images made with scanning equipment that produces an image of heat differences.

thermosphere highest layer of atmosphere extending from the mesopause to outer space.

Thornthwaite system climate classification based on moisture availability and of greatest use at the local level; climate types are distinguished by examining and comparing potential and actual evapotranspiration.

threshold condition within a system that causes dramatic and often irreversible change for long periods of time to all variables in the system.

thrust fault low-angle break with rocks on one side pushed over those of the other side by compressional forces.

thunder sound produced by the rapidly expanding, heated air along the channel of a lightning discharge.

thunderstorm intense convectional storm characterized by thunder and lightning, short in duration and often accompanied by heavy rain, hail, and strong winds.

tidal interval the time between successive high tides, or between successive low tides.

tidal range elevation difference between water levels at high tide and low tide.

tide periodic rise and fall of sea level in response to the gravitational interaction of the moon, sun, and Earth.

till sediment deposited directly by glacial ice.

till plain a broad area of low relief covered by glacial deposits.

tilted fault block crustal block between two parallel normal faults that has been uplifted along one fault and relatively downdropped along the other.

time zone Earth is divided into 24 time zones (24 h) to coordinate time with Earth's rotation.

tolerance ability of a species to survive under specific environmental conditions.

tombolo strip of wave- and current-deposited sediment connecting the mainland to an island.

topographic contour line line on a map connecting points that are the same elevation above mean sea level.

topography the arrangement of high and low elevations in a landscape.

tornado small, intense, funnel-shaped cyclonic storm of very low pressure, violent updrafts, and converging winds of enormous velocity.

tornado outbreak when a thunderstorm(s) produce more than one tornado.

tower karst high, steep-sided hills formed by solution of limestone or other soluble rocks in karst areas.

trace less than a measurable amount of rain or snow (i.e., less than 1 mm or 0.01 in.).

traction transportation process in moving water that drags, rolls, or slides heavy particles along in continuous contact with the bed.

trade winds consistent surface winds blowing in low latitudes from the subtropical highs toward the intertropical convergence zone; labeled northeast trades in the Northern Hemisphere and southeast trades in the Southern Hemisphere.

transform movement horizontal sliding of tectonic plates alongside and past each other.

transpiration transfer of moisture from living plants to the atmosphere by the emission of water vapor, primarily from leaf pores.

transportation movement of Earth materials from one site to another by gravity, water, wind, or glacial ice.

transported parent material rock fragments that form a soil and originated elsewhere and then were transported and deposited in the new location.

transverse dune a linear ridgelike sand dune that is oriented at right angles to the prevailing wind direction.

transverse stream a stream that flows across the general orientation or "grain" of the topography, such as mountains or ridges.

travertine calcium carbonate (limestone) deposits resulting from the evaporation in caves or caverns and near surface openings of groundwater saturated with lime.

tree line elevation in mountain regions above which cold temperatures and wind stress prohibit tree growth.

tributary stream channel that delivers its water to another, larger channel.

trophic level number of feeding steps that a given organism is removed from the autotrophs (e.g., green plant—first level, herbivore—second level, carnivore—third level, etc.).

trophic structure organization of an ecosystem based on the feeding patterns of the organisms that comprise the ecosystem.

Tropic of Cancer parallel of latitude at 23½°N; the northern limit to the migration of the sun's vertical rays throughout the year.

Tropic of Capricorn parallel of latitude at 23½°S; the southern limit to the migration of the sun's vertical rays throughout the year.

tropical region on Earth lying between the Tropic of Cancer (23½°N latitude), and the Tropic of Capricorn (23½°S latitude).

tropical climates climate regions that are warm all year.

tropical easterlies winds that blow from the east in tropical regions.

tropical monsoon climate climate characterized by alternating rainy and dry seasons.

tropical rainforest climate hot wet climate that promotes the growth of rainforests.

tropical savanna climate warm, semidry climate that promotes tall grasslands.

tropopause boundary between the troposphere and stratosphere.

troposphere lowest layer of the atmosphere, exhibiting a steady decrease in temperature with increasing altitude and containing virtually all atmospheric dust and water vapor.

trough elongated area or "belt" of low atmospheric pressure; also glacial trough, a U-shaped valley carved by a glacier.

trunk stream the largest channel in a drainage system; receives inflow from tributaries.

tsunami wave caused when an earthquake, volcanic eruption, or other sudden event displaces ocean water; builds to dangerous heights in shallow coastal waters.

tundra high-latitude or high-altitude environments or climate regions that are not able to support tree growth because the growing season is too cold or too short.

tundra climate characterized by treeless vegetation of polar regions and very high mountains, consisting of mosses, lichens, and low-growing shrubs and flowering plants.

turbulence chaotic, mixing flow of fluids, often with an upward component.

typhoon a tropical cyclone found in the western Pacific, the same as a hurricane.

ultisol soil that has a subsurface clay horizon; is low in bases and is often red or yellow in color.

unconformity an interruption in the accumulation of different rock layers; often represents a period of erosion.

uniformitarianism widely accepted theory that Earth's geological processes operate today as they have in the past.

unloading physical weathering process whereby removal of overlying weight leads to rock expansion and breakage.

uplift mechanisms methods of lifting surface air aloft, they are orographic, frontal, convergence (cyclonic), and convectional.

upper air westerlies system of westerly winds in the upper atmosphere, flowing in latitudes poleward of 20°.

upslope fog type of fog where upward flowing air cools to form fog that hugs the slope of mountains.

upwelling upward movement of colder, nutrient-rich, subsurface ocean water, replacing surface water that is pushed away from shore by winds.

urban heat island *see* heat island.

U.S. Public Lands Survey System a method for locating and dividing land, used in much of the Midwest and western United States. This system divides land into 6- by 6-mile-square *townships* consisting of 36 *sections* of land (each 1 sq mi). Sections can also be subdivided into halves, quarter sections, and quarter-quarter-sections.

uvala (valley sink) large surface depression resulting from coalescing of sinkholes in karst areas.

valley breeze air flow upslope from the valleys toward the mountains during the day.

valley glacier an alpine glacier that extends beyond the zone of high mountain peaks into a confining mountain valley below.

valley train outwash deposit from glacial meltwater, resembling an alluvial fan confined by valley walls.

variable one of a set of objects and/or characteristics of objects, which are interrelated in such a way that they function together as a system.

varve a pairing of organic-rich summer sediments and organic poor winter sediments found in exposed lake beds; because each pair represents 1 year of time, counting varves is useful as a dating technique for recent Earth history.

veering wind shift the change in wind direction clockwise around the compass; for example, east to southeast to south, to southwest, to west, and northwest.

vent pipelike conduit through which volcanic rock material is erupted.

ventifact rock displaying distinctive wind-abraded faces, pits, grooves, and polish.

verbal scale stating the scale of a map using words such as "one inch represents one mile."

vertical exaggeration a technique that stretches the height representation of terrain in order to emphasize topographic detail.

vertical rays sun's rays that strike Earth's surface at a 90° angle.

vertisol soil that develops in regions with strong seasonality of precipitation.

visualization a wide array of computer techniques used to vividly illustrate a place or concept, or the illustration produced by one of these techniques.

volcanic ash erupted fragments of volcanic rock of sand size or smaller (<2.0 mm).

volcanic neck vertical igneous intrusion that solidified in the vent of a volcano.

volcanism the eruption of molten rock matter onto Earth's surface.

volcano mountain or hill created from the accumulation of erupted rock matter.

V-shaped valley the typical shape of a stream valley where the gradient is steep.

warm front leading edge of a relatively warmer, less dense air mass advancing upon a cooler, denser air mass.

warping broad and general uplift or settling of Earth's crust with little or no local distortion.

wash (arroyo, barranca, wadi) an ephemeral stream channel in an arid climate.

water budget relationship between evaporation, condensation, and storage of water within the Earth system.

water mining taking more groundwater out of an aquifer through pumping than is being replaced by natural processes in the same period of time.

water table upper limit of the zone of saturation below which all pore spaces are filled with water.

water vapor water in its gaseous form.

wave base water depth equal to half the length of a given wave; at smaller depths the wave interacts with the underwater substrate.

wave crest the highest part of a wave form.

wave height vertical distance between the trough and adjacent crest of a wave.

wave period time it takes for one wavelength to pass a given point.

wave refraction bending of waves in map view as they approach the shore, aligning themselves with the bottom contours in the breaker zone.

wave steepness ratio of wave height to wavelength for a given wave.

wave trough the lowest part of a wave form.

wavelength horizontal distance between two successive crests of a given wave.

weather atmospheric conditions, at a given time, in a specific location.

weather radar radar that is used to track thunderstorms, tornados, and hurricanes.

weathering physical (mechanical) fragmentation and chemical decomposition of rocks and minerals at and near Earth's surface.

well artificial opening that reaches the zone of saturation for the purpose of extracting groundwater.

westerlies surface winds flowing from the polar portions of the subtropical highs, carrying fronts, storms, and variable weather conditions from west to east through the middle latitudes.

wet adiabatic lapse rate rate at which a rising mass of air is cooled by expansion when condensation is taking place. The rate varies but averages 5°C/1000 meters (3.2°F/1000 ft).

white frost a heavy coating of white crystalline frost.

wind air in motion from areas of higher pressure to areas of lower pressure; movement is generally horizontal, relative to the ground surface.

wind wave wave on a water body created when air currents push the water surface along.

windward location on the side that faces toward the wind and is therefore exposed or unprotected; usually refers to mountain and island locations.

xerophytic vegetation type that has genetically evolved to withstand the extended periods of drought common to arid regions.

yardang aerodynamically shaped remnant ridge of wind-eroded bedrock or partly consolidated sediments.

yazoo stream a stream tributary that flows parallel to the main stream for a considerable distance before joining it.

zone of ablation the lower elevation part of a glacier; where more frozen water is removed than added during the year.

zone of accumulation (glacial) the higher elevation part of a glacier; where more frozen water is added than removed during the year.

zone of aeration uppermost layer of subsurface water where pore spaces typically contain both air and water.

zone of depletion top layer, or *A* horizon, of a soil, characterized by the removal of soluble and insoluble soil components through leaching and eluviation by gravitational water.

zone of saturation subsurface water zone in which all voids in rock and soil are always filled with water; the top of this zone is the water table.

zone of transition an area of gradual change from one region to another.

zooplankton tiny animals that float and drift with currents in water bodies.

Index

D

H

I

J

K

L

M

T

X

Y

Z

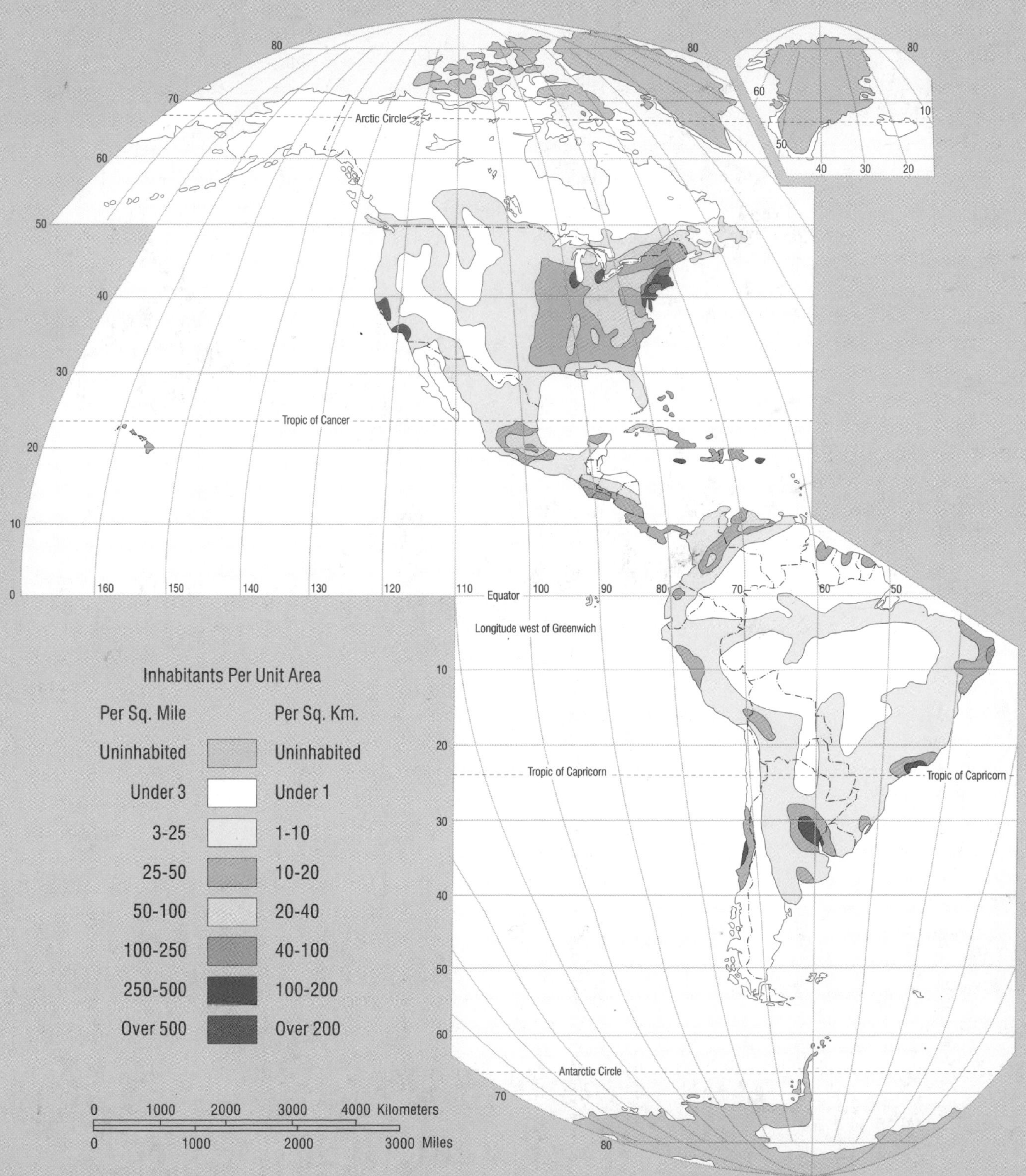
Arctic Circle
Tropic of Cancer
Equator
Longitude west of Greenwich
Tropic of Capricorn
Antarctic Circle
Inhabitants Per Unit Area
Per Sq. Mile
Per Sq. Km.
Uninhabited
Uninhabited
Under 3
Under 1
3-25
1-10
25-50
10-20
50-100
20-40
100-250
40-100
250-500
100-200
Over 500
Over 200
0 1000 2000 3000 4000 Kilometers
0 1000 2000 3000 Miles

World Map of Population Density